"Facts which at first seem improbable will, even on scant explanation, drop the
cloak which has hidden them and stand forth in naked and simple beauty."
—*Galileo Galilei, 1564–1642*

ENCYCLOPEDIA
of the
SOLAR SYSTEM

ENCYCLOPEDIA
of the
SOLAR SYSTEM

Edited by

PAUL R. WEISSMAN

Jet Propulsion Laboratory
California Institute of Technology
Pasadena, California

LUCY-ANN McFADDEN

Department of Astronomy
University of Maryland
College Park, Maryland

TORRENCE V. JOHNSON

Jet Propulsion Laboratory
California Institute of Technology
Pasadena, California

ACADEMIC PRESS

San Diego   London   Boston   New York   Sydney   Tokyo   Toronto

Jacket: An artist's view of the Solar System as observed from the Sun. Painting by Brad Greenwood.

Frontispiece: Launch of the Galileo spacecraft from the shuttle Atlantis, October 18, 1989. Galileo used gravity-assist flybys of Venus and the Earth to get to Jupiter, where it went into orbit on December 7, 1995. (NASA photo; courtesy of IMAX Corporation.)

Appendix: Launch of the Cassini Orbiter and Huygens Probe on Titan IV. (NASA photo)

Website: Academic Press maintains a WWW site for the Encyclopedia at *www.academicpress.com/solar/*. Author-recommended web resources for additional information, images, and research developments related to each chapter of this volume are available here.

This book is printed on acid-free paper. ∞

Academic Press
a division of Harcourt Brace & Company
525 B Street, Suite 1900, San Diego, California 92101-4495, USA
http://www.apnet.com

Academic Press Limited
24-28 Oval Road, London NW1 7DX, UK
http://www.hbuk.co.uk/ap/

Library of Congress Catalog Card Number: 98-84429

International Standard Book Number: 0-12-226805-9

PRINTED IN THE UNITED STATES OF AMERICA
98 99 00 01 02 03 MM 9 8 7 6 5 4 3 2

CONTENTS

· ·

CONTRIBUTORS   ix

FOREWORD   xiii

PREFACE   xv

GUIDE TO THE ENCYCLOPEDIA   xvii

The Solar System and Its Place in the Galaxy   1
Paul R. Weissman

The Origin of the Solar System   35
Patrick M. Cassen and Dorothy S. Woolum

The Sun   65
Jack B. Zirker

The Solar Wind   95
John T. Gosling

Mercury   123
Robert G. Strom

Venus: Atmosphere   147
Donald M. Hunten

Venus: Surface and Interior   161
James W. Head and Alexander Basilevski

Earth as a Planet: Atmosphere and Oceans   191
Timothy E. Dowling

Earth as a Planet: Surface and Interior   209
David C. Pieri and Adam M. Dziewonski

The Moon   247
Stuart Ross Taylor

Mars: Atmosphere and Volatile History   277
Fraser P. Fanale

Mars: Surface and Interior   291
Michael H. Carr

v

Phobos and Deimos   309
Peter C. Thomas

Atmospheres of the Giant Planets   315
Robert A. West

Interiors of the Giant Planets   339
Mark Scott Marley

Io   357
Dennis Matson and Diana Blaney

Titan   377
Athena Coustenis and Ralph Lorenz

Triton   405
William B. McKinnon and Randolph L. Kirk

Outer Planet Icy Satellites   435
Bonnie J. Buratti

Planetary Rings   457
Carolyn C. Porco

Planetary Magnetospheres   477
Margaret Kivelson and Fran Bagenal

Pluto and Charon   499
S. Alan Stern and Roger V. Yelle

Physics and Chemistry of Comets   519
Daniel C. Boice and Walter Huebner

Cometary Dynamics   537
Julio A. Fernández

The Kuiper Belt   557
Harold F. Levison and Paul R. Weissman

Asteroids   585
Daniel T. Britt and Larry A. Lebofsky

Near-Earth Asteroids   607
Lucy-Ann McFadden

Meteorites   629
Michael E. Lipschutz and Ludolf Schultz

Interplanetary Dust and the Zodiacal Cloud   673
Eberhard Grün

The Solar System at Ultraviolet
Wavelengths   697
Robert M. Nelson and Deborah L. Domingue

Infrared Views of the Solar System from
Space   715
Mark V. Sykes

The Solar System at Radio Wavelengths   735
Imke de Pater

Planetary Radar   773
Steven Ostro

Solar System Dynamics   809
Martin J. Duncan and Jack J. Lissauer

Chaotic Motion in the Solar System   825
Carl D. Murray

Planetary Impacts 845
Richard A. F. Grieve and Mark J. Cintala

Planetary Volcanism 877
Lionel Wilson

Planets and the Origin of Life 899
Christopher P. McKay and Wanda L. Davis

Planetary Exploration Missions 923
Louis Friedman and Robert Kraemer

Extra-Solar Planets: Searching for Other Planetary Systems 941
David C. Black

Appendix 957
Index 965

CONTRIBUTORS

Fran Bagenal
Planetary Magnetospheres
Department of Astrophysical, Planetary, and
Atmospheric Sciences
University of Colorado
Boulder, Colorado

Alexander Basilevsky
Venus: Surface and Interior
V. I. Vernadsky Institute for Geochemistry and
Analytical Chemistry
Russian Academy of Sciences
Moscow, Russia

David C. Black
*Extra-Solar Planets: Searching for Other
Planetary Systems*
Lunar and Planetary Institute
Houston, Texas

Diana Blaney
Io
Jet Propulsion Laboratory
California Institute of Technology
Pasadena, California

Daniel C. Boice
Physics and Chemistry of Comets
Department of Space Science
Southwest Research Institute
San Antonio, Texas

Daniel T. Britt
Asteroids
Lunar and Planetary Laboratory
University of Arizona
Tucson, Arizona

Bonnie J. Buratti
Outer Planet Icy Satellites
Jet Propulsion Laboratory
California Institute of Technology
Pasadena, California

Michael H. Carr
Mars: Surface and Interior
U.S. Geological Survey
Menlo Park, California

Patrick M. Cassen
The Origin of the Solar System
NASA Ames Research Center
Moffett Field, California

Mark J. Cintala
Planetary Impacts
NASA Johnson Space Center
Houston, Texas

Athena Coustenis
Titan
Observatoire de Meudon
Meudon, France

Wanda L. Davis
Planets and the Origin of Life
NASA Ames Research Center
Moffett Field, California

Deborah L. Domingue
The Solar System at Ultraviolet Wavelengths
Applied Physics Laboratory
Laurel, Maryland

Timothy E. Dowling
Earth as a Planet: Atmosphere and Oceans
Speed Scientific School
University of Louisville
Louisville, Kentucky

Martin J. Duncan
Solar System Dynamics
Physics Department
Queen's University
Kingston, Ontario, Canada

Adam M. Dziewonski
Earth as a Planet: Surface and Interior
Department of Earth and Planetary Sciences
Harvard University
Cambridge, Massachusetts

Fraser P. Fanale
Mars: Atmosphere and Volatile History
Planetary Geoscience Division
University of Hawaii
Honolulu, Hawaii

Julio A. Fernández
Cometary Dynamics
Dept. de Astronmía, Facultad de Ciencias
Universidad de la Republica
Montevideo, Uruguay

Louis Friedman
Planetary Exploration Missions
The Planetary Society
Pasadena, California

John T. Gosling
The Solar Wind
Los Alamos National Laboratory
Los Alamos, New Mexico

Richard A. F. Grieve
Planetary Impacts
Department of Energy, Mines and Resources
Ottawa, Canada

Eberhard Grün
Interplanetary Dust and the Zodiacal Cloud
Max-Planck-Institut für Kernphysik
Heidelberg, Germany

James W. Head
Venus: Surface and Interior
Department of Geological Sciences
Brown University
Providence, Rhode Island

Walter Huebner
Physics and Chemistry of Comets
Department of Space Science
Southwest Research Institute
San Antonio, Texas

Donald M. Hunten
Venus: Atmosphere
Department of Planetary Sciences
University of Arizona
Tucson, Arizona

Randolph L. Kirk
Triton
U.S. Geological Survey
Flagstaff, Arizona

Margaret Kivelson
Planetary Magnetospheres
Department of Earth and Space Sciences
University of California, Los Angeles
Los Angeles, California

Robert Kraemer
Planetary Exploration Missions
The Planetary Society
Pasadena, California

Larry A. Lebofsky
Asteroids
Lunar and Planetary Laboratory
University of Arizona
Tucson, Arizona

Harold F. Levison
The Kuiper Belt
Southwest Research Institute
Boulder, Colorado

Michael E. Lipschutz
Meteorites
Department of Chemistry
Purdue University
West Lafayette, Indiana

Jack L. Lissauer
Solar System Dynamics
NASA Ames Research Center
Moffett Field, California

Ralph Lorenz
Titan
Lunar and Planetary Laboratory
University of Arizona
Tucson, Arizona

Mark Scott Marley
Interiors of the Giant Planets
NASA Ames Research Center
Moffett Field, California

Dennis Matson
Io
Jet Propulsion Laboratory
California Institute of Technology
Pasadena, California

Lucy-Ann McFadden
Near-Earth Asteroids
Department of Astronomy
University of Maryland
College Park, Maryland

Christopher P. McKay
Planets and the Origin of Life
NASA Ames Research Center
Moffett Field, California

William B. McKinnon
Triton
Department of Earth and Planetary Sciences
Washington University
St. Louis, Missouri

Carl D. Murray
Chaotic Motion in the Solar System
Astronomy Unit
Queen Mary and Westfield College
University of London
London, England

Robert M. Nelson
The Solar System at Ultraviolet Wavelengths
Jet Propulsion Laboratory
California Institute of Technology
Pasadena, California

Steven Ostro
Planetary Radar
Jet Propulsion Laboratory
California Institute of Technology
Pasadena, California

Imke de Pater
The Solar System at Radio Wavelengths
Department of Astronomy
University of California, Berkeley
Berkeley, California

David C. Pieri
Earth as a Planet: Surface and Interior
Jet Propulsion Laboratory
California Institute of Technology
Pasadena, California

Carolyn C. Porco
Planetary Rings
Lunar and Planetary Laboratory
University of Arizona
Tucson, Arizona

Ludolf Schultz
Meteorites
Max-Planck-Institut für Chemie
Otto Hahn Institut
Mainz, Germany

S. Alan Stern
Pluto and Charon
Southwest Research Institute
Boulder, Colorado

Robert G. Strom
Mercury
Department of Planetary Sciences
University of Arizona
Tucson, Arizona

Mark V. Sykes
Infrared Views of the Solar System from Space
Steward Observatory
University of Arizona
Tucson, Arizona

Stuart Ross Taylor
The Moon
Research School of Physical Sciences
Australian National University
Canberra, Australia

Peter C. Thomas
Phobos and Deimos
Department of Astronomy
Cornell University
Ithaca, New York

Paul R. Weissman
The Solar System and Its Place in the Galaxy
The Kuiper Belt
Jet Propulsion Laboratory
California Institute of Technology
Pasadena, California

Robert A. West
Atmospheres of the Giant Planets
Jet Propulsion Laboratory
California Institute of Technology
Pasadena, California

Lionel Wilson
Planetary Volcanism
　　Environmental Sciences Division
　　University of Lancaster
　　Lancaster, England

Dorothy S. Woolum
The Origin of the Solar System
　　Department of Physics
　　California State University, Fullerton
　　Fullerton, California

Roger V. Yelle
Pluto and Charon
　　Department of Astronomy and Space Science
　　Boston University
　　Boston, Massachusetts

Jack B. Zirker
The Sun
　　National Solar Observatory
　　Sunspot, New Mexico

FOREWORD

. . . .

None of the other worlds in our solar system look like Earth.

When I was orbiting our planet in the Space Shuttle, I would often float over to a window and gaze down at the sparkling blue oceans, billowing white clouds, and rugged mountaintops of the Earth below. The signs of life are everywhere. A patchwork of fertile farmland lines the lazy Mississippi River, vast blooms of phytoplankton color the coastal waters of Southern Africa, and glittering coral reefs circle the islands of the Caribbean. At night, Earth's continents are outlined by twinkling city lights. Our planet is truly an oasis in space.

The view from space shows us how different Earth is from the other planets, but it also shows that many of the same processes that shape other worlds—wind and weather, impacts and volcanism—shape our planet, too. Through the Shuttle window, I watched an enormous dust storm grow until it blotted out all of Northern Africa. I saw the eroded remnants of an ancient impact crater in Eastern Canada, looked along a chain of volcanic islands in the Pacific Ocean, and gazed into a gaping rift valley in East Africa.

Imagine what it would be like to orbit an alien world: to watch a giant dust storm sweep across the ancient highlands of Mars, or look down on the slumping impact craters of Callisto, or see the gushing ice geysers of Triton. What would it be like to swing past Titan with Voyager 1, ride with Giotto past the icy heart of Halley's Comet, or wander with Sojourner across an ancient Martian floodplain? Though astronauts have never traveled beyond our own moon, humankind can now see the magnificent rings of Saturn, the tortured surface of Io, and the stormy atmosphere of Neptune through the eyes of robotic spacecraft.

Before the dawn of the Space Age, our instruments were confined to Earth's surface. Telescopes strained to gather light from worlds that were hundreds of millions of miles away. Even the most powerful produced fuzzy images of the planets, and only whetted scientists' appetites for more detailed observations. Then, in 1957 the Soviet Union launched the world's first satellite into orbit around Earth. Sputnik rocketed the world into the Space Age, and ignited a race between the United States and the Soviet Union for space superiority, the most public aspect of which was the race to the Moon. These new technical capabilities also stimulated the robotic exploration of our solar system, exploration that would continue long after Neil Armstrong left his bootprints in the lunar soil.

Today, spacecraft have visited every planet except Pluto. The Ulysses spacecraft circled above the poles of the Sun; Venera plunged headlong into the crushing carbon dioxide atmosphere of Venus; Vikings scooped up soil from the dusty red surface of Mars; and Galileo photographed barren, battered asteroids during its long journey to Jupiter. As these robot explorers radioed images and data back to Earth, they changed our view of the solar system forever.

The Encyclopedia of the Solar System appears at a propitious time. Not since 1609, when Galileo first turned his telescope to the heavens, has there been such a revolutionary change in our view of the solar system. Its pages are bursting with the knowledge gained as a result of data gathered by this fleet of robotic spacecraft and by a new generation of space- and ground-based telescopes. It takes us back 4.6 billion years to the origin of the solar system, then leads us on a tour that extends from the fiery interior of the Sun to the icy comets of the Kuiper Belt. It describes forces ranging from the solar wind that streams through all of interplanetary space, to the colossal impacts that affect every world.

Closer to home, this Encyclopedia makes it clear that our new perspective from space has revolutionized our understanding of Earth. Importantly, it also describes our continuing search for the answer to one of our most fundamental questions: Is life unique to our planet, or will it develop wherever conditions are right? Scientists are now asking whether primitive life might have begun in the hydrothermal vents of ancient Mars, or in the churning seas beneath the ice-covered surface of Europa. Our exploration is teaching us about Earth's origins and evolution; it may also teach us something about the origin of life itself.

Carl Sagan often pointed out how lucky we are to live at the very moment in human history when men and women are taking their first steps off the Earth. Every day for the last several years, astronauts have been living in an Earth orbit; every day, orbiting telescopes have been sending us images unobscured by Earth's atmosphere; and every day, distant spacecraft have been sending us information from faraway worlds. In this remarkable era of exploration, we are literally discovering new things about the universe every day.

The scientists whose work is described in this Encyclopedia have dedicated their careers to exploring the unknown. Their curiosity has led them to pose questions, propose theories, and conduct observations to help unravel mysteries that have intrigued scientists for centuries. This volume collects the contributions of the authors and, through them, hundreds of other scientists around the world. It represents our current state of knowledge on the origin, the evolution, and the fascinating components of our solar system. I invite you to join these scientists on their breathtaking journey. As you read their words, I encourage you to imagine, wonder, and question, just as they have.

Reach for the stars!

Sally K. Ride

PREFACE

· ·

"This is what hydrogen atoms can accomplish after four billion years of evolution."
—Carl Sagan, Cosmos, 1981

The quote above comes from the final episode of the public television series "Cosmos," which was created by Carl Sagan and several colleagues in 1981. Carl was describing the incredible accomplishments of the scientists and engineers who made the Voyager 1 and 2 missions to Jupiter and Saturn possible. But he just as easily could have been describing the chapters in this book.

This Encyclopedia is the product of the many scientists, engineers, technicians, and managers who produced the spacecraft missions which have explored our solar system over the past four decades. It is our attempt to provide to you, the reader, a comprehensive view of all we have learned in that 40 years of exploration and discovery. But we cannot take credit for this work. It is the product of the efforts of thousands of very talented and hard-working individuals in a score of countries who have contributed to that exploration. And it includes not only those involved directly in space missions, but also the many ground-based telescopic observers (both professional and amateur), laboratory scientists, theorists, and computer specialists who have contributed to creating that body of knowledge called solar system science. To all of these individuals, we say thank you.

Our goal in creating this Encyclopedia is to provide an integrated view of all we have learned about the solar system, at a level that is useful to the advanced amateur or student, to teachers, to non-solar system astronomers, and to professionals in other scientific and technical fields. What we present here is an introduction to the many different specialties that constitute solar system science, written by the worlds leading experts in each field. A reader can start at the beginning and follow the course we have laid out, or delve into the volume at almost any point and pursue his or her own personal interests. If the reader wishes to go further, the lists of recommended reading at the end of each article provide the next step in learning about any of the subjects cover

Our approach is to have the reader understand the solar system not only as a collection of individual and distinct bodies, but also as an integrated, interacting system, shaped by its initial conditions and by a variety of physical and chemical processes. The Encyclopedia begins with an overview chapter which describes the general features of the solar system and its relationship to the Milky Way galaxy, followed by a chapter on the origin of the system. Next we proceed from the Sun outward. We present the terrestrial planets (Mercury, Venus, Earth, Mars) individually with separate chapters on their atmospheres and satellites (where they exist). For the giant planets (Jupiter, Saturn, Uranus, Neptune) our focus shifts to common areas of scientific knowledge: atmospheres, interiors, satellites, rings, and magnetospheres. In addition, we have singled out three amazing satellites for individual chapters: Io, Titan, and Triton. Next is a chapter on the planetary systems most distant outpost, Pluto, and its icy satellite, Charon. From there we move into discussing the small bodies of the solar system: comets, asteroids, meteorites, and dust. Having looked at the individual members of the solar system, we next describe the different view of those members at a variety of wavelengths outside the normal visual region. From there we consider the important processes that have played such an important role in the formation and evolution of the system: celestial dynamics, chaos, impacts, and volcanism. Last, we look at three topics which are as much in our future as in our past: life on other planets, space exploration missions, and the search for planets around other stars.

A volume like this one does not come into being without the efforts of a great number of very dedicated people. We express our appreciation to the more than 50 colleagues who wrote chapters, sharing their expertise with you, the reader. In addition to providing chapters that captured the excitement of their individual fields, the authors have endured revisions, rewrites, endless questions, and unforeseen delays. For all of these we offer our humble apologies. To ensure the quality and accuracy of each contribution, at least two independent reviewers critiqued each chapter. The peer review process maintains its integrity through the anonymity of the reviewers. Although we cannot acknowledge them by name, we thank all the reviewers for their time and their conscientious efforts.

We are also deeply indebted to the team at Academic Press. Our executive editor, Frank Cynar, worked tirelessly with us to conceptualize and execute the encyclopedia, while allowing us to maintain the highest intellectual and scientific standards. We thank him for his patience and for his perseverance in seeing this volume through to completion. Frank's assistants, Daniela Dell'Orco, Della Grayson, Linda McAleer, Cathleen Ryan, and Suzanne Walters, kept the entire process moving and attended to the myriad of details and questions that arise with such a large and complex volume. Advice and valuable guidance came from Academic Press' director of major reference works, Chris Morris. Lori Asbury masterfully oversaw the production and copy editing. To all of the people at Academic Press, we give our sincere thanks.

Knowledge is not static. Science is a process, not a product. Some of what is presented in this volume will inevitably be out of date by the time you read it. New discoveries seem to come every day from our colleagues using Earth-based and orbiting telescopes, and from the flotilla of new small spacecraft that are out there adding to our store of knowledge about the solar system. In this spirit we hope that you, the reader, will benefit from the knowledge and understanding compiled in the following pages. The new millennium will surely add to the legacy presented herein, and we will all be the better for it. Enjoy, wonder, and keep watching the sky.

Paul R. Weissman
Lucy-Ann McFadden
Torrence V. Johnson

GUIDE TO THE ENCYCLOPEDIA

The *Encyclopedia of the Solar System* is a complete reference guide to this subject, including studies of the Sun, the Earth and the eight other major planets, the Moon and other natural satellites, planetary rings, comets, asteroids, meteorites, and interplanetary dust. Other entries discuss topics such as the dynamics of the solar system and planetary exploration missions. Each chapter in the Encyclopedia provides a scholarly overview of the selected topic to inform a broad spectrum of readers, from researchers to the interested general public.

In order that you, the reader, will derive the maximum benefit from the *Encyclopedia of the Solar System,* we have provided this Guide. It explains how the book is organized and how information can be located.

ORGANIZATION

The *Encyclopedia of the Solar System* is organized in a highly functional manner. It consists of 40 individual chapters that progress in sequence according to the physical arrangement of the solar system itself. That is, the encyclopedia begins with a summary chapter on the entire solar system, then follows with an chapter on the Sun, then the Solar Wind, then Mercury, Venus, Earth, the Moon, Mars, and so on. Following this are chapters on physical processes and on exploration. The final chapter of the book is Extra-Solar Planets: Searching for Other Planetary Systems.

Each chapter is a full-length narrative treatment of the subject at hand. Thus the Encyclopedias format allows readers to choose their own method for referring to the work. Those who wish specific information on limited topics can consult the A-Z Subject Index and then proceed to the desired topic from there. On the other hand, those who wish to obtain a full overview of a large subject can read the entire chapter on this subject from beginning to end; *e.g.,* The Sun. In fact, one can even read the entire Encyclopedia in sequence, in the manner of a textbook (or a novel), to obtain the ideal view of the complete subject of the solar system.

CHAPTER FORMAT

Each new chapter in the *Encyclopedia of the Solar System* begins at the top of a right-hand page, so that it may be quickly located. The authors name and affiliation are displayed at the beginning of the chapter. The chapter is organized according to a standard format, as follows:

- Title and Author
- Outline
- Glossary
- Defining Statement
- Body of the Chapter
- Cross-References
- Bibliography

OUTLINE

Each chapter in the Encyclopedia begins with an Outline that indicates the general content of the chapter. This outline serves two functions. First, it provides a brief preview of the text, so that the reader can get a sense of what is contained there without having to leaf through all the pages. Second, it serves to highlight important subtopics that will be discussed in the chapter. For example, the chapter Mars:Surface and Interior begins with the subtopic Mars Explorations.

The Outline is intended as an overview and thus it lists only the major headings of the chapter. In addition, extensive second-level and third-level headings will be found within the chapter.

GLOSSARY

The Glossary contains terms that are important to an understanding of the chapter and that may be unfamiliar to the reader. Each term is defined in the context of the particular chapter in which it is used. Thus the same term may appear as a Glossary entry in two or more chapters, with the details of the definition varying slightly from one chapter to another. The Encyclopedia includes approximately 500 glossary entries.

The following example is a glossary entry that appears with the chapter The Solar System and Its Place in the Galaxy.

> **Roche limit** The distance from a planet, within which another body will be disrupted because tidal forces from the planet exceed the self-gravity of the smaller body. For non-rotating bodies of equal density and zero strength, the Roche limit is about 2.2 planetary radii.

DEFINING STATEMENT

The text of each chapter in the Encyclopedia begins with a single introductory paragraph that defines the topic under discussion and summarizes the content of the chapter. For example, the chapter Planetary Radar begins with the following statement:

> Planetary radar astronomy is the study of solar system entities (the moon, asteroids, and comets, as well as the major planets and their ring systems) by transmitting a radio signal toward the target and then receiving and analyzing the echo. This field of research has primarily involved observations with Earth-based radar telescopes, but also includes certain experiments with the transmitter and/or the receiver on board a spacecraft orbiting or passing near a planetary object.

CROSS-REFERENCES

Chapters in the Encyclopedia have cross-references to other chapters. These cross-references appear within the text of the chapter, at the end of a paragraph containing material that is relevant to another chapter. The cross-references indicate related chapters that can be consulted for further information on the same topic, or for information on a related topic. For example, the chapter Titan has cross-references to Pluto and Charon, Triton, Planetary Impacts, and The Solar System at Radio Wavelengths.

BIBLIOGRAPHY

The Bibliography section appears as the last element in each chapter. This section lists recent secondary sources that will aid the reader in locating more information on the topic at hand. Review chapters and research papers that are important to a more detailed understanding of the topic are also listed here.

The Bibliography entries in this Encyclopedia are for the benefit of the reader, to provide references for further reading or research on the given topic. Thus they typically consist of a limited number of entries. They are not intended to represent a complete listing of all the materials consulted by the author or authors in preparing the chapter. The Bibliography is in effect an extension of the chapter itself, and it represents the authors choice as to the best sources available for additional information.

INDEX

The Subject Index for the *Encyclopedia of the Solar System* contains more than 4500 entries. Reference to the general coverage of a topic appears as a marginal entry, such as an entire section of a chapter devoted to the topic. References to more specific aspects of the topic then appear below this in an indented list.

ENCYCLOPEDIA WEBSITE

The *Encyclopedia of the Solar System* maintains its own editorial Web Page on the Internet at: **http://www.academicpress.com/solar/** This site gives information about the Encyclopedia project. It also features author-recommended links to other sites that provide information about the chapter topics of the Encyclopedia. The site will continue to evolve as more information becomes available.

THE SOLAR SYSTEM AND ITS PLACE IN THE GALAXY

I. Introduction
II. The Architecture of the Solar System
III. The Origin of the Solar System
IV. The Solar System's Place in the Galaxy
V. The Fate of the Solar System
VI. Concluding Remarks

Paul R. Weissman
*Jet Propulsion Laboratory,
California Institute of Technology*

GLOSSARY

Asteroid: Rocky, carbonaceous, or metallic body, smaller than a planet and orbiting the Sun. Most asteroids are in semistable orbits between Mars and Jupiter, but others are thrown onto orbits crossing those of the major planets.

Astronomical unit: The distance from the Sun at which a massless particle in an unperturbed orbit would have an orbital period of 365.2568983 days, equal to $1.4959787066 \times 10^{11}$ m, or about 9.2953×10^7 miles. Abbreviated AU, the astronomical unit is approximately the mean distance between the Earth and the Sun.

Comet: Body containing a significant fraction of ices, smaller than a planet and orbiting the Sun, usually in a highly eccentric orbit. Most comets are stored far from the planetary system in two large reservoirs: the Kuiper belt beyond the orbit of Neptune, and the Oort cloud at near-interstellar distances.

Eccentricity: Measure of the departure of an orbit from a perfect circle. A circular orbit has an eccentricity $e = 0$; an elliptical orbit has $0 < e < 1$; a parabolic orbit has $e = 1$; and a hyperbolic orbit has $e > 1$.

Ecliptic: Plane of the Earth's orbit around the Sun. The planets, most asteroids, and most of the short-period comets are in orbits with small or moderate inclinations relative to the ecliptic.

Heliocentric: Pertaining to a Sun-centered coordinate system.

Heliosphere: Cavity in the interstellar medium surrounding the solar system and dominated by the solar wind.

Inclination: Angle between the plane of the orbit of a planet, comet, or asteroid and the ecliptic plane, or between a satellite's orbit plane and the equatorial plane of its primary.

Jovian planet: Planet like Jupiter that is composed mostly of hydrogen, with helium and other gases, but possibly with a silicate/iron core; also called a gaseous planet. The Jovian planets are Jupiter, Saturn, Uranus, and Neptune.

Kuiper belt: Collection of some 10^9 to 10^{10} or more icy bodies in low-eccentricity, low-inclination orbits beyond Neptune, extending out possibly to about 10^3 AU.

Magnetosphere: Region of space around a planet or satellite that is dominated by its intrinsic magnetic field and associated charged particles.

Main sequence: When stars are plotted on a graph of their luminosity versus their surface temperature (or color), most stars fall along a line extending from high-luminosity, high-surface-temperature stars to low-luminosity, low-surface-temperature stars. This plot is known as the Hertzsprung–Russell diagram and the line is known as the "main sequence." Stars spend the majority of their lifetimes on the main sequence, during which they produce energy by hydrogen fusion occurring within their cores.

Meteoroid: Small fragment of an asteroid or comet that is in interplanetary space. When a meteoroid enters a planetary atmosphere and begins to glow from friction with the atmosphere, it is called a meteor. A fragment that survives atmospheric entry and can be recovered on the ground is called a meteorite.

Minor planet: Another term for an asteroid.

Oort cloud: Spherical cloud of some 10^{12} to 10^{13} comets surrounding the planetary system and extending out $\sim 10^5$ AU (0.5 parsec) from the Sun.

Orbit: Path of a planet, asteroid, or comet around the Sun, or of a satellite around its primary. Most bodies are in closed elliptical orbits. Some comets and asteroids are thrown into hyperbolic orbits, which are not closed, and so will escape the solar system.

Parallax: Apparent change in the position of a nearby star on the celestial sphere when measured from opposite sides of the Earth's orbit, usually given in seconds of arc.

Parsec: Distance at which a star would have a parallax of 1 second of arc, equal to 206,264.8 AU, or 3.261631 light-years; abbreviated as pc. One thousand parsecs are equal to a kiloparsec, which is abbreviated as kpc.

Perihelion: Point in the orbit of a planet, comet, or asteroid that is closest to the Sun.

Planet: Large body orbiting the Sun or another star, but not large enough to generate energy through nuclear fusion at its core. No formal definition of a planet exists and classifying exactly what is and is not a planet is often quite difficult. Some definitions demand that a planet should have an atmosphere, and/or a satellite, and/or be large enough to form itself into a sphere by self-gravity, and/or be able to gravitationally dominate its region of heliocentric space,

but there are counter examples to every one of these requirements.

Planetesimal: Small body formed in the early solar system by accretion of dust and ice (if present) in the central plane of the solar nebula.

Protostar: Star in the process of formation, which is luminous owing to the release of gravitational potential energy from the infall of nebula material.

Roche limit: Distance from a planet within which another body will be disrupted because tidal forces from the planet exceed the self-gravity of the smaller body. For nonrotating bodies of equal density and zero strength, the Roche limit is about 2.2 planetary radii.

Satellite: Body in orbit around a planet. A satellite was recently discovered orbiting an asteroid, and several other asteroid satellites are suspected to exist.

Secular perturbations: Long-term changes in the orbit of a body caused by the distant gravitational attraction of the planets and other bodies.

Semimajor axis: Half of the major axis of an elliptical orbit. Commonly taken to be the mean distance of the orbit of an object from its primary, though not precisely correct.

Solar nebula: Cloud of dust and gas out of which the Sun and planetary system formed.

Solar wind: Supersonic expansion of the Sun's outer atmosphere through interplanetary space.

Terrestrial planet: Planet like the Earth with an iron core and a silicate mantle and crust. The terrestrial planets are Mercury, Venus, Earth, and Mars.

Zodiacal cloud: Cloud of interplanetary dust in the solar system, lying close to the ecliptic plane. The dust in the zodiacal cloud comes from both comets and asteroids.

I. INTRODUCTION

The origins of modern astronomy lie with the study of our solar system. When ancient humans first gazed at the skies, they recognized the same patterns of fixed stars rotating over their heads each night. They identified these fixed patterns, now called constellations, with familiar objects or animals, or stories from their mythologies and their culture. But along with the fixed stars there were a few bright points of light that moved each night, slowly following similar paths through a belt of constellations around the sky. (The Sun and Moon also appeared to move through the same belt of constellations.) These wandering objects were the planets of our solar system. Indeed, the name "planet" derives from the Latin *planeta*, meaning "wanderer."

The ancients recognized five planets that they could see with their naked eyes. We now know that the solar system consists of nine planets (including the Earth), plus a myriad of smaller objects: satellites, rings, asteroids, comets, and dust. Discoveries of new objects, and new classes of objects, are continuing even today. Thus, our view of the solar system is constantly changing and evolving as new data and new theories to explain (or anticipate) the data become available.

The solar system we see today is the result of the complex interaction of physical, chemical, and dynamical processes that have shaped the planets and other bodies. By studying each of the planets and other bodies individually as well as collectively, we seek to gain an understanding of those processes and the steps that led to the current solar system. Many of those processes operated most intensely early in the solar system's history, as the Sun and planets formed from an interstellar cloud of dust and gas, 4.6 billion years ago. The first billion years of the solar system's history was a violent period as the planets cleared their orbital zones of much of the leftover debris from the process of planet formation, flinging small bodies into planet-crossing (and often planet-impacting) orbits or out to interstellar space. In comparison, the present-day solar system is a much quieter place, though all or most of these processes continue on a lesser scale today.

Our knowledge of the solar system has exploded in the past four decades as interplanetary exploration spacecraft have provided close-up views of all the planets except Pluto, as well as of a diverse collection of satellites, rings, asteroids, and comets. Earth-orbiting telescopes have provided an unprecedented view of the solar system, often at wavelengths not accessible from the Earth's surface. Ground-based observations have also continued to produce exciting new discoveries through the application of a variety of new technologies such as CCD (charge-coupled device) cameras, infrared detector arrays, adaptive optics, and powerful planetary radars. Theoretical studies have contributed significantly to our understanding of the solar system, largely through the use of advanced computer codes and high-speed, dedicated computers. Serendipity has also played an important role in many new discoveries.

Along with this increased knowledge have come numerous additional questions as we attempt to explain the complexity and diversity that we observe on each newly encountered world. The increased spatial and spectral resolution of the observations, along with *in situ* measurements of atmospheres, surface materials, and magnetospheres, has revealed that each body is unique, the result of the different combination of physical, chemical, and dynamical processes that formed and shaped it, as well as its different initial composition. Yet, even though each planet, satellite, and smaller object is now recognized to be very different from its neighbors, at the same time there are broad systematic trends and similarities that are clues to the collective history that the solar system has undergone.

We are also on the brink of an exciting new age of discovery with the detection of the first planet-sized bodies around nearby stars. Although the precise nature and origin of these extrasolar planets are still largely open questions, they are likely the prelude to the discovery of other planetary systems that may resemble our own.

A second astounding new discovery is the detection of possibly biogenic material in Martian meteorites (pieces of Mars rocks that were blasted off that planet by asteroid and/or comet impacts, and that have survived entry through the Earth's atmosphere). Although still very controversial, the detection of evidence of life evolving on a planet other than the Earth would suggest that life may also occur on other planets with the right physio-chemical resources and environment.

The goal of this chapter is to provide the reader with an introduction to the solar system. It seeks to provide a broad overview of the solar system and its constituent parts, to note the location of the solar system in the galaxy, and to describe the local galactic environment. Detailed discussion of each of the bodies that make up the solar system, as well as the processes that have shaped those bodies and the techniques for observing the planetary system, are provided in the following chapters of this Encyclopedia. The reader is referred to those chapters for more detailed discussions of each of the topics introduced here.

Some brief notes about planetary nomenclature will likely be useful. The names of the planets are all taken from Greek and Roman mythology (with the exception of Earth), as are the names of their satellites, with the exception of the Moon and the Uranian satellites, the latter being named after Shakespearean characters. The Earth is occasionally referred to as Terra, and the Moon as Luna, each the Latin version of their names. The naming system for planetary rings is different at each planet and includes descriptive names of the structures (at Jupiter), letters of the Roman alphabet (at Saturn), Greek letters and Arabic numerals (at Uranus), and the names of scientists associated with the discovery of Neptune (at Neptune).

Asteroids were initially named after Greek and Roman goddesses. As their numbers have increased, asteroids have been named after the family members of the discoverers, after observatories, universities, cities, provinces, historical figures, scientists, writers, artists, literary figures, and, in at least one case, the astronomer's cat. Initial discoveries of asteroids are designated by the year of their discovery and a letter code. Once the orbits of the asteroids are firmly established, they are given official numbers in the asteroid catalog; reliable orbits have been determined for about 8000 asteroids. The discoverer(s) of an asteroid are given the privilege of suggesting its name, if done so within 10 years from when it was officially numbered.

Comets are generally named for their discoverers, though in a few well-known cases, such as comets Halley and Encke, they are named for the individuals who first computed their orbits and linked several apparitions. Since some astronomers have discovered more than one short-period comet, a number is added at the end of the name to differentiate them, though this system is not applied to long-period comets. Comets are also designated by the year of their discovery and a letter code (a recently abandoned system used lowercase Roman letters and Roman numerals in place of the letter codes). The naming of newly discovered comets, asteroids, and satellites, as well as surface features on solar system bodies, is overseen by several commissions of the International Astronomical Union.

II. THE ARCHITECTURE OF THE SOLAR SYSTEM

The solar system consists of the Sun at its center, nine major planets, 63 known natural satellites (or moons), four ring systems, millions of asteroids (greater than 1 km in diameter), trillions of comets (greater than 1 km in diameter), the solar wind, and a large cloud of interplanetary dust. The arrangement and nature of all of these bodies are the result of physical and dynamical processes during their origin and subsequent evolution, and their complex interactions with one another. In studying the solar system, one of our primary goals is to understand those processes and to use that understanding to reconstruct the steps that led to the formation of the planetary system and its numerous components.

At the center of the solar system is the Sun, a rather ordinary main sequence star. The Sun is classified spectrally as a G2 dwarf, which means that it emits the bulk of its radiation in the visible region of the spectrum, peaking at yellow-green wavelengths. The Sun contains 99.85% of the mass in the solar system, but only about 0.5% of the angular momentum. The low angular momentum of the Sun results from the transfer of momentum to the accretion disk surrounding the Sun during the formation of the planetary system, and to a slow spin-down due to angular momentum being carried away by the solar wind.

The Sun is composed of hydrogen (75%), helium (23%), and heavier elements (2%). It produces energy through nuclear fusion at its center, with hydrogen atoms combining to form helium and releasing energy that eventually makes its way to the surface as visible sunlight. The central temperature of the Sun where fusion takes place is 15 million kelvins, whereas the temperature at the visible surface, the photosphere, is ~5800 K. The Sun has an outer atmosphere called the corona, which is visible only during solar eclipses, or through the use of specially designed telescopes called coronagraphs.

A star like the Sun is believed to have a typical lifetime of 9 to 10 billion years on the main sequence. The present age of the Sun (and the entire solar system) is estimated to be 4.6 billion years, so it is about halfway through its normal lifetime. The age estimate comes from radioisotope dating of meteorites.

A. DYNAMICS

The planets all orbit the Sun in roughly the same plane, known as the ecliptic (the plane of the Earth's orbit), and in the same direction, counterclockwise as viewed from the north ecliptic pole. Because of gravitational torques from the other planets, the ecliptic is not inertially fixed in space, and so dynamicists often use the invariable plane, which is the plane defined by the summed angular momentum vectors of all the planets.

To first order, the motion of any body about the Sun is governed by Kepler's laws of planetary motion. The laws of planetary motion are: (1) each planet moves about the Sun in an orbit that is an ellipse, with the Sun at one focus of the ellipse; (2) the straight line joining a planet and the Sun sweeps out equal areas in space in equal intervals of time; and (3) the squares of the sidereal periods of the planets are in direct propor-

TABLE I
Planetary Orbits[a]

Planet	Semimajor axis (AU)	Eccentricity	Inclination (°)	Period (years)
Mercury	0.38710	0.205631	7.0048	0.2408
Venus	0.72333	0.006773	3.3947	0.6152
Earth	1.00000	0.016710	0.0000	1.0000
Mars	1.52366	0.093412	1.8506	1.8807
Jupiter	5.20336	0.048393	1.3053	11.856
Saturn	9.53707	0.054151	2.4845	29.424
Uranus	19.1913	0.047168	0.7699	83.747
Neptune	30.0690	0.008586	1.7692	163.723
Pluto	39.4817	0.248808	17.1417	248.02

[a] J2000, Epoch: January 1, 2000

tion to the cubes of the semimajor axes of their orbits. The laws of planetary motion, first set down by J. Kepler in 1609 and 1619, are easily shown to be the result of the inverse-square law of gravity with the Sun as the central body, and the conservation of angular momentum and energy. Parameters for the orbits of the nine planets are listed in Table I.

Because the planets themselves have finite masses, they exert small gravitational tugs on one another, which cause their orbits to depart from perfect ellipses. The major effects of these long-term or "secular" perturbations are to cause the perihelion point of each orbit to precess (rotate counterclockwise) in space, and the line of nodes (the intersection between the planet's orbital plane and the ecliptic plane) of each orbit to regress (rotate clockwise). Additional effects include slow oscillations in the eccentricity and inclination of each orbit, and the inclination of the planet's rotation pole to the planet's orbit plane (called the obliquity). For the Earth, these orbital oscillations have periods of 19,000 to 100,000 years. They have been identified with long-term variations in the Earth's climate, known as Milankovitch cycles, though the linking physical mechanism is not well understood.

Relativistic effects also play a small but detectable role. They are most evident in the precession of the perihelion of the orbit of Mercury, the planet deepest in the Sun's gravitational potential well. General relativistic effects add 43 arc-seconds per century to the precession rate of Mercury's orbit, which is 574 arc-sec per century. Prior to Einstein's statement of general relativity in 1916, it was thought that the excess in the precession rate of Mercury was due to a planet orbiting interior to it. This hypothetical planet was given the name Vulcan and extensive searches were conducted for it, primarily during solar eclipses. No planet was detected.

A more successful search for a new planet occurred in 1846. Two celestial mechanicians, J. C. Adams and U. J. J. Leverrier, independently used the observed deviations of Uranus from its predicted orbit to successfully predict the existence and position of Neptune. Neptune was found by J. G. Galle on September 23, 1846, using Leverrier's prediction.

More complex dynamical interactions are also possible, in particular when the orbital period of one body is a small-integer ratio of another's orbital period. This is known as a "mean motion resonance" and can have dramatic effects. For example, Pluto is locked in a 2:3 mean motion resonance with Neptune, and although the orbits of the two planets cross in space, the resonance prevents them from ever coming within 14 AU of each other. Also, when two bodies have identical perihelion precession rates or nodal regression rates, they are said to be in a "secular resonance," and similarly interesting dynamical effects can result. In many cases, mean motion and secular resonances can lead to chaotic motion, driving a body into a planet-crossing orbit, which will then lead to it being dynamically scattered among the planets, and eventually either ejected from the solar system or impacted on the Sun or a planet.

Chaos has become a very exciting topic in solar system dynamics in the past twenty years, and has been able to explain many features of the planetary system that were not previously understood. It should be noted that the dynamical definition of chaos is not always the same as the general dictionary definition. In celestial mechanics the term chaos is applied to describe systems that are not perfectly predictable over time. That is, small variations in the initial conditions, or the inability to specify the initial conditions precisely, will lead to a growing error in predictions of the long-term behavior of the system. If the error grows exponentially, then the system is said to be chaotic. However, the chaotic zone, the allowed area in phase space over which an orbit may vary, may still be quite constrained. Thus, although studies have found that the orbits of the planets are chaotic, this does not mean that Jupiter may one day become Earth-crossing, or vice versa. It means that the precise position of the Earth or Jupiter in its orbit is not predictable over very long periods of time. Since this happens for all the planets, then the long-term secular perturbations of the planets on one another are also not perfectly predictable, and can vary.

On the other hand, chaos can result in some extreme changes in orbits, with sudden increases in eccentricity that can throw small bodies onto planet-crossing orbits. One well-recognized case of this occurs near mean motion resonances in the asteroid belt, which causes small asteroids to be thrown onto Earth-crossing orbits, allowing for the delivery of meteoroids to the Earth.

The natural satellites of the planets and their ring systems (where they exist) are governed by the same dynamical laws of motion. Most satellites and all ring systems are deep within their planets' gravitational potential wells and so they move, to first order, on Keplerian ellipses. The Sun, planets, and other satellites all act as perturbers on the satellite orbits. Additionally, the equatorial bulge of the planet, caused by the planet's rotation, also acts as a perturber on the satellite and ring particle orbits. Finally, the satellites raise tides on the planets (and vice versa) and these result in yet another dynamical evolution, causing the planets to transfer rotational angular momentum to the satellite orbits (in the case of direct, or prograde orbits; satellites in retrograde orbits lose angular momentum). As a result, satellites may slowly move away from their planets into larger orbits (or smaller ones in the case of retrograde motion).

The mutual gravitational interactions can be quite complex, particularly in multi-satellite systems. For example, the three innermost Galilean satellites (so named because they were discovered by Galileo in 1610)—Io, Europa, and Ganymede—are locked in a 4 : 2 : 1 mean motion resonance with one another. In other words, Ganymede's orbital period is twice that of Europa and four times that of Io. At the same time, the other Jovian satellites (primarily Callisto), the Sun,

TABLE II
Bode's Law
$[a_1 = 4/10, \; a_n = (3 \times 2^{n-2} + 4)/10]$

Planet	Semimajor axis (AU)	n	Bode's law
Mercury	0.387	1	0.4
Venus	0.723	2	0.7
Earth	1.000	3	1.0
Mars	1.524	4	1.6
Ceres	2.767	5	2.8
Jupiter	5.203	6	5.2
Saturn	9.537	7	10.0
Uranus	19.19	8	19.6
Neptune	30.07	9	38.8
Pluto	39.48	10	77.2

and Jupiter's oblateness perturb the orbits, forcing them to be slightly eccentric and inclined to one another, while the tidal interaction with Jupiter forces the orbits to evolve outward. These competing dynamical processes result in considerable energy deposition in the satellites, which manifests itself as volcanic activity on Io, as a possible subsurface ocean on Europa, and as past tectonic activity on Ganymede.

This last example illustrates a very important point in understanding the solar system. The bodies in the solar system do not exist as independent, isolated entities, with no physical interactions between them. Even these "action at a distance" gravitational interactions can lead to profound physical and chemical changes in the bodies involved. To understand the solar system as a whole, one must recognize and understand the processes that were involved in its formation and its subsequent evolution, and that continue to act even today.

An interesting feature of the planetary orbits is their regular spacing. This is described by Bode's Law, first discovered by J. B. Titius in 1766 and brought to prominence by J. E. Bode in 1772. The law states that the semimajor axes of the planets in astronomical units can be roughly approximated by taking the sequence 0, 3, 6, 12, 24, ..., adding 4, and dividing by 10. The values for Bode's Law and the actual semimajor axes of the planets are listed in Table II. It can be seen that the law works very well for the planets as far as Uranus, but then breaks down. It also predicts a planet between Mars and Jupiter, the current location of the asteroid belt. Yet Bode's Law predates the discovery of the first asteroid by 35 years, as well as the discovery of Uranus by 15 years.

The reason why Bode's Law works so well is not understood. It appears to reflect the increasing ranges of gravitational dominance of successive planets at increasing heliocentric distances. However, it has been argued that Bode's Law may just be a case of numerology and not reflect any real physical principle at all. Computer-based dynamical simulations have shown that the spacing of the planets is such that a body placed on a circular orbit between any pair of neighboring planets will likely be dynamically unstable. It will not survive over the history of the solar system unless protected by some dynamical mechanism such as a mean motion resonance with one of the planets. Over the history of the solar system, the planets have generally cleared their zones of smaller bodies through gravitational scattering. The larger planets, in particular Jupiter and Saturn, are capable of throwing small bodies onto hyperbolic orbits, which are unbound, allowing the objects to escape to interstellar space.

Thus, the comets and asteroids we now see in planet-crossing orbits must have been introduced into the planetary system relatively recently from storage locations either outside the planetary system or from protected, dynamically stable reservoirs. Because of its position at one of the Bode's law locations, the asteroid belt is a relatively stable reservoir. However, the asteroid belt's proximity to Jupiter's substantial gravitational influence results in some highly complex dynamics. Mean motion and secular resonances, as well as mutual collisions, act to remove objects from the asteroid belt and throw them into planet-crossing orbits. The failure of a major planet to grow in the asteroid belt is generally attributed to the gravitational effects of Jupiter disrupting the slow growth by accretion of a planetary-sized body in the neighboring asteroid belt region.

It is generally believed that comets originated as icy planetesimals in the outer regions of the solar nebula, at the orbit of Jupiter and beyond. Those proto-comets with orbits between the giant planets were gravitationally ejected, mostly to interstellar space. However, a fraction of the proto-comets were flung into distant but still bound orbits—the Sun's gravitational sphere of influence extends about 2×10^5 AU, or about 1 parsec. These orbits were sufficiently distant from the Sun that they were perturbed by random passing stars and by the tidal perturbation from the galactic disk. The stellar and galactic perturbations raised the perihelia of the comet orbits out of the planetary region. Additionally, the stellar perturbations randomized the inclinations of the comet orbits, forming a spherical cloud of comets around the planetary system and extending halfway to the nearest stars. This region is now called the Oort cloud, after J. H. Oort, who first suggested its existence in 1950. The current population of the Oort cloud is estimated at between 10^{12} and 10^{13} comets, with a total mass of about 40 Earth masses of material. About 80% of the Oort cloud population is in a dense core within $\sim 10^4$ AU of the Sun. Long-period comets (those with orbital periods greater than 200 years) observed passing through the planetary region come from the Oort cloud. Some of the short-period comets (those with orbital periods less than 200 years), such as Comet Halley, are long-period comets that have evolved to short-period orbits owing to repeated planetary perturbations.

A second reservoir of comets is the Kuiper belt beyond the orbit of Neptune, named after G. P. Kuiper, who in 1951 was one of the first to suggest its existence. Because no large planet grew beyond Neptune, there was no body to scatter away the icy planetesimals formed in that region. (The failure of a large planet to grow beyond Neptune is generally attributed to the increasing timescale for planetary accretion with increasing heliocentric distance.) This belt of remnant planetesimals may extend out several hundred AU from the Sun, perhaps even 10^3 AU, analogous to the disks of dust that have been discovered around main sequence stars such as Vega and Beta Pictoris (Fig. 1).

The Kuiper belt may contain many tens of Earth masses of comets. A slow gravitational erosion of comets from the Kuiper belt between 30 and 50 AU, due to the perturbing effect of Neptune, causes these comets to "leak" into the planetary region. Eventually some fraction of the comets evolve because of gravitational scattering by the Jovian planets into the inner planets region, where they can be observed as short-period comets. Short-period comets from the Kuiper belt are often called "Jupiter-family" or "ecliptic" comets, because most are in orbits that can have close encounters with Jupiter, and also are in orbits with inclinations close to the ecliptic plane. Based on the observed number of ecliptic comets, the number of comets in the Kuiper belt between 30 and 50 AU has been estimated at about 7×10^9 objects, with a total mass of about 0.1 Earth masses. Current studies suggest that the Kuiper belt has been collisionally eroded out to a distance of ~ 100 AU from the Sun, but that considerably more mass may still exist in orbits beyond that distance.

Although gravity is the dominant force in determining the motion of bodies in the solar system, other forces do come into play in special cases. Dust grains

FIGURE 1 Coronagraphic image of the dust disk around the star Beta Pictoris, discovered by the *IRAS* satellite in 1983. The disk is viewed nearly edge on and extends ~900 AU on either side of the star. The occulting disk at the center blocks out the view of the central star and of the disk within ~150 AU of the star. Infrared data show that the disk does not extend all the way in to the star, but has an inner edge at about 30 AU from Beta Pictoris. The disk interior to that distance may have been swept up by the accretion of planets in the nebula around the star. This disk is a likely analog for the Kuiper belt around our own solar system.

produced by asteroid collisions or liberated from the sublimating icy surfaces of comets are small enough to also be affected by radiation pressure forces. For submicron grains, radiation pressure is sufficient to blow the grains out of the solar system. For larger grains, radiation pressure causes the grains to depart from Keplerian orbits. Radiation pressure can also cause larger grains to spiral slowly in toward the Sun through two different mechanisms, known as the Poynting–Robertson and Yarkovsky effects.

Electromagnetic forces play a role in planetary magnetospheres where ions are trapped and spiral back and forth along magnetic field lines, and in cometary Type I plasma tails where ions are accelerated away from the cometary coma, achieving fairly high energies. Dust grains trapped in planetary magnetospheres and in interplanetary space also respond to electromag-

netic forces, though to a lesser extent than ions because of their much lower charge-to-mass ratios.

B. NATURE AND COMPOSITION

The solar nebula, the cloud of dust and gas out of which the planetary system formed, almost certainly exhibited a strong temperature gradient with heliocentric distance, hottest near the forming proto-Sun at its center and cooling as one moved outward through the planetary region. This temperature gradient is reflected in the compositional arrangement of the planets and their satellites versus heliocentric distance. Parts of the gradient are also preserved in the asteroid belt beween Mars and Jupiter and likely in the Kuiper belt beyond Neptune.

TABLE III
Physical Parameters for the Sun and Planets

Name	Mass (kg)	Equatorial radius (km)	Density (g cm^{-3})	Rotation period	Obliquity (°)	Escape velocity (km sec^{-1})
Sun	1.989×10^{30}	696,000	1.41	24.65–34 days	7.25[a]	617.7
Mercury	3.302×10^{23}	2,439	5.43	58.646 days	0	4.43
Venus	4.868×10^{24}	6,051	5.20	243.018 days	177.33	10.36
Earth	5.974×10^{24}	6,378	5.52	23.934 hr	23.45	11.19
Mars	6.418×10^{23}	3,396	3.93	24.623 hr	25.19	5.03
Jupiter	1.899×10^{27}	71,492	1.33	9.925 hr	3.08	59.54
Saturn	5.685×10^{26}	60,268	0.69	10.656 hr	26.73	35.49
Uranus	8.683×10^{25}	25,559	1.32	17.24 hr	97.92	21.33
Neptune	1.024×10^{26}	24,764	1.64	16.11 hr	28.80	23.61
Pluto	1.32×10^{22}	1,170	2.1	6.387 days	119.6	1.25

[a] Solar obliquity relative to the ecliptic.

The planets fall into two major compositional groups (Table III). The "terrestrial" or Earth-like planets are Mercury, Venus, Earth, and Mars, and are shown in Fig. 2. The terrestrial planets are characterized by predominantly silicate compositions with iron cores. This appears to result from the fact that they all formed close to the Sun, where it was too warm for ices to condense. Also, the modest masses of the terrestrial planets and their closeness to the Sun did not allow them to capture and retain hydrogen and helium directly from the solar nebula. The terrestrial planets all have solid surfaces that are modified to varying degrees by both cratering and internal processes (tectonics, weather, etc.). Mercury is the most heavily cratered because it has no appreciable atmosphere to protect it from impacts or weather to erode the cratered terrain, and also because encounter velocities with Mercury are very high that close to the Sun. Additionally, tectonic processes on Mercury appear to have been modest at best. Mars is next in degree of cratering, in large part because of its proximity to the asteroid belt. Also, Mars's thin atmosphere affords little protection against impactors. However, Mars also displays substantial volcanic and tectonic features, and evidence of erosion by wind and flowing water, the latter presumably having occurred early in the planet's history.

The surface of Venus is dominated by a wide variety of volcanic terrains. The degree of cratering on Venus is less than that on Mercury or Mars for two reasons: (1) Venus's thick atmosphere (surface pressure = 94 bars) breaks up smaller asteroids and comets before they can reach the surface and (2) vulcanism on the planet has covered over the older craters on the planet. The surface of Venus is estimated to be 600–800 million years in age. The Earth's surface is dominated by plate tectonics, in which large plates of the crust can move about the planet, and whose motions are reflected in features such as mountain ranges (where plates collide) and volcanic zones (where one plate dives under another). The Earth is the only planet with the right combination of atmospheric surface pressure and temperature to permit liquid water on its surface, and some 70% of the planet is covered by oceans. Craters on the Earth are rapidly erased by its active geology and weather, though the atmosphere provides protection only against very modest size impactors, on the order of 100 m diameter or less. Still, some 140 impact craters or their remnants have been found on the Earth's surface or under its oceans.

The terrestrial planets each have substantially different atmospheres. Mercury has a tenuous atmosphere arising from its interaction with the solar wind. Hydrogen and helium ions are captured directly from the solar wind, whereas oxygen, sodium, and potassium are likely the product of sputtering. In contrast, Venus has a dense CO_2 atmosphere with a surface pressure 94 times the pressure at the Earth's surface. Nitrogen is also present in the Venus atmosphere at a few percent relative to CO_2. The dense atmosphere results in a massive greenhouse on the planet, heating the surface to a mean temperature of 735 K. The middle and upper atmosphere contain thick clouds composed of H_2SO_4 and H_2O, which shroud the surface from view. However, it was recently discovered that thermal radia-

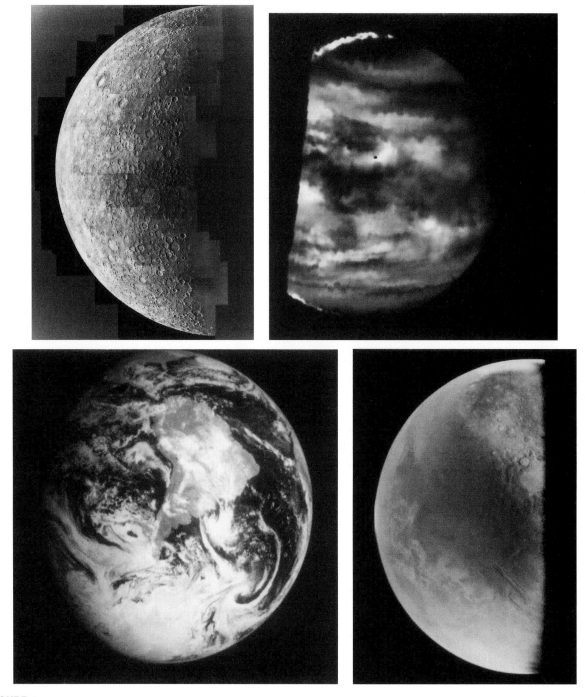

FIGURE 2   The terrestrial planets: the heavily cratered surface of Mercury as photographed by the *Mariner 10* spacecraft in 1974 (top left); clouds on the nightside of Venus, backlit by the intense infrared radiation from the planet's hot surface, as imaged by the *Galileo* NIMS instrument in 1990 (top right); South America and Antarctica as imaged by the *Galileo* spacecraft during a gravity assist flyby of the Earth in 1990 (bottom left); cratered and volcanic terrains on Mars, as photographed by the *Viking 1* spacecraft during its approach to the planet in 1976 (bottom right).

tion from the surface does penetrate the clouds, making it possible to view surface features through these infrared "windows."

The Earth's atmosphere is unique because of its large abundance of free oxygen, which is normally tied up in oxidized surface materials on other planets. The reason for this unusual state is the presence of life on the planet, which traps and buries CO_2 as carbonates and also converts the CO_2 to free oxygen. Still, the bulk of the Earth's atmosphere is nitrogen, 78%, with oxygen making up 21% and argon and water each about 1%. Various lines of evidence suggest that the composition of the Earth's atmosphere has evolved considerably over the history of the solar system, and that the original atmosphere was denser and had a much higher CO_2 content than the present-day atmosphere. Mars has a relatively modest CO_2 atmosphere with a mean surface pressure of only 6 millibars. The atmosphere also contains a few percent of N_2 and argon. Isotopic evidence and geologic features suggest that the past atmosphere of Mars may have been much denser and warmer, allowing liquid water to flow across the surface in massive floods.

The volatiles in the terrestrial planets' atmospheres (and the Earth's oceans) may have been contained in hydrated minerals in the planetesimals that originally formed the planets, and/or may have been added later from asteroid and comet bombardment as the planets dynamically cleared their individual zones of leftover planetesimals. It appears most likely that all of these reservoirs contributed some fraction of the volatiles on the planets.

The "Jovian" or Jupiter-like planets are Jupiter, Saturn, Uranus, and Neptune, and are shown in Fig. 3. The Jovian planets are also occasionally referred to as the "gas giants." They are characterized by low mean densities and thick hydrogen–helium atmospheres, presumably captured directly from the solar nebula during the formation of these planets. The composition of the Jovian planets is similar to that of the Sun, though more enriched in heavier elements. Because of this primarily gaseous composition and their high internal temperatures and pressures, the Jovian planets do not have solid surfaces. However, they may each have silicate–iron cores of several to tens of Earth masses of material at their centers.

The satellites of the Jovian planets are mostly icy bodies, predominantly water ice, with a few exceptions. One notable exception is Jupiter's innermost Galilean satellite, Io. However, Io has been heated tremendously over the history of the solar system by the tidal interaction noted in the previous section, and this can likely account for the loss of its volatile ices.

Because they formed at heliocentric distances where ices could condense, the giant planets may have initially had a much greater local density of solid material to grow from. This may, in fact, have allowed them to form ahead of the terrestrial planets interior to them. Studies of the dissipation of nebula dust disks around nearby solar-type protostars suggest that the timescale for the formation of giant planets is on the order of 10 million years or less. This is very rapid as compared with the ~100 million-year timescale currently estimated for the formation of the terrestrial planets (though questions have now been raised as to the correctness of that accretionary timescale). Additionally, the higher uncompressed densities of Uranus and Neptune (0.5 g cm^{-3}) versus Jupiter and Saturn (0.3 g cm^{-3}) suggest that the outer two giant planets contain a significantly lower fraction of gas captured from the nebula. This may mean that the outer pair formed later than the inner two giant planets, consistent with the increasing timescale for planetary accretion at larger heliocentric distances.

Because of their heliocentric arrangement, the terrestrial and Jovian planets are occasionally called the inner and outer planets, respectively, though sometimes the term inner planets is used to denote only Mercury and Venus, the planets interior to the Earth's orbit.

Pluto is an outlier to the system and is not easily classifiable as either a terrestrial or a Jovian planet. Rather, it bears the greatest resemblance to Triton, Neptune's large icy satellite that is slightly larger than Pluto, and to the icy planetesimals remaining in the Kuiper belt beyond the orbit of Neptune. (The retrograde orbit of Triton, the only one for a major satellite in the solar system, suggests that it may be an icy planetesimal that was captured from heliocentric space, and not formed coevally with Neptune in orbit around that planet.) Note also that Pluto does not readily fit into Bode's Law (see Table II). For these and other reasons, the designation of Pluto as a planet is often debated, and there are strong arguments both for and against the issue. As noted in the Glossary, the definition of a planet is an empirical one and often depends on the viewpoint of the observer. Because Pluto resides in the Kuiper belt, it is probably best thought of as the largest icy planetesimal to grow in that region of heliocentric space, rather than as a true planet. Pluto and its satellite Charon are shown in Fig. 4.

Pluto has a thin, extended atmosphere, probably methane and nitrogen, which is slowly escaping because of the low gravity of the planet. This puts it in a somewhat intermediate state between a freely out-flowing cometary coma and a bound planetary atmo-

FIGURE 3   The Jovian planets: the complex, belted atmosphere of Jupiter with the Giant Red Spot at the lower center, as photographed by *Voyager 1* during its approach in 1979 (top left); Saturn, its beautiful ring system, and three of its satellites, as photographed by *Voyager 1* in 1980 (top right); the featureless atmosphere of Uranus, obscured by a high-altitude methane haze, as imaged by *Voyager 2* in 1986 (bottom left); Neptune's atmosphere displays several large storm systems and a banded structure, similar to Jupiter, as photographed by *Voyager 2* in 1989 (bottom right).

sphere. Spectroscopic evidence suggests that methane frost covers much of the surface of Pluto, whereas its satellite Charon appears to be covered with water frost. Nitrogen frost has also been detected on Pluto. The density of Pluto is ~ 2 g cm^{-3}, suggesting that the rocky component of the planet accounts for about 60–70% of its total mass.

There has been considerable speculation on the possibility of a major planet beyond Pluto, often dubbed "Planet X." The search program that discovered Pluto in 1930 was continued for many years afterward but failed to detect any other distant planet, even though

the limiting magnitude was several times fainter than Pluto's visual magnitude of ~ 13.5. Other searches have been carried out, most notably by the *Infrared Astronomical Satellite* in 1983–1984. An automated algorithm was used to search for a distant planet in the *IRAS* data; it successfully "discovered" Neptune, but nothing else. Analyses of the orbits of Uranus and Neptune show no evidence of an additional perturber at greater heliocentric distances. Studies of the trajectories of the *Pioneer 10* and *11* and *Voyager 1* and *2* spacecraft have also yielded negative results. Analyses of the spacecraft trajectories allow one to set an upper

FIGURE 4 Pluto and its satellite Charon, as photographed by the *Hubble Space Telescope*. Pluto is the only planet that has not been imaged by a close spacecraft encounter.

limit on the unaccounted for mass within the orbit of Neptune of less than 3×10^{-6} solar masses ($M_\odot$), equal to about one Earth mass.

The compositional gradient in the solar system is perhaps best visible in the asteroid belt, whose members range from nickel–iron bodies in the inner belt, presumably the differentiated cores of larger asteroids that were subsequently disrupted by collisions, to volatile-rich carbonaceous bodies in the outer belt, which have never been melted or differentiated (Fig. 5). Thermally processed asteroids, including bodies like Vesta whose surface material resembles a basaltic lava flow, dominate the inner portion of the asteroid belt, at distances less than about 2.6 AU. At larger distances, out to the outer boundary of the main belt at about 3.3 AU, volatile-rich carbonaceous asteroids are dominant. The thermal gradient that processed the asteroids appears to be very steep and likely cannot be explained simply by the individual distances of these bodies from the forming proto-Sun. Rather, various special mechanisms such as magnetic induction, short-lived radioisotopes, or extreme solar flares have been invoked to try to explain the heating event that so strongly processed the inner half of the asteroid belt.

The largest asteroid is Ceres, at a mean distance of 2.77 AU from the Sun. (Note that Bode's law predicts a planet at 2.8 AU.) Ceres was the first asteroid discovered, by G. Piazzi on January 1, 1801. Ceres is 913 km in diameter, rotates in 9.08 hours, and appears to have a surface composition similar to that of carbonaceous chondrite meteorites. The second largest asteroid is Pallas, also a carbonaceous type with a diameter of 523 km. Pallas is also at 2.77 AU but its orbit has an unusually large inclination of 34.8°. Over 8000 asteroids have had their orbits accurately determined and have been given official numbers in the asteroid catalog; on the order of another 10^4 asteroids have been observed and have had preliminary orbits determined.

As a result of the large number of objects in the

asteroid belt, impacts and collisions are frequent. Several "families" of asteroids have been identified by their closely grouped orbital elements and are likely fragments of larger asteroids that collided. Spectroscopic studies have shown that the members of these families often have very similar surface compositions, further evidence that they are related. The largest asteroids, such as Ceres and Pallas, are likely too large to be disrupted by impacts, but most of the smaller asteroids have probably been collisionally processed. Increasing evidence suggests that many asteroids may be "rubble piles," that is, asteroids that have been broken up but not dispersed by previous collisions, and that now form a single but poorly consolidated body.

Beyond the main asteroid belt there exist small groups of asteroids locked in dynamical resonances with Jupiter. These include the Hildas at the 3 : 2 mean motion resonance, the Thule group at the 4 : 3 resonance, and the Trojans, which are in a 1 : 1 mean motion resonance with Jupiter. The effect of the resonances is to prevent these asteroids from making close approaches to Jupiter, even though many of the asteroids are in Jupiter-crossing orbits.

The Trojans are particularly interesting. They are essentially in the same orbit as Jupiter but they librate about points 60° ahead and 60° behind the planet in its orbit, known as the Lagrange L_4 and L_5 points. These are pseudostable points in the three-body problem (Sun–Jupiter–asteroid) where bodies can remain dynamically stable for extended periods of time. Some estimates have placed the total number of objects in the Jupiter L_4 and L_5 Trojan swarms as equivalent to the population of the main asteroid belt. Trojan-type 1 : 1 librators have also been found for Mars and the Earth (one each) and have been searched for at the L_4 and L_5 points of the other giant planets, though none has been detected. It is interesting that the Saturnian satellite Enceladus has two smaller satellites locked in Trojan-type librations in its orbit.

Much of what we know about the asteroid belt and the early history of the solar system comes from meteorites recovered on the Earth. It appears that the asteroid belt is the source of almost all recovered meteorites. A modest number of meteorites have been found that are from the Moon and from Mars, presumably blasted off of those bodies by asteroid and/or comet impacts. Cometary meteoroids are thought to be too fragile to survive atmospheric entry. In addition, cometary meteoroids typically encounter the Earth at higher velocities than asteroidal debris and thus are more likely to be fragmented and burned up during atmospheric entry. However, we may have cometary mete-

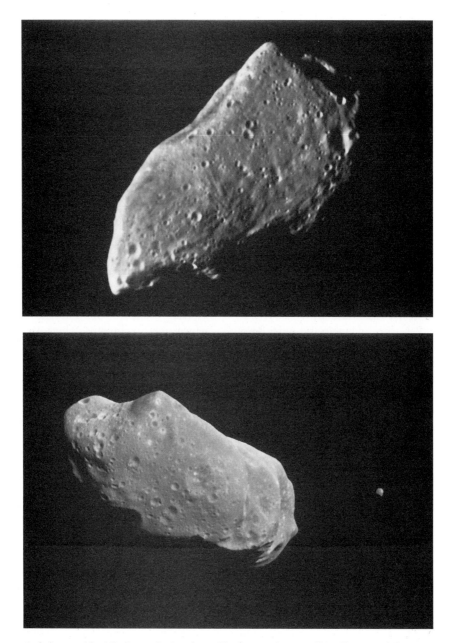

FIGURE 5   Three main belt asteroids: 951 Gaspra (top) and 243 Ida along with its small satellite Dactyl (bottom). All three asteroids are stony types and all exhibit heavily cratered surfaces. Gaspra is about 18 × 10 × 9 km in diameter, Ida is 54 × 24 × 15 km, and Dactyl is about 1.5 km in diameter. The asteroids were photographed by the *Galileo* spacecraft while it was en route to Jupiter, in 1991 and 1993, respectively. Ida's tiny satellite, Dactyl, was an unexpected discovery of two of *Galileo's* remote-sensing instruments, the Near Infrared Mapping Spectrometer and the Solid State Imaging system, during the flyby.

orites in our sample collections and simply not yet be knowledgeable enough to recognize them.

Recovered meteorites are roughly equally split between silicate and carbonaceous types, with a few percent being iron–nickel meteorites. The most primitive meteorites, that is, those that appear to show the least processing in the solar nebula, are the volatile-rich carbonaceous chondrites. However, even these mete-

orites show evidence of some thermal processing and of aqueous alteration, that is, processing in the presence of liquid water. Study of carbonaceous and ordinary (silicate) chondrites provides significant information on the composition of the original solar nebula, on the physical and chemical processes operating in the solar nebula, and on the chronology of the early solar system.

The other major group of primitive bodies in the solar system is the comets. Because comets formed farther from the Sun than the asteroids, in colder environments, they contain a significant fraction of volatile ices. Water ice is the dominant and most stable volatile. Typical comets also contain modest amounts of CO, CO_2, CH_4, NH_3, H_2CO, and CH_3OH, most likely in the form of ices, but possibly also contained within complex organic molecules. Organics make up a significant fraction of the cometary nucleus, as well as silicate grains. The American astronomer Fred Whipple described this icy-conglomerate mix as "dirty snowballs" (though the term "frozen mudball" may be more appropriate since the comets are more than 60% organics and silicates). It appears that the composition of comets is very similar to the condensed (solid) grains observed in dense interstellar cloud cores, with little or no evidence of processing in the solar nebula. Thus comets appear to be the most primitive bodies in the solar system. As a result, the study of comets is extremely valuable for those interested in learning about the origin of the planetary system and the conditions in the solar nebula 4.6 billion years ago.

Only one cometary nucleus, that of comet Halley, has been encountered by interplanetary spacecraft and imaged (Fig. 6). The nucleus was seen to be a highly irregular body, with dimensions of about 15 × 8 × 7 km. It has been suggested that cometary nuclei are weakly bound conglomerations of smaller dirty snowballs, assembled at low velocity and low temperature in the outer regions of the solar nebula. Thus, comets

FIGURE 6 The nucleus of Halley's comet, as photographed by the *Giotto* spacecraft in 1986. The nucleus is irregularly shaped with dimensions of 15 × 8 × 7 km. Jets of dust and gas are being emitted from active areas on the sunlit surface of the nucleus at left. (Copyright 1986 H. U. Keller, Max-Planck Institute for Aeronomie.)

may be "primordial rubble piles," in some ways similar to the asteroids, but with the difference that the "rubble" is primordially accreted macroscopic bodies in the solar nebula, rather than collisionally produced debris. A typical cometary nucleus is a few to ten kilometers in diameter.

Subtle and not-so-subtle differences in cometary compositions have been observed. However, it is not entirely clear if many of these differences are intrinsic or due to the physical evolution of cometary surfaces over many close approaches to the Sun. Because the comets that originated among the giant planets have all been ejected to the Oort cloud or to interstellar space, the compositional spectrum resulting from the heliocentric thermal profile is not spatially preserved as it has been in the asteroid belt. Although comets in the Kuiper belt are likely located close to their formation distances, physical studies of these distant objects are only just beginning. The data are not sufficient to reveal any compositional trends at the present time.

C. SATELLITES, RINGS, AND THINGS

The natural satellites of the planets, listed in Table IV, show as much diversity as the planets they orbit around. Among the terrestrial planets, the only known satellites are the Earth's Moon and the two small moons of Mars, Phobos and Deimos. The Earth's Moon is unusual in that it is so large relative to its primary (only Pluto's moon Charon is larger relative to its planet). The Moon has a silicate composition similar to the Earth's mantle and a very small iron core (Fig. 7).

It is now widely believed that the Moon formed as a result of a collision between the proto-Earth and another protoplanet about the size of Mars, late in the accretion of the terrestrial planets. Such "giant impacts" are now recognized as being capable of explaining many of the features of the solar system, such as the unusually high density of Mercury and the large obliquities of several of the planetary rotation axes. In the case of the Earth, the collision with another protoplanet resulted in the cores of the two planets merging, while a substantial fraction of the mantles of both bodies was thrown into orbit around the Earth where some of the material reaccreted to form the Moon. The tidal interaction between the Earth and Moon then slowly evolved the orbit of the Moon outward to its present position, at the same time slowing the rotation of both the Earth and the Moon. The giant impacts hypothesis is capable of explaining many of the features of the Earth–Moon system, including

TABLE IV
Orbital and Physical Parameters of Planetary Satellites

Name	Semimajor axis (10^3 km)	Orbital eccentricity	Orbital inclination (°)	Orbital period (days)	Mean radius (km)
Moon	384.40	0.0549	18–29	27.3216	1,737.4
Phobos	9.38	0.0151	1.08	0.319	$13 \times 11 \times 9.2$
Deimos	23.46	0.0003	1.79	1.262	$7.5 \times 6.1 \times 5.2$
J16 Metis	128.0	0.0	0.0	0.295	20
J15 Adrastea	129.0	0.0	0.0	0.298	10
J5 Amalthea	181.3	0.003	0.45	0.498	$131 \times 73 \times 67$
J14 Thebe	221.9	0.015	0.8	0.674	50
J1 Io	421.6	0.004	0.04	1.769	1,818
J2 Europa	670.9	0.010	0.47	3.552	1,560
J3 Ganymede	1,070	0.002	0.21	7.154	2,634
J4 Callisto	1,883	0.007	0.51	16.69	2,409
J13 Leda	11,094	0.148	26.70	238.7	5
J6 Himalia	11,480	0.163	27.63	250.6	85
J10 Lysithea	11,720	0.107	29.02	259.2	12
J7 Elara	11,737	0.207	24.77	259.6	40
J12 Ananke	21,200	0.169	147	631	10
J11 Carme	22,600	0.207	163	692	15
J8 Pasiphae	23,500	0.378	145	735	18
J9 Sinope	23,700	0.275	153	758	14
S18 Pan	133.6	0.0	0.0	0.575	10
S15 Atlas	137.6	0.0	0.0	0.602	$19 \times 17 \times 14$
S16 Prometheus	139.3	0.002	0.0	0.613	$74 \times 50 \times 34$
S17 Pandora	141.7	0.004	0.05	0.629	$55 \times 44 \times 31$
S11 Epimetheus	151.4	0.009	0.14	0.695	$69 \times 55 \times 55$
S10 Janus	151.5	0.007	0.34	0.695	$97 \times 95 \times 77$
S1 Mimas	185.5	0.020	1.53	0.942	199
S2 Enceladus	238.0	0.004	0.0	1.370	249
S3 Tethys	294.7	0.000	1.0	1.888	530
S14 Calypso	294.7	0.0	1.10	1.888	$15 \times 8 \times 8$
S13 Telesto	294.7	0.0	1.0	1.888	$15 \times 12 \times 8$
S4 Dione	377.4	0.002	0.02	2.737	560
S12 Helene	377.4	0.005	0.15	2.737	16
S5 Rhea	527.0	0.001	0.35	4.518	764
S6 Titan	1,222	0.029	0.33	15.945	2,575
S7 Hyperion	1,481	0.104	0.4	21.277	$180 \times 140 \times 112$
S8 Iapetus	3,561	0.028	14.72	79.330	718
S9 Phoebe	12,952	0.163	150	550.48	110

continues

the similarity in composition between the Moon and the Earth's mantle, the lack of a significant iron core within the Moon, and the high angular momentum of the Earth–Moon system.

Like most natural satellites, the Moon has tidally evolved to where its rotation period matches its revolution period in its orbit. This is known as "synchronous rotation." It results in the Moon showing the same face to the Earth at all times, though there are small departures from this because of the eccentricity of the Moon's orbit.

The Moon's surface displays a record of the intense bombardment that all the planets have undergone over the history of the solar system. Returned lunar samples have been age-dated based on decay of long-lived radioisotopes. This has allowed the determination of a chronology of lunar bombardment by comparing the sample ages with the crater counts on the lunar plains where the samples were collected. The lunar plains, or "maria," are the result of massive eruptions of lava during the first billion years or so of the Moon's history. The revealed chronology shows that the Moon

continued

Name		Semimajor axis (10^3 km)	Orbital eccentricity	Orbital inclination (°)	Orbital period (days)	Mean radius (km)
U6	Cordelia	49.75	0.000	0.14	0.335	13
U7	Ophelia	53.76	0.010	0.09	0.376	15
U8	Bianca	59.16	0.001	0.16	0.435	21
U9	Cressida	61.78	0.000	0.04	0.464	31
U10	Desdemona	62.66	0.000	0.16	0.474	27
U11	Juliet	64.36	0.001	0.06	0.493	42
U12	Portia	66.10	0.000	0.09	0.513	54
U13	Rosalind	69.93	0.000	0.28	0.558	27
U14	Belinda	75.26	0.000	0.03	0.624	33
U15	Puck	86.00	0.000	0.31	0.762	77
U5	Miranda	129.8	0.003	3.40	1.413	236
U1	Ariel	191.2	0.003	0.0	2.520	579
U2	Umbriel	266.0	0.005	0.0	4.144	585
U3	Titania	435.8	0.002	0.0	8.706	789
U4	Oberon	582.6	0.001	0.0	13.46	761
	S/1997 U1	7,169	0.082	140	580	30?
	S/1997 U2	12,214	0.509	153	1290	60?
N3	Naiad	48.23	0.000	0.0	0.294	29
N4	Thalassa	50.08	0.000	4.5	0.311	40
N5	Despina	52.53	0.000	0.0	0.335	74
N6	Galatea	61.95	0.000	0.0	0.429	79
N7	Larissa	73.55	0.000	0.0	0.555	104 × 89
N8	Proteus	117.6	0.000	0.0	1.122	208
N1	Triton	354.8	0.000	157	5.877	1,353
N2	Nereid	5,513	0.751	29	360.14	170
P1	Charon	19.40	0.0076	96.16	6.387	593

experienced a massive bombardment between 4.0 and 3.5 billion years ago, known as the Late Heavy Bombardment. This time period is relatively late as compared with the 100–200 million years required to form the terrestrial planets and to clear their orbital zones of most interplanetary debris. Similarities in crater size distributions on the Moon, Mercury, and Mars suggest that the Late Heavy Bombardment swept over all of the terrestrial planets. Recent explanations for the Late Heavy Bombardment have focused on the possibility that it came from clearing of cometary debris from the outer planets zones. However, the detailed dynamical calculations of the timescales for that process are still in process.

Like almost all other satellites in the solar system, the Moon has no substantial atmosphere. There is a transient atmosphere due to helium atoms in the solar wind striking the lunar surface and being captured. Argon has been detected escaping from the surface rocks and being temporarily cold-trapped during the lunar night. Also, sodium and potassium have been detected, likely the result of sputtering of surface materials due to solar wind particles (as on Mercury).

Unlike the Earth's Moon, the two natural satellites of Mars are both small, irregular bodies and in orbits relatively close to the planet. In fact, Phobos, the larger and closer satellite, orbits Mars faster than the planet rotates. Both of the Martian satellites have surface compositions that appear to be similar to carbonaceous chondrites in composition. This has resulted in speculation that the satellites are captured asteroids. A problem with this hypothesis is that Mars is located close to the inner edge of the asteroid belt, where silicate asteroids dominate the population, and where carbonaceous asteroids are relatively rare. Also, both satellites are located very close to the planet and in near-circular orbits, which is unusual for captured objects.

In contrast to the satellites of the terrestrial planets, the satellites of the giant planets are numerous and are arranged in complex systems. Jupiter has four major satellites, easily visible in small telescopes from Earth, and 12 known, lesser satellites. The discovery of the four major satellites by Galileo in 1610 (as a result of which they are known as the Galilean satellites) was one of the early confirmations of the Copernican theory of a heliocentric solar system. The innermost Galilean

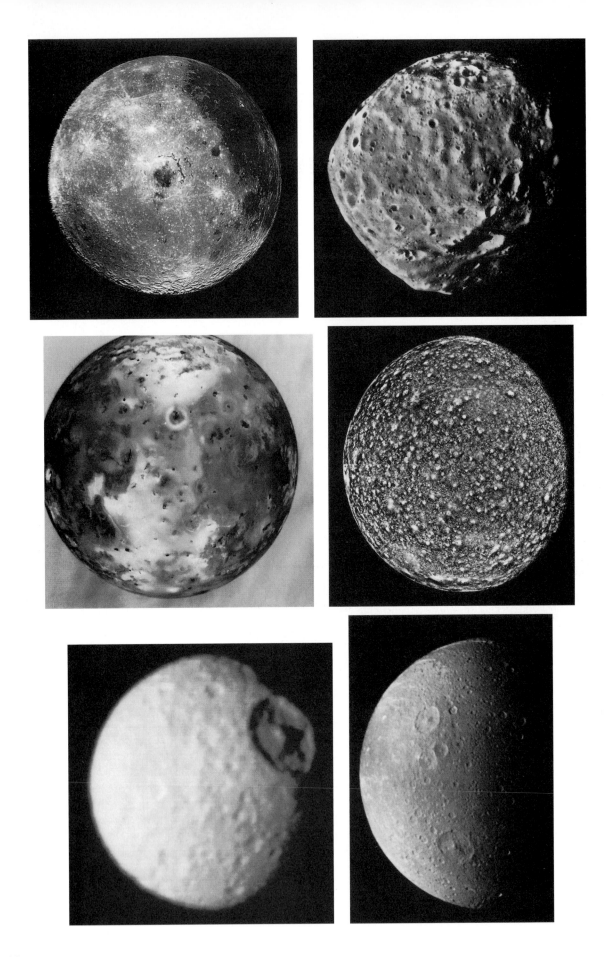

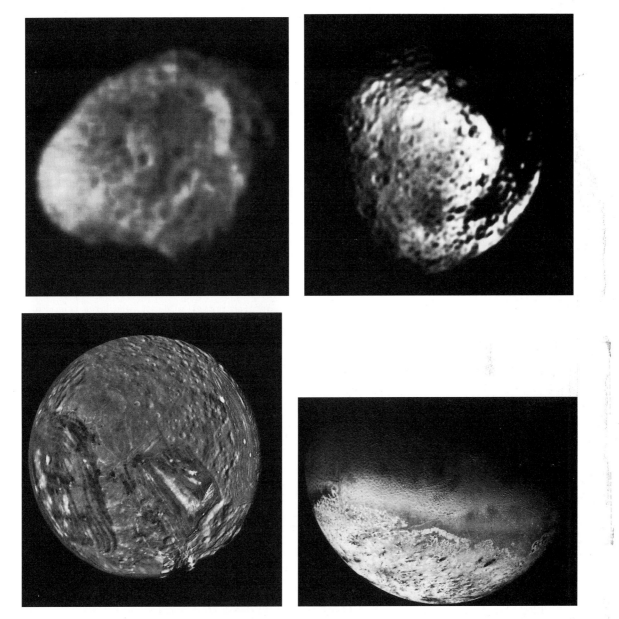

FIGURE 7 A sampling of satellites in the solar system: the heavily cratered surface of the Earth's Moon—at the center of the image is Mare Orientale, a large impact basin located on the east limb of the Moon as viewed from Earth (first page, top row, left); the larger of Mars's two moons, Phobos, is irregularly shaped and highly cratered (top row, right); the innermost Galilean satellite, Io, displays active vulcanism on its surface (first page, second row, left; the bright background is because the satellite was photographed against the disk of Jupiter); the outermost Galilean satellite, Callisto, displays a heavily cratered surface, likely dating back to the origin of the satellite system (first page, second row, right); one of Saturn's smaller satellites, Mimas, displays an immense impact crater on one hemisphere (first page, third row, left); the heavily cratered surface of Rhea, one of Saturn's intermediate-sized satellites, also showing some tectonic features (first page, third row, right); another Saturn satellite, Hyperion, is irregularly shaped and in chaotic rotation (second page, top row, left); Saturn's satellite Iapetus is black on one hemisphere and white on the other (second page, top row, right); Uranus's outermost major satellite, Miranda, has a complex surface morphology suggesting that the satellite was disrupted and reaccreted (second page, bottom row, left); Neptune's one large satellite, Triton, displays a mix of icy terrains (second page, bottom, right).

satellite, Io, is about the same size as the Earth's Moon and has active vulcanism on its surface as a result of Jupiter's tidal perturbation and the gravitational interaction with Europa and Ganymede (see previous section). The next satellite outward is Europa, a bit smaller than Io, which appears to have a thin ice crust overlying a possible liquid water ocean, also the result of tidal heating by Jupiter and the satellite–satellite interactions. Estimates of the age of the surface of Europa, based on counting impact craters, are very young, suggesting that the thin ice crust may repeatedly break up and re-form. The next satellite outward from Jupiter is Ganymede, the largest satellite in the solar system, even larger than the planet Mercury. Ganymede is another icy satellite and shows evidence of having been partially resurfaced at some time in its past. The final Galilean satellite is Callisto, another icy satellite that appears to preserve an impact record of comets and asteroids dating back to the origin of the solar system. As noted earlier, the orbits of the inner three Galilean satellites are locked into a $4:2:1$ mean motion resonance.

The lesser satellites of Jupiter include several within the orbit of Io and a number at very large distance from the planet. The latter are likely captured comets and asteroids. The orbits of the eight outer satellites are divided into two closely spaced groups, and this suggests that they may be fragments of larger objects that were captured and then somehow disrupted. The most likely disruption process is collision with another object, and it is such a random collision itself, occurring within the gravitational sphere of Jupiter, that could have resulted in the dynamical capture.

All of the close-orbiting Jovian satellites (out to the orbit of Callisto) appear to be in synchronous rotation with Jupiter. However, rotation periods have been determined for two of the outer satellites, Himalia and Elara, and these appear to be around 10 to 12 hours, much shorter than their ~250-day periods of revolution about the planet.

Saturn's satellite system is very different from Jupiter's in that it contains only one large satellite, Titan, comparable in size to the Galilean satellites, a number of intermediate-sized satellites, and a host of smaller satellites. Titan is the only satellite in the solar system with a substantial atmosphere. Clouds of organic residue in its atmosphere prevent easy viewing of the surface of that moon. The atmosphere is primarily nitrogen and also contains methane and possibly argon. The surface temperature on Titan has been measured at 94 K and the surface pressure is 1.5 bar.

The intermediate and smaller satellites of Saturn all appear to have icy compositions and have undergone substantial processing, possibly as a result of tidal heating. Again, orbital resonances exist between a number of the satellites and most are in synchronous rotation with Saturn. An interesting exception is Hyperion, which is a highly nonspherical body and which appears to be in chaotic rotation. Another moon, Enceladus, has a ring of material in its orbit that likely has come from the satellite, as a result of either a recent massive impact or active vulcanism on the icy satellite. Another satellite, Tethys, has two companion satellites in the same orbit, which oscillate about the Trojan libration point for the Saturn–Tethys system ahead and behind Tethys, respectively. Yet another particularly interesting satellite of Saturn is Iapetus, which is dark on one hemisphere and bright on the other. The reason(s) for this unusual dichotomy in surface albedos are not known.

Saturn has one very distant satellite, Phoebe, which is in a retrograde orbit and which is suspected of being a captured comet, albeit a very large one. Phoebe is not in synchronous rotation, but rather has a period of about 10 hours.

The Uranian system consists of five intermediate-sized satellites and a number of smaller ones. Again, these are all icy bodies. These satellites also exhibit evidence of past heating and possible tectonic activity. The satellite Miranda is particularly unusual in that it exhibits a wide variety of complex terrains. It has been suggested that Miranda and possibly many other icy satellites were collisionally disrupted at some time in their history, and the debris then reaccreted in orbit to form the currently observed satellites. Such disruption/reaccretion phases may have even reoccurred on several occasions for a particular satellite over the history of the solar system.

Two small, distant satellites of Uranus, S/1997 U1 and S/1997 U2, were discovered in late 1997 in retrograde, eccentric orbits around the planet. These are likely captured objects.

Neptune's satellite system consists of one large icy satellite, Triton, and a number of smaller ones. Triton is slightly larger than Pluto and is unusual in that it is in a retrograde orbit. As a result, the tidal interaction with Neptune is causing the satellite's orbit to decay, and eventually Triton will collide with the planet. The retrograde orbit is often cited as evidence that Triton must have been captured from interplanetary space and did not actually form in orbit around the planet. Despite its tremendous distance from the Sun, Triton's icy surface displays a number of unusual terrain types that strongly suggest substantial thermal processing and possibly even current activity. The *Voyager 2* spacecraft photographed what appears to be plumes from "ice volcanos" on Triton.

The lesser satellites of Neptune are mostly in orbits close to the planet. However, Nereid is in a very distant orbit and is likely a captured object.

Pluto's satellite Charon is the largest satellite relative to its primary in the solar system, being slightly more than half the size of the planet. The Pluto–Charon system is fully tidally evolved. This means that the planet and the satellite both rotate with the same period, 6.39 days, which is also the revolution period of the satellite in its orbit. As a result, the planet and the satellite always show the same faces to each other. Although both Pluto and Charon are icy bodies, their densities appear to be somewhat different: ~2 g cm^{-3} for Pluto versus ~1.7 g cm^{-3} for Charon (though the uncertainty on Charon's density is rather high). This suggests that the satellite may have a smaller rocky component than the planet.

In addition to their satellite systems, all the Jovian planets have ring systems. As with the satellite systems, each ring system is distinctly different from that of its neighbors (Fig. 8). Jupiter has a single ring at 1.8 planetary radii, discovered by the *Voyager 1* spacecraft. Saturn has an immense, broad ring system extending between 1.0 and 2.3 planetary radii, easily seen in a small telescope from Earth. The ring system consists of three major rings, known as A, B, and C ordered from the outside in toward the planet, a diffuse ring labeled D inside the C ring and extending down to the top of the Saturnian atmosphere, and several other narrow, individual rings.

Closer examination by the Voyager spacecraft revealed that the A, B, and C rings were each composed of thousands of individual ringlets. This complex structure is the result of mean motion resonances with the Saturnian satellites, as well as with small satellites embedded within the rings themselves. Some of the small satellites act as gravitational "shepherds," focusing the ring particles into narrow ringlets. Ground-based observers recently discovered nine small satellites, 10–20 km in radius, embedded in the F ring, a thin, single ringlet outside the main ring system.

The Uranian ring system was discovered accidentally in 1977 during observation of a stellar occultation by Uranus. A symmetric pattern of five narrow dips in the stellar signal was seen on either side of the planet. Later observations of other stellar occultations found an additional five narrow rings. *Voyager 2* detected several more fainter, diffuse rings and provided detailed imaging of the entire ring system. The success with finding Uranus's rings led to similar searches for a ring system around Neptune using stellar occultations. Rings were detected but were not always symmetric about the planet, suggesting gaps in the rings. Subse-

quent *Voyager 2* imaging revealed large azimuthal concentrations of material in one of the six detected rings.

All the ring systems are within the Roche limits of their respective planets, at distances where tidal forces from the planet would disrupt any solid body, unless it was small enough and strong enough to be held together by its own material strength. This has led to the general belief that the rings are disrupted satellites, or possibly material that could never successfully form into satellites. Ring particles have typical sizes ranging from micron-sized dust to centimeter- to meter-sized objects, and appear to be made primarily of icy materials, though in some cases contaminated with carbonaceous materials.

Another component of the solar system is the zodiacal dust cloud, a huge, continuous cloud of fine dust extending throughout the planetary region and generally concentrated toward the ecliptic plane. The cloud consists of dust grains liberated from comets as the nucleus ices sublimate, and from collisions between asteroids. Comets are estimated to account for about two-thirds of the total material in the zodiacal cloud, with asteroid collisions providing the rest. Dynamical processes tend to spread the dust uniformly around the Sun, though some structure is visible as a result of the most recent asteroid collisions. These structures, or "bands" as they are known, are each associated with specific asteroid collisional families.

Dust particles will typically burn up due to friction with the atmosphere when they encounter the Earth, appearing as visible meteors. However, particles less than about 50 μm in radius have sufficiently large area-to-mass ratios that they can be decelerated high in the atmosphere at an altitude of about 100 km, and can radiate away the energy generated by friction without vaporizing the particles. These particles then descend slowly through the atmosphere and are eventually incorporated into terrestrial sediments. In the 1970s, NASA began experimenting with collecting interplanetary dust particles (IDPs; also known as "Brownlee particles" because of the pioneering work of D. Brownlee) using high-altitude U2 reconnaissance aircraft. Terrestrial sources of particulates in the stratosphere are rare and consist largely of volcanic aerosols and aluminum oxide particles from solid rocket fuel exhausts, each of which is readily distinguishable from extraterrestrial materials.

The composition of the IDPs reflects the range of source bodies that produce them, and include ordinary and carbonaceous chondritic material and suspected cometary particles. Since the degree of heating during atmospheric deceleration is a function of the encounter velocity, recovered IDPs are strongly biased toward

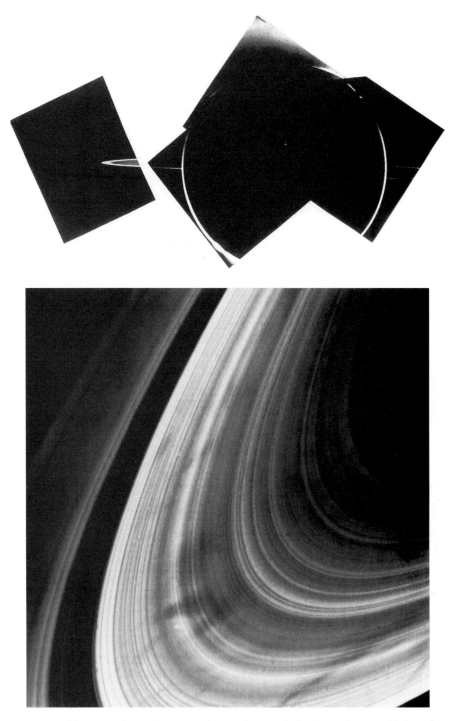

FIGURE 8   The ring systems of the Jovian planets: Jupiter's single ring photographed in forward-scattered light as the *Voyager 2* spacecraft passed behind the giant planet (top left); the multiple ringlets of Saturn's A and B rings (bottom left); the narrow rings of Uranus along with two "shepherd" satellites discovered by the *Voyager 2* spacecraft (top right); two of Neptune's rings showing the unusual azimuthal concentrations, as photographed by *Voyager 2* as it passed behind the planet; the greatly overexposed crescent of Neptune is visible at upper left in the image (bottom right).

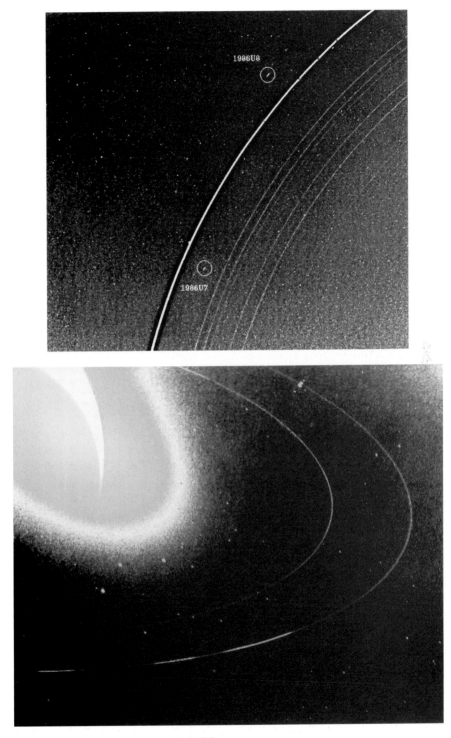

FIGURE 8 (*continued*)

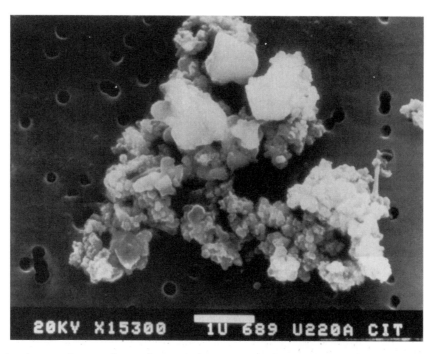

FIGURE 9   A scanning electron microscope image of a suspected cometary interplanetary dust particle (IDP). The IDP is a highly porous, apparently random collection of submicron silicate grains embedded in a carbonaceous matrix. The voids in the IDP may have once been filled with cometary ices. (Courtesy of D. Brownlee.)

asteroidal particles from the main belt, which approach the Earth in lower eccentricity orbits. Nevertheless, suspected cometary particles are included in the IDPs and represent our only laboratory samples of these icy bodies. The cometary IDPs show a random, "botroidal" (cluster-of-grapes) arrangement of submicron silicate grains (similar in size to interstellar dust grains), intimately mixed in a carbonaceous matrix. Voids in IDP particles may have once been filled by cometary ices. An example of a suspected cometary IDP is shown in Fig. 9.

Extraterrestrial particulates are also collected on the Earth in Antarctic ice cores, in meltponds in Greenland, and as millimeter-sized silicate and nickel–iron melt products in sediments. Recently, it has been shown that the IDP component in terrestrial sediments can be determined by measuring the abundance of ^{3}He. ^{3}He has normal abundances in terrestrial materials of 10^{-6} or less. The ^{3}He is implanted in the grains during their exposure to the solar wind. Using this technique, one can look for variations in the infall rate of extraterrestrial particulates over time, and such variations are seen, sometimes correlated with impact events on the Earth.

A largely unseen part of the solar system is the solar wind, an ionized gas that streams continuously into space from the Sun. The solar wind is composed primarily of protons (hydrogen nuclei) and electrons with some alpha particles (helium nuclei) and trace amounts of heavier ions. It is accelerated to supersonic speed in the solar corona and streams outward at a typical velocity of 400 km sec^{-1}. The solar wind is highly variable, changing with both the solar rotation period of 25 days and with the 22-year solar cycle, as well as on much more rapid timescales. As the solar wind expands outward, it carries the solar magnetic field with it in a spiral pattern caused by the rotation of the Sun. The solar wind was first inferred in the late 1940s based on observations of cometary plasma tails. The theory of the supersonic solar wind was first described by E. N. Parker in 1958, and the solar wind itself was detected in 1961 by the *Explorer 10* spacecraft in Earth orbit and in 1962 by the *Mariner 2* spacecraft while it was en route to a flyby of Venus.

The solar wind interaction with the planets and the other bodies in the solar system is also highly variable, depending primarily on whether or not the body has its own intrinsic magnetic field. For bodies without a magnetic field, such as Venus and the Moon, the solar wind impinges directly on the top of the atmosphere or on the solid surface, respectively. For bodies like the Earth or Jupiter, which do have magnetic fields, the field acts as a barrier and deflects the solar wind around it. Because the solar wind is expanding at super-

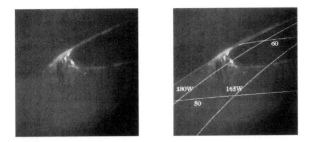

FIGURE 10 The auroral ring over the north polar region of Jupiter, as imaged by the *Galileo* spacecraft.

sonic speeds, a shock wave, or "bow shock," develops at the interface between the interplanetary solar wind and the planetary magnetosphere or ionosphere. The planetary magnetospheres can be quite large, extending out some 10 to 20 planetary radii upstream (sunward) of the Earth, and over 100 radii from Jupiter. Solar wind ions can leak into the planetary magnetospheres near the poles and these can result in visible aurora, which have been observed on both the Earth and Jupiter (Fig. 10). As it flows past the planet, the interaction of the solar wind with the planetary magnetospheres results in huge magneto-tail structures that often extend over interplanetary distances.

All the Jovian planets, as well as the Earth, have substantial magnetic fields and thus planetary magnetospheres. Mercury has a weak magnetic field but Venus has no detectable field. Mars has a patchy field, indicative of a past magnetic field at some point in the planet's history, but no organized magnetic field at this time. Nothing is known about Pluto's magnetic field. The *Galileo* spacecraft recently detected a magnetic field associated with Ganymede, the largest of the Galilean satellites. However, no magnetic field was detected for Europa or Callisto. The Earth's Moon has no magnetic field.

The most visible manifestation of the solar wind is cometary plasma tails, which result when the evolving gases in the cometary comae are ionized by sunlight and by charge exchange with the solar wind and then accelerated by the solar magnetic field. The ions stream away from the cometary comae at high velocity in an antisunward direction. Structures in the tail are visible as a result of fluorescence by CO^+ and other ions. Before the solar wind was suggested by Parker, its existence was inferred by L. Biermann based on his analysis of observations of cometary plasma (ion) tails.

At some distance from the Sun, far beyond the orbits of the planets, the solar wind reaches a point where the ram pressure from the wind is equal to the external pressure from the local interstellar medium flowing past the solar system. A shock will likely develop upstream (sunward) of that point and the solar wind will be decelerated from supersonic to subsonic. This shock is currently estimated to occur at about 90 ± 20 AU. Beyond this distance is a region still dominated by the subsonic solar plasma, extending out another 30–50 AU or more. The outer boundary of this region is known as the heliopause and defines the limit between solar system-dominated plasma and the interstellar medium. It is not currently known if the flow of interstellar medium past the solar system is supersonic or subsonic. If it is supersonic, then there must additionally be a "bow shock" beyond the heliopause, where the interstellar medium encounters the obstacle presented by the heliosphere. A diagram of the major features of the heliosphere is shown in Fig. 11.

The *Pioneer 10* and *11* and *Voyager 1* and *2* spacecraft, which are currently leaving the planetary region on hyperbolic trajectories, have been searching for the heliopause. These spacecraft are currently at distances ranging between 50 and 70 AU. There have been some indications from the Voyager plasma wave instruments that the spacecraft are approaching the heliopause but have not yet reached it. Based on the Voyager data, the heliopause is estimated to be at 110 to 160 AU from the Sun. The Voyager spacecraft are expected to continue to send measurements of this region of space until the year 2015, when they are each expected to be at about 130 AU from the Sun.

To many planetary scientists, the heliopause defines the boundary of the solar system, since it marks the changeover from a solar wind to an interstellar medium-dominated space. However, as already noted, the

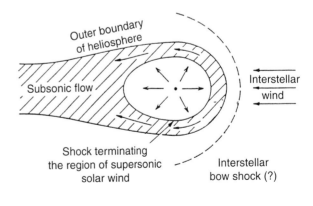

Outer boundary of heliosphere = "heliopause"
Region of subsonic flow = "heliosheath"

FIGURE 11 The major boundaries predicted for the heliosphere. (Reprinted with kind permission from Kluwer Academic Publishers, Axford, *Space Sci. Rev.* **78**, 9–14, Fig. 1, copyright © 1996.)

Sun's gravitational sphere of influence extends out much farther, to approximately 2×10^5 AU, and there are bodies in orbit around the Sun at those distances. These include the Kuiper belt, which may extend out to $\sim 10^3$ AU, and the Oort cloud, which is populated to the limits of the Sun's gravitational field.

III. THE ORIGIN OF THE SOLAR SYSTEM

Our knowledge of the origin of the Sun and the planetary system comes from two sources: study of the solar system itself, and study of star formation in nearby giant molecular clouds. The two sources are radically different. In the case of the solar system, we have an abundance of detailed information on the planets, their satellites, and numerous small bodies. But the solar system we see today is a highly evolved system that has undergone massive changes since it first condensed from the natal cloud, and we must learn to recognize which qualities reflect that often violent evolution and which truly record conditions at the time of solar system formation.

In contrast, when studying even the closest star-forming regions (which are about 140 pc from the Sun), we are handicapped by a lack of adequate resolution and detail. In addition, we are forced to take a "snapshot" view of many young stars at different stages in their formation, and from that attempt to generate a time-ordered sequence of the many different stages and processes involved. When we observe the formation of other stars we need also to recognize that some of the observed processes or events may not be applicable to the formation of our own Sun and planetary system.

Still, a coherent picture has emerged of the major events and processes in the formation of the solar system. That picture assumes that the Sun is a typical star and that it formed in a similar way to many of the low-mass protostars we see today.

The birthplace of stars is giant molecular clouds in the galaxy. These huge clouds of molecular hydrogen have masses of 10^5 to 10^6 solar masses, $M_\odot$. Within these clouds are denser regions or "cores" where star formation actually takes place. Some process, perhaps the shock wave from a nearby supernova, triggers the gravitational collapse of a cloud core. Material falls toward the center of the core under its own self-gravity and a massive object begins to grow at the center of the cloud. Heated by the gravitational potential energy of the infalling matter, the object becomes self-luminous, and is then described as a "protostar." Although

central pressures and temperatures are not yet high enough to ignite nuclear fusion, the protostar begins to heat the growing nebula around it. The timescale of the infall of the cloud material for a solar-mass cloud is about 10^6 years.

The infalling cloud material consists of both gas and dust. The gas is mostly hydrogen (77%) with helium (21%) and other gases. The dust is a mix of interstellar grains, including silicates, organics, and condensed ices. A popular model suggests that the silicate grains are coated with icy-organic mantles. As the dust grains fall inward, they experience a pressure from the increasing density of gas toward the center of the nebula. This slows and even halts the inward radial component of their motion. However, the dust grains can still move vertically with respect to the central plane of the nebula, as defined by the rotational angular momentum vector of the orginal cloud core. As a result, the grains settle toward the central plane.

As the grains settle, they begin to collide with one another. The grains stick and quickly grow from microscopic to macroscopic objects, perhaps meters in size (initial agglomerations of grains may look very much like the suspected cometary IDP in Fig. 9). This process continues and even increases as the grains reach the denser environment at the central plane of the nebula. The meter-sized bodies grow to kilometer-sized bodies, and these bodies grow to 100 km-sized bodies. These bodies are known as planetesimals. As a planetesimal begins to acquire significant mass, its cross section for accretion grows beyond its physical cross section because it is now capable of gravitationally deflecting smaller planetesimals toward it. These larger planetesimals then "run away" from the others, growing at an ever-increasing rate.

The actual process is far more complex than described here, and many details of this scenario still need to be worked out. For example, the role of turbulence in the nebula is not well quantified. Turbulence would tend to slow or even prevent the accretion of grains into larger objects. Also, the role of electrostatic and magnetic effects in the nebula are not understood.

Nevertheless, it appears that accretion in the central plane of the solar nebula can account for the growth of planets from interstellar grains. An artist's concept of the accretion disk in the solar nebula is shown in Fig. 12. In the inner region of the solar nebula, close to the forming Sun, the higher temperatures would vaporize icy and organic grains, leaving only silicate grains to form the planetesimals, which eventually merged to form the terrestrial planets. At larger distances where the nebula was cooler, organic and icy

FIGURE 12   Artist's concept of the accretion disk in the solar nebula, showing the orbiting planetesimals and the proto-Sun at the center. (Painting by William Hartmann.)

grains would condense and these would combine with the silicates to form the cores of the giant planets. Because the total mass of ice and organics may have been several times the mass of silicates, the cores of the giant planets may actually have grown faster than the terrestrial planets interior to them.

At some point, the growing cores of the giant planets became sufficiently massive to begin capturing hydrogen and helium directly from the nebula gas. Because of the lower temperatures in the outer planets zone, the giant planets were able to retain the gas and continue to grow even larger. The terrestrial planets close to the Sun may have acquired some nebula gas, but likely could not hold on to it at their higher temperatures.

Observations of protostars in nearby molecular clouds have found substantial evidence for accretionary disks and gas nebulae surrounding these stars. The relative ages of these protostars can be estimated by comparing their luminosity and color with theoretical predictions of their location in the Hertzsprung–Russell diagram. One of the more interesting observations is that the nebula dust and gas around solar-mass protostars seem to dissipate after about 10^7 years. It appears that the nebula and dust may be swept away by mass outflows, essentially superpowerful solar winds, from the protostars. If the Sun formed similarly to the

protostars we see today, then these observations set strong limits on the likely formation times of Jupiter and Saturn.

An interesting process that must have occurred during the late stages of planetary accretion is "giant impacts," that is, collisions between very large protoplanetary objects. As noted in Section II. C, a giant impact between a Mars-sized protoplanet and the proto-Earth is now the accepted explanation for the origin of the Earth's Moon. Giant impacts have similarly been invoked to explain the high mean density of Mercury, the retrograde rotation of Venus, the high obliquity of Uranus, and possibly even the formation of the Pluto–Charon binary. Although it was previously thought that such giant impacts were low-probability events, they are now recognized to be a natural consequence of the final stages of planetary accretion.

Another interesting process late in the accretion of the planets is the clearing of debris from the planetary zones. At some point in the growth of the planets, their gravitational spheres of influence grew sufficiently large that an encounter with a planetesimal would more likely lead to the planetesimal being scattered into a different orbit, rather than an actual collision. This would be particularly true for the massive Jovian planets, both because of their stronger gravitational fields and because of their larger distances from the Sun.

Since it is just as likely that a planet will scatter objects inward as outward, the clearing of the planetary zones resulted in planetesimals being flung throughout the solar system, and in a massive bombardment of all planets and satellites. Many planetesimals were also flung out of the planetary system to interstellar space, or to distant orbits in the Oort cloud. Although the terrestrial planets are generally too small to eject objects out of the solar system, they can scatter objects to Jupiter-encountering orbits where Jupiter will quickly dispose of them.

The clearing of the planetary zones has several interesting consequences. The dynamical interaction between the planets and the remaining planetesimals results in an exchange of angular momentum. Computer-based dynamical simulations have shown that this causes the semimajor axes of the planets to migrate radially. In general, Saturn, Uranus, and Neptune are expected to first move inward and then later outward as the ejection of material progresses. Jupiter, which ejects the most material because of its huge mass, migrates inward, but only by a few tenths of an astronomical unit.

This migration of the giant planets has significant consequences for the populations of small bodies in the planetary region. As the planets move, the locations of their mean motion and secular resonances will move with them. This will result in some small bodies being captured into resonances while others will be thrown into chaotic orbits, leading to their eventual ejection from the system or possibly to impacts on the planets and the Sun. The radial migration of the giant planets has been invoked in the clearing of both the outer regions of the main asteroid belt and the inner regions of the Kuiper belt.

Another consequence of the clearing of the planetary zones is that rocky planetesimals formed in the terrestrial planets zone will be scattered throughout the Jovian planets region, and vice versa for icy planetesimals formed in the outer planets zone. The bombardment of the terrestrial planets by icy planetesimals is of particular interest, both in explaining the Late Heavy Bombardment and as a means of delivering the volatile reservoirs of the terrestrial planets. Isotopic studies suggest that some fraction of the water in the Earth's oceans may have come from comets, though not all of it. Also, the recent discovery of an asteroidal-appearing object, 1996 PW, on a long-period comet orbit has provided evidence that asteroids may indeed have been ejected to the Oort cloud, where they may make up 1–3% of the population there.

IV. THE SOLAR SYSTEM'S PLACE IN THE GALAXY

The Milky Way is a large, spiral galaxy, about 30 kpc in diameter. Some parts of the galactic disk can be traced out to 25 kpc from the galactic center, and the halo can be traced to 50 kpc. The galaxy contains approximately 100 billion stars and the total mass of the galaxy is estimated to be about 4×10^{11} solar masses ($M_\odot$). Approximately 25% of the mass of the galaxy is estimated to be in visible stars, about 15% in stellar remnants (white dwarfs, neutron stars, and black holes), 25% in interstellar clouds and interstellar material, and 35% in "dark matter." Dark matter is a general term used to describe unseen mass in the galaxy, which is needed to explain the observed dynamics of the galaxy (i.e., stellar motions, galactic rotation) but which has not been detected through any available means. There is considerable speculation about the nature of the dark matter, which includes everything from exotic nuclear particles to brown dwarfs (substellar objects, not capable of nuclear burning) and dark stars (the burned-out remnants of old stars) to massive black holes. The galaxy is estimated to have an age of 10 to 15 billion years, equal to the age of the universe.

The Milky Way galaxy consists of four major structures: the galactic disk, the central bulge, the halo, and the corona (Fig. 13). As the name implies, the disk is a highly flattened, rotating structure about 15 kpc in radius and about 0.5–0.8 kpc thick, depending on which population of stars is used to trace the disk. The disk contains relatively young stars and interstellar clouds, arranged in a multi-arm spiral structure (Fig. 14). At the center of the disk is the bulge, an oblate spheroid about 3 kpc in radius in the plane of the disk, and with a radius of about 1.5 kpc perpendicular to the disk. The bulge rotates more slowly than the disk, and consists largely of densely packed older stars and interstellar clouds. It does not display spiral structure. At the center of the bulge is the nucleus, a complex region only 4–5 pc across, which appears to have a massive black hole at its center. The mass of the central black hole has been estimated at 2.6 million $M_\odot$.

The halo surrounds both of these structures and extends ~20 kpc from the galactic center. The halo has an oblate spheroid shape and contains older stars and globular clusters of stars. The corona appears to be a yet more distant halo at 60–100 kpc and consists of dark matter, unobservable except for the effect it has on the dynamics of observable bodies in the galaxy. The corona may be several times more massive than

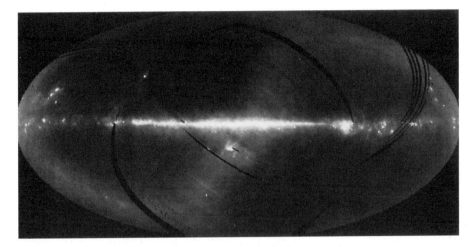

FIGURE 13   An image of the sky at infrared wavelengths as constructed from *IRAS* satellite data. The Milky Way galaxy is visible as the bright horizontal band through the image, with the galactic bulge at the center of the image. The fainter, S-shaped structure extending from lower left to upper right is the zodiacal dust cloud in the ecliptic plane. The plane of the ecliptic is tilted 62° to the plane of the galaxy. Dark gores are gaps in the data caused by incomplete scans by *IRAS*.

the other three galactic components combined. Many descriptions of the galaxy include the halo and the corona as a single component.

The galactic disk is visible in the night sky as the Milky Way, a bright band of light extending around the celestial sphere. When examined with a small telescope, the Milky Way is resolved into thousands or even millions of individual stars, and numerous nebulae and star clusters. The direction to the center of the galaxy is in the constellation Sagittarius (best seen from the Southern Hemisphere in June) and the disk appears visibly wider in that direction, which is the view of the central bulge.

The disk is not perfectly flat; there is evidence for warping in the outer reaches of the disk, between 15 and 25 kpc. The warp may be the result of gravitational

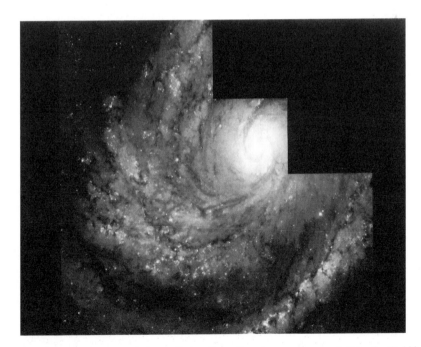

FIGURE 14   Messier 100, a large spiral galaxy in the constellation Coma Berenices, as photographed by the *Hubble Space Telescope*. The Milky Way galaxy may appear similar to this.

perturbations due to encounters with other galaxies, and/or with the Magellanic clouds, two nearby, irregular dwarf galaxies that appear to be in orbit around the Milky Way. Similarly, evidence has been building in recent years that the bulge is not an oblate spheroid, but rather appears to have a triaxial shape. This type of structure is observed in external galaxies and is referred to as a "bar"; such galaxies are known as barred spirals. In addition, the Milky Way's central bar appears to be tilted relative to the plane of the galactic disk. The nonspherical shape of the bulge and the tilt have important implications for understanding stellar dynamics and the long-term evolution of the galaxy.

Stars in the galactic disk have different characteristic velocities as a function of their stellar classification, and hence age. Low-mass, older stars, like the Sun, have relatively high random velocities and as a result can move farther out of the galactic plane. Younger, more massive stars have lower mean velocities and thus smaller scale heights above and below the plane. Giant molecular clouds, the birthplace of stars, also have low mean velocities and thus are confined to regions relatively close to the galactic plane. The disk rotates clockwise as viewed from "galactic north," at a relatively constant velocity of 160–220 km sec^{-1}. This motion is distinctly non-Keplerian, the result of the very nonspherical mass distribution. The rotation velocity for a circular galactic orbit in the galactic plane defines the Local Standard of Rest (LSR). The LSR is then used as the reference frame for describing local stellar dynamics.

The Sun and the solar system are located approximately 8.5 kpc from the galactic center, and 10–20 pc above the central plane of the galactic disk. The circular orbit velocity at the Sun's distance from the galactic center is 220 km sec^{-1}, and the Sun and the solar system are moving at approximately 17 to 22 km sec^{-1} relative to the LSR. The Sun's velocity vector is currently directed toward a point in the constellation of Hercules, approximately at right ascension 18^h 0^m and declination +30°, known as the solar apex. Because of this motion relative to the LSR, the solar system's galactic orbit is not circular. The Sun and planets move in a quasi-elliptical orbit between about 8.4 and 9.7 kpc from the galactic center, with a period of revolution of about 240 million years. The solar system is currently close to and moving inward toward "perigalacticon," the point in the orbit closest to the galactic center. In addition, the solar system moves perpendicular to the galactic plane in a harmonic fashion, with a period of 52 to 74 million years and an amplitude of ±49 to 93 pc out of the galactic plane. (The uncertainties in the estimates of the period and amplitude of

the motion are caused by the uncertainty in the amount of dark matter in the galactic disk.) The Sun and planets passed through the galactic plane about 2–3 million years ago, moving "northward."

The Sun and solar system are located at the inner edge of one of the spiral arms of the galaxy, known as the Orion or local arm. Nearby spiral structures can be traced by constructing a three-dimensional map of stars, star clusters, and interstellar clouds in the solar neighborhood. Two well-defined neighboring structures are the Perseus arm, farther from the galactic center than the local arm, and the Sagittarius arm, toward the galactic center. The arms are about 0.5 kpc wide and the spacing between the spiral arms is about 1.2–1.6 kpc. The local galactic spiral arm structure is illustrated in Fig. 15.

The Sun's velocity relative to the LSR is low as compared with other G-type stars, which have typical velocities of 40–45 km sec^{-1} relative to the LSR. Stars are accelerated by encounters with giant molecular clouds in the galactic disk. Thus, older stars can be accelerated to higher mean velocities, as noted earlier. The reason(s) for the Sun's low velocity are not known. Velocity-altering encounters with giant molecular clouds occur with a typical frequency of once every 300–500 million years.

The local density of stars in the solar neighborhood is about 0.11 pc^{-3}, though many of the stars are in binary or multiple star systems. The local density of binary and multiple star systems is 0.086 pc^{-3}. Most of these are low-mass stars, less massive and less luminous than the Sun. The nearest star to the solar system is Proxima Centauri, which is a low-mass ($M \simeq 0.1 M_\odot$), distant companion to Alpha Centauri, which itself is a double-star system of two close-orbiting solar-type stars. Proxima Centauri is currently about 1.3 pc from the Sun and about 0.06 pc (1.3 × 10^4 AU) from the Alpha Centauri pair it is orbiting. The second nearest star is Barnard's star, a fast-moving red dwarf at a distance of 1.83 pc. The brightest star within 5 pc of the Sun is Sirius, an A1 star ($M \simeq 2 M_\odot$) about 2.6 pc away. Sirius also is a double star, with a faint, white dwarf companion. The stars in the solar neighborhood are shown in Fig. 16.

The Sun's motion relative to the LSR, as well as the random velocities of the stars in the solar neighborhood, will occasionally result in close encounters between the Sun and other stars. Using the foregoing value for the density of stars in the solar neighborhood, one can predict that about 12 star systems (single or multiple stars) will pass within 1 pc of the Sun per million years. The total number of stellar encounters scales as the square of the encounter distance. This rate

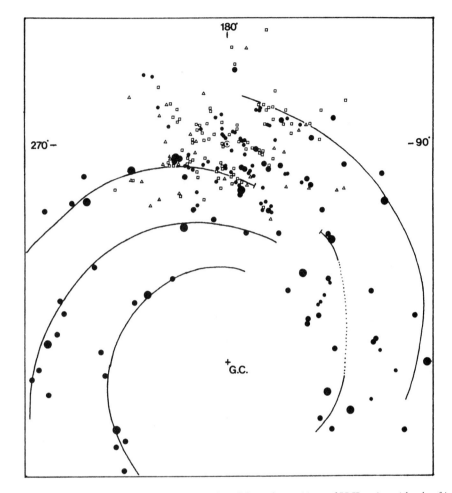

FIGURE 15 The spiral structure of the Milky Way galaxy as inferred from the positions of H II regions (clouds of ionized hydrogen) in the galaxy. The Sun and solar system are located at the upper center, as indicated by the ⊙ symbol. (Reprinted with kind permission from Kluwer Academic Publishers, Forbes and Shuter, *in* "Kinematics, Dynamics and Structure of the Milky Way," p. 221, Fig. 3, copyright © 1983.)

has been confirmed in part by data from the *Hipparcos* astrometry satellite, which measured the distances and proper motions of ~118,000 stars and which was used to reconstruct the trajectories of stars in the solar neighborhood.

Based on this rate, the closest stellar approach over the lifetime of the solar system would be expected to be at ~900 AU. Such an encounter would result in a major perturbation of the Oort cloud and would eject many comets to interstellar space. It would also send a shower of comets into the planetary region, raising the impact rate on the planets for a period of about 2–3 million years, and having other effects that may be detectable in the stratigraphic record on the Earth or on other planets. A stellar encounter at 900 AU could also have a substantial perturbative effect on the orbits of comets in the Kuiper belt and would likely disrupt the outer regions of that ecliptic comet disk.

Obviously, the effect that any such stellar passage will have is a strong function of the mass and velocity of the passing star.

The advent of space-based astronomy, primarily through Earth-orbiting ultraviolet and X-ray telescopes, has made it possible to study the local interstellar medium surrounding the solar system. The structure of the local interstellar medium has turned out to be quite complex. The solar system appears to be on the edge of an expanding bubble of hot plasma about 120 pc in radius, which appears to have originated from multiple supernovae explosions in the Scorpius–Centaurus OB association. The Sco–Cen association is a nearby star-forming region that contains many young, high-mass O- and B-type stars. Such stars have relatively short lifetimes and end their lives in massive supernova explosions, before collapsing into black holes. The expanding shells of hot gas blown off the

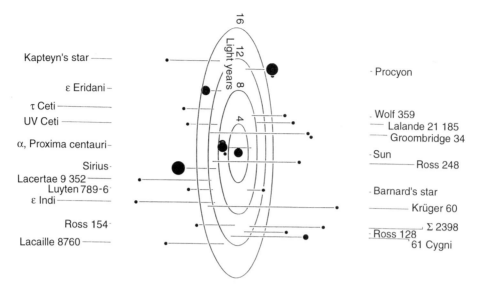

FIGURE 16 A three-dimensional representation of the stars in the solar neighborhood. Horizontal lines indicate the relative distance of the stars north (to the right) or south (to the left) of the celestial equator. The size of the dot representing each star denotes its relative brightness. [From Gilmore, G.F., *in* "Astronomy and Astrophysics Encyclopedia," S. P. Maron, Ed. Copyright © 1992 John Wiley & Sons, New York. Reprinted by permission of John Wiley & Sons, Inc.]

stars in the supernova explosions are able to "sweep" material before them, leaving a low-density "bubble" of hot plasma.

Within this bubble, known as the Local Bubble, the solar system is at this time within a small interstellar cloud, perhaps 2–5 pc across, known as the Local Interstellar Cloud. That cloud is apparently a fragment of the expanding shells of gas from the supernova explosions, and there appear to be a number of such clouds within the local solar neighborhood.

V. THE FATE OF THE SOLAR SYSTEM

Stars like the Sun are expected to have lifetimes on the main sequence of about 10 billion years. The main sequence lifetime refers to the time period during which the star produces energy through hydrogen fusion in its core. As the hydrogen fuel in the core is slowly depleted over time, the core contracts to maintain the internal pressure. This raises the central temperature and, as a result, the rate of nuclear fusion also increases and the star slowly brightens. Thus, temperatures throughout the solar system will slowly increase over time. Presumably, this slow brightening has already been going on since the formation of the Sun and solar system.

A $1 M_\odot$ star like the Sun is expected to run out of hydrogen at its core in about 10 billion years. As the production of energy declines, the core again contracts. The rising internal temperature and pressure are then able to ignite hydrogen burning in a shell surrounding the depleted core. The hydrogen burning in the shell heats the surrounding mass of the star and causes it to expand. The radius of the star increases and the surface temperature drops. The luminosity of the star increases dramatically and it becomes a red giant. Eventually the star reaches a brightness about 10^3 times more luminous than the present-day Sun, a surface temperature of 3000 K, and a radius of 100–200 solar radii. A distance of one hundred solar radii is equal to 0.46 AU, larger than the orbit of Mercury. Two hundred radii is just within the orbit of the Earth. Thus, Mercury and likely Venus will be incorporated into the outer shell of the red giant Sun and will be vaporized.

The increased solar luminosity during the red giant phase will result in a fivefold rise in temperatures throughout the solar system. At the Earth's orbit, this temperature increase will vaporize the oceans and roast the planet at a temperature on the order of ~1400 K or more. At Jupiter's orbit it will melt the icy Galilean satellites and cook them at a more modest temperature of about 600 K, about the same as current noontime temperatures on the surface of Mercury. Typical temperatures at the orbit of Neptune will be about the same as they are today at the orbit of the Earth. Comets in the inner portion of the Kuiper belt will be warmed sufficiently to produce visible comae.

The lowered gravity at the surface of the greatly expanded Sun will result in a substantially increased solar wind, and the Sun will slowly lose mass from its outer envelope. Meanwhile, the core of the Sun will continue to contract until the central temperature and pressure are great enough to ignite helium burning in the core. During this time, hydrogen burning continues in a shell around the core. Helium burning continues during the red giant phase until the helium in the core is also exhausted. The core again contracts and this permits helium burning to ignite in a shell around the core. This is an unstable situation and the star can undergo successive contractions and reignition pulses, during which it will blow off part or all of its outer envelope into space. These huge mass ejections produce an expanding nebula around the star, known as a planetary nebula (because it looks somewhat like the disk of a Jovian planet through a telescope). For a star with the mass of the Sun, the entire red giant phase lasts about 700 million years.

As the Sun loses mass in this fashion, the orbits of the surviving planets will slowly spiral outward. This will also be true for comets in the Kuiper belt and Oort cloud. Since the gravitational sphere of influence of the Sun will shrink as a result of the Sun's decreasing mass, comets will be lost to interstellar space at a greater rate from the outer edges of the Oort cloud.

As a red-giant star loses mass, its core continues to contract. However, for an initially $1 M_\odot$ star like the Sun, the central pressure and temperature cannot rise sufficiently to ignite carbon burning in the core, the next phase in nuclear fusion. With no way of producing additional energy other than gravitational contraction, the luminosity of the star plunges. The star continues to contract and cool, until the contraction is halted by degenerate electron pressure in the superdense core. At this point, the mass of the star has been reduced to about 70% of its original mass and the diameter is about the same as that of the present-day Earth. Such a star is known as a white dwarf. The remnants of the previously roasted planets will be plunged into a deep freeze as the luminosity of the white dwarf slowly declines.

The white dwarf star will continue to cool over a period of about 1 billion years, to the point where its luminosity drops below detectable levels. Such a star is referred to as a black dwarf. A nonluminous star is obviously very difficult to detect. There is some suggestion that they may have been found through an observing technique known as micro-lensing events. Dark stars provide one of the possible explanations of the dark matter in the galaxy.

VI. CONCLUDING REMARKS

This chapter has introduced the solar system and its varied members, viewing them as components of a large and complex system. Each of them—the Sun, the planets, their satellites, the comets and asteroids—is also a fascinating world in its own right. The ensuing chapters provide more detailed descriptions of each of these members of the solar system.

BIBLIOGRAPHY

Lewis, J. S. (1997). "Physics and Chemistry of the Solar System." Academic Press, New York.
von Steiger, R., Lallement, R., and Lee, M. A. (eds.) (1996). "The Heliosphere in the Local Interstellar Medium." Kluwer, Dordrecht.

THE ORIGIN OF THE SOLAR SYSTEM

I. A Theory of the Formation of the Solar System

II. Evidence from Astronomical Observations of T Tauri Stars

III. Evidence from the Analysis of Solar System Materials

IV. Important Unresolved Issues

Patrick Cassen
NASA Ames Research Center

Dorothy S. Woolum
California State University, Fullerton

GLOSSARY

Actinide: One of the group of elements of atomic number 89–103.

CAI (calcium-aluminum-rich inclusion): Pebble-sized particles in meteorites that are enriched in calcium, aluminum, and other refractory elements.

Chalcophile: Chemical tendency to concentrate in sulfide phases.

Chondrite: Meteorite that contains chondrules, or is chemically similar to such meteorites.

Chondrule: Millimeter-sized, spheroidal particles in meteorites that were once molten or partially molten droplets.

Corpuscular radiation: Atomic or subatomic particles streaming at high velocity.

Gravitational instability: Condition in which gravitational disturbances grow faster than they can be balanced by restoring forces.

Gravitational scattering: Alteration of a body's motion due to a close encounter with a massive body, such as a planet.

Lithophile: Chemical tendency to concentrate in silicate phases.

Obliquity: Angle between a planetary body's axis of rotation and the pole of its orbit.

Photospheric temperature: Temperature of the layer within the atmosphere of a star (or other astronomical object) at which it becomes optically thick; in effect, the observable temperature of the object.

Planetesimal: Small rocky or icy body formed in the solar nebula.

Pre-main sequence evolutionary tracks: Luminosity of a pre-main sequence star as a function of its photospheric temperature.

Pre-main sequence: Stage of stellar evolution prior to the onset of hydrogen fusion reactions.

Prograde: With rotational motion in the same sense as orbital motion; the opposite of retrograde.

Protoplanet: Body in orbit about a star, in the process of accumulating mass and destined to become a planet.

Protostar Cloud of gas undergoing gravitational contraction, destined to become a star.

Protostellar cloud: Cloud of interstellar gas from which a star is eventually formed.

Radio continuum: Radiation at radio frequencies (below about 10^5 MHz), with power smoothly distributed with frequency.

Radiochronometry: Method of determining an object's age using radioactive isotopes.

Refractories: Elements or phases that melt and/or evaporate at relatively high temperatures, typically above 1500 K; in contrast to volatiles.

Runaway accretion: Rapid growth of a single planetary object, relative to the growth of other bodies in its vicinity.

Siderophile: Chemical tendency to concentrate in metal phases.

Solar nebula: Disk of gas and dust around the young Sun from which the planets formed.

Spallation: Nuclear reaction in which the energy of the incident particle is so high that more than two or three particles are ejected from the target nucleus, changing its mass number and atomic number.

Spectral energy distribution: Power carried by radiation per unit wavelength (or frequency) as a function of wavelength (or frequency).

T Tauri star: Star whose spectral characteristics indicate it to be pre-main sequence, and of approximately solar mass.

Tidal evolution: Changes in the motions of an astronomical body due to forces associated with the changes in its shape, caused by the gravitational force of another body.

Since at least the time of René Descartes in the seventeenth century, the ordered arrangement of the planets has been interpreted to be a consequence of the *formation* of the solar system, that is, the result of physical processes acting in a systematic fashion to organize matter into a star and subordinate bodies. This perception, along with a Copernican predilection that planetary systems more or less like the solar system are not rare in our galaxy, impels the search for a quantitative theory of formation. Among the many hypotheses that have been proposed, the most fruitful invokes the collapse of tenuous interstellar matter to form a disk of gas and dust, from which

the Sun and other components of the solar system separated, under the action of dissipative forces. Thus it is held that the planets and other minor bodies of the solar system are contemporaneous by-products of the formation of the Sun itself. This view is supported by evidence from (1) astronomical observations of very young stars, (2) the examination of extraterrestrial material, (3) the developing theory of star formation, and (4) the recent discoveries of planets around other stars. However, a *predictive* model of the entire process is far from accomplished, and it is likely that no comprehensive theory will be regarded as compelling until the variety and properties of other planetary systems have been ascertained, and until theoretical conjectures are tested by application to circumstances other than those of our own solar system. [*See* THE SOLAR SYSTEM AND ITS PLACE IN THE GALAXY.]

I. A THEORY OF THE FORMATION OF THE SOLAR SYSTEM

A. OUTLINE

Most of the visible galaxy consists of stars. But it is known that between the stars, space is permeated by gas in a very rarefied state. Much of this gas occurs in clouds that are readily discerned by virtue of the fact that they contain very fine dust particles (interstellar grains) and are sufficiently thick to block the transmission of starlight; thus they appear as dark regions against the stellar background. It is in these interstellar clouds that new stars are born, and it is presumed that the Sun was born in such a cloud (Fig. 1). Although it is quite certain that stars form by the collapse of cloud material under its own gravity, the sequence of events that results in a star accompanied by a planetary system (or, for that matter, a multiple star system) is complicated and far from fully understood. Nevertheless, the essential features of such a sequence have been postulated, quantitative models based on the laws of physics are being developed, and their predictions are being tested by comparison with observations of stars that appear to be in their formative stages and the least evolved products of our own solar system. The main features of the theory adopted by most current researchers are outlined as follows.

Interstellar clouds in which star formation is observed to be occurring come in all shapes and sizes.

FIGURE 1   Images of the Eagle Nebula star-forming region in (a) visible light and (b) infrared light. Many of the sources of radiation seen in the infrared image are newly formed stars, still surrounded by the gas and dust from which they were (or are being) born. They can only be seen as "hot spots" because their visible light, which cannot penetrate the surrounding cloud, is absorbed and reradiated at the longer, infrared wavelengths that can escape the cloud. The stars that are seen in visible light are in front of the clouds. (Courtesy of Mark J. McCaughrean/NASA.)

They may contain as little as a few tens or as much as a million solar masses; they may be roughly globular, or they may be irregular or filamentary in appearance. But the individual stars seem to be forming in special regions where the density is anomalously high: perhaps 10^6 hydrogen (the dominant gas) molecules/cm^3 compared to a typical cloud density of 10^4 molecules/cm^3. These regions, called cloud cores, may have become denser than their surroundings by a gradual process, such as the diffusional loss of support provided by the magnetic field. Such diffusional loss might take 10^7 years. Or dense regions could be caused by a disturbance, such as a shock wave, created by a violent event within or near the cloud. In either case, the loss of support and increase in density are expected to lead to a state of gravitational instability (called Jeans instability), in which the forces that support the gas (thermal pressure, turbulent motions, and magnetic field) can no longer resist the increase in gravitational force that accompanies compression. The result is a dynamical collapse, in which the gravitational energy released is efficiently radiated away. If the core was not rotating, it would collapse from a size of about 0.1 parsec (3×10^{17} cm) to stellar dimensions (10^{11} cm), at which condition the density would have increased to the point where radiation can no longer easily escape, and a young star, supported by thermal pressure, is formed. However, if the core has even a very small amount of rotation (as will generally be the case), its angular momentum, which is nearly conserved during collapse, cannot be accommodated in an object as small as a single star. In the case of the solar system, most of the matter is presumed to have been initially contained in a disk surrounding a small stellar embryo (Fig. 2). The size of the disk was determined by the total angular momentum of the collapsed material. The star then grows by the inward migration of material within the disk, in response to the outward transport of angular momentum promoted by dissipative processes. In the solar system, the disk is called the primitive solar nebula.

The theory presupposes the outcome of collapse to be a single star and disk. This is by no means guaranteed; in fact, the contrary is to be expected because most stars are members of multiple systems. In this regard, the Sun is in the minority. To define the conditions that determine whether a single star or multiple system forms is a primary goal of current star formation research. Results to date suggest that single stars with disks form mainly from slowly rotating, centrally condensed, symmetrical progenitors. Rapidly rotating cloud cores, and those with relatively extended or irregular mass distributions, are thought to form multiple-star systems when they collapse.

The thermal state of the solar nebula was determined by a balance of the gravitational energy released by the accretion of material from it onto the proto-Sun, the illumination from the proto-Sun, and the rate at which energy was transported to and radiated away from its surfaces. Close to the forming Sun, the temperature was high enough to vaporize everything; far away it was cold enough so that even very volatile substances, such as carbon monoxide, could condense. The nebula gradually cooled as most of it was accreted to become the Sun. Its evolution may have been episodic, with short periods of rapid accretion interspersed with longer ones of relative quiescence. During this time, condensed dust and ices tended to coagulate, building from the submicron sizes characteristic of condensates or interstellar grains, to rocky or rock/ice objects centimeters or meters in size. As they did, they settled and accumulated in the midplane of the nebula, forming a thinner disk of particulates. Further growth occurred from this disk, probably by a snowball-like accumulation as solid objects bumped into each other, or less likely by another gravitational instability that had the effect of clumping material into planetesimal-sized objects. This stage of accumulation might have taken as little as 10^4 years, in the terrestrial planet region.

Once objects had grown to be several kilometers in size, their orbits around the Sun were disturbed mainly by their mutual gravitational attractions as they passed close to each other. This caused collisions among them, some of which resulted in further growth, and some in destruction by fragmentation. Eventually (within perhaps 10^7 years), a substantial population of lunar-sized objects grew, which coalesced through spectacular collisions to form the relatively few planets that now comprise the solar system. The planets probably attained almost all of their final masses within 10^8 years, although the cleanup of debris persisted for half a billion years. Since the rate of planetary growth depends on the frequency of embryo–planet encounters, it should have been faster close to the Sun, where the orbital periods are shorter and one might expect the density of material to be greater, than far from the Sun. Nevertheless, the outer planets apparently grew fast enough to gravitationally capture and retain large amounts of hydrogen and helium from the nebula, whereas this was apparently prevented from happening for the terrestrial planets.

In this picture, the asteroid belt is the result of a failed planet between Mars and Jupiter, its final stage of accumulation frustrated by the gravitational influence of Jupiter. Comets are presumed to be planetesimals produced by the first stage of accumulation be-

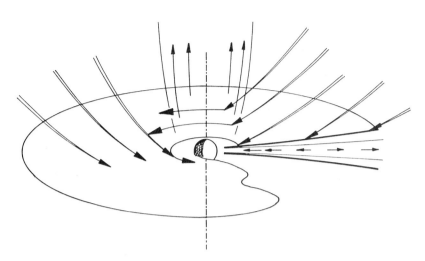

FIGURE 2 Conservation of angular momentum caused most of the material that collapsed to form the solar system to first form a nebular disk. The disk, known as the primitive solar nebula, was maintained by a near-balance of centrifugal force and gravity. Within the disk, processes that transfer angular momentum, such as turbulent friction or wave propagation, allowed material in the inner parts of the disk to transfer its angular momentum to material in the outer parts, so that the former moved inward to form the Sun, while the latter moved outward, absorbing the angular momentum now carried in the planets. Energetic processes operating near the region in which material was being accreted from the nebula onto the young Sun conspired to drive a flow of gas, called a stellar wind, to accelerate outward along the rotation axis. [Adapted from F. A. Podosek and P. Cassen (1994). Theoretical, observational, and isotopic estimates of the lifetime of the solar nebula. *Meteoritics* **29**, 6–25.]

yond the orbit of Jupiter. The orbits of those produced in the Uranus–Neptune region were made highly eccentric by encounters with the outer planets, and were then gradually circularized by extrasolar forces to produce the Oort Cloud. Those that formed beyond Neptune's orbit remained largely unperturbed and now constitute the Kuiper Belt. [*See* ASTEROIDS; KUIPER BELT.]

The theory outlined here has ancient roots in ideas expounded by Immanuel Kant and Pierre-Simon Laplace, dating back almost as far as Isaac Newton's discovery of the general law of gravitation. Both Kant and Laplace proposed nebular models (in 1755 and 1796, respectively) in which a hot gaseous nebula was centered on the Sun. In Kant's theory, the Sun was not part of the nebula; in Laplace's it was. In both, the nebula was rotationally flattened, and as it cooled and contracted, secondary concentrations (Kant) and rings (Laplace) ultimately resulted in the formation of planets. In later times, nebula theories fell into disfavor for a number of reasons. For example, the Laplace ring formation mechanism could be shown quantitatively to be unsound. But the primary reason for the abandonment of nebular theories was their failure, at the time, to explain the distribution of angular momentum in the solar system. With conservation of angular momentum during contraction, one expects the central condensation to acquire a very high angular velocity, a condition clearly not fulfilled by the Sun. The Jovian

planets account for the vast majority of the angular momentum of the observed solar system. No early nebular theory could quantitatively explain such an arrangement.

The majority of the other theories that were proposed in the last two centuries can be dubbed catastrophic. They involved either tidal encounters with other bodies (e.g., comets or another star) or direct collisions with stars or nebulae, in which filaments are formed around the proto-Sun from which the planets and their satellites coalesced. The probability for such occurrences in the lifetime of a normal star is very low, but of course that does not mean that it is impossible. If valid, such a theory would just imply that planetary systems around stars would be expected to be very rare. The key objections raised against such catastrophic theories are that much of the matter would fall back on the Sun and the rest of the hot filament would be tidally disrupted and thermally dissipated before it had a chance to condense and form planets. Associations of the proto-Sun with other stellar companions were also proposed to produce the circumsolar matter from which the planets formed. However, in these models, either gaseous filaments formed, in which case they suffer from the same flaws as noted here, or the equivalent of a solar nebula formed, in which case the model intersects other nebular theories.

The direct imaging of circumstellar disks by the *Hubble Space Telescope* (Fig. 3) culminates a phase

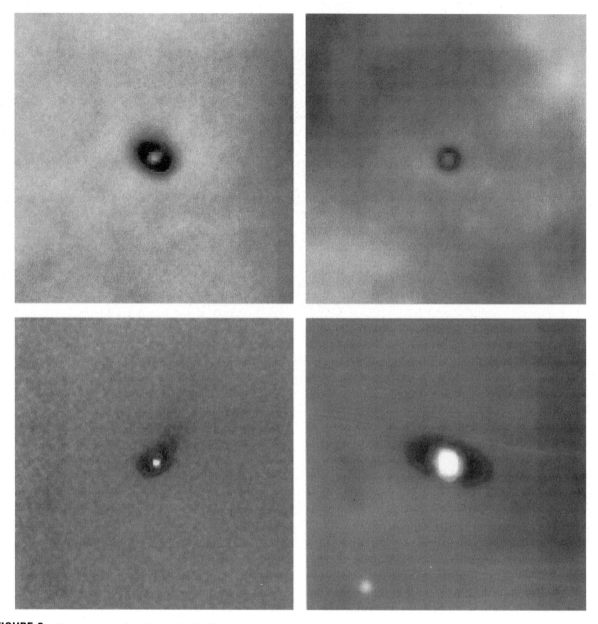

FIGURE 3   Images obtained with the *Hubble Space Telescope* provide dramatic support for the nebular hypothesis for the formation of the solar system. They reveal the presence of opaque disks of gas and dust, comparable in size to the solar system, surrounding young stars in the Orion star-forming region. These disks are seen in silhouette against a background of radiation from ionized gas; some are contained within envelopes of gas being ionized by nearby, bright O stars. (Courtesy of Mark J. McCaughrean and C. R. O'Dell/NASA.)

of solar system cosmogony in which competing hypotheses could differ in their fundamental premises. Even before the Hubble images were obtained, astronomical observations indicated that nebularlike disks around young stars are common (see Section II). Now, although many details remain poorly defined, virtually all modern theories of solar system origin embrace the nebular model, for it accounts for a number of prominent characteristics. The nearly circular, copla-

nar orbits of the planets, with angular momentum in the same direction as the Sun's, is a consequence of a common antecedent, the solar nebula. The disproportionate amount of angular momentum possessed by the planets in comparison with that of the Sun, once a major difficulty for nebular theories, is a natural result of dissipation in the nebula, as discussed in the following.

But some characteristics of the solar system have

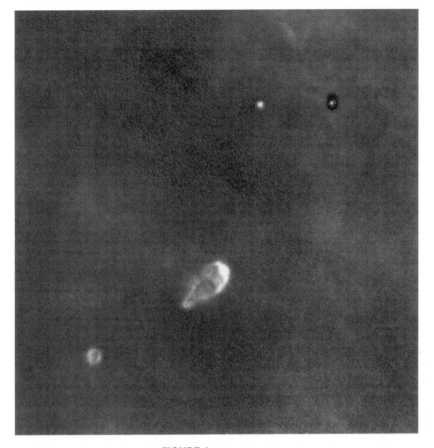

FIGURE 3 (*continued*)

yet to be explained by the theory. It has not been determined whether the regular spacing of the planetary orbits was established at the time of formation, or if it resulted from the gradual removal of objects in unstable orbits over the lifetime of the system. Nor is the generally prograde rotation of the planets an obvious consequence of the model; in fact, the explanation of this feature played a prominent role in debates about the origin of the solar system in the twentieth century. Current research indicates that prograde rotation is the expected outcome of the accumulation of solid material from neighboring orbits, but that the *rates* of planetary spin are not easily explained by the same process. The formation of two classes of planets, the rocky inner planets and the gas giants, once thought likely to be characteristic of planetary systems, may not be inevitable. A list of other important unresolved issues is contained in the final section. [*See* EXTRA-SOLAR PLANETS.]

The regular satellite systems of the outer planets, each similar in many respects to the solar system as a whole, are often considered to provide tests of origin theories. The satellites comprise approximately the same fraction of the mass of their primaries as the planets do of the Sun, their orbital motions are in the same direction as their primary's spin, and they tend to be regularly spaced. It is certainly expected that gaseous disks and accumulation processes of the type invoked for planet formation were also involved in satellite system formation. But an important difference is that the satellite systems are sufficiently compact that the mutual gravitational forces among the components are much stronger than in the planetary system. Therefore they are subject to much greater tidal evolution than the planets. In fact, the spin of most satellites is tidally coupled to that of their primaries, and the spacing of at least the major Jovian satellites has certainly been determined by tidal coupling. Neither condition holds for the planetary system in general.

B. PROTOSTELLAR COLLAPSE

A classic result of gravitational mechanics, derived by Sir James Jeans early in this century, is that a gas may become unstable and collapse under the force of its own gravitational field. It is this process that converts the tenuous gas of the galaxy into stars. The Jeans criterion for instability states that compressive perturbations of a gas (with uniform density ρ and speed of sound c_s) over distances greater than a critical length Λ are unstable, and induce collapse. The length Λ is given by

$$\Lambda = (\pi c_s^2 / G\rho)^{1/2}$$

In this formula, G is the gravitational constant. Thus thermal energy, which is proportional to c_s^2, helps to resist collapse, whereas high densities promote collapse. Of course real interstellar clouds do not have uniform density. The cloud cores from which stars form may be near a state of mechanical equilibrium, in which pressure gradients just balance gravity. The size R of such a self-gravitating structure can be estimated by setting the gravitational potential energy GM^2/R equal to the thermal energy Mc_s^2, to obtain $R \approx GM/c_s^2$. Combining this expression with the relation $M \approx \rho R^3$ yields $R \approx (c_s^2/G\rho)^{1/2}$, so R is always close to Λ, and the stability really depends on the particular distribution of matter within the core. For simple structures, like spheres, the precise stability criterion can be determined exactly. In some circumstances, instability may be induced by an external event, such as the passage of a shock wave or a disturbance of the local gravitational field.

Once a cloud core becomes unstable, its collapse is assured, because the infrared radiation produced by compression easily escapes, thereby preventing the buildup of thermal pressure. Theoretical analyses show that the collapse probably begins in the densest regions and propagates outward at the speed of sound. Material is accelerated inward toward a common center, slowly at first, but it eventually attains supersonic velocities. The collapse timescale is set by $R/c_s \approx GM/c_s^3$. For a one solar mass cloud core at a typical temperature of about 10 K, $R = 0.1$ parsec (3×10^{17} cm) and the timescale is about 10^6 years. If the cloud is supported by turbulent motions, in addition to thermal pressure, a greater mass will be supported within R, and the mass accumulation time will be correspondingly shorter. It is known that magnetic fields provide important forces in the support of molecular clouds, but their role in gravitational collapse has not been completely established. Theoretical models predict that they affect when instability occurs and how much material is involved, but suggest that they are of secondary importance once the collapse-inducing instability has begun to grow.

If the collapsing gas had no angular momentum, its inward supersonic motion would be arrested only at some point where its density was so great that radiation could no longer freely escape. The point at which this condition is fulfilled depends on both the density of the gas and the dominant wavelength of the radiation. Detailed calculations show that an object with a radius of about 2×10^{11} cm, or a few times the present solar radius, would be formed. But interstellar clouds do possess angular momentum; a typical angular velocity Ω, determined by the Doppler shifts of molecular emission lines at radio wavelengths, is about 10^{-14} sec^{-1}, which corresponds to a period of 2×10^7 years. Because this time is substantially longer than the collapse time, the angular momentum (per unit mass) j of the gas is retained during collapse. Centrifugal force opposes gravity in directions perpendicular to the rotation axis, but does not affect motion parallel to the rotation axis; therefore the gas tends to collapse to a *disk* rather than a spherical star. The centrifugal force on a unit mass of gas is j^2/r^3, where r is the distance from the rotational axis; it increases more rapidly with decreasing r than does the gravitational force $\approx Gm/r^2$ (where m is the total mass within the element's radius, which remains constant as the collapse proceeds). The forces balance where $r \approx j^2/Gm$. The gas at the edge of the collapsing region has $j \approx R^2\Omega = 10^{21}$ cm^2/sec, which yields, for $m = 1$ solar mass (and typical values of R and Ω), $r \approx 50$ AU. This radius is much greater than that to which the gas would fall in the absence of rotation, which means that most of the material will be supported in a disk by centrifugal force before it has had a chance to collapse all the way to stellar dimensions. It may not be a coincidence that 50 AU is about the size of our planetary system.

Because stars are observed to form typically in regions where there is a substantial oversupply of gas (the efficiency of star formation is low), the end of collapse is not believed to be caused simply by the depletion of the available reservoir. In fact, the theory of collapse says nothing about when the process ends, except to impose the implicit condition that material with sufficiently high angular momentum cannot collapse in the first place. This factor may indeed play a role in limiting the amount of material accumulated, if the angular momentum of the protostellar cloud increases outward to the point where such a condition prevails. On the other hand, it is known that stars young enough to be still embedded in the material from which they are forming have already produced a strong wind: an outward flow of material that, although apparently narrowly collimated about the axis of rotation, has sufficient momentum to react against the collapse. Although understanding of the origin and physics of these winds is very incomplete, it is likely that they play a role in preventing all the otherwise available material from being accreted.

C. EVOLUTION OF THE SOLAR NEBULA

Considerations of the kind described suggest that most of the material that formed the Sun first resided in a centrifugally supported disk, the primitive solar nebula. Yet we know that only a very small fraction of the mass was required to form the planets. This fraction can be estimated from the total mass of the planets, plus the mass of hydrogen and helium, the main light elements in which they are deficient, under the assumption that those gases must have been present in solar abundances even though they were not retained by the planets. The amount of material so determined is about $0.02 M_\odot$, the mass of the so-called "minimum mass nebula." (The Sun currently possesses more than 99% of the mass of the solar system, but less than 0.5% of its angular momentum.) Thus there must have been a profound redistribution of matter in order to form the Sun from the nebula, while at the same time leaving behind a small amount of material to form the planets, which retain most of the angular momentum. It is now known that such a redistribution of mass occurs whenever dissipative forces transfer angular momentum in the disk, as shown in the following.

Gas moves in nearly circular orbits in the disk. An exact description of its motion would involve a balance between the gravitational, centrifugal, pressure, viscous, and magnetic forces. But, to a good approximation, the first two of these dominate. The gravitational force depends on the distribution of matter throughout the disk, but can be specified in the usual way by a potential Φ, such that the radial force is $-\partial\Phi/\partial r$. The condition of centrifugal balance is then

$$j^2/r^3 = \partial\Phi/\partial r$$

and the energy of an element of gas (per unit mass) is

$$E = j^2/2r^2 + \Phi$$

Now consider two elements of gas (with the same mass, for simplicity) that interact with each other in some way that does not greatly disturb their circular motions, but transfers angular momentum between them. Allow energy to be dissipated, but require that the total angular momentum be conserved. The interaction may be due to viscous friction between adjacent elements, or

small, asymmetrical gravitational or magnetic forces that act over larger distances. In any case, the change in the total energy due to the interaction between element 1 and element 2 will be

$$dE = j_1 dj_1/r_1^2 - j_1^2 dr_1/r_1^3 + d\Phi_1$$
$$+ j_2 dj_2/r_2^2 - j_2^2 dr_2/r_2^3 + d\Phi_2$$

But $d\Phi = j^2 dr/r^3$ from the centrifugal balance relation, and conservation of angular momentum requires $dj_1 = -dj_2$. Therefore

$$dE = dj_1 (j_1/r_1^2 - j_2/r_2^2) = dj_1(\Omega_1 - \Omega_2)$$

The quantity Ω is the angular velocity. Now dE must be negative if energy is dissipated, so dj_1 and $(\Omega_1 - \Omega_2)$ must be of opposite signs. Hence the element with the lower angular velocity gains angular momentum, and must thus move outward, at the expense of that with the higher angular velocity, which must move inward. In the usual case, Ω decreases outward, so dissipation causes the outermost parts of a disk to expand and the innermost parts to move inward. It is presumed that this is how the Sun attained most of its mass from the primitive solar nebula.

The argument given here is independent of the details of the dissipative process that was responsible for the transport of angular momentum. Several mechanisms have been proposed, but their relative importances are unknown. They can be divided into two types: short-range processes, in which the interactions that transfer angular momentum occur between adjacent gas parcels, and long-range processes that can couple the nebula over large radius intervals. The former include turbulent friction, induced by hydrodynamic or hydromagnetic instabilities, and the propagation of rapidly dissipated waves. Recent work suggests that rotational effects in the nebula would have prevented purely hydrodynamic (nonmagnetic) turbulence from achieving the necessary angular momentum transport. Long-range processes include long-wavelength spiral structure induced by gravitational instabilities (analogous to the Jeans instability), which can occur if the disk is sufficiently cold and dense, and forces produced by a global magnetic field or coupling with the solar magnetic field. It is quite possible that more than one mechanism operated in the nebula at a time, or that different mechanisms prevailed at different stages of the nebula's evolution.

Whatever the mechanism, it probably caused an initially rapid influx of gas to the center of the disk, even while the interstellar cloud was still collapsing, to produce the proto-Sun. Then, over a much longer period, the rest of the nebula evolved, becoming cooler and less massive. This stage is estimated to last for perhaps a few to ten million years, based on the inferred range ages of young stars that exhibit evidence for circumstellar disks. Some material continued to accrete onto the Sun; a small fraction of it was retained in the form of planets and other orbiting bodies; and some (unknown fraction) was dispersed in the form of a stellar wind.

Something of the structure of the nebula during its evolutionary stages can be learned by including the pressure in the force balance. In the vertical direction, the gas must be supported by pressure against the vertical component of the Sun's gravity, which exceeded the nebula's self-gravity by the time the nebula mass was much less than that of the Sun's. This balance is expressed by the equation

$$\frac{\partial p}{\partial z} = -\rho \frac{GM}{(r^2 + z^2)^{3/2}} z \approx -\rho \Omega_K^2 z$$

where p and ρ are the gas pressure and density, M is the mass of the Sun, Ω_K is the Keplerian angular velocity, z is the distance above the midplane, and r is the radial distance to the Sun's center. From this equation, one can estimate that the thickness of the disk h will be

$$h \approx \sqrt{p/\rho \Omega_K^2} \approx c_s/\Omega_K$$

where c_s is now the sound speed in the nebula. Thus the nebula will be *thin* (i.e., $h/r \ll 1$) whenever $c_s \ll V_K \equiv r\Omega_K$. This condition is easily maintained because the large surface-to-volume ratio of a thin disk promotes efficient cooling.

The net radial force is the sum of the gravitational force and the pressure gradient:

$$\rho r \Omega_D^2 = \rho \frac{GM}{r^2} + \frac{\partial p}{\partial r}$$

where Ω_D is the angular velocity of gas in the disk. If the disk is thin, it is readily shown that the ratio of the radial pressure gradient to the gravitational force is of the order $(h/r)^2 \ll 1$, so the gravity is indeed balanced primarily by centrifugal force ($\Omega_D \approx \Omega_K$) as assumed earlier. Thus the statements that the disk (1) is nearly in centrifugal balance, (2) has supersonic azimuthal velocity, and (3) is thin are all equivalent. Nevertheless, the relatively small radial pressure gradient is important when considering the *relative* motions of gas and solid bodies in the nebula; solid bodies do not feel the pressure gradient, and therefore have a tendency to move at slightly greater azimuthal veloci-

ties than the gas. This velocity differential results in a frictional force between the gas and solids that causes the solids to lose angular momentum and drift inward, a potentially important effect in the evolution of protoplanetary material.

D. FORMATION OF PLANETESIMALS

It might be supposed that planets are formed in an analogous manner as stars, that is, by gravitational collapse due to a Jeans-type instability. However, the distinctly nonsolar composition of the planets argues that gravitational collapse of undifferentiated nebular material cannot be the whole story. In fact, current theory favors the idea that the planets of the solar system were built mainly by the gradual accumulation of solid material by collisions and, in the case of the gas giants, the subsequent attraction of hydrogen and helium from the nebula only after the solid constituent became sufficiently massive. It has been argued that the term "planet" should be reserved for objects formed in this manner, and that objects formed purely by gravitational collapse, regardless of their masses, should be distinguished from planets. The issue arises when considering the substellar mass companions of other stars, whose compositions and mode(s) of formation are unknown.

For planet building by gradual accumulation, the starting material must have been interstellar grains, and the fine dust and ice particles that condensed in the cooling nebula. These particles were mostly in the size range of 0.01 to 10 μm, small enough to be strongly (but not perfectly) coupled to the nebular gas through collisions with the gas molecules. They grew at a rate determined by how frequently they collided with each other and the efficiency with which colliding particles stuck together. Consider the *vertical motion* of a growing particle in the nebula. It is drawn downward by the vertical component of the star's gravity (again, assumed to be more important than the gravity of the nebula for most of the planet-building period), so this component of gravitational force is, for $z \ll r$,

$$m_p g_z \approx -m_p \frac{GM}{r^2} \cdot \frac{z}{r} = -m_p \Omega_K^2 z$$

where m_p is the mass of the particle. This force is balanced by the gas drag, which, for spherical particles smaller than the mean free path of the gas molecules, is

$$F_z = \frac{4\pi}{3} a^2 \rho V_z c_s$$

Here a is the particle radius and $V_z = dz/dt$ is the vertical velocity (negative for falling particles), assumed to be much less than c_s. Thus the vertical velocity is

$$V_z = -\frac{\rho_p}{\rho} \cdot \frac{a\Omega_K^2 z}{c_s}$$

where ρ_p is the density of the particle itself. If every grain encountered sticks to the growing particle, its rate of growth is given by

$$\frac{dm_p}{dt} = -\pi a^2 V_z \rho_{dust}$$

where ρ_{dust} is the spatial mass density of fine dust encountered as the particle settles. If ρ_{dust} is approximately constant, this equation can be integrated to yield the radius of the particle as a function of altitude z:

$$a = a_0 + (\Sigma_{dust}/8\rho_p) \cdot (1 - z/z_0)$$

where

$$\Sigma_{dust} = 2 z_0 \rho_{dust}$$

and the subscript 0 refers to the initial conditions.

What do these equations tell us about the earliest stages of the accumulation of solid material? First, it is necessary to estimate values for some of the parameters that appear in them. The earliest accumulates were probably poorly compacted, so ρ_p was plausibly less than that of rock, perhaps 1 g/cm³. If one assumes that Earth is composed of all of the available solid material between it and its neighbor planets, Σ_{dust} would be about 10 g/cm² in Earth's zone of the nebula. Under these conditions, the equation for a indicates that a very small dust grain would grow to a size of about 1 cm by the time it settled from high in the nebula to the midplane. A very rough estimate of the time τ required to do this can be obtained by approximating the equation for V_z with the relationship

$$dz/dt \approx h/\tau \approx \frac{\rho_p}{\rho} \cdot \frac{a\Omega_d^2 h}{c_s} \approx \frac{\rho_p a\Omega_d}{\rho}$$

or

$$\tau \approx \frac{h\rho}{\rho_p a\Omega_d} \approx \frac{\Sigma_{dust}}{2\eta\rho_p a\Omega_d}$$

where η is the mass ratio of solids to gas, about 10^{-2} for solar abundances. This formula yields a minimum

settling time of about 10^2 years for the particle to reach the midplane. Accurate integration of the vertical motion equation gives times about an order of magnitude longer than this estimate, because the particle spends a considerable fraction of the time at a size smaller than its final radius of about 1 cm.

The picture that emerges is one in which the small grains that were originally dispersed throughout the nebula gradually accumulate to pebble-sized objects that become concentrated in a layer at the midplane after some 10^3 orbital periods. Of course this picture is complicated by several factors not included in the simple analysis presented here. First, micron-sized particles that collide do not always stick; they can bounce off or destroy (i.e., fragment) each other. For relative velocities above about 1 km/sec, the precise value depending on the nature of the particles, fragmentation occurs rather than coalescence. For velocities less than about 10 m/sec, the particles stick. In between these two velocities, the particles will merely bounce, but not enough energy is dissipated in the collision to stick. The general consequence of imperfect sticking is to lengthen the time to accumulate to a given size by a factor proportional to the inverse of the sticking probability. Second, this analysis considers only a single growing particle in the presence of a background field of fine dust. Calculations that follow the coagulation of an ensemble of particles undergoing simultaneous evolution indicate that there is considerable dispersion in the size distribution at the midplane, as might be expected, but that most of the solid material is incorporated in a dust layer in about 10^3–10^4 orbital periods. Further, imperfect coupling to the gas causes particles to drift in the radial direction, as well as vertically, at rates that depend on their sizes. Indeed, these radial velocities can be 10^2–10^3 cm/sec, which, if sustained over the 10^3 years or longer taken to reach the midplane, would result in radial excursions of more than 1 AU.

Turbulence in the nebula can also have a profound effect on particle accumulation. It has been shown that even a small amount of turbulence is very effective in preventing very small particles from settling to the nebular midplane. On the other hand, it is now known that turbulence also produces strong concentrations of particles in certain size ranges (on the order of 0.1–1 cm, for plausible nebula parameters), which might greatly enhance their accumulation rates. The clusters so formed might then be the first objects to collect at the midplane.

Note that foregoing discussion presumes that all processes relevant to particle accumulation occur in the nebula. But it has also been suggested that particles might grow during the collapse phase, prior to their incorporation in the nebula, or that they could have been entrained in an energetic wind that originated close to the Sun and been transported great radial distances to the locations where they reentered the nebula and were finally incorporated into planetary bodies. In any case, the entire accumulation process should realistically be viewed in the context of an evolving system, in which the gas densities and motions, temperatures, and stability characteristics are themselves functions of time and position.

All of these complications severely limit our comprehension of the earliest phases of planet building. One of the most important challenges is to develop an understanding of the processes that prevail in the dense dust layer that is presumed to eventually collect at the midplane. The conditions in this region are expected to be quite different than in the rest of the nebula, and need to be understood in order to interpret evidence retained in the primitive meteorites. In this layer, the mass density of solids might have become as great as that of the gas. If the velocity dispersion σ of the particles remains low, so that $\sigma \Omega_K / \pi\, G \Sigma_{dust} \ll 1$, gravitational instabilities are predicted to rapidly (in a few years) produce clumping and further growth to greater than kilometer-sized bodies. For the parameter values considered earlier, σ must drop below about 10 cm/sec for this to occur. But the particles in such a dense dust layer can no longer be treated as isolated objects that interact only through binary collisions; they act collectively to modify the gas motions, as well as their own dynamical behavior. In particular, because they tend to have a greater orbital velocity than that of the gas (as mentioned before), their collective drag on the gas produces a turbulent boundary layer in which the velocity dispersion is maintained above the value that would permit gravitational instability. Therefore, further growth must proceed by collisions and cohesion. Once bodies have grown to a kilometer in size or greater, their motions are only slightly affected by gas drag, and gravitational interactions dominate their development.

E. FORMATION OF PLANETS

Although significant gaps remain in our understanding of the accumulation of solids to form 1- to 10-km-sized planetesimals, the key factors in the accretion of these planetesimals to form planets are more confidently delineated once gravitational interactions dominate. The processes are dynamical and the physics is well understood, in principle. Further, the problem is amenable to treatment by well-developed statistical methods.

The problem involves the consideration of a very large number of planetesimals. For example, even with perfect efficiency, building a planet out of 10-km-sized rocky bodies requires three hundred million such bodies to construct Earth and somewhere on the order of 4 billion to produce the inferred rock/ice core of Jupiter. The swarm of planetesimals that ultimately compose a planet start out in heliocentric Keplerian orbits of relatively low inclination and eccentricity, spread out over some radial distance in the protoplanetary disk. Somehow, they accrete to form the final planetary body.

By far the dominant force is the Sun's gravitational force. The Keplerian rotation results in orbital speeds that decrease monotonically with radial distance from the Sun, which produces a shearing motion in the planetesimal swarm. The largest perturbation to this organized heliocentric rotation is the mutual gravitational interactions among the planetesimals. Gravitational scattering tends to stir the swarm, increasing the inclinations and eccentricities of the orbits. In addition to randomizing the motion, it also results in a tendency toward equipartition of the kinetic energy associated with the random motion of the planetesimals; this means that all planetesimals tend to achieve the same kinetic energy associated with their random motion, regardless of the size (mass) of the body. Nongravitational perturbations also occur, the primary ones being inelastic collisions, which can result in accretion and/or fragmentation of the bodies, and gas drag. Both tend to damp the inclinations and eccentricities of the bodies.

The collision cross section of the planetesimals is enhanced over its geometrical (cross-sectional area) value as a result of gravitational attractions. The so-called gravitational focusing factor F_g is the ratio of the capture cross section to the geometrical cross section of the body and is

$$R_{grav}^2 / R_{geom}^2 = F_g = 1 + [V_{esc}/V_{rel}]^2$$

R_{geom} is simply the radius of the planetesimal and the escape velocity V_{esc} is defined by

$$V_{esc} = \sqrt{\frac{2GM_p}{R_{geom}}}$$

where M_p is the mass of the planetesimal. Thus, accretion is facilitated if V_{esc}/V_{rel} is large. Furthermore, if the largest bodies (which have the largest V_{esc}) also have the smallest V_{rel}, they can grow much more rapidly than their smaller neighbors. Equipartition of energy promotes such a condition, since massive bodies have lower velocities than less massive bodies with the same kinetic energy.

The evolution of the velocity distribution of the swarm is determined by the size distribution of the swarm. But the time evolution of the size distribution is determined by growth rates, which are themselves dependent on the collision rates and the velocity distribution. That is, size and velocity distributions are complex nonlinear functions of each other. If, for instance, the swarm evolves so as to achieve a small V_{esc}/V_{rel} ratio, F_g would be approximately equal to one. Then, if ρ_s is the mass density of the swarm and ρ_p is the planetesimal mass density,

$$dM_p/dt = \pi R_{geom}^2 F_g \rho_s V_{rel} = 4\pi R_{geom}^2 \rho_p \, dR_{geom}/dt$$

and dR_{geom}/dt is the same for all sized bodies, which means that smaller bodies would grow proportionally bigger in the same amount of time.

Numerical simulations, starting from plausible initial conditions, are necessary to follow the time evolution of the size and velocity distributions of the swarm. The results of the computer simulations yield two classes of outcomes. Under some circumstances, orderly growth occurs. In this case, V_{rel} becomes approximately equal to V_{esc} of the largest bodies, so F_g is of order unity for all bodies, which is like the situation referred to in the previous paragraph. The mass distribution falls steeply and smoothly at the high-mass end, and the maximum mass increases in time. Most of the mass resides in the most massive bodies (Fig. 4b). The second class of outcomes is considered to be more realistic. They exhibit runaway accretion, in which a few bodies grow much more rapidly than all the others (see Figs. 4c and 4d). This result can be achieved even with an initially uniform mass distribution, if the total mass is initially contained in a large number of small bodies, and if the collective effects of many gravitational perturbations (dynamical friction), which promote equipartition of energy, are included. Then Poisson statistics ensures that a few bodies experience many more collisions than the average, and therefore begin to grow faster than the rest; equipartition of energy keeps the V_{rel} of the largest bodies low, thereby allowing their growth to "run away." From a wide variety of simulations, it appears that once runaway accretion has started, no other effects considered to date (e.g., orbital resonances, three-body interactions) can prevent it and allow the small bodies to catch up with the runaway, at least in the terrestrial planet region of the protoplanetary disk.

Runaway accretion is terminated once the runaway has depopulated its accretionary zone of planetesimals. Further growth of the massive runaway, what might now be termed protoplanet-sized, is possible if it drifts radially to an undepleted zone owing to its interaction with the nebular gas, or if its accretionary zone is replenished with planetesimals by mutual gravitational scattering, by orbital evolution due to gas drag, or by gravitational perturbation of planetesimals by other protoplanets in neighboring accretion zones.

The width of the accretion zone for a planetesimal of mass M is proportional to its Hill sphere radius, which is a measure of the distance over which its gravitational influence dominates the solar influence. The Hill sphere radius is given by

$$R_H = a\{M_p/3M_\odot\}^{1/3}$$

where a is the semimajor axis of the orbit. Note that, as the planetesimal grows, the width of the accretion zone feeding it increases in proportion to dR_{geom}, whereas the planetesimal mass increases proportional to $R_{geom}^2 dR_{geom}$. After a certain period, the runaway runs out of planetesimals available for accretion and achieves a final mass that is proportional to

$$2\pi a[R_H]_{max}\Sigma_p$$

the mass of the protoplanetary disk annulus of width $[R_H]_{max}$. (Here, Σ_p is the surface mass density of the planetesimal disk.) There is some uncertainty in the proportionality constant, but most estimates vary between 3 and 4. If it is taken to be $2\sqrt{3}$,

$$M_{max} \approx 8 \cdot 3^{1/4}\pi^{3/2}\Sigma_p^{3/2}a^3/M_\odot^{1/2}$$

The surface density of planetesimals might be estimated by assuming that the solid components (rock, metal, and ice) of the planets were originally contained in small bodies distributed smoothly between the present planetary orbits. One would find that at 1 AU, Σ_p had to be about 10 g/cm^2 in order to provide the mass of Earth (5.98×10^{27} g), which is almost all rock and metal. According to this formula, the runaway mass would then be only about 0.02 Earth masses, so many runaway products (protoplanets) would have been required to merge to form the Earth subsequent to the termination of runaway accretion. If Σ_p were large enough to make $M_{max} = 1$ Earth mass, there would be a large excess of material in the terrestrial planet region.

After the termination of runaway accretion in the terrestrial planet region, one expects relatively close spacing of the protoplanets (e.g., R_H for Earth is 0.01

a

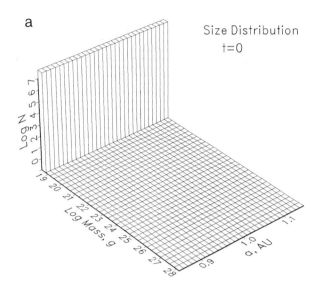

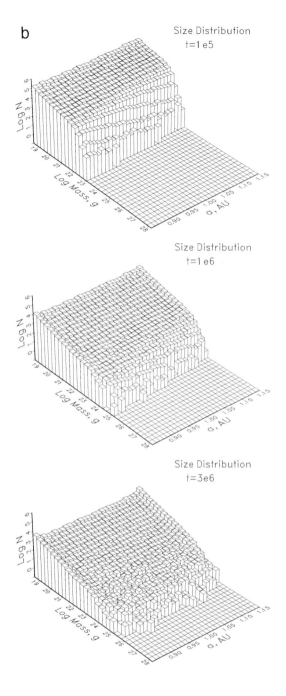

FIGURE 4. The results of simulations of planetary accretion: numbers of bodies as a function of mass interval and semimajor axis. (a) Initial conditions. At $t = 0$ the planetesimals have equal mass (4.8×10^{18} g) and the total mass within the 0.3-AU zone is 1.2 Earth masses. (b) The evolution in time, t (in years), of the distribution of masses for the case with no dynamical friction. Growth is "orderly" (see text) and only 10^{25}-g bodies grow in a million years. (c) The evolution in time of the distribution of masses for the case with dynamical friction. Runaway growth occurs and planet-sized bodies (10^{27} g) grow in a million years. (d) The evolution in time of the distribution of masses for the case with dynamical friction and gas drag, as would occur for accretion before the solar nebula was cleared. Runaway growth occurs. [From S. Weidenschilling *et al.* (1997). Accretional evolution of a planetesimal swarm. 2. The terrestrial zone. *Icarus* **128**, 429–455.]

AU). Such a system would not be dynamically stable for long; mutual gravitational perturbations would lead to crossing orbits and the gravitational scattering caused by close encounters would increase the relative velocities. They would not be damped, because the original large numbers of small bodies needed to establish equipartition of energy no longer exist. Significant eccentricities would develop (growing from 0.01 initially to 0.1–0.2 in the final stages), which would result in radial migration, crossing orbits, and violent inelas-

tic collisions. Simulations using widely varying initial conditions for the protoplanets all seem to result in 2–5 terrestrial planets within a total accretion time of about 100 million years (Myr), with or without consideration of gas drag and regardless of whether or not runaway accretion was obtained in the earliest accretionary stages (Fig. 5).

Further, nothing more than the minimum mass nebula is required to build the terrestrial planets in these accretionary timescales. That is, few planetesi-

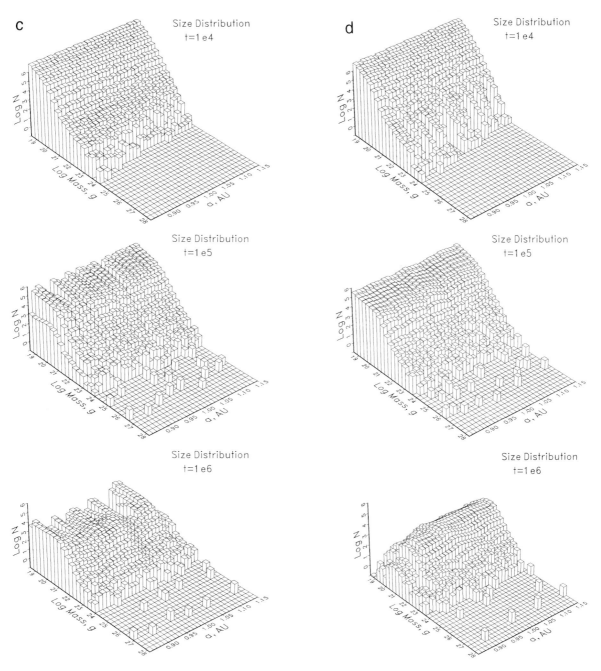

FIGURE 4 (*continued*)

mals are scattered out of the deep gravitational potential well of this zone of the disk. The consequences of this emerging picture of terrestrial planet formation are consistent with a number of observed inner solar system features. Frequent collisions with smaller bodies would have occurred in the final accretionary stage, a result consistent with the early cratering record read on planetary surfaces in the inner solar system. The expectation of occasional collisions with bodies of com-

parable size is consistent with the hypothesis of lunar formation, whereby a giant impact creates the Moon from a piece of Earth's mantle launched by the Earth-impacting body. A large impact in which part of the silicate crust of Mercury is lost may also explain why that planet has a large iron core, relative to its size. The obliquities of the planets, including the extreme case of Uranus, are also consistent with the occurrence of giant impacts in the final accretionary stages. Finally,

smooth extension of the solar system mass distribution or not remains to be discovered.

II. EVIDENCE FROM ASTRONOMICAL OBSERVATIONS OF T TAURI STARS

T Tauri stars are now known to be very young (ages between 10^5 and 10^7 years) stars of roughly solar mass. They were originally classified on the basis of observational characteristics such as emission lines, nonblackbody ultraviolet and infrared continuum radiation, variability, and association with dark clouds. These properties are now interpreted in terms of the theory of the formation of low-mass (up to about $2 M_\odot$) stars. In particular, the infrared radiation (as well as that at longer wavelengths) can be accounted for by radiation from the surface of a circumstellar disk and the ultraviolet radiation is believed to be associated with the accretion of material from the disk to the star. Thus, although even the nearest of these stars are too distant to spatially resolve characteristics that are unequivocally indicative of planet formation, evidence of their environments is contained in the spectral energy distributions of their radiation. It is therefore expected that observations of T Tauri stars at different stages of evolution provide information on the processes that occurred in the solar system during its early history.

Although not all T Tauri stars show evidence of disks around them, several different kinds of observations indicate that many do have disks. First, disks around stars in the Orion Nebula have been imaged directly by the *Hubble Space Telescope* (see Fig. 3). These disks are seen as dark silhouettes against the background field of bright emission from the hot gas of the nebula; some of them are roughly the size of the solar system. Other circumstellar disks have also been imaged by *Hubble* in scattered light. Second, for many T Tauri stars, radiation attributable to optically thin thermal emission from fine dust is seen at submillimeter and millimeter wavelengths. The total amount of dust can be determined from the intensity of the detected radiation. In a typical case, the amount of dust so inferred would be sufficient to substantially obscure the star at optical wavelengths, if the dust were distributed more or less uniformly around the star. But such obscuration is not usually observed, and it is therefore deduced that the dust is spatially confined. Because this result applies to many T Tauri stars, it is

concluded that the spatial distribution of dust is flat, that is, it forms a disk. Third, emission lines that are believed to be produced in a wind emanating from the star are inevitably Doppler-shifted to the blue, indicating that the gas producing them is coming toward the observer; red-shifted lines of this type are not observed. This observation is explained if the red-shifted lines that would be associated with gas flowing away from the observer were obscured by a circumstellar disk. Fourth, in some cases, line radiation from carbon monoxide, a standard tracer of interstellar gas, has been detected and mapped near T Tauri stars. The gas appears to be in a flattened configuration, and the spatial pattern of the magnitude of Doppler shift is consistent with Keplerian rotation about the star.

The detailed modeling of the spectral energy distributions of T Tauri stars constrains some of their properties (Fig. 6). If it is assumed that most of the mass of the disk resides at large distances, where the radio continuum radiation is optically thin, the total masses of the disks can be estimated. A necessary assumption is the fraction of material in fine, unaccumulated dust, since this is the only material directly detected. If this fraction is given by the solar abundance of condensable elements, the inferred total (mostly gas) masses of disks around T Tauri stars with detectable emission at radio wavelengths appear to lie in the range 10^{-3}–$1 M_\odot$, with many near $10^{-2} M_\odot$. Hence it seems that disks similar in mass to the minimum-mass solar nebula are not uncommon around young stars. Furthermore, the intensity of infrared radiation, which comes mostly from parts of the disk that are optically thick, places a strong constraint on the photospheric (effective) temperatures. In many cases, the disk temperature is well described by $T = T_1 r^{-q}$, where $\frac{1}{2} < q < \frac{3}{4}$ and $50\,\text{K} < T_1 < 400\,\text{K}$, if r is measured in AU. However, it is difficult to establish for any specific star whether the disk infrared radiation is dominated by stellar radiation that is absorbed and reemitted by disk material or represents an intrinsic disk energy source, such as that provided by accretion. Modeling the source of the ultraviolet radiation can be used to estimate the accretion rate. Derived values lie in the range 10^{-9}–$10^{-6} M_\odot$/year.

A difficulty in interpreting the characteristics of T Tauri stars and their disks in terms of a common evolutionary sequence, in which disks form early and slowly dissipate, is that correlations of disk properties with estimated age are frequently not compelling. For instance, there appear to be very young stars that show no evidence of possessing disks, while much older ones do; the masses derived from radio observations show

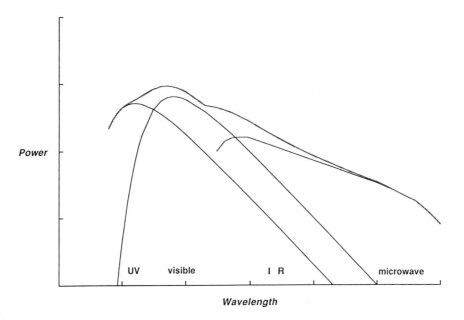

Power

UV   visible   I R   microwave

Wavelength

FIGURE 6   A schematic diagram of the energy radiated by a T Tauri star as a function of wavelength (the spectral energy distribution). The radiation comes from three components, each providing the dominant energy in a different wavelength region: the star itself (lower central curve), which radiates like a blackbody at a temperature of about 4500 K and provides the visible light; the disk (lower right curve), which radiates as the sum of a continuous distribution of cooler blackbodies, and dominates at infrared wavelengths; and the interaction region between the disk and the star (lower left curve), which radiates like a blackbody at about 10,000 K. The sum of these components is shown by the top curve. [Adapted from F. A. Podosek and P. Cassen (1994). Theoretical, observational, and isotopic estimates of the lifetime of the solar nebula. *Meteoritics* **29**, 6–25.]

no trend with age; and disk luminosity is not obviously correlated with age. The problem may be with the estimates of the ages themselves, which require a well-determined stellar luminosity *and* accurate theoretical pre-main-sequence evolutionary tracks. The latter may be substantially affected by disk accretion, which has not been taken into account until very recently. Other sources of error are variability, uncertainties about how far away the stars are, and unresolved stellar companions. The current best estimate for the maximum ages of stars with detectable, nebulalike disks is about 10 Myr.

Several other aspects of T Tauri stars may be important for understanding the early history of the solar system. The first is that many T Tauri stars are formed in dense clusters, in close proximity to other young, active stars. The effects of such dense stellar environments on nebular evolution and planet formation have not yet been systematically addressed. Second is the fact that these stars undergo episodes of intense brightenings, during which the luminosity of the system increases by as much as two orders of magnitude. There is evidence that such episodes are due to violent disk accretion events in which the temperature within the disk rises dramatically, so as to vaporize solid material to a much greater distance than would otherwise ob-

tain. It is reasonable to presume that the Sun and solar nebula went through such episodes, but attempts to examine the consequences for planet formation have only just begun. Another observation that might bear directly on the issue of planet formation is that some T Tauri disks appear to have gaps in them. This conclusion is drawn from spectral energy distributions in which there is a deficiency of near or midinfrared radiation (compared to a monotonic distribution). Although other interpretations are possible, such a deficiency would occur if material was "missing" from the disk at a particular radius (temperature). Theory predicts that the influence of a sufficiently massive companion (such as a Jupiter-sized planet) could produce such a gap. Thus it may be possible to indirectly detect the presence of planets if they coexist with disks. Finally, it is well established that T Tauri stars have extremely active chromospheres compared to the present-day Sun, and therefore are expected to be prodigious sources of nonthermal and corpuscular radiation. Though astrophysicists are confident that the Sun experienced a T Tauri phase, conclusive evidence from studies of planetary materials is missing. Excess spallation products and rare, highly radiation-damaged crystals found in one class of meteorites have been interpreted in terms of an exposure during this phase,

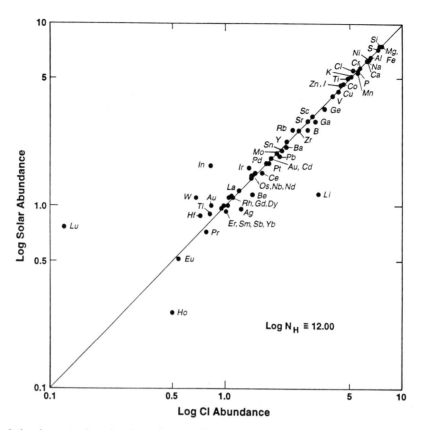

FIGURE 7 Elemental abundances in the solar photosphere are shown on a log–log plot versus those abundances measured in the CI carbonaceous chondrites. The abundances are normalized to 10^{12} hydrogen atoms: log $N_H = 12.00$. The remarkable 1:1 correspondence displayed for all but the most volatile elements is strong evidence for the creation of the CI meteorites out of unfractionated solar material, as well as for the essential homogeneity of the solar nebula. (Even some of the deviations are well understood. For instance, lithium in the Sun is low relative to CI abundances because lithium has been destroyed by nuclear reactions in the Sun.)

although it is presently not possible to dismiss more conventional exposures to cosmic rays in the regolith of the parent body for, albeit, unusually long periods of time.

III. EVIDENCE FROM THE ANALYSIS OF SOLAR SYSTEM MATERIALS

Most of the solid material of the solar system has been thoroughly altered by the physical and chemical processes that are ongoing in and on planet-sized bodies. But a vast amount of information about the earliest history of the solar system is still contained in those bodies small enough to have escaped such complete transformation: the meteorites, asteroids, and comets. Furthermore, even planetary bodies have not totally erased all record of their origins: clues can be found,

for example, in their compositions, the ages of certain constituents, and dynamical states. [*See* METEORITES.]

A. AGE AND COMPOSITION OF THE SOLAR SYSTEM

The key data regarding the age of the solar system and its overall composition come from the meteorites. Chondritic meteorites, in general, and the least altered carbonaceous chondrites in particular, are believed to be the most primitive objects to have been examined so far. This conclusion is based on the similarity of their composition to that of the Sun, their smooth heavy-nuclide abundance pattern, their ancient radiometric ages, and the evidence for their incorporation of both very short-lived extinct radionuclides and presolar material.

The chemical composition of certain carbonaceous chondrites (specifically those classified as CI) is largely

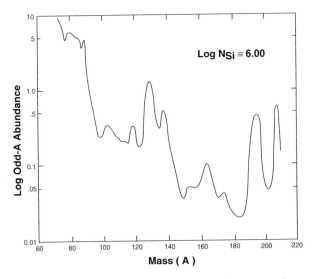

FIGURE 8 The log of the abundances, in CI carbonaceous chondrites, of the odd-atomic mass nuclides is plotted versus atomic mass (A) for $A > 70$. The abundances are normalized to 10^6 silicon atoms: log N_{Si} = 6.00. The abundance pattern is smooth, and its structure is understood in terms of the nucleosynthetic processes by which the elements heavier than iron were created primarily by the capture of neutrons. The fact that this pattern has been preserved in the CI meteorites indicates that little fractionation of the elements occurred between the creation of the elements and the formation of these meteorites, and that the abundances therefore represent the average cosmic abundances.

indistinguishable from that determined for the solar photosphere (Fig. 7), except for the very most volatile elements (and for light elements like lithium, which are depleted in the Sun by thermonuclear reactions). Thus they are believed to represent an accumulation of the essentially unaltered, primary mix of the solid material of the solar nebula and, for most elements, faithfully represent the average solar system abundances. Furthermore, their elemental compositions are generally similar to those determined in stars and the interstellar medium, in supernova ejecta and remnants, and in the galactic cosmic rays, so that one may even speak of "cosmic abundances." (Many astronomical exceptions exist, but they are usually understandable in terms of differences in stages of stellar evolution, the peculiarities of a particular astrophysical site, or, for cosmic rays, the effects of propagation.)

Another indication that CI carbonaceous chondrites are composed of primordial, unprocessed material is demonstrated by the fact that the odd atomic mass isotope abundances vary smoothly as a function of atomic mass, for those elements heavier than iron (Fig. 8). The smoothness of this distribution played an important part in the development of theories of nucleosynthesis. Indeed, the pattern is best explained

as the result of nucleosynthesis by neutron capture within stars and in explosive stellar environments. Subsequent chemical processing would have destroyed the smoothness of such a pattern, since it includes elements of widely disparate cosmochemical affinities (refractories, volatiles, siderophiles, chalcophiles, and lithophiles). Thus, a smooth abundance pattern suggests a lack of such processing since these elements were produced, mixed into the interstellar medium, and incorporated in carbonaceous chondrites.

The preservation of such a complete and ordered set of abundances in some meteorites has been extremely important for studies of the early history of the solar system. Its existence indicates that, despite evidence for a host of complexities, large regions of the solar nebula were basically elementally and isotopically homogeneous, and the accumulation of at least some solids was a rather gentle, chemically undiscriminating process. This idea is further supported by the unequivocal identification of rare chondritic materials with unique isotopic signatures, which demonstrate that they survived the formation of the solar system and all that that entails. The existence of this nearly homogeneous, primordial template can be exploited by examining deviations, which provide objective measures of the effects of, and clues to the nature of, more violent planet-building processes.

Although much of the material within carbonaceous chondrites has escaped severe chemical transformation, they do contain igneous inclusions that can be dated by radiometric means. In fact, measurement of the products of the decay of uranium isotopes (^{235}U and ^{238}U into ^{207}Pb and ^{206}Pb, respectively) yields an average crystallization age for the calcium–aluminum-rich inclusions (CAIs) in the Allende carbonaceous chondrite of 4.566 ± 0.002 billion years (Gyr), the oldest of any solar system material. The CAIs (Fig. 9) are chemically and mineralogically similar to predictions for objects in equilibrium with a high-temperature gas of solar composition, which suggests that they formed, or at least existed, in a hot phase of the nebula. Thus their age is plausibly associated with that of the earliest episode of solidification, taken to be the age of the solar system itself, as there does not appear to be a way to precisely date the time since the isolation of the protosolar cloud.

It is true that the oldest solids need not necessarily date back to the initial formation of the solar system. However, in addition to having the oldest ages, the chondrites also contain the products of short-lived radionuclides that are long since extinct. This suggests that these oldest solids may have formed close to the time of nucleosynthesis, which presumably occurred

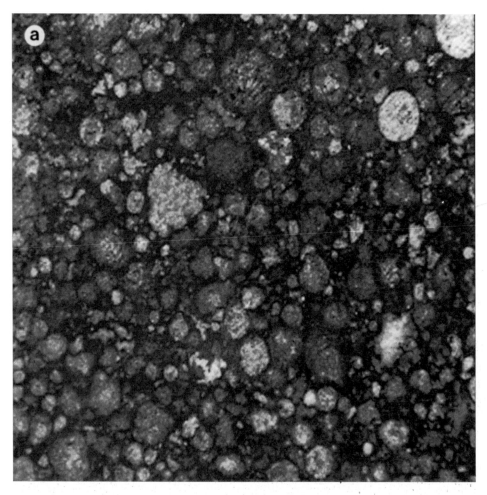

FIGURE 9   (a) A reflected light photograph of a piece of the Allende meteorite, cut to expose its interior structure. The meteorite has retained a high abundance of volatile elements, and yet also contains many light, irregular-shaped "pebbles" made of refractory material, the calcium- and aluminum-rich inclusions, referred to by the abbreviation CAIs. These inclusions are igneous (melted and recrystallized) stones, apparently made in the solar nebula and are the oldest objects to have had their formation ages accurately determined. The field of view is approximately 2 cm square. (Courtesy of Robert Gibb, California State University, Fullerton student.) (b) A fragment of an unusually large and rounded CAI in Allende. It is viewed in transmitted light in an optical microscope with crossed polarizers to reveal the interference colors, which allow the different minerals composing the inclusions to be distinguished. The original inclusion was about 1.6 cm in diameter. (Courtesy of Glenn J. MacPherson, Smithsonian Institution.)

in some stellar environment prior to the formation of the solar nebula. In particular, the identified products of extinct radionuclides include ^{26}Mg derived from ^{26}Al. The latter has a mean life of only 1.07 Myr, which is not much longer than the time estimated to form the solar system from the collapse of an interstellar cloud. It has been suggested, in fact, that the nucleosynthetic event that produced the ^{26}Al and the initiation of the collapse of the protosolar cloud were not unrelated events: perhaps the outflow from an evolved star or supernova, in which new elements were recently formed, compressed and destabilized a neighboring cloud, which became the Sun and its nebula.

It is significant that the ages of other bodies of the solar system are close to that of the oldest meteorites. The oldest rocks from the Moon that have been dated have ages (determined by Rb–Sr radiochronometry) of 4.45 and 4.48 Gyr, which indicates an interval of only some 100 Myr between the formation of the solar nebula and the crystallization of lunar crustal rocks, a time span within which the Moon must have formed. On the other hand, the best estimate for the ages of ordinary chondrites obtained from the ^{207}Pb–^{206}Pb chronometer, is 4.552 Gyr, or only several million years younger than the oldest CAIs. The age of Mars is inferred to be at least 4.5 Gyr, based on the crystalli-

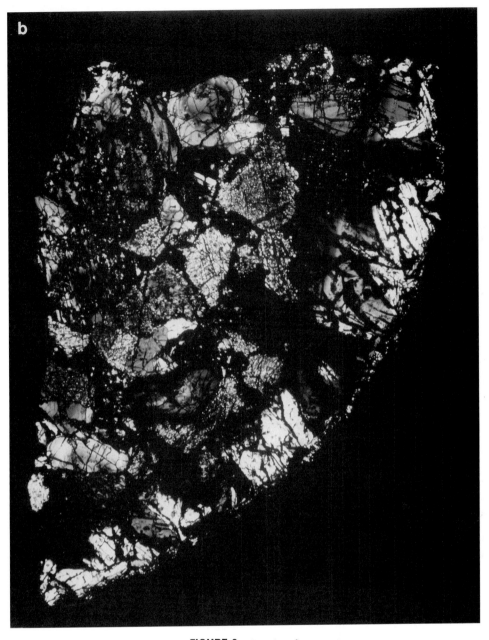

FIGURE 9   (*continued*)

zation age of a meteorite believed to come from that planet. The age of Earth has not been independently determined in such a manner (the oldest known minerals have an age of about 4.2 Gyr), but it can be shown to be very close to that of the meteorites if it is assumed that Earth shared the same reservoir of initial Pb isotopes, a likely circumstance. Thus there is compelling physical evidence that the solid bodies of at least the inner solar system formed within an interval that is only a tiny fraction of their ages, that is, they formed nearly contemporaneously.

B. OTHER TIMESCALES

Evidence regarding the duration of the nebular phase is derived from a number of isotopic measurements. Perhaps the most significant are those that indicate that a few CAIs were essentially free of ^{26}Al at the time they last solidified, despite the fact that they contain abundant aluminum. This result, taken together with the almost unvarying abundance level found in other CAIs and the inference that CAIs formed in the nebula, suggests that the nebula was present for at least a few million years, enough time for the ^{26}Al to decay to inconsequential levels. Another line of evidence that indicates that the nebula lasted for at least a few million years comes from Rb–Sr systematics. ^{87}Rb decays to ^{87}Sr; therefore, the initial ^{87}Sr/^{86}Sr ratio is higher in a sample obtained from a reservoir isolated later than in a sample from the same reservoir isolated earlier. Initial ^{87}Sr/^{86}Sr ratios found for some differentiated meteorites imply several million-year intervals of evolution in an environment with solar Rb/Sr, presumably the solar nebula, consistent with the available data for the absolute ages of these meteorites.

The igneous differentiation that produced the basaltic achondrites undoubtedly occurred on a parent body, that is, on a body much larger than the meteorite. Their Pb–Pb ages, referred to in the foregoing, thus indicate that igneous parent-body activity occurred within 20 Myr (perhaps even sooner) of the time of formation of the earliest solids detected, the Allende CAIs. Other indicators give similar results for other differentiated meteorites: enstatite chondrites show ^{129}I ages about 7 Myr younger than the Vigarano carbonaceous chondrite; pallasites (a type of stony-iron meteorite) give apparent ^{53}Mn ages of about 6 Myr younger than Allende; estimates for the differentiation timescales of iron meteorites (using ^{107}Pd) are typically less than a few tens of Myr; absolute Pb–Pb ages for zircons from a mesosiderite (another type of stony-iron) are within 15 Myr of the oldest CAIs. The discovery of evidence for the short-lived radionuclide ^{60}Fe in igneous meteorites suggests even shorter intervals between the formation of the earliest planetary products. Thus, the time required to build bodies large enough to produce planetary-type (igneous) differentiation was apparently less than 10 Myr. Precise Pb–Pb chronometry of ordinary (undifferentiated) chondrites yields comparable timescales for parent-body building.

Although there is an overall internal consistency among the timescales derived from the various radiochronometers, the measurements are difficult to make and their interpretations are not without controversy. Two factors can (and often do) render the derived results ambiguous: (1) the chronometer measures the time since isotopic closure, but the relation of this event to formation or other identifiable processes is often unclear; and (2) the distributions of some nuclides may have deviated from the uniform abundances that are thought to have generally characterized the nebula.

Evidence for isotopic heterogeneity of the nebula comes from a comparison of the abundances of the three stable isotopes of oxygen (atomic masses 16, 17, and 18) in meteorites with those in Earth and the Moon. Physical and chemical processing (e.g., evaporation or oxidation) can produce fractionation of isotopes that is *mass-dependent*, and terrestrial, lunar, and some meteoritic data are consistent with mass-dependent fractionation from a common reservoir such as the solar nebula. But some meteoritic data do not follow such a fractionation trend. Certain mineral assemblages, such as the CAIs in carbonaceous chondrites, contain oxygen isotope abundance ratios that can be explained as resulting from a mixture of an extremely ^{16}O-rich (perhaps pure) reservoir with another reservoir, such as one containing the average isotopic composition of the solar nebula. Furthermore, the oxygen isotope trends of the ordinary chondrites cannot be explained by simple mass-dependent fractionation either and appear to require a mixing with yet another isotopically distinct reservoir. Both of the anomalous reservoirs are presumed to be composed of solid material, since it is difficult to imagine how gaseous reservoirs could have avoided mixing with the ("normal") solar nebula. It should be emphasized, however, that even though the oxygen anomalies are of far greater magnitude than other isotopic anomalies, they are not obviously correlated with other anomalies, nor have their carrier phases even been identified. Therefore attempts have been made to find ways to selectively fractionate oxygen in a nebularlike environment. In fact, it is now known that such non-mass-dependent fractionations do occur in Earth's atmosphere and have been produced in laboratory experiments. It remains to relate these results to the meteoritic patterns.

Finally, it has been noticed that the inferred radioactive fractions ^{26}Al/Al, ^{53}Mn/Mn, ^{107}Pd/Pd, and ^{129}I/I are all roughly 10^{-4}, despite the fact that their mean lives vary by more than a factor of 20 (Table I). This had led to the suggestion that local (nebular or solar, not presolar) production of 10^{-4} (mass fraction) of the nebula may have been responsible for these extinct radionuclides. However, based on considerations of nucleosynthetic processes, it is deemed unlikely that all of these radionuclides could be produced in the nebula. Local production of ^{26}Al and ^{53}Mn by spallation reactions induced by solar energetic particles has been suggested, and would produce correlated effects; thus such hypotheses are testable, in principle. In a similar vein, a goal of studies of nucleosynthesis in stars is to

TABLE I
Extinct Radionuclides[a]

Nuclide	Mean life (million years)	Reference nuclide	Observed ratio
^{26}Al	1.07	^{27}Al	5.0×10^{-5}
^{60}Fe	2.2	^{56}Fe	1.6×10^{-6}
^{135}Cs	3.3	^{133}Cs	3
^{53}Mn	5.3	^{55}Mn	4.4×10^{-5}
^{107}Pd	9.4	^{108}Pd	2.0×10^{-5}
^{182}Hf	5.3	^{180}Hf	$8-80 \times 10^{-6}$
^{129}I	23.1	^{127}I	1.0×10^{-4}
^{92}Nb	50	^{93}Nb	2.0×10^{-5}
^{244}Pu	118	^{238}U	7×10^{-3}
^{146}Sm	149	^{144}Sm	7×10^{-3}

[a] From data summarized by A. G. W. Cameron (1993). Nucleosynthesis and star formation. *In* "Protostars and Planets III" (E. H. Levy and J. I. Lunine, eds.), pp. 47–73. Univ. Arizona Press, Tucson.

determine whether or not the measured and inferred abundances of all of the relevant radionuclides can be satisfactorily explained by remote production.

C. PHYSICAL CONDITIONS IN THE EARLY SOLAR SYSTEM

It may be that the type CI carbonaceous chondrites represent the pristine material of which the solid bodies of the solar system were formed. The fact that such unequilibrated material exists indicates that some fraction of the nebular solids was never substantially heated or otherwise radically processed. Data on the composition of the dust in the coma of Halley's comet, obtained from instruments on European and Russian spacecraft, indicate that virtually all major rock-forming elements in that comet occur at levels within a factor of two of solar/CI abundances. Although not all comets are necessarily the same, the measurement does reinforce the possibility that the vast majority of planet-forming material (now contained in the outer planets and comets) was never chemically or otherwise fractionated by nebular processes.

But the *ordinary* chondrites reveal another aspect of the nebular environment. These common meteorites are composed of spherical pebbles about 0.01–1 cm in size, their fragments, dust, and various other inclusions. The pebbles are the chondrules that lend their name to these important meteorite classes. They are found in all chondritic meteorites (except the CIs!), and, in some, constitute the bulk of the material. Laboratory analysis and experiments have proven that they were made by the heating of material (of chondritic composition, naturally) to the melting point or above (1800–2000 K), and then cooled at rates of 100–1000 K/hr; that is, they were liquid drops that cooled relatively rapidly, but were not quenched instantaneously. Their chemical fractionation patterns and mineralogy indicate that they are not the products of ordinary (planetary) igneous processes, but were made in a gaseous environment (e.g., the nebula). The short timescales that seem to be required (both heating and cooling) imply that their production was a *local* process; the times for overall nebula changes, or even transport times across nebula-scale gradients, would greatly exceed those inferred for chondrule production. Yet the prevalence of chondritic meteorites argues for a pervasive process. Thus, the chondrule-forming region was subject to energetic events that produced temperature excursions of at least several hundred degrees. The nature of these events is unknown and constitutes one of the primary unanswered questions regarding the origin of the solar system. Many possibilities have been proposed: lightning, flares caused by magnetic reconnection, gasdynamic shocks, collisions between solid bodies, high-velocity entry of dust balls into the nebula, ablation of drops from larger bodies, and exposure to intense solar radiation, among others. No explanation has gained favored status, either because of failings to account for chondrule properties or for the lack of testable predictions.

Estimates of the pressure in the chondrite-forming region of the nebula can be obtained from considerations of equilibrium chemistry and the contents of the most volatile elements in the meteorites. These estimates span a broad range, $10^{-6}-10^{-3}$ atm, but are all sufficiently low to imply direct gas-to-solid condensation in places where the nebula was hot enough to vaporize the common rock-forming elements. In fact, many aspects of chondritic chemistry and mineralogy appear to make sense in the context of fractionation during condensation as the nebula cooled through a series of equilibrium states, diminishing in temperature from one in which all material was vaporized to one of only a few hundred kelvins. But the picture is complicated by the coexistence, in the same meteorites, of distinctly low-temperature phases, apparently unaltered interstellar material (identified by its radical isotopic anomalies carried at the individual grain level), and refractories of igneous origin such as the CAIs. Clearly, these meteorites represent a mixture of nebular material that experienced a range of physical conditions that varied in both space and time. The histories of these objects have yet to be unraveled.

Meteorites are known to be magnetized, but the implications of this fact for the magnetic properties of the solar nebula are obscured by several factors: the possibility of contamination by terrestrial magnetization; disturbance of the natural remanence by chemical, mechanical, and thermal effects; ambiguities in the identification of the mineral carriers of the natural remanence; and uncertainties in how the natural remanence was acquired. Carbonaceous chondrites may give the most reliable estimates of primordial field strengths, 0.1–1 G, although even in this case the carrier phases have not been identified. Analyses of individual chondrules reveals that the magnetization is random in direction, which is interpreted to mean that it was acquired prior to accretion of the meteorite.

Energetic particle fluxes can affect meteoritic material in a number of ways: solar wind ions are implanted in exposed surfaces and can produce amorphous, radiation-damaged rinds on grains with prolonged exposure; solar flare Fe-group cosmic rays produce damaged crystal structure near the surfaces of mineral grains, which can be etched to reveal their tracks; galactic Fe-group cosmic rays produce tracks to greater depths; and both solar energetic and galactic cosmic ray protons can produce anomalous isotopic abundances by spallation. Tracks can also be produced internally by the energetic particles emitted by fissioning radionuclides (such as ^{238}U and ^{244}Pu). The effects of some of these processes can be used to diagnose conditions in the early solar system and other aspects pertinent to the accumulation of solid bodies. For instance, the time at which two meteoritic components came together can be deduced, in principle, if one of them possessed sufficient plutonium to produce measurable tracks in an adjacent, actinide-free, grain. Application of this technique is fraught with difficulties, but the resulting "compaction ages" (or, more appropriately, contact ages) are critically important for understanding primitive body chronology.

Furthermore, there are individual grains from some carbonaceous chondrites that contain spallation-produced ^{21}Ne at levels that seem to require precompaction exposure to especially high doses of energetic particles. Such an exposure might be acquired by prolonged residence in a parent-body regolith irradiated by current galactic cosmic ray (GCR) fluxes, or exposure for shorter times to a more intense flux, most plausibly from an active, T Tauri-like Sun. The issue is unresolved at present, but the latter appears more likely, since the equivalent regolith exposure time under the present GCR flux would greatly exceed both regolith model ages for asteroidal-sized bodies and the best estimates of relevant precompaction times. Therefore these grains (and others yet to be detected) may have recorded the only direct evidence that the Sun passed through an active T Tauri phase, as is presumed to have occurred.

D. PLANET-BUILDING PROCESSES

Meteorites display varying degrees of chemical and physical alteration, thereby providing clues to the stages by which planets were made. Metamorphism due to increasing pressure and temperature attests to the fact that bodies that never attained planetary size were nevertheless heated extensively, some to the point of igneous differentiation. Several sources of this energy have been suggested, among them collisional energy deposition, extinct radionuclides, and electromagnetic induction. The degree to which each contributed is debated. That hypervelocity collisions were pervasive throughout the planet-building stage is undisputed: all classes of meteorites show their effects (brecciation; deformation; mineralogical, chemical, thermal, and even isotopic alteration). If universally present in the relative abundances inferred for the aluminum-rich sites in which excess ^{26}Mg has been detected, the decay of ^{26}Al would provide enough energy to melt asteroid-sized bodies. The efficacy of electromagnetic induction heating depends on the strength of the magnetic field carried by the early solar wind, the absence of the solar nebula, and the physical characteristics (size and electrical conductivity) of the body; it remains a well-developed but untested hypothesis.

That the latest stages of planet building were pervaded by major collisions is attested to by the impact-scarred surfaces of planets and satellites (especially the atmosphereless, inactive objects like the Moon), and even the obliquities of the planets, which are best explained by collisions among the last planetary-sized bodies to be incorporated. Such an encounter is currently favored for the origin of the Moon. It has also been suggested that a major collision was responsible for stripping a differentiated Mercury of part of its silicate mantle, leaving an iron-rich planet, as is implied by its mean density. Finally, it is difficult to explain the elemental and isotopic compositions of the atmospheres of the terrestrial planets without invoking the influence of large-scale collisions, both those that resulted in a net gain of volatiles (e.g., comets bearing water) and those that would cause significant depletions. [*See* MERCURY.]

The compositions of the planets have frequently been interpreted in terms of the physical conditions of their origin. A useful concept was the idea that bulk composition, and therefore bulk density, was determined by the temperature distribution in the nebula, both being a reflection of the chemical equilibrium that was attained at the various planetary distances. The force of this argument has diminished with an increasing appreciation of the complexity of nebular evolution, accretionary processes, and planetary compositions. Certainly the incorporation of water as a major constituent of the outer planets, and its relative paucity on the terrestrial planets, is expected on the basis of nebular conditions that permit the condensation of ice only beyond the asteroid belt. But many compositional trends apparently require subtler interpretations.

For example, it is known that the gas-giant planets exhibit the following compositional features: (1) they contain greater-than-solar abundances, relative to hydrogen, of the major elements heavier than helium; (2) in particular, the major element carbon is enhanced over its solar abundance *in their atmospheres*; and (3) the enhancement of carbon increases systematically with distance from Jupiter to Neptune. From these facts, a model of the formation of the gas giants has emerged that begins with the collisional accretion of rock/ice bodies to form a core, accompanied by the gradual accretion of an atmosphere of nebular gas. As the mass of the atmosphere increases, it becomes dense enough to disrupt and dissolve accreting planetesimals; the constituents of the latter become part of the atmosphere, rather than the core. Accretion is finally terminated before the full complement of nebular gas has been accreted, because the planet has in some way become isolated from the nebula, or the nebula has been dissipated. The result is a planet with a rock/ice core and an atmosphere mostly of hydrogen and helium, but enhanced to some degree in heavy elements. Knowledge of the carbon abundance could be used to quantify this model, but to do so one must consider the following factors: (1) the total complement of heavy elements in the planet; (2) the ratio of condensed material to gas in the solar nebula; (3) the fraction of nebular carbon in gas (such as CO) and the fraction in condensable compounds (such as hydrocarbons); and (4) the fraction of planetesimal mass dissolved in the atmosphere. The first of these factors is constrained by the density of the planet and its gravitational moments (measured by accurately tracking the motions of passing spacecraft). The second is deter-

mined by the physical conditions in the nebula and constrained by cosmic abundances. Evaluations of the third factor rely on cosmochemical calculations and inferences about the compositions of outer solar system objects such as Pluto and the comets, as well as the meteorites. The fourth factor must be determined from models of the growth of the atmosphere and its physical interaction with accreting planetesimals. Thus, the compositions of the outer planets do not seem to be explained in terms of a simple, overriding principle; a broad range of complex nebular and planetary processes must be invoked.

The terrestrial planets offer similar challenges. The volatile element abundances in them are depleted relative to solar, and vary widely; but even potassium, only moderately volatile, is anomalous. It is not fractionated with respect to uranium by igneous processes, as demonstrated by similar K/U values from terrestrial samples of widely varying potassium concentrations, nor is K/U expected to be sensitive to atmospheric history, which has clearly varied from planet to planet. Yet K/U for Venus, Earth, and Mars (the latter inferred from meteorites believed to come from Mars) is substantially lower than that of CI meteorites (taken to be solar). There is no clear dependence on planetary mass. Moreover, K/U in eucrites, a kind of differentiated meteorite, is even lower than that of the planets, as is the lunar value. The fact that K is depleted relative to Si in primitive chondrites, again relative to CI, suggests the possibility that nebular fractionation processes played a role, although there is no clear dependence on heliocentric distance. Perhaps impacts, such as the one believed to be responsible for the origin of the Moon, affect K/U. [*See* PLANETARY IMPACTS.]

IV. IMPORTANT UNRESOLVED ISSUES

It is clear that a thorough understanding of the formation of the solar system has not yet been attained. In spite of its current popularity, it is doubtful that the theory described herein is correct in all respects. Some aspects are well established by observations and quantitative models, but others are very poorly understood. But it is difficult, or at least a highly subjective exercise, to identify those questions whose answers would be considered to complete our understanding. Nevertheless, the venerability of certain issues qualifies them immediately. The basic nature of chondrules, the major constituent of some of the most ancient material known, was recognized more than a hundred years ago; thousands have been analyzed down to the submicron level by hundreds of researchers in dozens of laboratories, yet the circumstances of their origin are unknown. On a grander scale, the conditions of star formation that lead to the birth of a single star surrounded by a protoplanetary disk have not yet been quantitatively and unequivocally distinguished from those that produce multiple stars, although considerable progress has been made in this area. These two problems bracket, in scale and discipline, a host of others whose resolutions would inarguably be considered major advances. Some of them are listed here.

1. What determined the mass of the Sun? Although it is widely believed that the onset of intense outflows disrupted the protostellar clouds from which stars are born, thereby halting accretion of interstellar gas, the mechanisms by which these outflows are generated, a quantitative theory of their behavior, their interactions with the protostellar environment, and the effects of other factors, such as rotation, are all incompletely understood at best.

2. What were the specific mechanisms responsible for the transport of angular momentum in the nebula and how efficiently did they act? Answers to these questions would determine or severely constrain the global nebula properties: the total mass available for making the planets and other bodies, the surface density as a function of time, the thermal structure of the nebula (which is related to the mass accretion rate from the nebula to the Sun), the duration of the nebula, and its stability with regard to its episodic behavior.

3. What is the true population of the Oort Cloud and Kuiper Belts? The bodies of these regions, long considered to be minor components of the solar system, are now recognized to be, potentially, a repository of a major fraction of angular momentum. Knowledge of their original numbers would provide a critical boundary condition on the solar system, namely, its primordial angular momentum.

4. What determined the masses of the giant planets? The fact that bodies exist that are squarely between rocks and stars is in some ways remarkable. Jupiter has perhaps six to ten times more mass of hydrogen and helium than rock and ice, but this is about five to ten times less than in a solar mixture. The idea of postformation loss of large amounts of hydrogen and helium is currently out of favor, but theories involving self-limiting mechanisms by tidal truncation (in which a planet exerts a gravitational torque on the surrounding nebula, thereby opening a gap in its feeding zone) or timely dispersal of the nebula are incomplete and difficult to test.

5. What happened to the nebular gas? This question, perhaps related to the preceding one, has attracted new attention because a venerable hypothesis, removal by a T Tauri wind, has been discredited, owing to recent observations of T Tauri stars suggesting that nebular accretion is necessary for a strong wind.

6. Can the nebular hypothesis satisfactorily explain the characteristics of planetary systems quite unlike those of the solar system? Recently discovered planetary-mass companions orbiting other stars have, in some cases, very small orbital radii or large eccentricities, in contrast to the well-spaced, orderly orbits of the solar system. These new discoveries hint at a previously unsuspected variety of planetary configurations, which will undoubtedly challenge aspects of formation theories based primarily on our local example.

7. What are the causes of the mass-independent isotopic heterogeneity observed in meteorites? This question is fundamental to understanding cosmochemical processes, the meaning of apparent radiometric ages, the nucleosynthetic prehistory of the solar system, and subsequent mixing processes in the nebula.

8. How were the oldest meteoritic inclusions (the calcium–aluminum-rich inclusions found mainly in carbonaceous meteorites) preserved to become mixed with components that were formed, or at least modified, independently, millions of years later? Theory predicts that such isolated nebular objects would drift into the Sun by orbital decay on timescales of a million years or less. Were they incorporated into larger objects that would resist orbital drift, to be broken out later and taken up in different bodies, or was there an outward flux or diffusion of nebular material that overcame inward orbital drift?

9. What processes were responsible for the patterns of chemical fractionation observed in the primitive meteorites, and the volatile abundances in the planets? The systematic depletion of siderophiles, the patterns of noble gas abundances, and other elemental and chemical variations imply the existence of fractionation processes, some systematic and some apparently chaotic, operating very early in the history of the solar system, in environments quite unlike well-studied terrestrial ones.

Most of the specific questions listed here bear heavily on those overriding ones that all studies of the solar system are related to, and that can only be properly answered by exploration. How common are planetary systems? To what degree are other planetary systems like ours? Do they contain Earth-like planets? Does life exist beyond our solar system? After centuries of thought, we are still left with more challenging questions than definitive answers. There appears little danger that we will soon eliminate the need for serious assaults on important solar system problems. [*See* EXTRA-SOLAR PLANETS: SEARCHING FOR OTHER PLANETARY SYSTEMS.]

BIBLIOGRAPHY

Black, D. C., and Matthews, M. S. (eds.) (1985). "Protostars and Planets II." Univ. Arizona Press, Tucson.

Kerridge, J. F., and Matthews, M. S. (eds.) (1988). "Meteorites and the Early Solar System." Univ. Arizona Press, Tucson.

Levy, E. H., and Lunine, J. I. (eds.) (1993). "Protostars and Planets III." Univ. Arizona Press, Tucson.

Lewis, J. (1995). "Physics and Chemistry of the Solar System." Academic Press, San Diego.

Taylor, S. R. (1992). "Solar System Evolution: A New Perspective." Cambridge Univ. Press, Cambridge, England.

Weaver, H. A., and Danly, L. (eds.) (1989). "The Formation and Evolution of Planetary Systems." Cambridge Univ. Press, Cambridge, England.

THE SUN

I. Overview
II. The Solar Interior
III. The Solar Atmosphere
IV. Solar Activity

J. B. Zirker
National Solar Observatory

GLOSSARY

Alfvén wave: Wave in a magnetic field, in which the field magnitude oscillates transversely to the direction of propagation.

C IV: Ion of carbon that has lost three of its initial six electrons.

Chromosphere: Region of the Sun characterized by temperatures between 6000 K and 20,000 K.

Continuum: That part of a spectrum that has neither absorption nor emission lines, but only a smooth wavelength distribution of radiant intensity.

Corona: Hot, tenuous outer atmosphere of the Sun.

Gaussian distribution: "Bell-shaped" curve that arises in statistical analysis.

Granule: Convection cell that surfaces in the photosphere.

Helioseismology: Study of the global oscillations of the Sun.

Heliosphere: Solar environment; the space filling the solar system.

Hydrostatic equilibrium: State of a gaseous or liquid medium in which the pressure at each level supports the weight of all the overlying material.

Kepler's laws: (a) The orbits of the planets are ellipses with the Sun at one focus; (b) the area swept out by the radius to a planet in equal time intervals is constant; (c) the square of a planet's period of revolution is proportional to the cube of its ellipse's major axis.

Kirchoff's law: The ratio of the radiant emissivity and absorptivity of a blackbody is equal to Planck's function.

Luminosity: Total radiative power emitted by a star.

Neutral line: Boundary between two regions of opposite magnetic polarity.

Neutrino: One of a family of three elementary particles, having neither mass nor charge.

Ohmic dissipation: Conversion of an electrical current to heat because of the resistance of the medium in which it travels.

Optical spectrum: Spectrum of a source that spans the visible wavelength range, approximately 380 to 850 nm.

Parsec: Astronomical unit of distance, equal to 3.26 light-years.

Photosphere: Radiant "surface" of the Sun, from which visible light escapes.

Power law: Mathematical relationship between two quantities in which one increases as a power (e.g., square or cube) of the other.

Spherical harmonics: Set of mathematical functions that comprise an orthogonal set, suitable for describing the distribution of any quantity over a sphere.

Spicule: Columnar dynamic structure of the solar chromosphere.

Supergranule: Convection cell, most easily detectable in the chromosphere, that is approximately 30 times as large as a granule.

Tokamak: Toroidal device used to investigate thermonuclear fusion in the laboratory.

Transition zone: Thin layer in the atmosphere that lies between the chromosphere and the corona.

I. OVERVIEW

The Sun is the central body of the solar system, a common main sequence star that is approximately 4.7 billion years old and that lies in a spiral arm of our galaxy, the Milky Way, at a distance of 8.5 kiloparsecs from the galactic center. To humankind, the Sun is literally the source of all life. To the astronomer, the Sun is an invaluable guide to the physics of other stars and a testing ground for astrophysical theories. Practically everything we know concerning the generation of stellar energy, the nucleosynthesis of the elements, the structure, dynamics, and evolution of stars, and stellar winds originated in studies of our nearest star. Spectroscopy, atomic theory, plasma physics, and climatology have all benefited from studies of the Sun.

The Sun and its planets were formed, astronomers think, about 5 billion years ago from an interstellar cloud of molecular hydrogen. After a few million years of gravitational contraction, the center of the Sun became hot enough to ignite the thermonuclear reactions that generate sunlight. List 1 summarizes some of the physical properties of the present Sun. [*See* THE ORIGIN OF THE SOLAR SYSTEM.]

The Sun is composed principally of hydrogen, with about 10% by number of helium and a few percent of all the other chemical elements (see List 2). The atoms are partially or fully ionized everywhere except in the cool surface layer known as the photosphere.

Solar energy is produced by the conversion of hydrogen to helium, at a central temperature of about 15 million kelvin (see List 3 for the reactions, which will be discussed in the next section). The radiant energy in the deep interior is in the form of gamma rays and X rays. This radiation flows toward the photosphere by a transport process that is similar to diffusion. As the energy flows outward, the gas temperature declines and the average energy of the photons also declines. At a distance of one quarter of the solar radius below the surface, the mode of energy transport changes from radiative diffusion to convection. In the convection zone (Fig. 1), energy is carried outward by the circulation of convection cells. Hot cells rise buoyantly until they reach the surface (the photosphere) where they radiate a portion of their heat to space. Then they cool and sink, only to repeat the process indefinitely. Since the convective motions are

LIST 1
Sun's Properties

Radius	695,970 km
Mass	1.989×10^{30} kg
Luminosity	3.85×10^{23} kW
Age	4.7×10^9 years
Temperature	
center	15.6×10^6 K
surface	6400 K
Rotation period	25 days (equator)
	35 days (latitude $\pm 70°$)

LIST 2
Chemical Composition of the Sun
(by number, relative to hydrogen)

	log N
Hydrogen	11.9
Helium	10.9
Carbon	8.2
Nitrogen	8.4
Oxygen	8.7
Neon	8.0
Magnesium	7.5
Silicon	7.4
Sulfur	7.0
Iron	7.7

LIST 3
Nuclear Reaction Chains

Proton–Proton Chain

$$p + p \rightarrow {}^2\text{H} + e^+ + \nu_e$$
$${}^2\text{H} + p \rightarrow {}^3\text{He} + \gamma \qquad 12.86 \text{ MeV}$$
$${}^3\text{He} + {}^3\text{He} \rightarrow {}^4\text{He} + p + p$$
or
$${}^3\text{He} + {}^4\text{He} \rightarrow {}^7\text{Be} + \gamma$$
$${}^7\text{Be} + e^- \rightarrow {}^7\text{Li} + \nu_e$$
$${}^7\text{Li} + p \rightarrow {}^8\text{Be} + \gamma \rightarrow {}^4\text{He} + {}^4\text{He} \qquad 17.35 \text{ MeV}$$

Carbon–Nitrogen Chain

$${}^{12}\text{C} + p \rightarrow {}^{13}\text{N} + \gamma$$
$${}^{13}\text{N} \rightarrow {}^{13}\text{C} + e^+ + \nu_e$$
$${}^{13}\text{C} + p \rightarrow {}^{14}\text{N} + \gamma$$
$${}^{14}\text{N} + p \rightarrow {}^{15}\text{O} + \gamma$$
$${}^{15}\text{O} \rightarrow {}^{15}\text{N} + e^+ + \nu_e$$
$${}^{15}\text{N} + p \rightarrow {}^{12}\text{C} + {}^4\text{He}$$

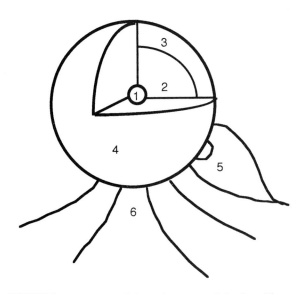

FIGURE 1 A diagram of the various parts of the Sun: (1) core, (2) radiative zone, (3) convection zone, (4) photosphere, (5) coronal streamer, (6) coronal hole. On this scale, the chromosphere and transition zone lie on the outer circle.

highly turbulent, they generate sound waves that propagate outward. These waves are thought to shock in the chromosphere, a zone that lies just above the photosphere.

The Sun has an extensive atmosphere, consisting of the chromosphere and the corona. The corona is normally hidden from the eye by the glare of the photosphere, but it is revealed as an array of nearly radial streamers during a total solar eclipse (Fig. 2) or in X-ray images of the Sun (Fig. 3). The corona owes its existence to nonradiative forms of energy, such as sound waves, Alfvén waves, and electric currents, that are generated by convective motions in and below the photosphere. This energy heats the tenuous solar atmosphere. As a result, the temperature rises sharply in a "transition zone" above the chromosphere and reaches 1 to 2 million kelvin in the inner corona. In some regions of the corona, principally the so-called "coronal holes," the plasma is able to expand freely into space as a solar "wind" that streams through and beyond the solar system.

The solar atmosphere is permeated with magnetic fields that are generated in or just below the convection zone and emerge through the photosphere as loops. Where particularly large loops cut through the photosphere, bipolar pairs of sunspots appear (Fig. 4). Surrounding and overlying the spots are magnetically active regions in which a variety of energetic transient events (e.g., flares) occur (Fig. 5; see also color insert). The Sun has an 11-year cycle of activity in which the number and

latitude distribution of sunspots (Fig. 6), the regularities of sunspot polarities, and the frequency of flares, among many other phenomena, vary systematically.

II. THE SOLAR INTERIOR

Our present knowledge of the solar interior is based mainly on theoretical models, which are constrained by observations of global quantities, such as the age, radius, and luminosity (total energy output) of the Sun. However, the study of solar oscillations (helioseismology) is also providing additional constraints and guidance for more refined models. Moreover, observations of the neutrino flux of the Sun provide a critical check on the accuracy of the so-called "standard model." In fact, much of the current research on solar models is driven by the need to satisfy both the neutrino data and the oscillation frequencies.

A. MODELS

Two general types of models are in use: static equilibrium models that describe the solar interior in its present state, and time-dependent models that trace the evolution of the Sun from an initial gas cloud to its present state. A successful evolutionary model yields a recognizable Sun within the accepted age (4.7 billion years) of development.

The standard model is a static model that incorporates the simplest assumptions and physics and still satisfies most of the global observations. It begins with these assumptions:

(a) The Sun is spherically symmetric and has a specified chemical composition and mass. The spectroscopy of comets provides the initial composition, which differs only slightly from that shown in List 2. The solar mass is determined most simply from Kepler's laws of planetary motion.

(b) The Sun neither loses nor gains mass during its evolution.

(c) The interior gas is static, except in localized shells where convection dominates energy transport.

(d) No diffusion occurs between zones of dissimilar composition.

The physics required for a model consists of the following elements:

(a) The production of energy by thermonuclear

FIGURE 2 Coronal streamers photographed at the total eclipse of 1983. (Courtesy of High Altitude Observatory.)

reactions that convert hydrogen to helium. The individual steps in this chain are summarized in List 3.

(b) The hydrostatic equilibrium of the gas, in which the weight of each spherical shell is supported by the difference of pressure between its upper and lower surfaces.

(c) The transport of energy by radiation, which requires a detailed calculation of the opacity of the solar gas as a function of density and temperature. Except in a central zone, where nuclear energy is released, the amount of radiant energy passing through each spherical shell is constant, that is, "radiative equilibrium" prevails.

(d) The transport of energy by convection, wherever it is more efficient than radiative transport. Convection is usually modeled by assuming that hot bubbles dump their excess heat after rising a distance of a few pressure scale heights.

(e) An equation of state that relates the pressure, temperature, and density. Throughout most of the interior, the perfect gas law is valid, but small corrections are required at the highest densities.

The basic structural equations conserve mass, momentum, and energy at each depth. Together with the equation of state, the rates of nuclear reactions, and the surface boundary conditions, they define a completely determined problem. An acceptable solution will re-

produce the observed solar radius and luminosity. The solution can be fine-tuned to some extent by varying the abundance of helium and the details of convective transport.

Figure 7 shows the run of temperature and density along a solar radius in the standard model. The transport of energy is radiative everywhere except in a convective shell approximately one quarter of a radius in thickness. One of the first successes of helioseismology was the verification of this thickness and the associated helium abundance.

The standard model is a triumph of economy, which accounts for the radius and luminosity, at a prescribed mass, with well-established physical principles. Prime credit for its success goes to H. Bethe, who first identified the essential nuclear processes in 1939.

B. NEUTRINOS

The proton–proton chain of nuclear reactions (see List 3) produces about 98.5% of the energy in the Sun, and the carbon–nitrogen cycle produces the remainder. In the process, four protons are converted to an alpha particle, the nucleus of helium, with intermediate products of deuterium (^2H), helium three (^3He), beryllium seven (^7Be), and lithium seven (^7Li). The chain has three branches, with different products and energy release. The principal chain is the burning of helium

FIGURE 3   An X-ray view of the Sun. The brightest areas are active regions. (Courtesy of Smithsonian Astrophysical Observatory.)

three to helium four, which produces about 86% of solar energy.

Notice that neutrinos (ν_e), elementary particles with essentially no mass and no electric charge, are produced at several points in the chain with maximum energies of 0.4, 0.8, and 1.4 million electron volts (MeV). In 1955, R. Davis, Jr., of the Brookhaven National Laboratory conceived an experiment to detect the high-energy neutrinos that reach the Earth. He based his technique on the conversion of a stable isotope of chlorine, ^{37}Cl, to a radioactive isotope of argon, ^{37}Ar, by the absorption of a neutrino. The capture rate is very slow since neutrinos interact very weakly with other atoms. Thus only a few argon atoms are produced each day in a target of several hundred tons. However, the radioactive argon can be collected with great efficiency by radiochemical means and yields a direct estimate of the high-energy neutrino flux. In 1965, Davis set up a 600-ton tank of perchloroethylene, a common dry-cleaning fluid, in the Homestake Mine in South Dakota, at a depth of 5000 feet. This arrangement was necessary to shield the chlorine detector from cosmic rays.

Davis's experiment soon indicated a discrepancy between the neutrino flux predicted by the standard model, about 7 SNU (Solar Neutrino Unit, defined as 10^{-36} captures per second per target atom), and the measured flux, about 2 SNU. The experiment has continued for over 25 years (it is still running) and, despite fluctuations in the data, has continued to show this discrepancy. After much effort to refine and calibrate the experimental technique, and further efforts to check the model, specialists have concluded that a real deficiency of a factor of three exists. This challenging result has been confirmed in recent years by four independent neutrino detectors: the GALLEX,

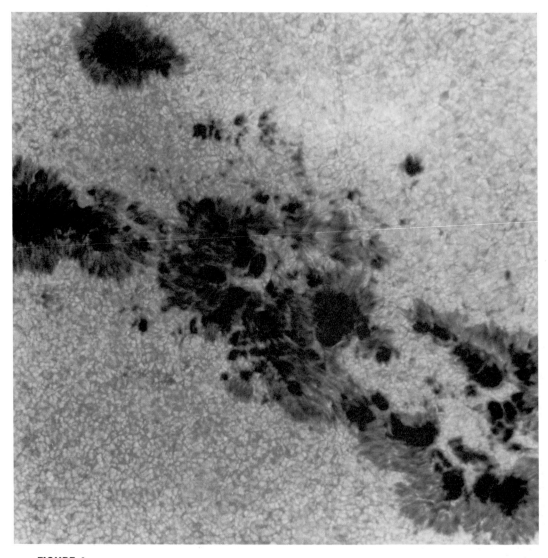

FIGURE 4   A large complex of sunspots. The small white features surrounding the dark spots are granules.

near Rome; the SAGE, in the Caucasus Mountains; and the Kamiokande I and II near Tokyo.

Many possible explanations for the neutrino deficit have been investigated, including revisions of the standard model and revisions in the properties of neutrinos. The favored idea at the moment, proposed by L. Wolfenstein, S. Mikheyev, and A. Smirnov, is that neutrinos "oscillate" among the three possible "flavors" of electron, muon, and tau-lepton neutrinos, as they pass through the Sun. Since each of the four devices mentioned here detect only the electron neutrinos, a deficit should be expected.

To check on this proposed explanation, several new detectors are being built. The Sudbury Neutrino Observatory, in Canada, will be able to distinguish elec-

tron neutrinos from other types in the decay of ^{8}Be. The Borexino machine, near Rome, will be especially useful in tracking low-energy electron neutrinos released in the capture of an electron by ^{7}Be. Super-Kamiokande will be able to detect neutrinos of any flavor when they scatter off electrons in the 700-ton water target. In efforts to shield their detectors, physicists have gone to extreme lengths. AMANDA has been lowered into a mile of ice at the South Pole, while NESTOR has been running since 1991 submerged deep in the Mediterranean Sea. Similarly, a Russian experiment lies deep in Lake Baikal. Within a few years, after these experiments have collected sufficient amounts of data, the neutrino deficit problem may be better understood, if not solved.

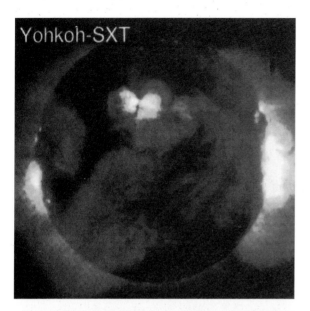

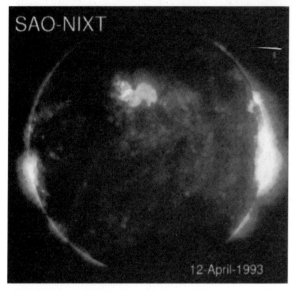

FIGURE 5 A solar flare observed simultaneously in soft X rays by the *Yohkoh* satellite and by a telescope (NIXT) aboard a rocket. (Courtesy of the *Yohkoh* team.) (See also color insert.)

C. HELIOSEISMOLOGY

In 1960, R. Leighton and his students at Cal Tech discovered that the visible surface of the Sun (the photosphere) is covered with patches a few thousand kilometers in size that oscillate vertically with a period of about 5 min. At any point on the surface, the oscillations build up in amplitude and then, within a few periods, decay, and the phase changes from one burst of oscillations to the next. The mean amplitude of the velocity signal, typically a few hundred meters per second, increases with altitude in the photosphere.

At first these oscillations were thought to be random pulses driven by granules, but in 1970, R. Ulrich at UCLA and independently C. Wolfe (at NASA/Goddard Space Center), J. Leibacher (now at the National Solar Observatory), and R. Stein (now at Michigan State University) suggested that the oscillations were evidence of standing sound waves in the solar convection zone. In short, the oscillations at the surface represent a complicated interference pattern of several million natural modes of vibration of the solar interior. Each model has a velocity amplitude of only a few centimeters per second, but their superposition raises the signal at any location to hundreds of meters per second.

The theory predicted a relation between the discrete frequency and the horizontal wavelength of each mode, a "dispersion relation." When plotted in a graph of frequency (ω) versus the reciprocal of the wavelength ($k = 1/\lambda$), the allowed modes fall along isolated ridges (Fig. 8). In 1975, the German astronomer F. L. Deubner confirmed this interpretation with a now classic set of observations that revealed the predicted ridges. The observed periods lie between 3 and 6 min, and the horizontal wavelengths range from a few thousand kilometers to the full circumference of the Sun.

1. Acoustic Oscillations

A drum or a horn produces a definite musical note as the result of the constructive interference of sound waves within a cavity of some definite size and shape. Only waves with certain wavelengths can fit exactly within the cavity, without any overlap. At any point in such a "standing wave," the air pressure and velocity oscillate sinusoidally. At "nodes," separated by distances of half a wavelength, the oscillations vanish.

The Sun contains no cavity with rigid walls, but gradients of temperature and density can reflect sound waves efficiently. The global oscillations discussed here are sound waves that are reflected from the bottom and top of the convection zone. Each allowed wave or mode can be assigned a set of three numbers that define the components of its wavelength along the radial, azimuthal, and polar directions. The ridges in the k–ω plane (see Fig. 8) are distinguished by the radial "quantum number," n, which specifies the number of nodes along a radius from the center of the Sun. The horizontal wavenumber k at any point is proportional to the "degree" l, the number of nodes along a meridian great circle. The azimuthal number, m, counts the number of nodes along the equator.

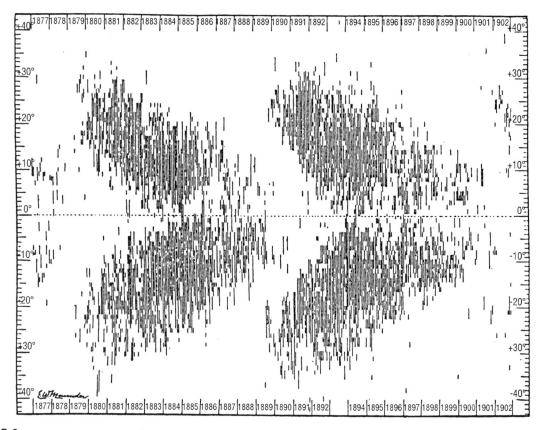

FIGURE 6   The "butterfly" diagram of the sunspot cycle. Spots of a new cycle appear at lower and lower latitudes. This is a copy of
E. W. Maunder's discovery diagram, published in 1904. (Courtesy of *Monthly Notices of the Royal Astronomical Society.*)

Sound waves in the convection zone propagate at
the local speed of sound, which increases with depth
because the temperature increases. As a result, sound
waves in the Sun are refracted as well as reflected. A
plane wave that descends into the Sun at an oblique
angle experiences a faster speed of sound at its deeper
edge. Thus this edge gains over the shallower edge,
and the plane wave turns back (refracts) toward the
surface. Such a wave will trace a series of arcs, reflecting
at the surface and refracting below it. If its horizontal
wavelength is a simple fraction of the solar circumference, the wave is resonant and can persist as a standing
wave. In general, waves of lower frequency (periods
around 6 min) penetrate more deeply into the interior
than waves of higher frequency.

Solar physicists derive the sound speed inside the
Sun in much the same way that seismologists determine the sound speed inside the Earth. Whereas seismologists observe many transient earthquakes, solar
physicists observe surface oscillations at many places
on the photosphere and for many 5-min periods. The
spatial distribution of the solar oscillations is fitted
with mathematical functions (spherical harmonics) that

assign the three quantum numbers (n, l, and m) to
each observed mode. The temporal fluctuations of the
velocity signal are Fourier-analyzed to derive the exact
frequency of every resolvable mode. These primary
data are then compared with the predicted frequencies
from a standard model of the interior and the comparison yields corrections to the model's radial temperature distribution. As longer and longer series of continuous observations become available, the modes'
frequencies are established with a precision of better
than one part in a billion.

2. Detection Schemes

In their quest for high-frequency precision, solar astronomers have resorted to a number of techniques.
Precision ultimately depends on the length of a continuous time series of observations, uninterrupted even
by the day–night cycle. These conditions can be met
at the South Pole during the austral summer, subject
only to occasional periods of bad weather. Several

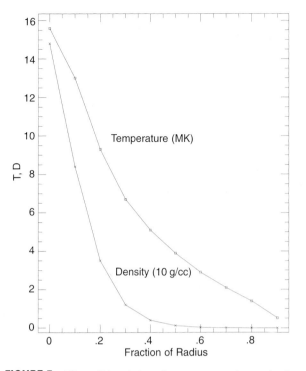

FIGURE 7 The radial variation of temperature and mass density in the solar interior as predicted by the "standard" solar model. The temperature scale is in millions of kelvin (MK) and the density scale is in tens of grams per cubic centimeter.

groups have observed the oscillations for as long as two weeks at the pole.

Alternatively, a network of observing stations has been used to observe the Sun continuously for several months. The Global Oscillation Network Group (GONG), a consortium of a hundred astronomers, is currently analyzing data from the most ambitious of these networks. GONG consists of six identical instruments distributed in longitude around the world. A map of photospheric velocities is obtained every minute from each site. These maps are collected every few weeks, calibrated at a center in Tucson, Arizona, and distributed to the consortium. GONG will operate continuously for at least three years to acquire perhaps the most complete set of observations attainable from Earth.

Finally, an oscillation detector is operating aboard the Solar and Heliospheric Observatory, a space mission launched in early 1996. The Sun is observable continuously from this spacecraft without any interference from weather or atmospheric scintillations, since it is located at the L1 Lagrangian point, a million kilometers from Earth in the direction of the Sun.

3. Some Results

The past decade of helioseismology, and especially the past year of GONG operation, has yielded several important results. We now know that helium nuclei represent about 24% of the mass of the interior, the heavy elements compose about 2.5%, and the remainder is hydrogen in the form of protons. Moreover, determinations of the sound speed along a radius yield the temperature profile of the interior. The agreement with theoretical models is very good.

Observations of the Sun's surface showed long ago that the solar equator rotates about 50% faster than the regions near the poles. Since the surface evidently does not rotate as a rigid body, how does the interior rotate? Helioseismology has provided an answer.

Sound waves propagate in a medium at a speed fixed by the temperature of the medium. If the medium itself is moving, the sound waves are carried along with it, and their observed speed is thus higher or lower depending on the wave's direction relative to the medium's direction. Thus, the observed frequencies of solar sound waves shift higher or lower, as a result of the Doppler effect, when the waves pass through layers that rotate at different speeds. Each original frequency is split into a pair that corresponds to waves traveling in the same or opposite direction as the rotation. Since short-period waves favor the surface and longer-period waves favor the deeper layers, it is possible to combine observations of rotational frequency splitting in such a way as to find the speed of rotation as a function of depth and latitude in the solar interior. Figure 9 illustrates some recent results. In the convection zone, rotation speeds tend to remain constant with depth, but vary with latitude. Just beneath the photospheric equator lies a belt of maximum speed. The radiative zone seems to rotate as a rigid body, somewhat slower than the equator at the surface.

4. Gravity Waves

Note that the derived rotation speeds are uncertain below the convection zone and missing entirely in the solar core, where few sound waves penetrate. Some astrophysicists expect that the Sun has a rapidly rotating core, a remnant of its original contraction from a gas cloud. To examine the deep interior, helioseismologists will have to study a different type of global oscillation, namely, gravity waves. In this type of oscillation, buoyancy, rather than pressure, supplies the restoring force. Gravity waves are predicted to have long periods (hours) and very small velocity amplitudes

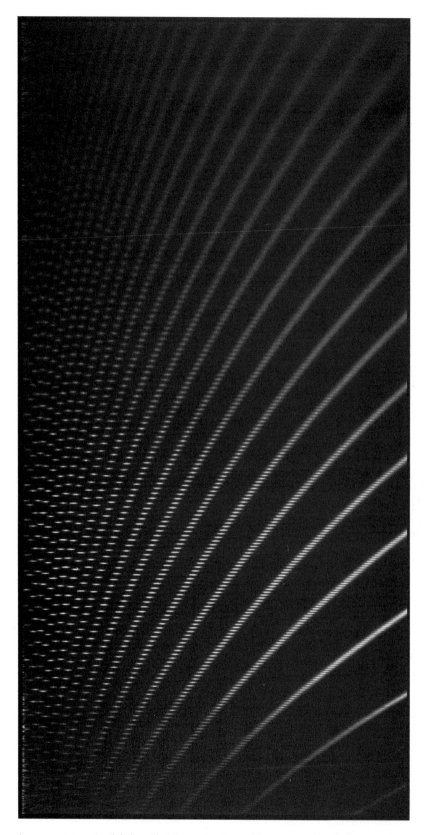

FIGURE 8 The acoustic power present in global oscillations, as functions of inverse wavelength (horizontal axis) and frequency (vertical axis). Each ridge is a family of modes that have a definite number of radial modes, the $n = 1$ mode lying lowest and farther to the right in the diagram. Each ridge is further subdivided (the small dashes) into modes with different numbers of meridional modes.

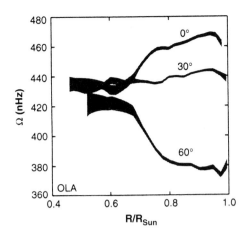

FIGURE 9 Rotation frequency as functions of radius and latitude, determined from helioseismic observations. (430 nHz corresponds to a rotation period of 27 days.) (Courtesy of *Science.*)

(millimeters per second). Since they are almost entirely confined to the deep interior, the detection of these oscillations will require very sensitive techniques. A number of researchers have claimed to detect gravity waves, but not convincingly. The study of these weak oscillations remains for the future.

5. Oscillations of Brightness

Global acoustic oscillations produce fluctuations in the brightness of the photosphere as well as of the next higher layer, the chromosphere. The lowest-degree oscillations, which correspond to the longest wavelengths, have been detected in the flux of integrated sunlight by active cavity radiometers aboard a series of satellites. NASA has attempted to place such detectors on every suitable satellite because of the potential importance of such measurements for the Earth's climate. The signal is only a few parts in a million of the steady solar output but is quite unambiguous.

6. The Future

The future successes of helioseismology are difficult to predict, but this new field of solar physics will undoubtedly play an important role in the investigation of the generation of solar magnetic fields. Studies of the solar cycle of magnetic activity will certainly benefit from helioseismic data. Moreover, astronomers are already attempting to find stellar analogs of solar oscillations, and indeed have found them in a class of white dwarfs.

III. THE SOLAR ATMOSPHERE

As Figs. 2 and 3 illustrate, the solar atmosphere is hardly a smooth, spherically symmetrical shell. Instead it consists of a complicated arrangement of interleaving structures that have spatial scales ranging from a few hundred kilometers to several hundred thousand kilometers. Each type of structure has its own characteristic temperature and density distribution. For a first crude description, however, it is useful to refer to several "layers," arranged along an outward radius, that correspond roughly to regions of increasing temperature and decreasing density. We have already referred to the names given to these layers: the photosphere, chromosphere, transition zone, and corona. In addition, one can distinguish between the active atmosphere, in the general vicinity of a sunspot group, and the quiet atmosphere, far from such a group.

A. THE SPECTRUM OF THE QUIET SUN

The solar spectrum is the source of much of our knowledge of the atmosphere. The photospheric continuum ("sunlight") has an approximate blackbody form with a color temperature of about 6000 K. Superposed on this continuum are the Fraunhofer absorption lines, which correspond to transitions in atoms of neutral and singly ionized metals, such as sodium, magnesium, and iron. The bands of many simple molecules, such as CO, CH, and CN, are also present in the spectrum. Over 20,000 atomic lines have been identified in the optical spectrum and their relative strengths, corrected for their atomic transition probabilities, yield the Sun's chemical composition. The relative concentrations of some of the more common elements are shown in List 2.

The optical spectrum of the chromosphere, originally observed at total eclipses at the limb of the Sun, is almost a "reversal" of the photospheric spectrum: each dark absorption line appears as a bright emission line. In addition, the chromosphere radiates strongly in the resonance lines of hydrogen, helium, magnesium, and other abundant elements, at wavelengths between 100 and 300 nm.

In the transition zone, the emission lines arise from multiply ionized abundant elements, such as C IV, N V, and O V. (C IV denotes the ion of carbon that has had three of its original electrons stripped off.) These lines appear mainly in the extreme ultraviolet, at wavelengths between 30 and 150 nm.

The spectrum of the nonflaring corona contains

resonance emission lines of highly ionized species, such as Fe IX to Fe XVI, at wavelengths between 10 and 50 nm. In addition, "forbidden" (quadrupole) transitions of ions such as Fe X, Fe XIV, and Ca XV appear in the optical coronal spectrum against a faint coronal continuum. The latter is produced by the scattering of photospheric light from free coronal electrons, and at radial distances beyond two solar radii, by scattering from interplanetary dust particles. [*See* INTERPLANETARY DUST AND THE ZODIACAL CLOUD.]

Solar flares produce the full range of plasma temperatures, from 10^4 to 10^7 K. The hottest regions briefly emit spectrum lines of such hydrogenlike ions as Ca XIX and Fe XXV.

The quiet Sun also emits a full spectrum of radio waves. The plasma density of the emitting region determines the characteristic wavelength. Thus the photosphere emits millimeter waves, the chromosphere emits centimeter waves, and the corona emits decimeter and meter waves. Active regions, sunspots, and flares emit polarized radio radiation with rapid time variations. [*See* THE SOLAR SYSTEM AT RADIO WAVELENGTHS.]

B. THE RADIAL TEMPERATURE AND DENSITY PROFILES

From the great number of spectroscopic observations collected over the past century, astronomers have constructed empirical one-dimensional models of temperature and density as functions of height through the atmosphere. Figure 10 shows averages for the quiet atmosphere. Note that the electron temperature passes through a minimum of about 4400 K, a few hundred kilometers above the visible surface and then rises steeply through the chromosphere and the transition zone to the million-degree corona. However, recent infrared observations of carbon monoxide molecules indicate that, over much of the solar surface where photospheric magnetic fields are weak, the temperature minimum is a mere 3000 K.

Except for the region near the temperature minimum, hydrogen is fully ionized. The atmosphere consists of a mixture of positive ions (principally protons) and electrons. It is electrically neutral and is called a "plasma." The radial distribution of ion number density at the poles differs from that at the equator.

The photosphere is sufficiently dense and opaque to ensure that thermodynamic equilibrium at the local temperature prevails in each volume. The spectrum of the radiation inside the photosphere resembles that of a blackbody and the ionization and excitation of the

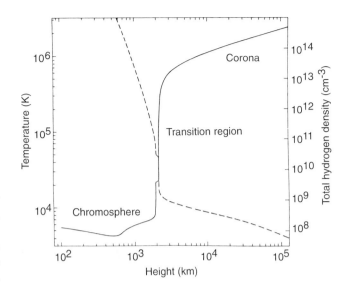

FIGURE 10   The variation with height of the temperature and proton density in a one-dimensional model of the solar atmosphere. (Courtesy of *Annual Reviews of Astronomy and Astrophysics.*)

elements are controlled by the Saha and Boltzmann laws, respectively. In the chromosphere, transition zone, and corona, the plasma is partially or wholly transparent to radiation, and collisions among particles are much less frequent. Therefore, thermodynamic equilibrium is not valid in these regions. A complex formalism, based on the assumption of a steady state of microscopic processes, has been developed over the past 30 years to analyze conditions in the more tenuous layers. The equilibrium ionization and excitation of the elements are governed by collisions among ions and electrons, at rates set by the local electron temperature and density. In this portion of the atmosphere, radiation is produced ultimately by collisions among particles and it emerges, without appreciable absorption, with a non-Planckian distribution.

C. THE ENERGY BALANCE OF THE ATMOSPHERE

According to current ideas, the steep rise in temperature through the atmosphere is the result of a balance between heating (by nonradiative energy that is generated in the convection zone) and cooling (by radiation to space). A comprehensive understanding of this energy balance and the resulting distribution of temperature and density has not been achieved as yet, but considerable progress has been made in studying the energy balance in individual layers.

At least three forms of nonradiative energy that heat the outer atmosphere have been suggested. Sound waves with periods between a few tens of seconds and a few hundred seconds can be generated as a by-product of turbulent convection below the photosphere. Waves with frequencies above a critical value (corresponding to about 300 s) can propagate vertically. As the waves rise into regions of rapidly decreasing density, their velocity amplitudes would increase accordingly, and when their amplitudes reached the local sound velocity, the waves would shock and dissipate their energy. The lower layers of the chromosphere are thought to be heated in this fashion. Propagating sound waves have been detected and studied by various means; their existence is well established.

Magnetohydrodynamic (MHD) waves, particularly Alfvén waves, have been suggested as agents for heating the upper chromosphere and corona. However, Alfvén waves are difficult to generate with the slow convective motions and, moreover, Alfvén waves do not dissipate readily. The best empirical evidence for the possible existence of such hydromagnetic waves in the atmosphere is the nonthermal broadening of spectrum lines in the low corona. The width of such lines is fixed in part by the distribution of thermal speeds of the radiating ions. However, observations show that the line widths are larger than simple thermal broadening can explain. The excess width has been attributed to excess nonthermal motions, among them MHD waves.

If sound waves shock in the chromosphere and if the evidence for MHD waves is weak, how is the million-degree corona heated? In recent years, E. N. Parker at the University of Chicago has argued effectively for heating by ohmic dissipation of electrical currents. He proposes that the currents arise from the slow twisting and braiding of coronal magnetic fields as the footpoints of the fields are shuffled by convective motions in and below the photosphere. The currents would be confined to extremely thin sheets (as narrow as 30 m) at places in the corona where the field direction changes discontinuously. As the convective motions continue to deform the coronal field, the current density rises in the sheets. Eventually they become unstable to a variety of plasma instabilities and dissipate, converting their electrical energy to heat and mass motions. During this process, magnetic fields with oppositely directed components can reconnect and release a fraction of their nonpotential energy. Such dissipative processes have been observed in Tokamaks and in three-dimensional numerical simulations, and are postulated to occur in flares (see Section IV). The result of such a field reconnection in the corona should be a tiny

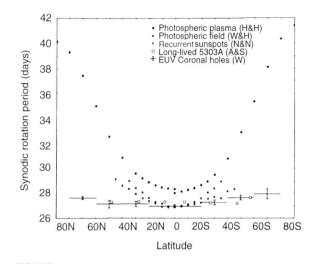

FIGURE 11 The rotation period of different parts of the Sun. The equator rotates faster than the poles and the corona rotates more rigidly than the photosphere. (Courtesy of Colorado Associated University Press.)

flare, a "nanoflare," that emits a burst of X rays and radio radiation. The corona would be heated, in Parker's scheme, by many small episodic injections of magnetic energy. To supply the corona's energy requirements, such nanoflares would have to occur in sufficient numbers. Counts of very small flares, observed with the *Yohkoh* satellite, fail to meet this criterion. The long-standing problem of coronal heating remains unsolved at the present time.

D. LARGE-SCALE MOTIONS

The rotation of the photosphere has been determined from daily maps of Doppler velocities (principally at the Mount Wilson Observatory) or from the displacements of individual features, such as sunspots or magnetic field patches. Although different methods yield slightly different results, the general features persist. The visible photosphere rotates from east to west but not as a rigid body would. Instead the equator rotates faster (in 25 days, relative to the distant stars) than the higher latitudes. At latitudes of ±75°, for example, the photosphere rotates in 34 days. This latitude-dependent effect is called the differential rotation (Fig. 11). The photosphere also flows slowly from the equator to the poles, at a speed of about 10 to 20 m/s.

As Fig. 11 shows, the corona (whose brightness is dominated by coronal streamers) tends to rotate more rigidly than the photosphere, but the details are complicated. A recent analysis of white light images shows that at latitudes less than 40°, and at distances from disc center less than 1.5 solar radii, the corona rotates differentially with a sidereal period of 26 days. At higher latitudes and greater heights the sidereal period is nearly constant at 34 days. Moreover, the corona decelerates at all latitudes during the rise of the sunspot cycle and accelerates after solar maximum. Similarly, X-ray images of coronal holes show quasi-rigid rotation over periods of several months.

So far, theorists have been unable to provide a detailed physical explanation for solar rotation. The photospheric rotation reflects the motions of the interior, which involve the interaction of convective flows and Coriolis forces. Recent results in helioseismology (see Section II) on the differential rotation inside the Sun should help to guide the theory. The quasi-rigid rotation of the corona is still more puzzling as it seems to involve the decoupling (reconnection) of coronal fields from surface fields at some intermediate height.

Three spatial scales of solar convective cells are recognized: the granulation (1000 km in size, with a lifetime of about 10 min), the mesogranulation (5000 km, lifetime of 8–10 h), and the supergranulation (30,000–40,000 km, lifetime of 24–48 h). The plasma at the center of supergranules rises into the photosphere at about 100 m/s, expands horizontally with a speed of about 400 m/s, and sinks down again at about 200 m/s. A fourth scale of convection, the so-called "giant cells," with sizes of several hundred thousand kilometers, have been postulated but not detected as yet.

E. SOLAR MAGNETIC FIELDS

As Fig. 3 illustrates, the solar atmosphere is highly structured because of its internal magnetic fields. In the photosphere, the gas pressure exceeds the magnetic pressure so the gas can concentrate and transport the field. Above the photosphere, the gas density and pressure decrease exponentially with a scale height of a few hundred kilometers, whereas the magnetic pressure falls off less rapidly. Thus, at a few hundred kilometers above the photosphere, the magnetic pressure greatly exceeds the gas pressure and the field shapes the gas into the structures we observe. Moreover, the plasma has a high electrical conductivity, and thus is constrained to flow along, not across, the magnetic field. As a result, the motions in the upper atmosphere are controlled indirectly by motions in the photosphere, with the magnetic field acting as the mechanical link.

Vector magnetic fields in the photosphere can be determined from the splitting and polarization of spectrum lines. Unfortunately, the fields in the corona are too weak to measure reliably by any known method. Astronomers therefore estimate coronal field strengths by extrapolating surface fields or, in special cases, from the polarization of radio bursts.

In Section III, F, we shall consider the properties of some important structures that make up the atmosphere, but first we describe the spatial organization of the fields.

Figure 12 shows a typical map of photospheric fields. Most of the magnetic flux is concentrated in the active regions at latitudes between 10 and 40°. Sunspots usually come in pairs of opposite polarity, with total magnetic fluxes on the order of 10^{22} maxwells. Smaller bipoles, called ephemeral active regions, with fluxes on the order of 10^{19} maxwells and lifetimes of a few hours, appear over a broader range of latitudes. Even smaller dipoles (the "intranetwork fields") emerge inside the supergranulation cells, uniformly over the entire Sun, and are swept to the cell boundaries within a day to form the coarse network.

In the quiet photosphere, away from active complexes, the horizontal flows of supergranulation cells redistribute the surface magnetic flux. A detailed magnetogram shows that the flux concentrates at the borders of these cells in a coarse network. These fields extend into the higher atmospheric layers, spreading laterally with height until, in the low corona, the pattern of a coarse network vanishes.

Observations at still higher spatial resolution show that the magnetic flux occupies a small fraction of the Sun's surface. The principal difference between an active region and the quiet Sun is not the strength of the magnetic fields but rather the "filling factor" of the magnetic flux, the fraction of the surface it covers. The flux emerges in the photosphere as narrow tubes as small perhaps as 100 km in diameter, with true field strengths ranging from a few hundred to about a thousand gauss. In the chromosphere and corona, these tubes flare out laterally to fill a much larger volume, with field strengths estimated from 1 to 10 G. The field at the center of a large sunspot can reach 3500 G.

Notice in Fig. 12 that, with few exceptions, all the western ("leading") halves of a bipole have the same polarity and that this polarity is different in the northern and southern hemispheres. In the next activity cycle of 11 years, the leading polarities in the two hemispheres will reverse. Each solar pole contains so-called "open" field lines, whose dominant polarity reverses near the maximum of the 11-year cycle. These

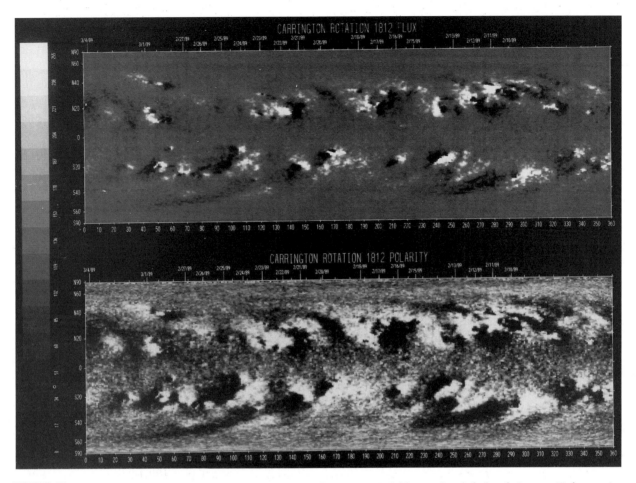

FIGURE 12   A magnetogram showing the strong and weak photospheric magnetic fields over the whole Sun during one 27-day rotation. Activity tends to concentrate in two bands at latitudes between 10° and 30°.

regularities are the so-called Hale–Nicholson laws (see Section IV).

As an active region decays (over a period of several months), differential rotation, meridian flow, and supergranule flows stretch out the region's flux into long bipolar regions at mid and high latitudes (see Fig. 12). As the bipolar field expands laterally and rises vertically, a helmet streamer (see Fig. 2) may develop in the overlying corona. As a result, the inner corona is organized into an array of streamers, separated by regions of open field lines, which are called "coronal holes." X-ray images of the corona (see Fig. 3) show that, in addition to these two basic structures, long coronal loops can connect different active complexes, even in different hemispheres.

Solar prominences form within a streamer, above the line in the photosphere that separates the opposite polarities of the weak bipolar region. They are flat sheets of plasma two orders of magnitude denser and

cooler than the surrounding corona. Prominences appear as dark filaments on the solar disk, when photographed in the hydrogen H alpha line (Fig. 13).

Quiescent filaments, which lie outside of active complexes, possess horizontal magnetic field components that obey a global rule. In the northern hemisphere, the axial fields of such filaments are directed to the *right* when viewed by an imaginary observer who stands in the positive polarity region adjacent to the filament. Such filaments are called *dextral*. In the southern hemisphere, the axial field components are directed toward the *left* when viewed from the neighboring positive polarity region. Such filaments are called *sinistral*. Most filaments in the northern hemisphere are dextral, most in the southern hemisphere are sinistral, and this regularity does not change from one 11-year solar cycle to the next, despite the change in polarities of the leading sunspots.

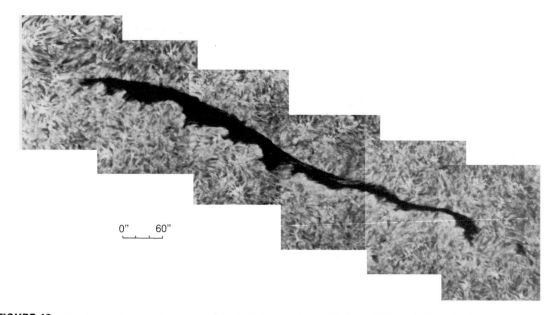

FIGURE 13   A quiescent filament photographed in the light of hydrogen H alpha (656.3 nm). Note the fine vertical structure.

F. STRUCTURES IN THE ATMOSPHERE

1. Sunspots

A sunspot is a region of high magnetic flux in the photosphere, the footpoint of a magnetic loop that extends into the corona. Spots are darker than the photosphere because they are one to two thousand kelvin cooler, and they are cooler because their magnetic fields inhibit convective heat transport (see Fig. 4).

The smallest spots are called pores. A pore appears as a simple dark patch, an umbra, and lacks the surrounding ring of striated brighter material, a penumbra, which characterizes a mature spot. Pores are typically smaller than 2500 km in diameter, usually persist for only 10 or 15 min, and then fade, presumably by submerging. A critical field strength of about 1500 G is required for a pore to form. Often a pore develops from a tiny magnetic patch within an intergranular lane.

A bipolar pair of spots first appears when a loop of magnetic flux rises from the interior and breaks through the photospheric surface. Some evidence suggests that the pair lies initially on the boundary of a supergranulation cell and the spots themselves form preferentially at the vertices of several cells. When formed, the spots in a pair begin to separate at about 0.5 km/s and the axis of the pair rotates over a few days until nearly parallel to the solar equator. Spot polarities in the north and south hemispheres obey the Hale–Nicholson laws, as previously mentioned.

A spot grows rapidly in area and darkness in its first few hours of life, by the merging of neighboring pores. Most spots live only a few days, but the active region in which they develop may last for many weeks. Small spots seem to decay by shrinking in area and magnetic flux, that is, by the submergence of their associated magnetic loop. Large spots decay by fragmenting or by shedding flux to their surroundings across an annular region called a "moat."

A large spot may have several umbrae within a single penumbra and grow to a diameter of 20,000 km. The umbra contains tiny bright regions, thought to be convection cells, and the penumbra consists of radial bright and dark filaments that terminate in a sharp boundary in the surrounding photosphere. The maximum umbral magnetic field increases with the area of the umbra and may reach 3000 G. The inclination of the field lines varies from nearly vertical in the umbra to nearly horizontal in the penumbra. The penumbral filaments are aligned along this horizontal field.

Above the penumbra, photospheric gas flows radially outward at speeds of a few kilometers per second and returns in the chromosphere to the spot's axis, in a circulation called the Evershed effect. Deeper in the penumbra, the filaments resolve into bright grains, some 250 km in diameter, that flow toward the umbra at speeds of about 0.5 km/s. Growing spots also exhibit penumbral waves that propagate radially outward at speeds of 10–20 km/s.

Observations made from space with precision bolometers show that the formation of a large spot actually reduces the Sun's total radiation by as much as 0.1% for as long as a week. This result implies that the spot acts as a barrier to the emerging energy flux in the photosphere. The energy not radiated from the umbra is stored somehow in the interior, and is released only later in the spot's history.

Very little is known at present about the roots of sunspots, that is, the structure of their magnetic fields below the photosphere. To account for the surface brightness of the umbra, some theorists have proposed that the umbral flux divides into a cluster of narrow filaments that are separated by field-free regions. A spot would then look like a Portuguese man-of-war jellyfish, with long tentacles penetrating the lower photosphere. In any case, spots do seem to be anchored in layers that rotate faster than the photosphere itself. Thus, small spots appear to move westward, relative to the surrounding photosphere, at a speed of about 100 m/s, whereas large spots move somewhat slower.

Sunspots are complicated structures and are not fully understood by any means. Their stability, energy balance, formation and decay, and complex patterns of motion are all remaining challenges for the theorist.

2. Active Regions

Sunspots are embedded in extended regions of high magnetic flux density, the active regions. Active regions are bright at all wavelengths, from X rays to radio waves (see Fig. 3), and are highly variable. They are the sites of the dramatic solar flares (see Section IV).

An active region is born as a small bipolar area that grows as new magnetic flux emerges and diffuses laterally into the surrounding quiet region. A region will grow in a week to a typical size of 10^5 km and total flux of 10^{22} maxwells, then slowly decay by diffusion and submergence.

The magnetic field strength outside the sunspots of an active region ranges from a few hundred to about a thousand gauss. The flux emerges from the photosphere in discrete flux tubes only a few hundred kilometers in diameter that cover only about 10% of the surface area. From the minimum to the maximum of the solar cycle, the magnetic flux of the Sun in all active regions varies by a factor of 15.

Active regions are composed of masses of loops and "fibrils" that connect opposite polarities and presumably outline the local magnetic fields. The loops terminate in chromospheric "plages" that are composed of smaller structures (faculae) and that are bright in X

rays. The longer loops extend upward into the corona and are visible in X-ray images (see Fig. 3).

The Japanese *Yohkoh* satellite (launched in 1991) derived the coronal temperature and electron density in active regions from measurements in soft X rays (1–3 keV). The temperature ranges from 4 to 10 million kelvin and the density from 10^9 to a few times 10^{10} cm^{-3}.

A "complex of activity" that consists of several bipolar active regions can form during the peak of the solar cycle. The complex may last for as long as nine months as new active regions emerge within it and old ones decay.

3. Chromospheric Spicules

Most of the plasma in the quiet Sun at temperatures between 6000 K and 15,000 K, the classic chromosphere, resides in spicules. These are thin, needle-shaped features with typical dimensions of 500 by 5000 km, a plasma density of 10^{11} cm^{-3}, and a lifetime of 10 to 15 min. Their magnetic fields are uncertain but are estimated to be some tens of gauss. They cover only a few percent of the solar surface and are clustered along the borders of the supergranules. Spicules rise with speeds of 10 to 30 km/s from the photosphere. Half of them fade when they reach maximum altitude, and half fall back to the photosphere. They carry an upward mass flux hundreds of times larger than that of the solar wind and have been suggested as the source of the wind's mass. Although several explanations have been offered for their origin, including chromospheric Alfvén waves or miniflares, none has full observational support.

4. The Transition Zone

Plasma that lies at temperatures between 10^4 K and 5×10^5 K composes the so-called "transition zone" between the chromosphere and corona. In the quiet Sun, most of this plasma resides above the borders of the supergranule cells, where most of the magnetic flux as well as the spicule clusters are located. The plasma at these transition temperatures is thought to lie in thin sheaths wrapped about the spicules and about their dense footpoints.

The spectrum of the zone is characterized by strong resonance lines of multiply ionized abundant species, such as C IV, N V, and Ne VII, that lie in the extreme ultraviolet between about 40 to 160 nm. Doppler shifts

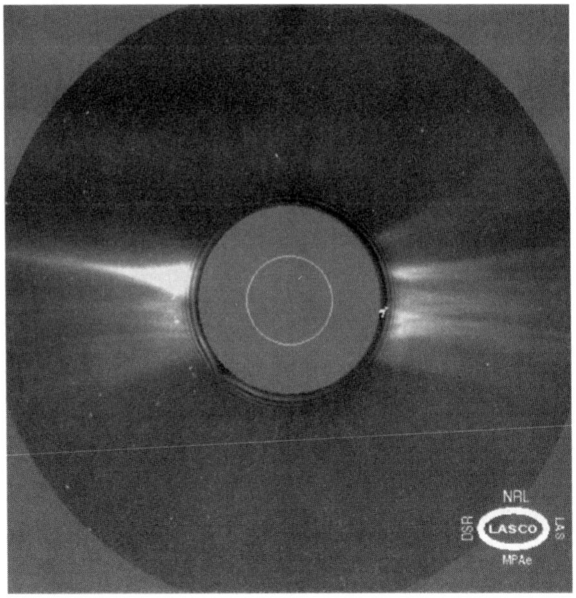

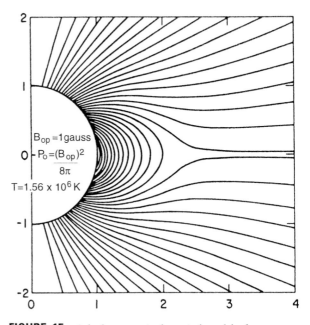

FIGURE 14 Coronal streamers imaged by the C2 coronagraph aboard the *SOHO* satellite. (Courtesy of the Lasco team.)

of such lines indicate rapid vertical motions, on the order of 10 km/s.

Spectra in the extreme ultraviolet (100 to 200 nm) obtained by a team at the Naval Research Laboratory show two types of relatively rare dynamic events. Explosive events with ejection speeds up to 400 km/s occur at an estimated rate of 24 s^{-1} over the whole Sun. Brief turbulent events with random motions on the order of 100 km/s occur near the borders of supergranule cells.

5. Coronal Streamers

Coronal streamers are the radial extensions of large bipolar magnetic regions that are the remnants of old active regions. Streamers have been observed from the *Solar and Heliospheric Observatory* (*SOHO*) satellite to extend as far as 30 solar radii into the heliosphere (Fig. 14). Their interfaces with the surrounding open corona are considered likely sources for the slow solar wind (speeds of less than 400 km/s). They are thought to merge with the heliospheric current sheet at large distances from the Sun.

The free electrons of streamers scatter and linearly polarize the white light of the photosphere. Since the scattering efficiency is independent of wavelength, the observed brightness of a streamer in white light can be analyzed to yield the number of electrons, per square

centimeter, in a column along each line of sight. As a streamer rotates past the observer and its orientation changes, an empirical three-dimensional density model of the streamer can be constructed. Despite their impressive appearance, helmet streamers are only a factor of two or three times denser than the surrounding corona.

White light observations show that a streamer is basically a long, fan-shaped coronal structure. The lower portion of a streamer consists of an arcade of closed loops that spans the underlying bipolar photospheric magnetic region. The arcade encloses a region of reduced density, the so-called "cavity."

In 1977, G. Pneuman and R. Kopp devised an influential three-dimensional hydromagnetic model of an equatorial belt of streamers that is embedded in symmetrical polar coronal holes (Fig. 15). The model approximates the situation near solar minimum when streamers cluster near the equator and the polar holes are at maximum size. In the model, the fast polar wind compresses the narrow blade of the streamer. A current sheet divides the oppositely directed open field lines in the upper part of the streamer and extends into space as the heliospheric current sheet. A magnetic neutral point lies at the intersection of the open and closed field systems.

FIGURE 15 A hydromagnetic dynamical model of a streamer. Magnetic field lines coincide with the streamlines of flow. The model approximates the situation near sunspot minimum when the streamers concentrate near the solar equator and the polar holes are at their maximum size. (Courtesy of *Solar Physics*.)

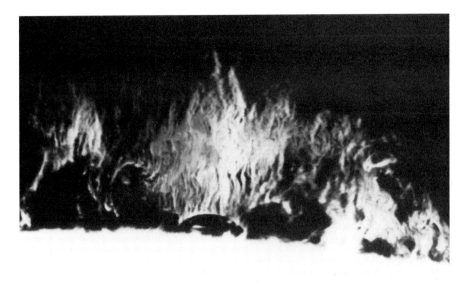

FIGURE 16   A quiescent prominence photographed in the light of H alpha (656.3 nm).

6. Prominences (Filaments)

In Section IV, we consider further the evolution of streamers during the solar cycle and their eruption as "coronal mass ejections."

Prominences are structures with chromospheric temperatures (6000 K to 15,000 K) that are embedded within the million-degree corona (Fig. 16). When viewed from above in the H alpha line (656.3 nm), they appear as dark "filaments." There are two main types: quiescent (in the quiet Sun) and active (inside or at the borders of active regions).

Quiescent prominences are long, thin, vertical sheets of dense plasma, with a fractal fine structure. A typical mature filament, as seen in H alpha, has a gross width of about 5000 km and a length of 10^5 km. Some filaments reach a solar radius in length. They lie along the neutral lines of the weak bipolar magnetic regions of the solar photosphere, within the cavity of an overlying streamer.

Active region prominences lie along the polarity inversion line of their bipolar regions. In comparison to quiescent prominences they are lower, are much narrower, and have stronger magnetic fields. Both types possess much fine structure, which may be fractal in nature and is thought to correspond to the internal magnetic fields.

Not much is known about the magnetic fields in and around quiescent filaments. Direct measurements indicate a horizontal field component of 2 to 30 G, directed at a small angle (typically 25°) to the plane of a quiescent prominence. In most quiescents, the normal component of the field crosses the vertical sheet of plasma in the direction *opposite* to that of a potential field that would span the underlying photospheric bipolar region. This result has suggested the Kuperus–Raadu model of magnetic field shown in Fig. 17. Here the filament sheet hangs in the bottoms of helical field lines that form the surrounding cavity.

When big quiescent prominences erupt (see Section IV), they often display helical motions that may indicate the presence of internal helical magnetic fields. Helical fields, with a strong axial component, have also been proposed for the thin active region filaments.

Quiescent prominences possess temperature transition zones between their cool centers and the hot coronal. As a result, they radiate the characteristic resonance lines of highly ionized abundant elements, particularly the C IV lines at 155 nm.

Prominences need a continuous energy supply to maintain their strong emission for many weeks. The sources of this energy are uncertain. Coronal X rays can be absorbed efficiently by the dense plasma, but are insufficient in themselves. Astronomers have searched for waves of various types, but although oscillations of different periods (typically several minutes) have been reported, clear evidence of wave dissipation is lacking. Theorists therefore postulate some form of magnetic or electrical current heating, similar to that postulated in the corona.

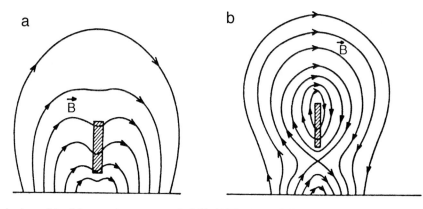

FIGURE 17 Two classic models of the prominence magnetic field. (a) The Kippenhahn–Schlüter model postulates a potential field arcade. (b) The Kuperus–Raadu model envisions a flux rope in which the prominence is suspended.

The mass of a large prominence (about 10^{15} g) seems to exceed the deficit of mass in the coronal cavity surrounding the prominence. Thus, during the formation of the prominence, some undetected process must supply its mass, presumably from the photosphere, where mass is plentiful. A continuous injection of plasma may also be required to maintain the prominence since there is some evidence that its plasma slowly descends to the photosphere along fine-scale vertical threads. Observations of the vertical motions of the plasma in filaments lead to a rather confused picture of the overall circulation. The speeds are generally only a few kilometers per second both up and down and somewhat larger in active region prominences.

Most large quiescent prominences end their lives by erupting dramatically, together with the overlying coronal streamer. A "coronal mass ejection" is then said to occur (see Section IV).

7. Coronal Holes

Eclipse observers had noted for many years the presence of faint (i.e., tenuous) regions between the more prominent streamers at the solar limb. In the mid-1960s, an association was noted between such dim regions and the onset of terrestrial magnetic storms. Theorists postulated that streams of "corpuscular radiation" emerged from these regions and interacted with the Earth's magnetic field a few days later. Other evidence, principally the orientation of comet tails, had pointed to a steady outflow of solar plasma, termed the "solar wind," but the source of the wind in the corona was not known. Shortly before the flight of *Skylab* (May, 1973–February, 1974), A. Krieger and colleagues traced a recurrent high-speed stream in the wind back to a dark equatorial coronal region that

appeared on their X-ray photographs. The *Skylab* mission confirmed that most of the fast streams (with speeds as high as 1000 km/s) originate in these dim regions, which were named "coronal holes." Larger holes have faster streams and the stream velocity is maximum near the centerline of the stream.

Figure 3 shows a large coronal hole in an X-ray image of the corona taken during the *Yohkoh* mission. Holes are also detectable as bright regions in images made in the spectrum line of chromospheric helium at 1083 nm.

Numerical extrapolations of photospheric magnetograms show that a hole's magnetic field lines open out into interplanetary space, allowing the coronal plasma to escape. Other regions, at the interface of two separate magnetic field regions, also possess open field lines and emit fast wind. Interplanetary magnetic field measurements, mapped back to the Sun, indicate a spatially averaged field strength of 6 to 12 G at the photosphere of a hole.

Continuous measurements of a polar hole were made in white light during *Skylab*. From these data, R. Munro and B. Jackson derived an empirical model of the hole's proton density and wind speed, assuming a typical proton flux through the hole (10^{14} cm^{-2} s^{-1}). Figure 18 shows their model. The density at two solar radii was only 5×10^5 cm^{-3}, a factor of at least three smaller than in a streamer at the same height. This particular hole had a severely nonradial geometry between 2 and 5 solar radii—its cross-sectional area increased with height seven times faster than a sphere. The computed wind speed reached sound speed (170 km/s) between 2.2 and 3.0 radii and a very rapid accleration to 450 km/s occurred between 3 and 5 radii. [*See* THE SOLAR WIND.]

A large fraction of the energy supplied at the base of a coronal hole (10^5 ergs cm^{-2} s^{-1}) escapes in the

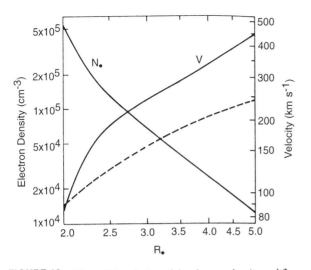

FIGURE 18 The radial variation of the electron density and flow speed in a coronal hole. The dashed line is the Parker wind solution for a coronal temperature of two million degrees kelvin. (Courtesy of *Astrophysical Journal*.)

form of the kinetic and gravitational energy of its wind stream. This large rate of energy loss poses a difficult theoretical problem, since E. N. Parker's classic explanation for the solar wind would require a coronal temperature of at least 2.5–3 million kelvin, whereas the holes have peak temperatures as low as 1.4 million kelvin. Somehow, energy or momentum must be deposited in the wind stream at radial distances of a few solar radii. Theorists have suggested that some form of hydromagnetic wave that propagates outward along the open field lines is responsible. Since Alfvén waves are detected in the wind near the Earth, they are considered prime candidates, but many problems arise concerning the production and dissipation of such waves. A complete understanding of the energetics of coronal holes and the wind within them remains for the future.

The size and location of coronal holes vary with the solar cycle. A few years before solar minimum, the poles of the Sun are covered by large coronal holes that occupy some 15% of their area. Occasionally the holes extend from a pole across the solar equator. Near solar maximum, the polar holes shrink in size, and smaller holes appear nearer the solar equator.

Like the rest of the corona, the holes rotate more rigidly (with periods of 26 to 29 days, independent of latitude) than the photosphere. Simulations of evolution of the coronal magnetic field, by Y. Wang and colleagues at the Naval Research Laboratory, suggest that this effect arises from transport of photospheric flux to higher latitudes and from reconnections of the coronal field.

IV. SOLAR ACTIVITY

Structures on the Sun change on timescales ranging from a few seconds to several months and their patterns evolve over a cycle of 22 years. We shall first discuss the solar magnetic cycle and then the more rapid transient events.

A. THE SUNSPOT CYCLE

1. Observations

The number and distribution of sunspots vary in a cycle of 8 to 13 years. At the beginning of the cycle, bipolar pairs of spots appear at latitudes above 60°. Then, as the cycle advances, increasing numbers of new spots appear at lower and lower latitudes. Thus, the latitude distribution of spots as a function of time mimics that of a butterfly's wings (see Fig. 6).

During a cycle, the magnetic polarity of all the leader (westernmost) spots in the northern solar hemisphere is the same, and is opposite to that of leader spots in the southern hemisphere. The polarity of leader spots reverses in each hemisphere at the beginning of the next cycle. Thus, the full magnetic cycle takes an average of 22 years. The leader spots of a new cycle possess the same polarity as the premaximum polar field of their hemisphere.

Similarly, the weak polar fields (2–10 G) reverse polarity close to the phase of sunspot maximum. However, oddly enough, the reversal at the two poles need not occur simultaneously! For example, the Sun possessed two negative poles during 1980.

The maximum number of sunspots in a cycle fluctuates by as much as a factor of two and this amplitude also seems to possess cycles, such as the 90-year Gleissberg cycle. In addition, periods of suppressed or negligible sunspot number have occurred during the past 7000 years, such as the Maunder minimum from 1645 to 1715. The Maunder minimum was accompanied by extraordinarily low temperatures throughout Europe and this correlation is perhaps the most convincing example of a solar influence on terrestrial weather.

As might be expected, the total magnetic flux in the photosphere varies with the 11-year cycle, by as much as a factor of five. During a cycle, the flux is constantly redistributed in latitude, from the intense concentrations in active regions to weaker large-scale regions at higher latitudes. Differential rotation, meridional flows, and supergranular flows combine to form the characteristic "chevron" distribution of weak flux (see Fig. 12).

Solar prominences (filaments) and coronal streamers also participate in the 11-year cycle. The mean latitude of prominences gradually increases from about 30° to about 60°. A few years after solar minimum, the upper latitude limit increases very rapidly. Newly forming quiescent prominences are said to "rush to the poles," where they form a nearly continuous "polar crown" at about 65° latitude.

At sunspot minimum, coronal streamers tend to cluster in a belt around the solar equator. Their current sheets (see Fig. 15) then divide the heliosphere into two hemispheres of opposite magnetic polarity. During the *Skylab* flight (1973), this effect was very well defined. The equatorial streamers formed the boundary between high-speed solar wind streams that emerged near the solar poles, flowed toward the Sun's equatorial plane, and were detected near Earth. At solar maximum, the streamers are well distributed in latitude.

2. Models

Empirical models of the solar magnetic cycle have been proposed by several astronomers, beginning in 1961 with H. Babcock at the Mount Wilson and Palomar Observatories. These models attempt to explain the salient regularities of the cycle (the butterfly diagram, the polar reversals, the reversal of sunspot polarities) with simple physical ideas, but without extensive mathematical calculations.

In Babcock's scenario (Fig. 19), the cycle begins with a large-scale dipole field that extends just below the photosphere (stage 1). Differential rotation in latitude wraps the meridional field lines around the solar equator (stage 2), producing a toroidal field of strong flux ropes. When a kink in a rope rises through the photosphere (because of its natural buoyancy) (stage 3), it emerges to make a small angle with the equator. Thus it forms a bipolar pair of sunspots, with the leader emerging at a slightly lower latitude than the follower. In this way, Hale's laws of sunspot polarity are fulfilled in each hemisphere. The active regions around sunspots expand into the corona. As a result, fields of opposite polarities from the opposite hemispheres can cancel (stage 4). This reconnection supposedly forms a disconnected loop that can float away into the corona. In this way, most of the amplified flux of one cycle is eliminated. A small remnant of following polarities is assumed to drift toward the nearest pole, reversing the polarity there.

Babcock's empirical model has several flaws, among them the assumption that most solar flux ends up in the corona or the solar wind. However, his basic ideas of surface flux transport and merging have survived. In 1964, R. Leighton demonstrated the role of supergranulation in a numerical model. N. Sheeley and several colleagues at the Naval Research Laboratory have simulated the evolution of surface flux over a full solar cycle, using observations of the emerging flux and combining the action of differential rotation, meridional flow, and supergranular flows. Their calculated evolution of the surface fields compares favorably with observations. In addition, Y. M. Wang and Sheeley were able to explain the observed rigid rotation of the corona. The transport of flux seems to be reasonably well understood. However, the *generation* of the field with all its regularities, by a subsurface solar dynamo, is still a goal of current research.

The basic physical ideas underlying a solar dynamo were developed by E. N. Parker in 1955. The essential ingredient in a dynamo model is the conversion of toroidal flux to poloidal flux and vice versa. (Poloidal fields lie along solar meridians, whereas toroidal fields lie parallel to the solar equator.) In Parker's model, small flux loops in a toroidal field rise buoyantly and twist as they rise because of the Coriolis force. The twist generates a poloidal component, parallel to the meridians. Poloidal field lines are assumed to reconnect to form a large-scale field from small cyclonic eddies, a process named the "alpha effect." Simultaneously the poloidal field is converted to a toroidal field by the action of latitude-dependent differential rotation. The two effects combine to produce waves of enhanced magnetic field that propagate slowly from the poles to the equator, and generate the "butterfly" latitude distribution of field strength (see Fig. 6).

Early dynamo models were kinematic: they postulated subsurface systematic motions without physical justification, but with some fine-tuning, they were successful in reproducing the butterfly diagram. Later models attempted to derive the driving motions from a theory of large-scale convection. Unfortunately, these dynamical models predicted a radial gradient of rotation speed with the wrong sign to explain why sunspots form at progressively lower latitudes. The empirical radial rotation gradient (see Fig. 9), derived from global oscillation observations, also contradicts the assumed gradient in some early dynamo models.

The resolution of these difficulties is a goal of current research. A favored approach is to shift the location of the solar dynamo from the interior of the convection zone to a thin layer at the inner boundary of

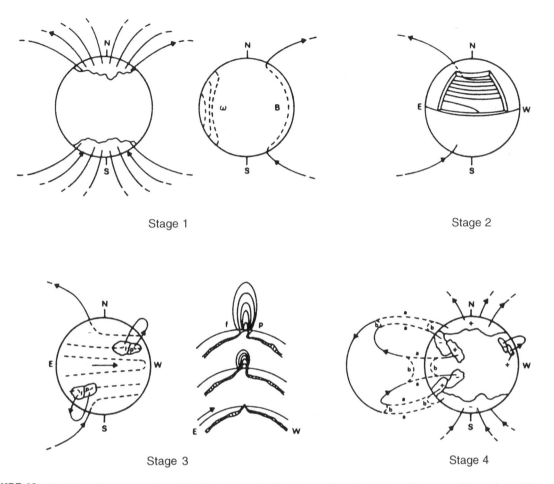

Stage 1   Stage 2

Stage 3   Stage 4

FIGURE 19   H. Babcock's conceptual model of the solar cycle. See the text for a description. (Courtesy of *Astrophysical Journal.*)

the zone. Models based on this idea and incorporating plausible physics do seem capable of maintaining the solar magnetic flux, but still have difficulties in transporting the amplified field to the surface.

B. SOLAR FLARES

1. Observations

A solar flare appears as sudden brightening of part of an active region (see Fig. 5). The region is heated to tens of millions kelvin, often within a few minutes, and may emit radiation over the full electromagnetic spectrum from gamma rays to long radio waves. In addition, beams of nonthermal electrons and protons can be accelerated to millions of electron volts and, along with masses of hot coronal gas, can be ejected into interplanetary space. The total energy of a large

flare (10^{31} ergs) is sufficient to power the entire United States for a hundred thousand years.

Some of this radiation and particle emission reaches the Earth and produces a variety of effects, including electric power grid surges, radio propagation anomalies, and auroras. In March, 1989, the entire electric power grid of eastern Canada was shut down by a powerful series of solar flares. Thus, flare research has a practical as well as an astrophysical aspect.

Since flares occur in active regions, their frequency follows the 11-year solar cycle. At sunspot maximum, scores of small flares and several large ones occur daily. The frequency with which flares occur, as a function of energy, is a power law with an index of -1.8. This means that flares with half the energy of another group of flares will occur about four times more often, on average. Flares last anywhere from a few minutes to many hours, usually depending on their area and total energy.

In general, flares tend to occur near the boundary between the opposite magnetic polarities of an active

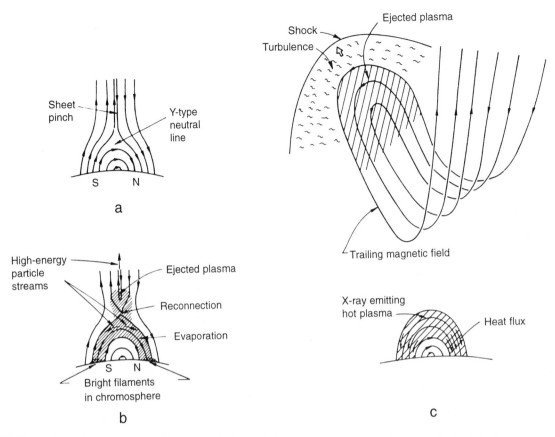

FIGURE 20   A diagrammatic sequence of the stages of a flare: (a) the preflare magnetic configuration, (b) the flash phase, (c) the thermal phase, following reconnection and ejection of hot plasma. (Courtesy of P. Sturrock, Colorado Associated University Press.)

region. The more complex the magnetic configuration of the region and the more rapidly new flux appears, the more likely a flare will occur. The simplest magnetic system that produces flares is a twisted magnetic loop that connects opposite magnetic polarity areas of the active region. A sunspot with opposite polarities within the same penumbra (a "delta" configuration) is especially prone to flaring.

The Japanese satellite *Yohkoh*, launched in August 1991, produced a vast quantity of high-quality observations of flares that are still being analyzed. These data are confirming many of the ideas developed since the flight of the *Solar Maximum Mission* (1980–1989), as well as revealing some surprises.

Although no single flare exhibits the full variety of flare phenomena, we shall attempt to describe the evolution of such a hypothetical flare, simply to give the reader an impression of the total event. The evolution of this ideal flare is illustrated schematically in Fig. 20. We can distinguish four phases: the preflare phase, the flash or impulsive phase, the thermal or gradual phase, and the postflare phase.

a. The Preflare Phase

In the preflare phase, which may last for minutes or tens of minutes, the active region shows signs of progressive heating. The brightness of many small sites increases and the region shows rapid fluctuations in H alpha, soft X rays, and microwaves. A filament that lies over the magnetic neutral line may be "activated," that is, may experience rapid changes in brightness and form, and may actually erupt just before the onset of the flare. Velocity oscillations have also been reported just before a flare.

Preceding the preflare phase is a long period in which magnetic flux continues to emerge and the field configuration is thought to become more sheared and twisted. One of the outstanding questions, still unanswered, is the conditions under which a flare is triggered.

b. The Flash Phase

The primary release of magnetic energy occurs in the low corona during the flash phase. The onset of the flare is marked by short pulses of hard X rays (photons with energy greater than 30 keV), microwaves, and possibly gamma rays (1- to 10-MeV photons). These are signatures of nonthermal electrons and protons that have been accelerated to high energy.

The X rays are the bremsstrahlung (German for "braking radiation," also known as free–free emission) of fast electrons colliding with charged particles in the coronal plasma. X rays may appear in bursts as short as a few milliseconds that are emitted from many small sites throughout the active region, or they may cluster into a longer burst lasting several minutes. The photon energy spectrum of hard X rays is generally a power law, sometimes extending to 300 keV, with indices between -2 and -5.

Gamma rays are energetic photons produced in part by the collision of fast protons with light nuclei. The most common flare gamma rays are the nuclear de-excitation lines at 4.4 MeV of carbon and at 6.1 MeV of oxygen. Protons with energy above 1 MeV can carry as much as 1% of the total flare energy, implying a rather efficient acceleration mechanism. *Yohkoh* observations of the gamma ray flare of October 27, 1991, showed that the carbon gamma rays originated at the footpoints of a flaring loop as fast charged particles slammed into the chromosphere.

Circularly polarized microwave emission, at 1- to 6-cm wavelengths, tracks the X-ray variations very closely, which suggests that both are produced by the same population of nonthermal electrons. The radio emission is thought to be the gyrosynchrotron radiation of fast electrons spiraling in the magnetic field of coronal loops. In fact, microwave images show that the emission occurs at the bottoms of loops.

Flares also commonly emit bursts of radiation at meter wavelengths. For example, Type III bursts at 5 to 600 MHz consist of plasma radiation, excited at progressively lower frequencies. They are caused by streams of electrons with energies above 40 keV that propagate outward along high loops at speeds that are a third the speed of light.

All of the nonthermal radiation described here arises from the sudden conversion of a significant fraction of the stored magnetic energy into beams of high-energy charged particles. The primary release is thought to occur at the tops of magnetic loops in tiny magnetic reconnection sites. The particles then collide with the static plasma and heat it rapidly. (*Yohkoh* observed "superhot" flares with temperatures above 30 million kelvin.) Heat conduction, shock waves, turbulent flows, and particle beams transport energy to locations much deeper and denser than the site of initial energy release. As these cooler layers are heated impulsively, they emit extreme ultraviolet radiation

(30–100 nm) at temperatures below a million kelvin. Very quickly, the flux of flare energy reaches the chromosphere, where H alpha is emitted at about 30,000 K.

Yohkoh images of a flaring magnetic loop, obtained on January 13, 1992, have confirmed much of this sequence of events. In particular, these observations showed that the high-energy particles are accelerated at or above the top of a flaring loop. Hard X rays emitted there indicate temperatures in excess of 20 million kelvin. The particles showered down the legs of the loop, heating it as high as 8 million kelvin and producing soft X rays. Additional hard X rays were emitted as the particles reached the feet of the loop.

c. The Gradual, Thermal Phase

As energy flows down the coronal loops, their ends in the chromosphere are heated and plasma explodes up the loops at speeds as high as 200 km/s, thereby increasing the coronal density. This process has been named (incorrectly) "evaporation." It has been followed in detail in spectra from *Yohkoh*.

The coronal plasma electron density reaches 10^{12} to 10^{13} cm^{-3}, a hundred times its normal density. Two parallel bright ribbons form in the H alpha chromosphere at the bottoms of the loops and separate at speeds of 5–100 km/s.

Within a few minutes, the magnetic and fast particle flare energy is thermalized. The temperature of the plasma rises as high as 20 million kelvin. The prime evidence for such high temperatures is the spectrum of emission lines in the soft X-ray range, 0.1–2 nm. Here are found the resonance lines of the hydrogenlike ions Fe XXV and Ca XIX, which can exist only at such high temperatures.

Later, as the plasma cools and expands, the degree of ionization of the heavy ions decreases gradually. The coronal plasma is still highly turbulent at this time, as indicated by the breadth of the emission lines. During this phase, which may last hours, the total mass and volume of the heated plasma rise gradually and then eventually fall. The original arcade of coronal loops rises, expands, and compresses the loops in the surrounding corona.

The expansion of the flare plasma into the overlying corona may produce a mass ejection. Often, a strong shock, moving at a speed of 500–1000 km/s, precedes the ejection and produces a Type II radio burst at meter wavelengths. The ejected mass can be as large as 10^{15} to 10^{16} g and carry a large fraction of the flare energy as kinetic energy. These flare-related coronal mass ejections have been detected in space. However, most ejections do not involve flares, but only the eruption of an unstable filament.

d. The Postflare Phase

Gradually, the flare plasma cools and much of it falls back to the chromosphere, as the active region returns to its approximate preflare state. Presumably, the region's nonpotential magnetic field has been reduced, but no direct information exists on the final (or initial!) configuration.

This relaxation process is sometimes accompanied by the formation of postflare coronal loops visible in H alpha and in the forbidden coronal emission lines typical of 2 to 4 million kelvin (Fe XIV and Ca XV). Coronal plasma seems to cool suddenly within a loop and flow downward. Higher loops appear successively, as though some process is gradually draining the upper corona.

An active region may emit a broadband circularly polarized continuum at meter wavelengths for hours or days following a flare. This Type IV radiation is thought to be synchrotron emission from energetic electrons trapped within a huge magnetic bottle.

2. Flare Models and Flare Physics

An enormous effort has been invested by solar physicists in attempts to understand the complex flare event. At present, general agreement exists on the basic physics involved in each phase, but detailed models are still goals for the future. The most obscure aspects of the flare phenomenon are the energy buildup, the triggering mechanism, and the acceleration of nonthermal charged particles. These are difficult to explain because they involve complex plasma physics at unresolvable spatial scales and large-scale magnetic field systems that are as yet unobservable. The later phases, involving the expansion and cooling of the flare plasma, are well documented and therefore better understood.

According to present ideas, flare energy is stored in the form of electric current systems or, equivalently, as nonpotential components of the active region's magnetic field. This energy accumulates as the footpoints of the magnetic field are shuffled by photospheric convective motions. A simple arcade of parallel coronal loops becomes sheared, and the individual loops may be twisted or braided. All this distortion creates increasingly large electric current densities within the coronal plasma.

The triggering of the flare may be driven by the intrusion of an opposite polarity field or may arise

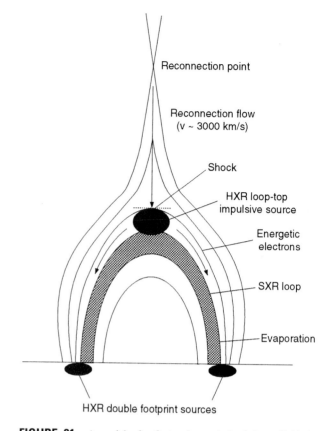

FIGURE 21 A model of a flaring loop, derived from *Yohkoh* observations. (Courtesy of the *Yohkoh* team.)

spontaneously as the field is distorted. Eventually, a current-driven instability may arise at some location. Several candidates have been proposed, among them the Buneman, ion–cyclotron, and ion–sound instabilities. When such an instability takes place, the resistivity of the plasma can increase by orders of magnitude and this allows magnetic field reconnection, in which nonpotential field components are destroyed and their energy is converted to charged particle kinetic energy via strong electric fields. Needless to say, this process is complicated and not well understood. To release a large flare energy, a large volume of magnetic flux must enter the reconnection zone and this requires some kind of external driving pressure or a propagating disturbance. Figure 21 is a model of a flaring loop, derived from *Yohkoh* observations of the flare of January 13, 1992.

Once the beams of charged particles have been accelerated, their subsequent evolution is somewhat easier to describe. Fast electrons produce hard X rays as bremsstrahlung when they impact the dense plasma at the footpoints of coronal loops. The so-called "thick-target" models of this process are successful in reproducing the energy spectrum and time variations of hard X-ray bursts.

The initial large-scale magnetic configurations in which a flare may occur are not known in any detail, although numerous plausible proposals abound. Vector magnetic observations do suggest that strong shearing of the field near the neutral line of the active region is probably a prerequisite, but these observations pertain to the photosphere and may not be definitive for the corona.

C. CORONAL MASS EJECTIONS

Figure 22 illustrates a typical "coronal mass ejection" as recorded in white light by a coronagraph aboard the *SOHO* satellite. These large disruptions of the solar corona were first observed at the solar limb in 1971 with a ground-based coronagraph by R. Hansen and associates. Later, they were detected further out from the Sun by coronagraphs aboard spacecraft such as the *Orbiting Solar Observatory 7*, the *Skylab* mission, the *Solwind* satellite, and the *Solar Maximum Mission*. Dramatic ejections were recorded on the solar disk in X rays by the *Yohkoh* satellite. We now recognize that such events occur about once daily at all phases of the solar cycle and that they represent the disruption (and subsequent re-formation) of a helmet streamer and its associated prominence.

Figure 22 shows the different parts of the ejection. A bright loop precedes a dim cavity, within which lies an erupting prominence. In front of the whole system is a fast-moving "forerunner." The total mass of such ejecta has been estimated at 2×10^{14} to 2×10^{16} g and the total energy at 2×10^{29} to 5×10^{31} ergs.

Most events are associated either with flares or with eruptive prominences, but a sizable percentage has no visible chromospheric counterpart. Flare-associated events have high and constant speeds (up to 1000 km/s). Prominence events, in comparison, accelerate slowly, from a few tens to about 300 km/s over a period of hours.

Yohkoh observations have shown that a disrupted helmet streamer begins to re-form within an hour. As its distorted field lines in the high corona begin to retract and simplify, the original internal arcade is rebuilt. Energy is then dumped into the arcade, which glows briefly in soft X rays.

Much effort has been expended toward explaining these events in physical terms. Models that employed flare-associated heating, with a "driver gas" propelling the large ejecta, were initially successful. But many arguments have been advanced to show that the flares associated with such events are consequences, not causes. The huge difference in size between the ejection and the flare and the fact that the eruption often begins before the flare starts imply that the flare is a secondary event.

An alternate explanation has been developed, particularly by B. C. Low at the High Altitude Observatory. In this scenario, the low-density "cavity" of a helmet streamer tends to rise buoyantly into the high corona. The tension in the magnetic arcade that surrounds the cavity tends to restrain the buoyant cavity, however. As the roots of the magnetic arcade spread over the solar surface, under the influence of supergranulation, the arcade's tension reduces, until finally the cavity breaks loose, tearing the helmet streamer apart and carrying the embedded prominence with it. In this picture the forerunner (see Fig. 22) may represent a supersonic shock wave.

However, the process is probably not this simple. The speeds of eruption are usually below the critical gravitational escape speed of about 550 km/s, so that most of the ejecta return somehow to the Sun. Moreover, J. Aly has pointed out that the energy of the fully opened magnetic field of a streamer *exceeds* that of its initial configuration, so that energy of some kind must be added to the system to enable it to erupt.

D. IRRADIANCE VARIATIONS

The total amount of energy emitted by the Sun (its luminosity) is thought to remain constant to a very high precision, over millions of years. During the period from 1902 to 1960, the Smithsonian Institution of Washington, D.C., carried out a careful series of measurements of the solar "constant." The program lasted as long as it did because variations of about 1% were found, with important implications for the Earth's climate. However, these measurements required corrections for absorption by the Earth's atmosphere of about the same size as the observed variations. The actual solar variations remained in doubt until recently.

New technology has laid the question to rest. A series of satellites (1980–1988) carried an extremely stable detector, the Active Cavity Radiometer Irradiance Monitor (ACRIM). This device, developed by Richard C. Wilson, measures the solar irradiance with a precision of 0.001% and is stable to within 0.05%.

The ACRIM measurements show that the amount of solar energy received by the Earth (the solar irradi-

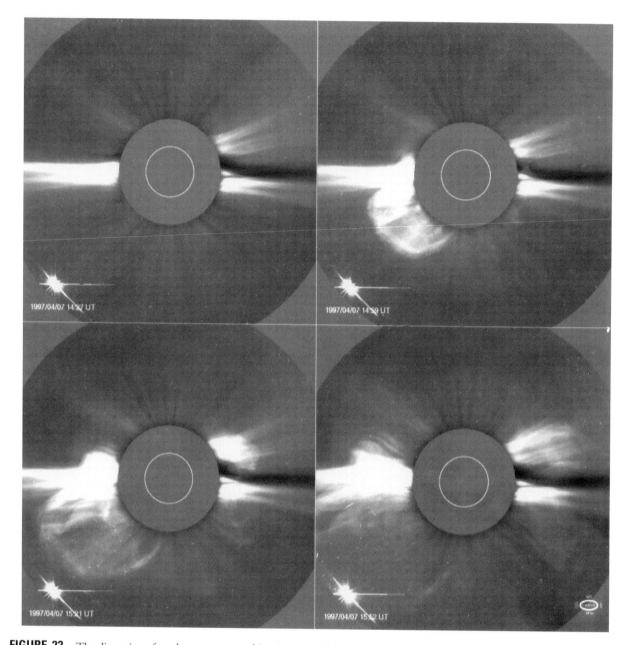

FIGURE 22 The disruption of a solar streamer, resulting in a coronal mass ejection, as recorded from a coronagraph aboard *SOHO*. A bright loop is preceded by a faint, fast "forerunner" shock. The loop encloses a solar prominence. (Courtesy of the *SOHO* team.)

ance) varies by about 0.1% in step with the 11-year sunspot cycle. At sunspot minimum, the irradiance is also at a minimum.

Daily dips in the irradiance of about 0.01% correspond to the emergence of large sunspots, and dips as large as 0.25%, lasting ten days, correspond to the passage of a large spot group across the solar disk. Evidently sunspots block the flow of energy from the interior, at least temporarily. On the other hand, bright faculae (hot magnetic clouds that lie just above the photosphere) contribute positively to the irradiance. By combining the observed brightness of different types of magnetic regions (spot, faculae, active regions, etc.), scientists can fit the observed irradiance variations in great detail. However, the physical processes that store and release the constant stream of solar energy remain controversial.

A LAST WORD

Solar research is a mature discipline. Most of the phenomena have been recognized for at least several decades, but some of the most difficult questions remain unanswered. How is the nonflaring corona heated? How is a flare triggered? How are magnetic fields generated in the interior? Why does the Sun have an activity cycle and why does it last 22 years?

A satisfactory "answer" in this context would be a complete physical theory that accords with the observations. As research progresses, however, more questions and deeper questions are constantly being raised. Thus there is a constant need for more detailed observations to constrain theory and to eliminate obsolete ideas.

In the near future such observations will come from several sources. The *SOHO* satellite is currently operating nicely, churning out exquisite global oscillation and coronal data, and the ground-based GONG network may continue for a complete solar cycle. The Japanese satellite *Solar B* is planned for launch soon after the year 2000. With such new information in hand, we can hope to see some of the long-standing problems in solar physics solved at last.

BIBLIOGRAPHY

Foukal, P. V. (1990). "Solar Astrophysics." John Wiley & Sons, New York.
Lang, K. R. (1995). "Sun, Earth and Sky." Springer-Verlag, New York/Berlin.
Phillips, K. J. H. (1992). "A Guide to the Sun." Cambridge Univ. Press, Cambridge, England.

THE SOLAR WIND

John T. Gosling
Los Alamos National Laboratory

I. Discovery

II. Statistical Properties in the Ecliptic Plane at 1 AU

III. Basic Nature of the Interplanetary Magnetic Field

IV. Stream Structure

V. The Heliospheric Current Sheet and Solar Latitude Effects

VI. Evolution of Stream Structure with Distance from the Sun

VII. Coronal Mass Ejections and Transient Disturbances

VIII. Variations with Distance from the Sun

IX. Termination of the Solar Wind

X. Kinetic Properties of the Plasma

XI. Heavy Ion Content

XII. Energetic Particles

XIII. Conclusions

GLOSSARY

Alfvén speed: Speed with which small-amplitude perturbations in the magnetic field propagate through a plasma.

Alpha particles: Helium nuclei having mass four times and charge twice that of a proton.

Coronal mass ejection: Solar event characterized by transient ejection of a large amount of new plasma and magnetic fields into the solar wind.

Coulomb interactions: Long-range interactions that occur between particles that are electrically charged.

Gyro radius: Radius of the orbit of a charged particle gyrating in a magnetic field.

Heliopause: Interface between the heliosphere and the interstellar plasma; the outer boundary of the heliosphere.

Heliosphere: Region of space containing plasma and magnetic fields of solar origin; a cavity carved in the interstellar medium by the solar wind.

Heliospheric current sheet: Surface in interplanetary space separating solar wind flows of opposite magnetic polarity; the interplanetary extension of the solar magnetic equator.

Interplanetary magnetic field: Remnant of the solar magnetic field dragged into interplanetary space by the flow of the solar wind.

Magnetic reconnection: Process by which basic magnetic field topology in a plasma is changed; pairs of magnetic field lines are broken and then rejoined in a new way.

Plasma: A gas of ionized (charged) particles.

Plasma beta: Ratio of gas pressure to field pressure within a plasma.

Shock: A discontinuous, nonlinear change in pressure commonly associated with supersonic motion in a gas or plasma.

Solar activity cycle: Cycle of ~11-year duration characterized by waxing and waning of various forms of solar activity, such as sunspots, flares, and coronal mass ejections.

Solar corona: Hot, tenuous outer atmosphere of the Sun from which the solar wind originates.

Stream structure: Pattern of alternating flows of low- and high-speed solar wind.

Termination Shock: A discontinuity in the solar wind flow in the outer heliosphere where the solar wind slows from supersonic to subsonic motion as it interacts with the interstellar plasma.

The solar wind is a plasma, that is, an ionized gas, that permeates interplanetary space. It exists as a consequence of the supersonic expansion of the Sun's hot outer atmosphere, the solar corona. The solar wind consists primarily of electrons and protons, but alpha particles and many other ionic species are also present at low abundance levels. At the orbit of Earth, 1 astronomical unit (AU) from the Sun, typical solar wind densities, flow speeds, and temperatures are on the order of 8 protons cm^{-3}, 470 km s^{-1}, and 1.2×10^5 K, respectively; however, the solar wind is highly variable in both space and time. A weak magnetic field embedded within the solar wind plasma is effective both in excluding some low-energy cosmic rays from the solar system and in channeling energetic particles from the Sun into interplanetary space. The solar wind plays an essential role in shaping and stimulating planetary magnetospheres and the ionic tails of comets.

I. DISCOVERY

A. EARLY INDIRECT OBSERVATIONS

In 1859, R. Carrington, a solar astronomer, made one of the first white light observations of a solar flare. He noted that a major geomagnetic storm began approximately 1.5 days after the flare, and tentatively suggested that a causal relationship might exist between the solar and geomagnetic events. Subsequent observations revealed numerous examples of apparent associations between solar flares and large geomagnetic storms. In the early 1900s, F. Lindemann suggested that this apparent relationship could be explained if large geomagnetic storms result from an interaction between the geomagnetic field and high-speed clouds of protons and electrons ejected into interplanetary space by solar activity. (Although we now know that the Sun does, in fact, frequently and sporadically eject clouds of charged particles into space, the phenomenon is not fundamentally a result of solar flares.) Early studies of geomagnetic activity also noted that some geomagnetic storms tend to recur at the ~27-day rotation period of the Sun as observed from Earth, particularly during declining years of the solar activity cycle. This observation led to the suggestion that certain regions of the Sun, commonly called M-regions (for magnetic), occasionally produce long-lived charged particle streams in interplanetary space. Further, because some form of auroral and geomagnetic activity is almost always present at high geomagnetic latitudes, it was suggested that charged particles from the Sun almost continuously impact and perturb the geomagnetic field. [See SUN.]

Observations of galactic cosmic rays (highly energetic charged particles that originate outside the solar system) in the 1930s also suggested that clouds of charged particles (and magnetic fields) are ejected from the Sun during intervals of high solar activity. For example, S. Forbush noted that the intensity of cosmic rays measured at Earth's surface often decreases suddenly during large geomagnetic storms, and then recovers slowly over a period of several days. Moreover, on a long-term basis, cosmic ray intensity varies in a cycle of ~11 years, but roughly 180° out of phase with the solar activity cycle. One possible explanation of these observations was that the cosmic rays from interstellar space are swept away from the vicinity of Earth by means of magnetic fields embedded in clouds of charged particles emitted from the Sun during large flares.

From a study of ionic comet tails in the early 1950s, L. Biermann concluded that there must be a continuous outflow of charged particles from the Sun. He based this conclusion on calculations that indicated that the orientations of ionic comet tails, which always point away from the Sun independent of the orbital inclination of the comets, could only be explained in terms of an interaction between material emitted from the comets and charged particles continuously flowing outward away from the Sun into interplanetary space. Biermann estimated that a continuous particle flux on the order of 10^{10} protons cm^{-2} s^{-1} was needed at 1 AU to explain the comet tail observations. He later revised his estimate downward to a value of $\sim 10^9$ protons cm^{-2} s^{-1}. (Direct measurements by spacecraft show that the average solar wind particle flux at 1 AU is actually $\sim 3 \times 10^8$ protons cm^{-2} s^{-1}.) [*See* PHYSICS AND CHEMISTRY OF COMETS.]

B. PARKER'S MODEL OF THE SOLAR WIND

Apparently inspired by these diverse observations and interpretations, E. Parker, in 1958, formulated a radically new theoretical model of the solar corona that proposed that the solar atmosphere is continually expanding into interplanetary space. Prior to Parker's work, most theories of the solar atmosphere treated the corona as static and gravitationally bound to the Sun. S. Chapman had constructed a model of a static solar corona in which heat transport was dominated by the thermal conduction of electrons. For a 10^6 K coronal temperature, the classic electron thermal conductivity is 8×10^8 ergs cm^{-1} s^{-1} deg^{-1}, and Chapman found that even a static solar corona must extend far out into interplanetary space. At the orbit of Earth, his model gave electron densities of 10^2 to 10^3 cm^{-3} and temperatures of $\sim 2 \times 10^5$ K. Parker realized, however, that such a static model leads to pressures at very large distances from the Sun that are seven to eight orders of magnitude larger than estimated pressures of the interstellar plasma. Because of this mismatch in pressures at large heliocentric distances, he reasoned that the solar corona could not be in hydrostatic equilibrium and must therefore be expanding. His consideration of the hydrodynamic (i.e., fluid) equations for mass, momentum, and energy conservation for a hot solar corona led him to unique solutions for the coronal expansion that depended on the value of the coronal temperature close to the surface of the Sun. The expansion produced low flow speeds close to the Sun, supersonic flow speeds (i.e., flow speeds greater than the speed with which sound waves propagate) far from the

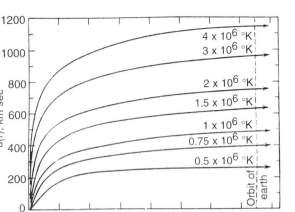

FIGURE 1 E. N. Parker's original solutions for solar wind flow speed as a function of heliocentric distance for different coronal temperatures. Subsequent work has demonstrated that the simple relationship between coronal temperature and solar wind speed illustrated here is incorrect. [From E. N. Parker (1963). "Interplanetary Dynamical Processes." Interscience, New York. Copyright © 1963. Reprinted with permission of John Wiley & Sons, Inc.]

Sun (Fig. 1), and vanishingly small pressures at large heliocentric distances. In view of the fluid character of the solutions, Parker called this continuous, supersonic, coronal expansion the "solar wind."

C. FIRST DIRECT OBSERVATIONS OF THE SOLAR WIND

Parker's theoretical ideas about the solar wind originally met with some resistance, but experimental confirmation was not long in coming. Several early Russian and American space probes in the 1959–1961 era penetrated interplanetary space and found tentative evidence for a solar wind. Firm proof was provided by C. Snyder and M. Neugebauer, who flew a plasma experiment on the *Mariner 2* space probe on its epic 3-month journey to Venus in late 1962. Their experiment detected a continual outflow of plasma from the Sun that was highly variable, being structured into alternating streams of high- and low-speed flows that lasted for several days each. Several of the high-speed streams apparently recurred at the rotation period of the Sun. Average solar wind proton densities (normalized for a 1 AU heliocentric distance), flow speeds, and temperatures during this 3-month interval were 5.4 cm^{-3}, 504 km s^{-1}, and 1.7×10^5 K, respectively, in essential agreement with the predictions of Parker's model. The *Mariner 2* observations also showed that helium, in the form of alpha particles, is present in the solar wind in variable amounts; the average alpha particle abundance relative to protons determined from the *Mariner 2* measurements was 4.6%, considerably lower than estimates of the helium abundance within the Sun itself. Finally, measurements made by *Mariner 2* confirmed the presence of an interplanetary magnetic field whose strength and orientation in the ecliptic plane were much as predicted by Parker (see Section III).

Despite the fundamental agreement of the observations with Parker's predictions, we still do not understand well the processes that heat the solar corona and accelerate the solar wind. Parker simply assumed the observational result that the corona is heated to a very high temperature, but he did not say how this is accomplished. Moreover, it is now known that thermal heat conduction is insufficient to power the coronal expansion. There now seem to be two main classes of models for heating the corona and accelerating the solar wind: heating and acceleration by waves generated by convective motions below the photosphere that propagate up into the corona; and bulk acceleration and heating associated with transient events in the solar atmosphere such as reconnection. Present observations are incapable of distinguishing between these and other possible alternatives.

II. STATISTICAL PROPERTIES IN THE ECLIPTIC PLANE AT 1 AU

Table I summarizes a number of statistical properties of the solar wind derived from satellite measurements in the ecliptic plane at 1 AU. The table includes mean values, standard deviations about the mean values (STD), most probable values, median values, and the 5–95% range limits for the ion number density (n), the flow speed (V_{sw}), the magnetic field strength (B), the alpha particle abundance relative to protons [A(He)], the proton temperature (T_p), the electron temperature (T_e), the alpha particle temperature (T_α), the ratio of the electron temperature to the proton temperature (T_e/T_p), the ratio of the alpha particle temperature to the proton temperature (T_α/T_p), the number flux (nV_{sw}), the sound speed (C_s), and the Alfvén speed (C_A) (the speed at which small-amplitude perturbations in the magnetic field propagate through the plasma). As noted previously, all the individual solar wind parameters exhibit considerable variability; we will see shortly that variations in solar wind parameters are often coupled to one another. It is of interest

that the average proton, alpha particle, and electron temperatures are not equal. Moreover, the proton temperature is considerably more variable than is the electron temperature, and the alpha particles almost always have a higher temperature than either the electrons or the protons. The alpha particle–proton temperature ratio scales roughly as the ratio of masses; that is, the alpha particles and the protons tend to have nearly equal thermal speeds and therefore temperatures that differ by a factor of about four. Finally, the solar wind flow speed is always greater than the sound speed and is almost always greater than the Alfvén speed, that is, the flow is usually both supersonic and superalfvénic.

III. BASIC NATURE OF THE INTERPLANETARY MAGNETIC FIELD

In addition to being a very good thermal conductor, the solar wind plasma is an excellent electrical conductor. Indeed, the electrical conductivity of the plasma is so high that the solar magnetic field is "frozen" into the solar wind flow as it expands away from the Sun. Because the Sun rotates, field lines in the equatorial plane of the Sun are bent into spirals (Fig. 2) whose inclinations relative to the radial direction depend on heliocentric distance and the speed of the solar wind. Each field line threads plasma emitted from a single point on the Sun. At 1 AU, the average field line spiral in the equatorial plane is inclined ~45° to the radial direction from the Sun.

In Parker's simple model, interplanetary magnetic field lines out of the equatorial plane take the form of helixes wrapped about the rotation axis of the Sun. These helixes are ever more elongated at higher solar latitudes and eventually approach radial lines over the poles of the Sun. The equations describing Parker's model of the interplanetary magnetic field far from the Sun are

$$B_r(r, \phi, \theta) = B(r_0, \phi_0, \theta)(r_0/r)^2$$
$$B_\phi(r, \phi, \theta) = -B(r_0, \phi_0, \theta)(\omega r_0^2/V_{sw}r)\sin\theta$$
$$B_\theta = 0$$

Here r, ϕ, and θ are radial distance, longitude, and latitude in a Sun-centered spherical coordinate system, B_r, B_ϕ, and B_θ are the magnetic field components in this coordinate system, ω is the angular velocity associated with solar rotation (2.9×10^{-6} radians s^{-1}), V_{sw} is the solar wind flow speed (assumed constant with distance

TABLE I
Statistical Properties of the Solar Wind at 1 AU

Parameter	Mean	STD	Most probable	Median	5–95% range
n (cm^{-3})	8.7	6.6	5.0	6.9	3.0–20.0
V_{sw} (km s^{-1})	468	116	375	442	320–710
B (nT)	6.2	2.9	5.1	5.6	2.2–9.9
A(He)	0.047	0.019	0.048	0.047	0.017–0.078
T_p ($\times 10^5$ K)	1.2	0.9	0.5	0.95	0.1–3.0
T_e ($\times 10^5$ K)	1.4	0.4	1.2	1.33	0.9–2.0
T_α ($\times 10^5$ K)	5.8	5.0	1.2	4.5	0.6–15.5
T_e/T_p	1.9	1.6	0.7	1.5	0.37–5.0
T_α/T_p	4.9	1.8	4.8	4.7	2.3–7.5
nV_{sw} ($\times 10^8$ cm^{-2} s^{-1})	3.8	2.4	2.6	3.1	1.5–7.8
C_s (km s^{-1})	63	15	59	61	41–91
C_A (km s^{-1})	50	24	50	46	30–100

from the Sun), and ϕ_0 is an initial longitude at a reference distance r_0 from Sun center.

Parker's model is in reasonably good agreement with suitable averages of the magnetic field measured in the ecliptic plane and at high latitudes over a wide range of heliocentric distances. However, the instantaneous orientation of the field often deviates substantially from the model field at all latidues. Moreover, there is evidence that interplanetary magnetic field lines commonly wander in latitude as they extend out into the heliosphere. This effect appears to be a result of motion of the footpoints of the field lines on the surface of the Sun associated with both differential solar rotation (the surface of the Sun rotates at different rates at different latitudes) and turbulent convective motions.

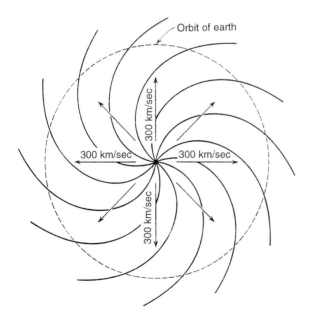

FIGURE 2 Configuration of the interplanetary magnetic field in the ecliptic plane for a uniform solar wind expansion of 300 km s^{-1}. [From E. N. Parker (1963). "Interplanetary Dynamical Process." Interscience, New York. Copyright © 1963. Reprinted with permission of John Wiley & Sons, Inc.]

IV. STREAM STRUCTURE

A. CORONAL STRUCTURE AND THE CORONAL MAGNETIC FIELD

Photographs taken at times of solar eclipse reveal that the corona is highly nonuniform, being structured by the complex solar magnetic field into a series of arcades, rays, holes (regions relatively devoid of material), and streamers. Some of this structure is evident in Fig. 3, which shows the solar corona as photographed in white light during the solar eclipse of March 18, 1988. Bright regions in the corona tend to be places where the field inhibits the coronal plasma from expanding outward, and dark regions (holes) are places where the solar wind expansion easily overpowers the coronal magnetic field. The magnetic field strength falls off sufficiently rapidly with height above the solar

FIGURE 3   The solar corona during the total eclipse of March 18, 1988, a year prior to solar activity maximum. The very large variation in coronal brightness as a function of height above the solar limb was attenuated by means of a radially graded filter placed at the focal plane of the camera. (Photograph courtesy of The High Altitude Observatory, National Center for Atmospheric Research.)

surface that it is incapable of containing the coronal expansion anywhere at altitudes above ~0.5–1.0 solar radii. The coronal expansion is what produces the "combed-out" appearance of coronal structures above those heights in the photograph.

B. HIGH-SPEED SOLAR WIND STREAMS

Given the great amount of coronal structure, it should not be surprising to learn that the solar wind is far from homogeneous. In fact, observations reveal that the solar wind in the ecliptic plane tends to be organized into alternating streams of high- and low-speed flows. Figure 4, which shows time histories of selected solar wind parameters at 1 AU for a 36-day interval in 1974, illustrates certain characteristic aspects of this stream structure that are particularly prevalent during the declining phase of the 11-year solar activity cycle. From top to bottom, the figure shows solar wind flow speed, the helium abundance relative to hydrogen, the proton density, and the azimuthal angle, ϕ_B, of the interplanetary magnetic field (relative to an inward-directed radial) plotted versus time. Three high-speed streams, which began on July 22, August 2, and August 19 and which persisted for a number of days each, are clearly evident in the figure. The July 22 and August 19 streams are actually the same stream encountered on successive solar rotations. For each stream the maximum speed exceeds 600 km s^{-1}; between streams the

speed falls to values below 350 km s^{-1}. Each high-speed stream is unipolar in the sense that ϕ_B is roughly constant throughout the stream. During the streams that began on July 22 and August 19, ϕ_B is approximately 135°, indicating that the field is directed outward away from the Sun along the interplanetary spiral. In contrast, during the intervening stream, ϕ_B is approximately 315° and the field is directed inward toward the Sun along the spiral. Sharp, long-lived reversals in field polarity occur at low speeds close to the leading edges of the high-speed streams, whereas more transient reversals occur elsewhere within the low-speed flows. The polarity reversals at the leading edges of the streams correspond to crossings of the heliospheric current sheet (discussed in more detail in the following section).

Variations in solar wind density are closely coupled to the field and flow structure. Particularly large and well-defined peaks in density occur in coincidence with the heliospheric current sheet crossings on July 23, August 2, and August 19. Smaller peaks in density occur in the low-speed solar wind that are loosely associated with more transient reversals in field polarity. The density tends to be lowest within the cores of the high-speed streams. Within the high-speed streams the helium abundance, A(He), is roughly constant at a value of about 4.5%, whereas within the low-speed flows A(He) is more variable, but tends toward lower values than within the cores of the high-speed streams. Relative minimums in A(He) occur at crossings of the heliospheric current sheet.

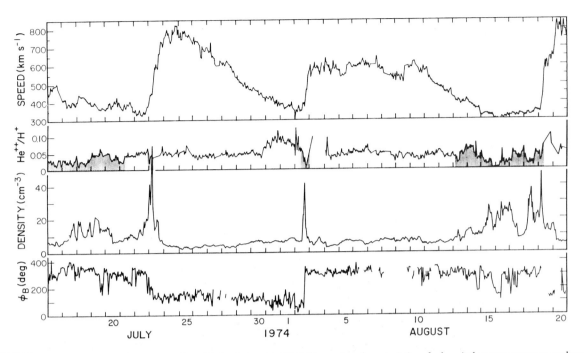

FIGURE 4 Selected solar wind parameters for a 36-day interval in 1974, illustrating characteristics of solar wind stream structure as observed at Earth's orbit. [From J. T. Gosling *et al.* (1981). *J. Geophys. Res.* **86**, 5438.]

C. THE CONNECTION BETWEEN CORONA STRUCTURE AND SOLAR WIND STREAM STRUCTURE

Figure 5 provides a schematic illustration of the connection between solar wind stream structure and coronal structure. Quasi-stationary high-speed streams originate in coronal holes (the dark regions in Fig. 3), which are large, nearly unipolar regions in the solar atmosphere. The coronal density is relatively low within coronal holes because the solar wind expansion there is relatively unconstrained by the solar magnetic

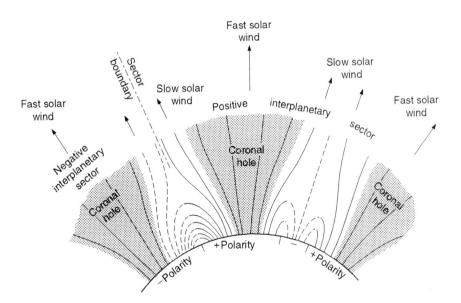

FIGURE 5 Schematic illustrating the relationship between coronal structure and solar wind stream structure. [From A. J. Hundhausen (1977). *In* "Coronal Holes and High Speed Wind Streams" (J. B. Zirker, ed.). Reprinted with permission of Colorado Associated University Press, Boulder.]

field. Low-speed flows, on the other hand, originate in the outer portions of coronal streamers (the relatively bright, combed-out structures in Fig. 3) that straddle regions of magnetic field polarity reversals. Close to the surface of the Sun within coronal streamers, the magnetic field is strong, field lines are entirely closed, the solar wind expansion is choked off, and coronal densities are high. At higher altitudes within streamers, the field is weaker and can be opened up by the coronal plasma pressure, producing the relatively dense, slow-speed flows at 1 AU characteristic of the region surrounding polarity reversals in the magnetic field.

D. SOLAR CYCLE EFFECTS

The corona continually evolves in response to the changing solar magnetic field associated with the advance of the ~11-year solar activity cycle. Near solar activity minimum and on the declining phase of the solar cycle, large coronal holes are found near the solar magnetic poles that often extend down to low heliographic latitudes. Thus quasi-stationary high-speed streams are common in the ecliptic plane at these times. Near solar activity maximum, however, strong magnetic fields choke off the coronal expansion over much of the Sun, and the solar wind flow in the ecliptic tends to be slower and more variable, often being disrupted by transient events associated with solar activity.

V. THE HELIOSPHERIC CURRENT SHEET AND SOLAR LATITUDE EFFECTS

A. RELATIONSHIP TO THE SOLAR MAGNETIC FIELD

We have seen that polarity reversals in the interplanetary magnetic field, which correspond to crossings of the heliospheric current sheet, map to the centers of coronal streamers. Coronal streamers, in turn, lie above regions in the lower solar atmosphere where the solar magnetic field reverses direction. On the declining phase of the solar activity cycle and near solar activity minimum, the Sun's large-scale magnetic field is approximately that of a dipole, similar to the Earth's. Regions where the Sun's field reverses direction from outward to inward and vice versa correspond approximately to the solar magnetic equator. Just as the Earth's

magnetic dipole is tilted relative to the Earth's rotation axis, so too is the solar magnetic dipole tilted with respect to the Sun's rotation axis. (The Sun's rotation axis, in turn, is tilted approximately 7° relative to ecliptic north.) However, the orientation of the solar magnetic dipole relative to the solar rotation axis is considerably more variable in time than is the orientation of the Earth's magnetic dipole relative to its rotation axis. As illustrated in the left portion of Fig. 6, near solar activity minimum the solar magnetic dipole tends to be aligned nearly with the rotation axis, whereas on the declining phase of the activity cycle it is generally inclined at a considerably larger angle relative to the rotation axis. Near solar maximum the Sun's field is not well approximated by a dipole.

B. THE BALLERINA SKIRT MODEL

Whenever the solar magnetic dipole and the solar rotation axis are closely aligned, the heliospheric current sheet in interplanetary space tends to coincide roughly with the solar equatorial plane. On the other hand, at times when the dipole is tilted substantially relative to the rotation axis, the heliospheric current sheet in interplanetary space becomes warped into an overall structure that resembles a ballerina's twirling skirt, as illustrated in the right portion of Fig. 6. Successive outward ridges in the current sheet (folds in the skirt) correspond to successive solar rotations and are separated in heliocentric distance by about 4.7 AU when the flow speed at the current sheet is 300 km s^{-1}. The maximum solar latitude attained by the current sheet in this simple picture is equal to the tilt of the magnetic dipole axis relative to the rotation axis.

C. SOLAR LATITUDE EFFECTS

On the declining phase of the solar activity cycle and near solar minimum, solar wind variability and stream structure are largely confined to a relatively narrow latitude band centered on the solar equator. This is illustrated in Fig. 7, which shows solar wind speed as a function of solar latitude as measured by the *Ulysses* space probe on the declining phase of the most recent solar cycle. *Ulysses* is nearly in a polar orbit about the Sun, reaching solar latitudes of ±80° in its ~6-year journey about the Sun. It is clear from the figure that the solar wind speed is highly variable at low heliographic latitudes, ranging from ~300 to ~850 km s^{-1} there, but is nearly constant at a value of 850 km s^{-1} at high latitudes. This latitude effect is a consequence

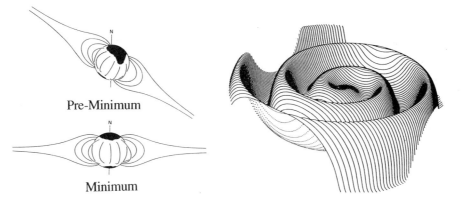

Pre-Minimum

Minimum

FIGURE 6 Right: schematic illustrating the configuration of the heliospheric current sheet in interplanetary space when the solar magnetic dipole is strongly tilted relative to the rotation axis of the Sun. The heliospheric current sheet separates interplanetary magnetic fields of opposite magnetic polarity and is the interplanetary extension of the solar magnetic equator. Left: schematic illustrating the changing tilt of coronal structure and the solar magnetic dipole relative to the rotation axis of the Sun as a function of phase of the solar activity cycle. [Adapted from J. R. Jokipii and B. Thomas (1981). *Astrophys. J.* **243**, 1115; and from A. J. Hundhausen (1977). *In* "Coronal Holes and High Speed Wind Streams" (J. B. Zirker, ed.). Reprinted with permission of Colorado Associated University Press, Boulder.]

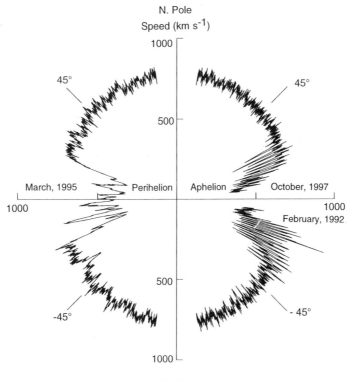

FIGURE 7 Solar wind speed as a function of heliographic latitude as measured by the *Ulysses* space probe from mid-February 1992 through mid-October 1997. Data shown in the left portion of the figure are centered on orbit perihelion at 1.4 AU, whereas data on the right are centered on orbit aphelion at 5.4 AU. The apparent difference in the latitude scale of structure in the flow speed in the left and right portions of the figure is an artifact associated with the fact that the spacecraft changed latitude very rapidly near perihelion but only very slowly near aphelion.

of two aspects of the solar wind already noted: (1) solar wind properties change rapidly with distance from the heliospheric current sheet, with flow speed increasing and density decreasing away from the current sheet; and (2) the heliospheric current sheet is commonly tilted relative to the solar equator, but is usually found within about ±30° of it. Neglecting transient events associated with solar activity (discussed later), both low- and high-speed flows, originating from coronal streamers surrounding the current sheet and from coronal holes, respectively, are observed at low latitudes as the Sun rotates. At high latitudes only high-speed flows from the polar corona holes are observed. The width of the band of solar wind variability changes as the tilt of the heliospheric current sheet changes; Fig. 7 indicates that the total band width varied from about 40° to about 70° during the phase of the solar activity cycle sampled by *Ulysses.* It is believed that the band of solar wind variability at solar activity maximum extends to significantly higher heliographic latitudes, perhaps almost to the poles.

VI. EVOLUTION OF STREAM STRUCTURE WITH DISTANCE FROM THE SUN

A. OBSERVATIONS OF EVOLVED STREAMS AT 1 AU

The high-speed solar wind streams shown in Fig. 4 are asymmetric in the sense that the speed rises more rapidly on the leading edges of the streams than it falls on the trailing edges. This and several other aspects of solar wind stream structure can be seen in Fig. 8, which shows the result of superposing data from 25 streams, keying on the peak in solar wind density at the leading edges of the streams. Note that on the average the density peaks as the speed rises and drops to low values as the speed falls. The plasma pressure also maximizes as the speed rises and decreases steadily throughout the trailing portion of the average stream. On the leading edge of the stream, the flow is deflected first in the sense of planetary motion about the Sun (i.e., so as to appear to be coming from east of the Sun) and then in the opposite direction shortly after the peak in plasma pressure. This pattern of variability, which often is highly repeatable from one stream to

the next, is the inevitable consequence of the evolution of a stream as it progresses outward from the Sun. Such evolution continues as a stream moves beyond Earth's orbit, producing dramatic changes in stream structure in the distant heliosphere.

B. KINEMATIC STREAM STEEPENING

The solar wind expansion from the Sun is variable in a spatial sense primarily because of the controlling influence of the highly structured solar magnetic field. As the Sun rotates with a period of 27 days as observed from Earth, alternately slow, then fast, then slow, and so on, plasma is directed outward along any radial line from the Sun. A snapshot of the speed of the solar wind as a function of heliocentric distance along a radial line in the equatorial plane in the inner heliosphere at some initial time t_0 might appear as shown in the left of the upper panel of Fig. 9. The dots in the snapshot identify different parcels of solar wind plasma moving at different speeds; these parcels originated from different positions on the Sun at different times and are therefore threaded by different magnetic field lines. At later times $t_0 + \Delta t$ and $t_0 + 2\Delta t$, the faster-moving plasma at the crest of the stream has overtaken and interacted with the slower plasma ahead, while at the same time it has outrun the slower-moving plasma from behind. Thus high-speed streams initially evolve with increasing heliocentric distance toward the sawtooth form illustrated on the right in the upper panel of Fig. 9. Material within the stream is rearranged as the stream steepens; parcels of plasma on the rising speed portion of the stream are compressed closer together, causing an increase in pressure there (lower panel of Fig. 9), whereas parcels on the falling speed portion of the stream are increasingly separated. The various parcels of gas cannot interpenetrate one another because they are threaded by different field lines.

C. STREAM DAMPING AND SHOCK FORMATION IN THE OUTER HELIOSPHERE

The steepening of a high-speed stream and the buildup in pressure on the rising speed portion of the stream are simple consequences of nonuniform coronal expansion and solar rotation. However, the pressure buildup produces forces (see lower panel of Fig. 9) that eventually limit the steepening of the stream and leads to novel features at large distances from the Sun. These forces

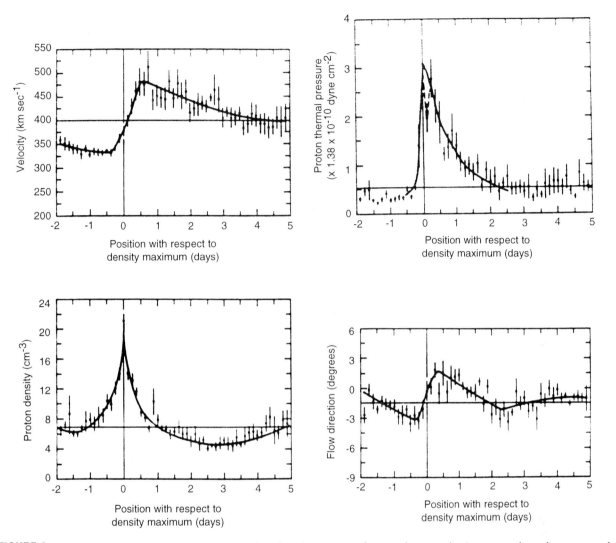

FIGURE 8 Average solar wind stream profiles at 1 AU for selected parameters (flow speed, proton density, proton thermal pressure, and bulk flow azimuth), obtained by superposing and averaging data for 25 different streams during 1965–1967. [Adapted from J. T. Gosling *et al.* (1972). *J. Geophys. Res.* **77**, 5442.]

act to accelerate the low-speed plasma on the leading edge of the stream and decelerate the high-speed plasma near the crest of the stream. The net result of the action of these forces is to limit the steepening of the stream and to transfer momentum and energy from the fast-moving plasma to the slow-moving plasma.

As long as the amplitude of a high-speed solar wind stream is not too large, the stream will gradually damp out with increasing heliocentric distance in the manner just described. However, when the stream amplitude, v_0, is greater than about twice the fast mode speed, C_f, the pressure on the rising speed portion of a stream increases nonlinearly as the stream steepens, and a pair of shock waves forms on either side of the high-pressure region (Fig. 10). [The fast mode speed is

the characteristic speed with which small-amplitude pressure signals propagate in a plasma. $C_f = (C_s^2 + C_A^2)^{0.5}$.] One of these shocks, known as a forward shock, propagates foward toward the trough of the stream and the other, known as a reverse shock, propagates backward toward and through the crest of the stream. Both of these shocks, however, are convected away from the Sun by the very high bulk flow of the wind. The major accelerations and decelerations associated with stream evolution now occur discontinuously at the shock surfaces, giving the stream speed profile the appearance of a double sawtooth wave, as shown in Fig. 10. The stream amplitude decreases and the compression region expands with increasing heliocentric distance as the reverse shock propagates back into the

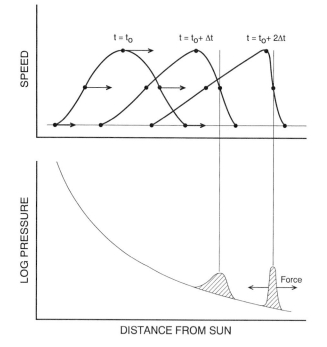

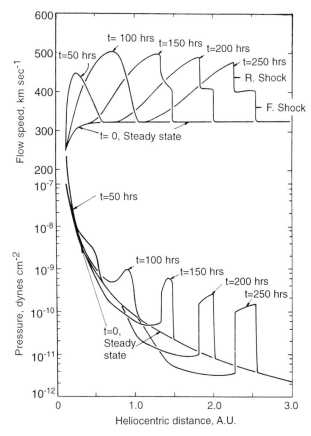

FIGURE 9 Above: schematic illustrating the kinematic steepening of a high-speed solar wind stream with increasing distance from the Sun. The steepening arises because plasma at the crest of the stream travels faster than plasma ahead of and behind the crest. Plasma on the leading edge of the stream becomes compressed during the steepening process, resulting in a buildup of pressure on the leading edge of the stream, as illustrated below. Forces associated with this buildup of pressure limit the steepening of the stream and transfer momentum from the fast-moving plasma to the slow-moving plasma ahead.

heart of the stream and the forward shock propagates into the lower-speed plasma ahead. Thus, damping of the stream occurs by removal of the fastest and slowest solar wind plasma. Observations indicate that few solar wind streams steepen sufficiently inside 1 AU to cause shock formation by the time the streams cross the Earth's orbit. Nevertheless, because C_f generally decreases with increasing heliocentric distance, virtually all large-amplitude solar wind streams steepen into shock wave structures at heliocentric distances beyond ~3 AU. At heliocentric distances beyond the orbit of Jupiter (~5.4 AU), a large fraction of the mass in the solar wind flow is found within compression regions bounded by shock waves on the rising portions of damped high-speed streams.

D. STREAM EVOLUTION IN TWO DIMENSIONS IN THE INNER HELIOSPHERE

So far we have considered stream evolution only along a fixed radius extending outward from the Sun in the

equatorial plane. The evolution of the stream is similar at all solar longitudes in the equatorial plane; however, the degree of evolution at any particular time is a function of longitude. When the coronal expansion is spatially variable but time stationary, a steady flow pattern such as sketched in Fig. 11 develops in the equatorial plane. This entire pattern corotates with the Sun and the compression regions are known as corotating interaction regions, or CIRs. It is worth emphasizing, however, that only the pattern rotates— each parcel of solar wind plasma moves outward nearly radially as indicated by the arrows. The region of high pressure associated with a CIR is nearly aligned with the magnetic field line spirals in the equatorial plane and the pressure gradients are thus nearly perpendicular to those spirals. Consequently, in the inner heliosphere the accelerations associated with the pressure gradients that form on the rising speed portion of a high-speed stream have transverse as well as radial

FIGURE 10 Snapshots of solar wind flow speed (above) and pressure (below) as functions of heliocentric distance at different times during the evolution of a large-amplitude, high-speed solar wind stream as calculated from a simple numerical model. [Adapted from A. J. Hundhausen (1973). *J. Geophys. Res.* **78**, 1528.]

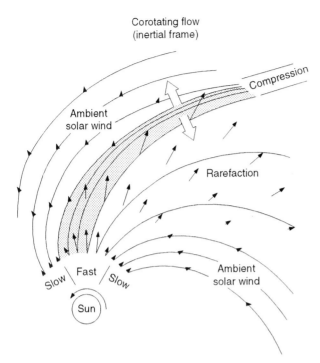

FIGURE 11 Schematic illustrating two-dimensional, quasi-stationary stream structure in the ecliptic plane in the inner heliosphere. The compression region on the leading edge of a stream is nearly aligned with the spiral magnetic field and the forces associated with the pressure gradients there have transverse as well as radial components. [From V. J. Pizzo (1978). *J. Geophys. Res.* **83**, 5563.]

components. In particular, the low-speed plasma near the trough of the stream not only is accelerated to a higher speed, as previously discussed, but also is deflected in the direction of solar rotation. In contrast, the high-speed plasma near the crest of the stream is both decelerated and deflected in the direction opposite to solar rotation. These transverse deflections are responsible for the systematic changes in flow direction observed near the leading edges of quasi-stationary, high-speed streams shown in Fig. 8.

E. TWO-DIMENSIONAL STREAM STRUCTURE IN THE OUTER HELIOSPHERE

In the outer heliosphere, CIRs become almost transverse to the radial direction. If the solar wind expansion is time stationary and there is only one high-speed stream in the equatorial plane, then in the outer heliosphere the forward shock eventually overtakes the reverse shock from the same stream on the previous rotation, as illustrated in the upper panel of Fig. 12. In this case, all the solar wind plasma in the equatorial plane becomes compressed at least once by the time it reaches a distance of about 20 AU from the Sun. When two or more streams are present in the equatorial plane, the forward and reverse shocks from the various streams intersect before the above closure can occur. For example, the lower panel of Fig. 12 shows a case of identical streams in the equatorial plane separated in solar longitude by 180° and for which all the plasma is compressed at least once by the time it reaches 10 AU.

The basic structure of the solar wind in the solar equatorial plane in the distant heliosphere thus differs considerably from that observed at 1 AU. Stream flow

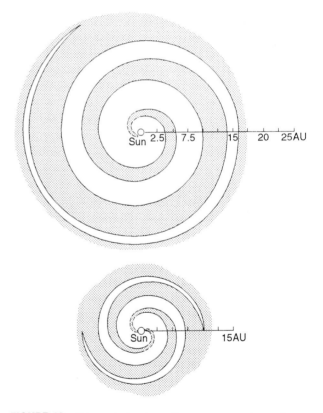

FIGURE 12 Schematic drawings of stream compression regions in the solar equatorial plane for a single solar wind stream (above) and for two identical streams separated by 180° in solar longitude (below). Beyond ~3 AU, the compression regions are bounded by forward–reverse shock pairs (solid lines) that diverge from one another, causing expansion of the compression regions. For the case of a single stream, the forward and reverse shocks from the same stream intersect one another at a distance of ~20 AU; for the case of two identical streams, the compression regions begin to overlap at ~10 AU. [Adapted from J. T. Gosling (1986). *In* "Magnetospheric Phenomena in Astrophysics" (R. I. Epstein and W. C. Feldman, eds.). Reprinted with permission of American Institute of Physics, New York.]

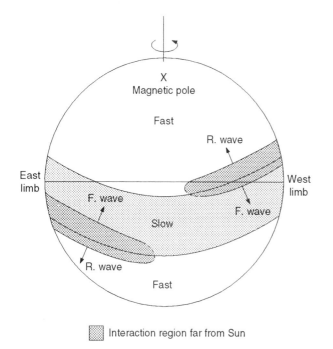

Interaction region far from Sun

FIGURE 13 A sketch illustrating the origin of north–south tilts of corotating interaction regions in the solar wind. The band of slow wind girdling the Sun at low latitudes is tilted relative to the heliographic equator in the same sense as the heliospheric current sheet (HCS). Solar rotation causes the fast wind to overtake the slow wind in interplanetary space far from the Sun along tilted interfaces between the slow and fast winds. The tilts are opposite in the northern and southern hemispheres. [Adapted from J. T. Gosling *et al.* (1993). *Geophys Res. Lett.* **20**, 2789.]

speed amplitudes are severely reduced and high-frequency velocity structures are damped out. The dominant structure in the solar equatorial plane in the outer heliosphere becomes the expanding CIR, where most of the plasma and field are concentrated. The CIRs are commonly bounded by shocks that eventually interact with one another.

F. STREAM EVOLUTION IN THREE DIMENSIONS

There is, of course, a three-dimensional aspect to stream evolution. In particular, the CIRs associated with stream evolution tend to have systematic north–south tilts that are opposite in the northern and southern solar hemispheres. The sketch shown in Fig. 13 illustrates how these tilts arise. A band of slow wind emanates from the Sun at low latitudes. This band is associated with coronal streamers that lie above the magnetic equator. Because the magnetic equator is commonly tilted relative to the heliographic equator,

so too is the band of slow wind. At higher latitudes a fast wind emanates from the coronal holes that surround the magnetic poles. Both of these flows are nearly radially outward from the vicinity of the Sun. As the Sun rotates, the fast wind overtakes the slow wind in interplanetary space along interfaces that are inclined relative to the solar equator in the same sense as is the band of slow wind. (At the other interfaces, where fast wind runs away from the slower wind, rarefactions are produced.) The interfaces, and the CIRs in which they are embedded (shaded in Fig. 13), have opposite north–south tilts in the northern and southern solar hemispheres. The forward and reverse waves bounding the CIRs always propagate roughly perpendicular to the interfaces. Thus, in addition to the motions already described, the forward waves in both hemispheres propagate toward and eventually across the equator, whereas the reverse waves in both hemispheres propagate toward higher latitudes. These motions can be discerned in plasma data by systematic north–south deflections of the flow, which are opposite in the northern and southern hemispheres, as a CIR sweeps over a spacecraft. As a result of these motions, forward shocks in the outer heliosphere are generally confined to the low-latitude band of solar wind variability, and the reverse shocks are commonly observed both within the band and poleward of it. However, the reverse shocks weaken as they propagate poleward and seldom reach latitudes more than about 15° above the low-latitude band.

G. QUASI-STATIONARY STREAMS AND RECURRENT GEOMAGNETIC ACTIVITY

A close relationship exists between coronal holes, high-speed solar wind streams, CIRs, and recurrent geomagnetic activity. It is now clear that the mysterious M-regions, hypothesized long before the era of satellite observations of the Sun and the solar wind, are to be identified with coronal holes, whereas the long-lived particle streams responsible for recurrent geomagnetic activity are to be identified with high-speed solar wind streams. Energy from the solar wind is transferred into the Earth's magnetosphere primarily via reconnection between the interplanetary magnetic field and the Earth's magnetic field at the magnetopause, the outer boundary of the Earth's magnetosphere. Magnetic reconnection favors oppositely directed magnetic fields. Since the Earth's field is generally directed northward in the outer dayside magnetosphere, dayside reconnection favors a southward-directed interplanetary magnetic field. The rate of reconnection, and hence the rate at which energy flows into the magnetosphere from the solar wind, depends on both the speed of the wind past the magnetosphere and the strength of the southward component of the interplanetary magnetic field, B_z. Both V_{sw} and B_z tend to be large within CIRs, where compression increases the strength of the interplanetary magnetic field, including any southward component present prior to compression. Thus, geomagnetic activity tends to peak during the passage of CIRs, which pass over Earth approximately every 27 days when the solar wind outflow is steady for long periods of time.

VII. CORONAL MASS EJECTIONS AND TRANSIENT DISTURBANCES

A. CHARACTERISTICS OF CORONAL MASS EJECTIONS

The solar corona evolves on a variety of time scales closely connected with the evolution of the coronal magnetic field. The most rapid and dramatic evolution in the corona occurs in events now known as coronal mass ejections, or CMEs (Fig. 14). CMEs originate in closed field regions in the corona where the magnetic field normally is sufficiently strong to constrain the coronal plasma from expanding outward. Typically these closed field regions are found in the coronal streamer belt that encircles the Sun and that underlies the heliospheric current sheet. The outer edges of CMEs often have the optical appearance of closed loops such as the event shown in Fig. 14. Whether or not these optical loops actually outline closed magnetic loops remains to be established. Indeed, the overall magnetic topology of CMEs in the corona and in interplanetary space is still poorly understood. Few, if any, CMEs appear to sever completely their magnetic connection with the Sun. During a typical CME, somewhere between 10^{15} and 10^{16} g of solar material is ejected into interplanetary space. Ejection speeds within about 5 solar radii of the solar surface range from less than 50 km s^{-1} in some of the slower events to greater than 2000 km s^{-1} in the fastest ones. The average CME speed at these heights is close to, but slightly less than, the overall 1 AU average ecliptic solar wind speed of ~470 km s^{-1}. Independent of their outward speeds, CMEs observed more than 0.5 solar radii above the limb of the Sun always continue to progress outward into interplanetary space, that is, they do not fall back into the Sun. Since observed solar wind speeds in the solar wind near 1 AU are never less than 280 km s^{-1}, the slowest CMEs must be accelerated further by the time they reach Earth's orbit. Usually it appears that the slowest CMEs are accelerated outward by the same forces that accelerate the normal solar wind. Some of the common characteristics of CMEs as observed by satellite-borne coronagraphs are provided in List 1.

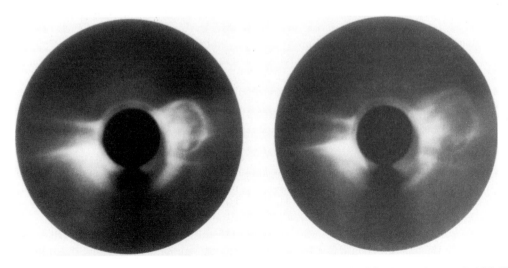

FIGURE 14 Two snapshots of a coronal mass ejection (CME) rising above the west limb of the Sun on August 10, 1973. Photographed with the coronagraph on *Skylab*. The Sun is occulted by a disk at the center of each photograph; the effective diameter of the disk is 1.5 solar diameters and the field-of-view is approximately 6 solar diameters. The snapshots were separated in time by 24 minutes. [Adapted from J. T. Gosling *et al.* (1974). *J. Geophys. Res.* **79**, 4581.]

LIST 1
Characteristics of Coronal Mass Ejection Events near the Sun

Mass ejected: 10^{15}–10^{16} g
Range of outward speed of leading edge: <50 km s^{-1}–>2000 km s^{-1}
Average outward speed of leading edge: $\sim$400 km s^{-1}
Average latitudinal width (as viewed from Sun center): $\sim$45°
Average longitudinal width—unknown, but events can cover $>180°$
Occurrence frequency: $\sim$3.5 events day^{-1} at solar activity maximum
$\qquad\qquad\qquad\qquad$ $\sim$0.1 events day^{-1} at solar activity minimum
Source region: Closed field regions (typically underlying heliospheric current sheet)
Associated solar activity: Eruptive prominences (common)
$\qquad\qquad\qquad\qquad\qquad$ Long-duration soft X-ray events (common)
$\qquad\qquad\qquad\qquad\qquad$ Hard X-ray events and optical flares (some of the time)
$\qquad\qquad\qquad\qquad\qquad$ Type II and IV radio bursts (the faster events)

B. ORIGINS, ASSOCIATIONS WITH OTHER FORMS OF SOLAR ACTIVITY, AND FREQUENCY OF OCCURRENCE

The processes that trigger CMEs and that determine their sizes and outward speeds are only poorly understood; there is presently no consensus on the physical processes responsible for initiating or accelerating these events. It is clear, however, that CMEs play a fundamental role in the long-term evolution of the solar corona. They appear to be an essential part of the way the corona responds to changes in the solar magnetic field associated with the advance of the $\sim$11-year solar activity cycle. Indeed, there is evidence to suggest that the release of a CME is one way that the solar atmosphere reconfigures itself in response to changes in the solar magnetic field. CMEs are commonly, but not always, observed in association with other forms of solar activity, such as eruptive prominences and solar flares. From a historical perspective one might be led to expect that large solar flares are the prime cause of CMEs; however, this is not the case. Many CMEs have no obvious associations with solar flares, and when CMEs and flares occur in association with one another the CMEs often lift off from the Sun before any substantial flaring activity has occurred. Moreover, the latitudinal and longitudinal extents of CMEs are commonly far greater than that of any associated flares. A typical CME extends about 45° in solar latitude close to the Sun, and some events have latitudinal widths exceeding 90°. Flares are usually confined to regions much smaller than this. When flares are observed in association with CMEs, they are usually found to one side of the CME span. Like other forms of solar activity, CMEs occur with a frequency that varies in a cycle of $\sim$11 years. It has been estimated that, on the average, the Sun emits about 3.5 CMEs day^{-1} near the peak of the solar activity cycle, but only about 0.1 CMEs day^{-1} near the minimum in solar activity.

C. INTERPLANETARY DISTURBANCES DRIVEN BY FAST CORONAL MASS EJECTIONS

The leading edges of the faster CMEs have outward speeds considerably greater than that associated with the normal solar wind expansion and drive transient shock wave disturbances in the solar wind. Indeed, fast CMEs are the prime source of transient solar wind disturbances throughout the heliosphere. Figure 15 shows calculated radial speed and pressure profiles of a simulated solar wind disturbance driven by a fast CME at the time the disturbance first reaches 1 AU. As indicated by the insert in the top portion of the figure, the disturbance was initiated at the inner boundary of the one-dimensional fluid calculation by abruptly raising the flow speed from 275 to 980 km s^{-1}, sustaining it at this level for 6 h, and then returning it to its original value of 275 km s^{-1}. The initial disturbance thus mimics a uniformly fast, spatially limited CME with an internal pressure equal to that of the surrounding solar wind plasma. A region of high pressure develops on the leading edge of the disturbance as the CME overtakes the slower ambient solar wind ahead. This region of higher pressure is bounded by a forward shock on its leading edge that propagates into the ambient solar wind ahead and by a reverse shock on its trailing edge that propagates backward into and through the CME. Both shocks are, however, carried away from the Sun by the convective flow of the solar wind, as in the case of the shocks associated

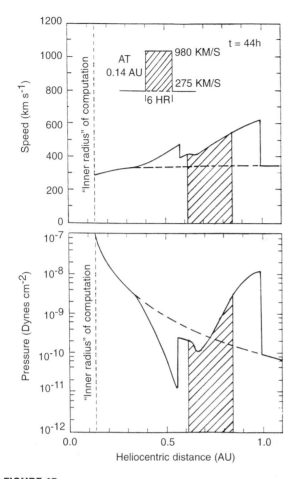

FIGURE 15 Solar wind speed and pressure as functions of heliocentric distance for a simple, one-dimensional gas-dynamic simulation of a CME-driven disturbance. The dashed line indicates the steady state prior to introduction of the temporal variation in flow speed imposed at the inner boundary of 0.14 AU and shown at the top of the figure. The hatching identifies material that was introduced with a speed of 980 km s^{-1} at the inner boundary, and therefore identifies the CME in the simulation. [Adapted originally from A. J. Hundhausen (1985). *In* "Collisionless Shocks in the Heliosphere: A Tutorial Review" (R. G. Stone and B. T. Tsurutani, eds.). American Geophysical Union, Washington, D.C.]

with steepened, corotating, high-speed streams. Both observations and more detailed calculations indicate that the reverse shock in CME-driven disturbances is ordinarily present only near the center of the disturbance.

The simple calculation shown in Fig. 15 is consistent with observations obtained near 1 AU in the ecliptic plane and illustrates to zeroth order the radial and temporal evolution of an interplanetary disturbance driven by a fast CME. The leading edge of the disturbance is a shock that stands off ahead of the CME. The ambient solar wind ahead of the CME is compressed, heated, and accelerated as the shock passes by, and the leading portion of the CME is compressed, heated, and slowed as a result of its interaction with the ambient plasma. Such compression leads to relatively high densities and strong magnetic fields in a broad region extending sunward from the shock to well within the CME. In the example illustrated, the CME slows from an initial speed of 980 km s^{-1} to less than 600 km s^{-1} by the time the leading edge of the disturbance reaches 1 AU. This slowing is a result of momentum transfer to the ambient solar wind ahead and proceeds at an ever slower rate as the disturbance propagates into the outer reaches of the heliosphere. Figure 16 displays selected plasma and magnetic field data from a shock wave disturbance driven by a CME and observed near 1 AU. The shock is distinguished in the data by discontinuous increases in flow speed, density, temperature, field strength, and pressure. As would be expected, the plasma identified as the CME has a higher flow speed than the ambient solar wind ahead of the shock. In this case it is also distinguished by an anomalously low temperature and a moderately strong and smoothly varying magnetic field strength.

D. CHARACTERISTICS OF CORONAL MASS EJECTIONS IN THE SOLAR WIND AT 1 AU

The identification of CMEs in solar wind plasma and field data is still something of an art. In this regard, shocks serve as useful fiducials for searching for plasma and/or field anomalies by which one can identify fast CMEs. List 2 provides a summary of plasma and field signatures that qualify as unusual compared to the normal solar wind, but that are commonly observed a number of hours after shock passage and that are often used to identify fast CMEs in the solar wind. Most of these anomalous signatures are observed elsewhere in the solar wind as well, where, presumably, they serve to identify those numerous, relatively low-speed CMEs that do not drive shock disturbances. Few CMEs at 1 AU exhibit all of these characteristics, and some of these signatures are more commonly observed than are others. It appears that CMEs are most reliably identified using signatures that reflect their unusual magnetic field topology (see next section). For example, the CME shown in Fig. 16 was identified in that way.

Measurements reveal that most CMEs expand as they propagate outward through the heliosphere. Such expansion can be a result of the CME's dynamic interaction with the ambient wind, or it can be a result of an initial high internal pressure or an initial front-to-rear speed gradient. CME radial thicknesses are

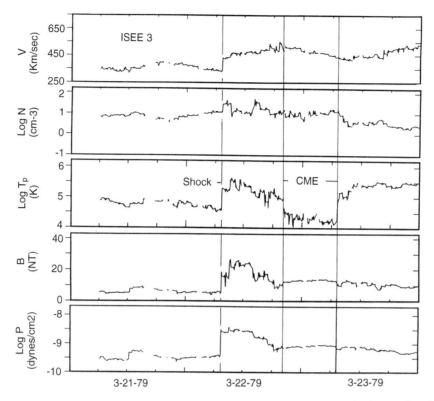

FIGURE 16 Selected plasma and magnetic field data from the *ISEE 3* spacecraft for a transient shock wave disturbance driven by a fast coronal mass ejection. From top to bottom, the quantities plotted are the solar wind flow speed, the log of the proton density, the log of the proton temperature, the magnetic field strength, and the log of the pressure (gas plus field). The CME has been identified by the presence of counterstreaming suprathermal electrons (not shown), but is also characterized by a higher speed than the ambient solar wind ahead of the shock, by an anomalously low temperature (for the observed speed), and by a moderately strong and smoothly varying magnetic field strength.

LIST 2
Characteristics of Coronal Mass Ejections in the Solar Wind at 1 AU

Common signatures: Counterstreaming (along the field) halo electrons

Counterstreaming (along the field) energetic (energy > 20 keV) protons

Helium abundance enhancement

Ion and electron temperature depressions

Strong magnetic field

Low plasma beta

Low magnetic field strength variance

Anomalous field rotation (flux rope)

Average radial thickness: 0.2 AU

Range of speeds: 300–1000 km s^{-1}

Single-point occurrence frequency: ~72 events year^{-1} at solar activity maximum

~8 events year^{-1} at solar activity minimum

Magnetic field topology: Predominantly closed magnetic loops rooted at both ends in Sun

Fraction of events driving shocks: ~1/3

Fraction of earthward-directed events producing large geomagnetic storms: ~1/6 variable; at 1 AU the typical CME has a radial width of ~0.2 AU and at Jupiter's orbit CMEs can have widths as large as 2.5 AU. Approximately one-third of all CMEs in the ecliptic plane have sufficiently high speeds relative to the ambient solar wind to drive interplanetary shocks at 1 AU, like the event shown in Fig. 16; the remainder do not have sufficiently high speeds to produce shock wave disturbances, and simply ride along with the rest of the solar wind. On the average, CMEs cannot be distinguished from the normal solar wind at 1 AU on the basis of either their speed or their plasma density. Near solar activity maximum CMEs account for about 15% of all solar wind measurements in the ecliptic plane at 1 AU, whereas near solar activity minimum they account for less than 1% of all the measurements. The Earth intercepts about 72 CMEs yr^{-1} near solar activity maximum and ~8 CMEs yr^{-1} near solar activity minimum. CMEs are much less common at high heliographic latitudes, particularly near activity minimum, when CMEs seem to be confined largely to the low-latitude band of solar wind variability.

E. MAGNETIC FIELD TOPOLOGY OF CORONAL MASS EJECTIONS AND THE PROBLEM OF MAGNETIC FLUX BALANCE

The solar wind expansion carries solar magnetic field lines out into interplanetary space to form the interplanetary magnetic field. In the normal solar wind, these field lines can usually be considered to be "open" in the sense that they connect to field lines of the opposite polarity only in the distant heliosphere very far from the Sun (Fig. 17, A). CMEs, on the other hand, originate in closed field regions in the solar corona not previously participating directly in the solar wind expansion, and CMEs thus open up new field lines by dragging "closed" field lines into interplanetaray space (Fig. 17, B). Indeed, it is the closed magnetic field topology common to most CMEs that most clearly distinguishes them from the ordinary solar wind. (It is becoming increasingly clear, however, that many CMEs in the solar wind also contain some open and, less often, disconnected field lines as well.) Because each CME adds new magnetic flux to interplanetary space, the interplanetary field strength would grow rapidly in the absence of other processes. If the closed field lines embedded within CMEs pinch off close to the Sun, that is, reconnect with themselves, to form disconnected plasmoids as indicated in Fig. 17, part C, no net increase in interplanetary field strength occurs. However, such reconnection requires a high degree

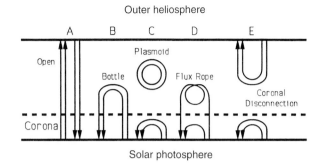

FIGURE 17 Schematic, not to scale, illustrating possible magnetic field topologies in the corona (below the dashed line) and in interplanetary space (above the dashed line). The topology on the left is that associated with the normal solar wind expansion, whereas that on the right is associated with magnetic reconnection of open field lines at the heliospheric current sheet. The central three drawings correspond to possible magnetic topologies associated with coronal mass ejections. [From D. J. McComas *et al.* (1992). *in* "Solar Wind 7" (E. Marsch and R. Schwenn, eds.). Pergamon Press, Oxford, England.]

of symmetry that is unlikely to occur. A more likely possibility is that the magnetic "legs" of CMEs commonly reconnect with near neighbors to form flux ropes (Fig. 17, D) that are partially disconnected from the Sun. In this case each CME adds less permanent magnetic flux to interplanetary space than it otherwise would. Many CMEs in the solar wind have an overall flux rope field topology, apparently because of such reconnection. It is interesting to note that sustained magnetic reconnection of this sort eventually produces both open and disconnected field lines within CMEs, as is sometimes observed. Finally, flux balance in interplanetary space can be maintained by the reconnection of previously open field lines to form U-shaped structures in interplanetary space that are magnetically disconnected from the Sun (Fig. 17, E) and that are subsequently convected to the outer reaches of the heliosphere by the flow of the solar wind. It is not presently clear what mix of these processes is actually responsible for maintaining a long-term balance of magnetic flux in interplanetary space; measurements do indicate, however, that all the interplanetary magnetic field topologies illustrated in Fig. 17, except possibly C, actually occur.

F. FIELD LINE DRAPING ABOUT FAST CORONAL MASS EJECTIONS

The closed field nature of CMEs effectively prevents any substantial interpenetration between the plasma

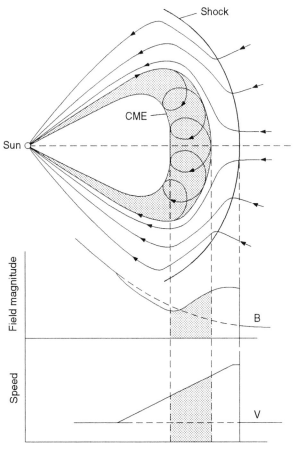

FIGURE 18 A hypothetical sketch of an interplanetary shock wave disturbance driven by a fast coronal mass ejection (above) and the corresponding radial variation of solar wind speed and field strength along the centerline of the disturbance (below). For simplicity, the ambient interplanetary magnetic field, which must drape around the CME as it forces its way outward through the solar wind, has been assumed to be radial. The sense and degree of draping actually observed in interplanetary space depend on a number of different factors (see text). [Adapted from J. T. Gosling *et al.* (1991). *J. Geophys. Res.* **96**, 7831.]

and, in addition, may produce a long magnetic tail behind a very fast CME in the outer heliosphere—somewhat akin to a cometary tail but pointing sunward rather than antisunward. On the other hand, draping does not affect the orientation of the field within the CME itself; there the field orientation is determined primarily by conditions and processes back at the Sun where the CME originates. As a final point of interest, Fig. 18 also illustrates that, just as the bow wave in front of a boat moving through water is considerably broader in extent than is the boat that produces it, so too is the shock in front of a fast CME somewhat broader in extent than is the CME that drives it. As a result, it is possible to encounter a CME-driven shock wave in interplanetary space without actually encountering the CME itself.

G. FAST CORONAL MASS EJECTIONS AND LARGE NONRECURRENT GEOMAGNETIC STORMS

High solar wind flow speeds and strong magnetic fields, often with strong southward components, are features common to solar wind disturbances driven by fast CMEs in the ecliptic plane. This is particularly true within the leading portions of these disturbances, where compression serves to elevate the field strength and draping often causes the field in the ambient plasma ahead of the CME to point out of the ecliptic plane (see Fig. 18). Consequently, solar wind disturbances driven by fast CMEs can be very effective in stimulating geomagnetic activity. Indeed, it is now known that all large, nonrecurrent geomagnetic storms occur during Earth-passage of solar wind disturbances driven by fast CMEs. CME-driven solar wind disturbances are thus the long-sought link between solar activity and geomagnetic activity. On the other hand, many CMEs are ineffective in a geomagnetic sense. Only about one in six CMEs passing Earth produces a large geomagnetic storm. Typically slow CMEs do not produce large geomagnetic storms because they lack the strong fields and high speeds necessary to stimulate significant activity in the Earth's magnetosphere.

VIII. VARIATIONS WITH DISTANCE FROM THE SUN

For a structureless solar wind, the solar wind speed remains approximately constant beyond the orbit of

within a CME and that in the surrounding solar wind. Thus the ambient plasma and magnetic field ahead of a fast CME must be deflected away from the path of the CME. Figure 18 illustrates that such deflections cause the ambient interplanetary magnetic field to drape about the CME. The degree of draping and the resulting orientation of the field ahead of a CME depend on the relative speed between the CME and the ambient plasma, the shape of the CME, the original orientation of the magnetic field in the ambient plasma, and position relative to the center-line of the CME. Draping can play an important role in reorienting the direction of the magnetic field ahead of a fast CME

Earth, the density falls off with heliocentric distance, r, as r^{-2}, and the magnetic field decreases with distance as described by the equations in Section II. The solar wind temperature also decreases with increasing heliocentric distance owing to the spherical expansion of the plasma; however, the precise nature of the decrease depends on particle species and the relative importance of such things as collisions and heat conduction (e.g., protons and electrons have different temperatures and evolve differently with increasing heliocentric distance). For an adiabatic expansion of an isotropic plasma, the temperature falls off as $r^{-4/3}$; for a plasma dominated by heat conduction, the temperature falls as $r^{-2/7}$.

Of course, as already noted, the solar wind is not structureless; rather it is characterized in the inner heliosphere by alternating streams of high- and low-speed flows of both spatial and temporal origin. The continual interaction of these flows with increasing heliocentric distance produces a radial variation of solar wind speed that differs considerably from that predicted for a structureless solar wind (see Fig. 1). High-speed flows decelerate and low-speed flows accelerate with increasing heliocentric distance as a result of momentum transfer from the high-speed flows to the low-speed flows on the rising speed portions of high-speed streams (see Sections VI and VII). Consequently, near the solar equatorial plane very far from the Sun (beyond ~15 AU), the solar wind flows at a nearly uniform speed close to 400 km s^{-1} most of the time (Fig. 19). Only rarely are significant speed perturbations observed at these very large distances from the Sun; these rare events are associated with disturbances driven by unusually fast CMEs that require a greater than normal distance to dissipate their momentum.

During the radial evolution of both quasi-stationary streams and transient disturbances, an ever greater fraction of the interplanetaary plasma and magnetic field becomes concentrated within the compression regions on the rising speed portions of high-speed streams; these compression regions are followed by large rarefaction regions relatively devoid of plasma and field. Thus, at low heliographic latitudes, the solar wind density and magnetic field strength tend to vary over a wider range in the outer heliosphere than near the orbit of the Earth, although the average density falls roughly as r^{-2} and the average magnetic field falls off roughly as predicted by the equations in Section II. On the other hand, plasma heating associated with the compression regions causes the solar wind temperature to fall off with increasing heliocentric distance more slowly than it otherwise would. Observations reveal that the actual temperature decrease for both

protons and electrons is somewhere between the adiabatic and conduction-dominated extremes.

IX. TERMINATION OF THE SOLAR WIND

Interstellar space is filled with a dilute gas of both neutral and ionized particles as well as a weak magnetic field. As space probes have not yet penetrated into the interstellar medium, the properties of the interstellar gas and the interstellar field in the vicinity of the heliosphere are poorly known. In the absence of the solar wind, the interstellar plasma would penetrate deep into the solar system. However, because of the magnetic fields embedded in these dilute plasmas, the interstellar plasma and the solar wind cannot interpenetrate one another. The result is that the solar wind blows a cavity in the interstellar plasma; the size and shape of the cavity depend on the momentum flux carried by the solar wind (which decreases with increasing heliocentric distance), the thermal pressure of the interstellar plasma (which is unknown), and the motion of the heliosphere relative to the interstellar medium. It is currently believed that the interface between the solar wind and the interstellar plasma, commonly known as the heliopause, occurs at a distance of 50 to 150 AU from the Sun in the direction of the Sun's motion through the interstellar medium.

The overall details of the solar wind interaction with the interstellar plasma are still speculative because of uncertainties in the properties of the local interstellar medium and because we lack direct observations of this interaction. Figure 20 shows what are believed to be the major elements of the interaction. The Sun and heliosphere are thought to be traveling at a speed of ~23 km s^{-1} relative to the interstellar medium. If this motion of the heliosphere relative to the interstellar medium exceeds the fast mode speed, C_f, in the interstellar plasma, then a bow shock stands in the interstellar plasma upstream of the heliosphere as shown in Fig. 20. This bow shock is analogous to the bow shocks that stand in the solar wind flow in front of planetary magnetospheres and ionospheres; it serves to initiate the slowing and deflection of the interstellar plasma around the heliosphere. The heliopause itself, which is the outermost boundary of the heliosphere, is a discontinuity that separates the interstellar plasma from the solar wind plasma. Sunward of the heliopause is a "termination" shock, where the solar wind flow is first affected by the presence of the interstellar plasma. The solar wind plasma is slowed (from supersonic motion to subsonic motion) as it crosses the termination

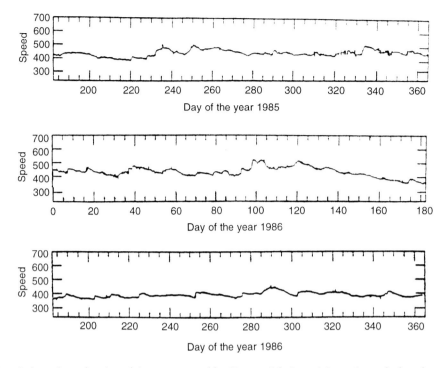

FIGURE 19 Solar wind speed as a function of time as measured by *Voyager 2* during a 1.5-year interval when the spacecraft was beyond 18 AU from the Sun. Because stream amplitudes are severely damped at large distances from the Sun, the solar wind speed there generally varies within a very narrow range of values. Compare with the speed variations evident in Fig. 4, obtained at 1 AU during a comparable period of the previous solar cycle. [Adapted from A. J. Lazarus and J. Belcher (1988). *In* "Proceedings of the Sixth International Solar Wind Conference" (V. J. Pizzo, T. E. Holzer, and D. G. Sime, eds.). National Center for Atmospheric Research, Boulder, Colorado.]

shock and is gradually turned by extended pressure gradients in the region between the termination shock and the heliopause so as to flow roughly parallel to the heliopause in the general direction of the flow of the external interstellar plasma. It is not known if the termination shock wraps entirely around the Sun, although it is commonly drawn that way. The shape of the heliosphere should be asymmetric as a result of its motion relative to the interstellar gas; it is compressed in the direction of that motion and is greatly elongated in the opposite direction.

It is unlikely that the solar wind's interaction with the interstellar gas is as smooth and uniform as drawn in Fig. 20 or is static in time because the momentum flux of the solar wind in the outer heliosphere is spatially and temporally variable. Present observations in the outer heliosphere suggest that the termination shock may be constantly in motion relative to the Sun owing to an ever changing solar wind momentum flux; it may never truly achieve an equilibrium position. Moreover, *Ulysses* measurements indicate that the momentum flux of the solar wind tends to be greater at high than at low heliographic latitudes, causing the heliopause to be farther from the Sun at high latitudes.

More direct information concerning the termination shock and the heliopause may be forthcoming in the years immediately ahead. The *Pioneer 11* and *Voyager 1* and *2* spacecraft, all of which carry a variety of particle and field experiments, are presently beyond the orbit of Neptune and are approaching the region where the termination shock is thought to reside. [*See* PLANETARY EXPLORATION MISSIONS.]

X. KINETIC PROPERTIES OF THE PLASMA

A. THE SOLAR WIND AS A NEARLY COLLISIONLESS PLASMA

Previous sections have emphasized that on a large scale the solar wind behaves like a compressible fluid and is capable of supporting relatively thin fluid structures such as shocks. As the solar wind is a very dilute plasma in which collisions are relatively rare, it is perhaps not obvious why the solar wind should exhibit this fluidlike

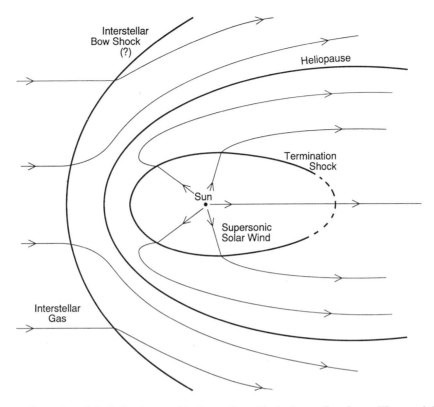

FIGURE 20 Schematic illustration of the heliosphere and its interaction with the interstellar plasma. The actual distance scale for this figure is unknown; the heliopause is probably ~100 AU from the Sun in the direction of the Sun's motion relative to the local interstellar medium. The arrows indicate the direction of the plasma flow.

behavior. For example, using values given in Table I, it is straightforward to show that the time between collisions for a typical solar wind proton at 1 AU is on the order of several days. (These collisions do not result from direct particle impacts such as colliding billiard balls, but rather from the long-distance "Coulomb" interactions characteristic of charged particles.) The time between collisions is thus comparable to the time for the solar wind to expand from the vicinity of the Sun to 1 AU; this is the basis for statements that the solar wind is a nearly collisionless plasma. An alternate, and perhaps better, statement is that the solar wind is a marginally collisional plasma.

Why does the solar wind behave like a fluid with so few particle collisions to effect fluidlike behavior? First, Coulomb collisions occur more freqently than noted in the foregoing when the solar wind temperature is low and the density is high. Second, the presence of the interplanetary magnetic field causes charged particles to gyrate about the field and thus they do not travel in straight lines between collisions. For typical conditions at 1 AU, solar wind electrons and protons have gyroradii of ~1.4 km and ~60 km, respectively, which are small compared to the scale size of structures

in the solar wind. Third, plasmas such as the solar wind are subject to a variety of instabilities that are triggered whenever the particle distribution function departs significantly from a Maxwellian (see next section). These instabilities produce collective interactions that mimic the effects of particle collisions. Finally, because the magnetic field is frozen into the solar wind, parcels of plasma originating from different positions on the Sun cannot interpenetrate one another.

B. KINETIC ASPECTS OF SOLAR WIND IONS

Collision rates in ordinary gases are usually very high, and such gases can usually be described by a single isotropic (i.e., the same in all directions) temperature, T. The range of individual particle speeds, v, present in an ordinary gas depends only on V_0 (the flow speed of the gas as a whole), T, and the particle mass, m. The problem of determining how individual particle speeds are distributed in a gas dominated by collisions, that is, of determining the so-called distribution function, $f(v)$, was first worked out by C. Maxwell in 1859, the same year that R. Carrington first observed a solar flare and noted its apparent association with a large geomagnetic storm. Maxwell showed that $f(v)$ $\sim \exp(-m(v - V_0)^2/2kT)$, where f is the number of particles per unit volume of velocity space and k is Boltzmann's constant (1.38×10^{-16} erg deg^{-1}). This form of $f(v)$ is often referred to as a Maxwellian distribution.

In contrast to the case for an ordinary gas, proton distribution functions in the solar wind are usually not isotropic because of the paucity of collisions and because the magnetic field provides a preferred direction in space. In the solar wind at 1 AU, the proton temperature parallel to the field is generally greater than is the temperature perpendicular to the field, on the average by a factor of ~1.4. Moreover, solar wind proton and alpha particle distributions often exhibit significant non-Maxwellian features such as the dou-

ble-peaked distributions illustrated in the left panel of Fig. 21. The second proton and alpha particle peaks are associated with a beam of particles streaming along the interplanetary magnetic field relative to the prime solar wind component at a speed comparable to the Alfvén speed. Measurements show that the relative streaming speed of such beams is usually comparable to or less than the local Alfvén speed, suggesting that the streaming speed is limited by a kinetic beam instability. Whenever the relative beam speed is higher than the Alfvén speed, the beam tends to disrupt and transfers its momentum and energy to the main solar wind component. Closer to the Sun, where the Alfvén speed is higher, relative streaming speeds between the beam and the main component can be as large as 100–200 km s^{-1}. Secondary proton beams are common in the solar wind in both low- and high-speed flows. The origin of these secondary beams is presently uncertain;

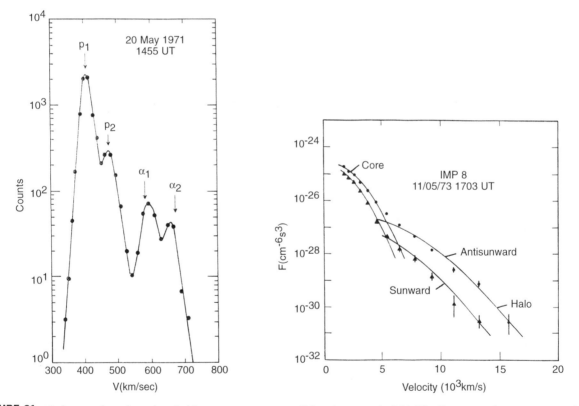

FIGURE 21   Left: a cut through a solar wind ion count spectrum parallel to the magnetic field. The first two peaks are protons and the second two peaks are alpha particles. (The velocity scale for the alpha particles has been artificially multiplied by a factor of 1.4.) Both the proton and alpha particle spectra show clear evidence for a secondary beam of particles streaming along the field relative to the main solar wind beam at about the Alfvén speed. [From J. R. Asbridge *et al.* (1974). *Solar Phys.* **37,** 451. Reprinted with permission of Kluwer Academic Publishers.] Right: a cut through a typical solar wind electron distribution parallel to the magnetic field in the high-speed solar wind. A break in the distribution at a speed of ~5×10^3 km s^{-1} for both sunward- and antisunward-moving electrons separates the core and halo populations. The bulk motion of the solar wind plasma is evident in the difference in the numbers of core electrons moving sunward and antisunward. There are far more halo electrons moving antisunward than sunward, because only one end of the field line at the spacecraft is connected to the hot solar corona. [From W. C. Feldman *et al.* (1975). *J. Geophys. Res.* **80,** 4181.]

it has been suggested that they may result from magnetic reconnection low in the solar corona and may play a fundamental role in the overall acceleration and heating of the solar wind.

C. KINETIC ASPECTS OF SOLAR WIND ELECTRONS

As illustrated in the right panel of Fig. 21, electron distributions in the solar wind contain two superimposed components, a relatively cold and dense thermal "core" population and a much hotter and more tenuous "halo" population. At 1 AU, the breakpoint between the core and the halo typically occurs at an energy of ~80 eV. This breakpoint moves steadily to lower energies as the core population cools with increasing distance from the Sun. Typically the core contains about 95% of all the electrons, and at 1 AU has a temperature of $\sim 1.3 \times 10^5$ K, whereas the halo contains about 5% of the electrons and has a temperature of $\sim 7.0 \times 10^5$ K. The core electrons generally are mildly anisotropic, with the temperature parallel to the field exceeding the temperature perpendicular to the field by a factor of ~1.1 on the average at 1 AU. However, the temperature anisotropy for core electrons varies systematically with density, such that at very low densities (<2 cm^{-3}) the temperature ratio often exceeds 2.0, whereas at very high densities (>10 cm^{-3}) the temperature ratio is often slightly less than 1.0. Such systematic variations of core electron temperature ratio with plasma density reflect the marginally collisional nature of the solar wind and the nearly adiabatic expansion of the core electrons in the spiral interplanetary magnetic field as the solar wind progresses outward from the Sun.

Solar wind electrons with halo energies ($>\sim 80$ eV) have speeds greater than 5×10^3 km s^{-1} and have far lower collision rates than do the colder core electrons. Beyond heliocentric distances of several solar radii, these electrons, which carry most of the solar wind heat flux, travel relatively unimpeded outward from the Sun along the interplanetary magnetic field. Because of their high speeds and nearly collisionless nature, the halo electrons serve as effective tracers of magnetic field topology in the interplanetary medium. At 1 AU, the flux of these hot electrons moving away from the Sun along the interplanetary magnetic field usually far exceeds the flux in the opposite direction (see Fig. 21). This nearly unidirectional flux arises because field lines in the normal solar wind are "open" (see Section VII, E) and are thus effectively connected to a hot generating source, the solar corona, at only one end. By way of contrast, field lines threading CMEs are most often connected to the Sun at both ends (see Sections VII, D and VII, E), and comparable fluxes of halo electrons are commonly observed in both sunward and antisunward directions within CMEs. Indeed, such counterstreaming fluxes of halo electrons are one of the most reliable signatures of CMEs in the solar wind (see List 2).

XI. HEAVY ION CONTENT

Although the solar wind consists primarily of protons (hydrogen), electrons, and alpha particles (doubly ionized helium), it also contains traces of ions of a number of heavier elements, reflecting the composition of the solar corona from which the solar wind originates. Table II provides estimates of the relative abundances of some of the more common solar wind elements summed over all ionization states. After hydrogen and helium, the most abundant elements are carbon and oxygen. The ionization states of all solar wind ions are "frozen in" close to the Sun because the characteristic times for ionization and recombination in the solar wind are large compared to the solar wind expansion time. Thus the ionization states measured in the solar wind far from the Sun are characteristic of the $1-2 \times 10^6$ K corona from which the solar wind originates. Commonly observed ionization states for some of the more abundant elements are provided in List 3. The relative abundances of these charge states change as the temperature in the corona from which the solar wind originates changes. Ionization state temperatures in the low-speed wind are typically in the range 1.4 to

TABLE II
Average Abundances of Elements in the Solar Wind

Element	Abundance relative to oxygen
H	1900 ± 400
He	75 ± 20
C	0.67 ± 0.10
N	0.15 ± 0.06
O	$= 1.00$
Ne	0.17 ± 0.02
Mg	0.15 ± 0.02
Si	0.19 ± 0.04
Ar	0.0040 ± 0.0010
Fe	$0.19 + 0.10, - 0.07$

LIST 3
Ionic Species Commonly
Observed in the Normal Solar Wind

H^+
He^{2+}
C^{5+}, C^{6+}
O^{6+}, O^{7+}, O^{8+}
Si^{7+}, Si^{8+}, Si^{9+}, Si^{10+}
Fe^{8+}, Fe^{9+}, Fe^{10+}, Fe^{11+}, Fe^{12+}, Fe^{13+}

1.6×10^6 K, whereas ionization state temperatures in the high-speed wind are typically in the range 1.0 to 1.2×10^6 K. Unusual ionization states, such as Fe^{16+} and He^{1+}, which are not common in the normal solar wind, are sometimes particularly overabundant within CMEs, reflecting the unusual coronal origins of those events.

The relative abundance values given in Table II are long-term averages. Observations reveal that these abundances can vary considerably as a function of time. Such variations have been extensively studied for helium (alpha particles), but are less well established for the heavier elements. Figure 22 shows a histogram of

measured alpha particle abundance values relative to protons. The most probable abundance value is ~4.5%, but the abundance ranges from less than 1% to values as high as 35% on occasion. The average helium abundance in the solar wind is about half that commonly attributed to the solar interior, for reasons presently unknown. Much of the variation in alpha particle abundance is related to the large-scale structure of the solar wind. The alpha particle abundance tends to be relatively constant within the cores of quasi-stationary, high-speed streams with an average value near 4.5%, but tends to be highly variable within low-speed flows. Particularly low (<2%) abundance values are commonly observed near magnetic field polarity reversals (see Fig. 4), including crossings of the heliospheric current sheet. Indeed, a distinct minimum in alpha particle abundance often occurs at the current sheet. Alpha particle abundance values greater than about 10% are relatively rare and account for less than 1% of all the measurements, At 1 AU, enhancements in alpha particle abundance above 10% occur almost exclusively within CME plasma. The physical cause of these variations is presently uncertain, although processes such as thermal diffusion, gravitational settling, and Coulomb friction in the solar corona all probably play roles.

XII. ENERGETIC PARTICLES

A proton moving with the average solar wind speed of 400 km s^{-1} has an energy of 0.84 keV, whereas an alpha particle moving with the same speed has an energy of 3.4 keV. Thus, by most measures, solar wind ions are low-energy particles. Interplanetary space is, nevertheless, filled with a number of energetic particle populations. Except in restricted regions of space, such as immediately in front of the planetary bow shocks or in the outer heliosphere close to the termination shock, the energetic particles found in the solar wind have insufficient energy densities to alter the bulk motion of the plasma or affect the overall structure of the interplanetary magnetic field. Thus, for the most part, energetic particles in the solar wind behave as test particles whose motions are guided and controlled by the interplanetary magnetic field.

Table III lists the main energetic particle populations observed in the solar wind, their approximate energy ranges, where these populations are observed within the heliosphere, the seed populations from which the energetic particle populations arise, and the sites where the seed particles are accelerated. With the

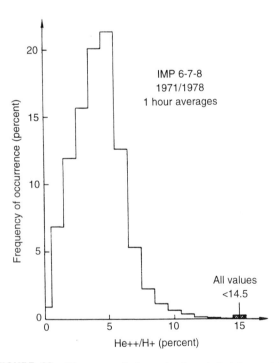

FIGURE 22 Histogram of observed solar wind alpha particle abundance relative to protons at 1 AU for the interval 1971–1978. [From G. Borrini *et al.* (1981). *J. Geophys. Res.* **86**, 4565.]

TABLE III
Energetic Particle Populations in the Solar Wind

Energetic particle population	Energy range (keV/nucleon)	Where observed	Seed population	Where particles are accelerated
Galactic cosmic rays	10^3–10^8	Everywhere (peak in outer heliosph.)	Stellar and galactic particles	Supernova remnants (?)—outside of heliosphere
Anomalous cosmic rays	10^3–10^5	Everywhere (peak in outer heliosph.)	Interstellar neutrals ionized in heliosphere	Termination shock—outer heliosphere
Corotating events	10^0–10^4	Field lines connected to corotating shocks	Solar wind ions	Forward/reverse shocks driven by corotating, high-speed streams—beyond ~2 AU
Gradual solar events	10^0–10^5	Field lines connected to CME-driven shocks	Solar wind ions	Interplanetary shocks driven by CMEs—solar wind
Impulsive solar events	10^0–10^5	Field lines connected to flare site	Chromospheric particles	Solar flares—at Sun
Bow shock particles	10^0–10^2	Field lines connected to bow shocks	Solar wind ions	Planetary and cometary bow shocks
Leaked particles	10^0–10^3	Field lines connected to magnetospheres	Magnetospheric particles	Planetary magnetospheres

exception of the nearly isotropic galactic cosmic rays, all of these energetic particle populations are produced by processes within the heliosphere itself. Both galactic cosmic rays and anomalous cosmic rays are observed throughout the heliosphere, but at greater intensities in the outer heliosphere than in the inner heliosphere. The other energetic particle populations are observed primarily on or near magnetic field lines that connect to their respective acceleration sites.

Shocks in nearly collisionless plasmas are particularly effective particle accelerators and are involved in producing at least four of the energetic particle populations listed in Table III. (The process by which a collisionless shock accelerates a small fraction of the ions it intercepts is reasonably well understood, although complex in detail.) The seed particles for the anomalous cosmic rays are interstellar particles that first enter the heliosphere as neutral particles, are subsequently ionized by solar ultraviolet radiation or by charge exchange with solar wind ions, and are then swept into the outer reaches of the heliosphere by the flow of the solar wind. It is presently believed that a small fraction of these newly ionized interstellar particles are accelerated to high energies as they interact with the termination shock. After acceleration, these ions, which are almost exclusively singly ionized H, He, N, O, and Ne, drift and diffuse back into the interior of the heliosphere to form the anomalous cos-

mic ray population. Solar wind ions act as the seed population for the other energetic particle populations that are shock-associated (corotating particle events, gradual solar energetic particle events, and bow shock particle events).

There are two energetic particle populations that are commonly associated directly with solar activity. Here these are called gradual solar energetic particle events and impulsive solar energetic particle events. Gradual events are observed near Earth about 10 times per year near solar activity maximum, typically last for at least several days, and generally are the most intense energetic particle events observed in interplanetary space. Although traditionally it has been thought that the particles in these events are accelerated in solar flares, more recent work shows that the acceleration occurs almost entirely in interplanetary space. That is, the energetic particles in these events are accelerated directly out of the solar wind by CME-driven shocks, as indicated in Table III. Impulsive events occur much more frequently (~1000 events per year near solar activity maximum) than do the gradual events, but they generally last for only a few hours and tend to be considerably weaker in intensity than the gradual events. In contrast to the particles in gradual events, the energetic particles in impulsive events are the result of (unknown) acceleration processes occurring close to the surface of the Sun.

XIII. CONCLUSIONS

Proof of the existence of the solar wind was one of the first great triumphs of the space age, and much has been learned about the physical nature of the wind in the intervening 35 years. Nevertheless, our understanding of the solar wind is far from complete. For example, we still do not know what physical processes heat and accelerate the solar wind or what determines its flow speed. We do not yet know if the low-speed wind arises primarily from quasi-stationary processes or from a series of small transient solar events. Likewise, the physical origins of coronal mass ejections are obscure; we do not fully understand why they occur or how they relate to the long-term evolution of the solar magnetic field and structure of the solar corona. We do not understand how a rough balance of magnetic flux is maintained in the solar wind in the presence of CMEs or how the magnetic topologies of CMEs evolve with time. In general, our ideas about the structure of the interplanetary magnetic field are still evolving and need testing with observations. Ideas about the termination of the solar wind in the outer heliosphere also remain to be tested, as do our perceptions about the three-dimensional structure of the heliosphere near solar activity maximum. The physical origin of variations in elemental abundances in the solar wind remains a mystery, as do temporal changes in the charge states of the heavier elements. The origin of double ion beams in the solar wind also remains unsolved, and we do not yet fully understand why different ionic species have different speeds and temperatures in the solar wind. Further analysis of existing data, new types of measurements, and fresh theoretical insights should lead to understanding in these and other areas of solar wind research in the years ahead.

BIBLIOGRAPHY

Barnes, A. (1992). *Res. Geophys.* **30**, 43–55.

Gosling, J. T. (1993). *J. Geophys. Res.* **98**, 18,937–18,949.

Grzedzielski, S., and Page, D. E. (eds.) (1990). "Physics of the Outer Heliosphere." Pergamon Press, Oxford, England.

Marsch, E., and Schwenn, R. (eds.) (1992). "Solar Wind 7." Pergamon Press, Oxford, England.

Marsden, R. G. (ed.) (1995). "The High Latitude Heliosphere." Kluwer, Boston, Mass.

Schwenn, R., and Marsch, E. (eds.) (1991). "Physics of the Inner Heliosphere. 1. Large-Scale Phenomena." Springer-Verlag, Berlin/Heidelberg, Germany.

Schwenn, R., and Marsch, E. (eds.) (1991). "Physics of the Inner Heliosphere. 2. Particles, Waves and Turbulence." Springer-Verlag, Berlin/Heidelberg, Germany.

Winterhalter, D., Gosling, J. T., Habbal, S. R., Kurth, W. S. and Neugebauer, M. (eds.) (1996). "Solar Wind 8," AIP Conference Proceedings, 382. American Institute of Physics, Woodbury, New York.

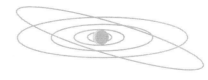

MERCURY

I. General Characteristics

II. Motion and Temperature

III. Atmosphere

IV. Polar Deposits

V. Interior and Magnetic Field

VI. Geology and Planet Evolution

VII. Origin

VIII. Conclusions

Robert G. Strom

University of Arizona

GLOSSARY

Antipodes: Opposite points on the surface of a sphere of a line through the center of the sphere.

Caloris Basin: Largest (1300-km diameter) well-preserved impact basin on Mercury viewed by *Mariner 10.*

Cold trapping: Process of trapping volatile material that would otherwise escape into space, by condensing it in exceptionally cold areas of a planet or satellite, usually the polar regions.

Graben: Long fault trough (valley) produced by subsidence between two inward-dipping boundary faults. It is the result of tensional stresses.

Hilly and lineated terrain: Broken-up surface of Mercury at the antipode of the Caloris impact basin.

Hot poles: Perihelion subsolar points on Mercury at the 0° and 180° meridians.

Intercrater plains: Oldest plains on Mercury that occur in the highlands and that were formed during the period of late heavy bombardment.

Lobate scarp: A long sinuous cliff (see *Thrust fault*).

Magnetopause: Outer boundary of a magnetosphere between the solar wind region and a planet's magnetic field region, where a strong thin current generally flows.

Magnetosheath: Region between a planetary bow shock and magnetopause in which the shocked solar wind plasma flows around the magnetosphere.

Magnetosphere: Magnetic cavity created by a planet where magnetospheric field lines have at least one end intersecting the planetary surface.

Obliquity: Inclination of the rotation axis of a planet to the plane of the planet's orbit around the Sun.

Period of late heavy bombardment: Earliest period in solar system history following planetary formation from about 4.5 to 3.8 billion years ago when the rate of meteoroid impact was very high compared to the present. The period of *early* heavy bombardment was the accretional formation of the planets that ended about 4.5 billion years ago.

Poynting–Robertson effect: Effect of the pressure of radiation from the Sun on small particles that causes them to spiral slowly into the Sun.

Regolith: Outermost fragmental layer on some airless planets and satellites that results from the fragmentation of rocks by repeated impacts of meteoroids.

Smooth plains: Youngest plains on Mercury with a relatively low impact crater abundance.

Tectonic framework: Global or large-scale pattern of fractures and folds formed by crustal deformation.

Thrust fault: Fault where the block on one side of the fault plane has been thrust up and over the opposite block by horizontal compressive forces.

Warm poles: Aphelion subsolar points on Mercury at the 90° and 270° meridians.

Mercury is the innermost and second smallest planet in the solar system (Pluto is smaller). It has no known satellites. The exploration of Mercury has posed questions concerning fundamental issues of its origin and, therefore, the origin and evolution of all the terrestrial planets. The data obtained by *Mariner 10* on its three flybys of Mercury on March 29 and September 21, 1974, and on March 16, 1975, remain our best source of detailed information on this planet. However, recent ground-based observations have provided important new information on the topography, radar, and microwave characteristics of its surface, discovered new constituents in its atmosphere, and helped constrain its surface composition. Mercury is often compared with the Moon because it superficially resembles that satellite. However, there are major differences that set Mercury apart from the Moon and, for that matter, all other planets and satellites in the solar system. The more we learn about this planet, the more we realize that Mercury is indeed a unique solar system body that provides insight into the origin and evolution of all the terrestrial planets. [*See* THE MOON.]

Mariner 10 imaged only about 45% of the surface at an average resolution of about 1 km, and less than 1% at resolutions between about 100 to 500 m (Fig. 1). This coverage and resolution are comparable to telescopic Earth-based coverage and resolution of the Moon before the advent of space flight. However, unlike the Moon in the early 1960s, only about 25% of the surface was imaged at sun angles low enough to allow adequate terrain analyses. As a consequence, there are still many uncertainties and questions concerning the history and evolution of Mercury. *Mariner 10* also discovered a magnetic field, measured the tem-

perature, and derived the physical properties of its surface.

On Mercury, the prime meridian (0°) was chosen to coincide with the subsolar point during the first perihelion passage after January 1, 1950. Longitudes are measured from 0° to 360°, increasing to the west. Craters are mostly named after famous authors, artists, and musicians, such as Dickens, Michelangelo, and Beethoven, whereas valleys are named for prominent radio observatories, such as Arecibo and Goldstone. Scarps are named for ships associated with exploration and scientific research, such as Discovery and Victoria. Plains are named for the planet Mercury in various languages, such as Odin (Scandinavian) and Tir (Germanic). Borealis Planitia (Northern Plains) and Caloris Planitia (Plains of Heat) are exceptions. The most prominent feature viewed by *Mariner 10* is named the Caloris Basin (Basin of Heat) because it nearly coincides with one of the "hot poles" of Mercury.

I. GENERAL CHARACTERISTICS

Mercury's diameter is only 4878 km, but it has a relatively large mass of 3.301×10^{23} kg. Because of its large mass in relation to its volume, Mercury has an exceptionally high mean density of 5.44 g/cm^3, second only to the density of the Earth (5.52 g/cm^3). The manner in which it reflects light is very similar to the way light is reflected by the Moon. The brightness (albedo) of certain terrains is greater than comparable terrains on the Moon. Mercury is covered with a regolith consisting of fragmental material derived from the impact of meteoroids over billions of years. Mercury's surface is heavily cratered, with smooth plains filling and surrounding large impact basins. Long lobate scarps traverse the surface for hundreds of kilometers, and large expanses of intercrater plains (the most extensive terrain type) fill regions between clusters of craters in the highlands. Also, a peculiar terrain consisting of a jumble of large blocks and linear troughs occurs antipodal to the Caloris Basin.

II. MOTION AND TEMPERATURE

Mercury has the most eccentric (0.205) and inclined (7°) orbit of any planet except Pluto. Its average distance from the Sun is 0.3871 AU (5.79×10^7 km). Because of its large eccentricity, however, the distance

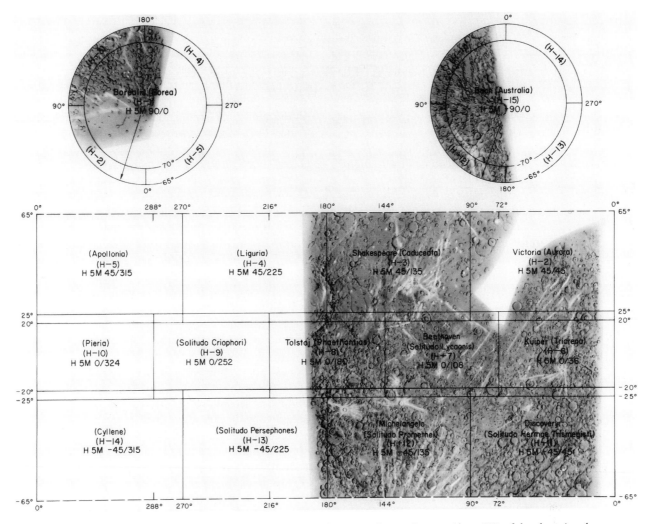

FIGURE 1 Shaded relief map of Mercury showing the quadrangle names and major features. About 55% of the planet is unknown.

varies from 0.3075 AU (4.6 × 10⁷ km) at perihelion to 0.4667 AU (6.98 × 10⁷ km) at aphelion. As a consequence, Mercury's orbital velocity averages 47.6 km/s, but varies from 56.6 km/s at perihelion to 38.7 km/s at aphelion. At perihelion the Sun's apparent diameter is over three times larger than its apparent diameter as seen from Earth. [*See* PLUTO AND CHARON.]

Mercury's rotation period is 58.646 Earth days and its orbital period is 87.969 Earth days. Therefore, it has a unique 3:2 resonant relationship between its rotational and orbital periods: it makes exactly three rotations on its axis for every two orbits around the Sun. This resonance was apparently acquired over time as the natural consequence of the dissipative processes of tidal friction and the relative motion between a solid mantle and a liquid core. As a consequence of this resonance, a solar day (sunrise to sunrise) lasts two

Mercurian years or 176 Earth days. The obliquity of Mercury is close to 0° and, therefore, it does not experience seasons as do Earth and Mars. Consequently, the polar regions never receive the direct rays of sunlight and are always frigid compared to torrid sunlit equatorial regions.

Another effect of the 3:2 resonance between the rotational and orbital periods is that the same hemisphere always faces the Sun at alternate perihelion passages. This happens because the hemisphere facing the Sun at one perihelion will rotate one and a half times by the next perihelion, placing it directly facing the Sun again. Because the subsolar points of the 0° and 180° longitudes occur at perihelion, they are called "hot poles." The subsolar points at 90° and 270° longitudes are called "warm poles" because they occur at aphelion. Yet another consequence of the 3:2 resonance is that an observer on Mercury (depending on location) would witness either a double sunrise or a double sunset, or the Sun would backtrack in the sky at noon during perihelion passage. Near perihelion, Mercury's orbital velocity is so great compared to its rotation rate that it controls the Sun's apparent motion in the sky as viewed from Mercury.

Although Mercury is closest to the Sun, it is not the hottest planet. The surface of Venus is hotter because of its atmospheric greenhouse effect. However, Mercury experiences the greatest range (day to night) in surface temperatures (635 K) of any planet or satellite in the solar system because of its close proximity to the Sun, its peculiar 3 : 2 spin orbit coupling, its long solar day, and its lack of an insulating atmosphere. Its maximum surface temperature is about 725 K at perihelion on the equator; hot enough to melt zinc. At night, the surface temperature drops to about 90 K. [*See* VENUS: ATMOSPHERE.]

III. ATMOSPHERE
. . . .

Although Mercury has an atmosphere, it is extremely tenuous with a surface pressure a trillion times less than Earth's. The number density of atoms at the surface is only 10^5 atoms/cm^3 for the known constituents (Table I). It is, therefore, an exosphere where atoms rarely collide; their interaction is primarily with the surface. *Mariner 10*'s ultraviolet spectrometer identified hydrogen, helium, and oxygen and set upper limits on the abundance of argon in the atmosphere. The hydrogen and helium are probably derived largely from the solar wind, although a portion of the helium may be of radiogenic origin and some hydrogen could result from the photodissociation of H_2O. The interac-

tion of high-energy particles with surface materials may liberate enough oxygen to be its principal source, but breakdown of water vapor molecules by sunlight could also be a possible source.

In 1985–1986, Earth-based telescopic observations detected sodium and potassium in the atmosphere. Both sodium and potassium have highly variable abundances (10^4–10^5 Na atoms/cm^3 and 100–10^4 K atoms/cm^3) near the surface on time scales of hours to years. Their abundances also vary between day and night by a factor of about 5, the day side being greater. Often bright spots of emission are seen at high northern latitudes or over the Caloris Basin. The temperature of the gas is about 500 K, but a hotter, more extended Na coma sometimes exists. Observed variations in the abundances of these elements are consistent with the photoionization time scale of 120 min for sodium and ~90 min for potassium. Photoionization of the gas will result in the exospheric ions being accelerated by the electric field in the planetary magnetosphere. Ions created on one hemisphere will be accelerated toward the planetary surface and recycled, but ions on the opposite hemisphere will be ejected away and lost. The total loss rate of sodium atoms is about 1.3×10^{22} atoms per second, so the atoms must be continuously supplied by the surface. The total fraction of ions lost to space from the planet is at least 30%. The atmosphere, therefore, is transient and exists in a steady state between its sources and sinks.

Although both sodium and potassium are probably derived from the surface of Mercury, the mechanism by which they are supplied is not well understood. The sodium and potassium in the Mercurian atmosphere could be released from sodium- and potassium-bearing minerals by their interaction with solar radiation, or by impact vaporization of micrometeoroid material. Both sodium and potassium show day-to-day changes in their global distribution.

If surface minerals are important sources for the exosphere, then a possible explanation is that their sodium/potassium ratio varies with location on Mercury. A possible explanation for some of the K and Na variations is Na and K ion implantation into regolith grains during the long Mercurian night (88 Earth days) and subsequent diffusion to the exosphere when the enriched surface rotates into the intense sunlight. At least one area of enhanced exospheric potassium emission apparently coincides with the Caloris impact basin, whose floor is highly fractured. This exospheric enhancement has been attributed to increased diffusion and degassing in the surface and subsurface through fractures on the basin floor, although other explanations may be possible.

TABLE I
Atmospheric Densities
on Mercury and the Moon

Species	Mercury (No. per cm³)	Moon (No. per cm³) Day	Moon (No. per cm³) Night
Hydrogen	23 (suprathermal) 230 (thermal)	< 17	—
Helium	6000	2000	40000
Oxygen	≤ 40,000	< 500	—
Sodium	20,000	70	
Potassium	500	16	
Argon	< 3 × 10⁷	1600	40000

IV. POLAR DEPOSITS

In 1991, high-resolution, full-disk radar images of Mercury were obtained from both the Arecibo and the linked Goldstone–Very Large Array radar facilities. The radar signals show very high reflectivities and polarization ratios centered on the poles of Mercury. The reflectivity and ratio values are similar to those of outer planet icy satellites and the residual polar water-ice cap of Mars. Therefore, Mercury's polar radar features have been interpreted to be water ice. The ice could be covered by several centimeters of regolith, potentially hiding it from optical view, and still be detected by radar, which can penetrate a few centimeters into the regolith. It has also been proposed that the radar characteristics are the result of scattering by inhomogeneities in the volume of elemental sulfur. In this case, it is proposed that sulfur volatilized from sulfides in the regolith was cold-trapped at the poles. The source of the sulfides may have been magmatic activity following initial accretion of material rich in troilite and pyrrhotite, or sulfur brought in by micrometeorites that constantly impact Mercury's surface.

Mariner 10 images of Mercury's polar regions show cratered surfaces where ice or sulfur could be concentrated in permanently shadowed portions of the craters. Radar studies have shown that the anomalies are indeed concentrated in the permanently shadowed portions of these polar craters (Fig. 2). The south polar radar feature is centered at about 88° south and 150° west and is largely confined within a crater (Chao Meng-Fu) that is 150 km in diameter, but a few smaller features occur outside this crater. In the north polar region, the deposits reside in about 25 craters down to a latitude of about 80°. Because the obliquity of Mercury is near 0°, it does not experience seasons and, therefore, temperatures in the polar regions should be <135 K. In permanently shaded polar areas, that is, the floors and sides of large craters, the temperatures should be less than 112 K, and water ice should be stable to evaporation on time scales of billions of years. If the deposits are water ice, then they could originate from comet or water-rich asteroid impacts that released the water to be cold-trapped in the permanently shadowed craters. Because comets and asteroids also impact the Moon, similar deposits would be expected to occur in the permanently shadowed regions of lunar craters.

The *Clementine* spacecraft orbited the Moon for about 3 months in 1994, and its bistatic radar experiment measured the magnitude and polarization of the radar echo versus bistatic angle. It discovered localized enhancements in the permanently shadowed regions of the lunar south pole. The enhancement has been interpreted as water ice.

In February 1998, the *Lunar Prospector* spacecraft made a positive identification of water ice in the regolith of permanently shadowed areas of craters at north and south polar regions. The amount of water ice appears to be between 0.5 and 1% of the regolith. This confirmation of water ice in permanently shaded lunar craters strongly suggests that the Mercurian polar deposits are also water ice derived from cometary and water-rich asteroidal impacts.

V. INTERIOR AND MAGNETIC FIELD

Mercury's internal structure is unique in the solar system and imposes severe constraints on any proposed origin of the planet. Mercury's mean density of 5.44 g/cm^3 is only slightly less than Earth's (5.52 g/cm^3), and larger than that of Venus (5.25 g/cm^3). Because of Earth's large internal pressures, however, its uncompressed density is only 4.4 g/cm^3 compared to Mercury's uncompressed density of 5.3 g/cm^3. This means that Mercury contains a much larger fraction of iron than any other planet or satellite in the solar system (Figs. 3 and 4). If this iron is concentrated in a core, then the core must be about 75% of the planet diameter, or some 42% of its volume. Thus, its silicate mantle and crust are only about 600 km thick. For comparison, Earth's iron core is only 54% of its diameter, or just 16% of its volume. [*See* THE EARTH AS A PLANET: SURFACE AND INTERIOR; PLANETARY MAGNETOSPHERES.]

Aside from Earth, Mercury is the only other terrestrial planet with a significant magnetic field. *Mariner 10* discovered that Mercury has a global intrinsic dipole magnetic field, but it is not well characterized because the spacecraft made only two passes through the field. The magnitude of the dipole moment is about 330 nT, or over 1000 times smaller than Earth's. Although weak compared to the Earth, the field has sufficient strength to hold off the solar wind, creating a bow shock, a magnetosheath, and a magnetosphere (Fig. 5). The magnetosphere strongly resembles a miniature version of Earth's magnetosphere. Because of the weaker field, however, Mercury occupies a much larger fraction of the volume of its magnetosphere than do other planets, and the solar wind actually reaches the surface at times of highest solar activity. The magnetopause subsolar distance is estimated to be about 1.35 ± 0.2 Mercury radii, and the bow shock distance

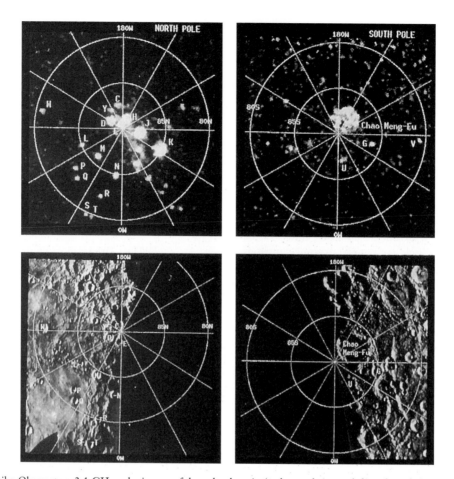

FIGURE 2 Arecibo Observatory 2.4-GHz radar images of the polar deposits in the north (upper left) and south (upper right) polar regions of Mercury. The resolution is 15 km (0.53°). The lower images are *Mariner 10* mosaics of the north (left) and south (right) polar regions. The letters in the radar images correspond to similarly lettered, permanently shadowed areas of craters in the *Mariner 10* images. The latitude–longitude grid on the *Mariner 10* images was derived from *Mariner 10* data, whereas the corrected grid on the radar images is derived from the positions of the radar features with respect to craters imaged by *Mariner 10*. (Courtesy of John Harmon, Arecibo Observatory, Puerto Rico.)

is about 1.9 ± 0.2 Mercury radii. Thus, the size of the magnetosphere is about 7.5 times smaller than Earth's. As a consequence, the equivalent region of intense radiation belts in the Earth's and the outer planet's magnetic fields is below the surface of Mercury. Therefore, the stably trapped charged particle environment of Mercury is probably very benign. Also because of the small size of Mercury's magnetosphere, magnetic events happen more rapidly and repeat more often than in Earth's magnetosphere.

Although other models may be possible, the maintenance of terrestrial planet magnetic fields is thought to require an electrically conducting fluid outer core surrounding a solid inner core. Therefore, Mercury's dipole magnetic field is taken as evidence that Mercury currently has a fluid outer core of unknown thickness. Thermal history models strongly suggest that Mercu-

ry's core would have solidified long ago unless there was some way of maintaining high core temperatures throughout geologic history. Proposed means of maintaining high temperatures are to either (1) provide more internal heat by enriching the core in uranium and thorium, (2) retain the heat longer by reducing the thermal diffusivity of the mantle, or (3) lower the melting point of the core material by adding some light alloying element. Most theoretical studies consider the addition of a light alloying element to be the most likely cause of a currently molten outer core. Although oxygen is such an element, it is not sufficiently soluble in iron at Mercury's low internal pressures. Metallic silicon has also been suggested, but sulfur is considered to be the most likely candidate. For a sulfur abundance in the core of less than 0.2%, the entire core should be solidified at the present time, and for an abundance

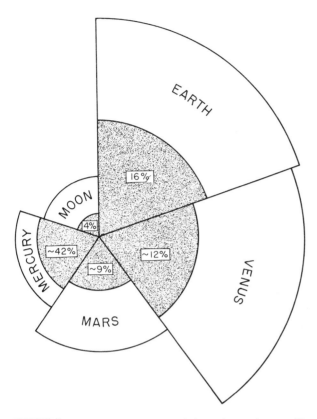

FIGURE 3 Comparison of terrestrial planet sizes and core radii. The percentage of the total planetary volume of the cores is also shown. The size of the Moon's core is not known, but the maximum possible size is shown.

of 7%, the core should be entirely fluid at present. Therefore, if sulfur is the alloying element, then Mercury probably contains between 0.2 and 7% sulfur in its core. As discussed in the following section, a possible sulfur abundance can be estimated from the planetary radius decrease derived from the tectonic framework.

VI. GEOLOGY AND PLANET EVOLUTION

Mercury has heavily cratered upland regions and large areas of younger smooth plains that surround and fill impact basins (Fig. 6). Infrared temperature measurements from *Mariner 10* indicate that the surface is a good insulator and, therefore, consists of a porous cover of fine-grained regolith. Earth-based microwave measurements indicate that this layer is a few centimeters thick and is underlain by a highly compact region extending to a depth of several meters. Mercury's heavily cratered terrain contains large areas of gently rolling intercrater plains, the major terrain type on the

planet. Mercury's surface is also traversed by a unique system of compressive thrust faults called lobate scarps. The largest well-preserved structure viewed by *Mariner 10* is the Calaoris impact basin, which is some 1300 km in diameter. Antipodal to this basin is a large region of broken-up terrain called the hilly and lineated terrain, probably caused by focused seismic waves from the Caloris impact.

A. GEOLOGIC SURFACE UNITS

The origin of some of the major terrains and their inferred geologic history are somewhat uncertain because of the limited photographic coverage and resolution, and the poor quality or lack of other remotely sensed data. In general, the surface of Mercury can be divided into four major terrains: (1) heavily cratered regions, (2) intercrater plains, (3) smooth plains, and (4) hilly and lineated terrain. Other relatively minor units have been identified, such as ejecta deposits exterior to the Caloris Basin.

1. Impact Craters and Basins

The heavily cratered uplands certainly record the period of late heavy meteoroid bombardment that ended

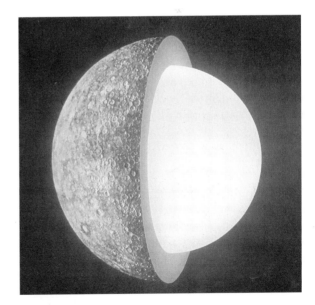

FIGURE 4 *Mariner 10* photomosaic together with an accurate artist's rendition of the size of Mercury's core compared to the silicate portion. The outer part of the core is still in a liquid state. [From R. G. Strom (1987). "Mercury: The Elusive Planet." Smithsonian Institution Press, Washington, D.C.]

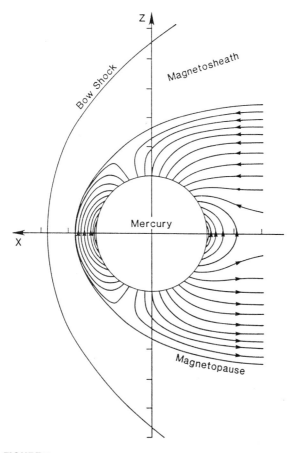

FIGURE 5 Magnetic field lines and a probable magnetopause and bow shock location for the Mercurian magnetic field. (From Vilas *et al.*, eds. (1998). *In* "Mercury." University of Arizona Press, Tucson.)

about 3.8 billion years ago on the Moon, and presumably at about the same time on Mercury. This period of late heavy bombardment occurred throughout the inner solar system and is also recorded by the heavily cratered regions on the Moon and Mars. The objects responsible for this bombardment may have been accretional remnants left over from the formation of the terrestrial planets and/or residual material left over from the formation of Uranus and Neptune and injected into the inner planet zone.

In the heavily cratered terrain on Mercury there is an increasing paucity of craters with decreasing crater diameter relative to heavily cratered terrain on the Moon. This paucity of craters is probably due to obliteration of the smaller craters by emplacement of intercrater plains during the period of late heavy bombardment. Below a diameter of about 20 km, the abundance of craters increases sharply. These craters may represent secondary impact craters from large craters or basins, or they could represent a group of objects called "vulcanoids" that were in long-lived or-

bits in Mercury's vicinity. These objects may have been accretional remnants left over from planet formation, or they could be fragments resulting from giant impacts that took place on Mercury near the end of its formation. The crater population superimposed on the smooth plains within and surrounding the Caloris Basin shows the same size distribution as the lunar highlands and the ridged plains of Mars over the same diameter range but at a much lower crater density. This suggests that, unlike the lunar maria, the Caloris smooth plains formed near the end of late heavy bombardment. [*See* MARS: SURFACE AND INTERIOR.]

Fresh impact craters on Mercury exhibit similar morphologies as those on the other terrestrial planets. Small craters are bowl-shaped, but with increasing size they develop central peaks, flat floors, and terraces on their inner walls. The transition from simple (bowl-shaped) to complex (central peak and terraces) craters occurs at about 10 km. At diameters between about 130 to 310 km, Mercurian craters have an interior concentric ring, and at diameters larger than about 300 km, they may have multiple inner rings. The freshest craters have extensive ray systems, some of which extend for distances over 1000 km. For a given rim diameter, the radial extent of Mercurian continuous ejecta is uniformly smaller by a factor of about 0.65 than that for the Moon. Furthermore, the maximum density of secondary impact craters occurs closer to the crater rim than for similar-sized lunar craters: the maximum density occurs at about 1.5 crater radii from the rim of Mercurian primaries, whereas on the Moon the maximum density occurs at about 2–2.5 crater radii. All of these differences are probably due to the larger surface gravity of Mercury (3.70 m/s^2) compared to the Moon (1.62 m/s^2).

Twenty-two multiring basins have been recognized on the part of Mercury viewed by *Mariner 10*. However, high-resolution radar images of the side not viewed by *Mariner 10* show several large circular features about 1000 km in diameter that may be impact basins. Based on ring tectonic theory, and the pattern and extent of grabens on the floor of Caloris, it is estimated that Mercury's lithosphere was thicker (>100 km) than the Moon's (25 to >75 km depending on location) at the end of late heavy bombardment. The 1300-km-diameter Caloris impact basin is the largest well-preserved impact structure (Fig. 7), al-

FIGURE 6 (a) Mercury as viewed by *Mariner 10* on its first approach in March, 1974. (b) Mercury's opposite hemisphere viewed by *Mariner 10* as it left the planet on the first encounter, and (c) the southern hemisphere viewed on the second encounter in September, 1974. (Courtesy of NASA.)

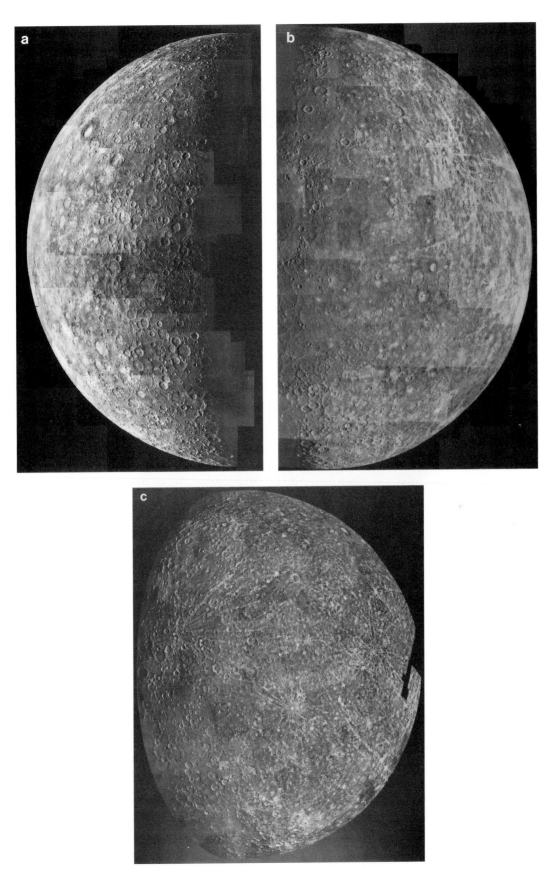

FIGURE 7 Photomosaic of the 1300-km-diameter Caloris impact basin showing the highly ridged and fractured nature of its floor. (Courtesy of NASA.)

though the much more degraded Borealis Basin is larger (1530 km). The floor structure of the Caloris Basin is like no other basin floor structure in the solar system. It consists of closely spaced ridges and troughs arranged in both concentric and radial patterns (Figs. 8a and 8b). The ridges are probably caused by compression, whereas the troughs are probably tensional graben that post date the ridges. This pattern may have been caused by subsidence and subsequent uplift of the basin floor. However, Earth-based radar observations indicate that the smooth plains surrounding the Caloris Basin are as much a 2.5 km lower than the surrounding terrain. This suggests than the crust has

been depressed by the weight of the smooth plains. Such a load could have induced concentric graben formation within the basin.

2. Hilly and Lineated Terrain

Directly opposite the Caloris Basin on the other side of Mercury (the antipodal point of Caloris) is the unusual hilly and lineated terrain that disrupts preexisting landforms, particularly crater rims (Figs. 9a and 9b). The Hills are 5 to 10 km wide and about 0.1 to 1.8 km high. Linear depressions that are probably tensional

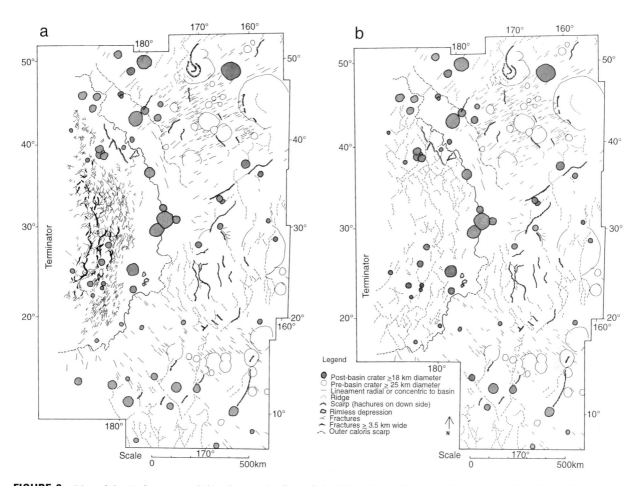

FIGURE 8 Map of the (a) fractures and (b) ridges on the floor of the Caloris Basin. The irregular continuous line follows the main rim of the basin, and the dash-dot line to the northeast of the main ring is a faint outer ring. The floor fractures and ridges have both radial and concentric components. [From R. G. Strom *et al.* (1975). *J. Geophys. Res.* **80**, 2478–2507.]

fault troughs form a roughly orthogonal pattern. Stratigraphic relationships suggest that the age of this terrain is the same as that of the Caloris Basin. Similar, but smaller, terrains occur at the antipodes of the Imbrium and Orientale impact basins on the Moon. The hilly and lineated terrain is thought to be the result of seismic waves generated by the Caloris impact and focused at the antipodal region (Fig. 10). Computer simulations of shock wave propagation indicate that focused seismic waves from an impact of this size can cause vertical ground motions of about 1 km or more and tensile failure to depths of tens of kilometers below the antipode. Although the lunar Imbrium Basin (1400 km diameter) is larger than the Caloris Basin, the disrupted terrain at its antipode is much smaller than that at the Caloris antipode. The larger disrupted terrain on Mercury may be the result of enhanced seismic wave focusing as a result of the large iron core.

3. Intercrater Plains

Mercury's two plains units have been interpreted as either impact basin ejecta or lava plains. The older intercrater plains are the most extensive terrain on Mercury (Figs. 11 and 12). They both partially fill and are superimposed by craters in the heavily cratered uplands. Furthermore, they have probably been responsible for obliterating a significant number of craters as evidenced by the paucity of craters less than about 40 km diameter compared to the highlands of the Moon. Therefore, intercrater plains were emplaced over a range of ages contemporaneous with the period of late heavy bombardment. There are no definitive features diagnostic of their origin. Because intercrater plains were emplaced during the period of late heavy bombardment, they are probably extensively brecciated and do not retain any signature of their original

FIGURE 9 (a) A portion of the hilly and lineated terrain antipodal to the Caloris impact basin. The image is 543 km across. (b) Detail of the hilly and lineated terrain. The largest crater in (b) is 31 km in diameter. (Courtesy of NASA.)

surface morphology. Although no landforms diagnostic of volcanic activity have been discovered, there are also no obvious source basins to provide ballistically emplaced ejecta. The global distribution of intercrater plains and the lack of source basins for ejecta deposits are indirect evidence for a volcanic origin. Additional evidence for a volcanic origin are recent *Mariner 10* enhanced color images showing color boundaries that coincide with geologic unit boundaries of some intercrater plains (Fig. 13). If intercrater plains are volcanic, then they are probably lava flows erupted from fissures early in Mercurian history. Crater densities on intercrater plains indicate ages between about 4 and 4.2 billion years.

4. Smooth Plains

The younger smooth plains cover almost 40% of the total area imaged by *Mariner 10*. About 90% of the regional exposures of smooth plains are associated with large impact basins, but they also fill smaller basins and large craters. The largest occurrence of smooth plains fills and surrounds the Caloris Basin (see Fig. 7) and occupies a large circular area in the north polar region that is probably an old impact basin (Borealis Basin). They are similar in morphology and mode of occurrence to the lunar maria. Craters within the Borealis, Goethe, Tolstoy, and other basins have been flooded by smooth plains, indicating that the plains are younger than the basins (Fig. 14). This is supported by the fact that the density of craters superimposed on the smooth plains that surround the Caloris Basin is substantially less than that of all major basins, including Caloris. Furthermore, several irregular rimless depressions that are probably of volcanic origin occur in smooth plains on the floors of the Caloris and the Tolstoy basins. The smooth plains' youth relative to the basins they occupy, their great areal extent, and other stratigraphic relationships suggest that they are

FIGURE 9 (*continued*)

volcanic deposits erupted relatively late in Mercurian history. *Mariner 10* enhanced color images show that the boundary of smooth plains within the Tolstoy Basin is also a color boundary, further strengthening the volcanic interpretation for the smooth plains. Based on the shape and density of the size/frequency distribution of superimposed craters, the smooth plains probably formed near the end of late heavy bombardment. They may have an average age of about 3.8 billion years as indicated by crater densities. If so, they are, in general, older than the lava deposits that constitute the lunar maria.

Two large radar-bright anomalies have been identified on the unimaged side of Mercury. One of these is similar to the radar signature of a fresh impact crater, but the other has a radar signature that is quite different from fresh impact craters on the Moon and Mercury. It has a structureless radar-bright halo (500 km diameter) and a radar-dark center (70 km diameter), which is similar to the radar signatures of large shield volcanoes on Venus and Mars. If this radar feature is indeed a shield volcano, then the geologic history, internal processes, and thermal history are much more dissimilar to those of the Moon than previously believed, because processes producing shield volcanoes did not occur on the Moon.

B. SURFACE COMPOSITION

Very little is known about the surface composition of Mercury. If the plains units (intercrater and smooth) are lava flows, then they must have been very fluid with viscosities similar to those of fluid flood basalts on the Moon, Mars, Venus, and Earth. The photometric characteristics are very similar to those of the Moon. However, at comparable phase angles and wavelengths in the visible part of the spectrum, Mercury appears to have systematically higher albedos than the Moon. Mercurian normal albedos range from 0.09 to 0.36 at 5° phase angle. The higher albedos are usually associated with rayed craters. However, the highest albedo (0.36) on *Mari-*

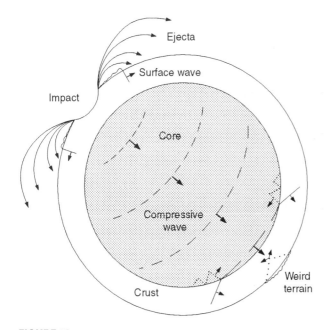

FIGURE 10 Diagrammatic representation of the formation of the hilly and lineated terrain by focused seismic waves from the Caloris impact. [From Schultz, P., and Gault, D. (1975). *The Moon* **12**, pp 159–177.]

ner 10 images is not associated with a bright rayed crater: it is a floor deposit in Tyagaraja Crater at 3° north latitude and 149° longitude. Whereas the lunar highlands/mare albedo ratio is almost a factor of 2 on the Moon, it is only a factor of 1.4 on Mercury. Furthermore, at ultraviolet wavelengths (58 to 166 nm), Mercury's albedo is about 65% lower than the Moon's at comparable wavelengths. These differences in albedo suggest that there are systematic differences in the surface composition between the two bodies.

Recalibration and color ratioing of *Mariner 10* images have been used to derive the FeO abundance, the opaque mineral content, and the soil maturity over the region viewed by *Mariner 10*. The probably volcanic smooth plains have a FeO content of <6 wt%, which is similar to that of the rest of the planet imaged by *Mariner 10*. Therefore, the surface of Mercury, may have a more homogeneous distribution of elements affecting color (e.g., more alkali plagioclase) than does the Moon. At least the smooth plains may be low iron or alkali basalts. Since the iron content of lavas is thought to be representative of their mantle source regions, it is estimated that Mercury's mantle has about the same FeO content

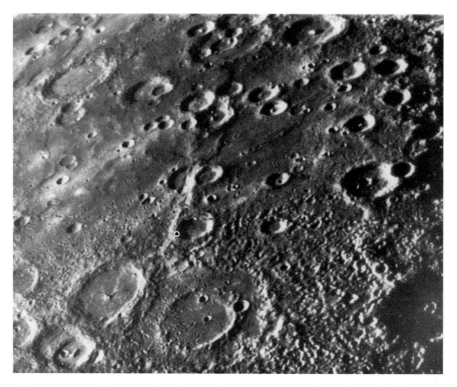

FIGURE 11 View of the intercrater plains surrounding clusters of craters in the mercurian highlands. Rodin, the large double-ring crater in the foreground, is about 250 km in diameter. (Courtesy of NASA.)

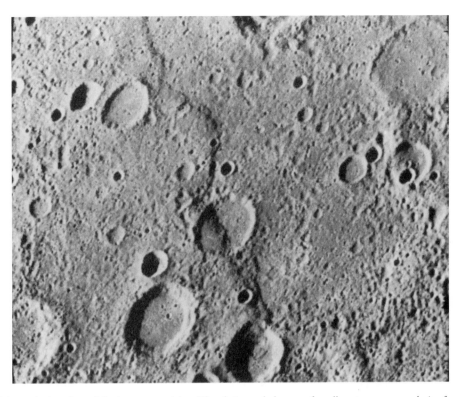

FIGURE 12 High-resolution view of the intercrater plains. The chains and clusters of small craters are secondaries from younger craters. The 90-km-diameter crater in the upper right-hand corner has been embayed by intercrater plains. The lobate scarp that diagonally crosses the image is a thrust fault. (Courtesy of NASA.)

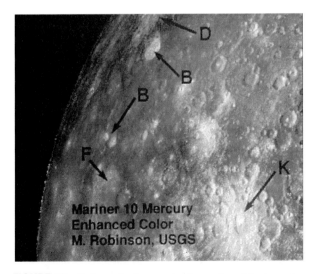

FIGURE 13 Enhanced color mosaic of a portion of the incoming side of Mercury as viewed by *Mariner 10*. The orange area at F has a sharp boundary that coincides with an intercrater plains boundary. The relatively dark blue unit at D is consistent with enhanced titanium content. The bright red unit at B may represent primitive crustal material, and Kuiper crater at K shows a yellowish color representing fresh material excavated from a subsurface unit that may have an unusual composition. (Courtesy of Mark Robinson, Northwestern University, Evanston, Ill.)

(<6 wt%) as the crust, indicating that Mercury is highly reduced with most of the iron in the core. In contrast, the estimated FeO contents of the mantle of the bulk Moon is 11.4%, of Venus and the Earth 8%, and of Mars ~18%.

There are some dark blue, low-albedo, and high-opaque mineral regions with diffuse boundaries that may be associated with fractures (see Fig. 13). These areas could be more mafic volcanic pyroclastic deposits. The bright rayed craters on Mercury have a very low opaque mineral index, which may indicate that the craters have excavated into an anorthositic crust. Color ratios of lunar and Mercurian crater rays also suggest that the surface of Mercury is low in Ti^{4+}, Fe^{2+}, and metallic iron compared to the surface of the Moon. This is consistent with Mercury's lower ultraviolet reflectivity and smaller albedo contrast. Earth-based microwave and mid-infrared observations also indicate that Mercury's surface has less FeO plus TiO_2, and at least as much feldspar as the lunar highlands. This has been interpreted as indicating that Mercury's surface is largely devoid of basalt, but it could also mean that the basalts just have a low iron content or are fluid alkali basalts. A comparison of Mercury's spectra with laboratory spectra of albite ($NaAlSi_3O_8$)

FIGURE 14   Photomosaic of the Borealis Basin showing numerous craters (arrows) that have been flooded by smooth plains. The largest crater is the Goethe basin, 340 km in diameter. (Courtesy of NASA.)

and anorthite ($CaAl_2Si_2O_8$) appears to show a spectral position intermediate between these minerals. This suggests that Mercury's surface may be rich in labradorite or bytownite [$(Na,Ca)(Al,Si)AlSi_2O_8$]. It has been suggested that eruption of highly differentiated basaltic magma may have produced alkaline lavas. On Earth there are low-viscosity alkali basalts that could produce the type of volcanic morphology represented by Mercury's plains. Mercury could be the only body in the inner solar system that has not experienced substantial high-iron basaltic volcanism and, therefore, may have undergone a crustal petrologic evolution different from that of other terrestrial planets.

C. TECTONIC FRAMEWORK

No other planet or satellite in the solar system has a tectonic framework like Mercury's, which consists of a system of compressive thrust faults called lobate scarps (Figs. 15 and 16). Individual scarps vary in length from about 20 km to over 500 km, and have heights from a few 100 m to about 3 km. They have a random spatial and azimuthal distribution over the imaged half of the planet and presumably occur on a global scale. Thus, at least in its latest history, Mercury was subjected to global compressive stresses. The only occurrences of features indicative of tensile stresses are localized fractures associated with the floor of the Caloris Basin and at its antipode, both of which are the direct or indirect result of the Caloris impact. No lobate scarps have been embayed by intercrater plains and they transect fresh as well as degraded craters. Few craters are superimposed on the scarps. Therefore, the system of thrust faults appears to postdate the formation of intercrater plains and formed relatively late in Mercurian history. This tectonic framework was probably caused by crustal shortening resulting from a decrease in the planet radius due to cooling of the planet. The amount of radius decrease is estimated to have been about 2 km based on the number of faults, their length, an average height of 1 km, a fault plane inclination of 25°, and extrapolated over the entire planet.

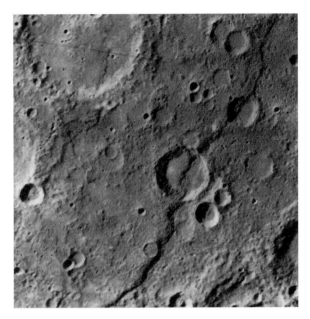

FIGURE 15 Photomosaic of Discovery scarp. This lobate scarp is a thrust fault over 2 km high and 500 km long. It cuts across two craters that are 55 and 35 km in diameter. (Courtesy of NASA.)

Also, there is apparently a system of structural lineaments consisting of ridges, troughs, and linear crater rims that have at least three preferred orientations trending in northeast, northwest, and north–south directions. The Moon also shows a similar lineament system. The Mercurian system has been attributed to modifications of ancient linear crustal joints formed in response to stresses induced by tidal spindown.

D. THERMAL HISTORY

All thermal history models of planets depend on compositional assumptions, such as the abundance of uranium, thorium, and potassium in the planet. Since our knowledge of the composition of Mercury is so poor, these models can provide only a general idea of the thermal history for certain starting assumptions. Nevertheless, they are useful in providing insights into possible modes and consequences of thermal evolution. Starting from initially molten conditions for Mercury, thermal history models with from 0.2 to 5% sulfur in the core indicate that the total amount of planetary radius decrease due to cooling is from about 6 to 10 km depending on the amount of sulfur (Fig. 17). About 6 km of this contraction is solely due to mantle cooling during about the first billion years before the start of inner core formation. The amount of radius decrease

FIGURE 16 The 130-km-long Vostok scarp transects two craters 80 and 65 km in diameter. The northwest rim of the lower crater (Guido d'Arezzo) has been offset about 10 km by thrusting of the eastern part of the crater over the western part. (Courtesy of NASA.)

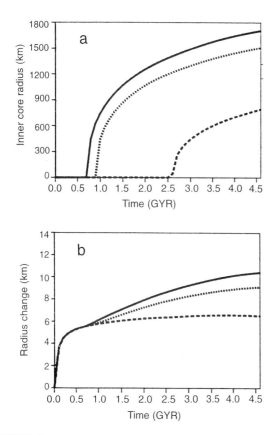

FIGURE 17 (a) A thermal history model for inner core radius as a function of time for three values of initial core sulfur content. The solid, dotted, and dashed lines are for sulfur contents of 0.2%, 1%, and 5%, respectively. (b) Model of the decrease in Mercury's radius due to mantle cooling and inner core growth for three values of initial core sulfur content as in (a). (From Vilas *et al.*, eds. (1998). *In* "Mercury." University of Arizona Press, Tucson.)

due to inner core formation alone is negligible for 5% sulfur and about 4 km for 0.2% sulfur.

If the 2-km-radius decrease inferred from the thrust faults was due solely to cooling and solidification of the inner core, then the core sulfur abundance is probably 2 to 3% and the present outer fluid core is about 500 or 600 km thick. In this case, inner core formation began about 3 billion years ago and, therefore, after the period of late heavy bombardment. This would imply that the observed tectonic framework began at about the same time, and that smooth and intercrater plains were emplaced before this event. Indeed, the geologic evidence indicates that at least the observed tectonic framework began to form relatively late in Mercury's history; certainly after intercrater plains formation and possibly after smooth plains formation. However, under initially molten conditions, the thermal history models indicate that the lithosphere has always been in compression, and that as much as a 6-

km-radius decrease occurred before inner core formation due to cooling of the mantle alone. If the smooth and intercrater plains are volcanic flows, then they had to have some way to easily reach the surface to form such extensive deposits. Early lithospheric compressive stresses would make it difficult for lavas to reach the surface, but the lithosphere may have been relatively thin at this time (<50 km). Large impacts would be expected to strongly fracture it, possibly providing egress for lavas to reach the surface and bury compressive structures. However, there is no geologic evidence for early compressive stresses, and one might expect that at least some thrust faults from this era would be partially preserved. Possibly there are some on the unimaged half of Mercury.

Other thermal history models with initially cool conditions, iron uniformly distributed throughout the planet, and heating by the decay of chondritic abundances of uranium and thorium indicate that core formation began 1.2 billion years after accretion and was complete by 1.8 billion years. This leads to extensive planetary melting and expansion that would fracture a thin lithosphere, providing egress for lavas (intercrater plains) to reach the surface. However, these events occur too late in history to account for the inferred surface ages, and the model does not account for the large iron core of Mercury. Furthermore, it is doubtful that Mercury formed in an initially cool condition.

E. GEOLOGIC HISTORY

Mercury's earliest history is very uncertain. If a portion of the mantle was stripped away, as invoked by most scenarios to explain its high mean density, then Mercury's earliest recorded surface history began after core formation and a possible mantle-stripping event (see Section VII). The earliest events are the formation of intercrater plains (≥4 billion years ago) during the period of late heavy bombardment. These plains may have been erupted through fractures caused by large impacts in a thin lithosphere. Near the end of late heavy bombardment, the Caloris Basin was formed by a large impact that caused the hilly and lineated terrain from seismic waves focused at the antipodal region. Further eruption of lava within and surrounding the Caloris and other large basins formed the smooth plains about 3.8 billion years ago. The system of thrust faults formed after the intercrater plains, but how soon after is not known. If the observed thrust faults resulted only from core cooling, then they may have begun after smooth plains formation and resulted in an inferred 2-km decrease in Mercury's radius. As the core continued to cool and the lithosphere thickened, compressive stresses closed off the magma sources and volcanism ceased near the end of late heavy bombardment. All of Mercury's volcanic events probably took place very early in its history, perhaps in the first 700 to 800 million years. Since that time, only occasional impacts of comets and asteroids have occurred. Today the planet may still be contracting as the present fluid outer core continues to cool.

VII. ORIGIN

The origin of Mercury and how it acquired such a large fraction of iron compared to the other terrestrial planets are not well determined. Chemical equilibrium condensation models for Mercury's present position in the innermost part of the solar nebula cannot account for the large fraction of iron that must be present to explain its high density. Although these early models are probably inaccurate, revised models that take into account material supplied from feeding zones in more distant regions of the inner solar system only result in a mean uncompressed density of about 4.2 g/cm³, rather than the observed 5.3 g/cm³. Furthermore, at Mercury's present distance, the models predict the almost complete absence of sulfur (100 parts per trillion of FeS), which is apparently required to account for the presently molten outer core. Other volatile elements and compounds, such as water, should also be severely depleted (<1 part per billion of hydrogen). [*See* THE ORIGIN OF THE SOLAR SYSTEM.]

Three hypotheses have been put forward to explain the discrepancy between the predicted and observed iron abundance. One (selective accretion) involves an enrichment of iron due to mechanical and dynamical accretion processes in the innermost part of the solar system, whereas the other two (postaccretion vaporization and giant impact) invoke removal of a large fraction of the silicate mantle from a once larger proto-Mercury. In the selective accretion model, the differential response of iron and silicates to impact fragmentation and aerodynamic sorting leads to iron enrichment owing to the higher gas density and shorter dynamical time scales in the innermost part of the solar nebula. In this model, the removal process for silicate from Mercury's present position is more effective than that for iron, leading to iron enrichment. The postaccretion vaporization hypothesis proposes that intense bombardment by solar electromagnetic and corpuscular radiation in the earliest phases of the Sun's evolu-

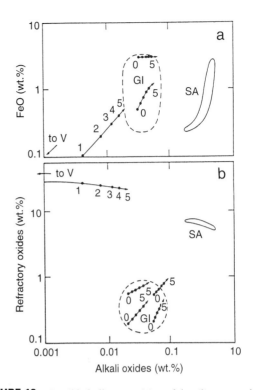

FIGURE 18 Possible bulk composition of the silicate mantle for the three models of Mercury's origin: selective accretion (SA), postaccretion vaporization (V), and giant impact (GI). The composition is parameterized for the FeO content, the alkali content (soda plus potash), and the refractory oxide content (calcium plus aluminum plus titanium oxides). The modifying effects of late infall of 0 to 5% of average chondritic meteorite material on several regolith compositions are indicated by arrows labeled 0 to 5. (From Vilas *et al.*, eds. (1998). *In* "Mercury." University of Arizona Press, Tucson.)

tion vaporized and drove off much of the silicate fraction of Mercury, leaving the core intact. In the giant impact hypotheses, a planet-sized object impacts Mercury and essentially blasts away much of the planet's silicate mantle, leaving the core largely intact. [*See* THE SUN; PLANETARY IMPACTS.]

Discriminating between these models is difficult, but may be possible from the chemical composition of the silicate mantle (Fig. 18). For the selective accretion model, Mercury's silicate portion should contain about 3.6 to 4.5% alumina, about 1% alkali oxides (Na and K), and between 0.5 and 6% FeO. Postaccretion vaporization should lead to very severe depletion of alkali oxides (~0%) and FeO (<0.1%) and extreme enrichment of refractory oxides (~40%). If a giant impact stripped away the crust and upper mantle late in accretion, then alkali oxides may be depleted (0.01 to 0.1%), with refractory oxides between about 0.1 to 1% and FeO between 0.5 and 6%. Unfortunately, our current knowledge of Mercury's silicate composition is ex-

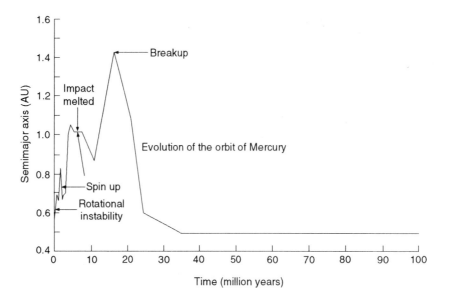

FIGURE 19   Results of a computer simulation of terrestrial planet evolution showing the change of "Mercury's" semimajor axis during its accretion. In this case, "Mercury's" semimajor axis spans the entire terrestrial planet region during the planet's growth. (From Vilas *et al.*, eds. (1998). *In* "Mercury." University of Arizona Press, Tucson.)

Provenance of planetesimals

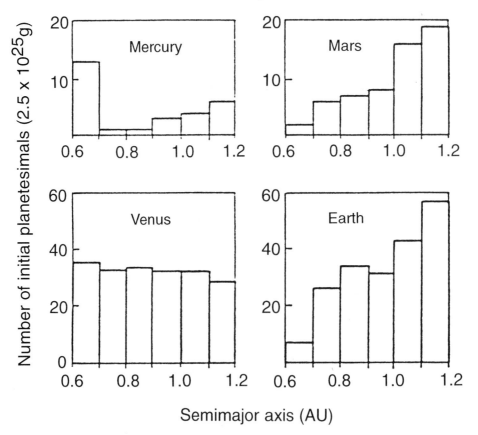

FIGURE 20   Results of a computer simulation of terrestrial planet evolution showing the region (semimajor axis) from which the terrestrial planets acquired their mass. In this simulation, "Mercury" acquires about half its mass from regions between 0.8 and 1.2 AU. (From Vilas *et al.*, eds. (1998). *In* "Mercury." University of Arizona Press, Tucson.)

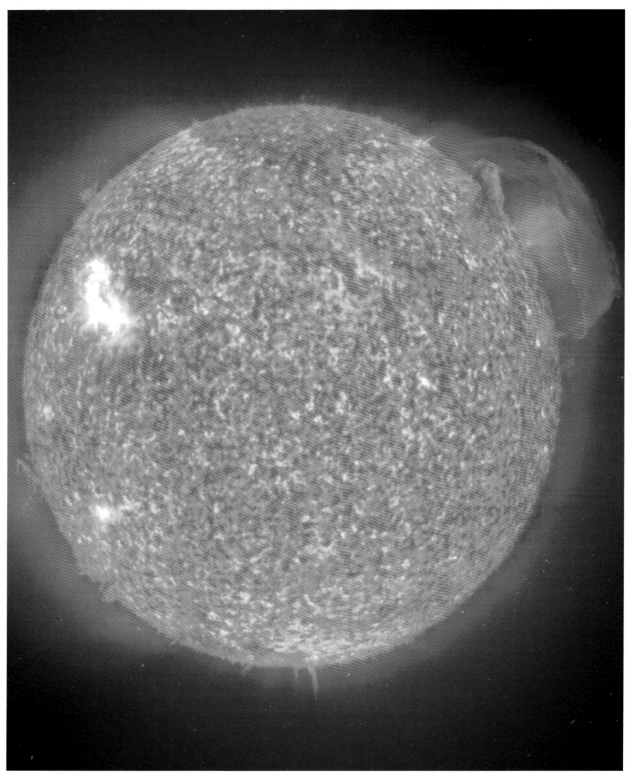

The erupting Sun—twisting magnetic fields propel this huge eruptive prominence about 150,000 km above the Sun's surface. The seething plasma of ionized gases is at a temperature of about 85,000 K and spans over 300,000 km. The space-based Solar Heliospheric and Oscillations satellite (SOHO) recorded this image from its vantage point in a Halo orbit. (SOHO-EIT Consortium, ESA, NASA)

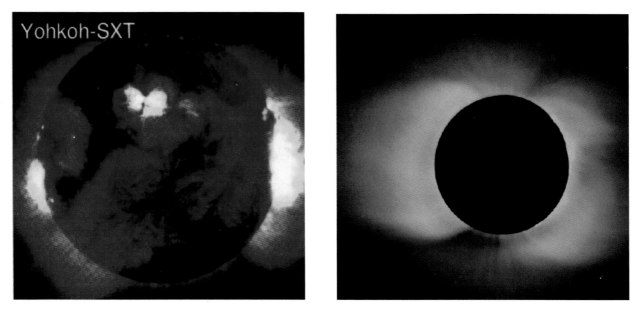

Above left, a solar flare observed in soft X-rays by the Yohkoh satellite. (Yohkoh team) Right, the sun's corona during a total eclipse, as observed by the High Altitude Observatory White Light Coronal camera from Chile.

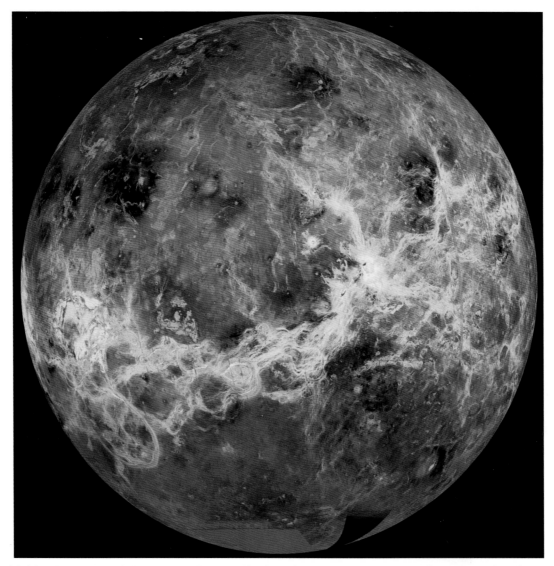

Global distribution of morphology on Venus from Magellan data. Globe represents view centered along equator of area between Theta Regio (bright area at left) and Atla Regio (bright area to right). (NASA photo)

Galileo Spacecraft image of the Earth-Moon system, taken 8 days after closest approach at a distance of 6.2 million kilometers. The Moon is moving from left to right in the foreground; it reflects only about one-third as much sunlight as Earth and hence appears dimmer. To improve visibility, color and contrast on both objects have been computer-enhanced. (NASA/JPL)

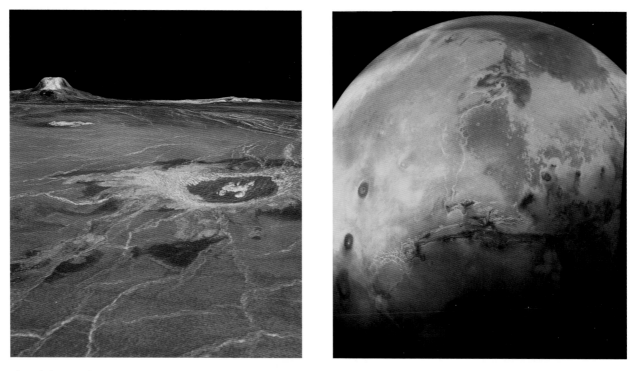

Above left, Magellan perspective views of regional units on Venus. The 3 km high shield volcano Gula Mons is seen at top left. The 53 km diameter crater Cunitz is seen superposed on the plains in the middle ground. Right, a distant view of Mars. In the center of the scene is the vast equatorial canyon system, Valles Marineris. To the left are three dark spots. These are large volcanoes in the Tharsis region. (NASA photos)

A composite of Voyager images showing Jupiter, Saturn, Uranus, and Neptune, scaled to their relative sizes. Earth and Venus are also shown scaled to their relative sizes.

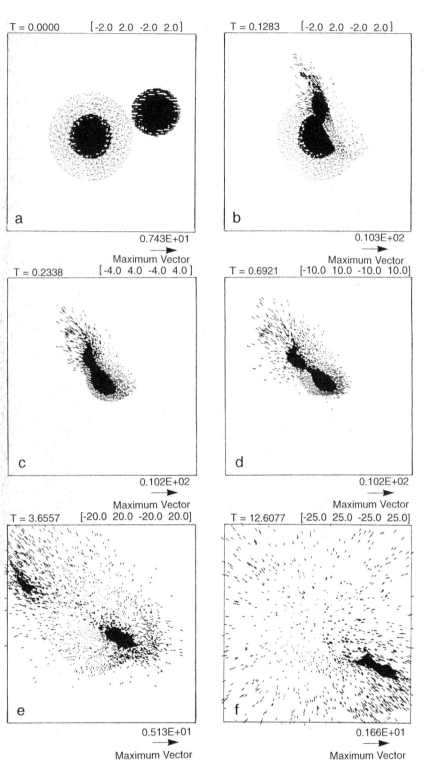

FIGURE 21 Computer simulation of a large, off-axis, 35-km/s impact with Mercury. In this simulation, the mantle separates from the core. A portion of the mantle must reaccrete to form the present-day Mercury. [from W. Benz *et al.* (1988). *Icarus,* **74,** 516–528.]

tremely poor, but near and mid-infrared spectroscopic measurements favor low FeO and alkali-bearing feldspars. If the tenuous atmosphere of sodium and potassium is being outgassed from the interior, as seems possible from recent observations, then the postaccretion vaporization model may be unlikely. Deciding between the other two models is not possible with our current state of ignorance about the silicate composition. Since the selective accretion hypothesis requires Mercury to have formed near its present position, then sulfur should be nearly absent unless the solar nebula temperatures in this region were considerably lower than predicted by the chemical equilibrium condensation model.

Support for the giant impact hypotheses comes from three-dimensional computer simulations of terrestrial planet formation for several starting conditions. Since these simulations are by nature stochastic, a range of outcomes is possible. They suggest, however, that significant fractions of the terrestrial planets may have accreted from material formed in widely separated parts of the inner solar system. The simulations indicate that during its accretion, Mercury may have experienced large excursions in its semimajor axis. These semimajor axis excursions may have ranged from as much as 0.4 to 1.4 AU owing to energetic impacts during accretion (Fig. 19). Consequently, Mercury could have accumulated material originally formed over the entire terrestrial planet range of heliocentric distances. About half of Mercury's mass could have accumulated at distances between about 0.8 and 1.2 AU (Fig. 20). If so, then Mercury may have acquired its sulfur from material that formed in regions of the solar nebula where sulfur was stable. Plausible models estimate FeS contents of 0.1 to 3%. However, the most extreme models of accretional mixing result in homogenizing the entire terrestrial planet region, contrary to the observed large systematic density differences. [See PLANETARY IMPACTS.]

The simulations also indicate that by-products of terrestrial planet formation are planet-sized objects up to three times the mass of Mars that become perturbed into eccentric orbits (mean e ~0.15 or larger) and eventually collide with the terrestrial planets during their final stages of growth. The final growth and giant impacts occur within the first 50 million years of solar system history. Such large impacts may have resulted in certain unusual characteristics of the terrestrial planets, such as the slow retrograde rotation of Venus, the origin of the Moon, the Martian crustal dichotomy, and Mercury's large iron core.

In computer simulations where proto-Mercury was 2.25 times the present mass of Mercury with an uncom-pressed density of about 4 g/cm^3, nearly central collisions of large projectiles with iron cores impacting at 20 km/s, or noncentral collisions at 35 km/s, resulted in a large silicate loss and little iron loss (Fig. 21). In the former case, although a large portion of Mercury's iron core is lost, an equally large part of the impactor's iron core is retained, resulting in about the original core size. At Mercury's present distance from the Sun, the ejected material reaccretes back into Mercury if the fragment sizes of the ejected material are greater than a few centimeters. However, if the ejected material is in the vapor phase or fine-grained ($\leq$1 cm), then it will be drawn into the Sun by the Poynting–Robertson effect in a time shorter than the expected collision time with Mercury (about 10^6 years). The proportion of fine-grained to large-grained material ejected from such an impact is uncertain. Therefore, it is not known if a large impact at Mercury's present distance would exclude enough mantle material to account for its large iron core. However, the disruption event need not have occurred at Mercury's present distance from the Sun; it could have occurred at a much greater distance (e.g., >0.8 AU; see Fig. 19). In this case, the ejected mantle material would be mostly swept up by the larger terrestrial planets, particularly Earth and Venus.

VIII. CONCLUSIONS

Mercury is the least known of all the terrestrial planets, but it is probably the only planet that holds the key to understanding details of the origin and evolution of all of these bodies. Because only half of the planet has been imaged at relatively low resolution, and because of the poor characterization of its magnetic field and almost complete ignorance of its silicate composition, there is little hope of deciding between competing hypotheses of its origin and evolution until more detailed information is obtained. A Mercury orbiter with a suitable complement of instruments could provide the data required to resolve these problems.

BIBLIOGRAPHY

Benz, W., Slattery, W., and Cameron, A. G. W. (1988). *Icarus* **74**, 516–528.

Harmon, J. K., Slade, M. A., Velez, R. A., Crespo, A., Dryer, M. J., and Johnson, J. M. (1994). *Nature* **369**, 213–215.

Potter, A. E., and Morgan, T. H. (1988). *Science* **247**, 675.

Robinson, M. S., and Lucey, P. G. (1997). *Science* **275,** 197–200.

Sprague, A., Kozlowski, R. W. H., and Hunten, D. M. (1990). *Science* **249,** 1140–1142.

Strom, R. G. (1984). "The Geology of the Terrestrial Planets" (M. H. Carr, ed.), NASA SP-469, pp. 13–55. NASA, Washington, D.C.

Strom, R. G. (1987). "Mercury: The Elusive Planet." Smithsonian Institution Press, Washington, D.C.

Various authors in a special issue on the *Mariner 10* encounter with Mercury. (1975). *J. Geophys. Res.* **80.**

Vilas, F., Chapman, C. R., and Matthews, M. S. (eds.) (1988). "Mercury." Univ. of Arizona Press, Tucson.

VENUS: ATMOSPHERE

I. Introduction

II. Lower Atmosphere

III. Middle and Upper Atmosphere

IV. Clouds and Hazes

V. General Circulation

VI. Origin and Evolution

Donald M. Hunten
University of Arizona

GLOSSARY

Adiabat: Process ocurring without exchange of heat with the surroundings. In an atmosphere, an adiabatic temperature gradient (about -10 K/km for Venus) is commonly found in regions of rapid vertical motion.

Bar: Unit of pressure, equal to 10^6 dynes/cm^2 or 10^5 pascals; the standard sea-level pressure of the Earth's atmosphere is 1.013 bars.

Catalytic cycle: Series of chemical reactions facilitated by a substance that remains unchanged.

Greenhouse effect: Heating of a planetary surface above the temperature that it would have in the absence of an atmosphere. The atmosphere transmits solar radiation in the visible, but impedes the escape of thermal infrared energy, thus creating the increased temperature.

Hydrostatic equation: Relationship that says pressure is equal to the weight of gas or liquid above the level of interest.

K or kelvin: Unit of absolute temperature; the freezing and boiling points of water are 273.16 K and 373.16 K, respectively.

Langmuir probe: Instrument used to measure electron and ion densities; the external sensor is usually a stiff wire; the current is measured as different voltages are applied.

Optical depth: Number of mean free paths for scattering or absorption; after an optical depth τ, radiation is reduced by a factor $e^{-\tau}$.

Refraction: Bending of a light ray as it traverses a boundary, for example, between air and glass or between space and an atmosphere.

Retrograde: Rotating clockwise viewed from the north, in the opposite sense to the orbital motions of the planets.

Scale height: Height range over which pressure falls by $1/e = 0.368$; equal to kT/mg, where k is Boltzmann's constant, T is temperature, m is the mean mass of the gas, and g is the acceleration of gravity.

Sidereal: Relative to the stars (rather than the Sun).

Stratosphere, mesosphere: Region (also called middle atmospheres) whose temperature is controlled by radiative balance. On Earth it extends from about 10 to 95 km, and on Venus from 65 to 95 km.

Thermosphere, exosphere: Outer parts of an atmosphere, heated by ionizing radiation and cooled by conduction. The exosphere is essentially isothermal and is also characterized by very long mean free paths.

Troposphere: Lowest region of an atmosphere, dominated by vertical mixing and often possessing clouds. On Earth it extends to 14 km (equatorial) and 9 km (polar), and on Venus to 65 km.

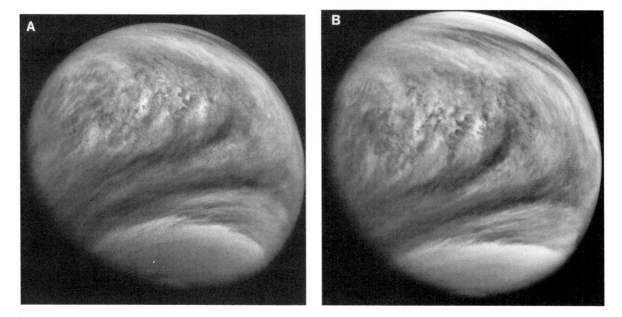

FIGURE 1 Two images of the cloud top region obtained from *Pioneer Venus Orbiter* 5 hours apart on May 28, 1990. The small but distinct westward movement of the large-scale cloud patterns can be seen. The original images have been processed to enhance the contrast and remove the limb darkening (which makes the edges considerably darker than the center). (Courtesy of Dr. L. Travis.)

Venus possesses a dense, hot atmosphere, primarily of carbon dioxide, with a pressure of 93 bars and a globally uniform temperature of 740 K at the surface. The surface is totally hidden at visible wavelengths by a cloud deck (really a deep haze) of concentrated sulfuric acid droplets that extends from 50 km altitude to a poorly defined top at 65 km (Fig. 1 and also Fig. 8). The clouds are thus located in the top part of the troposphere, which extends from 0 to 65 km. The middle atmosphere (stratosphere and mesosphere) extends from 65 to about 95 km, and the upper atmosphere (thermosphere and exosphere) from 95 km up. Although the rotation period of the solid planet is 243 Earth days (sidereal), the atmosphere in the cloud region rotates in about 4 days, and the upper atmosphere in about 6 days, all in the same retrograde direction.

I. INTRODUCTION

A. HISTORY

The study of Venus by Earth-based telescopes has been frustrated by the complete cloud cover. The presence of CO_2 was established in 1932, as soon as infrared-sensitive photographic plates could be applied to the problem. But establishment of the abundance was impossible because there was no way to determine the path length of the light as it scattered among the cloud particles. Moreover, it was assumed that nitrogen would also be abundant, as it is on Earth, and this gas cannot be detected in the spectral range available from the ground. Careful observation of the feeble patterns detectable in blue and near-ultraviolet images was able to establish the presence of the 4-day rotation at the cloud tops. These patterns are shown in the much more recent spacecraft images of Fig. 1. Radio astronomers, observing Venus's emission at the microwave wavelength of 3.15 cm, discovered in 1958 that it appears to be much hotter than expected, and this was confirmed by later results at other wavelengths. The most likely suggested explanation was that the radiation came from a hot surface, warmed by an extreme version of the greenhouse effect; but the required warming is so extreme that other hypotheses were debated. Spacecraft measurements, as will be described, finally settled the issue in favor of the greenhouse effect and showed that the pressure at the mean surface is 93 bars.

A large number of spacecraft experiments on 22 missions have been devoted to study of the atmosphere; it is better explored than that of any planet other than the Earth. United States missions, starting in 1962, were the flybys *Mariner 2, 5,* and *10* (which went on to Mercury); *Pioneer Venus Multiprobe* and *Orbiter* in 1978; the radar mapper *Magellan*; and the Jupiter-bound *Galileo*. Successful Soviet ones were *Venera 4–14*, which included entry and descent probes as well as flybys or orbiters, *Venera 15* and *16*, which were radar mappers, and *Vega 1* and *2*, which dropped both probes and balloon-borne payloads on their way to Halley's Comet. Early missions were devoted to reconnaissance, in particular to confirmation of the high surface pressure and temperature inferred from the microwave radio measurements. [*See* PLANETARY EXPLORATION MISSIONS.]

The composition of the clouds was another important question, but it was actually answered first from analysis of ground-based observations of the polarization of light reflected from the planet. Although such measurements were first made in the 1930s, the computers and programs to carry out the analysis did not exist until the middle 1970s. This analysis pinned down the refractive index and showed that the particles are spherical; these two properties eventually led to the identification of supercooled droplets of concentrated sulfuric acid (H_2SO_4). Measurements from the Pioneer Venus probes confirmed this composition and gave much greater detail on the sizes and layering of the haze.

B. MEASURING TECHNIQUES

Three principal techniques can be applied from Earth: spectroscopy, radiometry, and imaging. They can be used over a wide variety of wavelengths, from the ultraviolet to the shortest part of the radio spectrum. Spectroscopy, as mentioned earlier, was first applied in 1932 and led to the discovery of CO_2. Little more was done until the middle 1960s, when traces of water vapor were found and a tight upper limit was set on the amount of O_2. The development of Fourier spectroscopy permitted an extension further into the infrared, where CO, HCl, and HF were observed. Radiometry, and especially polarimetry, eventually led to the identification of the substance of the cloud particles. After the near-infrared "windows" were identified (see Section I, D), starting in 1983, spectroscopy of deeper parts of the atmosphere provided important further information. Visual studies, followed more recently by photography and infrared imaging, disclosed the 4-day rotation of the cloud tops and the 6-day period of a deeper region. Similar remarks apply to radio astronomical studies. Radiometry gave the data that finally led to the establishment of the high surface temperature, and millimeter-wave spectroscopy has led to the interesting results on CO discussed in Sec-

tion III, D. Until the early 1990s, all ground-based radio work used radiation from the whole disk, but modest spatial resolution is beginning to be available by interferometry (the technique of combining the signals from several antennas).

Many of the same techniques have been applied from flyby and orbiting spacecraft, but an important addition is the radio occultation experiment, which tracks the effect of the atmosphere on the telemetry carrier as the spacecraft disappears behind the atmosphere or reappears from behind it. On Venus, the regions observed in this way are the ionosphere and the neutral atmosphere from about 34 to 90 km. At greater depths, the refraction of the waves by the atmosphere is so great that the beam strikes the surface and never reappears. In addition to carrying several instruments for remote sensing, *Pioneer Venus Orbiter* (1978–1992) actually penetrated the upper atmosphere once per orbit, and took advantage of this by carrying a suite of instruments to make measurements *in situ*. Two mass spectrometers measured individual gases and positive ions; a Langmuir probe and a retarding potential analyzer measured electron and ion densities, temperatures, and velocities; and a fifth instrument measured plasma waves. Higher-energy ions and electrons, both near the planet and in the solar wind, were measured by a plasma analyzer, and important auxiliary information was provided by a magnetometer. In addition, the atmospheric drag on the spacecraft gave an excellent measure of the density as a function of height.

A large number of probes have descended part or all the way through the atmosphere, and the Vega balloons carried out measurements in the middle of the cloud region. All of them have carried an "atmospheric structure" package measuring pressure, temperature, and acceleration; height was obtained on the early Venera probes by radar, and on all probes by integration of the hydrostatic equation. Gas analyzers have increased in sophistication from the simple chemical cells on *Venera 4* to mass spectrometers and gas chromatographs on later Soviet and U.S. missions. In some, cases, however, there are suspicions that the composition was significantly altered in passage through the sampling inlets, especially below 40 km, where the temperature is high. A variety of instruments have measured the clouds and their optical properties. Radiometers observed the loss of solar energy through the atmosphere, and others have observed the thermal infrared fluxes. Winds were obtained by tracking the horizontal drifts of the probes as they descended, and the balloons as they floated. *Veneras 11–14* carried radio receivers to seek evidence of lightning activity.

TABLE I
Composition of the Venus Atmosphere

	Species	Mole fraction at 70 km	Mole fraction at 40 km
%	CO_2	96.5	96.5
	N_2	3.5	3.5
ppm[a]	He	~12	~12
	Ne	7	7
	Ar	70	70
	Kr	~0.2	~0.2
	CO	5170	45
	H_2O	≤1	45
	SO_2	0.05	~100
	H_2S	?	1
	COS		0.25
	HCl	0.4	0.5
	HF	0.005	0.005
	O_2	<0.1	0–20
%	D/H	1.6	1.6

[a] Parts per million.

C. COMPOSITION

The fact that carbon dioxide is indeed the major gas was established by a simple chemical analyzer on the *Venera 4* entry probe. The mole fraction was found to be about 97%, in reasonable agreement with the currently accepted value shown in Table I. The next most abundant gas is nitrogen; though it is only 3.5% of the total, the absolute quantity is about three times that in the Earth's atmosphere. The temperature profile is illustrated in Fig. 2, along with a sketch of the cloud layers.

Many of the strange properties of the atmosphere can be traced to an extreme scarcity of water and its vapor and the total absence of liquid water. On Earth, carbon dioxide and sulfuric, hydrochloric, and hydrofluoric acids are all carried down by precipitation, a process that is absent in the hot, dry lower atmosphere of Venus. All of them then react and are incorporated in geological deposits; the best estimates of the total amount of carbonate rocks in the Earth give a quantity of CO_2 almost equal to that seen in the atmosphere of Venus. Free oxygen is undetectable at the Venus cloud tops; one molecule in ten million could have been seen. There is, of course, plenty of oxygen in carbon dioxide, and dissociation by sunlight liberates it in copious quantities. It is readily detected (as is CO) by spacecraft instruments orbiting through the upper atmosphere, but is removed before it can reach the cloud level. Small quantities of O_2 are also found below the clouds, probably liberated by the thermal decom-

position of the cloud particles. All of these lines of evidence point to the action of a strong mechanism in the middle atmosphere that converts O_2 and CO back into CO_2. The observed HCl molecules are the key; they too are broken apart by solar radiation, and the free chlorine atoms enter a catalytic cycle that does the job. This chemistry is closely coupled to the sulfur chemistry (see Section IV) that maintains the clouds.

Carbon dioxide, aided by the other molecules listed in Table I, makes the lower atmosphere opaque to thermal (infrared) radiation; it is this opacity that makes the extreme greenhouse effect possible. Only a few percent of the incident solar energy reaches the surface, but this is enough. Venus is a remarkable and extreme example of the large climatic effects that can be produced by seemingly small causes. One chlorine atom in two and a half million can completely eliminate free oxygen from the middle atmosphere, and ozone

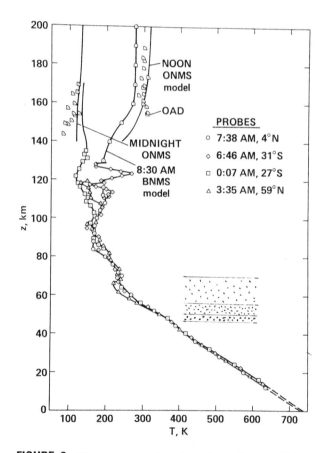

FIGURE 2 Temperature profiles from the surface to 200 km altitude, obtained by different experiments on the *Pioneer Venus Orbiter* and *Probes*. ONMS, orbiter neutral mass spectrometer; OAD, orbiter atmospheric drag; BNMS, bus neutral mass spectrometer. The cloud region with its three layers has been sketched in. [From D. M. Hunten *et al.* (eds.) (1984). "Venus." Univ. of Arizona Press, Tucson.]

has no hope of surviving in significant quantities. The temperature increase caused by the greenhouse effect is almost 500° C. The idea that the 30° seen on Earth could become 32° or 33° if its atmospheric content of CO_2 should double seems entirely probable to experts on Venus's atmosphere, and so does significant loss of ozone from release of chlorinated refrigerants. It thus seems that the obvious differences between Earth and Venus are all traceable to the differences in their endowments of water (vapor or liquid). Although origin and evolution are discussed in Section VI, a short preview is given here. It is plausible that both planets started out with similar quantities, but that the greater solar flux at Venus caused all its water to evaporate (a "runaway greenhouse"). Solar ultraviolet photons could then dissociate it into hydrogen (which escaped) and oxygen (which reacted with surface materials). Strong evidence in favor of this scenario is the extreme enhancement of heavy hydrogen (deuterium, or D), almost exactly 100 times more abundant relative to H than it is on Earth. Such a fractionation is expected because the escape of H is much easier than that of D. [See THE EARTH AS A PLANET: ATMOSPHERES AND OCEANS.]

D. NEAR-INFRARED SOUNDING

Study of the atmosphere below the clouds was revitalized in 1988 by the discovery of several narrow spectral windows in the near infrared, where the radiation from deep layers can be detected from above (Fig. 3). The two most prominent ones are at 1.74 and 2.3 μm (Fig. 4), and others are at 1.10, 1.18, 1.27, and 1.31 μm. As we have seen, at microwave radio wavelengths, radiation from the actual surface can escape to space. At other infrared wavelengths, the emission from the night side is characteristic of the temperature of the cloud tops, about 240 K. In the windows, the brightness, and therefore the temperature of the emitting region, is considerably higher. Images taken in a window reveal horizontally banded structures that appear to be silhouettes of the lowest part of the cloud (around 50 km) against the hotter atmosphere below (see Fig. 3 and Section IV). [See INFRARED VIEWS OF THE SOLAR SYSTEM FROM SPACE.]

Numerous absorption lines and bands allow inferences about the composition to levels all the way to the surface. One such spectrum is shown in Fig. 4. Each "window" allows the composition to be obtained at a different level; this is particularly important for water vapor, discussed in the next section. The measurement of carbonyl sulfide (COS) shown in Table I

was obtained by this analysis. This gas has resisted all attempts to measure it from entry probes, even though it has long been expected to be present. Other gases include CO, HF, HCl, and light and heavy water vapor, all in good agreement with prior results. These results are also included in Table I.

II. LOWER ATMOSPHERE

A. TEMPERATURES

It is convenient to regard the lower atmosphere as extending from the surface to about 65 km, the level of the visible cloud tops and also of the tropopause. This region has been measured in detail by many descent probes, with results in close agreement, and also by radio occultation. The temperature profile (see Fig. 2) is close to the adiabat, becoming noticeably less steep above the tropopause. As on Earth, the tropopause is a few kilometers lower at high latitudes than near the equator. The high surface temperature is maintained by the greenhouse effect, driven by the few percent of solar energy that reaches the surface. Converted to thermal infrared, this energy leaks out very slowly because of the opacity of the atmospheric gases at such long wavelengths. The molecules principally responsible are CO_2, SO_2, H_2O, and perhaps others. Quantitative calculations have shown that the greenhouse mechanism is adequate, and that the observed solar and infrared net fluxes can be reproduced. These models treat the temperatures as globally uniform, so that they can be restricted to considering vertical heat transport only.

The surface temperature is remarkably uniform with both latitude and longitude, largely because of a very long radiative time constant. A very slow atmospheric circulation is therefore adequate, but the details of how the nonuniform solar heating is converted to a uniform surface temperature are not understood. The "runaway greenhouse" that may have operated early in the history of the planet is discussed in Section VI.

B. WATER VAPOR

Table I shows rather uncertain quantities of H_2O, but there is no doubt that there is a major difference in the mole fractions below and above the clouds. This

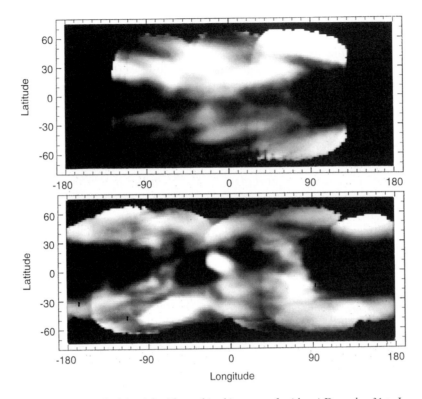

FIGURE 3   Near-infrared images (2.36 μm) of the night side combined into maps for (above) December 31 to January 7, 1991, and (below) February 7 to 15, 1991. Bright areas are thinner parts of the cloud through which thermal radiation from deeper layers can shine. [From D. Crisp *et al.* (1991). *Science* **253**, 1538–1541.]

is almost certainly because the concentrated sulfuric acid of the cloud particles is a powerful drying agent (see Section IV, C). A summary of the many attempts to measure the abundance below the cloud is given in Fig. 5. Direct measurements have been made by several mass spectrometers and gas chromatographs, but the amounts are so small and the results so divergent that there remain many questions. Indirect measurements come from radiation fluxes, which are strongly affected by the opacity of water vapor. The four Pioneer Venus probes carried infrared net flux radiometers (points labeled "7" in Fig. 5), and *Veneras 11* and *12* carried an instrument working with weaker absorptions in the near infrared (dashed line "2"). A major advantage of these measurements is that they cannot be affected by sampling errors, because they relate to the atmosphere far from the probe. It is likely that many of the divergences are due to the extreme difficulty of measuring such small quantities of a reactive molecule at the high temperatures of the lower atmosphere, but some of the variations may reflect real effects of latitude or height. Particularly puzzling has been the indication from the mass spectrometer on the *Pioneer Venus Large Probe* that the mole fraction falls off by nearly

a factor of 10 between 10 km altitude and the surface (Fig. 5, line "1"). It is likely that this result is incorrect; it is not supported by remote sensing of this region in the near-infrared windows.

The ratio of heavy to light hydrogen (D/H) (last line of Table I) was first measured on ions in the ionosphere and has been confirmed by the mass spectrometer just mentioned and by analysis of spectra taken from Earth in the near-infrared windows. In turn, the deuterium provides a valuable signature for distinguishing Venus water vapor in the mass spectrometer from any contaminants carried along from Earth. The likely enrichment process is discussed in Section VI.

III. MIDDLE AND UPPER ATMOSPHERE

A. TEMPERATURES

The middle atmosphere (stratosphere and mesosphere) extends from the tropopause at 65 km to the

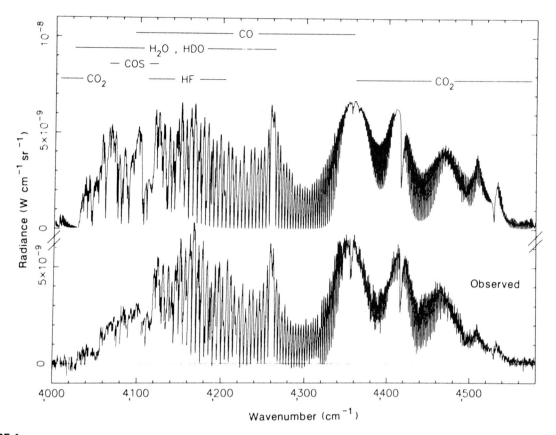

FIGURE 4   Near-infrared spectrum in the 2.3-μm window (bottom); the spectrum above it was calculated by making use of laboratory data for the six different molecules shown. [From B. Bézard *et al.* (1990). *Nature* **345**, 508–511.]

temperature minimum or mesopause at about 95 km (see Fig. 2). The upper atmosphere lies above this level. Here, temperatures can no longer be measured directly, but are inferred from the scale heights of various gases with use of the hydrostatic equation. On Earth and most other bodies, this region is called the "thermosphere" because temperatures in the outer layer, or "exosphere," are as high as 1000 K. The temperature is much more modest on Venus; the exospheric temperature is no more than 350 K on the day side. The corresponding region on the night side is sometimes called a "cryosphere" (cold sphere) because its temperature is not far above 100 K. Measurements of these temperatures by *Pioneer Venus Orbiter* are shown in Fig. 6. The large temperature difference translates into a pressure difference that drives strong winds from the day side to the night side, at all levels above 100 km.

On Earth, the exospheric temperature changes markedly with solar activity, being perhaps 700 K at sunspot minimum and 1400 K at maximum. The corresponding change at Venus is much more modest, per-

haps 50 K. Many of these differences are traceable to the fact that CO_2, the principal radiator of heat, is just a trace constituent of Earth's atmosphere but is the major constituent for Venus (and also Mars). Venus's slow rotation is responsible for the very cold temperatures on the night side, although the atmosphere does rotate substantially faster than the solid plane. [*See* MARS: ATMOSPHERE AND VOLATIVE HISTORY.]

B. IONOSPHERE

The principal heat source for the thermosphere is the production of ions and electrons by far-ultraviolet solar radiation. The most abundant positive ions are O_2^+, O^+, and CO_2^+. As part of these processes, CO_2 is dissociated into CO and O, and N_2 into N atoms. All of these ions, molecules, and atoms have been observed or directly inferred (Fig. 7). Some of the O^+ ions (with an equal number of electrons) flow around to the night side and help to maintain a weak ionosphere there. Venus lacks any detectable magnetic field, and the dayside ionosphere is

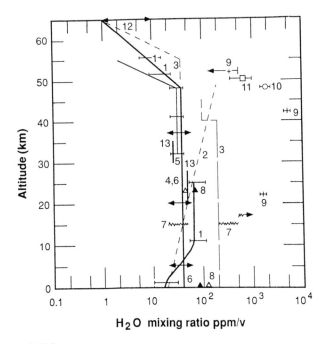

FIGURE 5 Water vapor mixing ratios or mole fractions from various experiments, mostly on Pioneer Venus and Venera probes [From T. M. Donahue and R. R. Hodges Jr. (1992). *J. Geophys. Res.* **97**, 6083–6091.]

therefore impacted by the solar wind, a tenuous medium of ions (mostly H^+) and electrons flowing from the Sun at about 400 km/sec. Electrical currents are induced in the ionosphere, and they divert the solar wind flow around the planet. The boundary between the two media, called the "ionopause," is typically at an altitude of a few hundred kilometers near the subsolar point, flaring out to perhaps 1000 km above the terminators and forming a long, taillike cavity behind the planet. [*See* SOLAR WIND.]

C. WINDS

The thermospheric winds carry the photochemical products O, CO, and N from the day side to the night side, where they are almost as abundant as they are on the day side. However, as Fig. 7 illustrates, all gases fall off much more rapidly on the night side because of the low temperature. They descend into the middle atmosphere in a region perhaps 2000 km in diameter and generally centered near the equator at 2 A.M. local time. This region can be observed by the emission of airglow emitted during the recombination of N and O atoms into NO molecules, which then radiate in the ultraviolet, and O_2 molecules, which radiate in the near infrared. The light gases hydrogen and helium are also carried along and accumulate over the convergent point of the flow; for these gases, the peak density is observed at about 4 A.M. These offsets are the principal evidence that this part of the atmosphere rotates with a 6-day period, a rotation that is superposed on the rapid day-to-night flow.

D. CHEMICAL RECOMBINATION

Oxidation of the CO back to CO_2 is much slower than the recombination of O and N atoms, but a very efficient process is required. This conclusion follows from Earth-based observations of a microwave (2.6-mm wavelength) absorption line of CO, from which a height distribution can be obtained from 80 to 110 km. It is found that the downward-flowing CO is substantially depleted on the night side below 95 km (as well as on the day side). The proposed solution involves reactions of chlorine atoms, as well as residual O atoms descending from the thermosphere. The chlorine acts as a catalyst, promoting reactions but not being consumed itself, and the reaction cycle works without the direct intervention of any solar photons other than the ones that produced the O atoms and CO molecules half a world away.

The availability of Cl atoms is assured by the observed presence of HCl at the cloud tops (Table I). On Earth, any HCl emitted into the atmosphere is

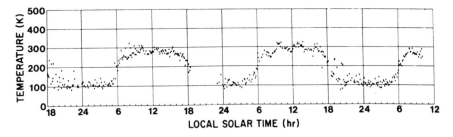

FIGURE 6 Diurnal variation of temperature in the upper thermosphere. Scale heights H were measured by the mass spectrometer on *Pioneer Venus Orbiter* and converted to temperatures by the formula $T = mgH/k$ (see the Glossary). The measurements sweep out the entire range of local solar time as Venus moves around the Sun about 2 1/2 times [From D. M. Hunten *et al.* (1984).]

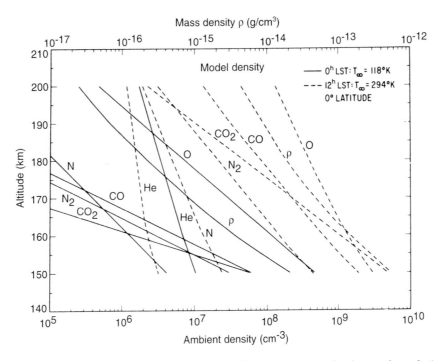

FIGURE 7 Daytime (dashed) and nighttime (solid) number densities of the major gases in the thermosphere obtained by fitting a large number of measurements by the mass spectrometer on *Pioneer Venus Orbiter*. [From D. M. Hunten *et al.* (1984).]

rapidly dissolved in water drops and rained out. Chlorine atoms reach the stratosphere only as components of molecules, such as the artificial ones CCl_4, CF_2Cl_2, and $CFCl_3$ and the natural one CH_3Cl, none of which dissolve in water. Once they have been mixed to regions above the ozone layer, they are dissociated by solar ultraviolet photons. Because liquid water is absent on Venus, the abundance of HCl is large to start with and it is not kept away from the stratosphere. Here again the atoms are released by solar ultraviolet. The chlorine abundance is nearly a thousand times greater than that on Earth, and Venus is an example and a warning of what chlorine can do to an atmosphere. The middle atmosphere is also the seat of important chemistry involving sulfur, which is discussed in the next section.

IV. CLOUDS AND HAZES

A. APPEARANCE AND MOTIONS

The clouds are perhaps the most distinctive feature of Venus. They do show subtle structure in the blue and near ultraviolet, illustrated in Fig. 1, which has been

processed to bring out the detail and flattened to remove the limb darkening. Although the level shown in the figure is conventionally called the "cloud top," it is not a discrete boundary at all. Similar cloud particles extend as a haze to much higher altitudes, at least 80 km; the "cloud top" is simply the level at which the optical depth reaches unity, and the range of visibility (the horizontal distance within which objects are still visible) is still several kilometers.

Study of daily images, first from Earth and later from spacecraft, reveals that the cloud top region is rotating with a period of about 4 days, corresponding to an equatorial east–west wind speed of about 100 m/sec. The speed varies somewhat with latitude; in some years, but not all, the rotation is almost like that of a solid body. Although there are not nearly as many near-infrared images like Fig. 3, they show a longer period consistent with the idea that the silhouettes are of the lower cloud, where entry probes have measured wind speeds of 70 to 80 m/sec.

B. CLOUD LAYERS

Several entry probes have made measurements of cloud scattering as they descended, but the most detailed results were obtained from Pioneer Venus and are

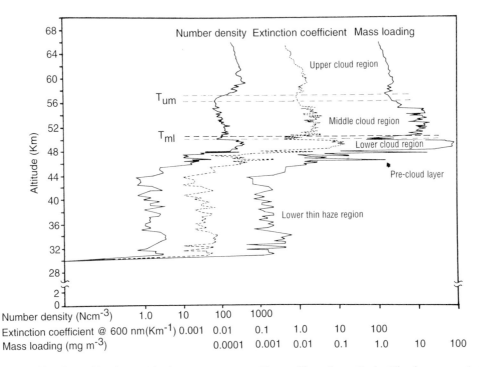

FIGURE 8   Cloud profiles obtained by the particle size spectrometer on *Pioneer Venus Large Probe*. The three curves indicate different properties: number of particles per cubic centimeter, extinction coefficient or optical depth per kilometer of height, and mass per cubic centimeter. [From D. M. Hunten *et al.* (1984).]

shown in Fig. 8. Three regions (upper, middle, and lower) can be distinguished in the main cloud, and there is also a thin haze extending down to 30 km. Size distributions are shown in Fig. 9; it is these, more than the gross properties of Fig. 8, that distinguish the regions. In the upper cloud, the one that can be studied from Earth or from orbit, most particles ("Mode 1") are about 1 μm in diameter and should really be considered a haze rather than a cloud; there are also larger ("Mode 2") particles with diameters around 2 μm. The same particles extend throughout the clouds, but the Mode 2 ones become somewhat larger in the middle and lower clouds, and a third population ("Mode 3"), greater than 6 μm in diameter, is also found. The existence of distinct modes is still not understood; the optical properties of all three are generally consistent with sulfuric acid, although there is some suspicion that the rare Mode 3 particles might be solid crystals.

C. CLOUD CHEMISTRY

A cloud particle of diameter 1 μm has a sedimentation velocity of 7.5 m/day at 60 km; this velocity varies as the square of the size. Though small, these velocities eventually carry the particles out of the cloud to lower

altitudes and higher temperatures, where they will evaporate. At still lower heights the hydrated H_2SO_4 must decompose into H_2O, SO_2, and oxygen, all of which are (at least probably) much more abundant beneath the clouds than above them (Table I). Atmospheric mixing carries these gases back upward. Nearly all the water vapor is absorbed by the cloud particles. Above the clouds, solar ultraviolet photons attack the SO_2, starting the process that converts it back to H_2SO_4. An important intermediate is the reactive free radical SO, and probably some elemental sulfur is produced. Ultraviolet spectra (pertaining to the region above the clouds) reveal the presence of the small amounts of SO_2 shown in Table I, but much less than has been measured below the clouds.

Sulfuric acid is perfectly colorless in the blue and near ultraviolet, and the yellow coloration that provides the contrasts of Fig. 1 must be caused by something else. Certainly the most likely thing is elemental sulfur, but yellow compounds are abundant in nature and the identification remains tentative. The photochemical models do predict production of some sulfur, but it is a minor by-product and the amount produced is uncertain. Probably the most likely alternative is ferric chloride, particularly for the Mode 3 particles in the lower cloud.

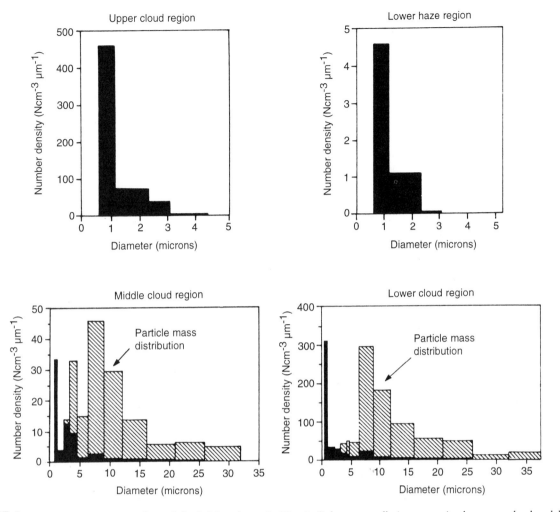

FIGURE 9 Particle size spectra at four of the heights shown in Fig. 8. Only very small sizes occur in the upper cloud and lower haze; the layers in between have three different populations or "modes" with sizes around 1 (or less), 3, and 7 μm. [From D. M. Hunten *et al.* (1984).]

D. LIGHTNING

Electromagnetic pulses have been observed by the entry probes *Venera 11, 12, 13,* and *14,* by *Pioneer Venus Orbiter,* and by *Galileo.* For many years it seemed that the most likely source was lightning, and many workers are convinced of its reality. However, some searches for the corresponding optical flashes have been negative, except for one ambiguous interval from *Venera 9.* A recent study from the Earth does seem to have turned up a few optical events. The negative results may simply be because the flashes are too faint, but another concern is that conditions on Venus do not seem propitious for large-scale charge separation. On Earth, lightning is seen during intense precipitation and in volcanic explosions. In thunderstorms, large drops are efficient at carrying charge of one sign away from the region where it is produced, and the gravitational force is large enough to resist the strong electric fields. This is not the case for small particles. There does not seem to be enough cloud mass on Venus to generate large, precipitating particles, although they are difficult to detect and may have been missed. As for volcanic explosions, most of them are driven by steam; on Venus, water is very scarce, and the 93-bar surface pressure means that, other things being equal, any explosion is damped by a factor of 93 compared with Earth. In spite of these concerns, lightning remains one of the more plausible explanations of the radio bursts, but it is important to seek others.

V. GENERAL CIRCULATION

Careful tracking of entry probes, notably the four of Pioneer Venus, has shown that the entire atmosphere is superrotating, with a speed decreasing smoothly from the 100 m/sec at the cloud top to near zero at 5–10 km. Winds in the meridional direction are much slower. Because the density increases by a large factor over this height range, the angular momentum is a maximum at 20 km. Small amounts of superrotation are observed in many atmospheres, especially thermospheres, but they are superposed on a rapid planetary rotation. (A familiar example is the midlatitude "prevailing westerlies" on the Earth.) In spite of a great deal of theoretical effort and a number of specific suggestions, there is still no accepted mechanism for the basic motion of the Venus atmosphere, nor is it given convincingly in any numerical general circulation model. What is needed is to convert the slow apparent motion of the Sun (relative to a fixed point on Venus) into a much more rapid motion of the atmosphere. There must also be a slow meridional (north–south) component, sometimes called a Hadley circulation, to transport heat from the equatorial to the polar regions.

There are no direct measurements above the cloud tops, but deductions from temperature measurements suggest a slowing of the 100 m/sec flow up to perhaps the 100-km level. At still greater heights the dominant flow is a rapid day–night one, first suggested on theoretical grounds and confirmed by the large observed temperature difference. But the flow is not quite symmetrical; maxima in the hydrogen and helium concentrations, and in several airglow phenomena, are systematically displaced from the expected midnight location toward morning. Possible explanations are a wind of around 65 m/sec or a wave-induced drag force that is stronger at the morning side than the evening.

VI. ORIGIN AND EVOLUTION

It is generally believed that the Sun, the planets, and their atmospheres condensed, about 4.6 billion years ago, from a "primitive solar nebula." The presumed composition of the nebula was that of the Sun, mostly hydrogen and helium with a small sprinkling of heavier elements. It is these impurities that must have condensed into dust and ice particles and accreted to form the planets. Evidently the Jovian planets were also able to retain a substantial amount of the gas as well, but the terrestrial planets and many satellites must have

been made from the solids. [See THE ORIGIN OF THE SOLAR SYSTEM.]

An intermediate stage in the accretion was the formation of "planetesimals," Moon-sized objects that merged to form the final planets. For the terrestrial planets (Mercury, Venus, Earth, Moon, and Mars), the number was probably about 500. These objects would not remain in near-circular orbits, and the ones in the inner solar system might end up as part of any of the terrestrial planets. One would therefore expect them to begin with similar atmospheric compositions, and indeed those of Venus and Earth have many interesting resemblances, as mentioned in Section I. The smaller bodies appear to have lost all or most of their original gas (or never possessed much in the first place).

Many of the differences between the atmospheres of Earth and Venus can be traced to the near-total lack of water on Venus. These dry conditions have been attributed to the effects of a runaway greenhouse followed by massive escape of hydrogen. A runaway greenhouse might have occurred on Venus because it receives about twice as much solar heat as the Earth. If Venus started with a water inventory similar to that of the Earth, the enhanced heating would have evaporated additional water into the atmosphere. Because water vapor is an effective greenhouse agent, it would trap some of the thermal radiation emitted by the surface and deeper atmosphere, producing an enhanced greenhouse warming and raising the humidity still higher. This feedback may have continued until the oceans were gone and the atmosphere contained several hundred bars of steam. (This pressure would depend on the actual amount of water on primitive Venus.) Water vapor would probably be the major atmospheric constituent, extending to high altitudes where it would be efficiently dissociated into hydrogen and oxygen by ultraviolet sunlight. Rapid escape of hydrogen would ensue, accompanied by a much smaller escape of the heavier deuterium and oxygen. The oxygen would react with iron in the crust, and also with any hydrocarbons that might have been present. Although such a scenario is reasonable, it cannot be proved to have occurred. The enhanced D/H ratio certainly points in this general direction, but could have been produced from a much smaller endowment of water (as little as 1%) than is in the Earth's oceans.

It used to be thought that Venus was a near twin of the Earth, perhaps a little warmer but perhaps able to sustain Earth-like life. It is still possible that the large divergences we now see could have arisen from different evolutionary paths; alternatively, the two planets may always have been very different.

BIBLIOGRAPHY

Bézard, B., de Bergh, C., Crisp, D., and Maillard, J.-P. (1990). *Nature* **345,** 508–511.

Bougher, S. W., Phillips, R. J., and Hunten, D. M. (eds.) (1997). "Venus II." Univ. of Arizona Press, Tucson.

Crisp, D., McMuldrough, S., Stephens, S. K., Sinton, W. M., Ragent, B., Hodapp, K.-W., Probst, R. G., Doyle, L. R., Allen, D. A., and Elias, J. (1991). *Science* **253,** 1538–1541.

Donahue, T. M., and Hodges, R. R., Jr. (1992). *J. Geophys. Res.* **97,** 6083–6091.

Fox, J. L., and Bougher, S. W. (1991). *Space Sci. Rev.* **55,** 357–489.

Hunten, D. M., Colin. L., Donahue, T. M., and Moroz, V. I. (eds.) (1984). "Venus." Univ. of Arizona Press, Tucson.

Krasnopolsky, V. I. (1986). "Photochemistry of the Atmospheres of Mars and Venus." Springer-Verlag, New York.

Russell, C. T. (ed.) (1991). "Venus Aeronomy." *Space Sci. Rev.* **55,** 1–489.

Yung, Y. L., and DeMore, W. B. (1982). *Icarus* **51,** 199–247.

VENUS: SURFACE AND INTERIOR

I. History of Exploration of Venus
II. Global Surface Morphology and Processes
III. Age of the Surface and Geological History
IV. Interior of Venus
V. Synthesis

James W. Head, III
Department of Geological Sciences, Brown University

Alexander T. Basilevsky
Russian Academy of Sciences, Moscow

GLOSSARY

Asthenosphere: Shallow region in a planetary interior that underlies the lithosphere and is characterized by weak, often partially molten rock, and in which seismic waves are attenuated. The Earth's asthenosphere lies at a depth of about 100–200 km.

Basalt: Igneous rock composed of solidified lava relatively rich in iron and magnesium in which the silica content is less than 53% by weight.

Convection: Transfer of heat in planetary interiors by vertical and lateral movement of mass.

Differentiation: Melting and chemical fractionation of a planet into layers, such as core, mantle, and crust.

En echelon: Parallel structural features offset like shingles on a roof as viewed edgewise.

Graben: Linear, fault-bounded trough in which the bounding faults dip inward and the fault movement is predominantly vertical, with the central floor moving down.

Hadley circulation: Global atmospheric circulation pattern in which the upward motion of air over tropical latitudes is offset by the downward motion at higher latitudes, producing trade winds that tend to blow toward the equator.

Inferior conjunction: Position of Venus when it is on a line between Earth and the Sun.

Lithosphere: Uppermost brittle thermal boundary layer of a planetary interior. On Earth the lithosphere averages about 100 km thickness.

Mantle: Spherical shell between the core and the crust in terrestrial planets.

Rift valley: Large topographic depression formed by extensional faulting; sometimes synonymous with graben, but faulting is usually much more complex than a single pair of bounding faults.

Synthetic aperature radar (SAR): Technique in which a radar image is obtained using signals transmitted and received by a small antenna mounted on a moving platform.

The last 35 years of exploration of Venus, the most Earth-like of the terrestrial planetary bodies, has revealed much about its surface and interior. Near-global, high-resolution gravity, topography, and altimetry data acquired by the *Magellan* mission have built on earlier spacecraft exploration by the United States and the Soviet Union to reveal a picture of the recent geologic history of Venus that lies in contrast to that of the Earth. Venus' surface is geologically young compared to the smaller terrestrial planetary bodies and is similar to that of the Earth. The nature of its surface and interior provide no evidence for lateral plate tectonics operating at the present. Instead, observed surface features apparently formed largely due to vertical crustal accretion and tectonics and illustrate significant changes in the relative importance of volcanism and tectonism as a function of time. This record may represent catastrophic and possibly episodic heat loss events. Why Venus is so similar to Earth in some ways and so different in others remains one of the major enigmas of comparative planetary sciences.

I. HISTORY OF EXPLORATION OF VENUS

Venus, at a distance of about 1.1×10^8 km (0.72 AU) from the Sun, is the most Earth-like of the terrestrial planets in terms of its radius of 6051 km (compared to Earth's 6378 km), its average density of 5240 kg/m³ (compared to Earth's of 5520 kg/m³), and its surface gravity 0.907 times that of Earth (8.87 m/sec²). It differs, however, in a number of significant ways. Its orbital period is 225 Earth days (compared to Earth's 365 days), its rotational period (sidereal) is 243 Earth days (compared to Earth's 24 hrs), it rotates in a retrograde manner (compared to Earth's prograde rotation), and its atmospheric composition is dominated by carbon dioxide (96%) and nitrogen (3+%), with trace amounts of sulfur dioxide, water vapor, carbon monoxide, argon, helium, neon, hydrogen chloride, and hydrogen fluoride (Earth's atomosphere is predominantly nitrogen and oxygen). In addition, its average surface temperature is 850 F (730°K), much hotter than on Earth, and its average surface atmospheric pressure is 90 times that of Earth (90 ± 2 bars).

Venus has occupied the interest of humans for centuries, at least since about 3000 B.C., when its erratic movement relative to the background of stars attracted the attention of the Babylonians. Observations by Galileo in 1610 revealed the lunar-like phases, and variable spots and markings were noted by later observers. In the 1700s, attempts to measure the Earth–Sun distance during the two transits of Venus across the Sun revealed important new information about Venus itself. Lomonosov, observing in St. Petersburg, correctly interpreted a gray halo seen surrounding the planet as

evidence of an atmosphere. This observation, combined with improved measurement of the diameter of Venus, led to the initial concept of Venus as the Earth's twin planet. Early brightness measurements and lack of detailed features suggested the presence of dense clouds, and the assumption that the clouds were made of water vapor conjured up images of steamy swamps on the surface.

In the prespace age part of the 20th century, observations established that the atmosphere was dominated by carbon dioxide rather than water. Continuing investigations of the composition of the clouds ultimately showed that they were composed of sulfuric acid rather than water vapor; these two factors, together with an emerging understanding of the high surface temperatures and pressures, began to change the perception of Venus as Earth's twin. Subsequent data from Earth-based observations and the *Venera*, *Pioneer–Venus Orbiter*, and *Vega* spacecraft provided more insight into the atmosphere of Venus.

During this early period, the nature of the surface of Venus and its rotation rate remained unknown. In the 1960s, early radar observations of Venus, first at the Jet Propulsion Laboratory and Lincoln Laboratory, and later at the Arecibo Observatory, led to the determination of the slow and retrograde rotation of Venus. The lack of information about surface features and topography meant, however, that Venus remained an uncertain and elusive member of the terrestrial family of planetary bodies to most planetary geologists and geophysicists. During this period, most of these scientists were occupying their time with exciting exploration of the Earth's seafloor and the surfaces and interiors of the Moon, Mars, and Mercury. In the early 1970s, hints at the geology of Venus began to emerge from the geochemical measurements of the Soviet *Venera 8* mission (indicating largely basaltic compositions) and from the panoramas of the surface from *Veneras 9* and *10*. At about the same time (1965–1975), Earth-based radar observations began to reveal information about the surface roughness, surface brightness, temperatures, and the physical nature of surface materials.

In the following decade, Arecibo and Goldstone Earth-based radar observations obtained near inferior conjunction provided images of portions of the surface in the resolution range of about 5–20 km. Discovery of features interpreted to be rift zones, volcanoes, and folded mountain belts began to provide some sense of the geology of the surface of Venus. Concurrently, synthesis of the results of the Apollo, Luna, Mariner, and Viking missions revealed that the smaller terrestrial planetary bodies (the Moon, Mars, and Mercury) are

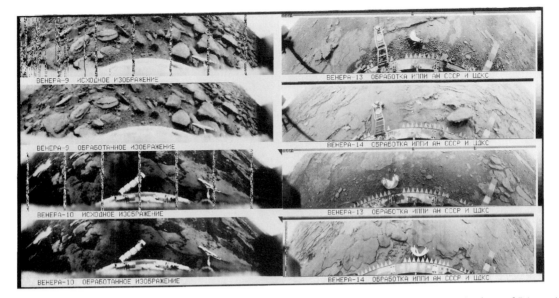

FIGURE 1   *Venera 9, 10, 13, and 14* lander images of the surface of Venus. (Courtesy of the Russian Academy of Sciences.)

characterized by ancient globally continuous lithospheres that have been little altered since the first half of solar system history. This is in stark contrast to the Earth at present, which is characterized by a series of laterally moving lithospheric plates, created at divergent boundaries and destroyed at convergent boundaries, producing a largely young and rapidly renewed surface. This new comparative planetology perspective for the terrestrial planets focused renewed interest in Venus as Earth's twin. Venus is most similar in size, density, and position in the solar system to Earth. Does this mean that its geological and geophysical characteristics are similar to the Earth and in contrast to those of the smaller terrestrial planetary bodies? Is it characterized by present-day Earth-like tectonics? Or could it represent some earlier stage in the evolution of Earth? [*See* EARTH AS A PLANET: SURFACE AND INTERIOR.]

In the late 1970s the Pioneer–Venus mission, using a radar altimeter, obtained further important information about the similarities and differences between Earth and Venus. A near-global topographic map with an average resolution of about 100 km was compiled, and although discrete highland areas were observed, the distribution of topography was clustered around an average value, in contrast to the nature of present Earth topography, which has two levels, continents and ocean basins. The geography of Venus became more well known, and the distribution of broad-scale geological features (rift valleys, linear mountain belts, topographic rises) began to emerge. Particularly productive during this time period were analyses of Earth-based imaging data, which showed that the observed

surface was relatively young, and correlations of the Earth-based radar images with Pioneer–Venus data, produced an increased knowledge of the geology of areas such as Beta Regio and Ishtar Terra. These analyses showed evidence for rifting and extensional deformation, volcanism and hot spots, and orogenic belts and compressional deformation. Continued acquisition of data from *Venera* landers (*Venera 13–14* and *Vega 1–2*; see Fig. 1) provided knowledge on the characteristics of the surface materials, showing them to be largely basaltic in composition. Doppler radio tracking of the *Pioneer–Venus Orbiter* provided gravity data over a large portion of Venus at resolutions of 300–2000 km and showed that there was a significant positive correlation of topography and gravity at long wavelengths. Analysis of the combined data sets began to raise specific questions about mantle dynamics and how they were related to surface processes and deformation patterns.

Major strides were made in 1983–1984 when the Soviet Union placed two spacecraft, *Venera 15* and *16*, in orbit around Venus. The *Venera 15/16* side-looking radar imager with 1- to 2-km resolution (see Fig. 2) revealed surface landforms and provided images over a large enough area to determine, by crater counting, approximately when the observed landforms and terrains formed. The northern quarter of the planet was imaged, showing this area to be largely a volcanic lava plain. The general smoothness of the plains and the large length of the lava flows visible on the images provided evidence for a basaltic composition of the lavas. About 10–15% of the area was shown to be a

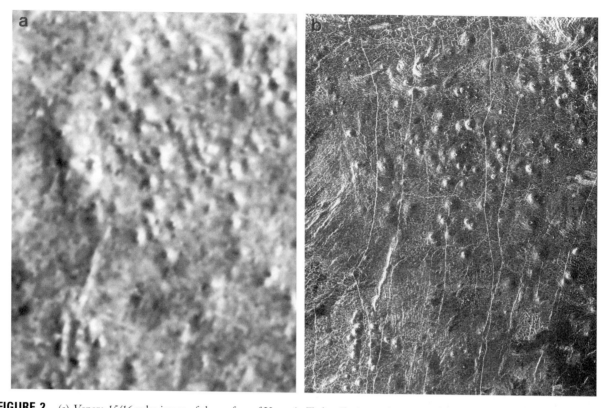

FIGURE 2   (a) *Venera 15/16* radar image of the surface of Venus in Tethus Regio, at about 1–2 km resolution, showing an abundance of small shield volcanoes ranging in diameter from ~2 to 12 km, and illustrating the high-resolution nature of data compared to earlier information. (Courtesy of the Russian Academy of Sciences.) (b) *Magellan* radar image of the same area, at about 120-m resolution. The increased resolution, combined with the complementary viewing geometry of *Venera* data, shows how important details of the small volcanoes and the stratigraphy and structure of surrounding plains are revealed. (NASA photo.)

particularly rough terrain formed by multiple intersections of ridges and grooves that originated through tectonic deformation. The rough terrain received a special name, "tessera," which in Greek means "tile." Comprehensive analysis of the tesserae morphology revealed evidence of earlier compressional and later extensional deformation responsible for the formatiom of tesserae from precursor rocks of unknown composition. In many cases it was clear that areas of tesserae are embayed by the surrounding volcanic plains. However, in some cases formation of tesserae due to the deformation of the observed plains was also considered a possibility, but the resolution of data was insufficient to provide confirmation. Several volcanoes superposed on regional plains were observed. Some of the volcanoes were associated with linear fracture zones cutting the regional plains, resembling the continental rift zones on Earth.

About 150 impact craters ranging from 8 to 140 km in diameter were identified within the ~20% of the planet imaged by *Venera 15/16*. Based on estimates of the crater-forming rate on Venus, the surface age was estimated to be about 500 million to 1 billion years. This estimate is, of course, the average surface age; it was clear from stratigraphic relations that some terrains and features are older than this average value whereas others are younger.

This age estimate, even being of relatively low accuracy, showed that the age of the dominant geologic formations on Venus was (1) significantly younger than those of the Moon and (2) comparable to, but slightly older than much of the surface of Earth (the ocean basins). The first point was predictable because the Moon, being much smaller than Venus and thus having a much larger surface area/volume ratio, cooled relatively quickly, thus halting observable tectonism and volcanism about 3 billion years ago. The second was expected by some planetary geologists who believed that Venus, with rift zones and compressional orogenic mountain belts observed in previous Earth-based data, might have an Earth-like, plate-tectonic style of geologic activity with horizontally moving lithospheric plates. Although the rifts and folded mountain belts were seen in much more detail, no conclusive evidence

FIGURE 3 Global distribution of morphology on Venus from *Magellan* data. Globe represents view centered along the equator of area between Thetis Regio (bright area at left) and Alta Regio (bright area to right). (NASA photo.) (See also color insert.)

was found in *Venera 15/16* data for a widespread plate tectonics style on Venus. In addition, no features formed by flowing water were found in the *Venera 15/16* images so the period of a water-rich environment hypothesized from the *Pioneer–Venus* measurements of the D/H ratio (ratio of deuterium, the hydrogen isotope of mass 2, and hydrogen, which can be used to infer the presence of water previously in the atmosphere) apparently took place earlier than 0.5–1 billion years ago.

The conclusions reached from analysis of *Venera 15/16* data were partly undermined by the fact that only one-fourth of Venus was surveyed; thus, one can always suspect that the other three-fourths of the planet might be very different. In addition, the resolution of *Venera 15/16* images was sufficient to observe various features and terrains and to make a first approx-

imate interpretation of what geologic processes were involved in their formation, but insufficient for more reliable geologic interpretaions, including establishing a time sequence of geologic events on the planet (see Fig. 2). Consequently, the next mission to Venus had to provide a higher resolution radar survey and global coverage of the planet. This was accomplished in 1990–1994 by the *Magellan* mission.

The *Magellan* mission, launched May 4, 1989, was designed to obtain near-global radar images of Venus' surface with a resolution of several hundred meters, a topographic map with 50 km spatial and 100 m vertical resolution, and gravity field data with 700 km resolution and 2–3 milligals accuracy. The scientific goals of the mission were to develop an understanding of the geological structure of the planet, including its density distribution and dynamics. *Magellan* arrived at

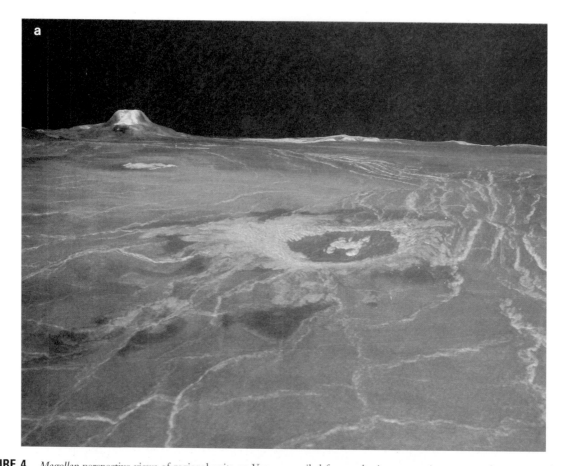

FIGURE 4 *Magellan* perspective views of regional units on Venus compiled from radar image mosaics superposed on topographic maps from altimetric data and viewed obliquely. (a) Looking northeast over regional volcanic plains showing the vast expanse of wrinkle-ridged plains in the foreground stretching to the horizon about 1300 km away where the 3-km-high shield volcano Gula Mons is seen. The 53-km-diameter crater Cunitz is seen superposed on the plains in the middle ground, about 215 km away. Vertical exaggeration on this and the following images is about 23 times. (See also color insert.) (b) Looking east toward Maat Mons, a large, 500-km-diameter shield volcano that rises about 8 km above mean planetary radius (MPR) in Alta Regio. In the foreground is a lava flow that has traveled about 600 km from the volcano, and together with the other bright volcano deposits, is superposed on the older regional plains. (c) Looking north over the highly deformed tesserae area known as Alpha Regio, about 1300 km in diameter. Note that the surrounding volcanic plains are much less heavily deformed and that they embay the tessera. The rim of the 330-km-diameter corona Eve is seen directly to the south–southwest of Alpha. (NASA photos.)

Venus on August 10, 1990, and much like its 16th century Portuguese namesake whose expedition first circumnavigated the Earth, began circling Venus. The spacecraft began mapping Venus from a near-polar elliptical orbit around the planet. Synthetic aperture radar (SAR), altimeter, and radiometer data were obtained, and the coherent X- and S-band radio subsystem was used for gravity field measurement by precision tracking of the spacecraft's orbit. During the first 8-month mapping cycle around Venus (one cycle is one complete revolution of the planet under the orbiting spacecraft), *Magellan* collected radar images, altimetry (surface topography), radiometry, and data on electrical characteristics of 84% of the planet's surface. Two further mapping cycles during the extended mission

from May 15, 1991 to September 14, 1992 were designed to image the south pole region, fill gaps from cycle 1, and collect stereo images. These cycles brought mapping coverage to 98% of the planet. Precision radio tracking of the spacecraft measured the gravitational field to reveal the internal mass distribution and permitted better interpretation of the forces responsible for the surface features. The fourth cycle was dedicated to this, followed by the fifth cycle during which aerobraking to circular orbit was achieved and higher-resolution global gravity measurements were obtained. At the end of cycle 6, the spacecraft was turned into the atmosphere to observe the behavior of molecules in the upper atmosphere, and the mission was terminated in October 1994, having ultimately achieved ra-

FIGURE 4   (*continued*)

dar mapping coverage over 98% and gravity data coverage over 95% of Venus. The Magellan mission, then, provided virtually global, high-resolution image and topography coverage, major advances in our knowledge of the physical properties of the surface, and much higher resolution global gravity data. Finally, data were at hand to fully compare Venus to Earth and the other terrestrial planetary bodies (Fig. 3; see also color insert).

In summary, following the initial acquisition of *Magellan* data, four tasks were at hand: (1) to provide a description and global inventory of the major features and geological processes observed, (2) to understand the ages of the surfaces and the stratigraphic and geologic sequence of events represented by them, (3) to understand the nature of the interior (structure and

dynamics), and (4) to link all of these into a coherent model for the evolution of Venus and its relation to the Earth and other terrestrial planetary bodies.

II. GLOBAL SURFACE MORPHOLOGY AND PROCESSES

A. *MAGELLAN* GLOBAL GEOLOGY AND ALTIMETRY

Magellan image and altimetry data (see Figs. 3 and 4; see also color insert for Fig. 4a) revealed that the sur-

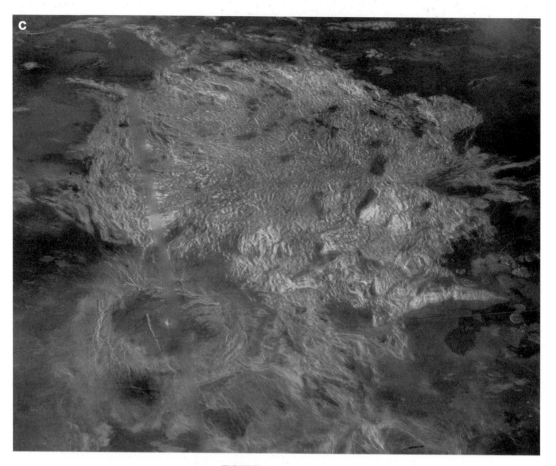

FIGURE 4   *(continued)*

face of Venus is dominated by an extensive mosaic of volcanic plains lying near the mean planetary radius (MPR) (see Fig. 5) and forming over 80% of the surface. Also seen are local areas of very highly deformed terrain (the tesserae revealed by earlier observations) lying above MPR and forming about 8% of the surface (see Figs. 4a and 6). Large rift valleys wind across the surface and converge on broad, several thousand kilometer-wide rises such as Alpha and Beta Regiones (see Figs. 3 and 7). Broad isolated shield volcanoes, generally less tall than those on Earth, dot the surface of the planet (see Fig. 4b). Broad ridges and arches extend for several thousand kilometers and suggest broad folding and shortening, whereas the surfaces of most plains are populated by extensive networks of wrinkle ridges, similar to those seen in the lunar maria. In Ishtar Terra, the linear mountain belts observed in previous data were revealed in beautiful detail. Hundreds of concentric ringed features called coronae are distributed across the surface (see Fig. 8), most with diameters less than 300 km, but several gargantuan ones exceed a thousand kilometers in diameter. Many

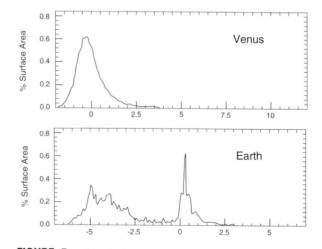

FIGURE 5   Altitude frequency distribution of topography for Venus and Earth shown as hypsometric curves [representing the percentage of surface area at different elevations relative to mean sea level on Earth or mean planetary radius (MPR) on Venus].

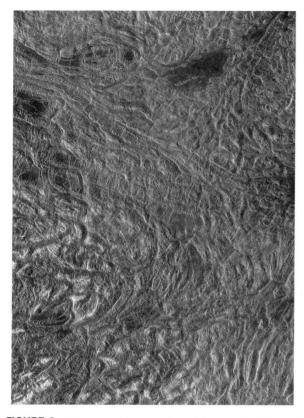

FIGURE 6   Tessera terrain. A portion of the complex deformed uplands of Alpha Regio (25.0°S, 3.0°) typical of the fabric of tesserae terrain. The crust is apparently initially folded into broad ridges 25–50 km apart as the crust is thickened by shortening and then faulted by extension as the crust relaxes in the waning stages of the intense deformation. Shear deformation also takes place. (NASA photo.)

coronae are isolated, but others form conspicuous chains thousands of kilometers long. Broadly speaking, *Magellan* data showed that Venus could be subdivided into three major catagories: (1) volcanic plains forming the vast majority of the planet (e.g., Lavinia, Niobe, Rusalka Planitiae), (2) intensely deformed tessera, forming upland plateaus (e.g., Fortuna, Aphrodite, and Tellus), and (3) broad topographic rises (e.g., Beta, Atla, Bell Regiones) with associated volcanoes and rifts.

Altimetry data from *Pioneer–Venus* and *Magellan* (see Fig. 5) demonstrated that the global altitude–frequency distribution of topography on Venus was unimodal and distinctly different from that of the bimodal Earth. Early investigators assessed the influence of oceanic load and the different thermal structure and showed that while these factors bring the two terrestrial peaks closer together, it does not erase the bimodality. One of the most significant factors accounting for the differences in the hypsometric curves

(those describing elevation differences from the mean) is the fact that the highlands of Venus (tessera) represent only about 8% of the surface area, compared to the 30% of Earth represented by continental regions. A second major factor is related to crustal thickness variations and their distribution between Venus and Earth. The contrast in average thickness difference between the crusts of the two terrains on Earth (ocean basins and continents) and Venus (plains and tessera) appears to be different. This is the most likely explanation for the separation of the two peaks on Earth relative to Venus. For example, adding 20 km of basaltic crustal thickness to the average thickness of the Venus highlands is about equal to adding 2 km in additional topographic elevation, or equivalent to 2 km additional separation between the Earth provinces and peaks relative to Venus. This is in contrast to the commonly held assumption that compositional differences between continents and ocean basins are responsible for the topographic variations between continental and oceanic crust.

FIGURE 7   Rift zone. Devana Chasma, 1170 km in length, is a major north–south trending rift valley in central Beta Regio (28.0°N, 283.0°), a region analogous to the East African rift on Earth, where the crust is being stretched and faulted. Here the 31-km-diameter crater Somerville, located near the center of the image, has been split by the relatively young faulting typical of these rift zones on Venus. (NASA photo.)

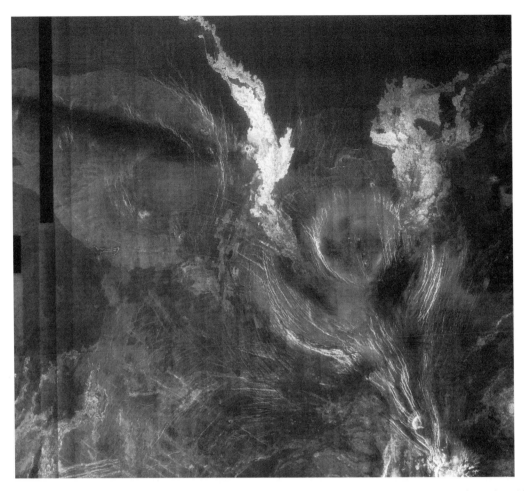

FIGURE 8   Coronae. *North of Gula Mons are two coronae; the western one is elongate (about 200 × 300 km) and is embayed by younger flows. The eastern one, about 225 km in diameter, contains prominent bright lava flows in excess of 200 km in length. Fractures leading away from the corona to the southeast are typical of those seen connecting coronae in chains. (NASA photo.)*

B. GEOLOGICAL FEATURES AND PROCESSES FROM *MAGELLAN* DATA

1. Volcanism

The extreme surface conditions on Venus (e.g., very high surface temperatures and pressures) are expected to have an influence on the ascent and style of eruption of magma and the morphology of the subsequent landforms. For example, the high atmospheric pressure on Venus reduces the amount of gas that comes out of solution as the molten rock rises toward the surface. This means that explosive eruptions like those typical of Hawaiian fire fountains or Mount St. Helens are not as likely to occur on Venus unless the gas content of the magma is much higher than typical of Earth.

In addition, the lack of gas bubbles in the upper crustal rocks means that the rising magma is commonly less dense than the rock through which it is rising, so it may more often travel all the way to the surface instead of stalling at a zone of neutral buoyancy within the upper crust. Magma would tend to ascend directly from greater depth and such eruptions would be characterized by relatively high total volumes and effusion rates. Thus, a higher number of more extensive flows is predicted. Because of the dense hot atmosphere, cooling of lava after it flows out on the surface will be first enhanced because of more efficient convective cooling and then inhibited because of the very high surface temperature. The net effect is that lava eruptions are predicted to flow to about 20% greater length than they would on Earth. These longer flows, together with predictions about the dynamics of magma ascent, suggest that volcanoes on Venus will be broader and flatter than those on Earth. [See PLANETARY VOLCANISM.]

In addition, prior to *Magellan* the range of rock-producing processes and products possible in the Venus environment during the production of primary magmas and remelting of crustal material was assessed; such analyses showed that a wide range of magma types was possible and that no magma type could be specifically excluded from consideration. Although predominantly basaltic compositions were predicted, variations due to water content, degrees of melting, and thermal gradient suggested that environments existed for more basic and more highly evolved, silica-rich magmas.

The global survey of *Magellan* data was used to document the characteristics, location, and dimensions of all major volcanic features on Venus and revealed over 1700 landforms and specific deposits in excess of 20 km diameter and countless thousands of smaller shield volcanoes. Among these were over 500 shield fields (concentrations of small volcanoes <20 km in diameter), over 230 intermediate volcanoes between 20 and 100 km diameter with a variety of morphologies, more than 140 large volcanoes in excess of 100 km diameter, more than 80 caldera-like structures independent of those associated with shield volcanoes (and typically 60–80 km in diameter), over 400 coronae (annulus of concentric ridges or fractures) and arachnoids (inner concentric, and outer radial network pattern of fractures and ridges), 43 novae (focused radial fracture patterns forming stellate patterns), 44 lava flood-type flow fields, and 43 sinuous lava channels (all of which are in excess of 10^2–10^3 km in length). The vast majority of volcanic features are apparently basaltic in nature, consistent with geochemical results from the Soviet *Venera* landers.

Two interesting variations on this theme were observed, providing evidence for some petrologic diversity. Over 150 biscuit-like, steep-sided domes (see Fig. 9) were observed to be distributed widely across the surface, and on the basis of their distinctive morphology were interpreted to represent eruptions of more evolved, silica-rich compositions, such as andesite, dacite, or rhyolite, perhaps similar to compositions represented by the *Venera 8* lander results interpreted to be different than typical basalts. However, other workers suggested that the unique Venus environment may lead basaltic eruptions to produce these features under special conditions. Several much larger deposits of viscous-appearing lavas provided evidence that more evolved magma may have been erupted on Venus, at least locally. A second variation was the presence of sinuous channels (see Fig. 10), which may represent much more fluid magma than typical of basalt. On the Moon such channels were interpreted to be due to very voluminous eruptions of fluid lava, and the resulting thermal erosion of the lavas into the substrate, to produce the channels. Channels on Venus are often much longer than even the spectacular ones on the Moon, the longest being Baltis Vallis, in excess of 6000 km. Some investigators think that these channels might represent the outpouring of vast quantities of very fluid magmas, with anomalously high iron or sodium content.

The range of morphologies observed by *Magellan* indicates that a spectrum of intrusive and extrusive processes have operated on Venus (see Figs. 9–11). Consistent with the predicted effects of the high surface atmospheric pressure, little evidence was found for extensive deposits or landforms due to explosive eruptions and the deposition of pyroclastic materials, although some impact-crater-related ejecta flows appear similar (see Fig. 12). The large size of many volcanic features (see Fig. 11) is evidence for the presence of very large magma reservoirs, also predicted by theory. The areal distribution, abundance, and size frequency distribution relationships of arachnoids, novae, large volcanoes, and coronae strongly suggest that they are the surface manifestation of mantle plumes or upwellings and that the different morphologies may represent variations in plume size and stage and thermal structure of the lithosphere.

Maps of the global distribution of volcanic features based on this census of *Magellan* data show that they are widespread, but not evenly distributed; however, there is little evidence for linear arrays of edifices or associations of styles and tectonic structure such as characterize lithospheric plate boundaries and hot-spot traces on Earth. Although globally widespread, features are less abundant in the highly deformed tesserae regions and much more abundant in the Beta–Atla–Themis (BAT) region, which covers about 20% of the planet (see Fig. 3). This region is unique in that it is the site of local concentrations of volcanic features with densities two to four times the global average, an interlocking network of rift and deformation zones, several broad rises several thousand kilometers in diameter with associated positive gravity anomalies and tectonic junctions, and evidence for volcanically embayed impact craters. Although the region as a whole is not anomalously older or younger than the rest of Venus, evidence shows that the most recent volcanic activity on the planet occurs here, and the presence of this series of concentrations suggests that the mantle in this region is anomalous.

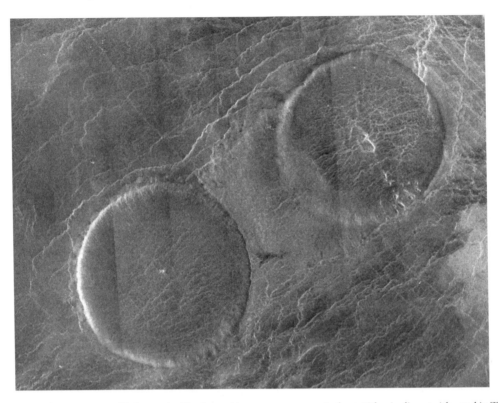

FIGURE 9   Domes. These two steep-sided, pancake-like domes (the westernmost one is about 62 km in diameter) located in Tinatin Planitia (12.0°N, 8.0°) are in contrast to the extensive surrounding smooth volcanic plains with wrinkle ridges and suggest that emplacement processes for the domes involved more viscous magma than those responsible for the smoother, more widespread plains. (NASA photo.)

Most of the types of volcanic features observed in the global census are seen both inside and outside the BAT area, but the large sinuous channels appear to be preferentially located in the lowlands and outside the BAT area. In addition, an observed deficiency of many volcanic features in several lowland areas of Venus may be due to an altitude-dependent influence of atmospheric pressure on eruption style that would favor lava floods (see Fig. 11) and sinuous channels at low elevations and edifices and magma reservoir-related features at higher elevations. In summary, volcanic edifices and structures on Venus are dominated by features interpreted to be related to mantle plumes and generally comparable to hot-spot volcanism on Earth in terms of numbers and rates of volcanism. However, the most areally extensive deposits on the surface are the volcanic plains occurring both between and below the observed edifices (see Fig. 4a).

There are several different varieties of volcanic plains. First, the appearance of plains-forming material varies in the sense of its homogeneity or inhomogeneity, thus forming varieties that can be described as smooth, mottled, digitate, lobate, etc. Second, plains also vary in the character and amount of superposed features, e.g., forming varieties that are fractured, ridged, domed, featureless, etc. Among the major plains types observed (see Fig. 13) are densely fractured plains, a terrain consisting of swarms of mostly subparallel grooves, which are interpreted to be fractures with a typical spacing of about 1 km. The undeformed precursor of this terrain is not observed but they appear to be plains of volcanic origin because, if the fractures are ignored, the terrain looks smooth and plains-like. Densely fractured plains make up only about 2–3% of the surface but occur throughout the planet, mostly in the form of patches tens to hundreds of kilometers across surrounded and embayed by younger plains. Some volcanic plains are less intensely deformed and typically have relatively broad (5–10 km wide) gently sloping ridges tens of kilometers long. The ridges are usually arranged en echelon and form belts. In addition to the broad ridges, these plains often have narrow ridges similar to the wrinkle ridges of the neighboring plains, as well as fractures. These fractured and ridged plains, occupying only about 1–3% of the surface, are usually present in the form of remnants tens to hundreds of kilometers across among the younger plains embaying them. Another variety of plains are those

FIGURE 10 Sinuous channel. This channel is only a few kilometers wide but its total length is in excess of several hundred kilometers, stretching across a major part of Helen Planitia (49.0°S, 271.0°). The structure, texture, and superposition relationships suggest that this channel and others like it may have been formed by thermal erosion, as hot lava cut into the underlying solid rock. One such channel, Baltis Valles, is in excess of 6000 km in length. (NASA photo.)

containing a population of small shield volcanoes less than 20 km in diameter (see Fig. 2). Between the abundant small shield volcanoes, these plains are smooth and relatively radar dark. These plains make up less than 5% of the surface but are very prominent because of the high density of small shields and localized source regions and their relatively distinctive stratigraphic occurrence. Although small shields can occur throughout the stratigraphic column and in association with a variety of features, their extreme concentration in this unit seems to call for some specific circumstances of formation.

The most abundant volcanic plains on Venus, composing about 70% of the surface and thus forming the background on which we see other features and terrains, are characterized by broad homogeneity in radar properties and by a general lack of volcanic structures and edifices. The plains are usually radar dark to

moderately dark, but sometimes have brighter members or mottled portions (see Figs. 4a and 4b). In addition to their homogeneity, one of the defining characteristics of these plains is the presence of wrinkle ridges superposed on their surfaces. The ridges are usually less than 1–2 km wide and a few tens of kilometers long and their trend often varies, sometimes intersecting to form a network. One of the few volcanic features associated with these plains are the river-like channel features a few kilometers wide and a hundred to a few thousand kilometers long (see Fig. 10). Wrinkle ridges criss-cross the channels. A final variety of plains has prominent lobate flow-like characteristics and is usually not characterized by superposed wrinkle ridges. These lobate plains tend to be associated with coronae, shields (see Fig. 4b), arachnoids, and faults.

The composition of the plains material is considered to be basaltic because of their morphology (see Fig. 11) and the *in situ* geochemical measurements made by the Soviet *Venera* and *Vega* spacecraft, which landed

FIGURE 11 Flow lobes. A spectacular flow field in the Lada Terra region (47.0°S, 25.0°) covering an area exceeding 500,000 km². The flows, traveling from the west, have breached a north–south trending ridge system and poured out into the surrounding area, embaying and burying older terrain. (NASA photo.)

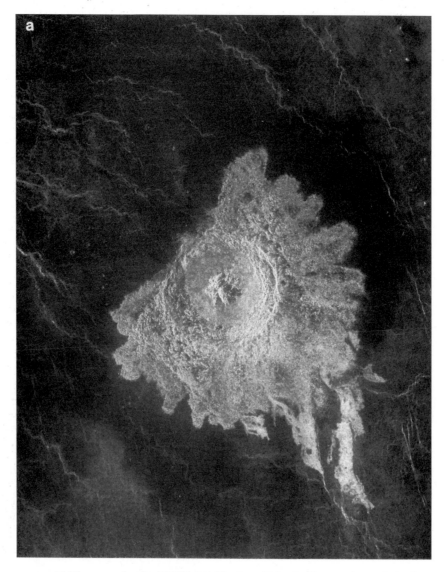

FIGURE 12 Impact craters. (a) The crater Aurelia (20.3°N, 331.9°), about 31 km in diameter, has well-developed terraced walls, a flat floor, and central peaks typical of a complex crater. Asymmetry of ejecta suggests that the projectile may have impacted obliquely from the northwest. (b) The crater Markham (4.1°S, 155.6°), about 69 km in diameter, shows spectacular crater ejecta outflows that extend many times the crater diameter away from the point of impact, are partly controlled by preexisting topography in the surrounding plains, and superficially resemble volcanic flows (see Fig. 11). (NASA photos.)

within plains with wrinkle ridges with subordinate patches of smooth and lobate plains.

What is the distribution of features in time? Stratigraphic evidence suggests to many that the majority of edifices and structures are superposed on the vast volcanic plains. The high density of features in the BAT area might mean that this is either an older area of more prolonged volcanism or an area of age similar to the rest of the planet, but with more concentrated volcanism. The areal distribution of impact craters shows that BAT is not significantly different from the crater retention age of the whole planet; volcanically embayed craters, however, are preferentially located in the BAT region, indicating that later volcanism is more abundant there and that it is probably the site of the most recent volcanism. In summary, there is evidence for changes in volcanic style, flux, and location with time; many workers believe that there was an early near-global formation of vast plains by flood volcanism and later localized centers of volcanism and much reduced flux concentrated in the BAT area. Other scientists think that the question of rate and volume of erupted material over the age of Venus (and hence time) is an unanswered question.

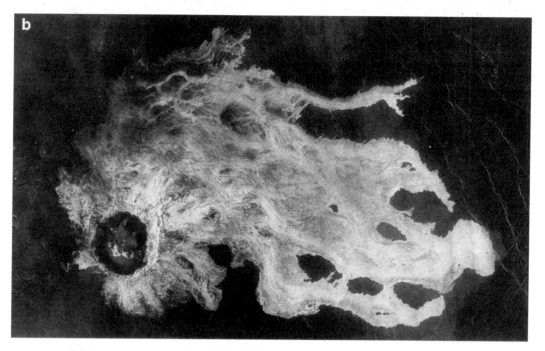

FIGURE 12 (*continued*)

2. Tectonism

Global *Magellan* data revealed the surface to have a wide variety of tectonic features, some similar to those on Earth and others different in their character and level of development. The vast majority of the surface is modified to some degree by tectonic activity, and superposition of tectonic events and volcanism is common. Deformation can be widespread with broadly distributed strain of low magnitude (e.g., the wrinkle-ridged plains) or it can be localized in zones of rifting and extension (see Fig. 7), or belts of intense shortening and folding. Portions of the surface are extremely highly deformed (tesserae terrain; see Figs. 4c and 6), suggesting either localized regions of very intense deformation or a change in deformation intensity as a function of time. The similarity of spacing and orientation of many features over hundreds to thousands of kilometers suggests that they may be linked to broader, deeper mantle processes. Although local strike-slip movement is observed, no large-scale lateral movement comparable to major strike-slip faults on Earth was noted. The presence of some tectonic features (e.g., Maxwell Montes) with extremely steep slopes (greater than 20–30° over distances of 10 km) suggested that they may be tectonically active today because the high surface temperatures would cause slopes to decrease over a very short period of geologic time; more recent work on the behavior of very dry rocks suggests, however, that the rocks may be stiffer, and these times may be longer.

Thus, the *Magellan* integrated global picture of tectonics on Venus does not resemble present-day active plate tectonics on Earth, where rigid lithospheric plates are deforming at their margins rather than within their interiors. Differences between Earth and Venus may be attributed to the presence of a hydrological cycle of erosion and water in the upper mantle on Earth, the absence of a mantle low-viscosity zone on Venus and the consequent more direct coupling of mantle convection and surface deformation, and the high surface temperature on Venus, which produces a shallower onset of ductile behavior of crustal rocks, producing a wider variety and abundance of crustal deformation features.

Individual features and deformational terrain types include tessera, fractures, graben, rifts, wrinkle ridges, deformation belts, orogenic belts, and coronae. The tesserae terrain is composed of areas of high radar backscatter, complex deformation patterns relative to other units, and topography standing higher than surrounding plains. It covers about 8% of the surface and is distributed nonrandomly, being preferentially located at equatorial and higher northern latitudes with a distinct paucity below about 30°S. Individual tesserae occurrences range in area from small patches up to the largest tessera, Ovda (see Fig. 6), with an area of about 2% of the surface area of Venus. Although tesserae exhibit a range of gravity signatures, many occurrences are interpreted to represent relatively shallow (crustal) levels of compensation. Tessera occurrences are virtually always embayed by younger plains. Only a small percentage of the length of all boundaries show no lava embayment and could be interpreted as tectonically active for long periods subsequent to initial tesserae formation. Several lines of evidence suggest the possibility that a widespread tessera-like basement, comprising at least 55% of the surface of Venus, is buried under a cover of later lava plains a few hundred meters to as much as 2–4 km thick.

Within the tesserae, a wide variety of deformational structures and patterns are observed, including those representing extension, compression, shear, and transpression (see Fig. 6). In some cases the apparently complex patterns can be resolved into single events, whereas in other cases multiple phases of deformation are more likely. Where relations can be determined stratigraphically, many workers believe that the earliest deformation within the tesserae is related primarily to crustal shortening and compression, followed by pervasive extensional deformation (a variety of graben) commonly oriented normal to the strike of earlier features. Some workers question whether all tesserae can be considered a single, global surface unit created by a single mechanism. The complexity of the deformation, however, means that there is controversy in its interpretation, and much more study is required before a concensus is reached.

Does tesserae terrain represent the end product of a long-term sequence of upwelling or downwelling followed by crustal deformation and tesserae formation that might be operating today? Tessera terrain as a geologic unit occupies the lowest portion of the stratigraphic column and no evidence for transitional stages between tesserae and volcanic rises and/or lowlands has been observed, providing evidence against this hypothesis and indicating that tesserae deformation was early in the preserved geologic history of Venus. No impact craters deformed by the earliest stage of tesserae deformation have yet been observed, suggesting that deformation sufficiently intense to eradicate earlier impact craters ceased relatively abruptly somewhat before ~300–750 million years ago; however, the starting time, and thus duration of tesserae formation, is unknown. On the basis of the very small number of on-tesserae craters deformed by later extensional deformation, this period probably did not last more than several tens of millions of years after the cessation of the earlier phase.

Tesserae terrain is undoubtedly the result of strong tectonic deformation of the crust but it is not clear what kind of material was deformed. No spacecraft have landed on tesserae so there is no direct data on the composition of tesserae material. It may represent the deformation of basaltic volcanic material similar to those that compose the adjacent younger plains. The fact that tesserae terrain is higher in altitude than the surrounding basaltic plains may mean that the tesserae material has a lower bulk density than the basalts and may be more evolved geochemically. The presence of geochemically evolved material elsewhere on Venus might be indicated by the presence of steep-sided volcanic domes.

Fractures and narrow graben (as opposed to the wider rift zones) are typical of many areas of Venus. Swarms of fractures are observed in tessera, in densely fractured plains, and are abundant in fractured and ridged plains. Fractures are usually much less abundant among younger materials, but commonly occur radiating away from such features as nova or varieties of coronae (see Fig. 8); such features have been interpreted as the surface manifestation of subsurface dikes emplaced radially from a magma reservoir. In some places, however, there are prominent fracture belts up to several thousands of kilometers long and a few hundred kilometers wide, composed of parallel and braided (anastomosing) small graben hundreds of meters to a few kilometers wide.

In other places, fractures form swarms and clusters of subparallel and anastomosing faults with feathering individual fractures. As a rule, these swarms and clusters are found in association with prominent topographic troughs (chasmata) and represent the distinctive rift zones that focus on broad highs such as Beta and Alta Regiones (see Fig. 7). In the huge rift zones of eastern Aphrodite the densely spaced subparallel fractures with quite consistent spacing form early parts of the rifts. They are criss-crossed by swarms of anastomosing fractures of different widths, and the floors of the widest of these latest fractures are the deepest depressions on Venus. Extensive, focused large-scale rift systems represent the latest stage of deformation; these cross-cut regional structures and units and are concentrated in areas (such as Beta and Atla Regiones) where gravity data strongly suggest that there is contemporaneous upwelling. Rifts represent minimal extension, often diverge from the center of rises, show associated large volcanic flows, and are the sites where coronae commonly show associated volcanism.

Compressional deformation is represented in broad ridges that form extensive ridge belts and mountain ranges, in the dense network of wrinkle ridges that populate the vast majority of the plains, and is thought by many workers to be seen in the fundamental fabric of the tessera. Local regions of relatively intense deformation are observed in the form of ridge belts (e.g., in Atalanta and Lavinia Planitiae), where the plains units are deformed into broad linear rises several tens of kilometers in width and hundreds of meters high. These features testify to the broad long-wavelength deformation of early plains, a phase of deformation in which modest shear was also involved. In the Ishtar Terra region, a series of mountain belts surround Lakshmi Planum, a plateau elevated several kilometers above the surrounding plains. First noted in Earth-based images, the mountain belts are hundreds of kilometers in length and several hundred kilometers wide; the tallest, Maxwell Montes, rises 11 km above MPR. The internal structure of the mountain belts is characterized by parallel folds and faults. The prominence of these mountain belts and their striking similarity to orogenic belts on Earth was one of the factors that led earlier workers to suspect that Venus might be characterized by plate tectonics. Subsequent deformational stages are represented by distributed wrinkle-ridge compressional deformation of the near-global volcanic ridged plains. The age of deformation of the ridged plains is interpreted by some to be immediately following their emplacement because only a tiny percentage of superposed craters are clearly cut by the compressional structures.

Another class of structure that is observed on Venus, coronae (see Fig. 8), are ring-like features characterized by a combination of tectonic and volcanic features and are typical of many areas of the plains. Coronae are elevated rings of concentric systems of grooves and ridges with diameters typically of 150–200 km, but several are more than a thousand kilometers across. Many hundreds of these features have been observed across the surface of Venus. One of the largest coronae is Heng-O, 1060 km in diameter (2°N, 355°E). A giant ring-like structure 2600 km in diameter, Artemis Chasma (35°S, 135°E), is considered a corona by some scientists. Typical coronae consist of several components, which may correspond to different episodes in coronae evolution. Most coronae contain a densely fractured terrain with radial and concentric patterns, forming rings or ring segments. A commonly younger structural component of coronae is a system of concentric ridges composing parts of the concentric annulus. These ridges deform the plains that merge gradually with the surrounding regional plains containing wrinkle ridges. The wrinkle ridges of the adjacent regional plains are structurally adjusted to the corona annulus. In some places, the latest structural component of coronae are concentric fractures that cut the regional plains and the ridges of the corona annulus. Patches of smooth and lobate volcanic plains undeformed by wrinkle ridges often occur within the corona interior and are seen radiating away from the corona margins in impressive arrays of petal-like lava flows.

This description demonstrates that formation of coronae was not a one-stage process, but consisted of several episodes. The life cycle of a typical corona may have started with the concentric-radial deformation, followed by emplacement of volcanic plains-forming material, which was eventually deformed by formation of concentric ridges. This latter episode usually occurred before or in conjunction with the emplacement of the regional plains. Next, plains-forming material was emplaced, producing patches of smooth and lobate plains. The latest deformation of the corona life cycle was emplacement of concentric fractures, which both predate and postdate the smooth and lobate plains. Many researchers interpret this sequence to be consistent with the formation of coronae by the impingement of rising plumes on the base of the lithosphere of Venus to create hot spots. In this scenario, the rising hot plume deforms the crust as it pushes up from below and spreads laterally beneath the surface. Pressure–release melting of the rising hot mantle material produces volcanism, resulting in the radiating flows. The latest stages of coronae evolution are believed to consist of concentric fractures and topographic subsidence, as well as burial by regional volcanic plains and dissection by regional tectonic events, such as rift zone formation. In an alternative interpretation, some scientists think coronae represent the impingement of diapirs, or ruptured, viscous material, from below the crust in a single event.

In summary, rates, styles, and areal distribution of deformation appear to vary over time; initial regional intense tesserae shortening interpreted by several workers gave way to more widespread apparently distributed extension, followed by widespread distributed modest compression, and most recently by rift systems of modest extension largely concentrated in areas of upwelling in the BAT region.

3. Impact Cratering

The global *Magellan* coverage revealed over a thousand impact craters ranging in diameter from less than 2–280 km, with variations in morphology depending on their sizes. Craters larger than 20–30 km in diameter show a sequence from craters with flat floors and central peaks (see Fig. 12a) to morphologies transitional to ringed basins, a sequence that is typical of complex craters on other planetary surfaces. Smaller craters are not typically bowl-shaped, as observed on other planets, but instead are typically of an irregular outline with irregular floors, or form groups or clusters of closely spaced craters. In addition, the density of smaller craters declines rapidly with decreasing diameter. Over 350 diffuse splotches with no central craters are observed on Venus. These latter trends are attributable to the effects of filtering of the projectiles (burnup, breakup, etc.) in the dense atmosphere of Venus and the formation of near-surface bursts (shock-induced splotches) and crater clusters. The atmospheric filtering effect on Venus is of tremendous significance. It has been estimated that about 98% of the craters between 2 and 35 km diameter that might have been formed on an airless Venus did not form because of the presence of its dense atmosphere!

In marked contrast to the characteristics of craters on other planets, most impact craters on Venus appear very pristine with few signs of age-dependent changes, largely because of the lack of abundant small impacts and the lack of a hydrologic cycle. Only about 38% of the population of craters appear even slightly modified, with about 4% embayed by volcanic deposits and 34% modified by tectonism. The only exception to the similar morphologies is a group of craters with associated dark parabolic features that are interpreted to represent the youngest part of the crater population. These dark parabaloids are interpreted to form by the lofting of ejecta to high altitudes, its dispersal, and its ultimate fallout down range. They compose about 10% of the total crater population, and by reason that this small percentage corresponds to a young age, their absolute age has been estimated to be not more than 50 Myr; thus, these craters and associated parabolic features may be used as stratigraphic markers for the most recent part of the geologic history of Venus. If, however, the age of other units were to be modified by new model calculations, the absolute age would scale accordingly. This does not change the fact that these features are relatively young. In fact, through the analysis of the relations of these parabolas with the surrounding geology, it has been possible to show that the youngest episodes of tectonic and volcanic activity in the rift zones of Alta Regio are younger than about 50 Myr.

Venus provides a unique laboratory to study the effects of gravity and the presence of an atmosphere on the impact cratering process and to test and validate principles obtained from laboratory experiments and observations on other planetary surfaces. Among the many important things learned is the significance of nonballistic ejecta emplacement due to the very large dynamic forces that act on the advancing ejecta curtain.

Strong response winds are induced by the outward expanding ejecta curtain, causing entrainment of ejecta and the production of ground-hugging debris flows, which spread out into the surrounding terrain to produce spectacular radial flow deposits (see Fig. 12b). [*See* PLANETARY IMPACTS.]

4. Eolian Processes

Eolian features, created by the interaction of atmospheric winds and surface materials, are mostly represented by surficial streaks and surficial patches. The streaks, over 3400 of which are seen, are typically tens of kilometers long and a few kilometers wide. Streaks often trail topographic obstacles, which implies that they were formed in response to wind turbulence around the obstacles, and sorting of particles in the low-velocity winds of Venus. Streaks tend to be oriented toward the equator, consistent with a Hadley model of atmospheric circulation. The surficial patches are radar dark areas of various form and size in local topographic lows and against or behind positive topographic obstacles. These deposits are typically associated with impact craters, suggesting that they are the results of redeposition or modification of debris produced by the impact. Dune fields were observed in two areas of Venus and in one place linear features (which may be yardangs, wind-sculpted linear ridges) were observed. Most Aeolian features are concentrated in the near-equatorial smooth plains, but they are widespread and have been observed at all latitudes and elevations. In the observed population of impact craters on Venus, the crater size distribution agrees well with theoretical models, which take into account the characteristics of the current atmosphere of Venus. This implies that for the recent past, the atmosphere of Venus was quite similar to that of the present. Therefore, the Aeolian resurfacing of Venus should have been approximately the same during all this time.

III. AGE OF THE SURFACE AND GEOLOGICAL HISTORY

A. AGE OF THE SURFACE AND IMPLICATIONS

Most planetary surfaces consist of a small number of major terrain types (e.g., continents and ocean basins on Earth, highlands and maria on the Moon) that have distinctly different ages and modes of formation. Initial considerations and observations of Venus led to the thought that there should be a wide variety of ages on the surface and that the highly deformed tesserae terrain might be some of the most ancient terrain. Thus, distinguishing the ages of different units on the basis of different size–frequency distributions was an effort that eagerly awaited the acquisition of high-resolution, global image data. One of the major surprises of *Magellan* was that early analyses indicated that the global distribution of impact craters could not be distinguished from a completely spatially random population. That a spatial distribution of impact craters on Venus' surface is indistinguishable from a random one shows that the surface has no large areas that are significantly different in absolute age. This meant that there was not a distinct bimodality or sequence of ages on different parts of the surface as observed on the Earth or on the smaller terrestrial planets. Although ongoing analyses may be able to ultimately demonstrate that the population is not completely spatially random, surface units do not appear to differ significantly in their age in ways similar to the lunar highlands and maria, for example.

In addition, the crater size–frequency distribution has been interpreted to mean that the global crater retention age is surprisingly young, about 300–750 Myr, more like the Earth than the smaller terrestrial planets. As the global picture emerged, it also became clear that the vast majority of superposed impact craters appeared pristine. Only a small number of impact craters had been modified by subsequent activity, with those interpreted to be cut by tectonism localized predominantly in areas of focused rifting, and those flooded by postformational volcanic activity localized predominantly in areas of volcanic rises. The unusual characteristics of the crater population suggested the radical idea that the extensive regional volcanic plains were emplaced geologically very rapidly and that subsequent volcanic and tectonic activity were at much lower rates and more localized. This surprising result initiated an intense debate that continues today about mechanisms of heat loss on the planet, the geological and internal evolution, and whether crater densities could be used on any of the geologic record to distinguish ages and to test observed stratigraphic relationships.

The density of craters on tesserae terrain was found to be about 50% higher than the average crater density on all Venus; if this estimate is statistically reliable it means that the surface age of tesserae is 50% more than the total surface age. Because tesserae cover such a small percentage of the surface (~8%) and because the number of craters is small, the statistics are poor and the possible error is large. Thus, the surface age of tesserae terrain is certainly greater than the average age of the total surface, but it may be greater by only a few million years or greater by a few hundred million years. The estimated surface age of tesserae terrain shows when this terrain was deformed and previous craters were destroyed. The age of the rocks forming the tesserae fabric is, of course, greater than the surface age, but how much greater is unknown.

Estimation of the surface age of the stratigraphically young lobate plains and large volcanoes is very important for understanding the geologic history of Venus. It was found that the age is about half of the average surface age of all Venus (approximately 150–250 million years). As is the case with tessera, the possible error bars are very large. The surface age of these volcanoes is actually the age of the lava emplacement and solidification, so in this case the surface age of the terrain and age of the material composing this terrain should be the same. Some lobate plains and large volcanoes (see Fig. 4b) seem to be very young. Some even postdate a specific variety of impact craters that have radar-dark parabolas; this ejecta disappears with time, just as the bright rays of Copernican-aged craters do on the Moon. The embayment and cutting of such young craters by a few deposits of lobate plains and young volcanoes show that geological activity has continued up to the geological present. Yet the average age of these lobate plains and large volcano deposits is significantly less than the average age of the whole population, which means that some other members of the population should be older than the average age. This, in turn, means that the duration of the period when lobate plains and large volcanoes formed was long.

If we consider that the regional plains deformed by wrinkle ridges occupy the majority of the surface (60–70%), while about half of the rest is older and another half is younger, then the average age of these regional plains has to be close to the average age of the surface of Venus (~300–750 million years). Could these regional plains have been emplaced over a short period of time, or were they deposited over an extended period of time at different rates in different places? In order to determine the answer to this question, and to investigate the sequence of units and any changes in styles of deformation in time and space, a global stratigraphic reconstruction of Venus is necessary.

B. STRATIGRAPHY AND GEOLOGIC HISTORY

Geological mapping of Venus has been underway for several years in the form of mapping of quadrangles at a scale of 1 : 5,000,000. In additon, other workers are examining the geological sequence of events in individual areas of scientific interest, and still others are compiling regional and near-global maps and stratigraphic sequences. In one such analysis, 36 different 1000×1000-km areas distributed randomly across the planet, and several larger regions, were studied and a series of broad stratigraphic units were identified (see Fig. 13a). Although there are differences of opinion about the description and sequence of units and there will certainly be an evolution of geological units through further study, this analysis, which appears applicable over the 40% of the surface examined so far, provides some insight into the emerging sequence of events and geologic and tectonic history.

The recognizable signatures of the main periods of the geologic history of Venus are represented by four major morphologic units: the areally dominant volcanic plains, large volcanoes, rift zones, and tesserae. Detailed stratigraphic analysis and cross-cutting relationships suggest to many workers that tesserae are older than plains whereas rift zones and large volcanoes are generally younger. The plains themselves can be subdivided into several morphologic units, which, as a rule, have quite clear embayment relations that allow workers to deduce their age sequence. The geologic history of Venus can be described as a sequence of superposed geologic units (see Fig. 13a) and then the time periods in which they may have formed can be assessed (see Fig. 13b).

1. The material composing tesserae terrain (Tt) is defined as areas whose morphology is determined by multiple intersections of ridges and grooves of deformational origin. Tesserae occupy about 8% of the surface, forming "islands" and "continents" rising above and being embayed by volcanic plains. The nature of tesserae rock material is unknown. It may just be highly deformed basaltic lavas generally similar to the lavas of the surrounding younger volcanic plains or it may be something geochemically more evolved, perhaps resembling granites of the Earth's continents. Several workers have described a sequence of deformational structures in tesserae [earlier compressional deformation (ridges) and later extensional structures (fractures/graben)], but the complexity of the terrain makes mapping results controversial. Different areas of tesserae may be different in the character of their rock material and in their mechanisms and exact time of formation. Most workers find that tesserae predates the formation of plains.

2. Material of densely fractured plains (Pdf) is characterized by systems of parallel, very densely spaced

faults that show deformation in an extensional and probably shear environment. If one ignores the superposed faults the terrain is regionally flat and is interpreted to be composed of lavas similar to younger plains. Remnants of densely fractured plains form islands embayed by and rising above the younger plains. Their total area is about 3% of the planet's surface.

3. Material of fractured and ridged plains (Pfr) is seen as broad (10–15 km wide) ridges, probably compressional folds, which are clustered into ridge belts up to thousands of kilometers in length. These ridges stand over younger adjacent plains, which are of basaltic composition, as determined by several Soviet landers. The total area of Pfr is also about 3% of the surface of Venus.

4. Material of shield plains (Psh, see Fig. 13b) consists of clusters of coalescing, gently sloping volcanic shields of 3–5 to 15–20 km in diameter. They typically overlap the fractured and ridged plains (Pfr) and are embayed by younger plains. The material of the shield plains is probably mostly basaltic lava. However, the only spacecraft landing within these plains (*Venera 8*) found high contents of potassium, uranium, and thorium, which are not typical of normal basalts. Shield plains occupy 5–10% of the planet's surface.

5. Material of plains with wrinkle ridges (Pwr) were shown by *Venera 9* and *10* and *Vega 1* and *2*, to be composed of basalts. The superposed wrinkle ridges are interpreted as narrow (1–2 km wide) compressional folds. The majority of these plains look homogeneously dark on *Magellan* images, but in some places relatively young lava flows are observed with higher radar reflectivity (relatively bright on images). Networks of wrinkle ridges deform both of these varieties of the plains. In many places Pwr plains are cut by very long channels (hundreds or even thousands of kilometers long) of 2–3 to 5–10 km width, probably formed by the flow of lava (see Fig. 10). These features are evidence of the extremely large lateral scale of volcanic floods forming these types of plains. The total area occupied by Pwr plains is 60–70% of the planet's surface.

6. Material of lobate plains (Pl) typically consists of fans of superposed lava flows emanating from a specific source (see Fig. 4b). The source is usually associated with rifts or with coronae that are circular or ovoidal structures typically a few hundred kilometers across sitting in many places among the plains. Large volcanoes are an extreme case of a lobate plains locality where the accumulation of lava flows produced prominent topography. The total surface area occupied by the lobate plains is 10–20% of the planet's surface.

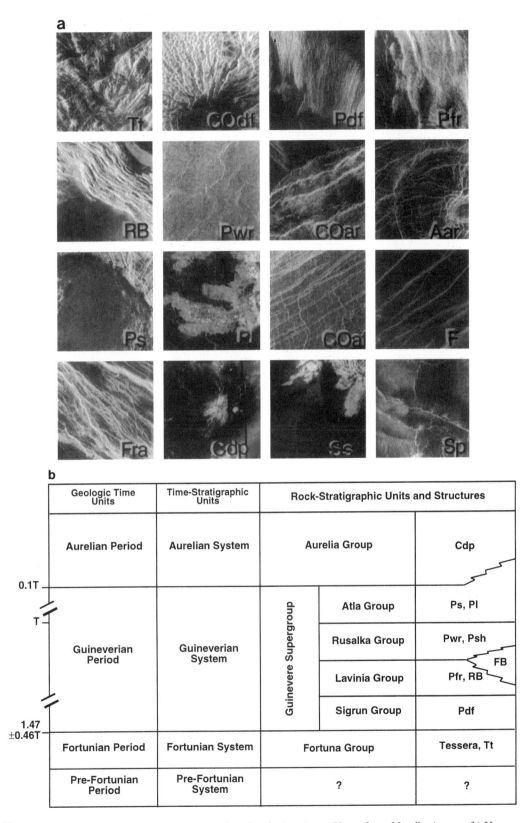

FIGURE 13 Stratigraphy and geologic history. (a) Examples of geologic units on Venus from *Magellan* images. (b) Venus stratigraphic units and global correlations (from Basilevsky and Head, 1995). T, the average age of the surface of Venus, is estimated to be most likely in the range of 300–750 million years.

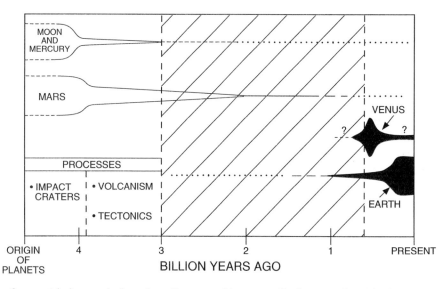

FIGURE 14 Ages of terrestrial planetary body surfaces. Represented in a generalized manner for each planetary body is the portion of presently exposed surface units that formed at a certain time. For example, for Earth, two-thirds of the presently exposed surface (the ocean floor) has formed in the last 200 million years, whereas only a very small part of the surface area dates from the middle of solar system history, and virtually none from the first third. In contrast, for the smaller bodies (the Moon, Mercury, and Mars), the vast majority of their surfaces date from the first third of solar system history. Venus' surface, similar to the Earth in relation to its average age, is different in terms of the major processes of geological evolution (no plate tectonics at present) and preserves virtually none of the first two-thirds of its history in presently exposed surface geological units. (Diagram by the authors.)

Together these units represent an interplay of volcanic lava extrusion and tectonic deformation. Both of these processes are related to activities in the interior of Venus. These materials also show on their surface the signatures of surficial geologic activities, such as wind-blown erosion and deposition, chemical weathering, down-slope mass wasting and impact cratering. The fact that these surficial processes and deposits influence only a small percentage of the surface of Venus and that volcanic and tectonic landforms hundreds of millions of years old look morphologically pristine means that during the observed part of the geologic history of Venus the significance of the internally driven volcanic and tectonic activities was much larger than the intensity of surficial activities. As a result, images of the surface of Venus may be used as textbook examples of many volcanic and tectonic landforms.

Using these observations, together with the size–frequency and density distribution of impact craters, we can attempt to reconstruct the geologic history of the planet (see Fig. 13b). In contrast to the Moon, Mars, and Mercury, no regional geologic units dating from the first 75–85% of the history of Venus remain at the surface (see Fig. 14), although some of the rocks in the tesserae are likely to date from earlier time periods. The geologic history of Venus recognizable on the *Magellan* images may actually span only the latest 5–10% of the planet's evolution. From compari-

sons with other bodies of the solar system, we know that about 4 billion years ago Venus underwent a period of heavy meteoritic bombardment, but no specific signature of that phase is now observed on the surface. From a general understanding of planetary formation and evolution we may infer that early in its history the interior of the planet was hotter than now, most likely resulting in more intensive mantle convection and surface volcanism. There is evidence that a more significant abundance of water might have occurred in the past, but because no related landforms are seen in *Magellan* images, this supposedly wetter period probably happened much earlier.

On the basis of regional and global stratigraphic analyses of the observed surface of Venus, we can outline the record of at most only the last 15–25% of the total history of Venus (see Fig. 14). The range in estimates of the mean crater retention age of the surface (from ~200 to over 1000 million years) means that it is best to describe the timing and duration of different events in terms of fractions of the mean surface age, T (see Fig. 13b). The beginning of the observed history of Venus was characterized by intensive tectonic deformation of global or semiglobal scale which formed the tesserae terrain. Termination of the compressional stage is estimated to have occurred at about 1.4T ago, whereas the extensional stage lasted for another 0.1–0.2T.

After tesserae formation, several stages of extensive volcanism occurred, burying vast areas of tesserae and forming what we see now as regional plains. The combined duration of the emplacement of these plains is estimated to be about 0.2–0.3T, with an implied average global rate of volcanism of a few km³/year. The solidified lavas of these initial plains were deformed in an extensional and/or shear environment, thus forming what we see now as remnants of densely fractured plains. Vast basaltic lava emplacement continued and these lavas were also deformed, but now the extensional/shear environment evolved into a compressional one. As a result, the new-formed lava deposits, as well as some of the earlier-formed materials, were deformed by ridges, forming what we see now as outcrops of fractured and ridged plains and ridge belts. Following this phase, a different style of emplacement of volcanics occurred in many places on Venus. Instead of vast volcanic plains, myriads of volcanic vents started to erupt lavas with a relatively low eruption rate to produce tens of thousands of small shield volcanoes forming what we see now as the shield plains. The reason for that change in the style of volcanism is unknown.

Following this, the style of volcanism changed again and vast volcanic floods covered about 90% of Venus' surface. These lavas were then almost totally deformed in the compressional environment, forming plains with wrinkle ridges, the most abundant plains unit on the planet. This dominance of widespread but small-scale compressional deformation may be explained qualitatively by cooling of the hot upwelling mantle material that formed the episodes of extensive volcanism. However, thorough theoretical modeling is necessary to understand if this mechanism is reasonable and what time period is necessary.

Regional plains-forming materials can be subdivided and are separated from each other, and from the underlying and overlying units, by unconformities formed, from oldest to youngest, by tesserae-forming deformation, dense fracturing, broad ridging, and finally wrinkle ridging. These tectonic episodes may be generally globally synchronous and represent successive episodes characterized by the dominance of compression, then extension, then again compression, and finally extension. Although evidence is accumulating for the near-synchronous nature of many of these phases, confirmation awaits complete global mapping.

The last global-scale tectonic episode, extensive wrinkle ridging, happened at about T (300–750 million years) ago, which was very close in time to the emplacement of the most areally abundant plains unit. This marked the transition to the present stage of the history of Venus, which is characterized by a predominance of regional rifting and related volcanism. The recent geologic activity of the planet is concentrated along linear zones with about 40,000 km total length occupying a few percent of Venus' surface. These are zones of extension and heavy fracturing of the crust through which basaltic magmas come from the interior to the surface, forming large volcanoes and areas of lobate plains. This stage appears to have lasted from about time T to the present, making it the longest time duration among the stratigraphic units considered, although the resulting tectonic and volcanic features and deposits cover only 10–20% of the surface of Venus.

These observations mean that the general intensity of tectonics and the flux of volcanism (a few tenths of a km³/year) in this latest period were much lower than that in earlier times. In summary, the morphologically observable part of the history of Venus was characterized by two key characteristics that stand in contrast to the comparable period of Earth history (approximately the Phanerozoic) when global geodynamic processes were dominated by plate tectonics:

1. Venus shows no signature of plate tectonics; instead, its global tectonic environment passed from an initial dominance of compression, through extension, then again compression, and finally extension, with the density of deformational structures and the strain rate declining with time.

2. The predominant component of volcanism on Earth during this time was extrusive volcanism at midoceanic ridges. In the beginning of that same period of time on Venus, plains-forming volcanism appears to have occurred at a rate comparable to volcanism at midocean ridges, but was emplaced in an entirely different style. For the last few hundred million years, Venus appears to have been dominated primarily by rift-associated volcanism emplaced at a production rate perhaps comparable to, or even lower than, present intraplate volcanism production rates on Earth.

IV. INTERIOR OF VENUS

A. INTRODUCTION

Present planetary interior structure can be viewed from two perspectives, that of internal compositional variations, usually configured in layers such as crust, mantle, and core, and that of variations in internal temperature and state, producing layers on Earth such as the litho-sphere, asthenosphere, and outer molten and inner solid core. Detailed assessment of the interior of a planet requires surface seismic networks and heat flow probes, which have not yet been emplaced on Venus. Thus, our present knowledge of the interior is based on its bulk density, information obtained from surface geochemical measurements, moment of inertia mea-surements, gravity measurements, inferences made from geologic structure, topography, and geologic his-tory, anologies with Earth, and assumptions about starting conditions.

The similar size and mean density of Earth and Venus, and their position in the same vicinity of the solar system, have been used to argue that their starting conditions were similar. Differences in self-compression (change in density as a funciton of depth) largely explain the fact that while Venus is 18% less massive than the Earth, its mean density is 5% less than that of the Earth. Even if Venus has the same internal structure, composition, and temperature with depth as that of the Earth, its mean density would still be almost 2% greater than that observed. How can this 3% difference from the Earth be explained? Some of the difference can be attributed to Venus' high sur-face temperature and its effect on interior convection temperatures and the corresponding influence on the density of convecting mantle material. Other differ-ences could be due to the fact that Venus is just not as much like the Earth as we think. For example, some of the difference could be due to differences in bulk composition or some could be due to basic differences in the internal thermal structure between Venus and Earth. Another factor could be the thickness of the crust. Basaltic crust on the Earth is typically 5–7 km thick on the two-thirds of the planet comprising the ocean floor. On Venus, if a basaltic crust is much thicker and exceeds the basalt–eclogite phase boundary at a depth where basalt increases in density, then a significant portion of the outer part of the planet could be denser than on the Earth. Through exploration of Venus, information has been sought to distinguish among these possibilities.

Another difference between Earth and Venus was revealed when a series of spacecraft measurements showed that Venus has essentially no intrinsic mag-netic field. If an intrinsic magnetic field exists at pres-ent, it must be no more than 10^{-5} that of Earth's. Although an immediately appealing idea is that the lack of a magnetic field may be due to Venus' very slow rotation rate, theory suggests that these factors are not related. Of course Venus may have had a mag-netic field in the past, but unfortunately the surface temperatures are currently above the point at which the record of magnetic fields would be preserved in rocks (the Curie point).

The gravity field of planets is an indicator of lateral variations in density related to surface topography, crustal structure and composition, and to any addi-tional mechanisms in the interior that form and main-tain density differences (e.g., broad convection, hot spots). On Earth, long-wavelength gravity anomalies are generally uncorrelated with topography and crustal density anomalies, being instead related to broad con-vective flow processes in the mantle. On Venus, the earliest *Pioneer–Venus* results (later confirmed in de-tail by *Magellan*) showed that Venus was markedly different than Earth; its long-wavelength gravity anomalies are extremely highly correlated with topog-raphy. Among the suggested reasons for this difference is that Venus does not have the equivalent of the Earth's asthenosphere (a partially molten layer at depth), which serves to partially decouple lithospheric plate movement from deeper mantle convection, or that the topography is supported by the finite strength of the interior. If the latter mechanism is not operating, then the implication is that the topography is geologi-cally very young and may be related to ongoing dy-namic thermal convection and heat loss mechanisms. Early detection and correlation of volcanism and rift-ing at Beta Regio, one of the most prominent positive gravity anomalies on Venus, provided impetus to the idea that dynamic convective processes (hot spots) were presently operating on Venus and were a major factor in heat loss.

B. CRUSTAL AND INTERIOR COMPOSITIONAL STRUCTURE

The Soviet *Venera* landers showed that the composi-tion of most plains units measured was consistent with tholeiitic basalts, similar to rocks observed on the Earth's ocean floor, and consistent with the volcanic surface unit morphologies observed in *Magellan* data (see Fig. 13). *Venera 8* measurements suggest that in some areas (possibly those associated with the distinc-tive steep-sided pancake domes; see Fig. 9) composi-tions may be more toward the granitic end of the spectrum (more silica rich). No measurements were made in the highlands and some workers have sug-gested that these areas, like the continents on Earth, may be more differentiated. Nonetheless, these data make it clear that the interior of Venus is differentiated and that a basaltic crust has been extracted from the mantle. The composition of the crust is thus thought to be largely basaltic, but how the composition might have changed with time is presently unknown, despite strong hints from stratigraphy that temporal variations have occurred.

Crustal thickness can be estimated by comparison of gravity and topography of different regions, and assumptions of average composition, and these have been used to estimate that the thickness of the crust averages 25–40 km and that there are local variations, particularly in tesserae regions, that probably exceed 50–60 km in thickness. How these thicknesses have changed with time, and indeed how they came about in the first place, are not well understood.

No direct information exists about the deep interior of Venus, but because of the high, Earth-like mean density, it is assumed by virtually all workers on the basis of comparisons to Earth that Venus has a differentiated core. Some investigators believe that the core may be frozen completely solid, whereas others suggest that core solidification has not yet commenced or is underway.

C. INTERNAL THERMAL STRUCTURE AND THERMAL EVOLUTION

In order to understand the present internal thermal structure and state without direct seismic and heat flow data, information must be gleaned from surface morphology and altimetry, gravity, topography, Soviet *Venera* lander surface measurements of composition and heat-producing elements, coupled with an understanding from laboratory measurements of the strength of rocks under Venus conditions, and geophysical modeling. Using *Magellan* data, estimates of the thickness of the effective elastic lithosphere have been obtained by comparing actual topographic profiles to flexural estimates assuming varying heat fluxes and by comparing gravity/topography relationships to flexural models as a function of wavelength. For coronae and other tectonic features the range of estimates of the thickness of the effective elastic lithosphere is 10–40 km and for gravity/topography relationships is 20–40 km. New data on very dry rocks suggest that the crust of Venus may be almost as strong as the mantle. The implications of these data for the thermal flux of Venus is that Venus is presently losing less heat than the Earth, perhaps because of the lack of the efficient terrestrial process of plate tectonics and subduction.

At present, the interior of the Earth is in a mantle convective regime dominated by the presence of a plate-tectonic-related mobile lithospheric lid, a lid that

is produced and moves laterally at high rates, flexes at plate margins, and is subducted back into the interior. The geologic record of Venus as revealed in *Magellan* data does not support the presence of plate tectonics at the present. Thus, for a variety of reasons, Venus is thought of as being in a mantle convective regime dominated by a sluggish or even a stagnant lithospheric lid. In addition, layering of the mantle on Venus is predicted to be even more likely than on Earth. Models of Venus mantle convection and heat loss suggest that Venus is not now in thermal equilibrium with Earth-like abundances of radioactivity, and thus there appears to be a present lack of balance between internal heat production and heat loss. Geophysicists are presently investigating whether Venus may undergo episodic, and different, phases and styles of heat loss and, if so, what the mechanism of heat loss might be or whether there might be some more fundamental differences between Earth and Venus in abundance of heat-producing elements.

V. SYNTHESIS

A. SUMMARY OF KEY SCIENTIFIC RESULTS FROM EXPLORATION OF VENUS

On the basis of the exploration of the surface and interior of Venus, most successfully using the global data set derived from the *Magellan* mission, we can cite the following key observations about the planet:

- Venus has a negligible intrinsic magnetic field, and its long-wavelength gravity anomalies are highly correlated with topography, unlike the Earth.
- *Magellan* high-resolution global images reveal that volcanism, tectonism, and impact cratering have been significant processes in the formation of surface structures and in its geologic history.
- The surface of Venus is covered mostly by volcanic materials. Volcanic surface features, such as vast lava plains, fields of small lava domes, and large shield volcanoes, are common.
- The typical signs of terrestrial plate tectonics (e.g., morphology and structure associated with seafloor spreading and continental drift) are not observed on Venus. Tectonics are dominated by a system of global rift zones and numerous broad, low domical structures called coronae, apparently produced by the upwelling of magma from the mantle.

- Despite Venus' dense atmosphere, evidence of substantial wind erosion is lacking, although limited wind transport of dust and sand is observed.

- The crust appears to be made predominantly of basalt. Some surface measurements and morphologic evidence suggest the presence of more silica-rich volcanic material. The crust appears to be forming from vertical crustal accretion, rather than the formation and lateral transport typical of the plate-tectonic system on Earth.

- The low density of impact craters and their size–frequency distribution is interpreted to mean that the average surface age for Venus is in the range of about 300–750 million years, less than about 10% of the total history of the planet. No evidence for surface units from the first 75–85% of the history of the planet is seen. Thus, the average age of the surface of Venus is very similar to that of the Earth (in contrast to that of the smaller terrestrial planetary bodies, the Moon, Mercury, and Mars). Yet the reason for this young age, and the processes involved, appears to be different for the two bodies.

- Although Venus has many features and characteristics similar to the Earth, it has fundamental differences in many others, including the morphology and distribution of its geologic features and their relationships and its current mechanism of heat loss.

B. MODELS FOR THE GEOLOGICAL AND GEOPHYSICAL EVOLUTION OF VENUS

What models for the evolution of Venus have been proposed on the basis of these observations and data? One of the keys to the understanding of Venus is the origin of the distinct tectonic fabric associated with tesserae and the factors responsible for the transition to geologic processes dominated by volcanism and more focused deformation (e.g., rift zones). Prior to *Magellan*, several general types of models for tesserae formation were proposed. In the first two models, tesserae terrain comes about through the normal evolution of crustal structure formed and modified by mantle convection patterns, but the models differ in the sign of the convection. Some proposed that proto-tesserae initially formed above large areas of mantle upwelling (hot spots or plumes) as regions of enhanced volcanism and crustal thickening. Subsequent to their formation, thermal decay and gravitational collapse and spreading

converted this distinctive, presumably Atla-like volcanic terrain into the highly deformed tesserae terrain. Others proposed that tesserae terrain forms over zones of long-term mantle downwelling, with the characteristic fabric resulting from coupling of mantle flow patterns and crustal deformation. Because *Magellan* observations indicate that the most likely regions of present downwelling (the lowlands such as Lavinia and Atalanta Planitiae) are not regions of tesserae, some workers proposed that these lowlands represent an initial evolutionary step in a process that eventually led to crustal thickening and formation of the high-standing, highly deformed tesserae terrain.

A third type of hypothesis for the evolution of Venus and the origin of tesserae terrain seeks to explain the *Magellan* observations in the context of uniformitarian thermal evolution and calls on the high surface temperature and the exponential relationship between temperature and rate of strain as a major factor in the evolution of the observed surface features. In this scenario, mantle convection is closely linked to the overlying lithosphere, and over most of the history of Venus a weak lower crust deforms readily, resulting in very high levels of surface strain and production of the observed tesserae deformation over most of Venus. At some point late in the thermal evolution, the heat flux declines to values that markedly decrease lower crust ductility, and rates of surface strain and deformation decrease rapidly. This model predicts that the change in strain rate should be sensitive to regional variations in thermal gradients and crustal thickness, with highland regions persisting in deformation long after the deformation rates in lowland regions have decreased to modest levels.

Although the deformation predicted in this hypothesis would be widespread, consistent with the predicted preplains distribution of tesserae, evidence for continuing deformation in the high-standing tesserae beyond the deformation in other tesserae regions, predicted by this model, has not yet been documented. In addition, areas of thermally enhanced topography (upwellings) are predicted to be the best candidates for continuing deformation and little evidence for associations of recent tesserae formation with areas thought to be sites of present upwelling is observed. Quantitative estimates of tectonic strain rates using the observed geologic record do not support the hypothesis that significant decreases in surface strain rates could be caused by a steady decline in heat flow over the last billion years. Thus, a steady temperature decline appears to be insufficient to cause the observed change in surface strain rate.

Another class of uniformitarian thermal evolution

model predicts evolutionary changes in mantle convection patterns. These models emphasize early very high rates of deformation related to a highly convective interior that has oscillatory properties, resulting in a highly deformable lithosphere capable of being incorporated into the convecting mantle. After a prolonged period of surface recycling into the mantle, the enhanced cooling results in diminished convective vigor to quasi-steady state, crustal and lithospheric stability, and a one-plate hot-spot-dominated planet about 500 Myr ago. These models do not make very specific predictions about the nature and evolution of landforms on Venus. Tessera terrain could be the remnants of this earlier phase of prolonged recycling. The vast outpouring of volcanic plains just after the end of tesserae formation is, however, not predicted, by a model of cooling and stabilization. No evidence on hand can definitively rule out these types of models, although some analyses favor episodic heat pulses, over monotonic thermal evolution, to explain the interpreted variations in strain rates in the observed geologic record.

The fourth category of hypothesis involves one or more periods of catastrophic resurfacing in the history of Venus. The analysis of the nature and distribution of impact craters using the global *Magellan* data base has shown that the crater population represents an average crater retention age of about 300–750 million years. These data, together with the very small number of craters that appear to be modified by volcanic activity, have led some workers to propose that the surface of Venus underwent a global volcanic and/or tectonic resurfacing event about 300–750 million years ago, completely eradicating the crater population, and that the present crater population is a production population with only a few craters modified by greatly reduced rates of volcanic activity. This is in contrast to an equilibrium resurfacing model, in which resurfacing takes place in a more stepwise manner. Several mechanisms have been proposed to account for the global resurfacing, including episodic plate tectonics and vertical crustal accretion and periodic overturn of the depleted mantle residuum.

On the basis of the characteristics of the geologic history outlined earlier, several of the proposed models seem plausible. For example, a mechanism of tesserae formation linked to global and perhaps catastrophic resurfacing at some point in the history of Venus is consistent with many of the observations. The near-simultaneous, widespread formation of the tesserae proposed by some workers, followed closely by large-scale volcanic flooding and embayment of the tesserae, is consistent with a discrete planet-wide event according to some interpretations. The distinctive differ-ence of styles of compensation between tesserae and volcanic rises also favors an early discrete mode of formation for tesserae terrain. In addition, the lack of a preserved geological record for the first 75–85% of the history of Venus, combined with the distinct change in intensity of deformation between the tesserae and post-tesserae terrains and the apparent low level of geologic activity for the last several hundred million years, could all be explained by a specific catastrophic event happening in the recent geologic history of Venus. This range of hypotheses is thus presently being explored in more detail by many scientists.

In the episodic plate tectonics model, periods of stable conductively thickening lithosphere result in small increases in mean interior temperatures, which lead to periods of enhanced mantle convection; these in turn initiate instabilities that lead to lithospheric foundering, periods of rapid lithospheric recycling and heat loss, and rapid resurfacing rates, prior to a return to stabilization, and another cycle. Although few specific predictions relative to the characteristics of terrain on Venus have been made by the proponents of this model, the tesserae could be remnants of the last episode of plate tectonics, perhaps representing crustal accretion and thickening adjacent to zones of subduction. There is little evidence for widespread regions representing remnant divergent plate boundaries and extensive intraplate fabric, or any asymmetries in crater ages that would be manifested in laterally spreading crust. However, insufficient data on the nature of crustal spreading in this environment is available to fully assess this hypothesis.

A variation on this hypothesis (although not episodic) has been proposed in which plate tectonics dominated the surface of Venus prior to about 800 Myr ago and the highly mobile crust and lithosphere created tesserae terrain. General planetary cooling caused thickening of the lithosphere and cessation of plate tectonics, followed quickly by global volcanic flooding (related to heat at the base of the lithosphere from a still rapidly convecting mantle), and then by hot spot volcanism as the planet finalized its evolution from a multiplate to a one-plate planet. This model is consistent with the general geologic history of Venus and the major characteristics of the tesserae terrain; further investigation is required in terms of geological mapping and the development of specific predictions related to the cratering and geologic record implied by the terminal phase of plate tectonics. Several other models suggest possible catastrophic global resurfacing/recycling episodes related to changes in mantle convection in the history of Earth and Venus due to phase changes, compositional stratification, and thermal evolution. However, none of these models make predictions sufficiently specific to compare to the Venus record.

Several investigators have proposed a model in which vertical crustal accretion on a one-plate planet results in a thickening basaltic crust and residual depleted mantle layer. Over time, positive compositional buoyancy decreases in significance and negative thermal buoyancy increases, resulting in net negative buoyancy for the depleted mantle layer. At this point, the depleted mantle layer founders, deforming and delaminating depleted mantle and overlying crustal material, and hot fertile mantle material ascends, undergoes pressure–release melting, and produces a phase of widespread surface volcanism. Following this overturn event, vertical crustal accretion continues at much reduced rates and the processes repeat at intervals of 300–750 Myr.

We believe that several aspects of the geological history of Venus are consistent with this model: (1) lack of preserved surface units from the first 75–85% of the history of Venus; (2) evidence for a vertically accreting crust over most of observed geologic time; (3) formation of tesserae terrain as the first major unit of the present stratigraphic column; (4) the sequence of deformation in tesserae terrain proposed by some workers (initial large-scale shortening to partly contemporaneous and subsequent extension related to relaxation); (5) a major change in the style and intensity of deformation (strain rate) following tesserae formation; (6) emplacement of widespread regional plains over the vast majority of the surface of Venus closely following the period of tesserae formation; and (7) a substantial decrease in volcanic flux and a change from large-scale regional plains emplacement to focused local sources (e.g., individual large volcanic edifices). Thus, on the basis of available model predictions and geological information, the vertical crustal accretion and depleted mantle layer overturn model appears to be compatible with a number of observations.

At present, this range of models is being tested against the results of global geologic mapping, and improved geophysical models of the interior and its thermal evolution. Nonetheless, we now have a much clearer picture of the nature of the surface of Venus in the last few hundred million years of its geologic evolution and its relation to Earth. Venus is a terrestrial planet very similar to Earth in size and mass, and thus comparisons offer a very promising way to understand some of the basic principles and trends in the evolution of terrestrial planetary bodies. During the observed part of Venus' history, Earth has been dominated by plate tectonic processes.

The emerging picture of Venus reveals a very different situation. The majority of what we observe appears to form by vertical processes, not lateral movement as on Earth. Vertical crustal accretion dominates and coronae and rift zones record vertical convective movement and limited lateral extension; mountain belts in Ishtar Terra may record local shortening. The general stratigraphic relations observed on Venus by many workers suggest the possibility of consecutive episodes of intense deformation, emplacement of extensive basaltic flood over a short period of time, and relative quiesence (see Fig. 13), in contrast to the approximately steady-state creation and destruction of lithospheric plates on Earth. No evidence of plate tectonic-style recycling is seen at present on Venus, yet there are important clues that Venus has changed its style of heat loss from the past to what we see today. In fact, some models predict that heat loss might have been not only catastrophic, but episodic. Observations of earlier Earth history show that there are major phases and events that have not been repeated in later geologic history (e.g., a major phase of continent building, a period of emplacement of iron-rich volcanic rocks known as komatiites, and a period of emplacement of feldspar-rich rocks known as anorthosites). Although plate tectonics is elegant in its global implications and simplicity, could the emerging picture of the surface and interior of Venus be telling us that an earlier part of Earth history may have been characterized by catastrophism?

On the basis of the successful exploration of Venus since the mid-1960s, future exploration of Venus (e.g., surface stations to make long-term seismic and heat flow measurements, balloons and landers to assess the atmosphere and global surface composition, and eventual sample return), combined with an emerging picture of the first two-thirds of the history of the Earth and other terrestrial planetary bodies, will surely result in great strides in understanding Venus and Earth in the coming decades.

ALSO SEE THE FOLLOWING ARTICLE

VENUS: ATMOSPHERE

BIBLIOGRAPHY

Barsukov, V. L., Basilevsky, A. T., Volkov, V. P., and Zharkov, V. N. (eds.) (1992). "Venus Geology, Geochemistry, and Geophysics: Research Results from the USSR." Univ. of Arizona Press, Tucson.

Basilevsky, A. T., and Head, J. W. (1995). Regional and global stratigraphy of Venus: A preliminary assessment and implications for the geologic history of Venus. *Planet. Space Sci.* **43**, 1523–1553.

Bougher, S. W., Hunten, D. M., and Phillips, R. J. (eds.) (1997). "Venus II: Geology, Geophysics, Atmosphere, and Solar Wind Environment." Univ. of Arizona Press, Tucson.

BVSP (Basaltic Volcanism Study Project) (1981). Basaltic volcanism on the terrestrial planets. Pergamon Press, New York.

Ford, J. P., Plaut, J. J., Weitz, C. M., Farr, T. G., Senske, D. A., Stofan, E. R., Michaels, G., and Parker, T. J. (1993). "Guide to Magellan Image Interpretation." Jet Propulsion Laboratory Publication 93–24.

Head, J. W., and Basilevsky, A. T. (1998). Sequence of tectonic deformation in the history of Venus: Evidence from global stratigraphic relationships. *Geology* **26**(1), 35–38.

Hunten, D. M., Colin, L., Donahue, T. M., and Moroz, V. I. (1983). "Venus." Univ. of Arizona Press, Tucson.

Magellan at Venus (1992). *J. Geophys. Res.* **97** (E8).

Magellan at Venus (1992). *J. Geophys. Res.* **97** (E10).

Roth, L. E., and Wall, S. D. (1995). "The Face of Venus." The Magellan Radar-Mapping Mission, NASA Special Publication 520.

Saunders, R. S., *et al.* (1992). Magellan mission summary. *J. Geophys. Res.* **97**(E8), 13067.

Solomon, S. C. (1993). The geophysics of Venus. *Physics Today* **46**, 49–55.

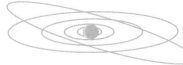

EARTH AS A PLANET: ATMOSPHERE AND OCEANS

I. Interplanetary Spacecraft Evidence for Life
II. Orbital Characteristics
III. Overview of Surface Characteristics
IV. Overview of Earth's Interior
V. Atmosphere and Oceans
VI. Magnetosphere
VII. Conclusions

Timothy E. Dowling
University of Louisville

GLOSSARY

Breccia: Rock composed of the sharp-edged fragments of previously formed rocks.

Chlorofluorocarbons (CFCs): Various compounds made with the halogens chlorine and fluorine. Their stability made them favored refrigerants until it was discovered that this also makes them efficient atmospheric ozone destroyers.

Coriolis acceleration: Component of the acceleration on a rotating planet that acts perpendicular to the motion and balances the horizontal pressure gradient in an atmosphere or ocean. It causes circulation around high- and low-pressure centers, and is strongest in the polar regions and weakest in the tropics.

ENSO: El Niño, "the child," and the Southern Oscillation. El Niño is the episodic appearance of warm water off the coast of South America, often at Christmas time, that devastates Peruvian fishing (usually one-fifth of the world's catch), causes drought conditions in Australia, and weakens the monsoon in India. The Southern

Oscillation is the historical name for the global (not just southern) atmosphere–ocean oscillation for which El Niño years are the extreme.

Lithosphere: Cold outer shell of a solid planet, including the crust and the upper part of the mantle, that does not act like a fluid in response to stress, even on million-year timescales. *Lithos* is Greek for "stone."

Mantle convection: Motion of mantle rock that is warm enough for stress to cause solid-state creep, which results in convective fluid behavior over million-year timescales.

Maxwellian distribution: Bell-shaped distribution of particle velocities in a gas at a given temperature.

Moment of inertia: Measure of resistance to angular acceleration (changes in spin rate). It is analogous to mass, which is a measure of resistance to translational acceleration. Units are mass times squared length.

Plate tectonics: Global dynamics of a lithosphere that is broken into plates, resulting in continental drift and plate boundaries marked by mountain building, volcanic activity, and earthquakes. Earth is the only body in the solar system that exhibits plate tectonics.

Specular reflection: Bright, mirror-like reflection.

Stratosphere: Region in an atmosphere overlying the troposphere that is strongly stabilized against convection by heating because of the absorption of ultraviolet radiation from the Sun. *Stratum* is Latin for "layer."

Troposphere: Bottom of an atmosphere where temperature falls off with height at close to the neutrally stable (adiabatic) lapse rate. Earth's troposphere contains 80% of the mass of its atmosphere and most of the water vapor, and consequently most of the weather. *Tropos* is Greek for "turning."

Volatiles: First chemical species to evaporate when temperature is raised.

Western boundary current: Strong ocean current that runs along the western edge of an ocean basin as a result of the much slower eastward group velocity of Rossby waves (planetary waves) relative to the westward group velocity. The Gulf Stream is a well-known example.

Earth orbits the Sun in the narrow distance range within which water occurs in all three of its phases (as solid ice caps, liquid oceans, and atmospheric water vapor), and this results in several unique characteristics. Large liquid-water oceans cover the majority of the planet's surface, such that the total amount of dry land is about equal to the surface area of Mars. The planet's lithosphere is broken into eight major plates and several minor ones that move relative to each other at a typical speed of a few centimeters per year; this phenomenon, called plate tectonics, does not occur anywhere else in the solar system. The plate boundaries are associated with regions of active mountain building, earthquakes, and volcanic activity. Like Mercury, Venus, and Mars, Earth has no ring system, but unlike its neighbors, it has a large satellite, the Moon, which is an airless body that preserves its impact-cratering record. Only about 140 impact craters have been identified on Earth's surface to date. Rapid erosion and the recycling of the planet's crust by plate tectonics combine to quickly remove the evidence of impact craters. The presence of intelligent life on Earth can be discerned from stable radio-wavelength

signals emanating from the planet that do not match naturally occurring signals, but do contain regular pulsed modulations that are the signature of information exchange. Global biological activity is indicated by the presence of atmospheric gases such as oxygen and methane that are in extreme thermodynamical disequilibrium, and by the widespread presence of a red-absorbing pigment (chlorophyll) that does not match the spectral signatures of any known rocks or minerals. The planet's atmosphere follows the general pattern of a troposphere at the bottom, a stratosphere in the middle, and a thermosphere at the top. There is the usual east–west organization of winds, but with north–south and temporal fluctuations that are larger than found in any other atmosphere. Earth has a strong magnetic field that is generated by fluid motions in its liquid iron–nickel core and has an active magnetosphere.

I. INTERPLANETARY SPACECRAFT EVIDENCE FOR LIFE

An ambitious but ever present goal in astronomy is to detect or rule out life in other solar systems, and in planetary science to detect or rule out life in our own solar system, apart from Earth. There are some weak indications that life in our solar system might have once existed, or even now exists, outside of Earth. We know that Mars had running water on its surface early in its history, because we can see fluvial channels on its surface in high-resolution images. We have a strong case that 12 of the thousands of meteorites that have been found on Earth's surface are from Mars; these are collectively called the SNC meteorites (pronounced "snick," the letters stand for the subclasses shergottite, nakhlite, and chassignite). One of them, ALH84001, which differs from the other 11 because it is much older, contains polycyclic aromatic hydrocarbons that may or may not be biological in origin, but are certainly intriguing. We believe ALH84001 comes from Mars because its oxygen isotopes match those of another SNC meteorite, EET79001, but do not match Earth rocks or non-SNC meteorites, and also because it contains shock-induced glass with air bubbles that hold noble gases that precisely match Mars's atmosphere, as determined by the *Viking* Landers, but not Earth's atmosphere. Farther out in the solar system, we know that present-day Europa, a satellite of Jupiter, has a smooth icy surface with cracks and flow features that resemble Earth's polar ice fields and suggest a liquid-water interior. [*See* MARS: SURFACE AND INTERIOR; METEORITES; OUTER PLANET ICY SATELLITES.]

However, to date we have no direct evidence for extraterrestrial life. This includes data from landers on Venus, Mars, and the Moon, and flyby encounters with 8 planets, a handfull of asteroids, a comet (Halley in 1986), and over 60 moons. Are the interplanetary spacecraft we have sent out capable of fullfilling the goal of detecting life? This question has been tested by analyzing data from the *Galileo* spacecraft's two flyby encounters with Earth, which, along with a flyby encounter with Venus, were used by the spacecraft's navigation team to provide gravity assists to send *Galileo* to Jupiter. The idea was to compare ground-truth information to what we can learn solely from *Galileo*. [*See* ATMOSPHERES OF THE GIANT PLANETS; IO; OUTER PLANET ICY SATELLITES.]

Galileo's first Earth encounter occurred on December 8, 1990, with closest approach 960 km above the Caribbean Sea; its second Earth encounter occurred on December 8, 1992, with closest approach 302 km above the South Atlantic. A total of almost 6000 images were taken of Earth by *Galileo's* camera system. Figure 1 shows the Earth–Moon system as seen by *Galileo*. (See also color insert.) Notice that the Moon is significantly darker than Earth. The spacecraft's instruments were designed and optimized for Jupiter, but nevertheless made several important observations that point to life on Earth. These strengthen the null results encountered elsewhere in the solar system. The evidence for life on Earth includes complex radio emissions, nonmineral surface pigmentation, disequilibrium atmospheric chemistry, and large oceans.

FIGURE 1   *Galileo* spacecraft image of the Earth–Moon system, taken 8 days after closest approach at a distance of 6.2 million km. The Moon is moving from left to right in the foreground; it reflects only about one-third as much sunlight as Earth and hence appears dimmer. To improve visibility, color and contrast on both objects have been computer-enhanced. (Courtesy of NASA/Jet Propulsion Laboratory.) (See also color insert.)

A. RADIO EMISSIONS

The only clear evidence obtained by *Galileo* for *intelligent* life on Earth was unusual radio emissions. Several natural radio emissions were detected, none of which was unusual, including solar radio bursts, auroral kilometric radiation, and narrowband electrostatic oscillations excited by thermal fluctuations in Earth's ionospheric plasma. The first unusual radio emissions were detected at 1800 UT and extended through 2025 UT, just before closest approach. These were detected by the plasma wave spectrometer (PWS) on the nightside, inbound pass, but not on the dayside, outbound pass. The signal strength increased rapidly as Earth was approached, implying that Earth itself was the source of the emissions. The fact that the signals died off on the dayside suggests that they were cut off by the dayside ionosphere, which means we can place the source below the ionosphere.

The unusual signals were narrowband emissions that occurred in only a few distinct channels and had average frequencies that remained stable for hours. Naturally occurring radio emissions nearly always drift in frequency, but these emissions were steady. The individual components had complicated modulations in their amplitude that have never been detected in naturally occurring emissions. The simplest explanation is that these signals were transmitting information, which implies that there is advanced technological life on Earth. In fact, the radio, radar, and television transmissions that have been emanating from Earth over the last century result in a nonthermal radio emission spectrum that broadcasts our presence out to interstellar distances. [*See* THE SOLAR SYSTEM AT RADIO WAVELENGTHS.]

B. SURFACE FEATURES

During its first encounter with Earth, the highest-resolution mapping of the surface by *Galileo's* solid-state imaging system (SSI) covered Australia and Antarctica with 1- to 2-km resolution. No usable images were obtained from Earth's nightside on the first encounter. The second encounter netted the highest-resolution images overall of Earth by *Galileo*, 0.3–0.5 km per pixel, covering parts of Chile, Peru, and Bolivia. The map of Australia from the first encounter includes 2.3% of Earth's total surface area, but shows no geometric patterns that might indicate an advanced civilization. In the second encounter, both the cities of Melbourne and Adelaide were photographed, and yet no geometric evidence is visible because the image resolution is only 2 km. The map of Antarctica, 4% of Earth's surface, reveals nearly complete ice cover and no signs of life. Only one image, taken of southeastern Australia during the second encounter, shows east–west and north–south markings that would raise suspicions of intelligent activity. The markings in fact were caused by boundaries between wilderness areas, grazing lands, and the border between South Australia and Victoria. Studies have shown that it takes nearly complete mapping of the surface at 0.1-km resolution to obtain convincing photographic evidence of an advanced civilization on Earth, such as roads, buildings, and evidence of agriculture.

On the other hand, many features are visible in the *Galileo* images that have not been seen on any other body in the solar system. The SSI camera took images in six different wavelength channels. A natural-color view of Earth was constructed using the red, green, and violet filters, which correspond to wavelengths of 0.670, 0.558, and 0.407 μm, respectively. The images reveal that Earth's surface is covered by enormous blue expanses that specularly reflect sunlight, which is easiest to explain if the surface is liquid, and that end in distinct coastlines. This implies that much of the planet is covered with oceans. The land surfaces show strong color contrasts that range from light brown to dark green.

The SSI camera has particular narrowband infrared filters that have never been used to photograph Earth before, and so they yielded new information for geological, biological, and meteorological investigations. The infrared filters allow the discrimination of H_2O in its solid, liquid, and gaseous forms; for example, clouds and surface snow can be distinguished spectroscopically with the 1-μm filter. False-color images made by combining the 1-μm channel with the red and green channels reveal that Antarctica strongly absorbs 1-μm light, establishing that it is covered by

water ice. In contrast, large regions of land strongly reflect 1 μm without strongly reflecting visible colors, which conflicts with our experience from other planetary surfaces and is not typical of igneous or sedimentary rocks or soil. Spectra made with the 0.73- and 0.76-μm channels reveal several land areas that strongly absorb red light, which again is not consistent with rocks or soil. The simplest explanation is that some nonmineral pigment that efficiently absorbs red light has proliferated over the planet's surface. It is hard to say whether an interstellar explorer might hypothesize that this is a biological mechanism for gathering energy from sunlight, but certainly *we* would recognize it on another planet as the signature of plant life. We know from ground truth that these unusual observations are caused by the green pigments chlorophyll *a* ($C_{55}H_{72}MgN_4O_5$) and chlorophyll *b* ($C_{55}H_{70}MgN_4O_6$), which are used by plants for photosynthesis. No other body in the solar system has the green and blue colorations seen on Earth. [*See* THE SOLAR SYSTEM AT ULTRAVIOLET WAVELENGTHS; INFRARED VIEWS OF THE SOLAR SYSTEM FROM SPACE.]

C. OXYGEN AND METHANE

Galileo's near-infrared mapping spectrometer (NIMS) detected the presence of molecular oxygen (O_2) in Earth's atmosphere with a volume mixing ratio of 0.19 ± 0.05. Therefore, we know that the atmosphere is strongly oxidizing. (It is interesting to note that Earth is the only planet in the solar system where one can light a fire.) In light of this, it is significant that NIMS also detected methane (CH_4) with a volume mixing ratio of $3 \pm 1.5 \times 10^{-6}$. Because CH_4 oxidizes rapidly into H_2O and CO_2, if thermodynamical equilibrium holds then there should be *no* detectable CH_4 in Earth's atmosphere. The discrepancy between observations and the thermodynamic equilibrium hypothesis, which works well on other planets (e.g., Venus), is an extreme 140 orders of magnitude. This fact provides evidence that Earth has biological activity and that it is based on organic chemistry. We know from ground truth that Earth's atmospheric methane is biological in origin, with about half of it coming from nonhuman activity like methane bacteria and the other half coming from human activity like growing rice, burning fossil fuels, and keeping livestock. NIMS also detected a large excess of nitrous oxide (N_2O) that is most easily explained by biological activity, which we know from ground truth comes from nitrogen-fixing bacteria and algae.

The conclusion is that the interplanetary spacecraft we have sent out to explore our solar system are capable of detecting life on planets or satellites, both the intelligent and primitive varieties, if it exists in abundance on the surface. On the other hand, if there is life on a planet or satellite that does not have a strong signature on the surface, as would probably be the case if Europa harbors life, then a flyby mission may not be adequate to decide the question. With regard to abundant surface life, we have a positive result for Earth and a negative result for every other body in the solar system.

II. ORBITAL CHARACTERISTICS

Earth orbits the Sun at a distance of only 108 times the diameter of the Sun. The warmth from the Sun that Earth receives at this distance, together with a 30 K increase in surface temperature resulting from the atmospheric greenhouse effect, is exactly what is needed for H_2O to appear in all three of its phases. This property of the semimajor axis of Earth's orbit is the most important physical characteristic of the planet that supports life. (One interesting consequence is that Earth is the only planet in the solar system where one can ski; this complements its unique fire-lighting property mentioned earlier.)

Orbiting the Sun at just over 100 Sun diameters is not as close as it may sound; a good analogy is to view a basketball placed just past first base while standing at home plate on a baseball diamond. For sunlight, the Sun-to-Earth trip takes 499 s or 8.32 min. Earth's semimajor axis, $a_3 = 1.4960 \times 10^{11}$ m = 1 AU (astronomical unit), and orbital period, $\tau_3 = 365.26$ days = 1 year, where the subscript 3 denotes the third planet out from the Sun, are used as convenient measures of distance and time. When the orbital period of a body encircling the Sun, τ, is expressed in years, and its semimajor axis, a, is expressed in AU, then Kepler's third law is simply $\tau = a^{3/2}$, without any proportionality constant. [See SOLAR SYSTEM DYNAMICS.]

A. LENGTH OF DAY

For all applications but the most demanding, the time Earth takes to turn once on its axis, the length of its day, is adequately represented by a constant value equal to 24 hours or 1440 minutes or 86,400 seconds. The standard second is the Système International (SI) second, which is precisely 9,192,631,770 periods of the radiation corresponding to the transition between two hyperfine levels of the ground state of the ^{133}Cs atom. When the length of day is measured with high precision, it is found that Earth's rotation is not constant. The same is likely to hold for any dynamically active planet. Information can be obtained about the interior of a planet, and how its atmosphere couples with its surface, from precise length-of-day measurements. Earth is the only planet to date for which we have achieved such accuracy, although we also have high-precision measurements of the rotation rate of pulsars, the spinning neutron stars often seen at the center of supernova explosions.

The most stable pulsars lose only a few seconds every million years and are the best-known timekeepers, even better than atomic clocks. In contrast, the rotating Earth is not an accurate clock. Seen from the ground, the positions as a function of time of all objects in the sky are affected by Earth's variable rotation. Because the Moon moves across the sky relatively rapidly and its position can be determined with precision, the fact that Earth's rotation is variable was first realized when a series of theories that should have predicted the motion of the Moon failed to achieve their expected accuracy. In the 1920s and 1930s, it was established that errors in the position of the Moon were similar to errors in the positions of the inner planets and, by 1939, clocks were accurate enough to reveal that Earth's rotation rate has both irregular and seasonal variations.

The quantity of concern is the planet's three-dimensional angular velocity vector as a function of time, $\vec{\Omega}(t)$. Since the 1970s, time series of all three components of $\vec{\Omega}(t)$ have been generated by using very long baseline interferometry (VLBI) to accurately determine the positions of quasars and laser ranging to accurately determine the positions of man-made satellites and the Moon, the latter with corner reflectors placed on the Moon by the Apollo astronauts. [See PLANETARY EXPLORATION MISSIONS.]

The theory of Earth's variable rotation combines ideas from geophysics, meteorology, oceanography, and astronomy. The physical causes fall into two categories: those that change the planet's moment of inertia (like a spinning skater pulling in her arms) and those that torque the planet by applying stresses (like dragging a finger on a spinning globe). Earth's moment of inertia is changed periodically by tides raised by the Moon and the Sun, which distort the solid planet's shape. Nonperiodic changes in the solid planet's shape occur because of fluctuating loads from the fluid components of the planet, namely, the atmosphere, the oceans, and, deep inside the planet, the liquid iron-nickel core. In addition, shifts of mass from earthquakes and melting ice cause nonperiodic changes. Over long timescales, plate tectonics and mantle convection significantly alter the moment of inertia and hence the length of day.

An important and persistent torque that acts on Earth is the gravitational pull of the Moon and the Sun on the solid planet's tidal bulge, which, because of friction, does not line up exactly with the combined instantaneous tidal stresses. This torque results in a steady lengthening of the day at the rate of about 1.4 ms per century and a steady outward drift of the Moon at the rate of 3.7 ± 0.2 cm yr^{-1}, as confirmed by lunar laser ranging. On the top of this steady torque, it has been suggested that observed 5-ms variations that have timescales of decades are caused by stronger, irregular torques from motions in Earth's liquid core. Calculations suggest that viscous coupling between the liquid core and the solid mantle is weak, but that electromagnetic and topographic coupling can explain the observations. Mountains on the core–mantle boundary with heights around 0.5 km are sufficient to produce the coupling and are consistent with seismic tomography studies, but not much is known about the detailed topography of the core–mantle boundary. Detailed model calculations take into account the time variation of Earth's external magnetic field, which is extrapolated downward to the core–mantle boundary. New improvements to the determination of the magnetic field at the surface are enhancing the accuracy of the downward extrapolations.

Earth's atmosphere causes the strongest torques of all. The global atmosphere rotates faster than the solid planet by about 10 ms^{-1} on average. Changes in the global circulation cause changes in the pressure forces that act on mountain ranges and changes in the frictional forces between the wind and the surface. Fluctuations on the order of 1 ms in the length of day, and movements of the pole by several meters, are caused by these meteorological effects, which occur over seasonal and interannual timescales. General circulation models (GCMs) of the atmosphere routinely calculate the global atmospheric angular momentum, which allows the meteorological and nonmeteorological components of the length of day to be separated. All the variations in the length of day over weekly and daily timescales can be attributed to exchanges of angular momentum between Earth's atmosphere and the solid planet, and this is likely to hold for timescales of several months as well. Episodic reconfigurations of the coupled atmosphere–ocean system, such as the El Niño–Southern Oscillation (ENSO), cause detectable variations in the length of day, as do changes in the stratospheric jet streams.

B. NEAR-EARTH OBJECTS

The most dramatic way to redistribute mass on a planet is to hit it with an asteroid or a comet. A fundamental lesson of planetary exploration has been that violent collisions are a natural, ongoing process everywhere in the solar system. Apart from what its inhabitants might muster, Earth has no special immunity from collisions. There are three families of asteroids that orbit the Sun near Earth. Collectively they are called the Aten–Apollo–Amor objects (AAAOs); the term "objects" is used because some are extinct comets. A few hundred AAAOs have been discovered to date. The Atens have semimajor axes less than $a = 1$ AU and are projected to number around 100; the Apollos are Earth-crossing objects and are projected to number 700 ± 300; the Amors are Mars-crossing objects and are projected to number 1500 ± 500. Because they are in or near planet-crossing orbits, these objects last only 10 to 100 million years before they are scattered out of the inner solar system or collide with a planet. [*See* NEAR-EARTH ASTEROIDS]

The AAAOs sample a wide range of the asteroid types seen in the main belt. It is likely that Apollo asteroids are the parent bodies of many meteorites found on Earth's surface. It turns out that there are about a dozen near-Earth asteroids that are less expensive to visit in terms of rocket fuel than the Moon. Calculations show that none of the known Apollo asteroids will collide with Earth anytime in the next century. However, Earth is also susceptible to being hit by long-period comets, which appear with little warning. Studies are currently under way to assess the hazards of a large impact. It is thought that the orbit of a potentially dangerous asteroid or comet could be safely diverted with present technology if at least a few years of lead time exists in which to plan a mission.

III. OVERVIEW OF SURFACE CHARACTERISTICS

A. IMPACT CRATERS

Whether or not impact craters are apparent on a solar system body is simply a matter of resurfacing rate. Jupiter's satellite Io has no visible impact craters because its active volcanism quickly covers them up. Europa, the next Jovian satellite out from Io, has only a few impact craters. On the other hand, our Moon is covered by impact craters that have collected during the 3 to 4 billion years since it was last volcanically active. All the airless, inactive satellites in the solar system are heavily cratered, as is every asteroid that has so far been imaged. [See PLANETARY IMPACTS.]

On Earth, no person is history has yet been killed by a meteorite, but houses and cars have been hit, and in 1911 the meteorite Nakhla fell in Egypt and killed a dog. (Ironically, Nakhla is the N in SNC, i.e., one of the Martian meteorites.) [See METEORITES.]

Earth has only about 140 impact craters identified on its surface to date, which is many fewer than Mars and Venus. Over 60% of Earth's impact craters are less than 0.2 billion years old, even though the impact rate has been reasonably steady for the last 3 billion years. Because water may be a liquid on Earth, it is the only solar system body that is literally soaking in a solvent, and this results in tremendous erosion rates. On airless bodies the cumulative number of craters larger than a given diameter, N, as a function of diameter, D, approximately follows an $N \propto D^{-2}$ power law. (One uses the cumulative, or integrated, number, rather than the number of craters between diameters D and $D + \Delta D$, say, because the resulting histogram is not sensitive to the binning size, ΔD.) On Earth, N follows the D^{-2} trend for its larger craters, but erosion reduces N to much smaller numbers for $D \leq 20$ km. Also, Earth's atmosphere burns up the smallest meteoroids, which are the vast majority by number. All impact craters on Earth with $D < 1$ km appear to have been formed by iron meteorites, which survive the stresses of atmospheric entry better than stony meteorites. Breakup by the atmosphere of a single object probably accounts for the Henbury crater field in Australia.

In addition to erosion, most of Earth's surface crust recycles through the action of plate tectonics, which destroys craters in a manner somewhat analogous to what a moving escalator would do to a neat pile of sand placed on one of its steps. There is one known crater under the shallow water of a continental shelf, but none on any deep ocean floor. Crater locations are biased toward the stable centers of continents, called cratons.

As on any body, Earth's craters fall into three categories. From the smallest to the largest, these are simple craters, complex craters, and basins. On less active bodies the morphological characteristics of these three crater types are well preserved, and one sees a general progression from the bowl-shaped simple craters to the uplifted central peak and slumping side walls of complex craters to the multiple concentric rings of basins.

The study of Earth's impact craters yields unique information about cratering processes, especially about the structure beneath crater floors, for which remote sensing cannot gather adequate data. On Earth, simple craters occur with diameters up to $D \sim 4$ km; all but a few have been lost to erosion. Meteor Crater (a.k.a. Barringer Crater) in Arizona is an example of a young simple crater, with a diameter of $D = 1.2$ km and a depth of 0.2 km. It was formed only 50,000 years ago (0.00005 billion years ago) by a 40- to 50-m iron meteoroid. Its rim is uplifed by 50 m and has overturned stratigraphy. Drilling on the floor has located a bowl-shaped layer of breccia that marks the transient crater depth of one-third the diameter, which is twice as deep as the present-day crater. When an impact takes place, the steep walls of the transient crater quickly slump inward and then get covered by a thin layer of fallback material, resulting in a final crater that is only half as deep as the transient crater.

Geological mapping of complex impact structures on Earth shows that the rim and the annular trough region between the rim and central peak both preserve the original surface rock material. On the other hand, the central peak is made of rock that has been lifted upward by as much as one-tenth of the final crater diameter, a depth of 10 km for a 100-km crater, thereby providing geologists with a view of rock types that are too deep to be reached by drilling. The initial impact happens so quickly and with such great force that the planet's surface momentarily acts like a fluid. The central uplift of a complex crater is like a freeze-frame of a movie of water into which a rock has been thrown. The central ring seen in larger complex craters resembles the same movie, but advanced a few frames. Erosion makes it difficult to find complex craters on Earth. Nevertheless, several have been identified by the telltale geology of their central peaks, including Carswell ($D = 39$ km), Siljan ($D = 55$ km), and Vredefort ($D = 140$ km). Recent work suggests that the diameter of Vredefort may be up to 300 km. The $D \sim 100$ km Manicouagan structure in Quebec, Canada, is particularly beautiful because of the annular lake that has formed in it, as shown in Fig. 2. (See also color insert.)

Not all large, circular structures on Earth are impact craters. One needs to find evidence of shocked material to establish formation by impact. During the impact event, a shock wave passes through Earth's crust that briefly raises the pressure in the rock to about 10 megabars (1000 GPa) for a duration of about 0.005 s. Detailed study of shock-metamorphosed minerals tells a researcher the magnitude of the impact pressure wave that has passed through the rock sample. The subsequent decompression of the rock does not release all

FIGURE 2   Space shuttle image of the Manicouagan impact basin, located at 51°23′N, 68°42′W, near Quebec, Canada. This structure was formed 210 ± 4 million years ago and is one of Earth's largest visible impact craters. Glaciers have eroded impact-brecciated rock and left a 70-km annular lake between more resistant igneous impact-melt rocks. (Courtesy of NASA Earth Data Analysis Center.) (See also color insert.)

of the compression energy: heating, melting, and vaporization occur. The Manicouagan structure in Quebec has 250-m cliffs that are striking in appearance because they have no stratigraphy—they are the cooled remnant of a single unit of impact melt.

Earth's fossil record is punctuated by episodes of mass extinction. A strong case has been made that the Cretaceous–Tertiary boundary (KT boundary; the letter "C" is used for Cambrian), which marks a 70–80% mass extinction that occurred 65 million years ago, the time when the dinosaurs disappeared, was caused by a large impact. The first clue came in 1980, when the Alvarez father and son team reported high levels of iridium, a common element in meteorites but a rare find in Earth's crust, at a KT boundary site in Italy. Similar findings were subsequently reported around the globe. The prime suspect is an approximately 65 million-year-old impact basin on Mexico's Yucatan Peninsula; the structure, called Chicxulub

("chik shu lub"), lays buried under 1.1 km of limestone and hence is not visible. However, it shows up in gravity data, and drilling has established the presence of impact melt and shock metamorphism at the site, which appears to be the largest impact structure ($D \sim$ 200–300 km) formed on Earth in the last 1 billion years. Spherules of impact glass, also of KT age, have been found around the region in Mexico and Haiti, and this provides spatial evidence linking the impact basin to the KT event. Theories for how the impact actually led to global mass extinctions vary, but most point out that sunlight would be diminished by the large cloud of debris thrown into the stratosphere, and that this would kill plant life and severely disrupt the food chain. Transient but severe climate change may also have contributed to the extinctions.

By far the most traumatic impacts on Earth occurred while it was first forming. Accretion of a terrestrial planet is thought to progress from smaller to larger bodies, until the final assembly is made by combining bodies with masses of about 10% of the mass of the present-day Earth. The leading theory for the formation of the Moon postulates that Earth's last major accretion event involved a collision between Earth and a Mars-sized impactor, and that the Moon coalesced out of ejected impactor material that was mixed with iron-poor crustal and mantle from Earth. We know that the details of the final assembly of a given planet must be important, because each planet has turned out significantly different. For example, Venus has no moon at all, and is also slowly spinning "backward" on its axis compared to Earth and most of the other planets. [See The Origin of the Solar System.]

B. PLATE TECTONICS

Photographs' from space show Earth's surface to be neatly partitioned by coastlines, mountain ranges, fault lines like the San Andreas Fault, island chains like the Hawaiian Islands, and island arcs like the Aleutian Islands. One does not describe the groupings of mountains on the other terrestrial planets as "ranges," which connote a preferred direction. Why is Earth's surface delineated in such a manner? The answer has to do with the planet's unique method of getting rid of its heat—plate tectonics.

Earth's cold outer shell, its lithosphere, is organized into 8 major plates and over 20 minor ones. The boundaries between plates are associated with transform faults, subduction zones, earthquakes, volcanos, and mountain ranges. Although there are a few localized examples of transform faults seen in images of other planetary bodies, only Earth exhibits global plate tectonics. Why this is the case is not entirely clear, but liquid water may play an important role. For plate tectonics to occur in the manner found on Earth, parts of the lithosphere have to eventually become negatively buoyant. One way to shut down plate tectonics is to make the lithosphere everywhere too light to be able to sink. In fact, this is the case for Earth's continents. The delicate balance of densities needed for plate tectonics to proceed may help to explain why it is absent on Venus. [See Venus: Surface and Interior.]

Plate motions can be directly measured by Global Positioning Satellite (GPS) techniques and are typically a few centimeters per year. The last 0.5 billions years have seen a dramatic change in Earth's appearance. By integrating the present-day plate motions backward in time and correlating geology, it is clear that 0.18 billion years ago the continents all fitted together into one supercontinent, named Pangaea.

The southern segment of Pangaea, called Gondwana, was a coherent unit earlier than this. During the time of Pangaea there was no Atlantic Ocean and the Pacific Ocean was enormous.

IV. OVERVIEW OF EARTH'S INTERIOR

What broke up Pangaea, and what drives plate motions today? Fundamentally, the only energy source strong enough is energy from the planet's interior. Earth's average surface heat flow is presently 75 mW m^{-2}, of which 70–75% is generated by radioactive decay throughout the interior and the balance is escaping heat left over from the original accretion of the planet. The radiogenic heat sources have dropped by about 50% in the last 3 billion years as the ^{235}U, ^{238}U, ^{232}Th, and ^{40}K isotopes decay. Today, about 60% of the internal heat drives the formation of new oceanic crust at midocean ridges. The physical evidence that firmly established plate tectonics was the discovery of mirror-symmetric stripes of magnetic anomalies along these midocean ridges. Earth's magnetic polarity flips irregularly in time, and as hot, new oceanic crust emerges, it cools below its Curie temperature and traps the current magnetic direction into the rock. As the ocean's crust moves away from the midocean ridges, it cools and thickens, and eventually is heavy enough to sink, or subduct, back into the mantle. Three-dimensional images of subducting slabs have been made using seismic tomography techniques. Figure 3 shows a seismic tomography cross section of Earth's mantle.

The concept of mantle convection is a central precept of geophysics. That a material can act like a solid on short timescales and a fluid on long timescales is part of our everyday experience—silly putty is a classic example: throw it down quickly on a table and it bounces back elastically, with no change in shape, but lean a book on it slowly and it flows plastically, deforming its shape like a fluid. Earth's mantle behaves like a fluid on long timescales because of solid-state creep, in which vacancies in the rock crystals slowly diffuse through the rock in response to stress.

Convection occurs when a fluid element is heated enough so that its buoyancy force overcomes the retarding forces of viscosity, which is the molecular-scale diffusion of momentum, and heat conduction, which is the molecular-scale diffusion of heat, and rises upward. The process occurs in the upside-down direction when a fluid element is cooled enough to sink, as can happen in the ocean, especially in the wintertime Labrador

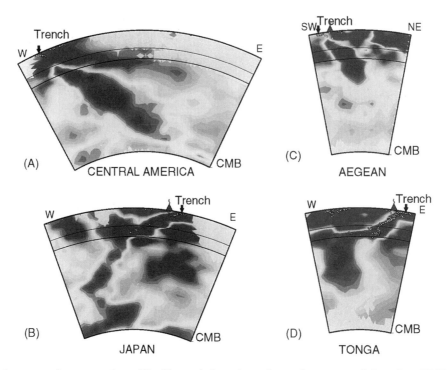

FIGURE 3   Seismic tomography cross sections of Earth's mantle from the surface to the core–mantle boundary (CMB). The blue and red colors indicate cold and hot temperature anomalies, respectively (gray indicates regions with insufficient data). The geographical orientation of each vertical cross section is indicated by the symbols E, W, SE, and NW. The angle of descent of the blue subducting slabs is as depicted with no vertical exaggeration. Thin lines depict the location of well-known discontinuities at 410 and 660 km depth. Small white circles indicate earthquake locations, which are seen to not extend deeper than the 660-km discontinuity. These results indicated that Earth's subducting slabs span the entire mantle. (Courtesy of Robert van der Hilst, Massachusetts Institute of Technology.)

Sea. The onset and strength of convection are predicted by a nondimensional number called the Rayleigh number, which measures the ratio of buoyancy force to the diffusive forces of viscosity and heat conduction. In 1935, the first quantitative measure of the viscosity of Earth's mantle was made by studying the elevations of ancient beach terrances in Scandinavia. There it was found that Earth's surface is still rebounding from the weight of the ice that was present during the last ice age. The rate of rebound yields a viscosity for the mantle that is 10^{23} times larger than the viscosity of liquid water. Nevertheless, the Rayleigh number for Earth's mantle turns out to be much larger than required for the onset of convection. That is to say, not only is Earth's mantle convecting, but it is doing so vigorously.

A. SEISMOLOGY

Earth's interior structure is known in detail from seismology, the analysis of seismic waves from earthquakes. Astronomers have recently applied analogous techniques to study the Sun's interior structure by observing sound waves on its surface, a technique known as helioseismology. The speed of seismic waves is a strong function of density. In 1909, Andrija Mohorovičić detected a large jump in wave speed that occurs at Earth's crust–mantle interface, which is called the Moho for short. Earth's crust accounts for only 0.5% of the planet's total mass. On average, the continental crust is about 35 km thick and the oceanic crust is about 5–10 km thick.

There are two basic types of seismic waves, shear waves and pressure waves. Shear waves have the important property that they do not pass through a liquid, and hence their absence at a particular seismic station is an indication that there was liquid along the path between the station and the earthquake location. Earth's liquid outer core was discovered in 1906 by this technique. The core–mantle boundary is at a depth of 2890 km, compared to the 6378-km equatorial radius of Earth. Earth's interior temperatures were higher in the past, and the entire core may have once been liquid, but now there is a solid inner core of radius 1228 km. Recent indications are that this inner core rotates slightly faster than the rest of the planet. The inner and outer core account for 1.7% and 30.8% of Earth's mass, respectively. The mantle accounts for 67%.

B. GRAVITY FIELD

Important information about planetary interiors comes from the analysis of their external gravity fields. Earth's gravity field is particularly well determined. Theories of mantle convection can be tested by the data, and even ocean circulations can be deduced. The highest-resolution global gravity model to date is complete to over 300 spherical harmonic terms. The biggest remaining uncertainties are for regions that do not have good ground coverage, including parts of Asia and the poles.

Temporal changes in Earth's gravity field due to tides are relatively well understood, and a major goal of present research is to refine measurements to the point where the nontidal temporal component can be discerned. Studies have shown that the degree 2 zonal harmonic, J_2, is affected by changes in atmospheric surface pressure. Further refinements will allow researchers to detect crustal movements associated with earthquakes, to improve the measurement of the viscosity of the mantle, to detect changes in Earth's fluid component, including shifts in the ocean circulation and in groundwater levels, to detect changes in ice fields and snowfields, and to test a range of theories about Earth's interior. Putting GPS receivers into low-Earth orbit is one way to improve measurements, as has been demonstrated by the *Topex/Poseidon* satellite experiment.

V. ATMOSPHERE AND OCEANS

· ·

Atmospheres are found on the Sun, eight of the nine planets, and three of the sixty-odd satellites, for a total count of a dozen. Each has its own brand of weather and its own unique chemistry. They can be divided into two major classes: the terrestrial-planet atmospheres, which have solid surfaces or oceans as their lower boundary condition, and the gas-giant atmospheres, which are essentially bottomless. Venus and Titan form one subgroup that is characterized by a slowly rotating planet with a rapidly rotating atmosphere. Mars, Io, Triton, and Pluto form a second terrestrial subgroup that is characterized by a thin atmosphere that is in large measure driven by vapor–pressure equilibrium with the atmosphere's solid phase on the surface. Both Io and Triton have active volcanic plumes. Earth's weather turns out to be the most unpredictable in the solar system. Part of the reason is that its mountain ranges frustrate the natural tendency for winds to settle into steady east–west patterns, and a second

reason is that its atmospheric eddies, the fluctuating waves and storm systems that deviate from the average, are nearly as big as the planet itself and as a result strongly interfere with each other. [*See* VENUS: ATMOSPHERE; IO; TITAN; TRITON; PLUTO AND CHARON.]

A. VERTICAL STRUCTURE

Earth may differ in many ways from the other planets, but not in the basic structure of its atmosphere. Planetary exploration has revealed that essentially every atmosphere starts at the bottom with a troposphere, where temperature decreases with height at a nearly constant rate up to a level called the tropopause, and then has a stratosphere, where temperature usually increases with height or, in the case of Venus and Mars, decreases much less quickly than in the troposphere. It is interesting to note that atmospheres are warm both at their bottoms and their tops, but do not get arbitrarily cold in their interiors. For example, on Jupiter and Saturn there is significant methane gas throughout their atmospheres, but nowhere does it get cold enough for methane clouds to form, whereas in the much colder atmospheres of Uranus and Neptune, methane clouds do form. Details vary in the middle-atmosphere regions from one planet to another, but each atmosphere is topped off by a high-temperature, low-density thermosphere that is sensitive to solar activity and an exobase, the official top of an atmosphere, where molecules float off into space when they achieve escape velocity. [*See* ATMOSPHERES OF THE GIANT PLANETS.]

In the first 0.1 km of a terrestrial atmosphere, the effects of daily surface heating and cooling, surface friction, and topography produce a turbulent region called the planetary boundary layer. Right at the surface, molecular viscosity forces the "no slip" boundary condition and the wind reduces to zero, such that even a weak breeze results in a strong vertical wind shear that can become turbulent near the surface. However, only a few millimeters above the surface, molecular viscosity ceases to play a direct role in the dynamics, except as a sink for the smallest eddies. The mixing caused by turbulent eddies can sometimes be represented as a viscosity with a strength that is a million times or more greater than the molecular viscosity.

B. GREENHOUSE EFFECT

Earth as a whole radiates with an effective temperature of 255 K, and therefore its flux peaks in the thermal infrared part of the spectrum. This effective temperature is 30 K colder than the average temperature on the surface, and quite chilly by human standards. What ensures a warm surface is the wavelength-dependent optical properties of the atmosphere. In particular, infrared light does not pass through the atmosphere as readily as visible light. The Sun radiates with an effective temperature of 5800 K and therefore its peak flux is in the visible part of the spectrum (or stated more correctly in reverse, we have evolved such that the part of the spectrum that is visible to us is centered on the peak flux from the Sun). The atmosphere reflects about 31% of this sunlight directly back to space, and the rest is absorbed or transmitted to the ground. The sunlight that reaches the ground is absorbed and then reradiated at infrared wavelengths. Water vapor (H_2O) and carbon dioxide (CO_2), the two primary greenhouse gases, absorb some of this upward infrared radiation and then emit it in both the upward and downward directions, leading to an increase in the surface temperature to achieve balance. Water vapor accounts for over 90% of the greenhouse effect on Earth, regardless of human activity, but steady increases in carbon dioxide due to human activity may have an effect on the global surface temperature. This is an ongoing area of research. On Mars, the primary atmospheric constituent is CO_2, which together with atmospheric dust causes a modest 5 K greenhouse effect. Venus has a much denser CO_2 atmosphere, which, along with atmospheric sulfuric acid and sulfur dioxide, absorbs essentially all the infrared radiation emitted by the surface, causing an impressive 500 K rise in the surface temperature. Interestingly, if all the carbon held in Earth's carbonate rocks were liberated into the atmosphere, Earth's greenhouse effect would approach that on Venus. [*See* MARS: ATMOSPHERE AND VOLATILE HISTORY; VENUS: ATMOSPHERE.]

C. TROPOSPHERE

Gravity causes the density of an atmosphere to fall off exponentially with height, such that Earth's troposphere contains 80% of the mass and most of the water vapor in the atmosphere, and consequently most of the clouds and stormy weather. Vertical mixing is an important process in the troposphere. Temperature falls off with height at a predictable rate because the air near the surface is heated and becomes light, and the air higher up cools to space and becomes heavy, leading to an unstable configuration and convection. The process of convection relaxes the temperature profile toward the neutrally stable configuration, called the adiabatic temperature lapse rate, for which the decrease of temperature with decreasing pressure (and hence increasing height) matches the drop-off of temperature that would occur inside a balloon that conserves its heat as it moves, that is, that moves adiabatically. The top of the troposphere is called the tropopause, which occurs in the altitude range 8–18 km on Earth depending on latitude. It is 10 km higher at the equator than at the poles. The cruising altitude of jet airplanes is purposely near the tropopause.

Water vapor in Earth's troposphere greatly accentuates convective activity because latent heat is liberated when moist air is raised above its lifting-condensation level, and this further increases the buoyancy of the rising air, leading to moist convection. Hurricanes are the best-organized and most dramatic examples of moist convection. Towering thunderstorms also get their energy from this process. Hurricanes occur only on Earth, because only Earth provides the necessary combination of high humidity and surface friction. Surface friction is required to cause air to spiral into the center of the hurricane, where it is then forced upward past its lifting-condensation level.

In contrast, windstorms like Jupiter's Great Red Spot and the hundreds of smaller ovals seen on Jupiter, as well as the dozens seen on Saturn and the couple seen on Neptune, are not hurricanes and do not require moist convection to drive them. Instead, they are simpler systems that are closely related to three types of long-lasting, high-pressure "storms," or coherent vortices, seen on Earth: blocking highs in the atmosphere and, in the ocean, Gulf Stream rings and Mediterranean salt lenses ("meddies"). Blocking highs are high-pressure centers that stubbornly settle over continents, particularly in the United States and Russia, thereby diverting rain from its usual path for months at a time. For example, the serious 1988 drought in the U.S. Midwest was exacerbated by a blocking high. Gulf Stream rings are compact circulations in the Atlantic that break off from the meandering Gulf Stream. The Gulf Stream is a river inside the Atlantic that runs northward along the eastern coast of the United States and separates from the coast at North Carolina, where it then jets into the Atlantic in an unsteady manner. Seen in three dimensions, the Gulf Stream has the appearance of a writhing snake. Similar western boundary currents occur in other ocean basins, for example, the Kuroshio Current off the coast of Japan and the Agulhas Current off the coast of South Africa. Jet streams in the atmosphere are a related phenomenon. When Gulf Stream rings form, they trap phytoplankton and zooplankton inside them, which are carried large distances. Over the course of a few months, the rings dissipate at sea, are reabsorbed into the Gulf Stream, or run into the coast, depending on which side of the Gulf Stream they formed. The ocean plays host to another class of long-lived vortices, Mediterranean salt lenses, which are organized high-pressure circulations that float under the surface of the Atlantic. They form when the extrasalty water that slips into the Atlantic from the shallow Mediterranean Sea breaks off into vortices. After a few years, they eventually wear down as they slowly mix with the surrounding water. The mathematical description of these long-lasting vortices on Earth is the same as that used to describe the ovals seen on Jupiter, Saturn, and Neptune. [*See* ATMOSPHERES OF THE GIANT PLANETS.]

D. STRATOSPHERE

The nearly adiabatic falloff of temperature with height in Earth's troposphere gives way above the tropopause to an increase of temperature with height. This results in a stable layer called the stratosphere. Observations of persistent, thin layers of aerosol and of long residence times for radioactive trace elements from nuclear explosions are direct evidence of the lack of mixing in the stratosphere. The temperature continues to rise with altitude in Earth's stratosphere until one reaches the stratopause at about 50 km. The source of heating in Earth's stratosphere is the photochemistry of ozone, which peaks at about 25 km. Ozone absorbs ultraviolet (UV) light, and below about 75 km nearly all this radiation gets converted into thermal energy. The Sun's UV radiation causes stratospheres to form in other atmospheres, but instead of the absorber being ozone, which is plentiful on Earth because of the high concentrations of O_2 maintained by the biosphere, other gases absorb the UV radiation. On the giant planets, methane, hazes, and aerosols do the job.

The chemistry of Earth's stratosphere is complicated. Ozone is produced mostly over the equator, but its largest concentrations are found over the poles, which implies that both dynamics and chemistry are important to the ozone budget. Mars also tends to have ozone concentrated over its poles, particularly over the winter pole. The dry Martian atmosphere has relatively few hydroxyl radicals to destroy the ozone. Some of the most important chemical reactions in Earth's stratosphere are those that involve only oxygen. Photodissociation by solar UV radiation involves the reactions $O_2 + h\nu \rightarrow O + O$ and $O_3 + h\nu \rightarrow O + O_2$, where $h\nu$ indicates the UV radiation. Three-body collisions, where a third molecule, M, is required to satisfy conservation of momentum and energy, include

$O + O + M \rightarrow O_2 + M$ and $O + O_2 + M \rightarrow O_3 + M$, but the former reaction proceeds slowly and may be neglected in the stratosphere. Reactions that either destroy or create "odd" oxygen, O or O_3, proceed at much slower rates than reactions that convert between odd oxygen. The equilibrium between O and O_3 is controlled by fast reactions that have rates and concentrations that are altitude dependent. Other reactions that are important to the creation and destruction of ozone involve minor constituents such as NO, NO_2, H, OH, HO_2, and Cl. An important destruction mechanism is the catalytic cycle $X + O_3 \rightarrow XO + O_2$ followed by $XO + O \rightarrow X + O_2$, which results in the net effect $O + O_3 \rightarrow 2O_2$. On Earth, human activity has led to sharp increases in the catalysts X = Cl and NO and subsequent sharp decreases in stratospheric ozone, particularaly over the polar regions. The Montreal Protocol is an international treaty signed in 1987, and amended in 1990 and 1992, that is designed to stop and eventually reverse the damage to the stratospheric ozone layer. The treaty calls for the phaseout of chlorofluorocarbons (CFCs), halons, and carbon tetrachloride by the year 2000 and methyl chloroform by the year 2005.

E. MESOSPHERE

Above Earth's stratopause, temperature again falls off with height, although at a slower rate than in the troposphere. This region is called the mesosphere. Earth's stratosphere and mesosphere are often referred to collectively as the middle atmosphere. Temperatures fall off in the mesosphere because there is less heating by ozone and emission to space by carbon dioxide is an efficient cooling mechanism. The mesopause occurs at an altitude of about 80 km, marking the location of a temperature minimum of about 130 K.

F. THERMOSPHERE

As is the case for ozone in Earth's stratosphere, above the mesopause, atomic and molecular oxygen strongly absorb solar UV radiation and heat the atmosphere. This region is called the thermosphere, and temperatures rise with altitude to a peak that varies between about 500 and 2000 K depending on solar activity. Just as in the stratosphere, the thermosphere is stable to vertical mixing. At about 120 km, molecular diffusion becomes more important than turbulent mixing, and this altitude is called the homopause (or turbopause). Rocket trails clearly mark the homopause—they are rapidly mixed below this altitude but linger relatively undisturbed above it. Molecular diffusion is mass dependent and each species falls off exponentially with its own scale height, leading to elemental fractionation that enriches the abundance of the lighter species at the top of the atmosphere.

For comparison with Earth, the structure of the thermospheres of the giant planets has been determined from *Voyager* spacecraft observations and the principal absorbers of UV light are H_2, CH_4, C_2H_2, and C_2H_6. The thermospheric temperatures of Jupiter, Saturn, and Uranus are about 1000, 420, and 800 K, respectively. The high temperature and low gravity on Uranus allow its upper atmosphere to extend out appreciably to its rings. [*See* ATMOSPHERES OF THE GIANT PLANETS.]

G. EXOSPHERE AND IONOSPHERE

At an altitude of about 500 km on Earth, the mean free path between molecules grows to be comparable to the density scale height (the distance over which density falls off by a factor of $e \approx 2.7128$). This defines the exobase and the start of the exosphere. At these high altitudes, sunlight can remove electrons from atmospheric constituents and form a supply of ions. These ions interact with a planet's magnetic field and with the solar wind to form an ionosphere. On Earth, most of the ions come from molecular oxygen and nitrogen, whereas on Mars and Venus most of the ions come from carbon dioxide. Because of the chemistry, however, ionized oxygen atoms and molecules are the most abundant ion for all three atmospheres.

Mechanisms of atmospheric escape fall into two categories, thermal and nonthermal. Both processes provide the kinetic energy necessary for molecules to attain escape velocity. When escape velocity is achieved at or above the exobase, such that further collisions are unlikely, molecules escape the planet. In the thermal escape process, some fraction of the high-velocity wing of the Maxwellian distribution of velocities for a given temperature always has escape velocity; the number increases with increasing temperature. An important nonthermal escape process is dissociation, both chemical and photochemical. The energy for chemical dissociation is the excess energy of reaction, and for photochemical dissociation it is the excess energy of the bombarding photon or electron, either of which is converted into kinetic energy in the dissociated atoms. A common effect of electrical discharges of a kilovolt or more is "sputtering," where several atoms can be ejected from the spark region at high velocities. If an ion is formed very high in the atmosphere, it can be swept out of a planet's atmosphere by the solar wind. Similarly at Io, ions are swept away by Jupiter's magnetic field. Other nonthermal escape mechanisms involve charged particles. Charged particles get trapped by magnetic fields and therefore do not readily escape. However, a fast proton can collide with a slow hydrogen atom and take the electron from the hydrogen atom. This charge–exchange process changes the fast proton into a fast, electrically neutral hydrogen atom.

Nonthermal processes account for most of the present-day escape flux from Earth, and the same is likely to be true for Venus. They are also invoked to explain the $62 \pm 16\%$ enrichment of the $^{15}N/^{14}N$ ratio in the Martian atmosphere. If the current total escape flux from thermal and nonthermal processes is applied over the age of the solar system, the loss of hydrogen from Earth is equivalent to only a few meters of liquid water, which means that Earth's sea level has not been effected much by this process. However, the flux could have been much higher in the past, since it is sensitive to the structure of the atmosphere. [*See* MARS: ATMOSPHERE AND VOLATILE HISTORY.]

H. VOLATILE INVENTORIES

Venus, Earth, and Mars have present-day atmospheres that are intriguingly different. The atmospheres of Venus and Mars are both primarily CO_2, but represent two extreme fates in atmospheric evolution: Venus has a dense and hot atmosphere, whereas Mars has a thin and cold atmosphere. It is reasonable to ask whether Earth is ultimately headed toward one or the other of these fates, and whether these three atmospheres have always been so different.

The history of volatiles on the terrestrial planets includes their origin, their interactions with refractory (nonvolatile) material, and their rates of escape into space. During the initial accretion and formation of the terrestrial planets, it is thought that most or all of the original water reacted strongly with the iron to form iron oxides and hydrogen gas, with the hydrogen gas subsequently escaping to space. Until the iron cores in the planets were completely formed and this mechanism was shut down, the outflow of hydrogen probably took much of the other solar-abundance volatile material with it. Thus, one likely possibility is that the present-day atmospheres of Venus, Earth, and Mars are not primordial, but have been formed by outgassing and by cometary impacts that have taken place since the end of core formation.

The initial inventory of water that each terrestrial

planet had at its formation is a debated question. One school of thought is that Venus formed in an unusually dry state compared with Earth and Mars; another is that each terrestrial planet must have started out with about the same amount of water per unit mass. The argument for an initially dry Venus is that water-bearing minerals would not condense in the high-temperature regions of the protoplanetary nebula inside of about 1 AU. Proponents of the second school of thought argue that gravitational scattering caused the terrestrial planets to form out of materials that originated over the whole range of terrestrial-planet orbits, and therefore that the original water inventories for Venus, Earth, and Mars should be similar.

An important observable that bears on the question of original water is the enrichment of deuterium (D) relative to hydrogen. A measurement of the D/H ratio yields a constraint on the amount of hydrogen that has escaped from a planet. For the D/H ratio to be useful, one needs to estimate the relative importance of the different hydrogen escape mechanisms and the original D/H ratio for the planet. In addition, one needs an idea of the hydrogen sources available to a planet after its formation, such as cometary impacts. The initial value of D/H for a planet is not an easy quantity to determine. A value of 0.2×10^{-4} has been put forward for the protoplanetary nebula, which is within a factor of two or so for the present-day values of D/H inferred for Jupiter and Saturn. However, the D/H ratio in Standard Mean Ocean Water (SMOW, a standard reference for isotopic analysis) on Earth is 1.6×10^{-4}, which is also about the D/H ratio in hydrated minerals in meteorites, and is larger by a factor of eight over the previously mentioned value. At the extreme end, some organic molecules in carbonaceous chondrites have shown D/H ratios as high as 20×10^{-4}. The enrichment found in terrestrial planets and most meteorites over the protoplanetary nebula value could be the result of exotic high-D/H material deposited on the terrestrial planets, or it could be the result of massive hydrogen escape from the planets early in their lifetimes through the hydrodynamic blow-off mechanism (which is the same mechanism that currently drives the solar wind off the Sun).

I. WEATHER

Planetary exploration has revealed that atmospheric circulations come in many varieties. Perhaps ironically, Earth is observed to have the most unpredictable weather of all. The goal of planetary meteorology is to understand what shapes and maintains these diverse circulations. The *Voyager* spacecraft provided close-up images of the atmospheres of Jupiter, Saturn, Uranus, and Neptune and detailed information on the three satellites that have atmospheres: Io, Titan, and Triton. The atmospheres of Venus and Mars have been sampled by entry probes, landers, orbiting spacecraft, and telescopic studies. Basic questions like why Venus's atmosphere rotates 60 times faster than does the planet, or why Jupiter and Saturn have superrotating equatorial jets, do not have completely satisfactory explanations. However, by comparing and contrasting each planet's weather, a general picture has begun to emerge.

Sunlight is the primary energy source for the winds on the terrestrial planets. Internal heat combines with sunlight to drive the circulations of the giant planets. Given that we know the nature of the energy source for the problem, and we know that the problem is governed by Newton's laws of motion, why then are atmospheric circulations difficult to understand? Several factors contribute to the complexity of observed weather patterns. In the first place, fluids move in an intrinsically nonlinear fashion that makes paper-and-pencil analysis formidable and often intractable. Laboratory experiments and numerical experiments performed on high-speed computers are often the only means for making progress on problems in geophysical fluid dynamics. In the second place, meteorology involves the intricacies of moist thermodynamics and precipitation, and we are only beginning to understand and accurately model the microphysics of these processes. And for the terrestrial planets, a third complexity arises from the complicated boundary conditions that the solid surface presents to the problem, especially when mountain ranges block the natural tendency for winds to organize into steady east–west jet streams. For oceanographers, even more restrictive boundary conditions apply, namely, the ocean basins, which strongly affect how currents behave. The giant planets are free of this boundary problem because they are completely fluid down to their small rocky cores. However, the scarcity of data for the giant planets, especially with respect to their vertical structure beneath the cloud tops, provides its own set of difficulties.

Most of the atmospheres in the solar system are rapidly rotating. On Jupiter the relative changes in wind speed of the east–west jet streams are on the order of 100 m s^{-1}, which corresponds to only about a 5-min deviation from the 9.92-h bulk rotation period of the planet. The winds on Earth are weaker, reaching 60 m s^{-1} in the jet streams. Earth rotates only 41% as fast as Jupiter, but this is still considered to be rapidly rotating. When planetary atmospheres are spun up in this manner, it is convenient to measure wind speeds relative to a noninertial reference frame that is rotating with the planet. For a terrestrial planet, the rotation rate of the solid surface provides a convenient reference frame. Because there is no solid surface on a giant planet except at its small rocky core, a problem arises. The problem is solved by measuring the winds relative to the rotation period of the planet's magnetic field, which is tied to the interior and indicates the bulk rotation (historically called the System III reference frame).

The use of a noninertial reference frame introduces new acceleration terms into the equations of motion: the centripetal acceleration and Coriolis acceleration. The centripetal acceleration naturally combines with the gravitational force, and the resultant force is usually referred to as simply the gravity. For rapidly rotating planets, the Coriolis acceleration is the dominant term that balances the horizontal pressure-gradient force in large-scale circulations, leading to geostrophic balance. Two other important effects of rapid rotation are the suppression of motions in the direction parallel to the rotation axis, called the Taylor–Proudman effect, and the coupling of horizontal temperature gradients with vertical wind shear, a three-dimensional relationship described by the thermal wind equation.

J. IRONIC UNPREDICTABILITY—AN ANECDOTE

The fickleness of Earth's weather compared to that of the other planets provides many fascinating scientific problems for meteorologists. Trying to live on such a planet presents Earth's inhabitants with practical problems as well. On the lighthearted side, there are common bromides such as "If you don't like the weather, wait 15 minutes," and "Everybody complains about the weather, but nobody does anything about it." On the serious side, lightning storms and tornados wreak havoc every year, and before the advent of weather satellites, hurricanes once struck populated coastlines without warning, causing terrible loss of life.

Even now, the tracks of hurricanes are notoriously difficult to predict. The point is best made with an example, and the following is a lighthearted anecdote from the author's personal experience: Perhaps he should have known better than to leave the windows of his apartment open on such a warm, breezy morning in the summer of 1991, but the apartment needed airing out, and the author was preoccupied with a desire to come up with a good way to illustrate to a group of distinguished terrestrial meteorologists that

the weather on Jupiter is more predictable than the weather on Earth. And so, he left the windows open, locked the door, and headed out to Boston's Logan airport to begin a ten-day trip to a symposium on "Vortex Dynamics in the Atmosphere and Ocean," which was to be held in Vienna. His preoccupation was not helped by the use of the singular "atmosphere" in the symposium's title, which, one could argue, should have been written with the plural "atmospheres." To be sure, Earth has its great vortices, like Gulf Stream rings, Mediterranean salt lenses, and atmospheric blocking highs, and even more powerful storms, like hurricanes, which are driven by moist thermodynamics (in fact, Hurricane Bob was at that moment slowly heading toward the Carolina coast). Yet Jupiter's Great Red Spot and the hundreds of other long-lived vortices found on the gas-giant planets are in many ways simple systems to study, and we have excellent observations of them from spacecraft like *Voyager* (and now *Galileo*).

After arriving at the conference, the author decided to make his case by pointing out that a *Voyager*-style mission to track hurricanes on Earth would most likely end in failure. This is because the *Voyager* cameras had to be choreographed 30 days in advance of each encounter to give the flight engineers time to sort through the conflicting requests of the various scientists and time to program the onboard computer. For the atmospheric working group, this constraint meant that success or failure depended on the accuracy of 30-day weather forecasts for the precise locations of the drifting Great Red Spot and other targeted features. On Earth, storms rarely last 30 days, and much less do they end up where they are predicted to be going a month in advance. The fact that the *Voyager* missions to Jupiter were a complete success, as were the subsequent Saturn, Uranus, and Neptune missions, illustrates in a practical way the remarkable predictability of the weather on the gas-giant planets relative to on Earth. [*See* PLANETARY EXPLORATION MISSIONS.]

Having made his point, on the road back to the Vienna airport after the conference the author was getting accustomed to the fact that taxis in German-speaking countries are Mercedes, when the driver explained that the announcer on the radio was saying that there had just been a coup in Moscow. This left him worried about his Russian colleagues, several of whom he had just met in the preceding week. On the flight back across the Atlantic, he was thinking about this when the Lufthansa pilot announced, with resignation in his voice, that because of thunderstorms the plane could not land in Boston and was being redirected to Montreal. After about two hours in Montreal, where the plane was nestled between several other waylaid international planes that were littered across the tarmac, the go-ahead was given to finish the trip to Boston. The landing was bumpy and the skyline was disturbingly dark, but there was a beautiful sunset that was framed with orange, red, and black clouds. It was only after getting off the plane that he first learned that Hurricane Bob had just hit Boston. Boston? Wasn't Bob supposed to hit the Carolina coast? It was difficult not to take this egregious forecasting error personally. On returning home, jet-lagged, the author discovered that his apartment was dark, the electricity was out, the windows were of course still open, the curtains, carpet, and furniture were completely soaked, and wall hangins and broken glass were strewn about the floor. The irony of the situation is not hard to grasp. *Voyager* would have returned beautiful, fair weather images of North Carolina and South Carolina, and would have completely missed the hurricane, which ended up passing through this writer's apartment a thousand kilometers north of the previous week's prediction.

VI. MAGNETOSPHERE

High above Earth's turbulent atmosphere, a different kind of storm rages, magnetic storms, in Earth's energetic magnetosphere. This is a highly time-variable region formed by Earth's strong magnetic field and shaped by the solar wind. Magnetospheres are also found above the four gas-giant planets, and the *Galileo* spacecraft has even discovered that Jupiter's satellite Ganymede has a magnetosphere. Magnetospheric phenomena are complex and involve the interaction of magnetic forces with plasmas and fluids. The general branch of physics that deals with this is called magnetohydrodynamics (MHD). The solar wind blows a magnetosphere back like a wind sock, forming a long magnetotail. The overall shape is like that of a comet. Jupiter's magnetotail is so long that Saturn passes through it, in fact, this occurred during the *Voyager* spacecraft encounters with Saturn. [*See* PLANETARY MAGNETOSPHERES; THE SOLAR WIND.]

Earth's magnetosphere has been extensively studied, but there is still much to learn. The Van Allen radiation belts were an important early discovery. It is now appreciated that different regions in the magnetosphere can be characterized by the nature of the plasma inhabiting them. For planets other than Earth, most of what we know comes from flyby missions. Unfortunately, these are limited to sampling only one chord

of a planet's inherently three-dimensional magnetosphere, and they also take only a snapshot. Therefore, separating spatial and temporal fluctuations is difficult. As such, it is interesting to see what can be learned from Earth from a flyby, in the same spirit as was discussed earlier with regard to the spacecraft evidence for life on Earth. [*See* PLANETARY MAGNETOSPHERES.]

The first flyby encounter with Earth's magnetosphere was carried out by the European Space Agency's *Giotto* spacecraft, whose primary mission was to encounter Comet Halley in 1986, which it did in a spectacular fashion, returning the first clear images of a comet's nucleus. The second was the *Galileo* spacecraft encounter in December, 1990. The following is a brief summary of what *Galileo* detected.

The encounter took place during an active geomagnetic period and recorded a rich set of observations. It is convenient to use the planet's radius, in this case denoted R_E, when discussing the geometry of a magnetosphere. Variations in the solar wind were monitored by the *IMP 8* spacecraft, which was about 38 R_E upstream of Earth and on nearly the Sun–Earth line as *Galileo* flew by. A host of ground-based magnetometers provided simultaneous observations that allowed the *Galileo* observations to be put into perspective after the fact. The flyby also gave the *Galileo* magnetometer team a valuable chance to calibrate its instruments.

Galileo approached Earth's magnetosphere from the tailward direction and crossed the magnetopause at 0304 UT, about 100 R_E downtail, as expected. This

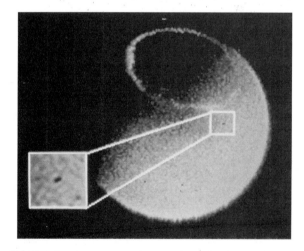

FIGURE 4 *Dynamics Explorer 1* ultraviolet image of Earth's northern auroral oval and dayglow, taken on October, 1981, from an altitude of 18,500 km. The dayglow emission is from atomic oxygen. Careful scrutiny reveals small atmospheric holes that are thought to be made by the atmospheric entry of tiny comets with masses of about 10^5 kg (about 100 tons). (Courtesy of Dr. L. A. Frank, University of Iowa.)

crossing was marked by a smoothing of magnetic field fluctuations. During the 2 hours after entry, the field strength was seen to increase slowly. The signature of a plasma bubble, or plasmoid, was encountered at 0500 UT. The center of a structure called a flux rope was apparently encountered at 0714 UT and showed features that resemble key aspects of numerical MHD simulations. Over the course of the first 16 hours, the spacecraft passed back and forth across the central neutral-plasma sheet. This sheet has a geometry that is affected by the dipole tilt of the planet's magnetic field, and interesting changes in all three components of the vector magnetic field were seen. Magnetic activity was observed that correlates with detections made at high-latitude ground stations. Further activity was detected over the auroral zone latitudes. A satellite image of Earth's northern aurora is shown in Fig. 4 Closest approach to Earth was at 2034 UT. When the magnetospheric bow shock was crossed, *Galileo* reported the expected drop in field strength. Overall, the experiment revealed a complicated correlation between ground-based detections of magnetic storm fluctuations and concurrent spacecraft observations made in the magnetotail.

VII. CONCLUSIONS

Viewing Earth as a planet is the most important change of consciousness that has emerged from the space age.

Detailed exploration of the solar system has revealed that it is a beautiful but violent place, and that our special planet has no special immunity to the powerful forces that continue to shape the solar system. The ability to remotely sense Earth's dynamic atmosphere, oceans, biosphere, and geology has grown up alongside our ever-expanding ability to explore distant planetary bodies. Everything we have learned about other planets influences how we view Earth. Comparative planetology has proven in practice to be a powerful tool for unraveling the most complicated of Earth's many secrets. The lion's share of understanding still awaits us, and in its quest we continue to be pulled outward.

BIBLIOGRAPHY

Beatty, J. K., and Chaikin, A. (eds.) (1990). "The New Solar System," 3rd Ed. Sky Publishing.

Geissler, P., Thompson, W. R., Greenberg, R., Moersch, J., McEwen, A., and Sagan, C. (1995). Galileo multispectral imaging of Earth. *J. Geophys. Res.* **100,** 16,895–16,906.

Grieve, R. A. F. (1993). Impact craters: Lessons from and for the Earth. *Vistas Astron.* **36,** 203–230.

Hartmann, W. K. (1993). "Moons and Planets," 3rd Ed. Wadsworth.

Hide, R., and Dickey, J. O. (1991). Earth's variable rotation. *Nature,* **253,** 629–637.

Kivelson, M. G., Kennel, G. F., McPherron, R. L., Russell, C. T., Southwood, D. J., Walker, R. J., Khurana, K. K., Coleman, P. J., Hammond, C. M., Angelopoulos, V., Lazarus, A. J., Lepping, R. P., and Hughes, T. J. (1993). The Galileo Earth encounter: Magnetometer and allied measurements. *J. Geophys. Res.* **98,** 11,299–11,318.

EARTH AS A PLANET: SURFACE AND INTERIOR

I. Introduction: How Does Earth Fit into the Scheme of Things and What is the Role of Geomorphology in Planetary Exploration?

II. Physiographic Provinces of Earth: Settings for Geomorphic Processes

III. Earth Surface Processes: How "Geomorphic Energy" is Expended within Various Geomorphic Settings

IV. Tools for Studying Earth's Deep Interior

V. Seismic Sources

VI. Earth's Radial Structure

VII. Earth in Three Dimensions

VIII. Earth as a Rosetta Stone

David C. Pieri
Jet Propulsion Laboratory,
California Institute of Technology

Adam M. Dziewonski
Harvard University

GLOSSARY

Drainage basin: Geomorphic entity that contains a drainage network. Typically a bowl-shaped catchment in humid areas, drainage basins in arid regions can be quite flat. Drainage patterns typically reflect the relative slopes (exponentially decreasing in the downstream direction) of the drainage basin in the systematic orientations of the joining or "junction" angles of their confluences, at all scales.

Geodesy: Global seismic tomography (GST) models of velocity anomalies and topography of major discontinuities (CMB, 660 km) can be compared with data on the Earth's rotation obtained by VLBI techniques.

Geomagnetism: Coincidence of the regions with the high rate of the secular variations and slow velocity anomalies (higher than average temperature) provide constraints on the thermal and mechanical coupling between the mantle and the core. The virtual geomagnetic pole paths, according to some publications, coincide with the two high velocity regions circumscribing the Pacific and, effectively, connecting the North and South Pole.

Geomorphology: Science of landscape analyses. Geomorphic investigations deal with the processes and time scales of landscape formation and degradation.

Gravity: Under the assumption that seismic anomalies are proportional to density perturbations, they provide constraints on the modeling of large wavelength gravity anomalies, the viscosity distribution in the mantle, and the ratio $\partial\ln\rho\backslash\partial\ln v$.

Mafic: Ferro-magnesian. Geological jargon that usually refers to basalts and other refractory igneous rock types.

Mantle convection: Distribution of seismic anomalies represents the current configuration of thermal and compositional heterogeneity advected by mantle flow, imposing a complex set of constraints on the possible modes of convection in the mantle.

Mineral physics: *In situ* ratio $\partial\ln v_s/\partial\ln v_p$ inferred for the lower mantle from GST is much higher than determined, at relatively low pressures, in the laboratory. Only recently has a generally accepted explanation of this fact been provided.

Orogenic, orogeny: Process of mountain building, with uplift generally occurring as a result of tectonic plate collisions.

Petrology and geochemistry: GST models have the potential to provide integral constraints on petrological and thermal models of the ridge systems. Velocity anomalies associated with continental shields confirm the hypothesis of "continental roots."

Physiographic: Referring to the physical appearance of the landscape.

Subaerial: Referring to landscapes, such as islands or continents, that are exposed to the air.

Terrane: A particular type of terrain. Generally used to denote the kind of terrain dominated or formed by a particular geomorphic process regime, such as a volcanic terrane or an aeolian terrane.

Tholeiitic: Referring to basaltic rocks generally found on the ocean floor, erupted from oceanic ridge zones or from shield volcanoes. Such rocks are considered in the mafic family.

Viscosity: Property of a fluid that resists flow; fluid dynamic stiffness or, in a sense, internal friction. For lava flows that typically have remarkably little excess energy above their solidus, viscosity can be the determining factor for the magnitude and morphology of lava flow fields associated with volcanoes and is often exponentially dependent on the core temperature of the flow.

I. INTRODUCTION: HOW DOES EARTH FIT INTO THE SCHEME OF THINGS AND WHAT IS THE ROLE OF GEOMORPHOLOGY IN PLANETARY EXPLORATION?

The surface of the Earth is perhaps the most geochemically diverse and dynamic among the planetary surfaces of our solar system. Uniquely, it is the only one with liquid water oceans under a stable atmosphere, and — as far as we now know — it is the only surface in our solar system that has given rise to life. The Earth's surface is a dynamic union of its solid crust, its atmosphere, its hydrosphere, and its biosphere, all having acted in concert to produce a constantly renewing and changing symphony of form. The unifying theme of the Earth's surficial system is water — in liquid, vapor, and solid phases — which transfers and dissipates solar, mechanical, chemical, and biological energy throughout global subaerial and submarine landscapes. The surface is a window to the interior processes of the Earth, as well as the putty which atmospheric processes continually shape. It is also the Earth's interface with extraterrestrial processes and, as such, has regularly borne the scars of impacts by meteors, comets, and asteroids, and will continue to do so. Until the advent of humans, all terrestrial organisms, much as the hapless two-dimensional Flatlanders, have perceived the Earth's global landscape as the limit of their respective universes.

Indeed, even human appreciation of Earth's role as a planetary body in the solar system was slow in coming. Until about 500 years ago, the prevailing Western view held that the Earth was most likely the center of the universe. Among Christians of that time, this view was taken as dogma, and so strongly held were these views among the establishment that even as prominent a physicist of the day as Galileo Galilei was not immune. He was charged and convicted of "suspicion of heresy" by the Roman Catholic Holy See. His crime in 1642 was the publishing of his analyses of telescopic observations in defense of the Copernican heliocentric theory versus the Aristotelian geocentric theory, which held that the Earth was the stationary center of the universe (Fig. 1).

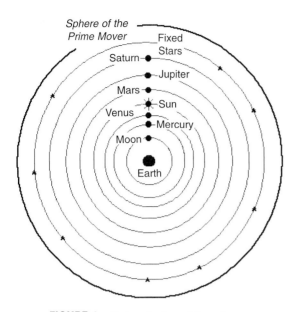

FIGURE 1 Ptolemeic view of the universe.

In many ways, during the latter half of the 20th century, we have undergone a more subtle, but analogous, evolution in thinking. This has occurred generally as a result of the exploration of our solar system with spacecraft and the detailed geomorphic analyses of the planetary surface images they returned. The spacecraft-based exploration of Mars has played a particularly pivotal role in this regard. Before the exploratory voyages of Russian and American spacecraft, the planets and their satellites were typically perceived as "out there" — strange, exotic entities with unimaginably hostile and alien environments — and not as "real" places. This was especially true after the initial U.S. *Mariner 5* and *7* observations of Mars (1965 and 1969) until the geologically provocative observations of *Mariner 9*. Before the 1971 *Mariner* mission, a post-Lowellian Mars was perceived as barren and lunar-like. As a result, the Earth seemed to many like the only reasonable past or present abode of life in the solar system.

During and after the *Mariner 9* mission, however, the situation changed dramatically: Mars was now viewed as a geologically active planet. *Mariner 9* images clearly showed massive volcanoes and deep canyons and, most importantly, very obvious evidence of the work of vast amounts of water in the landscape for substantial lengths of time. More detailed *Viking Orbiter* and *Lander* images later reinforced and amplified this view of Mars as a geologically active planet that was water rich at its surface during early epochs. Most recently, detailed surface imaging from the Mars *Pathfinder* lander and rover confirmed

the existence of apparent conglomerates in an area thought to be strewn with rocky debris from past catastrophic floods. Thus, our currently, rapidly evolving understanding of Martian geomorphology and environmental history has moved us markedly away from a geocentric "Earth-chauvinistic" point of view (very analogous to the 16th century Copernican vs. Ptolemeic arguments) toward a more expansive perspective that recognizes the possibility that relatively benign, life-engendering environments may occur on other planets. As promising as the early Martian surface environment appears, however, it was not sustainable, and the Earth stands out in our solar system as a planet with a surface characterized mainly by (a) open and deep oceans of liquid water complemented by subaerial land masses of comparatively modest extent and (b) a relatively temperate and stable climate under a benign and protective atmosphere. If we accept the view that Earth-type planets can exist elsewhere in the universe, they obviously will deserve special attention as potential incubators of life. [*See* MARS: ATMOSPHERE AND VOLATILE HISTORY.]

The search for other planetary systems and for life beyond the Earth will likely be one of humankind's transcendent pursuits of the 21st century. Even now, at the end of the 20th century, it is rapidly becoming clear, given recent Earth and space-based telescopic observations, that planetary systems may be relatively common throughout our galaxy. The proportion of these systems that harbor Earth-type "water planets" is, at present, still unknown. It is thought, however, that stars that resemble the sun in mass and spectral type (so-called "G-type" stars in the Hertzsprung-Russell Main Sequence classification scheme) may be more likely to have terrestrial planets and satellites in orbit around them. Such terrestrial bodies may be predominantly silicate in composition, and those of moderate age may have differentiated interiors with metallic cores (and thus possibly interior magneto-dynamos), which would generate biologically protective magnetic fields, such as the Earth's. Internally active planets stand a good chance of "sweating out" or degassing their original volatile inventory in volcanic eruptions, powered by internal radiogenic heating and gravitationally generated heat provided by the initial impact processes associated with planetary accretion. Such degassed volatiles (e.g., water vapor, H_2O; carbon dioxide, CO_2; methane, CH_4), along with possibly substantial additional volatile inventories imported by massive cometary impacts, form the original postaccretion atmospheres and oceans. If a newly accreted planet is roughly Earth-sized (as opposed to Mars-sized), it

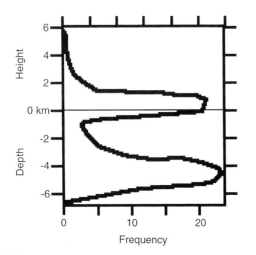

FIGURE 2 Hypsometric diagram of the Earth's topography. The bimodal seafloor–continental altitude dichotomy is obvious. The interface between the two, subject to tidal and climatic fluctuation stress, is thought to have provided, in part, stimuli for biological evolutionary adaptations.

will most likely (barring an improbable collision with another planet-sized body) retain those volatiles on or within its crustal layers over its lifetime, as will be the case with the Earth. [*See* EXTRA-SOLAR PLANETS.]

For the formation of the life forms familiar to us (although this may well be a biased sample), the ready availability of water, in some form, preferably liquid, seems to be essential. Likewise, the existence of an environmentally and temporally variable land–water interface may also be important in promoting the growth of tough, environmentally flexible, and persistent life. Thus, the formation of even modest land masses at a planetary surface that is otherwise dominated by the presence of liquid water may be key in pushing forward transcendent biological processes. For instance, the distribution of the earth's landscape altitudes, relative to the mean geoid, is bimodal — continental and sea floor (Fig. 2). Although limited in percentage of surface area coverage, the *interface* between the two modes is a relatively high-energy place called the littoral or tidal zone. Ocean tides in this zone generate frequent (twice daily) environmental stresses on its residents that profoundly encourage evolution and natural selection. [*See* PLANETS AND THE ORIGIN OF LIFE.]

Climatic changes, such as the great warm–wet epochs that induced inland seas to form over millions of years during various geologic eras on the Earth, probably serve a similar purpose at much longer time scales, as does the environmental stress associated with Ice Ages. Climatically cold periods not only result in

reductions of average continental air and ground temperatures at ice-sheet margins, but the massive sequestering of solid water on land areas as continental ice sheets radically changes sea levels and thus the extent and locations of crucial littoral zones. Additionally, critical accidents of continental geography sometimes amplify these effects, as in the case of the geologically recent, total desiccation of the Mediterranean Sea. Much of the understanding of the environmental history of the Earth is inextricably tied to the development and maturing of geomorphology as a geologic and geographic discipline.

Our own solar system has a variety of terrestrial planets and satellites in various hydrologic states with radically differing hydrologic histories. Some appear totally desiccated, such as the Moon, Mercury, and Venus. In some places, water is very abundant at the surface, such as on Earth and Europa (solid at the surface and possibly liquid underneath). In other places, such as Mars and Ganymede, it appears that water may have been very abundant in liquid form on the surface in the distant past. Also, in the case of Mars, water may yet be abundant in solid and/or liquid form in the subsurface today. Thus, for understanding geological (and, where applicable, *biological*) processes and environmental histories of terrestrial planets and satellites within our solar system, it is crucial to explore the geomorphology of surface and submarine landforms and the nature and history of the land–water interface where it existed. Such an approach and "lessons learned" from this solar system will also be key in future reconnaissance of extra-solar planets. [*See* MARS: SURFACE AND INTERIOR.]

II. PHYSIOGRAPHIC PROVINCES OF EARTH: SETTINGS FOR GEOMORPHIC PROCESSES

A. BASIC DIVISIONS

From a geographic and geomorphologic point of view, especially when seen from space, the surface of the Earth is dominated by its oceans of liquid water: approximately 75% of the Earth's surface is covered by liquid or solid water. The remaining 25% of subaerial land, the subject of nearly all historical geological and geomorphological study, lies mainly in its Northern

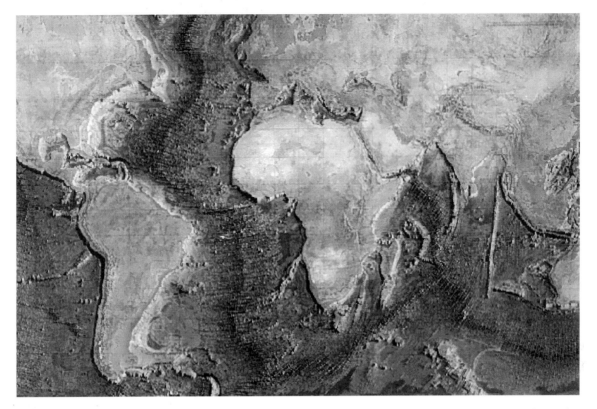

FIGURE 3   Section of the physiographic map of the Earth (Heezen and Tharp, 1997). Midoceanic ridges are among the Earth's most prominent and continuous landforms at this scale.

Hemisphere, where most of the world's population lives. The Southern Hemisphere is dominated by oceans, some subaerial continental and archipelago land masses (mainly parts of Africa, South America, southeast Asia, and Australia), and the large, mainly subglacial, island continent of Antarctica (Fig. 3).

Remarkably, despite the fact that geological and geographical sciences have been practiced on the Earth for about 200 years, it has only been during the last 40 or so that scientists have begun detailed mapping and geophysical explorations of the submarine land surface. Much of this research has been inextricably tied to post-World War II and Cold War American and Russian naval operations, particularly the development and operation of large atomic-powered submarines, which navigate in abyssal environments. From subsea remote-sensing technology, however, one of the most profound discoveries in the history of geological science has emerged: the paradigm of "plate tectonics." The extent, morphology, and dynamics of these massive tectonic plates were only realized after careful topographic and geomagnetic mapping of the intensely volcanic midoceanic ridges and their associated parallel-paired geomagnetic domains (Fig. 4).

Similar topographic mapping of the corresponding submarine trenches along continental or island-arc margins was equally revealing. The midoceanic ridges were found to be sites of accretion of new volcanically generated plate material, and the trenches the sites of deep subduction, where 50- to 60-Myr-old ocean crust is consumed beneath other overriding crustal plates. Tectonic plates represent the most fundamental and largest geomorphic provinces on Earth.

The Earth's crustal plates come in two varieties: oceanic and continental. Oceanic plates comprise nearly all of the Earth's ocean floors, and thus most of the Earth's crustal area. They are composed almost exclusively of ferromagnesian igneous rocks, mainly thoeleiitic basalts. Oceanic plates are created by volcanic eruptions along the apices of the Earth's midoceanic ridges: 1000-km-long sinuous ridges that rise from abyssal plains on either side, which are typically less than about 10 km thick (Fig. 5). Here, nearly continuous volcanic activity from countless submarine volcanic centers (far more than the 1000 or so active subaerial volcanoes) provides a steady supply of new basalt, which is accreted and incorporated into the interior part of the plate.

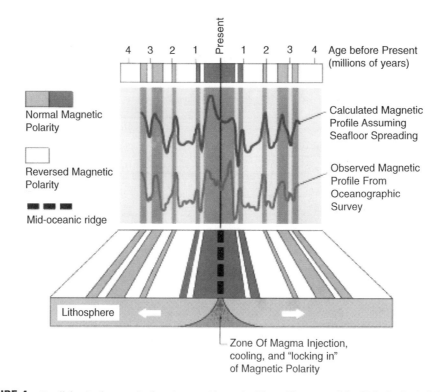

FIGURE 4   Parallel-paired magnetic domains at midoceanic ridges. (Courtesy of the U.S. Geological Survey.)

At plate edges, roughly the reverse occurs, where the outer, oldest plate margins are forced below overriding adjacent plate edges. Usually, when two oceanic plates collide, the resulting subduction zone forms an island arc along the trace of the collision. The islands, in this case, are the result of the eruption of more silicic andesitic magmas generated as part of the subduction process. The subducted plate margin is consumed along the axis of the resulting trench. Because the more silicic island arcs tend to be less dense and thus more resistant to subduction, they can be accreted onto plate margins and can thus increase the areal extent at the edges of oceanic plates or can enlarge the margins of existing continental plates. Trenches are the deepest part of the ocean: the Mariana Trench at −35,785 feet (below sea level) and the Tonga Trench at −35,326 feet are the two deepest. Oceanic plate material is typically less than about 60–100 Myrs old because that is roughly the survival time for thoeleiitic basalts between eruption at midoceanic ridges and subduction.

Continental plates tend to consist of much more silicic material, as compared with oceanic plates. Because of their lower density and the fact that they are isostatically compensated, they are much thicker than oceanic plates (30–40 km thick) and tend to "float" over the denser, more mafic (ferromagnesian) subja-

cent material in the Earth's upper mantle. When continental plates collide with oceanic plates, deep subduction trenches, such as the Peru–Chile trench along the west coast of South America result, occur, as the oceanic plate is forced under the much thicker and less dense continental plate. Usually, the landward side of the affair is marked by so-called *Cordillieran* belts of mountains, including andesitic-type volcanoes, which parallel the coastline. The Andes Mountains are an example of this type of tectonic arrangement.

When continental plates collide, a very different tectonic and geomorphic regime ensues. Here, equally buoyant and thick continental plates crush against each other, resulting in the formation of massive fold belts and towering mountains, as long as the tectonic zone is active (Fig. 6).

The Himalayan Range, driven by the tectonic collision of the Indian subcontinent with the southern part of Asia, is a classic example of this kind of phenomenon. Generally, the uplift rates are quite high in this type of collision (e.g., ~1 cm per year). Thus, a very high and steep topography is created (e.g., the Tibetan Plateau; Mt. Everest), often high enough to fundamentally alter atmospheric circulation, far beyond the familiar "rain shadow" effect, as is the case with both the Andes and the Himalayas. When aggregate stresses are ten-

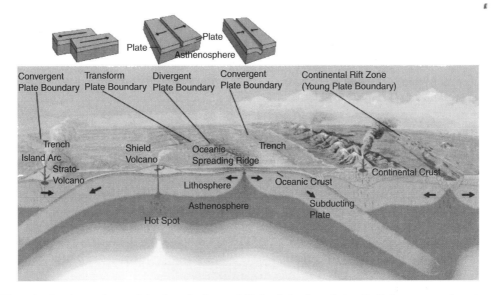

FIGURE 5 Tectonic plate interactions and the three fundamental kinds of plate boundaries. (Left) A convergent boundary caused by the subduction of oceanic material as it is overridden by another oceanic plate. (Center left) A subplate hot spot capped by a shield volcano (e.g., Hawaiian Islands). (Center right) A divergent plate boundary, in particular, a midoceanic spreading ridge. (Right) Another kind of convergent plate boundary, where the oceanic crust is being subducted by overriding continental crust, producing a chain of volcanic mountains (e.g., Andes Mountains). (Far right) A continental rift zone, another kind of divergent plate boundary (e.g., East African Rift). Finally, a transform plate boundary is shown at the upper middle of the scene, where two plates are sliding past each other without subduction. The three relationships are shown as block diagrams at the top of the figure. (Courtesy of the U.S. Geological Survey.)

sional rather than compressive, *extensional mountain ranges* can form, as tectonic blocks founder and rotate. The American Basin and Range Province is a good example of that type of mountain terrane. Another large subaerial extensional tectonic landform is the

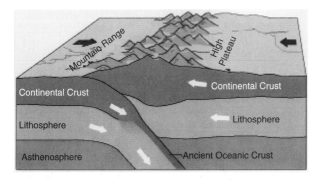

FIGURE 6 Continent–continent collision in cross section. Here continental crust and ancient oceanic crust are being subducted below an overriding plate of continental material. As a result, a towering mountain range arises (e.g., Himalayan Range). Such mountains remain rugged and topographically high as long as they are sustained by the plate collision process. If they are no longer being dynamically sustained, erosion rapidly takes its toll, and they are reduced to much more rounded and lower forms (e.g., Appalachian Mountains). (Courtesy of the U.S. Geological Survey.)

axial *rift valley* and associated inward-facing fault scarps, which form when aggregate tensional stresses tend to pull a continental plate apart (e.g., the East African Rift Valley).

All of the geomorphic provinces just discussed, with the exception of submarine abyssal plains, tend to be very dynamic, with lifetimes that are intrinsically short (100–200 Myr) relative to the age of the Earth (5 Byr). Some of the stable interior areas of continental plates, or *cratons,* do possess landforms and associated lithologic regions with ages comparable within a factor of two or three to the age of the Earth (2–3 Byr). The interior of the Canadian Shield and the Australian continent are two such special areas. Despite having been scoured repeatedly by continental ice sheets, the granitic craton of the Canadian Shield possesses a record of giant asteroidal and cometary impacts that are about 2 Byr old. For the most part, the oldest of these features, which were possibly once circular impact basins, have been distorted into quasi-linear sinuous features, hundreds of kilometers long, called *greenstone belts,* composed of highly ultramafic (i.e., very rich in refractory minerals) rocks. Near Sudbury, Ontario, the rich nickel mining district is exploiting highly deformed metal sulfide deposits that probably occurred as a result of a giant asteroidal impact eons ago (original impact basin size ~200 km diameter; age ~1.8 Byr). One of the Earth's largest impact features that retains a clearly circular character is the Manicouagan, Quebec, impact structure, which is over 100 km in diameter and probably occurred about 200 Myr ago. Ancient impact craters abound in Australia as well. These interior cratonic areas, in contrast to most of the rest of the Earth, which is mobile and active, provide a chance to view a part of the long sweep of the Earth's surface history. They are thus important, particularly in trying to understand how the environmental history of the Earth compares to that of the other terrestrial planets.

B. LANDFORM TYPES

1. Submarine Landforms

Geomorphically, *submarine oceanic basins* comprise the areally dominant landform of the Earth, but, ironically, they are probably less well explored than the well-imaged surfaces of Mars and Venus. Dominant features of oceanic basins are the oceanic ridge and rise systems, which have a total length of about 60,000 km (~1.5 times the equatorial circumference of the Earth), rise to 1–3 km above the average depth of the ocean, and can be locally rugged. In the Atlantic Ocean, oceanic rises exhibit a central rift valley that is at the center of the rise, whereas in the Pacific Ocean this is not always present. As mentioned earlier, new oceanic crust is created there, as the central rifts are almost always volcanically active. Ridge and rise systems can be offset by oceanic fracture zones, which are long (up to 3500 km) linear cracks in the oceanic crust. Submarine scarps associated with these features can have substantial relief (100 m to 4 km). Such fractures are the strain manifestations of intraplate stress fields.

Older crust within oceanic basins can have gently rolling abyssal hills, which are generally smoother than the ridge and rise systems. These may have been much more rugged originally, but are now buried beneath accumulated sediment cover. Perhaps the most areally dominant feature of ocean basins (with the largest ones occurring in the Atlantic Ocean) is the predominantly flat abyssal plains that stretch for thousands of kilometers, usually also covered with accumulated marine sediments. Generally very flat, in places they are punctuated by seamounts, which are conical topographic rises sometimes topped by coral lagoons, or which sometimes do not reach the oceans' surface. These features are subsea volcanoes associated with island arcs or with midplate hot spots. The famous Emporer Seamount

chain, the southeastern end of which terminates in the Hawaiian Islands, is a good example of the latter type. Here, a subcrustal mantle-plume has generated a stable hot spot over which the Pacific Plate has progressed steadily to the northwest, toward the Kurile–Kamchatka Trench and subduction zone. The current center of activity is under the Island of Hawaii and is driving eruptions at Kilauea and Mauna Loa volcanoes, as well as offshore at a new submarine volcanic center, called Loihi, now several hundred meters below the ocean surface. Eventually Loihi Volcano will be built by basaltic eruptions, layer by layer, into the newest of the Hawaiian Islands.

Oceanic margins represent another important, although more areally restricted, submarine landform province. "Atlantic style" continental margins (e.g., Fig. 7) tend to exhibit substantial ancient sediment accumulations and a shelf-slope-rise overall morphology, which probably represents submerged subaerial landscapes relict from the last Ice Age, when the sea level was lower (about 135 m below current sea level, worldwide). Continental shelves are usually less than about 100 km in width and have very shallow (~0.1°) topographic slopes. They typically end in a slope break that merges into the continental slope (~4° slope, about 50 km wide), which in turn merges into a gentle continental rise (~0.2° slope, about 50 km wide), which then typically transitions into an abyssal plain. Submarine canyons (also probably relict Ice Age features, e.g., Hudson Canyon) can deeply cut the continental shelf and slope and terminate in broad submarine sediment fan deposits at the seaward canyon outlet. "Pacific style" oceanic margins can be even more narrow. Along the margins of continents of the Pacific Rim, a short shelf and slope can terminate into deep submarine trenches, manifested by subduction zones (e.g., South America, Kamchatka), up to 10 km deep. Similar fore-arc submarine morphology is observed along the margins of Pacific island arcs (e.g., Aleutians and Kurile Is). Much shallower "back-arc" basins occur behind the arcs, on the overriding plate (e.g., Sea of Okhotsk).

2. Subaerial Landforms

The subject of classic geomorphological investigations, and historically far more well studied because they are where people on Earth live, are the *subaerial* landscapes. These terranes exist almost exclusively on continents; however, some important subaerial landscapes (particularly volcanic ones, e.g., Hawaii, Galapagos Islands) exist on oceanic islands. [In that vein, it is ironic that much of the historically seminal thinking that

FIGURE 7 Sonar reconstruction of the submarine margin offshore of the New Jersey coast. Numerous, deep submarine canyons cut by subsea avalanches occur off New Jersey, Delaware, and Maryland. Vertical exaggeration is 4:1, and depth is indicated by the scale bar at the bottom of the picture. (Courtesy of Bill Haxby and Lincoln Pratson, Lamont-Dougherty Earth Observatory, Columbia University.)

occurred in the development of geomorphic science was stimulated by observations of the landscape of England — a large island!] Most continental landscapes are predominately Cenozoic to late Cenozoic in age, because over that time scale (65 Myr or so), the combined action of plate tectonics, constructive landscape processes (e.g., volcanism and sedimentary deposition), and destructive landscape processes (e.g., erosion and weathering) have tended to rearrange, bury, or destroy preexisting continental landscapes at all spatial scales. Thus, while often retaining the *palimpsest* (i.e., imprint) of preexisting forms, subaerial landscapes on the Earth are constantly being reinvented.

Because the Earth's crust is so dynamic, one must realize from the planetary perspective that any geomorphic survey of the Earth's surface may be representative only of the current continental plate arrangement, and currently associated climatic and atmospheric circulation regimes. Plate tectonics is a powerful force in setting scenarios for continental geomorphology. For instance, during early Cenozoic times the global continental geography was characterized by the warm circum-global Tethys Sea and higher sea levels than now (possibly linked to higher rates of midoceanic spreading), which strongly biased the overall terrestrial climate toward the tropical range. The rearrangement of continental landmasses in the later Cenozoic closed the Tethys Sea, produced a circum-Antarctic ocean, and set up predominantly north–south circulation regimes within the Atlantic and Pacific Oceans. This global plate geography, combined

TABLE I
Classification of Terrestrial Geomorphological Features by Scale[a]

Order	Approximate spatial scale (km²)	Characteristic Units (with examples)	Approximate time scale of persistence (years)
1	10^7	Continents, ocean basins	10^8–10^9
2	10^6	Physiographic provinces, shields, depositional plains	10^8
3	10^4	Medium-scale tectonic units (sedimentary basins, mountain massifs, domal uplifts)	10^7–10^8
4	10^2	Smaller tectonic units (fault blocks, volcanoes, troughs, sedimentary subbasins, individual mountain zones)	10^7
5	10–10^2	Large-scale erosional/depositional units (deltas, major valleys, piedmonts)	10^6
6	10^{-1}–10	Medium-scale erosional/depositional units or landforms (floodplains, alluvial fans, moraines, smaller valleys and canyons)	10^5–10^6
7	10^{-2}	Small-scale erosional/depositional units or landforms (ridges, terraces, and dunes)	10^4–10^5
8	10^{-4}	Larger geomorphic process units (hillslopes, sections of stream channels)	10^3
9	10^{-6}	Medium-scale geomorphic process units (pools and riffles, river bars, solution pits)	10^2
10	10^{-8}	Microscale geomorphic process units (fluvial and eolian ripples, glacial striations)	

[a] From Baker (1986).

with greater ocean basin volume (linked to lower ridge spreading rates) and the onset of continental glaciation, lowered sea levels, exposing large marine continental self-environments to subaerial erosion. Our current global surface environment reflects a kind "oceanic recovery" after the last Ice Age, with somewhat higher sea levels. Thus, our current perception of the Earth's subaerial geomorphic landform inventory is strongly biased by our temporal observational niche in its environmental history. Hypothetical interstellar visitors who arrived here 50 Myr ago or may arrive 50 Myr in the future would likely have a much different perception because of this distinctive dynamic character.

Terrestrial subaerial landform suites are the classic landscapes studied in geomorphology. These are listed in Table I (adapted from Baker, 1986, and Bloom, 1998). Currently, on the Earth, globally dominant subaerial geomorphic regimes are related to the surface transport of liquid water and sediment due to the action of rainfall. Thus drainage basins dominate terrestrial landscapes at nearly all scales, from the continental scale to sub-100-m scales. These include currently active drainage basins in humid and semiarid climatic zones, to only occasionally active or relict drainages in arid zones. Drainage basin topographies and network topologies, however, are strongly influenced by the interplay of the orogenic aspects of plate tectonics (i.e., mountain building) and prevailing climatic regimes, including the biogenic aspects of climate (e.g., vegetative ground cover). Clearly, areas of rapid uplift (e.g., San Gabriel Mountains, California), have characteristically steep bedrock drainages, where gravitational energies are high enough to scour stream valleys, generally have parallel or digitate drainage patterns, have high local flood potentials, and respond strongly to local weather (e.g., spatial scales 10–100 km in characteristic dimension). At the other spatial extreme, major continental drainages (e.g., Amazon River, Mississippi River, Ob River), with highly dendritic overall pattern organization, are low average gradient systems that integrate the effects of a variety of climatic regimes at different spatial scales and tend to respond to mesoscale and larger climatic and weather events (e.g., 100- to 1000-km scale). Rainfall-derived terrestrial drainage patterns tend to evolve networks that strike a delicate balance among the uniform distribution of transport-related work throughout the system, drainage basin topography, and any system perturbations, such as

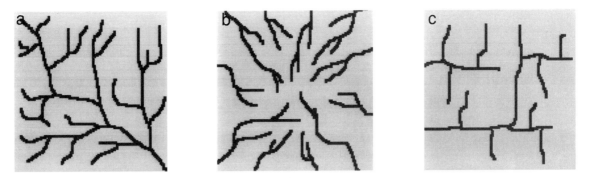

FIGURE 8 Three basic types of drainage patterns, which may be modified into endless variations. (a) Dendritic stream networks are typically expressed in horizontal sediments or uniformly resistant crystalline rock. Such stream networks are typically well evolved and overall probably approach a near-uniform distribution of erosional work. (b) Radial stream networks typically are seen on volcanoes, domes, and erosional knobs or residual topography and are usually the result of the dominance of a prominent central topographic feature with respect to runoff. (c) Trellis-shaped stream networks often occur on dipping or folded sedimentary volcanic or low grade meta-sedimentary rocks. Much of the nature of the underlying substrate and the basin topography is revealed in drainage patterns.

structural controls or tectonically driven base level changes (Fig. 8).

Arid regions make up substantial portions of some continents and their landforms are substantially sculpted by a combination of the rare high-intensity fluvial events, and much more persistent lower intensity eolian (i.e., wind-driven) processes, such as dune formation and sandblasting and weathering. Great sand seas and widespread sand dune fields are present in northern and southern Africa (e.g., Sahara, Kalahari), in the Middle East (e.g., the Arabian Desert), and in Asia (e.g., Tsar and Gobi Deserts). Often the underlying topography in these areas, however, is relict (e.g., a *palimpsest*) from a much more humid past climate regime (e.g., Nile Valley in Egypt and Sudan). The arid cratonic interior of the Australian continent is distinguished by its closed drainages into great ephemeral lakes (e.g., Lake Ayre) and by its tectonic stability. Like the Canadian Shield, it is a very old landscape, retaining the scars of numerous large meteorite impacts that are many tens of millions of years old.

Subaerial volcanic processes produce characteristic landforms in all terrestrial climate zones (Fig. 9). They tend to occur in belts, mainly at plate boundaries, with a few notable oceanic (e.g., Hawaiian Islands) and continental (e.g., San Francisco volcanic field, Arizona; Columbia and Snake River volcanic plains, Pacific Northwest; Deccan Traps, India), exceptions that occur within plate interiors. Although not as massive or as topographically high as their planetary counterparts (e.g., Martian volcanoes), they provide some of the most spectacular and graceful landforms on the Earth's surface (e.g., Mount Fujiyama, Japan; Mt. Kilamanjaro, Kenya). Our planet's central vent volcanic landforms range from the majestic strato-cone volcanic structures just mentioned to large collapse and resurgent caldrons or caldera features (e.g., Valles Caldera, New Mexico; Yellowstone Caldera, Wyoming; Campi Flegrei, Italy; Krakatau, Indonesia). More areally extensive and lower subaerial shield volcanoes, formed by more fluid lavas (and thus with topographic slopes generally less than 5) exist in the Hawaiian Islands, at Piton de la Fournaise (Reunion Island), in Sicily at Mount Etna (compound shield with somewhat higher average slopes, up to ~20°), and the Galapagos Islands (Equador), for example. Subaerial and submarine volcanoes occur on the Earth at nearly all latitudes. Indeed some of the world's most active volcanoes occur along the Kurile-Kamchatka-Aleutian arc, in subarctic to arctic environments, often with significant volcano–ice interaction. High-altitude volcanoes that occur at more humid, lower latitudes (e.g., Andean volcanoes like Nevado del Ruiz) can also have significant magma or lava–ice interactions. Volcanoes also occur in Antarctica, Mt. Erebus being the most active, with a perennial lava pond (Fig. 10). [*See* PLANETARY VOLCANISM.]

C. SUMMARY OF LANDFORMS

Overall, the Earth's geomorphic or physiographic provinces, as compared to those of the other planets in our solar system, are distinguished by their variety, their relative youth, and their extreme dynamism. Many of the other terrestrial-style bodies, such as the Moon, Mars, and Mercury, are relatively static, with landscapes more or less unchanging for billions of years. Although this may not have been the case early in their histories, as far as we can tell from spacecraft

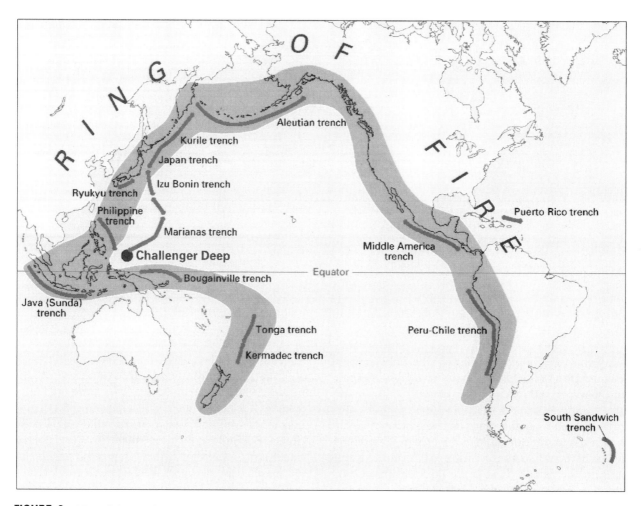

FIGURE 9   Map of the Pacific Ring of Fire, the most active belt of subaerial volcanism in the world. The distribution of volcanoes along the Ring of Fire follows oceanic plate boundaries, reflecting the release of energy to the Earth's surface by the process of plate subduction.

exploration, this is the case now. Other landscapes, such as those on Venus and Europa and a few of the other outer planets' satellites, appear younger and appear to be the result of very dynamic planet-wide processes, and possibly for Venus, a planet-wide volcanic "event." Currently these bodies appear relatively static, although this point may be credibly debated. In terms of dynamics, the Jovian satellite Io, of course, is an exception, with its vigorous ongoing volcanic activity. However, overall, it seems that the crusts of all of these bodies are currently somewhat less variegated than that of the Earth. Be aware, however, that this last statement may turn out to be just another example of "Earth chauvinism" and will be proved wrong once we eventually know the lithologies and detailed environmental histories of these bodies as well as we know the Earth's. [See VENUS: SURFACE AND INTERIOR.]

FIGURE 10   (a) View of remote Mt. Erebus Volcano, Antarctica. Erebus is located on Ross Island and is perennially active, with a persistent lava pond, first discovered by the Shackleton expedition in 1908. Erebus is the world's southernmost active volcano at latitude −77.54° and is 3794 m high. [Courtesy of the Mt. Erebus Volcano Observatory (MEVO) and the U.S. National Science Foundation (NSF).] (b) View of the active lava lake at Mt. Erebus, Antarctica. First Observed in 1908, it remains perennially active, sometimes tossing out anorthosite crystals, volcanic glass, and lava projectiles, or "bombs" up to 1 km from the vent. (Courtesy of the MEVO and NSF.)

FIGURE 10

III. EARTH SURFACE PROCESSES: HOW "GEOMORPHIC ENERGY" IS EXPENDED WITHIN VARIOUS GEOMORPHIC SETTINGS

The expenditure of energy in the landscape is what sculpts a planetary surface. Such energy is either "interior" (*endogenic*) or "exterior" (*exogenic*) in origin. The combined gravitational and radiogenic thermal energy of the Earth (endogenic processes) powers the construction of terrestrial landscapes. Thus, the Earth's main constructional landscape processes, plate tectonics and resulting volcanism, are endogenic processes.

Destructional processes, such as rainfall-driven runoff and streamflow, are essentially exogenic processes. That is, the energy that drives the evaporation of water that eventually results in precipitation, and the winds that transport water vapor, comes from an exterior source — the Sun (with the possible exception of very local, but often hazardous, weather effects near explosive volcanic eruptions, and endogenic energy source). In familiar ways, such destructional geomorphic processes work to reduce the "gravitational disequilibria" that constructive landscapes represent. For instance, the relatively low and ancient Appalachian Mountains, pushed up during one of the collisions between the North American and European continental landmasses, were probably once as tall as the current Himalayan chain. Their formerly steep slopes and high altitudes represented a great deal of gravitational disequilibria, and thus a great deal of potential energy that was subsequently expended as kinetic energy by erosive downhill transport processes (e.g., rainfall runoff and stream flow). Once the processes of continental collision ebbed and tectonic uplift ceased, continuing erosion and surface transport processes (such as rainfall, associated runoff, snowfall, and glaciation) over only a few tens of millions of years reduced the proto-Appalachian Mountains to their present gently sloping and relatively low-relief state.

Volcanic landforms provide myriad illustrations of the competition between destructive and constructive processes in the landscape. For example, Mt. Fujiyama, the most sacred of Japanese mountains, is actually an active volcano that erupts on the order of every 100–150 years. Its perfectly symmetrical conical shape is the result of volcanic eruptions that deposit material faster than it can be transported away, on average. If

FIGURE 11   Mt. Fuji, Japan. Perhaps the world's quintessential volcano, the perfect conical shape of Mt. Fuji has inspired Japanese landscape artists for centuries. It is considered a sacred mountain in Japanese tradition and thousands of people hike to its summit every year. Volcanologically, Mt. Fuji is termed a "strato-volcano" and rises to an altitude of 3776 m above sea level. It erupts approximately every 150 years, on average.

Fuji stopped erupting, it would become deeply incised by stream erosion and it would lose its classic profile over a geologically short time interval (Fig. 11).

A. CONSTRUCTIVE PROCESSES IN THE LANDSCAPE

Over the geologic history of the Earth, volcanism has been one of the most ubiquitous processes shaping its surface. Molten rock (lava) erupts at the Earth's surface as a result of the upward movement of slightly less dense magma. Its melting and upward migration are triggered by convective instabilities within the upper mantle. Volcanic processes very likely dominated the earliest terrestrial landscapes and competed with meteorite impacts as the dominant surface process during the first billion years of Earth history. With the advent of plate tectonics, multiphase melting of ultramafic rocks tended to distill more silicic lavas. Because silicate-rich rocks tend to be less dense than more mafic varieties, they tend to "float" and resist subduction, thus continental cores (cratons) were generally created and enlarged by island-arc accretion.

Most volcanism tends to occur on plate boundaries. Subaerial plate boundary volcanism tends to produce island arcs (e.g., Aleutian Islands; Indonesian archipelago) when oceanic plates override one another or subaerial volcanic mountain chains (e.g., Andes) underride more buoyant continental plates. Such volcanism tends to be relatively silicic (e.g., andesites), producing lavas with higher viscosities, thus tending to produce steeper slopes. Rough lava flows on these volcanoes tend to be classified as *aa* or *blocky* lavas. High interior gas pressures contained by higher viscosity magmas can produce very explosive eruptions, some of which can send substantial amounts of dust, volcanic gas, and water vapor into the stratosphere. Sometimes, for the largest class of explosive eruptions (e.g., Mt. Pinatubo, 1991), enough such material is injected so as to alter weather, short-term climate, and the chemical compositions of the global atmosphere (e.g., temporary 10% zonal depletion of stratospheric ozone after the Pinatubo 1991 eruption).

Another kind of volcanic activity tends to occur within continental plates. As is thought to have been widespread on the Moon, Mars, Venus, and Mercury, continental flood eruptions have erupted thousands of cubic kilometers of layered basalts. These are among the largest single subcontinental landforms on the Earth. Such lavas were mafic, of relatively low viscosity, and are thought to have erupted from extended fissure vents at very high eruption rates over relatively short periods (1–10 years). Recent work on the 100-km-long Carrizozo flow field in New Mexico, however, suggests that such massive deposits may have formed at much lower volume effusion rates over much longer periods than previously thought (10–100 years or more). The same may be true for lava flows of similar appearance on other planets.

Perhaps the most familiar kind of subaerial volcanism is the well-behaved, generally nonexplosive, Hawaiian-style low viscosity eruptions of tholeiitic basalts that form shield volcanoes. Such lavas tend to have relatively high volatile (e.g., H_2O) content, somewhat lower silica content (e.g., ~47–52% SiO_2) than adesitic lavas (~55–65% SiO_2), relatively high solidus temperatures (e.g., 1000–1160°C), and tend to erupt in long sinuous flows. Such flows can manifest themselves as very rough clinkery *aa* flows with well-defined central channels and levees or as very smooth, almost glassy,

pahoehoe flows that tend to form in diffused flow fields (Fig. 12).

These lavas are thought to be comparable to lavas observed in remote-sensing images of Martian central vent volcanoes (e.g., Alba Patera, Olympus Mons). Shield volcanoes on both planets tend to exhibit very low slopes (i.e., ≤5°). Active submarine basaltic volcanoes tend to occur along mid-oceanic ridges, also plate boundaries. Such volcanoes produce broad axially symmetric accumulations of mafic lavas. In deep water, lavas tend to form scalloped, pillow-like morphologies, and extrude like toothpaste, because of the vast cooling capacity of the surrounding seawater as compared to that of air and because dissolved gas tends to be contained by ambient hydrostatic pressures (typically thousands of atmospheres) (Fig. 13). Often the hot sulfide-rich waters circulating at such active submarine venting sites provide habitats for a wide variety of exotic chalcophile (sulfur-loving) biota found nowhere else.

Active plate tectonics — the kind that creates and consumes landscapes — appears to be a purely terrestrial process, at least within our solar system. Large plates with areas up to the order of 10% of the globe's area migrate slowly (<1 to ~10 cm/year). In the process, they collide, override (obduct), and underride (subduct) each other and generate a variety of landscapes at their interacting boundaries. Compressionally folded mountain ranges, such as the Himalayas, the Appalachians, the Zagros Crush Zone (Iran), and the Urals (western Russia), are the result of continental plate collisions. When an ocean plate underrides a continental plate at a steep angle, volcano-tectonic mountain ranges can result (e.g., Andes — see earlier). When such collisions occur at relatively shallow angles, a different kind of orogeny manifests itself, such as the Cordillieran orogeny that formed the basin and range and Cordillieran provinces of North American (e.g., Canadian and U.S. Rocky Mountains, Sierra Madre ranges of Mexico).

However, when continental plates start to pull apart, or rift, large tectonic valleys (i.e., horst and graben terranes), such as the East African Rift Valley or the Rio Grande Rift in New Mexico, tend to form, where large axial, normally faulted, rifts are the result of tensional stress in the continental crust. Such rift valleys are often the sites of active basaltic volcanism, engendered by the relatively thinned crust that results from the rifting process.

Often such tears in continental fabrics occur as three-cornered rifts. Once one of the arms of the rift fails, the overall stress field is reoriented so that the rifting process can proceed along two of the arms of

FIGURE 12 (a) Extruding toe of pahoehoe lava. This toe is about 30 cm wide. Note how it has erupted out of a crack in a previous toe and is flowing over yet another previous ropy-textured toe. (Courtesy of *Volcano World* Web Site, University of North Dakota, Professor Chuck Wood.) (b) Pahoehoe near the coast of Kilauea. Fields of Pahoehoe lava tend to form in a very complex intertwined fashion, and old cooled flows are often smooth enough to walk on in bare feet. (Photo by Steve Mattos, 1989, courtesy of *Volcano World* Web Site, University of North Dakota, Professor Chuck Wood.) (c) Advancing flow of incandescent aa lava. Generally, aa flows are very rough and meters to tens of meters thick. They form broad toes and lobes and can advance kilometers per day, as often happens during eruptions of large aa flows on Mauna Loa volcano in Hawaii (e.g., Mauna Loa 1984 eruption). (Photograph by R. W. Decker, U.S. Geological Survey, July 2, 1983, courtesy of *Volcano World* Web Site, University of North Dakota, Professor Chuck Wood.)

the tri-corner tear; the remaining "failed" arm is called an *aulacogen*. Because the thinned crust that comprises them typically represents a topographic low, such aulacogens become the locus for mainstreams of continental river drainage systems. The Mississippi and Congo Rivers are good examples of this phenomenon. When continental tears occur near ocean boundaries (coasts), narrow seas begin to open up. Current examples of this phenomenon are the Gulf of California and the Red Sea.

Clearly, the shape of the solid Earth surface can have profound effects on the dynamics at its interface with the global ocean and atmosphere. Major climatic regime shifts have occurred as a result of the opening and closing of ocean basins. Already mentioned here was the closing of the predominately east–west oriented Tethys Sea and the opening of the Atlantic Basin, associated with general continental cooling during the early Cenozoic. Such shifts in oceanic circulation also

have an effect on major atmospheric circulation modes and on heat transport by the atmosphere. More localized effects of plate tectonics include continental rain shadow zones created by orogenies driven by continental–continental plate collisions (e.g., the Himalayas) or by oceanic–continental plate subduction (e.g., Andes Range).

The transport of water across the land surface also has a hand in forming constructional landforms. Sediment erosion, transportation, and deposition can set the stage for a variety of landscapes, especially in concert with continental scale tectonic ("epirogenic") uplift. The Colorado Plateau in the southwestern United States is perhaps the best example of this type of landscape. The Grand Canyon of the Colorado River slices through the heart of the Colorado Plateau and exposes over 5000 vertical feet of sedimentary layers, the oldest of which date to the beginning of the Cambrian era (Fig. 14).

FIGURE 12 (*continued*)

FIGURE 13   Pillow lava forming. Here a lava "pillow" is being erupted under water off shore of the Island of Hawaii, during the eruption of Mauna Ulu Volcano in 1969. Under the confining hydrostatic pressure and the chilling effects of seawater, lava extrudes like toothpaste, in spastic bursting fits. The pillow shown here is approximately 1–3 m in characteristic dimension. (Photo by Gordon Tribble and courtesy of the U.S. Geological Survey.)

Water itself can form constructive landforms on the Earth. In its solid form, water can be thought of as another solid component of the Earth's crust, essentially as just another rock. Under the present climatic regime, the Earth's great ice sheets — Antarctica and Greenland — along with numerous valley glaciers scattered in mountain ranges across the world in all climatic zones, compose a distinct suite of landforms. Massive (up to kilometers thick) deposits of perennial ice form smooth, crevassed, plastically deforming layers of glacial ice, often with incorporated lithic sediments and clasts. Continental ice sheets depress the upper crust upon which they reside and can scour the subjacent rocky terrains to bedrock, as during the Wisconsin Era glaciation in Canada (i.e., last Ice Age in North America). Valley glaciers, mainly by mechanical and chemical erosion in concert, tend to carve out large hollows (*cirques*) in their source areas and have large outflows of meltwater at their termini. Sometimes large pieces of a marine glacier will calve off into the ocean and form *icebergs,* only the top one-eighth of which is visible in the air (due to the buoyancy of ice in water) (Fig. 15).

FIGURE 14   Classic view of the Grand Canyon of the Colorado River in Arizona. The massive layering records the local geologic history for at least the last 500 Myr. Comparable layering has also been observed recently in canyons on Mars. (Courtesy of the Grand Canyon Chamber of Commerce.)

B. DESTRUCTIVE GEOMORPHIC PROCESSES

Friction probably represents the largest expenditure of energy as geologic materials move through the landscape: friction of water (liquid or solid) on rock, friction of the wind, friction of rock on rock, or rock on soil. All of these processes are driven by the relentless force of gravity and generally express themselves as transport of material from a higher place to a lower one. Erosion (removal and transport of geologic materials) is the cumulative result, over time reducing the average altitude of the landscape and often resculpting or eliminating preexisting landforms of positive relief (e.g., mountains) and incising landforms of negative relief (e.g., river valleys or canyons). Overall, the source of potential energy for these processes (e.g., the height of mountain ranges) is provided by the tectonic activity of plates as they collide or subduct.

Subaerial landscapes on the Earth are most generally dominated by erosive processes, and subaqueous landscapes are generally dominated by depositional processes. Thus, from a planetary perspective, it is the ubiquitous availability and easy transport of water,

mostly in liquid form, that makes it the predominant agent of sculpting terrestrial landscapes on Earth. Based on the geologic record of ancient landscapes, it appears that this has been the case for eons on the Earth. Such widespread and constant erosion does not appear to have happened for such a long time on any other planet in the solar system, although it appears that Mars may have had a period of time when aqueous erosion was important and even prevalent. The imprint of that epoch, however, is now only a distant echo, given the billion-year stasis that the Martian landscape seems to have endured most recently.

Fluvial erosion and transport systems (river and stream networks) dominate the subaerial landscapes of the Earth, including most desert areas. Even in deserts where aeolian (wind-driven) deposits dominate the current landscape, the bedrock signature of ancient river systems, relict from more humid past climatic epochs, can be detected in optical and radar images taken from orbiting satellites. Surface runoff, usually due to the direct action of rainfall ("pluvial activity"), occurs in nearly all climatic zones (except the very coldest). The action of water flowing down gravita-

FIGURE 15   Tfhe Gulkana Glacier, Alaska on August 25, 1987. Note the meltwater stream at the glacier terminus at lower right. The longitudinal striations consist of rock that has been abraded as the sides of the glacier contact the confining valley walls. It is subsequently incorporated into the flowing body of ice. (Photograph by Bob Krimmel, courtesy of the U.S. Geological Survey.)

tional gradients tends to generate streamlines as small rivulets coagulate in kind of a quasi-random walk process, which generates bigger and bigger trunk streams, with many upstream branches. The form of the resulting ramified (branched) networks has generally been shown to correspond to the most uniform dissipation of work within the containing drainage basin.

On the Earth, such network forms resulting from this process tend to be scale-independent and take on a nearly fractal character. That is, network patterns tend to be replicated at nearly all scales, with their geometric relationships tending to be the same, no matter what the physical size of the network. This quasi-scale-independent character, in part, is the result of pluvial runoff occurring simultaneously on all available slopes, both within a particular drainage basin at the small end of the size scale ($\sim$1–10 km^2) during a

storm and in a time-averaged sense when considering the largest (continental scale) drainage basins. This occurs because, on the Earth, in general, the time between major basin inputs (i.e., major storms) is exceedingly short compared to time scales for the overall geomorphic alteration of major drainage networks. In desert areas, however, where much geomorphic work is accomplished during major downpours, which may be quite localized, this maxim may not be strictly true. The resulting ephemeral streams tend to be less morphometrically regular and more primitive in their network properties than the more finely and uniformly branched networks in humid climates.

In contrast to the situation on the Earth, the most visible and well-expressed Martian valley networks tend to be highly *irregular* in their network geometries, probably reflecting very restricted source areas and strong directional control by fractures and faults that

FIGURE 16 Sand dunes at Death Valley National Monument. In the great deserts of the world, sand sheets are the dominant morphology and are wind driven. In Death Valley, dunes are essentially trapped by surrounding deep topography; however, in open desert areas (e.g., Sahara or Arabian Peninsula), dune trains may stretch for tens or hundreds of miles. (Courtesy of the University of California at Berkeley.)

was not overcome easily by putative riverine formative processes in the past. In addition, they are distributed very sparsely and are primitive in their branching, very much like the canyon networks in the U.S. desert southwest. Thus, for most of its discernable history, the Earth's landscapes have been distinguished, overall, by well-integrated and complexly branched fluvial drainage networks.

Uniquely on the Earth (within this solar system at least) it is the competition between constant fluvial erosion and constant tectonic uplift (and in some land areas, frequent volcanic eruptions) that is the predominant determinator of the landscape's appearance. For instance, the present terrestrial landscape is not dominated by impact scars. Plate tectonic processes are, in part, responsible; however, fluvial erosion is probably the dominant factor for subaerial landscapes in this regard. Also, without constant tectonic reinforcement, rainfall would probably reduce a Himalayan-style, or Alpine range to Appalachian-style mountains within 10 Myr or so. On the Earth, when tectonic forces subside, constant fluvial erosion wins out and hilly landscapes are flattened.

Other erosive processes, independently or in con-cert with fluvial activity, also clearly play a role on the Earth. As mentioned already, the movement of water through the subsurface can often result in the undermining of surficial layers by mechanical transport of grains and chemical dissolution. Such "groundwater sapping" can result in a range of "sapping networks" from gulley sized within fresh landfill deposits to full-blown canyon networks deeply etched into consolidated sandstone. Such large sapping-generated networks can also form within layers of basalt, as seen on the east slopes of the Island of Hawaii, for instance. In addition, the chemical action of groundwater, in the absence of a mechanical transport component, can form *karstic terranes* — landscapes dominated by the presence of caves and sinkholes. Generally, karst terranes are found in limestone-dominated lithologies. Whereas groundwater sapping and karst formation on the Earth may be relatively less important than fluvial erosion, the opposite case may be true for Mars.

Another process regime that dominates arid and polar deserts on the Earth, and apparently is highly active, even today, on Mars, is that of aeolian erosion and transport. Named after Aeolus, the Roman god of the winds, aeolian processes include the wind-driven

FIGURE 17   (a) The famous Blackhawk landslide, Lucerne Valley, California. This is a head on view of a major prehistoric landslide from the north-facing side of the San Bernardino Mountains. The distal end of the side is the dark deposit that terminates at the lower end of the photo. The source of the slide lies within the foothills in the distance. (Courtesy of the U.S. Geological Survey.) (b) Computer rendering of the Blackhawk landslide (a). The vertical drop down the north slope of the San Bernardino Mountains is no more than 1200 m. The slide is visible at midfield and this view is orthogonal to the view in a. Vertical exaggeration is approximately 4:1. (Courtesy of the California State University at Fullerton.)

transport of fine material as well as erosion caused by the mechanical impact of that material when it is transported. Such fine lithics (or "dry" snow) can accumulate into familiar dune or drift morphologies. Where wind regimes are steady and powerful, as they are in most deserts, aeolian erosion can be a major agent of landscape change. As an illustration of the persistence and power of the wind, hundreds of windmills have been located in San Gorgonio Pass, near Palm Springs, California. Some of them are always spinning, as the winds in that area make electric power generation by windmills economically feasible. Rocks in this area are strongly polished on their exposed faces, even to the point of developing smooth facets on their windward sides.

Wind transport of fines is accomplished in a kind of kinetic energy cascade called "saltation." In this process, kinetic energy is transferred, nearly elastically, from one sand grain to another as they bounce along desert hardpan surfaces. Hard crystalline silicate sand grains saltate particularly well. Normally, such grains are well within a relatively static atmospheric boundary layer very close to the ground (~1 m or less). When grains are saltated, if the wind is above a certain threshold velocity depending on microterrain roughness and grain size and geometry, they bounce through the boundary layer into a higher velocity windstream. When they fall back through the boundary layer, they carry the kinetic energy of the windstream into energetic collisions with other static grains, which, in turn, saltates the new grains. The transfer of energy by saltation can occur in almost geometric, or explosive, proportions as a sand storm begins. Under these conditions, grain transport occurs in a saltation layer, which can vary in thickness. In the San Gorgonio Pass, for instance, telephone poles are armored with sheet metal for about 1 m above the ground, which is the depth of the typical saltation layer in that area. If the poles were not armored, their lower portions would be sandblasted to shreds relatively quickly.

In desert sand seas on the Earth, such as the Sahara, aeolian processes now dominate. Fleets of sand dunes move across the desert and take on morphologies dictated by the variability of the directionality of the wind as modulated by surrounding topography. Thus, "star dunes" (highly directionally variable winds forming

FIGURE 17 *(continued)*

equant star-like individual dune morphologies) and "seif dunes" (essentially linear dune forms under relatively unidirectional wind regimes) are contrasting end members of a morphologic continuum (Fig. 16).

Whereas sand dunes are found in relatively areally restricted deserts on Earth, on Mars fine dust and sand dune and drift morphologies appear to dominate the landscape and can reveal important information on current wind regimes and on the constitution of the fine material based on observations and models of terrestrial dune morphologies.

Aeolian, or fluvial, transport of fine material can only occur if a source of fine material is available to be transported. Another important terrestrial geomorphic process is *weathering* — the breakdown of consolidated material into constituent grains. Rock can be broken down in several ways. *Chemical weathering* can occur when natural acids (e.g., weak carbonic acid from CO_2 dissolved in rainwater) act on susceptible rocks, such as limestone (i.e., which is predominately formed by the accumulation of calcium carbonate in marine environments), or on calcareous (limy) sandstones. Once the carbonate has been dissolved, the remaining silicates (or other rock grains) are released into the environment. The *mechanical weathering* or breakdown of rock can occur due to the freeze-thaw expansion–contraction cycles of water within pores and cracks. The hydrostatic pressures exerted by freeze-thaw can easily overcome rock brittle strength thresholds at microscopic and macroscopic scales. The formation of salt crystals also exerts mechanical energy to break up rocks and can chemically weather rocks. Oxidation of minerals, particularly iron-containing minerals, is another form of chemical weathering. *Biological weathering* occurs through chemical weathering caused by biogenic acids, particularly in tropical areas. It can also occur mechanically, by bioturbation of soils and sediments, as well as by the physical pressure of root and stem turgor in cracks and fissures within solid rock. It is of significance that on the Earth, all three major forms of weathering are enhanced or enabled by the ubiquitous presence of water.

Perhaps some of the most dramatic forms of nonvolcanic landscape alteration that we see on the Earth today fall into the category that geomorphologists call *mass wasting*. Generally, the term mass wasting is applied to processes such as landslides, creep, snow and debris avalanches, submarine slides and slumps, volcano-tectonic sector collapses, and scour related to the action of glaciers. Mass-wasting processes tend to affect a relatively minor proportion of the Earth's surface at any given time, however, such as volcanic eruptions (with which they are often associated), when they occur near population areas, their effects can be devastating (Fig. 17).

Subaerial landslides are particularly prevalent in areas of deep weathering combined with high rainfall and high topographic relief. Thus, tropical areas, such as Indonesia, are particularly affected. It also happens that such areas are often tectonically and/or volcanically active, processes that tend to generate gravitationally unstable relief. Southern California is another prime example of a populated area affected by mass-wasting processes (e.g., landslides, slumps, creep), particularly when the ground becomes saturated by winter rains (e.g., Pacific Ocean El Niño conditions are particularly bad in this respect).

Massive gravitational instabilities are often generated within volcanoes, as newly erupted mass is added high on the volcanic edifice. Debris flows on steep volcano slopes are common in regions where relief and rainfall are high, such as Japan. Massive volcanic landsliding or sector collapse (total mechanical slope failure) can also occur. On the Hawaiian Island of Oahu, a major volcano sector collapse during Holocene time resulted in an enormous landslide, which radiated huge tsunami (seismic sea) waves throughout the Hawaiian chain and across the Pacific Basin.

Whereas prehistoric volcanic eruptions and landslides appear to be larger than any natural event recorded in human experience, meteorite or asteroidal impact undoubtedly has produced the largest geomorphic "events" that the Earth has experienced. Judging from the impact histories imprinted on the static crusts of other planets and satellites, the early Earth sustained direct and grazing impacts that probably had profound effects within subaerial and submarine landscapes, which may well have triggered major epochs of volcanic activity during Archean time and set the stage for the beginnings of plate tectonics. Nevertheless, traditional geomorphic thinking does not normally include impact process as a primary geomorphic landscape force, but it should. From a planetary perspective, massive early impacts were among the most fundamental (destructive) perturbations to the stability of ancient Earth's newly formed crust, as well as a periodic and massive perturbation (and profound threat) to the developing biosphere. The Cretaceous–Tertiary extinction of the dinosaurs, 65 million years ago, was just the most current and egregious object lesson.

Today we know that there are many asteroids in Earth-crossing and other orbits that could pose a serious threat to life on this planet. The geomorphic, atmospheric, and biological effects of an asteroid 10 to 100 km in diameter striking the Earth are almost unimaginable in their predicted ferocity and persistence. Geomorphically, the impact of a 10 km-diameter asteroid would have continental scale effects. Impact-generated shock waves would result in major seismic disturbances and ground deformation. The impact crater itself could be in the range of 100 km to many hundreds of kilometers in diameter. Ejecta could be deposited out to several crater diameters. Direct blast-caused devastation to the biosphere on a scale of at least hundreds of kilometers from the site would likely be total. Indirect biological devastation, due to debris lofted into the stratosphere, would be global in extent. Solar energy would be absorbed and reradiated or simply reflected from the top of the stratosphere, with very little reaching the surface for many months. This kind of impact-generated environmental change is implicated strongly in the extinction of the dinosaurs. [See NEAR-EARTH ASTEROIDS; PLANETARY IMPACTS.]

A sea impact, more probable by a factor of three, would perhaps be worse. All the primary effects mentioned for a land impact would probably occur, but there would be added features. Ocean-wide tsunami devastation far from the actual impact point would devastate coastal areas. Massive injections of water vapor into the stratosphere would tend to increase its opacity, thus making the ground surface even colder.

The only comparable events that have been experienced in written human history are massive volcanic eruptions, such as Krakatau or Tambora. The latter, in 1814, caused the famous "Year Without Summer" in North America during the following year. The Toba eruption in Indonesia (~75,000 years B.P.) was probably at least an order of magnitude more energetic than the Tambora 1814 event. Even these massive volcanic events are probably hundreds of times smaller than the impact of an asteroid 100 km in diameter. Such impacts rocked the Earth and other terrestrial planets and satellites repeatedly during their early history. On the Earth, plate tectonic activity has erased the evidence of these gargantuan impacts. On most other planetary surfaces in our solar system, however, the presence of impact basins hundreds to thousands of kilometers in diameter are testament to an ancient epoch of violence in their landscapes, unimaginable in human experience.

Impacts by smaller bodies are more common and thus pose a more immediate threat to humanity. The only recorded encounter with a large impacting body occurred on June 30, 1908, near the Stony Tunguska River in a swampy area of central Siberia. It did not leave a crater, but the associated air blast flattened more than 2100 km² of forest. Recent recovery by Italian scientists of particulate matter from the impact, trapped in tree resin whose age spans the impact event, suggests that the impacting body was probably a stony meteorite. [*See* METEORITES.]

IV. TOOLS FOR STUDYING EARTH'S DEEP INTERIOR

In comparison with other planets, the interior of the Earth can be studied in unprecedented detail. This is because of the existence of sources of energy, such as earthquakes or magnetic and electric disturbances. Seismic waves, for example, can penetrate deep inside the Earth, and the time they travel between the source (earthquake) and the receiver (seismographic station) depends on the physical properties of the Earth. The same is true with respect to electromagnetic induction, although observations are different in this case.

Observation and interpretation of seismic waves provide the principal source of information on the structure of the deep interior of the Earth. Both compressional (P waves) and shear (S waves) can propagate in a solid, only P waves in a liquid. Compressional waves propagate faster than shear waves by, roughly,

a ratio of $\sqrt{3}$. Velocities, generally, increase with depth because of the increasing pressure; hence the curved ray paths (Fig. 18).

At the discontinuities (which include the Earth's surface) waves may be converted from one type to another. Figure 18a shows P waves emanating from the source ("Focus"). The P waves can propagate downward (right part of the figure) and are observed as PP, PS, PPP, PPS, for example. They can also propagate upward, be reflected from the surface, and then observed as so-called "depth phases": pP, pPS. Depth phases are very important in determining the depth of focus.

Figure 18a shows rays in the mantle; there are also the outer core and inner core. The outer core is liquid and has distinctly different composition; the P-wave speed is some 40% lower than at the bottom of the mantle; also, there are no S waves. The inner core is solid, with a composition similar to that of the outer core. Figure 18b shows the rays (mostly P waves) that are reflected from the core–mantle boundary (CMB; a letter c is inserted, e.g., PcP) or that are transmitted through the outer core (letter K: PKP) or also through the inner core (letter I: PKIKP).

Figure 18a shows S-wave rays interacting with the CMB: reflected (ScS) or converted at the CMB into a P wave and then again reconverted into a S wave: SKS and SKKS. The latter indicates one internal reflection from the underside of the CMB.

Measurements of the travel times of the waves such as shown in Figure 18 have led to the derivation of models of the seismic wave speed as a function of depth. These, in turn, were used to improve the location of earthquakes and further refine the models. The first models were constructed early in the 20th century; the models published by Sir Harold Jeffreys in the late 1930s are very similar in most depth ranges to current ones. The upper mantle (the topmost 700 km) with its discontinuities and the inner core are exceptions.

In addition to the body waves, which propagate through the volume of the Earth, there are also surface waves, whose amplitude is the largest at the surface and decreases exponentially with depth. Surface waves are important in studying the crust and upper mantle and, in particular, their lateral variations, as the Earth is most inhomogeneous near the surface. There are Rayleigh waves with the particle motion in the vertical plane (perpendicular to the surface) and Love waves whose particle motion is in the horizontal plane (parallel to the surface). Surface waves are dispersed in the Earth because of the variation of the physical parameters with depth.

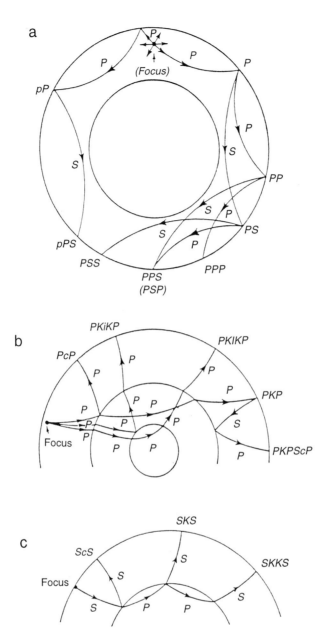

FIGURE 18 (a) Ray paths of the compressional waves (P) in the mantle, including their conversion to shear waves (S). (b) Ray paths of the P waves interacting with the outer and inner core. (c) Ray paths of the S waves interacting with the core; the S waves are converted into P waves in the outer core.

Very long period surface waves (~100 sec) are sometimes called "mantle waves," have horizontal wavelengths in excess of 1000 km, and maintain substantial amplitudes (and, therefore, sensitivity to the physical properties) down to depths as large as 600–700 km. Because of their long periods, mantle waves are attenuated relatively slowly and can be observed at the same station as they travel around the world several times along the same great circle (both in the minor and major arc direction).

Superposition of free oscillations of the Earth (known also as the normal modes) in the time domain will yield mantle waves. First specra of the vibrations of the Earth were obtained following the Chilean earthquake of 1960; the largest seismic event ever recorded on seismographs. The measurements of the frequencies of free oscillations lead to the renewed interest in the Earth's structure. In particular, they, unlike body waves, are sensitive to the density distribution and thus provided additional constraints on the mass distribution other than the average density and moment of inertia.

Sometime in the 1970s it became clear that further refinements in one-dimensional Earth models cannot be achieved, and perhaps do not make much sense, without considering the three dimensionality of the Earth structure. All three types of data described earlier are sensitive to the lateral heterogeneity. Travel times will be perturbed by slight variations of the structure along a particular ray path, compared to the prediction by a one-dimensional model. All we need is many observations of travel times along criss-crossing paths. Many millions of such data are available from the routine process of earthquake location; they are assembled from some 6000 stations around the world by the National Earthquake Information Center in Golden, Colorado, and by the International Seismological Centre in England. Surface waves, mantle waves, and periods of free oscillations in a three-dimensional Earth also depend on the location of the source and the receiver. Progress during the last decade in global seismographic instrumentation, in terms of the quality and distribution of the observatories and exchange and accessibility of the data, makes the required observations much more readily available.

V. SEISMIC SOURCES

Even though the field of seismology can be divided into studies of seismic sources (earthquakes, explosions) and of the Earth's structure, they are not fully separable. To obtain information on an earthquake, we must know what happened to the waves along the path between the source and receiver, and this requires the knowledge of the elastic and anelastic Earth structure. The reverse is also true: in studying the Earth structure, we need information about the earthquake; at least its location in space and time, but sometimes also the model of forces acting at the epicenter.

Most of the earthquakes can be described as a process of release of shear stress on a fault plane. Sometimes the stress release can take place on a curved surface or involve multiple fault planes: the radiation of seismic waves is more complex in these cases. Also, explosions, such as those associated with nuclear tests, have a distinctly different mechanism and generate P and S waves in different proportions, which is the basis for distinguishing them from earthquakes.

Figure 19 shows three principal types of stress release, sometimes also called the earthquake mechanism. The top part of Fig. 19a is a view in the horizontal plane of two blocks sliding with respect to each other in the direction shown by the arrows. Such a mechanism is called strike slip, and the sense of motion is left-lateral; there is also an auxiliary plane, indicated by a dashed line; a ground motion generated by a slip on the auxiliary plane (right lateral) cannot be distinguished from that on the principal plane. The bottom part of Fig. 19a is a stereographic projection of the sign of P-wave motion observed on the lower hemisphere of the focal sphere (a mathematical abstraction in which we encapsulate the point source in a small uniform sphere). The plus sign corresponds to compressive arrivals and minus sign to dilatational zones; quadrants with compressive arrivals are shaded.

The top part of Fig. 19b is a section in the vertical plane. In this case, the block on the right moves upward on a plane that dips at a 45° angle with respect to the block on the left; this mechanism is called thrust and is associated with compression in the horizontal plane and tension in the vertical plane and corresponds to the convergence of the material on both sides of the fault. Such processes are responsible for mountain building. The shaded central region in the bottom part of Fig. 19b, with the dilatational arrivals on the sides, is characteristic of the thrust, or reverse faulting, events. Figure 19c illustrates the opposite mechanism, in which tension is horizontal and compression vertical; this is called normal faulting and is associated with extension, which can lead to the development of troughs or basins. The "beach-ball" diagrams are commonly used as a graphic code to represent the tectonic forces. Some earthquakes are a combination of two different types of motion, e.g., thrust and strike slip; in this case the point at which the two planes intersect would be moved away from either the rim or the center of the beach-ball diagram.

The size of the earthquake is measured by magnitude. There are several different magnitude scales depending on the type of a wave whose amplitude is being measured. In general, magnitude is a linear function of the logarithm of the amplitude; thus a unit magnitude increase corresponds to a 10-fold increase in amplitude. Most commonly used magnitudes are the body-wave magnitude, m_b, and surface wave magnitude, M_S. The frequency of occurrence of earthquakes, i.e., a number of earthquakes per unit time (year) above a certain magnitude M, satisfies the Gutenberg–Richter law: $\log_{10} N = a \cdot M + b$. The value of a is close to -1, which means that there are, on average, 10 times more earthquakes above magnitude 5 than above magnitude 6. A new magnitude, M_W, based on the estimates of the released seismic moment [change of stress × fault area × offset (slip) on the fault] is becoming

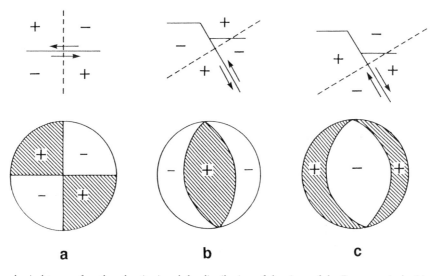

FIGURE 19   Three classical types of earthquakes (top) and the distribution of the signs of the P-wave arrivals: (a) strike slip (motion) in the horizontal plane); (b) thrust (or reverse) fault, vertical plane cross section; and (c) normal fault, vertical plane cross section.

increasingly popular; it is more informative for very large earthquakes, for which M_S may become saturated.

Figure 20 is a map of the principal tectonic plates, as defined in the plate tectonic theory. The direction of the arrows shows the motion of the plates; their length corresponds to the rate of motion. At a plate boundary where the arrows converge, we expect compression and, therefore, thrust faulting; one of the plates is subducted: hence the term "subduction zones." At a plate boundary where the arrows diverge, there is normal faulting and creation of a new crust: midocean ridges. For boundaries that slip past each other in the horizontal plane, also called the transform faults, there is strike-slip faulting.

Figure 21 shows the source mechanism of approximately 10,000 shallow earthquakes from 1976 through 1997 determined at Harvard University using the centroid-moment tensor (CMT) method; the center of each beach ball is at the epicenter—many earthquakes have been plotted on the top of each other. It is easy to see that thrust faulting is dominant at the converging boundaries (subduction zones), there are exceptions related to bending of the plates, plate motion oblique to the boundary and other causes. At midocean ridges,

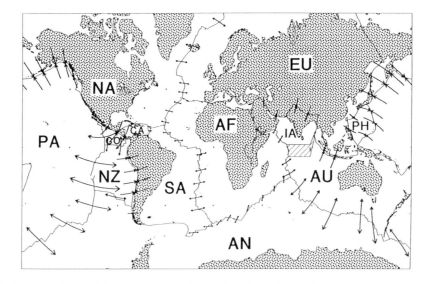

FIGURE 20   Principal tectonic plates and their motion in the absolute (hot spot) reference frame. The arrows represent the velocity vectors. [From Gordon, R. G., and Acton, G. (1989). Paleomagnetism and plate tectonics. *In* "Encyclopedia of Solid Earth Geophysics" (D. E. James, ed.), pp. 909–923. Van Nostrand and Reinhold, New York.]

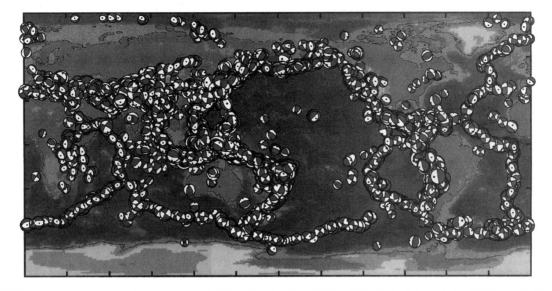

FIGURE 21 Source mechanisms of approximately 10,000 earthquakes from 1976 to 1997 obtained through the CMT analysis. The center of a beach ball is plotted at the epicenter. Only a small fraction of earthquakes are visible. Note the preponderance of earthquakes occurring on plate boundaries (Figure 20) and their mechanism corresponding closely to the type of the boundary (convergent, thrust faulting; divergent, normal faulting; transform, strike-slip faulting). Some earthquakes occur away from plate boundaries. They are particularly numerous in Asia and Africa along the west African rift system, but there are some in eastern North America and the center of the Pacific.

we see predominantly normal faulting, the faults where a midocean ridge is offset, show strike-slip faulting, in accordance with the plate tectonic theory. The exception is where the fault is complex. Along the San Andreas Fault, the most famous transform fault, we see many complexities that led to earthquakes other than the pure strike slip. For example, the Northridge earthquake of January 1994 was a thrust, and the Loma Prieta earthquake of October 1989 was half-thrust, half-strike slip. There are also earthquakes away from the plate boundaries. These are called intraplate earthquakes and their existence demonstrated the limits of the validity of the plate tectonic theory, as there should be no deformation within the plates. A very wide zone of deformaton is observed in Asia; the rare large earthquakes in eastern North America are sometimes associated with isostatic adjustment following the last glaciation.

There are also deep earthquakes, with the deepest ones just above 700 km depth; earthquakes with a focal depth from 50 to 300 km are said to be of an intermediate depth and are called "deep" when the focal depth is greater than 300 km. Intermediate and deep earthquakes are explained as occurring in the subducted lithosphere and are used to map the position of the subducted slab at depth. Not all subduction zones have very deep earthquakes; for example, in Aleutians, Alaska, and middle America the deepest earthquakes are above 300 km depth. The variability of the maxi-

mum depth and the mechanism of deep earthquakes have been attributed in the late 1960s to the variation in the resistance that the subducted plate encounters; more recent studies indicate more complex causes, often invoking the phase transformations (change in the crystal structure) that the slab material subjected to the relatively rapidly changing temperature and pressure may undergo.

VI. EARTH'S RADIAL STRUCTURE

A spherically symmetric Earth model (SSEM) approximates the real Earth quite well; the relative size of the three-dimensional part with respect to SSEM varies from several percent in the upper mantle to a fraction of a percent in the lower mantle and increases again above the CMB.

A concept of an SSEM, often referred to as an "average" Earth model, is a necessary tool in seismology. Such models are used to compute functionals of the Earth structure, and their differential kernels are needed to locate earthquakes and to determine their mechanism. Knowledge of the internal properties of the Earth is needed in geodesy and astronomy. Important inferences with respect to the chemical composition and physical conditions within the deep interior of the Earth are made using information on radial variations of the elastic and anelastic parameters and density.

An SSEM is a mathematical fiction that is very useful and often a necessary one, but a fiction nevertheless. This is most obvious at the Earth's surface, where one must face the dilemma of how to reconcile the occurrence at the same depth, or elevation, of water and rocks; the systems of equations governing the wave propagation in liquid and in solid are different. The commonly adopted solution is to introduce a layer of water whose thickness is such that the total volume of water in all the oceans and that calculated for the SSEM are equal. It is a reasonable decision, but it will be necessary to introduce corrective measures even when constructing the model, as practically all seismographs that record ground motion are located on land.

This article uses the preliminary reference Earth model (PREM) published in 1981 by Dziewonski and Anderson as an example. It has been derived using a large assembly of body-wave travel time data, surface wave dispersion and periods of free oscillations, collected through the end of 1970s. An effort to revise it is now under way: a large body of very accurate data has been assembled in the nearly 20 years since the publication of PREM. However, with the exception of the upper mantle, no substantial differences are expected.

Figure 22a shows the density, compressional velocity, and shear velocity in the model PREM. To illustrate the complexities in the uppermost 800 km of the model, its expansion is shown in Fig. 22b. In what follows, we shall give a brief summary of our knowledge and significance of the individual shells in the Earth structure.

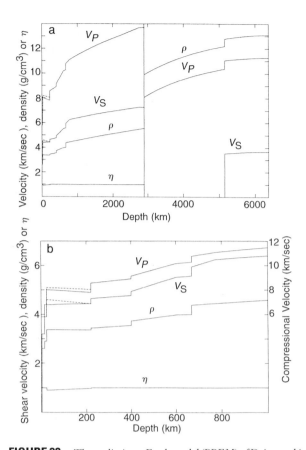

FIGURE 22 The preliminary Earth model (PREM) of Dziewonski and Anderson (1981) describing the compressional velocity (v_p), shear velocity (v_S), and density (ρ). (a) Model for the entire Earth and (b) an expansion of the uppermost 1000 km. In the top 220 km the model is characterized by transverse anisotropy, in which the waves propagating in the vertical (solid line) and horizontal (dashed lines) planes have different velocities. Parameter η, characterizing the propagation of P waves at intermediate angles, is unity in an isotropic medium and is about 0.95, just under the Moho. Below 220 km depth the model is isotropic.

A. CRUST

This is the most variable part of the Earth's structure, both in terms of its physical properties as well as history. Large areas of the Earth's surface are covered by soils, water, and the sediments. These provide support for life and economic activity. However, the vast proportion of what is called "the crust of the Earth" consists of crystalline rocks, mostly of igneous origin.

The primary division is between the continental and the oceanic crust. The former can be very old, with a significant fraction being older than 1.5 Ga. It is light, with an abundance of calcium, potassium, sodium, and aluminum. Its average thickness is 40 km, but varies substantially, from about 25 km in the areas of continental thinning due to extension (the basin and range province, for example) to 70 km under Tibet, in the area of continent — continent collision.

The oceanic crust is thin (7 km, on average, covered by some 4.5 km of the ocean), young (from 0 to 200 Ma), and somewhat more dense, with a greater abundance of elements such as magnesium and iron. It is created at the midocean ridges and is destroyed in subduction zones, with trenches being their surficial manifestation. The difference between oceanic and continental crusts is called by some the most important fact in Earth sciences, as it is related intimately to plate tectonics. The thinner, denser oceanic crust provides conditions more favorable for initiation of the subduction process.

Overall, crustal thickness follows the Airy's hypothesis of isostasy closely: thick roots under mountains and a thin crust under "depressed" areas — oceans. The seismic velocities in the crust increase with depth.

It is a subject of a debate whether this increase is gradual or the crust is layered; recently, the latter view has begun to prevail.

B. UPPER MANTLE: LITHOSPHERE AND ASTHENOSPHERE (25–40 km DEPTH)

The boundary between the crust and the upper mantle was discovered in 1909 by a Yugoslavian geophysicist Andreiji Mohorovicic. It represents a 30% increase in seismic velocities and some 15% increase in density. It is a chemical boundary with the mantle material primarily composed of a mineral olivine, being much richer in heavier elements, such as magnesium and iron.

The terms lithosphere and asthenosphere refer to the rheological properties of the material. The lithosphere, strong and brittle, is characterized by very high viscosity. It is often modeled as an elastic layer. It includes the crust and some 30 to 100 km of the upper mantle. The asthenosphere is hotter, its viscosity much lower, and in modeling is represented by yielding. Under loads, such as glacial caps, the lithosphere bends, whereas the asthenosphere flows.

The difference of rheological properties is explained by differences in temperature: the viscosity is an exponential function of temperature. The lithosphere is relatively cool; the transport of heat is mostly through conduction. The asthenosphere is hotter, and the convective processes are believed to become important.

Low viscosity of the asthenosphere is used to explain the mechanical decoupling between the plates (in the plate tectonic theory) and the underlying mantle. The depth of this decoupling varies with position: it is shallow near midocean ridges and increases as the plate cools with time and its lithosphere grows in thickness.

The continents, with its very old and cold shield regions, may be significantly different. If the hypothesis of the "tectosphere" is correct it may have roots that are 400 km deep and move as coherent units over long periods of the Earth's history. The depth of roots is still subject to a debate; it is true that plates that have a substantial proportion of the continental lithosphere have absolute plate velocities substantially lower than predominantly oceanic plates.

As the seismic velocities decrease with increasing temperature, the vertical gradient of seismic velocities in the transition between the lithosphere and the asthenosphere may become negative. This is called the "low velocity zone"; its presence creates a shadow zone in seismic wave propagation, making interpretation of data complex and nonunique. Also, the phenomenon of

anisotropy (propagation of waves with different speed depending on direction) complicates the modeling. For example, PREM is anisotropic down to a depth of 220 km; if isotropic modeling were used, which does not satisfy the data equally well, the low velocity zone would be very pronounced.

Measurements of attenuation of seismic waves led to the determination of models of Q (quality factor) for the shear and compressional energy. Anelastic dissipation of shear energy, due to grain boundary friction, is most important. Attenuation in the range of depths corresponding to the low velocity zone is several times stronger than in the lithosphere.

Somewhere below 200 km depth the velocities and Q begin to increase slowly: the effect of increasing pressure begins to dominate over the increase in temperature. The so-called "Lehmann discontinuity" is elusive and does not appear to be a global feature; this is one of the elements of PREM, where it is shown as a sudden increase in both P and S velocities, that will be changed in the next version of the reference Earth model.

C. TRANSITION ZONE (400–660 km DEPTH)

Knowledge of the composition of the transition zone is essential to the understanding of the composition, evolution, and dynamics of the Earth. In seismic models, this depth range has been known for a long time to have a strong velocity gradient; much too steep for an increase under pressure of the elastic moduli and density of a homogeneous material. It was first postulated in the 1930s that this steep gradient may be due to phase transformations: changes in the crystal lattice that for a given material take place at certain temperatures and pressures.

In the 1960s, when major improvement in seismic instrumentation took place, two discontinuities were discovered: one at 400 km and the other at 650 km (the current best estimate of the global average of their depth is 410 and 660 km, respectively). Their existence has been well documented by nearly routine observations of reflected and converted waves. There is still some uncertainty of how abrupt the velocity changes are: the 410-km discontinuity is believed to be spread over some 10–20 km, whereas the 660-km discontinuity appears to be abrupt.

This, in general terms, is consistent with the hypothesis that olivine is the principal constituent of the upper mantle. Laboratory experiments under pressures corresponding to depths up to 750 km show that olivine undergoes phase transformations to denser phases with higher seismic wave speeds. At pressures roughly corresponding to 400 km depth, the α-olivine transforms into β-spinel. The latter will transform to γ-spinel at about 500 km depth, with only a minor change in seismic velocities. Indeed, a seismic discontinuity at 520 km has been reported, although some studies indicate that in some parts of the world it may not be substantial enough to be detected. At 660 km γ-spinel transforms into perovskite and magnesio-wüstite.

Although olivine may be the dominant constituent, it is not the only one. The presence of other minerals complicates the issue. Also, there are other hypotheses of the bulk composition of the upper mantle: "piclogyte model," for example.

D. LOWER MANTLE (660–2890 km)

The uncertainties in the mineralogy of the upper mantle and the bulk composition of the Earth have created one of the most stubborn controversies in the Earth sciences: are the upper and lower mantle chemically distinct? A "yes" answer means that there has not been an effective mixing between these two regions throughout the Earth's history, implying that the convection in the Earth is layered. The abrupt cessation of seismic activity at about 660 km depth, coinciding with the phase transformation described earlier, and geochemical arguments — mostly with respect to isotopic composition of the midocean ridge basalts — are used as strong arguments in favor of the layered convection.

The whole mantle convection is favored by geodynamicists who develop kinematic and dynamic models of the mantle flow. For example, the geometry and motions of the known motions of the plates are much easier to explain assuming whole mantle circulation. Evidence has been presented for continuation of seismic velocity anomalies associated with the subducted slabs (which are much colder, and therefore faster, than the surrounding mantle) into depths exceeding 1000 km. There are, however, studies that dispute that, at least a simple model of the direct slab penetration, without a major modification as a result of the encountering the resistance expected in the encounter with the 660-km discontinuity. The whole mantle flow model is generally supported by the three-dimensional mapping of the seismic velocities in the mantle.

In the early 1990s a model of mantle avalanches was developed: the subducted material is temporarily accumulated in the transition zone as the result of an endothermic phase transformation at the 660-km discontinuity. Once enough material with the negative buoyancy collects, however, a penetration can occur in a flashing event, where most of the accumulated material sinks into the lower mantle. The calculations, originally performed in two-dimensional geometry, indicated the possibility of such events causing major upheavals in the Earth's history. However, when calculations were extended to three-dimensional spherical geometry, their distribution in space and time turned out to be rather uniform.

The computer models of the mantle convection are still tentative. There are many parameters that control the process. Some, such as the generation of the plates and plate boundaries at the surface, are difficult to model. Others, such as the variation of the thermal expansion coefficient with pressure- or temperature-dependent viscosity, are poorly known; even one-dimensional viscosity distribution with depth is subject to major controversies.

The lower mantle appears mineralogically uniform, with the possible exception of the uppermost and lowermost 100–150 km. There is a region of a steeper velocity gradient in the depth range of 660–800 km, which may be an expression of the residual phase transformations. Also, at the bottom of the mantle, there is a region of a nearly flat, possibly slightly negative gradient. This region, just above the CMB, known as "D"", is the subject of intense research. Its strongly varying properties, both radially and horizontally, are being invoked in modeling mantle convection, chemical interaction with the core, possible chemical heterogeneity (enrichment in iron), and as evidence for partial melting. The seismic velocities and density throughout the bulk of the lower mantle appear to satisfy the Adams–Williamson law, describing the properties of the homogeneous material under an adiabatic increase in pressure.

E. OUTER CORE (2981–5151 km)

The outer core is liquid: it does not transmit shear waves. This was discovered in the first decade of the 20th century through seismological evidence; the fact was suspected much earlier than that. Consideration of the average density and the moment of inertia pointed to a structure with a core that would be considerably heavier, possibly made of iron, judging from cosmic abundances. We know now that the core is mostly made of iron, with some 10% admixture of lighter elements, needed to lower its density. It has formed relatively early in the Earth's history in a melting event in which droplets of iron gravitationally moved toward the center.

The presence of a liquid with a very high electrical conductivity creates conditions favorable to self-excitation of a magnetic dynamo. The process is too complex to attempt an explanation here, but there is ample empirical evidence that such a process must be taking place. It is important to know that the magnetic field we observe at the surface is only a small fraction of the fields present in the core. Actually we see only one class of the field: the poloidal, whereas the toroidal field, possibly much stronger, is confined to the core.

Numerical models of the dynamo predicted several key phenomena observed at the surface: the primary dipolar structure with the alignment of the dipole axis close to the axis or rotation of the Earth, the westward drift of secular variations, and reversals of the polarity of the magnetic field. The later phenomenon is the cause of the magnetic anomalies on the ocean floor, which allowed estimating the rate of ocean spreading. The two most widely known models of dynamo are quite different in detail, with one by Glatzmaier and Roberts having the strongest field deep in the core, and one by Kwang and Bloxham being the strongest near the surface of the outer core.

Seismological data are consistent with the model of the core as that of a homogeneous fluid under adiabatic temperature conditions. As often near major discontinuities, there is some difficulty with pinning down the values near the end of the interval: just below the CMB and just above the inner core boundary (ICB).

F. INNER CORE (5251–6371 km)

An additional seismic discontinuity deep inside the core, which came to be called the inner core was discovered by Inge Lehmann in 1936. The fact that it is solid was postulated soon afterwards, but satisfactory proof awaited another 35 years, when observations and analysis of the free oscillations of the Earth showed that it indeed must have a finite rigidity.

It is believed that the inner core evolved during the history of the Earth, perhaps some 2 Ga ago. As the Earth was cooling, the temperatures at the Earth's center dropped below the melting point of iron (at the pressure of 330 GPa) and the inner core began to grow. The release of the gravitational energy associated with the precipitation of solid iron is believed to be the principal source of the energy driving the dynamo.

Seismologically, the inner core has been considered quite uninteresting, with a very small variation of the physical parameters across the region. This all changed in the mid-1980s when it was discovered that this region may be anisotropic, with the symmetry axis roughly parallel to the rotation axis. A deviation from that symmetry and an observation of temporal variation of travel times through the inner core brought forward an interpretation that the inner core rotates at a slightly (1°/year) higher rate than the mantle. This is being explained by the electromagnetic coupling with the dynamo field of the other core.

VII. EARTH IN THREE DIMENSIONS

Figure 23 is an example of results obtained using global seismic tomography (GST). It shows a triangular cut into a recent Earth model of the shear velocity anomalies in the Earth's mantle and shows only deviations from the average: if the Earth was radially symmetric, this picture would be entirely featureless. The surface is the top of the mantle (Mohorovicic discontinuity, or Moho) and the bottom is the core–mantle boundary. Seismic wave speeds higher than average are shown as darker shades, whereas slower than average are shown as light shades to white. Seismic velocities decrease with increasing temperature: the inference is that the light areas are hotter than average and dark are colder. Seismic wave speeds also vary with chemical composition, but there are strong indications that the thermal effect is dominant.

Density is also a function of temperature. Material hotter than average is lighter and, in a viscous Earth, will tend to float to the surface, whereas colder material is denser and will tend to sink. Thus our picture can be thought to represent a snapshot of the temperature pattern in the convecting Earth's mantle. In particular, the pixture implies a downwelling under South America and an upwelling originating at the core–mantle boundary under the central Pacific; sections passing through this anomaly indicate that this upwelling may continue to the surface.

A three-dimensional model of seismic velocities is obtained by solving an "inverse problem." Data can be observed travel time anomalies of seismic phases or, in a more complex application, entire seismograms consisting of various phases, and unknowns are the parameters describing the location and size of wave speed anomalies.

In practice, the medium must be represented by a finite number of parameters. In the example shown earlier, this is achieved by introduction of basis functions: spherical harmonics in geographical coordinates and Chebyshev polynomials in radius, both truncated at some order (degree). Evidence shows that the spec-

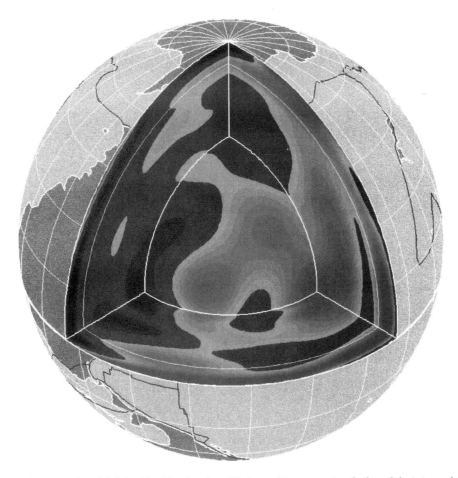

FIGURE 23   A three-dimensional model S12 of Su, Woodward, and Dziewonski, representing the lateral deviations of the shear velocities with respect to PREM. The sides represent a vertical cross section along three different profiles; the triangle at the center is a map at the core–mantle boundary. Faster than average velocities (caused by colder than normal temperature, presumably) are shown in dark and slower (hotter) in light shades. The scale is ±1.5%; significant saturation of the scale occurs in the upper mantle. Note the lateral and vertical consistency of the sign of the anomalies over large distances and depths. The mantle underneath South America is fast at nearly all depths, whereas the mantle under the central Pacific is slow.

trum of the lateral heterogeneity in the Earth is strongly dominated by very large wavelength features and that the aliasing caused by the truncation of the expansion is not a serious source of error. The alternative, frequently used, approach is to divide the medium into a three-dimensional array of cells, with velocity perturbation within each cell being constant. For the solution to be informative, there must be a sufficient coverage of the mantle volume with criss-crossing paths. This is necessary to locate the source of the observed travel time anomalies. If a particular region is sampled only in one direction, the observed anomaly could arise anywhere along the ray path.

The GST is limited by the distribution of globally detected earthquakes and by the locations of seismographic stations. There is not much that we can do about the distribution of seismicity, except that now and then an earthquake occurs in an unexpected place, so the coverage is expected to improve with time. Generally, the earthquake distribution is more even in the Northern Hemisphere. Much has been done in the last decade to improve the distribution and the quality of the seismographic stations, and recent results show considerably better resolution of the details in the top 200 km, for example. However, even using the available oceanic islands (which are very noisy, because of the wave action), there are oceanic areas with dimensions of several thousand kilometers where no land exists. A series of experiments by Japanese, French, and American seismologists have demonstrated that the establishment of a permanent or semipermanent network of ocean bottom high-quality seismographic stations is now a real, even though expensive, possibility.

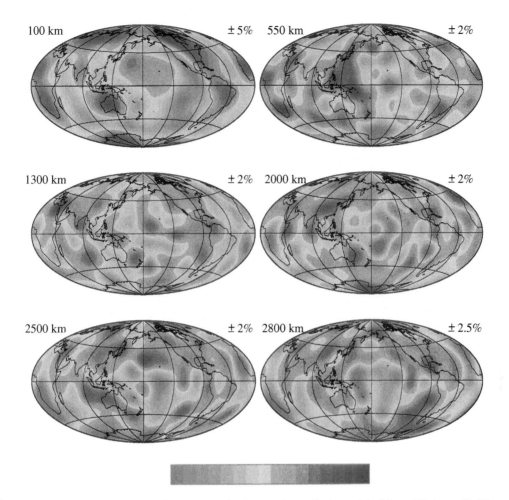

FIGURE 24 Maps of lateral variations of S velocities at six depths in a shear velocity model of Su and Dziewonski. The color range is shown at the bottom, and the range of the relative deviations is shown next to the individual panels.

Figure 24 is a collection of maps of the shear velocity anomalies from a recent model of the mantle by Su and Dziewonski (1997), built using a wide range of types of data. What is shown here is only one half of the model: the other half represents perturbations in the bulk sound velocity ($\nu_\phi = [\nu_P^2 - [\nu_S^2]]$); ν_ϕ and ν_S are independent of each other, whereas ν_S and ν_P both depend on the shear modulus. The nominal resolution of this model is about 100 km in depth and 1500 km horizontally near the surface.

At 100 km depth, the model agrees with the predictions of the plate tectonics and the thermal history of the continents. The stable continental areas (old and cool) are very fast (up to +7%), whereas material under the midocean ridges is much slower than normal (up to −5%). This negative anomaly decreases with the increasing age of the oceanic plate to become faster than average for ages greater than 100 Ma. The depth to which the anomalies associated with the midocean

ridges persists in the tomographic maps (~200 km) puzzles geodynamicists who think that midocean ridges are passive features.

The map at 550 km depth represents average shear velocity anomalies in the transition zone. The most characteristic feature is the fast anomalies in the western Pacific, under South America, the Atlantic, reaching to western Africa. In the western Pacific they can be associated with subduction zones, although they are much wider than an anomaly associated with a 100-km-thick slab. Studies of the topography of the 660-km discontinuity show that the areas of high seismic velocity are correlated with a depressed boundary, yielding credence to an interpretation that these anomalies are indicative of an accumulation (temporary, perhaps; see earlier discussion on the models of flow in the mantle) of the subducted material in the transition zone.

In the middle mantle (1300 km), the anomalies are

FIGURE 25   Low-pass filtered S-velocity model in a three-dimensional projection; the top 700 km of the structure is removed. The American fragment of the ring of high velocities circumscribing the Pacific and two large slow anomalies centered underneath the Pacific and Africa as clearly visible: see the two maps at the greatest depths in Fig. 24. The geodynamic significance of these largest scale anomalies has not yet been established, but they correlate well with the largest scale geoid anomalies.

not well organized. This contrasts with the results of "high-resolution" tomography (Grant, 1994; van der Hilst *et al.*, 1997), which in this depth range show two narrow high velocity features: one stretching from the Hudson Bay to Bolivia and the other from Indonesia to the Mediterranean. Even though elements of these two structures are present in the map shown, they are not equally well defined. Also, there are many other features of a comparable amplitude. This is also true with respect to models published by scientists at Berkeley and at Scripps, who used parameterization similar to that in Figure 24. Intensive efforts are made to understand the differences between the results of two different approaches to tomography.

The maps at 2000, 2500, and 2800 km depth show the evolution of the anomalies as the CMB is approached. The ring of high velocities circumscribing the Pacific basin is already visible at 2000 km; it strengthens considerably over the next 500 km and further increases even further toward the CMB. In the wavenumber domain of spherical harmonics, the spectrum of lateral heterogeneities is very red, being dominated by degree, 2 and 3. This is the dominant signal in the lower mantle, very clear in properly displayed data.

Figure 25 is a low-pass filtered, three-dimensional view of the anomalies in the lower mantle: the red is a 0.4% iso-surface and blue is +0.4%. We see the portion of the circum-Pacific ring of the fast anomalies and the two low velocity anomalies: one very concentrated under the Pacific and a more diffuse one under the Atlantic and Africa. The origin of this large-amplitude, very large wavelength signal has not yet been explained by geodynamic modeling, although an assumption that the velocity and gravity anomalies are correlated leads to a good prediction of the geoid field at the gravest harmonics.

It was believed since 1977, the time of publication of the first large-scale GST study, that three-dimensional images of lateral heterogeneity in the mantle will be an essential tool in addressing some of the fundamental problems in earth sciences. The results accumulated since then confirm that statement.

VIII. EARTH AS A ROSETTA STONE
• • • • • • • • • • • • • • • • • • • •

The Earth is unique among its partners in our solar system in that it has had liquid water oceans for most of its history, has a highly mobile crust, and a dynamically convecting interior. This combination means that the surface is and has been constantly driven by the movement of the interior, such that the oldest terrestrial subaerial landscapes are at most ≤10% of the age of the planet, and the oldest submarine landscapes are only a little more than 10% of that. Thus, the Earth not only has one of the most globally dynamic surfaces in the solar system, but its interior is also one of the most dynamic. (Only the tidally wracked and volcanically incessant surface of Io, Jupiter's innermost satellite, may be younger and more active, overall.) Driven by internal forces, the periodic conglomeration and separation of continental landmasses, causing opening and closing of oceans, and construction and destruction of mountain ranges profoundly impact the global climate. The environmental stresses caused by such reshuffling of the surface may themselves have influenced the progress of evolution on the planet—evolution that was possibly reset every 100 Myr or so by devastating asteroidal impacts. In the final analysis, the Earth is the only planetary body with which the human species has had intimate experience—for millenia. Thus, beyond being our home, the Earth is for us a crucial yardstick—a Rosetta stone—by which we will measure and interpret the processes, internal structure, and overall histories of other planets in this solar system and, someday, of other planets around other stars.

ALSO SEE THE FOLLOWING ARTICLE

EARTH AS A PLANET: ATMOSPHERE AND OCEANS

BIBLIOGRAPHY

Bagnold, R. A. (1940). "The Physics of Blown Sand and Desert Dunes," Metheuen & Co. Ltd., London.

Baker, V. R. (1986). Introduction: Regional landforms analysis. "Geomorphology from Space: A Global Overview of Regional Landforms" (N. M. Short and R. W. Blair, Jr., eds.), pp. 1–26. NASA Scientific and Technical Information Branch, Washington, DC.

Bloom, A. L. (1998). "Geomorphology: A Systematic Analysis of Late Cenozoic Landforms," 3rd Ed. Prentice Hall, Upper Saddle River, NJ.

Dziewonski, A. M., and Anderson, D. L. (1984). Seismic tomography of the Earth's interior. *Am. Sci.* **72** (5), 483–494.

Francis, P. W. (1993). "Volcanoes: A Planetary Perspective," Oxford Univ. Press, Oxford, UK.

Heezen, B., and Tharp, M. (1997). "Panoramic Maps of the Ocean Floor."

King, L. C. (1967). "Morphology of the Earth," 2nd Ed. Oliver and Boyd Ltd., Edinburgh.

Schumm, S. A. (1977). "The Fluvial System." Wiley, New York.

Short, N. M., and Blair, R. W., Jr. (eds.) (1968). "Geomorphology from Space: A Global Overview of Regional Landforms." NASA Scientific and Technical Information Branch, Washington, DC.

Snead, R. E. (1980). "World Atlas of Geomorphic Features." Robert E. Krieger Co., Huntington, NY, and Van Nostrand Reinhold, New York.

Su, W. J., and Dziewonski, A. M. (1997). Simultaneous inversion for 3-D variations in shear and bulk velocity in the mantle. *Phys. Earth Planet* **100** (1–4), 135–156.

Su, W. J., Woodward, A. M., and Dziewonski, A. M. (1994). Degree 12 model of shear velocity heterogeneity in the mantle. *J. Geol. Roy. Soc.* **99** (B4), 6945–6980.

van der Hilst, R. D., Widiyantoro, S., and Engdahl, E. R. (1997). Evidence for deep mantle circulation from global tomography. *Nature* **386**, 578–584.

THE MOON

I. Introduction

II. Physical Properties

III. Geophysics

IV. Lunar Surface

V. Lunar Structure

VI. Impact Processes

VII. The Maria

VIII. Lunar Highland Crust

IX. Lunar Composition

X. The Origin of the Moon

Stuart Ross Taylor

Australian National University, Canberra, and Lunar and Planetary Institute, Houston

GLOSSARY

Agglutinate: Common particle, usually about 60 μm, in the lunar soil, consisting of rock, mineral, and glass fragments bonded together by glass produced by meteorite impact.

Albedo: Percentage of incoming visible radiation that is reflected by the lunar surface.

Angular momentum: Property of rotating objects, usually expressed as mvr, where m is the mass, v is the velocity, and r is the distance from the center of rotation. The Earth and Moon have orbital angular momentum on account of their rotation around the Sun and spin angular momentum because of axial rotation.

Bouguer gravity: Free-air gravity, corrected for the attraction of the topography, so that it is dependent only on the internal density distribution.

Cumulate: Plutonic igneous rock composed of crystals accumulated by floating or sinking in the silicate melt, or magma.

Differentiation: Separation of chemical elements during crystallization into different mineral phases.

Europium anomaly and Eu*: Europium is divalent on the Moon and is mostly separated from the other trivalent rare earth elements because it is concentrated in plagioclase feldspar. The degree of enrichment or depletion is given by Eu/Eu*, where Eu is the measured abundance and Eu* is the abundance expected if Eu had the same relative concentration as the neighboring rare earth elements, samarium and gadolinium.

Gardening: Process of turning over the lunar soil or regolith by meteorite bombardment.

Mare (pl., maria): Latin word for sea, used first by Galileo to refer to the dark patches on the lunar surface, now known to be basaltic lava flows.

Mascons: Regions of the Moon of excess mass concentrations per unit area, identified by positive gravity anomalies and associated with basalt-filled multiring basins.

Moment of inertia: Quantity that is the measure of the density distribution within a planet, specifically the tendency for an increase of density with depth.

Oersted: Unit of magnetic intensity in the centimeter-gram-second system, equivalent to the gauss.

Planetesimals: Smaller bodies from which the terrestrial planets accreted. Asteroids are currently the best analogs of these now vanished objects.

P-wave velocity: Seismic body wave velocity associated with particle motion (alternating compression and expansion) in the direction of wave propagation.

Regolith: Blanket of loose surface material, produced by meteorite impact, that overlies bedrock.

Roche limit: Distance, about three planetary radii, within which a fluid satellite of zero tensile strength may be disrupted by tidal forces from the parent planet.

Solidus: Line or surface in a phase diagram below which the system is completely solid.

T he Moon is a unique satellite in the solar system, the largest relative to its planet. It is 1738 km in diameter, with a density of 3.344 g/cm³ (Earth density = 5.52 g/cm³), and has a mass 1/81 that of the Earth. Its orbit is inclined at 5.09° to the plane of the ecliptic. It rotates on its axis once every 27 days. The moment of inertia is 0.391, consistent with a small increase of density toward the center. The current consensus is that the Moon formed as a consequence of the collision with the Earth of a Mars-sized body about 4.5 billion years (b.y.) ago. The rocky mantle of the impactor spun out to form the Moon, while the core of the impactor fell into the growing Earth. This model explains the high spin of the Earth–Moon system, the strange lunar orbit, the low density of the Moon relative to the Earth, and the bone-dry and refractory composition of our satellite. The model also provides a source of energy to melt the early Moon. The geochemical and petrological evidence is clear that the Moon was molten and floated a crust of feldspar 4.44 billion years ago. This forms the present white highland crust. The interior crystallized into a sequence of mineral zones by about 4.4 billion years ago. Possibly a small metallic core formed. Heavy cratering of this crust, including many major basin-forming collisions, continued down to 3.85 billion years ago. Beginning about 4.3 billion years ago, and peaking between 3.8 and 3.2 billion years ago, partial melting occurred in the lunar interior, and basaltic lavas flooded the low-lying basins on the surface. This occurred mostly on the near side, where the crust is thinner. Activity ceased around 3.0 billion years ago, and the Moon has suffered only a few major impacts (forming, e.g., the young rayed craters such as Copernicus and Tycho) since that time.

I. INTRODUCTION

The Earth's Moon (Fig. 1) is a unique satellite in the solar system. None of the other terrestrial planets possesses comparable moons: Phobos and Deimos, the tiny satellites of Mars, are probably captured asteroids. The 60 or more satellites of the outer planets are mainly composed of low-density rock–ice mixtures, and either formed in accretion disks around their parent planets or were captured. None of them resembles the Moon and the origin of our unique satellite has been an outstanding problem. It is in plain sight, accessible even to naked-eye observation, yet it has remained until recently one of the most enigmatic objects in the solar system. [*See* PHOBOS AND DEIMOS; OUTER PLANET ICY SATELLITES.]

The Moon has played a pivotal role in human development. The axial tilt and 24-hour rotation period of the Earth may both be directly connected to lunar formation. Indeed the lunar tidal effects may have been crucial in providing an environment for life to develop. It is also possible that the Moon has stabilized the obliquity of the Earth, preventing large-scale excursions that might have had catastrophic effects on evolution. The other planets are so remote as to be only points of light, or enigmatic images in telescopes. Without the presence of the Moon, with its distinctive surface features and its regular waxing and waning phases to stimulate the human imagination, it might have taken much longer for us to appreciate the true nature of the solar system. In many other ways, such as the development of calendars, and in reminding us that there are other Earth-like bodies in the universe, the Moon has had a profound effect on the human race. One of the outstanding human achievements of the latter half of the twentieth century has been the exploration of the Moon, including the landing of astronauts on the lunar surface. A list of successful unmanned and manned spaceflight investigations of the Moon is given in Table I.

FIGURE 1   A composite full-Moon photograph that shows the contrast between the heavily cratered highlands and the smooth, dark basaltic plai_____ _____ Mare Imbrium is prominent in the northwest quadrant. The Apennine Mountains form the curved ridge southeast of Ma_____ _____ is the dark circular basalt patch at t_____

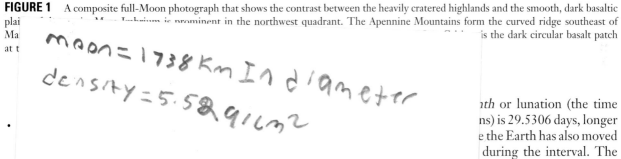

[handwritten annotation: moon = 1738 Km In diameter, density = 5.52 g/cm2]

_____nth or lunation (the time _____ns) is 29.5306 days, longer _____e the Earth has also moved _____ during the interval. The _____e equatorial plane of the _____e ecliptic (the plane of the _____un), but is closer to, but _____er. The axis of rotation of _____ly vertical to the plane of _____y at 1°32′ from the ecliptic _____ lunar orbit to the equato-_____es from 18.4° to 28.6°.

_____distance is 384,400 km or _____the distance varies from 363,000 to 406,000 km. The moon is closest to the Earth at *perigee* and farthest at *apogee*. The Moon is receding from the Earth, due to tidal interaction, at a rate of 3.74 cm/year.

A.

T_____
w_____
is_____
p_____
a_____
t_____
orbital speed of the Moon is 1.03 km/sec. _____ rotates on its axis once every 27.32166 days. This is the *sidereal month* and corresponds to the time taken for the average period of revolution of the Moon about the Earth.

TABLE I
Lunar Landing and Orbital Missions

(a) Pre-Apollo lunar landings

Spacecraft	Date	Landing site	Data returned
Ranger 7	August 1964	Mare Cognitum	Photographs
Ranger 8	Feb. 1965	Mare Tranquilitatis	Photographs
Ranger 9	March 1965	Crater Alphonsus	Photographs
Luna 9	Feb. 1966	Western Oceanus Procellarum	Photographs
Surveyor I	June 1966	Oceanus Procellarum north of Flamsteed	Photographs
Luna 13	Dec. 1966	Western Oceanus Procellarum	Photographs, soil physics
Surveyor III	April 1967	*Apollo 12* site at Oceanus Procellarum	Photographs, soil physics
Surveyor V	Sept. 1967	Mare Tranquilitatis, 25 km from *Apollo 11* site	Photographs, soil physics, chemical analysis
Surveyor VI	Nov. 1967	Sinus Medii	Photographs, soil physics, chemical analysis
Surveyor VII	Jan. 1968	Crater Tycho Ejecta blanket, North rim	Photographs, soil physics, chemical analysis

(b) Lunar orbital missions

Spacecraft	Date	Data returned
Orbiter I	Aug. 1966	Photographs
Orbiter II	Nov. 1966	Photographs
Orbiter III	Feb. 1967	Photographs
Orbiter IV	May 1967	Photographs
Orbiter V	Aug. 1967	Photographs
Galileo	Dec. 1990	Imaging
Clementine	Feb.–May 1994	Imaging, gravity, topography

(c) Apollo lunar landings

Apollo	Landing site	Latitude	Longitude	EVA (hr)	Traverse (km)	Date	Sample (kg)
11	Mare Tranquilitatis	0°67'N	23°49'E	2.24	—	July 20, 1969	21.7
12	Oceanus Procellarum	3°12'S	23°23'W	7.59	1.35	Nov. 19, 1969	34.4
14	Fra Mauro	3°40'S	17°28'E	9.23	3.45	Jan. 31, 1971	42.9
15	Hadley–Apennines	26°06'N	3°39'E	18.33	27.9	July 30, 1971	76.8
16	Descartes	8°60'N	15°31'E	20.12	27	April 21, 1972	94.7
17	Taurus–Littrow	21°10'N	30°46'E	22	30	Dec. 11, 1972	110.5

(d) Russian lunar sample missions

Luna	Landing site	Latitude	Longitude	Date	Sample (g)
16	Mare Fecunditatis	0°41'S	56°18'E	Sept. 1970	100
20	Apollonius highlands	3°32'N	56°33'E	Feb. 1972	30
24	Mare Crisium	12°45'N	60°12'E	Aug. 1976	170

(e) Russian lunar traverse vehicles

Vehicle	Landing site	Date	Traverse (km)
Lunokhod 1 (*Luna 17*)	Western Mare Imbrium	Nov. 1970	20
Lunokhod 2 (*Luna 21*)	Le Monnier crater, Eastern Mare Serenitatis, 180 km N of *Apollo 17* site	Jan. 1973	30

Tidal calculations have often been used to assess the history of the lunar orbit, but attempts to determine whether the Moon was once very much closer to the Earth, for example, near the Roche limit (≈18,000 km), which would place significant constraints on lunar origins, produce nonunique solutions. The problem is that the past distribution of land and sea is not known precisely. The continents approached their present dimensions only about 2 billion years (b.y.) ago in the Proterozoic era; oceans with small scattered land masses dominated the first half of Earth history, so that the extent of shallow seas, which strongly affect tidal dissipation, is uncertain. Work on tidal sequences in South Australia has shown that, in the late Precambrian (650 million years ago), the year had 13.1 ± 0.5 months and 400 ± 20 days. At that time, the mean lunar distance was 58.4 ± 1.0 Earth radii, so that during the Upper Proterozoic, the Moon was only marginally closer to the Earth.

Over 57% of the surface of the Moon is visible from the Earth, with variations of 6.8° in latitude and 8° in longitude. These variations in the lunar orbit are referred to as *librations* and are due to the combined effects of wobbles in the rotations of Earth and Moon.

The phases of the Moon as seen from the Earth are conventionally referred to as new moon, first quarter, full moon, and last quarter.

B. ECLIPSES

The presence of the Moon in orbit about the Earth close to the ecliptic plane produces two types of eclipses, so-called lunar and solar, that are visible from the Earth. *Lunar eclipses* occur at full Moon, when the Earth lies between Moon and Sun and so intercepts the light from the Sun. The Moon usually appears red or copper-colored during such events, since a portion of the red part of the visible solar spectrum is refracted by the Earth's atmosphere and faintly illuminates the Moon. When the Moon is partly shadowed, the border forms an arc of a circle, thus proving that the Earth has a spherical form. Typically there are two lunar eclipses a year, and they can be seen from all parts of the Earth where the Moon is visible.

In contrast, *solar eclipses,* in which the new Moon comes between the Earth and the Sun, are visible only from small regions of the Earth. Between two and five occur each year, but they reoccur at a particular location only once in every 300 or 400 years. The basic cause of the variability in eclipses is that the lunar orbit is inclined at 5.1° to the plane of the orbit of the Earth about the Sun (the plane of the ecliptic). For this reason, a solar eclipse does not occur at every new Moon. It is an extraordinary coincidence that, as seen from the Earth, the Moon and the Sun are very close to the same angular size of about 0.5° despite the factor of 389 in their respective distances, so that the two disks overlap nearly perfectly during solar eclipses. The Moon and Sun return to nearly the same positions every 6585.32 days (about 18 years), a period known to Babylonian astronomers as the *saros,* while other cycles occur up to periods of 23,000 years.

In past ages, eclipses were regarded mostly as ominous portents, and the ability to predict them gave priests, who understood their cyclical nature, considerable political power. There are many examples of the influence of eclipses on history, one notable example being the lunar eclipse of August 27, 413 B.C. This eclipse delayed the departure of the Athenians from Syracuse, resulting in the total destruction of their army and fleet by the Syracusans. Thus there is a certain irony that the word *eclipse* is derived from the Greek term for "abandonment."

C. ALBEDO

Albedo is the fraction of incoming sunlight that is reflected from the surface. Values vary from 5 to 10% for the maria to nearly 12 to 18% for the highlands. At full Moon, the lunar surface is bright from limb to limb, with only marginal darkening toward the edges. This observation is not consistent with reflection from a smooth sphere, which should darken toward the edge. This led early workers to conclude that the surface was porous on a centimeter scale and had the properties of dust. The pulverized nature of the top surface of the regolith provides multiple reflecting surfaces, accounting for the brightness of the lunar disk.

D. LUNAR ATMOSPHERE

The Moon has an extremely tenuous atmosphere of about 2×10^5 molecules/cm^3 at night and only 10^4 molecules/cm^3 during the day. It has a mass of about 10^4 kg, about 14 orders of magnitude less than that of the terrestrial atmosphere. The main components are hydrogen, helium, neon, and argon. Hydrogen and neon are derived from the solar wind, as is 90% of the helium. The remaining He and ^{40}Ar come from radioactive decay. About 10% of the argon is ^{36}Ar, derived from the solar wind.

E. MASS, DENSITY, AND MOMENT OF INERTIA

The mass of the Moon is 7.35×10^{25} g, which is 1/81 of the mass of the Earth. Although the Galilean satellites of Jupiter and Titan are comparable in mass, the Moon/Earth ratio is the largest satellite-to-parent ratio in the solar system. (The Charon/Pluto ratio is larger, but Pluto, an icy planetesimal, is less than 20% of the mass of the Moon and is called a planet only out of courtesy.) The lunar radius is 1738 ± 0.1 km, which is intermediate between that of the two Galilean satellites of Jupiter, Europa ($r = 1569$ km) and Io ($r = 1815$ km). The Moon is much smaller than Ganymede ($r = 2631$ km), which is the largest satellite in the solar system. [See Io; OUTER PLANET ICY SATELLITES; PLUTO AND CHARON.]

The lunar density is 3.344 ± 0.003 g/cm^3, a fact that has always excited interest on account of the Moon's proximity to the Earth, which has a much higher density of 5.52 g/cm^3. The lunar density is also intermediate between that of Europa ($d = 2.97$ g/cm^3) and Io, the innermost of the Galilean satellites of Jupiter, with a density of 3.57 g/cm^3. The other 60-odd satellites in the solar system are ice–rock mixtures and so are much less dense.

The lunar moment of inertia is 0.391 ± 0.002. This requires a slight density increase toward the center, in addition to the presence of a low-density crust (a homogeneous sphere has a moment of inertia of 0.400; the value for the Earth, with its dense metallic core that constitutes 32.5% of the mass of the Earth, is 0.3315).

F. ANGULAR MOMENTUM

The spin angular momentum of the Earth–Moon system is anomalously high compared with that of Mars, Venus, or the Earth alone. Some event or process spun up the system relative to the other terrestrial planets. However, the angular momentum of the Earth–Moon system (3.41×10^{41} g · cm^2/sec) is not sufficiently high for classic fission to occur. If all the mass of the Earth–Moon system were concentrated in the Earth, the Earth would rotate with a period of 4 hours. Yet even this rapid rotation is not sufficient to induce fission, even in a fully molten Earth.

G. CENTER OF MASS/CENTER OF FIGURE OFFSET

The mass of the Moon is distributed in a nonsymmetrical manner, with the center of mass lying 1.8 km closer to the Earth than the geometrical center of figure (Fig. 2). This is a major factor in locking the Moon into synchronous orbit with the Earth, so that the Moon always presents the same face to the Earth, although librations allow a total of 57% of the surface to be visible at various times.

Various explanations have been advanced to account for the offset of the center of mass from the center of figure. Dense mare basalts are commoner on the near side, but their volume is insufficient by about an order of magnitude to account for the effect. It has also been suggested that this offset could arise if the lunar core is displaced from the center of mass. However, such a displacement would generate shear stresses that could not be supported by the hot interior. Another suggestion is that a density asymmetry developed in the mantle during crystallization of the magma ocean, with a greater thickness of low-density Mg-rich cumulates being concentrated within the farside mantle. It is unlikely that such density irregularities would survive stress relaxation in the hot interior, unless actively maintained by convection. The conventional explanation for the center of figure/center of mass offset is that the farside highland low-density crust is thicker. It is massive enough and sufficiently irregular in thickness to account for the effect. The scarcity of mare basalts on the farside (Fig. 3) is consistent with a thicker farside crust. Lavas rise owing to the low density of the melt and do not possess sufficient hydrostatic head to reach the surface on the farside, except in craters in some very deep basins (e.g., Ingenii).

H. REMOTE SPECTRAL OBSERVATIONS

Spectral observations of the Moon from the Earth are limited to the visible and infrared portion of the electromagnetic spectrum between about 3000 and 25,000 Å. These studies identify plagioclase by a weak absorption band at 13,000 Å (1.3 μm) and pyroxene by two strong bands at about 9700–10,000 Å (0.97–1.0 μm), as well as olivine. This technique has enabled mapping of several distinctive mare basalt types on the lunar surface. In addition, mapping of pyroclastic glass deposits has been possible because of their characteristic absorption bands due to Fe^{2+} and Ti^{4+}. These features have also enabled the mapping of the FeO and TiO$_2$ contents of mare basalts and the amount of anorthosite in the lunar highland crust.

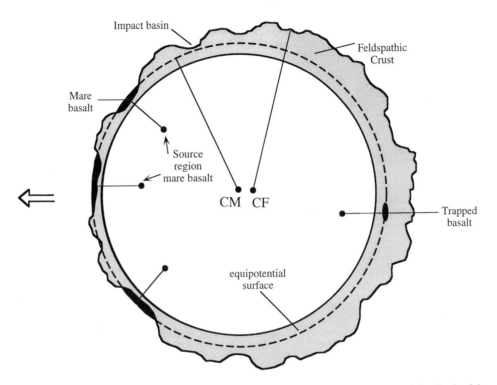

FIGURE 2 A cross section through the Moon in the equatorial plane that shows the displacement toward the Earth of the center of mass relative to the center of figure, due to the presence of a thicker farside, low-density feldspathic crust. It also illustrates that an equipotential surface is closer to the surface on the nearside. Magmas that originate at *equal depths* below the surface will have greater difficulty in reaching the surface on the farside, a problem exacerbated by the greater farside crustal thickness. However, not all flooding of lunar basins will be at the same level. Some magmas originate at different depths, whereas others come from different locations at different times. Others may extrude smaller or greater amounts of lava, leading to differences in the amount of basalt filling a particular basin. These factors will all contribute to filling of mare basins to differing depths not necessarily related to the equipotential surface.

III. GEOPHYSICS

A. GRAVITY

The young ray craters have negative Bouguer anomalies because of the mass defect associated with excavation of the crater and low density of the fallback rubble. Craters less than 200 km in diameter have negative Bouguer anomalies for the same reason (e.g., Sinus Iridium has a negative Bouguer anomaly of −90 mgal). Volcanic domes such as the Marius hills have positive Bouguer anomalies (+65 mgal), indicating support by a rigid lithosphere. The younger, basalt-filled circular maria on the nearside have large positive Bouguer anomalies, referred to as mascons (e.g., Mare Imbrium, +220 mgal). These are due to the uplift of a central plug of denser mantle material during impact followed by addition of dense mare basalt. The gravity signature of young, large, ringed basins, such as Mare Orientale, shows a "bull's-eye" pattern with a central positive

Bouguer anomaly (+200 mgal) surrounded by a ring of negative Bouguer anomalies (−100 mgal) with an outer positive Bouguer anomaly collar (+30–50 mgal).

The lunar highland crust is strong. High mountains such as the Apennines (7 km high), formed during the Imbrium collision 3.85 b.y. ago, are uncompensated and are supported by a strong cool interior. The older lunar highlands are in isostatic equilibrium and there are no Bouguer anomalies. Large impact basins on the thick farside crust are compensated. The absence of mascons associated with these large basins appears to be from two causes, both related to the thickness of the farside crust. Because of this, the rebound and uplift of a central dense plug of mantle rock are limited. The thickness of the low-density crust also inhibits the rise of basaltic lavas, so that, unlike on the nearside, the basins mostly remain unflooded, except for a few such as Mare Ingenii within the deep South Pole–Aitken Basin.

The gravity data are consistent with an initially molten Moon that cooled quickly and became rigid enough by 4.0 b.y. to support loads such as the circular moun-

FIGURE 3   The heavily cratered farside highlands. Note the scarcity of mare basalts. Mare Crisium is the dark circular patch of basalt on the northwest horizon. (Courtesy of NASA, *Apollo 16* metric frame 3023.)

tainous rings around the large, younger, ringed basins and the mascons. This places considerable restrictions on lunar themal models, as well as the amount of later melting in the lunar interior. Melting in the deep interior to produce the mare basalts had no effect on the strength of the crust (the volume of mare basalts is only about 0.1% of the whole Moon).

B. SEISMOLOGY

The lunar seismic signals have a large degree of wave scattering and a very low attenuation, so that during moonquakes the Moon "rings like a bell" owing to the absence of water and the very fractured nature of the upper few hundred meters. Observed moonquakes have been mostly less than 3 on the Richter scale; the largest recorded ones have a magnitude of about 4. Many are repetitive and reoccur at fixed phases of the lunar tidal cycle. The Apollo seismometers recorded the impacts of 11 meteorites with masses of more than one ton. The Moon is seismically inert compared to

the Earth, and tidal energy is the main driving force for the weak lunar seismic events. No seismic events are associated with the lunar grabens or rilles.

C. HEAT FLOW AND LUNAR TEMPERATURE PROFILE

Two measurements of lunar heat flow are available: 2.1 μW/cm^2 at the *Apollo 15* site and 1.6 μW/cm^2 at the *Apollo 17* site. It is interesting that these observed heat flows are close to Earth-based estimates from microwave observations. These values provide some constraints on the bulk lunar abundances of the heat-producing elements K, U, and Th. The lunar interior must have been stiff enough for the past 4.0 b.y. to account for the support of the mountain rings and the mascons. The most probable lunar temperature profile is shown in Fig. 4, which indicates temperatures of 800°C at a depth of 300 km. Unlike the Earth, which dissipates most of its heat by volcanism at the midocean ridges, the Moon loses its heat by conduction. Most

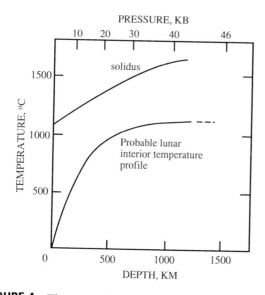

FIGURE 4 The present-day variation of lunar temperature with depth, showing that the temperature is well below that required for partial melting (solidus curve).

of its original internal heat has been lost, and differentiation has concentrated most of the K, U, and Th near the surface. The present heat flow could indicate lunar U values as high as 45 ppb or over twice the terrestrial abundances. A more conservative value of 30 ppb U is adopted here, based on petrological and geochemical constraints, as well as accommodating the high heat flow values. This uranium abundance is still 50% higher than the well-established terrestrial value of 20 ppb U. Although there has been considerable controversy over the reality of a higher than terrestrial lunar uranium abundance, it appears to be confirmed by the requirement from the *Clementine Mission* data for a higher than terrestrial lunar aluminum abundance. Both Al and U are refractory elements, not easily separated by nebular processes, and generally considered to be correlated in the terrestrial planets.

D. MAGNETIC FIELD

The lunar rocks contain a stable natural remnant magnetism. Apparently between about 3.6 and 3.9 b.y. ago there was a planetarywide magnetic field that has now vanished. The field appears to have been much weaker both before and after this period. The paleointensity of the field is uncertain, but perhaps was several tenths of an oersted. The most reasonable interpretation is that the Moon possessed a lunar dipole field of internal origin, all other suggested origins appearing less likely. The favored mechanism is that the field was produced by dynamo action in a liquid Fe core. A core 400 km

in diameter could produce a field of about 0.1 oersted at the lunar surface.

Localized strong magnetic anomalies are associated with patterns of swirls, as at Reiner Gamma. These swirls are antipodal to major impact basins and have been suggested to have formed by some focusing effect of the seismic waves that resulted from the large impacts that formed the basins. More work is clearly needed to substantiate this hypothesis and to understand the association of swirls and magnetic fields. Other remnant fields, with a field strength only about 1/100th of the terrestrial field, were measured at the Apollo landing sites.

IV. LUNAR SURFACE

The absence of plate tectonics, water, and life, and the essential absence of atmosphere, indicates that the present lunar surface is unaffected by the main agents that affect the surface of the Earth. Ninety-nine percent of the lunar surface is older than 3 b.y. and more than 80% is older than 4 b.y. In contrast, 80% of the surface of the Earth is less than 200 million years (m.y.) old. The major agent responsible for modifying the lunar surface is the impact of objects ranging from micron-sized grains to bodies tens to hundreds of kilometers in diameter.

Because of the effective absence of a lunar atmosphere, the lunar surface is exposed to ultraviolet radiation with a flux of about 1300 W/m². The absence of a magnetic field allows the solar wind (1–100 eV) and solar (0.1–1 MeV) and galactic (0.1–10 GeV) cosmic rays to impinge directly on the surface. The relative fluxes are 3×10^8, 10^6, and 2–4 protons/cm²/sec, respectively. The penetration depths of these particles extend to micrometers, centimeters, and meters, respectively.

The maximum and minimum lunar surface temperatures are about 390 K and 104 K. At the *Apollo 17* site, the maximum temperature was 374 K (111°C) and the minimum was 102 K (−171°C). The temperatures at the *Apollo 15* site were about 10 K lower. The conductivity of the upper 1–2 cm of the surface is very low (1.5×10^{-5} W/cm²). This increases about fivefold below 2 cm. A cover of about 30 cm of regolith is sufficient to damp out the surface temperature fluctuation of about 280 K to about ±3 K, so that structures on the Moon could be well insulated by a modest depth of burial. This in turn might produce difficulties in losing heat generated in buried structures. Impacts of micrometeoroids of about 1 mg mass could be expected about once a year on a lunar structure.

FIGURE 5 *Apollo 16* astronaut John Young and the lunar rover at Station 4 on the slopes of Stone Mountain, illustrating the nature of the lunar surface and the absence of familiar landmarks. Smoky Mountain in the left background, with Ravine crater (1 km in diameter) on its flank, is 9 km distant. (Courtesy of NASA, AS16-110-17960.)

The combination of strong sunlight, low gravity, awkward space suits, and absence of familiar landmarks makes orientation difficult on the lunar surface. All astronauts have commented on the difficulty of judging distance (Fig. 5).

A. REGOLITH

The surface of the Moon is covered with a debris blanket, called the regolith, produced by the impacts of meteorites. It ranges from fine dust to blocks several meters across. The fine-grained fraction is usually referred to as the lunar soil (Fig. 6). This is an unfortunate use of the term "soil," which has organic connotations, but the term is as thoroughly entrenched as astronomers' use of "metals" for all elements heavier than helium. Although there is much local variation, the average regolith thickness on the maria is 4 to 5 m, whereas the highland regolith is about 10 m thick.

Seismic velocities are only about 100 m/sec at the surface, but increase to 4.7 km/sec at a depth of 1.4 km at the *Apollo 17* site. The density is about 1.5 g/cm³ at the surface, increasing with compaction to about 1.7 g/cm³ at a depth of 60 cm. The porosity at the surface is about 50% but is strongly compacted at depth. The regolith is continuously being turned over or "gardened" by meteorite impact. The near-surface structure, revealed by core samples (the deepest was nearly 3 m at the *Apollo 17* site), shows that the regolith is a complex array of overlapping ejecta blankets, typically ranging in thickness from a few millimeters up to about 10 cm. These have little lateral continuity. Most of the regolith is of local origin: lateral mixing occurs only on a local scale, so that the mare–highland contacts are relatively sharp over a kilometer or so. The rate of growth of the regolith is very slow, averaging about 1.5 mm per million years or 15 Å per year, but it was more rapid between 3.5 and 4 b.y. ago.

Five components make up the lunar soil: mineral

FIGURE 6 The nature of the lunar upper surface is illustrated in this view of small pebbles being collected by a rake near the *Apollo 16* landing site in the Descartes highlands. Lunar sample 60018 was taken from the top of the boulder. (Courtesy of NASA, AS16-116-18690.)

bardment may well have produced megaregolith thicknesses in excess of 10 km. Related to this question is the degree of fracturing and brecciation of the deeper crust due to the large basin collisions. Some estimates equate this fracturing with the leveling off in seismic velocities (V_p) to a constant 7 km/sec at 20–25 km. In contrast to the highlands, bedrock is present at relatively shallow depths (tens of meters) in the lightly cratered maria.

B. TECTONICS

The dominant features of the lunar surface are the old heavily cratered highlands and the younger basaltic maria, mostly filling the large impact basins (see Figs. 1 and 3). There is a general scarcity of tectonic features on the Moon, in great contrast to the dynamically active Earth. Most of the lunar tectonic features are related to stresses associated with subsidence of the mare basins, following flooding with lava. There are no large-scale tectonic features, and the lunar surface acts as a single thick plate that has been subjected to only small internal stresses. Attention has often been drawn to a supposed "lunar grid" developed by tectonic stresses. However, the lineaments that constitute the "grid" are formed by the overlap of ejecta blankets from the many multiringed basins and have no tectonic significance.

Wrinkle ridges (or mare ridges) are low-relief, linear to arcuate, broad ridges that commonly form near the edges of the circular maria. They are the result of compressional bending stresses, related to subsidence of the basaltic maria from cooling.

Rilles, which are extensional features similar to terrestrial grabens, are often hundreds of kilometers long and up to 5 km wide. Unlike the wrinkle ridges, they cut only the older maria as well as the highlands and indicate that some extensional stress existed in the outer regions of the Moon prior to about 3.6 b.y. ago. They should probably be termed "grabens" so as to avoid confusion with the sinuous rilles, such as Hadley Rille, that are formed by flowing lava, presumably by thermal erosion. The set of three rilles, each about 2 km wide, that are concentric to Mare Humorum at about 250 km from the basin center are particularly instructive examples, showing a clear extensional relation to subsidence of the impact basin (Fig. 7).

C. LUNAR STRATIGRAPHY

The succession of events on the lunar surface has been determined by establishing a stratigraphic sequence

fragments, crystalline rock fragments, breccia fragments, impact glasses, and agglutinates. The latter are aggregates of smaller soil particles welded together by glasses. They may compose 25–30% of a typical soil and tend to an equilibrium size of about 60 μm. Their abundance in a soil is a measure of its maturity, or length of exposure to meteoritic bombardment. Most lunar soils have reached a steady state in particle size and thickness. Agglutinates contain metallic Fe droplets (typically 30–100 Å) produced by reduction with implanted solar wind hydrogen, which acts as the reducing agent, during melting of soil by meteorite impact.

A *megaregolith* of uncertain thickness covers the heavily cratered lunar highlands. This term refers to the debris sheets from the craters and particularly those from the large impact basins that have saturated the highland crust. The aggregate volume of ejecta from the presently observable lunar craters amounts to a layer about 2.5 km thick. The postulated earlier bom-

FIGURE 7   Three sets of curved rilles or grabens, each about 2 km wide, concentric to Mare Humorum, the center of which is about 250 km distant. The ruined crater intersected by the rilles is Hippalus, 58 km in diameter. The crater at bottom right, flooded with mare basalt, is Campanus, 48 km in diameter. (Courtesy of NASA, Orbiter IV-132-H.)

based on the normal geological principle of superposition. Geological maps based on this concept have been made for the entire Moon. Relative ages have been established by crater counting, and isotopic dating of returned samples has enabled absolute ages to be assigned to the various units. The formal stratigraphic sequence is given in Table II.

V. LUNAR STRUCTURE

A. LUNAR CRUST

The lunar highland crust averages 61 km in thickness but varies from an average of 55 km on the nearside to an average of 67 km on the farside (see Fig. 3). Yet there are wide variations in crustal thickness, with large areas on the farside crust exceeding 100 km in thickness. On the nearside, the crust reaches 95 km in thickness in the southern hemisphere (30°S, 30°E). The maximum relief on the lunar surface is over 16 km. The deepest basin (South Pole–Aitken) has 12 km relief. The crust composes 10% of lunar volume.

The mare basalts cover 17% of the lunar surface, mostly on the nearside (see Fig. 1). Although prominent visually, they are usually than 1 or 2 km thick, except near the centers of the basins. These basalts constitute only about 1% of the volume of the crust and make up less than 0.1% of the volume of the Moon.

Seismic velocities increase steadily down to 20 km. At that depth, there is a change in velocities within the crust that probably represents the depth to which

TABLE II
Lunar Stratigraphy

System	Age (billion years)	Remarks
Copernican	1.0 to present	The youngest system, which includes fresh ray craters (e.g., Tycho). Begins with the formation of Copernicus.
Eratosthenian	1.0–3.1	Youngest mare lavas and craters without visible rays (e.g., Eratosthenes).
Imbrian	3.1–3.85	Extends from the formation of the Imbrium Basin to the youngest dated mare lavas. Includes Imbrium Basin deposits, Orientale and Schrödinger multiring basins, most visible basaltic maria, and many large impact craters, including those filled with mare lavas (e.g., Plato, Archimedes).
Nectarian	3.85–3.92	Extends from the formation of the Nectaris Basin to that of the Imbrium Basin. Contains 12 large, multiring basins and some buried maria.
Pre-Nectarian	Pre-3.92	Basins and craters formed before the Nectaris Basin. Includes 30 identified multiring basins.

extensive fracturing, due to massive impacts, has occurred. At an earlier stage, this velocity change was thought to represent the base of the mare basalts, but these are now known to be much thinner. The main section of the crust from 20 to 60–80 km has rather uniform velocities of 6.8 km/sec, corresponding to the velocities expected from the average anorthositic composition of the lunar samples.

The *Galileo* spacecraft mission to Jupiter flew by the Earth–Moon system in December 1990 and imaged the western limb of the Moon, including the Orientale Basin, which was not well covered by the Apollo missions. The exterior ejecta deposits (the Hevelius formation) are not from the mantle of the Moon, but were excavated from the crust, and are dominantly anorthositic. The Hevelius formation is similar to typical *Apollo 16* highland soils. The Orientale ejecta also covered areas of mare basalt. The interior of the South Pole–Aitken Basin has a more mafic (Fe- and Mg-rich) composition relative to the more feldspathic lunar highlands. The floor of the basin is likely a mixture of crust and mantle. South Pole–Aitken is the deepest basin on the Moon and has probably excavated into the lunar mantle. It would be of great interest to study this area in detail, since no excavated mantle samples have ever been identified in the returned Apollo samples.

B. LUNAR MANTLE

The structure of the mantle appears to be relatively uniform (Fig. 8). The average P-wave velocity is 7.7 km/sec and the average S-wave velocity is 4.45

km/sec down to about 1100 km. There may be a minor discontinuity at 400–480 km with a slight change in seismic velocities, but this is not resolvable from the present data. The main foci for moonquakes lie deep within this zone at about 800–1000 km. The outer 800 km has a very low seismic attenuation, indicative of a volatile-free rigid lithosphere. Solid-state convection is thus extremely unlikely in the outermost 800 km. Below about 800 km, P- and S-waves become attenuated ($V_S = 2.5$ km/sec). P-waves are transmitted

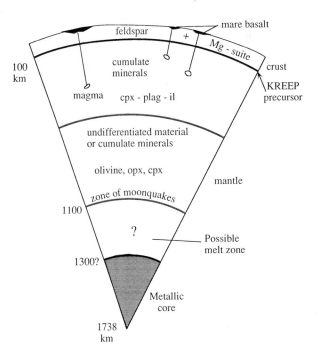

FIGURE 8 A diagram of the internal structure of the Moon.

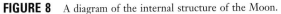

through the center of the Moon, but S-waves are missing, possibly suggesting the presence of a melt phase. It is unclear, however, whether the S-waves were not transmitted or were so highly attenuated that they were not recorded.

C. LUNAR CORE

The evidence for a metallic core is inconclusive but suggestive. Electromagnetic sounding data place an upper limit of a 400- to 500-km radius for a highly conducting core. The moment of inertia value of 0.391 ± 0.002 is low enough to require a small density increase in the deep interior, in addition to the low-density crust. In summary, a lunar core with radius about 400 km (4% of lunar volume) is consistent with the available data.

VI. IMPACT PROCESSES

A. CRATERS AND MULTIRING BASINS

One of the most diagnostic features of the lunar surface, which is in great contrast to the surface of the Earth, is the ubiquitous presence of impact craters at all scales, from micron-sized "zap pits" to multiring basins. The largest confirmed example is the South Pole–Aitken Basin (180°E, 56°S), 2500 km in diameter. The presence of the larger Procellarum Basin (3200-km diameter, centered at 23°N, 15°W) covering much of the nearside is questionable. Although the correct explanation for the origin of the lunar craters had already been reached by G. K. Gilbert in 1893 and R. B. Baldwin in 1949, this topic was the subject of ongoing controversy until about 1960, and the question still surfaces occasionally in popular articles. Since meteorites and other impacting bodies could be expected to strike the Moon at all angles, the circularity of the lunar craters was long used as an argument against impact and in favor of a volcanic origin. It was eventually realized that bodies impacting the Moon at velocities of several km/sec explode on impact and form a circular crater regardless of the angle of impact. The morphology of the craters resembles that of terrestrial explosion crates and is quite distinct from the landforms of terrestrial volcanic centers.

The smallest craters are simple bowl-shaped depressions, surrounded by an overturned rim and an ejecta blanket (e.g., Linné, 2450-m diameter, Fig. 9). With increasing size, more complex forms develop. At diameters greater than about 15–20 km, slump terraces appear on the crater walls. Central peaks formed by rebound appear at crater diameters greater than about 25–30 km (e.g., Copernicius, 93-km diameter, Fig. 10). Central peak basins, in which a fragmentary ring of peaks surrounds a central peak (e.g., Compton, 162-km diameter), develop in the size range 140–180 km. Larger craters develop internal concentric peak rings in place of the central peak (e.g., Schrödinger, 320-km diameter, Figs. 11 and 12). Such central peaks and peak rings may develop from fluidized waves during impact.

The ultimate form resulting from impact is the multiring basin, which may have six or more rings. The type lunar example is Orientale (Fig. 13). This structure is 920 km in diameter (about the size of France), with several concentric mountain rings having a typical relief of about 3 km with steep inward-facing scarps. These form in a few minutes following the impact of a body perhaps 50 km in diameter. The central portion has been flooded with mare basalt. Thirty such basins have been recognized on the Moon (Fig. 14), with another 14 probable. There is much controversy over the origin of multiring basins. One possibility proposes that the crust is fluidized by the impact and the rings form like ripples on a pond into which a stone has been dropped. The most likely explanation is that the mountain rings are fault scarps, formed by collapse into a deep transient crater formed by the initial impact.

The depth of excavation of the lunar basins decreases with increasing basin diameter. A transient cavity forms during the initial stage of the impact, but most excavated material comes from shallower depths. Thus no unequivocal lunar mantle material has been recognized in the returned samples from the lunar highland crust and the depths of excavation of the largest basins do not appear to have exceeded 60 km. Ejecta blankets incorporate much local material as they travel across the surface in a manner analogous to base surge. Apart from the ejecta blankets, numerous blocks from large impacts travel with sufficient velocity to produce secondary craters. These must be carefully distinguished from primary craters to avoid confusion in the dating of lunar surfaces by crater counting.

Shock pressures up to 100 GPa (1 GPa = 10 kbar) cause a variety of effects from the development of planar features in minerals (<10 GPa) to whole-rock melting (50–100 GPa). Above about 150 GPa, the rocks are vaporized. Vapor masses of a few times projectile mass and melt masses about 100 times the pro-

FIGURE 9   An oblique view of crater Linné, a simple bowl-shaped crater in northern Mare Serenitatis. The rim crest diameter is 2450 m. Note the ejecta blocks on the rim, dunelike features on the flanks, and secondary craters at 1–3 crater radii from the rim crest. Linné was famous in the nineteenth century as a "disappearing" lunar crater, since it was not seen by several observers. This was a consequence of observations at the limits of Earth-based telescopic resolution. (Courtesy of NASA, *Apollo 15* pan photo 9353.)

jectile mass may be formed. Impact melts compose 30–50% of all samples returned from the lunar highlands.

B. LUNAR CRATERING HISTORY AND THE LUNAR CATACLYSM

The intense cratering of the lunar highlands and the absence of a similar heavily cratered surface on the Earth were long recognized as due to an early "pre-geological" bombardment. In contrast, the lightly cratered basaltic mare surfaces, on which the cratering rate is about 200 times less, had escaped this catastrophe and were clearly much younger. The ages of the mare surfaces, dated from the sample return to be between 3.3 and 3.8 b.y., showed that the cratering flux was similar, within a factor of two, to that observed terrestrially. It also established that the intense cratering of the highlands occurred more than 3.8 b.y. ago. Most highland samples have ages in the range 3.8–4.0 b.y. The radiometric ages of the ejecta blankets from

the large collisions tend to cluster around 3.9 b.y., with the dates for the Imbrium collision being 3.85 and Nectaris 3.92 b.y. This is a surprisingly narrow range and indicates a rapid increase in the cratering flux before 3.8 b.y. This clustering has led to the concept of a "lunar cataclysm" or a spike in the collisional history at that time. The conventional, or noncataclysmic explanation, is that Imbrium and Orientale basins formed during the tail end of the accretion of the planets and so represent the final sweep-up of large objects. The problem with this scenario is that extrapolation from the rate at 3.8 b.y. back to 4.5 b.y. results in the accretion of a Moon several orders of magnitude larger than observed. Other arguments in favor of the cataclysm include the scarcity of impact melts older than 4 b.y. and the lead isotope data, which indicate a major resetting of the lead ages at 3.86 b.y. It is also probable that accretion of the Moon was essentially complete and that the Moon was at its present size 4440 m.y. ago, at the time of the formation of the feldspathic highland crust. Figure 15 shows a reconstruction of the lunar crater production rate with time.

FIGURE 10 Oblique view of crater Copernicus, 93 km in diameter, showing a central peak complex and well-developed slump terraces on the inner walls. (Courtesy of NASA, AS17-151-23260.)

VII. THE MARIA

The maria make up the prominent dark areas that form the features of "the man in the Moon" (Figs 1 and 16). After centuries of speculation during which the maria were thought to be composed of sediments, asphalt, and other unlikely materials, they were conclusively identified following the *Apollo 11* sample return in 1969 as formed of basaltic lavas. This conclusion had already been reached by earlier workers such as R. B. Baldwin and G. Kuiper and was strongly suggested by the data from the *Surveyor* landers. These vast plains cover 17% (6.4×10^6 km²) of the surface of the Moon, and they are exceedingly smooth, with slopes of 1:500 to 1:200 and elevation differences of only 150 m over distances of 500 km. This smoothness and the lack of volcanic constructional forms, which litter many terrestrial volcanic fields, remind one of plateau basalts on Earth and are probably due to several factors. These include a combination of high eruption rates and the low viscosity of the lunar lavas, which is about an order of magnitude lower than that of their terrestrial counterparts and is close to that of engine oil at room temperature. The lava flows (Fig. 17) are thin (10–40 m) and up to 1200 km long, a consequence

of the low viscosity and probable long duration of the eruption. Flow fronts are generally less than about 15 m in height. Occasional small volcanic domes and cones occur on the mare surface. The classic example is the region of the Marius Hills.

The maria are not all at the same level, and this is indicative of independent eruptions from diverse sources at differing depths in the interior. They are mostly subcircular in form owing to their filling of the multiring basins, originally excavated by impact. The lavas that fill the basins are the maria as in Mare Imbrium (see Figs. 1 and 14) or Mare Ingenii (see Fig. 16). The basins, as in the Imbrium Basin, were formed much earlier by impact. Thus the mare basalt fill is unrelated to the formation of the basin, but rather floods into the low-lying depressions much later. Some impact melt, distinct in composition to the lavas, may be formed at the time of the impact, but should not be confused with the basaltic mare lavas, which differ in both composition and age. Oceanus Procellarium (see Fig. 1) is the type example of an irregular mare, where the lavas have flooded widely over the highland crust. However, this mare may be filling parts of an old, large, and very degraded Procellarum Basin, 3200 km in diameter.

The mare lavas reach the surface because of the

FIGURE 11   The transition between central peak craters and peak-ring craters. The large central basin is Schrödinger (320 km in diameter), which has a well-developed peak ring. Antoniadi (135 km in diameter), southeast from Schrödinger, has both a central peak and a peak ring. The small crater immediately southwest of Antoniadi has a central peak only. (Courtesy of NASA, Orbiter IV-8M.)

density difference between the melt and that of the overlying column of rock. The scarcity of maria on the farside of the Moon (See Fig. 3) is due to the greater crustal thickness. An exception is part of the area of the deep depression of the South Pole–Aitken multiring basin (2500 km in diameter), on which is superimposed the Ingenii impact basin (650 km in diameter), now occupied in part by the lavas of Mare Ingenii (see Fig. 16). However, most of the South Pole–Aitken Basin, deeper than the nearside maria, is not flooded with lava. This argues for mantle heterogeneity and localized sources for the mare basalts, rather than some moonwide melting of the interior, with consequent flooding of lava to a uniform level.

Dark mantle deposits, which represent pyroclastic deposits formed probably by "fire fountaining" during lunar eruptions, occur, for example, around the southern borders of Mare Serenitatis. These pyroclastic deposits are composed mainly of glass droplets and fragments and can be distinguished from the ubiquitous glasses of impact origin by their uniformity, homogeneous composition, and absence of meteoritic contamination. Over 25 distinct compositions have been recognized. They commonly have a superficial coating of volatile elements such as Pb, Zn, Cl, and F, derived from volcanic vapors during the eruption. The dominant gas, however, is mostly CO. The source of these volatile elements is uncertain. They are rare in the lunar samples, and the Moon is generally thought to be strongly depleted in them. Possibly they come from

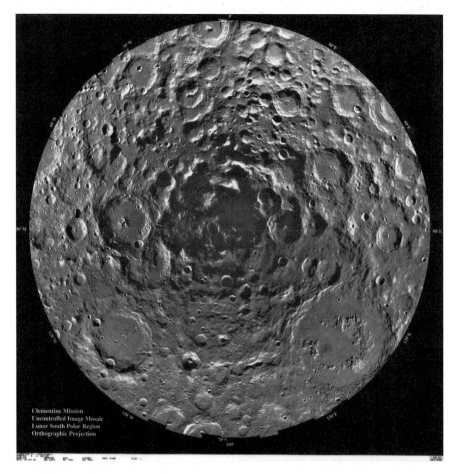

FIGURE 12   A mosaic of 1500 *Clementine* UV VIS images centered on the south pole, showing the heavily cratered south polar region of the Moon. The Schrödinger basin (320 km in diameter), which is the freshest peak-ring basin on the Moon, is at four o'clock. Schrödinger is slightly older than Orientale. Note the small volcanic cone in the bottom left-hand sector. It is possible that some ice (from cometary impacts?) is trapped in the permanently shadowed craters at the south pole (LPI *Clementine* press release.)

local cumulate sources, and thus do not imply an enrichment of the deep lunar interior in volatile elements. This remains an interesting question. Some areas of mare basalts appear to be covered by ejecta blankets from multiring basins; their presence is revealed by the haloes of ejected dark basalt from impact craters that have punched through the light-colored highland plains units of anorthositic composition to the underlying basalts.

Although they are prominent visually on the Moon, the maria typically form a thin veneer, mostly less than 1–2 km thick, except in the centers of the circular maria (e.g., Imbrium), where they may reach thicknesses of 3 to perhaps 5 km. The total volume of mare basalt is usually estimated at between 6 and 7 $\times$ 10^6 km^3 or about 0.1% of lunar volume. Cooling rates for mare basalts range from 0.1° to 30°C per hour, indicative of fast cooling in thin lava flows.

Sinuous rilles occur widely near the edges of the maria and are either lava channels or collapsed lava tubes. They have eroded into the surrounding lavas by a combination of thermal and mechanical erosion. The classic example is Hadley Rille (Fig. 18), visited by the *Apollo 15* mission. It is 135 km long and averages 1.2 km in width and 370 m in depth. Massive lava bedrock is exposed in the rille wall at the *Apollo 15* site. The sinuous rilles should not be confused with the straight or arcuate rilles, which are grabens of tectonic origin.

A. MARE BASALT AGES

The oldest ages for returned lunar mare basalts are from *Apollo 14* breccias; aluminous low-Ti basaltic clasts in these breccias range in age from 3.9 to 4.3 b.y. The oldest basalt from a visible maria is Apollo sample number 10003, a low-K basalt from Mare

FIGURE 13 Orientale is the classic example of a multiring basin. The diameter of the outer mountain ring (Montes Cordillera) is 930 km, about the size of France. Note the radial structures resulting from the impact. It is the youngest major impact basin on the Moon. This structure formed about 3800 m.y. ago in a few minutes following the impact of a planetesimal or asteroid about 50–100 km in diameter. Basalt has flooded the center of the Orientale Basin. The small circular patch of mare basalt northeast of Orinetale is Grimaldi. The western edge of Oceanus Procellarum fills the northeastern horizon. (Courtesy of NASA, Orbier IV-181M.)

Tranquilitatis with an age of 3.85 ± 0.03 b.y. This gives a younger limit for the age of the Imbrium collision, since the lavas of Mare Tranquilitatis overlie the Imbrium ejecta blanket.

The youngest dated sample is number 12022, an ilmenite basalt with an age of 3.08 ± 0.05 b.y. Low-Ti basalts are generally younger than high-Ti basalts. Stratigraphically younger flows, some of which appear to embay young ray craters, may be as young as 1 b.y. but are of very limited extent. The most voluminous period of eruption of lavas appears to have been between about 3.8 and 3.1 b.y. ago. Isotopic measurements show that the mare basalt source regions formed

at about 4400 m.y., and this age must represent the solidification of much of the magma ocean.

B. COMPOSITION OF THE MARE BASALTS

In comparison with terrestrial basalts, the silica contents of mare basalts are low (37–45%) and the lavas are iron-rich (18–22% FeO). The basic classification is chemical, with finer subdivisions based on mineral composition. The basalts are divided into low-Ti, high-Ti, and high-Al basalts. The low-Ti basalts include VLT (very-low-Ti), olivine, pigeonite, and ilmenite basalts. The high-Ti basalts include high-K, low-K, and VHT (very-high-Ti) basalts. The Clementine data suggest that there is a continuous variation in Ti contents. The major minerals are pyroxene, olivine (Mg-rich), plagioclase (Ca-rich), and opaques, mainly ilmenite. The basalts are highly reduced, with oxygen fugacities of 10^{-14} at 1100°C or about a factor of 10^6 lower than those of terrestrial basalts at any given temperature. Ferric iron is effectively absent, and 90% of Cr and 70% of Eu is divalent.

The lunar basalts are notably high in Ti, Cr, and Fe/Mg ratios and low in Ni, Al, Ca, Na, and K compared with terrestrial counterparts (Table III). They are depleted in volatile (e.g., K, Rb, Pb, Bi) and siderophile elements (e.g, Ni, Co, Ir, Au). The ratio of volatile (e.g., K) to refractory elements (e.g., U) is low. Thus lunar K/U ratios average about 2500 compared to terrestrial values of about 12,000. The rare earth elements (REE) display a characteristic depletion in divalent Eu (Fig. 19). This is one of the several pieces of evidence that the mare basalts come from a previously differentiated interior, rather than from a primitive undifferentiated lunar composition.

The differences in composition are mostly due to source region heterogeneity, with only minor evidence for near-surface fractionation. Variations in the amount of partial melting from a uniform source, subsequent fractional crystallization, or assimilation cannot account for the observed diversity. Some mare basalts are vesicular, evidence for a now vanished gas phase, usually thought to be CO.

C. ORIGIN OF THE MARE BASALTS

Mare basalts originate by partial melting, at temperatures of about 1200°C, deep in the lunar interior (see Fig. 8), probably at depths between 200 and 400 km. The basalts are derived from the zones and piles of cumulate minerals developed, at various depths, during

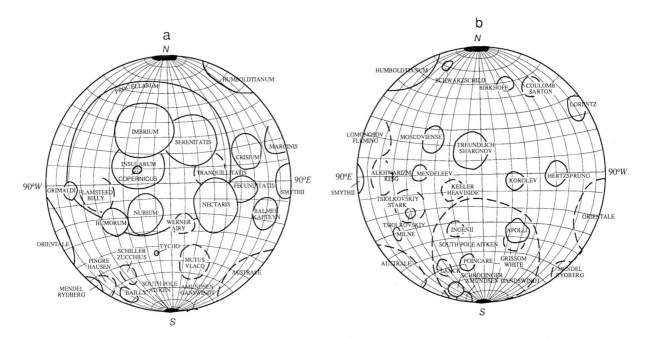

FIGURE 14 The distribution of major impact basins on (a) the nearside and (b) the farside of the Moon. (Courtesy of D. E. Wilhelms.)

crystallization of the magma ocean. The isotopic systematics of the mare basalts indicate that the source region had crystallized by 4.4 b.y. Partial melting occurred in these diverse mineral zones some hundreds of millions of years later due to the slow buildup of heat from the presence of the radioactive elements K, U, and Th. The melting was not extensive. Over 20 distinct types of mare basalt were erupted over an interval of more than 1 b.y., but the total amount of melt so generated amounted to only about 0.1% of the volume of the Moon. This forms a stark contrast to the state of the Moon at accretion, when it was probably entirely molten.

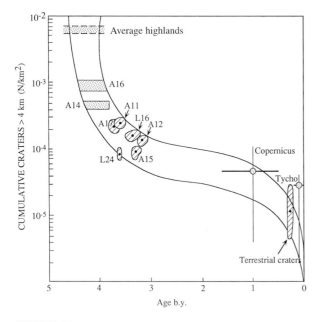

FIGURE 15 The production rate over geological time for lunar craters greater than 4 km in diameter. This illustrates the very high cratering rate prior to 3.8 b.y., but whether this represents the tail end of accretion or a spike ("cataclysm"), as preferred here, in the bombardment history is unclear. The terrestrial rate is for the past 200 m.y. [Updated from the Basalt Volcanism Study Prject (1981).]

VIII. LUNAR HIGHLAND CRUST

Most of the rocks returned from the highlands are polymict breccias, pulverized by the massive bombardment. However, some monomict breccias have low siderophile element contents. These are considered to be "pristine" rocks that represent the original igneous components making up the highland crust. Three principal pristine constituents make up the lunar highland crust, namely, ferroan anorthosites, the Mg suite, and KREEP.

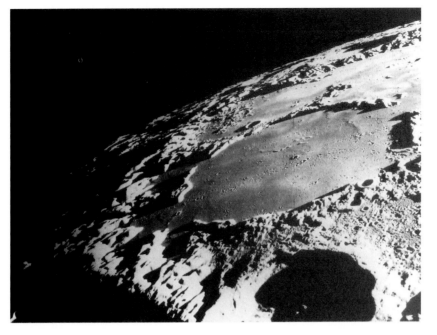

FIGURE 16   The distinction between maria and highlands is clearly shown in this view of the lunar farside. The large circular crater, filled with dark mare basalt, is Thomson (112 km in diameter), within the partly visible larger Mare Ingenii (370 km in diameter, 34°S, 164°E). The large crater in the highland terrain in the right foreground is Zelinskiy (54 km in diameter). The stratigraphic sequence, from oldest to youngest, is (a) formation of the white highland crust, (b) excavation of the Ingenii Basin, (c) excavation of Thomson Crater within the Ingenii Basin, (d) excavation of Zelinskiy Crater in the highland crust, (e) flooding of Ingenii Basin and Thomson Crater with mare basalt, and (f) excavation of small craters, including a probable chain of secondary craters, on the mare basalt surface. (Courtesy of NASA, AS15-87-11724.)

A. FERROAN ANORTHOSITE

Ferroan anorthosite is the single most common pristine highland rock type, making up probably 80% of the highland crust. The pristine clasts in lunar meteorites are mostly ferroan anorthosites. The major component (95%) is highly calcium-rich plagioclase, typically An_{95-97} with a pronounced enrichment in Eu (Eu/Eu* ~ 50). Low-Mg pyroxene is the next most abundant mineral, but the mafic minerals are only minor constitutents in this nearly monomineralic feldspathic rock. The anorthosites are typically coarsely crystalline with cumulate textures. An age of 4440 ± 20 m.y. has been obtained for the *Apollo 16* anorthosite 60025 and this is taken to represent the crystallization off ferroan anorthosites from the lunar magma ocean and the flotation of the feldspathic highland crust as "rockbergs." Alternatively, this date represents the "isotopic closure age" during cooling of the crust.

B. MG SUITE

The Mg suite comprises norites, troctolites, dunites, spinel troctolites, and gabbroic anorthosites. They are characterized by higher, and so more primitive, Mg/ Mg + Fe ratios compared to the ferroan anorthosites. They range in age from about 4.44 b.y. down to about 4.2 b.y., but typical ages are 100–200 m.y. younger than those of the ferroan anorthosites. The Mg suite is petrographically distinct from the older ferroan anorthosites and does not appear to be related directly to the crystallization from the magma ocean. It makes up a minor but significant proportion (perhaps 10%) of the highland crust and has two distinct and contradictory components in terms of conventional petrology. It is Mg-rich, and so primitive in terms of igneous differentiation, but also contains high concentrations of incompatible elements, typical or highly evolved or differentiated igneous systems. These characteristics point to an origin by mixing of these two distinct components.

The source of the highly evolved component is clearly KREEP. The source of the "primitive" Mg-rich component is less obvious. Conventional theories propose that the Mg suite arose as separate plutons that intruded the crust as separate igneous intrusions. However, all Mg suite rocks have parallel REE patterns, a characteristic compatible with mixing, but not expected to occur in separate igneous intrusions. Furthermore, it is of interest that the Mg suite contains Mg-rich orthopyroxene, a mineral that is lacking in

FIGURE 17   Mare basalt flows in southwestern Mare Imbrium. Flow thicknesses are in the range 10–30 m. The source of the flows is about 200 km southwest of crater La Hire (5 km in diameter), seen at right center. This crater is superimosed on Mount La Hire (about 30 km long at its base), a highland remnant that is partly submerged by the lavas. Note the prominent concentric wrinkle ridges on Mare Imbrium. (Courtesy of NASA, AS15-1556.)

mare basalts. Clearly the Mg suite originates in a location distinct from the source region of the mare basalts. During crystallization of the magma ocean, Mg-rich minerals (e.g., olivine and orthopyroxene) are among the first to crystallize and accumulate on the bottom of the magma chamber, in this case at depths exceeding 400 km. It is sometimes suggested that massive overturning has occurred to bring these within reach of the surface. However, the magma ocean had completed crystallization by 4400 m.y. with only the KREEP component remaining liquid until about 4360 m.y.; it was solid at the time of the formation of the Mg suite. There is no obvious source of energy for remelting early refactory Mg-rich cumulates. Subsequent melting to produce mare basalts took place in more differentiated cumulates and produced lavas with a different mineralogy, without the primitive characteristics of the Mg suite.

C. KREEP

KREEP is enriched in potassium, rare earth elements, and phosphorus, hence the name. It is commonly applied as an adjective to refer to highland rocks with an enhanced and characteristic trace element signature. KREEP originated as the final 2% or so melt phase from the crystallization of the magma ocean and is strongly enriched in those "incompatible" trace elements excluded from the major mineral phases (olivine, orthopyroxene, clinopyroxene, plagioclase, ilmenite) during crystallization of the bulk of the magma ocean. This residual phase was the last to crystallize, at about 4.36 b.y., and apparently pervaded the crust, with which it was intimately mixed by cratering. Its presence tends to dominate the trace element chemistry of the highland crust. Extreme REE enrichment up to 1000 times the chondritic (or solar nebula abundances) are

FIGURE 18 Hadley Rille, a typical sinuous rille, about 1 km wide, at the *Apollo 15* landing site, close to the base of the Apennine Mountains. (Courtesy of NASA, Lunar Orbiter IV-102H3.)

known (see Fig. 19). This extreme concentration of trace elements amounts to a significant part of the total lunar budget and so provides strong evidence for the magma ocean hypothesis.

D. KREEP BASALT

KREEP basalt, an enigmatic rock type with only a few undisputed examples, is highly enriched in incompatible elements (KREEP) but has a more primitive Mg/(Mg + Fe) ratio. This combination of primitive and evolved components suggests that they are derived, like the members of the Mg suite, from different sources and are probably impact melts. Possibly the Apennine Bench formation is composed to KREEP basalt. This formation appears to have formed close in time to the excavation of the Imbrium Basin.

E. BRECCIAS

Most of the rocks returned from the lunar highlands are breccias, usually consisting of rock fragments or clasts set in a fine-grained matrix. Lunar breccias are usually divided into monomict, dimict, and polymict breccias, consisting, respectively, of a single rock type, two distinct components, and a variety of rock types and impact melts. Polymict breccias, usually involving several generations of breccias, are the most common rock type returned from the lunar highlands. They are further subdivided into fragmental breccias, glassy

TABLE III
Elemental Abundances[a]

Oxide (wt%)	Cl	Earth mantle + crust	Bulk Moon	Highlands	Low-Ti basalt	High-Ti basalt
SiO$_2$	34.2	49.9	43.4	45.0	43.6	37.8
TiO$_2$	0.11	0.16	0.3	0.56	2.60	13.0
Al$_2$O$_3$	2.44	3.64	6.0	24.6	7.87	8.85
FeO	35.8	8.0	13.0	6.6	21.7	19.7
MgO	23.7	35.1	32	6.8	14.9	8.44
CaO	1.89	2.89	4.5	15.8	8.26	10.7
Na$_2$O	0.98	0.34	0.09	0.45	0.23	0.36
K$_2$O	0.10	0.02	0.01	0.03	0.05	0.05
Σ	99.2	100.1	99.3	100	100.4	99.5
Volatile elements						
K (ppm)	854	180	83	200	420	500
Rb (ppm)	3.45	0.55	0.28	0.7	1.0	1.2
Cs (ppb)	279	18	12	20	40	30
Moderately volatile element						
Mn (ppm)	2940	1000	1200	570	2150	2080
Moderately refractory element						
Cr (ppm)	3975	3000	4200	800	5260	3030
Refractory elements						
Sr (ppm)	11.9	17.8	30	130	101	121
U (ppb)	12.2	18	30	80	220	130
La (ppm)	0.367	0.55	0.90	2.0	6.0	5.22
Eu (ppm)	0.087	0.13	0.21	1.0	0.84	1.37
V (ppm)	85	128	150	30	175	50
Siderophile elements						
Ni (ppm)	16,500	2000	400	100	64	2
Ir (ppb)	710	3.2	0.01	—	0.02	0.04
Mo (ppb)	1380	59	1.4	5	50	50
Ge (ppm)	48.3	1.2	0.0035	0.02	0.003	0.003

[a] Element abundances in Cl chondrites (volatile-free = primitive solar nebula). Earth mantle + crust = primitive Earth mantle; bulk Moon; average lunar highland crust; Low-Ti mare basalt (12002) and High-Ti mare basalt (70215). Both of these latter samples are probable primary basaltic magmas. Data sources from Taylor (1982) and Hartmann *et al.* (1986).

melt breccias, crystalline melt breccias (or impact melt breccias), clast-poor impact melts, granulitic breccias, and regolith breccias.

F. THE MAGMA OCEAN

The geochemical evidence is clear that at least half and possibly the whole Moon was molten at accretion. This stupendous mass of molten rock is referred to as the "magma ocean" and a very energetic mode of origin of the Moon, such as the giant impact hypothesis, is required to account for it. The crystallization of such a body is difficult to constrain from our limited terrestrial experience. A reasoanble scenario is that

initial crystallization of olivine and orthopyroxene formed deep cumulates. As the Al and Ca content of the magma increased, plagioclase crystallized and floated in the bone-dry melt, forming "rockbergs" that eventually coalesced to form the lunar highland crust, dated at 4440 ± 20 m.y. The first-order variation in thickness from nearside to farside is probably a relic of primordial convection currents in the magma ocean. Excavation by large basin impacts has subsequently imposed additional substantial variations in crustal thickness.

Plagioclase was a very early phase to crystallize, since all lavas derived from the interior bear the signature of prior removal of Eu (and Sr) (see Fig. 19). Accordingly, the magma ocean was probably enriched in Ca and Al over typical terrestrial values, a conclusion

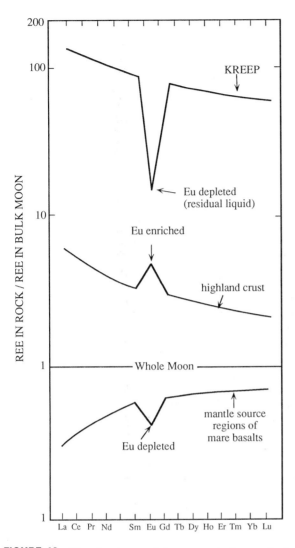

FIGURE 19 The abundances of the rare earth elements in the source regions of the mare basalts, the highland crust, and KREEP, relative to the bulk moon concentrations. These patterns result from the preferential entry of divalent europium (similar in radius to strontium) into plagioclase feldspar. This mineral floats to form the highland crust, and so depletes the interior of Eu. Mare basalts subsequently erupted from this region deep within the Moon bear the signature of this earlier depletion. KREEP is the final residue of the crystallization of the magma ocean. It is strongly depleted in Eu owing to prior crystallization of plagioclase and is enriched in the other rare earth elements, as well as other trace elements (e.g., K, U, Th, Ba, Cs, Zr, P) that are excluded from olivine, pyroxene, and ilmenite during the crystallization of the major mineral phases of the magma ocean.

reinforced by the *Clementine* and *Galileo* data. The implication is that the Moon was enriched in these and other refractory elements compared to our estimates of the terrestrial mantle. As crystallization of the magma ocean proceeded, those elements unable to enter olivine, ortho- or clinopyroxene, plagioclase, and ilmenite

were concentrated in the near-surface residual melt (KREEP), which eventually composed about 2% of the volume of the magma ocean. This appears to have pervaded and has been intimately mixed into the highland crust. Crystallization of the main phases was complete by 4400 m.y. and the final residue was solid by about 4360 m.y.

The crystallization sequence portrayed here was far from peaceful. During all this time, the outer portions of the Moon were subjected to a continuing bombardment, which broke up and mixed the various components of the highland crust. Perhaps coeval with these events was the intrusion into the crust of the Mg suite. Probably some local overturning of the deeper cumulate pile may have occurred, but such events did not homogenize the interior.

IX. LUNAR COMPOSITION

Estimates of the bulk composition of the Moon are given in Table III and compared to the terrestrial mantle abundances and to the bulk Earth. The most striking difference is in the abundance of iron. The Earth contains about 25% metallic Fe, the Moon less than about 2–3%. However, the bulk Moon contains 12–13% FeO, or 50% more than in current estimates of 8% FeO in the terrestrial mantle. The Moon is probably enriched in refractory elements such as Ti, U, Al, and Ca, a conclusion consistent with geophysical studies of the lunar interior, with the early appearance of plagioclase (An_{95-97}) during crystallization of the magma ocean, and with the limited heat-flow data indicating bulk lunar values of 30 or more ppb U (a refractory element). This conclusion is reinforced by the data from the *Galileo* and *Clementine* missions, which indicate that the highland crust is dominated by anorthosite. This requires that the bulk lunar composition contains about 5% Al_2O_3, compared with a value of about 3.6% for the terrestrial mantle. The Moon is bone-dry, no indigenous H_2O having been detected at ppb levels. It is strongly depleted to volatile elements (e.g., K, Pb, Bi) by a factor of about 50 compared to the Earth, or 200 relative to primordial solar nebula abundances, and is probably enriched in refractory elements (e.g., Ca, Al, Ti, U) by a factor of about 1.5 compared to the Earth. Along with its depletion in iron, the Moon also has a low abundance of siderophile elements that are depleted in order of their metal–silicate distribution coefficients. This observation suggests that they are segregated into a small lunar core.

A. LUNAR MINERALS

Only about 100 minerals have been identified in lunar samples, in contrast to the several thousand species that have been identified on Earth. This lunar paucity is due to the dry nature of the Moon and the depletion in volatile and siderophile elements. Extensive summaries of lunar mineralogy can be found in Frondel (1975) and in Heiken *et al.* (1991).

B. LUNAR METEORITES

Our understanding of the lunar crust has been aided by the discovery of lunar meteorites. Eleven specimens have been recovered from Antarctica and one from Western Australia. Most are from highland locations. From their feldspar-rich and KREEP-poor composition, they may be from the lunar farside; they are distinct from the nearside highland samples returned by *Apollo 14, 15, 16,* and *17* and *Luna 20.* However, their major element composition is close to that of estimates of the average highland crust. They confirm, as do the *Galileo* and *Clementine* missions, the essentially anorthositic nature of the lunar highland crust. [*See* METEORITES.]

C. TEKTITES

For many years, some scientists considered that tektites were derived from the Moon. However, the controversy over lunar versus terrestrial origin was settled in favor of the latter source by the first sample return from the Moon in 1969. It is firmly established from isotopic and chemical evidence that tektites are derived from the surface of the Earth by meteoritic or asteroidal impact. Because the debate still surfaces occasionally, readers interested in these glassy objects will find a useful review of the evidence for a terrestrial origin in Koeberl (1994).

X. THE ORIGIN OF THE MOON

In considering the origin of satellites, it is worth noting that only three of the giant planets, Jupiter, Saturn, and Uranus, possess substantial regular satellite systems. The capture of Triton was probably responsible for the destruction of any primordial satellite system of Neptune. Although these miniature solar systems around Jupiter, Saturn, and Uranus might have been expected to be similar, they are all quite distinct. Formation of satellite systems within our own solar system apparently does not lead to a uniform product. That a large element of chance has entered into the evolution of our present system has led to the realization of the inherent difficulties in constructing general theories of the origin of planetary systems. [*See* THE ORIGIN OF THE SOLAR SYSTEM.]

Hypotheses for the origin of the Moon must explain the high value for the angular momentum of the Earth–Moon system, the strange lunar orbit inclined at 5.1° to the plane of the ecliptic, the high mass relative to that of its primary planet, and the low bulk density of the Moon, much less than that of the Earth or of the other inner planets. The chemical composition revealed by the returned lunar samples added additional complexities to these classic problems, since the lunar composition is unusual by either cosmic or terrestrial standards. It is perhaps not surprising that previous theories for the origin of the Moon failed to account for this diverse set of properties and that only recently has something approaching a consensus been reached.

Hypotheses for lunar origin can be separated into five categories:

1. capture from an independent orbit;
2. formation as a double planet;
3. fission from a rapidly rotating Earth;
4. disintegration of incoming planetesimals; and
5. Earth impact by a Mars-sized planetesimal.

These are not all mutually exclusive, and elements of some hypotheses occur in others. For example:

1. Capture of an already formed Moon from an independent orbit has been shown to be highly unlikely on dynamic grounds. The hypothesis provides no explanation for its peculiar composition. In addition, it could be expected that the Moon might be an example of a common and primitive early solar system object, similar to the captured rock-ice satellites of the outer planets. This indeed had been the expectation of Harold Urey. It would be an extraordinary coincidence if the Earth had captured an object with a unique composition, in contrast to the many examples of icy satellites captured by the giant planets.

2. Formation of the Earth and the Moon in association as a double planet system immediately encounters the problems of differing density and composition of the two bodies. Various attempts to overcome the density problem led to co-accretion scenarios in which disruption of incoming differentiated planetesimals formed from a ring of low-density silicate debris. Popular models to provide this ring involved the breakup of differentiated planetesimals as they come within a Roche limit (about 3 Earth radii). The denser and tougher metallic cores of the planetesimals survived and accreted to the Earth, while their mantles formed a circumterrestrial ring of broken-up silicate debris from which the Moon could accumulate. This attractive scenario has been shown to be flawed since the proposed breakup of planetesimals close to the Earth is unlikely to occur. It is also difficult to achieve the required high value for the angular momentum in this model. Such a process might be expected to have been common during the formation of the terrestrial planets and so satellites should be common. The Moon, however, is unique.

3. In 1879, George Darwin proposed that the Moon was derived from the terrestrial mantle by rotational fission. Such fission hypotheses have been popular since they produced a low-density, metal-poor Moon. However, the angular momentum of the Earth–Moon system, although large, is insufficient by a factor of about four to allow for rotatioinal fission. If the Earth had been spinning fast enough for fission to occur, there is no available mechanism for removing the excess angular momentum following lunar formation. The lunar sample return provided an opportunity to test these hypotheses since they predict that the bulk composition of the Moon should provide some identifiable signature of the terrestrial mantle. However, in general they failed to account for significant chemical differences between the compositions of the Moon and that of the terrestrial mantle or to provide a unique terrestrial signature in the lunar samples. However, the Moon contains, for example, 50% more FeO, and has distinctly different trace siderophile element signatures. It also contains higher concentrations of refractory elements (e.g., Al, U) and lower amounts of volatile elements (e.g., Bi, Pb). The Moon and the Earth have distinctly different siderophile element patterns. The similarity in V, Cr, and Mn abundances in the Moon and the Earth is nonunique since CM, CO, and CV chondrites show the same pattern, probably due to volatile depletion in precursor planetesimals. These differences between the chemical compositions of the Earth's mantle and the Moon are fatal to theories that wish to derive the Moon from the Earth.

4. One proposed modification of the fission hypothesis uses multiple small impacts to place terrestrial mantle material into orbit. It is exceedingly difficult to obtain the required high angular momentum by such processes, since multiple impacts should average out. Most of these Moon-forming hypotheses should be general features of planetary and satellite formation and should produce Moon-like satellites around the other terrestrial planets. They fail to account for the unique nature of the Earth–Moon system and the very peculiar bone-dry composition of the Moon, nor do they account for the differences between the lunar composition and that of the terrestrial mantle.

These foregoing theories accounted neither for the lunar orbit nor for the high angular momentum, relative to the other terrestrial planets, of the Earth–Moon system, a rock on which all early hypotheses foundered.

A. THE SINGLE-IMPACT HYPOTHESIS

The single-impact hypothesis was developed by A. G. W. Cameron basically to solve the angular momentum problem, but, in the manner of successful hypotheses, it has accounted for other parameters as well and has become virtually a consensus. The theory proposes that during the final stages of accretion of the terrestrial planets, a body somewhat larger than Mars collided with the Earth and spun out a disk of material from which the Moon formed. This giant impact theory resolves many of the problems associated with the origin of the Moon and its orbit. The following scenario is one of several possible, although restricted, variations on the theme.

In the closing stages of the accretion of the terrestrial planets, the Earth suffered a grazing impact with an object of about 0.15 Earth mass (over 30% larger than Mars). This body is assumed to have differentiated into a silicate mantle and a metallic core. It came from the same general region of the nebula as the Earth, since the oxygen and chromium signatures of Earth and Moon are identical.

The impactor is disrupted by the collision and mostly goes into orbit about the Earth. Gravitational torques, due to the asymmetrical shape of the Earth following the impact, assist in accelerating material into orbit. Expanding gases from the vaporized part of the impactor also promote material into orbit. Following the impact, the mantle material is accelerated, but the core of the impactor remains as a coherent mass and is decelerated relative to the Earth, so that

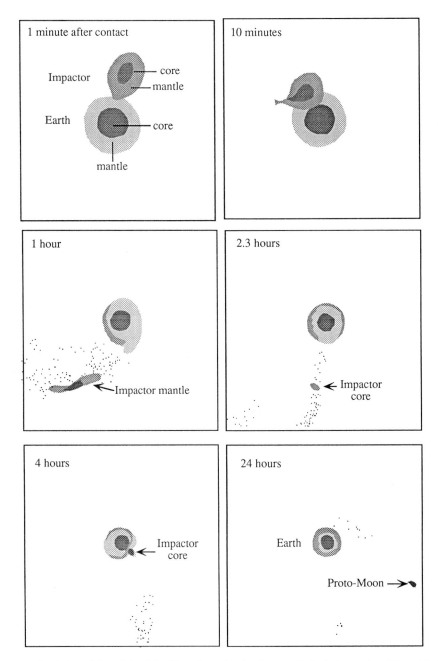

FIGURE 20   A computer simulation of the origin of the Moon by a glancing impact of a body larger than Mars with the early Earth. This event occurred about 4500 m.y. ago during the final stages of accretion of the terrestrial planets. Both the impactor and the Earth have differentiated into a metallic core and rocky silicate mantle. Following the collision, the mantle of the impactor is ejected into orbit. The metallic core of the impactor clumps together and falls into the Earth within about 4 hours in this simulation. Most terrestrial mantle material ejected by the impact follows a ballistic trajectory and is reaccreted by the Earth. The metal-poor, low-density Moon is thus derived mainly from the silicate mantle of the impactor. (Courtesy of A. G. W. Cameron.)

the core of the impactor accretes to the Earth within about 4 hours. A metal-poor mass of silicate remains in orbit.

In some variants of the hypothesis, this material immediately coalesces to form a totally molten Moon.

In others, it breaks up into several moonlets that subsequently accrete to form a partly molten Moon. This highly energetic event accounts for the geochemical evidence that indicates that at least half the Moon was molten shortly after accretion. Figure 20 illustrates several stages of a computer simulation of the formation of the Moon according to one version of the single giant impact hypothesis.

The giant impact event vaporized much of the material that subsequently was recondensed to make up the Moon. This effect explains such unique geochemical features as the extreme depletion of very volatile elements, the bone-dry nature of the Moon, and the enrichment of refractory elements in the Moon, in addition to providing an initially molten Moon.

This scenario is consistent with the planetestimal hypothesis for the formation of the planets from a hierarchical sequence of smaller bodies. Other evidence in support of the previous existence of large objects in the early solar system comes from the ubiquitous presence of heavily cratered ancient planetary surfaces, from the large number of impact basins with diameters up to 2000 km or so, and from the obliquities or tilts of the planets, all of which demand collisions with large objects in the final stages of accretion. The extreme example is that an encounter between Uranus and an Earth-sized body is required to tip that planet on its side. Clearly the Moon has a composition that cannot be made by any single-stage process from the material of the primordial solar nebula. The compositional differences from that of the primitive solar nebula, from the Earth, from Phobos and Deimos (almost certainly of CI carbonaceous chondritic composition), and from the satellites of the outer planets (rock–ice mixtures with the exception of Io) thus call for a distinctive mode of origin. [See Io.]

Unique events are difficult to accommodate in most scientific disciplines. The solar system, however, is not uniform. All eight planets (even such apparent twins as Venus and Earth) and over 60 satellites are different in detail from one another. All this diversity makes the occurrence of single events more probable in the early stages of solar system history.

BIBLIOGRAPHY

Basaltic Volcanism Study Prject (BSVP) (1981). "Basaltic Volcanism on the Terrestrial Planets." Pergamon, New York.

Belton, M. J. S., *et al.* (1992). Lunar impact basins and crustal heterogeneity: New western limb and far side data from *Galileo*. *Science* **255**, 570–576.

Cameron, A. G. W., and Benz, W. (1991). The origin of the Moon and the single impact hypothesis, IV. *Icarus* **92**, 204–216.

Frondel, J. W. (1975). "Lunar Mineralogy." John Wiley & Sons, New York.

Fuller, M. J., and Cisowski, S. M. (1987). Lunar paleomagnetism. *In* "Geomagnetism 2" (J. A. Jacobs, ed.), pp. 307–455. Academic Press, San Diego, Calif.

Hartmann, W. K., Phillips, R. J., and Taylor, G. J. (eds.) (1986). "Origin of the Moon." Lunar and Planetary Institute, Houston.

Heiken, G., Vaniman, D., and French, B. M. (1991). "The Lunar Sourcebook." Cambridge Univ. Press, New York.

Kaula, W. M., Drake, M. J., and Head, J. W. (1986). The Moon. *In* "Satellites" (J. A. Burns and M. S. Matthews, eds.), pp. 581–628. Univ. Arizona Press, Tucson.

Koeberl, C. (1994). Tektite origin by hypervelocity asteroidal or cometary impact; Target rocks, source craters and mechanisms. *Geol. Soc. Am. Spec. Paper* **293**, 133–151.

Lucey, P. G., Taylor, G. J., and Malaret, E. (1995). The abundance and distribution of iron on the Moon: Implications for crustal differentiation, structure, and the origin of the Moon. *Science* **268**, 1150–1153.

Nakamura, Y. (1983). Seismic velocity of the lunar mantle. *J. Geophys. Res.* **88**, 677–686.

Neumann, G. A., Zuber, M. T., Smith, D. E., and Lemoine, F. G. (1996). The lunar crust: Global structure and signature of major basins. *J. Geophys. Res.* **101**, 16,841–16,863.

Spudis, P. D. (1990). The Moon. *In* "The New Solar System" (J. K. Beatty and A. Chaikin, eds.), pp. 41–52. Cambridge Univ. Press, New York.

Taylor, L. A., Shervais, J. W., Hunter, R. H., Shih, C.-Y., Bansal, B. M., and Nyquist, L. E. (1983). Pre-4.2 AE mare basalt volcanism in the lunar highlands. *Earth Planet. Sci. Lett.* **66**, 33–47.

Taylor, S. R. (1982). "Planetary Science: A Lunar Perspective." Lunar and Planetary Institute, Houston.

Wilhelms, D. E. (1987). "The Geologic History of the Moon," U.S. Geol. Surv. Prof. Paper No. 1348. U.S. Geological Survey, Washington, D.C.

Zuber, M. T., Smith, D. E., Lemoine, F. G., and Neumann, G. A. (1995). The shape and internal structure of the Moon from the Clementine Mission. *Science* **266**, 1839–1843.

MARS: ATMOSPHERE AND VOLATILE HISTORY

Fraser P. Fanale
University of Hawaii at Manoa

I. Introduction

II. Mars's Volatile Inventory

III. Evidence for Early Epochal Climate Change

IV. Detailed Mechanisms of Long-Term Climate Change

V. Evidence for and Mechanisms of Ongoing Periodic Climate Change

VI. Summary

GLOSSARY

Albedo: Ratio of electromagnetic energy reflected by a surface to the amount of energy incident upon it. In reflectance spectroscopy, albedo is measured as diffuse reflectance in the visible wavelengths—conventionally at 0.56 μm.

Effused: Released from a magma by temperature or pressure release effects.

Exosphere: That level in a planet's uppermost atmosphere where thermally energized species can escape from the planet's gravitational field in an essentially collisionless path. Alternatively, that level where the mean free path is equal to the scale height.

Greenhouse effect: Heating of a planetary surface because the outgoing radiation at (long) thermal wavelengths is absorbed by certain gases such as CO_2 and radiated eventually back to the surface.

Hydrodynamic drag: Force exerted by a gaseous species, especially H, which, on escape from a planetary atmosphere, drags other heavier species with it.

Lobate ejecta: Ejecta from a crater that forms cohesive lobes resembling ejecta from a crater formed by dropping a rock into mud.

Mass fractionation: Preferential transfer of one element, compound, or isotope versus another because of differences in mass.

Mass wasting: Downslope movement of material caused largely by gravitational forces as opposed to fluvial erosion.

Megaregolith: Deep layer (perhaps over a kilometer deep) formed on a planetary surface by intense bombardment and accumulation of meteorite impact ejecta.

Morphology: Study of the shape of features on the surface of a planet and interpretation of same in terms of planetary history.

Obliquity: Angle that the axis of a planet makes with the plane of orbit, or the polar ''tilt'' of the planet.

Permafrost: Soil that is permanently frozen over two years to thousands of years; often used to describe "hard frozen" permafrost where the pores contain H_2O ice.

Thermokarst: Features on a planetary surface that are produced by melting or melting and freezing of ground ice.

Volatiles: Material such as H_2O and CO_2 that would be largely in the gaseous phase at magmatic or even surface temperatures. Volatiles may be sequestered in compounds that are not particularly volatile, such as $CaCO_3$ in rocks or soil.

I. INTRODUCTION

Much of the interest in Mars centers on the history of its surface and atmospheric volatiles such as H_2O and CO_2. One reason for this is that these volatiles and their history control cyclic climate change on Mars, the study of which casts new light on the cyclic climate change (glacial cycles) on Earth. An even more compelling reason is that, during their history, these volatiles controlled long-term changes in the delicate environment at the Mars atmosphere–surface interface. Therefore, understanding their history sheds light on the suitability of the past and present Mars environment for the possible instigation and subsequent sustenance of life.

First we must consider the inventory of Mars's surface and atmospheric volatiles. With regard to the magnitude and composition of this inventory, the first evidence to be discussed is that derived from the surface morphology of Mars (e.g., from fluvial features). Other lines of evidence are then discussed, including theoretical geochemical models and morphological observations that suggest vast amounts of current ground ice or hard frozen permafrost. Next, evidence is considered that the state of these volatiles was clearly different in earliest Martian history. This evidence suggests changes in the character of the fluvial features over Mars's life and also much greater erosion rates in history. Does this require a "warmer, wetter" Mars or are there alternative explanations?

To the extent that a "warmer, wetter" early Mars remains a possibility, a paradox is posed by the fact that most stellar astronomers insist that the Sun was much *weaker* in the early history of our solar system. Various possible solutions are discussed, the most prominent of which is that early Mars had a massive initial atmosphere that sustained a strong "greenhouse effect" on the atmosphere and surface environment, thereby overcompensating for the effect of the weak early Sun. The mechanism of such a putative greenhouse effect is discussed, along with various problems with this hypothesis that cast doubt on its validity. Several solutions to these problems are also considered.

Finally, morphological evidence is discussed suggesting that, in addition to the putative long-term epochal climate change, Mars continues to undergo cyclic climate change quite analogous in cause and effect to Earth's glacial cycles. The astronomical forcing function for these changes, the way in which volatiles migrate in response to it, and the resulting periodic effect on the Mars surface environment are considered.

II. MARS'S VOLATILE INVENTORY

A. MORPHOLOGICAL EVIDENCE

The morphology of Mars provides important evidence regarding the amounts of H_2O and CO_2 that were degassed to its surface and atmosphere and retained there. If adequate amounts were not present, then that alone could preclude a higher early atmospheric pressure and an atmospheric greenhouse effect as an explanation for climate change (see Section III). The most obvious evidence concerning the Mars volatile inventory is the present atmosphere itself. However, this represents a global basal CO_2 pressure of only ~7 mbar—about 1/1000th of the CO_2 inventory required for a significant greenhouse effect. Mars has both north and south polar caps. The south cap is composed primarily of CO_2, and at its temperature it is in equilibrium with the CO_2 atmospheric pressure. However, the cap is so thin that it represents an even lower contribution than the atmosphere itself. The next most important reservoir is inferred to be adsorbed CO_2 in the regolith. Although the morphology suggests the presence of a comminuted regolith, perhaps an average of ~1 km thick, it requires infusion of laboratory adsorption experiments measuring the ratio of the adsorbed phase versus the gas phase as a function of temperature and some assumptions to infer the amount of an adsorbed component. Our best estimates are that ~20–100× the atmospheric inventory is present as a "hidden ocean" of adsorbed CO_2 in that regolith. Even so, the resulting total (≤0.5 bar) falls far short of the

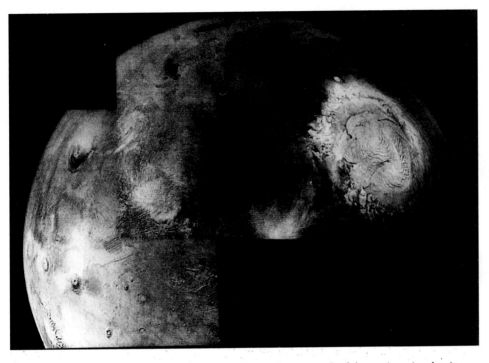

FIGURE 1 The north cap of Mars in spring. A thin cover of CO_2 ice still covers much of the cap in spring, but in summer the residual cap, which is composed of H_2O ice, is completely exposed. The cap is collared by layered terrain (see Fig. 2), which probably contains ~10 times the amount of H_2O as the cap itself.

CO_2 inventory needed to invoke an early greenhouse. Evidently, an invisible reservoir of CO_2 in the regolith (perhaps carbonates)—not yet firmly identified observationally—is required. This evidence will be discussed later.

In summary, the morphology itself provides little evidence for the presence of a massive early CO_2 inventory. In contrast, the morphology *does* provide valuable evidence for estimating the H_2O inventory. The most obvious evidence is the north cap, which is far more massive than the south cap and is composed of "dirty" H_2O ice (Fig. 1). This cap, however, contains only enough H_2O to cover the globe of Mars with 3–10 m of H_2O. A somewhat more inferential and less obvious, but more important, reservoir is the so-called "layered terrain" that underlies and collars both caps. Figure 2 shows layered terrain collaring the south cap. It is also of great importance in understanding ongoing climate change, as will be discussed later. The layered terrain could provide a useful lower limit to the H_2O inventory. The mass of this terrain is ~10× that of the ice cap. It may be that sublimation has left a lag deposit of ice-poor dust enclosing it (like a "Boston snowball"), which conceals its presence. Thus, the entire cap system may represent 10–30 m of H_2O if distributed over the surface of Mars. However, this assumes that it is mostly ice, whereas the ice-to-dust ratio is unknown at present.

The most important estimate of the H_2O inventory, however, comes from the large outflow channels (Fig. 3), which provide evidence constraining the total H_2O inventory. From the amount eroded by channels flowing to the Chryse Basin alone, one obtains an estimate of ~35 m of water if spread globally over the planet. The actual amount of H_2O that formed these channels was probably >35 m, since the amount of H_2O probably exceeded the amount of material eroded. Furthermore, it is likely that the Chryse channels formed there and not elsewhere simply because of the favorable conditions resulting from the Tharsis doming (which produced a hydrostatic head) and the higher heat flow (which resulted in a thinner permafrost cover), rather than because the Chryse area, which is about one-tenth of Mars's total area, just happened to be the only area that harbored H_2O. If so, the global estimate can be raised to ~>350 m of H_2O globally. This could be viewed as an estimate of the amount of water retained in the regolith at the time of outflow channel formation.

Given this H_2O estimate, if we assume a $CO_2 : H_2O$ ratio for Mars equal to that of Earth, we obtain a total degassed CO_2 estimate of ~3 bars. In fact this estimate

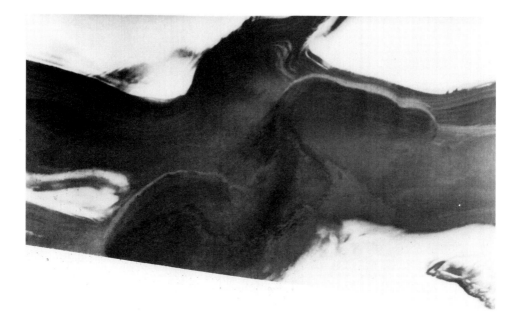

FIGURE 2   This "layered" terrain surrounds both caps and the layers consist of mixtures of H_2O ice and dust. If distributed globally, the H_2O in the layered terrain could represent as much as 35 m of H_2O.

is better expressed as "a few bars." Even the level of uncertainty implied by that expression does not take into account the dubious nature of the assumption that Mars and Earth had the same degassed $CO_2 : H_2O$ ratio. [*See* Earth as a Planet: Atmosphere and Oceans.]

B. OTHER LINES OF EVIDENCE

Obviously the morphological evidence is subject to much uncertainty. It is useful to compare these estimates of Mars's volatile inventory with other lines of evidence for two reasons: (1) to provide an independent check on the estimate of the H_2O inventory (~ -350 m globally) and (2) to provide some estimate of the degassed CO_2 inventory for which the morphology provides no direct evidence whatsoever.

Many theoretical geochemical models have been advanced that yield estimates of the degassed CO_2 and H_2O inventories on Mars. Though these models are based on many unlikely assumptions, they are entirely independent of those discussed earlier and therefore provide an interesting comparison.

The earliest of these models utilized observed calculations of elemental abundances in meteorites together with the ^{36}Ar abundance in the atmosphere (as a measure of planetary degassing) to estimate the degassed H_2O inventory. These models, although reasonable in principle, are mainly of historical interest. The reason

is that it has been subsequently established on the basis of isotopic evidence that certain meteorites called "SNC" meteorites are almost certainly derived from Mars. The vast body of geochemical knowledge derived from studies of these meteorites has allowed major revision of geochemical models. Perhaps the most convincing of these geochemical models predicts an enormous initial planetary H_2O inventory—several kilometers globally! Strangely, the same model predicts that very little water (tens of meters) survived and was degassed to the surface. The model predicts that nearly all the water was used up in oxidizing metallic iron and that the hydrogen thus produced escaped to space, dragging the rest of the degassed volatiles with it. It further estimates that of the small amount of H_2O retained in the interior, little was later degassed to the surface. The latter conclusion is based on the atmospheric ^{36}Ar and ^{40}Ar abundances. Thus these models predict very much lower remaining global H_2O inventories than are predicted by the visual evidence discussed earlier. [*See* Meteorites.]

However, we really have no way of knowing for certain whether the greater amount of oxidized iron in Mars in relation to Earth (as indicated by Mars's smaller metallic core in proportion to the planet's size) is the result of a water-destroying process or was an original condition of Mars-forming material. Second, the volatiles may have been added as a "late veneer" (e.g., comets) and may never have equilibrated with the interior.

FIGURE 3 An example of an outflow channel (as opposed to the valley networks—see text). The amount of H_2O discharged in these channels alone would correspond to 35 m of H_2O globally and 350 m globally if comparable (untrapped) H_2O reservoirs were typically present elsewhere on the planet.

The rare gas isotopic ratios seem to have been altered by major atmospheric loss to space, which occurred early in Mars history. The two leading candidate loss mechanisms are: (1) atmospheric "erosion" or loss caused by ejection of the atmosphere in high-velocity back jets from major early impact during and shortly after accretion and (2) "hydrodynamic drag" to space of the atmosphere caused early massive escape of H to space. Understanding the true significance of any early atmospheric escape episodes requires recognition of the possibility that all degassed rare gas would be entirely in the atmosphere and would be much more vulnerable to escape than H_2O and even CO_2, which might be sequestered in the regolith. Therefore all the models that use rare gases to estimate degassing efficiency for H_2O probably yield estimates that are far too low.

Given that the geochemical models, although quantitative in approach, are subject to great uncertainty as the result of early atmosphere loss processes, attention should be paid to the "outflow channel" estimate and to other approaches that, although less quantitative, deal with estimates of the *current* inventory and not the initial inventory. One such approach is based on an estimate of regolith thickness and current ground ice content. The surface of Mars is replete with evidence suggesting the presence of massive amounts of ground ice. One such type of evidence is the widespread occurrence of craters with cohesive lobate ejecta, which appear to require the presence of H_2O ice or water in the target material. An example is shown in Fig. 4. However, some investigations suggest that the same lobate morphology could be produced by atmospheric effects on the ejecta.

Another type of evidence is the widespread occurrence of features strongly resembling terrestrial "thermokarst" features, which result from single or multiple freeze–thaw cycles in an ice-rich soil. An example is shown in Fig. 5. However, on Mars the lateral scale of these latter features is so vast that it suggests that ice populates the pores of the Mars regolith down to depths of hundreds of meters or perhaps ~1 km. Still other types of features are the almost ubiquitous "softening" features, that is, the muting of the morphology of craters that seems to require a regolith strength so low that it in turn requires the presence of pore filling by H_2O ice. Such features are virtually absent within $\pm 30°$ of the equator. This further confirms the inference of an ice-filled regolith, because theoretical models based on diffusion rates suggest that the lower latitudes should be desiccated on a timescale of 10^8–10^9 years down to depths of tens to hundreds of meters.

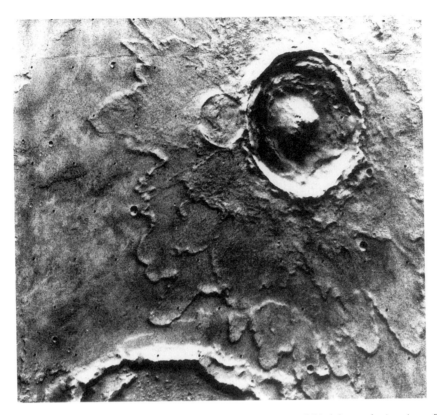

FIGURE 4   Lobate ejecta. Unlike Mercury and the Moon, many of Mars's craters exhibit lobate cohesive ejecta. This almost certainly indicates the presence (or former presence) of nearly ubiquitous deep, hard, frozen permafrost. In this case, the "muddy" nature of the ejecta is especially conspicuous and actually flows around an obstacle.

Thus several lines of evidence suggest that the pores of the regolith or megaregolith are filled with ice over a large portion of the Mars surface. The thickness of porous soil is difficult to estimate, but the large impact basins alone were capable of producing a 500-m global layer of unconsolidated material and most estimates of regolith thickness are ~1 km. If the regolith had an average porosity of 30%, which seems reasonable, an estimate of a water inventory of 30 m would result in excellent agreement with the estimate based on the outflow channels. However, the actual depth of the regolith and the porosity as a function of depth are not well known. Because the morphological estimates circumvent the uncertainties of rare gas loss to space that plague the geochemical models, the H_2O inventories based on morphological studies may be more reliable than those based on the more quantitative geochemical models.

Yet as mentioned earlier, a problem with the morphological estimates is that they provide virtually no evidence regarding the degassed and retained CO_2 inventory other than the estimate of 3 bars based on the dubious assumption that $CO_2 : H_2O$ was the same in the Mars volatile inventory as on Earth. Therefore it is very important to note that the "signature" (infrared absorption bands) of "carbonate" or at least some CO_2-containing compound has recently been detected in the mid-infrared spectrum of Mars' atmospheric dust. This is the first such detection of these bands and leads to an estimate of between 1% and 3% by mass of carbonate. If applied to the entire (1 km) regolith, this would lead to a total degassed and retained CO_2 inventory of 0.5–1.5 bar. It is encouraging to note that this is only a factor of two lower than the estimate derived by multiplying the morphologically derived H_2O inventory by Earth's $CO_2 : H_2O$ ratio. Also, the carbonate could be either enriched in the dust or the carbonate in the topmost surface material might have been partly depleted by sulfate aerosol leaching. Again, this is an estimate of current inventory and as such it circumvents the complexities of modeling a massive and highly fractionating atmospheric escape process.

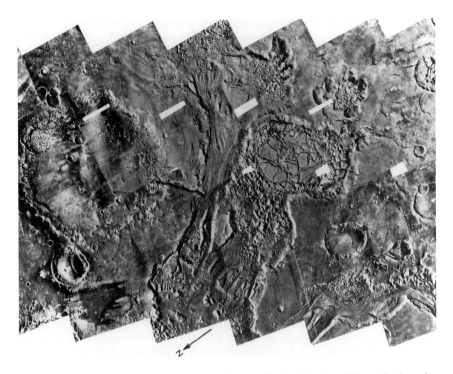

FIGURE 5 Karstlike topography on Mars. This is one of many landforms indicating freezing of deep soil, where the pores are largely filled with H_2O. Analogous terrain (but on a much smaller scale) is found in permafrost regions on Earth.

Thus, despite many uncertainties in morphologically derived volatile inventory estimates, and much lower estimates from geochemical models, the former coherently point to a current H_2O estimate of ~300 m globally and a current CO_2 estimate equivalent to 1–3 bars. A summary of these estimates is given in Table I. We will now consider the evidence for how astronomical influences, the greenhouse effect, and internal energy sources have operated on this inventory to produce (1) early climate change and (2) ongoing periodic climate change.

III. EVIDENCE FOR EARLY EPOCHAL CLIMATE CHANGE

Valley networks provide morphological evidence that a different erosional style operated on early Mars than that at present. These networks (Fig. 6) should not be confused with the large spectacular outflow channels (see Fig. 3) such as drain into Chryse Basin. The latter appear to be bursts from huge underground aquifers that were trapped beneath hard frozen permafrost and released when collapse occurred. Figure 7 shows a huge collapse at the head of one of these channels. The outflow channels therefore do not require a different surface climate; they conceivably could flow their great length even today beneath a thick ice cover.

Returning to the valley networks, these more closely resemble terrestrial river valleys. Most have tributaries and increase in size downstream. They tend to have V-shaped cross sections upstream but flat floors and steep walls in their lower reaches. Drainage of some sort and almost certainly by water is indicated. Since they are almost entirely confined to the heavily cratered terrain, which is a relic from ~3.8 billion years ago, it follows that *some* aspect of the Mars near-surface environment must have been very different at that time versus in middle to late Mars history. Thus these networks were originally regarded as the signature of a "warmer, wetter" early Mars surface climate. However, on closer examination, they do not possess the geometry that is observed in terrestrial river channels. Instead they are thought to possibly have been caused by mass wasting associated with underground flow. In this regard it should be noted that the heat production on Mars from radioactive decay would have been ~6–8 times higher than at present in earlier Mars history. In addition, accretionary heat may have been a major factor. These are internal factors not *directly*

TABLE I
Various Estimates of Mars's Volatile Inventory

	CO_2	H_2O (globally)
1. In atmosphere	~6 mbar	—
2. In south cap	<2 mbar	—
3. Adsorbed CO_2 on regolith	100–500 mbar	—
4. In north cap and layered terrain	—	10–30 m
5. From Chryse channel	—	>35 m
6. From Chryse channels assuming the Chryse region is not a unique repository of H_2O	—	>350 m
7. From "6," plus the assumption $CO_2 : H_2O$ (Mars) $= CO_2 : H_2O$ (Earth)	~3 bars	—
8. Some theoretical geochemical models[a]	—	<50 m
9. From ground ice estimates	—	300m
10. From carbonate infrared bands	~1–3 bars	—

[a] If one neglects rare gas/CO_2/H_2O fractionation during atmospheric loss (see text).

FIGURE 6   Valley networks resemble terrestrial river complexes more than the large outflow channels. They are found to be abundant in the oldest terrain, but are rare subsequently. They clearly indicate *some* very different conditions on early Mars, but they also somewhat resemble terrestrial groundwater "sapping" networks, so there is no consensus that rainfall was required for their formation.

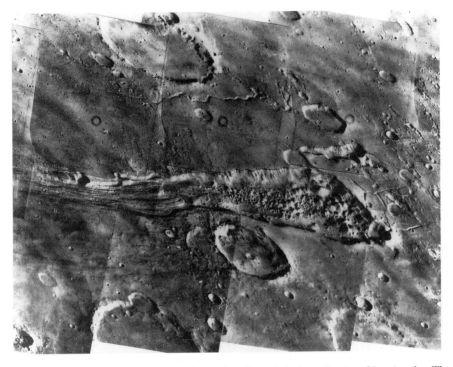

FIGURE 7 A huge collapse feature at the head of one of the outflow channels is about the size of Los Angeles. The discharge rate was clearly beyond any analogous event on Earth. Very important constraints can be placed on the Mars global water inventory from direct inspection of outflow channels, as well as from other approaches.

affecting atmospheric and surface temperature. The relative roles of genuine surface climate change versus internal heating are a matter of debate. Nonetheless, they are clearly evidence of a different near-surface regime early in Mars history.

A more convincing line of evidence suggesting different, specifically *surface* conditions is the absence of craters <15 km across in the earliest terrain. Even correcting for the longer time available for their erosion, it appears that erosion was much more effective on early Mars. Many investigators believe that this evidence requires very different surface conditions in the past, and may even require precipitation and runoff. However, a higher atmospheric pressure may have played an important role as well.

Having cited all of these "internal" factors as possible contributing factors in the creation of the early valley networks, we see that there are two basic alternatives: (1) the internal factors dominated and there is no need to suggest a warmer surface climate, and (2) the internal factors contributed, but the major role was played by a warmer surface environment. Even if we accept the first, we are still left with the problem of explaining the much greater early erosion rates. The issue is an important one because it relates to the possible origin and later extinction of life on Mars. As will be discussed at length in subsequent sections, if a major

role for a higher surface temperature is accepted, then a paradox is created: the Sun is thought to have been ~40% *less* luminous in early solar system history. Thus it is generally believed that a massive (several bar) CO_2 atmosphere with H_2O vapor would be required to *overcompensate* for the weak early Sun and produce warmer early conditions. Whether enough CO_2 was available and whether it could all be mobilized into the atmosphere at once are serious questions (see the following). This makes some supplementary role for internal factors seem attractive, but other contributing factors will be considered.

Despite all the alternative interpretations of the concentration of valley networks in early Mars history and the problems of the early greenhouse effect, there are two lines of evidence that actually distinguish between the two possibilities noted here and that tip the scales somewhat in favor of a largely climatic explanation: (1) On Earth, we have direct isotopic evidence for higher *surface* temperatures in the past. The existence of early limestone deposits, despite the supposedly weak early Sun, seems to require an atmospheric greenhouse effect. Thus, if we accept an exclusive role for internal effects on Mars, we are left in the position of having two nearly similar sets of evidence on Mars and Earth, and explaining one with internal factors and the other with external factors. We should keep this in mind when we consider the problems of a massive early CO_2 atmosphere on Mars. (2) A second argument is that, as discussed earlier, there is growing evidence for more intense erosion of large craters in Mars's past than the erosion rate consistent with preservation of subsequent craters. If substantiated, this effect could probably not be attributed to internal processes alone.

IV. DETAILED MECHANISMS OF LONG-TERM CLIMATE CHANGE

A. CO_2 GREENHOUSE

We have examined evidence for long-term (epochal) climate change and have established some approximate estimates of the available inventory of CO_2 and H_2O. Now we shall consider the ways in which the volatile inventories might have been operated on by "independent forces" in order to produce long-term climate change. The "forces" are of two basic types: surface and internal. The former include changes in solar output, changes in axial orientation of Mars, and changes in its orbit. The latter include changes in internal thermal gradient with time.

The canonical explanation of long-term climate change involves a CO_2- and H_2O-induced atmospheric "greenhouse effect." Without this effect, the temperature of a planet's surface is determined by the simple balance between radiation absorbed from the Sun and the thermal energy emitted by the surface and adsorbed by the atmosphere, which is proportional to the fourth power of the surface temperature. However, the radiative properties of an atmosphere, if present, can augment the surface temperature enormously. The augmentation is caused by absorption of surface radiation by the atmospheric gas and its reradiation back to the surface. Carbon dioxide is a relatively ineffective greenhouse gas with only one strong absorbing band—its 15-μm vibrational fundamental. Still, on Venus, the 100-bar CO_2 atmosphere raises the temperature of the surface by hundreds of degrees over what it would be without an atmosphere. On Mars, with its 7-mbar CO_2 atmosphere, the same effect currently amounts to only 7°C, and the low surface temperature likewise inhibits any significant greenhouse effect from H_2O vapor.

The idea of invoking a more massive early CO_2 atmosphere on Mars to produce a major greenhouse warming is also inspired by the fact that an early greenhouse effect seems to be required for Earth as well. Furthermore, the requirement is for a very effective greenhouse effect for both Mars and Earth, because of the fact that virtually all models for the evolution of our Sun indicate that solar luminosity in the earliest part of Earth and Mars history was actually *lower* by up to 40% than it is now! Thus, the problem of explaining why the oceans were apparently not thoroughly frozen in early Earth history (as attested by marine sediment) and why valley networks are abundant and erosion rates higher *only* in early Mars history is often referred to as the "weak early Sun paradox." [*See* THE ORIGIN OF THE SOLAR SYSTEM.]

We may approach this problem in three steps. First, we will examine how atmospheric radiative balance in a CO_2 atmosphere might warm Mars sufficiently under the most optimum, but not unphysical, of circumstances. Then we will list the reasons why serious doubt has been cast on this optimistic scenario. Finally, we will list possible alternative effects that might plausibly augment an insufficient greenhouse effect to help produce the possibly required climate change.

The predicted greenhouse effect in the simple case of a transparent CO_2 atmosphere is shown in Fig. 8. The surface temperature is shown as a function of CO_2 pressure for several values of solar luminosity. In this particular model, the calculations took into account an augmentation from an H_2O greenhouse effect with the assumption of an H_2O-saturated CO_2 atmosphere. It was also assumed in the case of the dashed lines that the albedo of the early surface was very low ($A_s = 0.1$), much lower than the present, and that it was subsequently raised by weathering. Assuming, according to solar models, that the luminosity of the Sun (S) was seven-tenths of the present value, the operative lines are those on the right of the figure. Thus, it would take 4–5 bars of CO_2 pressure to raise surface temperatures to the freezing point of H_2O, and the H_2O contribution would be minimal initially if Mars started out with a cold surface and a low but growing CO_2 pressure.

Although the model is based on sound physics, there are many reasons for doubting that this effect by itself could produce a "warm wet" Mars despite the lower solar luminosity. They are:

(1) The model does not include cloud formation. More recent models that include the obscuration effects of apparently inevitable CO_2 suggest that pressures considerably in excess of 5 bars of CO_2 may be required.

(2) Even the requirement of 4–5 bars is somewhat

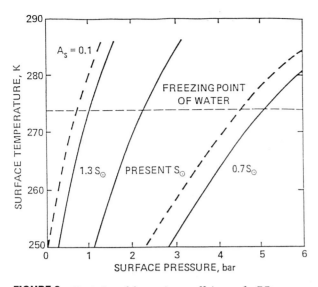

FIGURE 8 Depiction of the maximum efficiency of a CO_2 greenhouse on early Mars. Solid curves are for a current surface albedo of 0.215 and dashed curves are for the more favorable case of an albedo of 0.1. It is assumed that the early surface was darker owing to subsequent weathering. The two choices for solar luminosity are the present luminosity and 0.7 of the present luminosity, which is presumed for the "weak early Sun." These are maximum temperatures for any CO_2 pressure. In reality, clouds would possibly form, requiring considerably more CO_2 for any given temperature increment. Moreover, the actual CO_2 pressures may not have been as high as 4–5 bars. Therefore several supplementary warming effects discussed in the text are the subjects of current modeling.

higher than most independent estimates of total degassed CO_2, although these are subject to much uncertainty.

(3) Even keeping the CO_2 in the atmosphere is a problem in the first place. Most investigators agree that in the setting of a "warm wet" Mars, the CO_2 would be removed as carbonate on the timescale of 10^7–10^8 years—much shorter than the channel-forming epoch. This creates a requirement for rapid recycling of CO_2, which is plausible only because of the probably high early heat flow. However, most investigators agree that the paucity of ^{36}Ar and mass fractionation effects in the Mars atmosphere reflect massive atmospheric loss through a combination of impact erosion and hydrodynamic escape. Water could have been sequestered during this time and therefore been retained efficiently enough to explain the huge outflow channels and evidence of widespread ground ice much later (into the present) in Mars history. This cannot be said for the CO_2, which by definition had to be almost completely exposed in the atmosphere to such loss in order to operate as a greenhouse gas in the first place.

Thus it is clear to most investigators that the CO_2 and a cloudy atmosphere do not by themselves provide a satisfactory explanation for a warmer early Mars. Therefore, several other mechanisms are currently under investigation. These mechanisms are not to be regarded as alternatives to the CO_2 greenhouse but rather as supplements. The CO_2 greenhouse must have been a major factor on both Mars and Earth. As mentioned earlier, purely internal alternatives for Mars would not be satisfactory for the Earth, since on Earth, early *surface* climate change is unambiguously indicated. We will now consider the currently plausible "supplementary mechanisms."

B. CANDIDATE MECHANISMS TO AUGMENT THE CO_2 GREENHOUSE

Several gases are more effective greenhouse gases than CO_2 and may have been present in the early Martian atmosphere. Ammonia is one, but its lifetime in the face of photolysis is so short that it is an unlikely choice. Sulfur dioxide is a more interesting candidate. It is likely that early magmas on Mars may have been saturated with SO_2 and the effused SO_2 might even be equal to the effused CO_2. Further, even 1000 ppm SO_2 added to a CO_2 atmosphere could raise the surface temperature by an additional 20°C. This SO_2 augmentation has been and is still the subject of a great deal of theoretical radiative transfer modeling. Despite all this, no model has yet been published that convincingly suggests a major role for SO_2, although the issue has not yet been settled. The problem with SO_2 is that the atmospheric lifetime of SO_2 on Mars with respect to conversion to sulfate aerosol has been estimated to be <6 years, with most estimates <1 year. This requires an enormous influx of SO_2 to maintain a sufficient concentration in steady state. Still, SO_2 remains a plausible candidate worthy of further study.

Another candidate gas is CH_4. If metallic iron were in equilibrium with basaltic magmas, there would be more free H_2 effused than CO_2, so that on cooling, CH_4 could be synthesized, and CH_4 is a powerful greenhouse gas. Unfortunately, it seems unlikely that metallic iron would continue to be present in any surface magmas after hundreds of millions of years of volcanic activity. Even ordinary basaltic magmas can *indirectly* supply some methane. In such magmas, the Fe^{2+}/Fe^{3+} buffer controls the gas composition and CO_2 and CO are the dominant forms of carbon supplied. However, the equilibrium $H_2:H_2O$ ratio supplied at the melting point is predicted to be 1:140. Thus, if the H_2O/CO_2 ratio is high enough, enough free H_2

will be available to convert the CO_2 and CO partly to CH_4, which would be expected to occur if the gas reequilibrated at surface temperature. However, in the absence of an enormous rate of H_2 supply, a complication arises in that H_2 escapes rapidly and preferentially from the exosphere, so an intense degassing level is necessary to compensate for its escape. Preliminary calculations suggest that a high enough degassing rate to maintain a significant CH_4 concentration could continue for 10^7 to 10^8 years without supplying vastly greater amounts of other volatiles than can be accounted for. The model has not been widely accepted for two reasons:

(1) It is extremely difficult to estimate the exospheric temperature on early Mars, and the H escape rate and the model results depend very strongly on such estimates.
(2) The model assumes gas–gas chemical equilibrium, and it is not clear whether such equilibration could occur faster than photochemical displacements from molecular equilibrium.

Finally, recent models for the early Sun have called into question the premise on which the "paradox" is based. These models are preliminary and speculative, but considering the difficulties with the greenhouse model, they may prove to be significant.

V. EVIDENCE FOR
AND MECHANISMS OF
ONGOING PERIODIC CLIMATE CHANGE

Analogous to glacial cycles on Earth, Mars has experienced and is still experiencing *cyclic* climate change. The mechanism is the same for these two planets, namely, oscillations of the value of the inclination of the axis to the orbital plane (obliquity) and variations in the orbital eccentricity (departure from circularity). As indicated earlier, the layered deposits collaring the polar caps (see Fig. 2) are presumed to record these cycles of duration every 10^4–10^5 years. The periodicity is similar to the analogous cycles on Earth, but the magnitude of the wobble of the rotational axis—the "obliquity variation"—is several times greater for Mars than for Earth.

The primary mechanisms producing the layered effect are thought to be variations in atmospheric pressure caused by atmosphere–regolith exchange of adsorbed CO_2 and variations in atmospheric H_2O pressure. If the obliquity increased from its present nearly median value to its maximum axial tilt, the first response would be the vaporization of the CO_2 south pole caps. It is believed, however, that this would add only 1 or 2 mbar at most to the current 7 mbar pressure because the cap is so thin. The greatest part of the atmospheric pressure rise is thought to occur because of the penetration of higher surface temperatures to depth in the near-polar latitudes. Such a surface temperature rise near the poles could propagate to a depth of 500 m or more in 10^4 years. This would lead to CO_2 desorption as a result of the temperature rise, although some of that would be adsorbed by the (cooling) equatorial zone. In fact, a modest pressure rise—perhaps a factor of two—is expected. If, in the opposite part of the cycle, the obliquity decreases to its minimum value, that is, highest angle between the rotational axes and the orbital plane, the CO_2 cap would grow because its temperature would fall. On Mars, the

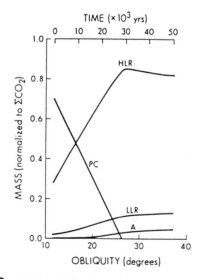

FIGURE 9 Quasi-periodic climate change on Mars. In addition to long-term climate change, quasi-periodic climate change, analogous to Earth's glacial cycles, also occurs on Mars with periods of 10^4–10^5 years. These are caused by variations in axial inclination on both planets. At Mars's lowest obliquity, the atmospheric (A) P_{CO_2} drops to perhaps as low as 1/100th of its present value and may reach twice its present value or more at highest obliquity. At lowest obliquity, the polar caps (PC) also become extremely cold and pump adsorbed CO_2 from both the high-latitude regolith (HLR) and low-latitude regolith (LLR). The amount of dust in the atmosphere is probably an extremely sensitive function of the atmospheric pressure, and its variation may account for the layered terrain collaring the polar caps (see text).

atmospheric pressure is always approximately equal to the CO_2 vapor pressure over the caps. Therefore, as the polar temperature falls, the global CO_2 pressure would fall rapidly until CO_2 was no longer the main atmospheric constituent. The atmosphere would simply serve as a conduit through which CO_2 was transferred from the regolith to a growing CO_2 cap. The entire complex process is pictured for a full cycle in Fig. 9. Starting with the present median situation, the CO_2 atmospheric pressure can vary downward by a factor of 100 and upward by a factor of about 2. It should be noted that recent celestial mechanical models suggest chaotic variations in obliquity and an obliquity range much greater than those depicted in the figure. This would allow for higher CO_2 pressures to be achieved at high obliquity and lower pressures at low obliquity.

The creation of the layered terrain can be directly related to the pressure variation. The present 7 mbar pressure is enough to raise periodic global dust storms. However, the ability to do so is very sensitive to the pressure in the mbar pressure range. If the pressure doubled, there would be almost constant dust storms, and if it dropped by a significant factor, there would be none at all. Thus, the degree of dust contamination of ice deposited at the poles would vary widely, causing albedo differences. Moreover, the thickness of each discernible layer corresponds quite well with the thickness expected for one obliquity cycle given the rate of deposition of polar material in the winter. The only discrepancy is that less than 25% of Martian history is immediately visible within the terrain, although we know that the current obliquity variation has been as intense or more intense through history. It is thought that this results from a continuing sinking of the column of layers and resulting continuing melting at the base of the column, resulting in the obscuration of the earlier deposited layers.

VI. SUMMARY

The two most important questions about Mars history from either an abiotic or exobiological standpoint are: (1) Did Mars experience a "warmer, wetter" climate and surface environment during the first billion years or so of its history? (2) What changes in the surface–atmosphere environment continued to occur until the present as the result of changes in solar insolation and periodic variations in the obliquity and orbital eccentricity of the planet?

In answer to the first question, several arguments have been delineated here that argue in favor of both affirmative and negative answers. Each of the following arguments or evidence could well be rephrased as questions, since most are still controversial given our current state of knowledge of Mars.

Arguments against a "warmer, wetter" early Mars are: (1) The concept of a "weak early Sun" has been only sporadically challenged. It appears that if Mars's surface was warmer early in its history, it would have to be so despite a solar luminosity only 70% of that at present. (2) Attempts to overcompensate for the weak early solar luminosity have relied mainly on a CO_2 greenhouse effect. However, the amounts of CO_2 required to be in the atmosphere are approximately 3–5 bars. Such amounts are very high and only marginally compatible with the best available estimates of Mars's total degassed CO_2 inventory. (3) If the CO_2 were kept in the atmosphere, it would be subject to impact erosion and loss to space over a period of time that probably included a large flux of impactors. In addition, efforts to show that enough CO_2 remains on Mars today, trapped as carbonates in the regolith, depend on preliminary spectroscopic detection and assay of carbonates in the dust and topmost regolith, which may not be typical of the unconsolidated regolith as a whole. The depth of the unconsolidated regolith is also largely a matter of speculation. (4) Attempts to circumvent these problems with small amounts of other, more effective greenhouse gases have difficulties. Water vapor could greatly augment the CO_2 greenhouse, but it would not be available in sufficient quantities if Mars's surface was initially cold. Sulfur dioxide, CH_4, and NH_3 are extremely effective greenhouse gases, but they are subject to rapid destruction by chemical reactions and photolysis. (5) The morphological evidence (the so-called "valley networks") that might appear to resemble river channels actually exhibit geometry more like terrestrial "sapping" channels caused by subsurface melting and collapse.

The arguments suggesting a warmer, wetter Mars are: (1) Even if the sapping channels were caused in part by internal processes, they clearly indicate at least a warmer, near-surface environment in early Mars history, which tapered off ~3.8 billion years ago despite the allegedly warming Sun. Abundant new surface flows probably suggest sporadic surface flow of water. (2) Other evidence suggests a higher atmospheric pressure or more abundant surface water: that the surface erosion rates were higher in early Mars history is indicated by the greater elimination rate of medium-sized craters (even when the greater exposure time for erosion is taken into account). Unlike the case of the valley networks, it is difficult to attribute these observations to internal processes alone. (3) Although it would appear that other species that must be better greenhouse gases than CO_2 could not be present in even the minor amounts required to augment the greenhouse effect, this is not necessarily true. For example, once the greenhouse effect got started, H_2O vapor would augment it in a major way. Sulfur dioxide is quickly destroyed on Earth following eruptions by H_2O vapor. However, if an early Mars volcano supplied SO_2 at a rate comparable to H_2O and if Mars's surface was cold, the ability to obtain H_2O vapor at a rate needed to supply SO_2 could be seriously questioned. NH_3 and CH_4 could be somewhat protected by constant dust storms. (4) Since the early heat flow on Mars was probably 6–8 times that of the present owing to greater radioactive decay and accretional heat, and there were *no* bodies of water, then CO_2 in the atmosphere would quickly accumulate. Though it could have been depleted by impact erosion, tentative measurements of carbonate abundance in the regolith and estimates of regolith depth suggest that the required CO_2 inventory could be mostly still present. The most likely alternative to stop CO_2 buildup would be removal by carbonate formation, which could operate over 10^7–10^8 years or longer and require possibly lake-sized bodies of water. Hence it is reasonable to suggest that Mars would automatically drive itself to a "damp" if not wet surface state. (5) On the much better understood Earth, we find evidence of early deposition of carbonates in bodies of water that should have been frozen if the weak early Sun was *not* overcompensated for by a substantial greenhouse effect. It may be awkward to attribute all of Mars's evidence for a different early erosional style to internal effect alone while still requiring a major early atmosphere greenhouse effect on our planet.

In answer to the second question, the ongoing period variations in obliquity and eccentricity of Mars are quite well established and the reality of more extreme chaotic variations is generally accepted. It is fairly clear that during most of Mars history, the obliquity becomes periodically low enough to cold trap most of the CO_2 from the atmosphere, as well as much of the adsorbed CO_2 on the regolith. During such times, the atmospheric CO_2 pressure falls to ~1 mbar or less, and during the times of lowest extreme obliquity, they may fall below the pressure of Ar and N_2. Dust storms probably do not occur then and the CO_2 cap that continues to limit the atmosphere pressure grows by a large factor, mostly at the expense of regolith-adsorbed CO_2, for which the atmosphere merely acts as a con-

duit. On the high-obliquity side, the atmospheric pressure may increase by only a factor of two or so, but dust storms may be much more frequent at present and perhaps continuous. There would be no permanent CO_2 cap if the obliquity were much greater than at present.

Though there are no great controversies over this interpretation, some mysteries remain. On Earth, the obliquity variation is much smaller, but the final thermal effect is thought to be the same as predicted for Mars. This is the result of complicated feedback effects involving the ocean and caps, which can augment the effect of redistributed solar insolation. We presume that these do not occur on Mars, but they may. We do not understand what the feedback effect of continuous dust storms or higher atmospheric pressure might be. Also, it is widely assumed that the layered terrain records the variations in a simple, understandable way. However, the precise reason for differences among the layers is not established. Finally, it appears that the layered terrain does not record more than a fraction of Mars history, perhaps because of subsidence and basal melting.

Thus, the mechanism responsible for the cyclic process that produced the layered terrain appears at present to be reasonably well understood and highly relevant to the understanding of Earth history as well. In contrast, the causes of clear differences in the environmental conditions in earliest Mars history versus middle and late Mars history remain a matter of intense controversy.

ALSO SEE THE FOLLOWING ARTICLE

MARS: SURFACE AND INTERIOR

BIBLIOGRAPHY

Carr, M. H. (1986). Mars: A water rich planet. *Icarus* **56**, 187–216.

Carr, M. H. (1996). "Water on Mars." Oxford Univ. Press, New York/Oxford.

Fanale, F. P., Postawko, S. E., Carr, M. H., Pollack, J. B., and Pepin, R. O. (1992). Mars: Epochal climate change and volatile history. *In* "Mars." Univ. Arizona Press, Tucson.

Kasting, J. F. (1991). CO_2 condensation and the climate of early Mars. *Icarus* **94**, 1–13.

Kieffer, H. H., and Zent, A. P. (1992). Quasi-periodic climate changes on Mars. *In* "Mars." Univ. Arizona Press, Tucson.

Pepin, R. O. (1987). Volatile inventories of the terrestrial planets. *Rev. Geophys.* **37**, 479–533.

Pollack, J. B. (1991). Kuiper Prize Lecture: Present and past climates of the terrestrial planets. *Icarus* **91**, 173–198.

MARS: SURFACE AND INTERIOR

I. Mars Exploration

II. General Characteristics

III. Impact Cratering

IV. Volcanism

V. Tectonics

VI. Canyons

VII. Erosion and Deposition

VIII. Poles

IX. Interior

X. Summary

Michael H. Carr

U.S. Geological Survey, Menlo Park

GLOSSARY

Chaotic terrain: Areas of the Martian surface where the ground has collapsed to form a surface of jostled blocks standing 1–2 km below the surrounding terrain.

Lithosphere: Rigid outer rind of a planet, as distinct from the underlying, more fluid asthenosphere.

Obliquity: Angle between the equatorial plane of a planet and the orbit plane, or the tilt of the rotational axis. It causes the planet to have seasons.

Outflow channels: Large channels that start full size and have few if any tributaries. They may be up to several tens of kilometers across and thousands of kilometers long, and are believed to have been formed by large floods.

Permafrost zone: Near-surface zone within which temperatures are always below 0°C. It may or may not contain ground ice.

Shield volcano: Broad volcano with a large summit pit formed by collapse and gently sloping flanks, built mainly from overlapping, fluid, basaltic lava flows.

SNC meteorites: Group of meteorites (Shergotty–Nakhla–Chassigny) believed to be derived from Mars because of their young ages, basaltic composition, and inclusion of gases with the same composition as the Martian atmosphere.

M ars, the outermost of the four terrestrial planets— Mercury, Venus, Earth, and Mars—is intermediate in size between the Earth and the Moon. The terrestrial planets all have solid surfaces and on these surfaces is preserved a partial record of how each planet has evolved. Successive events, such as volcanic eruptions or meteorite impacts, both create a new record and partly destroy the old record. The task of the geologist is to reconstruct the history of the planet from what is preserved at the surface. Both Mercury and the Earth's Moon appear to have become geologically inactive relatively early in their history, so most of the preserved record dates from very early in the history of the solar system, prior to 3 billion years ago. Recent results from the Magellan mission suggest that the geologic record on Venus is relatively young, most of the surface apparently having formed in the last half billion years. The record on Earth is also mostly young. Because the oceanic crust is constantly being subducted into the mantle by plate tectonics and replaced by younger crust at midoceanic ridges, nearly all the oceanic crust is less than 200 million years old. An older record is preserved only on the continents. But even here, the erosive action of water tends to destroy the record so that evidence from the first billion years of the Earth's history is preserved only in a few isolated places. In contrast, on Mars we appear to have a record that spans much of the history of the solar system. An ancient cratered terrain has survived from the first billion years and, as we shall see in the following, there are good reasons to believe that Mars is still volcanically active today and has been so throughout its history. We shall also see that the geologic activity has been diverse, parts of the surface having been variously affected by the action of volcanism, tectonism, wind, water, and ice.

Though we have viewed most of the surface of Mars and so have some basis for speculating about how it formed, our knowledge of the interior is very rudimentary. Much of our knowledge of the interior of the Earth is derived from seismic data on earthquakes. We have yet, however, to acquire similar data from Mars. We must infer what the interior is like indirectly from characteristics of the planet such as its moment of inertia, or by comparisons with the Earth and meteorites. Therefore, the following discussion will focus primarily on the surface.

I. MARS EXPLORATION

The modern era of Mars exploration began on July 14, 1965, when the *Mariner 4* spacecraft flew by the planet and transmitted to Earth 22 close-up pictures of the planet, with resolutions of several kilometers. Prior to that time we were dependent largely on telescopic observations, whose resolution at best is one to two hundred kilometers, and which reveal no topography but only surface markings. We knew from telescopic observations that Mars had a thin CO_2 atmosphere, polar caps that advanced and receded with the seasons, and surface markings that underwent annual

and secular changes, but geologic studies of the planet could realistically begin only when we acquired spacecraft data.

The *Mariner 4* pictures revealed an ancient surface that resembled the lunar highlands. These results were disappointing because it had been speculated that Mars, having an atmosphere and being larger than the Moon, might be more Earth-like than Moon-like. *Mariner 4* was followed by two more Mariner spacecraft in 1969 (Table I), which seemed to confirm Mars's lunarlike characteristics. However, our perception of Mars changed dramatically in 1972 when systematic mapping by the *Mariner 9* orbiter revealed the planet that we know today. As mapping progressed, huge volcanoes, deep canyons, enormous dry riverbeds, and extensive dune fields came into view, and a complex, variegated geologic history became apparent. Exploration of Mars continued in the 1970s as both the USSR and the United States sent landers to the surface and other orbital vehicles to the planet. Exploration in the 1970s culminated with the Viking mission, which successfully placed two landers on the surface and two other spacecraft in orbit. By the end of the Viking mission, almost all the surface had been photographed from orbit at a resolution of about 250 m/pixel and smaller fractions with resolutions as high as 10 m/pixel. In addition, the Viking landers had photographed in detail the two Viking landing sites and successfully carried out a variety of experiments directed mostly toward detecting life and understanding the chemistry of the soil.

In the early 1980s, our understanding of Mars was further enhanced when it became clear that we had samples of Mars in our meteorite collections here on Earth. A group of meteorites, called SNCs (pronounced "snicks"), were initially suspected to be of Martian origin because they were basaltic and had ages close to 1.3 billion years. These meteorites could not have come from the Earth because their oxygen isotope ratios are distinctively different from terrestrial ratios. The only plausible body that could have been volcanically active so recently was Mars. A Martian origin was later confirmed by finding gas included in the meteorites that has a composition identical to that of the Martian atmosphere, as measured by the *Viking* landers. The meteorites are believed to have been ejected from Mars by large impacts and subsequently captured by the Earth, after spending several million years in space.

Twelve of these meteorites had been identified in our collections at the time of writing this article. Eleven are 1.3 billion years old or younger. A twelfth, designated ALH84001, is 4.5 billion years old. The designa-

TABLE I
Mars Missions

Mission	Nation	Launch date	Fate
Mariner 4	US	11/28/64	Flew by 7/15/65; first close-up images
Mariner 6	US	2/24/69	Flew by 7/31/69; imaging and other data
Mariner 7	US	3/27/69	Flew by 8/5/69; imaging and other data
Mars 2	USSR	5/19/71	Crash landed; no surface data
Mars 3	USSR	5/28/71	Crash landed; no surface data
Mariner 8	US	5/8/71	Fell into Atlantic Ocean
Mariner 9	US	5/30/71	Into orbit 11/3/71; mapped planet
Mars 4	USSR	7/21/73	Failed to achieve Mars orbit
Mars 5	USSR	7/25/73	Into orbit 2/12/74; imaged surface
Mars 6	USSR	8/5/73	Crash landed
Mars 7	USSR	8/9/73	Passed by Mars
Viking 1	US	8/20/75	Lander on surface 7/20/76; orbiter mapping
Viking 2	US	9/9/75	Lander on surface 9/3/76; orbiter mapping
Phobos 1	USSR	7/7/88	Lost 9/2/88
Phobos 2	USSR	7/12/88	Mars and Phobos remote sensing
Mars Observer	US	9/22/92	Lost during Mars orbit insertion
Pathfinder	US	12/4/96	Landed 7/4/97; lander and rover data
Global Surveyor	US	11/7/96	Into orbit 9/11/97; imaging and other data

tion ALH indicates that it was found in the Allan Hills in Antarctica. The numbers 84001 indicate that it was the first meteorite found at that location in 1984. In 1996 it was tentatively suggested that carbonate globules within this meteorite, together with some disequilibrium mineral assemblages, polycyclic aromatic hydrocarbons (PAHs), and a number of different types of very small, segmented rods that resemble some terrestrial nanofossils, might be the result of biological activity. [*See* METEORITES.]

II. GENERAL CHARACTERISTICS

A. ORBITAL AND ROTATIONAL MOTIONS

The Martian day is almost the same as the Earth's day, but the year is almost twice as long (Table II). Because its rotational axis is inclined to the orbit plane, Mars, like the Earth, has seasons. But the Mars orbit has a significant eccentricity. This causes one pole, that tilted toward the Sun at perihelion, to have warmer summers than the other pole. At present the south has the warmer summers, but, because of a slow change in the direction of tilt of the rotational axis (precession), the hot and cold poles change on a 51,000-year cycle. The eccentricity also causes the seasons to vary significantly in length (see Table II). The magnitude of the tilt of the rotational axis is called the obliquity. At present the Mars obliquity is similar to the Earth's. Yet while the Earth experiences only minor changes in obliquity, the obliquity of Mars changes chaotically. It ranges mostly between 15° and 35°, but may occasionally reach 60°. Summer temperatures at the poles change dramatically during the obliquity cycle, being much warmer at high obliquities than at low obliquities.

B. SURFACE CONDITIONS

Mars has only a thin atmosphere that provides almost no thermal blanketing. As a result, temperatures at the surface have a wide diurnal range, controlled largely by the reflectivity of the surface and the thermal properties of the surface materials. Typically surface temperatures in summer at latitudes ±60° range from

TABLE II
Earth and Mars: General Characteristics Compared

	Earth	Mars
Mean equatorial radius (km)	6378	3396
Mass ($\times 10^{24}$ kg)	5.98	0.642
Mean distance from Sun (10^6 km)	150	228
Orbit eccentricity	0.017	0.093
Obliquity	23.5°	25.2°
Length of day	24 h	24 h 39 m 35 s
Length of year (Earth days)	365.3	686.9
Seasons (Earth days)		
Northern spring	92.9	199
Northern summer	93.6	183
Northern fall	89.7	147
Northern winter	89.1	158
Atmosphere	79% N_2, 21% O_2	95% CO_2, 3% N_2, 2% Ar
Surface pressure (mbar)	1000	7
Mean surface temperature (K)	288	215
Surface gravitational acceleration (cm/s^{-2})	981	371
Moons	1	2

180°K at night to 290°K at midday, but can range more widely if the surface consists of unusually low-density, fine-grained material. However, these temperatures are somewhat deceiving, because at depths of several centimeters below the surface, temperatures are at the diurnal mean of about 210–220°K. At the poles, in winter, temperatures drop to 150°K, at which point CO_2 condenses out of the atmosphere to form the seasonal cap. The surface pressure ranges from about 14 mbar in the deepest parts of the canyons to 0.3 mbar on top of the highest volcanoes, and changes annually as a result of the formation of the polar caps. At the two Viking landing sites, maximum daily winds were typically several meters per second for most of the year. However, during southern spring and summer, when dust storms are common in the southern hemisphere, wind gusts up to 40 m/sec were recorded. [*See* MARS: ATMOSPHERE AND VOLATILE HISTORY.]

The stability of water is of profound importance for understanding Martian geology. Under the conditions just described, the planet has a thick permafrost zone that extends down to depths of over a kilometer at the equator to a few kilometers at the pole. If water is present, it must exist as ice within this zone; liquid is stable only below the zone. Water ice is stable at the poles and has been detected at the north pole, under the seasonal CO_2 cap. At latitudes between roughly 40° and 80°, ice on the surface will tend to sublime

during summer, but it is permanently stable a few meters below the surface. At lower latitudes, ice is unstable at the surface and at all depths below the surface. Buried ice will be lost to the atmosphere at a rate controlled largely by the ability of the sublimed water to diffuse through the soil. Thus we should expect to find water ice at low latitudes only if it is prevented from subliming into the atmosphere because of impermeability of the overlying soil, or if water is being supplied, such as by volcanic activity, at a rate faster than it is being lost to the atmosphere. The stability conditions change slightly during the obliquity cycle. As discussed in following sections, geologic evidence suggests that conditions may have been very different in the distant past.

C. GLOBAL PHYSIOGRAPHY AND TOPOGRAPHY

The physiography of the surface has a marked north–south asymmetry. Much of the southern hemisphere is covered with heavily cratered terrain, and a large lobe of cratered terrain extends into the northern hemisphere between longitudes 30°W and 280°W. This terrain clearly dates from early in the planet's history, when impact rates were much higher than subsequently. By analogy with the Moon, the terrain is believed to have formed prior to 3.8 billion years ago. The 4.5-billion-year-old Martian meteorite ALH84001 is presumably from this terrain. Most of the cratered terrain stands at elevations of 1–4 km above the Mars datum. Exceptions are the floors of large impact basins such as Argyre and Hellas. (Elevations on Mars are referenced to the level at which the mean atmospheric pressure is 6.1 mbar.) Most of the rest of the planet is covered with sparsely cratered plains that formed subsequent to the postulated decline in impact rates around 3.8 billion years ago. At high northern latitudes, most of the plains are at elevations of 1–2 km below the datum. The cause of the dichotomy between cratered uplands and low-lying plains is unknown, but one suggestion is that the low-lying plains lie within the remnant of an enormous impact basin that formed at the end of planetary accretion.

Not all plains are at low elevations. Superimposed on the dichotomy between cratered uplands and low-lying plains are two bulges. The Tharsis bulge, centered on the equator at 100°W, is 5000 km across and 10 km high. The smaller Elysium bulge, centered at 30°N, 210°W, is 2000 km across and 4 km high. Both bulges have been sites of volcanic activity throughout much of the planet's history, and superimposed on them are large volcanoes (Fig. 1; see also color insert). The largest, which are in Tharsis, reach elevations of 27 km above the datum. The Tharsis bulge is also the center of a vast array of radial fractures that affect almost a third of the planet. To the east of Tharsis, just south of the equator, several enormous interconnected canyons are aligned along the faults radial to Tharsis. At the eastern end of the canyons are large areas of chaotic terrain. Out of these areas emerge enormous dry riverbeds that extend northward for thousands of kilometers down the regional slope to ultimately disappear in the low-lying northern plains. The channels are believed to have been cut during large floods.

The physiography of the poles is distinctively different from that of the rest of the planet. At each pole, extending out to the 80° latitude circle, is a stack of finely layered deposits a few kilometers thick. In the north they rest on plains, in the south on cratered upland. The small number of superimposed impact craters suggests that they are of a relatively young age, possibly only a few hundred million years old.

D. THE VIEW FROM THE SURFACE

We have close-up views of the Martian surface only at the two *Viking* landing sites at 22.3°N, 48°W and 48°N, 225.6°W, and the *Pathfinder* landing site at 19.3°N, 33.5°W. *Viking 1* landed on what appeared to be a level plain, featureless except for occasional ridges and impact craters (Fig. 2a). From the lander the surface resembles many rocky deserts of the Earth. The gently rolling landscape is yellowish brown and strewn with rocks in the centimeter to meter size range. The rocks are volcanic and were probably deposited in the area by distant large impacts. Between the rocks are drifts of fine-grained material, and locally bigger drifts cover the rocky debris. The drifts appear to be formed not from sand-sized, saltating particles like terrestrial dunes but from very fine-grained, wind-blown material that had been suspended in the atmosphere. The surface of the fine grained material is loosely cemented, probably by water-soluble salts such as sulfates, to form a crust. Impact craters are absent in the near field, although the rim of a large crater is visible on the horizon. Small craters are absent because small meteorites burn up or disintegrate in the thin atmosphere.

Viking 2 landed on a level, rock-strewn, but otherwise featureless plain. The strewn field is believed to be a lobe of ejecta from a 90-km-diameter impact crater situated 170 km to the west of the site. Although no chemical analyses were obtained from rocks at either Viking site, the rocks were believed to be basaltic on the basis of their reflectivity, the iron-rich nature of the weathered material, the composition of SNC meteorites, and the general basaltic nature of volcanic landforms on Mars.

The Viking landers measured the chemical composition of the soils at the two sites and performed a variety of chemical experiments on the soils, as part of their search for life. However, the analyses were incomplete. Two competing models for the soils are that (1) they consist largely of Fe-rich clays or (2) they resemble a poorly crystalline, partly hydrated, volcanic ash, called palagonite. The analyses were almost identical at the two sites, suggesting that what was analyzed was dust homogenized over the whole planet through repeated participation in global dust storms. The soils are oxidizing, 70–800 nmoles/g of O_2 being released upon humidification. Although the exact nature of the oxidants is unknown, they probably form as a result of (1) condensation on the surface of $OH \cdot HO_2$ and superoxides, formed by UV-induced photolysis of water in the atmosphere, and (2) UV-induced photolysis of water adsorbed on soil particles. Somewhat surprisingly, the soils contain no complex organic materials despite a continuous infall in meteorites. The organic materials are probably destroyed by the combined effects of the oxidants and UV radiation.

On July 4, 1997 the *Pathfinder* spacecraft landed in

FIGURE 1   Distant view of Mars. In the center of the scene is the vast equatorial canyon system, Valles Marineris. At its western end the canyon divides into a series of smaller, intersecting canyons to form a maze of depressions called Noctis Labyrinthis. Farther west, close to the left-hand margin of the picture, are three dark spots. These are large volcanoes in the Tharsis region. (See also color insert.)

southern Chryse Planitia, 850 km ESE of the *Viking 1* landing site. The landing was at the mouth of a large channel, Ares Vallis, and had been chosen in the hope of seeing evidence for the large flood that formed the channel; and because the flood probably deposited rocks in the area that would be accessible for analysis. The lander did see evidence of the flood, although it was rather subtle (Fig. 2b; see also color insert). Some of the rocks close to the lander were stacked together, leaning in the direction of the flood as indicated from the orbiter pictures, and broad linear depressions, presumably eroded by the flood, cross the site in the same direction. Horizontal markings on a distant hill may be terraces

cut by the flood. As expected, the scene was more rocky than the *Viking 1* site and most of the rocks were partly covered by red dust. The lander deployed a small rover that acquired close-up views and chemical analysis of some of the local rocks. The chemical analyses provided some of the most surprising results of the mission. Most of the local rocks were found to have silica contents more similar to those of terrestrial andesites rather than basalts. On Earth, andesites form mostly at convergent plate boundaries, but there is no morphologic evidence for plate boundaries on Mars, so the presence of andesites is puzzling and their origin remains unclear. In composition, the soils differ from the rocks mainly in having

FIGURE 2 (a) View of the surface of Chryse Planitia from *Viking Lander 1*. To the left are drifts of fine-grained, wind-deposited material. To the right, occasional outcrops of bedrock show through the rubbly surficial cover. (b) View looking to the west from the *Pathfinder* lander. The two peaks on the horizon are 0.8 km away. The large rock to the right is 4.8 m away and 1.4 m across. In the left half of the scene several of the rocks are leaning to the right, in the direction of the flood that is thought to have passed through the area and deposited most of the rocks in view. (See also color insert.)

more sulfur and chlorine, which could have been derived from volcanic gases.

III. IMPACT CRATERING

A. CRATERING RATES

All solid bodies in the solar system are subject to impact by asteroidal and cometery debris. The cratering rates are low. On Earth, in an area the size of the United States, a crater larger than 10 km across is expected to form every 10–20 million years, and one larger than 100 km across every 1 billion years. The rates on the other terrestrial planets are likely to be within a factor of two or three of these rates. As a consequence, any surface that appears heavily cratered must date back to a time when impact rates were much higher than at present. On the Moon, surfaces are either heavily cratered (lunar highlands) or sparsely cratered (maria), with no surfaces of intermediate crater density. This contrast arises because of the Moon's cratering history. Very early, cratering rates were high. Around 3.8 billion years ago they declined rapidly to roughly the present rate. Accordingly, surfaces that formed prior to 3.8 billion years ago are heavily cratered, and those that formed afterward are sparsely cratered. Mars appears to have experienced a similar cratering history, although we do not know when the decline occurred.

We assume that the cratering rate declined at the same time as on the Moon. Thus, as indicated earlier, the cratered uplands of Mars are thought to have formed around 3.8 billion years ago. Younger surfaces can be roughly dated from the number of superimposed impact craters by extrapolating recent lunar and terrestrial impact rates to Mars. [See THE MOON.]

B. CRATER MORPHOLOGY

Impact craters have similar morphologies on different planets. Small craters are simply bowl-shaped depressions with constant depth-to-diameter ratios. With increasing size, the craters become more complex as central peaks appear and the depth-to-diameter ratio decreases. At very large diameters, the craters become multiringed, relatively shallow basins and it is not clear which ring is equivalent to the crater rim of smaller craters. On Mars the transition from simple to complex crater takes place at 8–10 km in diameter, and the transition from complex crater to multi-ringed basin takes place between 130 and 150 km in diameter.

Although impact craters on Mars resemble those on the Moon, the patterns of ejecta are quite different. Lunar craters generally have continuous hummocky ejecta near the rim crest, outside of which is a zone of radial or concentric ridges, which merges outward into strings and loops of secondary craters. In contrast, the ejecta around most Martian craters, especially those in the 5- to 100-km-diameter size range, is disposed in discrete, clearly outlined lobes (Fig. 3). Various patterns are observed. The ejecta around craters smaller than 15 km in diameter is commonly enclosed in a single continuous lobate ridge or rampart, situated about one crater diameter from the rim. Around larger craters may be many large lobes, some superimposed on others, but all outlined by a peripheral rampart. Other craters have a distinct mound of ejecta immediately outside the rim, with more typical lobate ejecta outside the mound. The distinctive patterns of Martian ejecta have been attributed to two possible causes. The first suggestion, based on experimental craters formed under low atmospheric pressures, is that the patterns are formed by interaction of the ejecta with the atmosphere. The second, and generally preferred, explanation is that the ejecta contained water and had a mudlike consistency, and so continued to flow along the ground after ejection from the crater and ballistic deposition. This view is supported by the resemblance of the Martian craters to those produced by impacts into mud. The onset diameter for the lobate patterns increases with latitude, as expected if water-saturated material was ejected from below the permafrost zone. [See PLANETARY CRATERING.]

The previous discussion refers to fresh-appearing craters. Erosion rates at low latitudes for most of Mars's history are very low—typically 0.01–0.05 μm/yr, although rates may be higher locally. However, early in the planet's history, erosion rates were much higher. As a consequence, in the cratered uplands, craters range in morphology from fresh-appearing craters to barely discernible, rimless, circular depressions. In contrast, on volcanic plains in the equatorial regions, almost all craters appear fresh, although they may be billions of years old. Obliteration rates appear to have been higher at high latitudes. This has been attributed to ice-abetted creep of near-surface materials, but other factors may have contributed to the modification of craters, such as repeated burial and removal of wind-blown debris. Such a process has been invoked to explain the so-called pedestal craters that are particularly common at high latitudes. These craters are inset into a platform or pedestal that has about the same areal extent as the ejecta. The simplest explanation is that the region in which these craters are found was formerly covered with a layer of loose material that has since been removed by the wind except around craters, where the surface was armored by the ejecta.

IV. VOLCANISM
. . . .

Mars has had a long and varied volcanic history. Crystallization ages of SNC meteorites as young as 150 million years, and the scarcity of impact craters on some volcanic surfaces, suggest that the planet may still be volcanically active. If so, the rates are likely to be much lower than on the Earth. The large shield volcanoes, located primarily in the two provinces of Tharsis and Elysium, present the most spectacular evidence of volcanism. Shield volcanoes, such as those in Hawaii, are broad domes with shallow sloping flanks that form mainly by eruption of fluid, basaltic lava. Each has a summit depression formed by collapse, following eruptions on the volcano flanks. In contrast, volcanic cones tend to be smaller and have steeper flanks and a summit depression that is a true volcanic vent. Explosive activity accompanied by eruption of ash is common in the building of volcanic cones, and the lava tends to be more volatile rich, more siliceous, and more viscous than that which forms shields. On Mars, volcanic landforms suggestive of eruption of fluid lava are far more common than those suggestive of explosive activity and production of ash.

FIGURE 3 Impact craters on ridged, volcanic plains. Lobes of ejecta, each outlined by a low ridge, surround the crater. The linear ridges are believed to have formed as a result of compression caused by the Tharsis bulge. Picture is 250 km across.

In Tharsis, three large shield volcanoes form a northeast–southwest trending line, and an even larger volcano, Olympus Mons, stands 1500 km to the northwest of the line. Olympus Mons (Fig. 4) is 27 km high, and the main edifice is 550 km across; the three other shields are only slightly smaller. In contrast, the largest terrestrial shield volcano, Mauna Loa in Hawaii, is 120 km across and stands 9 km above the ocean floor. Each of the Tharsis volcanoes has a large summit caldera, that on Olympus Mons being 80 km in diameter. The flanks have a fine striated pattern caused by long linear flows, some with central leveed channels. The aligned Tharsis shields each have vents on their northeast and southwest flanks that have been the sources of flows that extend far across the surrounding volcanic plains. The main edifice of Olympus Mons is surrounded by a steep cliff, in places 6 km high. Beyond the cliff is the Olympus Mons aureole, consisting of several huge lobes of distinctively ridged terrain that extend from the base of the volcano out several hundred kilometers. The origin of the aureole is unknown, although it may result from gravitational collapse and outward movement of the peripheral parts of the edifice to form what is now the basal cliff. Throughout Tharsis and also in Elysium are numerous smaller shields that share some of the characteristics of the larger shields.

Many of the features of the shield volcanoes have close terrestrial counterparts, so the origin of the shields is one of the best-understood aspects of Martian geology. However, where terrestrial and Martian shield volcanoes differ most is in their size. The Martian shields are far larger than those on Earth, the summit calderas are much larger, and the individual flows are larger. Part of the cause of the large size of the Martian volcanoes is the lack of plate tectonics on Mars. The shield volcanoes in Hawaii are relatively short-lived. They sit on the Pacific plate, and the source of lava is in the mantle below the rigid plate. As a Hawaiian volcano grows, movement of the Pacific plate carries it away from the lava source so it becomes extinct within a few hundred thousand years. A trail

FIGURE 4   Olympus Mons. The volcano is 550 km across and 27 km high, with an 80-km diameter and multiple collapse pits at its summit. Parts of the aureole are visible at the top of the picture.

of extinct volcanoes across the Pacific attests to the long-term supply of magma from the mantle source presently below Hawaii. On Mars a volcano remains stationary over its magma source and will continue to grow as long as magma is available, so the volcano is correspondingly larger. Another factor probably contributes to the large size. For a volcano to continue to grow in height, the magma must have a density lower than that of the surrounding rocks so that it can rise into the magma chamber within the volcano and continue to do so as the magma chamber itself increases in elevation with the growth of the volcano. Whether this actually occurs will depend on the compositon of the lava and the density of the edifice. Shield volcanoes may have failed to grow on the Moon, for example, despite considerable volcanism, because the density of the molten lava was higher than the density of the near-surface rocks. Clearly, this was not the case for

Mars, since the volcanoes were able to grow to substantial heights.

Though the large shield volcanoes present the clearest evidence of volcanism, they may represent only a small fraction of the planet's volcanic activity. Surrounding the shield volcanoes in Tharsis and Elysium are extensive plains that are clearly volcanic as indicated by numerous flows superimposed one on another. These are probably mostly fissure-fed flows whose sources are now buried beneath the flows, although some of the flows originated from the shields, as already mentioned. Other plains, such as Lunae Planum and Hesperia Planum, are suspected of being volcanic but individual flows can rarely be seen (see Fig. 3). These may be analogous to flood basalts on Earth, which are the result of rapid effusion of very fluid lava that spread quickly over large areas. Local patches of intercrater plains, which resemble the Lunae

Planum plains, are common throughout the cratered uplands. Again, many of these plains may be volcanic, although direct evidence is lacking.

Not all Martian volcanism is the result of quiet effusion of fluid lava. Tyrrhena Patera (20°S, 252°W), an old volcano in the southern uplands, is surrounded by horizontal sheets of deeply eroded material that extend a few hundred kilometers away from the volcano. The most plausible interpretation is that the sheets are easily erodible ash flows that have subsequently been eroded by wind and/or water. The dissection of the flanks of other volcanoes, such as Ceraunius Patera (24°N, 97°W), Hecates Tholus (32°N, 210°W), and parts of Alba Patera (40°N, 110°W), and the blurring of flow features on the flanks of Elysium Mons have also been attributed to the presence of ash deposits. Clusters of small domes, such as those to the east of Hellas at 43°S, 239°W, may be indicative of eruption of more viscous lavas than on the shields. Finally, a number of features, particularly in the northwestern part of Elysium, have been attributed to water–ice interaction. As we shall see below, there are good reasons for believing that Mars has a water-rich crust, so interaction of the mantle-derived lavas with the ice-rich, near-surface rocks is expected. This supposition is supported by oxygen isotope data from the SNC meteorites, which show that the oxygen in the water in these basalts is isotopically distinct from the oxygen in the silicates. The water appears not to have been derived from the mantle but from some other source, presumably in the crust. [See PLANETARY VOLCANISM.]

V. TECTONICS

Most of the deformation of the Earth's surface is controlled by the movement of the large lithospheric plates with respect to one another. Linear mountain chains, transcurrent fault zones, rift systems, and oceanic trenches all form as a result of plate tectonics. On Mars there are virtually no plate tectonics, and so almost all the deformational features familiar to us here on Earth are rare or absent. The tectonics of Mars are dominated by the Tharsis bulge. The bulge is not isostatically compensated, that is, the extra mass of material in the bulge is not compensated for by lighter materials at depth. As a consequence, over the bulge is a large gravity anomaly. The origin of the bulge is unknown, but speculation ranges from sustained accumulation of volcanic materials to upward bowing of the crust as a result of mantle convection. Arrayed around the bulge is a vast system of roughly radial fractures that generally form closely spaced, parallel graben, indicating tensional deformation (Fig. 5). Throughout Tharsis, many of the fractures are buried by younger lava flows and are visible only in older islands that stand above the surrounding flows. To the north of Tharsis, the fractures are diverted around the large volcano Alba Patera to form a fracture ring. Also arrayed around the bulge, but particularly in Lunae Planum to the east, are circumferentially oriented wrinkle ridges, thought to have formed by compression (See Fig. 3). Both the fractures and the ridges appear to result from strain in the crust caused by the massive uncompensated load in Tharsis.

Some deformational features are not associated with the Tharsis bulge. Wrinkle ridges, suggestive of compression, are common on intercrater plains in the cratered uplands and on some plains that are far removed from Tharsis, such as Hesperia Planum and Syrtis Major. The causes of the compression are obscure. Some deformation, such as the arcuate faults around Isidis and Hellas, clearly result from accommodation of the crust to the formation or presence of the large basins. Circular fractures around large volcanoes such as Elysium Mons and Ascreus Mons have formed as a result of bending of the lithosphere under the volcanic load, and estimates can be made of the thickness of the lithosphere from the position of the fractures. Finally, large areas of the northern plains are cut by a polygonal pattern of fractures. The fractures divide the surface into a mosaic of blocks each a few to ten kilometers across. The cause of the fracture pattern has been the subject of considerable debate, and most observers have attributed the pattern in some way to the presence of abundant ground ice. Despite these examples, the variety of deformational features is rather sparse compared with the Earth because of the lack of plate tectonics. In particular, folded rocks of any type are likely to be rare.

VI. CANYONS

On the eastern flanks of the Tharsis bulge is a vast system of interconnected canyons. They extend just south of the equator from Noctis Labyrinthus, near the crest of the Tharsis bulge, eastward for about 4000 km until they merge with some large channels and chaotic terrain. The characteristics of the canyons change from west to east. Noctis Labyrinthus at the western end consists of numerous intersecting closed, linear depressions. The depressions are generally aligned parallel to faults in the surrounding plateau.

FIGURE 5   Intensely faulted ground in Tharsis. The faults are thought to have formed as a result of tension in the crust caused by the presence of the Tharsis bulge. Picture is 100 km across.

Farther east the depressions become deeper, wider, and more continuous to form roughly east–west-trending canyons (Fig. 6). Near the center of the system, three large canyons merge to form a depression 600 km across and in places over 6 km deep. Still farther east the canyons become shallower, fluvial features become more common both on the canyon floor and on the surrounding plateau, and finally the canyons end as the canyon walls become walls enclosing areas of chaotic terrain. In places, particularly in the central section, the canyons contain thick sequences of layered sediments.

The origin of the canyons is poorly understood. Straight sections of the canyon walls appear to be fault scarps, which suggests that the canyons are formed in part by downfaulting along faults radial to Tharsis. The importance of faulting is also suggested by graben parallel to the canyons. Some graben merge into lines of pits, which in turn merge to continuous depressions or canyons. However, processes other than faulting were clearly important in canyon formation. Most of the walls are dissected by gullies. In some places, branching side canyons have been cut into the adjacent plateau, and in other places the walls have collapsed in gigantic landslides. Thus, although most of the relief may have been created as a result of downfaulting, erosion has been important in enlarging the original structural depression.

The origin of the layered deposits within the canyons has been the subject of considerable debate. The consensus is that the deposits are water lain, and that at one time, and possibly several times, the canyons contained lakes. This is not surprising in that the canyons extend far below the depths at which the ground is expected to be frozen under present climatic conditions. Thus, groundwater below the permafrost zone could have seeped into the canyons and formed lakes, in which the observed sediments accumulated. Evidence of floods in the eastern ends of the canyons

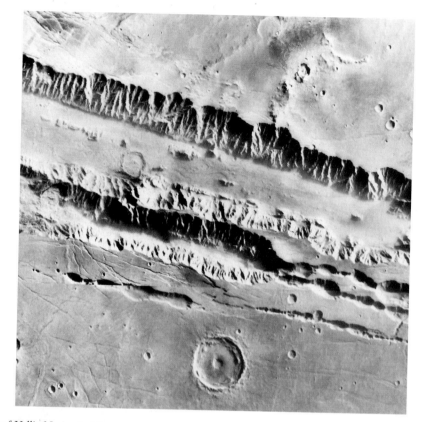

FIGURE 6 Section of Vallis Marineris. The upper canyon is 200 km across and 4 km deep. The lower canyon merges to the east with strings of closed depressions.

suggests that the lakes dissipated by draining catastrophically to the east, then northward, down large channels into the Chryse Basin.

VII. EROSION AND DEPOSITION

A. WATER

1. Outflow Channels

The role of water in the evolution of the Martian surface is one of the most intriguing problems of Martian geology. As already discussed, liquid water is unstable at the Martian surface: the temperatures are too low and the atmosphere too thin. Yet there is abundant evidence of water erosion, the most spectacular of which is the presence of large dry valleys, or outflow channels. Many such channels start in chaotic terrain

(Fig. 7). Individual areas of chaotic terrain may be as large as a few hundred kilometers across. Most are in a region south of the Chryse Basin and east and north of the canyons. The large channels near Chryse emerge full size from the chaotic terrain and extend down the regional slope for several hundreds to thousands of kilometers until they disappear in the northern plains. The channels, which may be up to several tens of kilometers across, generally have scoured floors and curvilinear walls and characteristically contain teardrop-shaped islands. Most have no tributaries. Not all the outflow channels start in chaotic terrain. Some in Elysium and around Hellas start adjacent to large volcanoes; others appear to start at faults.

Because they emerge full size and have no tributaries and because of their close resemblance to large terrestrial flood features, these channels are believed to have formed as a result of large floods. One possibility is that groundwater became trapped under high pressure beneath the thick permafrost zone and episodically burst out onto the surface. Eruptions of the trapped water could be triggered by meteorite impact, volcanic activity, or earthquakes. The size of the floods can be roughly estimated from the dimensions of the chan-

FIGURE 7   Chaos and channels south of the Chryse Basin. The areas of rubbly ground or chaotic terrain are the sources of large channels that extend northward down the regional slope. The ground is thought to have collapsed to form the rubble after removal of groundwater during channel formation. Picture is 750 km across.

nels. Many must have been enormous, with discharges in the range of 10^8 to 10^9 m^3/sec, 1000 to 10,000 times the discharge of the Mississippi. The discharges are so large compared with rates of freezing that flow could continue even under present climatic conditions. The fate of the water that flowed down the channels is uncertain. It presumably pooled in low-lying areas at the ends of the channels. Most of the large channels around Chryse terminate in the low-lying northern plains, where there are a variety of features suggestive of ice action. Terminal lakes in these areas would have quickly frozen under present climatic conditions, and possibly thick bodies of ice are still present at the site of the former lakes. Some channels end at low latitudes. In these cases, after freezing, the water would have slowly sublimed into the atmosphere and precipitated out at the poles.

2. Branching Valley Networks

Much of the ancient cratered uplands is dissected by branching valley networks that superficially resemble terrestrial river valleys (Fig. 8). In contrast to outflow channels, they have tributaries and increase in size downstream. They are also much smaller than outflow channels. The general consensus is that they formed

as a result of slow erosion by water erosion. Many have characteristics that suggest they formed by seepage of groundwater rather than by precipitation. It is unlikely that these valleys could form under present climatic conditions. The smaller headwater streams would rapidly freeze, thereby cutting off supply of water to the larger downstream valleys. Therefore, the networks have been taken as evidence that Mars was formerly much warmer and wetter than at present.

One possibility is that early Mars was much warmer and wetter because of the presence of a thick CO_2 atmosphere. Subsequently, but still early in the planet's history, Mars lost most of its thick atmosphere as the CO_2 reacted with the surface materials to form carbonates. As the atmosphere thinned, conditions changed to the cold, dry conditions that we observe today. Thus the old cratered upland is heavily dissected, but younger surfaces are not. There are, however, problems with this simple scenario. Although most of the dissected surfaces are old, occasional relatively young valley networks are also observed. Massive carbonate deposits have not been detected even though they should be close to the surface. Greenhouse warming by a CO_2 atmosphere on Mars is inefficient, particularly early in the planet's history when the energy output of the Sun is thought to have been smaller than at present. Because of these and other problems, some

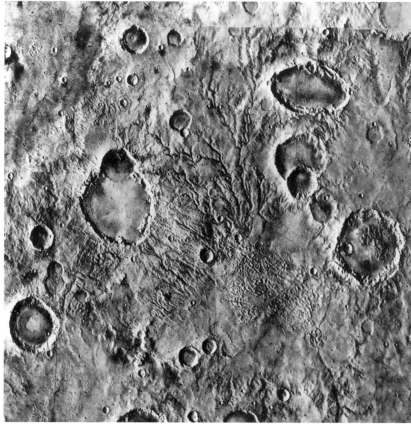

FIGURE 8   Branching valley networks in the southern uplands. The valleys in the center of the picture resemble terrestrial river valleys and are believed to be formed by slow erosion by running water. Picture is 300 km across.

radical alternative suggestions have been made, including massive episodic climate changes triggered by formation of the large floods. However, these suggestions raise more problems than they solve and the problem of Mars's climate history will remain unresolved until we get more information from the surface.

B. ICE

Although we have no direct evidence of the presence of ground ice, numerous surface features suggest that it may be common. The evidence is particularly strong in the 30–60° latitude belt. In this belt, many surface features such as ridges and crater rims appear, when viewed at high resolution, to be subdued and rounded, in contrast to their crisp, sharp appearance at low latitude. The rounding or softening has been attributed to the slow downhill movement or creep of near-surface materials as a result of the presence of ground ice. (As already indicated, ice is unstable at low latitudes, but stable below the surface at high latitudes.) Also in this

latitude band, debris flows extend 20–30 km away from cliffs and mountains. The simplest explanation is that debris shed from high ground contains ice that enables the eroded debris to flow. At low latitudes, because of the lack of ice, debris flows do not form; material eroded from cliffs remains adjacent to the cliff and protects it from further erosion. This process is particularly evident in what has been termed fretted terrain. These are sections of the plains/upland boundary in which flat-floored valleys extend deep into the uplands. Within these valleys, material has flowed away from the valley walls, down the valleys, and out over the adjacent plains. Again the simplest explanation is that the uplands material contains ice that facilitates flow of eroded debris.

A wide range of other observations, particularly in the low-lying northern plains, have been interpreted as the result of either ground ice or glaciers. These include polygonally fractured ground (analogous to arctic patterned ground?), closely spaced, curvilinear, parallel ridges and depressions (moraines?), local hollows (hollows left by removal of ice?), branching ridges (sites of former subglacial streams?), and striated ground (glacial scour?). The interpretation of many of these features is controversial, and we need more data before we can confidently determine whether the effects of ice action are trivial or widespread.

C. WIND

We know that the wind redistributes material across the Martian surface because we have observed dust storms and the changing pattern of surface markings that they cause. Given the violence of the dust storms, however, it is surprising that wind erosion is not more widespread. Wind appears mostly to move loose material around the surface; in most cases, wind erosion of the bedrock is trivial. Tails in the lee of craters are the most widespread indicators of wind action. Crater tails lighter than their surroundings are thought to be accumulations of bright material like those in the lee of obstacles at the Viking landing sites. Dark crater tails, and irregular dark patches in and beside craters, may be either areas scoured of loose material or areas where the loose material forms dunes or drifts. Around the north pole is a dark collar formed by a vast dune field in which individual crescentic dunes are clearly visible in high-resolution pictures. Farther south, in the northern plains, are a variety of features that suggest repeated deposition and removal of loose material by the wind.

Though the effects of wind erosion in most places are trivial, locally the effects may be substantial. This is particularly true where friable deposits are at the surface. In Southern Amazonis, and south of Elysium Planitia, thick, easily erodible deposits overlie the plains/upland boundary. Eroded into these deposits are arrays of curvilinear, parallel grooves that resemble terrestrial wind-cut grooves called yardangs. Wherever such wind erosion is observed, other evidence indicates that what is being eroded is a deposit that blankets the bedrock. Erosion of bedrock formations such as lava flows is minute. Wind may be ineffective as an erosive agent because of the lack of abrasive debris for the wind to move. On Earth, quartz sand is an effective erosive agent, and this may be rare on Mars. The dust lifted into the atmosphere during dust storms probably consists of soft, weathered materials, such as clays, that have little abrasive capacity.

VIII. POLES

During fall and winter, CO_2 condenses out onto the polar regions to form a seasonal cap that can extend as far equatorward as 40° latitude. In summer the CO_2 cap sublimes; that in the north sublimes completely. At this time, temperatures rise from the frost point of CO_2 (150°K) to the frost point of H_2O (200°K), and the amount of water vapor over the cap increases dramatically. At the south pole, the CO_2 cap did not completely sublime, when observed by the *Viking* orbiters, temperatures remained at 150°K, and no increase in water vapor was detected. These observations indicate that water ice is present under the CO_2 cap in the north. Because the CO_2 did not evaporate completely in the south, we cannot say whether or not a similar water ice cap is present in the south.

At both poles, evaporation of the CO_2 cap in summer reveals layered deposits several kilometers thick, and extending out to roughly the 80° latitude circle. Individual layers are best seen in the walls of valleys cut into the sediments, where layering is observed at a range of scales down to the resolution limit of our best pictures (Fig. 9). The frequency of impact craters on the upper surface of the deposits suggests a relatively young age, on the order of 10^8 years.

The poles act as a cold trap for water. Any water entering the atmosphere as a result of geologic processes, such as volcanic eruptions or floods, will ultimately be frozen out at the poles. The poles may also be a trap for dust, in that dust can be scavenged out of the atmosphere as CO_2 freezes onto the pole in the fall and winter. The layered deposits are, therefore, probably mixtures of dust and ice. The layering is thought to be caused in some way by periodic changes in the thermal and wind regimes at the poles, induced by variations in the planet's orbital and rotational motions. Precession and variations in eccentricity and obliquity are probably all important (see Section II, A). These cyclical motions affect the stability of H_2O and CO_2 at the poles and the dynamics of the atmosphere, hence the belief that they are in some way responsible for the observed layering.

IX. INTERIOR

Our perception of the interior of Mars is inferred indirectly from theoretical modeling, meteorites, comparisons with the Earth, and the shape, mass, and moment of inertia of the planet. Mars, like the Earth, is probably differentiated into crust, mantle, and core. Isotopic studies of SNC meteorites indicate that the differentiation took place at the end of accretion, around 4.5 billion years ago. Modeling of the condensation of the early solar nebula suggests that, during planetary

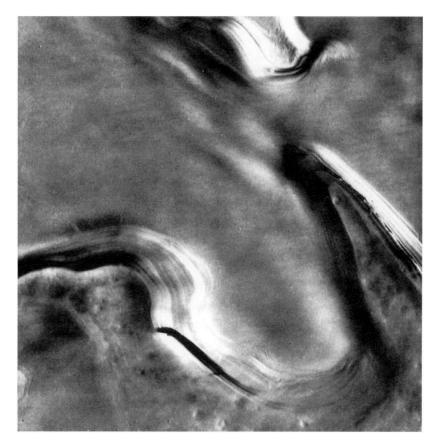

FIGURE 9 Layered terrain at the north pole. The layers are clearly visible along the escarpment in the upper half of the picture. The scarcity of impact craters indicates a relatively young age. Picture is 45 km across.

formation, Mars, being farther from the Sun, should have incorporated a higher proportion of moderately volatile elements, such as sulfur, sodium, and potassium, than did the Earth. If so, then Mars should have a sulfur-rich core, and elements such as copper that readily dissolve in sulfur-rich melts should have been largely scavenged from the mantle during core formation. This appears to be confirmed by SNC meteorites, for the mantle-derived melts from which the meteorites crystallized are depleted in such elements.

In principle, the size of the core can be determined from the mass of the planet and its moment of inertia. However, the moment of inertia is poorly determined, and we need to know the density of the mantle and the density of the core to accurately determine the core size. The mantle density can be estimated with some confidence, because it largely depends on the Fe/Fe + Mg ratio, which we know from SNC meteorites, but we have no way of determining the density of the core, which depends largely on the sulfur content. All we can say at present is that the possibilities range from a small S-poor core, consisting mostly of metallic iron, to a large S-rich core, consisting mostly of FeS. Determination of the size and composition of the core must await establishment of a seismic network.

We know even less about the crust of Mars. The Earth's crust is 30–40 km thick and granitic under the continents, and 5–10 km thick and basaltic under the oceans. The Moon's crust is anorthositic. The nature of the Martian crust is almost completely unknown. The planet may have a primitive crust that formed at the time of planetwide differentiation, 4.5 billion years ago, but if so we know neither its composition nor its thickness. This putative primitive crust has been pervasively injected by younger basaltic rocks, as indicated by numerous volcanic landforms and SNC meteorites. Isotopic data show that crustal material has not been mixed into the mantle as occurred within the Earth. The crust and mantle have remained chemically distinct since differentiation, except for injection of volcanic melts into the crust. The lithosphere, or rigid outer rind of the planet, includes both crust and upper mantle and is estimated to be 50–200 km thick, on the basis of the warping of the lithosphere under volcanic loads.

Modeling of the dynamics of the interior is hindered by lack of knowledge of the dimensions of the core and the compositions of the core and mantle. Recent modeling suggests, however, that the mantle may be convecting, with plumes of hot mantle rising from the core–mantle boundary. The complementary downward flow is by sheet-like movement over large areas. Tharsis and Elysium may be situated over two large plumes. Current average surface heat flow is estimated to be 40 mW/m^2, as compared with the Earth's 75 mW/m^2.

X. SUMMARY

. . . .

Mars is a geologically variegated planet on which have operated many of the geologic processes familiar to us here on Earth. It has been volcanically active throughout its history, the crust has experienced extensive deformation, largely as a result of massive surface loads, and the surface has been eroded by wind, water, and ice. Despite these similarities, the evolution of Mars and the Earth have been very different. The lack of plate tectonics on Mars has prevented the formation of linear mountain chains and the cycling of crustal materials through the upper mantle. Climatic conditions that inhibit the flow of liquid water across the surface have greatly limited the amount of water erosion, and the rates of volcanism, deformation, and erosion by wind and ice have been far lower than on Earth. As a result, a geologic record is preserved that spans almost the entire history of the planet. Considerable uncertainty remains about the interior of the planet, its climatic history, and the role of water in the evolution of the surface. These uncertainties will remain until we can perform more *in situ* observations at the surface and return a variegated set of samples to Earth for comprehensive analysis.

BIBLIOGRAPHY

Baker, V. R. (1982). "The Channels of Mars." Univ. Texas Press, Austin.
Beatty, J. K., and Chaikin, A. (1990). "The New Solar System." Sky Publishing, Cambridge, Mass.
Carr, M. H. (1981). "The Surface of Mars." Yale Univ. Press, New Haven, Conn.
Carr, M. H. (1996). "Water on Mars." Oxford Univ. Press, New York.
Hamblin, W. K., and Christiansen, E. H. (1990). "Exploring the Planets." Macmillan, New York.
Murray, B., Malin, M. C., and Greeley, R. (1980). "Earthlike Planets." Freeman, San Francisco.
Wilford, J. N. (1990). "Mars Beckons." Knopf, New York.

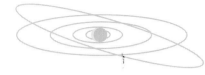

PHOBOS AND DEIMOS

I. Exploration

II. Orbits and Gravitational Environment

III. Surface Features

IV. Composition

V. Origin and History

Peter C. Thomas
Cornell University

GLOSSARY

Albedo: Ratio of the amount of light scattered by a surface to the amount incident upon it.

Ejecta: Material thrown out of an impact crater during its formation; nearly all ejecta is from the target, though some is from the projectile.

Regolith: Loose, fragmental material at the surface of a planet or satellite.

Phobos and Deimos are the two known natural satellites of Mars. They were discovered in 1877 by Asaph Hall at the U.S. Naval Observatory. Apart from the Earth's moon, they are the only natural satellites of the four terrestrial planets. Because of their small sizes and proximity to Mars, they are very difficult to study from the Earth; most of our knowledge of these objects comes from spacecraft data.

I. EXPLORATION

Phobos was first detected in spacecraft images by *Mariner 7* during its flyby of Mars in 1969. *Mariner 9* acquired the first detailed images in 1971–1972, and Viking in 1976–1980 made many flybys of both Phobos and Deimos at distances ranging from 50 km to a few thousand kilometers and mapped most of the satellites' surfaces at resolutions of 3–100 m. The *Phobos 2* spacecraft acquired some low-resolution images and spectroscopic data on Phobos in 1989.

II. ORBITS AND GRAVITATIONAL ENVIRONMENT

The Martian satellites are in nearly circular, equatorial orbits around Mars (Table I). Phobos orbits faster than Mars rotates, thus seen from the surface of Mars it would rise in the west. Because of its close orbit, it could only be seen from latitudes less than 68°. As Phobos is within the synchronous orbit distance, torques from the tidal bulge that it raises on Mars remove energy from Phobos, causing its orbit to gradually contract with time. This shrinkage of Phobos's orbit has been measured, and the observed rate implies that the satellite's orbit may intersect the surface of Mars within 30–100 million years. Depending on its internal strength, it might become disaggregated before then, forming something of a ring. Tidal effects cause Deimos's orbit to expand because it is outside the synchronous point, but at these much greater distances from Mars the tidal evolution rate is extremely low and has proved essentially impossible to measure. Deimos's orbit has probably changed very little over the age of the solar system.

Extrapolation of the orbits back in time faces many uncertainties, chief among them several possible resonances of Phobos's orbital period with the spin period of Mars, which would cause rapid changes in the eccen-

TABLE I
Physical Characteristics of Phobos and Deimos

	Phobos	Deimos
a (orbit semimajor axis, km)	9378	23,459
Period (hr)	7.65	30.3
e (orbital eccentricity)	0.01515 (±0.00004)	0.000196 (±0.000034)
i (orbital inclination, deg)	1.068	0.8965
Axes of fit ellipsoid (km)	13.05, 11.10, 9.30	7.8, 6.0, 5.1
Mean density (g/cm^3)	1.9 ± 0.1	1.8 ± 0.3
Surface gravity (cm/s^2)	0.3–0.6	0.2–0.3

tricity and inclination of Phobos's orbit. Other influences on how fast orbits change may include occasional tumbling, which has been suggested as a likely occurrence for all irregularly shaped satellites.

The Martian satellites have unusual gravitational environments. Rotational and tidal accelerations cause gravity to vary considerably over their surfaces, especially on Phobos, where the surface acceleration is as little as 0.3 cm/s^2 and as great as 0.6 cm/s^2. It is lowest near the points closest and farthest from Mars as they are subject to the greatest rotational and tidal forces and have the lowest component of self-gravity from the mass of Phobos. Escape velocity (to Mars orbit) from Phobos ranges from about 3 to 10 m/s; it is about 6 m/s over most of Deimos.

III. SURFACE FEATURES

The main surface features of Phobos are impact craters and long linear depressions called grooves. The craters are generally similar to impact craters on the Moon, which have been studied intensively by direct sampling and remote sensing. The largest crater on Phobos is Stickney (named for the wife of the discoverer), 9.6 km across, which is 90% the mean radius of the satellite.

The grooves are the most distinctive surface features on Phobos (Fig. 1). They are mostly 10–20 m deep and usually 100–200 m across and a few kilometers in length, though the largest are 100 m deep and nearly a kilometer wide; the longest extend over 15 km. They occur in patterns of nearly parallel members, with some sets of grooves crossing others. Their smooth slopes, chutes, and other lineations trending downslope, and overall morphology of arrayed and merging pits,

strongly suggest formation in loose materials (the regolith).

Theories of groove origins fall into two groups: disturbance of the regolith by fractures deeper in the body of Phobos, and effects on the regolith of ejecta from large impacts. The geometric pattern of the grooves is regarded as evidence for a fracture role in their formation; the presence of some raised rims on grooves is cited as evidence of an impact role. Fracturing might be due to the large crater Stickney, or older large impacts, or to influence of the tidal forces on Phobos. Attempts to match the groove pattern with expected stresses have not been completely successful, however. The origin of these features remains a puzzle, though it would appear they are at least in part the indirect results of the very large impacts on Phobos.

Both Phobos and Deimos have blocks up to 100 m across on the surface near likely source craters. Some of the blocks on Deimos are arranged in patterns that suggest they were larger pieces that broke apart on impact (Fig. 2). Well-defined, contiguous ejecta blankets are not visible on either satellite. There is some evidence of a hummocky ejecta blanket east of the crater Stickney (see Fig. 1), and some craters show bright raylike patterns indicative of effects of ejecta near the crater. The difficulty of finding obvious ejecta blankets may arise from blanketing by the fraction of crater ejecta that escapes temporarily into Mars orbit and then reimpacts the surface at fairly low velocities.

The most distinctive features on Deimos are long, tapered albedo features (Fig. 3). These are about 30% brighter than the surroundings, but are still quite dark (8% versus 6% albedo). They trend downslope and are apparently the result of loose material moving under the influence of gravity, which on Deimos is only about 1/3000th that on Earth. Thermal expansion and

FIGURE 1 View from *Viking Orbiter* of grooves and craters on Phobos. The hummocky materials in the lower right quarter of the image are probably material thrown out of the 10-km crater Stickney, which is just beyond the lower right corner.

contraction every Deimos "day" or micrometeorite impacts may provide the impetus for the motion of the material, but the gravity directs it downslope.

Deimos's largest well-defined crater is less than 3 km across, but an 11-km-wide depression near the south pole has been suggested as an impact scar. Although the surface is generally smoother and the craters are more eroded and filled in than on Phobos, there are nearly as many craters per unit area on Deimos as on Phobos.

Infrared thermal measurements have shown that both Phobos and Deimos have fine-grained loose material on their surfaces, but these data only tell about the upper few centimeters of the surface. The morphology of the grooves suggests that there is well over 100 m of loose material in many areas of Phobos.

Some craters on Deimos have been filled with over 10 m of debris. The regoliths are presumably the comminuted material produced by formation of impact craters. The visible crater population on Phobos is probably insufficient to produce the total amount of regolith, and thus one can conclude that the satellite has suffered many more impacts than are presently visible on the surface. Some of the indentations on the satellite may be the large, very degraded craters that

produced the debris. The smooth surface of Deimos also suggests that thick ejecta accumulations have buried substantial topography.

IV. COMPOSITION

The composition of the Martian satellites remains controversial. Compositon must be inferred from their mean densities and from the spectral characteristics of their surfaces. As shown in Table I, both satellites have mean densities slightly less than 2 g/cm³. Spectra of the satellites are difficult to obtain from Earth because of scattered light, and the only spacecraft designed to study these satellites, *Phobos 2*, failed after obtaining only a few low-resolution data. Nonetheless, the *Phobos 2* data and recent Hubble Space Telescope data have shown Phobos and Deimos to be redder than some previous analog materials of C-type objects. They may be more similar to D-type objects, or some other materials slightly brighter in the near infrared. If they are made of some other silicate materials as might be suggested by some spectral similarity to dark areas on the Moon, it would imply very high porosities. [*See* ASTEROIDS.]

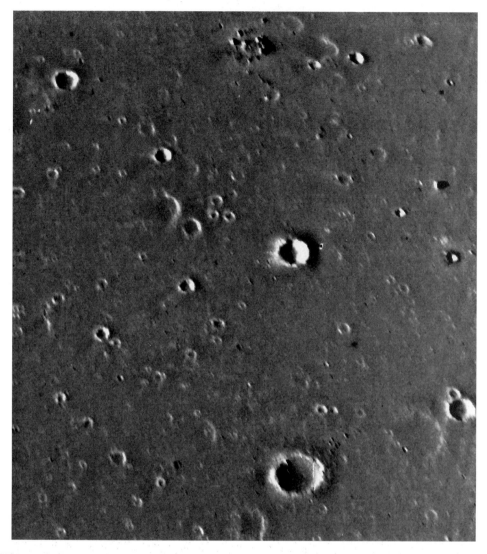

FIGURE 2 *Viking Orbiter* image of Deimos showing one of the ridge areas with many degraded craters and ejecta blocks from craters. The circle of blocks at the top may have originated when a larger block fell back to Deimos and fragmented.

The *Phobos 2* data also show there is some heterogeneity in the surface properties that may be related to excavation of different materials by impacts. The differences in color appear to be related to material ejected from crater Stickney; because the color pattern is not symmetric about the crater, it may be due to excavation of different materials within the large crater. [*See* PLANETARY IMPACTS.]

The low densities suggest that the satellites might have substantial porosity, especially if they contain any substantial fraction of mafic minerals. Even volatile-rich carbonaceous meteorites have densities over 2.2 g/cm³, and meteorites with fewer volatiles can have densities of well over 3 g/cm³. If the satellites were composed of the more volatile-rich material, porosities could be as low as 10–30%. But a composition of the denser chondritic material would require nearly 50% porosity for the whole satellite. Inclusion of ice, which could be stable at depth over geologic time, could reduce the bulk density. They do not show surface water of hydration. This latter characteristic makes it difficult to reconcile the bulk and surface properties, because the porosities of about 50% needed for the volatile-poor material may be unreasonably high. Thus, even though Phobos and Deimos appear to have surfaces that are similar to some meteorite and asteroidal surfaces, their bulk composition remains unresolved. [*See* METEORITES.]

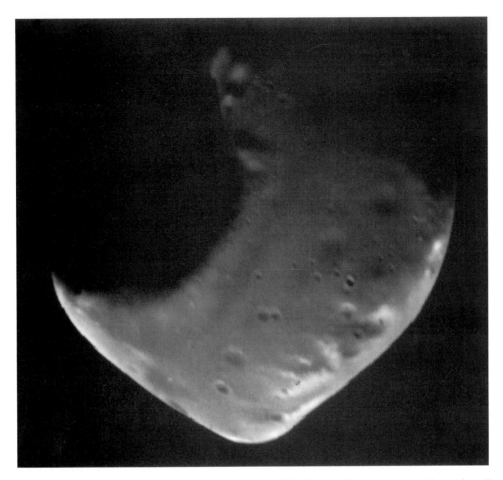

FIGURE 3 *Viking Orbiter* image of Deimos showing the irregular shape and brighter markings tapering away from ridges. Contrast between brighter and darker material is about 30%.

V. ORIGIN AND HISTORY

The ages of the surfaces of planets and satellites that are studied remotely are estimated from the number of impact craters that they have accumulated. This is a calculation made very difficult by imperfect knowledge of the rate of crater formation (the rate of impact of asteroids and comets of particular sizes) on objects other than Earth and the Moon. The high density of craters on Phobos and Deimos suggests that the surfaces are geologically old, but may still be substantially younger than the original accretion of the objects or even some impact-related fragmentation. By comparison to the Martian surface, it is likely that the surfaces of Phobos and Deimos record impacts occurring over more than half the age of the solar system. [*See* MARS: SURFACE AND INTERIOR.]

The origins of these satellites are not convincingly explained. Hints that they are composed of D-type or similar materials that apparently condensed farther from the Sun than did Mars have been used to support an origin in the asteroid belt, followed by capture into Mars orbit. In most caputre scenarios, asteroids passed close to Mars while it had a large protoatmosphere that could slow asteroidal objects sufficiently to capture into orbits. The objects that would survive in Mars orbit would be those captured just as the extended gas envelope was collapsing so that it would not last long enough to cause the satellite to spiral into Mars itself. The equatorial orbits of the satellites, and the presence of Phobos inside the synchronous orbit distance, with Deimos outside the synchronous orbit point, are very difficult to explain with capture mechanisms, however. If Phobos ever had a highly eccentric orbit (required in capture) it would have intersected that of Deimos and over short time scales would have collided with Deimos. Thus Phobos would have to be captured first, and its orbit circularized, before Deimos was captured. The short time scales for efficacy of protoatmosphere capture make it unlikely that Deimos could be captured if Phobos had been in orbit long enough to gain a circular orbit. Additionally, the distant, circular, virtually nonevolving orbit of Deimos by itself is difficult to explain by capture. Thus, although we cannot confidently say the Martian satellites formed in association with Mars, an origin in the asteroid belt remains very difficult to explain. A better knowledge of their compositions would go a long way toward showing a capture or *in situ* origin. [*See* MARS: ATMOSPHERE AND VOLATILE HISTORY.]

BIBLIOGRAPHY

Burns, J. A. (1992). Contradictory clues as to the origin of the Martian moons. *In* "Mars" (H. H. Kieffer, B. M. Jakosky, C. W. Snyder, and M. S. Matthews, eds.). Univ. Arizona Press, Tucson.

Burns, J. A. and Matthews, M. (1986). "Satellites." Univ. Arizona Press, Tucson.

Murchie, S., and Erard, S. (1996). Spectral properties and heterogeneity of Phobos from measurements by *Phobos 2. Icarus* **123,** 63–86.

Thomas, P., Veverka, J., Bell, J., Lunine, J., and Cruikshank, D. (1992). Satellites of Mars: Geologic history. *In* "Mars" (H. H. Kieffer, B. M. Jakosky, C. W. Snyder, and M. S. Matthews, eds.). Univ. Arizona Press, Tucson.

ATMOSPHERES OF THE GIANT PLANETS

I. Introduction

II. Chemical Composition

III. Clouds and Aerosols

IV. Dynamical Meteorology of the Troposphere and Stratosphere

V. Energetic Processes in the High Atmosphere

Robert A. West

*Jet Propulsion Laboratory,
California Institute of Technology*

GLOSSARY

Adiabatic temperature gradient: Rate of change of temperature with altitude in the part of the atmosphere (most of the atmosphere deeper than the altitude where the 2-bar pressure level is reached for the giant planets) where there is essentially no heating from sunlight. As gas parcels rise or fall owing to slight density differences, they will expand or compress with changing pressure, and the temperature change during expansion or compression is determined by the heat capacity of the gas, which in turn depends on the composition and (for molecular hydrogen) the relative amounts of *ortho*- and *para*-hydrogen.

Aerosol: In atmospheric physics, aerosol is a generic name for any particle (cloud, dust, haze) suspended in the air, although in the earth science community the term is usually restricted to apply to haze rather than cloud particles.

Aurora: Atmospheric glow caused by energetic particles from the magnetosphere that slam into the upper atmosphere, exciting the atoms they hit and causing them to radiate.

Chromophore: Any coloring material.

Deuterium/deuterated: Heavy form of the hydrogen atom, consisting of one proton and one neutron. A deuterated molecule (such as CH_3D, deuterated methane) has one or more deuterium atoms in place of hydrogen.

Homopause: Level in an atmosphere, above the stratosphere, at which gases cease being uniformly mixed and separate by diffusion, with the lighter elements diffusing upward.

Joule heating: Heating that occurs when a current flows through a resistive medium. It is a familiar phenomenon in everyday experience (it's the heat an electric toaster produces). In the high atmospheres of the giant planets, it may be an important process in heating the atmosphere to high temperature as currents of charged particles driven by magnetospheric electric fields collide with the neutral atmosphere atoms, which provide a resistance.

Latent heat: Heat that is released or absorbed during a phase change (vapor or liquid or ice to ice or liquid or vapor). Latent heat contributes to heating and cooling the atmosphere in regions where ice and liquid clouds form and dissipate. It also contributes to the heat capacity of a parcel of gas/cloud and therefore influences the adiabatic temperature gradient.

Meridional circulation: Motions of the atmosphere in the plane defined by the vertical and latitudinal coordinates. Atmospheric motions in the vertical and north/south directions participate in the meridional circulation.

Opacity: Measure of the ability of a substance (gas or aerosol particle) to block light. High-opacity materials block light very well. Same as optical depth.

Optical depth: Measure of the attenuation of the extinction of light along a path through the atmosphere, due to scattering and absorption by cloud particles or gas. The optical depth is given by $N\sigma l$, where N is the number density of particles, σ is the scattering or absorption cross section, and l is the path length traversed by the light beam.

Photolysis: Process that occurs when a molecule absorbs light of sufficiently high energy (usually ultraviolet light) and breaks apart.

Plasma: Fourth state of matter (the first three being solid, liquid, and gas) in which the material consists of ions (atoms stripped of one or more of their electrons) and electrons. High temperatures are required to maintain separation of the ions and electrons. The magnetospheres of the giant planets are filled with plasma (mainly protons and electrons from hydrogen). The electrically charged ions and electrons spiral along magnetic field lines in the magnetosphere and are dragged along with the magnetic field as it rotates with the planet.

Ram pressure: Pressure that the solar wind exerts on the magnetosphere of a planet as it flows around the magnetosphere at the flow velocity of the wind (several hundred kilometers per second). The magnetosphere acts as an obstacle, and pressure is exerted at the magnetopause, a shock front that forms a barrier between the solar wind and the magnetosphere.

Stratosphere: Upper layer of a planetary atmosphere, above the troposphere, characterized by a vertical temperature gradient that is stable against convection.

Troposphere: Deep levels of the atmosphere where the temperature gradient is governed by convection and is close to the adiabatic temperature gradient. The tropopause is defined to be the altitude or pressure level that corresponds to the top of the troposphere and the bottom of the stratosphere. It is very difficult to determine the altitude at which the temperature gradient departs from adiabatic and so a less rigorous but more easily determined location is often used. The tropopause is generally equated to the altitude of the temperature minimum near the 100-mbar pressure level.

Vorticity: Vector curl of the flow velocity. In dynamic meteorology, vorticity is often applied exclusively to the vertical component of the curl of the horizontal flow with respect to the planet, which, for zonal mean motion, is approximately equal to the negative latitudinal flow shear. Spots having cyclonic vorticity rotate counterclockwise in the northern hemisphere and clockwise in the southern hemisphere.

T he atmospheres of the giant planets—Jupiter, Saturn, Uranus, and Neptune—are very unlike those of the Earth, Mars, and Venus. They are composed mainly of hydrogen and helium, with some trace species, the most abundant of which are water, ammonia, and methane. They are cold enough to form clouds of ammonia and hydrocarbon ices, and extend deep into the interior of the planet, and indeed a significant fraction of the planet's mass may be responsible for the near-surface winds. The winds are primarily east–west (zonal) jets that alternate with latitude. Superimposed on the jets are spots of all sizes up to about three Earth diameters. Some of them, like Jupiter's Great Red Spot, are remarkably long-lived. At the highest altitudes, powerful auroras, as well as some still mysterious processes, heat the atmospheres to temperatures higher than current models can explain.

I. INTRODUCTION

To be an astronaut explorer in Jupiter's atmosphere would be strange and disorienting. There is no solid ground to stand on. The temperature would be comfortable at an altitude where the pressure is eight times that of Earth's surface, but it would be perpetually hazy overhead, with variable conditions (dry or wet, cloudy or not) to the east, west, north, and south. One would need to carry oxygen as there is no free oxygen,

TABLE I
Physical Properties of the Giant Planets

Property	Jupiter	Saturn	Uranus	Neptune
Distance from the Sun (Earth distance = 1[a])	5.2	9.6	19.2	30.1
Equatorial radius (Earth radius = 1[b])	11.3	9.4	4.1	3.9
Planet total mass (Earth mass = 1[c])	318.1	95.1	14.6	17.2
Mass of gas component (Earth mass = 1)	254–292	72–79	1.3–3.6	0.7–3.2
Orbital period (years)	11.9	29.6	84.0	164.8
Length of day (hours, for a point rotating with the interior)	9.9	10.7	17.4	16.2
Axial inclination (degrees from normal to orbit plane)	3.1	26.7	97.9	28.8
Surface gravity (equator–pole, m s^{-2})	(22.5–26.3)	(8.4–11.6)	(8.2–8.8)	(10.8–11.0)
Ratio of emitted thermal energy to absorbed solar energy	1.7	1.8	~1	2.6
Temperature at the 100-mbar level (K)	110	82	54	50

[a] Earth distance = 1.5×10^8 km, [b] Earth radius = 6378 km, [c] Earth mass = 6×10^{24} kg.

and special clothing to protect the skin against exposure to ammonia, hydrogen sulfide, and ammonium hydrosulfide gases, which form clouds and haze layers higher in the atmosphere. A trip to high latitudes would offer an opportunity to watch the most powerful, vibrant, and continuous auroral displays in the solar system. On the way one might pass through individual storm systems the size of Earth or larger, and be buffeted by strong winds alternately from the east and west. One might be sucked into a dry downwelling sinkhole like the environment explored by the *Galileo* probe. The probe fell to depths where the temperature is hot enough to vaporize metal and rock. It is now a part of Jupiter's atmosphere.

Although the atmospheres of the giant planets share many common attributes, they are at the same time very diverse. The roots of this diversity can be traced to a set of basic properties, and ultimately to the origins of the planets. The most important properties that influence atmospheric behavior are listed in Table I. The distance from the Sun determines how much sunlight is available to heat the upper atmosphere. The minimum temperature for all of these atmospheres occurs near the 100-mbar level and ranges from 110 K at Jupiter to 50 K at Neptune. The distance from the Sun and the total mass of the planet are the primary influences on the bulk composition. All the giant planets are enriched in heavy elements, relative to their solar abundances, by factors ranging from about 5 for Jupiter to 1000 for Uranus and Neptune. The latter two planets have a large fraction of elements (O, C, N, and S) that were the primary constituents of ices in the early solar nebula.

The orbital period, axial tilt, and distance from the

Sun determine the magnitude of seasonal temperature variations in the high atmosphere. Jupiter has weak seasonal variations; those of Saturn are much stronger. Uranus is tipped such that its poles are nearly in the orbital plane, leading to more solar heating at the poles than at the equator when averaged over an orbit. The ratio of radiated thermal energy to absorbed solar energy is diagnostic of how rapidly convection is bringing internal heat to the surface, which in turn influences the abundance of trace constituents and the morphology of eddies in the upper atmosphere. Vigorous convection from the deeper interior is responsible for unexpectedly high abundances of several trace species on Jupiter, Saturn, and Neptune, but convection on Uranus is sluggish. All of these subjects are treated in more detail in the sections that follow.

II. CHEMICAL COMPOSITION

This section is concerned with chemical abundances in the observable part of the atmosphere, a relatively thin layer of gas near the top (pressures between about 5 bars and a fraction of a microbar). To place the subject in context, some mention will be made of the composition of the interior. For more information on the interior [See INTERIORS OF THE GIANT PLANETS.]

The bulk composition of a planet cannot be directly observed, but must be inferred from information on its mean density, gravity field, and the abundances of constituents that are observed in the outer layers. The more massive planets were better able to retain the light elements during their formation, and so the bulk composition of Jupiter resembles that of the Sun. When the giant planets formed, they incorporated relatively more rock and ice fractions than a pure solar composition would allow, and the fractional amounts of rocky and icy materials increase from Jupiter through Neptune. [See THE ORIGIN OF THE SOLAR SYSTEM.] Most of the mass of the heavy elements is sequestered in the deep interior. The principal effects of this layered structure on the observable outer layers can be summarized as follows.

On Jupiter the gas layer (a fluid molecular envelope) extends down to about 40% of the planet's radius, where a phase transition to liquid metallic hydrogen occurs. Fluid motions that produce the alternating jets and vertically mix gas parcels may fill the molecular envelope but probably do not extend into the metallic region. Thus, the radius of the phase transition provides a natural boundary that may be manifest in the latitudinal extent of the zonal jets (see Section IV), whereas vertical mixing may extend to levels where the temperature is quite high. These same characteristics are found on Saturn, with the additional possibility that a separation of helium from hydrogen is occurring in the metallic hydrogen region, leading to enrichment of helium in the deep interior and depletion of helium in the upper atmosphere.

Uranus and Neptune contain much larger fractions of ice- and rock-forming constituents than do Jupiter and Saturn. A large water ocean may be present in the interiors of these planets. Aqueous chemistry in the ocean can have a profound influence on the abundances of trace species observed in the high atmosphere.

In the observable upper layers the main constituents are molecular hydrogen and atomic helium, which are well mixed up to the homopause level, where the mean free path for collisions becomes large enough that the lighter constituents are able to diffuse upward more readily than heavier ones. Other constituents are significantly less abundant than hydrogen and helium, and many of them condense in the coldest regions of the atmosphere. Figure 1 shows how temperature varies with altitude and pressure, and the locations of the methane, ammonia, and water cloud layers.

The giant planets have retained much of the heat generated by their initial collapse from the solar nebula. They cool by emitting thermal infrared radiation to space. Thermal radiation is emitted near the top of the atmosphere, where the opacity is low enough to allow infrared photons to escape to space. In the deeper atmosphere, heat is transported by convective fluid motions from the deep, hot interior to the colder outer layers. In this region, upwelling gas parcels expand and subsiding parcels contract adiabatically (e.g., with

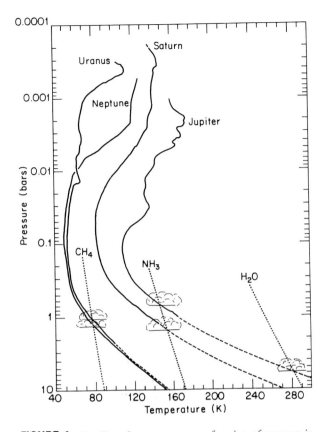

FIGURE 1 Profiles of temperature as a function of pressure in the outer planet atmospheres derived from measurements by the Voyager Radio Science experiment (solid curves). The dashed parts of the temperature profiles are extrapolations using the adiabatic lapse rate. At high altitudes (not shown), temperatures rise to about 1200 K for Jupiter, 800 K for Saturn and Uranus, and 300 K for Neptune. The dotted lines show vapor pressure curves divided by observed mixing ratios for water, ammonia, and methane. Condensate clouds are located where the solid and dotted curves cross. [From Gierasch and B. Conrath (1993). *J. Geophys. Res.* **98**, 5459–5469. Copyright American Geophysical Union.]

negligible transport of heat through their boundaries by radiation or conduction). Therefore temperature depends on altitude according to the adiabatic law $T = T_0 + C(z - z_0)$, where T_0 is the temperature at some reference altitude z_0, C is a constant (the adiabatic lapse rate) that depends on the gas composition, and z is altitude. The adiabatic lapse rate for dry hydrogen and helium on Jupiter is -2.2 K/km. On Uranus it is -0.8 K /km. The adiabatic lapse rate is different in regions where a gas is condensing or where heat is released as *ortho*-hydrogen and is converted to *para*-hydrogen. Both of these processes are important in the giant planet atmospheres at pressures between about 30 bars and 0.1 bar.

Hydrogen is the main constituent in the observable part of the giant planet atmospheres, but not until recently was it recognized as especially important for thermodynamics. The hydrogen molecule has two ground-state configurations for its two electrons. The electrons can have their spins either parallel or antiparallel, depending on whether the spins of the nuclei are parallel or antiparallel. These states are called the ortho and para states. Transitions between ortho and para states are slow because, unlike most molecules, the nuclear spin must change when the electron spin changes. At high temperature (about 270 K or higher), the ortho:para relative abundance is 3:1. At lower temperature, a larger fraction is converted to the para state. Heat release from conversion of *ortho*- to *para*-hydrogen can act in the same way as latent heat release from condensation. The relative fractions of *ortho*- and *para*-hydrogen are observed to be close to thermal equilibrium values in the giant planet atmospheres, leading to the question of how equilibrium is achieved. Catalytic reactions on the surfaces of aerosol particles are thought to be important in equilibrating the ortho and para states.

Temperature follows the adiabatic law at pressures deeper than about 2 bars. The atmospheric temperature would drop at the adiabatic rate to near absolute zero at the top of the atmosphere were it not for sunlight, which heats the upper atmosphere. Sunlight penetrates to pressure levels near 20 bars, depending on how much overlying cloud and haze opacity is present. The competition between convective cooling and solar heating produces a temperature minimum near the 100-mbar level (the tropopause). At pressures between about 100 and 0.1 mbar, the temperature is determined primarily by equilibrium between thermal radiative cooling and solar heating. At even lower pressures, other processes, including auroral heating, dump energy into the atmosphere and produce higher temperatures. More will be said about this in Section V.

The current inventory of observed gaseous species is listed in Table II. Molecular hydrogen and helium are the most abundant. Helium is in its ground state in the troposphere and stratosphere and therefore does not produce spectral lines from which its abundance can be determined. Its mixing ratio for Saturn, Uranus, and Neptune is inferred from its influence on the broad collision-induced hydrogen lines near the 45-μm wavelength, and from a combined analysis of the infrared spectrum and refractivity profiles retrieved from spacecraft radio occultation measurements. Helium on Jupiter is accurately known from measurements made by the *Galileo Probe*, which descended through the atmosphere. It is a little smaller than the mixing ratio inferred for the primitive solar nebula from which the planets formed. Helium is substantially depleted in

Saturn's upper atmosphere, consistent with the idea that helium is precipitating out in the metallic hydrogen region. For Uranus and Neptune the helium mixing ratio is close to the mixing ratio (0.16) in the primitive solar nebula. There is still some uncertainty in the helium mixing ratio, especially for Neptune, because additional factors, such as aerosol opacity and molecular nitrogen abundance, affect the shapes of the collision-induced spectral features and we do not have a completely consistent set of values for all of these parameters.

Mixing ratios of deuterated hydrogen and methane (HD and CH_3D) also provide information on the formation of the planets. Deuterium, which once existed in the Sun, has been destroyed in the solar atmosphere, and the best information on its abundance in the primitive solar nebula comes from measurements of the giant planet atmospheres. On Jupiter the deuterium mixing ratio is thought to be close to that of the primitive solar nebula. On Uranus and Neptune it is enhanced because those planets incorporated relatively more condensed material on which deuterium preferentially accumulated through isotopic fractionation. Isotopic fractionation (the enhancement of the heavier isotope over the lighter isotope during condensation) occurs because the heavier isotope has a lower energy than the lighter isotope in the condensed phase.

The elements oxygen, carbon, nitrogen, and sulfur are the most abundant molecule-forming elements in the Sun (after hydrogen), and all are observed in the atmospheres of the giant planets, mostly as H_2O, CH_4, NH_3, and (for Jupiter) H_2S. Water condenses even in Jupiter's atmosphere, at levels that are difficult to probe with infrared radiation (6 bars or deeper). A straightforward interpretation of Jupiter's spectrum indicated its abundance to be about a hundred times less than what is expected from solar composition. The *Galileo Probe* measurements indicated that water was depleted relative to solar abundance by roughly a factor of two at the deepest level measured (near 20 bars of pressure) and even more depleted at higher altitude. However, the probe descended in a relatively dry region of the atmosphere, analogous to a desert on Earth, and the bulk water abundance on Jupiter may well be close to the solar abundance. Water is not observed on any of the other giant planets because of the optically thick overlying clouds and haze layers. It is thought to form a massive global ocean on Uranus and Neptune based on the densities and gravity fields of those planets, coupled with theories of their formation.

Methane is well mixed up to the homopause level in the atmospheres of Jupiter and Saturn, but it condenses as ice in the atmospheres of Uranus and Nep-

TABLE II

Abundances of Observed Species in the Atmospheres of the Giant Planets

Constituent	Peak mixing ratio (by number) or upper limit			
	Jupiter	Saturn	Uranus	Neptune
Species with constant mixing ratio below the homopause				
H_2	0.86	0.97	0.82	0.79
HD	4×10^{-5}	4×10^{-5}		
He	0.14	0.03	0.15	0.18
CH_4	2×10^{-3}	2×10^{-3}		
CH_3D	3.5×10^{-7}	2×10^{-7}		
^{20}Ne	2×10^{-5}			
^{36}Ar	1×10^{-5}			
Condensable species (estimated or measured below the condensation region)				
NH_3	2.5×10^{-4}	2×10^{-4}		
H_2S	7×10^{-5}			
H_2O	6×10^{-4}			
CH_4			0.025	0.02–0.03
CH_3D			2×10^{-5}	2×10^{-5}
Disequilibrium species in the troposphere				
PH_3	5×10^{-7}	2×10^{-6}		
GeH_4	7×10^{-10}	4×10^{-10}		
AsH_3	2.4×10^{-9}	3×10^{-9}		
CO	2×10^{-9}	$1–25 \times 10^{-9}$	$<1 \times 10^{-8}$	1×10^{-6}
HCN			$<1 \times 10^{-10}$	1×10^{-9}
Photochemical species (peak values)				
C_2H_2	1×10^{-7}	3×10^{-7}	1×10^{-8}	6×10^{-8}
C_2H_4	7×10^{-9}			
C_2H_6	7×10^{-6}	7×10^{-6}	$<1 \times 10^{-8}$	$2 \times 10{-6}$
C_3H_4	2.5×10^{-9}			
C_6H_6	2×10^{-9}			

tune. Its mixing ratio below the condensation level is enhanced over that expected for a solar-composition atmosphere by factors of 2.3, 5.1, 35, and 40 for Jupiter, Saturn, Uranus, and Neptune, respectively. These enhancements are consistent with ideas about the amounts of icy materials that were incorporated into the planets as they formed. The stratospheres of Uranus and Neptune form a cold trap, where methane ice condenses into ice crystals that fall out, making it difficult for methane to mix to higher levels. Nevertheless, the methane abundance in Neptune's stratosphere appears to be significantly higher than its vapor pressure at the temperature that the tropopause would allow (and also higher than the abundance in the stratosphere of Uranus), suggesting some mechanism such as convective penetration of the cold trap by rapidly rising parcels of gas. This mechanism does not appear to be operating on Uranus, and this difference between Uranus and Neptune is symptomatic of the underlying difference in internal heat that is available to drive convection on Neptune but not on Uranus.

Ammonia is observed on Jupiter and Saturn, but not on Uranus or Neptune. Ammonia condenses as an ammonia ice cloud near 0.6 bar on Jupiter and at higher pressures on the colder outer planets. Ammonia and H_2S in solar abundance would combine to form a cloud of NH_4SH (ammonium hydrosulfide) near the 2-bar level in Jupiter's atmosphere and at deeper levels in the colder atmospheres of the other giant planets. Hydrogen sulfide was observed in Jupiter's atmosphere by the mass spectrometer instrument on the *Galileo Probe*. Another instrument (the nephelometer) on the probe detected cloud particles in the vicinity of the 1.6-bar pressure level, which would be consistent with the predicted ammonium hydrosulfide cloud. Evidence from thermal emission at radio wavelengths has been used to infer that H_2S is abundant on Uranus and Neptune. Ammonia condenses at relatively deep levels in the atmospheres of Uranus and Neptune and has not been spectroscopically detected. A dense cloud is evident at the level expected for ammonia condensation (2–3 bars) in near-infrared spectroscopic observations, but the microwave spectra of those planets are more consistent with a strong depletion of ammonia at those levels. An enhancement of H_2S relative to NH_3 could act to deplete ammonia by the formation of ammonium hydrosulfide in the deeper atmosphere. In that case, H_2S ice is the most likely candidate for the cloud near 3 bars.

Water, methane, and ammonia are in thermochemical equilibrium in the upper troposphere. Their abundances at altitudes higher than (and temperatures colder than) their condensation level are determined by temperature (according to the vapor–pressure law) and by meteorology, as is water in Earth's atmosphere. Some species (PH_3, GeH_4, and CO) are not in thermochemical equilibrium in the upper troposphere. At temperatures less than 1000 K, PH_3 would react with H_2O to form P_4O_6 if allowed to proceed to thermochemical equilibrium. Apparently the timescale for this reaction (about 10^7 s) is longer than the time to convect material from the 1000 K level to the tropopause. A similar process explains the detections of GeH_4 and CO. The detection of CO in the stratosphere of Neptune reinforces the notion that rapid convection is able to transport species from the deep interior to the stratosphere, although a small amount of the stratospheric CO may be produced by infalling meteorites or material blasted off Triton's surface by meteorite impact or transported to the atmosphere as ions along the magnetic field lines connecting the atmosphere to Triton.

Ammonia and phosphine are present in the stratospheres of Jupiter and Saturn, and methane is present in the stratospheres of all the giant planets. These species are destroyed at high altitudes by ultraviolet sunlight and by charged particles in auroras, producing N, P, and C, which can react to form other compounds. Ammonia photochemistry leads to formation of hydrazine (N_2H_4) and phosphine photochemistry leads to diphosphine (P_2H_4). These constituents condense in the cold tropospheres of Jupiter and Saturn and may be responsible for much of the ultraviolet-absorbing haze seen at low latitudes. Nitrogen gas and solid P_4 are other by-products of ammonia and phosphine chemistry. Solid phosphorus is sometimes red and has been proposed as the constituent responsible for the red color of Jupiter's Great Red Spot. That suggestion (one of several) has not been confirmed, and neither N_2H_4 nor P_2H_4 has been observed spectroscopically.

Organic compounds derived from dissociation of methane are present in the stratospheres of all the giant planets. The photochemical cycle leading to stable C_2H_2 (acetylene), C_2H_4 (ethylene), C_2H_6 (ethane), and C_4H_2 (diacetylene) is shown schematically in Fig. 2. The chain may progress further to produce polyacetylenes ($C_{2n}H_2$). These species form condensate haze

layers in the cold stratospheres of Uranus and Neptune. More complex hydrocarbon species (C_3H_8, C_3H_4) are observed in Jupiter's atmosphere primarily in close proximity to high-latitude regions, where auroral heating is significant. The abundant polar aerosols in the atmospheres of Jupiter and Saturn may owe their existence to the ions created by auroras in the upper atmosphere.

Hydrogen cyanide (HCN) is present in the stratospheres of Jupiter and Neptune, but for two very different reasons. On Jupiter, HCN was emplaced high in the stratosphere as a result of the 1994 impacts of comet Shoemaker-Levy 9. During the three years following the impacts, it was observed to spread north of the impact latitude (near 45°S), eventually to be globally distributed. It is expected to dissipate over the next decade or so. The most plausible explanation for Neptune's HCN is the same as that for some other disequilibrium species on Jupiter: rapid convection from the interior.

Quantitative thermochemical and photochemical models are available for many of the observed constituents and provide predictions for many others that are not yet observed. These models solve a set of coupled equations that describe the balance between the abundances of species that interact and include important physical processes such as ultraviolet photolysis, condensation/sublimation, and vertical transport. Current models heuristically lump all the transport processes into an effective eddy mixing coefficient, and the value of that coefficient is derived as part of the solution of the set of equations. As we gain more detailed observations and more comprehensive laboratory measurements of reaction rates, we will be able to develop more sophisticated models. Some models are beginning to incorporate transport by vertical and horizontal winds. Figures 3 and 4 show vertical profiles calculated from models for a number of photochemically produced species.

III. CLOUDS AND AEROSOLS

The appearance of the giant planets is determined by the distribution and optical properties of cloud and aerosol haze particles in the upper troposphere and stratosphere. Cameras on the *Voyager* spacecraft provided detailed views of all the giant planets, whose general appearances can be compared in Fig. 5. (See also color insert.) All of their atmospheres show a banded structure (which is difficult to see on Uranus) of color and shading parallel to latitude lines. These

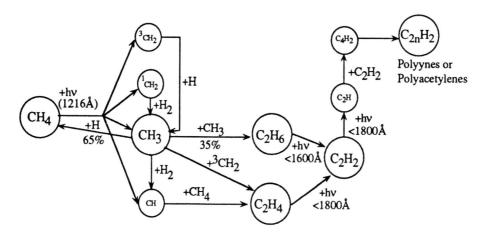

FIGURE 2 Summary of CH$_4$ (methane) photochemical processes in the stratospheres of the giant planets. Photodissocation by ultraviolet light is indicated by $+h\nu$ at the indicated wavelength. Methane photodissociation is the starting point in the production of a host of other hydrocarbons. [Revised by S. K. Atreya from Fig. 5-3 from J. B. Pollack and S. K. Atreya (1992). *In* "Exobiology in Solar System Exploration" (G. Carle *et al.*, eds.), NASA-SP 512, pp. 82–101.]

were historically named belts and zones on Jupiter and Saturn, with belts being relatively dark and zones relatively bright. Specific belts and zones were named in accordance with their approximate latitudinal location (Equatorial Belt, North and South Tropical Zones near latitudes ±20°, North and South Temperate Zones and Belts near ±35°, and polar regions).

The nomenclature should not be construed to mean that low latitudes are relatively warmer than high latitudes as they are on Earth and Mars. Nor is it true

that the reflectivities of these features remain constant with time. Some features on Jupiter, such as the North and South Tropical Zones, are persistently bright, whereas others, like the South Equatorial Belt, are sometimes bright and sometimes dark. On Jupiter there is a correlation between visible albedo and temperature, such that bright zones are usually cool regions and dark belts are usually warm near the tropopause. Cool temperatures are associated with adiabatic cooling of upwelling gas, and the correlation of cool

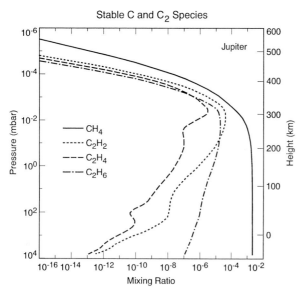

FIGURE 3 Vertical profiles of some photochemical species in Jupiter's stratosphere. The mixing ratios (horizontal axis) are plotted as a function of pressure. [From G. R. Gladstone *et al.* (1996). *Icarus* **119**, 1–52. Copyright by Academic Press.]

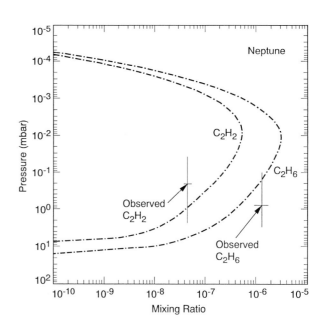

FIGURE 4 Vertical profiles of photochemical species in the Neptunian stratosphere. [From P. Romani *et al.* (1993). *Icarus* **106**, 442–463. Copyright by Academic Press.]

FIGURE 5 Voyager images of Jupiter, Saturn, Uranus, and Neptune, scaled to their relative sizes. Earth and Venus are also shown scaled to their relative sizes. (See also color insert.)

temperatures with bright clouds points to enhanced condensation of ice particles as condensable gases flow upward and cool. This correlation does not hold completely on Jupiter and almost not at all on the other giant planets. The mechanisms responsible for producing reflectivity contrasts and color remain largely mysterious, although a number of proposals have been advanced. These will be discussed in more detail.

Our understanding of aerosols and clouds is rooted in thermochemical equilibrium models that predict the temperature (and hence pressure and altitude) of the bases of condensate clouds. The cloud base occurs where the vapor pressure of a condensable gas equals its partial pressure. Model predictions for the four giant planets are shown in Fig. 6. The deepest cloud to form is a solution of water and ammonia on Jupiter and Saturn, with dissolved H_2S as well on Uranus and Neptune. At higher altitudes, an ammonium hydrosulfide cloud forms, and its mass depends on both the amounts of H_2S and NH_3 available and the ratio of S to N. At still higher altitudes, an ammonia or hydrogen sulfide cloud can form if the S/N ratio is less than or greater than 1, respectively. If the ratio is greater than 1, all the N will be taken up as NH_4SH, with the remaining sulfur available to condense at higher altitudes. This seems to be the situation on Uranus and Neptune, but the reverse is true for Jupiter and Saturn. Only the atmospheres of Uranus and Neptune are cold enough to condense methane, which occurs at 1.3 bars in Uranus and about 2 bars in Neptune. It is predominantly the uppermost clouds that we see at visible wavelengths.

Observational evidence to support the cloud stratigraphy shown in Fig. 6 is mixed. The *Galileo Probe* detected cloud particles near 1.6-bar pressure and sensed cloud opacity at higher altitudes corresponding to the ammonia cloud. With data only from remote-sensing experiments it is difficult to probe to levels below the top cloud, and the evidence we have for deeper clouds comes from careful analyses of radio occultations and of gaseous absorption lines in the visible and near infrared, and from thermal emission at 5, 8.5, and 45 μm. Contrary to expectation, spectra of the planets contain no features due to ice. The

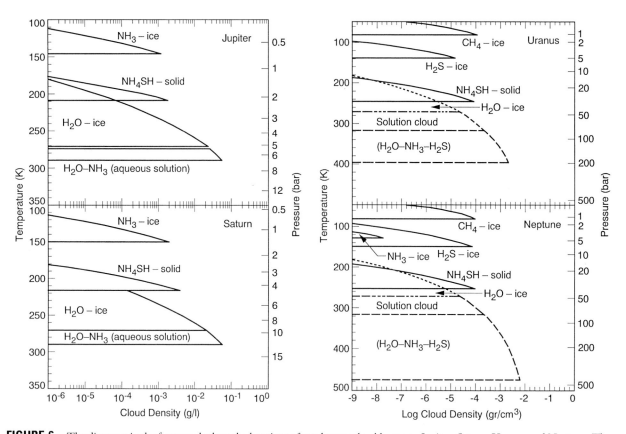

FIGURE 6   The diagrams in the four panels show the locations of condensate cloud layers on Jupiter, Saturn, Uranus, and Neptune. These figures indicate how much cloud material would condense at various temperatures (corresponding to altitude) if there were no advective motions in the atmosphere to move vapor and clouds. They are based on simple thermochemical equilibrium calculations, which assume, for Jupiter and Saturn, that the condensable species have mixing ratios equal to those for a solar composition atmosphere. [Figures for Jupiter and Saturn were constructed from models by S. K. Atreya and M. Wong, based on S. K. Atreya and P. N. Romani (1985). *In* "Planetary Meteorology" (G. E. Hunt, ed.), pp. 17–68. Cambridge Univ. Press, Cambridge, England. Those for Uranus and Neptune were first published by I. de Pater *et al.* (1991). *Icarus* **91**, 220–233. Copyright by Academic Press.]

Voyager radio occultation data showed strong refractivity gradients at locations predicted for methane ice clouds on Uranus and Neptune, essentially confirming their existence and providing accurate information on the altitude of the cloud base. Ammonia gas is observed spectroscopically in Jupiter's upper troposphere, and its abundance decreases with altitude above its cloud base in accordance with expectation. There is no doubt that ammonia ice is the major component of the visible clouds on Jupiter and Saturn, but it cannot be the only component and is not responsible for the colors (pure ammonia ice is white). In fact, all the ices shown in Fig. 6 are white at visible wavelengths. The colored material must be produced by some disequilibrium process like photochemistry or bombardment by energetic particles from the magnetosphere.

Colors on Jupiter are close to white in the brightest zones, gray yellow to light brown in the belts, and orange or red in some of the spots. The colors in Fig. 5 are slightly and unintentionally exaggerated owing

to the difficulty of achieving accurate color reproduction on the printed page. Colors on Saturn are more subdued. Uranus and Neptune are gray green. Neptune has a number of dark spots and white patchy clouds. Part of the green tint on Uranus and Neptune is caused by strong methane gas absorption at red wavelengths, and part is due to aerosols that also absorb preferentially at wavelengths longer than 0.6 μm.

Candidate materials for the chromophore material in outer planet atmospheres are summarized in Table III. All candidate materials are thought to form by some nonequilibrium process such as photolysis or decomposition by protons or ions in auroras, which acts on methane, ammonia, or ammonium hydrosulfide. Methane is present in the stratospheres of all the giant planets. Ammonia is present in the stratosphere of Jupiter. Ammonium hydrosulfide is thought to reside near the 2-bar level and deeper in Jupiter's atmosphere, which is too deep for ultraviolet photons to penetrate.

TABLE III
Candidate Chromophore Materials in the Atmospheres of the Giant Planets

Material	Formation mechanism
Sulfur	Photochemical products of H_2S and NH_4SH. Red allotropes are unstable.
H_2S_x, $(NH_4)_2S_x$, $N_2H_4S_x$	Photochemical products of H_2S and NH_4SH.
N_2H_4	Hydrazine, a photochemical product of ammonia, a candidate for Jupiter's stratospheric haze.
Phosphorus (P_4)	Photochemical product of PH_3.
P_2H_4	Diphosphine, a photochemical product of phosphine, a candidate for Saturn's stratospheric haze.
Products of photo- or charged-particle decomposition of CH_4	Includes acetylene photopolymers (C_xH_2), proton-irradiated methane, and organics with some nitrogen and/or sulfur. Confined to stratospheric levels where ultraviolet photons and auroral protons or ions penetrate.

There are two major problems in understanding which, if any, of the proposed candidate chromophores are responsible for the observed colors. First, no features have been identified in spectra of the planets that uniquely identify a single candidate material. Spectra show broad slopes, with more absorption at blue wavelengths on Jupiter and Saturn and at red wavelengths on Uranus and Neptune. All the candidates listed in Table III produce broad blue absorption. None of them can account for the red and near-infrared absorption in the spectra of Uranus and Neptune. Second, our understanding of the detailed processes that lead to the formation of chromophores is inadequate. Gas-phase photochemical theory cannot account for the abundance of chromophore material. It is likely that ultraviolet photons or charged-particle bombardment of solid, initially colorless particles like acetylene and ethane ice in the stratospheres of Uranus and Neptune or ammonium hydrosulfide in Jupiter's atmosphere breaks chemical bonds in the solid state, paving a path to formation of more complex hydrocarbons or inorganic materials that seem to be required. Additional laboratory studies are needed to address these questions. [See THE SOLAR SYSTEM AT ULTRAVIOLET WAVELENGTHS.]

Haze particles are present in the stratospheres of all the giant planets, but their chemical and physical properties and spatial distributions are quite different. Jupiter and Saturn have UV-absorbing aerosols abun-

dant at high latitudes and high altitudes (corresponding to pressures ranging from a fraction of a millibar to a few tens of millibars). The stratospheric aerosols on Uranus and Neptune do not absorb much in the UV and are not concentrated at high latitude. The polar concentration of UV-absorbing aerosols on Jupiter and Saturn suggests that their formation may be due to chemistry in auroral regions, where protons and/or ions from the magnetosphere penetrate the upper atmosphere and deposit energy. Association with auroral processes may help explain why UV absorbers are abundant poleward of about 70° latitude on Saturn, extend to somewhat lower latitudes on Jupiter, and show a hemispheric asymmetry in Jupiter's atmosphere. Saturn's magnetic dipole is nearly centered and parallel to Saturn's spin axis, but Jupiter's magnetic dipole is both significantly offset and tilted with respect to its spin axis, producing asymmetric auroras at lower latitudes than on Saturn. Other processes, such as the meridional circulation, also influence the latitudinal distribution of aerosols, so more work needs to be done to establish the role of auroras in aerosol formation.

Photochemistry is responsible for the formation of diacetylene, acetylene, and ethane hazes in the stratospheres of Uranus and Neptune. The main steps in the life cycle of stratospheric aerosols are shown in Fig. 7. Methane gas mixes upward to the high stratosphere, where it is photolyzed by ultraviolet light. Diacetylene,

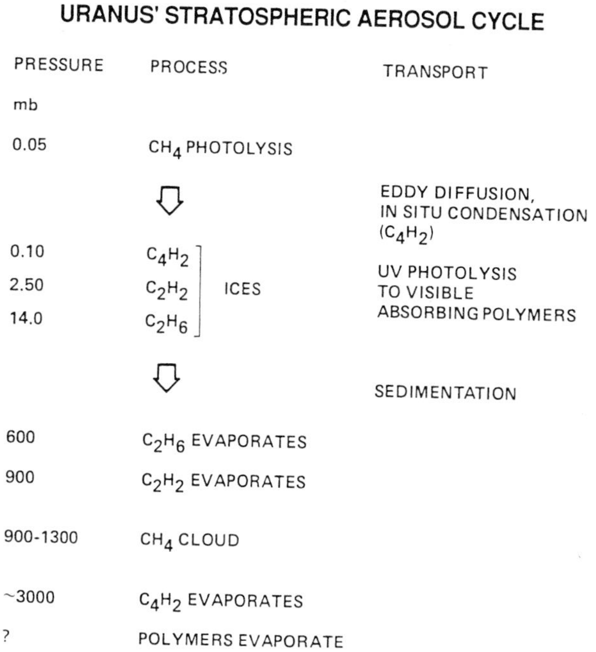

URANUS' STRATOSPHERIC AEROSOL CYCLE

PRESSURE mb	PROCESS		TRANSPORT
0.05	CH_4 PHOTOLYSIS		
	⇩		EDDY DIFFUSION, IN SITU CONDENSATION (C_4H_2)
0.10	C_4H_2		
2.50	C_2H_2	ICES	UV PHOTOLYSIS TO VISIBLE ABSORBING POLYMERS
14.0	C_2H_6		
	⇩		SEDIMENTATION
600	C_2H_6 EVAPORATES		
900	C_2H_2 EVAPORATES		
900–1300	CH_4 CLOUD		
~3000	C_4H_2 EVAPORATES		
?	POLYMERS EVAPORATE		

FIGURE 7 Life cycle for stratospheric aerosols on Uranus. [From J. Pollack *et al.* (1987). *J. Geophys. Res.* **92**, 15,037–15,066. Copyright American Geophysical Union.]

acetylene, and ethane form from gas-phase photo-chemistry and diffuse downward. Temperature decreases downward in the stratosphere, so ice particles form when the vapor pressure equals the partial pressure of the gas. On Uranus, diacetylene ice forms at 0.1 mbar, acetylene at 2.5 mbar, and ethane at 14 mbar. The ice particles sediment to deeper levels on a timescale of years and evaporate in the upper troposphere at 600 mbar and deeper. Polymers that form from solid-state photochemistry in the ice particles are probably responsible for the little ultraviolet absorption that does occur. They are less volatile than the pure ices and probably mix down to the methane cloud and below.

Photochemical models predict formation of hydrazine in Jupiter's stratosphere and diphosphine in Saturn's atmosphere. If these are the only stratospheric haze constituents, it is not apparent why the ultraviolet absorbers are concentrated at high latitude. As discussed earlier, auroral bombardment of methane provides an attractive candidate process for the abundant high-latitude aerosols on Jupiter and Saturn. However, we do not know enough to formulate a detailed chemical model of this process.

Thermochemical equilibrium theory serves as a guide to the location of the bases of tropospheric clouds, but meteorology and cloud microphysical processes determine the vertical and horizontal distribution of cloud material. These processes are too complex to let us predict to what altitudes clouds should extend, and so we must rely on observations. Several diagnostics are available to measure cloud and haze vertical locations. At short wavelengths, gas molecules limit the depth to which we can see. In the visible and near infrared are methane and hydrogen absorption bands, which can be used to probe a variety of depths depending on the absorption coefficient of the gas. There are a few window regions in the thermal infrared where cloud opacity determines the outgoing radiance. The deepest-probing wavelength is 5 μm. At that wavelength, thermal emission from the water-cloud region near the 5-bar pressure level provides sounding for all the main clouds in Jupiter's atmosphere. [*See* INFRARED VIEWS OF THE SOLAR SYSTEM FROM SPACE.]

The results of cloud stratigraphy studies for Jupiter's atmosphere are summarized in Fig. 8. There is spectroscopic evidence for the two highest tropospheric layers in Jupiter's atmosphere. There is also considerable controversy surrounding the existence of the water-ammonia cloud on Jupiter. The *Galileo Probe* descended into a dry region of the atmosphere and did not find a water cloud, but water clouds may be present in moister regions of the atmosphere that

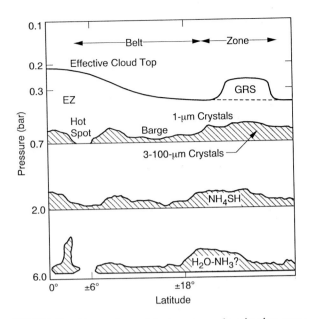

FIGURE 8   Observations of Jupiter at wavelengths that sense clouds lead to a picture of the Jovian cloud stratigraphy shown here. There has been no direct evidence for a water–ammonia cloud near the 6-bar pressure level, but it is likely that such a cloud exists from indirect evidence. The hot spots are named from their visual appearance at a wavelength of 5 μm. They are not physically much warmer than their surroundings, but they are deficient in cloudy material (see Fig. 9). [From R. West *et al.* (1986). *Icarus* **65**, 161–217. Copyright by Academic Press.]

are obscured by overlying clouds. There is evidence for a large range of particle sizes. Small particles (less than about 1-μm radius) provide most of the cloud opacity in the visible. They cover belts and zones, although their optical thickness in belts is sometimes less than in zones. Most of the contrast between belts and zones in the visible comes from enhanced abundance or greater visibility of chromophore material, which seems to be vertically but not horizontally well mixed in the ammonia cloud. The top of this small-particle layer extends up to about 200 mbar, depending on latitude. Jupiter's Great Red Spot is a location of relatively high-altitude aerosols, consistent with the idea that it is a region of upwelling gas.

Larger particles (mean radius near 6 μm) are also present, mostly in zones. This large-particle component appears to respond to rapid changes in the meteorology. It is highly variable in space and time and is responsible, together with the deeper clouds, for the richly textured appearance of the planet at 5-μm wavelength (Fig. 9). Some of the brightest regions seen in Fig. 9 are called 5-μm hot spots, not because they are warmer than their surroundings but because thermal radiation from the 5-bar region emerges with little attenuation from higher clouds. The *Galileo Probe*

FIGURE 9 At a wavelength of 5 μm most of the light from Jupiter is thermal radiation emitted near the 6-bar pressure level below the visible clouds. Places where the clouds are thin permit the deep radiation to escape to space, making these regions appear bright. Thicker clouds block the radiation and these appear dark. Jupiter's Great Red Spot is the dark oval just below the center. This image was taken with the NASA Infrared Telescope Facility. (Courtesy of J. Spencer.)

sampled one of these regions. The dark regions in the image are caused by optically thick clouds in the NH_4SH and NH_3 cloud regions. The thickest clouds are generally associated with upwelling, bright (at visible wavelengths) zones, but many exceptions to this rule are observed. Until we understand the chemistry and physics of chromophores we should not expect to understand why or how well albedo is correlated with other meteorological parameters.

Most of Jupiter's spots are at nearly the same altitude. Some notable exceptions are the Great Red Spot (GRS), the three white ovals just south of the GRS, and some smaller ovals at other latitudes. These anticyclonic features extend to higher altitudes, probably up to the 200-mbar level, compared to a pressure level of about 300 mbar for the surrounding clouds. Some of the anticyclonic spots have remarkably long lifetimes compared to the terrestrial norm. The GRS was recorded in drawings in 1879, and reports of red spots extend back to the seventeenth century. The three

white ovals in a latitude band south of the GRS formed from a bright cloud band that split into three segments in 1939. The segments shrunk in longitude over the course of a year, until the region (the South Temperate Belt) was mostly dark except for three high-albedo spots that remain to the present. Whereas anticyclonic ovals tend to be stable and long-lived, cyclonic regions constantly change.

Similar features are observed in Saturn's atmosphere, although the color is much subdued compared to Jupiter, and Saturn has nothing that is as large or as long-lived as the GRS. The reduced contrast may be related to Saturn's colder tropopause temperature. The distance between the base of the ammonia cloud and the top of the troposphere (where the atmosphere becomes stable against convection) is greater on Saturn than on Jupiter. The ammonia-ice cloud on Saturn is both physically and optically thicker than it is on Jupiter. Occasionally (about two or three times each century) a large, bright cloud forms near Saturn's equator. One well-observed event occurred in 1990, but its cause is unknown. It appears to be a parcel of gas that erupts from deeper levels, bringing fresh condensate material to near the top of the troposphere. It becomes sheared out in the wind shear and dissipates over the course of a year.

Uranus as seen by Voyager was even more bland than Saturn. Only a few low-contrast elongated clouds and a low-contrast banded pattern were seen in an otherwise featureless haze. Midlatitude regions on Uranus and Neptune are cool near the tropopause, indicating up-welling. But cloud optical thickness may be lower there than at other latitudes. The relation between cloud optical thickness and vertical motion is more complicated than the simple condensation model would predict.

Neptune's clouds are unique among the outer planet atmospheres. Voyager observed four large cloud features that persisted for the duration of the Voyager observations (months). The largest of these is the Great Dark Spot (GDS) and its white companion. Because of its size and shape, the GDS might be similar to Jupiter's Great Red Spot, but the GDS had a short life compared to the GRS.

There is no explanation yet of what makes the dark spot dark. The deepest cloud (near the 3-bar level) is probably H_2S ice, since ammonia is apparently depleted and NH_4SH would be sequestered at a deeper level. At higher altitudes there is an optically thin methane haze (near 2 bar) and stratospheric hazes of ethane, acetylene, and diacetylene. At high spatial resolution the wispy white clouds associated with the companion to the GDS and found elsewhere on the planet form and dissipate in a matter of hours. It was difficult to estimate winds from these features because of their transitory nature. Individual wisps moved at a different speed than the GDS and its companion, suggesting that these features form and then evaporate high above the GDS as they pass through a local pressure anomaly, perhaps a standing wave caused by flow around the GDS. Cloud shadows were seen in some places, a surprise after none was seen on the other giant planets. The clouds casting the shadows are about 100 km higher than the lower cloud deck, suggesting that the lower cloud is near 3 bar and the shadowing clouds near 1 bar, in the methane condensation region.

IV. DYNAMICAL METEOROLOGY OF THE TROPOSPHERE AND STRATOSPHERE
. . . .

Our understanding of giant planet meteorology comes mostly from Voyager observations, with observations from Galileo, the Hubble Space Telescope, and ground-based data adding to the picture. Although we have theories and models for many of the dynamical features, the fundamental nature of the dynamical meteorology on the giant planets remains puzzling chiefly because of our inability to probe to depths greater than a few bars in atmospheres that go to kilobar pressures and because of limitations in spatial and time sampling, which may improve with future missions to the planets.

Thermodynamic properties of atmospheres are at the heart of a variety of meteorological phenomena. In the terrestrial atmosphere, condensation, evaporation, and transport of water redistribute energy in the form of latent heat. The same is true for the outer planet atmospheres, where condensation of water, ammonia, ammonium hydrosulfide, and methane takes place. Condensables also influence the dynamics through their effects on density gradients. In the terrestrial atmosphere, moist air is less dense than dry air at the same temperature because the molecular weight of water vapor is smaller than that of the dry air. Because of this fact, and also because moist air condenses and releases latent heat as it rises, there can be a growing instability leading to the formation of convective plumes, thunderstorms, and anvil clouds at high altitudes. On the giant planets, water vapor is significantly heavier than the dry atmosphere and so the same type of instability will not occur unless a strongly upwelling parcel is already present. Some researchers proposed that the Equatorial Plumes on Jupiter and the elongated clouds on Uranus are the outer planet analogs to terrestrial anvil clouds.

Terrestrial lightning occurs most frequently over tropical oceans and over a fraction of the land surface. Its distribution in latitude, longitude, and season is indicative of certain properties of the atmosphere, especially the availability of liquid water. Lightning has been observed on the giant planets as well, either from imaging on the night side (Jupiter) or from signals recorded by plasma wave instruments. A somewhat mysterious radio emission from Saturn or its environment (the so-called Sudden Electrostatic Discharge events) has been interpreted as a lightning signature, although other interpretations have been proffered. The intensity and size of the lightning spots in the images imply that they are much more energetic than the average lightning bolt in the terrestrial atmosphere, and they occur in the water-ammonia cloud region as expected. The Galileo Probe did not detect lightning in Jupiter's atmosphere within a range of about 10,000 km from its location at latitude 6.5°N. [See THE SOLAR SYSTEM AT RADIO WAVELENGTHS.]

The heat capacity of hydrogen, and therefore the dry adiabatic lapse rate of the convective part of the atmosphere, depends on the degree to which the ortho/para states equilibrate. The lapse rate is steepest when equilibration is operative. The observed lapse rate for Uranus, as measured by the Voyager radio occultation experiment, is close to the "frozen" lapse rate—the rate when the relative fractions of ortho and para hydrogen are fixed. How can the observed relative fractions be near equilibrium when the lapse rate points to nonequilibrium? One suggestion is that the atmosphere is layered. Each layer is separated from the next by an interface that is stable and that is thin compared to the layer thickness. The air within each layer mixes rapidly compared to the time for equilibration, but the exchange rate between layers is slow or comparable to the timescale for conversion of ortho to para and back.

How can layers be maintained in a convective atmosphere? In the terrestrial ocean, two factors influence buoyancy: temperature and salinity. If the water is warmer at depth, or if the convective amplitude is large, the different timescales for diffusion of heat and salinity lead to layering. In the atmospheres of the outer planets, the higher molecular weight of condensables acts much as salinity in ocean water. Layering can be established even without molecular weight gradients. Layering in the terrestrial stratosphere and mesosphere has been observed. Layers of rapidly convecting gas occur where gravity waves break or where other types of wave instabilities dump energy. Between layers of rapid stirring are stably stratified layers with transport by diffusion rather than convection.

Some of the variety of the giant planet meteorology, as well as our difficulty to understand it, is nicely illustrated by observations of the wind field at the cloud tops. Wind vectors of all the giant planet atmospheres are predominantly in the east–west (zonal) direction (Fig. 10). These are determined by tracking visible cloud features over hours, days, and months. Jupiter has an abundance of small features and the zonal winds are well mapped. Saturn has fewer features, and they are of less contrast than those on Jupiter, but there is still a large enough number to provide detail in the wind field. Only a few features were seen in the Uranus atmosphere, and all but one of these were between latitudes 20°S and 40°S. The *Voyager 2* radio occultation provided an additional estimate for wind speed at the equator. Neptune has more visible features than Uranus, but most of them are transitory and difficult to follow long enough to gauge wind speed.

Figure 10 reveals a great diversity in the zonal flow among the giant planet atmospheres. Wind speed is relative to the rotation rate of the deep interior as revealed by the magnetic field and radio emissions. Jupiter has a series of jets that oscillate with latitude and are greatest in the prograde direction at latitude

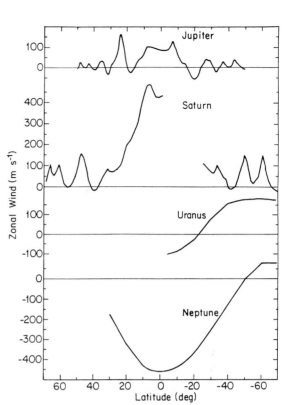

FIGURE 10 Zonal (east–west) wind velocity for the giant planets as a function of latitude. There are gaps where we lack contrasts suitable for tracking winds or where the atmosphere was in darkness (on the night side or shadowed by Saturn's rings). [From P. Gierasch and B. Conrath (1993). *J. Geophys. Res.* **98**, 5459–5469. Copyright American Geophysical Union.]

23°N, and near ±10°. The pattern of east–west winds is approximately symmetric about the equator except at high latitude. Saturn has a very strong prograde jet at low latitudes (within the region ±15°). It also has alternating but mostly prograde jets at higher latitudes, with the scale of latitudinal variation being about 10°. Uranus appears to have a single prograde maximum near 60°S, and the equatorial region is retrograde. Neptune has an enormous differential rotation, mostly retrograde except at high latitude. Various theories have been advanced to explain the pattern of zonal jets. None of them can account for the great variety among the four planets.

Some of the key observations that any dynamical theory must address include: (1) the magnitude, direction, and latitudinal scale of the jets; (2) the stability of the jets, at least for Jupiter and Saturn, where observations over long periods show little or no change; (3) the magnitude and latitudinal gradients of heat flux; and (4) interactions of the mean zonal flow with small spots and eddies. One of the controversies during the past two decades concerns how deep the flow extends into the atmosphere. It is possible to construct shallow-

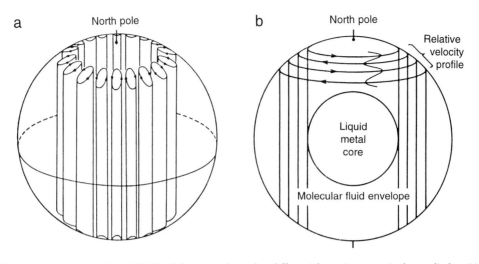

FIGURE 11 One model for the zonal wind fields of the giant planets has differential rotation organized on cylinders (a), exploiting the natural symmetry of a rotating deep fluid (b). [From F. Busse (1976). *Icarus* **29**, 255–260. Copyright by Academic Press.]

atmosphere models that have approximately correct jet scales and magnitudes. A shallow-atmosphere model is one in which the jets extend to relatively shallow levels (100 bars or less), and the deeper interior rotates as a solid body, or at least as one whose latitudinal wind shear is not correlated with the wind shear of the jets. The facts that the jets and some spots on Jupiter are very stable, that there is approximate hemispheric symmetry in the zonal wind pattern between latitudes ±60°, and that the Jovian interior has no density discontinuities down to kilobar levels suggested to some investigators that the jets extend deep into the atmosphere. A natural architecture for the flow in a rotating sphere with no density discontinuities is one in which the flow is organized on rotating cylinders (Fig. 11).

Apart from the stability and symmetry noted here, there is little evidence to suggest that the zonal wind pattern really does extend to the deep interior. The strength of the zonal jet at the location where the *Galileo Probe* entered (6.5°N) increased with depth, consistent with the idea of a deeply rooted zonal wind field on Jupiter. One way to test that hypothesis is to make highly precise measurements of the gravity field close to the planet. There are density gradients associated with the winds, and these produce features in the gravity field close to the planet. The largest signature is produced by Neptune's remarkable differential rotation. The *Voyager 2* spacecraft flew just above Neptune's atmosphere and provided the first evidence that the differential rotation cannot extend deep into the atmosphere. Gravity-field tests of the deep-wind hypothesis for the other giant planets are more difficult because the differential rotation is much weaker. No

spacecraft have come close enough to make the measurements.

What process maintains the zonal wind pattern? Voyager measurements shed some light on this question, but provided some puzzles as well. The ultimate energy source for maintaining atmospheric motions is the combination of internal thermal and solar energy absorbed by the atmosphere. Jupiter, Saturn, and Neptune all have significant internal energy sources, whereas Uranus has little or none. A measure of the amount of energy available for driving winds is the escaping radiative energy per square meter of surface area. Twenty times as much energy per unit area is radiated from Jupiter's atmosphere as from Neptune's, yet the wind speeds (measured relative to the interior as determined from the magnetic field rotation rate) on Neptune are about three times higher than those on Jupiter. Rather than driving zonal winds, the excess internal energy may go into driving smaller-scale eddies, which are most abundant to Jupiter.

What influence does the absorbed thermal radiation have? Most planets receive more solar radiation at their equator than at their poles. For Uranus the reverse is true. Yet the upper tropospheric and stratospheric temperatures on Uranus and Neptune are nearly identical, and the winds for both planets (as for Earth) are retrograde at the equator. According to one theory, deposition of solar energy may account for the fact that Uranus possesses very little internal energy today. Otherwise, it is hard to see how solar energy can be important for the tropospheric circulation of the giant planets.

What role do eddies have in maintaining the flow?

Measurements of the small spots on Jupiter and Saturn have allowed an estimate of the energy flow between the mean zonal wind and the eddy motions. For Jupiter the eddies at the cloud top appear to be pumping energy into the mean zonal flow, although that conclusion has been challenged on the grounds that the sampling may be biased. If further observation and analysis confirm the initial result, we need to explain why the jets are so stable when there is apparently enough energy in eddy motions to significantly modify the Jovian wind field. At the same time, other observations imply dissipation and decay of zonal winds at altitudes just above the cloud tops.

The relationship known to atmospheric physicists as the thermal wind equation provides a means of estimating the rate of change of zonal wind with height (which is usually impossible to measure remotely) from observations of the latitudinal gradient of temperature (which is usually easy to measure). One of the common features of all the outer planet atmospheres is a decay of zonal wind with height in the stratosphere, tending toward solid-body rotation at high altitudes. The decay of wind velocity with height could be driven by eddy motions or by gravity wave breaking, which effectively act as friction on the zonal flow.

Thermal contrasts on Jupiter are correlated with the horizontal shear and with cloud opacity as indicated by 5-μm images (see Fig. 9). Cool temperatures at the tropopause level (near 100 mbar) are associated with upwelling and anticyclonic motion, and warmer temperatures are associated with subsidence. Jupiter's Great Red Spot is an anticyclonic oval with cool tropospheric temperatures, upwelling flow, and aerosols extending to relatively high altitudes. Enhanced cloud opacity and ammonia abundance in cooler anticyclonic latitudes (mostly the high-albedo zones on Jupiter) are predicted in upwelling regions. The correlation is best with cloud opacity in the 5-μm region. At shorter wavelengths (in the visible and near infrared) there is a weaker correlation between cloud opacity and vorticity. Perhaps the small aerosols near the top of the troposphere, sensed by the shorter wavelengths but not at 5 μm, are transported horizontally from zone to belt on a timescale that is short compared to their rainout time (several months).

The upwelling/subsidence pattern at the jet scale in the upper troposphere penetrates into the lower stratosphere. We have relatively little information on the stratospheric circulation for the giant planets. Most of it is based on the observed thermal contrasts and the idea that friction is a dominant driver for stratospheric dynamics. We are beginning to appreciate the role of forcing by gravity or other dissipative waves. Figures

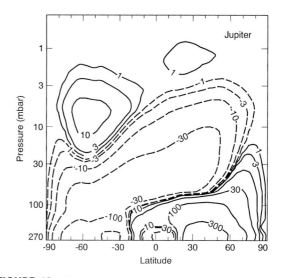

FIGURE 12 Contour lines trace the Jovian stratospheric long-term mean meridional circulation. The dashed contours follow the counterclockwise circulation and the solid contours follow the clockwise circulation. The numbers on the curves indicate the mass flow along the contour. [From R. West *et al.* (1992). *Icarus* **100**, 245–259. Copyright by Academic Press.]

12 and 13 show current estimates of the long-term mean meridional circulation for Jupiter and Uranus. The Uranus stratospheric circulation (see Fig. 13) is based on the frictional damping model and the observed thermal contrast as a function of latitude. The coldest temperatures in the lower stratosphere are at

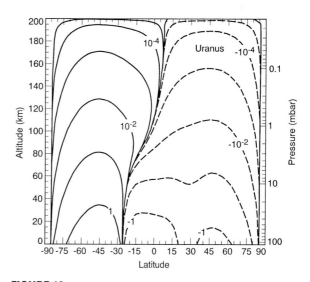

FIGURE 13 Uranian long-term average stratospheric meridional circulation. Gas rises at the bottom of the stratosphere near latitude 28°S and spreads to the north and south. The upwelling column migrates toward the equator at altitudes greater than about 60 km above the tropopause. (Courtesy of W. Wallace McMillan.)

midlatitudes, indicating upwelling there and subsidence at the equator and poles. A different pattern is expected if the deposition of solar energy controlled the circulation. Momentum forcing by vertically propagating waves from the deeper atmosphere is apparently more important than solar energy deposition.

The mean meridional circulation in Jupiter's stratosphere (see Fig. 12) differs from that predicted by the frictional damping model at pressure levels less than about 80 mbar. The zonal pattern of upwelling/sinking extends to about 100 mbar, giving way at higher altitude to a two-cell structure with cross-equatorial flow. There is also a hemispheric asymmetry. The high latitudes (poleward of 60°S and 40°N) are regions of sinking motion at the tropopause. Recent analysis of images from the Hubble Space Telescope indicate that the optical depth of the ammonia cloud decreases rapidly with latitude poleward of 60°S and 40°N and is well correlated with the estimated downward velocity. The descending dry air inhibits cloud formation. To produce the circulation shown in Fig. 12, there must be momentum forcing in the latitude range 40°S to 80°S and 30°N to 80°N at pressures between 2 and 8 mbar. Dissipation of gravity waves propagating from the deep interior is the most likely source of momentum forcing.

The long-term mean circulation as shown in Figs. 12 and 13 is slow and relatively unimportant at short time intervals (months to several years). Superimposed on the long-term mean are much faster processes such as horizontal eddy mixing, which can transport material in the north–south direction in days or weeks. The impacts of comet Shoemaker-Levy 9 on Jupiter in 1994 provided a rare opportunity to see the effects of eddy transport on small dust particles and trace chemical constituents deposited in the stratosphere immediately after impact. Particles spread rapidly from the impact latitude (45°S) to latitude 20°S, but there has been almost no transport farther toward the equator. Trace constituents such as HCN were observed to move across the equator into the northern hemisphere. [*See* PHYSICS AND CHEMISTRY OF COMETS.]

Long-term monitoring of the Jovian stratosphere has yielded some interesting observations of an oscillating temperature cycle at low latitudes. At pressures between 10 and 20 mbar, the equator and latitudes ±20° cool and warm alternately on time-scales of 2–4 years. The equator was relatively (1–2 K above the average 147 K) warm and latitudes ±20° were relatively (1–2 K below average) cool in 1984 and 1990. The reverse was true in 1986 and 1987. Changes in temperature must be accompanied by changes in the wind field, and these must be generated by stresses induced by wave forcing or convection. The similarities of the Jovian temperature oscillations to low-latitude temperature oscillations in the terrestrial atmosphere led some researchers to propose that the responsible mechanism is similar to that driving the quasi-biennial oscillation (QBO) on Earth: forcing by vertically propagating waves. The period of the oscillation is about 4 (Earth) years and so the phenomenon has been called the quasi-quadrennial oscillation or QQO.

The *Voyager* cameras provided much information about the shapes, motions, colors, and lifetimes of small features in the atmospheres of the giant planets. In terms of the number of features and their contrast, a progression is evident from Jupiter, with thousands of visible spots, to Uranus with only a few. Neptune has a few large spots that were seen for weeks and an abundance of small ephemeral white patches at a few latitudes. We do not have a good explanation for the contrasts and color, since the thermochemical equilibrium ices that form these clouds (NH_3, NH_4SH, H_2O, CH_4, and H_2S) are colorless. We need to know more about the composition, origin, and location of the colored material before we can understand how the contrasts are produced.

Fortunately, it is not necessary to understand how the contrasts are produced to study the meteorology of these features. One of the striking attributes of some of the clouds is their longevity. Jupiter's Great Red Spot has been observed since 1879 and may have existed much earlier. A little to the south of the GRS are three white ovals, each about one-third the diameter of the GRS. These formed in 1939–1940, beginning as three very elongated clouds (extending 90° in longitude) and rapidly shrinking in longitude. There are many smaller, stable ovals at some other Jovian latitudes. All of these ovals are anticyclonic and reside in anticyclonic shear zones. Since they are anticyclonic features, there is upwelling and associated high and thick clouds, and cool temperatures at the tropopause. Sinking motion takes place in a thin boundary region at the periphery of the clouds. The boundary regions are bright at 5-μm wavelength, consistent with relatively cloud-free regions of sinking. The Great Red Spot as revealed by *Galileo* instruments is actually much more complex, with cyclonic flow and small regions of enhanced 5-μm emission (indicating reduced cloudiness) in its interior.

Another attribute of many of the ovals is the oscillatory nature of their positions and sometimes shape. The most striking example is Neptune's Great Dark Spot, whose aspect ratio (ratio of shortest to longest dimension) varied by more than 20% with a period of about 200 hours, with a corresponding oscillation in orientation angle. Neptune's Dark Spot 2 drifted in latitude and longitude, following a sinusoidal law with amplitude 5° in latitude (between 50°S and 55°S) and 90° (peak to peak) in longitude. Other spots on Neptune and Jupiter, including the GRS, show sinusoidal oscillations in position. The Jovian spots largely remain at a fixed mean latitude, but the mean latitude of the GDS on Neptune drifted from 26°S to 17°S during the 5000 hours of observations by the *Voyager 2* camera. Ground-based observations in 1993 did not show a bright region at methane absorption wavelengths in the southern hemisphere, unlike the period during the *Voyager* encounter when the high-altitude white companion clouds were visible from Earth. The GDS may have drifted to the northern hemisphere and/or may have disappeared. Hubble Space Telescope images and ground-based images since the *Voyager* encounter show new spots at new latitudes.

Jupiter's Great Red Spot is often and incorrectly said to be the Jovian analog of a terrestrial hurricane. Hurricanes are cyclonic vortices. The GRS and other stable ovals are anticyclones. Hurricanes owe their (relatively brief) stability to energy generated from latent heating (condensation) over a warm ocean surface, where water vapor is abundant. Upwelling occurs in a broad circular region and subsidence is confined to a narrow core (the eye). The opposite is true for anticyclonic spots in the giant planet atmospheres, where subsidence takes pace in a narrow ring on the perimeter of the oval. The key to their stability is the long-lived, deep-seated background latitudinal shear of the jets. The stable shear in the jets provides an environment that is able to support the local vortices. Latent heat seems to play no role.

Atmospheric scientists have developed the groundwork for understanding the physics of vortices by studying the properties of idealized fluid flows. The simplest models assume that the atmosphere is like a shallow-water system, for example, an incompressible fluid bounded below by a hard surface. The depth of the fluid can be a function of latitude. Vorticity is conserved in a frictionless fluid, and the change of depth of the fluid with latitude can be determined by observations of the change in vorticity of individual vortices as they move in latitude (e.g., around the GRS, which acts as an obstacle to the flow). The simplest models consist of a single fluid layer. More complicated models have several layers.

A wide variety of phenomena can occur in such a system. Two important parameters are the Coriolis parameter beta = $[2\Omega \cos (\Theta)]/r$ and the radius of deformation $L_D = \sqrt{(gH)}/f$, where f is the local Coriolis parameter at the vortex location and H is the depth of the fluid layer. Beta is also approximately equal to

U/L^2, the ratio of the amplitude (U) of large-scale motions of length scale L squared. The three parameters that determine the properties of the vortices are the Rossby number, $R_0 = U/fL$, the ratio L_D/L, and a dimensionless beta parameter L/f. In different locations of the three-dimensional space defined by these three parameters, the vortices can be stable or unstable and can oscillate in aspect ratio and orientation, latitude and longitude by varying amounts. The theory has been remarkably successful in accounting for the range of phenomena observed, much of it alien to terrestrial meteorology. Nevertheless, the theory is an idealization and some aspects of the observations (e.g., the ejection of a jet of material from the GDS) are not accounted for.

V. ENERGETIC PROCESSES IN THE HIGH ATMOSPHERE

At low pressure (less than about 50 μbar), the mean free path for collisions becomes sufficiently large that lighter molecules diffusively separate from heavier ones. The level where this occurs is called the homopause. The outer planet atmospheres are predominantly composed of H_2 and He, with molecular hydrogen dissociating to atomic hydrogen, which becomes the dominant constituent at the exobase (the level where the hottest atoms can escape to space). This is also the region where solar EUV (extreme ultraviolet) radiation can dissociate molecules and ionize molecules and atoms. Ion chemistry becomes increasingly important at high altitudes. Some reactions can proceed at a rapid rate compared to neutral chemistry. Ion chemistry may be responsible for the abundant UV-absorbing haze particles (probably hydrocarbons) in the stratospheres of Jupiter and Saturn.

The high atmospheres of the giant planets are hot (1200 K for Jupiter to 300 K for Uranus and Neptune), much hotter than predicted on the assumption that EUV radiation is the primary energy source. Estimates prior to the *Voyager* observations predicted high-altitude temperatures closer to 250 K or less. One of the challenges of the post-Voyager era is to account for the energy balance of the high atmosphere. Possible sources of energy in addition of EUV radiation include (1) Joule heating, (2) currents induced by a planetary dynamo mechanism, (3) electron precipitation from the magnetosphere (and also proton and S and O ion precipitation in the Jovian auroral region), and (4) breaking inertia-gravity waves.

TABLE IV
Magnetic Field Parameters (Offset Tilted Dipole Approximation)

	Earth	Jupiter	Saturn	Uranus	Neptune
Tilt (degrees)	11.2	9.4	0.0	58.6	46.9
Offset (planetary radius)	0.076	0.119	0.038	0.352	0.485

Joule heating requires electric currents in the ionosphere that accelerate electrons and protons. It is a major source of heating in the terrestrial thermosphere. We do not have enough information on the magnetosphere to know how important this process or the others mentioned are for the giant planet atmospheres. The planetary dynamo current theory postulates that currents are established when electrons and ions embedded in the neutral atmosphere move through the magnetic field, forced by the neutral wind tied to the deeper atmosphere. Electric fields aligned with the magnetic field are generated by this motion and accelerate high-energy photoelectrons that collide with neutrals or induce plasma instabilities and dissi-

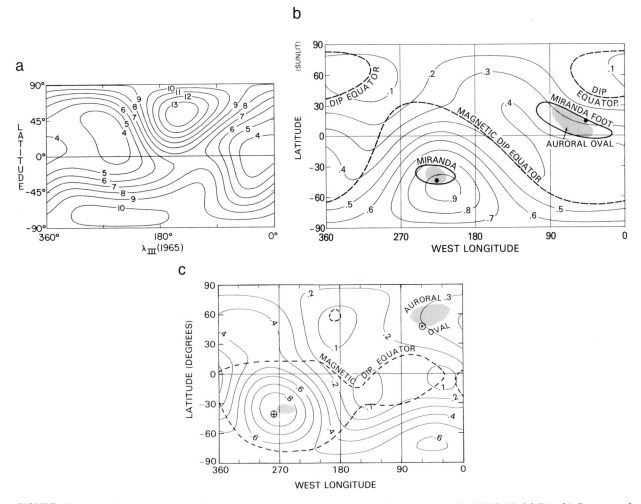

FIGURE 14 (a) Contours of magnetic field magnitude (gauss) on the surface of Jupiter (using the GSFC Model D$_4$). (b) Contours of constant magnetic field on the upper atmosphere of Uranus, along with the location of the auroral oval and the lines connected to the orbit of the satellite Miranda (Model Q$_3$). The magnetic dip equator is the location where the field lines are tangent to the surface. (c) Contours of constant magnetic field magnitude and pole locations (circled cross and dot) for Neptune (Model O$_8$). [From J. Connerney (1993). *J. Geophys. Res.* **98**, 18,659–18,679. Copyright American Geophysical Union.]

pate energy. Similar mechanisms are believed to be important in the terrestrial atmosphere.

Electron precipitation in the high atmosphere was one of the first mechanisms proposed to account for bright molecular hydrogen UV emissions. There is recent evidence for supersonic pole-to-equator winds in the very high atmosphere on Jupiter driven by auroral energy. These winds collide at low latitudes, producing supersonic turbulence and heating. Electron and ion precipitation outside of the auroral regions undoubtedly contributes to the heating, but the details remain unclear. The possible contribution from breaking planetary waves is difficult to estimate, but *Galileo Probe* measurements, details of the radio occultation profiles, and less direct lines of evidence point to a significant energy density in the form of inertia-gravity waves in the stratosphere and higher. How much of that is dissipated at pressures less than 50 μbar is unknown but could be significant to the energy budget of the high atmosphere.

The giant planets have extensive ionospheres. Like the neutral high atmospheres, they are hotter than predicted prior to the *Voyager* encounters. As for Earth, the ionospheres are highly structured, having a number of high-density layers. Layering in the terrestrial ionosphere is partially due to the deposition of metals from meteor ablation. The same mechanism is thought to be operative in the giant planet ionospheres. The Jupiter and Saturn ionospheres are dominated by the H_3^+ ion, whereas those of Uranus and Neptune are dominated by H^+.

Auroras are present on all the giant planets. Auroras on Earth (the only other planet in the solar system known to have auroras) are caused by energetic charged particles streaming down the high-latitude magnetic field lines. The most intense auroras on Earth occur when a solar flare disturbs the solar wind, producing a transient in the flow that acts on Earth's magnetosphere through ram pressure. As the magnetosphere responds to the solar wind forcing, plasma instabilities in the tail region accelerate particles along the high-latitude field lines.

The configuration of the magnetic field is one of the key parameters that determines the location of auroras. Jupiter's magnetosphere is enormous compared to Earth's. If its magnetosphere could be seen by the naked eye from Earth, it would appear to be the size of the Moon (about 30 arc minutes), whereas Jupiter's diameter is less than one arc minute. To a first approximation, the magnetic fields of Earth and the giant planets can be described as tilted dipoles, offset from the planet center. Table IV lists the strength, tilt, and radial offsets for each of these plan-

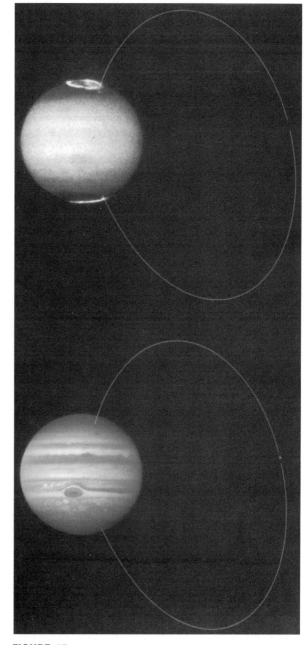

FIGURE 15 (Top) Image of Jupiter at ultraviolet wavelengths taken with the Wide Field and Planetary Camera 2 on the Hubble Space Telescope. Bright auroral ovals can be seen against the dark UV-absorbing haze and in the polar regions. Jupiter's north magnetic pole is tilted toward Earth, making it easier to see the northern auroral oval as well as some diffuse emission inside the oval. Small bright spots just outside the oval in both hemispheres are at the location of the magnetic field lines connecting to Io, depicted by a blue curve. Io is dark at UV wavelengths. (Bottom) Image taken a few minutes after the one above in a filter that samples the violet part of the spectrum just within the range that the human eye can detect. The Great Red spot appears dark at this wavelength and can just be seen in the top image as well. Io's small disk appears here along the blue curve, which traces the magnetic field lines in which it is embedded. (Courtesy of J. Trauger and J. Clarke.)

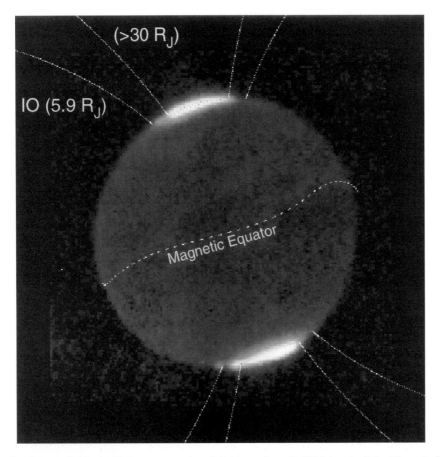

FIGURE 16   Auroral regions are bright in this image at wavelength 3.4 μm, where the H_3^+ ion emits light. Magnetic field lines connecting to Io and to the 30-Jupiter-radius equator crossing are shown. The brightest emissions are poleward of the $30R$, field line, which means the precipitating particles responsible for this emission come from more distant regions on the magnetosphere. [Reprinted with permission from J. Connerney *et al.* (1993). *Science* **262**, 1035–1038. Copyright 1993 American Association for the Advvancement of Science.]

ets. Earth and Jupiter have relatively modest tilts and offsets, Saturn has virtually no tilt and almost no offset, whereas Uranus and Neptune have very large tilts and offsets. Such diversity presents a challenge to planetary dynamo modelers. [*See* PLANETARY MAGNETO-SPHERES.]

The mapping of the magnetic fields onto the upper atmosphere determines where auroral particles intercept the atmosphere. Maps for Jupiter, Uranus, and Neptune are shown in Fig. 14, along with locations of field lines connected to the oribits of some satellites that may be important for auroral formation. The configuration for Saturn is not shown because contours of constant magnetic field magnitude are concentric with latitude circles owing to the field symmetry. Because of the large tilts and offsets for Uranus and Neptune, auroras on those planets occur far from the poles.

The Jovian aurora is the most intense and has received the most scrutiny. The remainder of this section will focus on what is known about it. It has been observed over a remarkable range of wavelengths, from X rays to the infrared, and possibly in the radio spectrum as well. Energetic electrons from the magnetosphere dominate the energy input, but protons and S and O ions contribute as well. Sulfur and oxygen k-shell emission seems to be the most plausible explanation for the X rays. Models of energetic electrons impacting on molecular hydrogen provide a good fit to the observed molecular hydrogen emission spectra. Secondary electrons as well as UV photons are emitted when the primary impacting electrons dissociate the molecules, and these secondaries also contribute to the UV emissions. Some of the UV emitted radiation is reabsorbed by other hydrogen molecules and some is absorbed by methane molecules near the top of the homopause. From the detailed shape of the spectrum it is possible to infer the depth of penetration of electrons into the upper atmosphere. In the near infrared (2–4 μm), emissions from the H_3^+ ion are prominent.

Attempts to account for all the observations call for more than one type of precipitating particle and more than one type of aurora.

Ultraviolet auroras from atomic and molecular hydrogen emissions are brightest within an oval that is approximately bounded by the closed field lines connected to the middle magnetosphere (corresponding to a region some 10–30 Jupiter radii from the planet) rather than the orbit of Io or open field lines connected to the tail. Weaker diffuse and highly variable UV emissions appear closer to the pole. They are produced by precipitation of energetic particles originating from more distant regions in the magnetosphere. There is also an auroral hot spot at the location where magnetic field lines passing through Io enter the atmosphere (the Io flux tube footprint). All of these features are evident in Fig. 15.

Io is a significant source of sulfur and oxygen, which come off its surface. The satellite and magnetosphere produce hot and cold plasma regions near the Io orbit, which may stimulate plasma instabilities. High spatial resolution, near-infrared H_3^+ images show emission from a region that maps to the last closed field lines far out in the magnetosphere (Fig. 16). This and evidence for auroral response to fluctuations in solar wind ram pressure indicate that at least some of the emission is caused by processes that are familiar to modelers of the terrestrial aurora. [*See* Io.]

Auroral emission is strongest over a small range of longitudes. In the north, longitudes near 180°, System III coordinates (which rotate with the magnetic field) show enhanced emission in the UV and also in the thermal infrared. The spectrum of the aurora in the UV resembles electron impact on molecular hydrogen, except the shortest wavelengths are deficient. This deficit can be accounted for if the emission is occurring at some depth in the atmosphere (near 10 μbar) below the region where methane and acetylene absorb UV photons. By contrast, the Uranian high atmosphere is depleted in hydrocarbons and does not produce an emission deficit.

Energy deposition at depth is also required to ex-plain the warm stratospheric temperatures seen in the 7.8-μm methane band. At 10 μbar of pressure, the hot spot region near longitude 180° appears to be 60–140 K warmer than the surrounding region, which is near 160 K. Undoubtedly such temperature contrasts drive the circulation of the high atmosphere. Auroral energy also contributes to anomalous chemistry. An enhancement is seen in acetylene emission in the hot spot region, whereas ethane emission decreases there. A significant part of the acetylene enhancement could be due simply to the higher emission from a warmer stratosphere, but a decrease in ethane requires a smaller ethane mole fraction.

Future work on the auroras of Jupiter and the other giant planets will focus on questions of which types of particles are responsible for the emissions, from what regions of the magnetosphere or torus do they originate, what are the acceleration mechanisms, and how does the deposited energy drive circulation and chemistry in the high atmosphere.

BIBLIOGRAPHY

Atreya, S. K., Pollack, J. B., and Matthews, M. S. (eds.) (1989). "Origin and Evolution of Planetary and Satellite Atmospheres." Univ. Arizona Press, Tucson.

Beaty, J. K., and Chaikin, A. (eds.) (1990). "The New Solar System," 3rd Ed. Sky Publishing, Cambridge, Mass.

Beebe, R. (1994). "Jupiter: The Giant Planet." Smithsonian Institution Press, Washington, D.C.

Bergstralh, J. T., Miner, E. D., and Matthews, M. S. (eds.) (1991). "Uranus." Univ. Arizona Press, Tucson.

Chamberlain, J. W., and Hunten, D. M. (1987). "Theory of Planetary Atmospheres: An Introduction to Their Physics and Chemistry," 2nd Ed. Academic Press, Orlando, Fla./San Diego.

Cruikshank, D. P. (ed.) (1995). "Neptune." Univ. Arizona Press, Tucson.

Gehrels, T., and Matthews, M. S. (eds.) (1984). "Saturn." Univ. Arizona Press, Tucson.

Rogers, J. H. (1995). "The Planet Jupiter." Cambridge Univ. Press, New York.

Wayne, R. P. (1991). "Chemistry of Atmospheres: An Introduction to the Chemistry of the Atmospheres of the Earth, the Planets, and Their Satellites, "2nd Ed. Oxford Univ. Press, New York.

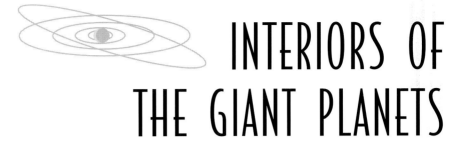

INTERIORS OF THE GIANT PLANETS

I. General Overview

II. Constraints on Planetary Interiors

III. Equations of State

IV. Interior Modeling

V. Planetary Interior Models

VI. Jovian Planet Evolution

VII. Future Directions

Mark S. Marley
New Mexico State University

GLOSSARY

Adiabatic compression: Compression of a gas without exchange of heat. Expansion or compression of rising or sinking air masses in Jovian planets is commonly assumed to be driven by adiabatic processes.

Bar: Unit of measure of atmospheric pressure, equal to the air pressure on Earth at sea level. Typical planetary interior pressures are measured in megabars (Mbar) or 10^6 bar.

Convection: Transport of energy by mass motion. In turbulent regions of planetary atmospheres and interiors, rising parcels of hot air and sinking parcels of cool air transport energy outward from the interior.

Critical point: Temperature and pressure for a given material above which there is no distinction between the liquid and gas phases.

Equation of state: Equation relating the pressure of a material to its temperature, composition, and density, typically derived from observational and theoretical considerations.

Heavy elements: All elements other than hydrogen and helium.

Ice: Mixture of water, ammonia, methane, and other volatile compounds in the interiors of Jovian planets, not literally in the form of condensed "ice."

Metallic hydrogen: High-pressure ($\gtrsim$1.4 Mbar) metallic form of hydrogen found in the interiors of Jupiter and Saturn.

Rock: Mixture of iron, silicon, magnesium, and other refractory elements found in the interiors of Jovian planets.

The giant, or Jovian planets—Jupiter, Saturn, Uranus, and Neptune—account for 99.5% of all the planetary mass in the solar system. An understanding of the formation and evolution of the solar system thus requires knowledge of the composition and physical state of the material in their interiors. But such information does not come easily. The familiar faces of these planets, such as the cloud-streaked disk of Jupiter, tell relatively little about what lies beneath. Knowledge of these planetary interiors must instead be gained from analysis of the mass, shape, and gravitational fields of the planets. The study of the behavior of planetary materials at high densities and pressures further provides the experimental and theoretical framework upon which planetary interior models are subsequently based. Interior models provide a window into the internal structure of these planets and shed light on processes that led to planet formation in the solar nebula.

I. GENERAL OVERVIEW

Several lines of observational evidence provide information on the composition and structure of the giant planets. The first and most easily obtained quantities are the mass (known from the orbits of natural satellites), radius (polar and equatorial radii), and rotation period (obtained originally from telescopic observations, now derived from remote and *in situ* observations of planetary magnetic fields). By the 1940s, these fundamental observations, coupled with the advances in understanding the high-pressure behavior of matter in the 1920s and 1930s, constrained the composition of Jupiter and Saturn to be predominantly hydrogen. Direct measurement of the planets' high-order gravity fields, interior rotation states, and heat flow, along with spacecraft and ground-based spectroscopic detection of atmospheric elemental composition, has since allowed the construction of more detailed interior models.

These models divide the giant planets into two broad categories. Jupiter and Saturn are predominantly hydrogen–helium gas giants with a somewhat enhanced abundance of heavier elements and dense cores. Uranus and Neptune are ice giants with hydrogen–helium envelopes and dense cores. The following description of Jupiter's interior, as illustrated schematically in Fig. 1, is qualitatively valid for Saturn and serves as a point of departure for understanding the interiors of Uranus and Neptune. Individual planetary interior structures are discussed in Section V.

The interior begins at the base of the outermost atmospheric envelope that we can see directly. The Jovian atmospheres consist of a gaseous mixture of molecular hydrogen, helium, methane, ammonia, and water. At 1 bar pressure (the pressure at sea level on Earth), the temperature in Jupiter's atmosphere is 165 K. Near this level, the ammonia condenses into clouds; the water condensation level is even deeper. In the colder atmospheres of Uranus and Neptune, methane also condenses into clouds. Deeper into the planet the pressure of the overlying atmo-

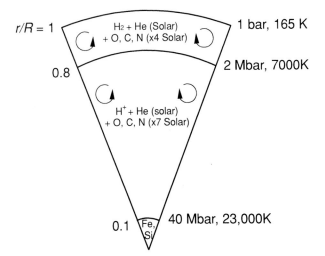

FIGURE 1 Highly schematic, idealized cross section of the interior structure of Jupiter. The numbers along the left refer to the relative radius ($r/R_{Jupiter}$) of the rocky core and the molecular-to-metallic hydrogen phase transition. On the right are listed the approximate temperatures and pressures at which these interfaces occur. Arrows indicate convection. H^+ is metallic hydrogen; Fe and Si suggest all rocky elements; likewise O, C, and N include all elements in more volatile ("icy") material. Numbers in parentheses give the enhancement of the particular element over its solar value. The real Jupiter is undoubtedly more complex. It is likely that interfaces are gradual and the composition of the various regions is inhomogeneous.

sphere compresses the gas, increasing its temperature and density. This process, adiabatic compression, is the same one responsible for the increase in temperature with decreasing altitude on Earth. One hundred kilometers beneath the cloud tops the temperature has reached 350 K.

As pressures and temperatures increase, the gas begins to take on the characteristics of a liquid. Since the critical point of the dominant constituent, molecular hydrogen, lies at 13 bars and 33 K, there is not a distinct gas–liquid phase boundary. By several hundred thousand bars the envelope closely resembles a hot liquid. This characteristic of the giant planets—they exist in the supercritical regime of their primary constituent—leads to their most fundamental property: these planets have essentially bottomless atmospheres.

Deeper into the planet the temperature and pressure continue to steadily increase. By 20,000 km beneath the cloud tops the temperature reaches 7000 K and the pressure is 2 Mbar. Recent experiments suggest that by this point hydrogen, previously present as molecules of H_2, has undergone a phase transition to a liquid, metallic state. Most of the mass of Jupiter consists of this metallic hydrogen:

protons embedded in a sea of electrons. Helium and other constituents exist as impurities in the hydrogen soup. For the remaining 50,000 km to Jupiter's core, the pressure and temperature continue to rise, reaching 200 Mbar and 20,000 K in the deep interior. Near the center of the planet the composition changes, perhaps gradually, from a predominantly hydrogen/helium mixture to a combination of rock and ice. The density of this rock and ice core is 10,000 to 20,000 kg m^{-3}, higher than the metallic hydrogen density of about 1000 kg m^{-3} (uncompressed water, like that which comes out of a tap, also has a density of 1000 kg m^{-3}).

Throughout most of the interior, the transport of energy by radiation is severely hampered by the high opacity of compressed hydrogen. Other constituents such as methane and water effectively block energy transport by radiation in those regions of the spectrum where the hydrogen is a less powerful absorber. Since conduction of heat by the thermal motion of molecules is also inefficient, convection is the dominant energy transport mechanism throughout the interior. However, there may be a thin zone in Jupiter's interior at temperatures of 1000 to 3000 K where energy transport by radiation indeed dominates. The rising and sinking convective cells in the interior move slowly, at velocities of just centimeters per second or less. Because of the continuous nature of the atmosphere, the wind patterns seen in the belts and zones of Jupiter and Saturn may have roots that reach into the deep, convective interior of the planet. Indeed the winds measured by the *Galileo* spacecraft's atmosphere probe continued to blow steadily at the deepest levels reached by the probe, about 20 bars.

The interior of Saturn is much like that of Jupiter. Saturn's lower mass and consequently lower pressures produce a smaller metallic hydrogen region. Uranus and Neptune lack a metallic hydrogen region; instead, at about 80% of their radius the abundance of methane, ammonia, and water increases markedly. In this region, temperatures of over 5000 to 10,000 K produce an ocean of electrically charged water, ammonia, and methane molecules, along with more complex compounds. Most of the mass of Uranus and Neptune exists in such a state. Deep in their interiors, all the planets likely have cores of primarily rocky material.

This picture of the interiors of the Jovian planets has been painstakingly pieced together since the 1930s. This chapter discusses the components of observation, experiment, and theory that are combined to reach these conclusions.

II. CONSTRAINTS ON PLANETARAY INTERIORS

A. GRAVITATIONAL FIELD

A variety of observations yield information about the makeup and interior structure of the Jovian planets. The mass of each of the four Jovian planets (Table I) has been known with some precision since the discovery of their natural satellites. The masses range from 318 times the Earth's mass ($M_\oplus$) for Jupiter to 14.5 $M_\oplus$ for Uranus. A second fundamental observable property is the radius of each planet measured at a specified pressure, typically the 1-bar pressure level. Radii are most accurately measured by the occultation technique, in which the attenuation of the radio signal from a spacecraft is measured as the spacecraft passes behind the planet. Jovian planet radii range from 11 times the Earth's radius ($R_\oplus$) for Jupiter to 3.9 $R_\oplus$ for Neptune. The combination of mass and radius allows calculation of mean planetary density, $\bar{\rho}$. Although a surprising amount can be learned about the bulk composition of a planet from just $\bar{\rho}$, more subtle observations are required to probe the detailed variation of composition and density with radius.

If the Jovian planets did not rotate, they would assume a spherical shape and their external gravitational field would be the same as that of a point of the same mass. No information about the variation in density with radius could be extracted. Fortunately, the planets do rotate and their response to their own rotation provides a great deal more information. This response is observed in their external gravitational field.

For a uniformly rotating body in hydrostatic equilibrium, the external gravitational potential, Φ, is

$$\Phi = -\frac{GM}{r}\left(1 - \sum_{n=1}^{\infty}\left(\frac{a}{r}\right)^{2n} J_{2n}P_{2n}(\cos\theta)\right)$$

G is the gravitional constant, M is the planetary mass, a is the equatorial radius, θ is the colatitude (the angle between the rotation axis and the radial vector **r**), P_{2n} are the Legendre polynomials, and the dimensionless numbers J_{2n} are known as the gravitational moments. The assumption of hydrostatic equilibrium means that the planet is in a fluid state, responding only to its rotation, and there are no permanent, nonaxisymmetric lumps in the interior. This assumption is believed to be quite good for the Jovian planets.

TABLE I
Observed Properties of Jovian Planets

Quantity	Jupiter	Saturn	Uranus	Neptune
M (kg)	1.8979×10^{27}	5.6833×10^{26}	8.679×10^{25}	1.024×10^{26}
a (km)	$71,492 \pm 4$	$60,268 \pm 4$	$25,559 \pm 4$	$24,764 \pm 20$
P_s (hours)	9.92492	10.65622	17.24 ± 0.01	16.11 ± 0.05
$J_2 \times 10^6$	$14,697 \pm 1$	$16,331 \pm 18$	$3,516 \pm 3$	$3,538 \pm 9$
$J_4 \times 10^6$	-584 ± 5	-914 ± 61	-31.9 ± 5	-38 ± 10
$J_6 \times 10^6$	31 ± 20	108 ± 50	—	—
q	0.08923	0.15485	0.02951	0.0261
Λ_2	0.1647	0.1055	0.1191	0.136
$\bar{\rho}$ (g cm^{-3})	1.334	0.69	1.27	1.64
Y	0.238 ± 0.007	0.06 ± 0.05	0.262 ± 0.048	0.26 ± 0.05
T_1 (K)	165	135	76	74

The gravitational harmonics are found from observations of the orbits of natural satellites, precession rates of elliptical rings, and perturbations to the trajectories of spacecraft. As a spacecraft flies by a planet, it samples the gravitational field at a variety of radii. Careful tracking of the spacecraft's radio signal reveals the Doppler shift due to its acceleration in the gravitational field of the planet. Inversion of these data yields an accurate determination of the planet's mass and gravitational harmonics (see Table I). In practice, it is difficult to measure terms of order higher than J_4, and the value of J_6 is generally quite uncertain. Progressively higher-order gravitational harmonics reflect the distribution of mass in layers progressively closer to the surface of the planet. Thus, even if they could be measured accurately, terms such as J_8 would not contribute greatly to understanding of the deep interior.

A planet's response to its own rotation is characterized by how much a surface of constant total potential (including the effects of both gravity and rotation) is distorted. The amount of distortion on such a surface of constant potential, known as a level surface, depends on the distribution of mass inside the planet, the mean radius of the level surface, and the rotation rate. The distortion, or oblateness, of the outermost level surface is measured from direct observations of the planet and is given by $\varepsilon = (a - b)/a$, where b is the polar radius. The equatorial and polar radii can be found from direct telescopic measurement or, more accurately, from observations of spacecraft or stellar occultations. Distortion of level surfaces cannot be described simply by ellipses. Instead the distortion is more complex and

must be described by a power series of shapes, as illustrated in Fig. 2. The most obvious distortion of a spherical planet (Fig. 2a) is illustrated in Fig. 2b. More subtle distortions are described by harmonic coefficients of ever increasing degree, as illustrated in Figs. 2c and 2d.

A nonrotating, fluid planet would have no J_{2n} terms in its gravitational potential. Thus the gravitational harmonics provide information on how the shape of a planet responds to rotating-frame forces arising from its own spin. Since the gravitational harmonics depend on the distribution in mass of a particular planet, they cannot be easily compared between planets. Instead a dimensionless linear response coefficient, Λ_2, is used to compare the response of each Jovian planet to rotation. To lowest order in the square of the angular planetary rotation rate, ω^2, $\Lambda_2 \approx J_2/q$, where $q = \omega^2 a^3/GM$. Table I lists the Λ_2 calculated for each planet. The Jovian planets rotate rapidly enough that the nonlinear response of the planet to rotation is also important and must be considered by computer models.

Since the gravitational harmonics provide information about the planet's response to rotation, interpretation of the harmonics requires accurate knowledge of the rotation rate of the planet. Before the space age, observations of atmospheric features as they rotated around the planet provided rotation periods. This method, however, is subject to errors introduced by winds and weather patterns in the planet's atmosphere. Instead, rotation rates are now found from the rotation rate of the magnetic field of each planet, generally as measured by the Voyager spacecraft (radio emissions arising from charged particles in Jupiter's magneto-

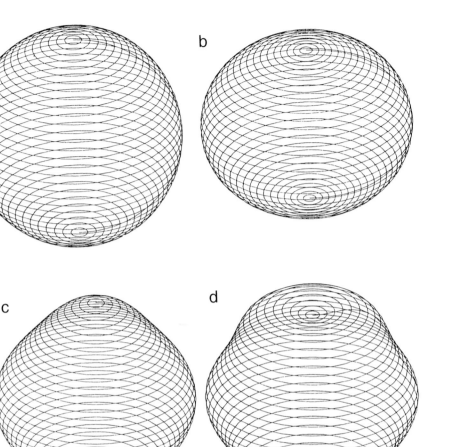

FIGURE 2   Illustration of the ways that a planet changes shape owing to its own rotation. A nonrotating planet (a) is purely spherical. Saturn's distortion due to its gravitional harmonic J_2 is shown approximately to scale in (b). The J_4 and J_6 distortions of Saturn are shown in (c) and (d), exaggerated by about 10 and 100 times, respectively.

sphere can be detected by radio telescopes on Earth). This approach assumes that the magnetic field is generated by convective motions deep in the electrically conducting interior of the planet and that the field's rotation consequently follows the rotation of the bulk of the interior.

B. ATMOSPHERE

The observable atmospheres of the Jovian planets provide further constraints on planetary interiors. First, the atmospheric temperature at 1 bar pressure, or T_1, constrains the temperature of the deep interior. The interior temperature distribution of the Jovian planets is believed to follow a specified pressure–temperature

path known as an *adiabat*. For an adiabat, knowledge of the temperature and pressure at a single point uniquely specifies the temperature as a function of pressure at all other points along the adiabat. Thus T_1 gives information about the temperature structure throughout the convective interior of the planet. Both the amount of sunlight that the atmosphere absorbs and the amount of heat carried by convection up from the interior of the planet to the atmosphere control T_1. For each planet, save Uranus, T_1 is higher than expected if the atmosphere were simply in equilibrium with sunlight. Thus each Jovian planet, save Uranus, exhibits a significant internal heat flow, ranging from 0.3 W m^{-2} at Neptune, to 2.0 W m^{-2} at Saturn, to 5.4 W m^{-2} at Jupiter.

Second, the composition of the observable atmosphere also holds clues to the internal composition.

This is because of the supercritical nature of the Jovian atmospheres. The principal component of the Jovian atmosphere, hydrogen, does not undergo a vapor–liquid phase change above 33 K. Since the planets are everywhere warmer than this temperature, the observed atmosphere is directly connected to the deep interior. Knowledge of the composition of the top of the atmospheres therefore provides some insight to the composition at depth. [*See* ATMOSPHERES OF THE GIANT PLANETS.]

The *Galileo* spacecraft atmosphere probe returned direct measurements of the composition of Jupiter's atmosphere. The composition of the remaining planetary atmospheres is inferred from spectroscopy. In planetary science, compositions are often stated relative to "solar" abundances. Solar abundances are the relative quantities of elements present in the solar nebular at the time of planetary formation. The solar abundances of hydrogen and helium are about 70% and 28% by mass, respectively. Oxygen, carbon, nitrogen, and the other elements make up the remainder. These elements are collectively called the "heavy elements" to distinguish them from hydrogen and helium. Measurements of the rate at which the atmospheric pressure decreases with height in these atmospheres require that hydrogen and helium must be the dominant components of the atmospheres of all four Jovian planets. Spectroscopy supports this conclusion and gives the relative abundance of hydrogen and helium. The helium mass fraction of each atmosphere, Y, is listed in Table I. The heavier elements are generally enriched in the Jovian atmospheres over their solar abundances.

C. MAGNETIC FIELD

All four Jovian planets possess a magnetic field. Jupiter's is large and complex; Saturn's is less complex and smaller. The magnetic fields of both Uranus and Neptune are very complex: they deviate substantially from a dipole and their field axes are tilted strongly with respect to their rotation axes. The only known mechanism for producing global planetary magnetic fields, the hydromagnetic dynamo process, requires nonuniform motion of a large electrically conductive region. Convection in the highly conductive interior of the Jovian planets is presumed responsible for formation of their fields. The level of complexity of each field plausibly relates to the depth of the electrically conducting region. Magnetic fields formed by relatively small, deep sources may be simpler and smaller than fields formed by large, shallow dynamos. [*See* PLANETARY MAGNETOSPHERES.]

III. EQUATIONS OF STATE

A. OVERVIEW

Beyond observations of the planets themselves, a second major ingredient in interior models is an *equation of state*, or EOS. An EOS is a group of equations, derived from laboratory observations and theory, that relate the pressure (P) of a mixture of materials to its temperature (T), composition (x), and density (ρ). Any attempt to model the interior structure of a giant planet must rely on an EOS. The construction of accurate equations of state is a primary activity in planetary interior modeling.

For an ideal gas, the well-known EOS is $P = nkT$. Here k is Boltzmann's constant and n is the number density of the gas. The composition of an ideal gas does not affect the pressure; only the number of molecules and atoms in a given volume, n, enters the equation. Under the conditions of high temperature and pressure found in the interiors of the giant planets, atoms and molecules interact strongly with one another, thus violating the conditions under which the ideal gas EOS holds. Additionally, the typical pressures reached in the interiors of the giant planets (tens to hundreds of megabars) are also amply sufficient to modify the electronic structure of individual atoms and molecules. This further adds to the challenge of understanding the EOS. In short, the properties of planetary materials at high pressures will differ substantially from those encountered in their low-pressure, and more familiar, forms. In practice, the behavior of planetary materials must be understood from both experiments and theory.

For pressures less than about 1–2 Mbar, depending on the material, shock wave experiments provide guidance in the construction of equations of state. In these experiments, a high-velocity projectile is fired into a container holding a sample of the material under study. The thermodynamically irreversible nature of shock compression causes both high temperatures and high pressures in the sample. High-speed measuring devices record the temperatures, pressures, and densities achieved during the brief experiments. A photograph of a shock tube at Lawrence Livermore National Laboratory, used extensively for planetary work, is shown in Fig. 3.

The temperatures and pressures reached in these experiments are the closest that terrestrial laboratories can come to reliably duplicating the conditions in the interiors of the Jovian planets. For Jupiter, the experi-

FIGURE 3 Photograph of the 60-foot-long, two-stage light-gas gun at Lawrence Livermore National Laboratory. This apparatus is used to obtain equation of state, shock temperature, and electrical conductivity data for planetary liquids (H$_2$, He, H$_2$O, NH$_3$, and the "synthetic Uranus" mixture). A projectile, fired from gun on left side of photo, travels down the barrel and impacts sample in container on extreme right.

ments model conditions about 90% of the way out from the planet's center. The experiments can equal pressures found at about 70% of Saturn's radius and 50% of Uranus and Neptune's.

Diamond anvils are used in another type of experiment to squeeze microscopically small samples of planetary materials to very high pressure. Although these experiments can reach higher pressures than the shock experiments, they are necessarily conducted at room temperature. Because of the low temperatures, these experiments are less applicable to the interiors of the Jovian planets.

B. HYDROGEN

For pressures less than about 1 Mbar, the behavior of molecular hydrogen, H$_2$, is understood fairly well from theory and the shock experiments. At higher pressures such as those encountered deeper in the interiors of Jupiter and Saturn, the hydrogen molecules are squeezed so closely together that they begin to lose their individual identities. Under these conditions, the hydrogen undergoes a phase transition to a metallic, pressure-ionized state commonly called metallic hy-

drogen. A shock wave experiment suggests that this transition occurs near 1.4 Mbar and 3000 K, however, more work is needed to fully understand this phase transition. At higher pressures, liquid metallic hydrogen consists of a dense mixture of ionized protons and electrons at temperatures over about 10,000 K. The EOS of liquid metallic hydrogen is understood well theoretically for pressures above about 10 Mbar, but the EOS is not well constrained from 1 to 10 MBar. A hydrogen phase diagram and temperature/pressure profiles for each giant planet are shown in Fig. 4. The detailed behavior of hydrogen near the phase transition itself, and even the exact location of the phase transition, is not known. Thus various simplifying assumptions must be made when considering these regions of giant planets. The EOS in this region is typically based on a mixture of theory and interpolation.

C. HELIUM

Helium has not been as well studied as hydrogen, but shock wave data do provide information to several hundred kilobars. Above that pressure, theory must guide models of the behavior of this element. Though

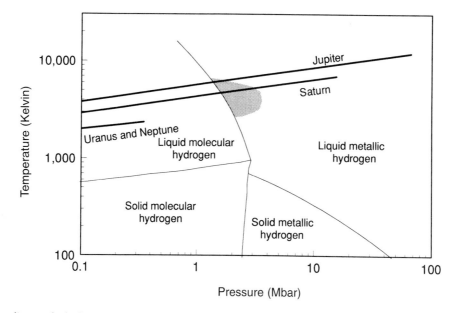

FIGURE 4   Phase diagram for hydrogen, the main constituent of Jupiter and Saturn. The approximate domains of liquid metallic hydrogen and molecular hydrogen are shown along with approximate interior temperature profiles for the Jovian planets. The shaded area indicates the approximate region in the interior of Saturn and, possibly, Jupiter where helium and metallic hydrogen cannot coexist in equilibrium. The locations of all phase boundaries are highly uncertain.

the equations of state of hydrogen and helium individually are reasonably understood, the behavior of mixtures of these two constituents is less well constrained. This is a serious theoretical void because the hydrogen–helium mixture composes most of the mass of Jupiter and Saturn and is an important component at Uranus and Neptune.

Current calculations of the behavior of hydrogen and helium mixtures show that helium is not soluble in hydrogen at all mass fractions and temperatures. At the temperatures predicted in the interior of Saturn, hydrogen and helium do not mix. According to this model, droplets of helium-rich material are constantly forming in the outer envelope of the planet. Because they are more dense than their surroundings, the drops fall to deeper, warmer levels of the envelope, where temperatures are high enough to again allow mixing. Thus at certain depths in Saturn's interior it is always raining helium. This remarkable conclusion is discussed in Section V, C in the context of the Saturn interior models.

D. ICES

The term "ices" is applied to mixtures of volatile elements in the form of water (H_2O), methane (CH_4), and ammonia (NH_3) in solar proportions, not necessarily present as intact molecules. Ices are a primary constit-

uent of Uranus and Neptune, but are less abundant in Jupiter and Saturn. As the planetary interior temperatures are over several thousand degrees Kelvin, they are present as liquids. Shock wave data on a mixture of water, isopropanol, and ammonia (dubbed "synthetic Uranus") have helped establish the equation of state of this material at pressures less than about 2 Mbar and temperatures less than about 4000 K. These experiments helped confirm that ices are a primary constituent of Uranus and Neptune. The shock wave data on this mixture show that at pressures exceeding ~200 kbar, the planetary ice constituents ionize to form an electrically conductive fluid. At pressures ≳1 Mbar, the ice constituents dissociate and the EOS becomes quite "stiff," meaning the density is not particularly sensitive to the pressure.

E. ROCK

The remaining planetary constituents are lumped into the category "rock." Rock is presumed to consist of a solar mixture of silicon, magnesium, and iron, with uncertain additions of oxygen and the remaining elements. Although the rock equation of state is not well known, it is also expected to be quite "stiff." The lack of a detailed rock EOS is not a serious limitation for planetary interior models, as the rock component is not a major fraction of the mass.

F. MIXTURES

Since all the planetary components, including gas, ice, and rock, are likely mixed throughout the interiors, equations of state of such mixtures are required for interior modeling. Hydrogen–helium mixtures, considered earlier, may not exist at all temperatures, pressures, and concentrations. The solubility of other mixtures, for example, rock or oxygen in metallic hydrogen, is less well known. From the limited data it appears that the planetary constituents other than hydrogen and helium do mix well under the temperature and pressure conditions typically found in planetary interiors. This is because delocalization of electrons at high pressure diminishes the well-defined intermolecular bonds present at lower pressures. Thus the separation of planetary materials into distinct layers of "pure" rock or ice is highly unlikely. If correct, such considerations also have important cosmogonic implications. For example, the rock cores of the planets likely did not "settle" from an initially well-mixed planet, but instead the gaseous components likely collapsed onto a preexisting rocky nucleus that formed in the protosolar nebula.

Since the EOS of all possible mixtures has not been studied, either experimentally or theoretically, approximations must be employed. One approximation, the additive volume law, weights the volumes of individual components in a mixture by their mass fraction. An implication of such approximations is that the computed densities of mixtures of rock, ice, and gas can be similar to that of pure ice. Thus it is not currently possible to differentiate between models of Uranus and Neptune with mantles of pure ice and models with mantles of a mixture of rock, ice, and gas.

IV. INTERIOR MODELING

In addition to an equation of state for the material in the interior of a planet, two more components are required to produce an interior model. The temperature and composition in the interior as a function of pressure, $T(P)$ and $x(P)$, must also be known. (These quantities are described as functions of pressure since the pressure increases monotonically toward the center of the planet.) The first of these ingredients, $T(P)$, is not difficult to find. If the Jovian planets are fully convective in their interiors, transporting internal heat to the surface by means of convection, the relation between temperature and pressure in their interiors is known as an adiabat. An adiabat has the property that

knowledge of a single temperature and pressure at any point allows specification of T as a function of P at any other point (assuming the material's EOS is known). Since the temperature and pressure in the convecting region of each Jovian atmosphere have been measured, a unique $T(P)$ relation for each planet can be found. If there is indeed a small radiative region in Jupiter's interior, a small correction is required.

More difficult to specify is the variation in composition through each planet, $x(P)$. The composition of each planet's atmosphere is known, but there is no guarantee that this composition is constant throughout the planet. Earth's core, for example, has a very different composition from the crust. For the Jovian planets, an $x(P)$ relation is typically guessed at, an interior model computed, and the results compared to the observational constraints. With multiple iterations, a variation in composition with pressure that is compatible with the observations is eventually found.

The combination of these three ingredients, an equation of state $P = P(T, x, \rho)$, a temperature–pressure relation, $T = T(P)$, and a composition–pressure relation, $x = x(P)$, completely specifies pressure as only a function of density, $P = P(\rho)$. Since the Jovian planets are believed to be fluid to their centers, the pressure and density are also related by the equation of hydrostatic equilibrium for a rotating planet:

$$\frac{\partial P}{\partial r} = -\rho(r)g(r) + \frac{2}{3} r\omega^2 \rho(r)$$

where g is the gravitational acceleration at radius r and ω is the angular rotation rate. This relation simply says that, at equilibrium, the pressure gradient force at each point inside the planet must support the weight of the material at that location. Combining the equation of hydrostatic equilibrium with the $P(\rho)$ relation finally allows determinations of the variation of density with radius in a given planetary model, $\rho = \rho(r)$.

The computed model must then satisfy all the observational constraints discussed in Section II. Total mass and radius of the model are easily tested. The response of the model planet to rotation and the resulting gravitational harmonics must be calculated and compared with observations. Figure 5, showing the relative contribution versus the depth from the center of the planet, illustrates the regions of a Saturn model that contribute to the calculation of the gravitational harmonics J_2, J_4, and J_6. Higher-degree modes provide information about layers of the planet progressively closer to the surface.

The construction of computer models that meet all the observational constraints and use realistic equa-

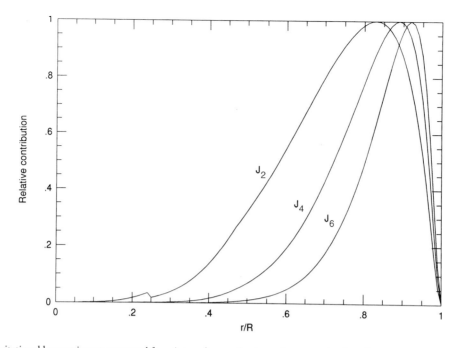

FIGURE 5   Gravitational harmonics are computed from integrals over density and powers of radius of a rotating planet. The curves illustrate the integrands for the harmonics J_2, J_4, and J_6 of a Saturn interior model. Higher-degree terms are proportional to the interior structure in regions progressively closer to the surface. All curves have been normalized to unity at their maximum value. The bump in the J_2 curve near 0.2 is due to the presence of the core.

tions of state requires several iterations, but the calculation does not strain modern computers. This effort is rewarded by models of the interiors of each of the Jovian planets. It should be emphasized that the overall modeling procedure is sufficiently difficult that a large range of interior models is typically not calculated. By necessity, each modeler begins with an ad hoc set of assumptions that limit the range of models that can be calculated. This inherent limitation of models should always be borne in mind when considering their results. The consensus for the structure of Jovian planet interior models is presented in the next section.

V. PLANETARY INTERIOR MODELS

A. GENERAL OVERVIEW

Even early "cosmographers" recognized that the giant planets of the solar system were distinct from the inner terrestrial planets. The terrestrial planets have mean densities of 4000 to 5000 kg m^{-3}, intermediate between the density of rocks and iron, whereas the giant planets have mean densities closer to that of water (1000 kg

m^{-3}), between 700 and 1700 kg m^{-3}. From this single piece of information, it is clear that the bulk composition of the giant planets must be substantially different from that of the terrestrial planets.

It has been known since the 1940s that if the interiors of Jupiter and Saturn are "cold," the primary component of these planets must be hydrogen. In this context, "cold" means that the densities throughout the interior must not deviate significantly from the values they would assume at the same pressures if the temperature was 0 K. The approximation is relevant because the behavior of substances at 0 K and high pressure can be calculated analytically. Hydrogen is then a likely dominant constituent because at the high pressures prevalent in the interiors of Jupiter and Saturn it would be a metallic fluid with a density of about 1000 kg m^{-3}, not the more familiar molecular gas. Since the density of "cold" metallic hydrogen is close to the bulk densities of Jupiter and Saturn, it was recognized as a plausible major constituent of these planets.

Mass–radius calculations provide a more compelling demonstration of the dominance of hydrogen in the interiors of Jupiter and Saturn. For a given composition, there is a unique relation between the radius of a spherical body in hydrostatic equilibrium and its mass. These relations can be calculated analytically for all elements at high pressure and zero temperature. Al-

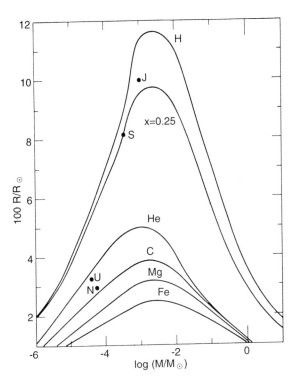

FIGURE 6 Mass–radius curves for objects of various compositions at zero temperature. Curve labeled $X = 0.25$ is for an approximately solar mixture of hydrogen and helium. Points J, S, U, and N represent Jupiter, Saturn, Uranus, and Neptune, respectively. Radius is in units of hundredths of a solar radius and mass is in units of solar masses ($1 R_\odot = 6.96 \times 10^5$ km and $1 M_\odot = 1.99 \times 10^{33}$ g). Jupiter and Saturn are clearly composed predominantly of hydrogen and helium; Uranus and Neptune must have a large complement of heavier elements.

though the interiors of Jovian planets are not at zero temperature, they are cool when measured on an atomic temperature scale. This is adequate for a qualitative calculation, but zero-temperature equations of state are insufficiently accurate for the calculation of detailed interior models.

Mass–radius curves for several likely planetary constituents are shown in Fig. 6. For low masses, the interior pressures are small compared to intermolecular forces and the volume of an object is just proportional to its mass, thus $R \propto M^{1/3}$. This is a realm with which we are familiar in daily life. At larger masses, the greater interior pressures ionize the material, liberating many electrons. In this regime, $R \propto M^{-1/3}$: when mass is added to an object, it shrinks. For intermediate masses where the curves meet, there is a region where the radius is not highly sensitive to the mass. At sufficiently high masses, the hydrogen in the core of the object will undergo fusion, the temperature will rise, and the zero-temperature relations shown in Fig. 6 are

no longer applicable. However, for planets and white dwarf stars, Fig. 6 is applicable. An important consequence of these considerations is that for any given composition, there is a maximum radius that a planet can have. For solar composition, the maximum radius is about 80,000 km for a planet with about three times Jupiter's mass.

The total mass and radius of each Jovian planet are plotted on Fig. 6 as well. This figure immediately proves that Jupiter must be composed primarily of hydrogen and helium. The maximum radii of planets composed of heavier, cosmically abundant elements are all much smaller. For example, only if Jupiter were very hot and very thermally expanded could carbon be a dominant constituent. But Jupiter's observed heat flux rules out a very hot ($>10^7$ K) internal state. Thus, Jupiter must primarily consist of a mixture of hydrogen and helium. Saturn's position on the graph implies a greater abundance of elements heavier than hydrogen, but still a primarily hydrogen bulk composition. Uranus and Neptune lie well below the mass–radius curve for hydrogen, thus revealing an appreciable component of heavier elements in their interiors.

Though the mass–radius relations clearly reveal the bulk composition of Jupiter and Saturn, they do not reveal information about the distribution of material inside the planet. It is here that the shape and gravitational harmonics enter the calculation. The response coefficient Λ_2 measures the response of the planet to its own rotation. For a uniform, hydrogen-rich material, $\Lambda_2 = 0.17$. Values smaller than 0.17 indicate a reduced gravitational response to rotation compared with that of the uniform composition hydrogen-rich planet. Such a reduced response results when more of the mass of the planet is concentrated in a dense core. Thus the smaller the Λ_2, the greater degree of central condensation.

Λ_2 varies (see Table I) from 0.16 for Jupiter to 0.11 for Saturn. The mass–radius relations show that the Jovian planets are not pure hydrogen and their Λ_2 values suggest that they are centrally condensed. Hence the heavier constituents are not uniformly distributed in radius, but are concentrated toward the center of each planet. Jupiter exhibits the least central condensation; Saturn and Uranus are most centrally condensed. Thus we begin to construct an elementary interior model.

Finally, the gravitational harmonics, J_2, J_4, and J_6, probe the detailed variation of the various planetary constituents. To simplify the interpretation of these harmonics, early interior models tended to employ three distinct compositional zones: an inner rocky core, an icy core surrounding the rock one, and a hydrogen/

helium envelope. More modern models allow the composition of various zones to vary gradually between layers and allow the outer envelopes to be enriched over solar abundance. The primary unknowns to be found from interior modeling are the size of the rocky/ icy core and the abundance of helium and other heavy elements in the envelope.

B. JUPITER

Jupiter contains more mass than that of all the other planets combined. Since its gravitational harmonics are also best known, Jupiter serves as a test bed for theoretical understanding of Jovian interiors. The observed physical characteristics of Jupiter are listed in Table I. The abundance of methane in Jupiter's atmosphere is about 3.5 times the solar abundance and the abundance of ammonia is about twice solar. Water does not show such enrichment, but is presumed to be present in and below a water cloud deck, beneath the observed ammonia clouds.

The general structure of Jupiter's interior was briefly described in Section I. Modern interior models attempt to specifically determine the degree of enrichment of heavy elements in the hydrogen/helium envelope of the planet. The atmospheric enrichment of methane and ammonia provides some indication that heavy element enrichment in the deeper interior may be expected. The Λ_2 value for Jupiter indicates that Jupiter is not homogeneous but is somewhat centrally condensed. More detailed modeling must determine how much material is segregated into a distinct core. The size and composition of Jovian planet cores and the amount of heavy element enrichment in the envelopes have bearing on the scenarios by which they are supposed to have formed.

The variation of density with radius for a typical Jupiter model is shown in Fig. 7. It should be emphasized that this is a single Jupiter model that is consistent with all available constraints. Other, equally valid interior models exist. Interior modeling of Jupiter clearly supports the presence of a dense core of heavy elements, likely consisting of both "rocky" and "icy" components. The mass of the core is model dependent, but lies between 5 and $10 M_\oplus$. The core is typically modeled as consisting of an inner rock core surrounded by an ice-rich outer core. Gravitational harmonics are not sensitive to the exact configuration, however, and a single, mixed core is also possible.

Surrounding the core is an envelope of hydrogen and helium. The temperature and pressure at the envelope–core interface are 39 Mbar and 23,000 K in this

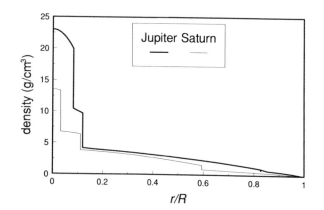

FIGURE 7 Density as a function of normalized radius for Jupiter and Saturn models. The transition from molecular to metallic hydrogen is responsible for the small density change near 60% of Saturn's and 85% of Jupiter's radii. The relative size of the presumed two-layer rock–ice core is about the same size for the two planets. Jupiter's core is more dense, however, owing to the much greater overlying pressure.

particular model. The gravitational harmonics require the envelope to be denser at each pressure level than a model that has only a solar mixture of elements. Thus the envelope must be enriched in heavy elements compared to a purely solar composition. The exact degree of this enrichment, expressed as the mass of material heavier than hydrogen and helium in the envelope, is somewhat uncertain, but is about $13 M_\oplus$. The total mass of heavy elements in Jupiter (core plus envelope) is thus about $20 M_\oplus$. If Jupiter had only a solar abundance of heavy elements, the equivalent value would be about $6 M_\oplus$. Other models find somewhat greater enrichment of heavy elements in the planet. The consensus is that, averaged throughout the planet, Jupiter is enriched in heavy elements over solar abundances by a factor of 3 to 5.

Jupiter's atmospheric abundance of helium, $Y = 0.238 \pm 0.007$, is less than the solar abundance of about 0.28. This depletion is likely an indication that the process of helium differentiation, described more fully in Section VI, may have recently begun on Jupiter. The interior models do not provide a sufficiently clear view into the interior structure to determine if this is the case. The inferred interior structure is, however, compatible with limited helium differentiation.

Hydrogen and helium compose about 90% of Jupiter's mass. Most of the hydrogen exists in the form of metallic hydrogen. Jupiter is the largest reservoir of this material in the solar system. Convection in the metallic hydrogen interior is likely responsible for the generation of Jupiter's magnetic field. The transition from molecular to metallic hydrogen takes place about

10,000 km beneath the cloud tops, compared to about 30,000 km at Saturn. The exceptionally large volume of metallic hydrogen is likely responsible for the great strength of Jupiter's magnetic field. The relative proximity of the electrically conductive region to the surface may explain why Jupiter's magnetic field is more complex than Saturn's.

The inferred interior structure of Jupiter is most consistent with the giant planet formation scenario known as nucleated collapse. In this scenario, a nucleus of rock and ice first forms in the solar nebula. When the nucleus has grown to about $5 M_\oplus$, the gas of the nebula collapses down upon the core, thus forming a massive hydrogen/helium envelope surrounding a rock/ice core. Planetesimals that accrete later in time cannot pass through the thick atmosphere surrounding the core. Instead they break up and dissolve into the hydrogen/helium envelope. This scenario accounts for both the core of the planet and the enrichment of heavy elements in the envelope.

C. SATURN

The observational constraints for Saturn are listed in Table I. Although Saturn has less than one-third of Jupiter's mass, it has almost the same radius. This is a consequence of the relative insensitivity of radius to mass for hydrogen planets in Jupiter and Saturn's mass range (see Fig. 6). Saturn's atmosphere, like Jupiter's, is enriched in methane and ammonia. One dramatic compositional difference is the abundance of helium at Saturn. The atmospheric abundance is a factor of four *lower* than the cosmic abundance. Since there is no known process by which Saturn could have accreted less helium than Jupiter, another process must be at work.

Saturn's interior is grossly similar to Jupiter's. It also has a core of rock and ice surrounded by a hydrogen-rich envelope. A sample Saturn model is shown in Fig. 7. Saturn's core mass is lower than Jupiter's and may be as small as $1 M_\oplus$. Temperatures inside Saturn are also cooler. In the model shown in Fig. 7, the temperature and pressure at the base of the metallic hydrogen envelope are 11,900 K and 13 Mbar. Like Jupiter, there is strong evidence that Saturn's envelope is enriched in heavy elements over solar abundance. The total mass of heavy elements in the envelope may be as large as $28 M_\oplus$; the expected solar component would be about $2 M_\oplus$. Heavy element enrichment in the envelope is much greater than at Jupiter, especially considering Saturn's lower total mass. This may be an indication that more condensed icy material was available to be incorporated into Saturn at its location in the solar nebula. Nevertheless, as at Jupiter, hydrogen and helium are the dominant component of Saturn's mass (~70%).

Saturn's low atmospheric helium abundance implies that the process of helium differentiation (see Section VI) has begun inside the planet. This process results in removal of helium from the outer molecular hydrogen envelope of the planet and enhancement of helium in the deep interior. Thus the helium fraction should increase with depth in Saturn's interior. The inferred density structure is consistent with this widely accepted explanation for Saturn's low atmospheric helium abundance. If helium is presumed to be uniformly depleted from the outer molecular envelope of the planet, it can be self-consistently accounted for in the deeper interior. The models lack the sensitivity to confirm that this is definitely the case, however.

D. URANUS AND NEPTUNE

Before the Voyager encounters, Uranus and Neptune were assumed to have similar interior structures. This assumption was well justified given their similar radii, masses, atmospheric compositions, and location in the outer solar system. Uranus and Neptune were modeled as having three distinct layers: an inner rocky core, a large icy mantle, and a methane-rich, hydrogen–helium atmosphere. Little more could be said with precision since their atmospheric oblateness and interior rotation rates were not accurately known.

Upon its arrival at Uranus in 1986 and Neptune in 1989, *Voyager 2* provided the measurements needed to constrain interior models and provide individual identities for each planet. Voyager observed the structure of the magnetic field of both planets and measured their rotation rates. In both cases the fields were off-center, tilted dipoles of similar strengths. Voyager also measured the higher-order components of the gravitational fields of both planets. The abundance of carbon in both atmospheres is about 30 times the solar value. Although Uranus and Neptune have similar radii and masses, the differences are such that the mean density of Neptune is 24% higher than the mean density of Uranus.

Voyager data revealed that, though similar, the interior structures of the two planets are not indentical. As with Jupiter and Saturn, the parameter Λ_2 provides information on the distribution of mass inside each planet. If Uranus and Neptune had a similar distribution of mass in their interiors, their Λ_2 parameters would be similar. As Table I shows, for Uranus $\Lambda_2 =$

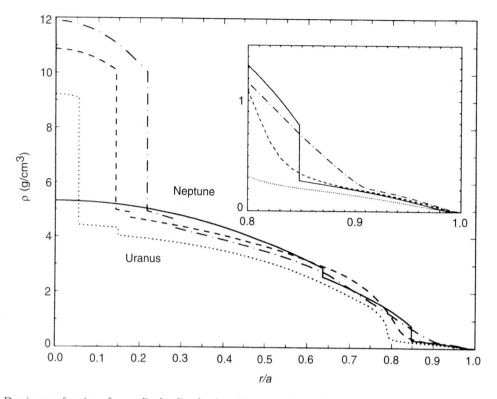

FIGURE 8 Density as a function of normalized radius for three Neptune and one Uranus interior models. The solid, dashed, and dot-dashed curves represent the range in possible Neptune models. Note the wide variety of acceptable core sizes, ranging from a model with no core to a model with a core extending to 20% of Neptune's radius. The dotted curve represents a single Uranus model. Because of Neptune's greater mass, it is everywhere denser than Uranus at the same relative radius. The inset shows the region of transition from a hydrogen-rich atmosphere to the icy mantle in more detail.

0.119, whereas for Neptune $\Lambda_2 = 0.136$. The larger value of Λ_2 for Neptune implies that Neptune is less centrally condensed than Uranus. Models show that this difference can be understood in terms of equal relative amounts of ice, rock, and gas that are simply distributed differently within the two planets. The two planets also follow virtually the same pressure–density law, another indication that they have very similar composition and structure.

Models (Fig. 8) of Uranus and Neptune's interior begin with a hydrogen-rich atmosphere that extends from the observable cloud tops to about 85% of Neptune's radius and 80% of Uranus'. The composition in this region does not vary significantly from the hydrogen-rich atmospheric composition. Near 0.3 Mbar and 3000 K ($0.85 R_{\text{Neptune}}$ and $0.80 R_{\text{Uranus}}$), the density rises rapidly to over 1000 kg m^{-3}. The density then increases steadily into the deep interior of both planets, where the pressure reaches 6 Mbar at 7000 K. The variation of density with pressure in this region is very similar to that found in the laboratory shock wave experiments on the artificial "icy" mixture known as "synthetic Uranus." The composition of this region

is thus undoubtedly predominantly "icy." However, since the density of rock/ice/gas mixtures can mimic the density of pure ice, the exact composition cannot be known with precision. Any hydrogen present in the deep interior would be in the metallic phase.

Interestingly, Uranus and Neptune models can be constructed that do not have cores. Other models with cores as large as $1 M_\oplus$ are also consistent with the available data.

The total mass of hydrogen and helium in Uranus and Neptune is about $2 M_\oplus$, compared to about $300 M_\oplus$ at Jupiter. Given the relatively small amounts of gas compared to ices in Uranus and Neptune, these planets are aptly termed "ice giants," whereas Jupiter and Saturn are indeed "gas giants."

Shock compression measurements show that the fluids of the hot, ice-rich region of Uranus and Neptune are expected to be substantially ionized and dissociated. The large electrical conductivities of such fluids, coupled with the modest convective velocities predicted for the interiors of Uranus and Neptune, can generate and sustain the observed magnetic fields of the planets. One possible explanation for the complexity of their magnetic fields is that the electrically conductive region of these planets is comparatively close (within about 4000 km) to the cloud tops, a consequence of the ionization behavior of water, ammonia, and methane. This is consistent with the trend in field complexity seen at Jupiter and Saturn.

Uranus and Neptune likely represent failed gas giant planets. The time to accrete solid objects onto the growing ice and rock planetary cores was much longer in the outer solar nebula than at the orbital distances of Jupiter and Saturn. Thus Uranus and Neptune took longer to grow. By the time the nebular gas was swept away, these planets had not yet grown massive enough to capture substantial amounts of hydrogen and helium gas from the nebula. Perhaps if the nebular gas had persisted for a longer time, Uranus and Neptune would have grown large enough to complete the capture of a hydrogen/helium envelope. In that case, these planets might now more closely resemble the current Jupiter and Saturn.

VI. JOVIAN PLANET EVOLUTION

The amount of energy radiated by each of the Jovian planets, except possibly for Uranus, is greater than the amount of energy that they receive from the Sun (see Table I). This internal heat source is too large to be explained by decay of radioactive elements in the rock cores of the planets. Temperatures, even in the deep interior, are far below the 10,000,000 K required for thermonuclear fusion. The source of the excess energy is gravitational potential energy that was converted to heat during the planets' formation and stored in their interiors. [See THE ORIGIN OF THE SOLAR SYSTEM.]

The potential energy of gas and solids in the solar nebula was converted to thermal energy when they were accreted onto the forming planet. Over time, the planets radiated energy into space and cooled, slowly losing their primordial energy content. Thus all four Jovian planets were initially warmer than they are now. During the early evolutionary stages the planets contracted as they cooled, thereby releasing even more gravitational potential energy. Today the planets all cool at essentially constant radius because the internal pressures depend only slightly on temperature. The coupled contraction and cooling is known as Kelvin–Helmholtz cooling.

Evolutionary models test whether Kelvin–Helmholtz cooling can account for the current observed heat flows of the Jovian planets. In these calculations, a series of sequentially cooler planetary interior

models is created with the last model representing the present-day planet. The time elapsed between each static model is calculated and thus the evolutionary age of the planet found.

Models predict that Jupiter should have cooled from an initially hot state (accompanied by an atmospheric temperature greater than about 340 K) to its current temperature in about 4.5 billion years. Since this is about the age of the solar system, the Kelvin–Helmholtz model is judged a success for Jupiter. For Saturn, however, the model is less successful. The models suggest that Saturn, with its current heat flow, should be about 2 billion years old. Since there is no reason to believe that Saturn formed 2.5 billion years later than Jupiter, another heat source must be adding to Saturn's Kelvin–Helmholtz luminosity. This leads to the hypothesis that differentiation of helium in the interior provides additional thermal energy to the planet.

The helium depletion hypothesis holds that as Saturn has cooled from an initially warmer state with the solar abundance of helium throughout, its interior reached the point (near 2 Mbar and 8000 K) at which hydrogen and helium no longer mix in all proportions. Like oil and water in salad dressing, the hydrogen and helium are separating into different phases.

As the helium-rich drops form in Saturn's envelope and fall to deeper, warmer layers of its interior, the helium eventually again mixes with hydrogen. Over time, this rainfall is depleting the supply of helium in the outer envelope and visible atmosphere and enriching the helium content deeper in the interior, close to the core. The overall planetary inventory of helium remains constant. This model is compatible with the observed depletion at Saturn. Jupiter, with a warmer interior and smaller helium depletion, has apparently only recently begun this process.

This process of helium differentiation liberates gravitational potential energy as the drops fall. Assuming that the observed atmospheric depletion of helium extends to the outer 20% of Saturn's mass, the potential energy released by the downward migration of helium drops is sufficient to explain Saturn's unusually high heat flow. No other process can simultaneously explain Saturn's anomalously high heat flow and the observed atmospheric depletion of helium.

The problem for Uranus and Neptune is somewhat different. The Kelvin–Helmholtz hypothesis predicts ages of the correct order of magnitude for Uranus and Neptune, but the ages are too large. In other words, the model predicts that these planets should have higher heat flows at the current time than they are observed to have. The problem is most severe for Uranus, which has no detectable heat flow. There are several possible resolutions to this contradiction.

One possibility is that gradients in the composition of Uranus with radius have served to impede convection in the deep interior. Composition gradients, for example, a gradual increase in the rock abundance with depth, can severely limit heat flow from the planet. In such a case, only the outermost layers could transport energy by convection to the atmosphere and cool effectively to space, thus producing a lower than expected heat flow. More of Neptune's interior than that of Uranus might be convective, thus explaining its higher current heat flow. Of course if this hypothesis were correct, then the existing interior models of these planets would have to be revised, since an initial assumption that the planets are fully convective would have been violated. Inhibition of convection in the deep interior by this mechanism has been proposed as one explanation for the strong nondipole component of both planets' magnetic fields.

VII. FUTURE DIRECTIONS

Models of Jovian planetary interiors have identified the mass of each planet's core and the approximate composition of their envelopes. These results have provided important constraints on the processes by which these planets form. In turn, formation models place limits on the mass, composition, and evolution of the solar nebula. Further progress, however, requires even tighter limits on the interior structure of these planets. Sufficiently detailed interior models may even provide constraints on the equation of state of hydrogen. Since Jupiter is the largest reservoir of metallic hydrogen in the solar system, it may potentially resolve issues such as the exact pressure of the transition between molecular and metallic hydrogen.

One might expect that future, more accurate measurements of each planet's gravitational harmonics would help to address questions such as these. The higher-order moments, however, are most sensitive to the density distribution in the outer 10 or 20% of the planetary radius. Thus, little additional information about the deep interior is likely to be forthcoming from such observational improvements. The higher-order harmonics do, however, provide some information about the state of rotation of the outer layers and may help address questions regarding the degree of differential rotation in the Jovian planets. For example, it is unknown if Jupiter rotates completely as a solid body, or if different cylindrical regions of its interior rotate at different rates.

Further improvements in delineating the equations of state of Jovian planetary components will help to clarify their interior structures. Better knowledge of the behavior of planetary constituents and their mixtures at high pressure will enable more accurate interior models to be constructed. Yet changes in understanding are unlikely to result from such improvements. Only significantly new and different sources of information offer the potential of providing fundamentally new insights into the interior structure of these planets.

Jovian seismology is one particularly promising new avenue of research into these planetary interiors. Much of our knowledge of the interior structure of the Earth arises from study of seismic waves that propagate through the interior of the planet. The speed and trajectory of these waves carry information about the composition and structure of the Earth's interior. During the collisions of the fragments of comet Shoemaker-Levy 9 with Jupiter, several experiments attempted to detect seismic waves launched by the impacts. If these waves had been detected, they would have provided a direct probe into the interior structure of Jupiter.

Another avenue for Jovian seismology is to detect resonant acoustic modes trapped inside Jupiter. The frequency of a given Jovian oscillation mode depends on the interior structure of the planet within the region in which the mode propagates. Thus measurement of the frequencies of a variety of modes would provide information on the overall interior structure of the planet. The study of such modes on the Sun, a science known as helioseismology, has revolutionized our knowledge of the solar interior. In 1992, French astronomers announced tentative detections of Jovian oscillations using two separate techniques. However, in both cases the observations and data analysis are difficult and interpretation of the results has been limited by the restricted number of observing nights on large telescopes. Future observational advances may allow unambiguous detection of Jovian oscillations.

As they would at Jupiter, oscillations of Saturn would perturb the external gravitational field of the planet. Though there is yet no way to detect such perturbations at Jupiter, this may be possible at Saturn. Saturn's rings are excellent detectors of faint gravitational perturbations and thus the possibility arises of using Saturn's rings as a seismometer. There is some evidence that certain wave features in Saturn's innermost C-ring may be produced by oscillation modes of the planet. Further spacecraft observations are required to confirm this hypothesis, however.

Definitive detection of oscillations of any Jovian planet would first serve to accurately determine the core size and rotation profile of the planet. Since such determinations would remove two sources of uncertainty surrounding the interior structure, more information could then be gleaned from the traditional interior model constraints. Seismology might also help to constrain more accurately the location of the transition from molecular to metallic hydrogen in Jupiter's interior. If so, seismology may untimately provide the tightest constraints on the hydrogen equation of state and interior structure of Jovian planets.

With the possible detection of giant planets around nearby stars by the Doppler radial velocity techniques, there may be more giant planets known outside of the solar system than are known inside it. If these planets are ultimately detected directly, they may someday provide further tests for our understanding of the interiors and evolution of giant planets. [See EXTRA-SOLAR PLANETS.]

BIBLIOGRAPHY

Atreya, S. K., Pollack, J. B., and Matthews, M. S. (eds.) (1989). "Origin and Evolution of Planetary and Satellite Atmospheres." Univ. Arizona Press, Tucson.

Beatty, J. K., O'Leary, B., and Chaikin, A. (eds.) (1990). "The New Solar System," 3rd Ed. Sky Publishing, Cambridge, Mass.

Bergstralh, J., Minor, E., and Matthews, M. S. (eds.) (1991). "Uranus." Univ. Arizona Press, Tucson.

Gehrels, T., and Matthews, M. S. (eds.) (1984). "Saturn." Univ. Arizona Press, Tucson.

Stevenson, D. J. (1982). *Annu. Rev. Earth Planet Sci.* **10**, 257–295.

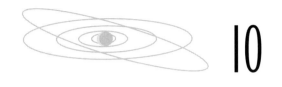

10

I. Overview and Significance

II. General Planetary Properties

III. Hot Lava Eruptions

IV. Heat Flow

V. Plume Eruptions

VI. Io's "Atmosphere"

VII. Future Perspective

Dennis L. Matson and
Diana L. Blaney
*Jet Propulsion Laboratory,
California Institute of Technology*

GLOSSARY

Albedo: Ratio between the amount of light reflected by an astronomical body and the amount of light received.

Ash: Fine particles of pulverized rock blown from an explosion vent. Less than 2 mm in diameter, ash may be either solid or molten when first erupted. The most common variety is vitric ash, glassy particles formed by gas bubbles bursting through liquid magma.

Caldera: Taken from the Portuguese word for kettle or cauldron, it is a volcanic depression, more or less circular or cirquelike in form, the diameter of which is many times larger than those of the included vents. Calderas are formed by subsidence or collapse and are distinguished from craters by both size and origin. Craters are formed by the volcanic ejection of material from a vent and rarely exceed a mile in diameter.

Column density: Measure of atmospheric density; namely, the number of a given species per unit area integrated over a defined column (e.g., line of sight to Io's surface).

Eruptive plume: Envelope of material that is thrown into the air by an eruption.

Geometric albedo: Ratio between the amount of light reflected by an astronomical body at 0° phase angle and the amount of light that would be reflected by a flat, diffusely reflecting disk of the same angular size.

Intrusion: Process of emplacement of magma in preexisting rock; also refers to igneous rock mass so formed within the surrounding rock.

Lava: Magma that has reached the surface through a volcanic eruption. The term is most commonly applied to streams of liquid rock that flow from a crater or fissure. It also refers to cooled and solidified rock.

Lava flow: Outpouring of lava onto the land surface from a vent or fissure; also, a solidified tonguelike or sheetlike body formed by outpouring lava.

Lava lake (pond): Lake (pond) of molten lava, usually in a volcanic crater or depression. The term refers to solidified and partially solidified stages, as well as to the molten, active lava lake.

Magma: Molten rock beneath the surface of the Earth.

Magma chamber: Subterranean cavity containing the gas-rich liquid magma that feeds a volcano.

Mantle: Zone of a planet below the crust and above the core, usually consisting of melted rock.

Phase angle: Angle between the Sun and the observer as seen from the target body.

Plinian eruption: Explosive eruption in which a steady, turbulent stream of fragmented magma and magmatic gases is released at a high velocity from a vent. Large volumes of tephra and tall eruption columns are characteristic.

Thermal anomaly: Area whose temperature differs from an expected value. For Io, the reference value is the temperature of the surface that would be in equilibrium with the amount of sunlight absorbed.

Thermal emission: Electromagnetic radiation produced by a body owing to its temperature.

I. OVERVIEW AND SIGNIFICANCE

Io is the innermost of the four satellites discovered by Galileo Galilei in 1610. It is the most geologically active of all the planets and satellites in the entire solar system, even exceeding the activity of Earth. Io is brilliantly colored with reddish, orangish, and white regions, offset by relatively small dark areas. With these shades of red, orange, and white, it looks much like a tomato and cheese pizza. In fact, this was exactly how Io was described by Voyager investigators when they first saw color pictures of its surface.

Typically, the surface is punctuated with a half-dozen or so plumes from geyserlike eruptions. Many volcanoes are also present. Some are currently erupting, and others show signs of recent activity. Calderas with lava lakes can be found easily. From time to time, large lava flows occur and radiate tremendous amounts of heat. Much of the surface is occupied by older, still cooling, flows. Supporting this huge thermal emission is the largest amount of heat coming out of the interior of any solid body in the solar system. A mosaic of Io made from *Galileo* spacecraft pictures is shown in Fig. 1.

Io also excels in the diversity and intensity of interactions with its environment. (1) Ionic, atomic, and molecular species escape from Io all of the time and form a torus around Jupiter. Notable examples, in terms of their optical detectability, are Na, K, S, O, and their ions. In addition to populating the torus, they diffuse throughout the Jovian magnetosphere. Some of them implant themselves in the surfaces of the other

satellites, causing optical alteration of these surfaces. (2) Io appears to be the source for fine dust particles that become charged and escape the Jovian system and travel as streams into interplanetary space. (3) Io is a key part of an electrical current circuit that, among other things, gives rise to aurora on Jupiter. (4) The emission of certain types of decametric radiation by the Jovian magnetosphere is due to disturbances induced in the magnetospheric plasma as it flows by Io. (5) Io is in an orbital resonance with Europa. This resonance is what causes Io's orbit to be slightly elliptical, which in turn gives rise to the tidal dissipation that produces the heat that drives Io's geologic activity. (6) The secular acceleration of Io along its orbit is a function of the tidal response characteristics of Jupiter's interior. [*See* ATMOSPHERES OF THE GIANT PLANETS; PLANETARY MAGNETOSPHERES; INTERPLANETARY DUST; THE SOLAR SYSTEM AT RADIO WAVELENGTHS.]

As a telescopic object, Io has long been noted for its peculiar optical properties. It is not like any other satellite. For more than half a century it has been known to have unusual colors. It is believed that this spectrum is due to a combination of yellowish and reddish forms of sulfur, sulfurous compounds, and white sulfur dioxide frost. At certain wavelengths in the infrared and in the ultraviolet, sulfur dioxide (SO_2) provides strong absorption features that also contribute to the uniqueness of Io's spectral reflectance. Sometimes, Io has excess emission at midinfrared wavelengths that is due to thermal emission from hot, erupting lava. Io's thermal emission (e.g., at a wavelength of 10 μm) falls off very, very rapidly when Io is eclipsed by Jupiter. This peculiar response gave rise to the erroneous idea that the surface was entirely covered by insulating fluff. In part, this effect is now believed to be due to the previously unappreciated relationships between the absorption of sunlight on warm lava and the wavelength at which the observation is made.

Io is a wonderful natural laboratory for studying volcanism, for it is the extreme example of volcanic activity in the solar system. Here volcanism can be seen in its purest form with the least amount of interference from nonvolcanic processes. Io is very frequently, if not continuously, experiencing eruptions of material onto its surface. This material comes in the form of fallout from plumes and as lava. Some eruptions are detected at infrared wavelengths by ground-based telescopes. Evidence of recent activity, as indicated by the enhanced level of thermal emission, is always detected in careful measurements at properly selected wavelengths. The tremendous amount of heat radiated by

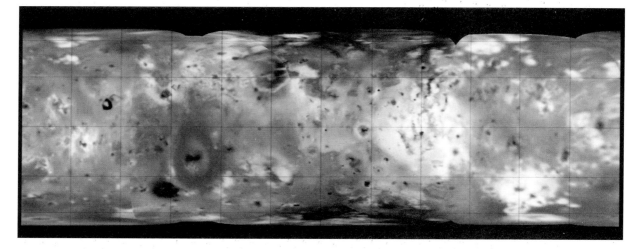

FIGURE 1 The false-color mosaic of Io was produced from images acquired in July and September of 1996 by the *Galileo* spacecraft. The area shown is 11,420 km in width. North is at the top. The grid lines in this cylindrical map projection are at 30° intervals of latitude and longitude. The yellowish and brown shades (shown here as mid-range gray tones) are due to sulfur or sulfurous materials, while the white and gray hues are deposits of SO_2 frost. The bright red materials (shown here as darker gray tones) and spots with low albedo ("black" spots) indicate areas of volcanic activity and are usually associated with high temperatures and recent surface changes. The color in the mosaic comes from near-infrared (756 nm), green, and violet filter images, which are composed as red, green, and blue. See website for color image. (P48496 _f.jpg, color global mosaic of Io.) (http://photojournal.jpl.nasa.gov/cgi-bin/PIAGenCatalogPage.pl?PIA00585)

the many cooling lava flows gives Io the highest heat flow of any planet or satellite, which poses a challenge for geophysical modelers. Where inside of Io does the heat originate? How is it brought to the surface? The mechanism for heat production is thought to be tidal dissipation. Tides are raised *on and in* Io by Jupiter. This interaction occurs because of Io's slight orbital eccentricity. The eccentricity is caused by an orbital resonance with Europa, the next satellite out from Jupiter [*See* PLANETARY VOLCANISM.]

II. GENERAL PLANETARY PROPERTIES

Our present knowledge of Io comes from: (1) the analysis of Io's light and radio emissions observed by ground-based telescopes and antennas; (2) the results from the flybys of the *Voyager 1* and *2* spacecraft; (3) the study of observations made by astronomical instruments in orbit about Earth; and (4) the results from the *Galileo* spacecraft in orbit about Jupiter. Some of the currently studied observational records go back 300 years or more. Others were obtained by the *Galileo* spacecraft in the late 1990s. Much of this

information can be distilled to simple facts and numbers. We list some of these in Table I.

In terms of global properties, Io is in many ways like Earth's Moon. For example, Io has a similar bulk density that indicates that it is mainly composed of silicate and other rock minerals. However, its surface properties are dramatically different because of the intensive heating of Io's interior by tidal dissipation and the ongoing volcanic eruptions. [*See* THE MOON.]

There has been a contentious, ongoing debate about the relative roles of silicate and other rock minerals versus those of sulfur and sulfurous minerals in the structures and processes seen on Io's surface. It was often said that "silicate minerals were not to be seen anywhere on Io." This debate has been recently resolved by the Earth-based observations of very high temperature lava; temperatures in the range of 1000–1500 K have been observed on about a dozen occasions. There is evidence from *Galileo* and ground-based telescope data for temperatures as high as 1800 K! These temperatures can only be explained by lava composed of molten rock. The coloration of the surface by various forms of sulfur *could still be* accomplished by relatively thin layers, analogous to paint or veneer. The idea that Io's surface *must* consist of global-scale sulfur deposits that are many tens of kilometers deep is now

TABLE I
Facts about Io

Orbital Information	
Semimajor axis	421,600 km (5.905 Jupiter radii)
Orbital period	1.769 days
Rotational period	Synchronous with orbit
Eccentricity (forced)	0.0041
Physical Properties	
Mass	$8.9319 \pm 0.0012 \times 10^{22}$ kg
	($1.215 \times$ Lunar; $0.0150 \times$ Earth)
Mean radius	1821.3 km
	($1.05 \times$ Lunar; $0.286 \times$ Earth)
Density	3.529 g/cm^3
	(Lunar = 3.3437 g/cm^3; Earth = 5.515 g/cm^3)
Gravity	1.80 m/s^2
	(Lunar = 1.62 m/s^2; Earth-9.82 m/s^2)
Global heat flow	>2.5 W/m^2
	(Lunar $\sim$ 0.0025 W/m^2; Earth = 0.087 W/m^2)
Radius of core	If pure Iron, 655 km (36% R$_{Io}$)
	If iron/iron sulfide mixture, 947 km (52% R$_{Io}$)
Magnetic field	Perhaps
Surface Properties	
Geometric albedo	0.7
Composition	Sulfur dioxide frost (everywhere, white deposits)
	Sulfur compounds (reddish yellow units)
	Rock (Unknown composition, small dark black areas)
Highest mountain	Euboea Mons, 13 km
Largest volcanic complex	Loki Patera
Plume heights	60–400 km
Plume velocity	0.5–1 km/sec

rejected. On the other hand, local accumulations of significantly thick sulfur deposits are not ruled out. In reality, both groups of minerals are important and are needed to understand Io's surface properties.

A. SURFACE PHYSIOGRAPHY

Io has relatively subdued topography when compared to other bodies its size. It has relatively few mountains. Vast expanses of volcanic plains constitute the most abundant landform. These can be seen in Figs. 2 and 5. Compared to other solar system bodies, these plains are very striking in that they show no impact craters. Instead, the plains are pockmarked by many volcanic calderas. One can often see dark material on the floors of these volcanic depressions, which are inferred to be warm, recently emplaced, lava. The height and steepness of the sides of many of these calderas imply that the walls are composed of material possessing considerable strength. This indicates the presence of rock rather than a subsurface composed of sulfur.

The 10-km height of Haemus Mons (seen in the lower portion of Fig. 2) also suggests that the underlying material must be rock. Sulfur is far too weak to be able to support the observed height of these mountains. The black material in the scene is probably lava that is still warm. From the widespread distribution of yellowish and reddish colors, it is obvious that as the lava cools it is quickly coated or modified by sulfur or sulfurous condensates.

Io's surface shows many local changes with time. Perhaps some of the easiest to see are changes in the shapes in Pele's plume deposits. Pele, named after the

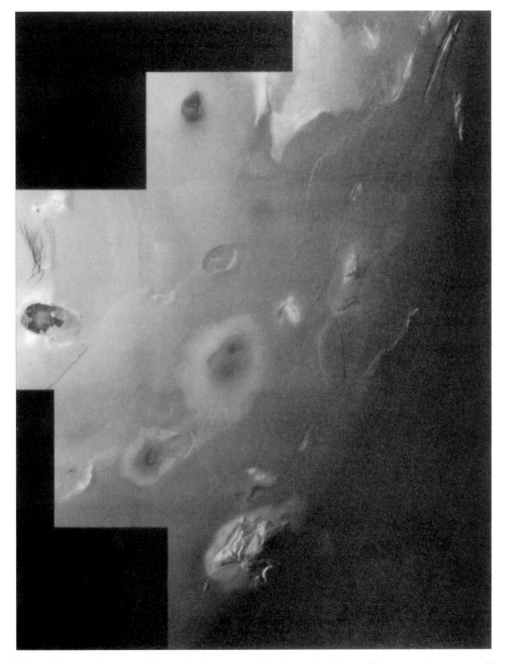

FIGURE 2 This is a *Voyager 1* mosaic of the south polar region. The low angle of the sunlight accentuates topographic relief. At the bottom of the picture is Haemus Mons, a huge, 10-km-high (32,000-ft) mountain. Other features are relatively flat, layered, volcanic plains, eroded plateaus, and bowl-shaped, volcanic craters and calderas. The dark, partially filled calderas at the far left is Creidne Patera. The patches of white are SO_2 frost. See website for color image. (http://photojournal.jpl.nasa.gov/cgi-bin/PIAGenCatalogPage.pl?PIA00327)

Hawaiian goddess of fire, is a large volcano that was active during the *Voyager* flybys. In Fig. 3 it is located near the center of Io's disk, and this *Galileo* picture clearly shows its large, and somewhat diffuse, plume deposit. The inset images show Pele's plume fallout zones for *Voyager 1* (upper inset) and *2* (lower inset).

Note the changes in shape, particularly of the outer edge.

Figure 4 shows a larger-scale comparison, with four views of the Jupiter-facing hemisphere of Io taken at different times. Surprisingly, the most obvious changes are between *Voyagers 1* (top right) and *2* (bottom left),

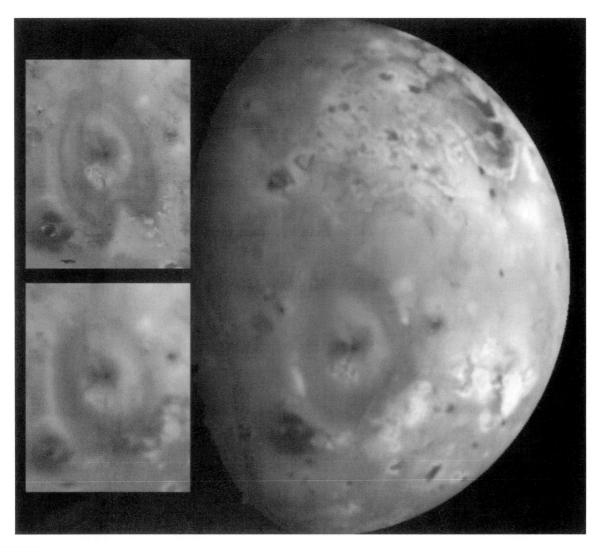

FIGURE 3 A comparative study of Pele plume deposits. The main image is a June 1996 *Galileo* picture with Pele slightly below the center of the disk. The upper inset is an image from *Voyager 1* (March 1979); the lower is from *Voyager 2* (July 1979). Note the changes in the shape of the outer bundary of the plume deposits. The differences in the forms of the deposits are thought to be due to the changing shape of the vent from which the plume emerges. At the time of the *Galileo* picture, the deposit is more circular (suggesting a more symmetric orifice) than during the time of either Voyager flyby. See website for color image. (11151996.jpg.) (http://photojournal.jpl.nasa.gov/cgi-bin/PIAGenCatalogPage.pl?PIA00718)

which were just four months apart. The eruptions at Surt and Aten Patera both produced darkened caldera floors and new, diffuse, pyroclastic deposits covering areas about 1400 km in diameter (about the size of the state of Alaska). In the *Galileo* image (lower left), these regions appear much more similar to *Voyager 1* conditions. This suggests that the plume deposits have mostly faded away and that the calderas have brightened with time. The Surt and Aten plume deposits have spectral properties similar to the plume deposits of Pele and also have faded, suggesting that Pele may have remained at least intermittently active over this longer time interval, whereas Surt and Aten are less frequently active.

Given all of the changes one can find in the pictures, it is a curious fact that on a hemispheric scale, changes of all types tend to average out. Astronomical measurements of Io over the last half-century have not seen significant changes in its global colors or level of reflectance. One can see this effect by viewing Fig. 4 at a distance, where upon it is difficult to see any difference between the three color images.

Io's surface is modified by many processes. Material coming from the interior can create a variety of landforms. Lava can form flows. With time, volcanic plains can be formed from coalesced flows. Lava lakes form, and some of them fill depressions and calderas. Entrained material can be carried with the SO_2 that issues

FIGURE 4 Surface changes in the Jupiter-facing hemisphere. The upper images are from the *Voyager 1* March 1979 flyby. Lower left is a picture from *Voyager 2* (July 1979). On the lower right is a *Galileo* image (June 27, 1996). North is up. Some of the large changes occurred over a period of four months between *Voyagers 1* and *2*. They were the eruptions of Surt (small dark spot in the lower left image, 45°N, in the 12 o'clock direction from the center) and Aten Patera (48°S at 5 o'clock), which darkened the caldera floors and laid down diffuse, pyroclastic deposits covering areas ~1400 km in diameter (comparable to the size of Alaska). At the time of *Galileo*, the Surt and Aten regions are more similar to the *Voyager 1*, preeruption image, suggesting that activity has stopped. See website for color image. (09091996 _full.jpg, resurfacing of the Jupiter-facing hemisphere of Io.) (http://photojournal.jpl.nasa.gov/cgi-bin/PIAGenCatalogPage.pl?PIA00712)

out of vents to form plumes and low-lying vent clouds. Eventually, the entrained material falls back to the surface and becomes part of the plume deposits. The SO_2 gas itself also comes back to the surface, where it eventually freezes out on cold, high-albedo surfaces. With all of this material coming from the interior and forming new landforms, there must be compensation in the form of some sort of general subsidence of the whole landscape as subsurface voids, formed by extraction of material, collapse. Little is known about subsidence on Io, or about the extent of any subsurface void space. We see mountains on Io, but little or noth-

ing is known about the process(es) by which they were formed.

However, it is clear from Io's colorful surface that different minerals are present, and have a story to tell, even if they constitute only a relatively small percentage of Io's total mass. Spectral reflectance measurements of Io show that there is a very deep absorption band in the infrared near a wavelength of 4 μm. This band, due to SO_2, is present everywhere on Io, with its strength varying from one location to another. It is strongest for the bright frost patches that cover much of Io's equatorial regions. Where frost is not evident, this band can be caused by SO_2 adsorbed on the local surface materials.

Researchers believe that sulfur and sulfur compounds are required to explain much of Io's overall spectrum. In particular, Io shows higher reflectance in the visual and infrared, it has a sharp drop-off in the blue, and a very low value throughout the ultraviolet spectral region. All of these are characteristic of the reflection of sunlight off of sulfur and sulfurous compounds. The total abundance of sulfur on Io's surface remains an open question. There may be only enough to paint or veneer the surface, or there may be enough to form some relatively thick deposits, on the order of a kilometer or more in thickness.

Notably absent from Io's spectrum is any indication of water. Neither is there reliable evidence for H_2S, a gas that is commonly found in association with SO_2 in terrestrial volcanic gases. It is unlikely that these compounds would be totally absent from Io; they are just not sufficiently abundant on the exposed surface to be detected. Larger quantities, if present, could be hidden from view. Sodium (Na) and potassium (K) and their compounds have not been detected on the surface of Io, but Na and K are seen in the magnetosphere around Io, and Na is seen escaping from Io. Thus these materials, and probably others, must be present either on the surface and/or in at least some of the eruptive plumes. Na, K, and some of their compounds are thought to be knocked off of Io and into the Jovian magnetosphere by collisions (sputtering) of charged particles with the surface and/or with the atmosphere (e.g., plumes).

B. INTERIOR OF IO

During a close flyby, a spacecraft is sensitive to the internal distribution of the mass inside a satellite. As the *Galileo* spacecraft flew by Io, small variations in the gravitational attraction on the spacecraft gave information about the distribution of mass. Io is differentiated. It has a core with a radius of about half that of the satellite. Thus, the core is approximately one-eighth of Io by volume. The composition of the core is either iron, iron sulfide, or a combination of both.

It is likely that a significant volume of Io's interior is molten. The presence of such melts has been inferred from the size of Io's heat flow, taken together with the thermal and tidal dissipation constraints of possible internal compositions.

It is not known if Io has an intrinsic magnetic field. During *Galileo's* single flyby, the magnetometer measured a magnetic signature that could be ascribed to an intrinsic dipolar field centered in Io. However, the electromagnetic environment near Io is very complex, and other effects could contribute to this signature. The known plasma currents and interactions, and the presence of the Io flux tube and its associated currents, can all introduce magnetic fields of their own. All of these and their magnitudes, as well as how they are oriented, are still not known.

III. HOT LAVA ERUPTIONS

The pictures obtained by the *Voyager 1* and *2* and *Galileo* spacecraft show many lava flows on the surface of Io. These flows can be seen readily in Fig. 5. In this region (2° to 65°N, long. 150° to 223°), there are rugged mountains several miles high, layered materials that form plateaus, and many irregularly shaped volcanic calderas. There are also many dark lava flows and bright SO_2 frost deposits, which have no discernible topographic relief at this scale. Some of the darkest lava flows may in fact be active or at least still warm. There are no impact craters. This implies that volcanism covers the surface with new material much more rapidly than the impacting flux of comets and asteroids can create observable craters.

Hot, fresh lava starts out dark. This interesting fact was established by the discovery of a correlation between thermal emission and albedo in Voyager data. Older flows, either as they cool or after they have become cold, show progressively higher albedos. In Fig. 5, one sees long, thin, dark flows (which may still be hot), some dark circular units (one of which is in a caldera), and many relatively broad lava flow fields. From the albedo markings on these, it appears that they have been laid down over time by a series of lava flows.

Lava eruptions on Io can now be detected by sensitive infrared radiometers attached to ground-based astronomical telescopes. The first such observation was made in 1978 by Fred Whitteborn and his colleagues of the NASA Ames Research Center. The event they recorded reached a temperature of approximately 600 K. At that time, it was not immediately clear to them (or to anyone else) what caused this excess thermal emission. However, they specifically cited the possibility of an eruption of lava onto the surface as one of the possible explanations for this strange observation. Careful monitoring of Io by William Sinton and his colleagues at Mauna Kea Observatory soon confirmed the existence of such outbursts. During their sightings, they saw temperatures as high as 700 K.

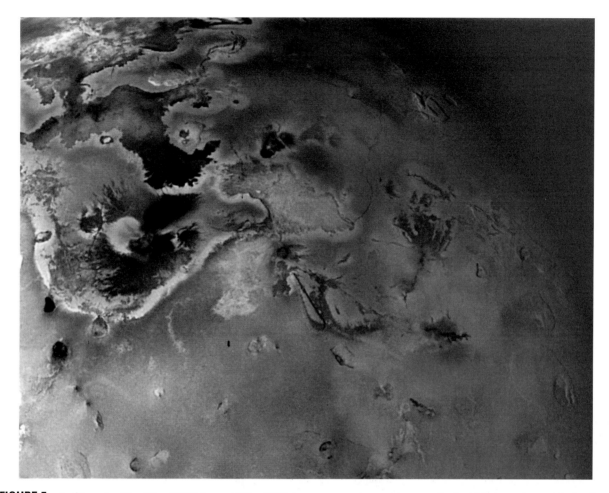

FIGURE 5   In this region (Lat. 2° to 65° N, Long. 150° to 223°) there are rugged mountains several miles high, layered materials that form plateaus, and an abundance of irregular-shaped, dark lava flows that occupy much of the scene. Many volcanic calderas are visible, some of them with dark, lava floors—presumably still warm. The white material in this *Galileo* image (November 6, 1996) is SO_2 frost. It can be seen outlining some of the lava flows and marking the bases of some escarpments. [01241997 _full.jpg, geologic landforms on Io (area three).] (http://photojournal.jpl.nasa.gov/cgi-bin/PIAGenCatalogPage.pl?PIA00537)

In 1986, a significant set of eruption temperatures was seen by Jet Propulsion Laboratory observers using an instrument at the NASA Infrared Telescope Facility on Mauna Kea. From the data collected that night, they estimated that the exposed lava reached temperatures of 1550 K! For the first time, there was no doubt that this had to be molten rock lava as opposed to melted sulfur. In 1990, they obtained the temperature of another eruption, which reached 1225 K. The temperatures of many "hot" sources on Io have now been measured. For example, John Spencer and his colleagues at Lowell Observatory have seen a number of eruptions that have exceeded temperatures of 1000 K. More recently, numerous sightings of similar events have been made by the spectrometers and cameras on the *Galileo* spacecraft in orbit about Jupiter. In addition to constraints on temperature, *Galileo* can provide

relatively precise information on the location of the lava. Some of these thermal anomalies were observed when Io was in Jupiter's shadow, and the lava can be imaged in its own thermal glow (Fig. 6).

The 1990 outburst is particularly important because it was observed at a number of different wavelengths and over a period of slightly more than 2 1/2 hours. The changes in flux with time at the different wavelengths permitted the first analysis of how these events evolve on Io. In this case the eruption started out small and hot. With time it became much larger and cooler. At first sighting, it had a temperature of 1225 K and an area of about 180 km^2. Two and a half hours later, it covered 3041 km^2 and showed an average temperature of 555 K. Over that time, the average rate of areal increase of the flow was 154,000 m^2 per second. For the purpose of illustration, if the lava is assumed to

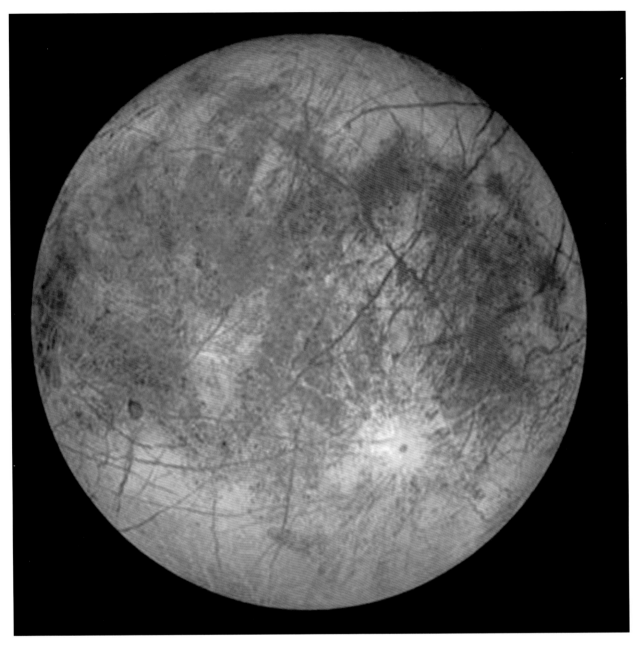

Views of three satellites of the giant planets. Above, Jupiter's ice-covered satellite Europa. Below left, Saturn's outermost large moon, Iapetus. Below right, Triton, a large satellite of Neptune. (NASA photos)

Above, a Voyager image of the main ring system of Saturn, showing spokes predominantly on the eastern half of the B ring, which is generally where they are most visible. Below, a Voyager false-color image of Saturn's rings. Subtle color variations due to differences in surface composition of the particles making up the rings are enhanced in this image produced by combining ultraviolet, clear, and orange frames.

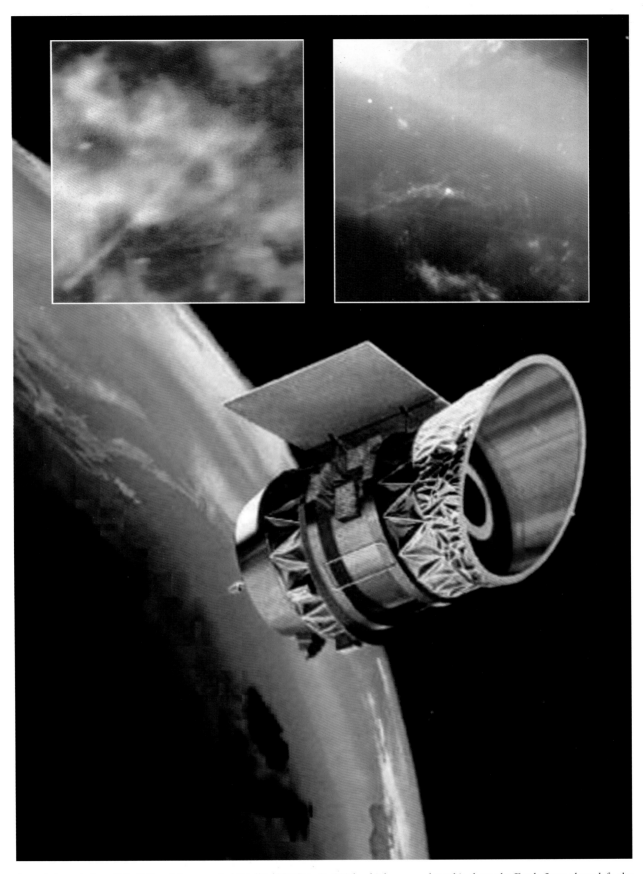

An artist's rendering of the Infrared Astronomical Satellite (IRAS) in its 900-km high, near-polar orbit above the Earth. Inset above left, the brightest of the "orphan trails" detected by IRAS, seen against a background of interstellar clouds. Orphan trails are probably associated with comets never before detected. Inset right, the sky as seen from space in the thermal infrared is filled with interstellar cirrus (the reddish clouds), rings of dust around the solar system arising from asteroid collisions (one of which is seen as the broad band extending diagonally across the top of the image), and contrail-like structures consisting of cometary debris—the birth of a meteor stream (see below the band).

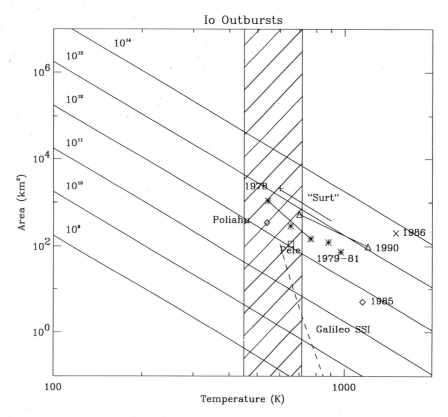

FIGURE 7 Area versus temperature data for Io's large eruptions. The crosshatched zone indicates temperatures at which a predominantly sulfur lava on Io could start to boil. Events observed to have temperatures above this zone must have lava composed of molten rock. Some events start at temperatures above this zone, but have been observed to enter it as they cool (e.g., points for 1990). These lavas are also molten rock, not sulfur. For temperatures in or below the zone, the lava in question could be composed mainly of molten sulfur. If the lava is pure sulfur, its boiling point would be ~450 K at a pressure of ~0.001 bar. The presence of other constituents in the lava could raise the boiling point.

IV. HEAT FLOW

A tremendously large amount of heat comes to Io's surface from its interior and is ultimately radiated to space. This high level of thermal emission provides conclusive proof that Io is the most volcanically active body yet discovered. In this section, we discuss how Io's heat flow is detected and measured.

All bodies emit radiation because of their temperature, and the physics of this thermal emission is well understood. As the surface of Io absorbs sunlight, it is warmed. As its temperature rises, so does the amount of thermal power it radiates. The total amount of emitted power and its distribution as a function of wavelength (i.e., its spectrum) can be calculated relatively accurately.

Volcanic processes provide additional heat to some parts of Io's surface. Temperatures in these places are higher, being heated by both solar and volcanic sources. Since we know the albedo of Io's surface, and some of its other thermophysical properties, we can estimate and subtract away that portion of thermal emission that is due to solar heating. What remains is the emission due to volcanic sources. We call such areas of elevated temperature *thermal anomalies*. (See Section VI, B for further discussion on thermal properties of Io's surface.) The largest thermal anomaly on Io is Loki Patera, which has been present since 1979. It is a complicated region consisting of a number of thermal anomalies and an eruptive plume that was seen by *Voyager 1*.

Astronomical observations of Io's thermal emission go back more than 30 years. However, only over the last dozen or so years has accurate monitoring of Io's excess thermal emission taken place. By observing this excess emission as Io rotates, it was discovered that Io has many thermal anomalies spread across its surface. On the basis of *Voyager 1* and *2* and *Galileo* data, there may be hundreds of thermal anomalies.

Consider a plot of the data for all of the thermal anomalies on Io (Fig. 8). The data have been binned by temperature interval, so that a point at a given temperature represents all of the area on Io that is within that temperature range. Now compare these data with points calculated using the lava flow model published by Michael Carr in 1986. In constructing his model, Carr took into account observations of terrestrial basalt flows and included successive series of eruptions as a feature. As these eruptions proceed, the composite flow, composed of both hot active and older cooler flows, becomes much wider. Thus, large areas can be covered. He also took into account that inactive flows are, to some extent, overrun by newer flows. From this ensemble of cooling flows, Carr was able to calculate the collective thermal emission spectrum, which he compared favorably to the data for Loki, the largest thermal anomaly on Io, and to other volcanic sites. In making a comparison with Carr's model, we are comparing *slopes*, not absolute values: the slope of the observational data are compared to the slope of the model points. Another caveat is that we are considering aggregate data for Io that really is divided up, more or less, among some 30 to 50 volcanic centers.

For the higher temperatures (i.e., those found at and near where the lava exists on Earth), Carr considered two different cases of vent geometry. Those model values constitute the intersecting curves at the lower right in Fig. 8. At high temperatures (i.e., 500 to 1000 K) the predictions of the model are very sensitive to the geometry of the vent. Unfortunately, this does not permit the model to predict details about Io. However, for temperatures of less than 500 K, the model is robust and the calculated slope is meaningful. The similarity in the slope between the observations and the model shows that cooling lavas on Io have the same distribution in temperature and size as do terrestrial flows. This also supports the hypothesis that cooling molten rock can account for all of the thermal anomalies on Io. While cooling, sulfur flows cannot be ruled out, they are not needed to explain the thermal data.

The hotter spots on Io, as seen in the outbursts, come and go with time and show much variability. They also occupy relatively little area on the surface. At the low-temperature end of the scale, the collected, cooler thermal anomalies occupy large areas and show relatively little change in total emission. On a global scale, the total emission from these thermal anomalies tends to be relatively stable from year to year. As comparison with the diagonal lines in Fig. 8 confirms, most of the thermally emitted power comes from the cooler anomalies.

The excess power radiated from Io is called *heat flow* and originates in its interior. It totals approximately 10^{14} W. When averaged over Io's entire surface, this amounts to about 2 W m^{-2}, which is a tremendously large heat flow, even for volcanic or geothermal areas on Earth. The heat flow for Earth is about 4.42×10^{13} W, and for the Moon about 7.58×10^{11} W. Thus, Io's present heat flow significantly exceeds that from Earth's interior and is more than 100 times the heat flow for the Moon, a body of approximately the same size and composition as Io. This large amount of heat flow is not produced by radioactive sources inside of Io, for it is a factor of about 200 greater than can be explained due to heating from decay of radioactive elements. The only process that can offer enough power is tidal dissipation, which occurs because Io is in a slightly eccentric orbit around Jupiter. As Io alternately finds itself closer and then more distant from Jupiter, the varying gravitational attraction causes tides in Io to first rise and then relax. During this continuous flexing of Io's rocks, mechanical energy is dissipated to produce heat. Once a portion of rock in Io's crust or mantle melts, it becomes more susceptible to the action of the tides. This increases the rate of heat production even more.

Unlike other rocky planets in which the preponderance of the heat flow reaches the surface by *conduction*, Io has a unique mode, *bulk transport*, for essentially all of its heat flow. Perhaps this is because Io has such a large amount of heat that must flow to a relatively small surface. This mode is also thought to have occurred on Earth billions of years ago, when Earth's heat flow was much higher than it is today. Assuming that all of the thermal anomalies on Io are created this way, Diana Blaney and colleagues have calculated that to sustain a heat flow of 2 W m^{-2}, the global resurfacing rate must be at least 1.3 cm per year averaged over all of Io's surface! This is very rapid, and is a good reason why impact craters are not to be found on Io. After a hit, the crater produced is quickly (in geologic time) buried by new deposits.

V. PLUME ERUPTIONS

Spacecraft navigators took pictures to use for determining the *Voyagers'* precise positions with respect to the satellites in the Jovian system. Such a picture was taken by *Voyager 1* on March 8, 1979, looking backward some 4.5 million km (2.6 million miles) at Io, three days after closest approach. This dramatic picture (Fig. 9) shows two simultaneously erupting plumes. The first can be seen contrasted against the black "sky" between 9 and 10 o'clock on the edge of

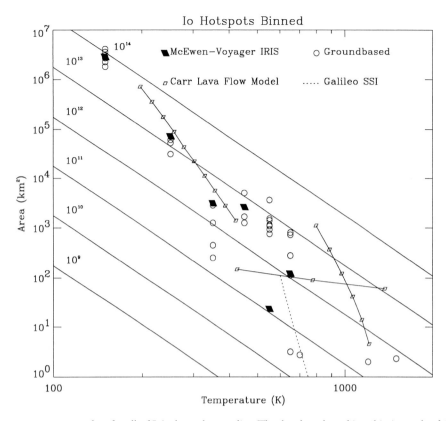

FIGURE 8 Area versus temperature data for all of Io's thermal anomalies. The data have been binned in intervals of 100 K and the areas summed. The sloped lines are isopower contours (ranging from 10^9 to 10^{14} W). For the whole of Io, it can be seen that more power is radiated by the lower-temperature anomalies than by the hotter ones. The agreement of the slope of the lower-temperature data with the slope of the Carr model is taken as an indication that lava flows composed of molten rock can explain all of Io's thermal anomalies. At the higher temperatures, the model predicts the areas to be dependent on the nature of the vent from which the lava effuses. Two different cases are shown.

the disklike image of Io. This plume is lofting fine particulate matter to altitudes of more than 260 km (150 miles). The second eruptive plume can be seen as a bright spot just on the nighttime side of the terminator (i.e., the demarcation between day and night). This eruption is raising a cloud of fine particles high enough above the surface that they catch the rays of sunlight. The "dark" hemisphere of Io is made visible because it is illuminated by light reflected from Jupiter. This picture is historic because it captured the first evidence of active eruptions other than on Earth.

Eventually nine plumes were discovered in the images taken by *Voyager 1* over the 6 1/2 days when it was near its closest approach. Four months later, eight of these plumes remained active and were seen by *Voyager 2*. Perhaps by coincidence, *Galileo* also obtained a picture that shows two active plumes (Fig. 10; see also color insert).

Plumes range in height from ~60 km to more than 400 km. Usually, they can be well simulated using ballistic trajectory models. Ejection velocities of 0.5 to

1.0 km per second are required to attain the observed heights. Ejection angles vary from the vertical, namely, 90°, down to approximately 55°. When the plume material returns to the ground, it produces a fallout deposit. In general, the shapes of the deposits are in the form of a circular ring. However, there are notable deviations from circularity, and these are believed to be due to the geometry of the vent from which the plume gas issues. Pele provides a good example of this deviation (see Fig. 3). Since plumelike deposits are seen in many places on Io where plumes are not currently active, most plumes may be short-lived or perhaps intermittent.

Eruptive plumes on Io are not the same as volcanoes. They are complex fluid phenomena more analogous to geysers on Earth. Plumes may, *or may not*, occur near lava eruptions in which molten rock comes to the surface from volcanoes. Ash plumes characteristic of Plinian eruptions (a type of violent, explosive eruption on Earth) have not been observed on Io.

The energy to erupt the plumes is derived from

FIGURE 9   This is the plume discovery picture taken by *Voyager 1* on March 8, 1979. Linda A. Morabito, an engineer at the Jet Propulsion laboratory in Pasadena, California, was the first person to see Io's plumes. This picture shows two plumes. One is on the edge of the disk and is seen in profile against the dark sky (at 9 to 10 o'clock with respect to the center of the disk). It is rising some 260 km (150 miles) above the surface. The other appears as a bright spot on the nightside of the morning terminator (i.e., the boundary between day and night, at 8 o'clock). It is reaching above the shadow of the local night to catch the rays of the rising sun. The nightside of Io can be seen because it is being illuminated by light reflected from Jupiter. (http://photojournal.jpl.nasa.gov/cgi-bin/PIAGenCatalogPage.pl?PIA00379)

heat below the surface. The observable properties such as temperature, shape, and velocity are dependent in a complex way on the subsurface features of the volcanic system from which they arise. In 1982, Susan Kieffer developed a model in which a volcanic system is divided into five regions: supply region, reservoir, conduit, vent, and plume. She used fluid dynamical and thermodynamic principles to study the motion of fluids moving through these regions. She found that Io's plumes are analogous to geysers. On Earth, the working fluid is water. On Io, the working fluid could be either liquid

sulfur, liquid SO_2, or both. She found that several different styles of eruption could involve SO_2 gas alone. With this model, she was able to explain the heights of the observed plumes.

The higher plumes require that SO_2 originate at relatively high temperature and pressure. Applied to Io, this means that SO_2 must start its journey at a significant distance below the surface (e.g., ~1 to 2 km) and come into contact with temperatures as high as 1400 K and above. Such temperatures are available from the subsurface migration of magma. In volcanic

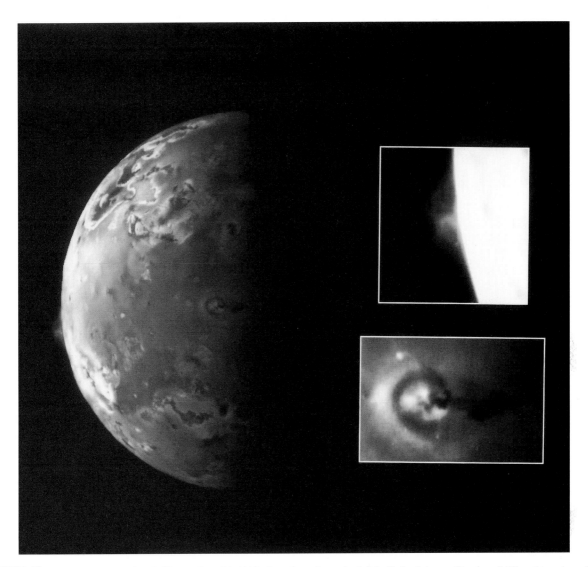

FIGURE 10 Two plumes seen by *Galileo* on June 28, 1997. One plume is on the bright limb of the satellite (at ~8:30) and is enlarged in the upper inset. It comes from a volcanic depression or caldera named Pillan Patera (after a South American god of thunder, fire, and volcanoes). This plume is ~140 km high (~86 miles) and was also detected by the *Hubble Space Telescope*. The second plume is near the terminator (the boundary where day meets night). It is enlarged and shown in the lower inset. The location of this eruption is at a volcano called Prometheus, named after the Greek fire god. Note the shadow cast by this plume. From measurement of the shadow, we know that this plume has a height of ~75 km (~45 miles). Prometheus was active at the times of the flybys of *Voyagers 1* and *2*. Is it possible that this plume has been erupting continuously for more than 18 years? By contrast, this picture is the first showing activity at Pillan Patera. (PIA00703.jpb) (http://photojournal.jpl.nasa.gov/cgi-bin/PIAGenCatalogPage.pl?PIA00703) (See also color insert.)

regions on Earth, such as Kilahea, Hawaii, there is about three times as much subsurface activity in terms of lava intrusions into the rock than there are eruptions of lava flows onto the surface. If the same holds true for Io, then its abundant surface activity suggests an even more active subsurface.

The thermodynamics that governs the eruption of SO_2 plumes on Io is shown in Fig. 11. Plume material starts out as a liquid at depth. Upon contact with a hot heat source, it is vaporized and pressurizes the

reservoir in which it resides. Once a path or conduit to the surface is found, the ascent of SO_2 liquid begins. As it rises up the conduit to the surface, it vaporizes and expands adiabatically (i.e., does not exchange heat with its environment), maintaining constant entropy as it cools. These paths plot as vertical lines in Fig. 11. The heavy, vertical lines show the eruption paths studied by Susan Kieffer. The high ends of the eruption paths mark the starting temperature of that eruption, which is at some depth below Io's surface. The highest-

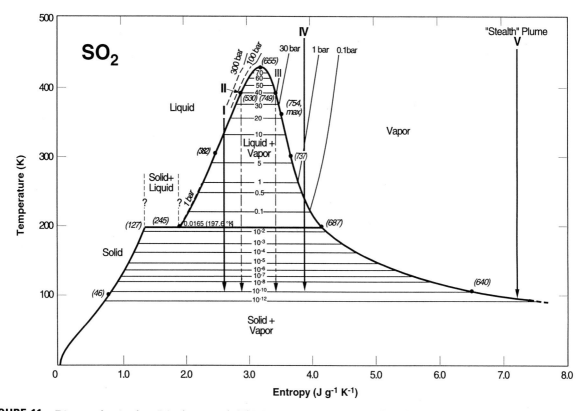

FIGURE 11 Diagram showing how Io's plumes work. This is a temperature–entropy phase diagram for SO$_2$. The heavy solid curves are phase boundaries, and thin lines are isobars. The heavy, vertical lines show possible eruption paths corresponding to different types of plumes. These paths are isentropic (i.e., the eruptions are assumed to be adiabatic). This condition is met when the eruption proceeds rapidly, with little or no heat exchange with the environment. An eruption starts at the high-temperature end of the isentrope (note that the "Stealth Plume" isentrope starts at 1400 K, well above the range plotted in this figure). In the starting reservoir the SO$_2$ is at both high pressure and temperature. It starts its trip up a conduit leading to the surface. As the SO$_2$ vapor expands, it also cools. This continues as the eruption continues down the isentrope. At some point, ~100–120 K, the expansion ceases. All of the available thermal energy has been converted into the kinetic energy of the uprushing SO$_2$ gas.

temperature paths (i.e., ~1500 K, and well above the scale of this figure) require a silicate lava to come into contact with a subsurface SO$_2$ reservoir.

The observed plumes are seen in reflected sunlight and can have two components—gas and particulate matter. The size of the particulate matter varies from a few hundredths of a micron to a few tenths of a micron in diameter. These particles arise either from the condensation of the gas, from the erosion of the vent walls, or as dissolved or colloidal materials that were suspended in the SO$_2$ liquid at depth. In Kieffer's diagram (Fig. 11), eruption paths I, II, and III permit condensation of particulate matter. In path IV, the eruption has proceeded mostly to completion before it enters the liquid and vapor field, and then the solid and vapor field. There is a relatively limited opportunity for the gas to condense and form particulate matter before all of the erupting material finds itself exiting the vent and in the very low pressures in the plume.

Along path V, which can start out as high as 1500 K, all of the expansion takes place before the solid field is encountered. Thus, no condensation of particulate matter is possible in this scenario. (See Section VI, C.)

The analogy of Io's plumes to geysers on Earth also extends to the need for replenishing the working fluid. On Earth, this is accomplished by groundwater recharging the geyser. On Io, the abundance and duration of plumes imply that SO$_2$ is somehow being recycled. It is amazing, but lava flows can effectively bury and recycle SO$_2$. A solid-state variant of the terrestrial recharging mechanism may supply the subsurface SO$_2$ needed by Io's plumes. Figure 12 illustrates the cycle. Thick SO$_2$ frost deposits on Io's surface are run over and buried by lava. The bottom of a lava flow is rapidly quenched by the cold SO$_2$. The thickness of SO$_2$ frost lost by sublimation is roughly equal to the thickness of the lava flow itself. The lava forms a cap over the SO$_2$ deposit, and subsequent lava flows do not cause

FORMING RESERVOIR V

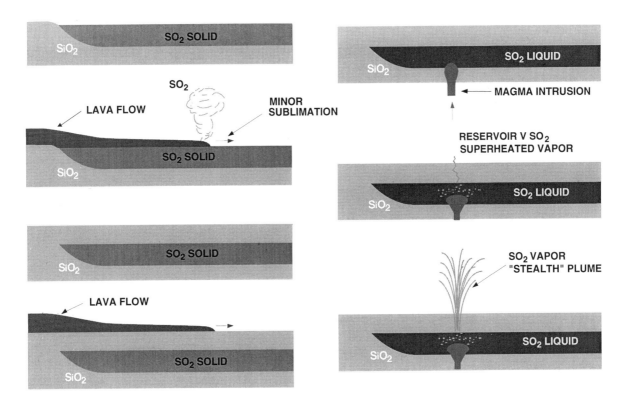

FIGURE 12 The SO$_2$ cycle. The active plumes draw upon SO$_2$ that is stored below the surface. If the eruptions are to continue, then these deep reservoirs of SO$_2$ must be replenished. This drawing illustrates a hypothetical process for producing "stealth" plumes. Starting at the upper left and proceeding downward: (1) First, SO$_2$ is collected as a frost deposit by cold-trapping gas from the atmosphere. (2) After the frost deposit has become some tens of meters thick, it is covered by a lava flow. The hot lava vaporizes some of the SO$_2$. (3) The remainder of the SO$_2$ is now buried. (4) Later, additioinal lava flows cover the earlier flows, burying the SO$_2$ deeper. This process happens many times until depths of 1 to 2 km are reached. When the pressure becomes high enough (see Fig. 11) the SO$_2$ will liquify. (5) Later, a lava intrusion encounters the buried SO$_2$. (6) Some of the SO$_2$ becomes a superheated vapor immediately upon contact with the 1400 K lava. Pressure increases and a conduit to the surface is forced open. (7) SO$_2$ travels up the conduit, expanding adiabatically along isentropic paths, as illustrated in Fig. 11.

any significant further loss of SO$_2$, they only bury the deposit deeper. Eventually, with many repeated flows, the SO$_2$ reaches a depth where it is liquefied by pressure. When an intrusion of lava comes into contact with the deeply buried SO$_2$ deposit, it creates superheated SO$_2$ vapor, which pressurizes the reservoir. Once the liquid/gas finds a route of escape to the surface, a plume is formed.

VI. IO'S "ATMOSPHERE"

Against the spectacular backdrop of Io's geologic activity, its atmosphere is very modest. Its ambient surface pressure is measured in nanobars to picobars, units that are a billion to a trillion times less than the pressure at Earth's surface. These pressures are comparable to good laboratory vacuums on Earth. Higher pressures on Io are found near the geyserlike plumes and in the vicinity of other types of vents, or areas where gas is emerging as a result of volcanic activity. SO$_2$ is the chief constituent of the atmosphere as well as a principal volatile on and below the surface. SO is also present, and there are probably other trace constituents yet to be discovered. Other gases (e.g., O$_2$) have been suggested as possibly significant but have not yet been detected.

There are two possible sources for the SO$_2$ in Io's atmosphere. The first is from the solid–vapor equilibrium between the atmosphere and frost deposits on

Io's surface. This would mean that the atmospheric pressure would be tied to the surface temperature of such frost deposits. During the night, the atmospheric pressure would drop dramatically as Io cooled. The other source for the atmosphere is the gas that issues from vents and plumes. In reality, both sources have roles to play.

A. THE VOLCANIC "ATMOSPHERE"

Atmospheric scientists have found Io to be a difficult subject. Until recently, there was only one observation that directly sensed SO_2 gas. In 1979 during the *Voyager* flyby, the IRIS instrument obtained near-infrared spectra showing SO_2 gas over Loki. This gas was part of the Loki plume. The next observation of SO_2 gas was made a dozen years later in 1990–1991. The rotational lines of SO_2 gas were detected at millimeter wavelengths by Emanuel Lellouch and his colleagues in France. These observations of Io were conducted with the 30-m IRAM (Institute de Radio-Astronomie Millimetrique) radio telescope and were the first direct ground-based detection of SO_2 gas. They found column densities in the range of $\sim 10^{18}$ to 10^{19} molecules cm^{-2} and detected a number of absorption lines that provided information needed to model the basic properties of Io's atmosphere. The Doppler shift of these lines indicated that most of the gas in Io's atmosphere is moving at relatively high speed, perhaps in the eruptive plumes. In addition, the observations showed that the column density of SO_2 gas was *not* well correlated with the SO_2 frost deposits on the surface. They also noted that there is no dependence on the abundance of SO_2 gas with time of day. Both of these facts indicate that gas pressure is not controlled by surface frost deposits.

Two years later, with the advent of the *Hubble Space Telescope*, direct measurements sensitive to SO_2 gas were possible at ultraviolet wavelengths. Ultraviolet measurements cannot be made from the surface of Earth because our atmosphere is opaque in this wavelength region. The *Hubble Space Telescope* discovered further evidence that Io's atmosphere was patchy, and in fact over most of the planet it is very thin and tenuous. At certain locations, volcanic gas increases the atmospheric density. At these places, the thickness of the atmosphere could be substantially higher as long as gas continues to be erupted from below the surface. [*See* THE SOLAR SYSTEM AT ULTRA-VIOLET WAVELENGTHS.]

In addition to spectroscopic evidence for SO_2 gas,

the Hubble Space Telescope investigators obtained ultraviolet images sensitive to its distribution. They were able to show that these images were consistent with the presence of SO_2 gas patches over the frost-poor region were Pele is located. Therefore, it appears that the bulk of Io's atmosphere is from volcanic sources. In regions very distant from volcanic sources of SO_2 gas, or at times when those sources are not operating, a very thin atmosphere may evaporate from warmer or even condense on colder SO_2 frost deposits.

B. THE VAPOR PRESSURE "ATMOSPHERE"

The most important control on the vapor pressure-driven atmosphere is the surface temperature. In 1988, Damon Simonelli and Josphef Veverka noted that the *Voyager* IRIS observations of the nightside of Io suggested temperatures that were about 35 K higher than would have been expected from a lunarlike surface. Unfortunately, the *Voyager* nighttime observations were at low spatial resolution. Also, the pointing knowledge (i.e., where the spacecraft was looking) for the IRIS experiment was too poor in 1982 to permit a more detailed analysis. Alfred McEwen and his colleagues obtained a similar result using better pointing information in 1992. Simonelli and Veverka suggested that this may be an indication that the heat was being transported to Io's nightside by some mechanism, possibly a solid-state greenhouse. In *the solid-state greenhouse effect*, sunlight passes through relatively translucent materials to be absorbed below the surface. Insolation thus absorbed produces internal heat, and this heat must be conducted to the surface before it can be radiated to space. This gives rise to warmer nighttime temperatures, which in turn would imply that the nighttime atmosphere of Io could be somewhat thicker than previously thought.

On the basis of multispectral radiometry obtained in 1994 with the IRTF telescope on Mauna Kea, Hawaii, Glenn Veeder and colleagues discovered that Io's high-albedo areas (which comprise about 80% of the surface) radiated only about half of the energy that they absobed from the Sun during the daytime. This meant that the remainder of the energy had to be released during the nighttime. Thus, nighttime temperatures are higher than previously expected, and temperatures during the daytime are lower. The result of such major reduction in the daytime surface temperature (i.e., from ~ 125 K to ~ 110 K at the equator) is to reduce the basal vapor pressure of an SO_2 atmosphere by about a factor of 1000. This means that evaporation from surface frost deposits plays only a minor role as a source of SO_2 supply to the atmosphere when significant volcanic sources are active. At times when there are no volcanic sources, it provides an SO_2 pressure of $\sim 10^{-10}$ to 10^{-9} bars at the subsolar point.

C. STEALTH PLUMES?

The amount of SO_2 gas needed to match the millimeter-wavelength observation would require 30–50 large plumes on Io. By comparison, *Voyager* saw only 10 such plumes. Has Io changed since the time of *Voyager*? Or did *Voyager* miss seeing something? Small plumes, for example, plumes of 10 km height, could have been missed, but they are also too small to satisfy the microwave data. *Voyager* detected the big, Prometheus-type plumes through the scattering of sunlight by small particles of SO_2 in the plume. However, if pure or nearly pure gas-phase plumes exist, then *Voyager* would not have seen them because they would have contained few if any particles with which to reflect sunlight. Such plumes have been dubbed "stealth" plumes because they would be hard to detect. Stealth plumes are extreme geyserlike eruptions that originate in a subsurface reservoir that is at high temperature and pressure (>1400 K, >40 bars). Susan Kieffer also studied the thermodynamics of such plumes in 1982 (see Fig. 11, path V). Geologically, these plumes would occur whenever a molten silicate intrusion encounters SO_2 liquid buried at depths of 1.5 km or deeper. Essentially the entire eruption occurs in the gas field and, in principle, particles cannot be formed through condensation. The *Galileo* spacecraft has detected several stealthlike plumes.

D. SOURCES, SINKS, AND TRANSPORT PROCESSES

New data and ideas have changed our perception of Io's atmosphere and how it interacts with the surface.

It is now clear that the geyserlike eruptive plumes and various forms of volcanic venting are very important as sources of atmospheric SO_2 gas. The global role of SO_2 frost as a source of atmospheric gas is relatively minor. At best, it can sustain an atmospheric gas column of only about 10^{15} molecules cm^{-2}. By comparison, the observations give gas columns that range from $\sim 10^{16}$ to 10^{18} molecules cm^{-2}.

Eruptive plumes, especially if a significantly large number of "stealthy" plumes are present, are a far more significant process than previously suspected. Ballistic transport and thermal escape of gas from Io remains a minor process. Because of the cold temperatures of surface frost deposits (e.g., 100–110 K), cold trapping by condensation on the surface acts as a voracious scavenger of SO_2 gas. Loss of atmospheric gas to space by a variety of processes is relatively small. However, the loss of these small amounts of gas and other materials into Io's torus is necessary to sustain the population of atoms and ions in the Jovian magnetosphere.

VII. FUTURE PERSPECTIVE

Our understanding of Io in the late 1990s is evolving very rapidly as a consequence of the unprecedented flow of new data. The *Galileo* spacecraft is in orbit about Jupiter and its Io intensive phase is yet to come. Ground-based astronomical telescopes and instrumented telescopes in Earth orbit are producing a stream of observations. Some results can be easily anticipated: (1) The rate of eruption of hot lava onto Io's surface will become accurately known. (2) The abundance of eruptive plumes, their locations, and their durations of eruption will be determined. (3) The atmosphere will become much better characterized, and (4) its interactions with the magneto-

TABLE II
Io Information on the World Wide Web

Organization	URL
Galileo Project, JPL	http://www.jpl.nasa.gov/galileo/
International Jupiter Watch Satellite Discipline	http://www.lowell.edu/users/ijw/
Space Telescope Science Institute	http://www.stsci.edu/
National Aeronautics and Space Administration	http://www.nasa.gov/

sphere will be better understood. (5) Io's internal structure will become better resolved. (6) The nature of magnetic fields and plasma at Io will be elucidated.

To help the reader keep up to date with these developments, a number of sites on the internet are suggested in Table II. By visiting these sites, you will be able to keep abreast of the unfolding Io story.

BIBLIOGRAPHY

Beatty, J. K., and Chaikin, A. (eds.) (1990). "The New Solar System," 3rd Ed. Sky Publishing, Cambridge, Mass.

Belton, M. J. S., West, R. A. and Rahe, J. (eds.) (1989). "Time-Variable Phenomena in the Jovian System," NASA SP-494. NASA, Washington, D.C.

Burns, J. A., and Matthews, M. S. (eds.) (1986). "Satellites." Univ. Arizona Press, Tucson.

Spencer, J. R., and Schneider, N. M. (1996). *Annu. Rev. Earth Planet. Sci.* **24**, 125–190.

TITAN

I. Introduction: Historical Context

II. The Voyager Mission to Saturn

III. The Atmosphere of Titan

IV. The Surface of Titan

V. Titan in the Solar System

VI. Space Missions: Infrared Space
Observatory and Cassini/Huygens

ATHENA COUSTENIS
Observatoire at Meudon, Paris

RALPH D. LORENZ
University of Arizona

GLOSSARY

Adiabatic lapse rate: Lapse rate is the drop of temperature with altitude, and frequently this quantity is close or equal to that for a parcel of air moving without transfer of energy ("adiabatic"). As a parcel moves upward, it expands as the atmospheric pressure falls. The expansion requires work to be done, this work coming from the internal energy of the air parcel. Thus the parcel cools as it rises.

Arcmin: Minute of arc, arc minute, or '. An angle equal to one-sixtieth of a degree, or sixty arc seconds (60"). The field of view of the Hubble Space Telescope Planetary Camera is 34". Titan as seen from Earth subtends an angle of about 1 arcsec.

Atmospheric layers: By analogy with Earth, Titan's atmosphere is subdivided into layers defined by the temperature variations with height (or pressure, through the hydrostatic law). On both Titan and Earth, the mean temperature profile is characterized by two temperature inversions (locations above which the temperature increases with altitude, whereas the opposite takes place beneath). The boundaries between regions are identified by terms ending with -pause (from the Greek "παύσης" = end). Convective instability exists in the lowest 40 km or so of the atmosphere (in other words, since the temperature decreases with altitude, the warmer air lying under colder air is unstable and so it rises while the colder air sinks), which is known as the *troposphere* ("turning-region"). The upper boundary here is the level where the overlying atmosphere is of such a low density that a substantial amount of radiative cooling to space can occur in the thermal infrared region of the spectrum. At this level, called the *tropopause*, radiation cools rising air so efficiently that the temperature tends to become constant with height and convection ceases. The *stratosphere* is the region where each layer is heated by radiation from the optically thick atmosphere below, and cooled by radiating to space; here density decreases monotonically with height, and therefore the layers do not try to move up or down through each other as in the troposphere. The temperature is also maximum at the upper boundary, which is known as the *stratopause*, due to the absorption and conversion of UV solar radiation by different gases and aerosols. Above the stratopause, the temperature declines again, reaching a minimum at the *mesopause*, where the second temperature inversion occurs, signifying the end of the *mesosphere*. The pressure at the mesopause is only a few microbars. With such low densities of gas above, energetic particles and solar photons in the extreme ultraviolet penetrate into the region, causing ionization and dissociation. The heating thus produced causes the temperature to increase rapidly with height, leading to the name *thermosphere* ("θερμό" = warm), the most extensive part of the atmosphere, in which the energy is transported by thermal conduction. A fairly small distance up into the thermosphere, diffusion takes over as the dominant process and the atmosphere starts to separate into its lighter and heavier components. For many practical purposes, this level (the *homopause*) may be considered to be the effective top of the atmosphere. The outermost part of the atmosphere, which extends to outer space from about 1500 km altitude on Titan, is also called the *exosphere* ("έξω" = out), since here light elements escape the planet's grip and are lost to space.

Dayglow: Fluorescent emission of radiation from the upper atmosphere, due to excitation by high-energy solar photons.

Deuterium: Heavy isotope of hydrogen, with a proton and neutron in its atomic nucleus. Deuterium (and hydrogen compounds like methane and water) participates in chemical reactions and physical processes such as diffusion and evaporation at different rates from normal hydrogen. The hydrogen-to-deuterium ratio (H/D) in an atmosphere is therefore an indicator of the separation processes that have occurred in it.

Dynamic inertia: Increase in radiative time constant due to mixing of the more massive deeper layers of an atmosphere. The thin atmosphere at high altitudes would be expected to respond rapidly to changes in sunlight, but if there is substantial vertical circulation, the changes will occur more slowly because mixing increases the effective mass of the layer under consideration.

Exobiology: Study of life beyond Earth. More broadly, the study of environments favorable for life and the origin and evolution of life in the universe.

Geometric albedo: Ratio of the total reflected flux in all directions by a planet to the incident solar flux received by a sphere of unit radius, divided by the phase integral. For a highly reflecting atmosphere or surface, the geometric albedo is in the range 0.65–0.80, depending on the phase function (see also *limb-darkening*).

Hadley circulation: Meridional (north–south) flow pattern in the atmosphere by which heat is transported away from warmer to colder regions. On Earth, this heated air rises at low latitudes and descends at high latitudes.

Infrared spectroscopy: Dispersal and measurement of infrared light intensity. Near-infrared (0.7 to 5 μm) spectroscopy is useful for mineral and ice identification; mid-infrared (5 to 30 μm) is more useful for measuring surface temperatures and gas abundances. Wavelength is often expressed as a wave number (e.g., a wave number of 10,000 cm^{-1} indicates that 10,000 wavelengths fit into 1 cm, so the wavelength is 1 μm; a wavelength of 10 μm therefore has a wave number of 1000 cm^{-1}).

Jeans escape: Process by which fast (energetic or hot) molecules of an atmosphere escape into space. The energy distribution of a gas at a given temperature has a hot tail, that is, a few atoms move faster than the rest. Therefore, at an altitude where collisions between molecules are rare, if the molecules in the hot tail move faster than the local escape velocity, they can escape to space. This process is fastest for hot atmospheres of light gases (hydrogen, helium) on bodies with low gravity.

Limb-darkening: Darkening of the edges of a planetary disk. This may be due to the scattering properties of the surface (e.g., if it is a strongly backscattering surface, like an icy one) or more usually to the presence of an optically thick atmosphere. It is often characterized by an exponent k, the Minnaert exponent, for a scattering law of the form $I = I_o\mu^k\mu_o^{k-1}$, where μ and μ_o are the cosines of the angle between the normal at a given point and the observer and the Sun, respectively, and I_o is the brightness of the center of the disk. $k = 0.5$ corresponds to a flat disk (rather like the Moon), and $k = 1$ corresponds to a Lambertian disk with strong limb-darkening. $k < 0.5$ corresponds to limb-brightening, typical of a scattering but optically thin region above an absorbing (dark) region in the atmosphere.

Mixing ratio: Fraction of the total volume (or the fraction of the number of molecules) of a given compound.

Mixing ratio scale height: If the mixing ratio of a given compound in an atmosphere changes with altitude, owing to its production or removal at different altitudes, its variation may be conveniently described by a function of the form $\exp(-h/H)$, where h is the change in altitude and H is the mixing ratio scale height, the vertical distance over which the mixing ratio changes by a factor e.

Oblateness: Flattening of a planet or satellite, usually due at least in part to its spin. The oblateness is the difference between the polar and equatorial radii, divided by the equatorial radius. Earth's oblateness is 1/298; Titan's atmosphere has an oblateness of ~1/250 at an altitude of 250 km, from which the rapid rotation of the upper atmosphere has been inferred.

Opacity: Capacity of an atmosphere to absorb (or sometimes scatter) radiation, also called optical depth. A beam of monochromatic radiation passing through an atmosphere with an optical depth of 1 will have its intensity reduced by a factor e (=2.718...), whereas an optical depth of 4 absorbs 99% of the radiation. Opacity is a function of wavelength, as well as the pressure, temperature, and composition of the region of the atmosphere under consideration.

Optically thin: Absorbing or scattering relatively little radiation, that is, "clear." The usual interpretation of this term is that the optical depth is less than 1.

Photolysis: Breaking of a chemical bond by a light photon (usually at ultravioleet wavelengths). The subsequent molecular fragments may recombine in a number of ways. Among easily photolyzed gases are methane, ammonia, and oxygen, but even "unreactive" gases such as nitrogen may react photochemically to produce other compounds (such as HCN on Titan, or nitrogen oxides on Earth).

Radio-occultation: Passing of a radio beam through a planet's atmosphere. Attenuation and refraction (bending) of the beam can be used to measure the density of electrons in the planet's ionosphere and the density of the gas in its atmosphere. The abrupt cutoff of the signal can also be used to make a precise measurement of the planet's surface radius.

Scale height: Vertical distance over which an atmospheric property (usually pressure or density) changes by a factor e. The property is described by a function of the form $\exp(-h/H)$, where h is the change in altitude and H is the

scale height. Four scale heights above the surface, the pressure is about 1/100th of its surface value. For an isothermal atmosphere, the pressure scale height is equal to $(c_p T/gM)$, where g is the local gravity, M the relative molecular mass, c_p the specific heat at constant pressure, and T the local temperature in K. Because all of these quantities vary with altitude, the scale height also varies with altitude, but it is often a convenient approximation to assume it to be constant. For Earth, the scale height is around 10 km, whereas for Titan's upper atmosphere the value is nearer 40 km.

Super-rotation: Rapid (prograde) rotation of the upper layers of the atmosphere, notable in both Venus and Titan. The atmosphere of Titan at a few hundred kilometers of altitude may make around 20 revolutions for every revolution of Titan's solid body.

Window: Spectral region that is relatively transparent, between two regions that have higher opacity. A window region can be important for remote sensing of a planetary surface and for limiting the extent of a greenhouse effect.

I. INTRODUCTION: HISTORICAL CONTEXT

Titan, Saturn's biggest satellite (the second largest satellite in our solar system), has attracted the eye of astronomers more and more since the beginning of the century. It has long been known to have a substantial atmosphere: in 1908, the Catalan astronomer Jose Comas Sola claimed to have observed limb-darkening on Titan (i.e., a stronger attenuation of the light reflected from Titan's limb rather than by its center, usually associated with a thick atmosphere). At the end of the twentieth century, we know that its massive atmosphere is most similar to Earth's among all the other objects in the solar system.

Titan was discovered on the night of March 25, 1655, when a novice Dutch astronomer, Christiaan Huygens, pointed his telescope at Saturn. He saw what he thought was a small star, 3 arcmin away from the planet. This object had been noticed before: Hevelius in Poland and Sir Christopher Wren in England had already seen it, but with star catalogues being scarce at the time, they had believed it to be just another star. Huygens, however, guessed that it was a satellite, and confirmed his guess a few days later. The exceptional size of the discovered satellite is the reason for the name it was given, upon a proposition by Sir William Herschel, who suggested names of gods associated with Saturn for naming its satellites.

After Sola's claims of having observed an atmosphere around Titan, Sir James Jeans decided in 1925 to include Titan and the biggest satellites of Jupiter in his theoretical study of escape processes in the atmospheres around solar system objects. His results showed that Titan could have kept an atmosphere, in spite of its small size and weak gravity, if low-temperature conditions (which he estimated between 60 and 100 K) prevailed. In this case, a gas of molecular weight higher or equal to 16 could not have escaped Titan's atmosphere since the satellite's formation. The constituents that could have been present in nonnegligible quantities in the mix of gas and dust particles that condensed to form the solar system and that, on the other hand, satisfy Jeans limit, are: ammonia, argon, neon, molecular nitrogen, and methane. The first, ammonia (NH_3), is solid at the estimated Titan temperature and therefore could not substantially contribute to its atmosphere. The last, methane (CH_4), is gaseous at this same temperature range and, unlike argon, neon, and molecular nitrogen, it exhibits strong absorption bands in the infrared.

The first detection of these bands was made in 1944, when Gerald Kuiper, of Chicago University (another Dutchman and a student of Jan Hendrik Oort), discovered spectral signatures on Titan at wavelengths longer than 0.6 μm, among which he identified two absorption bands of methane at 6190 and 7250 Å. By comparing his photographic observations with methane spectra taken at low pressures, Kuiper estimated the methane abundance to be 200 m-amagats. He then searched for similar behavior in the specta of other Saturnian satellites. In data he had obtained in 1952, he found—on the contrary—differences between Titan and the other satellites in the intensity observed in the ultraviolet and visible continuum. These differences, due to the presence of an atmosphere around Titan, are also responsible for the orange color of Titan.

In the years that followed, scientists struggled to understand Titan's atmosphere. In 1965 it was still hard to reach a consensus on a coherent value of the ground temperature from contradictory radio and infrared measurements, which ranged from 165 to 200 K. From 1972 to 1979, several scientists (Laurence Trafton, Barry Lutz, Tobias Owen, Uwe Fink, Harold Larson, and others) concentrated their efforts on an estimate of the methane abundance and of the pressure conditions in the atmosphere and at the surface from observations made in the 1- to 2-μm spectral region. The limb-dark-

ening was finally unambiguously observed in 1975, in accordance with an optically thick atmosphere. During this time, Laurence Trafton, from Texas University, conducted his own observations of the $3\nu_3$ methane band at 1.1 μm, in which he found an unexpected high intensity. This was indicative of either a methane abundance at least 10 times higher than that inferred by Kuiper or a broadening of the methane bands induced by collision with molecules of another as yet undetected gas, which would have to be quite abundant in the atmosphere. In either case, the intensity of the absorption band is a function of the methane abundance and of the local atmospheric pressure. By comparing the weak absorption bands of methane in Titan's visible spectrum with spectra of Jupiter and Saturn, where these bands have almost identical equivalent widths, in 1976 Barry Lutz and colleagues derived (from the pressure–abundance produced from Trafton's measurements) a 320 m-amagat abundance, leading to an estimate of the effective pressure on Titan of about 200 mbar. The immediate consequence of this result is that methane suddenly became just a minor atmospheric component, since even 1.6 km-amagat (Trafton's high value) could only produce a surface pressure of 16 mbar, which is an effective pressure of 8 mbar. (A km-amagat is a measure of the column density of a gas—one km-amagat corresponds to about 2.6×10^{24} molecules above every square centimeter.)

In 1975, Trafton had announced a tentative identification of the 3-0 S(1) quadrupolar line of hydrogen (H_2) in the spectrum of Titan, for which he had evaluated an abundance of 5 km-amagat. All efforts concentrated on the detection of NH_3 failed to produce more than low upper limits, suggesting that ammonia could either be photodissociated with subsequent production of H_2 or be restricted on the surface as ice.

By 1973, observations of the satellite's low albedo and of the positive polarization of the reflected light indicated the presence of a thick and cloudy atmosphere with cumuli present at high altitudes. Two sorts of aerosols were expected to coexist in Titan's atmosphere: solid methane clouds and a photochemical fog. Methane photolysis would produce a great variety of organic compounds and polymers.

Fredrick Gillett found evidence in Titan's thermal emission spectrum of not only methane, but also ethane (C_2H_6, at 12.2 μm), monodeuterated methane (CH_3D, at 9.39 μm), ethylene (C_2H_4, at 10.5 μm), and acetylene (C_2H_2, at 13.7 μm). Already, close examination of Titan's spectrum pointed out that the continuum of Titan's atmosphere decreased with frequency. This led to the assumption that it might be possible, at certain frequencies in the near infrared, to probe all the way down to the satellite's surface. The absence of any brightness increase up to wavelengths on the order of 2200 Å (1000 Å = 0.1 μm) argued in favor of an aerosol uniformly mixed at high altitudes. No indication was available, however, on the nature of the aerosols, but their presence in the atmosphere made all attempts at interpretation of spectroscopic observations extremely model dependent.

Two principal models prevailed in the pre-Voyager period. The first one, suggested by Robert Danielson in 1973 and refined by John Caldwell in 1977, favored methane as the main component (about 90%) and predicted surface conditions of $T = 86$ K for 20 mbar, as well as a temperature inversion in the higher atmospheric levels illustrated by the presence of emission features of hydrocarbon gases in the infrared spectrum of Titan. The second one, advocated by John Lewis (1971) and by Donald Hunten in 1977, was based on the assumption that ammonia dissociation should produce molecular nitrogen (transparent in the visible and infrared spectra) in large quantities and held that the surface temperature and pressure could be quite high (200 K for 20 bars). Independently of these two models, an explanation of the high ground temperatures observed was advanced, namely, a pronounced greenhouse effect, resulting essentially from H_2–H_2 pressure-induced opacity at wavelengths higher than 15 μm. This opacity blocks the thermal emission reflected by the surface, thus creating a heat-up of the lower part of the atmosphere.

Just prior to the *Voyager* encounter, Owen and Jaffe made radio observations with the newly completed Very Large Array and correctly obtained the emission temperature of the surface (near 94 K for a pressure close to 1.5 bar). They even suggested oceans of methane, but the paper failed to get the attention it deserved as it was published during the excitement of the *Voyager* encounter.

A year before *Voyager 1* arrived, an earlier, much less sophisticated (but cheaper!) spacecraft flew by Titan. *Pioneer 11* carried a rudimentary imager, a spin-scan radiometer, which produced a coarse image of Titan. The polarimetry of Titan from this instrument, however, was of high quality, and the data are used to this day.

II. THE VOYAGER MISSION TO SATURN

A. THE VOYAGER IMAGES

The *Voyager 1* spacecraft (launched in 1977) reached Titan on November 12, 1980, with closest approach distance to the center of Titan of only 6969 km (4394 km to the surface). *Voyager 2* flew by Titan nine months later but at a distance a hundred times greater (663,385 km), so that the bulk of our knowledge comes from *Voyager 1*. Titan's visible appearance is unexciting—a reddish brown ball, completely covered by thick haze that allowed no visibility of the surface (Fig. 1). The most obvious feature seen by *Voyager* was a difference in the brightness of the two hemispheres. This difference is on the order of 25% at blue wavelengths, and falls to a few percent in the ultraviolet and at red wavelengths. This so-called north–south asymmetry is probably related to circulation in the atmosphere pushing haze from one hemisphere to the other. The altitude of unity vertical optical depth is on the order of 100 km.

The asymmetry has been observed to reverse—when the *Hubble Space Telescope* (*HST*) first observed Titan in 1990, a little over a quarter of a Titan year after the *Voyager* encounters, the northern hemisphere was found to be brighter than the south. Whereas Voyager observed only up to red wavelengths, *HST* has since imaged Titan in the near infrared. At these wavelengths, the asymmetry is reversed, and indeed is somewhat stronger than in the visible (Fig. 2). This is due to the wavelength dependence of the atmospheric brightness (bright at short wavelengths due to Rayleigh scattering, dark in the near infrared due to methane absorption) and the haze (presumed to be dark in blue and bright at red and longer wavelengths, by analogy with "tholin" material generated in the laboratory). Tholin, according to Carl Sagan, comes from the Greek "θωλόν," meaning "muddy." The visible bright hemisphere has less haze than the darker one.

No imaging data yet exist of Titan in summer or winter, although photometry from 1970 to the present suggests that the hemispheric contrast varies smoothly. Limb-darkening is also strongly wavelength dependent. The disk at UV and violet wavelengths is fairly flat (Minnaert coefficient ~ 0.5, *see* Limb-darkening in Glossary), whereas it is near-Lambertian (coefficient ~ 1.0) at green and red wavelengths. In the near infrared, there is limb-brightening (Minnaert coefficient ~ 0.4).

Voyager also saw a dark ring above the north (winter) pole. This feature, termed the polar hood, extending from 70 to 90° north latitude, is most prominent at blue and violet wavelengths, and may be associated with lack of illumination in the polar regions during the winter (since the subsolar latitude goes up to 26.4°), and/or downwelling in global circulation.

In 1983, Kathy Rages and James Pollack (at NASA/ AMES) investigated the properties of the aerosols from

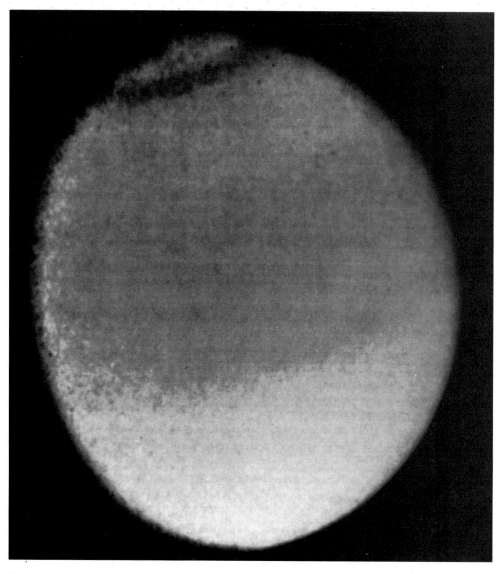

FIGURE 1 Titan as seen by *Voyager*. The bland appearance of Titan belies a complex world. The only features apparent in the images taken by *Voyager* were the detached haze layers (blue), the dark polar hood, and a difference in brightness between the two hemispheres. (Courtesy of NASA.)

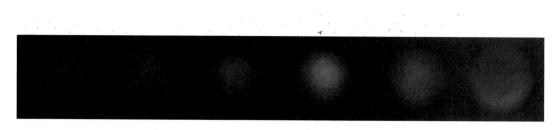

FIGURE 2 Titan at UV, visible, and near IR wavelengths from *Hubble Space Telescope*. At UV wavelengths (336 nm, left), the disk has fairly uniform brightness. As wavelength increases to blue (439 nm), the north–south albedo asymmetry increases and the disk becomes darker at the edges. By red wavelengths (673 nm), the asymmetry has disappeared. At near-infrared wavelengths (889 nm, right), the asymmetry has reversed and there is limb-brightening. (From Astronomy & Geophysics, Journal of the Royal Astronomical Society.)

high-phase-angle *Voyager* images and found the particle radius to be between 0.2 and 0.5 μm. These "smog" particles form a layer that enshrouds the entire globe of Titan and stretches to an altitude of about 200 km. These authors also estimated the optical thickness of the haze and the vertical distributions of the aerosols; they found a detached haze layer at 340–360 km in altitude with large, compact, irregular dark particles that may have many of the properties required of Titan's aerosols. The observations may also be consistent with an upwelling circulation, carrying particles upward from the summer/autumn hemisphere and allowing them to grow larger.

B. ATMOSPHERIC COMPOSITION

The nature of Titan's atmosphere emerged as a combination of the two pre-Voyager models. Molecular nitrogen (N_2, detected by the UV spectrometer) is by far the major component of the atmosphere (average of ~90%). Methane, the next most abundant molecule (2–8%), traces of hydrogen, and a host of organic gases, as well as oxygen compounds (see the following) were inferred from emission bands in infrared spectra.

The radio-occultation experiment provided a precise value of Titan's surface radius: 2575 ± 2 km, with a surface temperature of 94 ± 2 K and a pressure of about 1.44 bar. The main properties of Titan—as established at the time—are listed in Table I.

Under the surface conditions mentioned here, liquid methane and its principal by-product, ethane, are expected to exist and could even form an ocean, while in the troposphere, methane clouds (formed by condensation of methane gas) might cause rains. However, the degree of saturation in the lower atmosphere is unknown so the methane abundance is difficult to determine. Robert Samuelson and colleagues from Goddard Space Flight Center suggested a mean value of 3% above the condensation level and up to 6% below. The mean molecular weight value (from the radio-occultation experiment, which yields a profile of refractivity from which density and the temperature/molecular mass ratio can be determined) is around 28 amu (atomic mass units), confirming molecular nitrogen (N_2) as the dominant constituent, but it may be as high as 29.4 amu, implying the possible presence of a component heavier than nitrogen in Titan's atmosphere. Based on cosmological abundance, the presence of argon was suggested, but the gas has not yet been detected.

The infrared spectra taken by the Voyager InfraRed Interferometer Spectrometer (IRIS) cover the 200 to 1500 cm^{-1} spectra region with a spectral resolution of

TABLE I
Titan's Orbital and Body Parameters, and Atmospheric Properties Determined by Voyager (for atmospheric structure, see Table II)

Surface Radius	2575 km
Mass	1.35×10^{23} kg ($=0.022 \times$ Earth)
Mean Density	1880 kg m^{-3}
Distance from Saturn from Sun	1.23×10^9 m ($=20$ Saturn radii) 9.546 AU
Orbital Period around Sun	15.95 days 29.5 years
Obliquity	26.7°
Surface Temperature	94 K
Surface Pressure	1.44 bar
Composition (mole fractions):		
Nitrogen	N_2	90–97%
Argon	Ar	0–6%
Methane	CH_4	0.5–4%
Hydrogen	H_2	0.2%
Ethane	C_2H_6	1×10^{-5}
Acetylene	C_2H_2	2×10^{-6}
Propane	C_3H_8	5×10^{-7}
Ethylene	C_2H_4	1×10^{-7}
Diacetylene	C_4H_2	1×10^{-9}
Hydrogen Cyanide	HCN	1×10^{-7}
Carbon Monoxide	CO	$\sim 10^{-5}$
Carbon Dioxide	CO_2	1×10^{-8}
Water	H_2O	8×10^{-9}

4.3 cm^{-1}. (A wave number cm^{-1} is a reciprocal wavelength; see the Glossary). The data recovered by IRIS, under the responsibility of Rudy Hanel and colleagues from Goddard Space Flight Center, confirmed the presence of some simple hydrocarbons (such as methane, acetylene, ethylene, and ethane) and proved the existence of some more complex ones: diacetylene, methylacetylene, propane, and monodeuterated methane. Also, the signatures of three nitriles were found in the spectra: hydrogen cyanide (an important prebiotic molecule), cyanoacetylene, and cyanogen. Finally, carbon dioxide was identified at 667 cm^{-1}. By comparison with laboratory spectra, the abundances of these species were estimated.

The production of these compounds from methane photochemistry was examined in detail by Yuk Yung and colleagues (although earlier models existed), and their (and subsequent) models reproduce fairly well the relative abundance of some of these species. Molecular nitrogen, molecular hydrogen, carbon monoxide (CO), and argon (Ar) are expected to be uniformly mixed throughout the lower atmosphere. The production of organics in laboratory simulations was carried out by several groups: Carl Sagan and colleagues at Cornell University and Francois Raulin and colleagues at Paris University.

Methane should also be uniformly mixed globally, but since it may undergo phase changes in the troposphere, it may not be vertically constant. Thus, its mixing ratio is supposed to be constant above the condensation level, below which methane may be saturated and could form clouds. Such clouds may be necessary to account for the observed opacity between 200 and 400 cm^{-1}. The methane abundance can therefore increase with temperature down to the surface. After methane, the next most abundant hydrocarbons are ethane, acetylene, and propane. Unless we find ourselves in the presence of supersaturation conditions, each of these organics and the less abundant ones must condense at some level in the lower stratosphere and precipitate out, effectively reducing its gas-phase abundance to smaller amounts below this level.

Although no systematic quantitative study was performed at that stage, latitudinal variations were observed for some of the least abundant molecules, with an obvious enhancement of cyanoacetylene and cyanogen, as well as some hydrocarbons, at high northern latitudes, and seasonal effects were suggested as the origin of these variations.

C. TEMPERATURE STRUCTURE

The *Voyager 1* radio signal passed through Titan's atmosphere at two near-equatorial locations, with essentially identical results. The refraction of the signal measured at the ground gave two refractivity profiles, which were initially converted to density and temperature assuming a pure molecular nitrogen atmosphere (Fig. 3).

In the troposphere the temperature is bound between the surface value of ~94 K and a minimum of ~71 K at the tropopause (also called the "cold trap"). Therefore, it always remains higher than the condensation temperature of N_2, so nitrogen clouds do not form. On the other hand, condensation of methane is possible if the methane stratospheric mixing ratio exceeds 1.6%. The surface temperature and the tropospheric composition influence the properties of an ocean on the surface (since an ocean would be in thermodynamic equilibrium with the atmosphere). The surface pressure and temperature conditions suggest

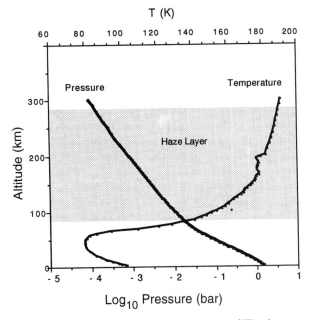

FIGURE 3 Temperature and pressure profiles of Titan's atmosphere. The atmosphere cools slowly with altitude, up to the tropopause at about 40 km altitude. Above this point, warming by the haze increases the temperature.

the presence of methane in liquid form. However, the maximum temperature lapse rate observed near the surface is compatible with the adiabatic value expected for a dry molecular nitrogen atmosphere, indicating—in principle—that methane saturation does not occur in this region. A lapse rate transition observed near 3.5 km can be interpreted in two different ways: as the mark of a boundary between a convective region near the surface and a higher radiative equilibrium zone, or as the bottom of a methane cloud deck.

The temperature profile in the lowest few kilometers excludes the presence of a global methane ocean on the surface. However, ethane is one of the main photochemical products of methane in Titan's atmosphere and also could accumulate as a liquid on the surface. An ethane–methane ocean would have a lower methane vapor pressure, so could be compatible with the observed lapse rate. This ethane ocean model, developed by Jonathan Lunine and others, was aesthetically appealing and was compatible with all the Voyager-era data.

The accuracy of the temperature degrades toward 200 km to ~10–15 K due to poor signal-to-noise and sensitivity to composition uncertainties. At this altitude, the temperature is around 180 K, substantially higher than the surface, due to the absorption of most of the sunlight by the haze.

In 1983, Robert Samuelson reproduced the temperature structure with an analytical radiative equilibrium model that was vertically homogeneous and considered the radiation balance in three broad spectral intervals (blue, red, and thermal). His results confirmed pre-Voyager models in which the temperature inversion observed on Titan was the result of strong stratospheric absorption of solar UV and penetration to near the surface of longer-wavelength solar visible radiation. Samuelson suggested the presence of enhanced opacity in the regions near 20 and 65 km, which he attributed to possible condensation clouds of methane and ethylene–ethane–propane, respectively. He also identified the region from 400 to 600 cm^{-1} as the thermal infrared "window" region, a spectral region of low optical depth that limits the greenhouse effect on Titan.

A more sophisticated radiative transfer model, using many more wavelengths and improved laboratory data on the properties of tholins and collision-induced absorption by N_2–CH_4, was developed by Christopher McKay and colleagues some years later. This model, by comparison with the visible spectrum of Titan and the Voyager temperature profile, placed constraints on the microphysics of the haze, the surface albedo, and the presence of methane clouds. This model suggested that a greenhouse warming (mostly due to methane and nitrogen) of 22 K, offset by cooling by the haze, kept Titan's surface 12 K warmer than the effective temperature of 82 K.

Above 200 km, the only supplementary data come from the UVS experiment during a solar occultation, which gives a density value of $2.7 \pm 0.2 \times 10^8$ cm^{-3} at 1265 km for a temperature of 186 ± 20 K, suggesting an average temperature of 165 K in the 200- to 1265-km altitude range. This same experiment allowed the detection of a methane mixing ratio of $8 \pm 3\%$ toward 1125 km and placed the homopause level at around 925 ± 70 km.

The thermal balance of the thermosphere of Titan was first studied by Friedson and Yung in 1984. They solved the equations of heat transfer and of hydrostatic equilibrium by using the UV measurements as boundary conditions. The heat budget of the atmosphere takes into account the sources of energy (essentially solar radiation and a contribution from electron precipitation in the magnetosphere) and the losses [cooling through emission of the minor components, mainly acetylene, in a non-local thermodynamical equilibrium (non-LTE) environment]. The heat is transported downward by molecular conduction. Friedson and Yung thus modeled the thermosphere profile down to the mesopause (~736 km and 110 K).

D. MAGNETOSPHERIC INTERACTION

Titan resides on the boundary of the Saturnian magnetosphere, so that sometimes it is within the magnetosphere and other times outside, depending on the level of solar activity. This leads to a widely varying particle environment. In addition, the relative directions of the magnetospheric flow and the solar illumination vary during Titan's orbit.

As far as could be determined from the Voyager 1 encounter, Titan has no intrinsic magnetic field—an upper limit of 5 nT was determined. Saturn's magnetic field lines are "draped" across Titan and the outflow of heavy ions (N_2^+, N_2CN^+, etc.) end up in Titan's ionosphere between 700 and 2700 km above the surface. The induced magnetotail has four lobes, due to a flow asymmetry.

Below 700 km altitude, galactic cosmic rays penetrate to the tropopause and cause some ionization. At these altitudes, meteoric ionization is also a contributor. These ionization sources are interesting in that they occur at broadly similar altitudes to those at which the photochemistry is occurring—since particle charge can strongly influence the coagulation of haze particles. The links between Titan's atmosphere and the space environment may be more manifest than those of other planets.

EUV excitation, and precipitation of magnetospheric electrons, excites nitrogen molecules in the upper atmosphere, leading to intense dayglow. Energetic nitrogen atoms can escape from Titan, although the loss amounts to less than 1% of the present atmosphere over the age of the solar system.

Hydrogen atoms can escape thermally (Jeans escape) from Titan—this makes the photochemistry irreversible, as hydrogen produced from methane is lost from Titan. The hydrogen is not sufficiently energetic to immediately escape from Saturn's gravity well, and so forms a toroidal cloud around Titan's orbit.

III. THE ATMOSPHERE OF TITAN
· ·

A. CHEMICAL COMPOSITION

Curiously, the bulk composition of Titan is more difficult to determine than the abundances of the trace constituents. Actual values for the major components vary roughly from 90 to 97% for molecular nitrogen and 0.5–4% for stratospheric methane (or 1.0 and

1.7% if argon is absent). The maximum methane abundance in the troposphere is 21% at the surface level (12% if there is no argon). From recent estimates, the molecular hydrogen abundance seems to be on the order of 0.2%.

Argon has not yet been detected, but it is important to establish, if not its actual abundance, then the lowest possible upper limit on the argon mixing ratio. Recent determinations of the mean molecular weight of Titan's atmosphere ($27.8 < m < 29.3$ amu) are uncertain enough to allow both extremes. Should the value of the mean molecular weight be higher than 28, as permitted by present uncertainties, an element heavier than nitrogen must be present in substantial amounts. At Titan temperature conditions, only argon is a plausible candidate. The first studies suggested an argon mole fraction of 0 to 27%. Argon resonance lines (which should occur at 1048 and 1067 Å in Titan's EUV dayglow, due to solar and photoelectron excitation processes) do not appear in the UV Voyager spectra. In a recent study, Darrell Strobel and colleagues concluded, from the analysis of the *Voyager 1* UVS solar occultation data and a spectrum of the north polar region dayglow obtained during the Titan flyby, that the upper limit on the tropopause argon mixing ratio, when combined with the mean molecular mass, is lower than 10%. More recently still, from thermal considerations using Voyager radio-occultation measurements, the upper limit of the argon mole fraction was set to 6%. A number of ground-based observations were dedicated to attempts to observe argon signatures: searches for H_2–Ar van der Waals molecules at 2.1 μm have been made, but were not successful owing to low spectral resolution. The only other possible approach to the problem from Earth is by observing the argon X-ray fluorescence at about 3 keV (and attempts are currently being made from space telescopes).

Ground-based observations of the $3\nu_2$ monodeuterated methane band at 1.6 μm confirm the value of $D/H = 1.5 \times 10^{-4}$ found from Voyager data analyses of the ν_6 CH_3D band at 8.6 μm and support the evidence for a deuterium enrichment in Titan's atmosphere with respect to the protosolar value and that of the giant planets ($D/H \sim 2$–3.4×10^{-5}). However, the enrichment factor is not known with enough precision to allow a firm determination of its origin. Several hypotheses have been suggested. A primitive nebula model was proposed in which two distinct deuterium reservoirs coexisted before the formation of the solar system: the main one in gaseous hydrogen (HD) and the second smaller one in deuterated isotopes of CNO compounds (such as methane, ammonia, and water)

trapped in ices or clathrates or adsorbed on grains. Or Titan's atmosphere could result from volatile degassing of grains originating from the second reservoir and having accreted to form the satellite. The fractionation mechanisms acting during or after the formation of the satellite could have contributed to the deuterium enrichment observed on Titan. Several processes have been suggested, including exchanges of deuterium between methane gas and cloud particles or the putative ethane–methane ocean, or even the icy crust underneath. An additional possibility could involve isotopic exchange catalyzed in the presence of metallic grains in the Saturnian nebula. Recent measurements from the ground and with the *Infrared Space Observatory* (*ISO*) are closer to a value of 7.5×10^{-5} for the D/H isotopic ratio in Titan, which is four times less than that observed in comets. According to works by Daniel Gautier and colleagues on the formation and evolution of the satellite, this observation would favor an interior origin of the atmosphere over a cometary one.

Carbon monoxide and carbon dioxide (CO_2) were the only two oxygen compounds found in Titan's atmosphere until recently. Their abundances are coupled and related to the probable presence of traces of water vapor (recently detected by ISO) in Titan's atmosphere. Their photochemistry has been studied extensively and the models in which the formation and destruction of CO and CO_2 are intimately related. When an independent source is provided in the form of hydroxyl radicals (OH, from an external water influx contained in chondritic or icy meteorites, consequently photodissociated to produce OH radicals in the high atmosphere), then CO can be formed from reactions of OH and CH_2 or CH_3. Carbon monoxide then moves downward and is destroyed near 500 km by action of OH with subsequent formation of carbon dioxide. On the other hand, through photolysis, various reactions, and mainly condensation, CO_2 can be removed to restore CO. The photochemical lifetime of carbon monoxide is very long (comparable to the age of Titan) and, as a consequence, it is expected to be uniformly mixed with the background N_2 atmosphere.

Water vapor was detected in Titan's atmosphere in 1998 by Athena Coustenis and colleagues, using ISO. The emission observed near 40 μm is compatible with a H_2O mole fraction of about 10 ppb near 400 km of altitude.

Carbon dioxide was identified in the *Voyager 1* IRIS spectra from the emission feature around 667 cm^{-1}, with an average mole fraction in Titan's stratosphere of about 15 ppb, representative of the 8-mbar level and assuming a constant mixing ratio above the condensation level.

Carbon monoxide was first detected in the near infrared from ground-based observations around 6350 cm^{-1}, and a constant mixing ratio of 5×10^{-5} was derived in the troposphere with a factor of 3 uncertainty (therefore in accordance with the Voyager value). Microwave measurements confirmed this value and it seems that CO is, as expected, uniformly mixed throughout the whole Titan atmosphere. Observations using heterodyne techniques suggested, however, that the carbon monoxide mixing ratio could be lower than previous estimates. When compared to the tropospheric value, this abundance suggests that CO may be depleted in the stratosphere. However, from very recent measurements, it appears that all data sets may be brought to agreement for a constant-with-height abundance of CO in the $1-5 \times 10^{-5}$ range.

Other trace constituents, such as cyanoacetylene, cyanogen, and dicyanoacetylene (the latter is observed as ice in emission at 478 cm^{-1} and has no equivalent gaseous band), were discovered in a special *Voyager* IRIS observational sequence, consisting of ~30 spectra, recorded at grazing incidence over Titan's north polar region in 1980. Later on these data yielded vertical distributions for most of the hydrocarbons and nitriles. The vertical distributions generally increase with altitude, confirming the prediction of photochemical models that these species form in the upper atmosphere and then diffuse downward in the stratosphere. Below the condensation level of each gas, the distributions are assumed to decrease following the respective vapor saturation law. However, the gradients in abundance represented by these profiles are not always in accordance with theoretical models. Titan is clearly not a simple place. Besides the variations as a function of altitude, some of the trace constituents discovered above the north polar region show enhanced abundances, probably due to seasonal effects.

Recent millimeter observations provided the stratospheric profile of hydrogen cyanide (HCN) on Titan. This vertical distribution, obtained from an analysis of the (1-0) line of hydrogen cyanide at 88.6 GHz, shows an increase with altitude. The mean mixing ratio scale height is ~47 km in the 100- to 300-km region and the HCN abundance around the 170-km level is ~3.3×10^{-7}. These results (disk-averaged) are generally consistent with the analyses of localized *Voyager* infrared measurements. However, the inferred vertical concentration gradient is much steeper and the abundance in the lower stratosphere smaller than predicted by photochemical models. [*See* THE SOLAR SYSTEM AT RADIO WAVELENGTHS.]

In 1992 came the first detection of a more complex nitrile than any of those detected by Voyager: CH$_3$CN

(acetonitrile) was observed by Bruno Bézard and colleagues in the millimeter range around 220.7 GHz at 0.1-MHz resolution, by use of the IRAM 30-m radio telescope at Pico-Veleta (Spain). The mixing ratio of acetonitrile is found to strongly increase with height in the stratosphere (by about two orders of magnitude between 200 and 400 km), attaining some 10^{-8}, in which case it should be easily detected by future space missions.

B. THERMAL PROFILES

Subsequent analyses of the first radio-occultation data in the 0- 200-km altitude range have used model N$_2$/CH$_4$/Ar atmospheres to study the allowable range of surface temperatures and compositions. Combined with the analysis of the emission observed in the ν_4 methane band at 7.7 μm, temperature profiles were derived from the upper stratosphere (about 450 km) down to the surface. In the upper atmosphere of Titan (0.1 to 10^{-2} nbar), Roger Yelle published in 1991 realistic models of the thermal profiles, including, among others, non-LTE (i.e., departing from local thermodynamical equilibrium conditions) heating/cooling in the rotation–vibration bands of the main hydrocarbons and HCN and aerosol heating. Two extra constraints are available for the temperature structure on Titan: first, the UVS measurement of the exospheric temperature, 186 ± 20 K at about 1265 km; and second, the one near 1 μbar (183 ± 11 K near 450 km) derived from the July 3, 1989, stellar occultation of Titan. The occultation of star 28 Sgr by Titan was observed from places as widely dispersed as Israel, the Vatican, and Paris.

This rare event provided information in the 250- to 500-km altitude range. A mean scale height of 48 km at 450 km altitude (~3-mbar level) was inferred. This allowed the mean temperature to be constrained at that level to between 149 and 178 K.

The temperature profile in the thermosphere was revisited by Emmanuel Lellouch and colleagues in 1990 using the emission observed by IRIS in the ν_4 band (yielding information in the 150- to 450-km zone), as well as the UV data. They found a "warm" stratospheric region (~175 K) up to 500 km, compensating for a cold mesopause (135 K) at 800 km. The most reliable thermal profile for Titan's atmosphere from the ground up to the exosphere takes into account all of the foregoing constraints (results of RSS and UVS, indications of the ν_4 CH$_4$ band, Yelle's models, and other measurements).

C. LATITUDINAL AND TEMPORAL VARIATIONS IN THE ATMOSPHERE

Ethane, acetylene, and propane are the three most abundant hydrocarbons after methane and were found to exhibit no noticeable compositional variations in latitude, but are homogeneously mixed in Titan's atmosphere, as seen by Voyager, from pole to pole. In contrast, the *Voyager 1* IRIS data suggest that ethylene and methylacetylene tend to increase (by a factor of 3–4) near the north polar region, while their mixing ratios remain constant from midlatitudes to the south pole. For the diacetylene mixing ratio, there is practically no difference between the equator and the south pole, whereas a factor of ~20 increase is observed near the north pole. Two of the nitriles (cyanoacetylene and cyanogen) are absent in the spectra corresponding to latitudes <~60°N. However, upper limits at a 3-σ level obtained from these selections indicate an increase of a factor of at least ~50 in cyanoacetylene and of 3–4 in cyanogen in the north polar region. Hydrogen cyanide is observed in emission at all latitudes in the IRIS data. Its abundance shows a steady increase from pole-to-pole by a factor of ~12 (a factor of ~4 from the equator to the north pole). Carbon dioxide shows a constant abundance of about 1.4 $\times$ 10^{-8} from the south pole to high latitudes, although its emission becomes extinct near the north pole.

The latitudinal variations of the gases have not been fully explained to date. The nitrile enhancement observed near the north pole was tentatively attributed by Yung in 1987 to a nitrile accumulation in that region during the winter when the polar region is in shadow. At the time of the *Voyager 1* flyby, the north polar region was just coming out of winter, having maximum nitrile abundances. Although the Voyager encounter occurred ~8 months after spring equinox, chemical time constants may have been large enough to prevent the UV flux from immediately depleting the strong nitrile concentrations observed. This could also explain the depletion in carbon dioxide, this molecule being produced mainly from water photolysis. Its formation may have then been inhibited in the north pole because of the low level of UV radiation.

Latitudinal temperature variations were found in Titan's stratosphere when the thermal profiles were derived from the Voyager data. A maximum temperature decrease of 17 K at the 0.4-mbar level (225 km in altitude) is observed between 5°S (the warmest region in the Voyager data) and 70°N, whereas the temperature drops by only 3 K from 5°S to 53°S. The stratospheric haze opacity at wave numbers larger than 600 cm^{-1} was found to show a north-to-south enhance-

ment of about 2.5. The coldest temperatures, found at high northern latitudes, are associated with enhanced gas concentration and haze opacity (as this may be caused by more efficient cooling) and/or dynamical inertia.

Voyager 2 infrared spectra (taken 9 months after *Voyager 1*, but from 170 times farther away) have been analyzed, even though the projected field of view of IRIS (0.25°) covered more than half the Titan diameter, allowing only two different regions to be sampled. The resulting composition and temperature (and their variations) were found to be broadly in agreement with *Voyager 1*, despite the poorer spatial resolution and signal-to-noise. The *Voyager 2* observations, taken only a tenth of a season on Titan away from *Voyager 1* observations, confirm the *Voyager 1* data and are consistent with the suggested cause for the temperature variations.

The latitudinal variations observed for the temperature and composition of Titan's stratosphere indicate that a full understanding of this atmosphere would consider seasonal and dynamical, as well as radiative, effects.

D. GENERAL CIRCULATION

The faintly banded appearance of Titan's haze suggests rapid zonal motions, that is, winds parallel to the equator—this general circulation pattern is seen in all the outer planets and on Venus. This impression is reinforced by the infrared temperature maps, which show very small contrasts in longitudinal direction, but rather larger ones of around 20° from equator to pole.

In the troposphere, the thermal contrasts are as small as 2 or 3 K, which suggests considerable meridional motions, since these would be much larger if heat were not being transported poleward by advection. The calculated radiative time constant is large enough so that the circulation should be symmetric about the equator, and the rotation of the solid body of Titan is slow enough so that a Hadley-type circulation may extend from equator to pole in each hemisphere.

There is some tentative evidence of this circulation—the detached haze layers observed by *Voyager 1* are consistent with haze being levitated by rising air in the summer/autumn (southern) hemispheere of Titan. At higher levels, pressures of around 50 mbar or less, the radiative time constant is short compared to the seasons and larger vertical velocities might be expected where the insolation is greatest. This is also the region where the zonal winds can become very large, leading to "super-rotation," as discussed in the next section.

E. ZONAL "SUPER-ROTATION" OF TITAN'S STRATOSPHERE

The upper atmosphere of Titan is believed to rotate faster than Titan's surface, a state inelegantly termed "super-rotation." It is what we would expect by analogy with Venus, which also has a thick, cloudy atmosphere and rapid zonal winds in the same sense as the rotation of the solid body. This state found some tentative confirmation in 1988 when some faint features were tracked in enhanced Voyager images from which wind speeds in the 28 to 99 m s^{-1} range were inferred. If real, these would correspond to air circulating around the equator in about 2 days. This may be contrasted with Titan's solid-body rotation rate of 16 days. [*See* VENUS: ATMOSPHERE AND OCEANS.]

The central flash observed during the 1989 stellar occultation by 28 Sgr allowed us to constrain an apparent oblateness of Titan at the 0.25-mbar level ($\sim$250 km of altitude), giving a value that may be as high as 0.014. The oblateness, if entirely due to the super-rotation of the atmosphere, would correspond to winds of about 180 m s^{-1} at 0.25 mbar. This is much faster than the surface rotation speed of 12 m s^{-1}, and presumably, although the measurement is ambiguous with regard to direction, in the same sense. The preliminary cloud-tracked winds were prograde, like Titan, and the surface is the only identifiable source of the atmospheric angular momentum, although the processes by which it is transferred and a balance maintained are not fully understood.

A sophisticated dynamic model of Titan's atmosphere by Frédéric Hourdin and colleagues is based on a general circulation model (GCM) of Earth, modified for Titan's radius of 2575 km, obliquity of 26°, and rotation rate of 16 days, as well as other factors such as the radiation balance (from McKay's model). One of the most striking results from this model is that, starting from rest, it spins up to produce a strong zonal super-rotation, with winds on the same order as the observed ones. A compact region of strong wind, known as a jet, appears at midlatitudes; a similar feature (and with about the same wind speed, around 120 m s^{-1}) is known to occur on Venus. As expected from simple arguments based on a calculation of the radiative time constant as a function of atmospheric density, the temperatures in the lower atmosphere of Titan remain nearly constant throughout the long Saturnian year of approximately 30 Earth years.

There have been attempts to explain the hemispherical asymmetry in Titan's brightness at visible wavelengths by a model that uses the general circulation to transport aerosol particles between the hemispheres. The initial two-dimensional circulation models (a single pole-to-pole Hadley cell, and the more complex circulation pattern from the Hourdin *et al.* GCM) changed the brightness as a function of latitude, but not quite in the manner observed by *Voyager* and *HST*.

F. CONDENSATES IN THE STRATOSPHERE

Most of the by-products of the nitrogen–methane photochemistry in the atmosphere of Titan saturate near the tropopause and fall through the troposphere to the surface in the form of liquids or solids. Ices are expected on the surface, but also in the lower stratosphere of Titan. Ices are but one form of particulate matter contained in planetary atmospheres. Both solids and liquids are formed by condensation processes, which imply the presence of seed nuclei as well as saturation of the gas or gases involved. One source for such nuclei is the photochemistry of gases high in the stratosphere or mesosphere. Polymerization of the photochemical by-products can give rise to submicrometer-size particles that in turn act as condensation centers.

Most of the features seen in emission in the Titan Voyager spectra are due to various gases, though some are due to condensates. Two of the condensates (ices of cyanoacetylene and dicyanoacetylene) have been positively identified by Samuelson and colleagues from their spectral signatures at 503 and 478 cm^{-1}, respectively. The latter feature lends itself well to analysis owing to its isolation from other features. Scattering and radiative transfer theory suggest a dicyanoacetylene cloud optical thickness between 0.04 and 0.15, depending on the vertical distribution of material, with a mean particle radius of $\sim$5 μm. Further analysis suggests that a time-independent downward vapor diffusion process from above (coupled with condensation at the cloud top and steady precipitation from the cloud bottom) is not consistent with the observed dicyanoacetylene cloud/vapor abundance ratio. This may mean that a time-dependent process (possibly associated with seasonal variations of temperature, chemistry, and dynamics) is operating in Titan's lower stratosphere. If such a process operates for dicyanoacetylene, it may operate for other volatiles as well, including water.

Although no emission lines of condensed ethane have been identified in Titan spectra, two independent studies of the spectral continuum between 200 and 600 cm^{-1} suggest its presence near or slightly above the tropopause. In one, IRIS limb spectra continua of the north polar hood were analyzed, yielding the wave number dependence of the aerosol cloud in the stratosphere. The ratio of opacity at high wavenumbers to that at low wavenumbers increases with decreasing depth, implying that condensates contribute to the aerosol continuum in the lower stratosphere. Later, a reanalysis of these data using scattering theory was performed and Samuelson and colleagues inferred that an ethane haze consisting of particles with mean radii ~10 μm could explain the data. Condensates other than ethane could also explain the wave number dependence of opacity, but were less likely from relative abundance considerations.

G. THE WEATHER NEAR THE SURFACE

The troposphere on Titan is about 40 km deep. The observed temperature structure is nearly in radiative balance, according to a model by McKay, so only a small amount of convection is present. The Voyager temperature profile indicates that the lowest few kilometers of the atmosphere are dry—the temperature gradient, or lapse rate, is shallower (more stable) than the dry adiabatic lapse rate, the rate given by the need for the buoyancy of a dry air parcel to just balance the vertical pressure gradient. On Titan, this adiabatic lapse rate, Γ, is about 1.4 K km^{-1}, compared to 6.5 km^{-1} for Earth (higher due to Earth's higher gravity). The observed lapse rate is about 0.7 K km^{-1}, somewhat steeper than the wet adiabatic lapse rate, so the implication is that the lowest few kilometers of atmosphere are not "wet," that is, they are unsaturated with respect to methane.

McKay and colleagues used their radiative balance model to investigate the effect of methane clouds on the temperature profile. They found that methane clouds might have at best a neutral effect or, at worst, a cooling effect. This model suggests the presence of patchy methane clouds limited below 30 km of altitude and containing >0.5-mm-size particles, which can be classified as rain. Compared with terrestrial clouds, Titan's methane clouds are predicted to have larger but fewer particles and more closely resemble terrestrial rain than terrestrial clouds.

Studies by Ralph Lorenz show that rain on Titan would fall very slowly, due to the thick atmosphere and low gravity (1.6 m s^{-1}, compared with 9 m s^{-1} on Earth). The smaller aerodynamic forces allow methane raindrops to grow to about 1 cm in diameter (compared with only about 0.6 cm on Earth). However, if the lower atmosphere is unsaturated, the slowly falling drops have time (assuming they are dropped from a cloud at 3 km) to evaporate before they reach the ground. Mountains could have their summits washed by rainfall, however.

The observational evidence for thick methane clouds on Titan is not strong. Certainly they cannot cover the whole disk, so if they exist, they must be in localized patches. Searches in *HST* and contrast-stretched Voyager images have suggested discrete cloud features (weather systems are a more appropriate analogy, since the features are hundreds of kilometers across), but no conclusive detection has been made.

If clouds do not exist, then the high concentrations of methane, and the low temperatures, in the tropopause imply supersaturation of methane to a remarkable degree. Recent studies suggest that low-latitude spectral continua of Titan can be fit best if additional opacity augments that of collision-induced absorption arising from various combinations of molecular nitrogen methane, and molecular hydrogen. This additional opacity can be due to either highly supersaturated methane vapor in the troposphere of Titan or ethane haze near the tropopause.

Finally, analyses of the IRIS spectral continuum between 200 and 600 cm^{-1} suggest radii of ~10 and 75 μm for ethane cloud particles in the north polar and equatorial regions of Titan's lower stratosphere, respectively. It is reasonable that ethane ice provides the residual continuum opacity required to fit the data, as ethane ice should be the dominant ice in the lowest regions of the stratosphere, although supersaturated methane could mimic the effect.

IV. THE SURFACE OF TITAN

A. NATURE OF THE SURFACE

As already mentioned, the surface of Titan is probably coated with atmospheric debris and condensates. Much of the outer part of the solid body of the satellite must, to be consistent with the observed mean density, consist of a thick layer of ice. Whether this is exposed at the surface, or completely or partially covered over with material precipitated out of the atmosphere, is one of the key questions that may be answered soon.

The nature and extent of the exchange of condensable species between the atmosphere and the surface, and the equilibrium that exists between the two, are other key topics.

The aerosol haze in the stratosphere has hindered the detection of the satellite's surface from either space missions (with the exception of the radio-occultation experiment on *Voyager* that gave the radius of Titan) or Earth-based observations, until very recently.

As already discussed, the haze and cloud materials, whatever they may be, are likely to precipitate onto the surface. Since at least some of these are expected to be liquids, and since the process of accumulation has gone on for a long time, the idea of an extensive ocean, like that suggested by Lunine in 1983, is attractive and has been in vogue for a long time. This "ethane–methane ocean" would serve as both the source of methane and the sink for photolysis.

However, the first remote-sensing technique to be used for sounding Titan's surface, radar, indicated that the surface may be nonuniform but mostly solid with at most small lakes.

Dewey Muhleman and colleagues used the National Radio Astronomy Observatory's Very Large Array in New Mexico combined as a receiver of the signal transmitted to Titan by the NASA Goldstone radio telescope in California. The echoes obtained by the signal sent from Earth show that Titan is not covered with a deep global ocean of ethane–methane, because such an ocean, devoid of suspended particulates and deeper than a few hundred meters, is a very poor reflector of radar signals (reflectivity of about 0.02). The mean radar cross section obtained varies between 5 and 15%, inconsistent with the presence of a significant body of liquid hydrocarbons in the hemisphere observed (eastern elongation: this corresponds to 90° longitude of the central meridian (LCM), as opposed to geographical longitude, which is about 210° then, assuming that Titan rotates synchronously with Saturn).

The highest value (about 0.15) is lower than that for the lowest of the Galilean satellites, Callisto, and much lower than that for Ganymede and Europa, which have reflective icy surfaces.

The detection of an echo on Titan was the first evidence against the global ocean model of the surface. The model held such prevalence that oceans rendered radar-reflective by suspended particulates or bubbles were contemplated, but it now seems clear that Titan's surface cannot be mostly covered in deep liquid hydrocarbons.

One region of Titan's surface seems brighter at radar wavelengths than the rest, and—at the risk of

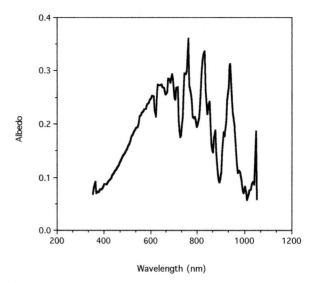

FIGURE 4   Titan's geometric albedo. The darkness at blue wavelengths indicates there is little gas above the absorbing haze. The smoothly increasing curve at visible wavelengths indicates the red color of the haze, while the dips correspond to methane absorption bands. The peaks between these bands are the windows through which the surface can be sensed.

overinterpreting a small set of difficult observations—shows polarization and spectral behavior characteristic of icy surfaces. The longitude at which this behavior is seen is that which is brightest in the near IR and was subsequently identified in Titan images (see later). Other Titan longitudes have a slightly lower radar cross section, and Titan remains the most distant object to have been detected by radar.

A Callisto-like surface, with ice rendered less reflective, perhaps by hydrocarbon lakes, is one interpretation of the radar data. However, this model has problems too, for example, it does not explain the polarization of the returning radar signal (different from that of the Galilean satellites).

It may be that the surface is actually "dry" and the hydrocarbon ocean is stored in a porous, uppermost few kilometers of methane clathrate or water ice, "bed rock." The surface in this model is porous enough (about 20% to a depth of about one-half kilometer, which is not unrealistic) to allow all the required ethane and methane to pass into a subsurface "aquifer" into which the liquid is stored. This model was suggested by David Stevenson of Caltech in 1992. One counterargument against such a porous surface is that hardened regolith upper layers enriched in organics (like "caliche" in the desert) may prevent ethane dribbling down into the regolith.

B. GEOMETRIC ALBEDO IN THE NEAR INFRARED

The near-infrared spectrum of Titan (0.7–5 μm), like that of the giant planets, is dominated by the methane absorption bands. At short (blue) wavelengths, light is strongly absorbed by the reddish haze particles (Fig. 4.) At red wavelengths, light is scattered by the haze, although the column optical depth is still high. In the near IR, the haze becomes more and more transparent (since the haze particles are smaller than the wavelength), although absorption by methane in a number of bands is very strong. Where the methane absorption is weak, clear regions or "windows" (situated near 4.8, 2.9, 2.0, 1.6, 1.28, 1.07, 0.94, and 0.83 μm) permit the sounding of the deep atmosphere and perhaps of the surface. In between these windows, contrary to what occurs on the giant planets, solar flux is not totally absorbed but scattered back through the atmosphere by stratospheric aerosols, especially at short wavelengths. The near-IR spectrum is thus potentially extremely rich in information on the atmosphere and surface of Titan.

Five different groups have independently demonstrated that the surface of Titan can be probed in this way, by measuring the geometric albedo in the 1- to 2.5-μm region. First came the observations of the giant planets and Titan by Uwe Fink and Harold Larson in 1979, using the Kitt Peak 4-m telescope. Dale Cruikshank and Jeffrey Morgan observed Titan in 1980 and looked for a 32-day variation in the albedo. These data were rebinned in longitude for a 16-day period by Keith Noll and Roger Knacke in 1993, and they gave a lightcurve for the 1- to 5-μm region windows. Titan's near-infrared spectrum was first used to investigate Titan's surface by Caitlin Griffith *et al.* in 1991. The first detailed radiative transfer models of the near-infrared spectrum produced by these authors indicated a surface albedo inconsistent with a global ocean, but left open the possibility for a cloud cover to partly account for the observed reflectivity in the atmospheric windows. Since 1990, these and other groups (Mark Lemmon *et al.*, Athena Coustenis *et al.*) have explored the infrared spectrum of Titan and have achieved full coverage of the satellite's orbit, unambiguously providing measurements of the surface reflectivity that show a change in Titan's albedo that is precisely correlated with Titan's rotation. Figure 5 shows a lightcurve for Titan.

The observations all agree in showing that the geometric albedo of Titan, measured over one orbit (16 days), shows significant variations indicative of a brighter leading hemisphere (facing Earth) and a darker trailing one. At conjunctions, that is, on the hemispheres facing Saturn and its opposite, the albedo was similar, of intermediate values between the maximum appearing near 120° LCM and the minimum near 230° LCM. Titan's surface then is heterogeneous, which is important information in itself because it rules out two of the previously cited possibilities: a global ocean or a uniform exposure of "dirty" water ice. No clouds are required—or observed. Recently, Griffith, Owen, and colleagues suggested the presence of an additional window near 2.9 μm.

C. SURFACE SPECTRUM

The Titan surface spectrum cannot be directly observed. However, using a computing code that includes such (admittedly uncertain) atmospheric properties as the haze production rate and the cloud opacity, it is possible to vary the surface albedo until the calculations reproduce the geometric albedo estimated from the observed Titan flux at different wavelengths in the near infrared. Such an effort was accomplished recently by Coustenis and colleagues, who modeled the surface spectrum from a large set of spectra taken with the Fourier Transform Spectrometer at the Canada–France–Hawaii Telescope on top of Mauna Kea in Hawaii. The inferred high absolute surface albedos argue for a mostly solid surface (in agreement with a previous study by Griffith *et al.* and with the radar experiment). The Titan surface spectrum, as modeled from observations in the near infrared of the satellite's albedo, holds the same shape with orbital phase and is not flat but shows higher values near 1 μm than at other wavelengths.

The depressions observed in the Titan spectrum near 1.6 and 2 μm (with respect to the albedo near 1 μm) are also found in Hyperion and Callisto data, where they are due to the water ice bands. The water ice bands are present in Titan's surface spectrum at all longitudes. The Hyperion spectrum is in general compatible with the Titan observations in the whole 1- to 2-μm region.

The existence of a second (or more) surface component is implied by the orbital variations. It could be spectrally neutral or not and mixed with water ice areally or intimately. Complex organics (tholins) show a neutral and fairly bright spectrum in the near IR, in agreement with high absolute albedos, but should be distributed uniformly with longitude. Hydrocarbon lakes or ices, silicate components, and other dark material are possible. Another possibility would be that the

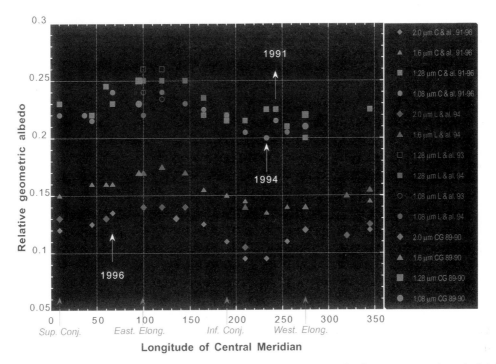

FIGURE 5 Titan's lightcurve at several near-IR window wavelengths. The maximum in brightness corresponds to the bright continent-sized area shown in Figs. 6 and 7. Different symbols correspond to various investigating teams.

orbital variations may be due to longitudinal differences in the ice morphology, that is, fresh or old, composed of big or small particles, and so on.

When the radar and IR data are taken together, Titan seems to be a unique object, unlike anything else in the solar system. Another technique is now employed to try and offer more constraints on the Titan surface problem.

D. IMAGING: CONTINENTS ON THE SURFACE?

In 1991, a new tool became available for the study of Titan's surface: high-resolution imaging with the possibility of resolving the disk (0.8 arcsec in angular diameter). The heterogeneity of Titan's surface, indicated in the near IR and radar lightcurves, was graphically revealed by 1994 observations of Titan's surface using the *Hubble Space Telescope*. Atmospheric turbulence typically prevents Titan from being observed from the ground, but *HST*, being above the atmosphere, is able to resolve the disk. Earlier *HST* observations had hinted at the ability to sense the surface (a number of *HST*'s filters sample the 940-nm window) but had been thwarted by the spherical aberration in the primary mirror. When the new Wide-Field Plane-

tary Camera (WFPC-2) was installed, which corrected this optical problem, Titan could be resolved at about 20 pixels across. In the filters used, a large proportion of light still comes from scattering by haze, but the near-full longitudinal coverage of the *HST* data set allows this to be determined and removed (since the haze seems to be longitudinally invariant).

A team led by Peter Smith of the University of Arizona produced maps of the surface in the 940-nm and 1070-nm windows, notably a large (2500 × 4000 km) bright region, at 110°E and 10°S, as well as a number of less bright regions (Fig. 6). A coarse map was also produced at red wavelengths (673 nm)—albeit somewhat blurred, since the haze optical depth is on the order of 3 at this wavelength, so each photon is reflected a number of times as it fights its way through the atmosphere. *HST* data taken in 1995 have confirmed the initial findings, and progress is being made in identifying spectrally distinct surface units, which may indicate regions of different composition. The contrast on the 940-nm and 1070-nm *HST* images was about 8%.

Images taken in 1994–1996 using the adaptive optics ADONIS camera at the 3.6-m ESO Telescope at Chile, under the leadership of Michel Combes of Paris Observatory, showed the same bright region at the equator and near 120° orbital longitude. A north–south hemispheric asymmetry is also apparent on Titan's

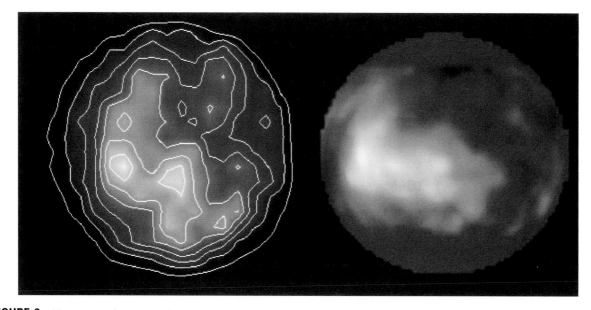

FIGURE 6 Titan images by *HST* and ADONIS—the large, bright, continent-sized feature is striking in both of these images of Titan's leading hemisphere. The *HST* picture (on the right) is a map of the 950-nm albedo reprojected onto a sphere, whereas the ADONIS image (on the left) is a 2-μm image with the estimated haze contribution removed and isophots plotted.

darker side. The adaptive optics system senses in real time the phase perturbations induced by the atmospheric turbulence on the incident wavefront reaching the telescope and applies opposite phase corrections with a thin deformable mirror in the pupil plane. Diffraction-limited (~0.13 arcsec resolution after deconvolution) images were thus obtained at 1.3, 1.6, and 2.0 μm with narrowband filters centered on the methane windows and in the wings of the absorption. At 2 μm, the light is less affected by haze scattering and absorption, so the image contrast was higher than in the *HST* images, around 30%.

Analysis of the 2.2-μm images (more sensitive to the atmospheric contribution than 2.0 μm) on the leading side of Titan showed a hemispheric asymmetry, and in particular a bright south limb, probably due to a strong aerosol concentration in this area. The 2.0-μm images showed an additional bright area near the equator. By subtracting the 2.2-μm images, the stratospheric contribution is extracted from the 2.0-μm images and therefore what remains must be attributed to the surface or to low cloud contribution. Following this procedure, the equatorial bright region could be seen to have two or three bright "peaks" and smaller, lower-contrast features could be seen near the poles on the leading hemisphere.

Spectroscopically resolved images, recorded with a circular variable filter (CVF) in adaptive optics, at 2.10 μm, in the wing of the H$_2$O ice band, are found

to be quite similar to the 2.0-μm images, when they were expected to show the strong absorption by liquid hydrocarbons (ethylene, ethane). This does not favor the presence of large (clean) hydrocarbon lakes in the dark regions (Fig. 7).

All images agree in showing the main bright equatorial region centered near 114° LCM and extending over 30° in latitude and 60° in longitude (see Fig. 6). The other bright spots are visible in the southwest region (near 25°S) and in the northern part (near 30°N). The exact nature of these features is still under investigation.

The location of the bright feature coincides with the longitudes where high albedo values were found by spectroscopy in the near infrared and strong radar echo signals returned. The bright spot might be due to differences in relief. Coustenis and colleagues have shown that elevating part of a uniformly bright surface (pushing the elevated region above thick methane absorption, hence making it brighter) cannot account for the observed difference in brightness. Some bright material is required, although this bright material may occur only on elevated terrain (e.g., methane rainfall might wash dark organic material on mountains.)

It is possible, on the basis of landforms found throughout the solar system, to make informed speculation on what Titan surface features might be like. Parallels may be drawn primarily with other icy satellites, but also with Venus, which, like Titan, has a dense, optically thick atmosphere.

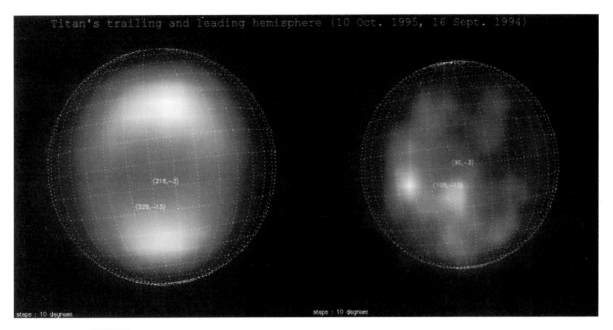

FIGURE 7 Titan's surface with ADONIS: bright (right) and darker (left) hemispheres at 2 μm.

E. LANDSCAPE SPECULATIONS

Titan's landscape may be unique in that it may well be very cratered, like virtually every other body in the solar system, but also have liquids present. Even without the latter factor, however, the study of cratering is an important probe of the resurfacing history of the body, allowing relative ages of regions to be determined. This is a vital window into both the body's interior, as processes such as crater relaxation indicate the mechanical properties of the crustal material, and to its atmospheric history, since small impactors are broken up, Tunguska-like, in a thick atmosphere. [*See* PLANETARY IMPACTS.]

The number of craters on Titan's surface is of course unknown, but simple interpolation or extrapolation of crater densities elsewhere in the Saturnian system suggests there should be ~200 craters larger than 20 km diameter in every million square kilometers, but recent resurfacing could reduce this number considerably. Airless satellites would have many more smaller craters, typically following an $N \propto D^{-2}$ type power law. However, Titan's atmosphere causes smaller impactors to slow before reaching the surface, so there is probably a turn-down in the crater number density below about 10 km, compared with a similar turn-down at 20 km on Venus. The crater density suggested here may underestimate (since the breakup of proto-Hyperion may have led to an enhanced impactor flux on Titan, owing to the orbital resonance

of the two satellites) or overestimate (if resurfacing has obliterated early craters) the actual number on Titan's surface.

The crater density at small sizes is also sensitive to Titan's climate history—if Titan had a thinner atmosphere in the past, abundant small impactors may have been able to reach the surface. The possibly widespread existence of liquids on the surface of Titan may make impacts into liquid-covered crust widespread—such impacts may leave subdued features on the seabed (termed "hydroblemes") and give rise to tsunami deposits.

Impact basins, even if not formed on liquid-covered surfaces, may be subdued because of the effects of viscous relaxation. Although Titan's present surface temperature is low enough that ice is very hard, the possible presence of ammonia hydrates in the crust may mobilize the ice. Further, the geothermal heat gradient makes the ice at depths of a few kilometers rather more mobile. The effect, as observed on other icy satellites, is to cause the floors of large craters to dome upward. Many such craters are observed to have small central pits as well, although the formation of these is less well understood. The interaction of a heavily cratered surface with large amounts of liquid may create some exciting and intriguing landscapes. Small bowl-shaped craters may fill to form simple lakes, while larger (>10 km diameter) craters may have a central peak that would form an island in a ring-shaped lake. Larger craters still may have a large island formed by the dome, with the central pit forming a pond—a "bullseye" lake.

Titan may well exhibit cryovolcanic activity, although this is entirely speculative at present. Studying volcanism on Titan (if *Cassini*—see later—yields evidence for it) is important, not only to understand the thermal history of Titan, but also to understand how volatiles, in particular methane, were delivered to the surface.

Titan's present environment is very placid: tidal currents are weak and rainfall—if it occurs—is soft, with the diurnal temperature contrasts small (and therefore winds are gentle). The solubility of ice in hydrocarbons is smaller than that of most rocks in water. Thus, except where the surface is more susceptible to erosion, due to organic deposits or perhaps water-ammonia ice, Titan's topography should not be significantly modified by erosion.

Some of these ideas about Titan's landscape may be actually observed by the *Huygens* lander in late 2004, when the Cassini mission will finally solve the mystery of the nature of the satellite's surface.

V. TITAN IN THE SOLAR SYSTEM

Titan seems to be an object intermediate between terrestrial planets (where the atmosphere consists of very minor constituents of the planet) and the giant planets (where the atmosphere is intimately related to the deep interior). With respect to the other satellites, the mean density of 1.86 g cm^{-3} and radius of 2575 km show that Titan lies somewhere between Ganymede and Callisto. These two satellites of Jupiter are believed to consist of rock (silicates and iron) and water ice (25–50% by mass). The similarities between the three satellites may suggest that they have similar interior properties, but, if Titan formed in lower temperatures than Ganymede and Callisto, it may have incorporated other ices having similar densities to water ice (ammonia-water and methane-nH$_2$O). All three satellites are silicate-enriched relative to cosmic abundances, which predict 52% water ice versus 40% rock by mass. Titan is unique in being the only satellite with an atmosphere that is thick compared to that of Earth. [See OUTER PLANET ICY SATELLITES.]

A. ORIGIN AND EVOLUTION OF TITAN

Titan's origin and evolution are explained in general terms by current models of the solar system as a whole,

but the details raise many questions. Models that start with the physical processes operating in the present-day environment (such as radiation-driven chemistry, atmospheric escape, photochemistry, surface-atmosphere coupling, etc.) and then attempt to "run the clock back" help to explain the origin and evolution of Titan, although we still lack many of the key observations that will allow this to be done properly.

Titan probably formed in the proto-Saturnian nebula, the disk of rock, ice, and gas that formed Saturn. In the early stages of Titan's accretion when it was small and its gravity was weak, the rock-ice planetesimals accumulated unmelted. In later stages, as the infalling planetesimals impacted with more energy on the growing satellite, they melted. At the end of accretion, therefore, Titan's interior was probably a solid rock-ice mix, overlain by the dense silicate components of later accretion, and further overlain by a liquid water mantle. A primordial atmosphere of water vapor plus ammonia would likely have been present. The ammonia was likely converted into nitrogen by photolysis and other processes early in Titan's history. Within the first few hundred million years, the ice crust atop the mantle would grow. In the interior, as radiogenic heating warmed the ice and silicates, the more mobile ice deformed and allowed the core to overturn—the dense rock layer falling to the center. Many models suggest that if ammonia is present, a liquid mantle persists to this day, perhaps 75 to 300 km below the surface.

The temperature of the proto-Saturnian nebula was likely lower than that of the proto-Jovian one, such that not only water, but also ammonia and methane and possibly nitrogen, may have been incorporated into the nascent Titan. A combination of devolatilization of nitrogen-containing ices condensed from the nebula (the clathrate hydrate model) and photolysis of ammonia in the early phases of accretion on primordial Titan appears to be the most plausible scenario for the evolution of its present atmosphere. The Galilean satellites, although of similar size, were probably too warm to incorporate these volatiles. Additionally, Titan is much less susceptible to atmospheric losses, by virtue of its distance from the Sun. This makes the atmosphere cooler, so Jeans escape is much slower. Second, because Jupiter is deeper in the solar gravity well than Saturn, impactors enter the Jovian system much faster than they do the Saturnian system. Also, as pointed out by Kevin Zahnle and colleagues, Titan is less deep in the Saturnian well than are the Galileans in the Jovian gravity well, so impactors (comets) on the Galileans have much higher velocities. They are therefore much better able to erode (or "blow off") the atmosphere, whereas for Titan the impactors move slowly enough that they typically add mass to the atmosphere rather than remove it.

Titan is also fortunate in that it lies mostly outside the Saturnian magnetosphere, inside which energetic particles would have stripped away the atmosphere. Also, in a classic "Goldilocks" problem of fine-tuning, Titan is warm enough to generate significant vapor pressure of volatiles (methane and nitrogen) yet not too warm to encourage Jeans escape. Further observations and modeling may bring new insights as to why Titan is unique in our solar system.

B. TITAN COMPARED TO EARTH AND VENUS

Because Titan has a substantial atmosphere with molecular nitrogen as the major component, the satellite is sometimes thought of as a kind of laboratory for scientists to study the chemical evolution of an atmosphere similar to that of the primitive Earth.

There are some key differences, however. The low temperature and gravity make Titan's scale height much larger, such that unlike the Earth (whose atmosphere may be thought of as a layer with the same relative thickness to Earth as an apple skin to an apple), Titan's atmosphere extends to a substantial fraction of the solid body's radius. [See EARTH AS A PLANET: ATMOSPHERE AND OCEANS.]

Second, the greater distance from the Sun, giving only about 1% of the sunlight that Earth receives, makes Titan much colder. This has a profound effect on the chemistry—oxygen compounds like water, carbon dioxide, and molecular oxygen have such low vapor pressures that they condense on Titan and are at most trace constituents of the atmosphere.

Interestingly, the totally different chemistry yields a temperature structure similar to that of Earth, subject to the different scale height and absolute temperature values. Just as ozone absorbs solar UV on Earth and leads to a warm stratosphere, the haze on Titan fulfills the same role. Similarly, water—a condensable greenhouse gas on Earth—has an analog in methane on Titan, also a major greenhouse gas and also condensable, perhaps to the extent of forming lakes on the surface. Carbon dioxide, a noncondensable greenhouse gas on Earth, has an abundance controlled by the balance between removal (by weathering of silicate rocks) and production (by burning and volcanic venting). On Titan, hydrogen has the same effect, produced by photochemistry and escaping into space.

Some other reasons for which conditions on present-day Titan do not replicate those on primordial Earth are the following (a) The major composition of Titan's haze (nitrogen–methane) is different from the Earth's (nitrogen-carbon monoxide–carbon dioxide). This could be due to the fact that the state of oxidation between the inner and the outer solar system atmospheres is very different. Carbon is present in its fully oxidized state (carbon dioxide) in all the terrestrial planets and in its fully reduced state in most of the outer solar system objects. This difference could reflect the ability of water to buffer the oxidation state of the carbon species in the atmospheres of the terrestrial planets, whereas the oxidation state of the carbon in the outer solar system atmospheres may be primordial. The present-day Earth is an extreme case in this regard due to the high abundance of molecular oxygen. This apparently reflects Earth's unique position in the solar system as an abode of life. (b) The infall of carbonaceous material (meteorites, comets, etc.) on a planet is smaller today than in the past. (c) Earth, being larger than Titan, has an atmosphere that has evolved significantly over its lifetime owing to volcanic and tectonic changes with time. Titan, by way of contrast, may have undergone early dramatic changes followed by tectonic quiescence. Subsequent evolution would be externally driven by photochemical processes, and perhaps by tectonic activity.

Regarding the question of life on Titan, an important drawback is that a major chemical link, oxygen, scarce because water remains trapped in a frozen state on the surface, is lacking in the atmosphere and cannot interact with the other components. So that, in spite of all the similarities between Titan and Earth (temperature structure, organic chemistry, condensation and rain in the lower atmosphere, possible liquid reservoir on the surface, etc), the cold temperatures (which slow the chemical reactions), the low sunlight input, and the lack of sufficient amounts of oxygen should forbid the miracle of life to happen in this distant satellite at present conditions. Nevertheless, Titan's environment, closer to Earth's than any other object in our solar system, does provide significant insight into the conditions that prevailed on Earth during its early history and into the chemistry that then gave rise to life.

Venus is another planet that is often compared to Titan and Earth because of its cloudy and superrotating deep atmosphere. The lower stratosphere of Titan is expected to be at least as stable against convection as that of Venus, and aerosol particles are just as abundant, yet cloud particles are larger. Particle radii of ~5 and ~10 μm are found, respectively, for condensed dicyanoacetylene and ethane in the north polar hood of Titan. On the other hand, ~1-μm-radius particles are the rule for the sulfuric acid clouds of Venus, al-

TABLE II
Atmospheres of Titan, Earth and Venus Compared

	Titan		Earth		Venus	
Bond albedo	0.29		0.36		0.77	
Cloud top (km)	10–30		—		~160	
Surface (equator)						
pressure (bar)	1.45		1.0		92	
temperature (K)	94		288		733	
Atmospheric structure	T(K)	Alt (km)	T(K)	Alt (km)	T(K)	Alt (km)
tropopause	71	40	217	15	250	90
stratopause	180	200	247	50	—	
mesopause	150	800	190	80	—	
thermopause	200	1500	1000	700	300	135
Surface gravity (m s^{-2})	1.35		9.81		8.87	
$r_{exobase}/r_{planet}$	1.6		1.08		1.03	

though admittedly somewhat larger particles ($r \sim$ 3.5–4 μm) may exist in Venus's lower cloud deck. The particle size difference between Venus and Titan is exacerbated if Titan's stratosphere at low latitudes is considered. There, clouds appear to contain particles of ethane ice with radii of ~75 μm or so.

An important difference is that Venus's cloud particles are bright and scatter most of the solar spectrum, whereas Titan's are dark (especially at blue wavelengths) and absorb it. While Venus's atmosphere falls off rapidly with altitude, the aerosol on Titan warms the stratosphere and gives a large scale height at high altitude, so the atmosphere tails off only slowly into space. [See VENUS: ATMOSPHERE.]

Titan, Earth, and Venus are compared in Table II.

C. TITAN AS A MEMBER OF THE SATURNIAN SYSTEM

The preceding discussion provides the reasons why the present-day atmospheres of Saturn and Titan are so different. As a gas-giant planet, cold and with an enormous gravitational field, Saturn has retained the light elements hydrogen and helium in abundance. If we were to imagine Titan's atmosphere removed and replaced with an unlimited amount of gas from Saturn, the heavy components, primarily ammonia and methane, would be retained and photolysis would eventually produce N_2 and higher hydrocarbons—including those that condense as aerosols—so we could end up with the atmosphere re-forming much as we see it

today. The main difference might be the contribution that came from the interior or from gases frozen on the surface. In terms of minor atmospheric composition then, Voyager and ground-based 5-μm data had shown that Saturn had methane (and its isotope mono-deuterated methane), ethane, acetylene, ammonia, phosphine, arsine, and germanium. Recent *Infrared Space Observatory* (*ISO*) observations showed that water, carbon dioxide, methylacetylene, and diacetylene also existed on Saturn, whereas it lacks ethylene, hydrogen cyanide, and other nitriles. Recently, *ISO* data have also produced water signatures on Titan. [*See* INFRARED VIEWS OF THE SOLAR SYSTEM FROM SPACE.]

In terms of bulk original composition, there is a gradient in composition observed with distance from Saturn, probably caused by the fact that the outer, cooler reaches of the nebula received more volatiles. This trend is seen much more graphically in the Jovian system. Titan is far larger than its sibling satellites, so in addition to retaining an atmosphere, its thermal history has been very different.

We know from high-resolution photometry and spectroscopy that Saturn's satellites contain water ice. Among them is Iapetus, one of the most mysterious objects in our solar system. The reason why Iapetus is so mysterious is because its two faces are radically different, one bright and the other very dark: its surface albedo varies from 0.4–0.5 on the trailing hemisphere (values typical for ice-covered objects) to 0.02–0.04 in the central parts of its leading hemisphere, a hemispheric asymmetry, as on Titan, but contrary. The shape and orientation of the dark material suggest that it is superimposed upon the bright terrain and that they are, in consequence, more recent. The mean density of the satellite is 1.16 g cm^{-3}, similar to that of the other icy satellites of Saturn and consistent with models predicting water ice as the main component of these bodies. The nature and origin of the dark material on Iapetus remain a mystery. The observed symmetry with respect to the direction of the orbital motion of the satellite strongly argues in favor of an external control, or even perhaps an external origin for this material. Studies of its composition, based on ground-based, low-resolution observations, show that it is very red in the visible and near infrared. A fit of the Iapetus spectrum was obtained recently with a mixture of HCN polymers, organic residuals, and water ice.

A recent study of UV Voyager archived images, in accordance with these models, suggests impacts of exogenic material from Phoebe (the outer most satellite, characterized by a retrograde orbit and with a surface albedo similar to but a little less red than that of Iapetus) on the surface of Iapetus, leading to ice evaporation, leaving a dark and red deposit on the ground. This deposit would be a mixture of Phoebe material and a preexistent nonvolatile constituent, uncovered by the impacts, similar to the D-type material found in the asteroids of this type, as well as on Hyperion.

Hyperion is another special satellite because of the complex aspect of its chaotic rotation and its irregular shape. Although its global composition is uncertain, we know that water ice is present on the surface. Its spectrum in the visible range and in the near infrared is compatible with that of Titan, showing a brighter leading and a darker trailing hemisphere. Hyperion is in resonance 4:3 with Titan and it has been suggested that Hyperion material was deposited in the past on Titan. Hyperion's spectrum is also similar to that of Iapetus, although its albedo is much higher (0.21 in the visible, lower than all satellite albedos but for those of Phoebe and Iapetus). Its red color and intermediate albedo value suggest the presence of an additional component (besides water ice) on the surface. A partial coverage by Phoebe-originating material has been suggested in this case also, as on Iapetus.

The observational data that we possess today have not provided a valid explanation of the composition of the surface of the satellites around Saturn or for the presence of an atmosphere around Titan. The diverse family of Saturn's satellites include a number of unique objects.

D. TITAN, TRITON, AND PLUTO

Titan is often compared to two other small objects of our solar system, the planet Pluto and Neptune's satellite Triton, because of their common atmospheric composition (Table III). Indeed, with Earth, these three objects are the only ones to possess an atmosphere essentially made of molecular nitrogen.

Titan and Triton have practically the same atmospheric composition, with molecular nitrogen and methane combining or breaking up under the action of photolysis and producing other organics, such as hydrocarbons and nitriles, which are responsible for the color of these two satellites: Titan is brown orange and parts of Triton are pink or pale yellow. But the two bodies are quite dissimilar in other aspects, partly due to the differences in temperature because their orbits are not at similar distances from the Sun. The atmosphere (with a pressure of about 16 μbar of N_2) is almost in vapor pressure equilibrium with the surface, heated to ~38 K. The bulk ratio rock/ice varies considerably, with Titan containing less rock, since it has a

TABLE III
Titan, Triton and Pluto Compared

	Titan	Triton	Pluto
Semi-major axis of orbit (AU)	9.54	30.06	39.44
Equatorial radius (km)	2575	1352	1208?
Surface gravity (m s^{-2})	1.35	0.78	0.60
Surface temperature (K)	94	38 ± 3	35–55
Bond albedo	0.29	0.7	0.2 − 1.
Surface Pressure (bar)	1.45	14 × 10^{-6}	3 × 10^{-5}?
Temperature at 1 mbar (K)	150–170	41	104 ± 21
T_{max} (K)	185 ± 20	102 ± 3	?
Atmospheric Composition	90–97% N_2 2–3% CH_4 0.2% H_2 10^{-6}–10^{-4}% CO	99% N_2 10^{-4}% CH_4 10^{-4}% H_2 CO?	99% N_2 5 × 10^{-3}% CH_4 CO?
Exobase height (km)	1500	930	?
$r_{exobase}/r_{planet}$	1.6	1.7	≫3?

smaller density. Since Triton was probably captured rather than formed in a planetary nebula, its formation conditions would have been vastly different from those of Titan. [See TRITON.]

By the same token, Pluto spectroscopic observations have revealed that principally N_2 and traces (on the order of a few percent) of CH_4 ice are also present on the planet's surface, along with carbon monoxide ice. Compared to Triton, Pluto has three and five times more methane and carbon monoxide ices, respectively. [See PLUTO AND CHARON.]

The surface temperature on the planet is approximately 40 K, which translates (under vapor pressure equilibrium conditions between the surface and the atmosphere) into a nitrogen atmospheric pressure on the surface of somewhere between 18 and 157 μbar. More recent predictions restrain this range from 10 to 30 μbar.

E. CLIMATE EVOLUTION: TITAN'S FATE

Titan's present surface temperature is elevated above the effective temperature by about 12 K, due to the competing influences of a greenhouse effect of ~20 K and an antigreenhouse (due to blocking of sunlight) effect due to the haze. The greenhouse is caused in part by collision-induced infrared opacity by nitrogen, methane, and also hydrogen. It is possible that in the past, methane may have been depleted in Titan's atmosphere, as its resupply may be episodic, while it is continuously destroyed by photolysis. In that event, the loss of the methane opacity could have led to cooling of the atmosphere to the point where nitrogen could have started to rain out. Albedo feedback due to the rainout, and the further loss of nitrogen opacity, could have led to further cooling, accompanied by a drop in atmospheric pressure, perhaps down to 100 mbar or less.

The feedbacks are complex and poorly understood at present. If methane were completely depleted, the production of both haze and hydrogen would stop. The haze residence time is on the order of 1000 years, so after this time, the sunlight reaching the surface would increase. Hydrogen is rather slower to escape, perhaps 2 million years, so after this time the surface would cool slightly.

If the methane were never depleted, for example, if it were buffered by lakes or an ocean of hydrocarbons on the surface, the evolution of the atmosphere would be less dramatic. Lakes would progressively drop in depth, as two molecules of less dense methane are used to make one of ethane. Further, nitrogen is less soluble in ethane than methane, so as the methane is used up, the lakes would expel nitrogen and increase the atmospheric pressure somewhat.

Titan's ultimate fate is of interest, both in its own right and in a more general exobiological context. The greenhouse effect makes Titan fairly sensitive to increases in tidal heating, for example, such that a Titan-like moon in orbit around a giant planet might be rather warmer than if it were itself a planet in a heliocentric orbit. If Titan were ~80 K warmer, its surface and near-subsurface might be liquid (the ammonia–water peritectic melting point is ~176 K). The coexistence of liquid water and the organics produced by photolysis allows the easy synthesis of prebiotic molecules such as amino acids. As the Sun becomes a red giant, its luminosity will increase by over an order of magnitude for several hundred million years.

However, as the atmosphere is warmed, it expands. Production of aerosol is tied to the absorption of UV at a given pressure (typically the aerosol production altitude in scattering models is 0.1 mbar) so that the altitude of haze production increases as the solar luminosity increases. Thus, for a fixed production rate, since the haze has much farther to fall, the column optical depth increases. This largely compensates for the increased luminosity. A small offsetting effect is the change in the solar spectrum—as the solar surface cools, more red light is produced, which penetrates deeper into the atmosphere than the present Sun. Detailed model study is required to examine these effects, but it seems likely that while Titan may get somewhat warmer when the Sun becomes a red giant, it will not become a balmy paradise.

F. OPEN QUESTIONS

Some questions of Titan's origin and evolution remain unanswered although some partial answers have recently been provided and are given in this chapter:

- What is the origin of nitrogen in Titan's atmosphere? Is it derived from a nebula containing molecular nitrogen or was it produced by chemical processing of ammonia?
- How much carbon monoxide was introduced into Titan and its early atmosphere?
- Was methane brought into Titan as frozen condensate or entrapped in clathrate?
- Why does Titan have an atmosphere, whereas the Galilean satellites do not?
- Could most of Titan's atmosphere have been added by impact of volatile-rich comets after accretion?
- What would be the effect of an ocean in contact with the atmosphere, as opposed to solid surface, on the evolution of Titan?

Precise knowledge of the abundances of the noble gases, such as argon and deuterium, in Titan's atmosphere may be diagnostic tests of the competing hypotheses on the origin of nitrogen and methane on Titan. The outstanding problem of the origin of atmospheric nitrogen on Titan (was it introduced as molecular nitrogen or NH_3?) could find an answer in measurements of the Ar/N_2 ratio in the present atmosphere: a large argon abundance suggests N_2, whereas a small one favors NH_3. Unfortunately, argon has not yet been detected on Titan.

A deuterium enrichment could originate from different procedures acting during or after the satellite's formation: fractionation processes involving deuterium exchange with cloud particles, ocean, or crust; isotopic exchange catalyzed by metal grains in the Saturnian nebula; outgassing of already D-enriched grains of interstellar origin, and so on. A precise measurement of the D/H value on Titan should tell us something about the ways in which hydrogen-containing compounds were incorporated in planets and satellites.

Water, recently detected in Titan's atmosphere, quickly dissociates into OH, which combines with methane photolysis products (such as CH_2 and CH_3) and produces CO and CO_2. The source for water on Titan can be found in the rings of Saturn, the meteorites, or comets. CO could also be directly captured in a clathrate hydrate directly from the solar nebula. However, all of these mechanisms involve numerous uncertainties that have not yet been quantified and that do not allow for a definite preference between the different photochemical models.

VI. SPACE MISSIONS: INFRARED SPACE OBSERVATORY AND CASSINI/HUYGENS

A. INFRARED SPACE OBSERVATORY

In November 1995, the *Infrared Space Observatory* was successfully put into orbit around Earth and has only recently returned some data of Titan in the infrared range. The data acquired on January 10, 1997, were mainly recorded by the Short Wavelength Spectrometer (SWS) in the grating mode and cover the range from roughly 2 to 50 μm. The spectral resolution is 0.2–1 cm^{-1}, about nine times better than that of IRIS/Voyager. The spectra show emission signatures of all of the expected minor constituents in Titan's stratosphere (hydrocarbons, nitriles, and CO_2) and the higher resolution allows us to resolve the bands and distinguish the various contributions. As a consequence, a better determination of the abundances and vertical distributions of these components is achieved on a disk-average scale. In December 1997 *ISO* SWS allowed for the first detection of water vapor lines in emission near 40 μm on Titan.

New data, taken by the Long Wavelength Spectrometer in the grating mode, but also by the SWS in the Fabry-Pérot mode, in June 1997, are currently under analysis and will provide information on the submillimeter part of Titan's spectrum. This has never been observed before, and important molecules—such as methane, carbon monoxide, hydrogen cyanide, and perhaps water—exhibit numerous rotational lines. These lines will provide precise measurements of the mixing ratios of these constituents.

B. CASSINI/HUYGENS

The *Cassini* mission is designed to address our principal questions regarding Titan. Although the mission's objectives span the entire Saturnian system, Titan is a target of particular interest (as for Voyager before it). During its four-year tour, the *Cassini* orbiter will make about 40 flybys of Titan, some as close as 1000 km (*Voyager 1* flew by at 4000 km from the surface), and direct visible, infrared, and radar instruments will collect data.

Additionally, the mission will see the deployment of the European-built *Huygens* probe. After release from the *Cassini* orbiter, this 300-kg probe will plunge into Titan's atmosphere and descend for 2.25 hours by parachute (Fig. 8). In addition to imaging the atmosphere and surface, the probe will take samples of the haze and atmosphere. These *in situ* measurements will complement the remote-sensing measurements to be made from the orbiter.

The orbiter carries a multimode radar, which will be able to completely penetrate the hazy atmosphere. The radar operates as a radiometer (to measure surface temperature or emissivity), scatterometer, and altimeter (to measure the reflectivity and topography along the orbiter groundtrack), and as a synthetic aperture imager. This latter mode, nearest closest approach, will image Titan's surface at 0.5- to 2-km resolution (i.e., about 3 times poorer resolution than *Magellan* at Venus) over about 1% of the surface for each flyby devoted to radar measurements. Thus perhaps 20–40% of Titan's surface will be imaged—in long thin strips—during the mission. Lower-resolution measurements will cover most of the surface. The imager on the orbiter carries filters tuned to the windows (e.g., 940 nm) in between the methane bands, and so, like *HST*, should be able to measure surface contrasts. Additionally, polarizers are carried that should be able to remove most of the light scattered by the haze at near 90° phase angle, so these measurements too will study the surface. The exact resolution achieveable will depend on the scene contrast and the haze optical depth at the time of the mission, as well as the image motion compensation that can be achieved, but may be better than 1 km. Other filters will be able to probe different altitudes in the atmosphere.

The Visual and Infrared Mapping Spectrometer (VIMS) instrument spans other spectral windows between 0.6 and 5 μm. This will allow spectral identification of surface materials with high (~500-m resolution) as well as resolved composition measurements. Looking at Titan's nightside, the instrument may be able to spot lightning or thermal emission from active cryovolcanism.

FIGURE 8   Artist's impression of the ESA *Huygens* probe entering the atmosphere. The *Cassini* orbiter spacecraft and Saturn are in the background. (Courtesy of NASA.)

The Composite Infrared Spectrometer (CIRS) instrument will allow the temperature to be profiled at different locations in the atmosphere, as well as take spatially resolved composition measurements (Fig. 9). These data will be valuable for verifying and refining models of photochemistry and atmospheric circulation. Direct measurement of the composition of Titan's atmosphere versus altitude will be made by the Gas Chromatograph Mass Spectrometer (GCMS) on the probe—because Titan's atmosphere has so many components, this will allow separation in two dimensions (by chromatography as well as mass spectroscopy). The GCMS will also analyze the pyrolysis products from the Aerosol Collector and Pyrolyzer (ACP), which sucks haze particles into the probe and traps them in a filter, which is subsequently baked in an oven to break down the haze macromolecules into smaller fragments that can be studied in the GCMS. The GCMS also has a heated inlet, so that if the probe survives impact with the surface, the volatile component of the surface material at the landing site can be determined.

The atmospheric composition at high altitude will

also be sampled directly, during the closest flybys by the orbiter. The Ion and Neutral Mass Spectrometer will analyze atomic and molecular composition at ~1000 km altitude. These, and spacecraft dynamics measurements, will allow direct comparison with the density profile measured by the entry deceleration of the *Huygens* probe.

The entry deceleration is measured by the Huygens Atmospheric Structure Instrument (HASI), which is the only probe instrument operating prior to parachute deployment on the probe: the deceleration is proportional to density, and from the density profile and hydrostatic equilibrium, a temperature profile of the upper atmosphere can also be derived. The temperature and pressure are measured directly from 170 km down as the probe descends by parachute. HASI also includes a Plasma Wave Analyzer (PWA), which will measure the electrical properties of the atmosphere (important in determining haze charging and coagulation physics), search for thunder and lightning, and, in the event of a successful landing, measure the dielectric properties of the surface material. The probe also carries a radar altimeter, which will estimate radar re-

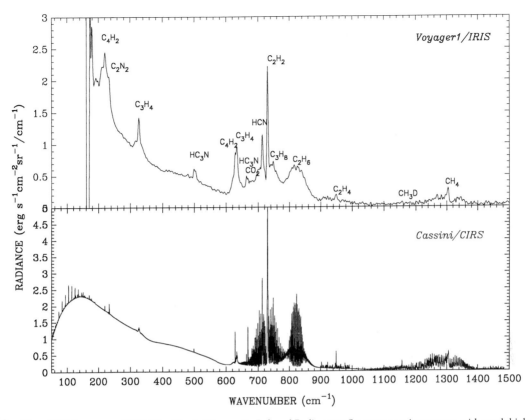

FIGURE 9 Titan IRIS data versus CIRS. The *Cassini* Composite Infrared Radiometer Spectrometer instrument, with much higher spectral resolution, will discriminate many more features in Titan's spectrum.

flectivity and surface topography. The radar altimeter, part of the probe system itself, passes its signal to the PWA for data processing.

The surface material is to be directly investigated by the Surface Science Package (SSP). This is a suite of sensors on the probe to measure (for a liquid surface) refractive index, density, and other properties, allowing a coarse identification. An acoustic sounder will measure the speed of sound in the atmosphere during descent (constraining temperature and relative molecular mass with high-altitude resolution) and place bounds on the depth of any liquid reservoir. Prior to impact, the sounder will also estimate the surface roughness at the landing site. In the event of a landing on a solid surface, an accelerometer and a penetrometer will measure the mechanical properties (e.g., particle size and stickiness) of the surface material. Tilt sensors will measure the probe's attitude during descent, and measure any wave motion on a liquid surface.

The probe's motion in Titan's atmosphere will be accurately measured along the probe–orbiter line of sight by the Doppler Wind Experiment (DWE) using an ultrastable oscillator that fixes the radio signal from the probe to a precise frequency, the Doppler shift which can be used to measure the line-of-sight velocity of the probe, and hence infer the wind field. It is possible that the probe signal may be measured on Earth as well as by the orbiter, allowing two-dimensional measurements.

The radio system on the orbiter will be used to study Titan in two ways: first, by tracking the spacecraft from Earth, to determine Titan's gravity field, which in turn constrains its internal structure (e.g., the size of a rock core, and perhaps the rigidity of the crust), and second, by multiple radio occultations. These will measure a temperature profile and indicate the extent of Titan's ionosphere.

The interaction of Titan with the Saturnian magnetosphere will be studied by the Magnetospheric Imaging Instrument (MIMI), the Cassini Plasma Spectrometer (CAPS), and the Planetary Radio Astronomy (PRA). The latter instrument will also search for radio emissions from lightning on Titan, although a similar search by *Voyager* failed to indicate any such emission.

The Cosmic Dust Analyzer (CDA) will measure the mass, velocity, and composition of particles in Titan's vicinity. These data will be valuable in understanding the origin of oxygen compounds in its reducing atmosphere, believed to be due to photochemical reactions involving meteoric water.

The *Cassini* mission promises to provide a wealth of data. Future ideas for Titan missions are not mature (after all, *Cassini's* arrival is still 7 years away), but a prominent concept is the use of an "aerobot," or intelligent balloon, to explore up close a variety of Titan locations.

BIBLIOGRAPHY

European Space Agency (1992). "Proceedings of the Symposium on Titan." Toulouse, France, September 1991, ESA SP-338. ESA, Noordwijk, The Netherlands.

Hunten, D., *et al.* (1984). "Titan in Saturn" (T. Gehrels and M. S. Matthews, eds.). Univ. Arizona Press, Tucson.

Lorenz, R. (1993). The surface of Titan in the context of the ESA Huygens Probe. *ESA J.* **17,** 275–292.

Spilker, L. (ed.) (1997). "Passage to a Ringed World: The Cassini–Huygens Mission to Saturn and Titan," NASA SP-533, Washington, D.C.

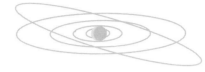

TRITON

I. Introduction
II. Discovery and Orbit
III. Pre-Voyager Astronomy
IV. *Voyager 2* Encounter
V. General Characteristics
VI. Geology
VII. Atmosphere and Surface
VIII. Origin and Evolution

William B. McKinnon
Washington University

Randolph L. Kirk
U.S. Geological Survey, Flagstaff

GLOSSARY

Albedo: Geometric albedo is the ratio of an object's brightness at zero phase angle to a perfectly diffusing disk with the same position and apparent size as the planet. Bond albedo is the fraction of the total incident light reflected by the body.

Caldera: Large basin-shaped volcanic depression more or less circular in form, the diameter of which is many times greater than that of any associated vent or vents.

Cryovolcanism: Volcanism where the volcanic materials are melted ices, such as water, ammonia, and methane, as distinguished from the common high-melting point volcanic materials of the terrestrial planets, such as basalt and rhyolite.

Diapir: A body of rock or ice that has moved upward due to buoyancy, attaining an inverted tear-drop or pear shape, and piercing and displacing the overlying layers.

Ekman layer: Idealized model of the planetary boundary layer in which the mean flow (as in an atmosphere or ocean) is modified near the ground by friction (either laminar or turbulent).

The resulting variation of flow speed and direction with height is described by the Ekman spiral.

Lithosphere: Stiff upper layer of a planetary body, generally including a chemically distinct crust and part of the upper mantle; the lithosphere is defined rheologically and so its thickness depends strongly on temperature.

Occultation: The passage of a celestial body across a line between an observer and another celestial object.

Phase angle: Angle at the surface of a planet between the source of light and the observer.

Regular satellite: Satellite with low orbital eccentricity and inclination.

Scale height: Vertical distance over which a given atmospheric parameter (e.g., pressure or density) changes by a factor of e.

Thermal wind: Wind shear developed in one direction due to a temperature gradient in an orthogonal direction.

I. INTRODUCTION

Triton is the major moon of the planet Neptune. It is also one of the most remarkable bodies in the solar

FIGURE 1   Photomosaic of the best Triton images, overlayed with a latitude–longitude grid. Centered on the Neptune-facing hemisphere at 15°N, 15°E, the resolution of the component images ranges from ~0.7 to 1.5 km/pixel. The latitude of the subsolar point on Triton at the time of the *Voyager* encounter was −45°, so the north polar region was in darkness. Because Triton's spin is tidally locked to Neptune, Triton's eastern hemisphere is also the leading hemisphere in its orbit. (Courtesy of Alfred McEwen, University of Arizona.)

system (Fig. 1). Its orbit is unusual, circular and close to Neptune, but highly inclined to the planet's equator (by 157°). Furthermore, Triton's sense of motion is retrograde, meaning it moves in the opposite direction to Neptune's spin (Fig. 2). Triton's history therefore must have been quite different from those of "regular" satellites, such as the moons of Jupiter, which orbit in a prograde sense in their primary's equatorial plane. The modern consensus is that Triton originally formed in solar orbit and was subsequently captured by Neptune's gravity.

Like nearly all solar system satellites, tides have slowed Triton's spin period to be coincident with its orbital period and shifted its spin axis to be perpendicu-

lar to its orbital plane. Consequently, one hemisphere of Triton permanently faces Neptune. The combination of Neptune's axial tilt (29.6°) and Triton's inclined orbit gives Triton a complicated and extreme seasonal cycle. In the distant geological past, tides associated with Triton's capture may have strongly heated and transformed its interior.

Although discovered soon after Neptune, little was learned about Triton until the modern telescopic era, and even so, most of the information we have was acquired during the *Voyager 2* encounter with Neptune in 1989. Triton is a relatively large moon (1352 km in radius), larger than all of the mid-sized satellites of Saturn and Uranus (200 to 800 km in

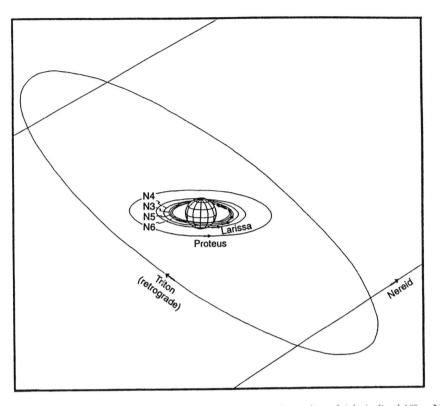

FIGURE 2 Orbits of Neptune's family of satellites. Shown is a perspective view along a line of sight inclined 18° to Neptune's equatorial plane. The innermost satellites are all relatively small and were not discovered until *Voyager 2* passed through the Neptune system. They orbit in Neptune's equatorial plane, while much more massive Triton circles outside them in an inclined, retrograde orbit. All the satellites have virtually circular orbits except for Nereid. The apparent crossing of Nereid's and Triton's orbits is an artifact of the projection. [From J. S. Kargel (1997). In "Encyclopedia of Planetary Sciences" (J. H. Shirley and R. W. Fairbridge, eds.). Chapman & Hall, London.]

radius), but not quite as large as the biggest icy satellites — the Galilean satellites and Titan (1570 to 2630 km in radius). It is a relatively dense world (close to 2 g cm^{-3}), rock rich, but with a substantial proportion of water and other ices. Ices comprise its reddish visible surface (Fig. 1; see also color insert), and the freshness of the ices cause Triton to be one of the most reflective bodies in the solar system (its total, or Bond albedo, is ≈0.85). This, combined with the satellite's distance from the Sun (30 AU), makes Triton's surface the coldest place yet measured in the solar system (38 K). Despite these frigid surface conditions, *Voyager 2* discovered a thin atmosphere of nitrogen surrounding the satellite (14 μbar surface pressure, where 1 bar is the approximate surface pressure of Earth's atmosphere). Triton's atmosphere is dense enough to support clouds and hazes and to transport particles across Triton's surface.

As with all solid planets and satellites, Triton's history is written into the geological record of its surface. Triton, however, is a geologically young body. Most of the approximately 40% of the satellite's surface that

was imaged by *Voyager* at sufficient resolution to tell is sparsely cratered. No heavily cratered terrains survive from early solar system times, an absence that may reflect an epoch of severe tidal heating. The geologic terrains that do survive are unique in the solar system. At least three major terrain types can be distinguished: smooth, walled, and terraced plains, which are thought to be due to icy volcanism; an enigmatic "cantaloupe" terrain, which may be due to diapirism; and a hemispheric-scale polar deposit or cap, which may be related to seasonal and climatic cycles of evaporation and redeposition of volatile ices. The polar cap is thought to be predominantly solid nitrogen. Other ices that have been identified on Triton are, in approximate order of abundance, H_2O, CO_2, CO, and CH_4. The cap is a site of present-day geological activity, in particular, the eruption of plumes or geysers of gas and fine particles.

The following sections describe Triton in greater detail, with emphasis on its geology, the interaction of its icy surface and the atmosphere (including the plumes), and its probable origin and violent early evolution.

II. DISCOVERY AND ORBIT

Acting on the mathematical prediction of Urbain Le Verrier, the planet Neptune was first identified at the Berlin Observatory on September 23, 1846. It was announced in England on October 1. On that day Sir John Herschel, son of the discoverer of Uranus, wrote to William Lassell, asking him to look for any satellites of the new planet "with all possible expedition," using his own 24-inch reflector. Lassell was a brewer by profession, owning a very successful establishment in Liverpool, and an astronomer by avocation. He made his own telescopes and was a keen visual observer. Herschel was no doubt seeking to ease some of the sting of Neptune's being found by continental astronomers, given that he was aware of the independent prediction of Neptune's position by John Couch Adams and the unsuccessful search for the planet from English soil. Lassell wasted no time, making his first observations on October 2, and on October 10, 1846, he discovered Triton. Over the next few months he established its orbital period (5 days and 21 hr), the orbit's angular size, and that Triton was moving in a retrograde sense with respect to the spin and orbital directions of nearly all other solar system bodies.

This orbit was clearly unusual, but given that Neptune's spin axis was unknown (its visible disk was too small to show bands or cloud features), one would be justified in assuming Neptune's rotation was similarly retrograde, and Triton's orbit no more unusual than the orbits of Uranus' satellites (two of which Lassell would discover). The gravitational effect of Neptune's oblateness, due to its spin, causes Triton's orbit to precess about Neptune's spin axis with a period of ~688 years. Once a long enough baseline of observations was established (over many decades), Neptune's obliquity of ~29° was determined, but it was not until the late 1920s that spectroscopic measurements showed that Neptune was indeed a prograde rotator.

Thus by 1930 it was established that Triton was a most unusual moon. It orbited at 14.3 Neptune radii, or R_N (using the modern value of 24,760 km for Neptune's equatorial radius). The orbit was circular inasmuch as this could be measured, but distinctly retrograde. It was also *alone*. No new satellites would be found until Gerard P. Kuiper located distant, eccentric Nereid (semimajor axis $\approx 223\ R_N$) in 1948 and *Voyager 2* identified six small regular satellites orbiting close to Neptune (within 4.75 R_N) in 1989 (Fig. 2). The contrast with the regular satellite systems of Jupiter, Saturn, and Uranus could hardly have been greater.

Onto this well-prepared stage came the discovery of the planet Pluto in early 1930 by Clyde Tombaugh. After a few months' observations, Pluto's orbit was determined to be so eccentric that Pluto actually crossed inside the orbit of Neptune for about 20 years of its 248-year orbital revolution. Although Pluto's orbit was also determined to be substantially inclined so that it did not actually intersect Neptune's, British astronomer R. A. Lyttleton realized differential precession of the orbits could cause them to intersect, either in the future or in the past. In 1936 he published a paper that theoretically explored the possibility that such an orbital configuration once did exist and that Pluto was in reality an escaped satellite of Neptune. [*See* PLUTO AND CHARON.]

One specific scenario Lyttleton addressed was that Triton and Pluto were once adjacent, regular, prograde satellites of Neptune. Tides raised on Neptune then caused the orbits of both of these bodies to expand (as does the orbit of our Moon), but at different rates. The orbit of the inner one moved too close to the orbit of the outer one, and a destabilizing series of gravitational encounters ensued. Triton's orbit was reversed, and Pluto was ejected from the Neptune system. As Triton's velocity change is the greater in this scenario (being reversed), the laws of mechanics demanded Pluto be the more massive body. This was in fact the presumption at that time because Pluto was supposed to have been sufficiently massive to have detectably perturbed Neptune's orbit according to some astronomers. We now know, of course, that Pluto is hardly massive at all, and in particular is substantially less massive than Triton; therefore, for this and a number of other reasons, planetary scientists have rejected Lyttleton's theory. The ghost of Lyttleton lives on in the popular imagination, though, and for most of this century it has contributed to the intrigue surrounding Triton.

The modern view of Triton's origin, and that of Pluto and other bodies in the deep outer solar system, is discussed later in this article.

III. PRE-VOYAGER ASTRONOMY

A. RADIUS AND MASS

Through the telescope Triton is a faint, 14th magnitude object, never more than 17 seconds of arc from Neptune (Fig. 3). Consequently, physical studies of the satellite from the ground have historically been

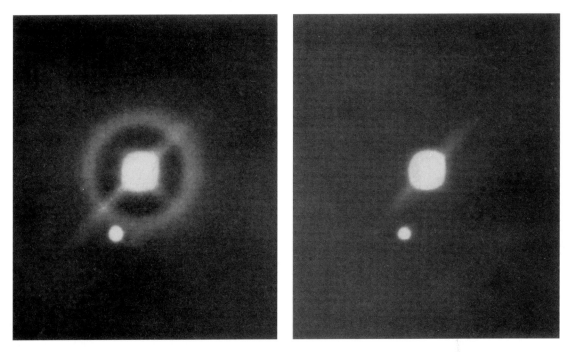

FIGURE 3 Photographic plates of Neptune and Triton taken by G. P. Kuiper in 1949 with the 82-inch telescope at the McDonald Observatory. [From G. P. Kuiper (1961). *In* "Planets and Satellites" (G. P. Kuiper and B. M. Middlehurst, eds.). University of Chicago Press.]

very difficult. Triton shows no visible disk. Only crude limits could be put on its size for many years. The most well-known was probably that of Kuiper, who in 1954 attached a "diskmeter" to the 5-m Hale Telescope atop Palomar Mountain and estimated Triton's radius to be 1900 km, but with a large uncertainty. Such a size implied that Triton was one of the largest moons in the solar system.

In the 1930s and 1940s astronomers attempted to measure the reflex motion of Neptune, i.e., Neptune's wobble as it and Triton orbit their common center of gravity. This was a common enough technique for determining the relative masses in binary star systems, but because the Triton/Neptune mass ratio was clearly rather small, detecting the slight oscillation of Neptune's position on a series of photographic plates would not be easy. Despite the apparent precision of the astrometric techniques, systematic error overwhelmed the observations. The numbers published implied Triton was very massive, possibly *the* most massive moon in the solar system. Kuiper's size estimate and the astrometric mass helped fuel the mystery of Triton for several decades.

B. INFRARED AND VISIBLE SPECTRA

The first real breakthrough occurred in 1978, when infrared detector technology had improved to the point

that the first compositional identifications could be made. Astronomers detected a methane (CH_4) band in Triton's infrared spectrum, at 2.3 μm, which they interpreted as evidence for a tenuous methane atmosphere. Given that Saturn's Titan had an atmosphere (then thought to be mostly CH_4) and that Triton was supposed to also be a large moon, this was a satisfying discovery. Within a few years, however, more absorption bands of methane were detected in the near infrared (Fig. 4). The relative depths of the new bands, plus their variability as Triton orbited Neptune, indicated that much (if not all) of the methane detected was in solid form, i.e., an ice on the surface of Triton, nonuniformly distributed in longitude. Ices on the surface implied that Triton might be a relatively bright, smaller world, rather than a darker, larger body of the same visual magnitude, but few if any astronomers could guess just how bright and reflective Triton would turn out to be. [*See* TITAN.]

Methane ice on the surface also offered a potential explanation for Triton's reddish visual color. Experiments had shown that when solid methane is irradiated by solar ultraviolet rays or bombarded by charged particles, such as could be supplied by the solar wind, by cosmic rays, or by neptunian magnetospheric particles (if such existed), the methane turns pink or red as hydrogen is driven off and the remaining carbon and hydrogen form various carbonaceous compounds. At Triton, however, it is important not to have too much radiation

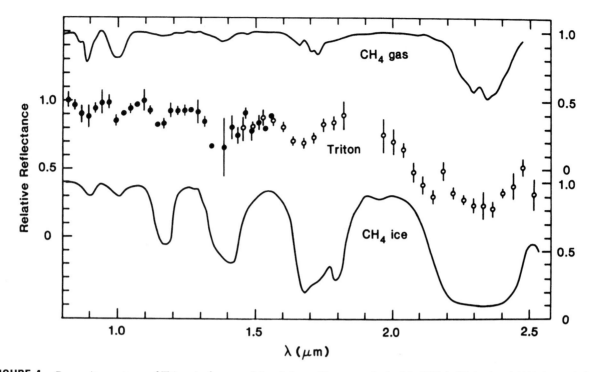

FIGURE 4 Composite spectrum of Triton in the range 0.8 to 2.5 μm. Data were obtained in 1982 (solid dots) and 1980 (open circles). For comparison, a calculated spectrum for cold (55 K) methane gas is shown, along with laboratory measurements of the diffuse reflectance from methane frost. Note how well the positions and relative strengths of the methane ice absorptions match the bands in Triton's spectrum. [From D. P. Cruikshank and R. H. Brown (1986). *In* "Satellites" (J. A. Burns and M. S. Matthews, eds.). Univ. of Arizona Press, Tucson.]

or charged particle bombardment, or *all* the hydrogen will be driven off and the carbon-rich residue turned black. Triton's persistent redness implies that the methane ice is refreshed on a relatively short time scale.

C. SEAS OF LIQUID NITROGEN?

An even more amazing discovery was made in the early 1980s. A single infrared spectral feature was found at 2.16 μm, since refined to 2.15 μm (see Fig. 4), a feature that could not be attributed to any of the usual spectral suspects (CH₄, H₂O, silicates, etc.). Nitrogen (N₂) *does* have an absorption at this wavelength, and as *Voyager 1* had recently determined the dominant atmospheric gas on Titan to be N₂ (not CH₄), finding nitrogen on Triton was not far-fetched. The amount of nitrogen gas required to account for the absorption was quite large, however, as nitrogen, a homonuclear diatomic molecule, is a very poor absorber of infrared light. In particular, the amount of N₂ was too large to be in vapor pressure equilibrium with surface N₂ ice at a temperature reasonable for Triton. Astronomers therefore concluded that in order to get the necessary pathlength for the absorption, the nitrogen had to

be in condensed form, either solid or liquid. Liquid nitrogen was the favored interpretation, and a fantastic vista emerged — a satellite covered with a global or near-global sea of liquid nitrogen, along with floating methane icebergs and/or methane-ice-coated islands, possibly eroded by eons of nitrogen rain.

The "problem" with liquid nitrogen is that it freezes at zero pressure at about 63 K. For Triton to have a global ocean at that temperature requires that (1) Triton absorb most of the sunlight striking it (have a low albedo) and (2) Triton's surface radiate infrared heat very inefficiently (have a low emissivity). The coexistence of solid methane and liquid nitrogen also requires specific thermodynamic conditions. For these and other reasons, planetary chemists offered a competing concept for Triton's surface, one in which both the nitrogen and the methane were solid and distributed nonuniformly (Fig. 5). Because of nitrogen's great volatility, it was argued that crystals of up to centimeter size could grow on Triton's surface over a season and so provide the pathlength for the 2.15-μm absorption. Methane, as in all the spectroscopic models, would be only a minor component; it dominates Triton's near-infrared spectrum (such as in Fig. 4) by virtue of the relative strengths of its absorptions.

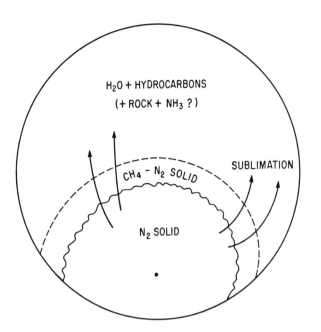

FIGURE 5 A pre-Voyager prediction for the state of Triton's surface. A subliming N_2 ice cap is centered on the illuminated south pole (dot). [From J. I. Lunine and D. J. Stevenson (1985). *Nature* **317**, 238–240.]

The difference between the two models for Triton's surface was far from academic. Condensed nitrogen on the surface meant that Triton had an atmosphere in vapor pressure equilibrium, one that was also predominantly N_2, as N_2 is much more volatile than CH_4. The vapor pressure is extraordinarily temperature sensitive (exponentially so), and because the state of N_2 on the surface (solid or liquid) also depends on temperature, the type of atmosphere Triton could have varied widely. For liquid nitrogen seas, Triton would possess an atmosphere of ~0.1 bar surface pressure, whereas for a solid surface the atmospheric pressure could be orders of magnitude less. The thicker the atmosphere the more dynamic the weather and the more vigorous the atmospheric response to seasonal and climatic changes.

It had already been recognized that Triton's seasonal cycle was complicated by the precession of its orbit. Its seasons vary in intensity and length, and in the decades before the *Voyager* encounter Triton was moving toward the peak of maximal southern summer (Fig. 6). Correspondingly, Triton's northern hemisphere was (and is) enduring prolonged darkness. The possibility of long-term cold traps at both poles, with strong seasonal atmospheric flows from pole to pole, was recognized. The illustration in Fig. 5 was in fact based in part on an analogy with Mars, with N_2 replacing CO_2 as the dominant, and condensible, atmo-

spheric constituent, and CH_4 replacing H_2O as the secondary, less volatile component. Specifically, a large cap of solid nitrogen was predicted for the south pole, sublimating languorously in the feeble summer sun.

D. SIMILARITIES WITH PLUTO

As Triton was coming into clearer astronomical focus in the 1980s, parallel developments were occurring for other outer solar system bodies, especially Pluto. Methane had been discovered on Pluto prior to Triton, and interpreted as a surface ice. By the mid-1980s it was clear Pluto's methane-dominated near-infrared spectrum bore a strong resemblance to that of Triton, although Pluto's methane absorptions were deeper; Pluto and Triton were similar in the photo-visual region as well, having similar red spectral slopes. Their common bond was reinforced by their similar visual magnitudes (Pluto and its moon together are only ~0.3 magnitudes fainter than Triton when referenced to a common distance and solar phase angle) and by Pluto being at a similar distance from the Sun as Triton in the 1980s (the interval 1979–1999 is the special time when Pluto's orbit crosses inside that of Neptune, making astronomical studies of the ninth planet easier).

Pluto's fundamental properties (mass and radius) were relatively well constrained by the time of the

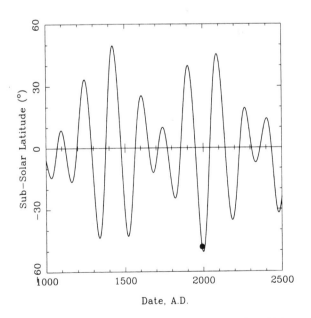

FIGURE 6 Seasonal excursion of the subsolar latitude on Triton. Dot shows the subsolar latitude at the time of the *Voyager* encounter. [From R. L. Kirk *et al.* (1995). *In* "Neptune and Triton" (D. P. Cruikshank, ed.). Univ. of Arizona Press, Tucson.]

Voyager 2 encounter with Neptune (and Triton). Pluto's relatively large satellite, Charon, had been discovered in 1978, which allowed determination of the mass of the Pluto–Charon system by means of Kepler's third law. Pluto–Charon turned out to have only one-fifth the mass of the Earth's moon. Charon's orbit is highly inclined, but by happy seasonal circumstance was about to go edge-on as seen from the Earth (as Saturn's rings are periodically seen to do). The mutual occultations and transits of Pluto and Charon began in 1985 and lasted several years. Careful monitoring of the system's light curve, plus observations of the occultation of a star by Pluto in 1988, established that Pluto's radius lay between 1150 and 1200 km. Pluto turned out to be a smallish, bright, more or less ice-covered world.

Pluto also turned out to be relatively dense as ice-rock bodies go, close to 2 gm cm^{-3}. The corresponding rock/ice ratio is about 70/30 and is, curiously, close to what is predicted for a body accreted in the deep cold reaches of the outer solar system. According to current thinking, the solar nebula at that distance from the Sun, when the Sun and planets were forming, was relatively cold and unprocessed. The outer nebula thus retained many of the chemical signatures of the interstellar gas and dust (molecular cloud) that was the ultimate source of the nebula. Specifically, carbon would be in the form of organic matter and carbon monoxide (CO) gas. CO is very volatile and does not condense as an ice until temperatures fall below ~20 K. This probably never occurred anywhere in the solar nebula, and therefore CO never accreted in bulk (although small amounts could be adsorbed on or enclathrated in water ice). The key point is that volatile CO ties up oxygen that could, in other circumstances, be liberated to form eminently condensible water ice. Therefore, bodies formed in the outer solar system, but not near a giant planet, should have relatively high rock/water-ice ratios. Even though water ice had not been detected (then) on Triton's surface, water ice is expected to be the volumetrically dominant ice in outer solar system bodies, as oxygen is the third most abundant element in the universe. [*See* ORIGIN OF THE SOLAR SYSTEM.]

In contrast, near a giant planet (specifically, in the subnebulae that probably formed around each) CO would be chemically converted to CH_4. The freed oxygen would then combine with abundant hydrogen to form water ice, and satellites that accreted in the subnebulae should have rock/ice ratios near or less than 50/50. If temperatures were cool enough for CH_4 condensation, then the rock/ice ratios would be lower still. This was one line of reasoning that pointed to

Pluto being an original solar-orbiting body and not an escaped satellite of Neptune.

Dynamical evidence against Pluto being an escaped satellite also accumulated. By the 1980s it was being argued that Triton and Pluto should be considered as two independent solar system bodies, with independent histories. The link between the two, in terms of brightness (and presumably size) and composition, was that they formed in the same region — the outer solar nebula near or beyond Neptune. Essentially, they are surviving examples of particularly large outer solar system planetesimals (or protoplanets). Pluto became locked in a dynamical resonance with Neptune, which preserved its peculiar orbital geometry, whereas Triton was later captured by Neptune's gravity. [*See* PLUTO AND CHARON.]

If the analogy with Pluto is correct, then Triton should also be rock rich. If Triton had a relatively bright, icy surface like Pluto, Triton's visible magnitude implied it would probably be somewhat larger (by about 15%), but its density would be similar to that of Pluto–Charon. Of course, the surface state and thus the size of Triton could not be pinned down before the *Voyager* encounter, but the consequences for Triton of being captured (as has been alluded to) were potentially spectacular. These include intense tidal heating and wholesale melting of the satellite. These will be discussed in greater detail later, but were appreciated by the planetary community on the eve of the *Voyager 2* encounter. With the observational and theoretical backdrop described in this section, and with the promise of resolution of fundamental questions and the revelation of novelty, anticipation was high.

IV. *VOYAGER 2* ENCOUNTER

Future history will no doubt record the Voyager project as one of humankind's great journeys of discovery. Originally conceived as a "grand tour" of all the giant planets and Pluto, the *Mariner*-class spacecraft that were eventually launched in 1977 (and renamed *Voyager*) were only designed to encounter Jupiter and Saturn. If they worked, however, a highly capable complement of remote-sensing instruments for the planets and satellites and *in situ* detectors for the magnetospheres and plasmaspheres would be carried into the outer solar system for the first time. Two spacecraft allowed for different encounter strategies, better satellite coverage, and modification of the second flyby to reflect discoveries made by the first.

The Jupiter encounters in 1979 were nothing short of spectacular. The wealth of images and data sent back about the Galilean satellites revealed bodies every bit as complex as the terrestrial planets and made planetary satellite studies the important discipline it is today. The *Voyagers'* flybys of Saturn in 1980 and 1981 were equally successful, but *Voyager 1* was targeted to pass close to Titan, a trajectory that sent it out of the ecliptic plane after it left Saturn. The trajectory of *Voyager 2* was carefully chosen to preserve the grand tour option, whereby each successive encounter would boost the spacecraft to a higher velocity and in just the right direction to reach the next giant planet, which were fortuitously arranged in the 1980s. That *Voyager 2* would reach Uranus, and then Neptune, was the decided wish of the Voyager scientists and engineers and of the planetary science community as a whole.

There was no guarantee *Voyager 2* would survive the 12-year trip from the Earth to Neptune, many years past its design life. Problems did develop. Radio communication was difficult, with one receiver out and its backup failing, and the articulated scan platform, upon which the remote sensing instruments were mounted, could no longer move as easily as before. Nevertheless, the Uranus encounter in 1986 was also successful, and in August 1988 *Voyager 2* sent back images of Neptune and Triton that were, for the first time, sharper than the best images taken by ground-based telescopes.

Each new *Voyager* encounter increased scientific and public awareness of the richness of the solar system. The *Voyager 2* flyby of Neptune and Triton in late August 1989 was going to be the last and proved to be perhaps the most exciting of all. It was certainly the best covered by the media. In any event, by July 1989 pictures of Triton (unresolved) were being sent back that were critical in adjusting the final trajectory of the spacecraft. Each image (of Triton, Neptune, etc.) was processed in real time through NASA's network of Deep Space tracking stations and then relayed to television monitors dispersed throughout the Jet Propulsion Laboratory in Pasadena (JPL), the nearby Caltech campus, and other locations around the world.

There was one last hurdle. In order to get to Triton, *Voyager 2* would have to pass very close to Neptune's north pole in order for Neptune's gravity to bend its trajectory southward (Fig. 7). This would be dangerously close (only 5000 km from the cloud tops) and in an unknown and potentially dangerous environment. To everyone's relief, *Voyager 2* made it past Neptune without incident just after midnight on August 25 (PDT), counting the more than 4 hr it took for *Voyager's* radio signals to reach Earth. Five hours later it

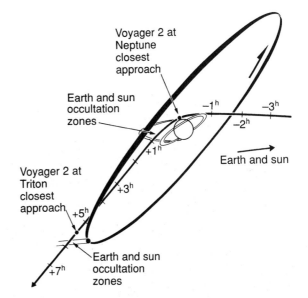

FIGURE 7 The trajectory of *Voyager 2* through the Neptune system. [From C. R. Chapman and D. P. Cruikshank (1995). *In* "Neptune and Triton" (D. P. Cruikshank, ed.). Univ. of Arizona Press, Tucson.]

passed within 40,000 km of Triton, sending back a sequence of mind-boggling images, one every 4 or 5 min, and storing others on its tape recorder for later transmission.

The images themselves deserve some comment. *Voyager's* vidicon (TV) cameras were not designed for the low light levels of Neptune. In order to compensate, exposure times could be lengthened, but then spacecraft motion would lead to smearing. The scan platform could not be counted on to move at just the right rate to compensate or, worse yet, it might seize up. JPL engineers therefore developed a procedure that allowed the entire spacecraft to "nod" at just the right rate to keep the images sharp during the long exposures at close encounters. The beautiful images returned form much of the basis for understanding, to the extent we do, Triton's geology and surface–atmosphere interactions.

V. GENERAL CHARACTERISTICS

Voyager 2 determined Triton to be smaller, brighter, and hence colder than anticipated (Table I). Essentially, its average geometric albedo of ≈0.7 is extreme even for an icy satellite and is exceeded in reflectivity only by Saturn's innermost icy satellites. Triton's global appearance was revealed during the approach

TABLE I
Properties of Triton

Radius, R	1352.5 km
Mass, M	2.140×10^{22} kg
Surface gravity, g	0.78 m sec^{-2}
Mean density, ρ	2065 kg m^{-3}
Percentage rock + metal by mass	65–70%
Distance from Neptune	354.8×10^3 km = 14.33 R_N
Distance from Sun	30.058 AU
Orbit period	5.877 days
Orbit period around Sun	164.8 years
Eccentricity	0.0000(16)
Inclination (present)	156.8°
Geometric albedo (average)	0.70
Bond albedo (average)	0.85
Surface temperature	38 K
Surface composition	N_2, H_2O, CO_2, CO, CH_4 ices
Surface atmospheric pressure	14 μbar
Atmospheric composition	N_2, minor CH_4
Tropopause height	8 km

sequence (Fig. 8). The view, mainly of the southern hemisphere, showed extensive bright polar materials, a bright equatorial fringe with streamers extending to the northeast, and darker low northern latitudes. Latitudes above 40°N were in shadow. The front page of the August 22, 1989, *New York Times* proclaimed: "Profile of Neptune's Main Moon: Small, Bright, Cold, and It's Pink." Radio tracking of the close approach of *Voyager* to the satellite yielded a very precise mass, which, when combined with the size, gave a very precise density of ≈2.065 g cm^{-3}. This density is essentially identical to that of the Pluto–Charon system, considering the uncertainty in the latter.

Once the radius and mass of a body are known, internal structural models can be created based on a set of plausible chemical components (for bodies formed in the outer solar system these would be rock, metal, ices, and carbonaceous matter). Such models provide context and to some extent guide interpretations of geological history. A calculation for Triton is illustrated in Fig. 9. Given that little direct information exists on the internal makeup of Triton, the model shown simply matches Triton's density and assumes the interior is hydrostatic (follows the fluid pressure–depth relation) and differentiated (the major chemical components are separated according to density). These last two assumptions are empirically consistent with

Triton's surface appearance, which indicates a prolonged history of melting and separation of icy phases. In the model, ice, structurally represented by the most abundant solar system ice (H_2O), forms a deep mantle around a rock + metal core. A metallic (Fe, Ni, and probably S) inner core is also shown. The proportions of rock and metal in the core are fixed to solar composition (carbonaceous chondrite) values, as relatively involatile rock and metal should have been completely condensed in the outer solar system. Melting and separation of metal from rock are justified by theoretical arguments for intense tidal heating in Triton's past and by the example of Ganymede, where the *Galileo* orbiter's discovery of a dipole magnetic field demands that such an inner metallic core exists.

Whether Triton is also a magnetized body depends on when its tidal heating ended, but *Voyager 2* passed too far away to tell. Triton is, however, a sufficiently rock-rich body that solid-state convection in its icy mantle should be occurring today, powered by the heat released by the decay of U, Th, and ^{40}K in its rocky core. Its icy mantle should also be warm enough (≳170–175 K) to mobilize lower melting point ices such as ammonia and methanol, which are among the minor ices a body formed in solar orbit might have accreted. If Triton did indeed form in solar orbit, it should have also accreted a large carbonaceous component, upwards of 10% by mass if comets such as Halley are a guide. Too little is known about the properties of this organic fraction to rigorously include it in structural calculations, such as Fig. 9, but it may have played an important role in Triton's physical and chemical evolution.

Voyager 2 confirmed the presence of nitrogen ice on Triton's surface. Specifically, a thin nitrogen atmosphere was detected with a surface pressure and temperature consistent with N_2 gas in vapor pressure equilibrium with N_2 ice (see Section VII). All of Triton's surface appears to be icy; even the darker northern hemisphere in Figs. 1 and 8 (see also color insert) has a geometric albedo of ~0.55. Triton's icy terrains are more forward scattering than other icy satellite surfaces; its phase integral (the ratio of Bond to geometric albedo) is greater than unity and only matched by that of Europa. Triton in fact exhibits no opposition surge (increase in brightness as zero phase is approached, e.g., the full Moon), indicating there is a relative dearth of fine particles in Triton's optical surface. Nitrogen is obviously very volatile, and theoretical models show nitrogen ice grains on Triton's surface can rapidly (over many decades) anneal and densify into a transparent glaze or sheet. It is thought that such a nitrogen glaze covers most of Triton, except perhaps for the

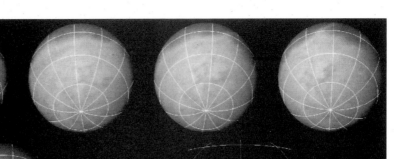

FIGURE 8 Triton approach sequence, overlaid with a latitude–longitude grid. Details on Triton's surface unfold dramatically as the resolution changes from about 60 km/pixel at a distance of 5 million km for the image in the upper left to about 5 km/pixel at a distance of 0.5 million km for the image in the lower right. Mainly looking at the southern hemisphere, Triton rotates retrograde (counterclockwise) over an observation period of 4.3 days. [Courtesy of NASA/Alfred McEwen, University of Arizona.]

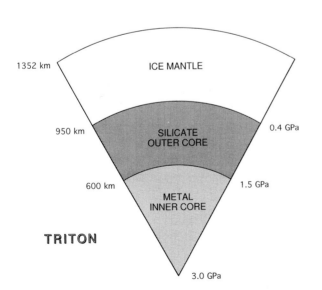

FIGURE 9 Internal structure model for present-day Triton.

bright equatorial fringe, which may be an unannealed frost deposit; the brightness of Triton's other units depends then on what lies underneath the glaze.

Triton's surface appearance also appears to be variable on short time scales, for a major change in the color of Triton occurred between 1977 and the *Voyager 2* flyby (Fig. 10). Triton has become remarkably less red, particularly at shorter wavelengths. Presumably, deposition of fresh nitrogen ice and frost have obscured more reddish surface ice in this interval. On any other moon, this would be a major event. On Triton, with its (presumably) active geology, extreme seasons, and sublimation, transport, and condensation of highly volatile ices in response to both, it seems an almost forgone conclusion that the satellite's global color, if not its overall brightness (and thus its surface temperature and atmospheric pressure), is not constant. Changes over time in Triton's methane spectral absorption signature have also been noted.

As mentioned earlier, the overall redness of the ice

(see color insert) is thought to be due to ultraviolet (UV) and charged particle processing of CH_4 (along with N_2), which can yield darker, redder chromophores: heavier hydrocarbons, nitriles, and other polymers. CH_4 exists as an atmospheric gas as well as a surface ice. *Voyager's* ultraviolet spectrometer solar occultation experiment determined the CH_4 mole fraction at the base of the atmosphere to be ~2 to 6 × 10^{-4}, near or at saturation for 38 K. Dark streaks and patches on the polar cap and elsewhere may be methane rich; if they are depleted of N_2 ice, they should be warmer than the global mean surface temperature, which is buffered by the latent heat of nitrogen condensation/sublimation.

The nature and chemistry of Triton's surface ices are being further refined by the latest ground-based spectroscopy. The spectral resolution of the newer instruments is remarkable (Fig. 11). In 1991 astronomers detected the spectral absorption of CO and CO_2 ice, along with CH_4 and N_2 ice, on Triton. The shapes of the absorption bands are so well determined that the abundances, grain sizes, and degree of mixing of various components can be modeled. It was determined that CH_4 ice is complexed in some way with the far more abundant N_2 ice (probably mixed in a solid solution), with an abundance relative to N_2 of 5 × 10^{-4}.

CO and N_2 form a relatively ideal solid solution, and if this is the case on Triton, the CO/N_2 abundance ratio is 1 × 10^{-3}. Only upper limits on the CO abundance for the atmosphere are available, but they are consistent with CO being a trace molecule in the ice and atmosphere. Most importantly, CO is a tracer of outer solar nebula or cometary chemistry (as discussed in Section III, D), but is not expected to form in giant planet protosatellite nebulae. The detection of CO directly supports a capture origin for Triton.

Because CO_2 is involatile on the surface of Triton, it is probably distributed as a discrete component. If true, approximately 10% of Triton's illuminated surface must be CO_2 ice (or be visible under a transparent N_2 glaze). This remarkable finding implies that dry ice is a major component of Triton's "bedrock." Water ice can also be seen unambiguously in the newest spectra for the first time (Fig. 11) and is apparently an even more important "bedrock" component.

The geology revealed by the *Voyager* encounter is as remarkable as it was unprecedented. The surface is almost wholly endogenic in nature. Intrusive and extrusive volcanism (calderas, flows, diapirs, etc.) dominates the landscape outside the polar terrain, with tectonic structures (mainly ridges) being decidedly subsidiary. Impact cratering is an even more minor

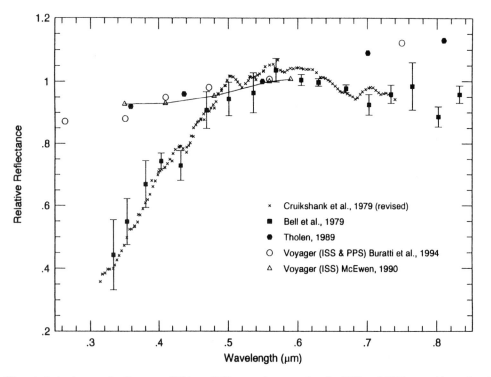

FIGURE 10 Historical visual spectral reflectance of Triton. Differences between data for 1977 and 1989 are evidence for changes on the surface of Triton. *Voyager* ISS and PPS refer to the imaging camera and the photopolarimeter, respectively. [From R. H. Brown *et al.* (1995). *In* "Neptune and Triton" (D. P. Cruikshank, ed.). Univ. of Arizona Press, Tucson.]

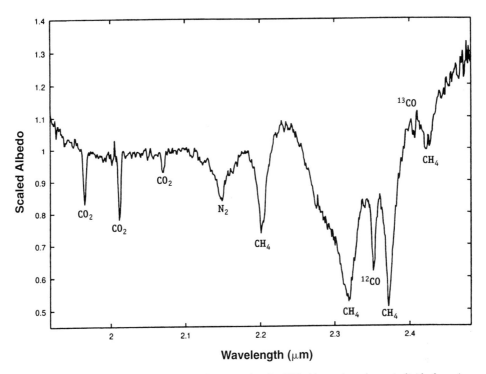

FIGURE 11 Near-infrared telescopic reflectance spectra of Triton taken in 1995. Absorptions due to individual species are indicated; the broad low at 2 μm is interpreted as due to H_2O ice. Data are in scaled reflectance, and the spectral resolution ($\lambda/\Delta\lambda$) is a remarkable 800 (cf. Figs. 4 and 10). [From R. H. Brown and D. P. Cruikshank (1997). *Annu. Rev. Earth Planet Sci.* **25**, 243–277.]

process. Triton's surface is geologically young and has apparently been active up until recent times. Triton's topography can be rugged, but does not exceed a kilometer or so in vertical scale (and usually no more than a few 100 m), due to the inherent mechanical weakness of most of the ices that comprise its surface and the long history of deformation the surface has undergone. Polar ices appear to bury much of this topography and so may in this sense constitute a true polar cap. It is usually assumed that this cap is mostly nitrogen, similar to the surface ice. Details of Triton's geology are pursued in the following section.

Triton's atmosphere is unique as well. It is too thin and cold for radiative processes to play a dominant role. Heat is transported by conduction throughout most of its vertical extent, which is by definition a thermosphere, up to an exobase of ~930 km, where the mean free path of N_2 molecules equals the pressure/density scale height. The thermospheric temperature is a nearly constant 102 ± 3 K above ~300 km altitude and is set by a balance between absorption of solar and magnetospheric energy in a well-developed ionosphere between ~250 and 450 km altitude and both radiation to space by a trace of CO and photochemically produced HCN below ~100 km and downward con-

duction to the cold, 38 K surface. The lowermost atmosphere is characterized by an interhemispheric, seasonal condensation flow. Turbulence near the ground forces the temperature profile to follow a convective, nitrogen-saturated lapse rate of ~ −0.1 K km^{-1} up to an altitude of ~8 km (as determined by observations of clouds, hazes, and plume heights, see Section VII), forming a troposphere (or "weather layer"). Unlike in the atmosphere of the Earth and other planets, there is no intervening radiatively controlled stratosphere between Triton's troposphere and thermosphere.

VI. GEOLOGY

Triton's surface, at least the 40% seen by *Voyager* at resolutions useful for geological analysis, can be roughly separated into three distinct regions or terrains: smooth, walled, and terraced plains, cantaloupe terrain, and bright polar materials. Each terrain is characterized by unique landforms and geological structures. Substantial variations within each terrain do occur, and the boundaries between each are in many locations gradational. But in general the classification of Triton's surface at any point is unambiguous. Certain geological structures are common to nearly all terrains, specifically, the tectonic ridges and fissures, and impact craters, naturally, can form anywhere.

Although Triton's surface is composed almost entirely of ices, many of the individual geological structures can be readily interpreted as variations of structures terrestrial planet geologists would find familiar, such as volcanic vents, lava flows, and fissures. The volcanic features in particular have inspired a designation "cryovolcanic" in order to distinguish them from those formed by traditional silicate magmatic processes. The physics and physical chemistry are fundamentally the same, however, whether one deals with silicate or icy volcanism. In addition, there are geological structures and features on Triton that are unusual and *not* readily interpretable in terms of terrestrial analogues. Some defy explanation altogether. Triton's terrains are explored one by one in the following sections, followed by a summary of its possible geological history.

A. UNDULATING, HIGH PLAINS

Plains units are found on Triton's eastern or leading hemisphere (referring to the sense of orbital motion, to the right in Fig. 1; see also color insert) and to the north of the polar terrain boundary (which shifts southward with increasing eastern longitude). Figure 12 shows a regional close-up of various plains near the terminator in the center of Fig. 1. To the bottom and right of the image are flat to undulating smooth plains centered around circular depressions or linear arrangements of rimless pits. These plains are relatively high-standing and bury preexisting topography, with edges that may be well defined or diffuse. There is little doubt that these high plains are the result of icy volcanism and that the various pits and circular structures are the vents from which this material emanated. In general, eruptions along deep-seated fissures or rifts often manifest as a series of vents, and the irregular, ~85-km-wide circular depression toward the lower left resembles a terrestrial volcanic caldera complex. This feature, Leviathan Patera (all the features on Triton have been given names drawn from the world's aquatic mythologies), sits at the vertex of two linear eruption trends. Toward the terminator (northeast), one of these trends is anchored by another caldera-like depression of similar scale.

Volcanic activity on the Earth often occurs in cycles, whereby magma formed by partial melting in the mantle rises due to buoyancy, accumulates at intermediate, crustal levels to form a magma chamber, and subsequently erupts; things are then quiescent until the magma is replenished and the cycle begins anew. The loss of magma volume often leads to collapse of the vent region over the magma chamber, forming a caldera. Cycles of eruption and collapse can create some complex forms, but calderas are generally composed of quasi-circular elements. The two paterae (from the Latin for saucer) in Fig. 12 are clearly of the caldera type in which renewed volcanism has occurred, as both are partially buried by younger icy lavas. Leviathan Patera in particular has a string of small rimless pits on its floor. Such pits can be formed by simple drainage, but they can also be formed by the venting of volcanic gases. An elongate (~15 × 25 km) depression at the eastern end of the other linear vent chain (Fig. 12, lower right) appears especially deep and is surrounded by a distinctive darker aureole. The pit depth and dark deposit are evidence for the venting of gas and entrained particles, which on Triton could be called a cryoclastic eruption (a variation on the more familiar pyroclastic).

On Earth the primary volcanic gases are water (steam) and CO_2 and often occur in sufficient quantities to make eruptions explosive if not catastrophic. Famous calderas formed in steam-driven cataclysms include Mono Lake in California and Krakatau. Triton's surface morphology does not indicate whether similarly violent volcanic events occurred there, but a role for volatile gases is probable. Volatile on Triton means something more volatile than the icy lavas themselves: nitrogen and methane are logical candidates as they are known to exist on Triton.

Compositions of the icy lavas are, strictly speaking, unknown. *Voyager 2* carried no remote-sensing instruments designed to determine compositions. The icy plain-forming lavas in Fig. 12 were clearly viscous enough to form thick enough deposits to bury preexisting topography. As this topography itself varies over a few hundred meters in elevation, the lavas must be at least this thick. The viscosities implied are certainly much greater than that of water or brine. Other chemical components must be involved, and these must be present in Triton's mantle, as the sizes of the paterae are measures of magma chamber depth (~70–100 km compared with a total mantle thickness of ~400 km; Fig. 9).

The favored composition for viscous lavas on the icy satellites has long been ammonia–water. As outlined by pioneering planetary chemist J.S. Lewis, ammonia (NH_3) is the chemically stable form of nitrogen in a low-temperature gas of solar composition and, when condensed, forms various hydrates with water ice, all of which have low melting points. Triton would not

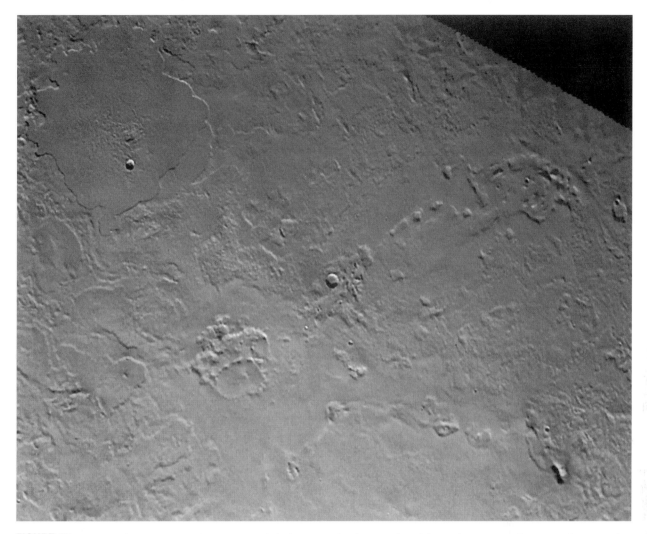

FIGURE 12 Young volcanic region on Triton. Toward the bottom and right, smooth undulating flows apparently emanate from complex caldera-like depressions and linear alignments of volcanic pits and vents, burying preexisting topography. At the upper left terraced plains surround an exceptionally level plain, Ruach Planitia. This region, 675 km across, is very sparsely cratered. (Courtesy of NASA/Paul Schenk, Lunar and Planetary Institute.)

have accreted much ammonia if it formed in solar orbit because N_2 would have been the dominant original form of nitrogen in the outer solar nebular for the same reasons CO and organic material were favored over CH_4 (see Section III, D), but it still would have acquired some NH_3 based on cometary compositions (up to a percent or two compared with water). A water-rich $NH_3–H_2O$ mixture (0 to 33 mole% NH_3) would be composed of frozen H_2O and ammonia dihydrate ($NH_3 \cdot 2H_2O$), which yields a lowest melting point (or eutectic) liquid at ~177 K at pressures typical of Triton's mantle. This melt (or cryolava) is ammonia rich (about 32%) and has a viscosity similar to some types of basaltic magma.

Comets also contain a host of other exotic, presumably interstellar, ices, some of which may have been important in Triton's geological history. For example, methanol (CH_3OH) pushes the minimum melting temperature of ammonia–water ice down to ~152 K, and the resulting lava is even more viscous, equivalent to certain types of silicic lavas on Earth. The range of viscosities available to liquids in the $H_2O–NH_3–CH_3OH$ system is certainly compatible with the appearance of the undulating smooth plains seen in Fig. 12. [*See* PHYSICS AND CHEMISTRY OF COMETS.]

The abundances of original ices may also have been altered, and new ices created altogether, during Triton's tidal heating epoch (see Section VIII). For example, *copious* NH_3 and CO_2 may have been produced chemically within Triton, provided there was a sufficient supply of nitrogen and carbon. Despite these exciting possibilities, neither NH$_3$ nor any complex, exotic ices have been discovered by ground-based spectroscopy (Section V).

B. WALLED AND TERRACED PLAINS

In the northwest corner of Fig. 12 lies an ~175-km-wide, remarkably flat plain, Ruach Planitia, that is bounded on all sides by a rougher plains unit that rises in one or more topographic steps (scarps) from the plain floor. It is one of four so-called walled plains identified on Triton; these are generally quasi-circular in outline, with typical relief across the bounding steps or scarps of ~150 to 200 m. Ruach Planitia and the other walled plains are the flattest places seen on Triton, which implies infill by a *very* fluid lava or other liquid. Clusters of irregular, coalesced pits toward the centers of these plains have been likened to eruptive vents or drainage pits.

The planitia themselves have been likened to calderas, but they are generally much larger than the nearby paterae and do not resemble them structurally. Specifically, there is no evidence for collapse at the periphery of any of the walled plains. Rather, the outline of the inward-facing scarps is indented and crenulate, with islands of the bounding plains occurring in the interior. If anything, the outlines of walled plains resemble eroded shorelines. How erosion occurred and under what environmental conditions on Triton is unclear. If the fluid that filled the planitia was responsible for the erosion, it does not explain the similar outline of the plains that overlap the eastern edge of Ruach Planitia (Fig. 12), which gives this area a terraced appearance and indicates that the rougher plains were laid down in layers. A distinct possibility is that the layers are composed, at least in part, of a more friable or volatile material, and that over time (or with higher heat flows) the layers disintegrated and the scarps formed by retreat. Similar processes of mass wasting, removal, and scarp retreat are believed responsible for the so-named etched plains of the martian south polar highlands and those on Io.

C. SMOOTH PLAINS AND ZONED MACULAE

Other plains units can be seen in Fig. 13, as well as the transition to the bright polar materials. At the top left is a hummocky terrain, composed of a maze of depressions and bulbous mounds. Stratigraphically, it is older than the volcanic plains to the north that over-

lap it, and appears older (more degraded) as well. The hummocky terrain gives way to a much smoother plains unit to the south. At the available resolution it is unclear whether this smoothness is due to volcanic flooding, volcanic or atmospheric condensate mantling, or some other form of degradation. These hummocky and smoother units are the most heavily cratered regions on Triton, equivalent in crater density (but not necessarily in age) to the lunar maria.

Among Triton's most perplexing geological features are the large zoned maculae (spots) close to the eastern limb in Fig. 13. Each such macula consists of a smooth, relatively dark patch or patches surrounded by a brighter annulus or aureole. The width of any given annulus tends to be relatively constant (20 to 30 km for the three major maculae in Fig. 13). The maculae betray almost no topographic expression, and so must vary in height across their extents by no more than a few tens of meters. The darkness and redness (see color insert) of the central patches imply the presence of carbonaceous material, which probably means some methane ice is present. The brightness of the annuli is similar to that of the bright terrain, so the simplest hypothesis is that they consist of similar ices (predominantly N$_2$).

The maculae could be considered a unique set of features on Triton, and perhaps irrelevant to the big picture there. The extreme eastern limb in Fig. 13 is composed of a mosaic of maculae, however, and much of the bright terrain in the rest of the image contains similar, although generally less distinct, features (see also Fig. 1). From this perspective the maculae can be considered outliers of the southern polar cap, which should be retreating at the season observed (late southern spring). Furthermore, another walled plain can be seen along the middle left edge of the frame. Its eastern rim is incomplete and breaks down into a region of small mesas. If this planitia were filled with bright ice, it would passably resemble, in plan and in albedo, the bright terrains to the south, especially those near the boundary with the smoother plans. The resemblance would be further improved if the planitia are bowed upwards, for which there is independent topographic evidence (see Fig. 12). It may be that the maculae are in fact planitia underneath and that the mysterious erosive process that has cut back the planitia scarps has operated more extensively on Triton.

D. CANTALOUPE TERRAIN, RIDGES, AND FISSURES

The entire western half of Triton's nonpolar surface in Fig. 1 is termed cantaloupe terrain, as it appears

FIGURE 13   Southeastern limb of Triton showing (from bottom) hummocky terrain, smooth terrain, and bright polar terrain. A prominent bulbous ridge zigzags across the top, and distinct bright-surrounding-dark albedo features of uncertain origin, termed maculae (spots), are seen at the right and, more faintly, along the limb and in the bright terrain. The largest crater on Triton, the 27-km-diameter, central-peaked Mozamba, is to the right of the largest prominent macula, Zin. (Courtesy of NASA/Paul Schenk, Lunar and Planetary Institute.)

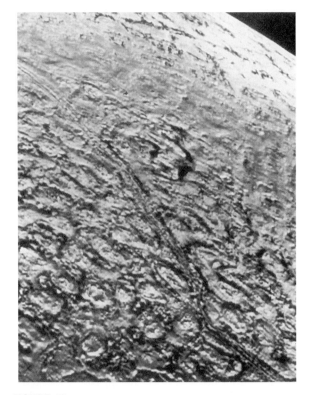

FIGURE 14 Cantaloupe terrain at the bottom and polar terrain at the top in this high-resolution *Voyager* image taken from a distance of 40,000 km. Each cantaloupe "dimple" is about 25–35 km across. A tectonic ridge and fissure set runs through the cantaloupe terrain, probably formed by the extension of Triton's icy crust. Toward the south (upper right), smooth materials, and beyond them, brighter ice, appear to mostly bury cantaloupe and fissure topography. (Courtesy of NASA/JPL.)

covered by large dimples and criss-crossed by prominent quasi-linear ridges (see color insert.). Much of the terrain displays a well-ordered structural pattern: at high resolution the dimples become a network of interfering, closely spaced, elliptical and kidney-shaped depressions, termed cavi (Fig. 14). Unlike impact craters, the cavi are of roughly uniform size, ~25 to 35 km in diameter, and do not overlap or crosscut. They are clearly endogenic, or internal, in origin, but the leading explanation is diapirism (not volcanism).

Diapirism is triggered by a gravitational instability involving a less dense material rising through overlying denser material. The required buoyancy may be thermal or compositional. Probably the best known terrestrial examples of diapirs are salt domes, in which a layer of salt rises as a series of individual blobs, or diapirs, through overlying dense sedimentary strata. In one region of extreme dryness, the Great Kavir in central Iran, the salt diapirs breach the surface, rotating and pushing the overlying strata to the side. The shapes, close spacing, and interference relations of the diapirs of the Great Kavir in fact bear a significant resemblance to the cavi. That the cavi are depressed (although many contain central mounds) has been interpreted as being due to the maturity of the diapiric spreading.

The implications of a diapiric origin for cantaloupe terrain are that Triton possesses distinct crustal layering and, based on the spacing of the cavi, that the overlying denser layer or layers is ~20 km thick. This crustal layer could simply be a weaker ice (possibly ammonia rich) that responded to heating from below or it may be ice truly denser than the ammonia–water ices presumably below (e.g., CO_2 ice).

Numerous small pits occur throughout the cantaloupe terrain and are found on the materials between cavi, which are often organized into subdued, short, sinuous ridges (Fig. 14). These make impact crater identification nearly impossible at the available resolutions.

Triton's surface is crosscut by a system of ridges and fissures, which are best expressed in the cantaloupe terrain (Fig. 1; see also color insert). The ridges occur in a variety of forms: pairs of low, parallel ridges bounding a central trough (essentially a trough with raised rims), ~6–8 km across crest to crest; similar but wider ridge-bounded troughs with one or more medial ridges (one, Slidr Sulcus, can be seen in Fig. 14); and single, broad, bulbous ridges (e.g., Fig. 13). The fissures, which are less numerous, appear to be simple, long, narrow valleys only 2–3 km wide. All of these fundamentally tectonic features appear to result from the extension of Triton's surface. The medial ridges may be due to subsequent dike-like intrusions of icy material. The bulbous appearance of the ridge in Fig. 13 may be due to overflow of such injected ice, which may also be a source for much of the adjacent smooth plains. Topographic profiles across the ridges show typical relief of a few hundred meters, with flanking troughs. Ridges on Triton bear more than a passing resemblance to so-called triple bands on Europa, seen as ice ridges by the *Galileo* orbiter (a matter that warrants further study). [*See* OUTER PLANET ICY SATELLITES.]

E. BRIGHT POLAR TERRAINS

Most of Triton seen by *Voyager* is actually bright terrain of one type or another, but the imagery is generally not of sufficient quality for geological analysis. Interpretations are further confused by the numerous dark streaks, plumes, and clouds. Nevertheless, the bright terrains represent substantial, not superficial, deposits. The view in Fig. 14 looks across the edge of the cantaloupe terrain, into a band of subdued or mantled cantaloupe-like topography, and then into brighter materials beyond. Cantaloupe-like topographic elements and sections of a linear ridge appear engulfed by bright ice, probably up to a few hundred meters in thickness. The important questions are whether the bright ice thickness increases into the interior of the bright materials in the distance and does it become sufficiently deep to qualify as a true polar cap.

Low-resolution imagery shows that quasi-circular elements can be made out at many locations well within the bright materials. Ridges also cross into the bright terrains, and one bright lineament is seen close to the south pole. The implication is that much of the polar topography is incompletely buried. However, there are extensive bright, featureless regions as well (up to several 100 km across), which indicate either complete burial at these locations or obscuration by clouds. Overall thickness of the bright polar ice is therefore probably less than 1 km, but even if not organized as a uniform ice cap or sheet, a thick deposit of a volatile ice such as N_2 could be warm and deformable enough at its base to flow laterally. Although not literally a polar cap, much of the bright polar terrains may behave as if glaciated.

F. GEOLOGICAL HISTORY

It is notable that the volcanic province in Fig. 12 is one of two similar ones, with the second occurring to the southeast and together stretching across 1000 km of Triton's surface. The alignments of volcanic vents in both provinces suggest extension and rifting of Triton's relatively strong icy outer shell, or lithosphere. The volcanic plains in Fig. 12 are also very sparsely cratered (the largest crater visible is 16 km across), much less cratered than, say, the lunar maria. Estimates of the rate at which comets bombard Triton made by the late planetary geologist Eugene M. Shoemaker are necessarily uncertain, but suggest that these provinces are no more than a billion years old and possibly much less. A broad region of Triton's sublithospheric mantle was thus hot and partially molten late in solar system history.

The high volcanic plains postdate most of the other terrains on Triton. They stratigraphically overly the terraced plains to the west and the hummocky plains to the east. The terraced plains grade into and appear to superpose the cantaloupe terrain. The relative age of the cantaloupe terrain cannot be determined by traditional crater counting methods because (as noted earlier) no reliable crater counts can yet be made there, but cantaloupe terrain nevertheless appears to be the stratigraphically oldest unit on Triton. The linear ridges clearly postdate the cantaloupe terrain, yet some ridges fade into the terraced plains to the east and another is discontinuous as it crosses the hummocky and smooth plains near the equator to the east (Figs. 1 and 13); no ridges cut the high volcanic plains.

The eastern hummocky and smooth plains comprise the most heavily cratered region on Triton, and when due account is taken of the concentration of cometary impacts on Triton's leading hemisphere, they appear to be about twice as old as the high volcanic plains to the north and northwest. The cantaloupe terrain, then, must be even older. The hummocky terrain may be a degraded version of cantaloupe terrain. Indeed, cantaloupe terrain has been suggested to underlie much of Triton's surface (e.g., cantaloupe-like topography extends well south into the bright region of the trailing hemisphere).

The youngest surfaces on Triton, naturally, involve the mobile materials of the bright terrains. These probably include the zoned maculae of the eastern hemisphere. The geological substrate on which the bright materials reside may of course be older. The walled plains themselves are locally the youngest stratigraphic units. Ruach Planitia and a larger planitia immediately to the west are less cratered than the high volcanic plains, albeit with a large statistical uncertainty. The filling of these walled plains may thus represent the most recent volcanic activity on the hemisphere of Triton seen by *Voyager*.

VII. ATMOSPHERE AND SURFACE

A. ATMOSPHERE

Triton is one of only seven solid bodies in the solar system with an appreciable atmosphere, and one of only four in which the major component of the atmosphere also condenses onto the surface. (Of the others, Io's tenuous SO_2 atmosphere is least well known; the CO_2 atmosphere of Mars has been studied extensively and differs in that most of the material is in the atmosphere rather than on the surface; and Pluto is probably similar to Triton, although surprises undoubtedly await future missions to that body.) Triton's atmosphere is composed primarily of nitrogen. The complicated oscillation of the subsolar latitude with time drives an exchange of N_2 and trace species between the atmosphere and surface frost deposits in the two hemispheres that is equally complicated and as yet not fully understood. Internal heating (which is comparatively important because of Triton's extreme distance from the Sun and large proportion of rocky materials containing radioactive elements) and even glacier-like creep of solid nitrogen caps may also play important roles in the interaction of atmosphere and surface. [*See* Io; Mars: Atmosphere and Volatile History.]

As described in Section III, spectroscopic evidence prior to the *Voyager 2* encounter indicated that nitrogen existed on Triton in condensed form. *Voyager* showed Triton to be much smaller, brighter, and colder than had been guessed. Surface temperatures could be inferred from the visible reflectivity as well as measured directly by the infrared interferometer spectrometer (IRIS). Occultations (passage of the spacecraft or a star behind Triton) observed by the ultraviolet spectrometer (UVS) and radio science subsystem (RSS) probed different parts of the atmosphere, revealing its temperature and density, from which pressure and composition could be deduced. These investigations revealed a consistent picture of a surface and lowermost atmosphere at about 38 K. The pressure at the surface is only 14 microbars, indicating that the gas is in equilibrium with solid nitrogen at the same temperature. The thermal structure of the lower atmosphere is not well constrained, but the temperature probably reaches a minimum at about 8 km height, above which it increases to about 100 K in the upper atmosphere because of heat deposited from space and conducted downwards. In meteorological parlance, Triton's thermosphere directly overlays its troposphere.

The *Voyager* images and occultation data revealed a variety of condensates in the lower atmosphere. Most of the atmosphere contains a diffuse haze that can be seen against the background of space at Triton's limbs, and which probably consists of hydrocarbons produced by the action of sunlight on trace gases such as methane. Discrete clouds were also seen at the limbs (Fig. 15) and against the unlit part of the satellite beyond the terminator, where they formed east–west trending "crescent streaks" roughly 10 km wide, a few hundred kilometers long, and 1 to 3 km above the surface. One crescent streak may have been as much as 400 km long and 5 km high and appeared to move eastward at a speed of 13 m sec^{-1}. At the limbs, clouds could be distinguished from haze by being optically thicker and localized both in height (10 km or less) and in horizontal extent (patchy and mainly concentrated at mid

FIGURE 15 Limb cloud over Triton's south polar cap near $-30°$ S. The brightness has been increased to better show the cloud. Mainly situated between ~3 and 6 km altitude, this cloud is interpreted as being caused by nitrogen condensation. (Courtesy of NASA/JPL.)

to high southern latitudes, where they cover a third of the limb). The sharper upper boundary to the clouds suggests that they consist of condensed nitrogen rather than involatile solids like the haze.

The crescent streaks provide clues to atmospheric motion by their east–west orientation and the apparent eastward motion of the largest, highest cloud. Further clues come from markings on the surface. Over 100 dark "streaks" were seen in the southern hemisphere, mainly between latitudes of 15° and 45° S. The streaks range from 4 to over 100 km in length and many are fan-shaped. The vast majority extend to the northeast from their narrow end (presumably the origin point); a smaller number are directed westward. These streaks are extremely similar to "wind tails" that are common on Mars and are seen on the Earth and Venus as well. On these other bodies, wind tails are created by deposition (or sometimes erosion) of loose material by localized eddies downwind of topographic features. It was initially difficult to understand how wind tails could form on Triton, however, because the atmosphere is so thin that even the slightest tendency for dust grains to stick to one another would prevent their being lifted by the wind.

The interpretation of the surface streaks as wind created was nevertheless strengthened by the discovery, shortly after closest encounter, that some of the streak-like features were actually atmospheric phenomena. Stereoscopic viewing of images obtained from varying angles as *Voyager 2* passed by Triton (Fig. 16) revealed that, although the majority of the streaks were on the surface (or at least too low to measure their altitude, less than 1 km), at least two had an altitude of roughly 8 km. These features were subsequently named Mahilani Plume (48°S 2°E, with a very narrow, straight cloud 90–150 km long) and Hili Plume (57°S 28°E, actually a cluster of several plumes with broadly tapering clouds up to 100 km long). Thus, it is clear that winds on Triton *do* transport suspended material, but the question is *how* the material becomes suspended.

The plumes were entirely unexpected, and explaining their vigorous activity became a major focus of research, as described later. What is clearest is that they complete a coherent picture of winds on Triton at the time of the encounter. Unlike most surface streaks, both plume clouds extend westward from their apparent sources (the plumes proper—narrow, possibly unresolved vertical columns linking the horizontal plume clouds with the surface). Images of Mahilani appear to show kilometer-sized "clumps" within the cloud moving westward at 10–20 m sec^{-1} and elongation of the cloud from 90 to 150 km at a similar speed. Thus, putting all these descriptions together (crescent streak clouds, dark surface streaks, and plume tails), the wind is northeast nearest the surface, eastward at intermediate altitudes, and westward at 8 km, the top of the troposphere.

This is precisely the circulation pattern predicted at the time of encounter, the height of summer in the southern hemisphere. Heating by sunlight is presently causing solid nitrogen in the south to sublimate (evaporate); meanwhile in the colder north, the atmosphere is precipitating. Because of the rotation of Triton once every 5.877 days, however, the wind does not blow directly from south to north to make up the difference. Instead, gas is transported northward only in a thin skin of atmosphere near the surface (the Ekman layer) in which the flow is northeastward. The atmosphere above the 1-km-thick Ekman layer circulates from west to east. The westward flow at the altitude of the plumes can be explained if Triton's atmosphere is slightly warmer over the equator than at the south pole (perhaps because the equator is darker), in which case the temperature gradient will drive a thermal wind that causes the eastward flow to weaken and eventually change to westward flow with increasing altitude.

Basic properties of the plumes can be inferred from the images. The plume clouds do not settle out visibly (no more than the ~1-km resolution of the best images) over their length, so the suspended particles must be smaller than about 5 μm. From this particle size and the width and contrast of the clouds—about 5% darker when seen against Triton—one can further infer the amount of solids: about 10 kg sec^{-1} must be discharged if the material is dark or twice as much if it is bright. (Bright material in a cloud would appear relatively dark against Triton's very bright surface, although not as dark as intrinsically dark material. However, bright particles deposited from such a cloud would not show up as a dark streak on the surface.) The cloud moves horizontally at the wind speed, 10–20 m sec^{-1}, but the vertical velocity in the plume must be significantly faster because the plumes are not blown visibly askew by the wind. The columns may be just barely resolved in the best images, which are unfortunately missing alternate columns of image pixels because of data compression on the spacecraft. Thus the plumes may be 2 km across or perhaps smaller. The source area must have similar (or smaller) dimensions. Little or no structure is visible in the columns, although a "sheath" of descending material around the plume has been described by some authors. The active lifetime of the plumes can be estimated at a few Earth years: shorter, and *Voyager* would have been unlikely to see any plumes active; longer, and active plumes should have been more numerous compared with surface streaks.

B. PLUME MODELS

Numerous attempts have been made to model the plumes in order to answer the questions of where the particulates, the gas suspending them, and the energy to drive the gas flow originate. Most models have taken their cue from the presence of the active plumes (and surface streaks) at mid to high southern latitudes at a season when the sun was almost directly overhead (Fig. 17), and assumed that the plumes are somehow solar powered. It is also possible, however, that Triton's internal heat drives the plumes and that their location is determined not by the Sun but by a local enhancement of this heat source (i.e., by cryovolcanic activity) or by the thickness of the nitrogen "cap," the equivalent area of the northern hemisphere being hidden in darkness during the encounter.

It is conceivable that the plumes are purely an atmospheric phenomenon. One early suggestion was that the plumes are dust devils, localized regions of spinning and ascending hot atmosphere formed above patches on the surface that are bare of N_2 frost and that can therefore be heated by the sun to higher temperatures than their frosty surroundings. This model was inspired by early analyses of *Voyager* occultation data that seemed to show that the atmosphere was much warmer than the surface frosts. Later analyses disputed this conclusion, weakening the dust devil hypothesis. Despite their name, Tritonian dust devils would also have difficulty picking up dust from the surface and

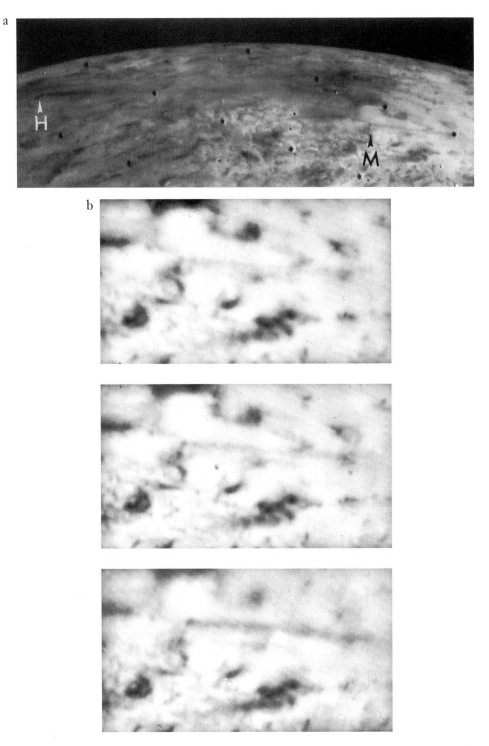

FIGURE 16   (a) *Voyager 2* image of the southern polar region of Triton in which geyser-like eruptions were discovered. Here the plumes are viewed obliquely with Hili (H) and Mahilani (M) plumes marked. (b) Highly magnified images of Mahilani plume on Triton taken from increasingly oblique angles and at increasing resolution (top to bottom). The images have been projected onto a spherical surface with a viewing geometry similar to that at the top. The increasing parallax from top to bottom makes the plume "stem" appear to grow taller. (Courtesy of NASA/Alfred McEwen, University of Arizona.)

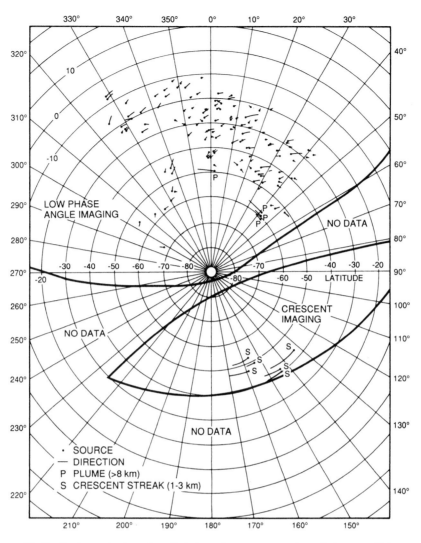

FIGURE 17 The geographic distribution and orientation of wind streaks, crescent streaks, and plumes on Triton as seen by *Voyager*. The latitude and longitude of each feature source are plotted as dots; tails indicate streak or plume length and direction. [From C. J. Hansen *et al.* (1990). *Science* **250**, 421–424.]

becoming visible because their winds are not strong enough. However, it has been suggested that if the hot areas on the ground were not only nitrogen free but contained methane frost, they would give off clouds of methane gas. Being lighter than nitrogen, this methane would ascend. Although little dust would be picked up, part of the methane might recondense in the atmosphere, making the rising plume visible. Falling back onto the ground, the methane frost would initially be pure, transparent, and indistinguishable from the nitrogen underneath, but over time it would darken from exposure to radiation, explaining the surface wind streaks. Although this model ingeniously solves the problem of how such a gently rising plume picks up enough solids to become visible, there is also the objec-

tion that the plume would be blown sideways. An additional objection is that deposits of methane ice are unlikely to form on Triton because solid methane tends to be incorporated into the crystal structure of the much more abundant nitrogen frost. A final variation on these types of plume model suggests that nitrogen rather than methane is ascending and condensing. Of the same composition as the rest of the atmosphere, the plume in this model would be buoyant only because it is warm. Condensation during its ascent could release enough heat to accelerate the plume substantially, but the nitrogen must somehow start off fast enough to pick up dust and to avoid being blown sideways by winds near the base of the plume.

How could a plume of nitrogen gas get started?

One possibility is that they are geysers. Like geysers on Earth, which consist of water and water vapor, those on Triton would be eruptions of volatile material that has been heated underground. Whereas the water in terrestrial geysers starts as a liquid and partially boils, however, Tritonian geysers would start as hot gas that would partially condense as it expanded to the ambient pressure. This expansion could drive a gas flow powerful enough to pick up dust and form the observed plumes. Solar-powered nitrogen geysers have been studied in some detail. The pieces of the model are as follows.

1. Plumes as Jets

The energy needed to drive the plumes is determined by how much gas is involved and how fast it has to be erupted. The worst case assumption is that the nitrogen does not condense as it rises. Instead of becoming buoyant and accelerating, it is denser than its surroundings because of any inert dust entrained in it and the small amount of N_2 (several percent by mass) that crystallizes immediately upon eruption. The plume is therefore slowed both by gravity and by interaction with the atmosphere around it. How high it will rise depends on both the size of the eruption and its speed, and can be calculated based on laboratory simulations. As an example, a jet with a diameter of 20 m, a velocity of 230 m sec^{-1}, and 5% solids by mass will reach the observed altitude of 8 km on Triton. The plumes might be this small, but they could be as big as 1–2 km in diameter, in which case they could be somewhat slower. As discussed, plumes erupting more slowly could also reach 8 km if condensation continues after eruption. There is little problem for the model of the plume ascending too vigorously because it will be stopped at about 8 km by the increasing atmospheric temperature (buoyancy) above this altitude.

2. Eruption Velocity and Temperature

Both the initial velocity of the gas and the amount of solid nitrogen that will condense can be calculated from the initial and final temperatures and the thermodynamic properties of nitrogen. The example just given (5% solids, 230 m sec^{-1}) is attained for nitrogen expanding freely (no change in entropy) and cooling from 42 to 38 K. Thus, the subsurface gas must be heated about 4 K to power the geyser to the right altitude. We also learn from this calculation that the 10- to 20-kg sec^{-1} of solids estimated to be feeding

the plumes is accompanied by as much as 400 kg sec^{-1} of gas. Given the latent heat of sublimation of nitrogen, about 100 MW of power is needed to convert solid to gas at this rate.

3. Temperature of a Solid-State Greenhouse

The "greenhouse effect" usually describes heating of the Earth's atmosphere (or that of another planet) when sunlight at visible wavelengths penetrates the atmosphere before being absorbed, but longer wavelength thermal radiation is absorbed by the atmosphere and cannot escape to space as easily. A similar effect can take place in a transparent solid, e.g., nitrogen ice on Triton. The amount of solar energy is not great at Triton's distance from the Sun, but nitrogen is an excellent thermal insulator and the deeper the sunlight is absorbed the warmer the subsurface will get. A 6-m-thick layer of clear nitrogen ice over a dark subsurface layer would actually melt at the base, whereas even a 4-m layer would blow itself apart because the hot ice would produce gas at a pressure higher than the weight of the solid above. (This cannot be how plumes originate, however, because the production of gas would cease very quickly as chunks of the ruptured layer cooled.) Heating by 4 K can be achieved with a greenhouse layer only 1–2 m thick.

4. Subsurface Energy Transport

What happens after sunlight is absorbed below Triton's surface and before hot gas is erupted? As just estimated, 100 MW are needed to heat the gas in a typical plume. This is the amount of power deposited by sunlight on a region of Triton about 10 km in diameter, much bigger than the 1- to 2-km size of the plume sources. We can therefore conclude that gas (or energy to produce gas by sublimation) is stored over time and then released quickly, is transported horizontally from the larger area to the geyser, or both. Somewhat counterintuitively, gas is not mainly "stored" in voids in the nitrogen ice, but is produced on demand from hot ice, whereas heat transport is mainly carried by flowing gas rather than ordinary thermal conduction. Nitrogen ice can give off more than 100,000 times its own volume of gas as it cools just 4 K. If there are voids in the solid nitrogen, this gas will flow to colder areas and recondense, warming them by releasing its latent heat. Depending on the size of such void spaces, the gas flow can transport energy hundreds of times more efficiently than conduction. Not only could flow between meter-sized blocks of solid readily supply a geyser, but when a path to the surface was first opened, eruption would be vigorous at first and decline over a period of about a year, roughly the estimated lifetime of the plumes. Energy transport by the production of gas, its flow through pores, and recondensation at colder points are known on Earth: "heat pipes" containing a condensible gas (with a wick to return the liquid to the hot end) conduct heat better than metal and are used for baking potatoes from the inside out and for controlling the temperature of spacecraft, including *Voyager*! How a suitably fractured layer of nitrogen ice, overlain by a clear, gas-tight greenhouse layer, might form on Triton is discussed in the next section.

The idea of solar-powered geysers thus seems extremely promising, although much work remains to take the separate pieces that have been modeled so far and make sure that they fit together. Internally powered geysers (more similar to their terrestrial counterparts) have not been studied nearly as thoroughly, but several possibilities exist. As discussed later, the nitrogen "polar caps" on Triton may be so thick near their center (over a kilometer) that they begin to melt at the base. Liquid N_2 finding its way to the surface could erupt as a boiling geyser, with more than enough energy to power the plumes. Gases other than nitrogen could also be erupted from deeper in Triton's water–ice mantle, driven by internal heating. If ammonia is present along with water ice, some melting of the mixture at temperatures of 176 K or above is possible. Dissolved gases might drive a geyser-like eruption of this liquid, but the amount of such gases is not enough to explain the plumes unless most of the liquid can be left behind. Thus, this type of eruption may be a better explanation for possible cryovolcanic landforms in Triton's northern hemisphere than for plumes in the south.

C. POLAR CAP AND CLIMATE

We turn now from the plumes to a consideration of how Triton's surface frosts and atmosphere change over time. Here, too, the *Voyager* images yielded a surprise: at the height of southern hemisphere summer (Fig. 6), most of the southern hemisphere was covered with a bright deposit (a polar cap), but the visible portion of the northern, winter hemisphere was darker. Models of the redistribution of N_2 frost with the seasons can be constructed with varying degrees of complexity, but a fundamental expectation is that the summer hemisphere should have less of a polar cap than the winter one!

The basic physics of seasonal frost-distribution models is as follows.

1. The whole atmosphere and all frosted areas are at very nearly the same temperature. If a frosted area were colder, more nitrogen would condense there and release of latent heat would raise the temperature. Conversely, a warm frost area would be cooled by sublimation. Winds would quickly even out the atmospheric pressure and temperature.

2. At this fixed temperature, sublimation occurs where frosts are exposed to the sun and condensation where the average input of solar energy is less. Sublimation/condensation rates can be calculated from the amount of sunlight absorbed at each point on Triton.

3. Bare (unfrosted) areas can be warmer than the atmosphere and frosts (if they are dark and/or well exposed to the sun) but they cannot be colder, or frost would immediately condense on them.

Using the albedo of the surface as measured by *Voyager,* models indicate that frost in most of the southern hemisphere is currently subliming, thinning the surface deposits. Nitrogen is presumably being deposited in the northern hemisphere and in a few of the brightest areas of the south where little sunlight is absorbed, but what about the long run? By assuming that frost has some given albedo and that the surface underneath has some other albedo, one can model the redistribution of nitrogen over long periods. A layer of nitrogen frost about a meter thick is moved back and forth as the sun shines on one hemisphere and the other, and the pressure and temperature of the atmosphere change as well. Most notably, such models predict that all nitrogen deposited in the southern hemisphere the last time it was winter there would have resublimated before *Voyager* arrived. Correspondingly, the northern hemisphere should be extensively frosted.

How can these predictions be reconciled with observation? If the frost was actually darker than the surface, then the *Voyager* images would indicate frost in the north but not the south, but there is no good reason to expect nitrogen frost to be dark. Frost might not be deposited in the north until after the *Voyager* encounter if the surface there retains heat exceptionally well. Alternatively, the frost might deposit mainly in shadows and on north-facing slopes where *Voyager* could not see it, or it could be glassy and transparent, hence invisible. There is some evidence for the last possibility from laboratory observations of condensing nitrogen, calculations of the rate at which loose frost grains would merge or anneal into a dense, transparent layer, and even from observations of the light-scattering properties of Triton's equator. These suggestions would each explain the dark, apparently frost-free northern hemisphere, but the bright "cap" in the south must be explained as well. Perhaps it is a much thicker deposit of nitrogen that never completely sublimes away (this is certainly the impression one gets geologically). Although nitrogen frost may be very transparent when first annealed, changing temperatures will make the residual cap expand and contract, fracturing it and making it appear bright. Thus, we are led to the idea of a clear, uncracked (i.e., gas-tight) seasonal frost layer over a thick, fractured permanent cap: precisely the kind of layering hypothesized earlier to explain the plumes as solar-powered geysers.

What controls the size of the residual cap, and why is one not seen in the north? A good candidate is solid-state creep, or flow, of the thick nitrogen deposit, similar to the flow of glaciers and spreading of polar caps on the Earth and Mars. Models based on terrestrial polar caps, combined with estimates of the rate at which solid nitrogen would flow, suggest that the permanent cap is about a kilometer thick at the center and could easily extend to 45° latitude or even beyond if the total amount of nitrogen on Triton were great enough. Cap spreading also prevents the eventual disappearance of the seasonal frosts predicted by the models discussed earlier. Because the pole always receives less sunlight than the edges of the seasonal frost deposits, more frost will be deposited at the pole than at the edges, and after each cycle of redistribution the seasonal frosts will be thicker but smaller in extent. Spreading eventually offsets this process. There may be a northern as well as a southern permanent cap. If this northern cap extends less than 45° from the pole, it would lie in the dark portion of Triton unseen by *Voyager.* The southern permanent cap might be larger because of hemispheric differences in the heat released from Triton's interior or it might also extend only 45°, in which case the bright deposits extending almost to the equator have still to be explained. Some of this bright material may be nitrogen "snow" that condenses in the atmosphere into grains that are too big to anneal on a seasonal time scale into a transparent layer. It should be apparent from this discussion that, as with the plumes, we seem to have many pieces of the puzzle of the polar caps (and perhaps a few spurious pieces of unrelated puzzles), but they have yet to be assembled into a final picture of Triton's surface–atmosphere interaction.

Additional clues to the behavior of volatiles on Triton are being gathered from Earth-based spectroscopic measurements, and some constraints on the models described earlier can be expected. For example, spectral detection of carbon dioxide and water ice on Triton has proved that at least some areas are bare of nitrogen because CO_2 and H_2O are almost completely involatile and do not mix with N_2. In contrast, CH_4 and CO, which have also been detected, are volatile and are either dissolved in the solid nitrogen or mixed with it grain by grain. The 2.15-μm spectral feature by which nitrogen was first identified is also being monitored. The shape and position of this spectral feature change with temperature. To date, the 38 K temperature measured by *Voyager* has been confirmed, but the passing of Triton's southern summer may bring changes. If the temperature dips to about 35 K (or slightly warmer in the presence of CH_4 and CO), solid nitrogen will undergo a phase transition to a different solid form that will radically change the absorption feature. This phase transition also changes the volume of the solid. Widespread fracturing of the hypothesized clear nitrogen layer may occur as a result, leading to a sudden overall brightening and a decrease of spectral contrast. Finally, future occultations of stars by Triton (such as occurred in 1993, 1995, and 1997) will allow monitoring of its atmospheric pressure and dynamics, which are also expected to change with time. Similar monitoring of Pluto will provide a second case against which to test theoretical models.

VIII. ORIGIN AND EVOLUTION

Triton and Pluto turn out to be remarkably similar in size, in density, and, with the detection of N_2 and CO ice on Pluto in 1992, in surface and atmospheric compositions as well. There is little doubt that they share a common heritage. Moreover, they are not isolated in the outer solar system. An entirely new reservoir of minor planets has been found orbiting near and beyond Neptune: the Kuiper Belt. The first Kuiper Belt object was found in 1992 and, as of this writing, over 60 have been discovered. Although none of these are as large as Pluto or Triton, they are not small objects either: for an assumed albedo of a few percent (appropriate to small outer solar system asteroids and known comet nuclei), their diameters range between ~100 and 800 km. The density or chemical makeup of these objects is not yet known, but it is most probable that they are smaller members of an outer solar system population whose largest *known* examples are Pluto and Triton. [*See* KUIPER BELT.]

The link among Triton, Pluto, and the Kuiper Belt is strengthened by what is known of the orbital dynamics of this region. For example, a number of Kuiper Belt objects share the same dynamical resonance with Neptune that Pluto occupies (this orbital resonance prevents encounters between Neptune and Pluto and is one of the strong arguments against the Pluto-as-escaped-satellite hypothesis). In this sense, Pluto and its companion "Plutinos" are more like the Trojan or Hilda groups of asteroids (which are locked in orbital resonances with Jupiter), only that Pluto–Charon is the clearly dominant member of its group.

Dynamical calculations show that Pluto and its companions were probably swept into this orbital resonance as Neptune's orbit expanded early in the solar system's history. This happened as Neptune finished its accretion, in the process clearing the heliocentric space around it of any leftover planetesimals or protoplanets. During this time the flux close to Neptune of bodies orbiting near and beyond Neptune would have been quite high, and even today Neptune continues to deplete the inner Kuiper Belt population, the short-period comets being one result. It is perhaps not surprising then that Neptune should have had a catastrophic encounter with at least one escapee from the Kuiper Belt: Triton. [*See* COMETARY DYNAMICS.]

Satellite capture does not occur easily. Generally, objects passing near a planet leave with the same speed that they came in with. Even complicated trajectories called temporary gravitational captures (enjoyed by Comet Shoemaker–Levy 9) are just that, temporary. To be permanently captured, a cosmic body must lose energy (velocity) by running into or through something. In Triton's case, it could have collided with another stray body just passing by Neptune, but the probability of this having happened is quite low. Because Triton orbits close to Neptune, in the region usually occupied by regular satellites, it is much more likely that it ran into a regular satellite or its precursor protosatellite disk.

Triton's orbit once captured would have been very elliptical ($e \sim 0.99$), stretching from near Neptune to up to several 1000 R_N away. If Triton was captured by passage through a protosatellite disk, then its orbit would continue to shrink and become more circular because of the gas drag encountered during each subsequent passage through the disk. This is actually a problem for disk capture models because the disk must dissipate before gas drag causes the satellite's orbit to decay completely and merge with the planet. For this reason, and because of the difficulty of permanently capturing a body as massive as Triton by gas drag alone, the leading mechanism for Triton's capture is

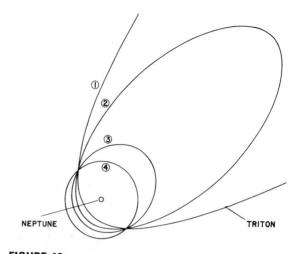

FIGURE 18 Schematic tidal evolution of Triton's orbit, subsequent to capture by Neptune. A very extended elliptical orbit (1) shrinks and circularizes (2 through 4). In doing so, Triton courses through much of the space close to Neptune and would have profoundly disturbed, if not destroyed, any original satellite system. Nereid, Neptune's outermost moon, may be a remnant of this original system. Triton would have been spectacularly heated and melted during this orbital evolution. [From W. B. McKinnon *et al.* (1995). *In* "Neptune and Triton" (D. P. Cruikshank, ed.), pp. 807–877. Univ. of Arizona Press, Tucson.]

presently collision with a preexisting satellite. Such a collision may or may not have markedly damaged Triton, as Triton is substantially larger than any of the uranian satellites, which probably provide the best examples of what Neptune's original satellite system could have been like. In any event, the inclination of the postcapture orbit depends on the initial encounter geometry and is essentially random; Triton could have ended up either prograde or retrograde.

After capture by collision, Triton's orbital evolution would have been strongly influenced by tides. Every time Triton reapproached Neptune, Neptune's gravity would raise a tidal bulge on Triton. The periodic rise and fall of the bulge would dissipate energy as heat, which would be extracted from the energy of Triton's orbit. Because the tidal couple between Triton's bulge and Neptune would be (on average) radial, no change in Triton's orbital angular momentum would occur. Based on these constraints, and ignoring for the moment any further encounters with original satellites, Triton's orbital configuration after capture would evolve as depicted in Fig. 18. The important point is that early on Triton's periapse (the closest point to Neptune in its orbit) would lie as low as half its present semimajor axis. Triton's tidal evolution probably took several 100 million years, so there would have been sufficient time for Triton's orbit to evolve through and interact with any preexisting satellites.

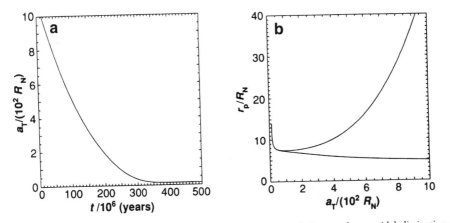

FIGURE 19   (a) Example evolution of Triton's semimajor axis, a_T, as a function of time, t, due to tidal dissipation within Triton. (b) Evolution of Triton's minimum and maximum periapse distance, r_p (the closest point to Neptune in its orbit), as a function of semimajor axis due to the combined influence of semiannual solar perturbations and tidal dissipation. The periapse distance oscillates between the two curves shown. [Adapted from P. Goldreich *et al.* (1989). *Science* **245**, 500–504.]

This point is emphasized in Fig. 19b, which includes the periodic effects of *solar* tides on Triton's evolving orbit. When Triton's orbit was very large and eccentric, its periapse would have fluctuated and may have periodically been as low as 5 R_N! Triton would have had ample opportunity for further collisions with Neptune's original satellites, possibly accreting them in the process; it may also have scattered original satellites into distant orbits, caused them to crash into Neptune, or perhaps even ejected them from Neptune altogether. There is now nothing left of Neptune's original system (if it indeed existed) other than the inner satellites and Nereid. The inner satellites all lie within 5 R_N, however, which is perfectly consistent with this capture scenario. Nereid may also be a survivor of this orbital mayhem. Little is known about this distant moon, save its size (~340 km in diameter) and reflectivity (~20%), but these meager facts make it more akin to a regular satellite than a dark captured asteroid or comet.

The end state of Triton's orbital evolution is an extremely circular orbit (Fig. 18). As such, the orbital energy potentially dissipated by tides within Triton represents an absolutely enormous reservoir, about 10^4 kJ kg^{-1}. It is sufficient to completely melt all the ice, rock, and metal within Triton 10 times over. The magnitude of Triton's temperature change, however, depends on the heating *rate*. Two such models are illustrated in Fig. 20. Tidal heating after capture in either model is at first modest, as the satellite spends most of its time far from Neptune. As its semimajor axis shrinks and its orbital period decreases, the average heating rate begins to rise. The epoch of greatest heating occurs when the relative change in the semimajor axis is the greatest (because orbital energy is inversely proportional to semimajor axis), roughly when the semimajor axis drops below 100 R_N. Because the orbit can only evolve as fast as the tides can convert orbital energy to heat, the response of Triton to tidal flexing is crucial. If Triton responds as a dissipative elastic sphere, then the semimajor axis drops continuously (Fig. 19a) and the tidal heating rises and then falls smoothly as the orbit becomes more circular (Fig. 20, elastic sphere model). The calculations in Figs. 19a and 20 are actually for two different elastic sphere models, but are shown to represent a range of possible time scales.

A dissipative elastic sphere is clearly an idealized and oversimplified model for Triton. Triton is in reality a complex rock, metal, organic matter, water–ice, and volatile ice body. The volatile ices especially should be melted and mobilized within Triton early in its history (e.g., ammonia), with or without tidal heating. A partially molten body is a particularly dissipative body, so when capture occurs and tidal heating begins, heat concentrates in the partially liquid regions. This causes more melting, which makes the body more dissipative, which results in greater tidal heating. Thus, within a few hundred million years after capture, Triton in all probability went through an episode of runaway melting. This is illustrated schematically in Fig. 20, where in the model labeled thin shells Triton melts spontaneously when enough energy has been accumulated to do so (in reality the runaway occurs much earlier). Thereafter Triton is a nearly totally molten, but still dissipative body. Its tidal heating curve rises and falls sharply over the course of ~100 million years.

During this epoch of extreme tidal heating Triton's

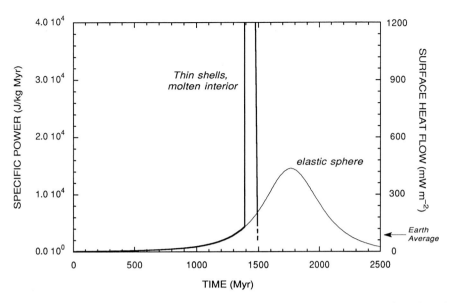

FIGURE 20 Power dissipated per unit mass and surface heat flow for Triton as its postcapture orbit shrinks and circularizes. Two models are shown. One assumes that Triton remains a uniform, undifferentiated sphere, whereas the second allows for melting. In both cases the time scales are longer than in the calculations in Fig. 19a due to updated parameters for Triton, but the periapse variations as a function of semimajor axis in Fig. 19b are unchanged. The thin shells model is more realistic than the elastic sphere, but even here the meltdown of Triton has been artifically suppressed. In reality a thermal runaway probably occurs much earlier. [From W. B. McKinnon *et al.* (1995). *In* "Neptune and Triton" (D. P. Cruikshank, ed.), pp. 807–877. Univ. of Arizona Press, Tucson.]

heat flow is an amazing $\sim$2–4 W m^{-2}, equal or greater than that measured today from Io. Its surface temperature is governed by this flux and corresponds to a black body temperature of 80–90 K. During and after this epoch there would likely have been large chemical exchanges between the global oceanic mantle with its dissolved volatiles and the hot rock core below. Much of Triton's volatiles may have been driven into a massive atmosphere. Atmospheric components plausibly include CO, CH$_4$, CO$_2$, and NH$_3$, or even H$_2$ (from photolysis of methane or ammonia or as a minor component in Triton's original ice). Conservative assumptions yield an atmospheric greenhouse with surface temperatures well above 100 K; more extreme possibilities allow for surface temperatures greater than 200 K.

A most intriguing aspect of raising a massive greenhouse atmosphere by tidal heating is that it may persist well after the tidal heating input has tapered off and Triton's interior has begun to freeze. It may only collapse after enough of it has been lost to space due to solar UV heating driven to hydrodynamic escape, which could have taken in excess of 1 billion years. While the atmosphere existed it would have kept Triton's surface warmer and enhanced the geological mobility of the satellite's surface layers. Unfortunately, there are as yet no definitive indicators of the atmosphere's former presence (e.g., ancient aeolian or fluvial features or peculiar crater shapes). If a thick atmo-

sphere existed, Triton's continued geological activity has obscured the evidence.

Regardless, once tidal heating ended, Triton's interior should have begun to freeze. It would probably have taken a few 100 million years to do so, but even today such freezing would not be complete. Triton's ice mantle is probably warm enough, due to radiogenic heating from the core, that any ammonia- and methanol-rich fluids are stable, and Triton's inner core of alloyed iron, nickel, and sulfur should likewise be warm enough (more than $\approx$1250 K) to allow for a eutectic liquid mixture of those elements.

The possible persistence of cryomagmas in Triton's mantle due solely to radiogenic heating has raised the question as to whether any of the geological observations in Section VI actually *demand* that Triton was massively tidally heated. Certainly, solar-powered plume models do not require Triton to be internally active at all. Triton's surface, however, is so peculiar (in the sense of being unique or special). Furthermore, the extent and intensity of the geological activity recorded there are only seen on satellites that are undergoing active and substantial tidal heating (Io, Europa, and Enceladus). While no ironclad argument can be made, Triton's geology and chemistry in all likelihood indicate that it did indeed experience massive tidal heating.

The proof of Triton's history and provenance requires further exploration of this extraordinary body. For example, determination of the compositions of Triton's icy lavas, and terrains in general, would be key constraints. Detailed exploration of the Neptune system by spacecraft is also a technically and fiscally feasible proposition, given recent and projected technological advances. Instruments and electronics are being increasingly miniaturized, thereby requiring smaller and less costly launch vehicles. Missions to Triton can also take advantage of innovative flight strategies, such as using solar electric propulsion *in the inner solar system* to boost the spacecraft to high speeds and cut travel time, and aerobraking in the Neptune atmosphere to go into initial Neptune orbit. Thereafter a complement of advanced instruments can be trained on Triton during repeated encounters, filling out our picture of this amazing satellite.

As for Triton's ultimate future, as a retrograde satellite its orbit is actually decaying due to tides it raises on Neptune. In the 1960s it was estimated that Triton would closely approach Neptune and be torn apart by tides in a geologically short time. Present estimates imply less peril: Triton's orbit will probably shrink by no more than 15% over the next 5 billion years, giving Triton plenty of time for further geological and atmospheric adventures.

BIBLIOGRAPHY

Beatty, J. K., *et al.* (1998). "The New Solar System, 4th Edition." Sky Publishing, Cambridge, MA.

Cruikshank, D. P. (ed.) (1995). "Neptune and Triton." Univ. of Arizona Press, Tucson.

Greeley, R., and Batson, R. (1997). "NASA Atlas of the Solar System." Cambridge Univ. Press, Cambridge, UK.

Littmann, M. (1990). "Planets Beyond: Discovering the Outer Solar System." Wiley & Sons, New York.

"Moons and Rings" (Voyage through the Universe Series) (1991). Time-Life Books, Alexandria, VA.

Morrison, D., and Owen, T. (1996). "The Planetary System." Addison-Wesley, Reading MA.

Rothery, D. (1992). "Satellites of the Outer Planets." Clarendon Press, Oxford.

Smith, B. A., and the Voyager Imaging Team (1989). Voyager 2 at Neptune: Imaging science results. *Science* **246**, 1422–1449.

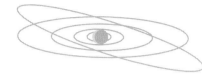

OUTER PLANET ICY SATELLITES

I. Summary of Characteristics

II. Formation of Satellites

III. Observations of Satellites

IV. Individual Satellites

Bonnie J. Buratti

Jet Propulsion Laboratory,
California Institute of Technology

GLOSSAY

Bond albedo: Fraction of the incident radiation reflected by a planet or satellite. The bolometric Bond albedo is this quantity integrated over all wavelengths.

Carbonaceous (C-type) material: Carbon–silicate primordial material rich in simple organic compounds. C-type material is spectrally flat and exists on the surfaces of several outer planet satellites.

D-type material: Primordial, low-albedo material believed to be rich in organic compounds. It is redder than C-type material.

Differentiation: Melting and chemical fractionation of a planet or satellite into a core and mantle.

Geometric albedo: Ratio of the brightness at a phase angle of zero degrees (full illumination) compared with a diffuse, perfectly reflecting disk of the same size.

Greenhouse effect: Heating of the lower atmosphere of a planet or satellite by the absorption of visible radiation and subsequent reradiation in the infrared.

Ionosphere: Outer portion of an atmosphere where charged particles are abundant.

Lagrange points: Equilibrium points in the orbit of a planet or satellite around its primary.

Magnetosphere: Region around a planet dominated by its magnetic field and associated charged particles.

Opposition effect: Surge in brightness as a satellite becomes fully illuminated to the observer.

Phase angle: Angle between the observer, the satellite, and the Sun.

Phase integral: Integrated value of the function that describes the directional scattering properties of a surface.

Primary body: Celestial body (usually a planet) around which a satellite, or secondary, orbits.

Regolith: Surface layer of rocky debris created by meteorite impacts.

Roche limit: Distance (equal to 2.44 times the radius of the primary) at which the tidal forces exerted by the primary on the satellite equal the internal gravitational forces of the satellite.

Synchronous rotation: Dynamical state caused by tidal interactions in which the satellite presents the same face toward the primary.

An outer planet icy satellite is any one of the celestial bodies in orbit around Jupiter, Saturn, Uranus, Neptune, or Pluto. They range from large, planetlike, geologically active worlds with significant atmospheres, such as Triton, to tiny irregular objects tens of kilometers in diameter. These bodies are all believed to have as major components some type of frozen

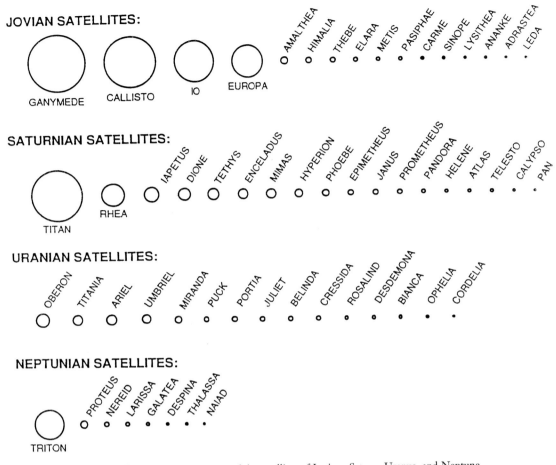

FIGURE 1   The relative sizes of the satellites of Jupiter, Saturn, Uranus, and Neptune.

volatile, primarily water ice, but also methane, ammonia, nitrogen, carbon monoxide, carbon dioxide, or sulfur dioxide existing alone or in combination with other volatiles. The outer five planets have among them a total of 60 known satellites. There undoubtedly exist many more undiscovered small satellites in the outer solar system. The relative sizes of the satellites of these four planets are illustrated in Fig. 1. Table I is a summary of their characteristics. This chapter covers the satellites of Jupiter, Saturn, Uranus, and Neptune, exept Io, Triton, and Titan. [*See* Io; Titan; Triton; Pluto and Charon.]

I. SUMMARY OF CHARACTERISTICS

A. DISCOVERY

None of the satellites of the outer planets was known before the invention of the telescope. When Galileo turned his telescope to Jupiter in 1610, he discovered the four large satellites in the Jovian system. His observations of their orbital motion around Jupiter in a manner analogous to the motion of the planets around the Sun provided important evidence for the acceptance of the heliocentric (Sun-centered) model of the solar system. These four moons—Io, Europa, Ganymede, and Callisto—are sometimes called the Galilean satellites.

In 1655, Christian Huygens discovered Titan, the giant satellite of Saturn. Later in the seventeenth century, Giovanni Cassini discovered the four next largest satellites of Saturn. More than one hundred years would pass before the next satellite discoveries were made: the Uranian satellites Titania and Oberon and two smaller moons of Saturn. As telescopes acquired more resolving power in the nineteenth century, the family of satellites grew (see Table I). The smallest satellites of Jupiter and Saturn, and all the small satellites of Uranus and Neptune (except Nereid), were discovered during flybys of the Pioneer and Voyager spacecraft (see Table II).

TABLE I
Summary of the Properties of the Outer Planet Satellites

Satellite	Distance from primary (10³ km)	Revolution period (days) R = retrograde	Orbital eccentricity	Orbital inclination (degrees)	Radius (km)	Density (g/cm³)	Visual geometric albedo	Discoverer	Year discovered
Jupiter									
J15 Adrastea	128	0.30	0.0	0.0	10		<0.1	Jewitt *et al.*	1979
J16 Metis	129	0.30	0.0	0.0	20		<0.1	Synnott	1979
J5 Amalthea	181	0.49	0.003	0.4	131 × 73 × 67		0.05	Barnard	1892
J14 Thebe	222	0.67	0.015	0.8	50		<0.1	Synnott	1979
J1 Io	422	1.77	0.004	.04	1,818	3.53	0.6	Galileo	1610
J2 Europa	671	3.55	0.010	0.5	1,560	2.99	0.6	Galileo	1610
J3 Ganymede	1,070	7.15	0.002	0.2	2,634	1.94	0.4	Galileo	1610
J4 Callisto	1,883	16.69	0.007	0.5	2,409	1.85	0.2	Galileo	1610
J13 Leda	11,094	239	0.148	26.7	5			Kowal	1974
J6 Himalia	11,480	251	0.163	27.6	85		0.03	Perrine	1904
J10 Lysithea	11,720	259	0.107	29.0	12			Nicholson	1938
J7 Elara	11,737	260	0.207	24.8	40		0.03	Perine	1904
J12 Ananke	21,200	631R	0.17	147	10			Nicholson	1951
J11 Carme	22,600	692R	0.21	163	15			Nicholson	1938
J8 Pasiphae	23,500	735R	0.38	145	18			Melotte	1908
J9 Sinope	23,700	758R	0.28	153	14			Nicholson	1914
Saturn									
S18 Pan	134	0.57	0.0	0.0	10	—	—	Showalter	1990
S15 Atlas	138	0.60	0.000	0.0	19 × 17 × 14		0.4	*Voyager*	1980
S16 Prometheus	139	0.61	0.002	0.0	74 × 50 × 34		0.6	*Voyager*	1980
S17 Pandora	142	0.63	0.004	0.05	55 × 44 × 31		0.6	*Voyager*	1980
S10 Janus	151	0.69	0.007	0.14	97 × 95 × 77	0.65	0.6	Dollfus	1966
S11 Epimetheus	151	0.69	0.009	0.34	69 × 55 × 55	0.65	0.5	Fountain and Larson	1978
S1 Mimas	186	0.94	0.020	1.5	199	1.4	0.8	Herschel	1789
S2 Enceladus	238	1.37	0.004	0.0	249	1.2	1.0	Herschel	1789
S3 Tethys	295	1.89	0.000	1.1	530	1.2	0.8	Cassini	1684
S14 Calypso	295	1.89	0.0	1.1	15 × 8 × 8		0.6	Pascu *et al.*	1980
S13 Telesto	295	1.89	0.0	1.0	15 × 12 × 8		0.9	Smith *et al.*	1980
S4 Dione	377	2.74	0.002	0.02	560	1.4	0.55	Cassini	1684
S12 Helene	377	2.74	0.005	0.15	16		0.5	Laques and Lecacheux	1980
S5 Rhea	527	4.52	0.001	0.35	764	1.3	0.65	Cassini	1672
S6 Titan	1,222	15.94	0.029	0.33	2,575	1.88	0.2	Huygens	1655
S7 Hyperion	1,481	21.28	0.104	0.4	180 × 140 × 112		0.3	Bond and Lassell	1848
S8 Iapetus	3,561	79.33	0.028	14.7	718	1.2	0.4–0.08	Cassini	1671
S9 Phoebe	12,952	550.4R	0.163	150	110		0.06	Pickering	1898
Uranus									
U6 Cordelia	49.7	0.33	0.0005	0.14	13			*Voyager 2*	1986
U7 Ophelia	53.8	0.38	0.010	0.09	15			*Voyager 2*	1986
U8 Bianca	59.2	0.43	0.001	0.16	21			*Voyager 2*	1986
U9 Cressida	61.8	0.46	0.0002	0.04	31		~0.04	*Voyager 2*	1986
U10 Desdemona	62.7	0.47	0.0002	0.16	27		~0.04	*Voyager 2*	1986
U11 Juliet	64.4	0.49	0.0006	0.06	42		~0.06	*Voyager 2*	1986
U12 Portia	66.1	0.51	0.0002	0.09	54		~0.09	*Voyager 2*	1986
U13 Rosalind	69.9	0.56	0.00009	0.28	27		~0.04	*Voyager 2*	1986
U14 Belinda	75.3	0.62	0.0001	0.03	33			*Voyager 2*	1986
U15 Puck	86.0	0.76	0.00005	0.31	77		0.07	*Voyager 2*	1985
U5 Miranda	130	1.41	0.003	3.4	236	1.2	0.35	Kuiper	1948
U1 Ariel	191	2.52	0.003	0.0	579	1.6	0.36	Lassell	1851
U2 Umbriel	266	4.14	0.005	0.0	585	1.5	0.20	Lassell	1851
U3 Titania	436	8.71	0.002	0.0	789	1.7	0.30	Herschel	1787
U4 Oberon	583	13.46	0.001	0.0	761	1.6	0.22	Herschel	1787
1997 U1	7775	654	0.2	146	20?	?	?	Gladman *et al.*	1997
1997 U2	8846	795	0.34	154	40?	?	?	Gladman *et al.*	1997
Neptune									
N8 Naiad	48.2	0.29	0.000	0.0	29			*Voyager 2*	1989
N7 Thalassa	50.1	0.31	0.0002	4.5	40			*Voyager 2*	1989
N5 Despina	52.5	0.33	0.0001	0.0	74		0.05	*Voyager 2*	1989
N6 Galatea	62.0	0.43	0.0001	0.0	79			*Voyager 2*	1989
N4 Larissa	73.6	0.55	0.000	0.0	104 × 89		0.06	*Voyager 2*	1989
N3 Proteus	117.6	1.12	0.0004	0.0	208		0.06	*Voyager 2*	1989
N1 Triton	354.8	5.87R	0.000015	157	1,353	2.08	0.73	Lassell	1846
N2 Nereid	5,513	360.1	0.751	29	170		0.14	Kuiper	1949

The natural planetary satellites are generally named after figures in world mythologies who were associated with the namesakes of their primaries. They are also designated by the first letter of their primary and an Arabic numeral assigned in order of discovery: Io is J1, Europa J2, and so on. When satellites are first discovered but not yet confirmed or officially named, they are known by the year in which they were discovered, the initial of the primary, and a number assigned consecutively for all solar system discoveries, for example, 1980 J27. Official names for all satellites are assigned by the International Astronomical Union.

After planetary scientists were able to map geologic formations of the satellites from spacecraft images, they named many of the features after characters or locations from Western and Eastern mythologies.

B. PHYSICAL AND DYNAMICAL PROPERTIES

The motion of a satellite around the center of mass of itself and its primary defines an ellipse with the primary at one of the foci. The orbit is defined by three primary orbital elements: (1) the semimajor axis, (2) the eccentricity, and (3) the angle made by the intersection of the plane of the orbit and the plane of the primary's spin equator (the angle of inclination). The orbits are said to be regular if they are in the same sense of direction (the prograde sense) as that determined by the rotation of the primary, and if their eccentricities and inclinations are low. The orbit of a satellite is irregular if its motion is in the opposite (or retrograde) sense of motion, if it is highly eccentric, or if it has a high angle of inclination. The majority of the outer planets' satellites move in regular, prograde orbits. Many of the satellites that move in irregular orbits are believed to be captured objects.

Most of the planetary satellites present the same hemisphere toward their primaries, a state that is the result of tidal evolution. When two celestial bodies orbit each other, the gravitational force exerted on the nearside is greater than that exerted on the farside. The result is an elongation of each body to form tidal bulges, which can consist of either solid, liquid, or gaseous (atmospheric) material. The primary tugs on the satellite's tidal bulge to lock its longest axis onto the primary–satellite line. The satellite, which is said to be in a state of synchronous rotation, keeps the same face toward the primary. Since this despun state occurs rapidly (usually within a few million years), most natural satellites are in synchronous rotation.

The satellites of the outer solar system are unique worlds, each representing a vast panorama of physical processes. The small satellites of Jupiter and Saturn are irregular chunks of ice and rock, perhaps captured asteroids that have been subjected to intensive meteoritic bombardment. Several of the satellites, including the Saturnian satellite Phoebe and areas of the Uranian satellites, are covered with C-type material, the dark, unprocessed, carbon-rich material found on the C class of asteroids. The surfaces of other satellites such as Hyperion and the dark side of Iapetus contain D-type primordial matter (named after the D class of asteroids), which is spectrally red and believed to be rich in organic compounds. Both D- and C-type material are common in the outer solar system. Because these materials represent the material from which the solar system formed, understanding their occurrence and origin will yield clues on the state and early evolution of the solar system. Iapetus presents a particular enigma: one hemisphere is 10 times more reflective than the other. [See ASTEROIDS.]

Before the advent of spacecraft exploration, planetary scientists expected the icy satellites to be geologically dead worlds. They assumed that heat sources were not sufficient to have melted their mantles to provide a source of liquid or semi liquid ice or ice–silicate slurries. Reconnaissance of the icy satellite systems of the four outer giant planets by the two *Voyager* spacecraft uncovered a wide range of geologic processes, including currently active volcanism on Io and Triton. At least two additional satellites (Europa and Enceladus) may have current activity. The medium-sized satellites of Saturn and Uranus are large enough to have undergone internal melting with subsequent differentiation and resurfacing. Among the Galilean satellites, only Callisto lacks evidence for periods of such activity after formation.

Recent work on the importance of tidal interactions and subsequent heating has provided the theoretical foundation to explain the existence of widespread activity in the outer solar system. Another factor is the presence of nonice components, such as ammonia hydrate or methanol, which lower the melting point of near-surface materials. Partial melts of water ice and various contaminants—each with its own melting point and viscosity—provide material for a wide range of geologic activity. The realization that such partial melts are important to understanding the geologic history of the satellites has spawned an interest in the rheology (viscous properties and resulting flow behavior) of various ice mixtures and exotic phases of ices that exist at extreme temperatures or pressures. Conversely, the types of features observed on the surfaces provide clues to the likely composition of the satellites' interiors.

Because the surfaces of so many outer planet satellites exhibit evidence of geologic activity, planetary scientists have begun to think in terms of unified geologic processes that function throughout the solar system. For example, partial melts of water ice with various contaminants could provide flows of liquid or partially molten slurries that in many ways mimic terrestrial or lunar lava flows formed by the partial melting of mixtures of silicate rocks. The ridged and grooved terrains on satellites such as Ganymede, Enceladus, Tethys, and Miranda may all have resulted from similar tectonic activities. Finally, explosive volcanic eruptions occurring on Io, Triton, Earth, and possibly Enceladus may all result from the escape of volatiles released as the pressure in upward-moving liquids decreases. [See PLANETARY VOLCANISM.]

II. FORMATION OF SATELLITES

A. THEORETICAL MODELS

Because the planets and their associated moons condensed from the same cloud of gas and dust at about the same time, the formation of the natural planetary satellites must be addressed within the context of the formation of the planets. The solar system formed 4.6 ± 0.1 billion years ago. This age is derived primarily from radiometric dating of meteorites, which are believed to consist of primordial, unaltered matter. In the radiometric dating technique, the fraction of a radioactive isotope (usually rubidium, argon, or uranium), which has decayed into its daughter isotope, is measured. Since the rate at which these isotopes decay has been measured in the laboratory, it is possible to infer the time elapsed since formation of the meteorites, and thus of the solar system. [See THE ORIGIN OF THE SOLAR SYSTEM.]

The Sun and planets formed from a disk-shaped rotating cloud of gas and dust known as the protosolar nebula. When the temperature in the nebula cooled sufficiently, small grains began to condense. The difference in solidification temperatures of the constituents of the protosolar nebular accounts for the major compositional differences of the satellites. Since there was a temperature gradient as a function of distance from the center of the nebula, only those materials with high melting temperatures (e.g., silicates, iron,

aluminum, titanium, and calcium) solidified in the central (hotter) portion of the nebula. Earth's Moon consists primarily of these materials. Beyond the orbit of Mars, carbon, in combination with silicates and organic molecules, condensed to form the carbonaceous material found on C-type asteroids. Similar carbonaceous material is found on the surfaces of the Martian moon Phobos, several of the Jovian and Saturnian satellites, regions of the Uranian satellites, and possibly Triton and Charon. In the outer regions of the asteroid belt, formation temperatures were sufficiently cold to allow water ice to condense and remain stable. Thus, the Jovian satellites are primarily ice–silicate admixtures (except for Io, which has apparently outgassed all its water). On Saturn and Uranus, these materials are predicted to be joined by methane and ammonia, and their hydrated forms. For the satellites of Neptune and Pluto, formation temperatures were low enough for other volatiles, such as nitrogen, carbon monoxide, and carbon dioxide, to exist in solid form. In general, the satellites that formed in the inner regions of the solar system are denser than the outer planets' satellites, because they retained a lower fraction of volatile materials.

After small grains of material condensed from the protosolar nebula, electrostatic forces caused them to stick together. Collisions between these larger aggregates caused meter-sized particles, or planetesimals, to be accreted. Finally, gravitational collapse occurred to form larger, kilometer-sized planetesimals. The largest of these bodies swept up much of the remaining material to create the protoplanets and their companion satellite systems. One important concept of planetary satellite formation is that a satellite cannot accrete within the Roche limit, the distance at which the tidal forces of the primary become greater than the internal cohesive forces of the satellite.

The formation of the regular satellite systems of Jupiter, Saturn, and Uranus is sometimes thought to be a smaller-scaled version of the formation of the solar system. A density gradient as a function of distance from the primary does exist for the regular system of small, inner Neptunian satellites and for the Galilean satellites (see Table I). This implies that more volatiles (primarily ice) are included in the bulk composition as the distance increases. However, this simple scenario cannot be applied to Saturn or Uranus because their regular satellites do not follow this pattern.

The retrograde satellites are probably captured asteroids or large planetesimals left over from the major episode of planetary formation. Except for Titan and Triton, the satellites are too small to possess gravitational fields sufficiently strong to retain an appreciable atmosphere against thermal escape. At least one satellite (Ganymede) has a magnetic field.

B. EVOLUTION

Soon after the satellites accreted, they began to heat up from the release of gravitational potential energy. An additional heat source was provided by the release of mechanical energy during the heavy bombardment of their surfaces by remaining debris. The satellites Phobos, Mimas, and Tethys all have impact craters caused by bodies that were nearly large enough to break them apart; probably such catastrophes did occur. The decay of radioactive elements found in silicate materials provided another major source of heat. The heat produced in the larger satellites was sufficient to cause melting and chemical fractionation; the dense material, such as silicates and iron, went to the center of the satellite to form a core, while ice and other volatiles remained in the crust. A fourth source of heat is provided by tidal interactions. When a satellite is being tidally despun, the resulting frictional energy is dissipated as heat. Because this process happens very quickly for most satellites (~10 million years), another mechanism involving orbital resonances among satellites is believed to cause the heat production required for more recent resurfacing events. Gravitational interactions tend to turn the orbital periods of the satellites within a system into multiples of each other. In the Galilean system, for example, Io and Europa complete four and two orbits, respectively, for each orbit completed by Ganymede. The result is that the satellites meet each other at the same point in their orbits. The resulting flexing of the tidal bulge induced on the bodies by their mutual gravitational attraction causes significant heat production in some cases. [See PLANETARY IMPACTS; SOLAR SYSTEM DYNAMICS.]

Some satellites, such as the Earth's Moon, Ganymede, and several of the Saturnian satellites, underwent periods of melting and active geology within a billion years of their formation and then became quiescent. Others, such as Io and Triton, and possibly Enceladus and Europa, are currently geologically active. For nearly a billion years after their formation, the satellites all underwent intense bombardment and cratering. The bombardment tapered off to a slower rate and presently continues. By counting the number of craters on a satellite's surface and making certain assumptions about the flux of impacting material, geologists are able to estimate when a specific portion of a satellite's surface was formed. Continual bombardment of satellites causes the pulverization of both rocky and icy surfaces to form a covering of fine material known as a regolith.

Many scientists expected that most of the craters formed on the outer planets' satellites would have disappeared owing to viscous relaxation. The two *Voyager* spacecraft revealed surfaces covered with craters that in many cases had morphological similarities to those found in the inner solar system, including central peaks, large ejecta blankets, and well-formed outer walls. Recent research has shown that the elastic properties of ice provide enough strength to offset viscous relaxation. Silicate mineral contaminants or other impurities in the ice may also provide extra strength to sustain impact structures.

Planetary scientists classify the erosional processes affecting satellites into two major categories: endogenic, which includes all internally produced geologic activity, and exogenic, which encompasses the changes brought by outside agents. The latter category includes the following processes: (1) meteoritic bombardment and resulting gardening and impact volatization; (2) magnetospheric interactions, including sputtering and implantation of energetic particles; (3) alteration by high-energy ultraviolet photons; and (4) accretion of particles from sources such as planetary rings.

Meteoritic bombardment of icy bodies acts in two major ways to alter the optical characteristics of the surface. First, the impacts excavate and expose fresh material (cf. the bright ray craters on Ganymede). Second, impact volatilization and subsequent escape of volatiles result in a lag deposit enriched in opaque, dark materials. The relative importance of the two processes depends on the flux, size distribution, and composition of the impacting particles, and on the composition, surface temperature, and mass of the satellite. For the Galilean satellites, older geologic regions tend to be darker and redder. Both the Galilean and Saturnian satellites tend to be brighter on the hemispheres that lead in the direction of orbital motion (the so-called "leading" side, as opposed to the "trailing" side); this effect is thought to be due to preferential micrometeoritic gardening on the leading side. The Uranian satellites show no similar dichotomy in albedo, but their leading sides do tend to be redder, possibly due to the accretion of reddish meteoritic material on that hemisphere.

For satellites that are embedded in planetary magnetospheres, their surfaces are affected by magnetospheric interactions in three ways: (1) chemical alterations; (2) selective erosion, or sputtering; and (3) deposition of magnetospheric ions. In general, volatile components are more susceptible to sputter erosion than refractory ones. The overall effect of magnetospheric erosion is thus to enrich surfaces in darker, redder opaque materials. A similar effect is believed to be caused by the bombardment of UV photons, although much fundamental laboratory work remains to be done to determine the quantitative effects of this process. [See PLANETARY MAGNETOSPHERES.]

III. OBSERVATIONS OF SATELLITES

A. TELESCOPIC OBSERVATIONS

1. Spectroscopy

Before the development of interplanetary spacecraft, all observations from Earth of objects in the solar system were obtained by telescopes. One particularly useful tool of planetary astronomy is spectroscopy, or the acquisition of spectra from a celestial body.

Each component of the surface or atmosphere of a satellite has a characteristic pattern of absorption and emission bands. Comparison of the astronomical spectrum with laboratory spectra of materials that are possible components of the surface yields information on the composition of the satellite. For example, water ice has a series of absorption features between 1 and 4 μm. The detection of these bands on three of the Galilean satellites and several satellites of Saturn and Uranus demonstrated that water ice is a major constituent of their surfaces. Other examples are the detections of SO_2 frost on the surface of Io, methane in the atmosphere of Titan, nitrogen and carbon dioxide on Triton, and carbon monoxide on Pluto.

2. Photometry

Photometry of planetary satellites is the accurate measurement of radiation reflected to an observer from their surfaces or atmospheres. These measurements can be compared to light-scattering models that are dependent on physical parameters, such as the porosity of the optically active upper surface layer, the albedo of the material, and the degree of topographic roughness. These models predict brightness variations as a function of solar phase angle (the angle between the observer, the Sun, and the satellite). Like the Earth's Moon, the planetary satellites present changing phases to an observer on Earth. As the face of the satellite

becomes fully illuminated to the observer, the integrated brightness exhibits a nonlinear surge in brightness that is believed to result from the disappearance of mutual shadowing among surface particles. The magnitude of this surge, known as the "opposition effect," is greater for a more porous surface.

One measure of how much radiation a satellite reflects is the geometric albedo, p, which is the disk-integrated brightness at "full moon" (or a phase angle of zero degrees) compared to a perfectly reflecting, diffuse disk of the same size. The phase integral, q, defines the angular distribution of radiation over the sky:

$$q = 2 \int_0^\pi \Phi(\alpha) \sin \alpha \, d\alpha$$

where $\Phi(\alpha)$ is the disk-integrated brightness and α is the phase angle.

The Bond albedo, which is given by $A = p \times q$, is the ratio of the integrated flux reflected by the satellite to the integrated flux received. The geometric albedo and phase integral are wavelength dependent, whereas a true (or bolometric) Bond albedo is integrated over all wavelengths.

Another ground-based photometric measurement that has yielded important information on the satellites' surfaces is the integrated brightness of a satellite as a function of orbital angle. For a satellite in synchronous rotation with its primary, the subobserver geographical longitude of the satellite is equal to the longitude of the satellite in its orbit. Observations showing significant albedo and color variegations for Io, Europa, Rhea, Dione, and especially Iapetus suggest that diverse geologic terrains coexist on these satellites. This view was confirmed by images obtained by the *Voyager* spacecraft.

Another important photometric technique is the measurement of radiation as one celestial body occults, or blocks, another body. Time-resolved observations of occultations yield the flux emitted from successive regions of the eclipsed body. This technique has been used to map albedo variations on Pluto and its satellite Charon and to map the distribution of infrared emission—and thus volcanic activity—on Io.

3. Radiometry

Satellite radiometry is the measurement of radiation that is absorbed and reemitted at thermal wavelengths. The distance of each satellite from the Sun determines the mean temperature for the equilibrium condition that the absorbed radiation is equal to the emitted radiation:

$$\pi R^2 (F/r^2)(1 - A) = 4\pi R^2 \varepsilon \sigma T^4$$

$$T = \left(\frac{(1 - A)F}{4\sigma \varepsilon r^2} \right)^{1/4}$$

where R is the radius of the satellite, r is the Sun–satellite distance, ε is the emissivity, σ is Stefan–Boltzmann's constant, A is the Bond albedo, and F is the incident solar flux (a slowly rotating body would radiate over $2\pi R^2$). Typical mean temperatures in degrees Kelvin for the satellites are: the Earth's Moon, 280; Europa, 103; Iapetus, 89; the Uranian satellites, 60; and the Neptunian satellites, 45. For thermal equilibrium, measurements as a function of wavelength yield a blackbody curve characteristic of T: in general, the temperatures of the satellites closely follow the blackbody emission values. Some discrepancies are caused by a weak greenhouse effect (in the case of Titan), or the existence of volcanic activity (in the case of Io).

Another possible use of radiometric techniques, when combined with photometric measurements of the reflected portion of the radiation, is the estimate of the diameter of a satellite. A more accurate method of measuring the diameter of a satellite from Earth involves measuring the light from a star as it is occulted by the satellite. The time the starlight is dimmed is proportional to the satellite's diameter.

A third radiometric technique is the measurement of the thermal response of a satellite's surface as it is being eclipsed by its primary. The rapid loss of heat from a satellite's surface indicates a thermal conductivity consistent with a porous surface. Eclipse radiometry of Phobos, Callisto, and Ganymede suggests that these objects all lose heat rapidly.

4. Polarimetry

Polarimetry is the measurement of the degree of polarization of radiation reflected from a satellite's surface. The polarization characteristics depend on the shape, size, and optical properties of the surface particles. Generally, the radiation is linearly polarized and is said to be negatively polarized if it lies in the scattering plane and positively polarized if it is perpendicular to the scattering plane. Polarization measurements as a function of solar phase angle for atmosphereless bodies are negative at small phase angles; comparisons with laboratory measurements indicate that this is characteristic of complex, porous surfaces consisting of multisized particles.

TABLE II
Major Flyby Missions to the Outer Planetary Satellites

Mission	Objects	Encounter dates
Pioneer 10	Jovian satellites	1979
Pioneer 11	Jovian satellites	1979
	Saturnian satellites	1979
Voyager 1	Jovian satellites	1979
	Saturnian satellites	1980
Voyager 2	Jovian satellites	1979
	Saturnian satellites	1981
	Uranian satellites	1986
	Neptunian satellites	1989
Galileo	Jovian satellites	1996–1998

In 1970, ground-based polarimetry of Titan that showed it lacked a region of negative polarization led to the correct conclusion that it has a thick atmosphere.

5. Radar

Planetary radar is a set of techniques that involve the transmittance of radio waves to a remote surface and the analysis of the echoed signal. Among the outer planets' satellites, the Galilean satellites and Titan have been observed with radar. [See PLANETARY RADAR.]

B. SPACECRAFT EXPLORATION

Interplanetary missions to the planets and their moons have enabled scientists to increase their understanding of the solar system more in the past 20 years than in all of previous scientific history. Analysis of data returned from spacecraft has led to the development of whole new fields of scientific endeavor, such as planetary geology. From the earliest successes of planetary imaging, which included the flight of a Soviet Luna spacecraft to the farside of the Earth's Moon to reveal a surface unlike that of the visible side, devoid of smooth lunar plains, and the crash landing of a United States Ranger spacecraft, which sent back pictures showing that the Earth's Moon was cratered down to meter scales, it was evident that interplanetary imaging experiments had immense capabilities. Table II summarizes the successful spacecraft missions to the outer planetary satellites. The *Cassini/Huygens* spacecraft was launched in 1997 for a four year study of the Saturnian system.

The return of images from space is very similar to

the transmission of television images. A camera records the level of intensity of radiation incident on its focal plane, which holds a two-dimensional array of detectors. A computer onboard the spacecraft records these numbers and sends them by means of a radio transmitter to Earth, where another computer reconstructs the image.

Although images are the most spectacular data returned by spacecraft, a whole array of equally valuable experiments are included in each scientific mission. For example, a gamma-ray spectrometer aboard the lunar orbiters was able to map the abundance of iron and titanium on the Moon's surface. The *Voyager* spacecraft included an infrared spectrometer capable of mapping temperatures; an ultraviolet spectrometer; a photopolarimeter, which simultaneously measured the color, intensity, and polarization of light; and a radio science experiment that was able to measure the pressure of Titan's atmosphere by observing how radio waves passing through it were attenuated.

The *Pioneer* spacecraft, which were launched in 1972 and 1973 toward an encounter with Jupiter and Saturn, returned the first disk-resolved images of the Galilean satellites. By far the greatest scientific advancements were made by the *Voyager* spacecraft, which returned thousands of images of the satellite systems of all four outer planets, some of which are shown in Section IV. Color information for the objects was obtained by means of six broadband filters attached to the camera. The return of large numbers of images with resolution down to a kilometer has enabled geologists to construct geologic maps, to make detailed crater counts, and to develop realistic scenarios for the structure and evolution of the satellites.

Further advances are being made by the *Galileo* spacecraft, which was launched in 1990 and began obtaining data at Jupiter in 1996. The mission consists of a probe that explored the Jovian atmosphere and an orbiter designed to make several close flybys of the Galilean satellites. The orbiter contains both visual and infrared imaging devices, an ultraviolet spectrometer, and a photopolarimeter. The visual camera is capable of obtaining images with better than 20-m resolution. The only other currently approved mission to the outer planet satellites is the *Cassini* mission to Saturn, due to be launched in 1997. The *Cassini* spacecraft will contain a probe to study the atmosphere and surface of Titan and an orbiter to perform an in-depth study of Saturn, its rings, and satellites. Its instruments include a camera, an imaging spectrometer, infrared and ultraviolet spectrometers, and a suite of fields and particles experiments. [*See* PLANETARY EXPLORATION MISSIONS.]

IV. INDIVIDUAL SATELLITES

A. THE GALILEAN SATELLITES OF JUPITER

1. Introduction and Historical Survey

When Galileo trained his telescope on Jupiter, he was amazed to find four points of light that orbited the giant planet. These were the satellites Io, Europa, Ganymede, and Callisto, planet-sized worlds known collectively as the Galilean satellites. Analysis of telescopic observations over the next 350 years revealed certain basic features of their surfaces. There was spectroscopic evidence for water ice on the outer three objects. The unusually orange color of Io was hypothesized to be due to elemental sulfur. Orbital phase variations were significant, particularly in the cases of Io and Europa, which indicated the existence of markedly different terrains on their surfaces. Large opposition effects observed on Io and Callisto suggested that their surfaces were porous. The density of the satellites decreases as a function of distance from Jupiter (see Table I). Theoretical calculations suggested that the satellites had differentiated to form silicate cores and (in the case of the outer three) ice crusts. The *Voyager 1* and *2* and *Galileo* missions to Jupiter (see Table II) revealed the Galilean satellites to be four unique geological worlds.

2. Europa

When the Voyager spacecraft encountered the second Galilean satellite, Europa, they returned images of bright, icy plains crisscrossed by an extensive network of darker fractures. Higher-resolution images obtained by the *Galileo* spacecraft show possible "dirty geysers" situated along the fractures (Fig. 2). The existence of only a handful of impact craters suggests that geologic processes were at work on the satellite until a few hundred million years ago or less. Activity may even be current, although no direct evidence exists for it. Europa is very smooth: the only prominent topographic relief are ridges with a height of a few hundred meters. Some regions appear to be formed by shifting plates of ice (see bottom of Fig. 2).

Part of Europa is covered by a darker mottled terrain (see left side of Fig. 2). Dark features also include hundreds of brown spots of uncertain origin and larger areas, which may be the result of silicate-laden water

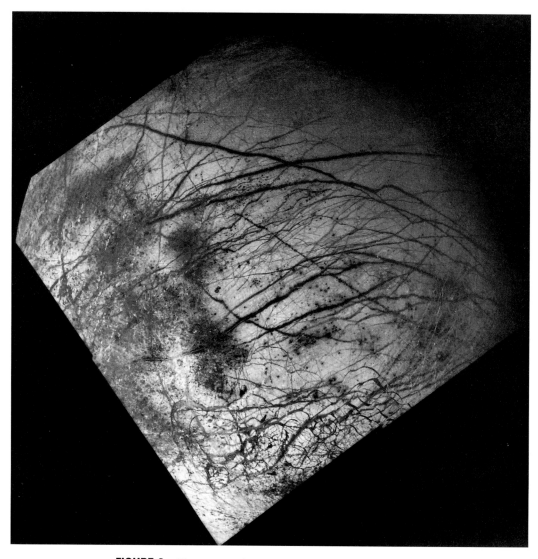

FIGURE 2 Photomosaic of Europa assembled from *Galileo* images.

rising to the surface. The reddish hue of Europa is believed to be due in part to contamination by sulfur from Io.

The cracks, which planetary geologists have called lineae, are a few kilometers wide and 10–15% lower in albedo than the ambient terrain. Central bright stripes down the centers of some lineae may be fresh ice extruded from below. The darker albedo of the lineae is due to compositional differences: these areas are contaminated with silicate minerals or ices darkened by irradiation. The mechanism for the formation of lineae is probably some form of tidal interaction and subsequent heating and refreezing of the ice crust. Calculations show that Europa may still have a liquid mantle. Although some scientists have dis-

cussed the possibility of a primitive life-form teeming in the mantle, there is no evidence that life does exist there.

The Hubble Space Telescope discovered a very thin atmosphere of molecular oxygen on Europa. Its surface pressure is only about 100-billionth that of the terrestrial atmosphere. This oxygen is formed by impacts to Europa's surface by charged particles, micrometeorites, and sunlight.

3. Ganymede

The icy moon Ganymede, which is the largest Galilean satellite, also shows evidence for geologic activity. Its

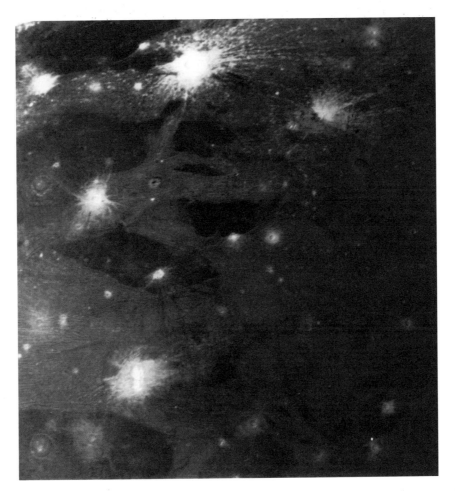

FIGURE 3   Voyager image of Ganymede, showing the darker, more cratered terrain and the brighter grooved terrain.

dark, heavily cratered terrain is transected by more recent, brighter grooved terrain (Fig. 3). Although they show much diversity, the grooves as seen in *Voyager* images are typically one-third to one-half kilometer high and 10 km or less in width. *Galileo* images show that much smaller scale ridges exist as well. The grooves were emplaced during several episodes between 3.5 and 4 billion years ago. Their formation may have occurred after a melting and refreezing of the interior, which caused a slight crustal expansion and subsequent faulting and flooding by subsurface water.

The grooved terrain of Ganymede is brighter because the ice is not as contaminated with rocky material that accumulates over the eons from impacting bodies. The satellite is also covered with relatively fresh bright craters, some of which have extensive ray systems. In the cratered terrain, there appear outlines of old, degraded craters, which geolo-

gists call palimpsests. The polar caps of Ganymede are brighter than the equatorial regions; this is probably due to the migration of water molecular released by evaporation and impact toward the colder high latitudes.

The Hubble Space Telescope discovered that Ganymede has a thin atmosphere of molecular oxygen similar to Europa's. Hubble also obtained preliminary evidence for polar aurorae, which are visible brightenings due to the impact of charged particles on the atmosphere. During its flyby of Ganymede in 1996, the *Galileo* spacecraft discovered a magnetosphere around the satellite, a finding that implies that it has a magnetic field. The existence of a magnetic field suggests that Ganymede has differentiated into a metallic core, a mantle, and a crust, consistent with analysis of the gravitatioinal pull that Ganymede exerted on the *Galileo* spacecraft as it flew by. *Galileo* also discovered an ionosphere.

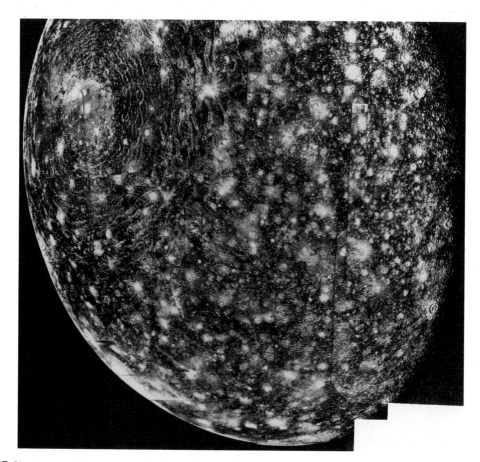

FIGURE 4 *Voyager 1* photomosaic of Callisto. The Valhalla impact basin, which is 600 km wide, dominates the surface.

4. Callisto

Callisto is the only Galilean satellite that does not show evidence for extensive resurfacing at any point in its history. It is covered with a relatively uniform, dark terrain saturated with craters (Fig. 4). There is, however, an absence of craters larger than 150 km and very small craters. In a process known as viscous relaxation, ice slumps and flows over periods of billions of years, and Callisto is apparently not able to maintain the structure of a large crater for as long as material that is primarily rock. One type of feature unique to Callisto is the remnant structures of numerous impacts. The most prominent of these, the Valhalla Basin, is a bright spot encircled by many fairly regular rings (see Fig. 4). Callisto may be contaminated by C-type or D-type material that accreted onto its surface during cometary or asteroidal impacts, or from ambient dust. High-resolution *Galileo* images show what appears to be loose particulate matter overlaying some regions of the surface (Fig. 5). The leading hemisphere of Callisto appears to be fluffier than the trailing side, possibly due to more intense meteoritic bombardment.

5. The Small Satellites of Jupiter

Jupiter has 11 known small satellites, including three discovered by the *Voyager* mission. They are all probably irregular in shape (see Table I). Within the orbit of Io are at least three satellites: Amalthea, Adrastea, and Metis. Amalthea is a dark, reddish, heavily cratered object reflecting less than 5% of the visible radiation it receives; the red color is probably due to contamination by sulfur particles from Io. Little else is known about its composition except that the dark material may be carbonaceous.

Adrastea and Metis, both discovered by *Voyager,* are the closest known satellites to Jupiter and move in nearly identical orbits just outside the outer edge of the thin Jovian ring, for which they may be a source of particles. Between Amalthea and Io lies the orbit of Thebe, also discovered by *Voyager.* Little is known about the composition of these satellites, but they are most likely primarily rock–ice mixtures. The three inner satellites sweep out particles in the Jovian magnetosphere to form voids at their orbital positions.

Exterior to the Galilean satellites, there is a class

FIGURE 5 *Galileo* image of Callisto, showing the fine, dustlike deposits found in certain regions.

of four satellites moving the highly inclined orbits (Lysithea, Elara, Himalia, and Leda). They are dark objects, reflecting only 2 or 3% of incident radiation, and may be similar to C- and D-type asteroids.

Another family of objects is the outermost four satellites, which also have highly inclined orbits, except they move in the retrograde direction around Jupiter. They are Sinope, Pasiphae, Carme, and Ananke, and they may be captured asteroids.

B. THE SATURNIAN SYSTEM

1. The Medium-Sized Icy Satellites of Saturn: Rhea, Dione, Tethys, Mimas, Enceladus, and Iapetus

The six largest satellites of Saturn are smaller than the Galilean satellites but still sizable—as such they represent a unique class of icy satellite. Earth-based telescopic measurements showed the spectral signature of ice for Tethys, Rhea, and Iapetus; Mimas and Enceladus are close to Saturn and difficult to observe because of scattered light from the planet. The satellites' low densities and high albedos (see Table I) imply that their bulk composition is largely water ice, possibly combined with ammonia or other volatiles. They have smaller amounts of rocky silicates than the Galilean satellites. Resurfacing has occurred on several of the satellites. Most of what is presently known of the Saturnian system was obtained from the *Voyager* flybys in 1980 and 1981. The six medium-sized icy satellites are shown to relative size in Fig. 6.

The innermost medium-sized satellite Mimas is covered with craters, including one (named Herschel) that is as large as a third of the satellite's diameter (see upper left of Fig. 6). The impacting body was probably nearly large enough to break Mimas apart; such disruptions may have occurred on other objects. There is a suggestion of surficial grooves that may be features caused by the impact. The craters on Mimas tend to be high-rimmed, bowl-shaped pits; apparently surface gravity is not sufficient to have caused slumping.

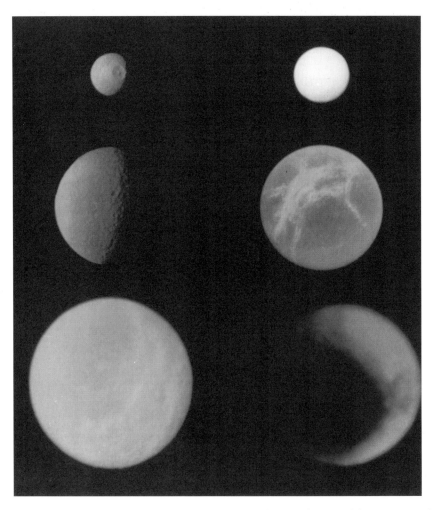

FIGURE 6 The six medium-sized icy Saturnian satellites. They are, left to right from the upper left, Mimas, Enceladus, Tethys, Dione, Rhea, and Iapetus.

The next satellite outward from Saturn is Enceladus, an object that was known from telescopic measurements to reflect nearly 100% of the visible radiation incident on it (for comparison, the Moon reflects only about 11%). The only likely composition consistent with this observation is almost pure water ice, or other highly reflective volatile. When *Voyager 2* arrived at Enceladus, it transmitted pictures to Earth that showed an object that had been subjected, in the recent geologic past, to extensive resurfacing; grooved formations similar to those on Ganymede were evident (Fig. 7). The lack of impact craters on this terrain is consistent with an age less than a billion years. It is possible that some form of ice volcanism is presently active on Enceladus. The heating mechanism is believed to be tidal interactions, perhaps with Dione. About half of the surface observed by *Voyager* is extensively cratered and dates from nearly 4 billion years ago.

A final element to the enigma of Enceladus is the possibility that it is responsible for the formation of the E-ring of Saturn, a tenuous collection of icy particles that extends from inside the orbit of Enceladus to past the orbit of Dione. The position of maximum thickness of the ring coincides with the orbital position of Enceladus. If some form of volcanism is presently active on the surface, it could provide a source of particles for the ring. An alternative source mechanism is an impact and subsequent escape of particles from the surface. [*See* PLANETARY RINGS.]

Tethys is covered with impact craters, including Odysseus, the largest known impact structure in the solar system. The craters tend to be flatter than those on Mimas or the Moon, probably because of viscous relaxation and flow over the eons under the stronger gravitational field of Tethys. Evidence for resurfacing episodes is seen in regions that have fewer craters and

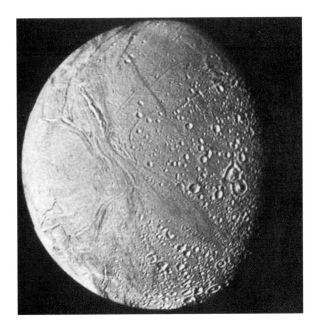

FIGURE 7 *Voyager 2* photomosaic of Enceladus. Both heavily cratered terrain and recently resurfaced areas are visible.

higher albedos. In addition, there is a huge trench formation, the Ithaca Chasma, which may be a degraded form of the grooves found on Enceladus.

Dione, which is about the same size as Tethys, exhibits a wide diversity of surface morphology. Most of the surface is heavily cratered (Fig. 8), but gradations in crater density indicate that several periods of resurfacing occurred during the first billion years of its existence. The leading side of the satellite is about 25% brighter than the trailing side, possibly due to more intensive micrometeoritic bombardment on this hemisphere. Wispy streaks (see Figs. 6 and 8), which are about 50% brighter than the surrounding areas, are believed to be the result of internal activity and subsequent emplacement of erupting material. Dione modulates the radio emission from Saturn, but the mechanism for this phenomenon is unknown.

Rhea appears to be superficially very similar to Dione (see Fig. 6). Bright wispy streaks cover one hemisphere. However, there is no evidence for any resurfacing events early in its history. There does seem to be a dichotomy between crater sizes—some regions lack large craters whereas other regions have a preponderance of such impacts. The larger craters may be due to a population of larger debris more prevalent during an earlier episode of collisions.

When Cassini discovered Iapetus in 1672, he noticed that at one point in its orbit around Saturn it was very bright, but on the opposite side of the orbit it nearly disappeared from view. He correctly deduced that one hemisphere is composed of highly reflective material and the other side is much darker. *Voyager* images show that the bright side, which reflects nearly 50% of the incident radiation, is fairly typical of a heavily cratered icy satellite. The other side, which is centered on the direction of motion, is coated with a material with a reflectivity of about 3–4% (Fig. 9).

Scientists still do not agree on whether the dark material originated from an exogenic source or was endogenically created. One scenario for the exogenic deposit of material entails dark particles being ejected from Phoebe and drifting inward to coat Iapetus. The major problem with this model is that the dark material on Iapetus is redder than Phoebe, although the material could have undergone chemical changes after its expulsion from Phoebe to make it redder. One observation lending credence to an internal origin is the concentration of material on crater floors, which implies an infilling mechanism. In one model, methane erupts from the interior and is subsequently darkened by ultraviolet radiation.

Other aspects of Iapetus are unusual. It is the only large Saturnian satellite in a highly inclined orbit, and it is less dense than objects of similar albedo. This

FIGURE 8 The heavily cratered face of Dione is shown in this *Voyager 1* image. Bright wispy streaks are visible on the limb of the satellite.

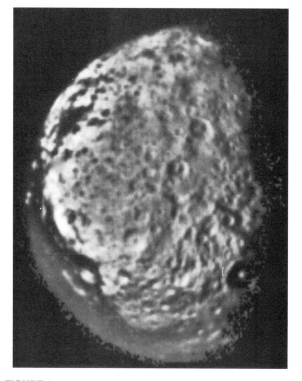

FIGURE 9 *Voyager 2* image of Iapetus, showing both bright and dark terrains.

latter fact implies a higher fraction of ice or possibly methane or ammonia in its interior.

2. The Small Satellites of Saturn

The Saturnian system has a number of unique small satellites (Fig. 10). Telescopic observations showed that the surface of Hyperion, which lies between the orbits of Iapetus and Titan, is covered with ice. Because Hyperion has a visual geometric albedo of 0.30, this ice must be mixed with a significant amount of darker, rocky material. Its composition may be similar to D-type asteroids. Although Hyperion is only slightly smaller than Mimas, it has a highly irregular shape (see Table I). This suggests, along with the satellite's battered appearance, that it has been subjected to intense bombardment and fragmentation. There is also good evidence that Hyperion is in chaotic rotation—perhaps a collision within the last few million years knocked it out of a tidally locked orbit. [*See* CHAOTIC MOTION IN THE SOLAR SYSTEM.]

Saturn's outermost satellite, Phoebe, a dark object (see Table I) with a surface composition probably similar to that of C-type asteroids, moves in a highly inclined, retrograde orbit, suggesting that it is a captured object. *Voyager* images show definite variegations con-

FIGURE 10   The small satellites of Saturn. They are, clockwise from the top, Atlas, Pandora, Janus, Calypso, Helene, Telesto, Epimetheus, and Prometheus.

sisting of dark and bright (presumably icy) patches on the surface. Although it is smaller than Hyperion, Phoebe has a nearly spherical shape.

Three types of unusual small satellites have been found in the Saturnian system: the shepherding satellites, the co-orbitals, and the Lagrangians. All of these objects are irregularly shaped (see Fig. 10) and probably consist primarily of ice. The three shepherds, Atlas, Pandora, and Prometheus, are believed to play a key role in defining the edges of Saturn's A- and F-rings. The orbit of Saturn's innermost satellite, Atlas, lies several hundred kilometers from the outer edge of the A-ring. The other two shepherds, which orbit on either side of the F-ring, not only constrain the width of this narrow ring but may cause its kinky appearance.

The co-orbital satellites Janus and Epimetheus, which were discovered in 1966 and 1978, exist in an unusual dynamical situation. They move in almost identical orbits at about 2.5 Saturn radii. Every four years the inner satellite (which orbits slightly faster than the outer one) overtakes its companion. Instead of colliding, the satellites exchange orbits. The four-year cycle then begins over again. Perhaps these two satellites were once part of a larger body that disintegrated after a major collision.

Three other small satellites of Saturn orbit in the Lagrangian points of larger satellites: one is associated with Dione and two with Tethys. The Lagrangian points are locations within an object's orbit in which a less massive body can move in an identical, stable orbit. They lie about 60° in front of and in back of the larger body. Although no other known satellites in the solar system are Lagrangians, the Trojan asteroids orbit in two of the Lagrangian points of Jupiter.

C. THE SATELLITES OF URANUS

1. The Medium-Sized Satellites of Uranus: Miranda, Ariel, Umbriel, Titania, and Oberon

The rotational axis of Uranus is inclined 98° to the plane of the solar system; Earth-based observers currently see the planet and its system of satellites nearly pole-on. The orbits of Ariel, Umbriel, Titania, and Oberon are regular, whereas Miranda's orbit is slightly inclined. Figure 11 is a telescopic image of the satellites typical of the quality attainable before the advent of spacecraft missions.

Theoretical models suggest that the satellites are

FIGURE 11   Telescopic view of Uranus and its five satellites obtained by Ch. Veillet on the 154-cm Danish-ESO telescope. Outward from Uranus they are: Miranda, Ariel, Ubriel, Titania, and Oberon. (Photograph courtesy of Ch. Veillet.)

composed of water ice, possibly in the form of methane clathrates or ammonia hydrates, and silicate rock. Water ice has been detected spectroscopically on all five satellites. Their relatively dark visual albedos, ranging from 0.13 for Umbriel to 0.33 for Ariel (see Table I), and gray spectrum indicate that their surfaces are contaminated by a dark component such as graphite or carbonaceous material. Another darkening mechanism that may be important is bombardment of the surface by ultraviolet radiation. The higher density of Umbriel implies that its bulk composition includes a larger fraction of rocky material than the other four satellites. Heating and differentiation have occurred on Miranda and Ariel, and possibly on some of the other satellites. Models indicate that tidal interactions may provide an important heat source in the case of Ariel.

Miranda, Ariel, Oberon, and Titania all exhibit large opposition surges, indicating that the regoliths of these bodies are composed of very porous material, perhaps resulting from eons of micrometeoritic "gardening." Umbriel lacks a significant surge, which suggests that its surface properties are in some way unusual. Perhaps its regolith is very compacted, or it is covered by a fine

FIGURE 12   The five major satellites of Uranus, shown to relative size based on *Voyager 2* images. They are, from the left, Miranda, Ariel, Umbriel, Titania, and Oberon.

dust that scatters optical radiation in the forward direction.

The *Voyager 2* spacecraft encountered Uranus in January 1986 to provide observations of satellites that have undergone melting and resurfacing (Fig. 12). Three features on Miranda, known as ''coronae,'' consist of a series of ridges and valleys ranging from 0.5 to 5 km in height (Fig. 13). The origin of these features is uncertain: some geologists favor a compressional folding interpretation, whereas others invoke a volcanic origin or a faulting origin. Both Ariel, which is the geologically youngest of the five satellites, and Titania are covered with cratered terrain transected by grabens, which are fault-bounded valleys. Umbriel is heavily cratered and it is the darkest of the satellites, both of which suggest that its surface is very old, although the moderate-resolution images obtained by Voyager cannot rule out heating or geologic activity. Some scientists have in fact interpreted small albedo variegations on its surface as evidence for melting events early in its history. Oberon is similarly covered with craters, some of which have very dark deposits on their floors. On its surface are situated faults or rifts, suggesting resurfacing events (Voyager provided ambiguous, medium-resolution views of the satellite). In general, the Uranian satellites appear to have exhibited more geologic activity than the Saturnian satellites and Callisto, possibly because of the presence of methane, ammonia, nitrogen, or additional volatiles.

There is some evidence that Umbriel and Oberon, as well as certain regions of the other satellites, contain D-type material, the organic-rich primordial constituent that seems to be ubiquitous in the outer solar system. D-type material is seen in the dark, red D-type asteroids, on the dark side of Iapetus, on Hyperion, and on specific areas of the larger satellites.

2. The Small Satellites of Uranus

Voyager 2 discovered 10 new small satellites of Uranus, including two that act as shepherding satellites for the outer (epsilon) ring of Uranus (see Table I). All of these satellites lie inside the orbit of Miranda. Images of two satellites, Puck and Cordelia, provided sufficient resolution to directly determine their radii (see Table I). The sizes of the other bodies were derived by making the assumption that their surface brightnesses are equal to those of the other inner satellites and estimating the projected area required to yield their observed integral brightnesses. Puck appears to be only slightly nonspherical in shape. It is likely that the other small satellites are irregularly shaped. The satellites' visual geometric albedos range from 0.04 to 0.09, which is slightly higher than that of Uranus's dark ring system. No reliable color information was obtained by *Voyager 2* for any of the small satellites, although their low albedo suggests that they are C-type objects.

D. THE SATELLITES OF NEPTUNE

1. Introduction

Neptune has eight known satellites: one is the large moon Triton and the remaining seven are small, irregularly shaped bodies (see Table I). The small satellites can be divided into two categories: the six inner bodies, which move in highly regular, circular orbits close to Neptune (<5 planetary radii), and the outer satellite Nereid, which moves in an eccentric orbit bringing it from 57 to 385 planetary radii from Neptune. Nereid's orbit is by far the most eccentric of any known natural satellite. Triton has an appreciable atmosphere, seasons, and currently active geologic processing.

Only Triton and the outer satellite Nereid were known before the reconnaissance of Neptune by the *Voyager 2* spacecraft in 1989. Nereid was discovered in 1949 by Gerard P. Kuiper at McDonald Observatory in Texas. In keeping with the theme of water and oceans for the Neptunian system, the satellite was named after the sea nymphs known in Greek mythology as Nereids. Larissa was probably detected in 1981 when it blocked a star that astronomers were measuring for possible occultations by planetary rings. Because it was not subsequently observed or tracked, it was not classified as a satellite until its existence was confirmed by *Voyager* observations. Reliable ground-based observations of Nereid were limited to estimates of its visual magnitude. Telescopic observations reported near the time of the Voyager encounter suggesting that one side of Nereid was significantly brighter than the other were not confirmed by *Voyager* images or by later telescopic observations.

2. Orbital and Bulk Properties

The six inner satellites were all discovered within a few days during the *Voyager* encounter with Neptune in August 1989. They were given names of mythical nautical figures by the International Astronomical Union. For four of these satellites (Proteus, Larissa, Galatea, and Despina), as well as Nereid, Voyager images provided sufficient resolution to determine their dimensions (see Table I). All five bodies are irregularly shaped. The sizes of Thalassa and Naiad were derived by making the assumption that their albedos are equal to those of the other inner satellites. The size of the satellites increases with the distance from Neptune. Proteus is the largest known irregular satellite in the solar system. The satellite has probably not been subjected to viscous relaxation; rather its mechanical properties have been determined by the physics of water ice, with an internal temperature below 110 K.

Spacecraft tracking of the six inner satellites, and ground-based observations of Nereid, provided accu-

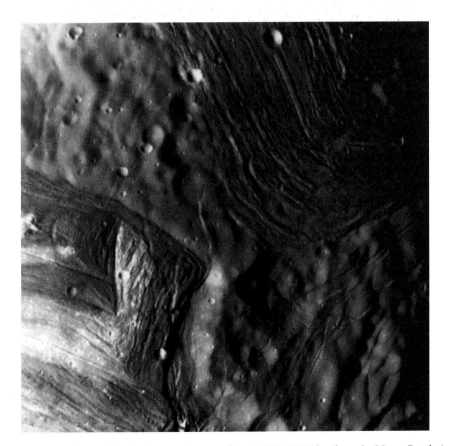

FIGURE 13 Image of Miranda obtained by the *Voyager 2* spacecraft at 30,000–40,000 km from the Moon. Resolution is 560 to 740 m. Older, cratered terrain is transected by ridges and valleys, indicating more recent geologic activity.

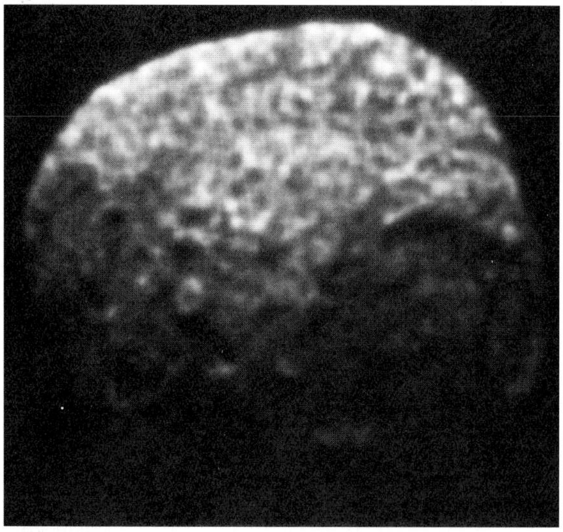

FIGURE 14 The best Voyager image of Proteus, with resolution is 1.3 km/pixel.

rate orbit determinations, which are listed in Table I. All the small inner satellites except Proteus orbit inside the so-called synchronous distance, which is the distance from Neptune at which the rotational spin period equals the Keplerian orbital period. However, Proteus has been tidally despun so that its rotational period equals its orbital period. Voyager observations suggest that Nereid is in nonsynchronous rotation, although its rotational period has not yet been determined. Theoretical calculations suggest that it may be in chaotic rotation, like Hyperion.

The masses of the satellites were not measured directly by Voyager. Limits may be obtained by assuming reasonable values for their bulk densities. These values range from 0.7 g/cm³, corresponding to water ice with a bulk porosity of about 30%, to 2 g/cm³, corresponding to water ice with a significant fraction of rocky material. If the satellites were formed from captured material, it is believed that the higher density is more reasonable. In any case, the small satellites have less than 1% of the mass of Triton. The ring system of Neptune contains only a very small amount of mass, possibly one-millionth of the small satellites' combined masses.

3. Appearance and Composition

Figure 14 depicts the best Voyager images obtained for Proteus, with a resolution of 1.3 km/pixel. The large feature—possibly an impact basin—has a diameter of about 250 km. Close scrutiny of this image reveals a concentric structure within the impact basin. Possible ridgelike features appear to divide the surface. The regions of Proteus outside of the impact basin show signs of being heavily cratered.

The best image of Larissa was obtained at a resolution of 4.2 km/pixel and that of Nereid at a resolution of 43 km/pixel. Neither image has sufficient resolution to depict surface features. In 1988, 1989, and 1991, several ground-based observers reported large lightcurve amplitudes (up to factors of four), which they interpreted as significant albedo variegations on Nereid. Although there is some uncertainty (a factor of two at most) in comparing Voyager to gound-based observations because the orientation of Nereid's spin axis is unknown, a lightcurve produced for Nereid from Voyager images over a 12-day period shows an amplitude of less than 15%.

Analysis of calibrated, integral Voyager measurements of the four inner satellites reveals that their geometric albedos are about 0.06, in the Voyager clear filter with an effective wavelength of about 480 nm. The integral brightness of Nereid is almost three times that of Proteus, which is slightly smaller; its geometric albedo is therefore ~0.20.

The limited spectral data obtained by Voyager suggest that Proteus, Nereid, and Larissa are gray objects. The dark albedos and spectrally neutral character of the inner satellites suggest that they are carbonaceous objects, similar to the primitive C-type asteroids, possibly the Uranian satellite Puck, the satellites of Mars, and several other small satellites. Nereid, however, with its markedly higher albedo, probably has a surface of water frost contaminated by a dark, spectrally neutral material. It is more similar to the differentiated satellites of Uranus than to the dark-C type objects. It is also similar in albedo, size, and color to Phoebe, the outer planet of Saturn, which moves in an inclined, retrograde orbit, suggesting that it is a captured object.

4. Origins and Evolution

The irregular orbit of Nereid indicates that it is probably a captured object. Jupiter, Saturn, and Neptune thus all appear to have outer, captured satellites.

The evolution of the inner satellites was likely punctuated by the capture of Triton. Initially, the inclinations and eccentricities of the satellites would have been increased by the capture, and subsequent collisions would have occurred. The resulting debris would then have reaccreted to form the present satellites. Models of the collisional history of the satellites suggest that with the exception of Proteus they are much younger than the age of the solar system. The heavily cratered surface that appears in the one resolvable *Voyager* image of these bodies (see Fig. 14) does suggest that they have undergone vigorous bombardment.

The only satellite that has been shown to have a dynamical relationship with the rings of Neptune is Galetea, which confines the ring arcs. The orbits of the satellites have probably evolved under the influence of tidal evolution and resonances. For example, the inclination of Naiad is possibly due to its escape from an inclination resonance state with Despina.

ACKNOWLEDGMENTS

Portions of this work were performed at the Jet Propulsion Laboratory, California Institute of Technology, under contract with the National Aeronautics and Space Administration.

BIBLIOGRAPHY

Beatty, J. K., and Chaikin, A. (eds.) (1990). "The New Solar System," 3rd Ed. Sky Publishing, Cambridge, Mass.

Belton, M. J. S., and the Galileo Science Teams (1996). *Science* **274**, 377–413.

Bergstralh, J., and Miner, E. (eds.) (1991). "Uranus." Univ. Arizona Press, Tucson.

Burns, J., and Matthews, M. (eds.) (1986). "Satellites." Univ. Arizona Press, Tucson.

Gehrels, T. (ed.) (1984). "Saturn." Univ. Arizona Press, Tucson.

Hartmann, W. K. (1983). "Moons and Planets," 2nd Ed. Wadsworth, Belmont, Calif.

Morrison, D. (ed.) (1982). "The Satellites of Jupiter." Univ. Arizona Press, Tucson.

Stone, E., and the Voyager Science Teams (1989). *Science* **246**, 1417–1501.

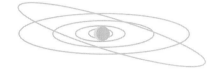

PLANETARY RINGS

I. Introduction

II. Sources of Information

III. Overview of Ring Structure

IV. Ring Processes

V. Ring Origins

VI. Prospects for the Future

Carolyn C. Porco
Department of Planetary Sciences,
University of Arizona

GLOSSARY

Occultation: Obscuration of a body brought about by the passage of another body in front of it. Occultations of spacecraft telemetry signals can arise from passage of the spacecraft behind the rings as seen from the Earth; occulations of stars by the rings can be observed from a moving spacecraft; and occultations of stars by the rings, brought about the motion of the planet against the sky, can be observed from the Earth.

Opposition: Planet beyond the orbit of the Earth is at opposition when it is 180° away from the Sun as seen from the Earth.

Optical depth: Measure of the integrated extinction of light along a path through a medium, such as the disk of particles forming the rings. Normal optical depth refers to the extinction along a path perpendicular to the ring plane.

Plasma: Ionized medium in which electrons have been stripped from neutral matter to make a gas of charged ions and electrons.

Precession: Systematic change in the orientation of an eccentric orbit due to the oblateness of the primary; the slow, periodic, and conical motion of the rotation axis of a spinning body.

I. INTRODUCTION

Planetary rings are those strikingly flat and circular appendages embracing all the giant planets in the outer solar system: Jupiter, Saturn, Uranus, and Neptune. Like their cousins, the spiral galaxies, they are formed of many bodies, independently orbiting in a central gravitational field, and evolved to a state of equilibrium where their random motions perpendicular to the plane are very small compared to their orbital motions. In Saturn's rings, for example, orbital velocities are 10's km/sec while various lines of evidence indicate random motions as small as millimeters per second. The ratio of vertical to horizontal dimensions of the rings is consequently extreme: $1:10^7$. Rings, in general, find themselves in the Roche zone of their mother planet, that region within which the tidal effects of the planet's gravity field prevent similar-sized ring particles from coalescing under their own gravity into a larger body. Ring particles can vary from micron-sized powder to objects as big as houses, and the arrangement of their orbits around the planet can take on a bewildering variety, most of which is still unexplained. [*See* SOLAR SYSTEM DYNAMICS.]

II. SOURCES OF INFORMATION

A. PLANETARY SPACECRAFT

Despite the fact that rings have been observed from the surface of the Earth ever since Galileo Galilei discovered Saturn's rings in 1610, the study of planetary rings did not emerge as the rich field of scientific investigation it is today until the *Voyagers* made their his-

FIGURE 1   The November 1995 sun crossing of Saturn's ring plane observed by the *Hubble Space Telescope*. The image has been computer enhanced to remove the background light from the planet and to enhance the visibility of the rings. [From P. D. Nicholson *et al.* (1996). *Science* **272**, 26.]

toric tours of the outer solar system in the 1980's. Not even the *Pioneer* spacecraft, the first human artifacts to pass through the realms of Jupiter and Saturn in the mid to late 1970s, hinted at the enormous array of phenomena to be found within these systems.

Voyager 1 arrived first at Jupiter in March 1979, followed by *Voyager 2* four months later. After its encounter with Saturn in November 1980, *Voyager 1* was placed on a trajectory that took it out of the solar system; *Voyager 2* encountered Saturn in August 1981 and then sailed on to reach Uranus in January 1986, and Neptune, its last planetary target, in August 1989. Each spacecraft was equipped with a suite of instruments collectively capable of covering a wide range of wavelength and resolution. Over 27,000 images of the planetary ring systems in the outer solar system were acquired by the *Voyager* cameras over a large range in phase angle and at resolutions that are impossible to obtain from the ground. Also, occultations by the rings of bright stars observed from the moving spacecraft with ultraviolet detectors, and occultations by the

rings of the spacecraft telemetry radio signals observed from the Earth, produced maps of the radial architecture of the rings at spatial scales of ∼100 m. In addition to these remote-sensing observations, *in situ* measurements were made of charged particles, plasma waves, and, indirectly, impacts of micron-sized meteoroids as each spacecraft flew through the ring region of each planet. These data sets contributed in varying degrees to the picture that ultimately emerged of the unique character and environment of the rings surrounding the giant planets.

The *Galileo* spacecraft followed on the heels of the *Voyagers* when it was placed in orbit around Jupiter in December 1995. Images of the jovian ring system are few — 29 in number — but have improved resolution and image quality over those obtained by *Voyager*. *Voyagers'* findings of three distinct ring components, and the intimate relationship between these components and the satellites orbiting within them, have been confirmed by *Galileo*, while new ring features have been discovered.

FIGURE 2 The August 10, 1995 Earth crossing of Saturn's ring plane observed by the *Hubble Space Telescope*. [From P. D. Nicholson *et al.* (1996). *Science* **272**, 26.]

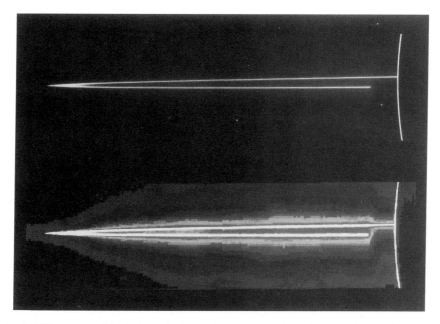

FIGURE 3 A mosaic of *Galileo* images of the jovian main ring and a computer-enhanced version showing the toroidal-shaped "halo" ring interior to the main ring.

FIGURE 4   A *Galileo* image of the outermost jovian gossamer ring.

B. EARTH-BASED OBSERVATIONS

In the decade following the last *Voyager* planetary encounter in 1989, computers have become more powerful, earth-based telescopic facilities and instrumentation have become more sophisticated and sensitive, and the *Hubble Space Telescope* (HST) was placed in orbit around the Earth. These advances have proven invaluable for furthering the study of planetary rings.

1. Stellar Occultations

In a remarkable and rare event, Saturn's rings occulted a very bright near-infrared star, 28 Sagittarii, on July 3, 1989. A voluminous amount of high quality data was collected from a wide geographical net of observatories extending from Chile to Arizona, from Hawaii to Texas. Although the resolution of the raw data was limited to ~20 km, the radial positions of sharp-edged ring features could be located to a precision of ~2 km, rivaling some of the best data collected by *Voyager*. These high-caliber results, in combination with *Voyager* imaging and occultation data taken 9 years earlier, permitted the refinement of the Saturn pole position, the absolute locations of ring features, and the precession rates of nonaxisymmetric features throughout the rings. [*See* SOLAR SYSTEM DYNAMICS.]

Observations of stellar occultations by planetary rings were also observed by the high-speed photometer on the HST before that instrument was removed to make way for HST's corrective optics. Three Saturn ring occultations and two Uranian ring occultations were observed between October 1991 and April 1993, and the results have been used to study the narrow

eccentric rings of both Saturn and Uranus and to further improve the knowledge of Saturn's rotational pole.

2. Ring Plane Crossings

Ring plane crossings, those times when the plane containing a planet's rings sweeps over the Earth or the Sun as the planet moves in its orbit, are unique observational opportunities: at such instances, either the viewer or the source of illumination are exactly in the ring plane. In a ring crossing by Earth, one can view the illuminated rings exactly edge on to determine their thickness or to detect any nonplanar ring components. It is also a time when the entire equator plane of the planet collapses down into a single line as seen from Earth, making it a convenient time to search for diaphanous rings and for faint satellites with near-zero orbital inclination to the equator plane. In a crossing by the Sun, the rings are momentarily unilluminated by direct sunlight; it becomes possible to examine phenomena normally washed out by the bright rings.

Sun crossings occur exactly twice per planetary orbital period, i.e., once every ~15 years for Saturn. Earth crossings can occur multiple times near opposition due to the relatively slow and prograde–retrograde–prograde apparent motion of outer planets like Saturn as seen from the Earth during these times. Earth and solar crossings occur close in time due to the near alignment of the lines of intersection of the orbits of Saturn and Earth with Saturn's ring plane, a circumstance resulting from the small mutual inclination (~2.5°) between the orbits of Earth and Saturn. Consequently, during Earth crossings, the rings are

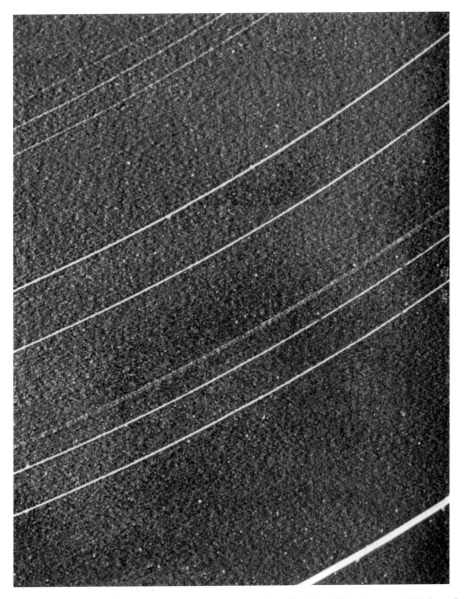

FIGURE 5 A mosaic of *Voyager* images showing the 9 narrow uranian rings. The faint 10th ring, not visible here, lies nearly midway between the two outermost rings.

only barely illuminated by the sun. Saturn ring plane crossings of 1966 and 1980 saw the discoveries of five rather small saturnian satellites (Janus, Epimetheus, Telesto, Calypso, and Helene), as well as Saturn's diffuse E ring. [*See* OUTER PLANET ICY SATELLITES.]

The most recent ring plane crossing events occurred during 1995 and 1996. The Sun crossing occurred during the interval of November 17–21, 1995 (Fig. 1). During this particular ring plane crossing epoch, Earth crossings occurred on three occasions: May 22, 1995, August 10, 1995, and February 11, 1996 (Fig. 2). These events were well imaged by astronomers on the ground

and from the *Hubble Space Telescope*. Findings include refinement in the precession of Saturn's pole and a clearer understanding of the nature of the particles comprising the diffuse E and G rings. The discovery of an orbital displacement in the expected position of one of the innermost saturnian satellites, Prometheus, is the most puzzling and as yet unexplained result.

C. NUMERICAL STUDIES

Recent advances in the speed and design of desk-top computers have taken the study of ring systems into the

FIGURE 6 A *Voyager* backlighted image of the uranian rings showing an extensive sheet of icy, powder-sized particles filling in the ring region. The linear streaks are stars smeared by the spacecraft motion during the exposure.

numerical arena where many dynamically important factors that are not easily treated by analytical methods can be efficiently explored. Numerical methods have been used to simulate a myriad of ring processes, including the collisional and gravitational interactions among orbiting ring particles, the effects of micrometeoroid impacts onto the rings, the behavior of small charged ring particles under the influence of rotating magnetic fields, and the evolution of debris resulting from a catastrophic disruption of a satellite orbiting close to the rings.

III. OVERVIEW OF RING STRUCTURE

Rings are characterized by an enormous variety of structural detail, only some of which has been attributed successfully to known physical processes, either external or internal to the rings (Section IV). Looking across all four ring systems, we find diffuse tenuous rings, like the rings of Jupiter (Figs. 3 and 4). These rings almost certainly have their origin in the release of dust from each of the four small satellites — Adrastea, Metis, Amalthea, and Thebe — which are colocated with the rings. The densest ring of Jupiter, the main ring, has very small normal optical depth, $\tau_N \sim 10^{-6}$ or less, in tiny ($<10~\mu$m) particles; the optical depth may be even smaller for large (>1 mm) particles. It has a relatively sharp outer edge, suspiciously coincident with the orbit of Adrastea; the satellite Metis orbits within the main ring, creating a depression in ring brightness. A $\sim$13,000-km vertically thick toroidal ring, or halo, lies interior to the main ring. The faint outermost gossamer ring extends out to $\sim$222,000 km, about 3 jovian radii. Unlike the rings around Saturn, Uranus, and Neptune, the jovian ring particles appear to be mainly silicate or carbonaceous in composition. They are also likely contaminated by sulfur emanating from the volcanoes on Io. [*See* Io.]

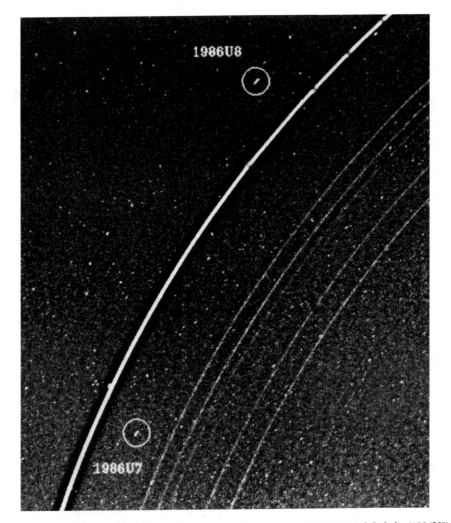

FIGURE 7   The outermost ε ring, shepherded by the small uranian satellites Cordelia (1986U7) and Ophelia (1986U8). The ε ring varies in width. Its average width of ~60 km makes it the widest of the uranian rings. The satellites are streaked by their orbital motion during the exposure. [*See* OUTER PLANET ICY SATELLITES.]

Ten, narrow, sharp-edged continuous rings are found encircling Uranus (Figs. 5 and 6). Most of these are eccentric and some are inclined to Uranus' equator plane by a few parts in 10⁴. Since a ring of colliding debris left to itself would spread in radius, narrow rings and sharp edges require some kind of confining action to keep them so. It is clear that for the outermost ring, ε, the gravitational perturbations to ring orbits from the two small ($r \lesssim 15$ km) satellites on either side of it provide the required mechanism (Fig. 7). The ring's own internal self-gravity is probably responsible for maintaining its eccentric shape and rigid precession. Some of the other Uranian rings also appear, in part, to be similarly controlled by either of the ε shepherding satellites. However, a satisfactory explanation has not yet been found for all the eccentric Uranian rings, some of which (like the middle α and β rings) appear

to have insufficient mass to counter the inward drag forces exerted on the rings by the distended uranian atmosphere.

Extensive sheets of icy powder, like fine snow, conspicuous when backlit by the Sun, fill in the ring systems of Uranus and, possibly, Neptune. Broad (>1000 km) and narrow (<500 km) continuous but diaphanous ($10^{-4} < \tau_N < 10^{-2}$) rings encircle Neptune (Fig. 8), one of which contains the set of discrete, clustered, and denser-than-average arc segments for which Neptune has become famed (Figs. 9 and 10). Neptune's ring system is peculiar in that it shares the Roche region around the planet with a set of four relatively large satellites, ranging in diameter from 60 to 160 km. One of these is Galatea, which is responsible, for the most part, for maintaining the arcs. These satellites, which collectively are nearly as massive as Mimas,

FIGURE 8 A *Voyager* backlighted view of Neptune's ring system showing the main continuous rings; the arcs have not been captured in this mosaic. Only the small powder-sized material can be seen easily in this geometry.

are a Saturn-like ring system waiting to happen (Section V.)

Finally, the rings of Saturn, containing as much mass as the 200 km radius saturnian satellite, Mimas, are home to almost all the ring phenomena described earlier and more: empty gaps in the rings whose widths vary with longitude, narrow uranian-like rings, time-variable markings called "spokes," systematic variations in particle number density and ring height that spiral around the rings, and so on. Saturn's are the only rings whose composition is known with certainty: they are made predominantly of water ice, whereas rocky material seems most likely at Jupiter, and mixtures of ammonia and methane ices coated with carbon are plausible constituents of the rings of Uranus and Neptune. The main saturnian rings are the classical components seen from Earth: A, B, and C (Fig. 11). The narrow F ring immediately outside the main rings was discovered by *Pioneer* (Fig. 12); the innermost D ring (Fig. 13) and the tenuous G ring were not clearly identified as rings until *Voyager* arrived there. The very broad E ring, whose particle number density peaks at the orbit of Enceladus, was discovered from the ground; its nature, and that of the G ring, has been delineated with increasing accuracy most recently by Earth-based observations made during the ring plane crossing events in 1995 (Fig. 14).

Dimensions of the major components in each of the four ring systems are given in Table I.

IV. RING PROCESSES

The fact that certain architectural details are common to all ring systems speaks of common physical processes operating within them. To date, only a subset of planetary ring features can be confidently explained. The physical processes believed to be responsible for the creation of ring features can be broken down into two categories: external and internal.

A. EXTERNAL CAUSES OF RING STRUCTURE

In a disk system of colliding particles on Kepler orbits, angular momentum is naturally transferred outward across the disk, and its rate of flow is related to both the vertically integrated mass per unit area (the surface mass density), Σ, and the kinematic viscosity, ν. (The latter is a parameter, akin to friction, that arises when one uses a fluid model to describe the collisional behavior of ring particles. Such a model has been successfully

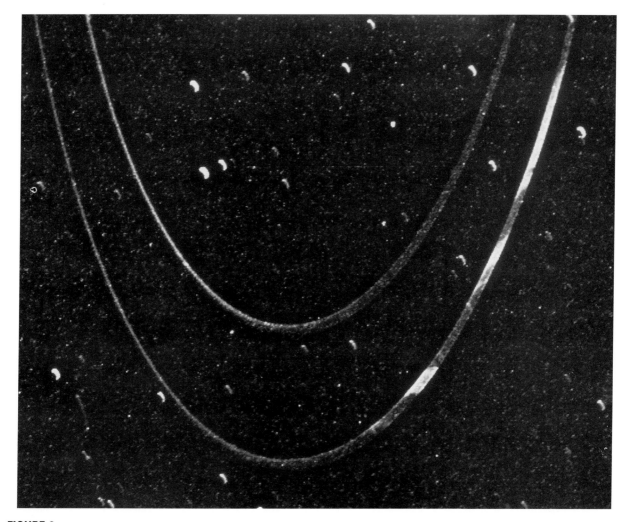

FIGURE 9 A *Voyager* image of the two outermost neptunian rings. Orbital motion is clockwise. The direction of motion of the spacecraft during the exposure, indicated by the trailed stars, has smeared the arcs to look much wider than their natural widths of ~15 km.

applied to certain ring regions.) The degree to which external perturbations on ring particle orbits will create visible disturbances in a featureless disk system depends on the ring's natural ability to keep up with the rate of change in angular momentum imposed on it by the external perturbation. If the angular momentum is removed or deposited by external means at a rate that is less than the ring's ability to transport it away from the excitation region, then the ring response (if any) will take the form of a wave. If the rate of removal or deposition is greater, then the rings will respond by opening a gap, i.e., the particles themselves must physically move, carrying angular momentum with them, to accommodate the external driving force.

Operating over sufficiently long time scales, external perturbations can create a staggering variety of features in planetary rings. Satellites are the most common source of external disturbances. The two outermost major ring elements comprising Saturn's rings — the very dense B ring and the intermediate optical depth A ring — contain many examples of features caused by external perturbations of satellites. For example, the 320-km-wide Encke gap in the outer A ring is believed to be maintained against collisional diffusion by the gravitational perturbations of the 20-km-diameter satellite, Pan, orbiting within it (Fig. 15); radial oscillations of characteristic azimuthal wavelength ~0.7° seen in the edges of this gap are also attributable to this small satellite. Throughout the rings can be found spiral density and bending waves, in which variations in either the particle number density or the height of the rings above the equator plane of the planet take on a tightly wound spiral pattern. Such waves are created by the action of gravitational

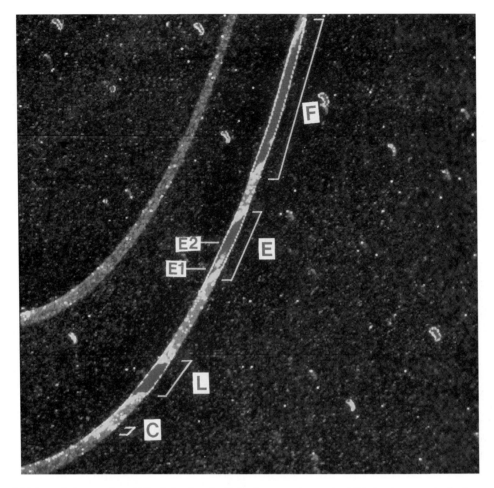

FIGURE 10 A high resolution portion of Fig. 9 showing details in the arc region made visible by the smear in the image. The positions of the arcs are indicated. The official names are Courage (C), Liberté (L), Egalité (E), and Fraternité (F). The Egalité "arc" is composed of two arcs. [From C. C. Porco *et al.* (1995). *In* "Neptune and Triton" (D. P. Cruikshank, ed.). Copyright © 1995 by The Arizona Board of Regents. Reprinted by permission of the University of Arizona Press.]

resonances due to satellites, such as the F ring shepherds, Janus/Epimetheus (the coorbital satellites of Saturn), and Mimas, which are capable of exciting either radial or vertical motion in the ring particles' orbits. The coorbitals and Mimas are responsible for the two strongest resonances within Saturn's rings, which create the seven- and two-lobed patterns of radial oscillations observed in the sharp outer edges of the A and B rings, respectively.

At Uranus, it is clear that the particles within the ε ring are shepherded in their movement around the planet by the gravitational perturbations of two small satellites on either side of it, Cordelia and Ophelia. At Neptune, the eccentricities of the arc particle orbits are excited by a gravitational resonance with the satellite Galatea into a coherent 30-km amplitude radial distortion that travels through the arcs at the orbital speed of the satellite. This particular resonance also seems

capable of confining the arcs both in radius and azimuth — one satellite doing double duty — although it alone may not be sufficient to explain the observed configuration of arcs. Small, kilometer-sized bodies embedded within the ring or arcs might assist Galatea in arc confinement as well as slow the rapid retreat of the arcs from the satellite. (Unfortunately, satellites this size are well below the detection limit in *Voyager* images.)

Other sources of external driving can arise from the planet. The magnetic field may affect ring particles if they have sufficiently large charge-to-mass ratios. Jupiter's halo ring particles are small enough that the charges they accumulate can allow electromagnetic forces to dominate gravity. The inner and outer boundaries of Jupiter's toroidal ring seem to be delineated by the perturbations of the magnetic field. If ring particles find themselves orbiting within a distended

FIGURE 11   A *Voyager* image of Saturn's main rings showing the main components, A, B, and C. The C ring is the innermost and most optically thin component. The B ring is the middle and densest component. The A ring is the outermost component whose middle region is nearly devoid of features. It is separated from the B ring by the Cassini Division. The very narrow F ring is seen beyond ring A. The globe of Saturn is extremely overexposed to bring out the details in the rings, a circumstance that enhances the shadow cast by the rings across the equatorial region of the planet.

planetary atmosphere, as at Uranus, the resulting drag forces can gradually affect their orbits, leading to a drift, either inward or outward, with time.

A planet undergoing acoustic oscillations may also be a source of resonant gravitational perturbations on ring particles. Acoustic oscillations, which have been observed in the Sun, arise from variations in the internal density of a fluid planet that can take on coherent patterns and a variety of frequencies as they circumnavigate the planet. These rotating density variations produce commensurate variations in the planet's external gravity field and can, in principle, be strong enough to resonantly excite ring particle eccentricities and inclinations. This theory makes predictions about the attributes of oscillation-produced ring features that cannot be tested against currently available data on Saturn's rings, but will be tested against the very high resolution ring observations that can be expected when *Cassini* begins a thorough examination of the Saturn system in 2004 (Section VI).

Yet another possibility for externally creating ring structure arises from the redistribution of mass and angular momentum caused by meteoroid bombardment of the rings. Saturn's rings present a large surface area — twice that of Saturn itself — to the hail storm of interplanetary debris raining down on them. The total mass falling onto the rings in the age of the solar system may be greater than the mass of the rings themselves; this process is therefore likely to be a major contributor to ring erosion and modification. Its effects have been studied with numerical computer methods. Results show that the rate of change in orbital distance of affected ring particles can be several centimeters per yer. At this rate, the entire C ring of Saturn will fall into the planet in $\sim 10^8$ years. Because the ejecta from each impact are distributed preferentially in one direction, meteoroid bombardment provides a mechanism for altering radial structure, especially when the initial radial distribution of mass is grossly nonuniform as near an abrupt and large change in optical depth. The shapes of the inner edges of the A and B rings and features near them can be roughly explained by this process and may take as little as $\sim 10^7$ to 10^8 years to evolve to their currently observed configurations.

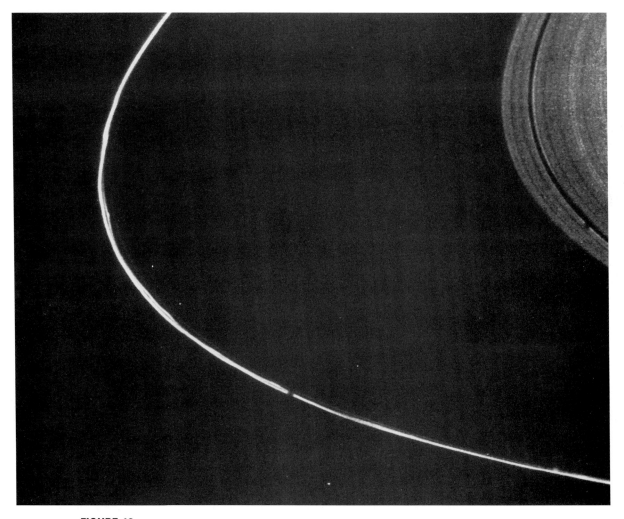

FIGURE 12 A high-resolution *Voyager* image of the F ring showing its stranded and braided structure.

These results offer the hope that other structural features in the rings of Saturn and those of the other giant planets may be explainable by this process. However, its effectiveness depends sensitively on the amount of external material impacting the rings, a quantity that is presently very uncertain and will be investigated by the *Cassini* mission.

The impacts of micrometeoroids onto Saturn's rings have been proposed as the first step in the production of spokes, those ghostly features in the B ring that come and go while revolving around Saturn (Fig. 16; see also color insert). Spokes are almost certainly powder-sized ice particles that have been lifted off bigger ring particles; the elevation mechanism is believed to involve the electrical charging of these small particles by transient plasma clouds created during an impact. The fact that the contrast and areal coverage of spokes vary with a period equal to that of Saturn's magnetic field supports the suggestion that electromagnetic forces operating on initial charged particles play a role in spoke creation and development.

B. INTERNAL CAUSES OF RING STRUCTURE

Two physical concepts underlie the internal workings of ring systems: the availability of free mechanical energy due to the difference in orbital velocities across the rings, and the viscosity and energy dissipation that arises from the presence and the inelastic nature of collisions among ring particles. Collisions tap the free mechanical energy available due to the differential orbital velocity, leading to random motions among the particles. The energy in these random motions is dissipated during collisions due to the chipping, cracking, compaction, and sound propagation through the parti-

FIGURE 13 A *Voyager* image of the very optically thin saturnian D ring, which falls interior to ring C.

cles. Equilibrium between these two processes, achieved after several orbital periods, yields a monotonically decreasing relation between the random velocity, $\bar{v}$, and optical depth, τ. That is, in equilibrium each ring region characterized by a particular optical depth (or surface mass density) has a unique random velocity. At low τ, the random velocities and ring thickness tend to be large; at high τ, the reverse is true. In this case, the classic relation for $\Sigma\nu$, the only locally changing quantity in the expression for the flow of

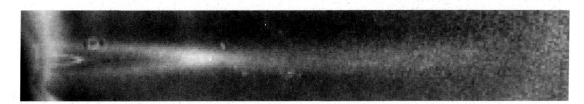

FIGURE 14 A *Hubble Space Telescope* image processed to reduce the scattered light from the planet, showing the G ring (inner) and the very broad (outer) E ring of Saturn. North is down. The image was taken during the solar ring plane crossing period of November 1995. Diagonal features in the lower left corner and the "donut" to the upper right of the G ring are artifacts. The brightest portion of the E ring is coincident with the orbit of Enceladus, but the remainder of the E ring extends well beyond that. [Reprinted with permission from P. D. Nicholson *et al.* (1996). *Science* **272**, 26. Copyright 1996 American Association for the Advancement of Science.] [*See* OUTER PLANET ICY SATELLITES.]

TABLE I
Locations of Major Ring Components

Planet	Ring component[a]	Radial location and extent
Jupiter (1 R_J = 71,492 km)	Halo (toroidal)	92,000–122,500 km
	Main	122,500–128,940 km
	Gossamer	181,000–222,000 km
Saturn (1 R_S = 60,330 km)	D	67,000–74,500 km
	C	74,500–92,000 km
	B	92,000–117,580 km
	Cassini division	117,580–122,200 km
	A	122,200–136,780 km
	F	140,200 km
	G	166,000–173,000 km
	E	181,000–483,000 km
Uranus (1 R_U = 26,200 km)	6	41,837 km
	5	42,235 km
	4	42,571 km
	α	44,718 km
	β	45,661 km
	η	47,176 km
	γ	47,627 km
	δ	48,300 km
	λ	50,024 km
	ε	51,149 km
Neptune (1 R_N = 25,225 km)	Galle	41,000–43,000 km
	Le Verrier	53,200 km
	Lassell/Arago	53,200–57,500 km
	Adams	62,932 km

[a] The saturnian F ring has multiple narrow strands; the ring's core is ~1 km wide. The uranian rings are all <12 km wide, except the ε ring, which varies from ~20 to ~100 km in width. The uranian η ring has a diffuse component that extends ~55 km beyond the ring. Arago is a narrow feature that forms the outside edge of Lassell. Adams and Le Verrier are ~100 km wide, except for the arcs in the Adams ring, which are ~15 km wide.

angular momentum across the rings, is a peaked function of optical depth. This implies that contiguous ring regions differing widely in optical depth can stably exist side by side by supporting the same value of angular momentum flow.

The mechanism responsible for this outcome was regarded as a possibility for explaining the large degree and variety of ring structure in Saturn's B ring (Fig. 17) until laboratory measurements on ice particles indicated that the particle inelasticity was higher, and consequently the rings were thinner and the particle number density larger, than originally believed. When it was recognized that very dense rings and highly inelastic collisions violate the principles of kinetic theory on which much of ring theory was based, it became necessary to introduce a new effect into the theory: the transport of angular momentum (via sound waves)

across a tightly packed system of orbiting particles. The result of adding this effect, which becomes important in high τ regions, was to change the variation of $\Sigma\nu$ with ring optical depth from a peaked function to a monotonically increasing function. On the basis of this conclusion, it seemed impossible for ring regions of differing optical depth, and therefore differing natural angular momentum flow, to exist stably side by side. Other possibilities, however, have been suggested to explain the fine-scale structure within a dense ring like the Saturn B ring. These suggestions include adjacent narrow ring regions alternating in behavior between a liquid and a solid and the possibility that density waves may be driven to the point of instability in very dense ring regions.

V. RING ORIGINS

Three distinct scenarios are considered plausible for the origin of rings, although each meets with some difficulty when trying to explain all observed ring systems simultaneously. (1) Rings may be the inner unaccreted remnants of the circumplanetary nebulae that ultimately formed the satellite systems surrounding each planet; (2) they may be the debris that resulted from one or more satellites, tidally evolving inward into the Roche zone, being completely disrupted by cometary or meteoroid impacts, and evolving quickly into a ring system, replete with small embedded satellites (Fig. 18); or (3) they may be the result of the disruption of an icy planetesimal in heliocentric orbit that strayed too close to the planet and was simultaneously captured and torn apart by planetary tides, such as Comet Shoemaker-Levy 9 (SL-9), and evolved into a ring/satellite system (Fig. 19).

Several lines of circumstantial evidence indicate that planetary rings may be young and not primordial. The most reliable evidence rests on the calculation of the rate of separation expected in the orbits of satellites and ring particles locked in gravitational resonance, e.g., the predicted recession of the Saturnian F ring shepherds from the A ring due to their resonant interactions with A ring particles. Simple inverse extrapolation of these rates brings the satellites and rings into physical coincidence roughly 10^6 years ago. Others, such as the estimates for the lifetime of all rings against erosion by micrometeoroid impacts, yield similar time scales. On the basis of these arguments, an ancient origin for any of the present ring systems seems unlikely.

The second and third possibilities are somewhat

FIGURE 15   Four computer-enhanced *Voyager* images showing four different views of Pan as it orbits within the Encke gap.

more appealing at first glance. The large number of satellites presently orbiting each of the giant planets, and the ever-increasing discoveries of icy planetesimals found in the Kuiper Belt (a suspected source of planet-crossing bodies), indicate sufficient fodder for ring creation. The fate of Comet SL-9 lent dramatic support to the idea of a ruptured planetesimal origin for rings, and interpretation of the crater populations on the surfaces of outer solar system satellites suggests that satellite disruption must have been a common event in the past. [*See* KUIPER BELT.]

However, all three scenarios run aground in one way or the other, especially in the case of Saturn. Some theoretical models, and several parameters, used in calculating both the orbital changes in satellites and rings due to resonant interactions, and in calculating the rate of meteoroid impacts onto the rings, are only poorly known. Thus, the best arguments against ancient rings are still inconclusive. Also, the possibility of creating Saturn's massive ring system in the recent past from satellite disruption is rather low, as Mimas-sized bodies within the Roche zone are nonexistent now and probably were rare in the past. (However, the opposite is the case at Jupiter, Uranus, and Neptune, where each has sufficient mass even today in ring-region satellites to create the present ring systems.) Finally, the disruption of a planetesimal large enough to create Saturn's rings is not likely at Saturn: the low material density of Saturn itself makes breakups, ala SL-9, rather unlikely, and the frequency with which large icy planetesimals cross the orbit of Saturn is too low to make tidal disruption and capture a plausible outcome.

So while any of the proposed mechanisms are plausible for making the ring systems of Jupiter, Uranus, and Neptune, with youthful rings having the edge in these three cases, the origin of Saturn's massive ring system remains an unsolved mystery.

VI. PROSPECTS FOR THE FUTURE

Despite improvements in ground-based observing facilities and instrumentation that can be expected in the future, the next most spectacular advance in the study of rings will certainly come when the *Cassini* spacecraft, launched in October 1997, arrives at Saturn in the summer of 2004. *Cassini* is equipped with instruments capable of remotely sensing the rings from the

FIGURE 16 A *Voyager* image of the main ring system showing spokes predominantly on the eastern half of the B ring, which is generally where they are most visible. By the time the ring particles have rotated around to the western side of the rings, most of the elevated particles comprising the spokes have settled down to the ring plane again, making the spokes nearly invisible. (*See* also color insert.)

ultraviolet through the microwave, as well as making *in situ* measurements of the number and nature of micrometeoroids impacting the rings. Results expected to come from *Cassini* include the discovery of satellites well below the *Voyager* detection limit ($r \sim 6$ km), both internal and external to the rings; the rate of recession of resonance-producing satellites and concomitantly a better estimate for the age of the rings; high resolution maps of the rings' composition and radial structure; the detailed behavior and dynamics of time-variable features; and a better understanding of the mechanisms responsible for the variety of still-unexplained ring architecture.

The ring systems of today offer invaluable insights into the processes operating in primordial times in the flattened system of gas and colliding debris that ultimately formed the solar system. Yet almost all the results on the internal workings of Saturn's rings that will come from *Cassini* — the collisional frequency and elasticity of ring particles, the kinematic viscosity, and self-gravity — will be made on the basis of inference, as direct imaging of ring particles and their interactions will be impossible from the trajectory that *Cas-*

sini will follow through the Saturn system. [*See* THE ORIGIN OF THE SOLAR SYSTEM.]

For this reason, it is likely that in the not-too-distant future we will dispatch, to follow in the wake of *Cassini*, a small spacecraft capable of hovering over the rings of Saturn or orbiting within one of the large ring gaps. Views of the rings from this vantage point will capture ring particles — big and small — in the acts of colliding, chipping, breaking, coalescing, and so on. With observations like these, planetary scientists will be given an unprecedented opportunity to determine precisely how much energy and angular momentum are lost or exchanged during each collision and to peer into the past, as it were, and observe some of the processes at work in the solar nebula disk.

Finally, sometime during the first half of the next century, an orbiter should arrive at the planet Neptune. Time spent monitoring Neptune's inner satellites and rings will inform a deeper understanding of that timeless ballet danced by planetary rings and their satellite companions. It will give earthlings a chance to witness a collection of bodies that may one day find themselves ripped asunder and wreathing beautiful blue Neptune

FIGURE 17 A high-resolution *Voyager* image of Saturn's B ring. The smallest visible features are about 4 km across. This is the densest portion of the B ring in which there are no gaps.

FIGURE 18 An artist's rendition of a satellite being catastrophically disrupted by a meteoroid or cometary collision, and on the verge of becoming a ring system. Painting by William K. Hartmann.

FIGURE 19   An artist's rendition of a large icy planetesimal, the size of a typical Kuiper Belt object, approaching Saturn. Painting by William K. Hartmann. [*See* THE KUIPER BELT.]

in a ring system that rivals Saturn's. Future generations of planetary ring explorers have much to look forward to.

BIBLIOGRAPHY

Beatty, J. K., and Chaikin, A. (eds.) (1990). "The New Solar System," 3rd Ed. Sky Publishing Corporation and Cambridge Univ. Press, Cambridge, MA.

Durisen, R. H., Bode, P. W., Cuzzi, J. N., Cederbloom, S. E., and Murphy, B. W. (1992). *Icarus* **100**.

French, R. G., Nicholson, P. D., Porco, C. C., and Marouf, E. A. (1991). "Uranus" (J. T. Bergstrahl, E. D. Miner, and M. S. Matthews, eds.). Univ. of Arizona Press, Tucson.

Marley, M. S., and Porco, C. C. (1993). *Icarus* **106**.

Nicholson, P. D., and Dones, L. (1991). *Rev. Geophys.* **29**.

Ockert-Bell, M. E., Burns, J. A., Daubar, I. J., Thomas, P. C., Veverka, J., Belton, M. J. S., and Klaasen, K. P. (1998). *Icarus*. Submitted for publication.

Porco, C. C., and Goldreich, P. (1987). *Astron. J.* **93**.

Porco, C. C., Nicholson, P. D., Cuzzi, J. N., Lissauer, J. J., and Esposito, L. W. (1995). "Neptune and Triton" (D. Cruikshank, ed.). Univ. of Arizona Press, Tucson.

PLANETARY MAGNETOSPHERES*

I. What Is a Magnetosphere?

II. Types of Magnetospheres

III. Planetary Magnetic Fields

IV. Magnetospheric Plasmas

V. Dynamics

VI. Interactions with Moons

VII. Conclusions

M. G. Kivelson[1] and F. Bagenal[1,2]

[1]University of California, Los Angeles, and
[2]University of Colorado

GLOSSARY

Alfvén speed: Speed of propagation of disturbances in a magnetized plasma that bend a magnetic field without changing its magnitude.

Contact surface: In the vicinity of a comet, the surface that separates outflowing cometary plasma from the slowed solar wind that is approaching the comet.

Coulomb collision: Charged particles interact at large distances through the Coulomb force. The interaction is called a Coulomb collision because it produces a change of particle momentum similar to that in a conventional collision.

Flux transfer event: Localized spatial region in which magnetic reconnection links the solar wind magnetic field to a planetary magnetic field, producing a configuration that transports flux from the dayside to the nightside of the planet.

Geomagnetic activity: Disturbances in the magnetized plasma of a magnetosphere associated with fluctuations of the surface field, auroral activity, reconfiguration and changing flows within the magnetosphere, strong ionospheric currents, and particle precipitation into the ionosphere.

Gyrofrequency: Frequency of the circular motion of a charged particle perpendicular to a magnetic field.

Gyroradius: Radius of the (circular) orbit of a charged particle traced by its component of motion perpendicular to a magnetic field.

Heliopause: Interface between outflowing solar wind plasma and interstellar plasma.

Heliosphere: Volume of space around the Sun within which the solar wind is largely confined by interstellar plasma.

Ionopause: Surface separating ionospheric plasma and the solar wind in the vicinity of an unmagnetized planet.

Ionosphere: Highly ionized upper atmosphere of a planet bound by collisional or gravitational forces.

---

* UCLA Institute of Geophysics and Planetary Physics Publication No. 4857.

Lorentz force: Total electromagnetic force on a charged particle.

Magnetic storm: Prolonged interval of intense geomagnetic activity, often lasting for days.

Magnetosphere: Volume of space carved out of the solar wind in the vicinity of a magnetized planet within which plasmas are linked magnetically to the planet.

Maxwellian distribution: Distribution of particle velocities for a gas in thermal equilibrium.

Plasmoid: Region within a magnetosphere in which plasma is confined by a magnetic structure that is not directly linked to the planet.

Reconnection: Process in which the magnetic configuration changes as if two field lines were broken and reconnected in a new configuration. This can occur when two plasmas containing oppositely directed magnetic fields flow toward each other.

Ring current: Current carried by energetic particles that flows at radial distances beyond a few planetary radii in the near-equatorial regions of a planetary magnetosphere.

Solar wind: Magnetized, highly ionized plasma that flows radially out from the solar corona at supersonic and super-Alfvénic speed.

Space weather: Variable level of geomagnetic activity controlled by the conditions in the solar wind.

Substorm: Elementary disturbance of the magnetosphere that produces geomagnetic activity.

Thermal diffusion: Heat transport resulting from a temperature gradient in a solid body.

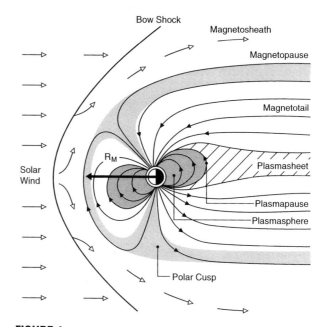

FIGURE 1   Schematic illustration of the Earth's magnetosphere. The Earth's magnetic field lines are shown as modified by the interaction with the solar wind. The solar wind, whose flow speed exceeds the speeds at which perturbations of the field and the plasma flow directions can propagate in the plasma, is incident from the left. The pressure exerted by the Earth's magnetic field excludes the solar wind. The boundary of the magnetospheric cavity is called the magnetopause, its nose distance being R_M. Sunward (upstream) of the magnetopause, a standing bow shock slows the incident flow and the perturbed solar wind plasma between the bow shock and the magnetopause is called the magnetosheath. Antisunward (downstream) of the Earth, the magnetic field lines stretch out to form the magnetotail. In the northern portion of the magnetotail, field lines point generally sunward, whereas in the southern portion, the orientation reverses. These regions are referred to as the northern and southern lobes, and they are separated by a sheet current flowing generally dawn to dusk across the near-equatorial magnetotail in the plasmasheet. Low-energy plasma diffusing up from this ionosphere is found close to Earth in a region called the plasmasphere, whose boundary is the plasmapause.

I. WHAT IS A MAGNETOSPHERE?

The term magnetosphere was coined by T. Gold in 1959 to describe the region above the ionosphere in which the magnetic field of the Earth controls the motions of charged particles. The control by the planetary magnetic field extends many Earth radii into space, but finally terminates near 10 Earth radii in the direction toward the Sun. At this distance, the magnetosphere is confined by a low-density, magnetized plasma called the solar wind, which flows radially outward from the Sun at supersonic speeds. Qualitatively, a planetary magnetosphere is the volume of space from

which the solar wind is excluded by a planet's magnetic field. (A schematic illustration of the terrestrial magnetosphere is given in Fig. 1, which shows how the solar wind is diverted around a surface called the magnetopause that encloses the Earth.) This qualitative definition is far from precise. Most of the time, solar wind plasma is not totally excluded from the region that we call the magnetosphere. Some solar wind plasma finds its way in and indeed many important dynamical phenomena result from intermittent direct links between the solar wind and the plasmas governed by a planet's magnetic field. Moreover, unmagnetized planets in the flowing solar wind carve out cavities whose properties are sufficiently similar to those of true magnetospheres to allow us to regard them as such. Moons embedded in the flowing plasma of a planetary magnetosphere create interaction regions that resemble the regions surrounding unmagnetized planets. If a moon is sufficiently strongly magnetized, it may carve out a true magnetosphere completely contained within the magnetosphere of the planet.

Magnetospheric phenomena are of both theroretical and phenomenological interest. Theory has benefited from the data collected in the vast plasma laboratory of space in which different planetary environments provide the analogue of different laboratory conditions. Furthermore, magnetospheric plasma interactions are important to diverse elements of planetary science. For example, plasma trapped in a planetary magnetic field can interact strongly with the planet's atmosphere, heating the upper layers, generating neutral winds, and ionizing the neutral gases. Energetic ions and electrons that precipitate into the atmosphere can modify atmospheric chemistry. Interaction with plasma particles can contribute to the isotopic fractionation of a planetary atmosphere over the lifetime of a planet. Impacts of energetic charged particles on the surfaces of planets and moons can modify surface properties, changing their albedos and spectral properties. The motions of charged dust grains in a planet's environment are controlled by electrodynamic forces as well as by gravity. Recent studies of dusty plasmas show that the electrodynamic forces may be critical in determining the role and behavior of dust in the solar nebula as well as in the present-day solar system.

In Section II, the different types of magnetospheres and related interaction regions are introduced. Section III presents the properties of observed planetary magnetic fields and discusses the mechanisms that produce such fields. Section IV reviews the properties of plasmas contained within magnetospheres, describing their distribution, their sources, and some of the currents that they carry. Section V covers magnetospheric dynamics, both steady and "stormy." Section VI addresses the interactions of moons with planetary plasmas. Section VII concludes the article with remarks on plans for future space exploration.

II. TYPES OF MAGNETOSPHERES

A. THE HELIOSPHERE

The solar system is dominated by the Sun, which forms its own magnetosphere referred to as the heliosphere.

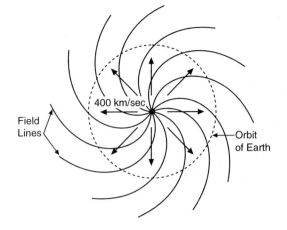

FIGURE 2 The spiral orientation of the average solar wind magnetic field. The solar wind is shown flowing radially outward from the Sun at 400 km s^{-1} and distorting the interplanetary magnetic field into a spiral configuration.

The size and structure of the heliosphere are governed by the motion of the Sun relative to the local interstellar medium, the density of the interstellar plasma, and the pressure exerted on its surroundings by the outflowing solar wind that originates in the solar corona. The corona is a highly ionized gas that is so hot that it can escape the Sun's immense gravitational field and flow outward at supersonic speeds. The solar wind is threaded by magnetic field lines that map back to the Sun. A useful and picturesque description of the field contained within a plasma relies on the idea that if the conductivity of a plasma is sufficiently large, the magnetic field is frozen into the plasma, and field lines can be traced from their source by following the motion of the plasma within which it is frozen. Because the roots of the field lines remain linked to the rotating Sun (the Sun rotates about its axis with a period of approximately 25 days), the field lines twist in the form of an Archimedean spiral as illustrated in Fig. 2.

The solar wind speed is greater than the Alfvén speed [$v_{A} = B/(\mu_{0}\rho)^{1/2}$], the speed at which rotational perturbations of the magnetic field propagate along the magnetic field in a magnetized plasma. Here B is the magnetic field magnitude, μ_{0} is the magnetic permeability of vacuum, and ρ is the mass density of the plasma. The outflow of the solar wind flow along the direction of the Sun's motion relative to the interstellar plasma is terminated by the forces exerted by the interstellar plasma. Elsewhere the flow is diverted within the boundary of the heliosphere. Thus the Sun and the solar wind are (largely) confined within the heliospheric cavity; the heliosphere is the biggest of the solar system magnetospheres. [*See* SOLAR WIND.]

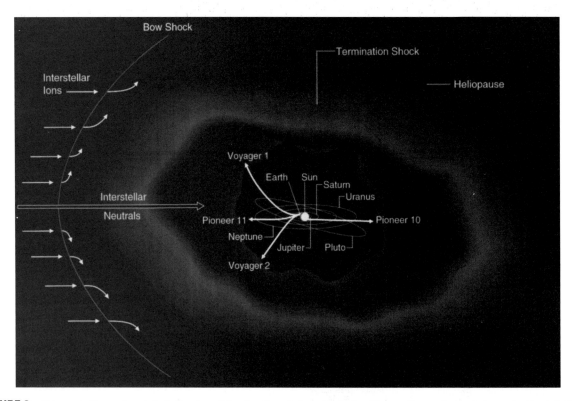

FIGURE 3 Schematic illustration of the heliosphere. The direction of plasma in the local interstellar medium relative to the Sun is indicated and the boundary between solar wind plasma and interstellar plasma is identified as the heliopause. An internal shock, referred to as the termination shock, is shown within the heliopause. Such a shock, needed to slow the supersonic solar wind inside of the heliopause, is a new feature in this type of magnetosphere. Beyond the heliopause, the interstellar flow is diverted around the heliosphere and a shock probably exists beyond the diverted flow.

Our knowledge of the heliosphere beyond the orbits of the giant planets is principally theoretical, but our expectations regarding its properties are founded on information that is well established. We know that the solar wind density decreases as the inverse square of the distance from the Sun and we are confident that, as the plasma becomes sufficiently tenuous, the pressure of the interstellar plasma impedes its further expansion. The solar wind must slow down across a shock (the termination shock) before reaching the boundary that we call the heliopause, which excludes the interstellar plasma. (The different plasma regimes are schematically illustrated in Fig. 3.) The distance to the heliopause is estimated as roughly 100 AU (AU is an astronomical unit, equal to the mean radius of Earth's orbit or about 1.5×10^8 km), a region not yet probed by spacecraft. Although we have not yet crossed the heliopause, we are confident that it is there because our sound theoretical arguments are supported by tantalizing hints that the boundary lies not far beyond our most distant spacecraft. One type of hint is given by bursts of radio noise that do not weaken with distance from known sources within the solar system. The termination shock is a plausible source of such noise. Another hint comes from extremely energetic singly charged particles that could have been ionized and accelerated near the termination shock. Both processes are consistent with a shock at roughly 100 AU. *Voyager 1* is now more than 69 AU from the Sun. In the first decade of the twenty-first century, we expect that it will encounter the termination shock and provide direct evidence for the models that have been developed.

B. MAGNETOSPHERES OF THE UNMAGNETIZED PLANETS

Earth has a planetary magnetic field that has long been used as a guide by travelers such as scouts and sea voyagers. However, not all of the planets are magnetized. Table I summarizes some key properties of some of the planets, including their surface mag-

TABLE I
Properties of the Solar Wind and Scales of Planetary Magnetospheres

	Mercury	Venus	Earth	Mars	Jupiter	Saturn	Uranus	Neptune	Pluto
Distance, a_{planet} (AU)[a]	0.31–0.47	0.723	1[b]	1.524	5.2	9.5	19	30	30–50
Solar wind density (amu cm^{-3})[b]	35–80	16	8	3.5	0.3	0.1	0.02	0.008	0.008–0.003
Radius, R_{Planet} (km)	2,439	6,051	6,373	3,390	71,398	60,330	25,559	24,764	1,170 ($\pm$33)
Surface magnetic field B_0 (gauss = 10^{-4} T)	3×10^{-3}	$<2 \times 10^{-5}$	0.31	$<10^{-4}$	4.28	0.22	0.23	0.14	?
R_{MP} (planetary radii)[c]	1.4–1.6R_M	—	10R_E	—	42R_J	19R_S	25R_U	24R_N	?
Observed size of magnetosphere (km)	1.4R_M	—	8–12R_E	—	50–100R_J	16–22R_S	18R_U	23–26R_N	?
	3.6×10^3	—	7×10^4	—	7×10^6	1×10^6	5×10^5	6×10^5	?

[a] 1 AU = 1.5×10^8 km.
[b] The density of the solar wind fluctuates by about a factor of 5 about typical values of $\rho_{sw} \sim [(8 \text{ amu cm}^{-3})/a_{planet}^2]$.
[c] R_{MP} is calculated using $R_{MP} = (B_0^2/2\mu_0\rho u^2)^{1/6}$ for typical solar wind conditions of ρ_{sw} given above and $u \sim 400$ km s^{-1}.

netic field strengths. Neither Mars nor Venus has a planetary magnetic field. The nature of the interaction between an unmagnetized planet and the supersonic solar wind is determined by the electrical conductivity of the body. If conducting paths exist across the planet's interior or ionosphere, then electric currents flow through the body and into the solar wind, where they create forces that slow and divert the incident flow. The diverted solar wind flows around a region that is similar to a planetary magnetosphere. Mars and Venus have ionospheres that provide the required conducting paths. The barrier that separates planetary plasma from solar wind plasma is referred to as an ionopause. The analogous boundary of the magnetosphere of a magnetized planet is called a magnetopause. Earth's Moon, with no ionosphere and a very low conductivity surface, does not deflect the bulk of the solar wind incident on it. Instead, the solar wind runs directly into the surface, where it is absorbed. The absorption leaves the wake, the region immediately downstream of the Moon in the flowing plasma, devoid of plasma, but the void fills in as solar wind plasma flows toward the center of the wake. The different types of interaction are illustrated in Fig. 4.

The magnetic structure surrounding Mars and Venus has features much like those found in a true magnetosphere surrounding a strongly magnetized planet. This is because the interaction causes the magnetic field to drape around the planet. The draped

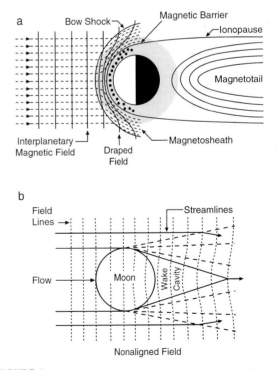

FIGURE 4 Schematic illustrations of the interaction regions surrounding (a) a planet like Mars or Venus, which is sufficiently conducting that currents close through the planet or its ionosphere, and (b) a body like the Moon, which has no ionosphere and low surface and interior conductivity. [Adapted from J. G. Luhmann (1995). *In* "Introduction to Space Physics" (M. G. Kivelson and C. T. Russell, eds.). Cambridge Univ. Press, reprinted with permission.]

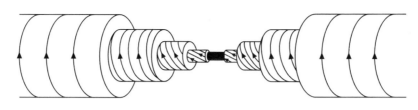

Interior Structure of Flux Rope

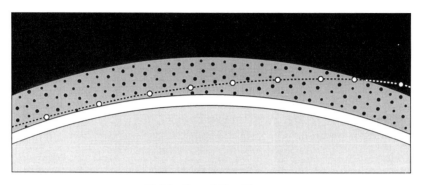

Distribution of Flux Ropes

FIGURE 5 Schematic illustration of a flux rope, a magnetic structure that has been identified in the ionosphere of Venus (shown as black dots within the ionosphere) and extensively investigated (a low-altitude pass of the *Pioneer Venus Orbiter* is indicated by the dashed curve). The rope (see above) has an axis aligned with the direction of the central field. Radially away from the center, the field wraps around the axis, its helicity increasing with radial distance from the axis of the rope. Structures of this sort are also found in the solar corona and in the magnetotails of magnetized planets.

field stretches out downstream (away from the Sun), forming a magnetotail. The symmetry of the magnetic configuration within such a tail is governed by the orientation of the magnetic field in the incident solar wind, and that orientation changes with time. For example, if the interplanetary magnetic field (IMF) is oriented northward, the east–west direction lies in the symmetry plane of the tail and the northern lobe field (see Fig. 1 for the definition of lobe) points away from the Sun while the southern lobe field points toward the Sun. A southward-oriented IMF would reverse these polarities, and other orientations would produce rotations of the symmetry axis.

Much attention has been paid to magnetic structures that form in and around the ionospheres of unmagnetized planets. Magnetic flux tubes of solar wind origin pile up at high altitudes at the dayside ionopause, where, depending on the solar wind dynamic pressure, they may either remain for extended times, thus producing a magnetic barrier that diverts the incident solar wind, or penetrate to low altitudes in localized bundles. Such localized bundles of magnetic flux are often highly twisted structures stretched out along the direction of the magnetic field. Such structures, referred to as flux ropes, are illustrated in Fig. 5.

C. INTERACTIONS OF THE SOLAR WIND WITH ASTEROIDS, COMETS, AND PLUTO

Asteroids are small bodies (<1000 km radius and more often only tens of kilometers) whose signatures in the solar wind were first observed by the *Galileo* spacecraft in the early 1990s. Asteroid-related disturbances are closely confined to the regions near to and downstream of the magnetic field lines that pass through the body, and thus the interaction region is fan-shaped as illustrated in Fig. 6 rather than bullet-shaped like Earth's magnetosphere. Unlike Earth's magnetosphere, there is no shock standing ahead of the disturbance in the solar wind. The signature found by *Galileo* in the vicinity of the asteroid Gaspra suggested that the asteroid is magnetized at a level similar to the magnetization of meteorites, but the measurements were made at locations so remote from the body that its field was not measured directly. Several additional missions are collecting data close to other asteroids and we may soon have better determinatioins of internal fields if they are present. [*See* ASTEROIDS.]

Comets are also small bodies. The spectacular appearance of an active comet, which can produce a glow

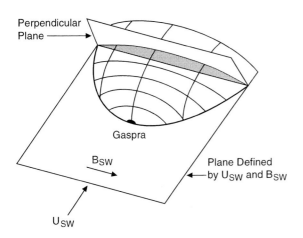

FIGURE 6 Schematic illustration of the shape of the interaction between an asteroid and the flowing solar wind. The disturbance spreads out along the direction of the magnetic field downstream of the asteroid. The disturbed region is thus fan-shaped, with greatest spread in the plane defined by the solar wind velocity and the solar wind magnetic field. The curves bounding the intersection of that plane with the surface and with a perpendicular plane are shown.

over a large visual field extending millions of kilometers in space on its approach to the Sun, is somewhat misleading for comet nuclei are no more than tens of kilometers in diameter. It is the gas and dust released from these small bodies by solar heating that we see spread out across the sky. Some of the gas released by the comet remains electrically neutral, with its motion governed by purely mechanical laws, but some of the neutral matter becomes ionized either by photoionization or by exchanging charge with ions of the solar wind. The newly ionized cometary material is organized in interesting ways that have been revealed by spacecraft measurements in the near neighborhood of comets Halley and Giacobini-Zinner. Figure 7 shows schematically the types of regions that have been identified. Of particular interest is that the different gaseous regions fill volumes of space that are many orders of magnitude larger than the actual solid comet. The solar wind approaching the comet first encounters the expanding neutral gases blown off the comet. Passing through a shock that begins to slow the flow, the solar wind encounters ever-increasing densities of newly ionized ions, referred to as pick-up ions. Energy is extracted from the solar wind as the pick-up ions are swept up, and the flow slows further. Still closer to the comet in a region referred to as the cometopause, a transition in composition occurs as the pick-up ions of cometary origin begin to dominate the plasma composition. Close to the comet, at the contact surface, ions flowing away from the comet carry enough mo-

mentum to stop the flow of the incident solar wind. [*See* PHYSICS AND CHEMISTRY OF COMETS.]

Pluto is also a small body even though it is classified as a planet. Pluto's interaction with the solar wind has not yet been observed, but it is worth speculating about what that interaction will be like in order to test our understanding of comparative planetology. The solar wind becomes tenuous and easily perturbed at large distances from the Sun (near 30 AU), and either escaping gases or a weak internal magnetic field could produce a magnetospherelike interaction region that is many times Pluto's size. At some phases of its 248-year orbital period, Pluto moves close enough to the

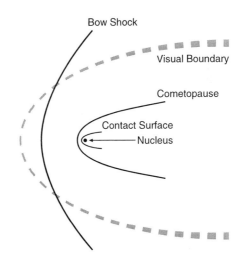

FIGURE 7 Schematic illustration of the magnetic field and plasma properties in the neighborhood of a comet. The length scale is logarithmic. The nucleus is surrounded by a region of dense plasma into which the solar wind does not penetrate. This region is bounded by a contact surface. Above that lies an ionopause or cometopause bounding a region in which ions of cometary origin dominate. Above this, there is a transition region in which the solar wind has been modified by the addition of cometary ions. As ions are added, they must be accelerated to become part of the flow. The momentum to accelerate the picked-up ions is extracted from the solar wind, and thus in the transition region the density is higher and the flow speed is lower than in the unperturbed solar wind. The region filled with cometary material is very large, and it is this region that imposes the large scale size on the visually observable signature of a comet. Spacecraft observations suggest that the shock bounding the cometary interaction region is weak as the effects of ion pickup serve to slow the flow to near the critical sound and Alfvén speeds, reducing the need for a shock transition. Similar to Venus-like planets, the solar wind magnetic field folds around the ionopause, producing a magnetic tail that organizes the ionized plasma in the direction radially away from the Sun and produces a distinct comet tail with a visual signature. The orientation of the magnetic field in the tail is governed by the solar wind field incident on the comet, and it changes as the solar wind field changes direction. Dramatic changes in the structure of the magnetic tail are observed when the solar wind field reverses direction.

Sun for its surface ice to sublimate, producing an atmosphere and possibly an ionosphere. Models of Pluto's atmosphere suggest that the gases would then escape and flow away from the planet. If the escape flux is high, the solar wind interaction would then resemble that of a comet rather than of Venus or Mars. Pluto's moon, Charon, may serve as a plasma source within the magnetosphere, and this could have interesting consequences of the type addressed in Section VI in relation to the moons of Jupiter and Saturn. For most of its orbital period, Pluto is so far from the Sun that its interaction with the solar wind is more likely to resemble that of the Moon, with absorption occurring at the sunward surface and a void developing in its wake. It is unlikely that a small icy planet will have an internal magnetic field that is large enough to produce a magnetospheric interaction region, although recent observations of magnetic fields of asteroids and some of the Galilean satellites are challenging our ideas about magnetic field generation in small bodies. [See PLUTO AND CHARON.]

D. MAGNETOSPHERES OF MAGNETIZED PLANETS

In a true magnetosphere, the scale size is set by the distance, R_{MP}, along the planet–Sun line at which the sum of the pressure of the planetary magnetic field and the pressure exerted by plasma confined within that field balances the dynamic pressure of the solar wind. (The dynamic pressure is ρu^2, where ρ is the mass density and u is its flow velocity in the rest frame of the planet. The thermal and magnetic pressures of the solar wind are small compared with its dynamic pressure.) Assuming that the planetary magnetic field is dominated by its dipole moment and that the plasma pressure within the magnetosphere is small, one can estimate R_{MP} as $R_{MP} \approx R_P(B_0^2/2\mu_0\rho u^2)^{1/6}$. Here B_0 is the surface equatorial field of the planet and R_P is its radius. Table I gives the parameters needed to evaluate R_{MP} for the different planets and shows the vast range of scale sizes in terms of both the planetary radii and absolute distance. The large scale sizes of the magnetospheres of the outer planets arise both because R_P is large and because the solar wind density is so small at the orbits of the outer planets that the solar wind pressure is not effective in balancing a strong magnetic field.

Within a magnetosphere, the magnetic field differs greatly from what it would be if the planet were placed in a vacuum. The field is distorted, as illustrated in Fig. 1, by currents carried on the magnetopause and in the plasma trapped within the magnetosphere. Properties of the trapped plasma and its sources are discussed in Section IV. An important source of magnetospheric plasma is the solar wind. Figure 1 makes it clear that, along most of the boundary, solar wind plasma would have to move across magnetic field lines to enter the magnetosphere. The Lorentz force of the magnetic field opposes such motion. However, shocked solar wind plasma of the magnetosheath easily penetrates the boundary by moving along the field in the polar cusp. Other processes that enable solar wind plasma to penetrate the boundary are discussed in Section V.

III. PLANETARY MAGNETIC FIELDS

Because the characteristic timescale for thermal diffusion is greater than the age of the solar system, the planets tend to have retained their heat of formation. At the same time, the characteristic timescale for diffusive decay of a magnetic field in a planetary interior is much less than the age of the planets. Consequently, primordial fields and permanent magnetism are small and the only means of providing a substantial planetary magnetic field is an internal dynamo. For a planet to have a magnetic dynamo, it must have a large region that is fluid, electrically conducting, and undergoing sufficiently vigorous convective motion. The deep interiors of the planets and many larger satellites are expected to contain electrically conducting fluids: terrestrial planets and the larger satellites have differentiated cores of liquid iron alloys; at the high pressures in the interiors of the giant planets Jupiter and Saturn, hydrogen behaves like a liquid metal; for Uranus and Neptune, a water–ammonia–methane mixture forms a deep conducting "ocean." The fact that some planets and satellites do not have dynamos tells us that their interiors are stably stratified and do not convect. Models of the thermal evolution of terrestrial planets show that as the object cools, the liquid core ceases to convect and further heat is lost by conduction alone. In some cases, such as the Earth, convection continues because the nearly pure iron solidifies out of the alloy in the outer core, producing an inner solid core and releasing gravitational energy that can drive continued convection in the liquid outer core. The more gradual cooling of the giant planets allows convective motions to persist. [See INTERIORS OF THE GIANT PLANETS.]

Of the nine planets, six are known to generate mag-

TABLE II
Planetary Magnetic Fields

	Mercury	Earth	Jupiter	Saturn	Uranus	Neptune
Magnetic moment/M_{Earth}	4×10^{-4}	1[a]	20,000	600	50	25
Surface magnetic field at dipole equator (gauss)	0.0033	0.31	4.28	0.22	0.23	0.14
Maximum/minimum[b]	2	2.8	4.5	4.6	12	9
Dipole tilt and sense[c]	$+14°$	$+10.8°$	$-9.6°$	$0.0°$	$-59°$	$-47°$
Obliquity[d]	$0°$	$23.5°$	$3.1°$	$26.7°$	$97.9°$	$29.6°$
Solar wind angle[e]	$90°$	$67–114°$	$87–93°$	$64–117°$	$8–172°$	$60–120°$

[a] $M_{Earth} = 7.906 \times 10^{25}$ gauss cm^{-3} = 7.906×10^{15} tesla m^{-3}.
[b] Ratio of maximum surface field to minimum (equal to 2 for a centered dipole field).
[c] Angle between the magnetic and rotation axes.
[d] The inclination of the equator to the orbit.
[e] Range of angle between the radial direction from the Sun and the planet's rotation axis over an orbital period.

netic fields in their interiors. Exploration of Venus has revealed an upper limit to the degree of magnetization comparable to the crustal magnetization of the Earth, suggesting that its core is stably stratified and that it does not have an active dynamo. The question of whether Mars does or does not have a weak internal magnetic field was disputed for many years because spacecraft magnetometers had measured the field only far above the planet's surface. The first low-altitude magnetic field measurements were made by *Mars Global Surveyor* in 1997. It is now known that the planetary magnetic field of Mars cannot have a surface field strength greater than 1/6000th of Earth's surface field. Pluto has yet to be explored. Though models of Pluto's interior suggest that it is probably differentiated, its small size makes one doubt that its core is convecting and any magnetization is likely to be remanent. Earth's moon has a negligibly small planet-scale magnetic field, though localized regions of the surface are highly magnetized. The magnetic fields of Jupiter's large moons are discussed in Section VI.

The characteristics of the six known planetary fields are listed in Table II. Assuming that each planet's magnetic field has the simplest structure, a dipole, we are able to compare the equatorial field strength (B_0) and the tilt of the dipole axis with respect to the planet's spin axis. Note that although the net magnetic moment of the planets varies by nearly eight orders of magnitude, the surface fields are on the order of a gauss = 10^{-4} T (except for Mercury), indicating that the strength of the dynamo scales with planetary size. The degree to which this dipole model is an oversimplification of more complex structure is indicated by the ratio of maximum to minimum values of the surface field.

This ratio has a value of 2 for a dipole. The larger values, particularly for Uranus and Neptune, are indications of strong nondipolar contributions to the planets' magnetic fields. Similarly, the fact that the magnetic axes of these two planets are strongly tilted (Fig. 8) also suggests that the dynamos in the icy giant planets may be significantly different from those of the aligned, dipolar planetary magnetic fields.

Recall that the size of a planet's magnetosphere (R_{MP}) depends only on the ambient solar wind density and the planet's radius and magnetic field, since the solar wind speed is approximately constant with distance from the Sun. Thus we expect a planet with a strong magnetic field to have a large magnetosphere, and even the weak fields of Uranus and Neptune produce moderately sized magnetospheres in the tenuous solar wind of the outer solar system. Table I shows that the measured sizes of planetary magnetospheres generally agree quite well with the theoretical R_{MP} values. Jupiter is the only notable exception, where the plasma pressure inside the magnetosphere is sufficient to further "inflate" the magnetosphere. This makes the magnetosphere of Jupiter a hugh object—about 1000 times the volume of the Sun, with a tail that extends at least 6 AU in the antisunward direction, beyond the orbit of Saturn. If the Jovian magnetosphere were visible from Earth, its angular size would be twice that of the Sun, even though it is at least four times farther away. The magnetospheres of the other giant planets are smaller (although large compared with the Earth's magnetosphere), having a similar scale of about 20 times the planetary radius, comparable to the size of the Sun. With only a weak magnetic field and being embedded in the denser solar wind close to

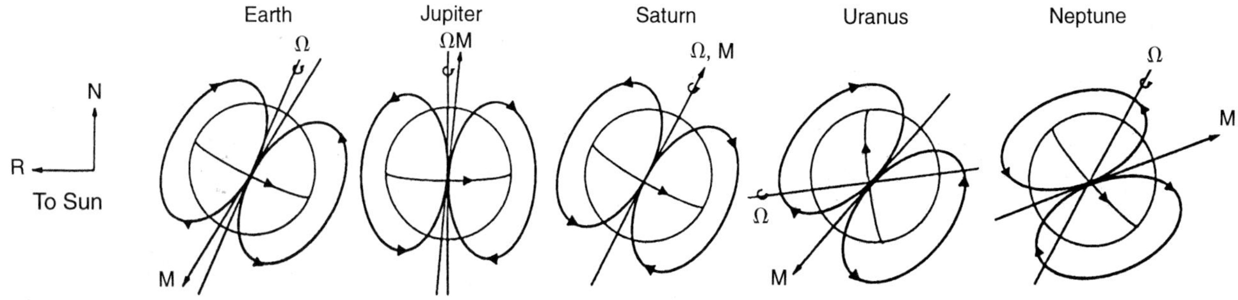

FIGURE 8   Orientation of the planets' spin axis (Ω) and their magnetic fields (**M**) with respect to the ecliptic plane (horizontal). The larger the angle between these two axes, the greater the magnetospheric variability over the planet's rotation period. The variation in the angle between the direction of the solar wind (close to radial from the Sun) and a planet's spin axis over an orbital period is an indication of the degree of seasonal variability.

the Sun, Mercury has a very small magnetosphere. Figure 9 compares the sizes of several planetary magnetospheres.

Although the size of a planetary magnetosphere depends on the strength of a planet's magnetic field, the configuration and internal dynamics depend on the field orientation (illustrated in Fig. 8). The orientation of a planet's magnetic field is described by two angles (tabulated in Table II): the tilt of the magnetic field with respect to the planet's spin axis and the angle between the planet's spin axis and the solar wind direction, which is generally within a few degrees of being radially outward from the Sun. Since the direction of the spin axis with respect to the solar wind direction varies only over a planetary year (many Earth years for the outer planets), and the planet's magnetic field is assumed to vary only on geological timescales, these two angles are constant for the purposes of describing the magnetospheric configuration at a particular epoch. Earth, Jupiter, and Saturn have both small dipole tilts and small obliquities. This means that the orientation of the magnetic field with respect to the solar wind does not vary appreciably over a planetary rotation period and that seasonal effects are small. Thus Mercury, Earth, Jupiter, and Saturn have symmetric and quasi-stationary magnetospheres, each exhibiting only a small wobble at the planetary rotation period owing to their ~10° dipole tilts (or a barely detectable wobble in the case of Saturn). In contrast, the large dipole tilt angles of Uranus and Neptune mean that the orientation of their magnetic fields with respect to the interplanetary flow direction varies considerably over a planetary rotation period, resulting in highly asymmetric and time-variable magnetospheres. Furthermore, Uranus's large obliquity means that the configuration of its magnetosphere will have strong seasonal changes over its 84-year orbit.

IV. MAGNETOSPHERIC PLASMAS

A. SOURCES OF MAGNETOSPHERIC PLASMAS

Magnetospheres contain considerable amounts of plasma, electrically charged particles in equal proportions of positive charge on ions and negative charge on electrons, from various sources (Fig. 10.) The main source of plasma in the solar system is clearly the Sun. The solar corona, the upper atmosphere of the Sun (which has been heated by some as yet undetermined process to temperatures of 1–2 million kelvin), streams away from the Sun at a more or less steady rate of 10^9 kg s^{-1} in equal numbers (8×10^{35} s^{-1}) of electrons and ions. The boundary between the solar wind and a planet's magnetosphere, the magnetopause, is not entirely plasma-tight. Wherever the interplanetary magnetic field has a component antiparallel to the planetary magnetic field, magnetic reconnection (discussed in Section V) is likely to occur and solar wind plasma will enter the magnetosphere across the magnetopause. Solar wind material is identified in the magnetosphere by its energy and characteristic composition of protons (H$^+$) with ~4% alpha particles (He^{2+}) and trace heavy ions, many of which are highly ionized.

Second, although ionospheric plasma is generally cold and gravitationally bound to the planet, a small fraction has sufficient energy to escape up magnetic field lines and into the magnetosphere. In some cases, field-aligned potential drops accelerate ionospheric ions and increase the escape rate. Ionospheric plasma has a composition that reflects the composition of the planet's atmosphere (e.g., abundant O$^+$ for the Earth and H$^+$ for the outer planets).

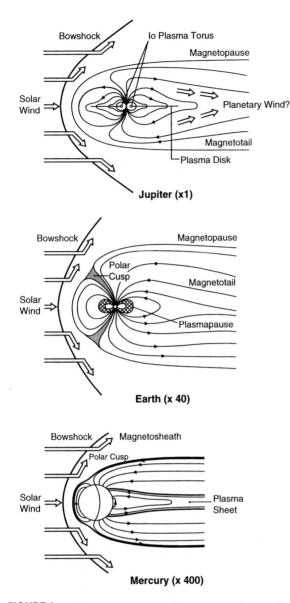

FIGURE 9 Schematic comparison of the magnetospheres of Jupiter, Earth, and Mercury. Relative to the Jupiter schematic, the one for Earth is blown up by a factor of 40 and that for Mercury is blown up by a factor of 400. The planetary radii are given in Table I.

Third, the interaction of magnetospheric plasma with any natural satellites or ring particles that are embedded in the magnetosphere can generate significant quantities of plasma. Magnetospheric plasma flowing by a satellite can ionize the outermost layers of its atmosphere, thus providing a major source of plasma. Energetic particle sputtering of the satellite surface or atmosphere produces ions of lower energy than the incident energy directly, and can create an extensive cloud of neutral atoms that are eventually ionized, possibly far from the satellite. The distributed

sources of water-product ions (totaling ~2 kg s^{-1}) in the magnetosphere of Saturn suggest that energetic particle sputtering of the rings and icy satellites is an important process. Although the sputtering process, which removes at most a few microns of surface ice per thousand years, is probably insignificant in geological terms, sputtering has important consequences for the optical properties of the satellite or ring surfaces. [See PLANETARY RINGS.]

Table III summarizes the basic characteristics of plasmas measured in the magnetospheres of the planets that have detectable magnetic fields. The composition of the ionic species indicates the primary sources of magnetospheric plasma: satellites in the cases of Jupiter, Saturn, and Neptune and the planet's ionosphere in the case of Uranus. In the magnetospheres where plasma motions are driven by the solar wind, solar wind plasma enters the magnetosphere, becoming the primary source of plasma in the case of Mercury's small magnetosphere and a secondary plasma source at Uranus and Neptune. At Earth, both the ionosphere and the solar wind are important sources. Earth's Moon remains well beyond the region in which sputtering or other plasma effects are important. In the magnetospheres where plasma flows are dominated by the planet's rotation (Jupiter, Saturn, and within the Earth's plasmasphere), the plasma is confined by the planet's strong magnetic field for many days so that substantial densities are accumulated.

B. ENERGY

Plasmas of different origins can have very different characteristic temperatures. Ionospheric plasma has a temperature on the order of ~10,000 K or ~1 eV, much higher than the temperature of the neutral atmosphere from which it was ionized (<1000 K) but much lower than the ~1 keV temperature characteristic of plasmas of solar wind origin, which are heated as they cross the bow shock and are subsequently thermalized. Plasmas from satellite sources extract their energy from the planet's rotation through a complicated process. When the neutrals are ionized, they experience a Lorentz force as a result of their motion relative to the surrounding plasma; this force accelerates both ions and electrons, which then begin to gyrate about the magnetic field at a speed equal to the magnitude of the neutral's initial velocity relative to the flowing plasma. At the same time, the ion is accelerated so that its bulk motion (the motion of the instantaneous center of its circular orbit) moves at the speed of the incident plasma, close to corotation with the planet near the

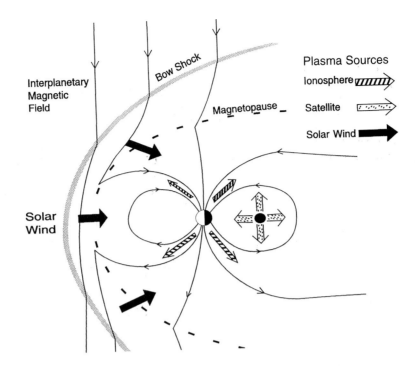

FIGURE 10   Sources of magnetospheric plasma. Plasma from the solar wind leaks through the magnetopause (black arrows), escapes from a planet's ionosphere (hatched arrows), and/or is produced in the magnetospheric interaction with a satellite (shaded arrows).

large moons of Jupiter and Saturn. Because the electric field pushes them in opposite directions, the new ion and its electron separate after ionization. Hence a radial current develops as the ions are "picked-up" by the magnetic field. This radial current is linked by field-aligned currents to the planet's ionosphere, where the Lorentz force is in the opposite direction to the planet's rotation. Thus the planet's angular momen-

tum is tapped electrodynamically by the newly ionized plasma.

In the hot, tenuous plasmas of planetary magnetospheres, collisions between particles are very rare. By contrast, in the cold, dense plasmas of a planet's ionosphere, collisions allow ionospheric plasmas to conduct currents and cause ionization, charge exchange, and recombination. Thus cold, dense, collision-dominated

TABLE III
Plasma Characteristics of Planetary Magnetospheres

	Mercury	Earth	Jupiter	Saturn	Uranus	Neptune
Maximum density (cm^{-3})	~1	1–4000	>3000	~100	3	2
Composition	H$^+$	O$^+$, H$^+$	O^{n+}, S^{n+}	O$^+$, H$_2$O$^+$, H$^+$	H$^+$	N$^+$, H$^+$
Dominant source	Solar wind	Ionosphere[a]	Io	Rings, Dione Tethys	Atmosphere	Triton
Strength (ions s^{-1}) (kg s^{-1})	?	2×10^{26} 5	>10^{28} 700	10^{26} 2	10^{25} 0.02	10^{25} 0.2
Lifetime	Minutes	Days[a] Hours[b]	10–100 Days	30 days Years	1–30 Days	~1 day
Plasma motion	Solar wind driven	Rotation[a] Solar wind[b]	Rotation	Rotation	Solar wind +rotation	Rotation (+solar wind?)

[a] Inside plasmasphere.
[b] Outside plasmasphere.

plasmas are expected to be in thermal equilibrium, but, surprisingly, even hot, tenuous plasmas in space are generally found not far from equilibrium, that is, their particle distribution functions are observed to be approximately Maxwellian (though the ion and electron populations often have different temperatures). This fact is remarkable because the source mechanisms tend to produce particles with an initially very narrow range in energy, and timescales for equilibration by means of Coulomb collisions that are usually much longer than transport timescales. The approach to equilibrium is achieved by interaction with waves in the plasma. Space plasmas support many different types of plasma waves and these waves can grow in the presence of such energy sources as non-Maxwellian distributions of newly created ions. Interactions between plasma waves and particle populations not only bring the bulk of the plasma toward thermal equilibrium but also accelerate or scatter particles at higher energies.

Plasma detectors mounted on spacecraft that fly through a magnetosphere provide detailed information about the particles' velocity distribution, from which bulk parameters such as density, temperature, and flow velocity are derived. Plasma properties are determined only in the vicinity of the spacecraft. Data from planetary magnetospheres other than Earth's are limited in both duration and spatial coverage, so there are considerable gaps in our knowledge of the changing properties of the many different plasmas in the solar system. Some of the most interesting space plasmas, however, can be remotely monitored by observing emissions of electromagnetic radiation. Dense plasmas, such as Jupiter's plasma torus, comet tails, Venus's ionosphere, the solar corona, and so on, have collisionally excited line emissions at optical or UV wavelengths. These radiative processes, particularly at UV wavelengths, can be significant sinks of plasma energy. Figure 11 shows an image of optical emission from the plasma that forms a ring deep within Jupiter's magnetosphere near the orbit of its moon, Io (see Section VI). Observations of these emissions serve to monitor the temporal and spatial variability of the Io plasma torus. Similarly, when magneteospheric particles bombard the planets' polar atmospheres, various auroral emissions are generated from radio to X-ray wavelengths and these emissions can also be used for remote monitoring of the system. Thus, our knowledge of space plasmas is based on combining the remote sensing of plasma phenomena with spacecraft measurements that provide "ground-truth" details of the particles' velocity distribution and of the local electric and magnetic fields that interact with the plasma.

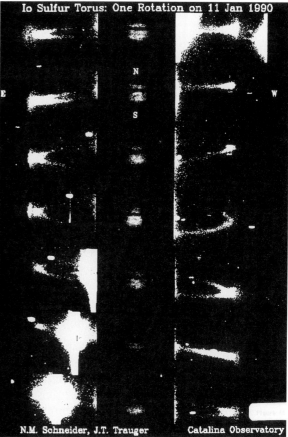

FIGURE 11   Optical emission from S^+ ions in the Io plasma torus in Jupiter's magnetosphere. These images were obtained with a ground-based telescope over Jupiter's 10-hour rotation period and demonstrate the changing aspect of the ring of ionized material that is trapped in Jupiter's tilted magnetic field. [Reprinted with permission from Nick Schneider and John Trauger (1995). *Astrophys. J.* **450**, 450–462.]

C. ENERGETIC PARTICLES

Significant populations of particles at keV–MeV energies, well above the energy of the thermal population, are found in all magnetospheres. The energetic particles are largely trapped in long-lived radiation belts (summarized in Table IV) by the strong planetary magnetic field. Where do these energetic particles come from? Since the interplanetary medium contains energetic particles of solar and galactic origins, an obvious possibility is that these energetic particles were "captured" from the external medium. In most cases, the observed high fluxes are hard to explain without identifying additional internal sources.

Compositional evidence supports the view that some fraction of the thermal plasma is accelerated to high energies, either by tapping the rotational energy

TABLE IV
Energetic Particle Characteristics in Planetary Magnetospheres

	Earth	Jupiter	Saturn	Uranus	Neptune
Phase space density[a]	20,000	200,000	60,000	800	800
Plasma beta[b]	<1	>1	>1	~0.1	~0.2
Ring current, ΔB[c] (nT)	10^{-23}	200	10	<1	<0.1
Auroral power (W)	10^{10}	10^{14}	10^{11}	10^{11}	<10^8

[a] The phase space density of energetic particles (in this case 100 MeV/gauss ions) is measured in units of $c^2(cm^2 \text{ s sr MeV}^3)^{-1}$ and is listed near its maximum value.

[b] The ratio of the thermal energy density to magnetic energy density of a plasma, $\beta = nkT/(B^2/\mu_0)$. These values are typical for the body of the magnetosphere. Higher values are often found in the tail plasma sheet and, in the case of the Earth, at times of enhanced ring current.

[c] The magnetic field produced at the surface of the planet due to the ring current of energetic particles in the planet's magnetosphere.

of the planet, in the cases of Jupiter and Saturn, or by acceleration in the distorted magnetic field in the magnetotail of Earth, Uranus, and Neptune. In a nonuniform magnetic field, the ions and electrons drift in opposite directions around the planet, producing an azimuthal electric current: the ring current. If the energy density of the energetic particle populations is comparable to the magnetic field energy, the ring current produces a magnetic field that significantly perturbs the planetary magnetic field. Table IV shows that this is the case for Jupiter and Saturn, where the high particle pressures inflate and stretch out the magnetic field and generate a strong ring current in the magnetodisk. Although Uranus and Neptune have significant radiation belts, the energy density of particles remains small compared with the magnetic field and the ring current is very weak. In Earth's magnetosphere, the strength of the ring current is extremely variable, as discussed in Section V. Relating the magnetic field produced by the ring current to the kinetic energy of the trapped particle population (scaled to the dipole magnetic energy external to the planet), we find that while the total energy content of magnetospheres varies by many orders of magnitude and the sources are very different, it appears that the particle energy builds up to only 1/1000th of the magnetic field energy in each magnetosphere. Earth, Jupiter, and Saturn all have energetic particle populations close to this limit. The energy in the radiation belts of Uranus and Neptune is much less than this limit, perhaps because it is harder to trap particles in nondipolar magnetic fields.

Where do these energetic particles go? Most appear to diffuse inward toward the planet. Loss processes for energetic particles in the inner magnetospheres are ring and satellite absorption, charge exchange with neutral clouds, and scattering by waves so that the particles stream into the upper atmospheres of the planets, where they can excite auroral emission and deposit large amounts of energy, at times exceeding the solar energy input.

V. DYNAMICS

Magnetospheres are ever-changing systems. Changes in the solar wind, in plasma source rates, and in energetic cosmic ray fluxes can couple energy, momentum, and additional particle mass into the magnetosphere and thus drive magnetospheric dynamics. Sometimes the magnetospheric response is direct and immediate. For example, an increase of the solar wind dynamic pressure compresses the magnetosphere. Both the energy and the pressure of field and particles then increase even if no particles have entered the system. Sometimes the change in both field and plasma properties is gradual, similar to a spring being slowly stretched. Sometimes, as for a spring stretched beyond its breaking point, the magnetosphere responds in a very nonlinear manner, with both field and plasma experiencing large-scale, abrupt changes. These changes can be identified readily in records of magnetometers (a magnetometer is an instrument that measures the magnitude and direction of the magnetic field), in scattering of radio waves by the ionosphere or emissions of such waves from the ionosphere, and in the magnetic field configuration, plasma conditions and flows, and energetic particle fluxes measured by a spacecraft moving through the magnetosphere itself.

Auroral activity is the most dramatic signature of magnetospheric dynamics. Records from ancient days include accounts of the aurora (the lights flickering in the night sky that inspired fear and awe), but the oldest scientific records of magnetospheric dynamics are the measurements of fluctuating magnetic fields at the surface of the Earth. Consequently, the term geomagnetic activity is used to refer to magnetospheric dynamics of all sorts. Fluctuating magnetic signatures with timescales from seconds to days are typical. For example, periodic fluctuations at frequencies between ~1 mHz and ~1 Hz are called magnetic pulsations. In addition, impulsive decreases in the horizontal north–south component of the surface magnetic field (referred to as the H-component) with timescales of tens of minutes occur intermittently at latitudes between 65° and 75° often several times a day. The field returns to its previous value typically in a few hours. These events are referred to as substorms. A signature of a substorm at an ~70° latitude magnetic observatory is shown in Fig. 12. The H-component decreases by

hundreds to 1000 nT (the Earth's surface field is 31,000 nT near the equator). Weaker signatures can be identified at lower and higher latitudes. Associated with the magnetic signatures and the current systems that produce them are other manifestations of magnetospheric activity, including particle precipitation and auroral activation in the polar region and changes within the magnetosphere previously noted.

The auroral activity associated with a substorm can be monitored from above by imagers on spacecraft and the dramatic intensification of the brightness of the aurora as well as its spatial extent can be accurately determined. Figure 13 shows an image of the aurora taken by the UltraViolet Imager on the *Polar* spacecraft on April 6, 1996. This image was taken 15 minutes after the onset of a substorm on the nightside of the Earth. Note that the intense brightness is localized in a high-latitude band surrounding the polar regions. This region of auroral activity is referred to as the auroral oval. Only during very intense substorms does the auroral region move far enough equatorward to be visible over most of the United States.

Twenty-seven-day variations in the level of geomagnetic activity are imposed by the periodicity of equatorial solar rotations viewed from the moving Earth, and are linked to active locations on the Sun that are the source of high-speed solar wind streams. (The rotation period of the equatorial region of the Sun is 25 days, but Earth advances in its orbit during that time, and this explains the longer period observed from Earth.) Geomagnetic activity also varies over an 11-year cycle as the number of sunspots waxes and wanes and the structure of high- and low-speed streams in the solar wind changes. The intensity of geomagnetic activity is governed not only by the speed of the solar wind but also by the orientation of the magnetic field embedded in the solar wind. When the magnetic field of the solar wind points strongly southward and remains so for prolonged intervals, geomagnetic activity may become particularly intense and persist at a high level for days. Then the disturbance is called a magnetic storm. During a magnetic storm, the ring current (see Section IV, C) energy increases and correspondingly the current that it carries grows larger. The signature of the increased current is a depression of the northward horizontal field at the surface of the Earth at low and middle latitudes. During a magnetic storm, the magnetic fluctuations at midlatitude stations may increase much above the tens of nT typical of substorm signatures. Sensitive spacecraft operating systems may be damaged or destroyed by the increase of energetic particle fluxes and radio communications can be disrupted. Great magnetic storms in which mid-

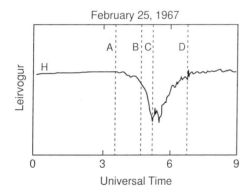

February 25, 1967

FIGURE 12 The variation of the H-component of the surface magnetic field of the Earth at an auroral zone station at 70° magnetic latitude plotted versus universal time in hours during a 9-hour interval that includes a substorm. (A) The beginning of the growth phase, during which the magnetosphere extracts energy from the solar wind, and the electrical currents across the magnetotail grow stronger. (B) The start of the substorm expansion phase, during which currents from the magnetosphere are diverted into the auroral zone ionosphere and act to release part of the energy stored during the growth phase. Simultaneously plasma is ejected down the tail to return to the solar wind. (C) The end of the substorm onset phase and the beginning of the recovery phase, during which the magnetosphere returns to a stable configuration. (D) The end of the recovery phase.

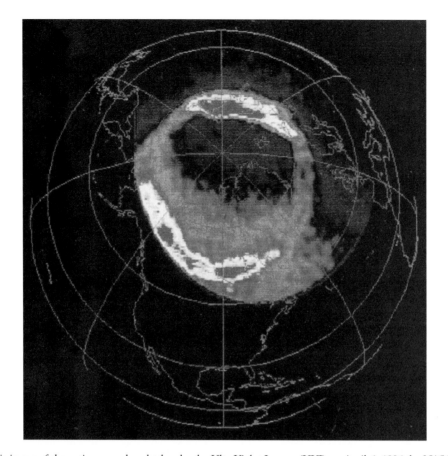

FIGURE 13   This image of the entire auroral oval taken by the UltraViolet Imager (UVI) on April 6, 1996, by NASA's *Polar* spacecraft shows intense, separated aurora on both the dayside of the Earth (foreground) and on the nightside. It is the final image of a sequence started 57 minutes earlier that shows continuous, evolving activity on the dayside and the sudden onset of a relatively weak auroral substorm on the nightside that developed within 15 minutes to the extent seen in this image. Much stronger storms are expected to be observed during periods when the Sun is more active. The UVI camera field of view is roughly circular and does not encompass the entire globe of Earth on which the image is superposed. Faint circular traces mark latitudes of 0, 20, 40, 60, and 80 degrees. The UltraViolet Imager was designed and built at NASA's Marshall Space Flight Center and the University of Alabama at Huntsville. The principal investigator is Dr. George E. Parks of the University of Washington. (Courtesy of NASA Photo, National Aeronautics and Space Administration, Goddard Space Flight Center, Office of Public Affairs, Greenbelt, Md. 20771.)

latitude perturbations become as large as 500 nT occur only a few times in each solar cycle and may produce auroras as far south as Rome (geographic north latitude: 41°54'00") or even Mexico City (geographic north latitude: 19°24'00").

The fundamental role of the magnetic field in the solar wind may seem puzzling. It is the orientation of the interplanetary magnetic field that is critical and at Earth it is normally tilted southward when substorm activity is observed. The issue is subtle. Magnetized plasma flowing through space is frozen to the magnetic field. The high conductivity of the plasma prevents the magnetic field from diffusing through the plasma and, in turn, the plasma particles are bound to the magnetic field by a "$\mathbf{v} \times \mathbf{B}$" Lorentz force. How, then,

can a plasma ion or electron move from a solar wind magnetic field line to a magnetospheric field line?

The coupling arises through a process called reconnection, which occurs when plasmas bound on flux tubes with oppositely directed fields approach each other sufficiently closely. The weak net field at the interface may be too small to keep the plasma bound on its original flux tube and the field connectivity can change. Newly linked field lines will be bent at the reconnection location. The curvature force at the bend accelerates plasma away from the reconnection site. At the dayside magnetopause, for example, solar wind magnetic flux tubes and magnetospheric flux tubes can reconnect in a way that extracts energy from the solar wind and allows solar wind plasma to penetrate the magnetopause. A diagram first drawn in a French café by J. W. Dungey in 1961 (and reproduced frequently thereafter) provides the framework for understanding the role of magnetic reconnection in magnetospheric dynamics (Fig. 14). Shown in the diagram on the left are southward-oriented solar wind field lines approaching the dayside magnetopause. Just at the nose of the magnetosphere, the northern ends of the solar wind field lines break their connection with the southern ends, linking instead with magnetospheric fields. Accelerated flows develop near the reconnection site. The reconnected field lines are dragged tailward by their ends within the solar wind, thus forming the tail lobes.

The reconnection process transports magnetic flux from the dayside of the Earth to the nightside. The path of the foot of the flux tube crosses the center of the polar cap, starting at the polar edge of the dayside auroral zone and moving to the polar edge of the nightside auroral zone as shown schematically in Fig. 14a. Ultimately that flux must return and this process is also shown, both in the magnetotail, where reconnection is shown closing a flux tube that had earlier been opened on the dayside, and in the polar cap (Fig. 14b), where the path of the foot of the flux tube appears at latitudes below the auroral zone, carrying the flux back to the dayside. In the early stage of a substorm (between A and B in Fig. 12), the rate at which mag-netic flux is transported to the nightside is greater than the rate at which it is returned to the dayside. This builds up stress in the tail, reducing the size of the region within the tail where the magnetic configuration is dipolelike and compressing the plasma in the plasma sheet (see Fig. 1). Only after reconnection starts on the nightside (at B in Fig. 12) does flux begin to return to the dayside. Complex magnetic structures form in the tail as plasma jets both earthward and tailward from the reconnection site. In some cases, the magnetic field appears to enclose a bubble of tailward-moving plasma called a plasmoid. At other times, the magnetic field appears to twist around the earthward- or tailward-moving plasma in a flux rope (see Fig. 5). Even on the dayside magnetopause, twisted field configurations seem to develop as a consequence of reconnection and, because these structures are carrying flux tailward, they are called flux transfer events.

The diversity of the processes associated with geomagnetic activity and their complexity and the limited data on which studies of the immense volume of the magnetosphere must be based have constrained our ability to understand details of substorm dynamics. However, both new research tools and anticipated practical applications of improved understanding have accelerated progress toward the objective of being able to predict the behavior of the magnetosphere during a substorm. The new tools of the 1990s and into the

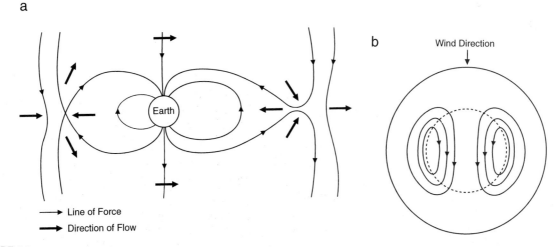

a

b

FIGURE 14 Adapted from the schematic view of reconnection sketched by J. W. Dungey in 1961. (a) A noon–midnight cut through the magnetosphere showing, from left to right, in addition to two dipolelike field lines (rooted at two ends in the Earth): a solar wind field line with plasma flowing earthward; a newly reconnected pair of field lines, one of solar wind origin and one dipolelike field line, with plasma flowing toward the reconnection point from two sides near the midplane and accelerated both north and south away from the reconnection point; two reconnected field lines with one end in the solar wind and one end in the Earth flowing over the polar caps; two field lines about to reconnect in the magnetotail carried by plasma flow toward the midplane of the diagram; and a newly reconnected field line moving farther away from the Earth in the solar wind. (b) A view down on the northern polar cap showing flow lines moving from day to night near the center, above the auroral zone, and returning to the dayside at latitudes below the auroral zone. [Adapted from J. W. Dungey (1961). *Phys. Rev.* **6,** 47.]

next century include a fleet of spacecraft in orbit around and near the Earth (*Wind, Polar, Geotail,* and several associated spacecraft) making coordinated measurements of the solar wind and of different regions within the magnetosphere, better instruments making high time-resolution measurements of particles and fields, spacecraft imagers covering a broad spectral range, ground radar systems, and networks of magnetometers. The anticipated applications relate to the concept of forecasting space weather much as we forecast weather on the ground. An ability to anticipate an imminent storm and take precautions to protect spacecraft in orbit, astronauts on space stations, and electrical systems on the surface (which can experience power surges during big storms) has been adopted as an important goal by the space science community, and improvements in our understanding of the dynamics of the magnetosphere will ultimately translate into a successful forecasting capability.

Dynamical changes are also expected in the magnetospheres of the other planets. In passes through Mercury's magnetosphere, the *Mariner* spacecraft observed substorms that lasted for minutes. Substorms or related processes should also occur at outer planets, but the timescale for global changes in the system is expected to increase as its size increases. For a magnetosphere as large as Jupiter's, the equivalent of a substorm is likely to occur only once every week or two as contrasted with several each day for Earth. Until December 1995, when *Galileo* began to orbit Jupiter, no spacecraft had remained long enough within a planetary magnetosphere to monitor its dynamical changes. Before the end of the century, the data from Jupiter should be sufficiently complete to provide insight into the changes that are occurring in that giant planet's magnetosphere.

The importance of Jupiter's internal source of heavy ions and the role of centrifugal forces should produce dynamical responses different in detail from those familiar at Earth. Plasma loaded into the magnetosphere near Io may ultimately be flung out down the magnetotail, and this process may be intermittent, possibly governed both by the strength of internal plasma sources and by the magnitude of the solar wind dynamic pressure that governs the location of the magnetopause for any fixed level of internal plasma pressure. Various models have been developed to describe the pattern of plasma flow in the magnetotail as heavily loaded magnetic flux tubes dump plasma on the nightside, but the evidence on which the models are based is incomplete.

Processes directly driven by the solar wind may also be important. It is known that a considerable amount of solar wind plasma enters the magnetosphere. One way to compare the solar wind as a plasma source with Io, which deposits ions into the magnetosphere at a rate of a few $\times 10^{28}$ s^{-1}, is to estimate how much plasma can enter the magnetosphere as a result of dayside reconnection. If the solar wind near Jupiter flows at 400 km s^{-1} with a density of 0.5 particles cm^{-3}, it carries $\sim 10^{31}$ particles s^{-1} onto the circular cross section of a magnetosphere with $>50R_J$ radius. If reconnection is approximately as efficient as it is at Earth, where a 10% efficiency is often suggested, and if a significant fraction of the solar wind ions on reconnected flow tubes enter the magnetosphere, the solar wind source could be important and, as at Earth, the solar wind may contribute to the variability of Jupiter's magnetosphere. As *Galileo* continues to map out the nightside of the magnetosphere, it will help clarify the nature and sources of magnetospheric dynamics at Jupiter.

VI. INTERACTIONS WITH MOONS

Embedded deeply within the magnetosphere of Jupiter, the four Galilean moons (Io, Europa, Ganymede, and Callisto, whose properties are summarized in Table V) are immersed in magnetospheric plasma that corotates with Jupiter (i.e., flows once around Jupiter in each planetary spin period). Io is itself the principal source of the plasma in which it is embedded, providing approximately 1 ton per second of ions to Jupiter's magnetosphere, thus creating the Io plasma torus alluded to in Section IV. The plasma sweeps ahead of the moons whose Keplerian orbital speeds are slow compared with the speed of local plasma flow. Plasma interaction regions develop around the moons, with details depending on whether or not the moon has an internal magnetic field. At Saturn, Titan, shrouded by a dense atmosphere, is also embedded within the magnetosphere. These interaction regions differ greatly from the model planetary magnetosphere that we illustrated in Fig. 1. An important difference in the nature of the interaction region is that the plasma flows onto the moons at a speed that is smaller than either the sound speed or the Alfvén speed, so the flow can easily be deflected upstream of the moon and no bow shocks are formed. The ratio of the thermal pressure to the magnetic pressure is typically small in the plasma that flows onto the moons and this minimizes the changes associated with the interaction. Nonetheless, significant changes in both magnetic field and plasma properties have been observed near Ganymede and Titan (both roughly as large as Mercury) and near

TABLE V
Properties of the Galilean Moons of Jupiter

Moon	Distance to orbit (R_J)	Rotation period (Earth days)[a]	Radius (km)	Radius of core (moon radii)[b]	Mean density (kg m^{-3})	Surface B at dipole equator (nT)	Approx. ave. B_{Jup} (nT)
Io	5.9	1.77	1818	0.25 to 0.5	3530	≤1300	−1800
Europa	9.4	3.55	1560		2990	Small	−400
Ganymede	15	7.15	2634	0.25 to 0.5	1940	750	−70
Callisto	26	16.7	2409		1851	Small	−15

[a] Jupiter's rotation period is 9 hr 55 min, so corotating plasma moves faster than any of the moons.

[b] Core densities can be assumed in the range from 5150 to 8000 kg m^{-3}. This corresponds to maximum and minimum core radii, respectively.

[c] The magnetic field of Jupiter oscillates in both magnitude and direction at Jupiter's rotation period of 9 hr 55 min. The average field over a planetary rotation period is southward-oriented, that is, antiparallel to Jupiter's axis of rotation. Neither the orbits nor the spin axes of the moons are significantly inclined to Jupiter's equatorial plane, so we use averages relative to the spin axis of Jupiter to represent the average along the spin axis of the moon.

Io (roughly as large as Earth's Moon). Near Titan, significant magnetic field perturbations were observed by *Voyager 1*. It appears that the field of Saturn drapes around the moon's ionosphere much as the solar wind field drapes to produce the magnetosphere of Venus. [*See* TITAN.]

Near Ganymede, both the magnetic field and the plasma properties depart dramatically from their values in the surroundings. The changes occur because the moon has an internal magnetic field strong enough to stand off the flowing plasma and carve a bubblelike magnetospheric cavity within Jupiter's magnetosphere, as illustrated in Fig. 15. Similarly, in its December 1995 pass near Io, *Galileo* found that both the magnetic field and the plasma properties were substantially different than in the surrounding torus, partly because the moon is a prodigious source of new ions, which greatly changes the plasma properties in Io's immediate vicinity. There is also reason to think that Io, like Ganymede, has an internal field, but there are some difficulties in knowing whether the measurements can be interpreted purely in terms of currents flowing in the plasma, recognizing that large currents are produced by ions picked up in the vicinity of Io. The answer to this dilemma is likely to become clear only after another close spacecraft pass by Io. [*See* IO.]

VII. CONCLUSIONS

We have described interactions between flowing plasmas and diverse bodies of the solar system. The interaction regions all manifest some of the properties of

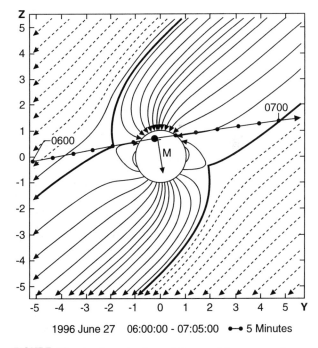

1996 June 27   06:00:00 - 07:05:00   ●—● 5 Minutes

FIGURE 15 A schematic view of Ganymede's magnetosphere embedded in Jupiter's magnetospheric field in a plane that is normal to the direction of corotation flow. Dashed lines are attached to Jupiter at both ends. Solid lines are attached to Ganymede at one or both ends. The thick line separates field lines doubly attached to Jupiter from the Ganymede-linked field lines and is the equivalent of the magnetopause in a planetary magnetosphere. The projection of the trajectory of *Galileo's* first pass in the vicinity of Ganymede is illustrated and this model represents the measured magnetic field quite well.

magnetospheres. Among magnetospheres of magnetized planets, one can distinguish (1) the large, symmetric and rotation-dominated magnetospheres of Jupiter and Saturn; (2) the small magnetosphere of Mercury, where the only source of plasma is the solar wind that drives rapid circulation of material through the magnetosphere; and (3) the moderate-sized and highly asymmetric magnetospheres of Uranus and Neptune, whose constantly changing configuration does not allow substantial densities of plasma to build up. The Earth's magnetosphere is an interesting hybrid of the first two types, with a dense corotating plasmasphere close to the planet, and tenuous plasma, circulated by the solar wind-driven convection, in the outer region. All of these magnetospheres set up bow shocks in the solar wind. The nature of the interaction of the solar wind with nonmagnetized objects depends on the presence of an atmosphere that becomes electrically conducting when ionized. Venus and Mars have tightly bound atmospheres so that the region of interaction with the solar wind is close to the planet on the sunward-facing side, with the interplanetary magnetic field draped back behind the planet to form a magnetotail. Bow shocks form in front of both of these magnetospheres. Comets cause the solar wind field to drape much as at Venus and Mars; they produce clouds extended over millions of kilometers. The interaction of the solar wind with the cometary neutrals weakens or eliminates a bow shock. Small bodies like asteroids disturb the solar wind without setting up shocks. Within the magnetospheres of Saturn and Jupiter, the large moons interact with the subsonic magnetospheric flow, producing unique signatures of interaction with fields that resist draping. No shocks have been observed in these cases.

The complex role of plasmas trapped in the magnetosphere of a planetary body must be understood as we attempt to improve our knowledge of the planet's internal magnetic field, and this means that the study of magnetospheres links closely to the study of intrinsic properties of planetary systems. Although our understanding of the dynamo process is still rather limited, the presence of a planetary magnetic field has become a useful indicator of the properties of a planet's interior. As dynamo theory advances, extensive data on the magnetic field may provide a powerful tool from which to learn about the interiors of planets and large satellites. For example, physical and chemical models of interiors need to explain why Ganymede has a magnetic field whereas its neighbor of similar size, Callisto, does not, and why Uranus's and Neptune's magnetic fields are highly nondipolar and tilted whereas Jupiter's and Saturn's fields are nearly dipolar and aligned.

Continued exploration of the plasma and fields in the vicinity of planets and moons is needed to reveal features of the interactions that we do not yet understand. We do not know how effective reconnection is in the presence of the strong planetary fields in which the large moons of Jupiter are embedded. We have not learned all we need to know about moons as sources of new ions in the flow. We need many more passes to define the magnetic fields of some of the planets and all of the moons, because single passes do not provide constraints sufficient to determine more than the lowest-order properties of the internal fields. Temporal variability of magnetospheres over a wide range of timescales makes them inherently difficult to measure, especially with a single spacecraft. Spurred by the desire to understand how the solar wind controls geomagnetic activity, space scientists combine data from multiple spacecraft (e.g., *Polar*, *Wind*, and *Geotail*) and from ground-based instruments to make simultaneous measurements of different aspects of the Earth's magnetosphere, coordinated in the International Solar-Terrestrial Program. [*See* PLANETARY EXPLORATION MISSIONS.]

As it orbits Jupiter, the *Galileo* spacecraft is mapping out different parts of the Jovian magnetosphere, monitoring changes and measuring the interactions of magnetospheric plasma with the Galilean satellites. In the next decade, several spacecraft will explore Mars, and *Cassini* will make extensive measurements of Saturn's magnetosphere and define the roles played by the moons and rings. There is great interest in returning to Mercury with an orbiting spacecraft that would be able to characterize the mysterious magnetic field of this planet. And finally, Pluto beckons as the only unexplored planet, the most distant of the solar family. It is sure to interact with the solar wind in an interesting way. As new technologies lead to small, lightweight instruments, we look forward to missions of the new millennium that will determine if Pluto or Charon have magnetic fields and help us understand the complexities of magnetospheres large and small throughout the solar system.

BIBLIOGRAPHY

Bagenal, F. (1992). Giant planet magnetospheres. *Annu. Rev. Earth Planet. Sci.* **20**, 289.

Cheng, A. F., and Johnson, R. E. (1989). Effects of magnetosphere interactions on origin and evolution of atmospheres. *In* "Origin and Evolution of Planetary and Satellite Atmospheres" (S. K. Atreya, J. B. Pollack, and M. S. Matthews, eds.), Univ. Arizona Press, Tucson.

Kivelson, M. G., and Russell, C. T. (eds.) (1995). "Introduction to Space Physics." Cambridge Univ. Press, Cambridge, England.

Luhmann, J. G. (1986). The solar wind interaction with Venus. *Space Sci. Rev.* **44**, 241.

Luhmann, J. G., Russell, C. T., Brace, L. H., and Vaisberg, O. L. (1992). The intrinsic magnetic field and solar wind interaction of Mars. *In* "Mars" (H. H. Kieffer *et al.*, eds.). Univ. Arizona Press, Tucson.

Russell, C. T., Baker, D. N., and Slavin, J. A. (1988). The magnetosphere of Mercury. *In* "Mercury" (Vilas, Chapman, and M. S. Matthews, eds.). Univ. Arizona Press, Tucson.

Van Allen, J. A. (1990). Magnetospheres, cosmic rays and the interplanetary medium. *In* "The New Solar System" (Beatty and Chaikin, eds.). Sky Publishing, Cambridge, Mass.

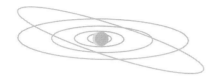

PLUTO AND CHARON

I. Historical Background

II. Pluto's Orbit and Spin

III. The Mutual Events

IV. Pluto's Surface Properties and Appearance

V. Pluto's Interior and Bulk Composition

VI. Pluto's Atmosphere

VII. Charon

VIII. The Origin of Pluto and Charon

S. Alan Stern
Southwest Research Institute

Roger V. Yelle
Boston University

GLOSSARY

Albedo: Fraction of incoming light reflected from a surface. An albedo of 0 is completely absorbing, whereas an albedo of 1 is completely reflecting.

Arcsecond: One arcsecond is 1/3600th of a degree.

Astronomical unit (AU): Mean distance of the Earth from the Sun; 149.6×10^6 km.

Ecliptic: Plane of the Earth's orbit about the Sun.

Equilibrium vapor pressure: Ambient pressure of the gas phase over a condensed phase when the gas and condensed phase are in thermodynamic equilibrium (i.e., when the rate of condensation from gas to ice equals the rate of sublimation from ice to gas). In effect, vapor pressure is a measure of the amount of gas that an ice layer at a specified temperature will evolve in a closed container (or planetary atmosphere). Vapor pressures are extremely sensitive functions of temperature and are also related to the composition and structure of a given ice.

Hydrodynamic escape: Limiting case of atmospheric escape that occurs when the escape rate is so rapid that the atmosphere at high altitudes reaches an outward velocity comparable to the sound speed. This occurs if the thermal energy of the gas molecules becomes comparable to the gravitational binding energy. Hydrodynamic escape allows the upper atmosphere of a planet to escape wholesale, as opposed to the usually slower processes of Jeans-type thermal leakage or solar wind ion pickup.

Ice dwarf: Term given to the small (500–3000 km diameter) miniplanets believed to have been created in large numbers during the formation of the giant planets and later scattered to the Sun's cometary reservoir or ejected from the solar system by close encounters with the forming giant planets. Pluto, Charon, and Triton may be remnants of this population.

Insolation: Flux of sunlight at all wavelengths falling on a body. For the Earth, this amounts to a flux of 1.4×10^6 ergs cm^{-2} s^{-1}.

Jeans escape: Escape process first characterized in 1926 by James Jeans, in which sufficiently energetic atoms and molecules in the high-energy tail of the exosphere's thermal Maxwellian leak away from a planetary atmosphere

Obliquity: Angle between a body's equatorial and orbital planes; equivalently, the angle between a body's rotational pole and the pole of its orbit.

Phase function: Curve describing the change in brightness of a body as a function of the angle between the observer, the body, and the Sun. This angle is usually referred to as the phase angle and is expressed in astronomical magnitudes per degree of change.

Refractory: Any substance that is generally inert and produces only an insignificant vapor pressure at a given temperature. Many common minerals, as well as some heavy organics, are examples of refractory materials.

Rotational lightcurve: Plot depicting the variation in brightness of an object as a function of time as it rotates on its axis. This variation can be caused by nonsphericity (i.e., shape effects) or albedo markings; for objects as large as Pluto and Charon, albedo markings usually dominate.

Space ultraviolet: That part of the ultraviolet electromagnetic spectrum that can only be observed from space because the Earth's atmosphere is opaque at those wavelengths; it is commonly thought of as the region below 3000 Å.

Stellar occultation: When a planet or asteroid passes in front of a star and is briefly hidden from view. Such events can be used to probe the size and also the atmospheric structure of the planet (or asteroid) doing the occulting.

Volatile: Any substance that outgasses or produces a significant vapor pressure at a given temperature. Ice is a volatile on Earth (270–300 K), but involatile in the outer solar system ($T < 100$ K). By contrast, the ices of CH_4, CO, and N_2 are volatile throughout the planetary region wherever $T > 30$ K.

Pluto is the ninth planet; for most of each orbit, it is the farthest known planet from the Sun. Pluto is in an elliptical, 248-year orbit that ranges from 29.5 to 49.5 astronomical units (AU) from the Sun. Its satellite Charon is close enough to Pluto in size that the pair are widely considered to be a double planet. Recent advances in astronomical techniques, a unique set of mutual eclipses between Pluto and Charon, the occultation of a bright star by Pluto, and results obtained by the Hubble Space Telescope (*HST*) have resulted in a rapid expansion of knowledge about the Pluto–Charon system. Both Pluto and Charon are rich in ices, but their surface compositions, albedos, and colors are different. Also, unlike Charon, Pluto is

known to possess distinct surface markings, polar caps, and an atmosphere. Major questions under study include the structure of Pluto's atmosphere, the individual densities of Pluto and Charon, and the origin of the binary.

I. HISTORICAL BACKGROUND

A. OVERVIEW

Pluto was discovered in February 1930, at Lowell Observatory in Flagstaff, Arizona. This discovery was made by Clyde Tombaugh (1906–1997), an observatory staff assistant working on a search for a long-suspected perturber of the orbits of Uranus and Neptune. That search, which was first begun in 1905 by the observatory's founder, Percival Lowell, never located the large object originally being searched for, and it is generally believed that the positional discrepancies of Uranus and Neptune that prompted that search were flawed. Still, the search for Lowell's "Planet X" resulted in the discovery of the tiny planet Pluto.

Within a year of Pluto's discovery, its orbit was well determined. That orbit is both eccentric and highly inclined to the plane of the ecliptic, compared to the orbits of the other planets (Table I). However, no quantitative discoveries about Pluto's physical properties were made until the early 1950s. This lack of information was largely due to the difficulty of observing Pluto with the scientific instruments available in the 1930s and 1940s. Between 1953 and 1976, techno-

TABLE I
Pluto's Heliocentric Orbit[a]

Orbital element	Value
Semimajor axis, a	39.44 AU
Orbital period, P	247.688 yr
Eccentricity, e	0.254
Inclination, i	17.14°
Long. ascending node, Ω	110.29°
Long. perihelion, ω	223.94°
Perihelion epoch, T	5.1 September 1989 UT

[a] Note: Osculating elements on JD 2449000.5 referred to the mean ecliptic and equinox of J2000.0.

logical advances in photoelectric astronomy made possible several important findings. Among these were the discovery of Pluto's ~6.39-day rotation period, the discovery of Pluto's reddish surface color, and the discovery of Pluto's high axial tilt, or obliquity.

Between 1976 and 1989, the pace of discoveries increased more dramatically. In rapid succession, there was the discovery of methane on Pluto's surface; the detection of Pluto's satellite Charon; the prediction, detection, and then the study of a set of mutual eclipse events between Pluto and Charon that occurs once every 124 years; and the occultation by Pluto of a bright star, confirming the presence of an atmosphere. In addition, the 1989 *Voyager 2* encounter with the Neptune system gave us detailed insights into the object believed to be Pluto's closest analog in the solar system, Triton, thereby showing how complex and scientifically interesting Pluto would likely be under close scrutiny by spacecraft. [See TRITON.]

In the 1990s, it has been discovered that Pluto's surface consists of a complex mixture of low-temperature volatile ices, that this surface displays large-scale bright and dark units, and that Pluto's atmosphere consists primarily of nitrogen gas, with only a minor amount of carbon monoxide and only a trace of methane. Additionally, Pluto's context in the solar system has become much better defined through the discoveries of many smaller objects in the region of the solar system beyond Neptune called the Edgeworth–Kuiper Belt. [See Kuiper Belt.]

B. THE DISCOVERY OF CHARON

Charon (pronounced correctly as "Kharon," but more colloquially known as "Sharon") was discovered by J. W. Christy and R. S. Harrington on a series of photographic plates made in 1978 at the U.S. Naval Observatory's Flagstaff Station. Interestingly, these images were taken less than four miles from Lowell Observatory, where Pluto had been discovered 48 years before.

Charon was apparent on the 1978 Naval Observatory images as a bump or elongation in Pluto's apparent shape. The recognition that the bump was in fact a close-in satellite was made when it was recognized that this bump regularly revolved with Pluto in a 6.39-day period, which matched Pluto's rotation period. Ironically, Charon's bump or elongation of Pluto was seen occasionally on plates made in the 1960s, but had not been recognized to be a satellite. This was because the elongation of Pluto's image by Charon was attributed to turbulence in the Earth's atmosphere causing

a distortion of Pluto's pointlike image (the two are <1 arcsecond apart, and blended together by atmospheric seeing). What Christy and Harrington recognized in 1978 was that although Pluto was distorted, none of the stars in the photographs were. This led them to look for a periodicity in the elongations, and hence the discovery that the elongation was due to an object that circled Pluto.

In the first few months after Charon's discovery, Christy and Harrington determined that Charon's orbit is (1) synchronous with Pluto's rotation and (2) highly inclined to the plane of the ecliptic. During that same year, 1978, Leif Andersson was the first to recognize that Pluto's orbit motion would cause Charon's orbital plane to sweep through the line of sight to the Earth for a period of several years every half Pluto orbit, or 124 terrestrial years. Mutual eclipses (also called mutual events) would then begin occurring every 3.2 days (half Charon's orbit period). These eclipses were predicted to progress over a period of 5 to 6 years, from shallow, partial events to central events lasting up to five hours, then to recede again to shallow grazing events. It was widely recognized that such a series of mutual eclipses and occultations would be scientifically valuable events. These mutual events began occurring in 1985 and ended in 1990. Figure 1 illustrates the mutual event geometry. These events (described in Section III) yielded a wealth of data on both Pluto and Charon.

Sensitive searches for other satellites of Pluto have been made since Charon's discovery, but no objects have been found. From these searches it has been estab-

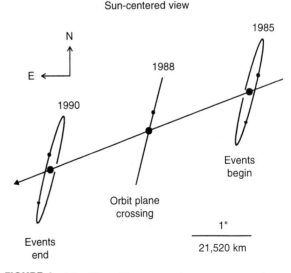

Sun-centered view

FIGURE 1 The Pluto–Charon mutual event geometry that occurred in the 1980s. (Provided with permission by R. L. Marcialis.)

lished that Pluto probably does not have any other satellites brighter than an astronomical *V* magnitude of 22. This magnitude constraint corresponds to bodies (with Charon-like geometric albedos, i.e., $A \sim 0.4$) as small as 37 km in radius.

II. PLUTO'S ORBIT AND SPIN

A. PLUTO'S HELIOCENTRIC ORBIT

Relative to the other known planets, Pluto's orbit is unusually eccentric (eccentricity $e = 0.25$), highly inclined (inclination $i = 17°$), and large (semimajor axis $a = 39.4$ AU). Pluto's orbit period is 248 years, during which the planet ranges from inside Neptune's orbit (Pluto's perihelion is near 29.7 AU) to nearly 49.5 AU. The fact that Pluto's perihelion is closer to the Sun than Neptune's orbit is quite unusual: No other known planet in the solar system crosses the orbit of another. The significant change in heliocentric distance as Pluto moves around the Sun causes the surface insolation on Pluto and Charon to vary by factors of three, which has important implications for the atmosphere (see Section VI). The Pluto–Charon barycenter passed its once-every-248-year perihelion at 05.1 ± 0.1 September 1989 UT; this will not occur again until A.D. 2236.

As of 1996, orbit integrations using osculating elements were able to predict Pluto's position to 0.5 arcsecond accuracy over timescales of a decade.

As noted, Pluto's perihelion lies inside Neptune's orbit. In the mid-1960s, it was discovered through computer simulations that Pluto's orbit librates in a 2:3 resonance with Neptune, which prevents mutual close approaches between the objects. This discovery has been verified by a series of increasingly longer and more accurate simulations of the outer solar system that now exceed 2×10^9 years. Pluto and Neptune never closely approach one another, because the argument of Pluto's perihelion (i.e., the angle between the perihelion position and the position of its ascending node) librates (i.e., oscillates) about 90° with an amplitude of approximately 23°. This ensures that Pluto is never near perihelion when it is in conjunction with Neptune. Thus, Pluto is "protected" because Neptune passes Pluto's longitude only near Pluto's aphelion, never allowing Neptune and Pluto to come closer than ≈ 17 AU. Indeed, Pluto approaches Uranus more closely than Neptune, with a minimum separation of ≈ 11 AU, but still too far to significantly perturb its orbit.

In the late 1980s, it was discovered that Pluto's orbit displays a high degree of sensitivity to initial conditions, called chaos by modern dynamicists. This discovery of a formal kind of chaos in Pluto's orbit does not imply that Pluto will suddenly undergo some near-term orbital change (or for that matter that it recently, by astronomical standards, did so). However, it does mean that Pluto's orbit is unpredictable on long timescales and could rapidly evolve in the very distant future. The timescale for this dynamical unpredictability has been established by Jack Wisdom and co-workers at MIT to be 2×10^7 years. [*See* Chaotic Motion.]

B. PLUTO'S LIGHTCURVE, ROTATION PERIOD, AND POLE DIRECTION

As indicated earlier, since the mid-1950s, Pluto's photometric brightness has been known to vary regularly with a period of about 6.39 days; more precisely, this period is 6.387223 days. Despite Pluto's faintness as seen from Earth, its period was easily determined using photoelectric techniques beccause the planet displays a large lightcurve amplitude, 0.35 magnitudes at visible wavelengths, which is equivalent to 38%.

Since at least 1955, it has also been known that Pluto's lightcurve is exhibiting an increase in its amplitude with time. Although the 6.387223-day period is identical to Charon's orbit period, Charon's photometric contribution is too small to account for the lightcurve's amplitude. This is turn implies that the lightcurve period is caused by surface features on Pluto. Figure 2 shows the shape of the combined Pluto–Charon lightcurve and its evolution over the past few decades.

The first study of Pluto's polar obliquity (or tilt relative to its orbit plane) was reported in 1973. By assuming that the variation of the lightcurve amplitude from the 1950s to the early 1970s was caused by a change in the aspect angle from which we see Pluto's spin vector from Earth, it was then determined that Pluto has a high obliquity, that is, $90 \pm 40°$. In 1983, additional observations allowed the obliquity to be refined to $118.5 \pm 4°$. Even more recently, the results of the Pluto–Charon mutual events (or eclipses, see Section III) have given a very accurate value of $122 \pm 1°$. Pluto's corresponding pole position lies near declination $-9°$, right ascension 312° (equator and equinox of 1950).

It is important to note, however, that torques on the Pluto–Charon binary cause Pluto's obliquity to oscillate between ~105 and ~130° with an ~3.7 × 10⁶-year period. Thus, although Pluto presently reaches perihelion with its pole vector nearly normal to the Sun and roughly coincident with the orbit velocity vector, this configuration is only coincidental. The pole position executes a 360° circulation with a 3.7 × 10⁶-year precession period.

C. CHARON'S ORBIT AND THE SYSTEM MASS

The discovery that Charon orbits Pluto with a cycle time equal to Pluto's rotation period immediately implied that the system has reached spin-orbit synchronicity: an unprecedented situation in all the solar system. Although there are no data to the contrary, the strict demonstration of *complete* tidal evolution, in which Charon's rotational period is equal to both its orbital period and Pluto's day, awaits observational confirmation. However, this synchronicity is widely expected because standard tidal theory shows that Charon's internal rotation will synchronize to its orbit period (as is the case for many planetary satellites) 10 to 100 times faster than Pluto's rotation period becomes locked to the satellite's orbit period, which is known to have already occurred.

Table II gives a solution to Charon's orbital elements obtained from various data. This fit relies on a semimajor axis determination of $a = 19,636 \pm 8$ km derived from ground-based and Hubble Space Telescope data; it is statistically indistinguishable from ground-based results obtained in the mid-1980s of $a = 19,640 \pm 320$ and $a = 19,558 \pm 153$ km.

Based on Charon's known orbital period and the

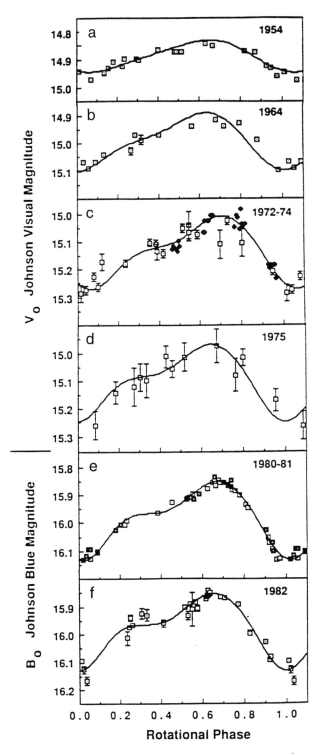

FIGURE 2 The evolution of Pluto–Charon's lightcurve over the past several decades. [Adapted from R. L. Marcialis (1988). *Astronom. J.* **95**, 941.]

TABLE II
Charon's Orbit[a]

Orbital element	Value
Semimajor axis, a	$19,636 \pm 8$ km
Orbital period P	6.387223 ± 0.00002 days
Eccentricity, e	0.0076 ± 0.003
Inclination, i	$96.2 \pm 0.3°$
Long. perihelion, ω	$222.99 \pm 0.5°$
Mean anomaly, M	$34.84 \pm 0.35°$

[a] From D. J. Tholen and M. W. Buie (1997). *In* "Pluto and Charon" (S. A. Stern and D. J. Tholen, eds.). Univ. Arizona Press, Tucson.

19,636-km semimajor axis, the system's (i.e., combined Pluto + Charon) mass is $1.47 \pm 0.002 \times 10^{25}$ g. This is very small, just 2.4×10^{-3} M_{Earth}.

Data from the mutual events showed that unless Charon's orbit has a very special orientation relative to Earth, Charon's orbital eccentricity is very low. Recently, *HST* observations have shown that Charon's orbital eccentricity is nonzero, with a best-estimated present value of 0.0076. The fact that the orbit is not precisely circular indicates that some disequilibrium forces have disturbed it from the exact value of zero expected from tidal evolution. It is most likely that the disturbance causing this is generated by occasional close encounters between the Pluto–Charon system and one of the 100-km or larger-diameter bodies now known to orbit with Pluto in the Edgeworth–Kuiper Belt (cf. Section VIII). [*See* KUIPER BELT.]

III. THE MUTUAL EVENTS

A. BACKGROUND

The realization after Charon's discovery that mutual eclipses between Pluto and Charon would soon occur opened up the possibility of studying the Pluto–Charon system with the powerful data analysis techniques developed for eclipses between binary stars. Initial predictions by Leif Andersson indicated that the events could begin as early as 1979. As Charon's orbit pole position was refined, however, the predicted onset date moved to 1983–1986 (this was fortuitous, since knowledge of the pole could have changed to indicate that the events had already just ended in the mid-1970s!)

After a multiyear effort by several groups to detect the onset of these events, the first definitive eclipse detections were made on 17 February 1985 by Richard Binzel at McDonald Observatory, and were confirmed during an event 3.2 days later on 20 February 1985 by David Tholen at Mauna Kea. These first, shallow events (~0.01–0.02% in depth) revealed Pluto and Charon grazing across one another as seen from Earth.

The very existence of these eclipses proved the claim (by 1985 widely accepted) that Charon was in fact a satellite, rather than some incredible topographic high on Pluto. The mutual eclipses persisted until October 1990, and dozens of events were observed. Important results from the 1985–1990 mutual events included reconstructed surface "maps" of Pluto and Charon;

individual albedos, colors, and spectra for each object; and improvements in Charon's orbit. First, however, was the opportunity to use event timing to accurately determine the radii of Pluto and Charon.

B. RADII AND AVERAGE DENSITY OF PLUTO AND CHARON

Prior to the mutual events, the radii of Pluto and Charon were highly uncertain. Since Pluto and Charon remained unresolved in terrestrial telescopes (their apparent diameters are both <0.1 arcsecond), direct measurements of their diameters were not available. A well-observed, near-miss occultation of Pluto in 1965 had constrained Pluto's radius to be <3400 km, but no better observations were available until the mutual events. However, circumstantial evidence that Pluto was smaller than 3400 km was inferred from the combination of Pluto's *V* astronomical $\approx$ 14 magnitude and the 1976 discovery of CH_4 frost (see Section IV, C), which exhibits an intrinsically high albedo when prepared in a laboratory under low-temperature conditions. The small system mass determined after the discovery of Charon in 1978 strengthened this inference, but Pluto's radius was still uncertain within the bounds 900–2200 km.

The first concrete data to remedy the situation came when a stellar occultation by Charon was observed on 7 April 1980. The 50-second length of the star's disappearance, observed by a 1-m telescope at Sutherland, South Africa, gave a clear lower limit for Charon's radius of 600 km. This result was later refined to 601.5 ± 0.8 km.

As noted earlier, the best radius measurements for Pluto and Charon have resulted from fits of mutual event lightcurves. The analysis of such data yielded radii solutions of 1151 ± 6 and 593 ± 13 km for Pluto and Charon, respectively. Although these are the best available data, they disagree at the level of a few percent with results from a stellar occultation by Pluto itself (see Section VI). This discrepancy is not fully understood but is related to uncertainties in the exact value of Charon's orbital semimajor axis, Pluto and Charon limb darkening, and other quantities. For now, it is only possible to conclude that Pluto's true radius is between 1145 and 1200 km, and that Charon's radius is between 600 and perhaps 640 km.

The two striking implications of the small radii and like masses are (1) that Pluto and Charon form a true example of a double planet (with the system barycenter outside of Pluto) and (2) that Pluto indeed is a very small planet—even smaller than the seven largest planetary satellites (the Moon, Io, Europa, Ganymede, Callisto, Titan, and Triton).

Based on the radii and the total mass of the binary, it is possible to derive a system-average density of 1.95 ± 0.1 g cm^{-3}, where the large error bar is dominated by the uncertainty in the radii of Pluto and Charon. Any density of 1.8 g cm^{-3} or higher implies that the system is compositionally dominated by rocky material, probably hydrated chondrites, as opposed to ices. This result and its implications will be discussed in more detail in Section V.

IV. PLUTO'S SURFACE PROPERTIES AND APPEARANCE

Pluto's surface properties have been studied since the 1950s. During that span of time, photometric, spectroscopic, and polarimetric techniques have been applied, and the explorable wavelength regime has expanded from the ground-based window to the reflected infrared (IR) and the space ultraviolet. Thermal-IR and millimeter-wave measurements of the Pluto–Charon binary have also been made.

A. ALBEDO AND COLOR

Two of the most basic photometric parameters that one desires to know for any solid body are its albedo and color. Accurate knowledge of Pluto's albedo was obtained only after the onset of the mutual events, because until then Pluto's radius was unknown, and there was no definitive way of removing Charon's contribution.

The very first report of eclipse detections revealed a factor of two difference in depth between partial eclipses of Charon and Pluto, indicating that Pluto's geometric albedo is substantially higher than Charon's.

Now that the eclipse season is complete, a more complete data set is available for analysis. Comprehensive models for the analysis of mutual event lightcurve data simultaneously solve for the individual radii of Pluto and Charon, the individual albedos, and Charon's orbital elements. The modeling of these parameters is complicated by solar phase angle effects, the presence of shadows during eclipse events, and instrumental and timing uncertainties. To derive the albedo lightcurve for Pluto alone, its albedo at the longitude of the best superior eclipses (in which Charon was completely hidden) must first be determined; albedos

at other rotational epochs are then derived from this anchor point, assuming that Charons' rotational lightcurve contributes only a small constant to the combined Pluto + Charon lightcurve.

The assumption that a constant Charon contribution can be removed is not unreasonable, because (1) its geometric cross section is small (one-fourth of Pluto's) and (2) its eclipsed hemisphere has a geometric albedo only about 50–60% of Pluto's. However, recent HST results have shown that Charon does vary somewhat in brightness ($\approx 8\%$) as it rotates on its axis.

Analysis of large sets of mutual event data in the way just described has found that Pluto's maximum, disk-integrated, B-bandpass (~ 4360 Å) geometric albedo is 0.61. Rotational variations cause this albedo to range from values as low as 0.44 to values as high as 0.61 as Pluto rotates.

Information on Pluto's color comes from both unresolved photometry of the binary and the mutual events. As reported in Section I, Pluto's visible-bandpass color slope has been known to be red since the 1950s. Recent analysis of pre-mutual event photometry yields B-V and U-B color differences of 0.84 and 0.31, respectively, for Pluto + Charon. There is only weak evidence that this value has changed since the 1950s when photoelectric measurements were first made. Eclipse data have revealed that the B-V color of Pluto itself is very close to 0.85 astronomical magnitudes. By comparison, this color is much less red than the refractory surfaces of Mars (B-V = 1.36) and Io (B-V = 1.17), and slightly redder than its closest analog in the solar system, Triton (B-V = 0.72).

B. SOLAR PHASE CURVE

The photometric behavior of a planet or satellite as it changes in brightness on approach to opposition can be used to derive surface scattering properties, and therefore its microphysical properties. Knowledge of the complete solar phase curve is also required to transform geometric albedos into bolometric Bond albedos. *HST* observations give linear phase coefficients for Pluto and Charon of 0.029 ± 0.001 and 0.866 ± 0.008 magnitudes/deg, respectively.

Pluto's maximum solar phase angle (ϕ_{max}) as seen from Earth is just $\approx 1.9°$. Therefore, no measurements of the large-angle scattering behavior have been possible. Without measurements at large phase angles, no measurements of Pluto's phase integral q or Bond albedo A can be made. However, some improvement in estimates of q and A could become possible if the *Cassini* spacecraft obtains Pluto's phase curve from

Saturn orbit, where $\phi_{max} \approx 18°$. However, what is really needed are flyby spacecraft measurements of Pluto at high phase angles. For the present, the best available phase integral to use for Pluto is probably Triton's (Pluto and Triton also have similar linear phase coefficients). Triton's q has been measured by *Voyager 2*, giving $q = 1.2$ (at green wavelengths) to 1.5 (at violet wavelengths). If Pluto is similar, then its surface may have Bond albedos ranging from 0.3 to 0.7.

C. SURFACE COMPOSITION

Progress in understanding Pluto's surface composition required the development of sensitive detectors capable of making moderate spectral resolution measurements in the infrared, where most surface ices show diagnostic spectral absorptions. Although this technology began to be widely exploited as early as the 1950s in planetary science, Pluto's faintness (e.g., 700 times fainter than the Jovian Galilean satellites) delayed compositional discoveries about it until the mid-1970s.

The first identification of a surface constituent on Pluto was the discovery by Dale Cruikshank, Carl Pilcher, and David Morrison in 1976 of CH_4 ice absorptions between 1 and 2 μm (a wavelength of 1 μm = 10,000 Å). Cruikshank *et al.* made this discovery using infrared photometers equipped with customized, compositionally diagnostic filters. In their report, they also presented evidence against the presence of strong H_2O and NH_3 absorptions in Pluto's spectrum. Confirmation of the methane detection came in 1978 and 1979, when both additional CH_4 absorption bands and true IR spectra of Pluto became available.

In mid-1992, another breakthrough occurred when Toby Owen and a number of colleagues made observations using a new, state-of-the-art IR spectrometer at the UK Infrared Telescope (UKIRT) on Mauna Kea. These data revealed the presence of both N_2 and CO ices on Pluto. These molecules are much harder to detect than methane because they produce much weaker spectral features. However, their presence on Pluto indicates that the surface is chemically more heterogeneous and more interesting than had previously been established. Because N_2 and CO are orders of magnitude more volatile (i.e., have higher vapor pressures) than CH_4, their presence also implies that they likely play an important role in Pluto's atmosphere.

Rotationally resolved spectra of Pluto's CH_4 absorption bands have been reported by a number of groups. Their studies showed that Pluto's methane is present at all rotational epochs, but the band depths are correlated with the lightcurve so that the minimum absorption occurs at minimum light. Mutual event spectroscopy has now demonstrated that Charon is not the cause of this variation, since Charon's surface is devoid of detectable CH_4 absorptions (see Section VII). This important discovery suggests that Pluto's dark regions could contain reaction products resulting from the photochemical or radiological conversion of methane and N_2 to complex nitriles and higher hydrocarbons.

We are thus left with the following picture of Pluto's surface composition: CH_4 appears to be rotationally ubiquitous, but with its surface coverage more widespread in regions of high albedo. CO and N_2 have also been detected, but have not yet been mapped with rotational phase. In the bright areas of the planet, where these ices are thought to mainly be located, N_2 dominates the surface abundance, and the CO is more abundant than the previously known (but more spectroscopically detectable) CH_4. Pluto's strong lightcurve and red color demonstrate that at least one other widespread, probably involatile surface constituent exists. This may be either rocky material or, more naturally, hydrocarbons resulting from radiation processing of the CH_4 due to long-term exposure to ultraviolet sunlight. Whether the frost we are seeing is a surface veneer or the major component of Pluto's crust is unclear.

D. SURFACE TEMPERATURE

Results from the *Infrared Astronomical Satellite* (*IRAS*) indicated that Pluto's surface temperature was in the range of 55 to 60 K around perihelion. However, more recently, it has come to be appreciated that the situation in Pluto's surface is likely more complicated. One line of evidence for this conclusion comes from millimeter-wave measurements of Pluto's Rayleigh–Jeans blackbody spectrum. Such measurements, reported first by Altenhoff Wilhelm and collaborators, and then later by Alan Stern, Michel Festou, and David Weintraub, and independently confirmed by David Jewitt, indicate that a significant fraction of Pluto's surface is significantly colder than 60 K, most likely in the range of 35 K to 42 K. As described in the 1993 Stern *et al.* report, although the surface pressure of N_2 is not well known, it must be less than $\approx$60 microbars. This is consistent with an N_2 ice temperature of $\approx$40 K, assuming vapor pressure equilibrium between the N_2 ice and the atmosphere. This provides additional evidence that Pluto's surface cannot be uniformly at 50–60 K, but must have colder regions, from which the N_2 atmosphere is presumably generated. [See INFRARED VIEWS OF THE SOLAR SYSTEM FROM SPACE.]

A second, later-coming line of evidence comes from high-resolution spectroscopy of the temperature-sensitive 2.15-μm N_2 ice absorption band, which Kimberley Tryka and her co-workers found indicates a surface temperature of about 40 K for the widespread nitrogen ices on Pluto. Because the *IRAS* measurements made in the early 1980s are not flawed, it is now thought likely by many planetary scientists that Pluto's surface temperature varies from place to place, with ≈ 40 K regions where N_2 ice is sublimating and ≈ 55–60 K regions where N_2 ice is not present in great quantities. If this is correct, then the strong temperature contrasts across Pluto's surface imply important complications in the dynamics of both Pluto's atmosphere and the way in which volatiles migrate over the surface over time.

E. SURFACE APPEARANCE AND MARKINGS

Because Pluto is less than 0.1 arcsecond across as seen from Earth, its disk could not be resolved until the advent of the Hubble Space Telescope. Still, evidence for surface markings has been available since the mid-1950s, when lightcurve modulation was first detected. Because Pluto is large enough to be essentially spherical (and, indeed, mutual event and stellar occultation data show that it actually is), the distinct variation in this lightcurve must be related to large-scale albedo features.

From the lightcurve in Fig. 2, it can be seen that Pluto's surface must contain at least three major longitudinal provinces. Information on the latitudinal distribution of albedo can be gained by observing the evolution of this lightcurve as Pluto moves around its orbit while the pole position remains inertially fixed, assuming, of course, that the surface albedo distribution is time invariant.

The most complete mapping products obtained from photometric data inversions (variously using rotational lightcurves and mutual event lightcurves) have been obtained by two teams. The first team, led by Marc Buie of Lowell Observatory, has used both mutual event lightcurves and rotational lightcurve data compiled from 1954 to 1986 to compute a complete map of Pluto. The second group, consisting of Eliot Young and Richard Binzel of MIT, numerically fit a spherical harmonic series to

each element of a finite element grid using the Charon transit mutual event lightcurve data as the model input. Because Young and Binzel used only mutual event data, their map is limited to the single hemisphere of Pluto that Charon eclipses. Because the two groups use different data sets and different numerical techniques, their results are complementary and serve to check one another on the Charon-facing hemisphere that they share in common.

These two maps are shown in Fig. 3. Although there are differences between them, it must be remembered that each map has intrinsic noise. The common features of these maps are (1) a very bright south polar cap, (2) a dark band over midsouthern latitudes, (3) a bright band over midnorthern latitudes, (4) a dark band at high northern latitudes, and (5) a northern polar region that is as bright as the southern cap. Even better results, possibly including some color information, are expected as these models are improved and a more complete set of mutual event lightcurve data is included in them.

In 1990, the Hubble Space Telescope imaged Pluto, but owing to its then-severe optical aberrations, these images (obtained by R. Albbrecht and a team of collaborators) cleanly separated Pluto and Charon but did not reveal significant details about the surface of Pluto. After HST was repaired by an astronaut crew in late 1993, its optics were good enough to resolve crude details on Pluto's surface. And in mid-1994, the first actual images of Pluto were obtained that revealed significant details about its surface. These images were made by Alan Stern, Marc Buie, and Laurence Trafton using the Faint Object Camera (FOC) of the Hubble Space Telescope. The 20-image HST data set is longitudinally complete, rotationally resolved, and obtained at both blue and ultraviolet wavelengths. The various images that HST obtained were combined to make blue and ultraviolet (UV) maps of the planet, such as the one shown in Fig. 4. The HST images and derived maps reveal that Pluto has (1) a highly variegated surface, (2) extensive, bright, asymmetric polar regions, (3) large midlatitude and equatorial spots, and (4) possible linear features that are hundreds of kilometers in extent. The dynamic range of albedo features across the planet detected at the FOC's resolution in both the 410- and 278-nm bandpasses exceeds 5 : 1. Although more sophisticated HST map inversions are planned, and new HST images will no doubt be obtained in the future, the existing HST maps already provide important inputs to modelers interested in volatile transport and comparative studies of Pluto and Triton.

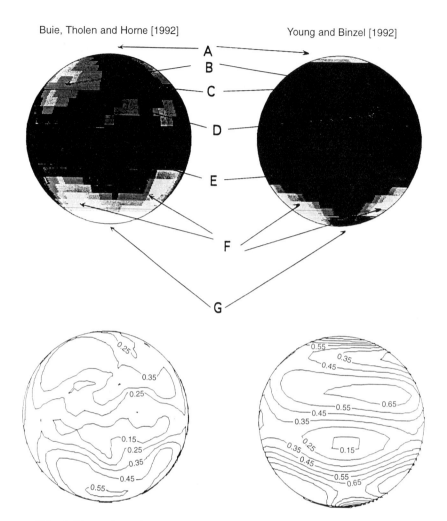

FIGURE 3 Two maps of Pluto's Charon-facing hemisphere. The map on the left was derived by M. Buie, K. Horne, and D. Tholen using both mutual event and lightcurve data. The map on the right was derived by E. Young and R. Binzel from their mutual event data. Although the fine details of these maps differ, their gross similarities are striking. [For additional details, see R. P. Binzel *et al.* (1997). *In* "Pluto and Charon" (S. A. Stern and D. J. Tholen, eds.). Univ. Arizona Press, Tucson.]

V. PLUTO'S INTERIOR AND BULK COMPOSITION

A. DENSITY

As noted earlier, knowledge of Charon's semimajor axis and the radii of Pluto and Charon allows us to specify the average Pluto + Charon (or system) density to be 1.95 ± 0.08 g cm^{-3}. This value is strikingly close to Triton's density, which *Voyager 2* showed to be 2.054 ± 0.03 g cm^{-3}. However, one must keep in mind that the system density is an average between the two

bodies, and does not (necessarily) represent either body alone.

To determine the separate densities of Pluto and Charon, one must either obtain precise astrometric measurements that detect the barycentric wobble between Pluto and Charon or remove Charon's effect by modeling. Since 1992, both HST and ground-based measurements have been gathered to address the mass ratio, and therefore the relative masses and densities of Pluto and Charon. These are very difficult measurements and, as of early 1997, it is not clear whether Pluto and Charon have nearly identical densities near the system average or whether they might have somewhat different densities. In particular, it is possible within the current errors reported for these measure-

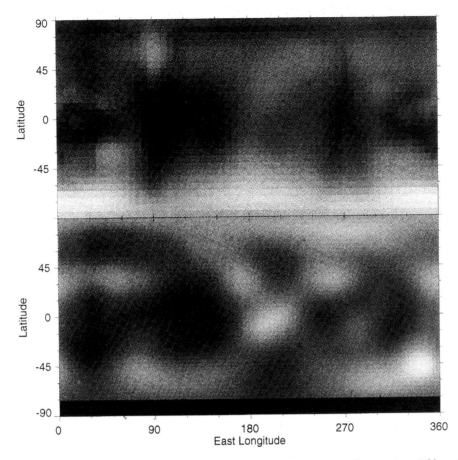

FIGURE 4 Direct-imaging maps of Pluto obtained from Hubble Space Telescope images made in 1994 at visible and UV wavelengths. [Adapted from S. A. Stern *et al.* (1997). *Astronom. J.* **113**, 827.] Note: these maps are not scaled to true brightnesses; Pluto's surface is highly reflective.

ments that Charon could have a density perhaps as low as 1.8–1.9 g cm⁻³, which, if true, would indicate that it has a significantly higher proportion of ices in its interior than does Pluto.

B. BULK COMPOSITION AND INTERNAL STRUCTURE

The discovery that the system average density is near 2 g cm⁻³ was a major surprise resulting from the mutual events. Many scientific papers had previously predicted values closer to the density of water ice (~1 g cm⁻³), or even lower. Thus, contrary to earlier thinking, the Pluto–Charon binary is dominated by material denser than water ice.

Based on a system density near the 2 g cm⁻³ system average, and the fact that Pluto must be within 10% of the system average (or else Charon's density would be outside the observationally allowed range of 1.8–2.3 g cm⁻³), a three-component model for Pluto's bulk composition and internal structure can be derived. In such a model, Pluto's bulk density is assumed to consist of three of the most common condensates in the outer solar system: water ice ($\rho = 1.00$ g cm⁻³), "rock" ($2.8 < \rho < 3.5$ g cm⁻³, depending on its degree of hydration), and methane ice ($\rho = 0.53$ g cm⁻³). More complicated models employing multiple rock components or other volatiles (e.g., CO or N₂ ice) are unlikely to provide additional insight, because they introduce more unknowns than constraints.

From three-component models, it has been found that Pluto's rock fraction is in the range of 60 to 80%, with preferred values close to 70%. By comparison, the large (e.g., $R > 500$ km) icy satellites of Jupiter, Saturn, and Uranus have typical rock fractions in the range of 50 to 60% by mass. Only Io, Europa, and Triton rival Pluto in terms of their computed rock content.

Pluto's high rock (i.e., nonvolatile) mass fraction is in contrast to the ≈50:50 rock:ice ratio predicted for objects formed from solar nebula material according to many nebular chemistry models and our present-day understanding of the nebular C/O ratio. This high rock fraction indicates that the nebular material from which Pluto formed was CO- rather than CH_4-rich. As such, roughly half of the available nebular oxygen should have gone into CO, rather than H_2O formation, which in turn would lead to a high rock : ice ratio.

There are two possible ways out of the apparent nebular chemistry dilemma imposed by Pluto's high rock fraction. One is that Pluto's estimated radius of 1150 km may indeed be too small; a value near 1200 km, as suggested by some stellar occultation models, would solve the problem. Alternatively, William McKinnon and Damon Simonelli have independently suggested that a giant impact may have induced volatile loss from an already-differentiated Pluto, which may have raised Pluto's rock fraction and somewhat (perhaps 20%) to reach its present value. As we discuss in Section VIII, such an impact is thought to be responsible for the formation of the Pluto–Charon binary.

The gross internal thermal structure of Pluto depends on several factors, virtually all of which are unknown. These include material viscosities in the interior, the internal convection state, the actual rock fraction and radioisotope content, and the internal density distribution (i.e., most fundamentally, the differentiation state). It would appear likely that Pluto's deep interior reaches temperatures of at least 100–200 K, but not much higher. Whether or not Pluto is warm enough to exhibit convection in its ice mantle depends on both the internal thermal structure and the radial location of water ice in the interior.

Based on the results just given and laboratory equations of state, Pluto's central pressure can be estimated to lie between 0.6 and 0.9 GPa (gigapascals) if the planet is undifferentiated, or 1.1–1.4 GPa if differentiation has occurred. As such, the high-pressure water ice phase, Ice VI, is expected in the deep interior if the planet has not differentiated. If differentiation has occurred, a higher-pressure form of water ice called Ice II may be present, but only near the base of the convection layer. If Pluto did differentiate, then its gross internal structure may be represented by a model like that shown in Fig. 5.

VI. PLUTO'S ATMOSPHERE

A. ATMOSPHERIC COMPOSITION

The existence of an atmosphere on Pluto was strongly suspected after the discovery of methane on its surface in 1976, largely because at the likely surface temperatures on Pluto (~40–60 K), enough methane should be in the vapor phase to constitute a significant atmosphere. This argument was supported by the high reflectivity of Pluto's surface, which suggested some kind of resurfacing—most plausibly due to volatile laundering through an orbitally cyclic atmosphere. Nonetheless, there was no definitive evidence for an atmosphere until the late 1980s.

The definitive proof of Pluto's atmosphere came from the occultation of a 12th magnitude star by Pluto in 1988, by providing the first direct observational evidence for an atmosphere. The best measurements of the occultation were obtained by Robert Millis and James Elliot, and their various MIT, Lowell Observatory, and Australian co-workers. These teams used both NASA's mobile *Kuiper Airborne Observatory* (which contained a 36-inch-diameter telescope) and ground-based telescopes to observe the occultation event. They discovered that light from the star was diminished far more gradually than it would be from an airless body. The apparent extinction of starlight observed during the occultation was caused by atmospheric refraction (i.e., the degree of bending of the starlight by the atmosphere), which varies with height. The rate at which the refractivity of the atmosphere varies with altitude depends on the ratio of atmospheric temperature (T) to atmospheric mean molecular weight (m). The Pluto occultation data imply $T/m = 3.7 \pm 0.7$ K/g at and above an altitude of 1215 km. If the atmosphere were composed entirely of methane ($m = 16$ g/mole), the implied atmospheric temperature would be 60 K, whereas an N_2 or CO atmosphere ($m = 28$ g/mole) would be at a temperature near 106 K.

From the stellar occultation data alone it is impossible to separately determine the mean atmospheric molecular weight and temperature of Pluto's atmosphere. However, theoretical calculations of the atmospheric temperature made by Roger Yelle and Jonathan Lunine of the University of Arizona indicated a value of 106 K in the upper atmosphere, under a variety of assumed compositions. This value is relatively large compared with the surface temperature (~40–50 K) because the efficiency at which the atmosphere radiates and cools is very small at the cold temperatures on Pluto. An upper atmospheric temperature near 106 K implies that the atmospheric mean molecular weight is close to 28 g/mole. This is consistent with either N_2 or CO gas, or both. In the future, it may eventually be possible to measure this temperature directly through extremely high resolution spectroscopy, if emission lines in Pluto's atmosphere are discovered.

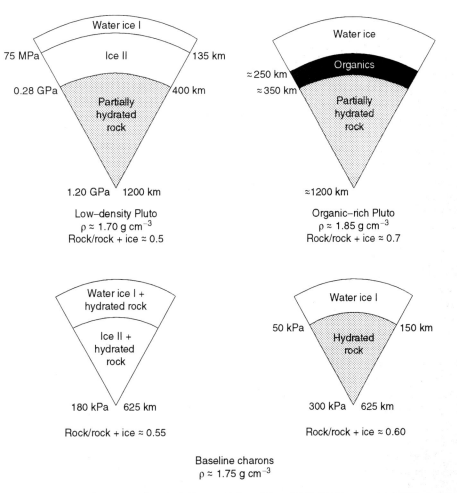

FIGURE 5 Typical interior structural models for Pluto and Charon. [Adapted from W. B. McKinnon *et al.* (1997). *In* "Pluto and Charon" (S. A. Stern and D. J. Tholen, eds.). Univ. Arizona Press, Tucson.]

The recent detection of N_2 ice absorption features on Pluto's surface (cf. Section IV), coupled with the discovery that Triton's atmosphere also consists predominantly of N_2 and only a trace of CO, suggests that Pluto probably has an N_2-dominated atmosphere. Nevertheless, if the high-temperature (106 K) atmospheric model is correct, then at least a few percent methane is thought to be required, because methane (which is efficient at atmospheric heating) is thought to be responsible for the elevated atmospheric temperatures. Thus, although the presence of a major atmospheric constituent heavier than methane seems well established, it is very difficult to make quantitative predictions about atmospheric composition with the presently available data.

The line of argument supporting a nitrogen-dominated atmosphere with only a minor amount of methane was significantly strengthened in 1994 when Leslie Young and her colleagues detected CH_4 gas in Pluto's

atmosphere for the first time. This discovery, which was made possible by sensitive, high-resolution IR spectroscopy of the 2.3-μm CH_4 band system, revealed a total methane mixing ratio of probably <1%, and perhaps as little as 0.1% in the atmosphere.

B. ATMOSPHERIC STRUCTURE

The 1988 occultation data exhibit interesting behavior at altitudes below 1215 km, as is shown in Fig. 6. The starlight, which was decreasing gradually at higher altitudes, dropped suddenly to a value close to zero below this level; this is called "lightcurve steepening." The drop is still not as sudden as would be expected from the setting of a star behind the limb of an airless planet, however, and two explanations have been proposed for this change in the characteristics of the occultation lightcurve.

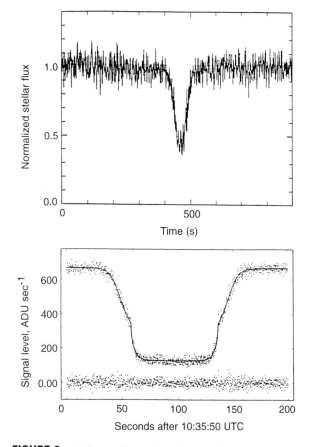

FIGURE 6 Stellar occultation data showing the refractive signature of Pluto's atmosphere and the steepening of the lightcurve around the half-light level that is discussed in the article. The upper panel is a ground-based data product; the lower panel was obtained from the *Kuiper Airborne Observatory*. [Adapted from Elliot *et al.* (1989). *Icarus* **77**, 148.]

In the first model, the steepening is caused by the presence of aerosol hazes in the lower atmosphere. Condensation clouds can be ruled out as an explanation for the aerosol layer because of the temperature structure of the atmosphere. Because reproducible albedo features have been seen on Pluto's surface, any such aerosol layer must be transparent when viewed from above, but relatively opaque when viewed horizontally. The aerosols must also extend around most of the planet since the steepening of the occultation lightcurve was seen in both immersion and emersion. It has been suggested that the aerosols could be "photochemical smog" similar to the aerosols discovered on Titan and Triton (and a distant cousin to the air pollution in the industrial basins on Earth, such as Los Angeles). [*See* TITAN.]

In the second model, the sudden drop in the brightness of starlight below 1215 km is caused by a gradient in atmospheric temperature. Such a gradient is expected from theoretical modeling (see earlier) because atmospheric temperatures are expected to be higher than surface temperatures. Changes in atmospheric temperature cause a variation of refractivity with height in the atmosphere that could be manifested as the accelerated diminution of starlight seen in the occultation.

The haze layer and temperature gradient explanations imply differences in the way that the color of starlight changes during an occultation. Future occultations may help decide between the two explanations if simultaneous observations can be made at two or more well-separated wavelengths.

If the temperature gradient explanation is correct, Pluto's surface radius is likely near 1206 ± 11 km. If the haze layer explanation is correct, Pluto's surface radius is more difficult to determine, but is probably closer to 1180 km. In either case, the occultation implies a radius that is a few percent larger than the nominal mutual event solution (1151 km); in the case of the haze model, the radius cannot be much less than 1180 km or else the haze would be so thick so to completely obscure the surface. Clearly, there is a discrepancy between the radii determined from the occultation and those derived from the mutual events, which future research will have to resolve.

One interesting feature of Pluto's atmosphere worth mentioning is the very rapid rate at which it escapes to space. Because of Pluto's low mass and consequently weak gravitational binding energy, sufficiently energetic molecules at the top of the atmosphere are able to escape the gravitational pull entirely. This can result in a condition called hydrodynamic escape, in which the high-altitude atmosphere achieves an internal thermal energy greater than the planetary gravitational potential energy acting on the atmosphere. Estimates indicate that the present escape rate is so high that atmosphere may be lost to the escape process (thus requiring replenishment from sublimating surface ices) on timescales as short as 30–200 years. Relatively speaking, the atmosphere of Pluto is escaping at a rate far greater than any other planetary atmosphere in the solar system. However, in saying this, one must also note that models indicate that <2% of Pluto's total mass has been lost to space from such escape since the planet's formation.

A second interesting feature of Pluto's atmosphere is its strong orbital-seasonal effects. These effects are driven by the fact that the strength of solar heating varies by a factor of almost four around Pluto's orbit, which in turn causes the vapor pressures of N_2, CO, and CH_4 to vary by factors of hundreds to thousands.

Therefore, unlike any other planet, Pluto's atmosphere is thought to be essentially seasonal, with the perihelion pressure being many many times the aphelion pressure. Indeed, some models predict that by A.D. 2010, just 20 years after perihelion, Pluto's atmosphere will largely condense onto the surface, a condition called atmospheric collapse.

VII. CHARON

Pluto's only known satellite, Charon, was discovered in 1978. Charon's radius of 600 km is about half of Pluto's, implying that its mass is most likely between 10 and 14% of Pluto's. By comparison, typical satellite : planet mass ratios are 1000 : 1 or greater, and even the mass ratio of the Moon to the Earth is only 81 : 1.

Relative to Pluto, very little is known about Charon. The first important fact came from a stellar occultation in 1980. That event, observed from South Africa, revealed a lower limit to Charon's radius of 601 km. Much of what is known has been discovered from the mutual events. Most of the remainder has been learned using HST.

The mutual events resulted in several key discoveries. These included (1) the fact that Charon's average visible surface albedo is 30–35%, much lower than Pluto's, and (2) that Charon's visible surface color is quite neutral, unlike Pluto's clearly reddish tint.

Another major set of advances that resulted from the eclipse events was the first set of constraints on Charon's basic surface composition. These came from the subtraction of spectra made just prior to eclipse events from those made when Charon was completely hidden behind Pluto. The resulting "net" spectrum thus contains the Charon-only signal. As shown in Fig. 7, this technique has been applied in both the visible (0.55–1.0 μm) and infrared (1–2.5 μm) bandpasses. The visible light data show that Charon's surface does not display the prominent CH_4 absorption bands that Pluto does, indicating that Charon's surface has little or (more likely) no substantial methane on it. Additionally, there is no evidence for strong absorptions due to a number of other possible surface frosts, including CO, CO_2, H_2S, NH_3, N_2, or NH_4HS, on Charon. The IR spectra of Charon do show that Charon does, however, display clear evidence of water ice absorptions, which Pluto does not. It is tempting to speculate (as some authors have) that Charon may have lost its volatiles through the escape of a primordial atmosphere or by heating resulting from its formation in a giant impact.

Since the launch of the Hubble Space Telescope, it has been possible to routinely separate Charon's light from Pluto's, and to learn Charon's phase coefficient, UV albedo, and rotational lightcurve. Most notably among these, Marc Buie and Dave Tholen have determined that Charon displays a small but significant lightcurve variation near 8% as it rotates on its axis.

Because the only identified surface constituent of Charon is water ice, which is not volatile at the expected 50–60 K surface radiative equilibrium temperature at perihelion, one does not expect Charon to have an atmosphere. The fact that CH_4 is not present on the surface supports this expectation. However, absence of evidence is not the same as evidence of absence, and one eagerly awaits more stringent constraints on other potential surface volatiles, including CO, N_2, and the condensates of noble gases. However, one published interpretation of the 1980 Charon stellar occultation claims there is some evidence for a weak atmospheric refraction signal. To definitively resolve the issue of Charon's atmosphere, either a better-observed stellar occultation event or a spacecraft flyby is required.

In summary, based on the scant data available, one concludes that Charon's surface albedo, color, and composition are more like those of the five major Uranian satellites than Pluto. Indeed, Uranus's satellite Ariel, with a radius of 580 ± 5 km, a geometric albedo of 0.4, very subtle surface color, strong water ice absorption bands, and no detectable atmosphere, makes a nice first-order model for Charon's outward appearance. However, Ariel's mean density is near 1.65 g cm^{-3}, about 25–35% higher than Charon's, implying that Ariel and the other Uranian satellites likely have a higher rock-to-volatile ratio than does Charon. Table III compares some basic facts about Pluto and Charon. [See OUTER PLANET ICY SATELLITES.]

VIII. THE ORIGIN OF PLUTO AND CHARON

Any scenario for the origin of Pluto must of course provide a self-consistent explanation for the major attributes of the Pluto–Charon system. These include: (1) the existence of the binary's exceptionally low, ~8 : 1 planet : satellite mass ratio; (2) the synchronicity of Pluto's rotation period with Charon's orbit period; (3) Pluto's inclined, elliptical, Neptune-resonant orbit; (4) the high axial obliquity of Pluto's spin axis and Charon's apparent alignment to it; (5) Pluto's small mass (~10^{-4} of Uranus's and Neptune's); (6) Pluto's

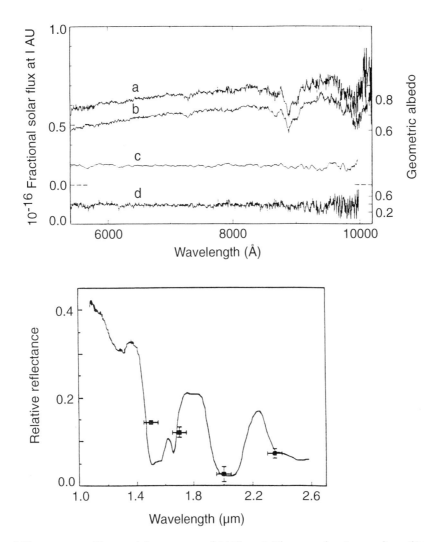

FIGURE 7 Pluto and Charon spectra. Top panel shows spectra of: (a) Pluto + Charon made prior to eclipse; (b) Pluto-only after second contact with Charon hidden; (c) Charon-only smoothed to 80-Å resolution resulting from the subtraction of (a) − (b); and (d) the raw Charon-only spectrum resulting from the subtraction of (a) − (b). Notice that the strong methane absorption bands present in Pluto's spectrum are not detected in the Charon-only spectrum. [This panel adapted from U. Fink and M. DiSanti (1988). Bottom panel shows the Marcialis *et al.* (1987) detection of water ice in Charon's reflectance spectrum (data points) against a laboratory spectrum of water ice at 55 K; adapted from R. L. Marcialis *et al.* (1987). *Science,* **237,** 1349.]

high rock content—the highest among all the outer planets and their major satellites; and (7) the dichotomous surface compositions of Pluto and Charon. This formidable list of constraints on origin scenarios is very clearly dominated by Charon's presence, the unique dynamical state of the binary, and the low mass of Pluto/Charon compared to other planets. In the past few years, much progress has been made in understanding Pluto's likely origin and its context in the outer solar system. This work began with theoretical consideration in the late 1980s and early 1990s, and was advanced considerably by the recent discoveries of numerous 100- to 400-km objects in the Edge-

worth–Kuiper Belt, where Pluto also resides. [*See* Kuiper Belt.]

A. THE ORIGIN OF THE BINARY

Several scenarios have been examined for the origin of the Pluto–Charon binary. These include coaccretion in the solar nebula, mutual capture via an impact between proto-Pluto and proto-Charon, and rotational fission. Gravitational capture of Pluto by Charon without physical contact is not dynamically viable.

TABLE III
Pluto and Charon Comparison

Parameter	Pluto	Charon
Rotation period	6.387223 days	6.387223 days
Radius	1150–1215 km	600–640 km
Density	≈ 2.1 g cm^{-3}	≈ 1.3 g cm^{-3}
Perihelion V_0	13.6 magnitude	15.5 magnitude
Mean B geometric albedo	0.55	0.38
Rotational lightcurve	38%	8%
B–V color	0.85 mag	0.70 mag
V–I color	0.84 mag	0.70 mag
Known surface ices	CH_4, N_2, CO	H_2O
Atmosphere	Confirmed	Doubtful

The formation of Pluto and Charon together in a subnebular collapse is also not considered realistic because of their small size; standard planetary formation theory suggests that bodies in the Pluto and Charon size class formed via solid-body accretion of planetesimals. Similarly, the rotational fission hypothesis is unlikely to be correct, because the Pluto–Charon system has too much angular momentum per unit mass to have once been a single body.

The more likely explanation for the origin of the Pluto–Charon binary is an inelastic collision between two bodies, Pluto and proto-Charon, which were on intersecting heliocentric orbits. A similar scenario has been proposed for the origin of the Earth–Moon binary, based in part on its relatively high mass ratio (81:1) and high specific angular momentum. In the collision theory, Pluto and the Charon-impactor formed independently by the accumulation of small planetesimals, and then suffered a chance collision which dissipated enough energy to permit binary formation. [See THE MOON.]

An important qualitative difference between the Pluto–Charon and Earth–Moon giant impacts is that the relative collision velocities, and hence the impact energies of the Pluto–Charon event, were much smaller. This enormously reduced the thermal consequences of the collision. Thus, whereas the Earth may have been left molten by the Mars-sized impactor necessary to have created the Moon, the proto-Charon impactor would probably have raised Pluto's global mean temperature by no more than 50 to 75 K. This would have been insufficient to melt either body, but may have been sufficient to induce the internal differentiation of either. It would have also produced a substantial short-lived, hot, volatile atmosphere with intrinsically high escape rates. Such an escaping atmosphere could have interacted with the Charon-forming orbital debris, and also perhaps affected Pluto's present-day volatile content.

Although the giant impact scenario for the formation of the Pluto–Charon system is now widely accepted, we must point out that no realistic simulations of this putative event have been carried out. In effect, the giant impact hypothesis has come to be accepted because it is the only scenario that remains at all viable given the various constraints—most particularly the high specific angular momentum of the binary. Clearly, this is an area ripe for future work.

B. THE ORIGIN OF PLUTO ITSELF

The presence of volatile ices, including methane, nitrogen, and carbon monoxide on Pluto and water ice on Charon, argues strongly for their formation in the outer solar system. The average density and consequent high rock content of these two bodies also argue for formation from the outer solar nebula, rather than from planetary subnebular material. As described earlier, it is thought that the two objects (or more precisely Pluto and a Charon-progenitor) formed independently and subsequently collided, thus forming the binary either through direct, inelastic capture or through the accretion of Charon from debris put in orbit around Pluto by the impact.

The first widely discussed theory for Pluto's origin was R. A. Lyttleton's 1936 suggestion, which was based on the fact that Pluto's orbit is Neptune-crossing. In Lyttleton's well-remembered scenario, Pluto was formerly a satellite of Neptune that was ejected via a close encounter between itself and the satellite Triton. According to Lyttleton, this encounter also reversed the orbit of Triton. Variants on the "origin-as-a-former-satellite-of-Neptune" hypothesis were later proposed. However, all of these scenarios were dealt a serious blow by the discovery of Charon, which severely complicates the Pluto-ejection problem by requiring either: (1) Charon to also be ejected from the Neptune system in such a way that it enters orbit around Pluto or (2) Charon to be formed far beyond Neptune where Pluto currently orbits and then captured into orbit around Pluto (presumably by a collision). Other strong objections to scenarios like Lyttleton's also exist. First among these is the fact that any object ejected from orbit around Neptune would be Neptune-crossing and therefore subject to either accretion or rapid dynamical demise. It is implausible that such an object would be transferred to the observed 2:3 Neptune:Pluto resonance, because stable 2 : 3 libration orbits are dynamically disconnected in orbital phase space from orbits intersecting Neptune. Further, because Pluto is less massive than Triton by about a factor of two, it is impossible for Pluto to reverse Triton's orbit to a retrograde one, as is observed. Further still, Pluto's rock content is so high that it is unlikely that Pluto formed in a planetary subnebula. Of course, none of these facts was known until decades after Lyttleton made his original (and then quite logical) suggestion that Pluto might be a former satellite of Neptune.

The heliocentric formation/giant collision scenario described here for the origin of the binary can account for most of the major attributes of the system, including the elliptical, Neptune-crossing orbit, the high axial obliquities, and the 8 : 1 mass ratio. Further, the present tidal equilibrium state would naturally be reached by Pluto and Charon in 10^8–10^9 years after the binary's formation—a small fraction of the age of the solar system.

Still, such a scenario begs two questions. First, why is Pluto so small? and second, how could Pluto and the Charon-progenitor, alone in over 10^3 AU^3 of space, "find" each other in order to execute a mutual collision? That is, the giant impact hypothesis still fails to explain (1) the existence of Pluto and Charon themselves, (2) the very small masses of Pluto and Charon compared to the gas giants in general, and Neptune and Uranus in particular, (3) the fact that the collision producing the impact was highly unlikely, and (4) the binary's position in the Neptune resonance. [See ATMOSPHERES OF THE GIANT PLANETS.]

In 1991, Alan Stern of the Southwest Research Institute suggested that the solution to (1)–(3) lies in the possibility that Pluto and Charon were members of a large population (300 to 3000) of small ($\sim 10^{25}$ g) "ice dwarf" planets present during the accretion of Uranus and Neptune in the 20 to 30-AU zone. Such a population would make likely the Pluto–Charon collision, as well as three otherwise highly unlikely occurrences in this region: the capture of Triton into retrograde orbit, and the tilting of Uranus and Neptune. Similar conclusions based on different considerations were reached by William McKinnon of Washington University in the 1980s.

According to Stern's work, the vast majority of the ice dwarfs were either scattered (with the comets) to the Oort Cloud or ejected from the solar system altogether by perturbations from Neptune and Uranus. Only Pluto–Charon and Triton remain in the 20- to 30-AU zone today, specifically because they are trapped in unique dynamical niches that protect them against loss to such strong perturbations.

If this hypothesis is correct, it would imply that Pluto, Charon, and Triton are important "relics" of a very large population of small planets, dubbed ice dwarfs, which by number (but not mass) dominate the planetary population of the solar system. As such, these three bodies would no longer appear as isolated anomalies in the outer solar system and would be genetic relations from an ancient, possibly heterogeneous ice dwarf ensemble, and therefore worthy of more intense study.

C. THE CONTEXT OF PLUTO IN THE OUTER SOLAR SYSTEM

When the existence of the ice dwarf population was first suggested, the solar system beyond Neptune appeared to be inhabited only by Pluto and the numerous comets scattered out of the planetary region during the accretion of the giant planets.

Since late 1992, however, our concept of the outer solar system has evolved considerably, owing to a rapid set of discoveries of faint (i.e., 22–25th astronomical magnitude), largish bodies orbiting between 30 and 50 AU in what is known as the Edgeworth–Kuiper Belt (Fig. 8). The first such objects were detected by David Jewitt and Jane Luu using the University of Hawaii's 2.2-m telescope on Mauna Kea. [See Kuiper Belt.]

As of this writing in mid-1998, over 65 miniplanets with diameters of 100 to 500 km have been discovered. Many of these are apparently in the 2 : 3 mean motion resonance with Neptune that Pluto also occupies. Because this population of objects has been discovered after searches covering only a tiny fraction of the ecliptic sky, it is estimated that many times the discovered population exists. Similarly, in 1995, a team of observers led by Anita Cochran reported the discovery of evidence for much smaller objects the size of conventional comets in this region of the solar system. These objects were discovered at much fainter astronomical magnitudes near $V = 28.5$ using the HST. Based on these discoveries, current models of the population of the region between 30 and 50 AU from the Sun now indicate that some 40,000 or more 100- to 500-km-diameter objects and perhaps several billion comets 1 to 20 km in diameter reside there. The total mass of bodies currently in the 30- to 50-AU zone amounts to roughly $0.1 M_\oplus$.

Interestingly, collisional evolution models that have been developed by Don Davis, Paolo Farinella, and Alan Stern have provided strong evidence that the 100- to 500-km-diameter objects seen in the Edgeworth–Kuiper Belt could not have grown there in the age of

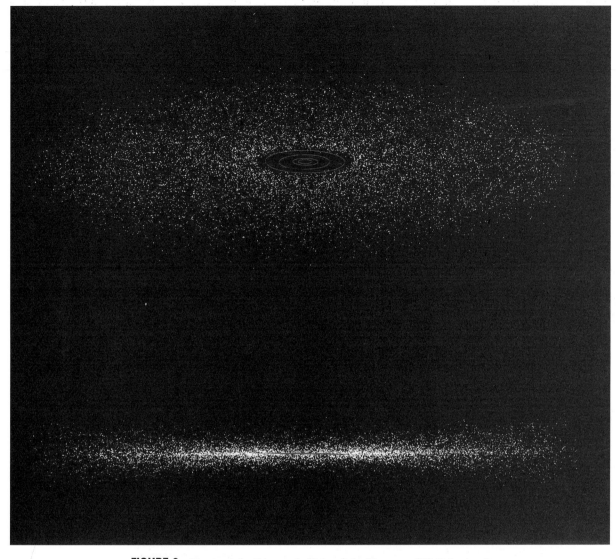

FIGURE 8   Pluto and the Edgeworth–Kuiper Belt. (Courtesy of H. F. Levison.)

the solar system, unless the mass of the primordial Edgeworth–Kuiper Belt was many times higher—in the range of 10 to perhaps 50 Earth masses. Very recently, Stern and his collaborator Joshua Colwell have shown that a combination of collisional and dynamical erosion of the 30- to 50-AU zone was sufficient to deplete the mass of this region down to its present-day value.

Both the discovery of the rapidly expanding cohort of objects found in the 30- to 50-AU zone and the circumstantial evidence that this region of the solar system was much more heavily populated when the solar system was young finally provide a context for Pluto (and the putative Charon-progenitor as well).

We now see that Pluto did not form in isolation, and does not exist so today. Instead, Pluto is simply one of a large number of significant miniplanets that grew in the region beyond Neptune when the solar system was young. Pluto's presence there today is in large measure due to its location in the stable 2 : 3 resonance with Neptune. The question now has moved from why a small planet like Pluto formed in isolation, to why a large population of objects hundreds and thousands of kilometers in diameter formed in the 30- to 50-AU zone without progressing to the formation of a larger planet there. Perhaps the answer lies in the perturbing influence of Neptune.

BIBLIOGRAPHY

Binzel, R. P. (1990). Pluto. *Sci. Am.* **252,** (6), 50–58.

Stern, S. A. (1992). The Pluto–Charon system. *Annu. Rev. Astron. Astrophys.* **30,** 185–233.

Stern, S. A., and Mitton, J. (1998). "Pluto and Charon: Ice-Dwarfs on the Ragged Edge of the Solar System." John Wiley & Sons, New York.

Stern, S. A., and Tholen, D. J. (eds.) (1997). "Pluto and Charon." Univ. Arizona Press, Tucson.

Tombaugh, C. W., and Moore, P. (1980). "Out of the Darkness: The Planet Pluto." Stackpole Books, Harrisburg, Penn.

Whyte, A. J. (1980). "The Planet Pluto." Pergamon, Toronto.

PHYSICS AND CHEMISTRY OF COMETS

I. Introduction

II. Cometary Nucleus

III. Cometary Coma

IV. Cometary Plasma Tails

V. Cometary Dust Tails

VI. Molecular and Elemental Abundances
in Comets

VII. Meteoroid Streams

VIII. Hazards of Cometary Impacts

Daniel Boice and Walter Huebner
Southwest Research Institute

GLOSSARY

Albedo: Fraction of incident light reflected by an object.

Aphelion: Point in a heliocentric orbit farthest from the Sun.

Apparition: Archaic term referring to the perihelion passage of a comet. Possibly related to the diffuse or "ghostly" appearance of the comet.

Astronomical unit (AU): Mean distance between the Earth and the Sun, equal to 149.598 million km.

Coma: Roughly spherical, outflowing atmosphere of a comet, consisting of gases evolved from sublimating ices and dust grains entrained in the escaping gases. The visible cometary coma typically extends tens to hundreds of thousands of kilometers from the nucleus (in the inner solar system). Together, the coma and the nucleus form the comet's "head."

Dust tail: A broad, curved diffuse tail composed of fine dust grains blown back from the coma by solar radiation pressure, extending tens of millions of kilometers from the nucleus (in the inner solar system). Also known as a Type II tail. Comet tails do not usually appear until the comet is within about 2 AU of the Sun.

Hydrogen halo: Extension of the cometary coma consisting of hydrogen atoms produced from the photodissociation of water molecules, extending hundreds of thousands to millions of kilometers from the nucleus.

Kuiper belt: Belt of some 10^9 or more comets in low inclination, low eccentricity orbits beyond the orbit of Neptune. The Kuiper Belt is believed to be the source of Jupiter-family short-period comets.

Meteoroid stream: Trail of large dust particles (radius $>100 \ \mu m$) left in the orbit of a comet. When meteoroid streams intersect the Earth's orbit they produce visible meteor showers.

Nucleus: Solid part of the comet that is the source of all cometary activity, typically a few to tens of kilometers in diameter. The nucleus is a conglomerate of frozen gases (predominantly water ice), organic grains, and silicate grains, intimately mixed at very fine scales.

Oort Cloud: Cloud of some 10^{12} to 10^{13} comets surrounding the planetary system and extending out to near-interstellar distances. The Oort Cloud is the source of long-period comets and possibly Halley-type comets.

Perihelion: Point in a heliocentric orbit closest to the Sun.

Plasma tail: A narrow straight tail of a comet composed of ionized molecules from the coma, being accelerated away from the coma by interactions with the solar wind and the solar magnetic field, extending tens of millions of kilometers (in the inner solar system). Also known as a Type I tail.

I. INTRODUCTION

The word "comet" originates from the Greek, *kometes,* meaning "long-haired" because of a comet's appearance as a hairy star or a star with a tail. The term "comet" usually refers to the large-scale phenomenon visible in the sky rather than its disproportionately small source, the nucleus. The proper body of a "comet" is its solid, icy-conglomerate nucleus. When the orbit of a nucleus brings it into the inner solar system, sunlight heats the ices in the nucleus and the visible phenomenon known as a comet occurs. Cometary nuclei originated during the formation of the solar system, some 4.6×10^9 years ago.

Cometary nuclei are minor but very numerous bodies in the solar system. They are composed of frozen gases (mostly H_2O but also CO, CO_2, H_2CO, CH_3OH, and small amounts of others) and organic and silicate dust. When a nucleus is heated by sunlight as it approaches the Sun, the ices sublimate, producing a diffuse atmosphere or coma, surrounded by a very large hydrogen halo (visible only in the ultraviolet). The coma consists of both evolved gases and dust grains entrained in the escaping gases. Interaction of the dust with the radiation pressure of sunlight and the molecules with the solar wind and the solar magnetic field

then produce a dust tail and a plasma tail, respectively (see Fig. 1).

Comets are distinguished by their moderate to high orbital eccentricities and inclinations. The orbital eccentricity of a comet can range from a moderately eccentric ellipse to almost parabolic to even hyperbolic for comets perturbed strongly by giant planets. Similarly, orbits can be inclined at any angle to the ecliptic, including retrograde orbits that go around the Sun in a direction opposite to that of the Earth and other planets. One distinguishes between long-period and short-period comets: the dividing period is arbitrarily set at 200 years, as comets with periods greater than that have generally been observed only once in modern times.

Long-period comets move in highly elliptical, nearly parabolic orbits that take them far beyond the orbits of the planets, close to the edge of the solar system at aphelion. This means that long-period comets spend most of their orbital period of thousands to millions of years at distances of thousands to tens of thousands of astronomical units (AU) from the Sun, a substantial fraction of the distance to the nearest star. The distribution of the directions of the orbital aphelia of long-period comets on the sky is close to isotropic, forming an approximately spherical distribution around the Sun.

The spherical distribution of orbits, along with the peculiar distribution of the orbital energy of long-period comets led the Dutch astronomer Jan Oort to propose in 1950 that the source of long-period comets was a vast cloud of comets surrounding the planetary system and extending to near-interstellar distances. That cloud is now called the Oort Cloud and has an estimated population of 10^{12} to 10^{13} comets. [*See* COMETARY DYNAMICS.]

Short-period comets are usually divided into two subgroups: Halley-type comets with orbital periods between 20 and 200 years and Jupiter-family comets with periods less than 20 years. Orbits of Jupiter-family comets have inclinations that are generally close to the plane of the ecliptic, less than about 40°. Orbits of Halley-type comets have random inclinations, like long-period comets. Both Jupiter-family and Halley-type orbits are unstable and chaotic because of frequent close encounters with Jupiter and the other giant planets.

The most recent cometary catalog (1997) lists orbits for 936 observed comets. Approximately 80% of those are long-period comets observed on only one apparition. About 18% are Jupiter-family short-period comets, several observed on tens of perihelion passages, and the remainder are Halley-type comets.

FIGURE 1   Comet Hale-Bopp (1995 O1) photographed by Dennis diCicco on April 5, 1997, with a Schmidt camera. The head (consisting of the coma and the unresolved nucleus) is seen at the left of the photograph and the plasma tail is the narrow, filamentary feature pointing toward the upper right corner. The broad dust tail is bright near the comet head and fades as it curves toward the bottom right corner. (Courtesy of Dennis diCicco, *Sky & Telescope Magazine.*)

Early records of comets are preserved in cuneiform tablets dating back to the Babylonian era. The Greeks (about 500–300 B.C.) began the debate about the true nature of comets. Two schools of thoughts emerged. One considered comets as atmospheric phenomena, whereas the other treated them as true celestial bodies, beyond the Earth's atmosphere. Aristotle was a strong proponent of the first school of thought and wrote extensively about the effects of comets on meteorology. The Roman philosopher Seneca defended the other school of thought, arguing by logic that comets are in the realm of stars. Observations of comets were not limited to the Middle East and Europe. From about 300 B.C. until the Renaissance, the Chinese kept the most reliable records of comets. Ancient Chinese annals contain observations of many comets, including several apparitions of Comet Halley, the oldest in 240 B.C. In the Western hemisphere, appearances of comets are found in Aztec records.

With the onset of astronomy in Europe in the Middle Ages, questions about the nature and origin of comets led to extensive observations. European astronomers adopted the school of thought that comets were "atmospheric exhalations." From observations of the parallax of the bright comet of 1577, Tycho Brahe determined that comets were not phenomena within the Earth's atmosphere but were farther away than the Moon, i.e., comets were interplanetary bodies. His pupil, Johannes Kepler, thought that comets followed straight line paths and concluded in 1619 that they originated outside the solar system. However, Johannes Hevelius suggested parabolic orbits in 1668. Isaac Newton initially argued against a parabolic orbit for the sungrazing comet of 1680, preferring the hypothesis of two independent comets, one for the inbound and one for the outbound leg. However, Newton later showed that the orbit of the comet could indeed be fit by a parabola.

The first catalog of cometary orbits was compiled by Edmond Halley in 1705. Halley noticed that the comets of 1531, 1607, and 1682 had very similar orbits and were spaced at ~76-year intervals. Halley proposed that these were the same comet returning on a regular basis and successfully predicted its return in 1758. Halley's prediction was a brilliant application of Newton's theory of gravity published in the *Principia*.

Unfortunately, Halley did not live to see his famous prediction come true, as he died in 1742.

Comets are designated by the International Astronomical Union (IAU) according to the following nomenclature. After a comet discovery (or recovery) is confirmed, a designation is assigned consisting of the names of the discoverers of the comet (up to three codiscoverers are allowed) and the year of discovery or recovery, followed by an alphanumeric code that indicates the half-month during which the comet was found and the order in which it was found in that half-month. For example, Comet Hale-Bopp, 1995 O1, was discovered independently by Alan Hale and Thomas Bopp in 1995, in the second half of July (letter "O"), and was the first comet discovered in that half-month.

In the case of the recovery of a previously seen periodic comet, the recoverer's name is generally not used as the comet already bears the name of the original discoverer(s) (an exception is made if the comet is considered "lost" and then is independently rediscovered). When the orbit of the recovered comet has been determined sufficiently accurately, the comet is designated as a periodic comet with the names of the discoverers (as in the original designation), a number indicating the historical order in which the comet was recognized to be periodic, a prefix of "P/," and an additional number in case the observer has discovered more than one periodic comet. Thus, Comet 10P/Tempel 2 was originally discovered by William Tempel in 1873, was the 10th comet to be recognized as periodic, and was the second periodic comet found by Tempel. Comets 1P/Halley and 2P/Encke are exceptions to the previous naming convention. They were named in honor of the extensive work done on their orbits by Edmond Halley and Johann Encke, respectively.

In addition to the intrinsic interest that the physics and chemistry of comets hold for researchers, the study of comets is important from two other aspects. Comets are icy remnants left over from the accretion of the cores of outer planets. According to this scenario, comets were scattered gravitationally by the giant planets to distant orbits in the Oort Cloud where stellar and galactic perturbations lifted their perihelia out of the planetary region. Further galactic and stellar perturbations randomized the orbits into a spherical cloud. Comets remain there, in a frozen state of storage, until gravitational perturbations from passing stars, interstellar molecular clouds, or the tidal forces of the galaxy bring them back into the inner solar system or eject them into interstellar space. Many comets likely also reside in low eccentricity, low inclination orbits beyond the orbit of Neptune in a disk-like distribution called the Kuiper Belt. Those comets were never ejected because no large planet grew beyond Neptune (Pluto is too small to eject comets and is now regarded by many astronomers as the largest member of the Kuiper Belt).

Comets in these two storage reservoirs are probably the least modified objects in the solar system, preserved in a state similar to their formation and relatively unaltered by the processes that have changed the larger planets and their satellites closer to the Sun. Therefore, cometary studies can provide important clues to the chemical composition of the solar nebula and the physical processes that occurred in the early stages of the formation of our solar system.

Another important aspect of cometary studies involves the question of the origins of life on Earth. The discovery of organic material in the dust of Comet P/Halley by the *Vega* and *Giotto* spacecraft raised the intriguing possibility that prebiotic chemical reactions may have taken place on the dust particles in comets, perhaps even in interstellar space before these particles were incorporated into cometary nuclei in the solar nebula. Current studies suggest that the early atmosphere on Earth was inhospitable to the evolution of the organic material necessary for life. A way out of this dilemma is the possibility that prebiotic molecules may have been delivered to the Earth by impacting comets. Thus, important clues to understanding the origins of life on Earth may be found by investigating the organic material in comets.

II. COMETARY NUCLEUS

The nucleus of a comet is a solid body with a typical dimension of a few kilometers, consisting of intimately mixed ices (frozen gases) and refractory components (dust). The concept of the nucleus as an icy conglomerate or "dirty snowball" was originally proposed by the American astronomer Fred Whipple in 1950. As a comet approaches the Sun the increasing solar insolation raises surface temperatures on the nucleus, resulting in sublimation of the water ice. The evolving gases flow off the low mass nucleus at thermal speeds, carrying with them organic and silicate grains embedded in the icy-conglomerate mix.

One of the important effects of this gas and dust overflow is the generation of nongravitational forces on comets. It was recognized in the 19th century that the time of return of several short-period comets was not precisely as predicted by gravitational theory. Comets such as P/Encke and P/Halley returned to perihelion both early and late, often by several days.

One of the key points of Whipple's icy-conglomerate model was the ability to explain nongravitational forces as the result of a "rocket force" resulting from the sublimation of ices on the sunward side of the nucleus. As evolving gases depart the nucleus, Newton's third law of action–reaction causes them to provide an impulse in the opposite direction.

The generation of a secular acceleration or secular deceleration in cometary orbits results from a combination of two processes. First, nuclei are rotating and the thermal lag between maximum insolation and maximum gas emission adds a transverse component to the otherwise radial (i.e., sunward) force term. Depending on whether the nucleus rotation is prograde or retrograde, the transverse acceleration can either add or detract from the comet's orbital velocity. Second, comets are not active uniformly and often display more activity on one leg of their orbit versus the other (i.e., pre- versus postperihelion). This perihelion asymmetry leads to an integrated radial acceleration over a complete perihelion passage that is nonzero, and thus can change the period of the orbit.

The dependence of nongravitational forces in comets on the sublimation rate of water ice was confirmed by Brian Marsden and colleagues in the early 1970s. They showed that typical nongravitational forces were on the order of a few times 10^{-5} relative to the Sun's gravitational attraction for long-period comets and a few times 10^{-6} for short-period comets.

Determination of nongravitational forces for observed long-period comets is more difficult as comets are only seen on one apparition. However, Marsden's nongravitational force model solved an important problem with regard to long-period comets. It was observed that some long-period comets appeared to have slightly hyperbolic orbits, seeming to approach the solar system from interplanetary space. However, when nongravitational forces were included in the orbit solutions, many of these supposedly hyperbolic comets were shown actually to be in elliptical orbits coming from the Oort Cloud. Thus, the nongravitational force model helped confirm the Oort Cloud as the source of long-period comets.

Whipple's "dirty snowball" model was confirmed dramatically by images returned by the *Vega* and *Giotto* spacecraft which flew past Comet Halley in 1986. The nucleus, shown in Fig. 2, was seen to be an irregularly shaped body, with dimensions of about $16 \times 8 \times 7$ km. Active areas emitting dust jets into space were sprinkled randomly over the nucleus, covering about 30% of the sunlit surface of the nucleus; no active areas were seen on the nightside. The remainder of the surface appeared to be inactive.

FIGURE 2 A composite image of the nucleus of Comet P/Halley as photographed by the camera onboard the *Giotto* spacecraft on March 14, 1986. The Sun is located toward the left, 29° above the horizontal. Note the elongated and irregular shape of the nucleus, which has dimensions of $16 \times 8 \times 7$ km. The heterogeneity of the irregular surface is well illustrated; several surface features can be seen, including active regions and hills. The smallest features that can be resolved are about 100 m across. (Courtesy of H. U. Keller, Max-Planck-Institut für Aeronomie.)

One surprise in the images of the Halley nucleus was the low reflectivity or albedo of the surface. The Halley nucleus reflected only about 3.5% of the incident sunlight. This low albedo was not expected for what was generally thought of as an icy body. However, it can be understood as the result of intimately mixed, carbon-rich, organic grains that presumably have very low albedos.

Although the *Vega* and *Giotto* spacecraft flybys were too fast and too far away to measure the gravitational deflection of their trajectories by the cometary nucleus, the mass of the Halley nucleus was determined indirectly based on the comet's nongravitational forces. The orbit of P/Halley is very well determined, having been seen on 30 perihelion passages stretching over more than 2200 years. The nongravitational forces appear to have been remarkably constant over that time. By comparing the nongravitational forces with the observed rate of mass loss from the comet (gas and dust production) over the complete perihelion passage in 1986, it was possible to derive estimates of the total nucleus mass.

The resulting mass estimates, by a number of different researchers, ranged from 1 to 6×10^{14} kg. Comparing this mass range with the derived volume of the nucleus obtained from spacecraft images resulted in estimates for the bulk density of the Halley nucleus between 200 and 1200 kg m^{-3}. This implied that the nucleus might be a fairly porous body with significant voids or possibly a moderately compacted snowball.

An independent means of measuring the density of a cometary nucleus was used in the study of Comet Shoemaker-Levy 9, which was disrupted tidally by a close approach to Jupiter in 1992 and subsequently impacted the planet in 1994. Shoemaker-Levy 9 was discovered in March 1993, after the disruption event, and was found to consist of a string of 21 individual cometary nuclei. Quite surprisingly, it was found that Shoemaker-Levy 9 was in a highly elliptical orbit around Jupiter; such temporary captures of periodic comets by Jupiter are occasionally observed. By integrating the orbits of the fragments backward in time it was shown that they had disrupted when they were previously closest to Jupiter, only about 0.3 Jupiter radii above the cloud tops of the Jovian atmosphere. Such a close approach is within the Roche limit of the planet where its gravity can tear a smaller body apart.

Researchers showed that if one modeled the cometary nucleus as many smaller dirty snowballs held together only by their own self-gravity, then the tidal forces from Jupiter could easily pull the "cometesimals" apart. However, as the disrupted pieces moved away from Jupiter, their self-gravity would cause them to reassemble themselves into multiple nuclei. It was shown that the number of resulting nuclei was inversely proportional to the bulk density of the original cometary nucleus. The 21 fragments of Comet Shoemaker-Levy 9 implied a bulk density of the original nucleus between 500 and 1100 kg m^{-3}. The precise value depended on the rotation state of the original nucleus, which was unknown. Again, this result suggested that the nucleus might be a fairly porous body or a moderately compacted one.

Another outcome of the study of the disruption of Comet Shoemaker-Levy 9 was estimates of the strength of the cometary nucleus. Because the tidal forces on the cometary nucleus could be calculated very precisely during its passage close to Jupiter, it was shown that the binding strength of the individual cometesimals was only about 10^2–10^2 N m^{-2}, several orders of magnitude less than that for solid ice or for rock. Similar strength values have been obtained from studies of sungrazing comets that disrupted when they passed within the Roche limit of the Sun.

This and other evidence have led researchers to suggest that cometary nuclei may be weakly bound agglomerations of many smaller icy cometesimals. This modification to Whipple's model has been referred to as the "fluffy aggregate" or the "primordial rubble pile" model (Fig. 3). It helps explain not only the behavior of Comet Shoemaker-Levy 9 but also many other phenomena associated with comets.

For example, comets are occasionally observed to disrupt spontaneously, i.e., pieces break off the nucleus for no apparent reason (except for tidal disruption events such as Shoemaker-Levy 9 where the reason for the disruption event is quite apparent). These secondary nuclei have typical lifetimes of a few days to a few weeks, slowly moving away from the main nucleus due to nongravitational jetting forces (in a few rare cases, such as P/Biela in 1853, the comet returned as a double comet). Disruption events are understood more easily if cometary nuclei are weakly bound aggregates or rubble piles.

Rubble pile models may also help explain cometary outbursts, where comets suddenly release large amounts of dust and gas, increasing their visual brightness by several magnitudes. If cometesimals were to break off the nucleus, exposing fresh volatile ices beneath them to solar illumination, sudden increases in gas and dust production would be possible.

Although comets are the most pristine bodies in the solar system, they have undergone some physical processing during their histories. Short-lived radionuclides such as ^{26}Al may have warmed the interiors of comets soon after their formation in the solar nebula. Long-period cometary nuclei likely suffered some collisional evolution while they were still orbiting the Sun in the giant planets zone, prior to their ejection to the Oort Cloud. Similarly, short-period nuclei have been evolved collisionally in the Kuiper Belt, before they were fed into the inner planets region.

Both Oort Cloud and Kuiper Belt comets have been irradiated by solar protons and galactic cosmic rays during their long storage in their respective dynamical reservoirs. The effect of these energetic particles is to sputter away volatiles and to form refractory organics in the topmost meter of the nucleus surface. Thus, comets may be covered by a nonvolatile crust before they ever enter the planetary region.

While the comets are in cold storage in the Oort Cloud and the Kuiper Belt they sweep up interstellar matter, perhaps providing a "frosting" of extra-volatile molecules on their surfaces. Such a frosting has been

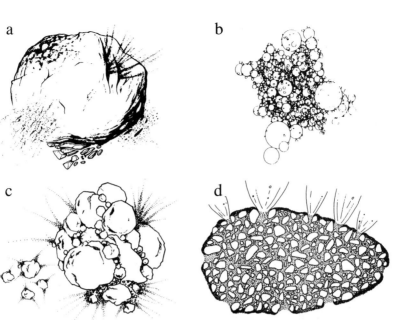

FIGURE 3 Four artist's concepts of suggested models for the structure of cometary nuclei: (a) the icy conglomerate model, (b) the fluffy aggregate model, (c) the primordial rubble pile, and (d) the icy-glue model. Evidence has continued to mount that the fluffy aggregate and the primordial rubble pile are the most likely representations of nucleus structure. All but model (d) were suggested prior to the Halley spacecraft encounters in 1986.

suggested as a possible explanation for the anomalous brightness of dynamically new comets from the Oort Cloud on the inbound legs of their orbits. However, the surfaces of comets will also be "sandblasted" by interstellar dust grains encountering the nuclei at velocities of several tens of kilometers per sec^{-1}. Estimates of this process suggest that sandblasting will remove the veneer of volatile molecules as quickly as it can form.

Comets will also be heated occasionally by stars passing through the Oort Cloud. Calculations show that the surfaces of all the comets in the Oort Cloud will be warmed to about 40 K at some time in their histories. Because this is comparable to (or less than) the expected formation temperatures of the comets in the solar nebula, such warmings will probably not have much of an effect.

Once a comet enters the planetary region, the heating by the Sun dominates all other physical processes. As the nucleus warms, the initially amorphous water will transform to crystalline ice, releasing trapped volatiles in the process. Also, the warming of subsurface layers of the nucleus will mobilize volatile ices such as CO and CO_2, producing activity at relatively large distances from the Sun. As the comet moves closer to the Sun, the water ice in the topmost layers will sublimate and carry dust grains into the coma, but perhaps

leaving behind large grains too heavy to lift off the nucleus surface. Such large grains may form a lag deposit or crust on the nucleus surface, effectively cutting off further activity. If gas pressure continues to build beneath the surface, it may fail catastrophically, resulting in a sudden outburst from the comet as the overlying crust is blown away.

Most of the processes just described only affect the surface layers of comets, likely only the first few meters. Beneath that depth, comets may remain quite cold with little or no physical change since their formation 4.6 billion years ago.

Cometary nuclei continually lose mass because they are too weak gravitationally to bind the gases evaporating and the dust escaping from active areas on the nucleus. As a result, the typical lifetime of a comet is limited. Estimates of the physical lifetimes of cometary nuclei range from tens to thousands of returns. Comet 2P/Encke has been observed on a total of 57 returns to date and is likely much older than that. Two different end states for nuclei are possible: (1) the comet may sublimate all of its ices and disintegrate, leaving nothing but a meteoroid stream of nonvolatile particles in its orbit or (2) the comet's active areas may become completely mantled with dust, forming a regolith of refractory material and preventing further losses. Such dormant or extinct objects may then have an asteroidal appearance and be indistinguishable from primitive-type asteroids from the main belt.

III. COMETARY COMA

The coma of a comet develops when the comet approaches the Sun. Solar radiation, mostly visible light, sublimates the volatiles (up to about 90% water), which escape into the near vacuum of space because of the low gravitational binding of the nucleus. The evolving gases are accelerated away from the cometary surface, reaching terminal velocities equal to thermal speeds. Dust grains are entrained in the outflowing gas and are similarly accelerated, with the smallest, submicron and micron-sized grains achieving velocities similar to that of the gas. Larger grains decouple from the expanding gas flow and do not achieve as high velocities.

Idealistic models of the coma outflow depict radial expansion into space from a uniformly active nucleus surface. However, because there is little or no activity on the nightside of the nucleus, some of the flow must be diverted hydrodynamically toward the nightside. The situation becomes even more complex in more realistic models where active areas are scattered randomly on the nucleus surface, and only detailed hydrodynamic codes can model the outflow. The numerous jets visible in the photographs of the nucleus of Comet Halley in 1986 are illustrative of this complexity (Fig. 4).

From roughly a dozen or more different parent molecules, a plethora of chemical species is formed in the expanding coma by solar photolytic processes, secondary electron processes, and gas-phase chemical reactions. Photolytic processes depend mainly on solar ultraviolet light to form radicals and ions. The highly reactive radicals and ions interact mainly with the neutrals via fast ion-neutral reactions in the inner coma, where pressures are high enough to make molecular collisions frequent.

Dissociation of water produces OH and H. Because of its high thermal velocity, the hydrogen expands more rapidly, forming a huge hydrogen halo around the coma, with typical dimensions of several times 10^5 km. The hydrogen halo is visible only in the ultraviolet and thus is not accessible with Earth-based instruments. One of the early successes of the Space Age was the confirmation of the existence of a hydrogen coma surrounding Comet Tago-Sato-Kosaka (1969 T1) by Arthur Code and colleagues using the *OAO-2* satellite.

FIGURE 4   Dust jets emitted from the Halley nucleus, taken by the camera onboard the *Giotto* spacecraft on March 14, 1986. The Sun is at the left and slightly above the horizontal. (Courtesy of H. U. Keller, Max-Planck-Institut für Aeronomie.)

IV. COMETARY PLASMA TAILS

Cometary plasma or ion tails are produced when molecules and radicals in the coma are ionized by charge exchange with solar wind particles, and then become subject to Lorentz forces resulting from the solar magnetic field. These forces accelerate the ions rapidly in the anti-sunward direction, producing the narrow tails often seen in comets. Typically, ion tails, also known as Type I tails, have a bluish color resulting from CO^+ ions.

Before cometary plasma physics matured as an important discipline of space physics, theories were advanced by several pioneering researchers to explain the bright plasma tails seen in comets. In a historical context, the modern study of comet–solar wind interaction began with the statistical investigation of the directions of cometary ion tails by the German astrophysicist Ludwig Biermann in 1951. From the aberration angle of about 3° relative to the radial direction from the Sun, Biermann deduced that a "solar corpuscular radiation" must exist in order to sweep away cometary ions at high speed. The radial velocity of these solar charged particles (i.e., the solar wind) was deduced to be on the order of a few times 100 km sec^{-1}.

However, to facilitate the momentum coupling between the solar plasma and the cometary ions by Coulomb collisions (interactions between charged particles via the electrostatic force), a very high plasma number density, far exceeding the limit set by solar coronal white light measurements, had to be invoked. To overcome this difficulty, Hannes Alfvén advocated the presence of a magnetic field in interplanetary space so that the comet ionosphere would drape the field lines into a magnetic tail. In this classical picture, cometary ions are channeled into the anti-sunward direction by the magnetic field.

Expanding coma ions and electrons eventually interact with the solar wind (the high-speed stream of charged particles from the Sun). On the sunward side, the coma plasma is decelerated by the opposing solar wind and the flow of the plasma is diverted to the anti-solar direction, forming the so-called ion (or plasma) tail. Other structures and boundaries may exist in the plasma environment, including a bow shock or wave, a magnetic field pileup region, a region of plasma stagnation, a central magnetic field-free cavity close to the nucleus, and a boundary called the ionopause separating the magnetic field-free cavity from the plasma stagnation region. [See THE SOLAR WIND.]

Plasma tails often display considerable structure, including disconnection events where the tail of the comet is seen to break away and the comet then grows a new plasma tail. These events are believed to be caused by sector boundary crossings where the magnetic field of the Sun reverses polarity. An example of a disconnection event involving Comet Halley in 1986 is shown in Fig. 5.

V. COMETARY DUST TAILS

Dust particles are released from the nucleus and entrained in the rapidly expanding coma gas. As the grains move sunward, their motion is opposed by the solar radiation pressure force, which decelerates the grains and then accelerates them in the anti-sunward direction. This motion is referred to as the "fountain model" because of its resemblance to water ejected vertically (i.e., sunward) out of a fountain and then falling back to earth while spreading horizontally at the same time.

Initially, the dust grains can be thought of as moving in the same orbit as the comet, except for the differential motion induced by their initial acceleration by coma gas drag. However, as the smaller grains (typical radii $\approx 1-5$ μm) are accelerated anti-sunward by solar radiation pressure, they move in orbits with slowly increasing semimajor axes. As a result, their Keplerian velocities decrease slowly relative to that of the com-

etary nucleus and they lag behind the nucleus and coma in their orbits. This produces the broad curving dust tails seen for many comets.

Because the effectiveness of the radiation pressure force is a function of the area-to-mass ratio of individual dust grains, dust particles are segregated spatially according to size. The smallest grains are accelerated more rapidly in the anti-sunward direction, whereas the larger grains drift away from the nucleus at relatively low velocities. Also, because dust production from the nucleus can be periodic or sporadic due to the existence of discrete active areas on the nucleus surface, discrete structures can be produced in the otherwise diffuse dust tails. These are known as synchrones and represent bursts of particles produced over a short span of time, but with a wide range of particle sizes.

In extreme cases, large grains in the coma and dust tail can produce an apparent "anti-tail," a sunward pointing spike-like structure. These are actually an illusion that requires a very specific geometry so that the large grains lagging behind the comet in its orbit appear to be projected in the sunward direction. The most famous case of an anti-tail was that for Comet Arend-Roland 1956 R1. The anti-tail was visible as the Earth passed through the orbital plane of the comet.

The grains in cometary dust tails eventually become part of the zodiacal cloud of interplanetary particles. Comets are estimated to contribute about two-thirds of the interplanetary dust complex, with the remainder coming from asteroid collisions and cratering events. Samples of these grains have been recovered in the Earth's atmosphere by high-flying research aircraft (Fig. 6). Because they are small, they can cool quickly as they decelerate in the Earth's atmosphere, and thus survive entry intact. These interplanetary dust particles (IDPs) are the only samples of cometary material currently available for study on Earth. [See INTERPLANETARY DUST AND THE ZODIACAL CLOUD.]

VI. MOLECULAR AND ELEMENTAL ABUNDANCES IN COMETS

As noted in Section I, the study of comets is important because of the cosmochemical record of the primordial solar nebula that is frozen in cometary volatiles and dust grains. Analyses of ground-based observations of cometary comae and *in situ* measurements by the spacecraft that flew by Comet Halley in 1986 have provided us with a wealth of information on elemental,

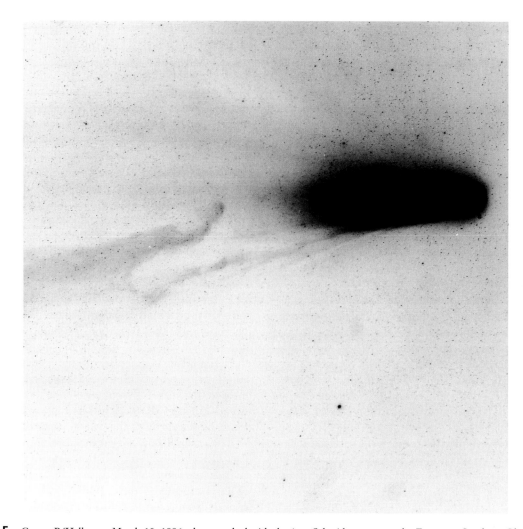

FIGURE 5 Comet P/Halley on March 10, 1986, photographed with the 1-m Schmidt camera at the European Southern Observatory in La Silla, Chile. The comet was at a heliocentric distance of 0.85 AU (postperihelion) and 1.05 AU from the Earth. This photograph, printed as a negative, was taken 1 day after the closest approach of the *Vega 2* spacecraft. The narrow plasma tail, seen on the right, has a filamentary structure near the comet head. Just above the center of the figure, a plasma condensation (or disconnection event) stands out. The broad, diffuse dust tail is seen curving to the left. (Courtesy of the European Southern Observatory.)

molecular, and isotopic abundances in comets. These data provide insights into the chemical and physical processes in the solar nebula, 4.6 billion years ago.

Chemical species that have been detected in comets are listed in Table I. They have been identified by their spectra in the ultraviolet (UV), visible, infrared (IR), and radio ranges of the spectrum and in a few cases by *in situ* mass spectrometry during the spacecraft flybys of Comet Halley. Question marks in Table I label those species that have been identified only once. In addition, the metals Na, K, Ca, V, Mn, Cr, Fe, Co, Ni, and Cu are associated with sublimation (vaporization) or sputtering of the dust, detected mostly in sungrazing comets (except for Na, which is detectable in

most comets at $r < 0.7$ AU). However, in the very active Comet Hale-Bopp, sodium was observed as a rapidly moving, narrow component on one edge of the dust tail and as a slower moving, broad, and diffuse component in the dust tail at larger heliocentric distances ($r > 1$ AU).

Mass spectrometer identifications are somewhat problematic because of the overlap of masses from different species and the modest resolution of the Halley flyby instruments. For example, a peak in mass channel 28 contains CO, N_2, and C_2H_4 as possible main constituents. In some cases, isotopes can be inferred, such as H_2DO in channel 20. There are similar problems with other mass channels. Although H_2O is

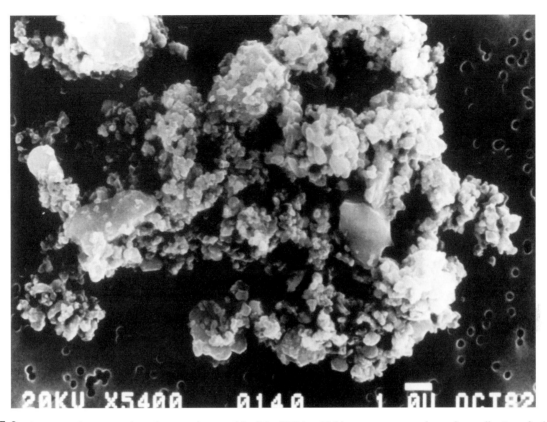

FIGURE 6 A suspected cometary interplanetary dust particle. The IDP is a highly porous, apparently random collection of submicron silicate grains embedded in a carbonaceous matrix. The voids in the IDP may have once been filled with cometary ices. (Courtesy of D. Brownlee, University of Washington.)

so abundant that most of it must come from the nucleus (although icy grains in the coma may provide a second, extended source), it is not known what fraction of it and other species come from spatially distributed sources in the coma; these sources are associated with the dust. For example, water ice may be stored and shielded from the Sun in sintered, porous dust particles entrained in the gas escaping from the nucleus. Protonated species, such as CH_5^+, NH_4^+, and H_3O^+, can only be produced by ion–molecule reactions.

It is interesting to compare the cometary species in Table I with identified gas-phase molecules detected in the interstellar medium, as presented in Table II. All of the neutral cometary molecules (except CO_2, S_2, and the neutral cometary radicals NH_2 and NH) are also identified interstellar molecules. CO_2 and S_2 are symmetric species that have no dipole moment and, therefore, no pure rotational spectra by which most interstellar molecules are identified. However, S_2 does have a fine structure transition, but searches for it in the interstellar medium have turned out negative so far. From the presence of CO_2H^+ (protonated CO_2) in the interstel-

lar medium, one can conclude that CO_2 is present. Solid interstellar CO_2 (ice or frost) has also been identified. Interstellar NH has been identified in the UV in diffuse clouds with background stars. Because NH and NH_2 are light molecules, their rotational transitions are in the submillimeter region of the spectrum, for which observational capabilities have been developed only recently. There is no reason to expect that NH_2 is not present in the interstellar medium. Thus all molecules in comets, with the exception of S_2, also appear to be present in the interstellar medium.

In contrast to neutral molecular species, many cometary radicals and ions have not been identified in the interstellar medium. This is not surprising. Photolytic processes are the primary mechanisms for dissociation and ionization, and the solar radiation field is very different from the interstellar radiation field.

The molecular similarities seem to indicate a close affinity between comets and interstellar clouds. This is not to say that comets might have formed at the very low densities in the interstellar medium. Rather, interstellar molecules appear to have survived in the

TABLE I
Chemical Species Identified in Comets

Identification by radio, microwave, IR, visual, and UV spectra

H	C	O	S	CO_2	HDO	CHO	DCN	HNC	CO	CS	NH	OH
C_2	$^{12}C^{13}C$	CH	$H_4C_2O_2$	^{13}CN	$H^{13}CN$	$HC^{15}N$	OCS	SO_2	S_2	SO		
C_3	NH_2	H_2O	HCOOH	C_2H_2	H_2S	H_2CS	HNCO	CH_4	HCO	CN	HC_3N	Na
NH_3	H_2CO	HCN	CH_3OH	CH_3CN	HC_3N	NH_2CHO	C_2H_6					
C^+	CO^+	CH^+	CN^+	HCO^+	CO_2^+	H_2O^+	H_2S^+	N_2^+	H_3O^+			

Identification by mass spectra

Mass	Ions			Neutrals		
1	H^+					
12	C^+					
13	CH^+					
14	CH_2^+	N^+				
15	CH_3^+	NH^+				
16	O^+	CH_4^+	NH_2^+			
17	OH^+	NH_3^+	CH_5^+			
18	H_2O^+	NH_4^+		H_2O		
19	H_3O^+					
23	Na^+					
28				CO	N_2?	C_2H_4?
30				H_2CO		
31	H_3CO^+					
35	H_3S^+					
36	C_3^+					
37	C_3H^+					
39	$C_3H_3^+$					
44				CO_2		

outer parts of the solar nebula where densities were high enough to condense and form comets. The list of interstellar molecules may be taken as a guide to identify new species from observed, but still unassigned, cometary spectral lines. It is also unlikely that large, porous, cometary-type dust particles exist in the interstellar medium. It appears that large cometary dust particles are composed of small interstellar dust particles sintered into aggregates, a process that occurred during or possibly after comet formation.

Assimilating the available comet observations, model results, and the arguments presented earlier, the best estimate for the average icy composition of cometary nuclei is listed in Table III. Large variations for all species except H_2O may be expected from comet to comet and perhaps even from one active area to another on the same cometary nucleus. However, it is also possible that most comets have very similar compositions and the observed differences are the result of different observing techniques (e.g., radio versus UV versus visible versus IR spectra), different observing conditions (e.g., heliocentric and geocentric distance, pre- versus postperihelion), and the physical evolution of the near-surface cometary materials. Data come from many sources and many allegedly "normal" comets, although the *in situ* P/Halley measurements strongly influence the relative abundances. As was suspected for many years, it is now clear that water is the dominant species of ice.

TABLE II
Identified Interstellar Molecules in the Gas Phase

H_2	C_2	CO	CS	$NaCl^a$	HCl	SiO	SiS
$AlCl^a$	KCl^a	PN	AlF^a	SiN^a	SiH^b	HF^b	
H_2O	SO_2	H_2S	OCS	HNO	C_3^a	HCN	C_2O
HNC	SiC_2^a	NH_2	N_2O	$MgNC^a$	$MgCN^a$	$NaCN^a$	
NH_3	HC_2H	C_3O	$HNCO$	$HNCS$	C_3S	H_2CO	H_2CS
$c\text{-}C_3H$	$l\text{-}C_3H$	$HCCN$	H_2CN				
CH_4	SiH_4^a	C_5^a	HC_3N	C_4Si^a	$OHCHO$	H_2CNH	CH_2CC
H_2NCN	CH_2CO	C_3H_2	$HCCNC$	$HNCCC$			
H_3CNC	CH_3CN	NH_2CHO	CHC_2HO	$H_2C_2H_2^a$	CH_3OH	CH_2C_3	CH_3SH
HC_5N	CH_3C_2H	CH_3CHO	CH_2CHCN	H_3CNH_2	$c\text{-}CH_3OCH_2$		
CH_3C_3N	CH_3OOCH	CH_3COOH	CH_3C_3N	C_7H^a	C_2H_6		
HC_7N	CH_3C_4H	CH_3CH_2CN	CH_3CH_2OH	CH_3OCH_3	C_8H^a		
$CH_3C_5N^b$	$(CH_3)_2CO$						
HC_9N							
$HC_{11}N$							
$(C_2H_5)_2O$							
CH^+	SO^+	CO^+					
HCO^+	HN_2^+	HOC^+	HCS^+	H_2D^{+b}	H_3^+		
$HCNH^+$	H_3O^+	$HOCO^+$					
CH_3NH^+							
OH	CH	CN	NO	NS	NH	SO	CP^a
SiC^a							
HCO	C_2S	C_2H	CH_2				
C_3N							
C_4H	CH_2CN						
C_5H							
C_6H							

[a] Detection only in the envelopes around evolved stars.
[b] Claimed but not yet confirmed.

TABLE III
Abundance Estimates for Ices in Comet Nuclei

H_2O	0.85
CO	0.04
CO_2	0.03
H_2CO	0.02
CH_3OH	0.02
N_2	0.01
H_2S, HCN, NH_3, CH_4, CS_2, C_2H_2, H_2CO_2, C_2H_4, etc.	0.03

Cometary nuclei are inhomogeneous and there appear to be compositional differences between comets. To be specific, all variations are associated with detections in the coma. Because the coma may be the result of gas emission from active areas, which are only a small part of the surface of a comet nucleus, caution must be exercised in the interpretation of these variations. Some researchers suggest that the variations are due to gross compositional differences of the nuclei, possibly related to the place and time of comet formation. However, others argue that the differences may be only local, due to the inhomogeneous nature of a cometary nucleus.

The best known variation is the mass ratio of dust-to-gas release rates, χ. The measured dust-to-gas ratio in Comet Halley, the only one for which there are *in situ* data, was 2. Some comets appear to be very dusty ($\chi > 1$), whereas others appear to be almost dust free ($\chi < 0.1$). Again, such a simple interpretation must be viewed with caution. Such measurements, based solely on remote sensing, are subject to instrumental biases as well as the varying behavior of comets with heliocentric distance. Also, a dust-free comet may actually contain a very large mass (but small number) of submillimeter-sized, aggregated dust particles that are inefficient light scatterers, making their detection very difficult. However, such an argument only shifts the emphasis from dust-rich versus dust-poor comets to comets with small versus large dust particles.

It is inconceivable that comets formed without dust. Dust particles are condensation nuclei for the frozen gases of a comet. Thus, "dust-free" comets may contain very large, aggregated dust particles, whereas dust-rich comets contain larger numbers of virtually unprocessed (not sintered) dust from the interstellar medium. Comet Hale-Bopp falls into this latter category. Almost nothing is known about the fraction of organic to silicate dust in comets other than P/Halley, where they comprise roughly equal mass fractions.

Based on observations of CO in the UV and CO^+, the main plasma tail ion in the visible range of the spectrum, the ratio of CO to H_2O, is also variable. Some comets have a very weak plasma tail or none at all, whereas others, such as Comets Morehouse and Humason, have very dominant CO^+ tails. Finally, there is the variation of relative abundances of some minor species such as CN and C_2. In most comets, the ratio of production rates for C_2/CN varies from 1.2 to 1.5. However, in about 30% of comets, C_2 and C_3 are depleted by factors up to 20.

P/Giacobini-Zinner is a typical example of a comet with C_2 and C_3 depletion. Observations of Comet Yanaka (1988 Y1) at 0.91 AU heliocentric distance showed strong emissions of oxygen and NH_2, but no trace of either C_2 or CN, the two spectroscopic species that are among the strongest in most comets. However, because these observations were restricted to the innermost part of the coma, they may indicate that C_2 and CN come from dust particles that had not yet released these species as the dust flowed outward.

A comparative study of 85 comets found that most comets are very similar to each other in chemical composition. However, there are compositional groupings of comets based on observed C_2 and C_3 abundances in the coma. This study found that a majority of Jupiter-family short-period comets, (i.e., comets from the Kuiper Belt) appear to be significantly depleted in the parent molecules that lead to C_2 and C_3, suggesting that compositional differences are related to the place of comet formation. Alternatively, the Jupiter-family comets are much older (i.e., have made more perihelion passages) than long-period and Halley-type comets in the study, and the compositional differences in their comae may reflect physical and chemical processing of the near-surface layers of cometary nuclei. This study also reported that most CN and some C_2 are produced from grains in the coma rather than from ices in the nucleus, whereas virtually no NH comes from grains.

The coexistence of oxidized and reduced species in comets should be noted. Interstellar clouds are rich in the very stable molecules CO and N_2. It is only natural to assume that the solar nebula was also rich in these species. However, the regions of the solar nebula where the giant planets formed were probably rich in CH_4, NH_3, C_2H_2, and C_2H_6 because the temperatures were low enough and the densities high enough to convert CO to CH_4 and other hydrocarbons, and N_2 to NH_3. It is now believed that the abundances of CH_4 and NH_3 may be low in comets. This conclusion was confirmed by data from Comets Hyakutake and Hale-Bopp, as well as by Comet P/Halley. If the low CH_4 and NH_3 abundances can be confirmed in other comets, then comets must have formed outside the subnebulae of the giant planets, probably in the trans-Neptunian region of the solar nebula.

Investigations of the dynamical evolution of comet orbits suggest three classes of comets from two dynamical reservoirs: (1) The Oort Cloud, at about 45,000 AU in a spherical distribution around the Sun, which is the source of long-period comets, (2) the Kuiper Belt in the plane of the ecliptic beyond Neptune, which is the source of Jupiter-family short-period comets, and (3) the Halley-type comets, far out of the ecliptic plane and in part in retrograde orbits, appear to be Oort Cloud comets captured into the inner solar system by gravitational interactions with planets. [*See* THE KUIPER BELT.]

It is quite possible that the ratios of CH_4/CO and NH_3/N_2 vary from comet to comet. The ratio of reduced-to-oxidized species may be a measure of metamorphosis toward chemical equilibrium, related to the place of origin of comets. Oort Cloud comets may be more similar to the interstellar composition (i.e., richer in oxides), whereas some short-period comets may be more similar to other solar system bodies (i.e., containing more reduced species). This would suggest an apparent inconsistency because P/Halley is a short-period comet while being rich in oxides. However, P/Halley is likely a captured Oort Cloud comet and

TABLE IV
Relative Element (Number)
Abundances of Comets and the Sun

Element	Ice	Dust	Ice and dust ($\chi = 1$)	Ice and dust ($\chi = 2$)	Sun
H	0.5929	0.4810	0.5464	0.5273	0.92048
He	—	—	—	—	0.07835
C	0.0570	0.1934	0.1137	0.1370	0.00030
N	0.0154	0.0100	0.0132	0.0122	0.00008
O	0.3347	0.2114	0.2834	0.2621	0.00061
Ne	—	—	—	—	0.00008
Na	—	0.0024	0.0010	0.0014	—
Mg	—	0.0238	0.0099	0.0140	0.00002
Al	—	0.0016	0.0007	0.0009	—
Si	—	0.0439	0.0183	0.0258	0.00003
S	—	0.0171	0.0071	0.0101	0.00001
Ca	—	0.0015	0.0006	0.0009	—
Ti	—	0.0001	—	0.0001	—
Cr	—	0.0002	0.0001	0.0001	—
Mn	—	0.0001	—	0.0001	—
Fe	—	0.0124	0.0052	0.0073	0.00004
Co	—	0.0001	—	0.0001	—
Ni	—	0.0010	0.0004	0.0006	—
Total	1.0000	1.0000	1.0000	1.0000	0.9996

thus should share the same compositional traits as long-period comets.

Analyses of the elemental abundances of the gas-phase species and the organic particles of Comet P/Halley concluded that the relative abundances of most elements in Comet P/Halley are the same as those of the Sun, but with several notable exceptions: noble gases, hydrogen, and nitrogen are depleted. Table IV presents the relative elemental abundance in comets (primarily based on P/Halley) and compares them to those of the Sun. The important quantities to compare are the ratios of $C : N : O : S$, which are discussed next.

To understand hydrogen, nitrogen, and noble gas discrepancies, it must be realized that all matter in the Sun is bound gravitationally. Thus the Sun reflects closely the elemental abundances of the solar nebula. Save for the very small fraction escaping in the solar wind (10^{-4} of the solar mass, if the solar wind was constant over the age of the solar system), even the most volatile elements at the high temperatures of the

Sun are bound by its gravitational field. The solar abundance of hydrogen is larger than the abundance of all other elements combined. The high abundance of hydrogen in the Sun must have existed also in the solar nebula. The excess of hydrogen not bound chemically to other elements formed H_2 at the low temperature of the outer solar nebula. Comets apparently formed at a temperature high enough to prevent condensed H_2 from being incorporated. Any H_2 that was trapped in the ice has escaped, as it is a very small and mobile molecule. Thus, comets show a very large depletion of hydrogen relative to the other elements when compared to solar abundances.

The reason for the nitrogen discrepancy is similar but slightly different. Nitrogen is less reactive than carbon or oxygen and tends to form N_2. At low temperatures, N_2 is too volatile to condense if comets formed at temperatures between 25 and 50 K. Carbon, which is also underabundant in the gas (i.e., ice) phase, can condense as hydrocarbon grains. The conversion of CO and N_2 in the solar nebula or on interstellar grains to form CH_4, which is also very volatile, and NH_3 was incomplete (except in the regions of the giant planet subnebulae). Ices of the precometary grains probably also underwent some chemical processing due to UV photolysis or other energetic events in the nebula. Oxygen is bound in water and silicates. Thus the temperature in the nebula region where comets formed was high enough to prevent condensation of CO and N_2, and only small amounts of these gases were trapped during the formation of cometary ices. It should be noted that gases can be released in the interior of the nucleus by processes other than sublimation of icy constituents. Amorphous ice can trap significant quantities of guest molecules at low temperature, which are then released upon warming as the amorphous ice transforms first to cubic and subsequently to hexagonal crystalline ice at temperatures between about 120 and 160 K.

The solar ratio of N : C is approximately 1 : 3. Thus, if the probability of trapping N_2 is about the same as for CO, the ratio of N_2 : CO in comets should be about 1 : 6. Considering that carbon is very reactive whereas nitrogen is not, some CO may have been converted to condensable organic materials, which may make N_2 : CO in comets closer to 1 : 5. N_2 is difficult to detect by ground- or space-based spectroscopy of comets. In addition, N_2 shares the same mass channel as CO, making in situ mass spectroscopic identification very difficult. N_2^+ has been observed in some comet plasma tails, and the $N_2^+ : CO^+$ ratio measured in Halley and other comets was less than 10^{-4}. It is not clear if this value can be applied directly to the nucleus composi-

TABLE V
Isotopic Ratios in Comets and Other Reservoirs

Species	Solar system	Interstellar matter	Comets
D/H	1 to 2 × 10⁻⁵	1.5 × 10⁻⁵	3.2 × 10⁻⁴ᵃ
D/H		10⁻⁴ to 10⁻²ᵇ	1.9 to 3.5 × 10⁻⁴ᶜ
$^{12}C/^{13}C$	89	43 ± 4ᵈ	95 ± 12ᵉ
$^{12}C/^{13}C$		65 ± 20ᶠ	70 to 130ᵍ
$^{12}C/^{13}C$		12 to 110ᵇ	10 to 1,000ʰ
$^{14}N/^{15}N$	272	≈400ᶠ	>200ᵉ
$^{16}O/^{18}O$	498	≈400ᶠ	493ᵃ
$^{24}Mg/^{25}Mg$	7.8		Variableʰ
$^{25}Mg/^{26}Mg$	0.9		<2ʰ
$^{32}S/^{34}S$	22.6		22ʰ
$^{56}Fe/^{54}Fe$	15.8		15ʰ

ᵃ From *in situ* mass spectrometry of Comet Halley coma.
ᵇ Range of observed values in dense ISM clouds.
ᶜ From radio wavelength spectra of HDO in Comet Hyakutake.
ᵈ From visual spectra.
ᵉ From ground-based observations of CN bands in Comet Halley.
ᶠ From radio astronomical data.
ᵍ From ground-based observations of C_2 spectra.
ʰ From *in situ* mass spectrometry of Comet Halley dust.

tion, so the relative abundance of N:C remains rather uncertain.

In the gas phase, carbon is also underabundant in the coma. However, when the carbon contained in organic particles is added, assuming the mass ratio dust-to-gas in the coma to be $\chi = 1$ or 2, then the relative carbon abundance in Comet P/Halley is restored to the solar value. Oxygen is sufficiently abundant in water ice so that the organic contribution to the oxygen reservoir is negligibly small. Because nitrogen is also underabundant in organic particles, its relative abundance in comets remains below the solar value. It should be mentioned that some sulfur is also contained in the organic particles and its relative abundance in comets is very close to the solar value. Phosphorus, an element important for the origins of life, however, has not been detected in comets, but its relative solar abundance is so low that it may have simply eluded detection.

Measurements of isotopic ratios in comets from both ground-based spectroscopy and *in situ* spacecraft measurements of Comet Halley show good agreement with solar system values (Table V), strongly suggesting that comets formed out of the same compositional mix as the rest of the solar system. One substantial difference exists for the deuterium:hydrogen ratio that is enriched considerably over values found for the atmospheres of the giant planets. However, the cometary value is in good agreement with enrichments seen in meteorites and in dense cloud cores in the interstellar medium. This suggests that comets obtained their deuterium and hydrogen from condensed (i.e., solid) phases, whereas giant planets collected their hydrogen directly from the nebula gas.

The D:H ratio measured in Halley (3.2 ± 0.3 × 10⁻⁴) is twice that of terrestrial ocean water, which has a D:H value of 1.56 × 10⁻⁴. It had been thought that cometary impacts early in the solar system's history may have contributed a substantial fraction of the Earth's oceans. However, the substantial enrichment in Halley means that the Earth's oceans had to have been diluted by a less enriched source of water, perhaps hydrated silicates in meteorites if comets contributed to creating Earth's oceans.

VII. METEOROID STREAMS

In 1866, Giovanni Schiaparelli found that the orbit of the Perseid meteoroid stream, seen annually as a meteor shower in mid-August, was similar to the orbit of Comet 109 P/Swift-Tuttle (1862 O1). This was the first definite connection established between comets and meteor showers. Since then, over a dozen comets have been identified as sources of meteor showers. These include the associations of Comet 1P/Halley with the ε Aquarids (May) and Orionids (August), Comet 55P/Tempel-Tuttle with the Leonids (November), and Comet 2P/Encke with the Taurids (October–December).

The gradual erosion of a comet during its passage through the inner solar system leads to the formation of a trail of solid particles strewn along its orbit. The trail consists of large ($r > 100 \, \mu m$) nonvolatile grains lifted off the surface of the nucleus by the evolving gases. Because of the relatively large size of the grains, they are not blown away as easily by radiation pressure forces as are the far more numerous small grains entrained in the escaping coma gases. Thus, the large grains remain in roughly the same orbit as the parent comet, but drift slowly away from it along the orbit due to their nonzero ejection velocities, due to small, slow changes in their orbits caused by radiation pressure forces.

If the trail's orbit crosses that of the Earth, these cometary particles may encounter the Earth. They enter the atmosphere at high speeds and are heated by friction to high temperatures, vaporizing the material and giving rise to visible meteors. Because of the concentration of dust grains in the comet's orbit, many enter the atmosphere at the same time, producing a meteor shower. Typical meteor velocities range from about ~15–25 km sec^{-1} for meteors associated with Jupiter-family short-period comets, such as P/Encke, to ~50–70 km sec^{-1} for Halley-type comets, such as P/Swift-Tuttle. Meteors typically burn up at altitudes between 80 and 120 km above the ground.

Occasionally, a meteoroid stream contains larger cometary fragments that can produce spectacular fireballs and bolides that explode in the Earth's atmosphere. Typical observed meteor rates for larger showers such as the Perseids and Geminids, for a single observer location, are on the order of 60 per hour. This can be compared with the rate of sporadic (random) meteors of about 10–15 per hour.

The association between comets and meteor showers is important because it provides the opportunity to gain further insight into the nature of comets. Observations of meteors yield information on the composition and density of cometary dust, cometary orbital evolution, the ages of comets in their present orbits, and lower limits on the masses of cometary nuclei. Meteor showers often show variations in magnitude from year to year, usually peaking near or soon after the time of perihelion passage of the parent comet. The cometary debris is generally distributed inhomogeneously along the orbit.

The best known case of this is the Leonid meteor shower associated with 55P/Tempel-Tuttle, which peaks every 33 years with spectacular showers. Observed Leonid meteor rates for a single observer location have been known to approach 50,000 per hour. P/Tempel-Tuttle passed perihelion in February 1998, and spectacular showers may occur in November 1998 or 1999, although the peak rate is predicted to be a more modest 5000 per hour.

Cometary meteoroid streams were observed directly for the first time by the Infrared Astronomical Satellite (IRAS) in 1983–1984. Also detected by IRAS was a small asteroid, 3200 Phaeton, with an orbit that matched that of the Geminid meteor shower (December). Phaeton is thought to be an extinct cometary nucleus, either depleted of its reservoir of volatile ices or covered with a refractory regolith, preventing the release of volatiles from its interior. It is possible that up to half of the Earth-crossing asteroids may be defunct comets.

Eventually, after the parent comet becomes inactive or disintegrates entirely, a meteoroid stream will disperse due to gravitational perturbations by the planets. The time scale for this is about 10^4 years. Cometary grains may suffer a number of fates: they may become part of the sporadic (random) meteors that are continually seen throughout the year, slowly being swept up by the planets and their satellites, or their orbits may decay slowly due to radiation pressure forces until they approach too close to the Sun and are vaporized. Finally, they may be ejected to interstellar space by encounters with the giant planets.

VIII. HAZARDS OF COMETARY IMPACTS

The fact that some cometary orbits intersect the Earth's orbit leads to the question as to whether there is a possibility of a collision, an idea first suggested by Halley in 1705. Ample evidence shows that the Earth, like other members of the inner solar system, has undergone many such collisions in its history. More than 140 impact craters caused by asteroid and comet bombardment have been found on the Earth, and many more have undoubtedly been erased through erosion and tectonic activity. Impacts of this magnitude release cataclysmic amounts of energy. [*See* PLANETARY IMPACTS.]

For example, a small stony asteroid about 60 m in diameter and traveling at 15 km sec^{-1} has a kinetic energy equivalent to about 15 megatons of TNT. Such an object entered the Earth's atmosphere and exploded over central Siberia on June 30, 1908. Called the Tunguska event, it knocked down trees in a radial pattern in a heavily forested area up to 40 km from the center of the blast, but no crater was found. The object apparently exploded in the atmosphere at an altitude of about 8 km. Although the type of the impactor has not been firmly established, studies of the event suggest that this was a small stony asteroid and not a comet. Small cometary nuclei are expected to explode higher in the atmosphere than similarly sized asteroidal fragments. However, larger nuclei, about 350 m in diameter or larger, are expected to survive entry intact and will impact the Earth's surface.

In 1996, Comet Hyakutake, 1996 B2, was discovered only about 8 weeks before it passed within 0.102 AU of the Earth. Hyakutake was a typical long-period comet with an estimated nucleus diameter of 1–3 km. If it had struck the Earth, it would have created a crater perhaps 20–50 km in diameter. In 1983, Comet IRAS-Araki-Alcock, 1983 H1, was discovered only 15 days prior to passing 0.031 AU from the Earth. In fact, the comet was not recognized as a comet until only 8 days before closest approach! With an estimated nucleus radius of ~4 km, Comet IRAS-Araki-Alcock could have formed a crater 100–150 km across! The climatic disturbance caused by the dust placed into the atmosphere by such a large impact could have had global consequences.

Fortunately, the probability of impact by comets on the Earth is relatively small. A typical long-period comet crossing the Earth's orbit has an impact probability of $\sim 2.2 \times 10^{-9}$ per orbit. The probability of impact of a typical short-period comet is about three times that value (per perihelion passage). However, the energy imparted by a short-period comet impact is generally less than for a long-period comet because the encounter velocities with the Earth are typically less. Comet are estimated to account for only about 10% of the impactors on the Earth, with asteroids accounting for the other 90%.

BIBLIOGRAPHY

Huebner, W. F. (ed.) (1990). "Physics and Chemistry of Comets." Springer-Verlag, Berlin.

Newburn, Jr., R. L., Neugebauer, M., and Rahe, J. (eds.) (1991). "Comets in the Post-Halley Era." Kluwer, Dordrecht.

Yeomans, D. (1991). "Comets, a Chronological History of Observation, Science, Myth and Folklore." Wiley, New York.

COMETARY DYNAMICS

I. Statistics of Comet Orbits

II. Dynamical Evolution of Long-Period Comets

III. The Oort Cloud

IV. The Jupiter Family

V. The Trans-Neptunian Belt

VI. Dynamical Aspects of Cometary Origin

VII. Summary

Julio A. Fernández

Facultad de Ciencias, Montevideo, Uruguay

GLOSSARY

Apparition: Observation of a comet from Earth when it passes close to perihelion.

Centaurs: Objects orbiting the Sun beyond Jupiter and crossing or approaching one or more of the other Jovian planets; their orbits are very unstable.

Edgeworth–Kuiper belt (or Kuiper belt): Reservoir of low-inclination comets, presumably of primordial origin, located beyond Neptune.

Gaussian distribution: Probability distribution of a continuous random variable whose form is that of a bell symmetric about its central value.

Halley-type comets: Generally defined as comets with orbital periods smaller than 200 years and Tisserand parameters (with respect to Jupiter) smaller than 2.

Jupiter-family comets: Generally defined as comets with orbital periods smaller than 20 years and Tisserand parameters (with respect to Jupiter) greater than 2.

Long-period comets: Comets with orbital periods longer than 200 years.

New comets: Comets coming from the Oort cloud, most of them presumably for the first time.

Nongravitational forces: Jet forces acting on the comet nucleus arising from the nonisotropic sublimation of the cometary ices.

Oort cloud: Extended comet reservoir of spherical structure stretching to 10^4–10^5 AU.

Orbital energy: Sum of the kinetic and potential energies that is proportional to the reciprocal of the semimajor axis.

Original orbit: Orbit that a comet has (with respect to the barycenter of the solar system) before entering the planetary region.

Osculating orbit: Instantaneous orbit of a perturbed body that changes with time.

Random walk: Process that proceeds by random steps, either positive or negative; its cumulative effect is given as the square root of the sum of the squares of the number of steps.

Tisserand parameter: Invariant of the motion of an infinitesimal particle (comet) in the circular, restricted three-body problem Sun–comet–perturbing planet (in general Jupiter).

C ometary dynamics deals with the study of cometary motion based on Newton's laws of mechanics. The knowledge of the type of orbits followed by comets and what forces perturb their motion is essential in learning about their origin and later evolution, in particular how comets can be transferred to the inner planetary region, where they become observable. Dynamical studies of comets are of fundamental importance in the framework of theories of solar system origin, since these primitive bodies are considered to be residues of planetary formation. In this regard, the recent discovery of a trans-Neptunian population of bodies, probably the residual outer edge of the protoplanetary disk, is particularly relevant since it may be the source of the Jupiter-family comets. [*See* THE KUIPER BELT.]

I. STATISTICS OF COMET ORBITS

Comets have fascinated people since immemorial times owing to their sudden and sometimes spectacular apparitions. Yet, it was not until 1577 that the question of their celestial nature was definitively settled when the parallax of a very bright comet could be measured by several astronomers—Tycho Brahe among them— who recognized unambiguously that it was farther away than the Moon. Newton's law of universal gravitation was first used by Edmond Halley to show that comets moved around the Sun in elliptic orbits, generally so eccentric that they could be taken as parabolas. Halley built the first catalog of cometary orbits with 24 entries and noticed that three of them—comet apparitions of 1531, 1607, and 1682 (this one observed by himself)—probably corresponded to three different passages of the same comet orbiting the Sun in about 76 years. Comet Halley, as it was later named, was the first one to be recognized as periodic.

The record of cometary apparitions is quite sparse until the seventeenth century, with most of them coming from Chinese sources. The earlier references are usually so vague that they do not allow one to estimate fairly good orbits. Records since then have steadily improved, adding to the naked-eye discoveries those made by telescopic means. Furthermore, modern comet discoveries are usually followed by accurate astrometric measurements, which are essential for the computation of their orbits. The orbital periods (P) of comets are found to span from a few years to several million years (Myr). Comets with the shortest periods are called periodic, whereas the rest are called nonperiodic or long-period (LP). The limiting period is set rather arbitrarily at 200 years, based on the fact that comets with $P > 200$ yr have so far been observed only once. To a large extent, this is an observational artifact due to the lack of systematic observations of comet apparitions for more than a couple of centuries or so. Periodic comets are sometimes divided into Halley-type (HT) and Jupiter-family (JF) comets, with the boundary usually at $P = 20$ yr (though a few HT comets are found to have $P < 20$ yr). A better distinction between HT and JF comets based on another dynamical parameter, the Tisserand parameter, will be provided in Section IV.

Instead of the orbital period, we can talk about the energy E of a comet, which is the sum of the kinetic and potential energies. The orbital energy is proportional to the reciprocal of the semimajor axis ($1/a$), and from now on we shall refer to the latter (with opposite sign) as equivalent to E, namely, $E \equiv -1/a$. Orbital energies E are negative for elliptic orbits and positive for hyperbolic ones; $E = 0$ for a parabolic orbit. For elliptic orbits it is also possible to relate E to the orbital period P through Kepler's third law, $P = a^{3/2} = -E^{-3/2}$, where P is given in years, a in astronomical units (AU), and E in AU^{-1}.

There are nearly 940 comets with computed orbits recorded up to the present (end of 1997), of which about 80% are LP comets and the rest are periodic. The orbits of a few observed comets covering a wide range of orbital periods are shown in Fig. 1. Among the LP comets, there is a special group of sungrazers, known as the Kreutz family, that probably come from a single parent comet tidally disrupted by the Sun (see Section II, D). The discovery rate of Earth-crossing LP comets has remained nearly constant for about a century, which suggests that these are almost completely known. By making allowance for some missed comets, we can estimate that the rate of passages of LP comets in Earth-crossing orbits brighter than absolute total magnitude $H_T \simeq 10.5$ is about 3 yr^{-1}. (H_T measures the brightness of the comet nucleus plus surrounding coma, assuming ideally that the comet was located at 1 AU from Earth and the Sun and at an angle Sun–comet–Earth of 0°, and that the brightness decreases with the heliocentric distance r as r^{-4}.)

According to a resolution adopted by the International Astronomical Union (the main organization of professional astronomers of all the world) in August 1994, each discovered comet is given a designation consisting of the year of discovery, an uppercase code letter identifying the half-month of discovery during that year (A = Jan. 1–15, B = Jan. 16–31, ..., Y = Dec. 16–31, I being omitted), and a consecutive numeral to indicate the order of discovery announcement during that half-month. This designation is prefixed, generally by "P/" or "C/", according to whether the comet is periodic or not. The prefix "D/" is used for a few periodic comets that no longer exist or are deemed to

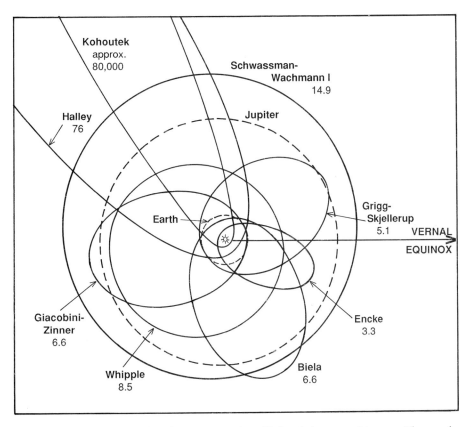

FIGURE 1 Orbits of some selected comets. Below their names are the orbital periods expressed in years. The vernal equinox denotes the position of the Sun, as seen from Earth, at the time of the northern spring equinox, about March 21.

have disappeared. In addition to this official designation, the tradition of naming comets after their discoverer(s) is maintained. For instance, the discovery of a comet was announced independently by Alan Hale and Thomas Bopp on July 23, 1995, being the first in the second half of July. Follow-up observations showed that it was an LP comet. Accordingly, its given name was C/1995 O1 (Hale-Bopp). If a periodic comet is observed to return, the prefix P/ is preceded by a sequential number. For instance, the first comet shown to be periodic by Edmond Halley in 1682 now receives the official designation 1P/Halley.

It is interesting to analyze the distributions of orbital inclinations (i) of the different dynamical classes. For LP comets, the observed i distribution fits a sine law quite well (Fig. 2a), though showing some excess of retrograde orbits (i.e., with $i > 90°$). The closeness to a sine law indicates that the orbital planes of LP comets are more or less randomly distributed. By contrast, HT comets move predominantly in direct orbits (Fig. 2b), whereas all JF comets so far discovered move in direct orbits with most of their orbital planes lying close to the ecliptic plane (Fig. 2c).

An important orbital parameter is the perihelion distance (q), which is the distance of closest approach to the Sun. For LP comets, the frequency distribution of q shows a maximum near Earth's orbit ($q \simeq 1.0$–1.1 AU). It falls steeply closer to and farther away from the Sun (Fig. 3). The latter can be explained in terms of an observational selection effect, since most distant comets can pass undetected. As for the LP comets coming close to the Sun, they may disintegrate or fade away after a few passages. For instance, when we limit the sample of LP comets to near-parabolic orbits, say, for semimajor axes $a > 10^4$ AU (orbital periods $P > 10^6$ years), the q distribution turns out to be roughly uniform, at least in the inner planetary region. Most of these comets are on their first perihelion passage (see Section III, B), so they are still little affected by solar radiation.

So far, the overwhelming majority of the discovered LP comets have perihelion distances smaller than a few AU. Only 10 have perihelia beyond Jupiter's orbit (>5.2 AU), the farthest one being for comet C/1991 R1 (McNaught-Russell), which has $q = 7.0$ AU. In addition, there are two periodic comets in the trans-

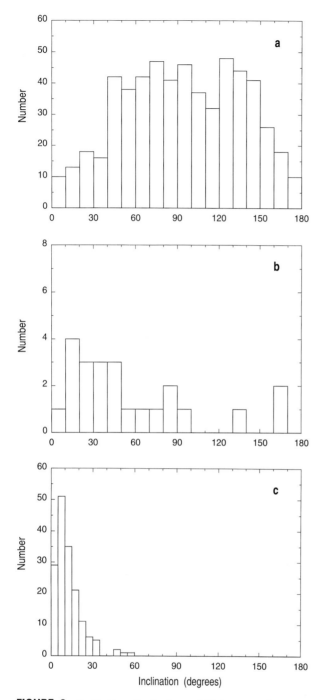

FIGURE 2 Inclination distributions of (a) long-period comets discovered after 1758 (the Kreutz family of sungrazing comets has been considered as a single comet, as well as C/1988F1 and C/1988J1, which move in similar orbits), (b) intermediate-period comets with $20 < P < 200$ yr, and (c) short-period comets with $P \leq 20$ yr.

Jovian region: 29P/Schwassmann-Wachmann 1 and the famous one that collided with Jupiter in 1994, D/1993 F2 (Shoemaker-Levy 9), and several Centaurs and trans-Neptunian objects that we will refer to later.

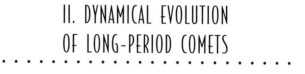

II. DYNAMICAL EVOLUTION OF LONG-PERIOD COMETS

A. OSCULATING AND ORIGINAL ORBITS

In principle, a comet orbit can be determined from three accurate astrometric positions (in practice, several more) spanning several days. First, a parabolic solution is tried to fit the observations. Once good astrometric positions covering a period of several weeks are available, an elliptic (or sometimes hyperbolic) solution is tested as a better approximation to the actual orbit. Since the observed comet is continuously perturbed by the planets, its orbit will change with time. Therefore, what we obtain from the astrometric observations is the *osculating orbit,* that is, the orbit the comet has at a particular instant. The epoch adopted for the osculating orbit is usually near perihelion, where most comet observations are performed.

When we deal with an LP comet, we are interested in obtaining its *original orbit,* that is, the orbit it had before entering the planetary region. The original orbit is referred to the barycenter of the solar system. Knowledge of original orbits is vital to assess the place from where LP comets come. To obtain the original orbit of an observed comet, we integrate its osculating orbit obtained at a given epoch backward in time until the comet is so far away from the planets (say, heliocentric distances ~ 50 AU) that their perturbations on the comet can be assumed to be negligible. To this purpose, the equations of motion have to be solved, namely,

$$\frac{d^2\vec{r}}{dt^2} = -\frac{GM_\odot \vec{r}}{r^3} + \nabla R \qquad (1)$$

where G is the gravitational constant, $M_\odot$ is the Sun's mass, $\vec{r}$ is the Sun–comet radius vector expressed in the heliocentric frame of reference, and R is the disturbing function. The latter describes the perturbations of the planets of masses m_i on the comet's orbit and is given by

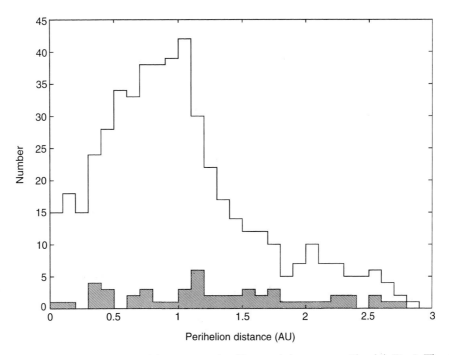

FIGURE 3   Distribution of perihelion distances of the same sample of long-period comets considered in Fig. 2. The shaded histogram is for near-parabolic comets with original semimajor axes > 10^4 AU.

$$R = G \sum_i m_i \left(\frac{1}{d_i} - \frac{xx_i + yy_i + zz_i}{r_i^3} \right) \qquad (2)$$

where d_i, r_i are the planet–comet and the planet–Sun distances, respectively, and (x, y, z), (x_i, y_i, z_i) are the heliocentric coordinates of the comet and the planets, respectively. In many cases, a better solution can be obtained for the original orbit if an extra term that takes into account what is known as the nongravitational force (i.e., a jet reaction on the comet nucleus produced by the sublimating gases) is introduced in Eq. (1). We will describe this force in Section II, C.

The frequent classification of a comet orbit as a "parabola" (i.e., eccentricity $e = 1$ or energy $E = 0$) reflects our poor knowledge of the orbit. The energy (or reciprocal semimajor axis) of an LP comet is very close to zero, so the computation of its original value $(1/a)_{orig}$ requires a large number of accurate astrometric measurements. Improvements in the observation of comets at large heliocentric distances, added to the use of fast computers that can handle the perturbations of all the planets, have facilitated the accurate determination of $(1/a)_{orig}$ for a large number of LP comets.

B. PLANETARY PERTURBATIONS

The orbital parameter of an LP comet that is most perturbed by the planets is generally the orbital energy or reciprocal semimajor axis ($1/a$). The other orbital parameters—inclination (i), argument of perihelion (ω), longitude of the ascending node (Ω), and the perihelion distance (q)—experience only minor changes. Therefore, we can conceive the dynamical evolution of an LP comet on successive passages through the planetary region as a random walk in energy space, where each step $\varepsilon = \delta(1/a)$ corresponds to the energy change during a passage. The basic assumption is that two successive perturbations ε in the comet's energy are independent of each other, which is justified on the basis that the comet moving in an orbit of long orbital period will encounter a planetary configuration completely different from the previous one. We note that since comets can either gain or lose energy, ε can be negative or positive, respectively.

Through the random walk in energy space, some LP comets will pass from initial near-parabolic orbits (orbital energies $\simeq 0$) to periodic orbits (orbital energies $\lesssim -0.03$ AU^{-1}). Dynamically "young" comets (i.e., those having no more than a few passages by the planetary region) will tend to concentrate around near-zero energies, since they have not had time yet to evolve toward larger (negative) binding energies (i.e., small semimajor axes). LP comets with larger binding energies are regarded as dynamically "old" and may have accumulated tens, hundreds, or thousands of passages through the planetary region.

Let us define $\Psi(\varepsilon)$ as the probability that a comet experiences an energy change ε after a perihelion passage. Numerical integrations of fictitious comets have shown that $\Psi(\varepsilon)$ can be approximated by a Gaussian $[= (1/\sigma\sqrt{2\pi})e^{-\varepsilon^2/2\sigma^2}]$ or a double-exponential $[= (1/\sqrt{2}\sigma)e^{-\sqrt{2}|\varepsilon|/\sigma}]$ distribution with long tails of large values of ε due to strong planetary perturbations during close encounters. In the previous distributions, σ is the standard deviation. For randomly oriented LP comets in Jupiter-crossing orbits, the typical energy change $\bar{\varepsilon}$ (which can be taken as the standard deviation of the ε distribution) is about 7×10^{-4} AU^{-1}. For more distant comets, $\bar{\varepsilon}$ decreases very quickly because they do not approach Jupiter, the greatest perturber. There is also a dependence on the comet's inclination. In general, it is somewhat larger for direct orbits ($i < 90°$) than for retrograde ones ($i > 90°$). Comets in retrograde orbits will meet the planets at larger relative velocities, so they will on the average be less perturbed. Values of $\bar{\varepsilon}$ as a function of q and for different ranges of orbital inclinations are shown in Fig. 4.

The average number of revolutions required to pass from an initially parabolic orbit to an orbit with a certain energy E_P will be on the order of

$$\langle n \rangle = (E_P/\bar{\varepsilon})^2 \tag{3}$$

Taking $E_P = -0.03$ AU^{-1} and $\bar{\varepsilon} = 7 \times 10^{-4}$ AU^{-1} for a Jupiter-crossing LP comet, we get a number of 1850 revolutions to reach a periodic orbit.

Let us now define $\nu(1/a, t)d(1/a)$ as the number of LP comets passing perihelion per year with reciprocal semimajor axes in the range $(1/a, 1/a + d(1/a))$ at a certain time t. The total number of comets in this range of $(1/a)$ will thus be given by

$$N(1/a, t)d(1/a) = a^{3/2}\nu(1/a, t)d(1/a) \tag{4}$$

where $a^{3/2}$ is the comet's orbital period in years and a is expressed in AU. The increase in the number of comets of a certain $1/a$ per year will be given by $\partial N(1/a, t)/\partial t$. The diffusion equation will be expressed as

$$\partial N(1/a, t)/\partial t = \int_{-\infty}^{+\infty} \nu(1/a - \varepsilon, t)\Psi(\varepsilon)d\varepsilon \\ - \nu(1/a, t)d(1/a) \tag{5}$$

where the first term on the right-hand side represents the number of comets acquiring the reciprocal semimajor axis $(1/a)$ per year and the second term represents those changing from $(1/a)$ to other values.

From Eq. (5) it is possible to show that comets

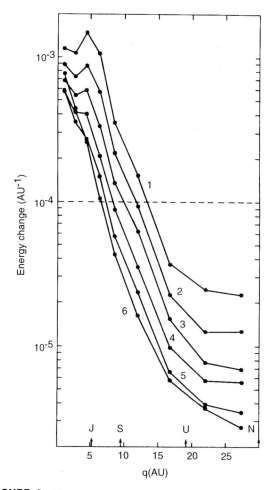

FIGURE 4 Typical energy change per perihelion passage as a function of a comet's perihelion distance and for different inclination ranges: curve 1 for $0 < i < 30°$, ..., curve 6 for $150° < i < 180°$. [From J. A. Fernández (1981). *Astron. Astrophys.* **96**, 26. Used with permission.]

diffusing inward from a field of parabolic comets ($1/a = 0$) would reach a uniform $(1/a)$ distribution after a large number of passages ($t \to \infty$).

Most LP comets will be ejected to interstellar space during their random walk in energy phase space (i.e., they acquire positive orbital energies). In fact, from an initial population of N_0 comets on near-parabolic orbits, the number $N(n)$ still gravitationally bound to the solar system after n perihelion passages is

$$N(n) \simeq \tfrac{1}{2}N_0 n^{-1/2} \tag{6}$$

For instance, if a comet in an initially parabolic orbit requires an average number of 1850 revolutions to reach a periodic orbit, as estimated before, Eq. (6) suggests that only one in ~90 comets will survive to reach such a dynamical state.

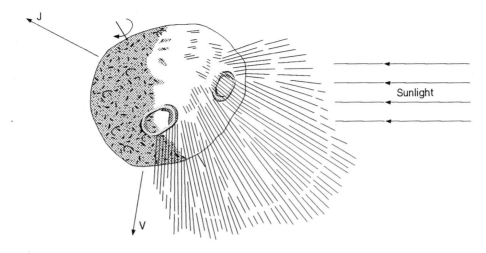

FIGURE 5 The action of nongravitational forces on comets. The sublimating gases from the nucleus give rise to a net jet force $\vec{J}$ in the opposite direction to the maximum outgassing. Because of the thermal inertia, the zone of maximum outgassing on a rotating nucleus will lag with respect to the subsolar point.

C. NONGRAVITATIONAL FORCES

In 1819, the German astronomer Johann Franz Encke discovered that the comet with the shortest orbital period so far known (3.3 yr) departed significantly from a purely gravitational motion. Comet 2P/Encke, as it is now named, showed a shortening of about 2.5 hr in its orbital period per orbit. Secular decreases of orbital periods were also found for other periodic comets like 3D-A/Biela and 16P/Brooks 2. The first idea was that a resisting medium affected the orbital motion of small, light bodies like comets. However this explanation could not apply to comets like 8P/Tuttle, which showed a secular increase of P, or 21P/Giacobini-Zinner, which initially showed an increase and later a decrease. In 1950, the American astronomer Fred Whipple found a satisfactory explanation for nongravitational forces in terms of the jet reaction produced by the nonisotropic, sublimating gases from the cometary nucleus. The thermal inertia on a rotating nucleus will cause the region of maximum outgassing to shift toward the nucleus "afternoon" hemisphere, giving rise to a net force $\vec{J}$ deviating from the radial direction to the Sun (Fig. 5). Therefore, in general there will be radial, transverse, and normal nongravitational components (J_r, J_t, J_n) acting on the comet nucleus. In the more realistic situation of a comet nucleus with a few active areas scattered on a dust-mantled surface, the sublimating gases may follow a rather more complex pattern than the one depicted in Fig. 5, but the physical principles are the same.

The main nongravitational effect that can be detected in a periodic comet observed at previous apparitions is a delay or advance in the time of the perihelion passage with respect to that derived from purely gravitational theory, which means a change ΔP in its orbital period P. For instance, for the last few apparitions, 1P/Halley has arrived at its perihelion with an average delay of $\Delta P \simeq 4.1$ days. The change ΔP can be expressed in terms of the radial and transverse nongravitational components by means of the Gauss equation

$$\Delta P = \frac{6\pi}{\eta^2 a} \int_0^P \left[\frac{e \sin f}{(1 - e^2)^{1/2}} J_r + \frac{a(1 - e^2)^{1/2}}{r} J_t \right] dt \quad (7)$$

where $\eta = 2\pi/P$ is the mean motion, e the eccentricity, and f the true anomaly, which is defined as the angle between the radius vector Sun–comet and the perihelion direction.

For a comet with a symmetric lightcurve (outgassing) with respect to perihelion, only the transverse component J_t will have some effect in the change ΔP. In this case, the term of Eq. (7) containing J_r integrated over the whole orbital period P will vanish, and only the term containing J_t will remain. However, most comet lightcurves are moderately or highly asymmetric, so the integral of the term containing J_r will not vanish; indeed, it may become dominant.

The evaluation of the nongravitational force is more difficult for LP comets, since none of them has been observed on a second apparition to check for an advance or delay in the time of perihelion passage. Nevertheless, nongravitational terms have been fitted to the equations of motion of several LP comets, leading to more satisfactory orbital solutions.

D. SUNGRAZERS

The recent discovery of more than 20 sungrazing comets by the coronograph on board the ESA/NASA *Solar and Heliospheric Observatory* (*SOHO*) spacecraft has renewed interest in this class of comets. The *SOHO* sungrazers add to the previous 23 comets discovered visually and by the *SOLWIND* and *Solar Maximum Mission* satellites. They all have similar orbital elements and belong to a family of comets (known as the Kreutz family) that are probably the fragments of a single large parent comet that was tidally disrupted by the Sun. The injection of a large comet from the Oort cloud into a sungrazing orbit appears to be a low probability event. Furthermore, some of the Kreutz members have orbital periods $P \lesssim 10^3$ yr, suggesting that the progenitor was already on a dynamically evolved orbit. It is hard to explain how the progenitor could have withstood several sungrazing passages without being tidally disrupted already on its first passage. The study of the dynamical evolution of LP comets with small perihelion distances (say, $q \lesssim 2$ AU) and inclinations close to 90° has shed light on this puzzle. It has been shown that long-term secular perturbations by the planets, involving the so-called Kosai resonance (where the dynamics is characterized by the libration of the comet's argument of perihelion ω around 90° or 270°), cause correlated changes in the perihelion distance, eccentricity, and inclination of such comets. As a result, their perihelion distance can drop to very small values during part of the dynamical evolution. It is then possible that the progenitor of the Kreutz family was not originally a sungrazer, but evolved to such a state after several passages that led to its tidal disruption by the Sun.

Alternatively, the Kreutz family progenitor may have been thrown from the Oort cloud into an initially sungrazing orbit; and then evolved rapidly to a small semimajor axis due to the powerful nongravitational forces which must be present on a comet with such a small perihelion distance.

III. THE OORT CLOUD
· ·

A. OBSERVATIONS

As discussed in Section II, B, LP comets subject to planetary perturbations will tend to acquire a uniform distribution of their orbital energies (or reciprocal semimajor axes). However, in 1950 the Dutch astronomer Jan Hendrik Oort found a dramatic excess of comets with original near-parabolic orbits (say, with $0 < (1/a)_{\mathrm{orig}} < 10^{-4}$ AU^{-1}), which showed up as a spike in an otherwise smooth $(1/a)$ distribution (Fig. 6). Since the typical energy change by planetary perturbations is several times larger than the binding energies of the comets falling in the spike (cf. Section II, B), Oort argued that they were entering the planetary region for the first time, and for this reason called them "new." It is very likely that after a single passage most new comets will be either ejected to interstellar space or transferred to much less eccentric elliptic orbits. This finding led Oort to the conclusion that a huge swarm of $\sim 10^{11}$ comets surrounds the solar system at distances of 10^4–10^5 AU. This structure, called the Oort cloud, is the source of LP comets.

Not all "new" comets will necessarily be new in a strict sense, since a small fraction of them (varying with q) will return to the Oort cloud and then come back again to the planetary region. In the present discussion, new comets will be those coming from the Oort cloud, that is, having original near-parabolic orbits with $0 < (1/a)_{\mathrm{orig}} < 10^{-4}$ AU^{-1}, no matter if sometime before they passed through the planetary region.

There are a few comets with original slightly hyperbolic orbits, but they may be due to observational errors and/or nongravitational forces that were not considered in the orbital computations. Confirming this assumption, the sample of dynamically young comets with large perihelion distances—where nongravitational forces are probably less effective—shows very few hyperbolic members. Since comets approaching from interstellar space would come on strongly hyperbolic orbits, their lack is firm evidence that the observed LP comets belong to the solar system.

B. INJECTION OF NEW COMETS

Oort cloud comets are subject to the action of external perturbers, in particular passing stars, molecular clouds, and galactic tides. The quasi-steady-state supply of new comets is due to the action of distant passing stars (typically, distances greater than $\sim 5 \times 10^4$ AU) and galactic tides. By contrast, close stellar passages and penetrating encounters with molecular clouds may cause sudden enhancements in the flux of new comets, which have been called "comet showers" (see Section III, F).

The perihelion distances of Oort cloud comets will change with time owing to the action of external perturbers. To be injected into the planetary region (and

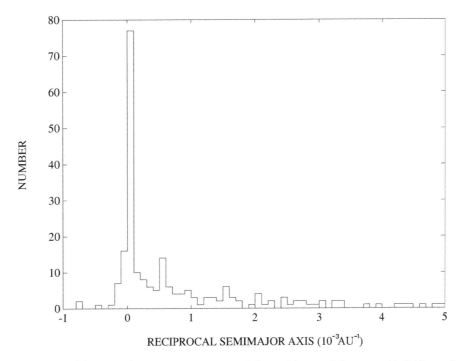

FIGURE 6 Distribution of the original inverse semimajor axes of observed long-period comets with $(1/a)_{orig} < 5 \times 10^{-3}$ AU^{-1}.

become potentially observable), an Oort cloud comet will have to experience a drastic reduction of its perihelion distance q, on the order of $\Delta q \sim q$. For a near-parabolic comet, the change in the perihelion distance, Δq, can be derived from the approximate relation

$$\Delta q/q \sim 2\Delta v_T/v_T \tag{8}$$

where $v_T \sim (2GM_\odot q)^{1/2}/r$ is the transverse component of the comet's velocity at a distance r. Therefore, the condition $\Delta q \sim q$ can be accomplished by a perturbation of its transverse velocity $\Delta v_T \sim v_T$.

If planetary perturbations were neglected, the diffusion of perihelion distances by external perturbers of small-q Oort cloud comets would lead to a uniform q distribution. However, the giant planets will remove Oort cloud comets to different degrees. Thus, most new comets crossing Jupiter's orbit will be removed from the Oort cloud after a single passage owing to its powerful gravitational field. There will be an ever-increasing fraction of comets returning to the Oort cloud for larger q as planetary perturbations become weaker. As a consequence, the frequency of passages of new comets should sharply increase in the outer planetary region, as new comets can have the opportunity to "bounce" back and forth between the planetary region and the Oort cloud before being definitely removed by either planetary perturbations or external perturbers. This is what is actually found in numerical simulations (Fig. 7). For new comets in Jupiter crossing orbits, the q distribution should be nearly uniform in agreement with the observed one (cf. Fig. 3). On the other hand, Fig. 7 shows that the frequency of passages

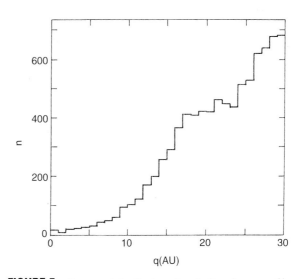

FIGURE 7 Computed distribution of perihelion distances of hypothetical "new" comets injected into the planetary region, obtained from Monte Carlo studies that included both planetary and stellar perturbations. [From J. A. Fernández (1982). *Astron. J.* **87**, 1318. Used with permission.]

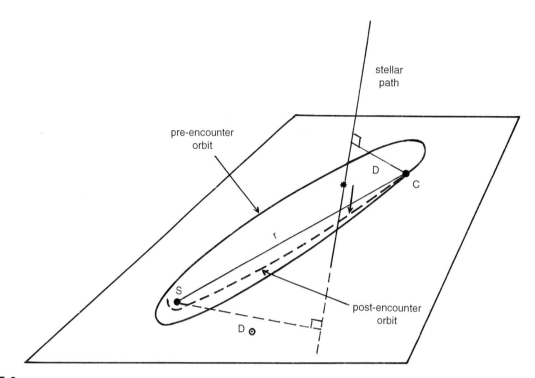

FIGURE 8 Geometry of a stellar encounter. The comet can be perturbed in such a way that it is injected into the planetary region. S, Sun; C, comet.

of new comets may be about an order of magnitude greater at Neptune's distance. Jupiter therefore appears to provide a very efficient dynamical barrier to Oort cloud comets in preventing all but a tiny fraction of them from reaching the inner planetary region.

C. STELLAR PERTURBATIONS

Let us now evaluate the magnitude of the perturbation caused by a passing star of mass M and relative velocity V on a comet located at a heliocentric distance r (Fig. 8). The usual approach is to assume that the comet is at rest in a heliocentric frame of reference during the stellar encounter and that the star's path is only slightly perturbed by the Sun's gravity, so it can be taken as a straight line without loss of accuracy. Under these simplifying assumptions, the impulsive change in the comet's velocity will simply be

$$\overrightarrow{\Delta v} = \overrightarrow{\Delta v_c} - \overrightarrow{\Delta v_\odot} \tag{9}$$

where $\overrightarrow{\Delta v_c}$ and $\overrightarrow{\Delta v_\odot}$ are the impulses received by the comet and the Sun from the passing star, respectively. They are given by

$$\overrightarrow{\Delta v_c} = \frac{2GM}{VD}\frac{\overrightarrow{D}}{D} \tag{10}$$

$$\overrightarrow{\Delta v_\odot} = \frac{2GM}{VD_\odot}\frac{\overrightarrow{D_\odot}}{D_\odot} \tag{11}$$

where D and $D_\odot$ are the distances of closest approach of the star to the comet and the Sun, respectively.

During an orbital revolution of period P, a comet will be perturbed by many stars. Let $s(D_\odot)dD_\odot = 2n_* D_\odot dD_\odot$ be the rate of stellar passages with impact parameters in the range $(D_\odot, D_\odot + dD_\odot)$. n_* is the stellar flux in the solar neighborhood, which is about 10 stars Myr^{-1} passing through a circle of one-parsec (pc) radius at an average encounter velocity with the Sun of $V = 30$ km s^{-1} (1 pc is the distance at which the radius of Earth's orbit is seen to subtend an angle of 1 arc-second: 1 pc = 206,265 AU). The cumulative change in the orbital velocity of the comet during P, Δv_{cum} will be expressed as

$$\Delta v_{\text{cum}}^2 = P \int_{D_m}^{D_M} \Delta v^2 s(D_\odot) dD_\odot \tag{12}$$

where $D_m = (2n_* P)^{-1/2}$ is the minimum distance of closest approach of a star to the Sun expected during

P. D$_M$ is the maximum distance of a passing star that may have some dynamical influence, and it can be taken as infinity without too much error.

Passing stars are the most efficient perturbers of the inner portions of the Oort cloud, say, for distances $\lesssim 10^4$ AU. Even if comets started out their dynamical evolution in orbits close to the ecliptic plane, a full randomization of their orbital planes should have been achieved for semimajor axes $a \gtrsim 3000–4000$ AU as a result of very close stellar passages.

D. GALACTIC TIDAL FORCES

Tides from the Galaxy have a significant influence on the shape and extent of the Oort cloud. A simple calculation shows that the potential of the galactic disk dominates over the potential of the galactic nucleus so that, as a first approximation, we can neglect the effect of the latter. The galactic disk can be approximately modeled as a homogeneous disk of density ρ in the midplane of the Galaxy, so its potential can be simply expressed as

$$U = U_0 + 2\pi G\rho z^2 \tag{13}$$

where U_0 is a constant and z is the distance to the galactic midplane. A value of $\rho = 0.15 M_\odot$ pc^{-3} has been derived for the solar neighborhood from the comparison of different gravitational potential models for the Galaxy with the observed velocity dispersions of tracer stars. During the course of its galactic motion, the solar system oscillates about the galactic midplane, with a half-period of $T_z = \sqrt{\pi/G\rho} \sim 34$ Myr, reaching a maximum height of about 70 pc.

From this potential, the tidal force of the galactic disk acting on a comet at a galactic latitude ϕ is

$$(dU/dz)_c - (dU/dz)_\odot = 4\pi G\rho r \sin \phi\, \hat{z} \tag{14}$$

where r is the Sun–comet distance and $\hat{z}$ is the unit vector perpendicular to the galactic plane. We note that $r \sin \phi = z_c - z_\odot$. From Eq. (14), the change in the transverse velocity v_T of an Oort cloud comet at a time-average heliocentric distance $\langle r \rangle = 1.5a$, during an orbital revolution P, can be easily derived, leading to

$$(\Delta v_T)_{\text{tide}} = 3\pi G\rho a P \cos \alpha \sin 2\phi \tag{15}$$

where α is the angle between the orbital plane and the plane perpendicular to the galactic disk containing the radius Sun–comet. By introducing the value of $(\Delta v_T)_{\text{tide}}$

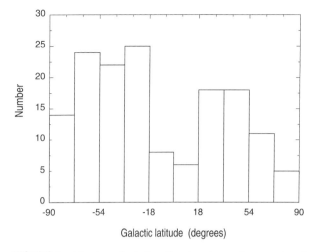

FIGURE 9 Number of aphelion directions in equal-area strips of the celestial sphere parallel to the galactic equator, taken from a sample of 151 "young" comets with $a_{\text{orig}} > 500$ AU.

into Eq. (8), we find that $(\Delta q)_{\text{tide}} \propto a^{7/2}$, which indicates that the effectiveness of vertical galactic tides in the injection of Oort cloud comets into the planetary region will increase very rapidly with the comet's semimajor axis. Indeed, for semimajor axes greater than $\sim 2 \times 10^4$ AU, the vertical galactic tide becomes dominant as compared to stellar perturbations.

Equation (15) shows that the velocity change will be maximum for midgalactic latitudes ($\phi = 45°$) and negligible near the galactic poles and the galactic equator. Since there is a correspondence between the change in the comet's transverse velocity and the change in its perihelion distance, the injection rate of Oort cloud comets will follow a similar pattern with galactic latitude. This is in agreement with the observed distribution of the aphelion directions (i.e., the farthest points from the Sun) of the LP comets, which shows concentrations in both galactic hemispheres at midgalactic latitudes (Fig. 9). A striking numerical imbalance in the north–south aphelion distribution may be partly due to inhomogeneities in the distribution of aphelion points (see Section III, F), and partly to the uneven distribution of comet observers in the Northern and Southern Hemispheres.

We note that the overall effect of the tidal force of the galactic disk on a comet's orbital energy is negligible. This is because its perturbing action along the outgoing leg of the comet's orbit roughly compensates for that along the incoming leg, so that the net effect is very small, typically on the order of (q/a). Therefore, Oort cloud comets can be rather frequently injected into the planetary region but rarely ever ejected by this force.

E. CATASTROPHIC ENCOUNTERS WITH GIANT MOLECULAR CLOUDS

Under the action of stellar perturbations, Oort cloud comets will, on average, gain energy. Once the root-mean-square change of the comet's velocity reaches the escape velocity, it will be lost to interstellar space, that is, when

$$\Delta v_{rms} = v_{esc} = \left(\frac{2\,GM_\odot}{r}\right)^{1/2} \simeq \left(\frac{4\,GM_\odot}{3a}\right)^{1/2} \quad (16)$$

where again we assume that the comet is at the time-average heliocentric distance $\langle r \rangle = 1.5a$. Conversely, from Eq. (16) we can derive the semimajor axis a_{esc} for which comets will achieve escape velocities over the age of the solar system (4.6 billion yr). For stellar perturbations, this condition is fulfilled for $a_{esc} \simeq 1.3 \times 10^5$ AU, so comets with initial semimajor axes $a > a_{esc}$ would have been lost to interstellar space by now.

But stellar perturbations do not define the outer boundary of the outer Oort cloud. Penetrating encounters with giant molecular clouds (GMCs), though rare, may have catastrophic consequences for the outer portions of the Oort cloud. The previous stability limit based only on stellar perturbations can be considerably shortened if the action of molecular clouds is properly taken into account. The most devastating effects will occur when the solar system encounters a large GMC of a mass $\sim 5 \times 10^5 M_\odot$ and radius ~ 20 pc. There might have been between 1 and 10 such encounters during the solar system's lifetime. Simple estimates show that penetrating encounters with GMCs will constrain the stability boundary of the Oort cloud to about one-fourth the boundary imposed by stellar perturbations, that is, to about 3×10^4 AU. This value can be compared with the observed maximum separations between members of wide binary stars in the Galaxy, which are coincidentally on the order of a few times 10^4 AU.

The fact that we define a stability boundary does not mean that the Oort cloud is empty beyond it, since the outer portions will be continuously replenished with comets from the "stable" core that gain energy under the action of strong external perturbers. This is similar to the process of thermal escape of gaseous molecules from the outer layers of a planetary atmosphere, which is kept in a quasi-steady state by the continuous supply of molecules from the lower atmosphere. In this regard, we note that nearly 40% of the original semimajor axes of new comets have $a > 3 \times 10^4$ AU, indicating that they come from the outer Oort cloud. Nevertheless, we should expect a significant drop in the number of comets in the outer portions of the Oort cloud as they have ever-decreasing dynamical lifetimes.

F. TEMPORAL VARIATIONS IN THE INJECTION RATE OF NEW COMETS: COMET SHOWERS

Oort cloud comets have been thermalized by external perturbers over the age of the solar system, for semimajor axes $a \gtrsim 10^4$ AU. But not all the directions of the velocity vectors of thermalized comets are possible. Oort cloud comets entering the inner planetary region will be quickly removed by planetary perturbations, since they experience typical energy changes greater than their binding energies in the Oort cloud. There will be a region in velocity phase space known as the loss cone, whose axis is the solar direction. Oort cloud comets will diffuse into the loss cone under the action of external perturbers, will be injected into the planetary region, and then removed from the Oort cloud by planetary perturbations. Results for the efficiency in filling the loss cone by different external perturbers show a slight predominance of galactic tides, which is reflected in the galactic pattern of the distribution of aphelion points on the sky (cf. Fig. 9). For a certain $a > a_{fill}$, the diffusion speed of Oort comets into the loss cone is so fast as compared to losses by planetary perturbations, that it will be kept filled at all times. The combined effect of stellar perturbations and the vertical galactic tidal force gives $a_{fill} \sim 3 \times 10^4$ AU. Consequently, the outer portions of the Oort cloud for $a > a_{fill}$ will provide a steady supply of Oort cloud comets into the inner planetary region.

A random perturber—either a very close stellar passage or a molecular cloud—will affect the inner portions of the Oort cloud, where the loss cone is essentially empty. As a result, the loss cone will be suddenly refilled, so the zone of the Oort cloud with filled loss cones will temporarily expand inward. The result will be a sharp enhancement in the supply of Oort cloud comets to the inner planetary region, which we define as a comet shower.

The intensity of a comet shower will depend on the perturber and on the degree of central condensation of the Oort cloud. For a heavily concentrated Oort cloud, a close stellar passage at $D_\odot \sim 10^4$ AU can trigger a comet shower with a frequency of perihelion passages ~ 100 times the background comet flux during its phase of highest intensity. The same effect can be achieved by a penetrating encounter with a GMC. Closer stellar passages at, say, ~ 5000 AU, might trigger showers ~10^3 times more intense, depending again on the degree of central condensation of the Oort cloud. Stellar encounters at ~5000 AU should be expected at average time intervals on the order of 1–2×10^8 yr. A star penetrating deeply into the core of the Oort cloud will strongly perturb those comets that happen to pass nearby. Since $D \ll D_\odot$, the comets will receive impulses $\Delta v \propto 1/D$ [cf. Eqs. (9) and (10)]. Therefore, the shorter the distance D, the greater the perturbation, so comets injected into the planetary region might have their aphelion points clustered in a sky area along the stellar path. Indeed, some anomalous clusterings of aphelion points of dynamically young comets have been regarded as signatures of past close stellar passages.

Monte Carlo simulations that include both stellar and galactic tidal perturbations illustrate very nicely the production of comet showers at average intervals of several times 10^7 yr (Fig. 10). For $a = 10^4$ AU, several spikes (showers) punctuate the otherwise quiescent comet flux, indicating the sporadic action of close stellar passages on comets in the inner core (Fig. 10a). For $a = 2 \times 10^4$ AU, the fluctuations are more frequent, although less conspicuous, as more stars are able to perturb comets with larger a (Fig. 10b). Finally, the temporal variation in the injection rate becomes too "noisy" for $a = 3 \times 10^4$ AU (Fig. 10c). The lack of conspicuous peaks in the latter can be interpreted in terms of Oort cloud comets with loss cones already filled, thus precluding the production of substantial variations in the rate of comets injected into the planetary region from those distances.

Whether compelling evidence of past comet showers can be found in the impact cratering record is not clear yet. Studies of impact crater ages are not conclusive in this respect. The major problem seems to be that the terrestrial cratering rate is dominated by Earth-crossing asteroids rather than comets. The latter may only produce ≲10% of all observed terrestrial craters. Simple calculations show that a comet shower must be at least about 50–300 times more intense than the background comet flux to show up in crater statistics. This might occur only at intervals of some 10^8 yr, which is longer than many dated craters so far available. Studies of crater ages on the Moon—once we return there—might be more promising at detecting age clusterings attributable to past intense showers.

As mentioned, the observed aphelion distribution of new and dynamically young comets follows a pattern reflecting the influence of the galactic disk potential with the highest concentration of aphelion points at midgalactic latitudes. This strongly suggests that most

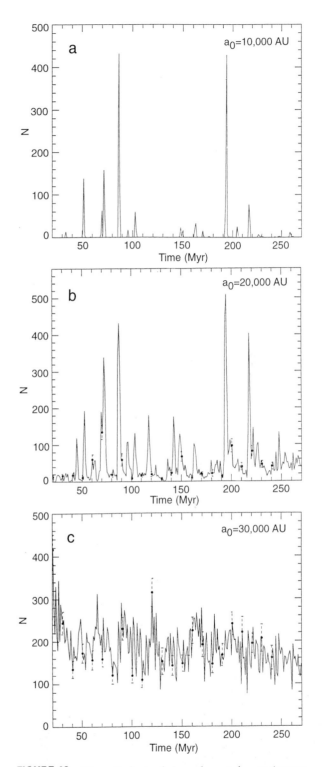

FIGURE 10 Monte Carlo simulations of comet showers (comets injected with $q < 2$ AU) triggered by stellar perturbations on Oort cloud comets with semimajor axes: (a) 10,000 AU, (b) 20,000 AU, and (c) 30,000 AU. [From J. Heisler (1990). *Icarus* **88**, 104. Used with permission.]

new comets are currently being deflected into the inner planetary region mainly by the tidal force of the galactic disk. Since this is a quasi-steady-state perturber, the frequency of comet passages may be close to its background level at present. Enhanced comet fluxes produced during close encounters with stars or molecular clouds—events apparently uncorrelated with galactic structure—might erase or severely weaken the galactic signature. Nevertheless, as we argued before, a small fraction of the overall aphelion sample appears to follow a pattern (clustering) independent of the galactic signature, which might reflect close stellar passages in the recent past.

IV. THE JUPITER FAMILY

Jupiter family comets move in very unstable orbits subject to frequent perturbations by Jupiter. The first JF comet recognized as such was D/1770 L1 (Lexell), whose period was found to be only 5.6 yr. As mentioned earlier (cf. Section II, C), the comet with the shortest orbital period so far discovered is 2P/Encke with $P = 3.3$ yr. There have been 168 JF comets discovered through the end of 1997. The discovery rate of JF comets has been steadily increasing. Though no JF comets discovered prior to 1892 had $q > 2$ AU, deep sky surveys have led to the discovery of a growing number of distant JF comets. At present about half of the observed population have $q > 2$ AU. This suggests that a large fraction of the population of distant JF comets ($q \gtrsim 2$ AU) has still to be discovered.

There are some JF comets found to librate temporarily around mean motion resonances, for instance, 9P/Tempel 1, though they will tend to decouple after some time (a mean motion resonance corresponds to a commensurability between the orbital period of the perturbing planet—in this case Jupiter—and that of the body). The dynamical control of Jupiter can be seen as a spatial concentration of JF comets in its vicinity at a given time. Furthermore, low-velocity, long-lasting encounters of comets with Jupiter may lead to their capture as temporary satellites. This is particularly the case with 39P/Oterma, 74P/Smirnova-Chernykh, 82P/Gehrels 3, and 111P/Helin-Roman-Crockett, all of which have the distinctive properties of being near the 3:2 resonance with Jupiter and Tisserand parameters $T \sim 3$, indicating the possibility of low-velocity encounters with Jupiter [see Eqs. (17) and (18)].

The study of the random-walk process of LP comets in energy phase space (see Section II, B) led to the idea that JF comets could be the end products of the dynamical evolution of LP comets, involving perhaps hundreds to thousands of revolutions. Yet, despite its dynamical feasibility, the question as to whether this mechanism is sufficient to maintain a steady-state population of JF comets is still open. A more serious objection is that the inclination distribution of comets captured from an isotropic, near-parabolic Oort cloud flux should include a significant fraction of JF comets in retrograde orbits, which is in contradiction with their observed flat i distribution (cf. Fig. 2c).

The Tisserand invariant derived from Jacobi's integral of the circular, restricted three-body problem provides another useful criterion for distinguishing among different source regions of JF comets. For a body moving on an orbit with orbital elements q, a, and i, the Tisserand invariant T is given by

$$T = \frac{1}{a} + 2\sqrt{2q\left(1 - \frac{q}{2a}\right)}\cos i \qquad (17)$$

which is valid under the assumptions that the perturbing planet (Jupiter) has a circular orbit of unit radius and that other planets do not perturb the body. As Fig. 11 shows, the Tisserand parameters of most JF comets with $q < 1.3$ AU have values in the range $2.5 < T < 3$. By contrast, HT and LP comets have Tisserand invariants $T < 2$. Consequently, should JF comets come from the capture of near-parabolic comets, they would have $T < 2$, since T is (more or less) conserved during the capture process. This argues against a field of near-parabolic comets from the Oort cloud as a suitable source.

Bearing in mind the dynamical objections expressed here, it may still be arguable that JF comets—at least most of them—come from near-parabolic comets captured by Jupiter. However, a disk or belt of comets beyond Neptune, possibly the remnants of the formation of the outer planets, may be a more suitable source (see next section). [See THE KUIPER BELT.]

Comets bound in short-period orbits are subject to physical loss mechanisms, such as splitting, disintegration, and deactivation, that will finally remove them from the population of active comets. Dust particles left behind on the nucleus surface after the sublimation of volatile ices can form an insulating dust mantle that chokes off further sublimation, so that the nucleus may become inert and acquire an asteroidal appearance. Several authors have suggested that Apollo and Amor (AA) asteroids, whose orbits cross or approach Earth's orbit, may be defunct cometary nuclei. Yet most AA asteroids move in orbits of long dynamical stability, whereas the orbits of JF comets are usually very unstable. Most AA asteroids are well inside Jupiter's orbit

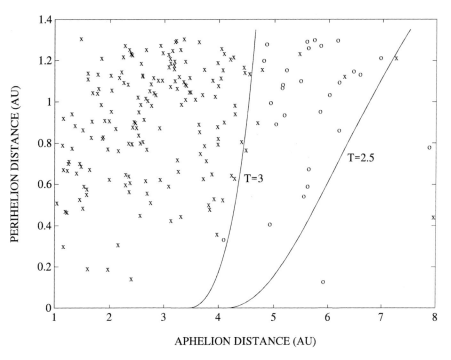

FIGURE 11 Plot of aphelion (Q) versus perihelion distances (q) for Apollo–Amor asteroids (x) and Jupiter-family comets with $q < 1.3$ AU (o). The solid curves are the loci in the (Q, q) plane, giving Tisserand parameters of $T = 2.5$ and $T = 3$ in the planar case.

(see Fig. 11) or avoid close encounters with Jupiter. Nevertheless, a few AA asteroids move in typical cometary orbits, and even an object of asteroidal appearance on a near parabolic orbit—1996PW—has recently been discovered. Whether these objects are bona fide asteroids scattered by Jupiter to highly eccentric orbits or comets is not known yet. [See ASTEROIDS; NEAR-EARTH ASTEROIDS.]

The encounter velocity U of a body with respect to Jupiter (assumed to be on a circular orbit) can be expressed in terms of the Tisserand invariant T as

$$U = (3 - T)^{1/2} \qquad (18)$$

which shows that encounters with Jupiter are possible only in the case $T < 3$. As discussed, the great majority of JF comets have $T < 3$, whereas the opposite holds for most AA asteroids (i.e., encounters with Jupiter are not possible), in accordance with the previous discussion on orbital stability. As Fig. 11 shows, the great majority of AA asteroids are located to the left of the curve $T = 3$.

The dynamical distinction between JF comets and AA asteroids should not be understood as a definitive proof in favor of distinct source regions for these two classes of bodies. For instance, if JF comet nuclei can build permanent dust mantles, they may have much longer physical lifetimes than expected for ice-exposed nuclei, which might allow some of them to decouple from Jupiter's orbit. Several possible mechanisms are able to shorten aphelion distances of inactive comets, such as nongravitational forces and perturbations by the terrestrial planets, facilitated perhaps by Jovian commensurabilities. At the end, it is probable that some AA asteroids are indeed defunct cometary nuclei, whereas the rest are bona fide asteroids, transferred from the asteroid belt to the region of the terrestrial planets through chaotic motion induced around mean motion and secular resonances (i.e., the resonances between the precession rates of the orbits of asteroids and of Jupiter).

V. THE TRANS-NEPTUNIAN BELT

Nearly a half century ago, the Irish astronomer Kenneth Essex Edgeworth and the Dutch-American astronomer Gerard P. Kuiper argued independently that icy planetesimals formed beyond Neptune could not grow to massive bodies given the long collisional timescales at such distances. Consequently, trans-Neptunian planetesimals would have been left unaccreted in a ring stretching between ~30 and 50 AU with a

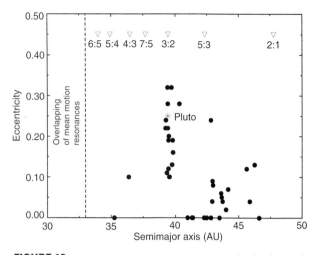

FIGURE 12 Semimajor axes versus eccentricities for the observed Edgeworth–Kuiper belt bodies. The orbital parameters were taken from Minor Planet Electronic Circulars 1996-Q06 and 1996-W03. The semimajor axes of the main mean-motion resonances, as well as Pluto's position in the parametric plane, are indicated.

structure resembling that of the asteroid belt. Such a trans-Neptunian population remained in the realm of theoretical speculation for several decades, despite several sky surveys. Until 1992, the binary system Pluto–Charon—discovered in 1930 and 1978, respectively—remained as the only known trans-Neptunian object. Pluto was more regarded as the ninth planet rather than a massive member of the trans-Neptunian population. Then another trans-Neptunian object—1992QB₁—was discovered by David Jewitt and Jane Luu with the 2.2-m telescope on Mauna Kea, Hawaii. This discovery was followed by more than 50 other trans-Neptunian objects in the short time span of five years. The trans-Neptunian population has been called the Kuiper Belt or, more properly, the Edgeworth–Kuiper (E–K) belt. [*See* KUIPER BELT.]

Preliminary computations of the orbits of the discovered trans-Neptunian objects indicate that ~15 of them (i.e., about 40%) move in Pluto-like orbits, that is, in the 3:2 mean-motion resonance with Neptune. This result is not surprising at all, since this mean-motion resonance (and others) provides a mechanism of protection from Neptune encounters. This explains why Pluto has survived over the solar system lifetime despite the fact that its orbit crosses that of Neptune, since this resonance prevents Pluto from getting closer than ~18 AU from Neptune. The locations of the discovered E–K Belt objects in the (eccentricity, semimajor axis) plane are shown in Fig. 12. [*See* PLUTO AND CHARON.]

All of the discovered trans-Neptunian objects are of relatively large size, with estimated diameters ranging

between 100 and 700 km if low geometric albedos are assumed. From the discovered objects, it is possible to estimate a population of 70,000 objects with diameters greater than 100 km in a belt between 30 and 50 AU. The inferred trans-Neptunian population turns out to be about two orders of magnitude more massive than the asteroid belt. Like the asteroid belt, the trans-Neptunian belt is also very flat, most inclinations are smaller than 10°, and only two so far have inclinations greater than 30°. The orbits of some E–K belt bodies are still quite uncertain, so they have been preliminarily placed on circular orbits (eccentricity $e = 0$).

Numerical investigations by several researchers show that the dynamical evolution in the E–K belt proceeds rather quickly for $a \lesssim 45$ AU, so it is expected that the primordial population of the E–K belt in the range $30 \lesssim a \lesssim 45$ AU has been heavily eroded, except for a few stability zones around some mean-motion resonances with Neptune. Secular resonances may play a key role in exciting high eccentricities in E–K belt bodies on initially near-circular orbits. Once in more eccentric orbits, such bodies can evolve toward Neptune-crossing orbits. Figure 13 depicts how an E–K belt body initially on a near-circular orbit at $a \sim 40$ AU evolves until it becomes a Neptune-crosser. First, it tends to evolve in eccentricity and perihelion distance, keeping its semimajor axis more or less constant [represented by a diagonal in the (q, e) plane]. When it becomes a Neptune-crosser, strong perturbations during close encounters with Neptune (it is in a nonresonant orbit) produce large changes in a and e, keep-

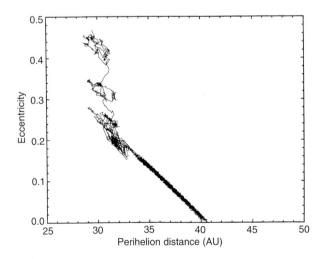

FIGURE 13 Dynamical evolution in the parametric plane (q, e) of a hypothetical Edgeworth–Kuiper belt body initially on a circular orbit located in the invariable plane. Lines of constant semimajor axis are diagonal. [From M. J. Holman and J. Wisdom (1993). *Astron. J.* 105, 1987. Used with permission.]

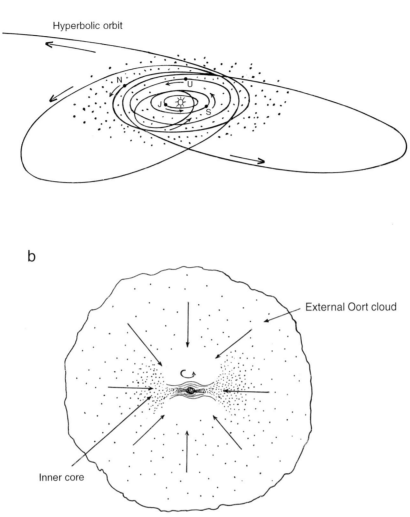

FIGURE 14 Two possible scenarios of comet origin: (a) comets were the residues of the outer planets' accretion from where they were scattered to the Oort cloud by gravitational perturbations, mainly by Neptune; (b) comets formed in the outer parts of the primitive solar nebula as it collapsed, stretching perhaps from the edge of Neptune's accretion zone to some 10^4 AU.

ing in this case q more or less constant [a vertical line in the (q, e) plane]. It is therefore not surprising to find a significant fraction of the discovered objects in the 3:2 resonance, since it provides a protection mechanism against close encounters with Neptune; it is expected that other bodies might remain trapped in other low-order resonances such as 4:3, 5:4, 6:5, and 5:3. The overlapping of mean-motion resonances often causes large-scale chaos, and this occurs in a zone close to Neptune ($a \lesssim 33$ AU) that is already strongly perturbed by close encounters with the planet. Thus, this inner edge of the E–K belt should have been cleared very early in the history of the solar system. Outside the stable libration regions around mean-mo-

tion resonances, the motion is highly chaotic and non-resonant bodies with semimajor axes $a \lesssim 45$ AU are expected to evolve to Neptune-crossing orbits on timescales smaller than the solar system's age.

At heliocentric distances greater than ~45 AU, the trans-Neptunian population is stable, and is expected to have remained almost undisturbed over the solar system's history. Consequently, this distant population might be very large. The fact that only four among the discovered trans-Neptunian objects have $a > 45$ AU is an artifact of the increasing difficulty of detecting more distant, fainter bodies.

Once E–K belt objects become Neptune-crossers in nonresonant orbits, their further evolution proceeds rather quickly. It is expected that about one-third of them will become Jupiter-crossers on timescales of ~10^7 yr. From the point of view of the transport rate of bodies from the E–K belt to the inner planetary region, as well as from the dynamical characteristics of the resulting population of JF comets (in particular, the resulting strong concentration of inclinations toward small values), the E–K belt fulfills the requirement as the main source of JF comets.

If JF comets come indeed from the E–K belt, there should be a large number of stray bodies in the outer planetary region, in transit from the outer to the inner planetary region. Actually, there have already been several discoveries of this class of objects, known now as Centaurs, starting with Chiron in 1977 and culminating with 1997 CU$_{26}$ in an orbit between Saturn and Uranus.

VI. DYNAMICAL ASPECTS OF COMETARY ORIGIN

There has been an almost two-centuries-old debate on whether comets formed in the solar system or in interstellar space. The latter theory can be traced back to Pierre-Simon de Laplace, who suggested that the Sun moved through an interstellar field of comets, capturing those members with low relative velocities. Almost at the same time, Joseph-Louis Lagrange proposed the theory that comets and meteors formed from eruptions of the giant planets, an idea that had been postulated by Johannes Hevelius in the sixteenth century.

The theory of interstellar comet origin had the appeal of explaining the isotropic distribution of their orbital planes. Yet the failure to observe comets with markedly hyperbolic orbits has been a strong argument against an interstellar origin (cf. Section III, A). This cleared the way for the opposite view that comets formed in the solar system, either as a by-product of planet formation or in the outer portions of the collapsing cloud of gas and dust that formed the Sun and the planets (known as the solar nebula). [See THE ORIGIN OF THE SOLAR SYSTEM.]

As the solar nebula contracted, it spun faster to conserve angular momentum, leading to a flattened structure. The material of low angular momentum eventually settled at the center, giving rise to the proto-Sun. This was surrounded by the material of higher angular momentum, which formed the planetary disk. Dust particles within the planetary disk accumulated into kilometer-sized bodies, that we call planetesimals,

either by gravitational instabilities or by mutual collisions. The formed planetesimals may have continued growing by mutual collisions into asteroid-sized bodies and, finally, protoplanetary objects that were the forerunners of the planets. Once the outer protoplanets acquired strong gravitational fields, they scattered the remaining unaccreted planetesimals from their influence zones. Most were ejected to interstellar space, but a fraction was stored in the Oort cloud. Given the low temperatures estimated for the outer portions of the solar nebula, it is likely that such planetesimals—or "cometesimals"—would consist of a mixture of different kinds of frozen volatiles or "snows" and dust particles in a way resembling the physicochemical structure of comet nuclei.

From a dynamical viewpoint, a probable comet source might be the outermost part of the planetary region, close to Neptune's orbit (Fig. 14a). Unaccreted cometesimals initially confined in Neptune's influence zone may have been scattered by its gravitational perturbations. The relatively smooth diffusion in energy phase space may have made it possible for most cometesimals to fall into the narrow range of near-zero energies [$0 < (1/a) \lesssim 10^{-4}$ AU^{-1}] that corresponds to Oort cloud comets. Once cometesimals reached such orbits, perturbations by passing stars and galactic tides played a fundamental role in removing cometary perihelia from the planetary region, thus preventing the scattered cometesimals from being further perturbed and finally ejected by the planets. By contrast, Jupiter's perturbations are so strong that cometesimals under its gravitational control probably jumped directly from elliptic orbits [$(1/a) \gtrsim 10^{-4}$ AU^{-1}] to hyperbolic ones [$(1/a)$ negative]. To a lesser extent, the same objection applies to Saturn. Uranus' contribution of scattered cometesimals to the Oort cloud may have been partially hindered because they fell first under the gravitational control of Jupiter and Saturn. (Neptune's efficiency was less affected by competing perturbations from these two largest planets owing to its greater distance.)

An alternative—though not necessarily contradictory—view of the formation of the Oort cloud is that comets formed during the earliest phases of the collapsing solar nebula, either *in situ* or in the inner core (Fig. 14b), thus avoiding the very costly process of dynamical transport from the planetary region to the Oort cloud. The cometary mass locked in such a primordial inner core might be very large. Dynamical or observational constraints may set a very loose upper limit of hundreds to thousands of Earth masses at present. Members of such a core might be very slowly released to the outer or classic Oort cloud and to the planetary region by the very rare, strong perturbations caused by very close stellar passages or penetrating encounters with GMCs.

There is still the possibility that the early galactic environment in which the solar system formed was quite different from the present one. In this regard, we have to bear in mind that most stars seem to form in clusters within molecular clouds. If the solar system indeed formed in such a dense environment like most stars, the much stronger external perturber field would have produced a much more tightly bound comet cloud. The natal molecular cloud and the star cluster would have dissipated shortly after the formation of the solar system, thus preventing the formed Oort cloud from being disrupted. This scenario offers a natural explanation for the formation of a tightly bound core of the Oort cloud, without having to invoke an *in situ* formation.

VII. SUMMARY

Cometary dynamics has developed greatly during the last few decades, thanks to the enhanced capability of computer power, new theoretical tools incorporated celestial mechanics such as the theory of chaos, a better knowledge of the external forces acting on Oort cloud comets, and various search programs for minor bodies leading to the discovery of new classes, such as the E–K belt objects and Centaurs. This field shows a great vitality with many challenging aspects still to be addressed, although we expect that the main features depicted here will stand the test of time.

In summary, the present comet reservoir surrounding the solar system may stretch from the vicinity of Neptune's orbit to 10^5 AU, passing from a flat structure—the Edgeworth–Kuiper belt—to a spherical one—the Oort cloud. Comets can be transferred to Earth's vicinity from both the Oort cloud and the E–K belt through different dynamical routes. In particular, the latter may be the source of the JF comets. From the foregoing discussion, we can define three regions of the Oort cloud (Fig. 15): (1) An inner core, with semimajor axes $a \lesssim 3000\text{–}4000$ AU, that might be concentrated toward the ecliptic plane if comets formed in the protoplanetary disk and were later scattered by the planets. Its mass should be constrained to a few tens of Earth masses by the

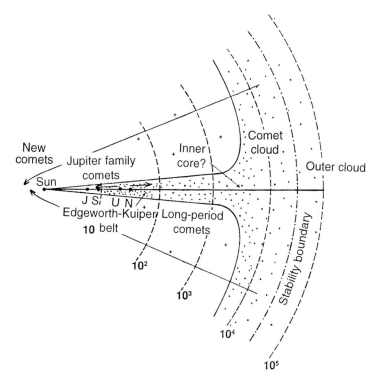

FIGURE 15   Diagram depicting the possible spatial distribution of comets around the solar system if they formed in the protoplanetary disk. A comet disk might still be present, stretching to a few 10^3 AU; its inner edge close to Neptune's orbit may be identified with the Edgeworth–Kuiper belt. Beyond a few 10^3 AU, the action of external perturbers should have by now randomized the orbital planes of comets, leading to a spherical structure. The core of a few 10^3 AU in radius should contain evolved long-period comets (crosses) and perhaps an as yet undetected more tightly bound population of the Oort cloud. The dashed lines indicate distances to the Sun in AU.

limited capacity of Neptune and the other giant planets to remove matter from their influence zones. On the other hand, its mass could be considerably larger (up to several hundred Earth masses) if comets formed *in situ.* The flattened inner portions of the core closer to the planetary region may be identified with the Edgeworth–Kuiper belt, which may be the source of the Jupiter-family comets. Note, however, that the E–K belt comets were likely formed *in situ* in their current orbits, unlike the Oort cloud comets which were likely ejected from the outer planets region. The inner core constitutes a reservoir for the outer portions of the Oort cloud. (2) A spherical, stable Oort cloud for $4000 \lesssim a \lesssim 30{,}000$ AU, containing comets that are dynamically stable over periods comparable to or larger than the age of the solar system (4.6×10^9 yr). (3) An outer Oort cloud, for $a \gtrsim 30{,}000$ AU, whose population has dynamical lifetimes shorter than the solar system age; it therefore needs to be continuously replenished, with comets coming from the inner regions of the Oort cloud.

SEE ALSO THE FOLLOWING ARTICLE

PHYSICS AND CHEMISTRY OF COMETS; CHAOTIC MOTION IN THE SOLAR SYSTEM

BIBLIOGRAPHY

Bailey, M. E., Clube, S. V. M., and Napier W. M. (1990). "The Origin of Comets." Pergamon, Oxford, England.

Carusi, A., and Valsecchi, G. B. (eds.) (1985). "Dynamics of Comets: Their Origin and Evolution," IAU Coll. No. 83. Reidel, Dordrecht.

Ferraz-Mello, S. (ed.) (1992). "Chaos, Resonance and Collective Dynamical Phenomena in the Solar System," IAU Symp. No. 152. Kluwer, Dordrecht.

Newburn, R. L., Neugebauer, M., and Rahe, J. (eds.) (1991). "Comets in the Post-Halley Era," IAU Coll. No. 121. Kluwer, Dordrecht.

Smoluchowski, R., Bahcall, J. N., and Matthews, M. S. (eds.) (1986). "The Galaxy and the Solar System." Univ. Arizona Press, Tucson.

Weissman, P. R. (1995). The Kuiper belt. *Annu. Rev. Astron. Astrophys.,* **33,** 327–357.

Yeomans, D. K. (1991). "Comets. A Chronological History of Observation, Science, Myth, and Folklore." Wiley Science Editions, New York.

THE KUIPER BELT

I. Historical Perspective

II. Basic Orbital Dynamics

III. Observations of the Kuiper Belt and Related Structures

IV. Physical Observations of Trans-Neptunian Objects

V. Long-Term Stability of Orbits in the Kuiper Belt

VI. Sizes of Trans-Neptunian Objects and the Total Mass of the Kuiper Belt

VII. Ecliptic Comets and the Scattered Disk

VIII. Formation of the Kuiper Belt and the Scattered Disk

IX. Concluding Remarks

Harold F. Levison
Southwest Research Institute

Paul R. Weissman
Jet Propulsion Laboratory, California Institute of Technology

GLOSSARY

Albedo: Fraction of incident light reflected from a surface. An albedo of 1.0 indicates a surface that is totally reflecting; an albedo of zero indicates a surface that is totally absorbing.

Astronomical unit (AU): Formally, the distance from the Sun to a massless object in an unperturbed circular orbit that has an orbital period of 365.2568983 days, equal to $1.4959787066 \times 10^{11}$ m. The AU is commonly taken to be the average distance of the Earth from the Sun, and is the unit of distance used by planetary astronomers.

Chaotic motion: Dynamical situation where the error in the prediction of the long-term motion of a body grows exponentially over time. This exponential growth leads to an inability to predict the long-term behavior of the body. Chaotic motion can be confined to fairly narrow regions of space so that the orbit of the body will not

change much over time. However, this is unusual for objects in the solar system. Most often chaotic motion will lead to sudden and drastic changes in the orbit of a body.

Centaur: Body in a heliocentric orbit with an average distance from the Sun that lies between the orbits of Jupiter and Neptune. Typically the orbits of these objects also cross one or more of the orbits of the giant planets Saturn, Uranus, and Neptune. Most Centaurs have likely been perturbed out of the Kuiper Belt and/or the scattered disk, and some fraction of them will evolve to and appear as Jupiter-family comets. Centaurs are part of the ecliptic comets population.

Eccentricity: Measure of the departure of an elliptical orbit from circularity. The eccentricity, e, is equal to $(1 - b^2/a^2)^{1/2}$, where a and b are the major and minor axes of the elliptical orbit. Circular orbits have $e = 0$, elliptical orbits have

$0 < e < 1$, radial and parabolic orbits have $e = 1$, and hyperbolic orbits have $e > 1$.

Ecliptic: Plane of the Earth's orbit around the Sun. The planets, most asteroids, most of the short-period comets, and the Kuiper Belt are all in orbits with small or moderate inclinations (or tilts) relative to the ecliptic.

Ecliptic comet: Object that has escaped the Kuiper Belt or the scattered disk and is being scattered inward to an orbit crossing one or more of the planets. The designation of such objects as comets assumes an icy composition because of their likely formation far from the Sun in the colder regions of the solar nebula. Ecliptic comets include Centaur objects and Jupiter-family comets.

Inclination: Measure of the orientation of an orbit. The inclination is the angle between the plane of the orbit and some reference plane, usually the ecliptic.

Invariable plane: Plane passing through the barycenter (center-of-mass) of the solar system and perpendicular to the total angular momentum vector of the solar system.

Jupiter-family comet: Active comet in a low to moderate inclination orbit with a semimajor axis less than that of Jupiter's orbit. Most Jupiter-family comets are in orbits that cross or can closely approach Jupiter's orbit. Jupiter-family comets are a subset of the ecliptic comets.

Kuiper Belt: Collection of a billion or more bodies in low eccentricity, low inclination orbits beyond Neptune, extending out possibly as far as 1000 astronomical units.

Longitude of the ascending node: Measure of the orientation of an orbit. The *nodes* of an orbit are the points where the orbit crosses some reference plane, usually the ecliptic. The *ascending node* is where the nodal crossing is in the upward direction (where "upward" is defined by some reference direction, typically the angular momentum vector of the Earth's orbit). The *longitude of the ascending node* is the angle between the location of the ascending node and some standard direction in the reference frame, usually the direction of the vernal equinox.

Mean motion resonance: Dynamical situation where the ratio of the orbital periods of two orbiting objects can be expressed as the ratio of two small integers. Mean motion resonances can

lead to strong changes in the orbit of one or both of the bodies or can actually enhance orbital stability, depending on the precise nature of the resonance.

Perihelion: Location in an orbit of closest approach to the Sun.

Planetesimal: Small body formed in the early solar system by accretion of dust and ice (if present) near the central plane of the solar nebula.

Precession: Slow, smooth increase or decrease of an angle.

Scattered disk: Collection of icy planetesimals in high eccentricity orbits in the ecliptic plane beyond Neptune. Objects in the scattered disk may be escapees from the Kuiper Belt and/or may be scattered Uranius–Neptune planetesimals.

Secular resonance: Dynamical situation where there is a commensurability among the frequencies associated with the rates of precession of the argument of perihelion or of the regression of the nodal line of two bodies. Secular resonances can lead to very large changes in the orbit of one or more of the bodies.

Semimajor axis: One-half of the major axis of an elliptical orbit. The semimajor axis is commonly thought of as the average distance between an object and the body it is in orbit about.

Trojan asteroids: Asteroids locked in a 1:1 mean motion resonance with Jupiter and librating about the L_4 and L_5 Lagrange points, 60° ahead and behind Jupiter in its orbit. Trojan-type orbits are also possible for objects in 1:1 resonances with other planets. A Trojan-type librating asteroid has been found for Mars.

Vernal equinox: Direction of the Sun as viewed from the Earth as it crosses the celestial equator moving northward. Denotes the beginning of spring in the northern hemisphere.

I. HISTORICAL PERSPECTIVE

Since its discovery in 1930, Pluto has traditionally been viewed as the last vestige of the planetary system—a lonely outpost at the edge of the solar system, orbiting beyond Neptune with a 248-year period. Pluto receives very little light from the Sun (being almost 40 times farther from the Sun on average than the Earth) and thus it is very cold. The view was that it was a distant, isolated and unfriendly place, with nothing of substance beyond it.

Pluto itself has always appeared to be an oddity among the planets. Traditionally, the planets are divided into two main groups. The first group, the terrestrial planets, formed in the inner regions of the solar system where the material from which the planets were made was too warm for water and other volatile gases to be condensed as ices. These planets, which include the Earth, are small and rocky. Farther out from the Sun, the cores of the planets grew from a combination of rock and condensed ices, and captured significant amounts of nebula gas. These are the Jovian planets, the giants of the solar system; they most likely do not have solid surfaces. Then there is Pluto, unique, small (its radius is only ~1170 km), and made of a mix of rock and frozen ices.

The planets formed in a disk of material that originally surrounded the Sun. As the Sun formed from the collapse of its parent molecular cloud, it faced a problem. The cloud had a slight spin and as it collapsed, the spin rate had to increase in order to conserve angular momentum. The cloud could not form a single star with the amount of angular momentum it possessed, so it shed a disk of material that contained very little mass (as compared with the mass of the Sun), but most of the angular momentum of the system. As such, the planets formed in a narrow disk structure; the plane of that disk is known as the invariable plane. Then there is Pluto, unique, having an orbital inclination of 15.6° with respect to the invariable plane.

The orbits of the planets are approximately ellipses with the Sun at one focus. As the planets formed in the original circumsolar disk, they tended to evolve onto orbits that were well separated from one another. This was required so that their mutual gravitational attraction would not disrupt the whole system. (Or to put it another way, if our system had not formed that way we would not be here to talk about it!) Then there is Pluto, unique, having an orbit that crosses the orbit of its nearest neighbor, Neptune.

So, the historical view was that Pluto was an oddity in the solar system. Unique for its physical makeup and size as well as its dynamical niche. This view, however, changed in September 1992 with the announcement of the discovery of the first of a population of small (compared to planetary bodies) objects orbiting beyond the orbit of Neptune, in the same region as Pluto. Since that time over 60 objects with radii between about 50 and 500 km have been discovered. Given the very small area of the sky that has been searched so far, these 60+ objects imply that there are

approximately 70,000 similar objects occupying this region of space, between 30 and 50 AU from the Sun. There are almost certainly many more smaller ones. As discussed in more detail in the following sections, these objects likely have a similar physical makeup to that of Pluto and many have similar orbital characteristics. Thus, Pluto has been transformed from an oddity, to the founding member of what is perhaps the most populous class of objects in the planetary system.

The discovery of the Kuiper Belt, as it has come to be known, represents a revolution in our thinking about the solar system. First predicted on theoretical grounds and later confirmed by observations, the Kuiper Belt is the first totally new class of bodies to be discovered in the solar system since the first asteroid was found on New Year's day in 1801. Its discovery is on a par with the discovery of the solar wind and the planetary magnetospheres in the 1950s and 1960s, and it has radically changed our view of the outer solar system.

Speculation on the existence of a trans-Neptunian disk of icy objects dates back to the late 1940s. In the 1940s and early 1950s, Kenneth Edgeworth and Gerard Kuiper independently considered the structure of our planetary system. They noticed that if one were to grind up the giant planets and smooth out their masses to form a disk, this disk would have a very smooth distribution, with a density that decreased as the distance from the Sun. That is until Neptune, at which point there is an apparent edge, beyond which there was thought to be nothing except tiny Pluto. Edgeworth and Kuiper suggested that perhaps this edge was not real. Perhaps the disk of planetesimals that formed the planets extended past Neptune, but that the density was too low or the formation times too long to form large planets. If so, they argued, these planetesimals should still be there in nearly circular orbits beyond Neptune. Unfortunately, Edgeworth's contribution was overlooked until recently, and thus this disk has come to be known as the Kuiper Belt.

The idea of a trans-Neptunian disk received little attention for many years. Objects in the hypothetical disk were too faint to be seen with the telescopes of the time, so there was no way to prove or disprove their existence. Comet dynamicists showed that the lack of detectable perturbations on the orbit of Halley's comet limited the mass of such a disk to no more than 1.3 Earth masses ($M_\oplus$) if it was 50 AU from the Sun.

However, the idea was resurrected in 1980 when Julio Fernandez proposed that a cometary disk beyond Neptune could be a possible source reservoir for the short-period comets (those with orbital periods <200 years). Subsequent dynamical simulations showed that a comet belt beyond Neptune is the most plausible source for the low inclination, Jupiter-family comets. This work led observers to search for Kuiper Belt objects, the first one being discovered by David Jewitt and Jane Luu in August 1992. Since then, over 60 Kuiper Belt objects with estimated radii $\gtrsim 50$ km (assuming a typical cometary nucleus albedo of 0.04) have been found by ground-based searches, and ~ 30 comet-sized (radii $\lesssim 10$ km) objects have been detected using the Hubble Space Telescope. (Because the Kuiper Belt is believed to be the source of the Jupiter-family short-period comets, typical cometary albedos of 4% are assumed for the surfaces of Kuiper Belt objects.)

Two populations of objects in the solar system are related to the Kuiper Belt. The first consists of a large number ($\sim 10^6$) of (presumably) icy objects on planet-crossing orbits interior to Neptune's orbit. These objects are known as ecliptic comets and they most likely originated in the Kuiper Belt. Ecliptic comets include two distinct subpopulations: Centaurs and Jupiter-family comets. Known Centaurs are 20- to 100-km-radius objects with semimajor axes beyond the orbit of Jupiter, most of which appear to be inactive (the one known exception is the first and largest Centaur discovered, 2060 Chiron). Jupiter-family comets are much smaller objects a few kilometers in diameter on Jupiter-crossing orbits but with semimajor axes interior to Jupiter. Most of the known Jupiter-family comets are active comets as active comets are much brighter and hence easier to detect than inactive ones.

Both Centaurs and Jupiter-family comets are just the brightest members of the ecliptic comets, i.e., the most easily discovered. Active Jupiter-family comets are bright because they are relatively close to the Earth and because they produce active cometary comae. Centaurs are bright because they are relatively big.

The second related population of objects is known as the scattered comet disk. It was suggested by a number of theoretical researchers in the 1980s, predicted by numerical models in the 1990s, and then discovered by observers in 1996. The scattered disk occupies the same volume of physical space as the Kuiper Belt, but as discussed later, it has a different dynamical character and a different origin.

This article discusses the available information on the Kuiper Belt, scattered disk, and Centaurs (for a discussion of the Jupiter-family comets.) Observational information comes in the form of (i) orbits determined for the discovered objects, (ii) sizes of the objects estimated from their brightnesses, and (iii) physical observations (e.g., broad-band colors) of the objects. In addition, there is a vast body of theoretical studies of the trans-Neptunian region based primarily on computer-based simulations. These simulations have provided considerable insight into the motion, spatial distributions, size distribution, and evolution of bodies in this region, as well as their relationship to other solar system populations. [See COMETARY DYNAMICS; PHYSICS AND CHEMISTRY OF COMETS.]

Why study the Kuiper Belt and the scattered disk? It is rare in the history of astronomy that a whole new region of the solar system is discovered that needs to be understood. For that reason alone, the Kuiper Belt deserves study. Indeed, as the following sections illustrate, the structure of the orbits of objects in the trans-Neptunian region is much more complex, and therefore more interesting, than would have been dreamed of only a few years ago.

It has also become clear that not only is this region interesting because of its novelty and complexity, but also because it has supplied (and most likely will continue to supply) us with important clues about the formation of the solar system. The discovery of dust disks around nearby stars by the IRAS satellite in 1983–1984 suggests that Kuiper Belts may be a natural consequence of star and planet formation, thus providing us with a tool for studying solar system formation around other stars.

II. BASIC ORBITAL DYNAMICS

Much of the story of the Kuiper Belt to date involves the distribution of the orbits of its members. This section presents a brief overview of the important aspects of the dynamics of small bodies in the solar system. [See SOLAR SYSTEM DYNAMICS.]

The most basic problem of orbital dynamics is the two-body problem: a planet, say, orbiting a star. In this case, the orbit of the planet is constrained to lie in a single plane. The orbit's trajectory is an ellipse with the Sun at one of the foci. Energy, angular momentum, and the orientation of the ellipse are conserved quantities. The semimajor axis, a, of the ellipse is a function of the orbital energy. The eccentricity, e, of the ellipse is a function of the energy and the angular momentum. For a particular semimajor axis, the angular momentum is a maximum for a circular orbit, $e = 0$. These two-body orbits are known as Kepler orbits.

In the real solar system there are nine planets and many smaller bodies, each acting to gravitationally perturb the orbits of the others. However, the Sun is much more massive than any of the planets. As a result, the orbits of objects in the solar system can be viewed as slightly perturbed Keplerian orbits about the Sun (unless the object in question gets particularly close to another perturber, such as a planet). Such an orbit is characterized by an instantaneous Keplerian orbit, which is described by its semimajor axis and eccentricity, as well as by three angles that describe the orientation of the orbital ellipse in space. The first, known as the inclination, i, is the angle between the angular momentum vector of the orbit and some reference direction for the system. In our solar system, the reference direction is usually taken as the angular momentum vector of the Earth's orbit (which defines the ecliptic plane) but is sometimes taken to be the angular momentum vector of all the planetary orbits combined (which defines the invariable plane).

The point where the orbit passes through the reference plane in an "upward" direction is called the ascending node. Here the reference plane is the plane that is perpendicular to the reference angular momentum vector and "upward" is in the direction along the reference angular momentum vector. The second orientation angle of the orbit is the angle between the ascending node and some reference direction in the reference plane, as seen from the Sun. In our solar system, the reference direction is usually taken to be the direction toward the vernal equinox. This angle is known as the longitude of the ascending node, Ω.

The third and final orientation angle is the angle between the ascending node and the point where the orbit is closest to the Sun (known as perihelion), as seen from the Sun. It is called the argument of perihelion, ω. Another useful angle, known as the longitude of perihelion, $\tilde{\omega}$, is defined to be $\omega + \Omega$.

The first-order gravitational effect of the planets on one another is that each applies a torque on the other's orbit, as if the planets were replaced by rings of material distributed smoothly along their orbits. This torque causes both the longitude of perihelion, $\tilde{\omega}$, and the longitude of the ascending node, Ω, to precess. For a given planet the precession of $\tilde{\omega}$ is typically dominated by one frequency. The same is true for Ω, although the dominant frequency is different. The periods associated with these frequencies range from 4.6×10^4 to 2×10^6 years in the outer planetary system. This is much longer than the orbital periods of the planets (which are all less than 250 years).

The orbit of a small object in the solar system, when it is not being strongly perturbed by a close encounter with a planet or is not located near a resonance (see later), is usually characterized by a slow oscillation in e and i and a circulation (i.e., continuous change) in $\tilde{\omega}$ and Ω. The variation in the eccentricity is coupled

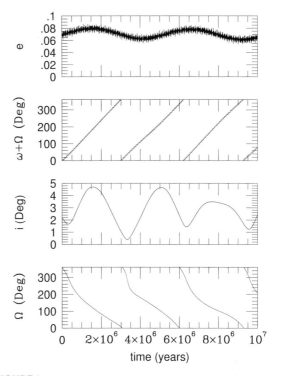

FIGURE 1 Temporal evolution of the orbit of the first Kuiper Belt object found, 1992 QB$_1$. As described in the text, the eccentricity, e, and inclination, i, oscillate, whereas the longitude of the ascending node, Ω, and the longitude of perihelion $\tilde{\omega} = \omega + \Omega$ circulate.

with the $\tilde{\omega}$ variation and the variation in the inclination is coupled with the Ω variation. Figure 1 shows this behavior for the first discovered Kuiper Belt object, 1992 QB$_1$.

The behavior of objects that are in a resonance can be very dramatic. Two types of resonances are known to be important in the Kuiper Belt. The most basic is known as a mean motion resonance. A mean motion resonance is a commensurability between the orbital period of two objects, i.e., the ratio of the orbital periods of the two bodies in question is a ratio of two (usually small) integers. Perhaps the most well-known and important example of a mean motion resonance in the solar system is the one between Pluto and Neptune.

As noted earlier one of the unique aspects of Pluto's orbit is that when Pluto is at perihelion, it is closer to the Sun than Neptune. Normally, this configuration would soon lead to close encounters between the two planets that would eventually scatter Pluto (i.e., perturb it to a very different orbit) and, eventually, most likely out of the solar system. (Because Pluto is much less massive than Neptune, conservation of energy and angular momentum would lead to Pluto's orbit undergoing radical changes, and not Neptune's.) However,

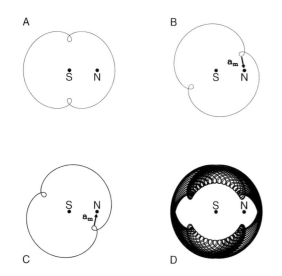

FIGURE 2 Mechanics of the 2 : 3 mean motion resonance between Neptune and Pluto. The double-lobed curve represents the orbit of Pluto as seen in a coordinate frame that rotates at the average speed of Neptune. Thus, Neptune is almost stationary. As drawn, the frame is rotating counterclockwise. Locations of the Sun and Neptune are shown as dots and are labeled S and N, respectively. (A) The orbit of an object exactly at the resonance. The gravitational perturbations of Neptune cancel out due to the symmetry in the geometry. Thus, this orbit is stable. (B) If the symmetry is broken then there is a net acceleration due to Neptune. Here, the strongest perturbation (a_m) is at the upper lobe. Pluto is leading Neptune here, so the net acceleration will decrease Pluto's semimajor axis and it will start to precess clockwise. (C) The strongest perturbation is in the lower lobe. Pluto will start to precess counterclockwise. (D) The orbit of an object that *librates* in the resonance.

close encounters do not occur because Pluto is locked in a mean motion resonance where it goes around the Sun twice every time Neptune goes around three times. This resonance is known as the 2 : 3 mean motion resonance.

Figure 2A illustrates how the 2 : 3 resonance works. The trajectory of Pluto is shown in a frame that rotates with Neptune. The figure is constructed so that in a nonrotating frame, all the planets would move in a counterclockwise direction. However, in the rotating frame shown, Neptune (N) and the Sun (S) are fixed; their locations are shown as dots. Pluto moves in a clockwise direction when further from the Sun than Neptune and moves in a counterclockwise direction when closer to the Sun. In Fig. 2A, an object with Pluto's eccentricity and exactly at Neptune 2 : 3 mean motion resonance would have a trajectory that is a double-lobed structure oriented so that the loops (which are the object's perihelion passages) are always 90° from Neptune. The configuration shown in Fig. 2A will remain fixed only if the object is *exactly* at the

location of the resonance. This is not the case for Pluto. In Pluto's case, the double-lobed structure will slowly precess in the rotating frame, possibly leading to close encounters with Neptune. What makes resonances like this one important is that there are stabilizing forces that lock nearby orbits into the resonances. In other words, the resonances are stable.

How is the resonance stabilized in Pluto's case? Suppose the semimajor axis of Pluto was slightly larger than the exact location of the resonance. Pluto would be moving slightly too slowly to maintain the configuration in Fig. 2A. In other words, it would lag behind an object that was exactly at the resonance and Pluto's double-lobed trajectory would precess clockwise, as shown in Fig. 2B. Because of the symmetry of the exact resonance (Fig. 2A) the forces on Pluto by Neptune average out over long periods of time. This is not true for the new configuration. Indeed, Pluto would receive its largest acceleration (a_m) from Neptune when in or near the upper lobe in Fig. 2B. At this point, Pluto is leading Neptune in their orbits and thus Pluto will be slowed down, decreasing its semimajor axis. Remember, this scenario started out with Pluto's semimajor axis being too large. As Pluto's semimajor axis decreases, its precession rate in the figure decreases, goes through zero (when its semimajor axis is equal to that of the resonance), and starts to move the other way (when its semimajor axis is less than that of the resonance).

Now Pluto's semimajor axis is smaller than the resonance and the double-lobed trajectory is precessing in the counterclockwise direction. The trajectory moves through the point where the lobes are at 90° to the Sun–Neptune line and evolves into a configuration similar to the one shown in Fig. 2C. In this case, Pluto gets its largest acceleration when it is near perihelion and is trailing Neptune in their orbits (near the lower lobe of the trajectory). Thus, Pluto's velocity is increased, increasing its semimajor axis. As Pluto's semimajor axis increases, its precession rate in the figure decreases, goes through zero (when its semimajor axis is equal to that of the resonance), and starts to move the other way (when its semimajor axis is greater than that of the resonance). The end result is that the double lobe structure oscillates or "librates" back and forth in a frame that rotates with Neptune, as shown in Fig. 2D. Therefore, this orbit is stable.

Mean motion resonances of small bodies are identified by the planet involved and by the two integers that define that ratio of orbital periods. The two integers are separated by a colon, the one associated with the small body appears first. For example, Pluto orbits the Sun twice (2) every time Neptune orbits the Sun

three (3) times. So, Pluto is said to be in Neptune's
2 : 3 mean motion resonance.

The other type of resonance that is important in
the Kuiper Belt is called a secular resonance. A secular
resonance occurs when there is a commensurability
between the frequencies associated with the precession
rates of either $\tilde{\omega}$ or Ω of two objects. A small body in
a secular resonance can undergo drastic changes in its
eccentricity and inclination. Indeed, it has been shown
that secular resonances can drive objects into the Sun
(i.e., $e \to 1$).

To understand how such extreme behavior can
come about, imagine a system with two planets and
one small body. For simplicity, assume that the orbits
of these three objects are in the same plane. The principal effect of the two planets on one another is to cause
their orbits to precess. The orbit of the small body
will also precess. Figure 3A shows the long-term trajectories of these objects in a fixed frame. Figure 3B shows
the same system in a frame that rotates with the precession rate of the small body. Note that the orbit of the
secular small body (the outermost orbit) is now an
ellipse. The trajectories of the two planets are still, on
average, axisymmetric and thus the small body experiences no long-term torques.

However, if the small body is in a secular resonance
with one of the planets, it precesses at the same rate
as the planet. For example, Fig. 3C shows a system
where the small body (outermost orbit) is in a secular
resonance with the inner planet. The system is shown
in a frame that rotates with the small body's precession
rate. In this case, the long-term trajectory of the planet
is a fixed ellipse and is no longer axisymmetric. Thus
the small body feels a significant long-term torque,
which can lead to a significant change in the eccentricity (which is related to the angular momentum) of the
small body.

There are actually two types of secular resonances.
The first, which was discussed earlier, is a resonance
between the precession rates of the longitudes of perihelion. As discussed, this can lead to changes in eccentricity. These resonances are identified by the Greek
letter ν with a numbered subscript that indicates the
resonant planet (one for Mercury through nine for
Pluto). In the Kuiper Belt, the perihelion secular resonance with Neptune, or ν_8, is most important. The
other type of secular resonance occurs when the small
body's nodal precession rate is the same as for a planet.
This type of resonance can cause significant changes
in the inclination of the orbit of the small body. These
resonances are identified by ν_{1x}, where x is the number
of the resonant planet. For example, the nodal resonance with Neptune is ν_{18}.

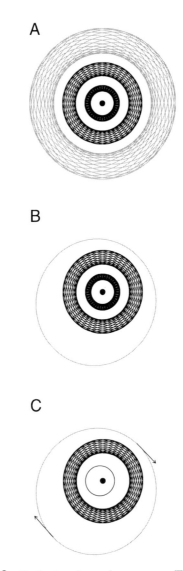

FIGURE 3 Mechanics of a secular resonance. Three orbits are
shown in each panel. The inner two are planets, which are shown
as solid lines. The outer orbit (dashed line) is for an object that is
very small compared to either of the planets. The orbits of each
object are ellipses, and the ellipses are precessing due to the mutual
gravitational effects of the planets. (A) The orbits of the objects
over a period of time that is long compared to the precession time
of the orbits. Here, we are looking in a fixed, nonrotating reference
frame. Each orbit sweeps out a torus of possible positions. (B) The
same as A, except that we are looking in a frame that rotates at the
precession rate of the small outer body. Thus, its orbit is again an
ellipse. This panel shows the geometry if no secular resonance exists.
Note that the trajectories of the planets look axisymmetric. Therefore, there is no net torque on the outer small object. (C) Same as
B except that the outer object is in a secular resonance with the
inner planet, i.e., both orbits precess at the same rate. As a result,
the outer object no longer sees an axisymmetric gravitational perturbation from the inner planet. Indeed, it feels a significant torque.

The dynamical structure of the Kuiper Belt and the scattered disk has been sculpted by a combination of mean motion and secular resonances and by the evolution of these resonances during the formation of Uranus and Neptune. In order to illustrate this point, we must first discuss what is known of the objects in the Kuiper Belt and scattered disk. Observational data are reviewed in the next section.

III. OBSERVATIONS OF THE KUIPER BELT AND RELATED STRUCTURES

The greatest areal search for trans-Neptunian objects is that by Clyde Tombaugh, which covered the entire northern sky photographically to a visual magnitude (V, yellow-green in color) of 15. Tombaugh succeeded in discovering Pluto in 1930. Charles Kowal searched ~16% of the sky photographically, generally near the ecliptic, to a visual magnitude of ~20, and discovered the first outer solar system, planet-crossing object (other than Pluto and recognized comets), 2060 Chiron, in 1977. Chiron's orbit crosses Saturn's with a perihelion of 8.45 AU and an aphelion of 19.0 AU, just inside the orbit of Uranus. As of January 1998, six additional outer solar system, large, planet-crossing objects have been discovered. These objects are known as Centaurs.

Orbital data and estimated radii, R, for the seven Centaurs are given in Table I. The fact that these objects are on orbits that cross the orbit of one or more planets and the fact that they are not trapped in resonances im-

TABLE I
Characteristics of Known Centaurs[a]

Name	a (AU)	e	i (deg)	R (km)
Chiron	13.6	0.38	6.9	150
Pholus	20.2	0.57	24.7	120
Nessus	24.5	0.52	15.7	40
1995 GO	18.0	0.62	17.6	50
1994 TA	16.9	0.30	5.4	20
1995 DW2	24.9	0.24	4.2	50
1997 CU26	15.7	0.17	23.4	150

[a] As of January 1, 1998.

plies that they will, in time, suffer a close encounter with one of these planets. Such encounters will lead to drastic changes in the orbits of these objects. Indeed, these objects will bounce around the solar system, gravitationally scattering off one planet after another until they finally reach their ultimate fate of (i) being thrown out of the planetary system altogether, (ii) impacting one of the planets, or (iii) hitting the Sun. On average, it will take between ~10^5 and 10^8 years for these objects to reach their final states with a median lifetime of 5×10^7 years. The maximum inclination among the seven known Centaurs is 25° for 5145 Pholus, suggesting that their source reservoir is likely in the ecliptic plane. As will be discussed in more detail later, the most likely source is the Kuiper Belt.

The first successful detection of an object beyond the orbit of Neptune (other than Pluto and its satellite Charon) was by David Jewitt and Jane Luu. Using a CCD camera on the 2.2-m University of Hawaii telescope, they searched ~1 square degree of sky to a visual magnitude of 24 and found object 1992 QB$_1$ in August 1992. The discovery image is shown in Fig. 4. At the time 1992 QB$_1$ was at a heliocentric distance of 41.2 AU. The object had an R (red) magnitude of 23.5, was reddish in color (see next section), and stellar in appearance with no evidence of cometary coma. The radius of this object can be estimated from the following formula:

$$R^2 = 4.53 \times 10^5 \, r^2 \, \Delta^2 p^{-1} \, 10^{-0.4V},$$

where R is the radius in kilometers, r is its heliocentric distance in AU, Δ is its distance from the Earth in AU, p is the fraction of incoming light that the object reflects (known as the geometric albedo), and V the object's visual magnitude. If 1992 QB$_1$ has a typical cometary albedo of 0.04 (the albedo of 1992 QB$_1$ and other Kuiper Belt objects have not yet been measured), then it has a radius of ~120 km. Subsequent observations allowed Brian Marsden to determine an orbit for 1992 QB$_1$ with a semimajor axis of 43.8 AU, an eccentricity of 0.088, an inclination of 2.2°, and an orbital period of 290 years. The perihelion distance of 40.0 AU is well beyond the orbit of Neptune. Long-term dynamical simulations suggest that orbits like that of 1992 QB$_1$ are stable over the age of the solar system.

The second trans-Neptunian object, designated 1993 FW, was discovered in March 1993 at a heliocentric distance of 42.1 AU. 1993 FW is similar in size to 1992 QB$_1$ (possibly slightly larger), but less red in color and, again, stellar in appearance. A subsequent orbit solution found $a = 43.9$ AU, $e = 0.041$, $i = 7.7°$,

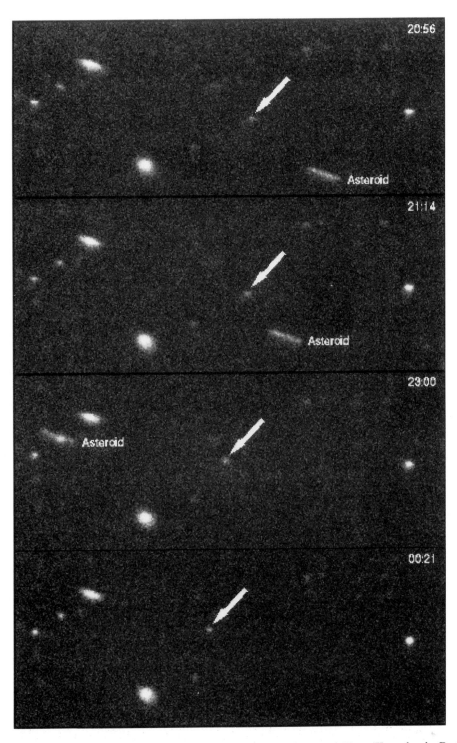

FIGURE 4 Discovery images of 1992 QB₁ with the University of Hawaii 2.2-m telescope on Mauna Kea, taken by David Jewitt and Jane Luu. The Kuiper Belt object is about R magnitude 23.5. Numbers at the upper right of each image refer to the time of day. The arrow indicates 1992 QB₁, which moves slowly through the field of view. (Courtesy of David Jewitt and Jane Luu.)

and $P = 291$ years. Again, this orbit is expected to be stable over the age of the solar system. Thus, it appeared that the Kuiper Belt had a dynamical structure similar to what most researchers expected at the time, i.e., Kuiper Belt objects were in random, low-inclination, low-eccentricity orbits about the Sun.

A real surprise came with the discovery of the next four objects. These objects were significantly different in that their heliocentric distances at discovery were substantially closer to the Sun, in a region where the orbits cannot be stable unless protected by some dynamical mechanism. The four objects, 1993 RO, 1993 RP, 1993 SB, and 1993 SC, were found at heliocentric distances ranging from 32.3 to 35.4 AU. Interestingly, all four objects were approximately 60° from Neptune in the sky, suggesting a possible Trojan-type (1:1 mean motion resonance) dynamical relationship. However, further observations allowed Brian Marsden to show that all but 1993 RP (for which there are insufficient observations and which is now considered lost) are trapped in the 2:3 mean motion resonance with Neptune, similar to the orbit of Pluto.

Continued ground-based searches have now discovered a total of 60 trans-Neptunian objects, which are listed in Table II, in order of discovery. The columns in the table are the semimajor axis, eccentricity, inclination, orbital period, and an estimated radius, based on an assumed cometary albedo of 0.04. Those objects marked with a superscript c have been discovered too recently to have accurate orbits determined. Those marked with a superscript b do not have accurate orbits and are presumed lost. Many of these are assumed to be on circular orbits if they are beyond 40 AU and on orbits locked in mean motion resonances inside of 40 AU.

The largest object discovered so far appears to be 1996 TO$_{66}$ with a radius of ~500 km. This is only slightly smaller than Pluto's satellite Charon, which has a radius of ~600 km. For comparison, Pluto has a radius of ~1170 km. The next three largest trans-Neptunian objects are similar in size with radii of ~300 km. The smallest known object is 1993 RP at ~50 km.

In addition to large objects, approximately 30 objects have been detected in the Kuiper Belt with V magnitudes between 28.6 and 27.8 (radii between 5 and 10 km assuming an albedo of 0.04) using the Hubble Space Telescope's Wide Field Planetary Camera 2. These objects were detected using statistical techniques and are not currently recoverable. Thus, only very limited information about them is available.

IV. PHYSICAL OBSERVATIONS OF TRANS-NEPTUNIAN OBJECTS

Physical studies of Kuiper Belt and Centaur objects have been limited because of the intrinsic faintness of these distant objects. Obtaining meaningful physical observations with good signal to noise is extremely difficult at magnitudes fainter than $V = 20$ and requires considerable dedication of telescope resources as well as very careful data reduction and calibration. Observations to date (January 1998) have been limited to broad-band spectral measurements of about one-third of the Centaur and Kuiper Belt objects, infrared spectra of a few objects, and light curve measurements of a few objects. Coverage is better for Centaur objects than for Kuiper Belt objects because the former are closer and, hence, generally brighter.

Broad-band spectral observations in the visible region of the spectrum have demonstrated that both Centaurs and Kuiper Belt objects apparently fall into two distinct groups characterized by their $B - V$ (blue minus visible) and $V - R$ (visible minus red) colors. One group consists of objects that are essentially gray, with colors similar to or only slightly redder than the sunlight that illuminates them. The other group consists of objects that display some of the reddest colors observed for any small bodies in the solar system.

This dichotomy in Centaur and Kuiper Belt object colors is illustrated in Fig. 5, which shows $B - V$ and $V - R$ colors for 13 objects, which include four Centaurs, eight Kuiper Belt objects, and one scattered disk object, 1996 TL$_{66}$. The gray objects in the figure are similar in color to C- and D-type asteroids in the main asteroid belt (both C and D are carbonaceous asteroid types); and include three Centaurs, two Kuiper Belt objects, and one scattered disk object. The red objects include one Centaur (5145 Pholus) and six Kuiper Belt objects. For three other distant objects, one Centaur and two Kuiper Belt objects, only $V - R$ colors have been determined. Of these, both Kuiper Belt objects appear to be members of the gray group, and the Centaur appears to be a member of the red group. None of them as yet appears to fall between the two populations. Also, the gray group includes 2060 Chiron, the first Centaur discovered, which is not among those shown in Fig. 5. [See ASTEROIDS.]

Gray colors have generally been identified with the presence of carbon-rich surface materials and are prevalent throughout the asteroid belt, although predominantly in the outer main belt. Red colors are more common in the outer planets region and have generally

TABLE II
Characteristics of Known Trans-Neptunian Objects[a]

Name	a (AU)	e	i (deg)	R (km)	Name	a (AU)	e	i (deg)	R (km)
1992 QB$_1$	44.3	0.08	2.2	120	1995 WY$_2$	45.7	0.04	1.7	120
1993 FW	43.5	0.05	7.8	120	1995 YY$_3$	39.4	0.22	0.4	60
1993 RO	39.6	0.20	3.7	80	1996 KV$_1$[b]	43.0	0.04	8.4	120
1993 RP[b]	39.3	0.1	2.6	50	1996 KW$_1$[b]	46.6	0	5.5	120
1993 SB	39.5	0.32	1.9	80	1996 KX$_1$[b]	39.5	0.10	1.5	60
1993 SC	39.9	0.19	5.1	120	1996 KY$_1$[b]	39.5	0.10	30.9	80
1994 ES$_2$	45.5	0.12	1.1	100	1996 RQ$_{20}$	44.4	0.12	31.6	120
1994 EV$_3$	42.8	0.05	1.7	120	1996 RR$_{20}$	40.1	0.19	5.3	120
1994 GV$_9$	43.5	0.06	0.6	120	1996 SZ$_4$	39.8	0.25	4.7	80
1994 JS	42.3	0.22	14.1	80	1996 TK$_{66}$	43.0	0	3.3	120
1994 JV[b]	35.3	0	18.1	120	1996 TL$_{66}$	84.7	0.59	24.0	300
1994 JQ$_1$	44.0	0.05	3.8	120	1996 TO$_{66}$	43.6	0.12	27.3	480
1994 JR$_1$	39.4	0.12	3.8	100	1996 TP$_{66}$	39.7	0.34	5.7	150
1994 TB	39.8	0.32	12.1	120	1996 TQ$_{66}$	39.7	0.13	14.6	150
1994 TG[b]	42.3	0	6.8	120	1996 TR$_{66}$	42.6	0.22	12.3	100
1994 TH[b]	40.9	0	16.1	120	1996 TS$_{66}$	44.2	0.13	7.4	190
1994 TG$_2$[b]	42.4	0	2.2	120	1997 CQ$_{29}$[c]	44.4	0.07	2.9	150
1994 VK$_8$	42.8	0.03	1.5	150	1997 CR$_{29}$[c]	42.0	0	20.2	150
1995 DA$_2$	36.2	0.07	6.6	80	1997 CS$_{29}$	43.7	0.01	2.3	300
1995 DB$_2$	46.3	0.13	4.1	100	1997 CT$_{29}$[c]	42.9	0.05	1.0	300
1995 DC$_2$	43.8	0.07	2.3	120	1997 CU$_{29}$	43.3	0.03	1.5	150
1995 FB$_{21}$[b]	42.4	0	0.7	100	1997 CV$_{29}$[c]	44.2	0.10	7.8	120
1995 GH[b]	42.9	0.09	22.9	120	1997 CW$_{29}$[c]	39.4	0.08	19.0	150
1995 GA$_7$[b]	39.5	0.1	3.5	100	1997 QH$_4$[c]	44.4	0.07	12.5	150
1995 GY$_7$[b]	41.3	0	0.9	100	1997 QJ$_4$[c]	39.6	0.12	16.4	100
1995 HM$_5$	39.4	0.25	4.8	80	1997 RT$_5$[c]	42.3	0	12.8	120
1995 KJ$_1$[b]	43.5	0	2.7	150	1997 RY$_6$[c]	41.4	0	12.8	120
1995 KK$_1$[b]	39.5	0.2	9.3	60	1997 RX$_9$[c]	41.9	0	30.6	80
1995 QY$_9$	40.1	0.27	4.8	100	1997 SZ$_{10}$[c]	39.6	0.20	12.7	60
1995 QZ$_9$	39.8	0.15	19.5	100	1997 TX$_8$[c]	39.4	0.19	8.4	80

[a] As of January 1, 1988.
[b] Observations are too incomplete to calculate an accurate orbit. Object presumed lost.
[c] Object has been discovered too recently to have an accurate orbit.

been taken to be indicative of the presence of complex hydrocarbons on the icy surfaces of the objects.

Primitive solar system materials in the outer planets region are expected to be initially gray or black due to the high fraction of carbon and simple hydrocarbons mixed with relatively transparent ices. Laboratory experiments have shown that irradiation of such materials with UV photons from stars and cosmic rays will pro-duce more complex hydrocarbons and turn the material reddish in color. However, continuing irradiation of the material will eventually produce elementary carbon and refractory organic residue that will once again turn gray.

Various hypotheses have been put forward to explain the different colors of Centaur and Kuiper Belt objects. Conceivably, objects may be reddened by irra-

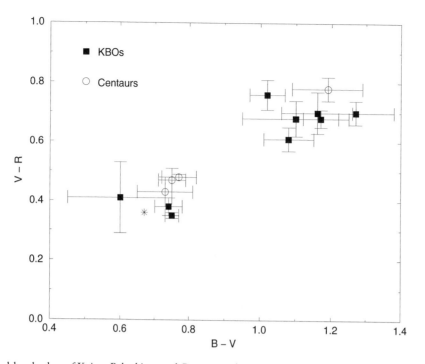

FIGURE 5 The broad-band colors of Kuiper Belt objects and Centaurs as determined by Stephen Tegler and William Romanishin. Squares and circles refer to Kuiper Belt objects and Centaurs, respectively. *B*, *V*, and *R* refer to magnitudes measured through blue, green-yellow ("visual"), and red filters, respectively. Error bars represent 1σ uncertainties. The asterisk represents the color of the Sun. Redder colors are to the upper right in the diagram. Note the two distinct populations. (Reprinted with permission from *Nature*. Copyright Macmillan Magazines Limited.)

diation, but then resurfaced due to impacts, which excavate primitive gray material from the bottoms of craters and spreads it across the surface, "repainting" the objects gray. However, it is not clear that individual impacts on such large objects could excavate sufficient material to cover the entire surface area. Additionally, the fact that there appears to be two distinct color groups with little evidence of intermediate color objects suggests that whatever may change the colors of Centaurs and Kuiper Belt objects acts quickly compared to the length of time that they retain their colors, and acts over the complete surface. This might be consistent with a sudden resurfacing event due to a major impact, but is not consistent with the slow reddening of the new materials that would subsequently take place.

Alternatively, because the first Centaur, 2060 Chiron, is known to display cometary-like outbursts and is gray in color, it has been suggested that the generation of a coma may spread gray material originating at depth over the surface of a formerly red object. Arguing against this idea is the fact that many of the Kuiper Belt objects are far from the Sun in regions where surface temperatures would be too low to produce significant cometary activity, unless ex-

tremely volatile ices were involved. Also, the first two Centaurs found, 2060 Chiron and 5145 Pholus, have very similar perihelion distances, 8.45 and 8.68 AU, respectively, yet one is gray and active and the other is red and without apparent activity.

Of course, the dichotomy in colors may simply represent some intrinsic difference in the formation mechanism for the two groups of objects. However, the red and gray colors are found equally, to first order, for both Centaurs and Kuiper belt objects; statistics are not yet good enough to detect significant population differences. No correlations have been detected between colors and orbital elements (semimajor axis, eccentricity, inclination) or absolute magnitudes (i.e., physical radius).

The range of colors seen for Centaurs and the Kuiper Belt objects is also seen in long- and short-period comet nuclei. Comets do not display the same clear dichotomy in colors, but that may be a result of the contamination of the cometary observations by dust comae, as well as by the poor signal to noise often present in cometary observations.

Broad-band spectral measurements have been extended into the infrared for about half a dozen Centaur and Kuiper Belt objects. These all show essentially

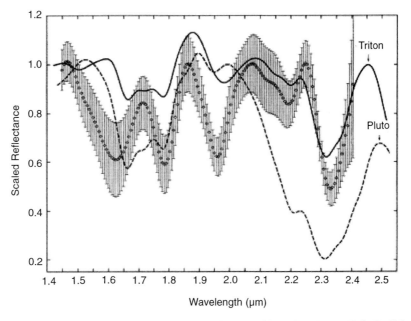

FIGURE 6 Dots and error bars represent the spectrum of 1993 SC as determined by Robert Brown, Dale Cruikshank, Yvonne Pendleton, and Glenn Veeder. Solid and dotted curves represent spectra of Neptune's satellite Triton and Pluto, respectively. The absorption features in the spectra have been identified with methane and possibly nitrogen ices. (Courtesy of *Science*.)

gray colors in the infrared, regardless of whether the objects appear gray or red in the visible region of the spectrum.

A moderate resolution spectrum in the near-infrared (IR) has been obtained for 1993 SC, one of the brightest of the Kuiper Belt objects (Fig. 6). The spectrum shows absorption features at 1.62, 1.79, 1.95, 2.20, and 2.32 μm and is similar in many ways to reflection spectrum in the near-IR for both Triton and Pluto (which are also shown in Fig. 6). The features at 1.79, 2.20, and 2.32 μm have been interpreted previously as evidence for methane (CH_4) in solid solution with nitrogen (N_2) ice on Triton's surface. The identification of these bands with N_2 ice on 1993 SC is far less certain and may represent pure methane ice alone. A small inflection in the spectrum of 1993 SC at 2.15 μm is also similar to one seen on Triton and is attributed to N_2 ice. However, the issue is not so clear for 1993 SC, and the error bars on the 1993 SC measurments are such that this feature may not be real. Other spectral features on 1993 SC are suggestive of simple hydrocarbon ices such as ethane (C_2H_6), ethylene (C_2H_4), and acetylene (C_2H_2).

If nitrogen ices are indeed present in sufficient quantity on the surfaces of Kuiper Belt objects, then that would have important implications for the surface temperatures, and hence the albedos of these objects. Up until now it has generally been assumed that Kuiper

Belt objects have very low albedos, similar to that for carbonaceous asteroids and cometary nuclei. Such albedos imply that subsolar temperatures on 1993 SC at its current heliocentric distance of 34.3 AU should be on the order of 65 K. At that temperature methane and other simple hydrocarbon ices should be stable, but N_2 should have a fairly high sublimation rate. Thus, one would not expect to see stable N_2 ice unless the albedo of 1993 SC was considerably higher than 0.04, which would then result in lower equilibrium surface temperatures and, hence, lower sublimation rates. Note that both Triton and Pluto have fairly high mean albedos, $A = 0.85$ and 0.55, respectively. These make the presence of N_2 ice more reasonable on these bodies, although the surface temperature of Pluto is controversial and N_2 may be confined to the near-polar regions on that planet. Conceivably, any N_2 ice on 1993 SC, if it is there, may also be confined to the polar regions of that object. The few measured albedos for Centaur or Kuiper Belt objects include 2060 Chiron, which has $A \approx 0.14$, Pholus with $A \approx 0.04$, Pluto with $A \approx 0.55$, and Charon with $A \approx 0.38$.

If there are indeed stable N_2 ices on the surfaces of Kuiper Belt objects, then it may be that the albedos of Kuiper Belt objects are considerably higher than 0.04. This would mean that the objects are considerably smaller than current estimates. Estimated radii are proportional to $A^{-1/2}$. Obviously, this would have substantial implications for estimates of the total mass of material in the Kuiper Belt and for what the Kuiper Belt is telling us about the origin of the planetary system. However, more observations are clearly needed in order to clarify whether N_2 ice is really there. If there is no N_2 ice, then there would be no problem with the currently assumed low albedos for Kuiper Belt objects.

The spectrum of 1993 SC strengthens the argument that there is a common link among Pluto, Triton, and Kuiper Belt objects. However, the similarities may only be skin deep because these spectra only measure surface compositions and do not necessarily reflect the bulk composition of these objects. If Kuiper Belt objects did form under the same conditions as Triton and Pluto, then their bulk composition will be approximately two-thirds rock and complex organics, and about one-third volatile ices, the latter being predominantly water ice. Methane and nitrogen ices on Triton and Pluto, and perhaps 1993 SC, may only be thin veneers of surface frost.

Another recently obtained infrared spectrum shows evidence for water ice on the surface of the Centaur object 1997 CU_{26}. Water ice or frost on the surfaces of Centaurs and KBO's may also result in higher albedos, and hence may imply smaller diameters and masses for these bodies.

Several attempts have been made to obtain light curves for Centaurs and Kuiper Belt objects, which would presumably be indicative of their rotation periods. Rotation periods have been determined for several Centaurs, including 2060 Chiron, 5.92 hr; 5145 Pholus, 9.98 hr; and 1995 GO, 8.93 hr. Among Kuiper Belt objects, attention has again focused on 1993 SC, one of the brightest Kuiper Belt objects. However, no evidence has been found for a systematic variation in the brightness of 1993 SC, to a level of about 12%. This could suggest that the object is close to spherical with little in the way of albedo features on its surface or that it is currently being viewed pole on and that the same hemisphere is being observed at all times. Further observations of 1993 SC and other Kuiper Belt objects are required to resolve this question. The rotation periods of both Pluto and Charon are 6.387 days, but because they have been mutually, tidally despun, their original periods were likely shorter.

Physical observations of Centaur and Kuiper Belt objects should improve as more of the new, large-aperture telescopes come on line and as the associated detectors and instrumentation improve over time. At present we have only the tantalizing clues described in this article as to the nature of these fascinating objects.

V. LONG-TERM STABILITY OF ORBITS IN THE KUIPER BELT

One of the fundamental difficulties with studying astronomy is that the systems being studied change on time scales that are very long compared to the length of time that astronomers have been observing them. The solar system is about 4.6 billion years old, the precession periods of the planets are thousands to millions of years, and yet modern science has only existed for a few hundred years. For example, we have studied Pluto for only one-quarter of its orbital period! We have only seen it close to the perihelion of its fairly eccentric orbit.

However, one of the fundamental goals of astronomy is to determine where the solar system came from and where is it going. Perhaps the most important modern tool in accomplishing this goal is the computer. In the last few decades a new branch of astronomy has developed known as numerical experimentation, which consists of running experiments on the computer. The advantage of this method is that it is possible to run an experiment on the computer that would take much too long to perform in normal (as opposed to cyber) space.

Here, we are interested in the long-term dynamical behavior of objects in the Kuiper Belt. We are interested in asking questions such as: "If there was an object in some specific orbit 4 billion years ago, would it still be there today?" "Will Pluto still be in a stable orbit 4 billion years from now?" Clearly, we cannot wait around for the answer. So, exacting computer algorithms have been developed that simulate the motion of objects along their orbits, taking into account the gravitational effects of the Sun, planets, and sometimes smaller objects. Using these algorithms, billions of years can pass in cyber space while only a few weeks or months pass in real space.

So let us consider the Kuiper Belt. As described briefly in Section I, modern work on the Kuiper Belt was prompted by a mystery that existed in the early 1980s concerning the origin of a certain type of comet, known as Jupiter-family comets. These comets are short lived as compared to the age of the solar system, with lifetimes of only $\sim 10^5$–10^6 years. Thus, there must be a "reservoir" serving as a source for these comets, continually replenishing the transient population on planet-crossing orbits. Also, Jupiter-family comets have small orbital inclinations compared to most comets; the median inclination of the Jupiter family is $\sim 10°$. Dynamical simulations showed that the only possible way to get such a flat inclination distribution for these comets was for them to have originated in a flattened disk. At the time (about 1988), the only possible place to hide such a disk was beyond the orbit of Neptune. Thus, to many researchers in the field, the existence of Jupiter-family comets was the first evidence for the existence of the Kuiper Belt.

However, there was a major problem with the Kuiper Belt as the source of Jupiter-family comets. Recall that the dynamical lifetimes of these comets are short. So, if the Kuiper Belt is the source of these comets, then whatever mechanism removes comets from the Kuiper Belt and feeds them into the inner solar system needs to be active today. Initially, Julio Fernandez suggested that lunar-sized bodies in the Kuiper Belt itself would scatter smaller bodies on to planet-crossing orbits. Later, Michael Torbett showed that such massive objects were not necessary because orbits in the Kuiper Belt just outside the orbit of Neptune were formally chaotic.

Usually, if two objects are placed very close to one another with the same initial velocities, the relative separation of the two bodies will increase linearly with time. However, in certain cases, the two bodies will separate exponentially. This kind of behavior is one of the formal definitions of chaos and has two very important consequences. First, it leads to unpredictability, i.e., it is impossible to predict the dynamical behavior of objects for long periods of time. Second, objects on chaotic orbits may undergo drastic and unexpected changes in their orbits [*See* CHAOTIC MOTION IN THE SOLAR SYSTEM.]

The behavior of the eccentricity of an object in an initially nearly circular (small e) orbit in the Kuiper Belt is shown in Fig. 7. The eccentricity remains small for a long period of time, in this case for over 200 million years. The orbit looks very well behaved for this length of time. Then, at about 2.5×10^7 years, the object evolves very quickly on to a high eccentricity orbit. The eccentricity is so large that the object can and does encounter Neptune, which gravitationally kicks it out of the Kuiper Belt.

The most complete study of the long-term behavior of objects in the Kuiper Belt has been performed by Martin Duncan, Harold Levison, and collaborators. They followed the orbital evolution of thousands of test particles for the age of the solar system, studying a wide range of initial semimajor axes, inclinations, and eccentricities. Each particle was followed until it suffered a close encounter with Neptune. Once comets encounter Neptune, they will evolve rapidly ($\sim 10^7$ to $\sim 10^8$ years) into the inner planets region or be ejected to the Oort Cloud or to interstellar space. This issue is described in more detail in Section VII.

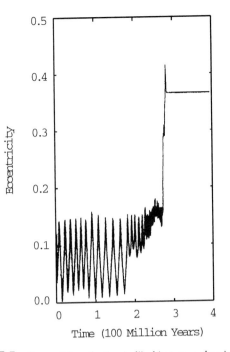

FIGURE 7 Eccentricity of a "typical" object on a chaotic orbit in the Kuiper Belt as a function of time. At first, the object appears to be well behaved as the eccentricity slowly oscillates up and down. Then the behavior quickly changes and the eccentricity gets large. So large, in fact, that the object then suffers a close encounter with Neptune, which gravitationally scatters the object out of the Kuiper Belt. It is important to note that this sudden increase in eccentricity is not a result of some large perturbation that the object experienced, such as an encounter with a massive object. It is a result of the slow buildup of small perturbations of the planets. (Courtesy of *Astronomical Journal*.)

The results of these simulations are shown in Fig. 8. The vertical strips indicate the length of time required for a particle to encounter Neptune as a function of its initial semimajor axis and eccentricity. Strips that are colored light gray represent objects that survive for the length of the simulation, 4×10^9 years, the approximate age of the solar system. As can be seen in Fig. 8, the Kuiper Belt can be expected to have a complex structure, although the general trends are readily explained. Objects with perihelion distances less than ~ 35 AU (shown as a solid curve) are unstable, unless they are near, and presumably librating about, a mean motion resonance with Neptune. Indeed, the results in Fig. 8 show that many of the Neptunian mean motion resonances (shown as vertical lines) are stable for the age of the solar system. Objects with semimajor axes between ~ 40 and 42 AU are unstable. This is presumably due to the presence of three overlapping secular resonances that occur in this region of the solar system: two with Neptune and one with Uranus.

Indeed, secular resonances appear to play a critical

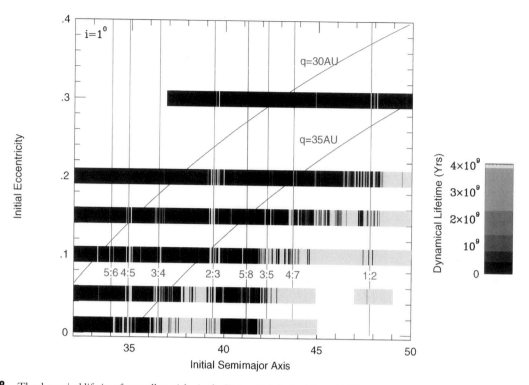

FIGURE 8   The dynamical lifetime for small particles in the Kuiper Belt derived from 4 billion year integrations. Each particle is represented by a narrow vertical strip, the center of which is located at the particle's initial eccentricity and semimajor axis (initial orbital inclination for all objects was 1°). The shade of each strip represents the dynamical lifetime of the particle. Strips colored light gray represent objects that survive for the length of the integration (4×10^9 years). Dark regions are particularly unstable on these time scales. For reference, locations of the important Neptune mean motion resonances are shown and two curves of constant perihelion distance, q, are also plotted. (Courtesy of *Astronomical Journal.*)

role in ejecting particles from the Kuiper Belt. This can be seen in Fig. 9, which shows some of the results of a 1 billion year integration. Here, the strips show the length of time required for a particle to encounter Neptune as a function of its initial semimajor axis and inclination. These particles all had initial eccentricities of 0.01. Also shown are the locations of the Neptune longitude of perihelion secular resonance (in dark gray) and the Neptune longitude of the ascending node secular resonance (in light gray). It is important to note that much of the clearing of the Kuiper Belt occurs where these two resonances overlap. This includes the low inclination region between 40 and 42 AU. The Neptune mean motion resonances are also shown.

As mentioned earlier, many of the Neptune mean motion resonances are stable at low inclinations. However, it can be seen in Fig. 9 that these resonances are often unstable at high inclinations. This instability is again most likely due to secular resonances. Figure 9 shows that the ν_8 secular resonance converges on the 3 : 4 and 2 : 3 mean motion resonances at large inclinations. Numerical integrations show that the unstable

orbits in these regions of phase space are chaotic and temporarily librate about both the local mean motion resonance and the nearby secular resonance, confirming that the resonance overlap is the cause of the instability.

It is interesting to compare the numerical results to the current best orbital elements of the known Kuiper Belt objects. This comparison is made in Fig. 10. Locations of all the Kuiper Belt objects with well-established orbits are shown as filled circles in Fig. 10. Stable regions as indicated by the numerical integrations in Figs. 8 and 9 are shown as gray areas. The main conclusion from this comparison is that the orbits of objects inside of ~41 AU have sufficiently high eccentricities that they must be in Neptune mean motion resonances to be stable. Orbits of objects outside of this region tend to have lower eccentricities and are not in obvious mean motion resonances (although there does appear to be two objects in the 3 : 5 resonance). It is interesting to note that the transition from resonant to nonresonant orbits occurs near the location of the overlapping secular resonances at 40–42 AU.

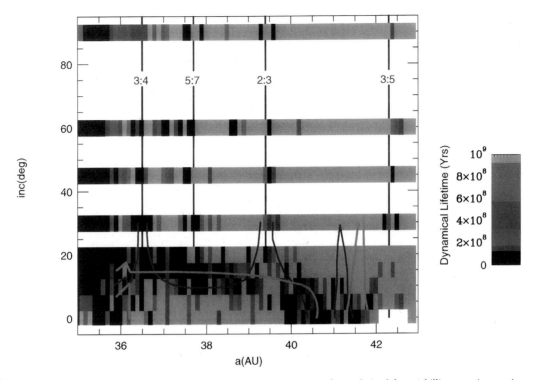

FIGURE 9 The dynamical lifetime for test particles with an initial eccentricity of 0.01 derived from 1 billion year integrations. This plot is similar to Fig. 8. In addition, two curves show the locations of the Neptune longitude of perihelion secular resonances (ν_8) and the Neptune longitude of the ascending node secular resonances (ν_{18}). Vertical lines show the location of the important Neptune mean motion resonances.

Looking at Fig. 10, it is surprising that there are, as yet, no known objects in nonresonant orbits with semimajor axes between 36 and 39 AU, despite the fact that dynamical simulations indicate that most orbits in this region are stable, provided their initial eccentricities are less than ~0.05. In order to interpret this result, let us consider what the integrations are saying about what we should expect in the Kuiper Belt. Integrations were performed by (i) initially taking the giant planets in their current orbits and with their current masses, and massless test particles in a smooth distribution of initial orbits, and (ii) integrating the orbits of the planets and test particles under the gravitational effects of the planets. Thus, if the 36–39 AU region indeed proves to be unpopulated, then some mechanism other than the long-term gravitational effects of the planets in their current configuration is likely required to have cleared it. Three mechanisms that may have accomplished this come to mind.

- A hypothetical early outward migration of Neptune would cause its mean motion resonances to sweep through this region, thereby sweeping most objects into the mean motion resonances. The migration of Neptune was first demonstrated by Julio Fernandez

and Wing Ip, and the capture mechanism was later proposed by Renu Malhotra to explain Pluto's current orbit. The orbital element distribution predicted by this model is shown in Fig. 11A. Note that the predicted orbital element distribution matches that for real objects in the Kuiper Belt fairly well. In general, this mechanism predicts that the mean motion resonances will be overpopulated relative to a more uniform initial distribution. One interesting consequence of this model is that in order to get eccentricities of 0.32 in Neptune's 2:3 mean motion resonance requires that Neptune must have had an initial semimajor axis of 25 AU (compared to a current value of 30 AU). One possible problem with this model is that it predicts that we should see a significant number of objects in orbits in the 1:2 mean motion resonance. A significant fraction of these objects should have large enough orbital eccentricities to get close enough to the Sun to have been discovered. Unfortunately for this mechanism, none have yet been found.

- Harold Levison, Alan Stern, and Martin Duncan have shown that the location of the perihelion secular resonance with Neptune would have been much closer to the Sun if the Kuiper Belt was massive in

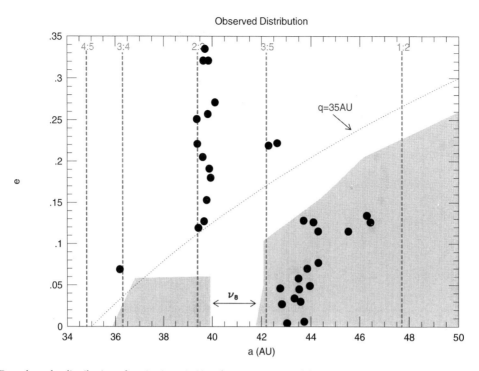

FIGURE 10 Dots show the distribution of semimajor axis (*a*) and eccentricity (*e*) of the known Kuiper Belt objects that have been observed at more than one opposition. Dashed lines show the locations of Neptune's important mean motion resonances. Notice that inside of the ν_8 secular resonance, which is marked in the figure, all objects are in mean motion resonances. Shaded regions show the locations of stable nonresonant orbits (see Fig. 8). Objects that lie outside these regions and that are not in a Neptune's mean motion resonance are not stable for the age of the solar system. The dotted curve is a curve of constant perihelion distance (*q* = 35 AU).

the past. Indeed, if the Kuiper Belt initially contained about 10 Earth masses ($M_\oplus$) of material between 30 and 50 AU, then the ν_8 resonance would have been at 36 AU. This amount of mass is significantly larger than what is in the Kuiper Belt today (see Section VI on the size distribution of objects), but is not inconsistent with what may have been there at earlier times (see Section VIII on the formation and evolution of the Kuiper Belt). As the Kuiper Belt eroded to its current mass through a combination of physical collisions and dynamical ejection, the ν_8 secular resonance swept outward, exciting the eccentricities of the orbits of objects along the way. Only those objects in orbits between the initial and final positions of the secular resonance would be effected. In addition, only those objects that were not in orbits in mean motion resonances would have been removed from the Kuiper Belt due to encounters with Neptune. Figure 11B shows the orbital element distribution predicted by this mechanism. It predicts correctly that objects in orbits inside of 41 AU will be locked in mean motion, whereas those outside will not be. However, it does not predict the large eccen-

tricities seen for the orbits of objects in the 2 : 3 mean motion resonance.

- Some process may have pumped up the eccentricity and inclination of the orbits of objects in the region between 36 and 39 AU to values greater than $e \sim 0.05$ and/or $i \sim 10°$ where the dynamical lifetimes are short. One method for exciting random motions in a disk is by mutual gravitational encounters between objects in the disk. In this vein, it is interesting to note that the escape velocity of the largest known Kuiper belt object (assuming the sizes in Section III and a density of 1 g cm^{-3}) is approximately 400 m/sec, which is about 10% of its heliocentric orbital velocity. Thus, if there were enough of these large objects initially for the Kuiper Belt to be relaxed dynamically by mutual gravitational scattering, then the Kuiper Belt objects would have typical orbital eccentricities of about 10%, just enough to depopulate the region between 36 and 39 AU, except in resonances where high eccentricity orbits are stable. Unlike the previous hypothesis, this mechanism predicts that mean motion resonances will not be overpopulated relative to a more uniform initial distribution.

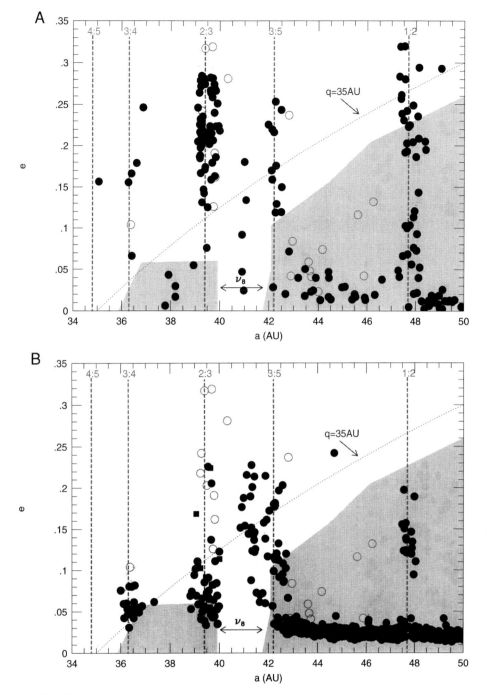

FIGURE 11 Same as Fig. 10 except that black dots show the distribution of semimajor axis (*a*) and eccentricity (*e*) predicted by two models of Kuiper Belt formation. Open circles show the orbits of known Kuiper Belt objects that have been observed at more than one opposition. (These objects are shown as black dots in Fig 10.) (A) The distribution predicted by the migrating planet model. (B) The distribution predicted by the massive Kuiper Belt model.

These three explanations are not exhaustive. Physical collisions and gas drag may also have played an important role. Nonetheless, these three models make very different predictions about the distribution of the orbits of objects in the Kuiper Belt. Thus, it seems likely that further observations will help resolve whether any of the mechanisms described earlier have played an important role.

VI. SIZES OF TRANS-NEPTUNIAN OBJECTS AND THE TOTAL MASS OF THE KUIPER BELT

As described briefly in Section I, the disk out of which the planetary system accreted was created as a result of the Sun shedding angular momentum as it formed. As the Sun condensed from a molecular cloud, it left behind a disk of material (mostly gas with a little bit of dust) that contained a small fraction of the total mass but most of the angular momentum of the system. It is believed that the initial solid objects in the protoplanetary disk were boulder sized, on the order of meters to tens of meters in diameter. These objects formed larger objects through a process of accretion, whereby bodies collided and stuck to one another. The boulders accreted to form mountain-sized bodies, which in turn accreted to form asteroids and comets, which in turn accreted to form planets (or the cores of giant planets which then accreted gas directly from the solar nebula). Understanding this process is one of the main goals of astronomy today.

There are few clues in our planetary system about this process. We know that the planets formed and we know how big they are. Unfortunately, the planets have been so altered by internal and external processes that it is difficult to learn anything about their formation. Luckily, we also have the asteroid belt, the Kuiper Belt, and the scattered disk. These structures contain the best clues to the planet formation process because they are regions where the process started, but for some reason, did not run to completion (i.e., a large planet). Thus, the size distribution of objects in these regions may show us how the processes progressed with time and (hopefully) what stopped them. Unfortunately, this type of information has been contaminated in the asteroid belt due to the collisional breakup of many large bodies. Thus, the Kuiper Belt and the scattered disk are perhaps the best places to learn about the accretion process.

The best clue for understanding planet formation in the trans-Neptunian region is the size distribution of the objects found there. This distribution is usually given in the form of a power law of the form

$$n(R)\, dn \propto R^{-q}\, dR, \qquad q > 0$$

where n is the number of objects with radius, R, between R and $R + dR$. The slope of this distribution, q, contains important clues about the physical strengths, masses, and orbits of the objects involved in the accretion process.

For example, there are two extremes to the accretion process. If two large, strong objects collide at low velocities then the amount of kinetic energy in the collision is small compared to the amount of energy holding the objects together. In this case, the objects merge to form a larger object. If two small, weak objects collide at high velocities then the energy in the collision will overpower the gravitational and material binding energies. In this case the objects break apart, forming a large number of much smaller objects. In realistic models of the Kuiper Belt with a range of sizes and velocities (where small objects usually move faster than large objects due to the equal partition of energy), we expect small objects to fragment and large objects to grow. This produces a size distribution with $q \sim 3$ at small sizes and a much steeper slope at large sizes where accretion is important.

By culling information from a variety of sources, including both observed Jupiter-family comets and Kuiper Belt objects, it is possible to attempt to construct a preliminary estimate of the size distribution of icy planetesimals formed in the outer solar system. Much of what follows is crude, and the results are not intended to be a finished product. On the contrary, the results should be viewed as a forum for a comparison of the known theoretical and observational constraints on the sizes of observed long- and short-period comets and Kuiper Belt objects.

In all, there are currently five major constraints on the size distribution of objects between 30 and 50 AU from the Sun (which is the dynamically active region of the Kuiper Belt). They are:

1. As described earlier, the current interest in the Kuiper Belt was prompted by the suggestion that Jupiter-family comets originated there. If so, it is possible to estimate that total number of comet-sized ($1 < R < 10$ km) objects in the Kuiper Belt from the number of known Jupiter-family comets. The total number of Jupiter-family comets (both active and extinct) is expected to be

$$N_{\rm J} = N_{\rm K} p_e f_{\rm J} L_{\rm J},$$

where $N_{\rm K}$ is the current number of comets in the Kuiper Belt, p_e is the mean probability that any comet between 30 and 50 AU will escape the Kuiper Belt per year, $f_{\rm J}$ is the fraction of those comets that become Jupiter-family comets once they leave the Kuiper Belt, and $L_{\rm J}$ is the dynamical lifetime of a Jupiter-family comet. With the exception of $N_{\rm K}$, all other variables have been estimated based on either observational or theoretical work. Thus, given the observed number of Jupiter-family comets, there are approximately 7×10^9 objects with radii between 1 and 10 km in the Kuiper Belt between 30 and 50 AU from the Sun.

2. In 1994, Anita Cochran and collaborators used the Hubble Space Telescope to detect Halley-sized objects in the Kuiper Belt. They detected approximately 30 objects with V magnitudes between 28.6 and 27.8. Assuming an albedo of 0.04, this magnitude range corresponds to radii between 5 and 10 km. These observations covered only about 4 square arc minutes of sky. If one assumes a uniform inclination distribution of objects between 0 and 12°, the HST observations imply that there are $\sim2 \times 10^8$ comets in this size range in the Kuiper Belt. This should be viewed as a lower limit to the number between 30 and 50 AU because the objects detected were all closer than 40 AU.

3. It has been found that Pluto's moon Charon has a significant orbital eccentricity (near $e = 0.003$). This eccentricity is surprisingly large because the time it takes for tides to damp the eccentricity to zero, $\sim9 \times 10^6$ years, is short compared to the age of the solar system. It is possible to place crude estimates on the total number of large objects in the Kuiper Belt because Kuiper Belt objects passing between Pluto and Charon can excite the eccentricity of Charon's orbit. Under the assumption that this mechanism is the only one of importance, preliminary results suggest that there are between 3×10^6 and 3×10^8 Kuiper Belt objects with radii between 20 and 330 km, between 30 and 50 AU. Because other unrecognized mechanisms may also be important, this estimate should be viewed as an upper limit. [See PLUTO AND CHARON.]

4. The searches that discovered the Kuiper Belt objects listed in Table II can also be used to estimate the total number of objects in the Kuiper Belt. In order to accomplish this, a consistent set of observations is required. Such an analysis is possible on the 15 objects that were discovered in the Mauna Kea and Cerro-Tololo surveys, which covered 3.9 and 4.4 deg² of sky, respectively. The Mauna Kea survey found that there are ~3.8 objects per square degree

brighter than a limiting magnitude of $R \sim 24.2$. The Cerro-Tololo survey found ~1.4 deg^{-2} brighter than $R \sim 23.2$. Combining these numbers, there must be $\sim7 \times 10^4$ objects with $R > 50$ km and $\sim2.6 \times 10^4$ objects with $R > 80$ km between 30 and 50 AU from the Sun. Because these searches were only along the ecliptic, and Kuiper Belt objects have since been found in orbits inclined up to 30°, these estimates are likely only lower limits.

5. Charles Kowal's search that discovered Chiron (the first Centaur) in 1977 covered 6400 deg² photographically to $V \sim 20$. This search should have found any object as large or larger than Pluto currently near the ecliptic and within 60 AU of the Sun, although Kowal's longitude coverage was not complete. Thus, there is probably only one such object within 50 AU, Pluto.

The constraints just described are displayed in Fig. 12, which plots the cumulative number of objects larger than radius R between 30 and 50 AU from the Sun, as a function of R. These constraints are shown as filled circles in the plot. It should be noted that some of these numbers are very uncertain. For example, point 1 can be off by as much as an order of magnitude because of uncertainties in the physical lifetimes of comets and in the current orbital element distributions of objects in the Kuiper Belt. In addition, the location of this point on the abscissa, which represents the radius of the smallest detectable Jupiter-family comet in the inner solar system, is also unknown. Its value is thought to be near 1 km, but is uncertain by at least a factor of two.

However, the data do show that the size distribution is not a simple power law. The circles in Fig. 12 show that $q = 3$ fits the constraints fairly well for objects with $R \lesssim 10$ km. At $R > 10$ km the circles indicate a slope of about $q = 4.5$. Thus, the size distribution of objects in the Kuiper Belt might best be described by a broken power law of the form

$$n(R)\, dn \propto \begin{cases} R^{-q_1}\, dR & \text{if } R \le R_0 \\ R^{-q_2}\, dR & \text{if } R > R_0 \end{cases},$$

where R_0 is the radius where the power law changes slope. The solid black curve in Fig. 12 shows such a distribution with $R_0 = 10$ km, $q_1 = 3$, and $q_2 = 4.5$.

Interestingly, models of the accretion of objects in the outer solar system and of the size distribution of long-period comets find similar broken power laws. In addition, an analysis of the size distribution of the Kuiper Belt objects discovered to date by ground-based

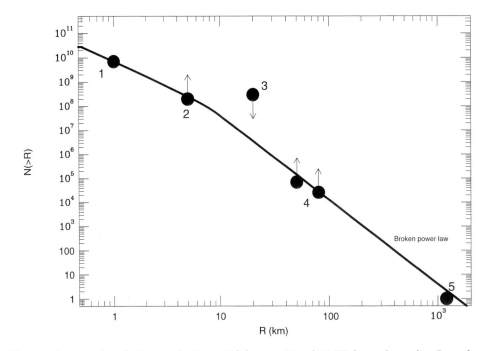

FIGURE 12 The cumulative number of objects in the Kuiper Belt between 30 and 50 AU, larger than radius R as a function of R. Filled circles indicate the known constraints as listed in the text, and arrows indicate upper/lower limits.

telescopes, those indicated by points 4 and 5 in Fig. 12, found a slope of $q_2 = 4$, quite close to the value predicted in Fig. 12. The broken power law shown in Fig. 12 suggests that Kuiper Belt objects with radii less than 10 km are evolved collisionally, but that larger objects are most likely the results of accretion.

Given the limited number of objects that have been discovered in the Kuiper Belt to date, any conclusions about the size distribution in the belt must be regarded as speculative. Nevertheless, the Kuiper Belt is clearly a rich area for future studies of the growth of planetesimals in the solar nebula. It is hoped that the discovery of additional objects, along with accurate photometric measurements from which their sizes can be inferred, will help shed more light on this problem.

Finally, it is possible to integrate under the broken power law shown in Fig. 12 in order to estimate the total mass in the Kuiper Belt between 30 and 50 AU. Such an integration with limits between $R = 1$ km and 1200 km (the approximate radius of Pluto) and assuming a density of 1 g cm^{-3} shows that the total mass between 30 and 50 AU is 0.13 $M_{\oplus}$.

As with many scientific endeavors, the discovery of new information tends to raise more questions than it answers. Such is the case with the just-described mass estimate. Edgeworth's and Kuiper's original arguments for the existence of the Kuiper Belt were based on the idea that it seemed unlikely that the disk of

planetesimals that formed the planets would have ended abruptly at the current location of the outermost known planet. An extrapolation into the Kuiper Belt (between 30 and 50 AU) of the current surface density of nonvolatile material in the outer planets region predicts that there should originally have been about 30 $M_{\oplus}$ of material there. However, as stated earlier, our best estimate is over 200 times less than that figure!

Were Kuiper and Edgeworth wrong? Is there a sharp outer edge to the planetary system? The answer appears to be no to both questions. Evidence has mounted that supports the idea of a massive primordial Kuiper Belt. Models of collisional processes have shown that the Kuiper Belt is currently eroding away due to collisions. These models predict that the Kuiper Belt will lose a significant fraction of its objects in the next few times 10^8 years. Because the Kuiper Belt is currently losing mass due to collisional erosion, it clearly had to be more massive in the past.

These collisional evolution models have also shown that it is not possible for objects with radii greater than about 30 km to form in the current Kuiper Belt, at least by two-body accretion, over the age of the solar system. The current surface density is too low to accrete bodies larger than this size. However, the models show that objects the size of 1992 QB$_1$ could have grown in a more massive Kuiper Belt if the mean orbital eccentricities of the accreting objects were very small. A Kuiper Belt of at least several Earth masses is required in order for 100-km-sized objects to have formed. It is interesting to note that according to this model, 100-km-sized objects had to have formed before Uranus and Neptune grew to their current masses because the presence of those planets would pump the eccentricities of the orbits of the initial objects too high for QB_1-sized objects to accrete.

The same applies even more strongly to the accretion of Pluto and Charon. For those two bodies to have grown to their current sizes in the trans-Neptunian region, there must have originally been a far more massive Kuiper Belt. Otherwise, Pluto and perhaps even Charon would have had to have accreted in a denser region of the solar nebula (presumably closer to the Sun) and then been transported to their current orbit by gravitational scattering. There is no known mechanism that could accomplish this.

The idea of a historically massive Kuiper Belt also solves two other mysteries. First, as discussed in Section IV, objects in the Kuiper Belt seem to fall into two distinct classes. Inside of 41 AU, they are all in mean motion resonances with Neptune and have eccentricities larger than 0.1. Beyond 41 AU they tend to be in nonresonant orbits and most have eccentricities less than 0.1. If the Kuiper Belt were massive in the past, the ν_8 secular resonance would have been closer to the Sun than it is today. As the Kuiper Belt's mass declined, the ν_8 resonance would have swept outward, thereby exciting the orbital eccentricity of objects that were between its initial and its current position. The orbits of these objects would be unstable unless they were in a mean motion resonance. This mechanism thus predicts that most objects inside of the current ν_8 resonance should be on high eccentricity orbits in Neptune mean motion resonances, whereas objects beyond this region should be in low eccentricity, nonresonant orbits. This is precisely what is observed. Thus, an early massive Kuiper Belt could help explain the current orbital element distributions of known Kuiper Belt objects.

The second potentially solved mystery has to do with Neptune's eccentricity. The orbital eccentricity of the other three giant planets range from 0.047 to 0.054. Neptune's orbital eccentricity is 0.009, much smaller. Recent analytic models by William Ward and Joseph Hahn have shown that Neptune's eccentricity could have been decreased through gravitational interactions with a massive Kuiper Belt.

With these arguments in mind, it is possible to build a strawman model of the Kuiper Belt, which is depicted in Fig. 13. There are three distinct zones in the Kuiper Belt in this model. Region A is a zone where the dy-namical perturbations of the outer planets have played an important role in determining the Kuiper Belt's heliocentric structure. In this region the planets have tended to pump up the eccentricities of the orbits of objects. About half of the objects in this region have been removed dynamically from the Kuiper Belt by being thrown into Neptune-crossing orbits. The remaining objects have been perturbed to larger orbital eccentricities, resulting in encounter velocities between objects that are large enough that accretion has stopped and collisional erosion has become important. Thus, we expect that a significant fraction of the mass in region A has been removed by collisions. Indeed, the solid curve in this region of Fig. 13 shows an estimate of the current Kuiper Belt surface density, whereas the dotted curve shows an estimate of the initial surface density, extrapolated from that of the outer planets. If these curves are at all representative of reality, the current surface density in this region has been depleted by a factor of 10^3. The outer boundary of region A is somewhere between 50 and 65 AU.

Region B in Fig. 13 is a zone where collisions are important in shaping the structure of the Kuiper Belt, but where the gravitational effects of the planets are not important. Without the effects of the planets, the orbital eccentricities of objects in this region will most likely be small enough that collisions lead to accretion of bodies, rather than erosion. Large objects could have formed here, so that the size distribution of objects may be significantly different than in the region immediately interior to it. However, the surface density may not have changed very much in this region over the age of the solar system. The outer boundary of region B is very uncertain, but will likely be beyond 100 AU. Region C is a zone where collision rates are low enough that the surface density of the Kuiper Belt and the size distribution of objects in it have remained virtually unchanged over the history of the solar system.

It is only natural to draw parallels between the Kuiper Belt and the dust disks discovered around main sequence stars by the IRAS satellite. These disks extend out to ~900 AU (in the case of β Pictoris) or more from the central star. The most likely source of the dust in the IRAS disks is colliding and sublimating comets. Infrared excesses have been found to be common around solar-type stars and it would be unusual if the Sun did not have a similar dust disk.

The amount of dust in the Kuiper Belt between 30 and 100 AU is constrained to be less than 0.3 $M_\oplus$, based on estimates of collisionally produced dust in the belt and upper limits on zodiacal dust infrared emission by the COBE satellite. This estimate is 10^{-2} and 10^{-4} times the luminosity of the two best

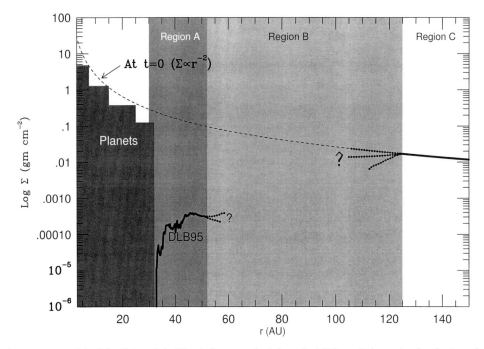

FIGURE 13 A strawman model of the Kuiper Belt. The dark area at the left marked "Planets" shows the distribution of solid material in the outer planets region. Notice that it follows a power law with a slope of about −1.5 versus heliocentric distance. The dashed curve is an extension into the Kuiper Belt of the power law found for the outer planets and illustrates the likely initial surface density distribution of solid material in the solar nebula. The dashed curve has been scaled so that it is twice the surface density of the outer planets, under the assumption that planet formation was 50% efficient. The Kuiper Belt is divided into three regions. In region A, the Kuiper Belt has been shaped by collisions and by dynamical perturbations by the planets. Most of the material in region A has been eroded away (both dynamically and physically) because of the large orbital eccentricities excited by planetary perturbations. The solid curve shows a model of the mass distribution in region A by Martin Duncan, Harold Levison, and collaborators. In region B, collisons have been important, but gravitational perturbations by the planets have not. Orbital eccentricities were small enough that accretion was the dominant process in region B. In region C, neither the gravitational effects of the planets nor collisions have been important. The Kuiper Belt has remained virtually unchanged in this region over the age of the solar system. Dotted curves illustrate the unknown shape of the surface density distribution in region B. (From Weissman and Levison (1997). *In* "Pluto and Charon," (D. J. Tholen and S. A. Stern, eds.). University of Arizona Press, Tucson.)

extra-solar dust disks, Vega and β Pictoris, respectively. Both the Vega and the β Pectoris disks are much younger than our solar system, as they are in orbit around stars that are more massive than the Sun and thus have shorter lifetimes. Thus, they have not had as much time to erode as the Kuiper Belt about our solar system. Still, a more massive Kuiper Belt, as discussed earlier, could be present at larger solar distances.

VII. ECLIPTIC COMETS AND THE SCATTERED DISK

As described in Section I, the current renaissance in Kuiper Belt research was prompted by the suggestion that Jupiter-family comets originated there. Thus, as part of the research intended to understand the origin of these comets, a significant amount of research has gone into understanding the dynamical behavior of objects that are on orbits that can encounter Neptune. These studies show that once objects (either escapees from the Kuiper Belt or leftover planetesimals from the Uranus–Neptune zone) encounter Neptune, gravitational encounters can spread them throughout the planetary system. The distribution of these objects as predicted by numerical integrations by Martin Duncan and Harold Levison is shown in Fig. 14.

Figure 14 therefore shows the predicted distribution of the objects that make up the ecliptic comet distribution described earlier. Those that get close to the Sun are the Jupiter-family comets. It is somewhat surprising that about a third of the objects in the simulations spend at least some of their time as Jupiter-family comets. The Jupiter-family comets that we see today are all small, $R \lesssim 10$ km. However, if our understanding of the size distribution of these objects is correct

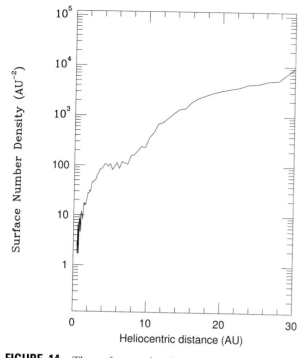

FIGURE 14 The surface number density (on the plane of the ecliptic) of ecliptic comets as determined from numerical integrations. There are approximately 10^6 comets in this population.

(see Section V), we should expect to see a 100-km-sized Jupiter-family comet about 0.4% of the time. What a show that would be!

Those ecliptic comets between Jupiter and Neptune are the Centaurs (only the largest of which are observable). The simulations predict that there are $\sim 10^6$ ecliptic comets currently in orbits between the giant planets.

Most ecliptic comets do not last forever. They either get scattered out of the planetary system by a close encounter with a planet or hit something (a planet or the Sun) in a relatively short amount of time: a few tens of millions of years. However, Levison and Duncan's integrations showed that a small fraction, about 1%, are scattered into long-lived, eccentric orbits outside the orbit of Neptune, although mainly inside of one to two hundred AU from the Sun. The long-range gravitational effects of the planets, particularly resonances with the planets, can lift the orbits of the objects beyond the outer edge of the planetary system (i.e., Neptune's orbit), protecting them from close encounters with the planets.

An example of this type of behavior is shown in the dynamical simulation in Fig. 15. This hypothetical object initially underwent a random walk in semimajor axis due to encounters with Neptune. At about $7 \times$

10^7 years it was trapped temporarily in Neptune's 3:13 mean motion resonance for about 5×10^7 years. It then performed another random walk in semimajor axis until about 3×10^8 years, when it was trapped in the 4:7 mean motion resonance, where it remained for 3.4×10^9 years. Notice the increase in the perihelion distance near the time of capture into this resonance. While trapped, the particle's eccentricity became as small as 0.04. After leaving the 4:7 resonance, the object was trapped temporarily in Neptune's 3:5 mean motion resonance for $\sim 5 \times 10^8$ years and then went through a random walk in semimajor axis for the remainder of the simulation.

Many of the objects that had their perihelion distances raised by resonances can survive in these eccentric orbits for the 4.6 billion-year age of the solar system. Thus, Duncan and Levison's simulations predicted the existence of a scattered disk in the outer solar system.

The simulations just described indicate that objects in the scattered disk will tend to have more eccentric and/or inclined orbits than those in the Kuiper Belt. Two recently discovered trans-Neptunian objects may be the first members of the scattered disk to be found. The first, 1996 TL$_{66}$, was discovered in October 1996 by Jane Luu and colleagues and is estimated to have a semimajor axis of 84 AU, a perihelion of 35 AU, an eccentricity of 0.58, and an inclination of 24°. Such a high eccentricity orbit could only result from gravitational scattering by a giant planet, in this case Neptune.

A second possible scattered disk object, 1996 RQ$_{20}$, was discovered in September 1996 by Eleanor Helin and colleagues. Recent observations indicate its semimajor axis is 44 AU, its eccentricity is 0.11, and its inclination is 32°. The high inclination of this object suggests that it may be a member of the scattered disk.

VIII. FORMATION OF THE KUIPER BELT AND THE SCATTERED DISK

Given the advances in both our observational and our theoretical understanding of the trans-Neptunian region, it is now possible to construct a scenario of how the Kuiper Belt and the scattered disk formed. The following scenario is by no means unique and there are variations that could reproduce the observed structures just as well. In addition, many of the stages of this scenario have yet to be modeled in detail and thus are still very uncertain.

In the outer solar system, Jupiter and Saturn formed

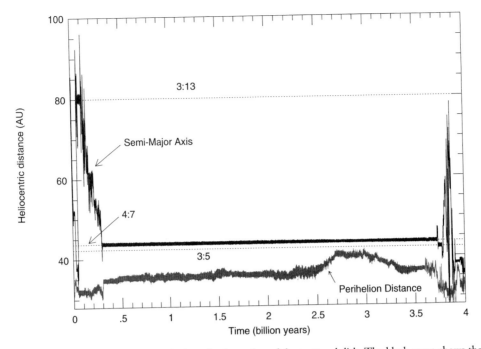

FIGURE 15 The temporal behavior of the orbit of a long-lived member of the scattered disk. The black curve shows the behavior of the comet's semimajor axis. The gray curve shows the perihelion distance. The three dotted curves show the location of the 3:13, 4:7, and 3:5 mean motion resonances with Neptune. This object initially underwent a random walk in semimajor axis due to encounters with Neptune. At about 7×10^7 years it was trapped temporarily in Neptune's 3:13 mean motion resonance for about 5×10^7 years. It then performed a random walk in semimajor axis until about 3×10^8 years, when it was trapped in the 4:7 mean motion resonance, where it remained for 3.4×10^9 years. Notice the increase in the perihelion distance near the time of capture into the resonance. While trapped in this resonance, the particle's orbital eccentricity became as small as 0.04. After leaving the 4:7 resonance, it was trapped temporarily in Neptune's 3:5 mean motion resonance for $\sim 5 \times 10^8$ years and then went through another random walk in semimajor axis for the remainder of the simulation.

early on when the gaseous component of the solar nebula, mainly hydrogen and helium, was still present. As the gas in the solar nebula dissipated, there was a large population of small, icy bodies in low-inclination, nearly circular orbits about the Sun outside the orbit of Saturn and extending outward perhaps as far as several hundred astronomical units. These objects gradually grew into larger planetesimals by mutual collisions. As the planetesimals increased in mass, they began to gravitationally perturb one another into more eccentric and more inclined orbits about the Sun. The growth of planets was more rapid inside of ~40 AU, leading to the formation of Uranus and Neptune. Outside of 40 AU (i.e., in the Kuiper Belt) the bodies remained relatively small.

As the bodies that eventually became Uranus and Neptune grew in mass, they began to gravitationally scatter the remaining nearby planetesimals both outward into the trans-Neptunian region and inward toward Jupiter and Saturn. Usually, the objects scattered outward returned to the Uranus–Neptune region where they would again be scattered. However, Jupiter and Saturn were so massive that they could efficiently

eject many of these planetesimals out of the planetary system, thereby acting as a sink for these objects. As a result, there was a net flux of objects toward Jupiter and Saturn. This inward transport of mass required an outward migration of the orbits of Uranus and Neptune in order to conserve the total angular momentum and energy of the system. As Neptune moved outward, it may have trapped planetesimals external to it (i.e., in the Kuiper Belt) into mean motion resonances. Once the planetesimals were trapped in the resonances, Neptune's heliocentric migration (and other dynamical processes) would have pumped up the eccentricity of the orbits of these objects to values as large as 0.3, leading to much of the present Kuiper Belt structure described in Section V.

As discussed earlier, the objects scattered outward by Uranus and Neptune can be relatively long-lived. After their first encounter with Neptune, about 1% of the objects survive for the age of the solar system, stored in an extended scattered disk beyond Neptune. How much mass could there be in the scattered disk? There are ~32 Earth masses of material in Uranus and Neptune. Because planet formation is not 100%

efficient, it is not unreasonable to expect that a similar amount of mass was initially scattered outward by Neptune (the efficiency factor is not known to within a factor of a couple of orders of magnitude). So crudely, we may expect as much as a few tenths of an Earth mass of material in the scattered disk. This value is consistent with current estimates of the number of scattered disk objects in the solar system.

IX. CONCLUDING REMARKS

Because the study of the trans-Neptunian region is still in its infancy, much of the formation scenario is speculative and much of it is likely to change rapidly in the coming years (or even months) as this field matures and more data become available. One reason for this is that only a very small fraction ($\lesssim 0.06\%$) of the Kuiper Belt and scattered disk objects with radii >50 km have been discovered so far. However, it has already been determined that the structure of the trans-Neptunian region is much more complex, and there-fore more interesting, than would have been dreamed of only a few years ago. It has also become clear that not only is this region interesting because of its novelty and complexity, but also because it has supplied (and most likely will continue to supply) us with important clues to the formation of the solar system.

Pluto and its satellite Charon occupy an important position in this new scenario. They are no longer just a pair of odd objects on the fringes of the planetary system. We now recognize that they are the largest known members of a huge population of objects that grew in the outer reaches of the solar nebula beyond Neptune. Other Pluto-sized objects may still orbit in the more distant regions of the Kuiper Belt and/or the scattered disk.

The Kuiper Belt also provides us with a link to dust disks detected around many main sequence stars. It suggests that those stars likely had accretion disks and may have even formed planetary systems like our own closer in to the stars.

The study of the Kuiper Belt is one of the most active frontiers of solar system research at present, and it will be interesting to see what new discoveries and revelations are made in the coming years.

ASTEROIDS

I. Introduction
II. Locations and Orbits
III. Composition and Characteristics
IV. Mystery, Promise, and Peril

Daniel T. Britt and Larry A. Lebofsky
University of Arizona

GLOSSARY

Albedo: Reflectivity of an object's surface at zero phase relative to a perfectly diffuse standard surface under the same illumination and distance conditions.

Astronomical unit (AU): Average distance between the Earth and the Sun. This is approximately 149,600,000 km or 96,000,000 miles.

Charge-Coupled Device (CCD): Solid-state detector that is extremely sensitive to photons. Used with computers, CCDs have achieved unprecedented levels of low-light astronomical observations.

Eccentricity: The factor that an asteroid's orbit deviates from circular.

Inclination: Tilt of an object's orbit with respect to the solar system ecliptic.

Kirkwood gaps: Zones of the asteroid belt that have been depleted of objects by orbital resonance with Jupiter.

Kuiper Belt: Zone beyond the orbit of Neptune but within approximately 1000 AU that is the probable source for short-period comets. Approximately 40 Kuiper Belt objects have been discovered.

Light-curve inversion: An asteroid light curve is the graph of an object's brightness versus time. Because asteroids are usually not perfect spheres, some faces that are larger than others, and when they point toward the observer the apparent brightness of an asteroid increases.

Peaks of the curve provide a measure of that asteroid's rotation rate, and the shape of the light curve can be "inverted" statistically to model the object's shape.

Oort Cloud: Theoretical source and storage area for long-period comets. This area is believed to be a spherical cloud of comets between 1000 and 100,000 AU.

Radiometric modeling: The thermal emission of an asteroid provides an estimate of an asteroid's surface temperature and albedo. A dark asteroid, for instance, would absorb more of the visible sunlight because it has a low albedo, but would radiate that additional energy in thermal wavelengths, showing a warmer surface temperature. Combined data on thermal "temperature" and visible reflectance can provide the albedo of an object and an estimate of its size.

Reflectance spectroscopy: Study of the physical and mineralogical properties of materials over the wavelength range of reflected electromagnetic radiation. Light interacts with the atoms and crystal structure of materials producing a diagnostic set of absorptions and reflectances.

Regolith: Layer of unconsolidated rocky fragments and debris that cover the bedrock of an object. On asteroids the regolith is produced by meteorite and micrometeorite impacts.

Semimajor axis: Average distance of an object from the Sun.

Speckle interferometry: Most observations of

asteroids occur from ground-based observatories and must look through the turbulent atmosphere of the Earth. Interferometry attempts to minimize the effect of turbulence by taking many rapid images and recombining them into a single clear image with the atmosphere removed.

I. INTRODUCTION

A. WHAT ARE ASTEROIDS?

Asteroids (or more properly, minor planets) are small, naturally formed solid bodies that orbit the Sun, are airless, and show no detectable outflow of gas or dust. Figure 1 shows images of the only three asteroids that humankind has seen in detail: 243 Ida, 951 Gaspra, and 253 Mathilde. The difference between asteroids and other naturally formed Sun-orbiting bodies, planets and comets, is largely historical and to some extent arbitrary. To the ancient Greeks and other peoples, there were three kinds of bright objects populating the heavens. The first and most important group was the

stars, or *astron* in Greek, which are fixed relative to each other. The English word *star* is probably an Old English and Germanic derivation of the Indo-European base word stêr, which provided the source of the Greek *astron* and the Latin *astralis*. Terms for the study of stars were based on the Greek root, i.e., astronomy or astrophysics. The second group of objects was planets or Greek *planetos* meaning "wanderer" as the planets were not fixed but wandered against the background of the stars in predictable paths. For the ancients these included the Sun, Moon, Mercury, Venus, Mars, Jupiter, and Saturn. The final group were the comets or *kometes* meaning "long-haired" because of their long tails or comas and their unpredictable paths and appearances.

Asteroids were not known to the ancients, and the first asteroid, 1 Ceres, was not discovered until 1801 by the Sicilian astronomer Giuseppe Piazzi. He was "searching" in the gap between Mars and Jupiter for what theorists at the time speculated would be the location of a "missing planet." 1 Ceres was thought initially to be this new planet, but its apparent small size seemed to rule out the category of planet. Because these new objects were planet-like in their sun-centered orbits, but star-like in that they were unresolvable points of light in a telescope, Piazzi applied the disused Greek root for a single star, "aster," to this new addition to the celestial population. Although

Mathilde   **Gaspra**   **Ida**

FIGURE 1   The three asteroids that have been imaged by spacecraft flyby: 243 Ida, 951 Gaspra (*Galileo*), and 253 Mathilde (*NEAR*). The relative sizes of these objects are to scale with the portion of 253 Mathilde shown here being 59 × 47 km. (Courtesy of Johns Hopkins University/Applied Physics Laboratory.)

asteroids share many of the characteristics of planets (Sun-centered orbits, seemingly solid bodies), the primary distinction is that they are simply much smaller than the known planets and are much dimmer when viewed from the Earth.

Similarly, the distinction between asteroids and comets is also based on their observational qualities rather than on any inherent difference in physical properties or composition. Comets are characterized by their coma, or cloud, of sublimating gas and expelled dust. This gives them their characteristic diffuse "fuzzy" halo and long streaming tail. Compared to the fuzzy look of comets, an asteroid is a "star-like" sharp point of light. However, comets only become "cometary" when they enter the inner solar system and are heated sufficiently by the Sun to evaporate their volatile materials. The point at which frozen comets begin to sublimate and turn cometary can vary, but for most comets this is approximately 4 AU or four times the distance between the Earth and the Sun. A number of asteroids with semimajor axes greater than 4 AU may be composed of the same collection of volatile ices, dust, and carbonaceous organics as comets. Because their orbits are less elliptical than currently active comets, they never travel close enough to the Sun to appear cometary. They are "solid" bodies only because their surfaces stay cold enough to keep their gases frozen. Active comets probably evolve into asteroids as they lose their ices from repeated trips into the inner solar system. As the gases boil off, what remains is a crust of dust and rock. There are several objects currently that are classified as asteroids but have shown brief comas, indicating that they are possibly comets at the end of their active lives.

In the final analysis, asteroids are defined by what they are not: They move against the celestial background so they are not stars. Of the objects that "wander," asteroids are either not large enough to be planets or are not actively shedding gas and dust to be a comet.

B. DISCOVERIES, NUMBERS, AND NAMES

Because asteroids appear as relatively small and dim points of light moving slowly against the stellar background, the problem of finding and identifying an object as an asteroid is fundamentally a question of precise "bookkeeping." The field of view seen through a telescope at any one moment is filled with literally hundreds of points of light and just one or two may be asteroids. The asteroid may move a small amount relative to the stars during the course of a night's observations, but the trick is to know the relative positions of all the viewed stars precisely enough to know when one of the points of light is out of place. Today the viewing through a telescope is done by extremely sensitive charge-coupled-devices (CCD) that feed their digital data directly to computers to do the "bookkeeping" of the stars and known asteroids. In the days when Giuseppe Piazzi discovered 1 Ceres, all observations were done with an eye to the viewfinder, and the bookkeeping was done by hand drawings of the star fields. Discoveries were made by visually comparing each point of light in the telescope field with a chart that was drawn on a previous observation. With these methods it is not surprising that only four more asteroids were found in the 45 years after Piazzi found 1 Ceres.

The application of photography to astronomy revolutionized the search for asteroids in the last half of the 19th century and the early part of the 20th century. A photographic plate has a number of advantages over the naked eye and a cold hand drawing in the dark: First a photographic plate is essentially an instant and precise local star chart. Second, they are far more light sensitive than the human eye and sensitive to wavelengths of light not visible to the eye. Finally, with long exposures and telescopes that move slowly to compensate for the Earth's rotation, stars will be fixed as bright dots on the plate and asteroids will appear as streaks as they move relative to the stars. With this equipment, finding asteroids was simply a matter of taking consistent exposures and looking for streaks. Modern searches have replaced the photographic plate with the CCD and the computer. CCDs are more sensitive to light and are linked directly with computers that can screen out the sources of "noise" such as stars, known asteroids, and the increasing collection of "space junk" accumulating in near-Earth orbit. As of this writing there are about 8500 numbered asteroids, and new numbered asteroids are being discovered at the rate of over 30 per month.

Asteroids, like planets, comets, and moons, are named. A newly discovered asteroid is given a six-character temporary "name" based on the date of discovery. The first four characters are the year of discovery, followed by a letter indicating which half-month of the year the discovery took place. The final character is a letter assigned sequentially to the asteroids discovered in the half-month in question. Thus asteroid 1998 CE would be the fifth asteroid discovered in the first 2 weeks of February in 1998. If a half-month has more than 24 discoveries, then the letter sequence starts over with additional numerical characters added as a subscript. The 25th object discovered in the first half of February would be 1998 CA_1. If an asteroid is observed long enough to determine its orbit accurately, the ob-

TABLE I
Diameters of the 20 Largest Asteroids

Asteroid name and number	Asteroid class	Semimajor axis	Diameter (km)
1 Ceres	C	2.767	940
4 Vesta	V	2.362	576
2 Pallas	B	2.771	538
10 Hygeia	C	3.144	430
704 Interamnia	D	3.062	338
511 Davida	C	3.178	324
65 Cybele	P	3.429	308
52 Europa	C	3.097	292
87 Sylvia	P	3.486	282
451 Patientia	C	3.063	280
31 Euphrosyne	C	3.156	270
15 Eunomia	S	2.644	260
324 Bamberga	C	2.683	252
3 Juno	S	2.670	248
16 Psyche	M	2.922	246
48 Doris	C	3.112	246
13 Eugenia	C	2.576	244
624 Hector	D	5.201	232
24 Themis	C	3.133	228
95 Arethusa	C	3.068	228

ject is given a permanent number and a name. The numbers are not assigned in order of discovery, but sequentially by order of orbit determination. Once an asteroid has a number, its discoverer has the right to suggest a name for the object. Asteroids are unique in that the naming has few, if any, rules and can be named after persons living, dead, or imaginary; mythological characters or creatures; or, in several cases, pets.

C. SIZES AND SHAPES

The primary defining characteristic of an asteroid is its size, i.e., "smaller than a planet." Table I gives a listing of the diameters of the 20 largest asteroids. Asteroids drop rapidly, with the largest asteroid 1 Ceres being almost twice as large as the next smaller. The sizes decrease rapidly from this point so there are only five asteroids with diameters greater than 400 km and only three with diameters between 400 and 300 km. The asteroid population starts to become relatively abundant only below 300 km in diameter.

This size–frequency distribution where the number of objects increases exponentially as the size decreases linearly is called a "power law" size distribution. In this case the number of fragments N with mass greater than m is shown by m raised to the negative power q:

$$N(>m) \propto m^{-q}.$$

Asteroids, by observation, follow an approximately -3 power law exponent. This is consistent with an initial population of strong, solid bodies that have been ground down by repeated impacts with each other over the age of the solar system. Most asteroids today are probably simply fragments of larger, midsized parent bodies that collided with other similarly sized asteroids and shattered themselves into much smaller fragments. This power law is seen not only in the sizes of asteroids, but in the sizes of the craters on the Moon, Mars, and the moons of Jupiter and Saturn. On all these bodies, the distribution of crater sizes shows that the asteroid population follows a power law. Interestingly, the crater sizes on these different objects do not follow the same power exponent. It appears that the asteroid size–frequency distribution is somewhat different in different parts of the solar system. There is no currently accepted explanation for this observation, but rather is another of the suprising questions that continually arise from the exploration of the solar system. Given the conditions in the asteroid belt today, only the largest asteroids are large enough to have survived from the beginning of the solar system.

The power law predicts and observations confirm that by far the most common asteroids are the smallest. Asteroid search programs using powerful telescopes, extremely sensitive CCD sensors, and state-of-the-art software regularly find asteroids in near-Earth space with diameters of only 5–10 m. The primary limitation on our ability to find asteroids is their size. Smaller objects reflect less light and, after a point, a small object is not observable because the light it reflects drops below the limiting sensitivity of the telescopic system trying to detect it. The good news is that we have probably discovered and tracked all asteroids in the main asteroid belt and near-Earth space larger than 20 km. The bad news is that there are thousands of small asteroids, some in Earth-crossing orbits, several kilometers in diameter, that remain undiscovered and potential threats to Earth. [*See* NEAR-EARTH ASTEROIDS.]

Because most asteroids are probably collisionally produced fragments of larger asteroids, it should not be a surprise that they are not perfect spheres. The three asteroids that have been imaged directly all show very irregular shapes. The collection of shapes can be increased by using radar imaging of near-Earth asteroids and these are equally as irregular. Figure 2 shows a radar image of the asteroid 1620 Geographos. These shapes can be characterized as triaxial ellipsoids, which are objects that have different dimensions on each of their principle axes. Whereas a sphere can be characterized by one dimension, its diameter, a triaxial ellipsoid requires dimension for each axis x, y, and z, and some axes can be radically different from the others. In the case of Ida the long dimension in Fig. 1 is approximately three times greater than the short dimension. Shapes and diameters can also be approximated by several other methods. The most common are light-curve inversion, radiometric modeling, and speckle interferometry.

Star/asteroid occultations provide a direct measurement of an asteroid's shape and a rare opportunity for amateur astronomers to become involved in significant scientific research. The principle is simple: Because asteroids are much smaller than the Earth when they

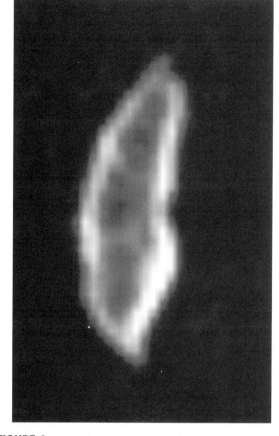

FIGURE 2 Radar image of the asteroid 1620 Geographos. The long axis is 5.1 km and the narrow axis visible in this image is a maximum of 1.8 km wide. Although this object is an extreme example, all the asteroids "seen" so far by either spacecraft or radar are very irregular in shape. (Courtesy of NASA/JPL/Caltech.)

pass in front of or "occult" the essentially parallel light from a star, the pattern and the timing of the occultation will be directly proportional to the asteroid's size and shape. Essentially the asteroid is creating a "shadow" in the starlight projected on the Earth. Observers in different parts of the world simply time the disappearance of the occulted star by a common clock and reconstruct their "chords" or time-tagged observations of the star winking in and out behind the asteroid. With modern equipment available to amateurs such as CCD detectors, computer-driven imaging systems, precise broadcast time, and Global Positioning System locators, these measurements can be taken with very high accuracy and provide an excellent "snapshot" of the two-dimensional shape of the asteroid at the moment of occultation.

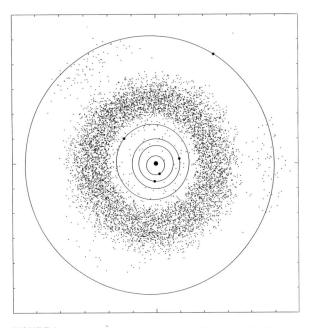

FIGURE 3 Positions projected onto the ecliptic plane for October 4, 2000 of the planets Mercury through Jupiter and 7722 numbered asteroids with accurately known orbits. (Copyright 1998 Institute for Remote Exploration.)

II. LOCATIONS AND ORBITS

A. ZONES, ORBITS, AND DISTRIBUTIONS

Asteroids orbit in almost any region of the solar system, but as shown in Fig. 3, by far the largest concentration is found in the "belt" between 1.8 and 4.0 AU. Figure 4 graphs asteroid populations relative to their semimajor axis (an orbit's average distance from the Sun) and highlights some of the peculiar structures of the asteroid belt. The first impression is of concentrations of asteroids separated by gaps where the asteroid population is either thin or nonexistent. These gaps are called *Kirkwood gaps* and are located where the orbital period of asteroids are simple fractions of the period of Jupiter such as one-fourth, one-third, one-half, or two-fifths. These gaps are formed by the steady influence of Jupiter's huge gravitational attraction. An asteroid with an orbital period of a simple fraction of Jupiter's 11.86-year period will be in resonance with Jupiter and have a close approach in the same place in its orbit over and over again. For example, for an asteroid with a 6-year period, the resonance will occur every other orbit. Jupiter's pull will impart some energy to the asteroid's

orbit, forcing a steady change or perturbation that will reoccur every 12 years. Asteroids that are not in resonance with Jupiter are still affected by its gravitational perturbations, but because these interactions occur at random time intervals and at random locations of the orbit, they will on average cancel out and not change the orbit. The effect of a Jupiter resonance is increased eccentricity of an asteroid's orbit. This does not change its "average" distance from the Sun, but will make the perihelion move closer to the Sun and the aphelion farther from the Sun. Over time increased eccentricity will cause the asteroid to become planet crossing, and close encounters with the inner planets or Jupiter will alter its orbit drastically. The end result is either a collision with a planet or ejection from the solar system. For asteroids, orbital life in Kirkwood gaps is short, but exciting.

Moving a little farther out in the main belt are the Hilda asteroids at 4.0 AU. The Hildas demonstrate the difficulties of generalizing on the subject of orbital dynamics. In the main belt, orbital resonances with Jupiter produce gaps in the asteroid population, but the Hildas are in a two-thirds resonance with Jupiter, which in this case produces a concentration of asteroids.

The Trojan asteroids lie in the orbit of Jupiter but reside in dynamically stable zones 60° ahead and behind Jupiter. These are the "Lagrangian points," named after the French theorist J. L. Lagrange who suggested that two objects can share the same orbit indefinitely as long as the objects form an equilateral triangle with the Sun. A total of 416 Jupiter Trojans have been discovered, and, in theory, any planet can have a Trojan swarm of asteroids sharing its orbit, but to date only one other planet (Mars with asteroid 5261 Eureka) has been shown to possess an asteroid at its Lagrangian points.

The next major group of asteroids are the Centaurs, found between Saturn and Uranus. These may not be asteroids at all but rather large comets perturbed inward from the Kuiper Belt. Because Centaurs orbit deep in the outer solar system, they have little chance to warm sufficiently to show cometary activity, so they are considered asteroids until proven otherwise. However, even at these extreme distances some of the seven Centaurs discovered to date have shown brief periods of cometary activity.

The final outward group is the Kuiper Belt. Objects populating space beyond the orbit of Neptune but inside about 1000 AU. Again these objects are probably cometary. In fact the existence of the Kuiper Belt was first suggested in 1949 as a source area for short-period comets. The first object was discovered in 1992 (1992

A mosaic of Hubble Space Telescope images depicting the temporal evolution of the Shoemaker-Levy 9 G impact. The "string of pearls" comet (seen above) broke into 21 fragments, all of which impacted Jupiter in mid-July of 1994. Below, the progression of views from lower left to upper right show: an impact plume 5 minutes after contact; the impact site after 1.5 hours; after 3 days and subsequent fragment impacts; after 5 days. (Images courtesy of R. Evans, J. Trauger, H. Hammel and HST Comet Science Team, National Space Science Data Center, and NASA)

Whole-rock samples of common meteorite types (approximate longest dimension, in cm). Top left: Whitman, H5 (6 cm); top right: Allende; C3V (8 cm)—note 1 cm chondrule in center; center left: Springwater pallasite (18 cm); center right: Sioux Co. eucrite (8 cm); bottom left: Sanderson IIIB medium octahedrite (13 cm)—note large FeS inclusions. (All Arizona State University) Bottom right: nearly complete fusion crust, the Noblesville H chondrite, which fell on August 31, 1991. (NASA Johnson Space Center)

Large meteoroids: Above, a fireball (shown by arrow) of an 80 m object with estimated mass of 1 MTonne, that apparently skipped out of the atmosphere, observed on August 10, 1972 moving left to right over Grand Teton National Park. (Photo courtesy of Dennis Milon) Below, the 1–km diameter Meteor Crater in Arizona formed by the explosive impact of the Canyon Diablo IA octahedrite meteoroid about 50 ka ago. (Photo courtesy of Allan E. Morton)

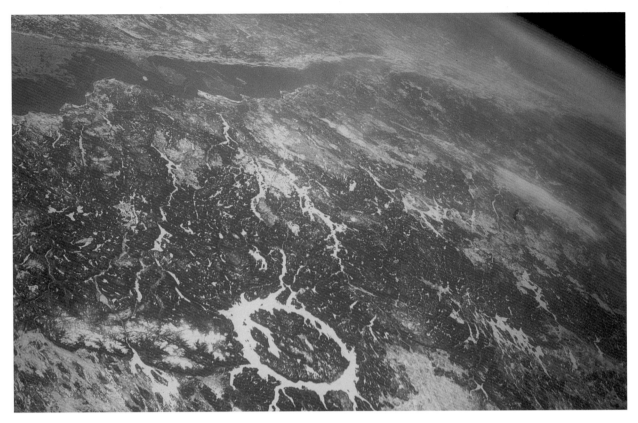

Above, a Space Shuttle image of the Manicouagan impact basin, located near Quebec, Canada. This structure was formed 210 ± 4 million years ago and is one of Earth's largest visible impact craters. Glaciers have eroded impact-brecciated rock and have left a 70 km annular lake between more resistant igneous impact-melt rocks. (NASA)

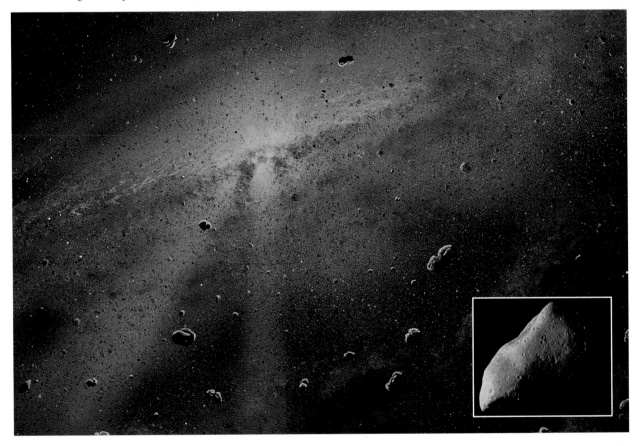

Painting of the formation of the asteroids in the solar system. Asteroid 951 Gaspra is seen in the inset at lower right, from a picture obtained by the Galileo spacecraft during its approach to the asteroid on October 29, 1991. (NASA/JPL)

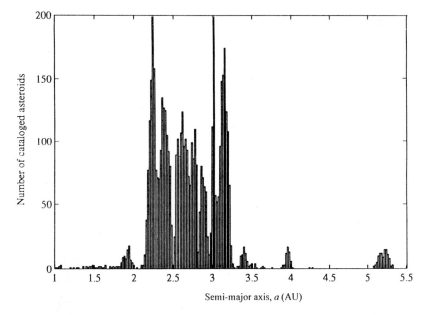

FIGURE 4 Asteroid populations relative to their semimajor axis. Kirkwood gaps are visible as areas of depleted populations within the main belt. For instance, the 3:1 resonance with Jupiter is at 2.5 AU and shows a deep "well" of missing asteroids. (Courtesy of Charles Kowal, "Asteroids: Their Nature and Utilization." Wiley, New York.)

QB1) with a semimajor axis of 44 AU and an estimated diameter of several hundred kilometers. Given the size of the 60 Kuiper Belt "asteroids" discovered so far, there are probably hundreds of thousands of objects larger than a kilometer populating this belt. [*See* KUIPER BELT OBJECTS.]

Inward from the main asteroid belt are asteroids that cross the orbits of the inner planets: the Amor, Apollo, and Aten asteroids. Amor asteroids are Mars crossers, but do not reach the orbit of the Earth. Apollos cross Earth's orbit but their semimajor axis is always ≥1 AU. This differentiates them from Atens, which also cross the Earth's orbit but have a semimajor axis inside of Earth's orbit. These objects are collectively called near-Earth asteroids (NEAs). They are relatively small objects, with the largest known NEA being Amor 1036 Ganymed with a diameter of 38.5 km. NEAs are also subject to a power law distribution, so as the population increases their sizes drop rapidly. There are currently approximately 200 known NEAs and it is estimated that the total population of NEAs larger than a kilometer in diameter is at least 1500. Compared to the size and population of the main belt, 1500 kilometer-sized asteroids seem a trivial assembly except that these objects can and, in the course of geologic time, do collide frequently with Earth. Compositional data indicate that NEAs are drawn from every zone of the asteroid belt with dark, probably primitive, asteroids originating in the outer belt coexisting with differentiated asteroids drawn from the inner belt. Sources for NEAs are probably Kirkwood gaps. NEAs may have started as fragments of larger asteroids that were thrown into Jupiter resonance by collisions with other large asteroids.

B. FAMILIES AND SATELLITES

As discoveries of asteroids accumulated in the early part of this century, astronomers noted that it was common for several asteroids to have very similar orbital elements and that asteroids tended to cluster together in semimajor axis, eccentricity, and inclination space. In 1918 K. Hirayama suggested that these clusters were "families" of asteroids. Figure 5 shows a plot of orbital inclination vs semimajor axis with some of the major families identified. These families are probably the result of the collisional breakup of a large parent asteroid into a cloud of smaller fragments sometime in the distant past. Time and the gravitational influence of other solar system objects have gradually dispersed the orbits of these fragments, but not enough to erase the characteristic clustering of families. Hirayama suggested five families, and this number has been increased greatly by the work of generations of orbital dynamists.

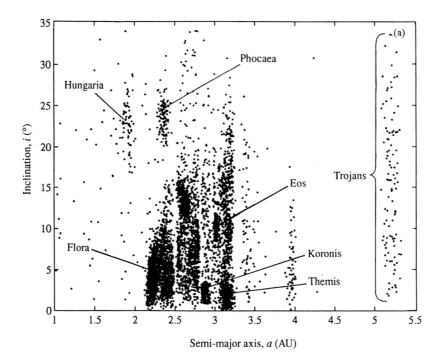

FIGURE 5 Plot of orbital parameters of numbered asteroids in semimajor axis vs inclination space. Major asteroid families and planets are identified in the plot. (Courtesy of Charles Kowal, "Asteroids: Their Nature and Utilization." Wiley, New York.)

The study of asteroid families provides a probe into the internal structure and composition of the original parent. Particularly for differentiated asteroids, families can provide a glimpse at geologic units that are usually deeply hidden in the interiors of planets. A family may have members that represent the metallic core, a metal-rock transition zone called the core–mantle boundary, a collection of dense, iron-rich units in the mantle, and others in the crust of the former planetesimal.

For decades in asteroid science, the question of the possibility of asteroids having satellites was a hotly debated issue. Theorists postulated the possibility or impossibility with mathematical arguments, and observers reported tantalizing hints of possible companions to asteroids. The hints included "secondary events" during occultations (the star blinking out an extra time) and unusual light-curve patterns that could be explained by an eclipsing satellite. As is so often the case, arguments continued right up to the day that a picture, shown in Fig. 6, of the asteroid 243 Ida and its satellite Dactyl was produced. It is theorized that Dactyl was formed from material that was thrown off of Ida during an impact. This debris may have been ejected strongly enough to escape the surface but not to escape the gravitational attraction of Ida. Because the Ida/Dactyl system was only the second asteroid

ever encountered by spacecraft, this result indicates that asteroids can have satellites and that satellites may be common.

C. ASTEROIDS AND METEORITES

There are several strong lines of evidence (which will be reviewed in the next section) that meteorites come from the asteroid belt. As such this material represents "cheap" (compared to the cost of a space mission to sample an asteroid) samples of the highly diverse, numerous, and effectively inaccessible asteroid population. The advantage of having these samples delivered to us is that they can be subjected to the level of sophisticated and highly detailed analysis that can only be done in a laboratory. This gives us access to priceless details on the earliest period of solar system formation, the mineralogy of the asteroid belt, and the processes that have shaped asteroids over the life of the solar system. However, the scientific price we pay for these cheap samples is that we lose what geologists call *provenance,* information on the location, structural setting, and context of the sample. That information is vital to understanding how a meteorite and its parent asteroid fit into the story of the origin and evolution of the solar system. Meteorites do not automatically provide

FIGURE 6   243 Ida and its satellite, Dactyl. (Courtesy of NASA/ JPL/Caltech.)

D. DELIVERY OF METEORITES

There are a number of lines of evidence that show the ultimate source region for meteorites is the asteroid belt. The strongest evidence is the direct observations of three recovered ordinary chondrites that were photographed falling by camera networks. The network observations allowed for not only the recovery of the fall, but also the determination of the object's orbit. In each case the aphelion lay in the asteroid belt. Other evidence includes the similarity of meteorite reflectance spectra to several classes of asteroids. The existence of xenoliths (pieces of other meteorite types included in meteorite breccias) in meteorites requires that the source region have the mineralogical diversity found in the asteroid belt. The solar–wind-implanted gases found in gas-rich regolith meteorites indicate that implantation took place in regions consistent with the location of the asteroid belt.

There are several factors that bias the population of meteorites arriving on Earth and therefore limit our sample of the asteroid belt. First, the dynamic processes that deliver meteorites from the asteroid belt to Earth are probably biased strongly toward sampling relatively narrow zones in the asteroid belt. Calculations suggest that the vast majority of meteorites and planet-crossing asteroids originate from just two resonances in the belt, the 3:1 Kirkwood gap and the so-called ν_6. Both of these zones are in the inner asteroid belt where the asteroid population is dominated by material that is differentiated or highly metamorphosed.

A second factor is the relative strength of the meteorites. Many meteorites begin the process of evolving into an Earth-crossing orbit by being ejected at high velocity from the parent body by a major impact. To survive the stress of impact and acceleration without being crushed into dust, the meteorite must have substantial cohesive strength. At the opposite extreme, the almost completely metallic iron meteorites and the mostly metallic stony irons have such great cohesive strength that it would be difficult to break off pieces for ejection. Strength is also an important selection factor for meteorites that survive the deceleration and heating of atmospheric entry. Atmospheric entry usually involves a variety of thermal and dynamical stresses that typically break up most stony meteorites from one large individual into showers of much smaller stones. These selection effects introduce biases of unknown magnitude into the link between asteroids and meteorites that limit the usefulness of the meteorite collection as a representative sample of the asteroid belt. It is very likely that the meteorites available to us represent

the location and taxonomic class of their parent bodies. The very fact that a meteorite is literally "in our hands" indicates that some violent event occurred that may have fragmented and perhaps destroyed the parent body. The best that can be done is to link individual asteroid spectral classes with meteorite compositional groups. This is done by comparing laboratory visible and near-infrared reflectance spectra of meteorites to telescopically obtained spectra and albedo of asteroids. This task is somewhat speculative as most meteorites have never been on the surface of an asteroid, asteroid surface conditions are unknown, and the effects of the space environment on the surface materials are poorly understood. All spectral matches between asteroids and meteorites, including the ones detailed here, should be viewed with healthy skepticism. [*See* METEORITES.]

only a small fraction of the asteroids and that most asteroids either cannot or only rarely contribute to meteorite collections.

E. RELATION TO COMETS

As discussed in the first section, comets were known to the ancients as "long-haired" wanderers whose paths, appearance, and apparitions were mysterious and unpredictable. Comets are characterized by compositions that include a frozen but highly volatile collection of compounds such as water, carbon monoxide, carbon dioxide, methane, ammonia, hydrogen cyanide, methyl alcohol, and many others that were probably formed in the cold outer reaches of the solar system. For comets to be stable over the life of the solar system, they must be "stored" in orbits well beyond Jupiter so that solar radiation will not warm and volatilize their inventory of frozen gases. These storage yards of comets include the Oort Cloud beyond the orbit of Pluto and the already discussed Kuiper Belt between the orbits of Uranus and Neptune. However, dynamic events such as the gravitational pull of a passing star perturb the orbits of enough comets that every year a few are sent into the inner solar system. The resulting orbits are typically highly inclined and very eccentric so that a comet crosses the orbits of several planets on its course into the inner solar system. [*See* COMETARY DYNAMICS; PHYSICS AND CHEMISTRY OF COMETS.]

The difference between comets and asteroids are primarily two factors: First is the obvious physical appearance of the object in the inner solar system. Comets inside the orbit of Jupiter throw out spectacular gas and dust tails whereas asteroids have no detectable gas emissions. Comets are literally burning themselves out by shedding a major fraction of their mass in the gas that is being boiled off by the heat of the Sun. The second factor is the characteristic short-lived high inclination, high eccentricity orbit. An orbit that repeatedly crosses the orbits of the largest planets in the solar system is not going to last very long in astronomical terms. One standard for discriminating between asteroids and comets is the basis of their orbits. Dynamists use the Tisserand invariant (a mathematical expression that relates oribtal motions in a restricted three-body problem). The rule of thumb is that asteroids have Tisserands less than 3 and comets have Tisserands greater than 3. The fate of comets is to either fall completely apart into a stream of dust from the stresses of losing gas (this is probably the source of the spectacular meteor showers that sometimes grace the night skies) or to encounter a planet. The planetary

encounter will (1) destroy the comet by collision (such as Shoemaker-Levy 9 and Jupiter, for example), (2) eject it from the solar system, or (3) circularize the orbit into one similar to an asteroid's. Here is where the difference between comets and asteroids blurs. If a comet has a lucky encounter with a planet, its orbit can become much like those of asteroids and will be likely to survive the age of the solar system. There are several objects such as 4015 Wilson-Harrington, which straddle the fence between asteroids and comets. This object was originally observed in 1948 as a comet. It was subsequently lost and in 1978 rediscovered as asteroid 1979VA. Only later was it noted that it was highly probable that the 1948 comet was the "asteroid" discovered in 1979. This case emphasizes that comets that lose their volatiles become asteroid-like, particularly if they have orbits that have been circularized by one or more encounters with planets. A number of asteroids may have compositions rich in volatile materials, but either orbit too far in the outer solar system for volatilization to take place or have surfaces that are covered with a thick lag of dust and rock that insulates the volatile core from observable degassing. Either way, this demonstrates the compositional and evolutionary heterogeneity of asteroids and the continuing supply of surprises in store for those who study asteroids.

III. COMPOSITION AND CHARACTERISTICS

A. TELESCOPIC OBSERVATIONS OF COMPOSITION

Our understanding of the composition of asteroids rests on two pillars: the detailed study of meteorite mineralogy and geochemistry and the use of remote-sensing techniques to analyze asteroids. As discussed in a previous section, meteorites provide an invaluable but limited sample of asteroidal mineralogy. To extend this sample to what are effectively unreachable objects, remote sensing uses a variety of techniques to determine asteroid composition, size, shape, rotation, and surface properties. The best available technique for the widespread study of asteroid composition is visible and near-infrared reflectance spectroscopy using ground-based and Earth-orbiting telescopes. Reflectance spectroscopy is fundamentally the analysis of the "color" of asteroids over the wavelength range of $0.2-3.6$ μm. An experienced rockhound limited to the three colors of the human eye can identify a surprisingly wide variety of rock-forming minerals. The important silicate olivine is green, pyroxenes are dark blue, and important copper minerals such as azurite are colored vividly. These colors are a fundamental diagnostic property of the mineralogy. The atoms of a mineral's crystal lattice interact with light and absorb specific wavelengths depending on its structural, ionic, and molecular makeup, producing a unique reflectance spectrum. The reflectance spectrum is essentially a set of colors, but instead of three colors our remote-sensing instruments "see" very precisely in 8, 52, or even several thousand colors. What can be seen are very precise details of the mineralogy of the major rock-forming minerals olivine, pyroxene, and feldspar; the presence of phyllosilicates, organic compounds, and hydrated minerals; and the abundance of free iron and opaque minerals.

In addition to a spectroscopic inventory of minerals, telescopic measurements yield several other critical pieces of information. The albedo, or fundamental reflectivity of the asteroid, can be determined by measurements of the visible reflected light and the thermal emission radiated at longer wavelengths. A dark asteroid will absorb much more sunlight than it reflects, but will heat up and radiate that extra absorbed energy at thermal wavelengths. Ratioing reflected and emitted fluxes at critical wavelengths provides an estimate of an asteroid's albedo. Reflectance measured at a series of phase angles can be used to model the photometric properties of the surface material and estimate physical properties such as surface roughness, surface soil compaction, and the light-scattering properties of the asteroidal material. Phase angle measurements can also characterize the polarization of the surface and provide insight into the texture and mineralogy of the surface.

B. TAXONOMY AND DISTRIBUTION OF CLASSES

The basic knowledge of asteroids is limited primarily to ground-based telescopic data, usually reflectance spectra in the visible and near-infrared wavelengths and albedos that are indicative of composition. These data are the basis of asteroid taxonomy. Asteroids that have similar color and albedo characteristics are grouped together in a class denoted by a letter or group of letters. Asteroids in particularly large classes tend to be broken into subgroups with the first letter denoting the dominant group and the succeeding letters denoting less prominent spectral affinities or subgroups.

Asteroid taxonomy has developed in tandem with the increase in the range and detail of asteroid observational data sets. Early observations were often limited in scope to the larger and brighter asteroids, and in wavelength range, to filter sets used for stellar astronomy. As observations widened in scope and more specialized filter sets and observational techniques were applied to asteroids, our appreciation of the variety and complexity of asteroid spectra also increased. The asteroid classification system has evolved to reflect this complexity, and the number of spectral classes has increased steadily. Table II gives a listing of the expanded "Tholen" asteroid classes and the current mineralogical interpretation of their reflectance spectra.

After the first asteroid taxonomy was proposed it was clear that the new asteroid classes were not distributed uniformly throughout the asteroid belt. The S class dominated the inner asteroid belt whereas the C class was far more abundant in the outer asteroid belt. The most populous taxonomic classes (E, S, C, P, and D classes) peak in abundance at different heliocentric distances. In addition, the distribution of each class is roughly Gaussian in shape and tends to range over about 1 AU. Figure 7 shows the distribution of taxonomic classes. What does this remarkable heliocentric distribution mean? If we assume that the bulk of spectral and albedo differences between asteroid classes reflects real differences in mineralogy, then what we are seeing are rough compositional zones in the asteroid belt. These zones probably are the result of a combination of effects that include the primordial distribution of mineralogy, the subsequent thermal evolution of the asteroid belt, and the sum of dynamic processes that scatter and pulverize asteroids.

The dominant cause of this taxonomic zonation may be an echo of the conditions in the early solar nebula. During the earliest period of its formation the solar system may have been a flattened whirlpool of hot gases and dust. As the nebular cooled, the temperature, pressure, and chemical state of the nebular gas controlled the mineralogy of the condensing grains. According to models of solar system condensation, the high to moderate temperature silicate minerals would tend to dominate the inner solar system, whereas lower temperature carbonaceous minerals would be common in the cooler, outer regions of the solar system. The transition between moderate and low temperature nebular condensates is apparently what is seen in the taxonomic zonation of the asteroid belt. The innermost major group of asteroids is the E class, which are thought to be composed of the iron-free silicate enstatite, indicating formation under relatively reducing conditions. This class peaks in abundance at about

TABLE II
Meteorite Parent Bodies

Asteroid class	Inferred major surface minerals	Meteorite analogues
Z	Organics + anhydrous silicates? (+ice??)	None (cosmic dust?)
D	Organics + anhydrous silicates? (+ice??)	None (cosmic dust?)
P	Anhydrous silicates + organics? (+ice??)	None (cosmic dust?)
C (dry)	Olivine, pyroxene, carbon (+ice??)	"CM3" chondrites, gas-rich/blk chondrites?
K	Olivine, orthopyroxene, opaques	CV3, CO3 chondrites
Q	Olivine, pyroxene, metal	H, L, LL chondrites
C (wet)	Clays, carbon, organics	CI1, CM2 chondrites
B	Clays, carbon, organics	None (highly altered CI1, CM2??)
G	Clays, carbon, organics	None (highly altered CI1, CM2??)
F	Clays, opaques, organics	None (altered CI1, CM2??)
W	Clays, salts????	None (opaque-poor CI1, CM2??)
V	Pyroxene, feldspar	Basaltic achondrites
R	Olivine, pyroxene	None (olivine-rich achondrites?)
A	Olivine	Brachinites, pallasites
M	Metal, enstatite	Irons (+EH, EL chondrites?)
T	Troilite?	Troilite-rich irons (Mundrabilla)?
E	Mg-pyroxene	Enstatite achondrites
S	Olivine, pyroxene, metal	Stony irons, IAB irons, lodranites, windonites, siderophyres, ureilites, H, L, LL chondrites

2 AU. The next group out is the S class (which includes as many as seven compositionally distinct subclasses), which peaks in abundance at 2.3 AU. This class is thought to be rich in the moderate temperature silicates olivine and pyroxene and may also contain large amounts of free metal. The mix of mineralogies in the S class suggests more oxidizing conditions in this region of the solar nebula. The C class, which peaks in abundance at 3 AU, shows a major transition in asteroid mineralogy. Compared to the S class, the C asteroids probably contain less free metal, more oxidized silicates, important low temperature carbon minerals, and significant amounts of volatiles such as water. The P asteroids, which peak in abundance at about 4 AU, and the D asteroids, peaking at 5.2 AU, dominate the outer asteroid belt. Although there are no direct meteorite analogues for these asteroid types, their spectra may indicate mineralogies rich in low temperature materials such as carbon compounds, complex organics, clays, water, and volatiles. The P and D classes are probably transitional objects between the rocky asteroids of the main belt and the volatile-rich comets in the Kupiter Belt and the Oort Cloud. The general trend of the taxonomic zonation supports the predictions of some solar system condensation models and may provide a guide to the chemical and mineralogical variations within the solar nebula.

Several processes since the formation of the asteroids from the solar nebula may have blurred the taxonomic imprint from the original condensation. Appar-

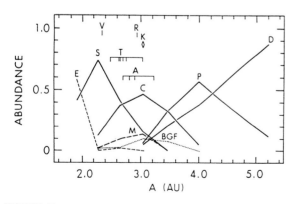

FIGURE 7 Distribution of taxonomic classes from Bell *et al.* (1989). (Courtesy of University of Arizona Press.)

ently a radially dependent thermal event heated much of the asteroid belt soon after accretion. Evidence from meteorites shows that some parent asteroids were completely melted (basaltic achondrites, irons, stony irons), some asteroids were strongly metamorphosed (ordinary chondrites), and some were heated only enough to boil off volatiles and produce aqueous alteration (CI and CM carbonaceous chondrites). This event seems to have been much more intense in the inner asteroid belt and strongly effected the E, S, A, R, V, and M class asteroids. The dynamic interaction of asteroids with each other and the planets, particularly Jupiter, has altered the original orbital distribution of asteroids and cleared whole sections of the belt. The net result probably has been to expand the original compositional zones and to produce orbital overlaps of zones that once may have been distinct from each other.

C. ASTEROID COMPOSITIONS

Any interpretation of asteroid composition must be an extrapolation of our limited and biased sample of meteorites and our extensive, but often not very informative, collection of meteorite and asteroid reflectance spectra to a much larger and more complex population. Any systematic attempt to suggest composition for the many asteroid classes will include large uncertainties and will need to be treated for the foreseeable future as a work in progress. Table 2 gives the current best mineralogical interpretations and possible meteoritic analogues for the asteroid classes.

To explain the compositional meaning of asteroid reflectance spectra, we will treat the asteroid belt as a series of zoned geologic units, starting at the outer zones of the main belt and working inward toward the Sun. The asteroid outer belt is dominated by the P and D classes. Spectra of these types are dark, red to very red, generally anhydrous (lacking absorptions associated with water or hydrated clays), and relatively featureless. Figure 8 shows spectra of representative members of these classes. No direct analogues for these asteroid types exist in meteorite collections. The analogues cited most commonly are cosmic dust or CI carbonaceous chondrites that are enriched in organics. However, the spectral characteristics of these asteroids are difficult to duplicate with material that is delivered to the inner solar system. P and D asteroids are probably composed of primitive materials that have experienced different geochemical evolution than cosmic dust or CI chondrites. The Z class was proposed for the trans-Jovian asteroid 5145 Pholus, which has the reddest visible and near-infrared reflectance spectrum

observed so far. Once again, this object does not have a meteorite analogue but is probably dominated by primitive material such as complex organics. There is a general increase in "redness" with increasing heliocentric distance in the outer belt asteroids going from the moderately red P asteroids, through the redder Ds, to the much redder Z asteroid.

The dark inner belt asteroids of the B, C, F, and G classes are characterized by relatively featureless flat spectra as shown in Fig. 8. The proposed analogues for most of these asteroids are the dark CI and CM carbonaceous chondrite meteorites, and the spectral differences between the asteroid types are thought to represent varying histories of aqueous alteration or thermal metamorphism. The CI carbonaceous chondrites are rich in water, clay minerals, volatiles, and carbon. They probably represent primitive material that has been mildly heated and altered by the action of water. CMs are less water rich, more heated, and have less carbon. These two meteorite types are most analogous to the "wet" C asteroids. Spectra of B, F, and G asteroids also show hydration features similar to CI and CM carbonaceous chondrites, but differences in their spectral slope and UV absorptions suggest these asteroids underwent varying degrees of aqueous alteration. About 40% of the C types are anhydrous, suggesting that the C class may be a collection of objects with roughly similar reflectance spectra but greatly varying mineralogy. Analogues for the "dry" Cs include CM carbonaceous chondrites that have been devolatilized and/or ordinary chondrites that have been darkened by regolith processes.

Sunward of 3 AU, differentiated bright asteroids become much more common. This zone was strongly affected by the early solar system heating event and contains those classes most likely to represent differentiated and metamorphosed meteorites. Perhaps the best asteroid/meteorite spectral matches are the V class asteroids with basaltic achondrite meteorites. Spectrally, V types are interpreted to be a differentiated assemblage of primarily orthopyroxene with varying amounts of plagioclase, which makes them very close analogues to the basaltic howardite/eucrite/diogenite (HED) association of meteorites. Their strong spectral features are shown in Fig. 9. The petrology of these meteorites indicates that they are basaltic partial melts originating on asteroids that underwent extensive heating and differentiation. Thus, these meteorites probably represent the surface melts and upper-crustal rocks of a differentiated asteroid. In addition, similarities in petrology, chemical trends, and oxygen isotopes suggest that all of the HED meteorites are closely related and probably come from a single parent body. A dozen

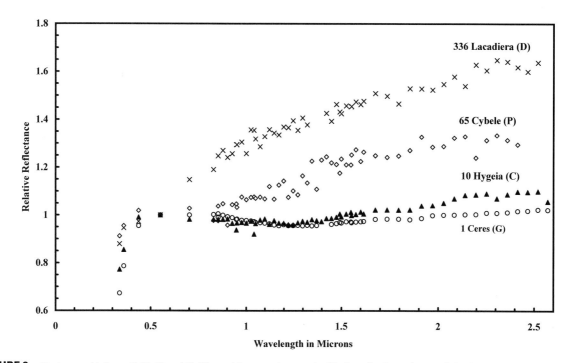

FIGURE 8   Dark asteroid classes P, D, G, and C. These objects are characterized by low albedos and very subdued spectral absorption features.

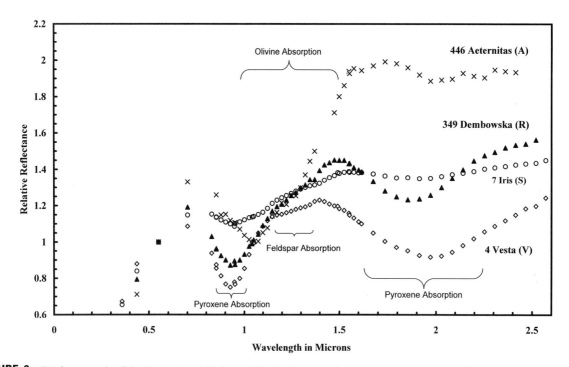

FIGURE 9   Bright asteroids of the V, A, S, and R classes. The bright asteroids are examples of the uses of reflectance spectroscopy in identifying mineral composition. Spectral absorptions of the major rock-forming minerals are labeled. Pyroxene has two absorptions, a sharp feature near 1.0 μm and a broader, shallower feature near 2.0 μm. Olivine has a strong absorption between 1.0 and 1.3 μm, and feldspar has a weak "shoulder" at 1.3 μm.

small V class asteroids have been identified in the main Asteroid Belt in orbits that stretch from near Vesta to near the 3 : 1 resonance that is thought to supply planet-crossing asteroids. These objects are almost certainly ejecta fragments from a major impact on Vesta. This scenario is supported by the *Hubble Space Telescope* identification of a large impact crater near Vesta's South pole, which may be the source of the small V class asteroids. Three V class asteroids have also been identified in earth-crossing orbits. These objects have almost identical V class spectra, have deep, fresh-appearing spectral features that suggest they are relatively young rocky fragments, and are small enough (less than 4 km in diameter) to be fragments of the crust of a disrupted differentiated asteroid. Interestingly, their orbits are very similar, suggesting that they were all part of a larger fragment that was disrupted collisionally while in an earth-crossing orbit. This may be a case where the study of meteorite chemistry and spectra have come together with remote sensing and orbital dynamics to describe the origin, evolution, and current location of a major meteorite parent body.

The A class asteroids are thought to represent the next lower zone of a differentiated asteroid. These asteroids are interpreted to be nearly pure olivine and may be derived from the mantle of extensively differentiated parent bodies. The best meteorite analogues for this interesting asteroid type are the extremely rare brachinites. The 200-g meteorite Brachina is the only non-Antarctic member of this type. This demonstrates what some workers have termed "the great dunite shortage." If some asteroids are differentiated then there should be, along with the identified fragments of crust (V types) and core (M types) material, a substantial amount of olivine-rich mantle material similar to terrestrial dunite. However, A type asteroids are relatively rare and dunite-like meteorites are very rare. Other suggested meteorite analogues for the A type asteroids are the somewhat more abundant pallasite meteorites, but they are also a suggested analogue for some S type asteroids. Another possible mantle-derived asteroid is the R class, which is a single-member class made up of the asteroid 349 Dembowska. Analysis of its reflectance spectra suggests a mineralogy that contains both olivine and pyroxene and may be a partial melt residue of incomplete differentiation. Unfortunately, the meteorite collection contains no potential analogues for this mineralogy.

A more common asteroid class is the M class, which has the spectral characteristics of almost pure iron–nickel metal. Asteroids in this class are thought to be a direct analogue to the metallic meteorites. This

material may represent the cores of differentiated asteroids, and examples of M class spectra are shown in Fig. 10. However, there is a great deal of geochemical variety in the iron meteorite population. The 13 different classes of iron meteorites suggest origins from a number of different parent bodies and/or a variety of geochemical conditions. Recent observations have added new complications to the simple M = metal picture. Six M class asteroids have been shown to have hydrated minerals on their surfaces. The spectral characteristics of M asteroids, moderate albedo and a featureless red-sloped spectrum, can also be characteristics of some clay-rich silicates. This raises the possibility that the "wet" M asteroids are assemblages of clays, like the CI carbonaceous chondrites, but without the carbon-rich opaques that darken CIs. A new class, the W (or "wet") class, has been coined to classify these unusual objects.

The T asteroids may be related to the metal-rich Ms. The Ts are inner asteroid belt objects that have featureless spectra, a strong red continuum slope, low albedo, and no hydration features. Their spectra are similar to the mineral troilite (FeS) that is most common as an accessory mineral in iron meteorites. Although there are no pure troilite meteorites, some iron meteorites are as much as 35% troilite by volume. Theoretical work suggests that the metallic cores of asteroids would tend to concentrate troilite along radial dendrites during their cooling and crystallization. Troilite-rich zones would be much weaker than the bulk metal, and collisions would tend to split the core along these troilite-rich zones. This could create metallic core fragments with surfaces enriched in troilite.

The E class asteroids are another example of the perils of extrapolation from limited information to a convenient meteorite analogue. Looking at the spectrum of the "type" asteroid for the E class, 44 Nysa, it was easy to assume that these asteroids were excellent analogues for bright, red-sloped, but spectrally featureless enstatite achondrites. The only problem was that enstatite meteorites are entirely anhydrous, and 44 Nysa was observed to be strongly hydrated. Although some E class asteroids are probably composed of the same differentiated enstatite assemblages as the enstatite achondrites, about half of the observed Es are hydrated and cannot be composed of anhydrous enstatite. "Wet" E asteroids such as Nysa may be related to the W asteroids and have surfaces rich in hydrated silicate clays.

Perhaps the most complex class of asteroids is the S class. Several authors have suggested breaking up this class into three or seven subclasses. Figure 11 shows the seven subclasses of the Gaffey classification

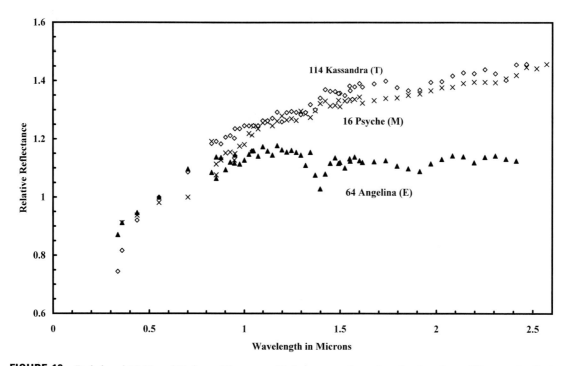

FIGURE 10   Red-sloped M, T, and E classes. These asteroids lack strong absorptions, but have large differences in albedo.

scheme. The intricacies of the subclasses are beyond the scope of this work, but it is enough to know that S class spectra, on average, indicate subequal amounts of olivine and pyroxene and a substantial metallic component. A standard analogue for S asteroids are the pallasites that are assemblages of large olivine grains set in a matrix of iron–nickel metal. This means that some S asteroids may represent the core–mantle boundary regions of differentiated asteroids where the silicates (principally olivine) of the mantle are in direct contact with the metallic core. However, the S class is very large and includes a number of objects that do not conform to this standard interpretation. A number of S asteroids are rich in pyroxene and may represent a larger cross section of the mantle and a lower crust of an asteroid. Meteorite analogues for this "mantle" material include ureilites, lodranites, brachinites, siderophyres, and winonites. Some S asteroids have lower metal contents and may be the parent bodies of ordinary chondrite meteorites. Others (principally the Eos family) have already been split out to form the K class. The K type asteroids have lower albedos and flatter spectral features than most S asteroids, and the CV and CO carbonaceous chondrites are interpreted as their analogues.

A long-running controversy in asteroid science has been the identification of the asteroidal source of ordi-

nary chondrite meteorites. Ordinary chondrites are by far the largest meteorite type, accounting for approximately 80% of observed meteorite falls, but so far only a few asteroids have been identified as the Q class, thought by many scientists to be the ordinary chondrite analogue. A number of S class asteroids have spectral absorption bands roughly similar to those or ordinary chondrites, but S asteroids typically have a strong red continuum slope that is not seen in ordinary chondrites. A number of explanations for the lack of ordinary chondrite parent bodies have been put forward: (1) At least some S asteroids are ordinary chondrites, but regolith processes enrich the metal content of the regolith and increase their apparent spectral red slope, (2) ordinary chondrite parent bodies are in the asteroid belt, but 4.6 billion years of collisions have ground them down to sizes that are too small to observe with current telescopes, and (3) regolith processes can darken ordinary chondrites, so their parent asteroids actually have spectra of the dark C type asteroids. Whatever the answer, this subject promises to be a source of lively debate for some time.

In general, the differentiated asteroids of the V, A, R, S, and M classes may represent examples of a transect from the crust to the core of differentiated asteroids and, as such, can tell us a great deal about the geochemical evolution of a differentiated body. In this

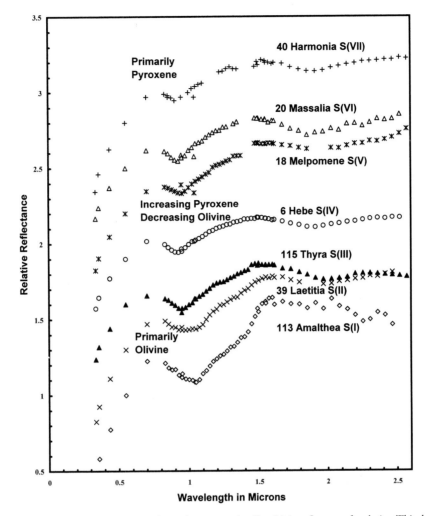

FIGURE 11   The seven "Gaffey" subtypes of S asteroids. Each spectrum is offset 0.3 in reflectance for clarity. This demonstrates the "grab-bag" nature of the S class with the extremes of the subclasses having mineralogies that range from pure olivine in subtype S(I) to pure orthopyroxene in subtype S(VII), with the intervening subtypes being mixtures of varying proportions of these two minerals. The general trend is for decreasing olivine and increasing pyroxene as the subtype number increases.

scenario, the V class asteroids would be the surface and crustal material. The A asteroids would be from a completely differentiated mantle, whereas R asteroids would represent a mantle that experienced only partial differentiation. The S asteroids would be material from some region in the mantle or the core–mantle boundary. Finally, M class materials represent samples of the metallic cores of these asteroids.

D. ASTEROID DENSITY, POROSITY, AND ROTATION RATES

A fundamental physical property of an asteroid is its density. To first order, asteroid density is related to its composition and should be similar to the densities

of analogue meteorites and constituent minerals. However, as is often the case, first-order expectations are often frustrated by unexpected direct measurements. The primary complication is porosity. Asteroids appear to have significant porosity. Asteroids appear to be significantly underdense relative to their suggested meteorite analogues. Meteorite analogues themselves have moderate porosities, suggesting that if the proposed analogues are correct, asteroids commonly have very large porosities. Studies of the collisional dynamics of the asteroid belt have suggested a history of intense collisional evolution and that only the largest asteroids retain their primordial masses and surfaces. Asteroids below 300 km in diameter will have been shattered by energetic collisions with some objects reaccreting to form gravitationally bound rubble piles and the remainder ejected into smaller fragments to be further shattered or fragmented. It may be that most asteroids are shattered heaps of loosely bound rubble with significant porosity in the form of large fractures, vast internal voids, and loose-fitting joints between major fragments. Add the microporosity found in meteorites to this macroporosity, which is likely the remnant of the original accretion of the meteoritic material, and it should be expected that the average asteroid will have a very large porosity. Additional data to support this rubble-pile model of asteroids came from the *NEAR* spacecraft's flyby of 253 Mathilde. This object has six identified impact craters that are larger than the size necessary to shatter the asteroid. The only way that Mathilde could have survived these repeated huge impacts is if it were already a shattered rubble pile that dissipates much of the energy of large impacts in the friction of the pieces of rubble grinding against each other.

Additional data on the collisional history and density of the asteroid belt can be seen in the rotation rates of asteroids. For objects that have not been disrupted catastrophically by collisions, rotation rates are probably set by the accretion conditions of the solar nebula and would tend to be relatively slow (one or two revolutions per day). For small asteroids that are fragments of catastrophic impacts, rotation rates are set by the conditions of angular momentum partitioning during the collision and tend to be much more rapid (more than five revolutions per day). Medium-sized rubble piles lie between these two groups. Interestingly, the distribution of rotation rates as a function of size varies with taxonomic type for the major C, S, and M classes. The density of an object affects the transfer of angular momentum such that, on average, less dense objects will rotate slower for a given size range than denser objects. In relative terms the distribution of rotation rates confirm the general structure of the compositions of these classes: C asteroids are apparently less dense than S asteroids, which are in turn less dense (on average) than M asteroids.

E. SURFACES OF ASTEROIDS

As shown in Figs. 1 and 6, the surfaces of asteroids appear cratered, lined with fractures, and covered with regolith. As discussed in earlier sections, asteroids are affected strongly by collisional disruption and have a complex history of impact fracturing and fragmentation. Objects in the size range shown in Figs. 1 and 6 are probably formed as disrupted fragments from larger objects and are likely rubble piles themselves.

Because asteroids are far too small to retain an atmosphere that could offer some protection from the exposure to space, the surfaces are exposed to an extremely harsh environment. There are a range of processes associated with exposure to the space environment: high levels of hard radiation, high-energy cosmic rays, ions and charged particles from the solar wind, impacts by micrometeorites, impacts by crater-forming objects, and finally impacts by other large asteroids that could destroy the parent asteroid. The overall result of these processes is threefold: (1) large impacts shatter the parent asteroid, creating substantial internal fracturing, porosity, and an extremely rough and irregular surface; (2) small impacts and micrometeorites create a regolith that blankets the asteroid in a fine soil of debris from the bedrock; and (3) micrometeorites, radiation, and solar wind produce chemical and spectral alteration in the regolith soil and exposed bedrock that "weathers" the surface of the asteroids.

The "space weathering" of asteroids has been a hotly debated subject in asteroid science. For years a number of leading asteroid scientists did not believe that regoliths even existed on small asteroids, much less that the regoliths were affected by space weathering. However, of the three small asteroids viewed by spacecraft, all have significant regoliths. On two of these asteroids, 234 Ida and 951 Gaspra, regoliths appear to have been altered by space-weathering processes, although just how this alteration affects asteroidal material is poorly understood. In previous sections, spectra of meteorites and asteroids were compared and matched, but what was not discussed was the fact that most asteroid spectra do not actually have direct matches to meteorite spectra. The typical asteroid spectrum has some troubling differences from the most likely meteorite analogue spectrum. The S class asteroids have a strong spectral red slope that is not seen at all in the population of likely meteorite analogues. Spectra of CI and CM carbonaceous chondrites are typically redder than C asteroids. Ordinary chondrites have spectral charcteristics, particularly their neutral continuum slope, that are seen only in very small, very rare Q class asteroids. In the asteroid population, general spectral trends appear to be associated with the age of an asteroid's surface. The red continuum slope of S class asteroids declines in magnitude with asteroid size. This effect appears to be related to the age of the asteroidal surface, with younger, less altered surfaces tending to be less red. This effect is seen in the meteorite population. Meteorites that have evidence of residing on the surfaces of asteroids have strong spectral differences from meteorites that were not exposed to space. The topic of space weathering continues to be hotly debated and researched and these long lists of unknowns are typical of the lure of asteroids.

IV. MYSTERY, PROMISE, AND PERIL

A. ASTEROIDS AND EXTINCTION

It is a rare but exciting event in science when a single idea by a small group of scientists ignites an entirely new field of study and redefines the scientific debate. That is exactly what happened to such diverse fields as impact physics, asteroid observations, and paleontology after Alvarez and colleagues hypothesized that the iridium anomaly found in Cretaceous–Tertiary (K/T) boundary sediments was the mark of an impact event that destroyed the dinosaurs. The K/T impactor was probably a 10-km-diameter asteroid or comet traveling at about 20 km a second. An object that massive would create an explosion of 60,000,000 megatons and carve a crater over 200 km in diameter. The idea that an impact caused the extinction of the dinosaurs is still the subject of lively debate, but some basic facts are not in dispute. Asteroids hit the Earth all the time and when they do, they cause major explosions. Military satellites built for detecting nuclear explosions and missile launches regularly detect explosions from small asteroids hitting the upper atmosphere. The rate is approximately 1 kiloton-sized explosion every month! These explosions are so high up, so fast, and usually in such remote areas that it is extremely rare that anyone on the ground notices.

Asteroid impacts are a consistent and steady-state fact in the solar system. One just has to look at the extensively cratered surface of any solid body to realize that impacts happen. To some extent, the fact that the Earth has active geological processes that erase the scars of impact craters rapidly and a thick atmosphere that filters out the smaller impactors has lulled us into a false sense of security. The real question is not whether asteroids hit the Earth, but rather how often does it happen. Because impactors are subject to the same power law distribution of sizes that is seen in the asteroid belt, small impacts will be more frequent and large "species-killing" impacts will be much rarer. However, as those who live near dormant volcanoes should realize, rare events on human time scales can be common and frequent events on geologic time scales. Asteroid

impacts on Earth can be predicted in statistical terms by looking at the crater populations on the other planets and by estimating the populations and orbits of Earth-crossing asteroids and comets. From these data Chapman and Morrison estimated a time vs cratering energy curve shown in Fig. 12. It is reasonable to expect kiloton-sized explosions in the upper atmosphere about once a month, megaton-sized explosions at ground level (Tuguska impact in Russia) every few hundred years, 20 megaton blasts (meteor crater in Arizona) every 50,000 years, 100,000 megaton devastation every 300,000 years (these are impacts by one of the 1500 km-sized asteroids in near-Earth orbits), and K/T-sized impacts about every 100,000,000 years. There is plenty of evidence in the geologic and fossil record for repeated major impacts, some of which are associated with mass extinctions. For instance, there were five mass extinctions during the last 600 million years, about what would be predicted by a purely impact-driven extinction model. The bottom line is that asteroid impacts should be treated as one of the steady-state processes that result from a dynamic solar system. Although the chances of a cratering event like the one that dug the almost 1-mile-diameter Meteor Crater in Arizona happening on any random day are small, the probability is 100% that it will happen sometime. The only question is when.

B. TELESCOPIC SEARCHES AND EXPLORATION

No one doubts that having a Meteor Crater-sized impact near your hometown would severely depress property values. When faced with predictable dangers, it is sensible to take precautions. In the same way that people who live on the Gulf coast of North America track hurricanes and people who live in tornado-prone Oklahoma build houses with cellars, is seems a reasonable precaution to identify, track, and study the asteroids in near-Earth space. Several programs do just this by searching the sky with large telescopes using CCDs. In the United States the Spacewatch program in Arizona and the NEAT (Near Earth Asteroid Tracking) program in Hawaii are prolific discoverers of asteroids. The problem with these search and discovery programs is that a few dedicated individuals essentially run them on a shoestring. Only recently has NASA started a program to coordinate and fund NEA searches. This is still a long way from the proposed "Spaceguard" program that would find and track 90% of all the kilometer-sized NEOs for modest amounts of money.

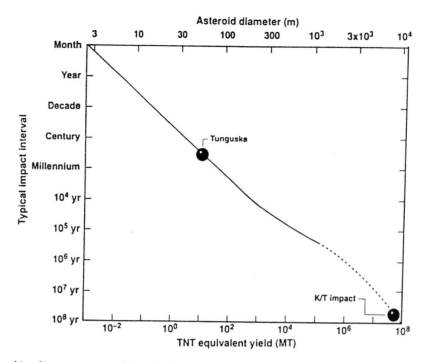

FIGURE 12   Relationship of impact energy and time for the Earth. Small TNT equivalent yields occur frequently, whereas larger yields are much less common, but do occur regularly throughout geological time. (Courtesy of Chapman and Morrison (1994). *Nature* **367**, 35.)

C. SPACECRAFT MISSIONS TO ASTEROIDS

Although telescopic studies are by far the most prolific source of data on asteroids, critical science questions on asteroid composition, structure, and surface processes can only be addressed by spacecraft missions getting close to these objects. The range of spacecraft encounters includes flybys, rendezvous, and sample return missions, which provide information of ever-increasing detail and reliability. We have seen the results of three flybys, two by the *Galileo* spacecraft (243 Ida and 951 Gaspra) on its way to Jupiter and one by the *Near Earth Asteroid Rendezvous* (*NEAR*) spacecraft (253 Mathilde). *NEAR* is the first dedicated asteroid mission and is heading for a January 10, 1999 rendezvous with the asteroid 433 Eros. The plan is for *NEAR* to orbit Eros for one year and map its morphology, elemental abundances, and mineralogy. The long orbit is primarily to allow its X-ray/gamma ray spectrometer (XGRS) enough time to operate (the XGRS determines elemental abundances by counting the energy spectrum of X-rays and γ-rays emitted by the materials). Other instruments include an imaging camera, a near-infrared reflectance spectrometer, a laser range finder, and a magnetometer.

The next mission to asteroids is *Deep Space 1* (*DS1*), which will launch in October, 1998 and include flybys of asteroids. *DS1* carries a multiwavelength reflectance spectrometer and an advanced magnetometer, but this is primarily a technology demonstration to test the new solar–electric propulsion ion drive system.

The Japanese space agency is developing an ambitious *Muses-C* asteroid sample return mission. It is planned to launch in January, 2002 and rendezvous with asteroid 4660 Nereus in September, 2003. The spacecraft will deposit a soft lander on the surface equipped to transfer samples back to the spacecraft. The lander will also carry a small rover supplied by NASA that will traverse the asteroid and collect scientific data from a number of sites. After 2 months, the spacecraft will return to Earth, landing a recovery capsule via parachute in January, 2006.

D. ASTEROIDS AS RESOURCES

This article has described how NEAs can be a serious threat to the Earth and its inhabitants, but asteroids can also be a major source of benefit to humankind. Asteroids have two characteristics that make them uniquely valuable for civilization in general and space exploration in particular. First, they have a wide range of compositions that include a number of critical elements and compounds. Low temperature primitive asteroids probably have complex carbon compounds and organics as well as hydrated clays. These materials could be mined and processed for a variety of products to support space exploration. These include fuel, water, and oxygen, arguably the most valuable commodities to spacefarers. In addition to supporting space stations and colonies directly, asteroids can be valuable sources of commercial ores. An ore in geologic terms is simply a concentration of an element or mineral that is high enough per unit mass to be extracted profitability. Asteroids are full of useful elements, and probably the most useful are metallic asteroids that have concentrated a number of valuable elements during their formation. For example, consider the iron asteroid that created Meteor Crater in Arizona. Samples of that object, called the Cañon Diablo iron, are about 90% iron and 7% nickel with some standard "trace" elements such as gold at 1.57 grams per metric ton and platinum at 6.3 grams per metric ton. This may not seem like much, except that 6 g/ton is considered high-grade platinum ore anywhere on Earth. If the Cañon Diablo parent asteroid was 100 m in diameter, as suggested by the energies required to dig the meteor crater, then it was carrying about 200,000 troy ounces of gold and 1,300,000 troy ounces of platinum. At current prices, over $600 million dollars of valuable metals were wasted digging a picturesque hole in the Arizona plateau country (i.e., the Meteor Crater). This "cost" in lost minerals is for just one very small asteroid and just two of its range of trace elements. Iron meteorites routinely have compositions that qualify them as rich ores of cobalt, chromium, gallium, germanium, iridium, and a number of other rare/valuable elements.

The second advantage that asteroids have is that they and their potential resources are already in space, outside the Earth's deep gravity well. The often-quoted price to move something off the surface of the Earth and into space is $10,000 per kilogram. Asteroids are already in space and can potentially provide the raw material for a huge range of space constructions from space stations to solar power satellites to interplanetary spacecraft. It is easy to be overwhelmed by a vision of asteroids moved into Earth's orbit by electromagnetic mass drivers and 15-km-long solar power plants, and forget the enormous technical, economic, and engineering problems that stand in the way of making these dreams reality. Asteroids are out there, they are still largely unexplored, and, in the case of NEAs, still largely undiscovered. They have great potential for both building and destruction, for benefit and catastrophe.

ALSO SEE THE FOLLOWING ARTICLE

NEAR-EARTH ASTEROIDS

BIBLIOGRAPHY

Asteroids II. Binzel, R. P., Gehrels, T., and Matthews M. S. (eds.) (1989). "Asteroids" (R. P. Binzel, T. Gehrels, and M. S. Matthews, eds.), Vol. II. Univ. of Arizona Press, Tucson.

Gehrels, T. (ed.) (1994). "Hazards due to Comets and Asteroids." Univ. of Arizona Press, Tucson.

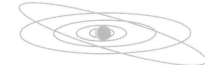

NEAR-EARTH ASTEROIDS

I. Significance
II. Origins
III. Impact Hazards
IV. Population
V. Physical Properties
VI. Future Studies
VII. Extraterrestrial Resources

Lucy-Ann McFadden
University of Maryland, College Park

GLOSSARY

Bode's rule: Empirical relation from which the distance of planets from the Sun in astronomical units can be derived: $D = 0.4 + 0.3 \times 2^n$, where $n = 1$ for Earth and increases by one for each planet, including asteroid 1 Ceres as a planet.

Chaos/chaotic motion: Term used to refer to nonlinear dynamics whereby the motion is not directly and specifically predictable in terms of either time or position.

Charge-coupled device (CCD): Electronic detector made from silicon used to quantitatively measure the light flux from an astronomical object.

Conservation of angular momentum: Fundamental physical law requiring that the quantity of angular momentum (p) be constant for objects subject to forces resulting in a circular motion at constant speed: $p = mvr$, were m is mass, v is velocity, and r is radius of the circle of motion.

Electromagnetic radiation: Particles and waves reflected or emitted from matter.

Ephemerides: List of positions of an object as a function of time.

Lithophile: Material made of elements that are commonly found in rocks, such as Si, O, Al, Ca, and Fe; derived from the Greek, meaning "rock-loving."

Magnitude: Astronomical term of brightness in which a difference of 5 magnitudes is a factor of 100 in brightness. All magnitudes are scaled to the flux of Alpha Lyrae, which is designated as magnitude 0.

Phase angle: Angle between the Sun, a given object, and the observer with the object at the vertex.

Schmidt telescope: Telescope with high image quality made of a spherical mirror and a corrector plate. The result of this configuration is that objects in a large field of view are in focus. This is the telescope of choice for survey studies.

Telemetry: Technique of using a stream of electronic signals to carry information from one site to another.

Triaxial ellipsoid: Solid produced by the rotation of an ellipse around three axes.

NEAR-EARTH ASTEROIDS reside in the vicinity of Earth near 1.0 AU (the mean distance between Earth and the Sun). We define this group of asteroids as any object orbiting the Sun with a perihelion of $q < 1.3$ AU, the aphelion of Mars. Aphelia, Q, lie within a sphere of radius 4.2 AU, defined by Jupiter's perihelion. Among this broad group is a subset distinguished by their approaching the orbit of Earth, called Amors. These asteroids have a semimajor axis $a > 1.0$ AU and perihelion $1.017 < q < = 1.3$ AU, between the aphelion of Earth and the perihelion of Mars (Fig. 1a). Those that actually cross Earth's orbit, called Apollos, have $a > 1.0$ AU and $q < = 1.017$ AU, Earth's aphelion (Fig. 1b). The third subset orbits mostly inside the orbit of Earth. These are called Atens and have $a < 1.0$ AU and $Q > 0.983$ AU, Earth's perihelion. Most of the near-Earth asteroids originated in the Main Asteroid Belt, although some of them might have evolved into their current orbits from the reservoir of short-period comets extending beyond Jupiter and into the outer solar system. The range of composition and physical characteristics of near-Earth asteroids spans those found among the Main Belt asteroids. Comparisons to comets are continually sought but are limited by observational constraints attributed mainly to their different physical states. Knowledge of these bodies is necessary to assess the collision hazard to Earth. The *Near-Earth Asteroid Rendezvous* (*NEAR*) mission is the first designed to orbit an asteroid. 433 Eros is its target. Plans are for a year of data collection in 1999 to study its geology, chemistry, and internal and magnetic structure, providing the first *in situ* remote-sensing measurements of a near-Earth asteroid. Sample return mssions and hopefully economic development of asteroids to promote human activities on Earth and in space are prospects for the future.

I. SIGNIFICANCE

A. REMNANTS OF THE EARLY SOLAR SYSTEM

From a scientific point of view, near-Earth asteroids are studied for the same reason as those from the Main Belt, they are remnants of the early solar system (Fig. 2; see also color insert). As such, they contain information that has been lost in the planets through large-scale, planetary processes such as accretion, tectonism, volcanism, and metamorphism. Knowledge of the asteroids as unprocessed material from the early solar nebula is critical to piecing together a scenario of the formation of the solar system. [*See* THE ORIGIN OF THE SOLAR SYSTEM.]

We are reasonably certain that most near-Earth asteroids derived from the Main Belt, the region between Mars and Jupiter. This region is a dividing point in the solar system, where the planets that formed closer to the Sun, the terrestrial planets, are dominated by rocky, lithophile material. Beyond the asteroid belt, the planets are dominated by gases. Near-Earth asteroids might have originated elsewhere in the solar system, such as the cometary reservoirs lying at great distances from the Sun. Knowing about material from these reservoirs reveals information about the low temperatures and resulting chemistry in the very outer

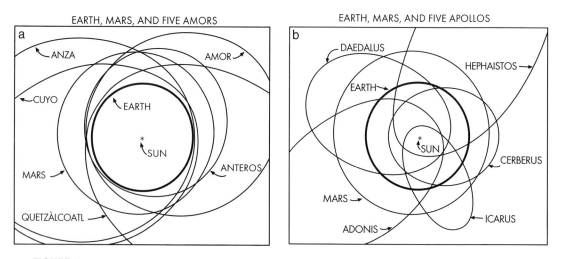

FIGURE 1   (a) Amor asteroids approach Earth, but do not cross its orbit. (b) Apollo orbits cross that of Earth.

FIGURE 2   Formation of the asteroids in the solar system, painting by Bill Hartmann. (Reprinted with permission.) (See also color insert.)

regions of the solar system. The objective of scientific study of the near-Earth asteroids is to determine whether or not these hypotheses are valid and which near-Earth asteroids, if any, might be derived from which regions of asteroidal and cometary reservoirs.

B. PROXIMITY TO EARTH

A practical reason for studying near-Earth asteroids is that they come closer to Earth than Main Belt asteroids and thus provide a more advantageous observing opportunity for scientific study. Near-Earth objects, with a low orbital inclination and small semimajor axis, are also accessible targets for spacecraft. The cost and hence accessibility of an asteroid from an Earth-launched rocket are related to the energy required to direct the spacecraft into the asteroid's orbit. Once free of Earth's gravity field, the largest demand on fuel, small changes in velocity, keep fuel requirements low and within cost limitations. Hence, the proximity of near-Earth asteroids to Earth makes them opportune targets for study from both the ground and space.

C. INVENTORY AND HAZARD ASSESSMENT

Near-Earth asteroids have been the popular subject of discussion in books, newspapers, and network television shows. The recognition that as asteroid probably caused the extinction of the dinosaurs in a geological episode known as the Cretaceous–Tertiary Boundary event has highlighted the potential for destruction

should an energetic collision occur again (Fig. 3). Furthermore, as scientists analyze the energy involved in collisions, we realize that they are tremendous and larger than anything created by human activities (nuclear weapons) or naturally occurring phenomena on Earth such as volcanoes or earthquakes. Scientists are alarmed at the results of computer simulations that consider the interactions of colliding asteroids with various Earth systems both natural and civilized. Coupled with these computer simulations is the very real phenomenon of the collision of comet Shoemaker–Levy 9 with Jupiter, which was observed worldwide through telescopes in 1993. We now recognize the very real and possibly devastating hazard posed to Earth if hit by a high-energy asteroid or comet. [*See* EARTH AS A PLANET; PLANETARY IMPACTS.]

Although chips, hand-sized rocks, and large boulders, all called meteorites, are continuously landing on Earth, and astronomers find house-sized objects occasionally passing between Earth and the Moon, knowledge of the near-Earth asteroids, their locations, and physical and chemical characteristics is more than just an idle curiosity. This leads us to a third reason to consider the nature of near-Earth asteroids, namely, to inventory and assess their hazard potential to the Earth. [*See* METEORITES.]

D. RESOURCE POTENTIAL

The fourth aspect of these asteroids' significance is related to the future of space exploration and exploitation. As humans extend their activities beyond Earth and as we consider what we might do in the future in space, one quickly realizes that moving materials from Earth is hard work and very costly. If we move beyond Earth, are there materials in space that are useful and can be used more efficiently *in situ*, rather than transporting material from Earth (Fig. 4)? The answer is probably yes, but locating the material and designing procedures for using them is a large project that is just beginning. Part of our efforts to understand the nature of near-Earth asteroids will contribute to such endeavors.

II. ORIGINS

In the widely accepted scenario of the formation of the solar system, gas and dust collapse into a disk-shaped nebula from which the planets and the smaller

FIGURE 3 "Dinosaurs' Demise," painting by Don Davis. (Reprinted with permission.)

planetesimals form. Planets grow after seeding conditions begin and molecules form aggregates, which then form clumps that continue to grow into objects large enough to be called planetesimals. This process starts with grains about 1 μm in size, which behave as discrete particles sweeping up smaller grains and growing in size. Both electromagnetic and gravitational forces come into play to overcome the destructive forces of erosion from particle collisions. Planet growth is gravitationally controlled and is called accretion. Asteroids are planetesimals that were prevented from growing to the size of the major planets by pervasive erosional forces that counteract accretion, the net effect being to keep the asteroids small.

We know that the early formation of Jupiter was a major force in establishing the Main Asteroid Belt, as well as in the failure of a larger planet forming between Mars and Jupiter at 2.8 AU. This location is the expected distance for a planet according to Bode's rule. The details of the Main Belt formation are not well known, as this all happened about 4.5 billion years ago, long before any human was around to observe it. Since the beginning, gravitational interactions between planets and asteroids, and asteroids and asteroids, have resulted in perturbations of some asteroid orbits. On occasion, these perturbations result in the orbit, over time, evolving into one crossing a planet. More often than not, the sum of gravitational perturbations ejects the body from the solar system entirely. But in this chapter we are interested in those asteroids that successfully evolved into near-Earth orbits.

FIGURE 4 Astronaut mining an asteroid, painting by Pamela Lee. (Reprinted with permission.)

A. RELATIONSHIP TO MAIN BELT ASTEROIDS

If we are to understand the relationship of near-Earth asteroids to Main Belt asteroids, one must understand that not all asteroids are the same. We have known since early asteroid studies in the 1940s that they come in different colors (see Section V). Techniques to study both reflected and emitted electromagnetic radiation from the asteroids were developed and used to derive information about their mineral and chemical composition. Two scientists in the late 1970s, Jonathan Gradie and Edward Tedesco, recognized that there is a relationship between the apparent composition of the asteroids and their distance from the Sun. This finding was very exciting because it represented observational support for a model predicted by John Lewis in which the solar nebula was in a state of equilibrium when it formed. The composition changes as a function of temperature and hence distance from the Sun. Therefore, we do not expect all asteroids to have the same composition and, furthermore, the exact nature of that material holds clues to the temperature at which the material formed and where it formed relative to the Sun. Given that asteroids retain information about the conditions in the early solar system, such information is valuable as scientists piece together the scenario leading to the formation of our solar system and look for evidence of the existence of other solar systems. Studies of the composition of near-Earth asteroids (NEAs), carried out by this author in her dissertation research, led to the conclusion that such composition spanned the range found among the Main Asteroid Belt, thus establishing that many or most of the NEAs are derived from the Main Belt. Furthermore, physical information derived from NEAs can be reasonably considered to apply to Main Belt asteroids.

Have near-Earth asteroids always been planet-crossing objects? The answer is no, based on statistical analysis of the evolution of many asteroid orbits over the age of the solar system. Of all the small bodies in the solar system, only Main Belt asteroids can survive unperturbed throughout the age of the solar system, 4.56 billion years. The lifetime of a planet-crossing body against subsequent gravitational perturbations is relatively short. With powerful computer workstations and fast numerical integration codes, estimates of the lifetimes of NEAs have become possible. Planet-crossing asteroids have short lifetimes, on the order of 10 million years or less. Within this time frame, they will either collide with a planet or evolve to another trajectory that is not planet-crossing. Bear in mind that this time interval applies to the average of the entire population and does not refer to the exact lifetime of any particular asteroid. It turns out that the orbital evolution of a specific asteroid cannot actually be determined owing to nonlinear and hence unpredictable motions within the solar system that prevent one from determining the specific evolutionary path of an object's orbit. [See CHAOTIC MOTION IN THE SOLAR SYSTEM.]

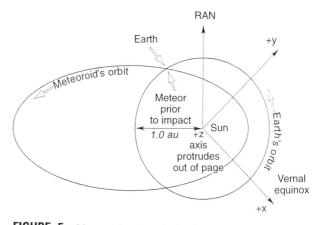

FIGURE 5 Meteoroid orbit calculated just before entry into Earth's atmosphere on February 1, 1994. [From Tom McCord *et al.* (1995). *J. Geophys. Res.* **100**, 3246.]

Recognizing the short lifetime of NEAs, we want to know specifically where they came from. Which part of the asteroid belt? Did they all derive from the Main Belt or are some of them fragments from larger planets, or remnants of once active comets? Asking this question is necessary to place any physical and chemical information available about the near-Earth asteroids into its proper context. What region of the solar system are we learning about in our study of NEAs? We will present this answer to the extent known in Section V.

B. RELATIONSHIP TO METEORITES

Just as we ask, "Which Main Belt asteroids are related to any given near-Earth asteroid?" we are also driven to ask, "Which meteorites landing on Earth are related to any individual or subset of the NEAs?" If we can answer that question, then we can make a very rich connection between all the geochemical, isotopic, and structural information that we have from laboratory studies of meteorites and the near-Earth asteroids.

On February 1, 1994, a meteoroid entered Earth's atmosphere and was detected at an altitude of approximately 54 km by an Earth-orbiting satellite with a downward-looking telescope and silicon detector system. The energy of the impact was derived from the observed brightness. The mass and size of the meteoroid can be estimated from the observed velocity just before entry and its energy. The orbit of the meteoroid just prior to impact with Earth was derived from the meteoroid's position in space and its velocity vector at impact (Fig. 5). Unfortunately, the meteoroid burned up in the atmosphere, and nothing hit the ground. We therefore know the mass and size of an object in an Earth-crossing orbit, but have no chemical or mineralogical data on it.

In October 1993, another meteoroid did survive its passage through the atmosphere and was recovered from the trunk of a car. The orbit of that fragment was derived not from military sensors, but from information collected from eyewitnesses and their video cameras that were mostly being used to record a high school football game at the time. [See METEORITES.]

The determination of meteorite orbits serves as a constraint on the mechanisms that result in meteorite delivery to Earth. Again, with the aid of numerical computer simulations, we believe there are certain regions of the asteroid belt that produce most of the meteorites. Small changes in the velocity of fragments from asteroid–asteroid collisions near regions in the Main Belt that are resonant with Jupiter's orbit, such as the 3:1 Kirkwood Gap, produce changes in the fragments' orbital eccentricity. These orbital changes result in an Earth-intersecting orbit.

C. RELATIONSHIP TO COMETS

Relationship to comets? Comets are predominantly icy and dusty objects that come from the outer reaches of the solar system. Their orbital periods are long, their eccentricity is high, and they may have large or small inclinations. What is their relationship to NEAs? In the 1950s, Ernst Öpik concluded that comets must be a partial source of near-Earth asteroids because he could not produce the number of observed meteorites from the asteroid belt alone via his calculations. Öpik's work was made without benefit of knowledge of the resonances in the asteroid belt and their role in supplying meteorites. But the hypothesis continues to merit consideration as knowledge of comets and asteroids increases and we continue to simulate the dynamical evolution of interacting small bodies under the gravitational influences of the planets. [See COMETARY DYNAMICS; PHYSICS AND CHEMISTRY OF COMETS.]

The idea that comets can evolve into asteroid orbits has always seemed far-fetched to me. How could a comet, with an extremely elliptical orbit, shrink into an orbit like that of a near-Earth asteroid, which rarely extends beyond Jupiter? Changing an orbit requires significant addition or subtraction of energy. When computational simulations of orbital perturbations acting on comets can be shown to produce such orbits, I perceive just how long a time period of 4.5 billion years really is, as the magnitude of the perturbations at any given time is small. Given such analyses, observing

astronomers (as opposed to theoretical astronomers, who make calculations and predictions for the observers) seek evidence for these masquerading comets among the planet-crossing asteroids. For the sake of studying comets, consider what an advantage this would be to study the nucleus, the gas-free portion, of a comet at relatively close range to Earth without the obscuring gaseous and dusty coma? In 1983, George Wetherill predicted that 20% of the near-Earth asteroid population is made of extinct comet nuclei.

Are there hints that any particular NEA was once a comet? There are some suggestive lines of evidence and some compelling but not completely persuasive arguments, but no positive proof that comets masquerade as asteroids. We are not completely sure what the identifying feature of an extinct, totally gas-free comet might be. Employing the scientific method, we make our best guess at the identifying feature, go look for it, and continue our efforts to study the nature of the solid nucleus of comets from objects that we know to be comets.

Let me review the suggestive arguments linking some NEAs to comets. They fall into two categories: arguments based on dynamical evidence and those based on physical evidence.

1. Tisserand Invariant

If we look at the orbital parameters of asteroids and comets, they separate out readily when plotting eccentricity versus semimajor axis (Fig. 6). The boundary between the two types of objects is defined by the gravitational interactions between a small body, the Sun, and Jupiter and the law of conservation of angular momentum. These interactions lead to an approximation that the relationship between semimajor axis, eccentricity, and inclination is a constant and smaller than that value for most comets. The same relationship,

$$a_J/a + 2\sqrt{(a/a_J(1 - e^{-2})} \cos i = C$$

when solved for parameters characteristic of asteroid orbits, gives a value larger than the constant. This value, called the Tisserand invariant, is less than 3 for comets and greater than 3 for asteroids, represented by the solid line in Fig. 6. We reason from this that objects found in the cometary region with no evidence of coma or tail might be extinct or dormant comets, however, such objects are not usually in planet-crossing orbits and are not considered to be near-Earth objects. But the population of planet-crossing bodies

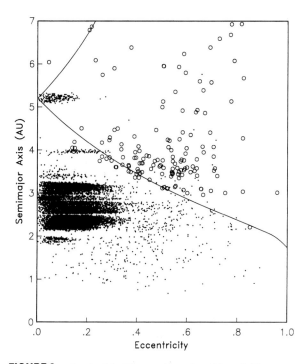

FIGURE 6 Graph of the Tisserand invariant. The solid line represents the Tisserand invariant with a value of 3. (Graph provided by Jeff Bytof, Jet Propulsion Laboratory, Pasadena, Calif.)

near Earth and close to this boundary might evolve across it over time and may have in the past been comets. Computer simulations have shown that objects lying close to the Tisserand invariant boundary can exhibit evidence of chaotic motion and thus cross the boundary within short time intervals.

2. Computer Simulations of Short-Period Comet Orbits

When computer simulations of hypothetical short-period comets, including the gravitational forces of the seven planets, Venus through Neptune, are run for time intervals up to 225,000 yr, the probability of such an object evolving into a near-Earth asteroid orbit with $Q \leq 4.2$ AU is less than 1 in 360, or 0.0028. Note that the number of test objects in this study carried out by N. W. Harris and M. E. Bailey was 360, hence the probability is defined relative to that number. When nongravitational forces are considered, that being the jetting effect of cometary outgassing, the probability increases. But the efforts to accurately model the effects of cometary activity are not fully developed. Thus, although some scientists think that as many as 50% of the NEA population might be extinct comets, there is evidence based on simulations of the dynamical

evolution of short-period comets that the contribution may be smaller, say, 20% or less.

3. Evidence from Physical Studies

Gas and dust are the primary indicators of a comet. Bearing in mind that all NEAs appear as point sources through a telescope and have no obvious indication of an extended gas or dust coma or cometary tail, my colleagues Alan Chamberlin and Rita Schulz and I set out to see if there might be residual evidence of cometary activity seen as faint gaseous emission features. Meanwhile, my colleagues Jane Luu and Dave Jewitt looked for evidence of dust comae. So far we have observed only upper limits on these materials. Measuring an upper limit means that we have detected no evidence of a sought-after material to the limits detectable by our measuring instruments. In the case of 3200 Phaethon, those limits are $<9 \times 10^{22}$ molecules/s for CN gas compared to 1.4×10^{27} molecules/s for comet Halley at its maximum level of activity. No dust coma was found on Phaethon, with less than three-thousandths of a percent of the surface active, compared to 13% of Halley's surface that was active.

2212 Hephaistos has an orbit similar to that of comet P/Encke, but numerical simulations of its orbit backward in time show that such an orbit could also evolve from the Main Belt.

D. METEOR SHOWER ASSOCIATIONS

Meteor showers occur when many particles and fragments of less than a gram enter the upper atmosphere of Earth and burn up. The heat, generated primarily by friction, is seen as a streak of light when viewed from the ground. Periodic episodes of meteor activity are called meteor showers and correspond to Earth's passage through a cluster of dust fragments orbiting the Sun in their own orbit. In many instances, the orbital elements of a known comet match that of the meteor shower, and hence it is reasonable to assume that the dust grains were once part of that comet. For this reason, any near-Earth object with orbital elements close to those of a meteor shower might reasonably be suspected to be an extinct comet.

The near-Earth asteroids 2101 Adonis and 2201 Oljato have orbits similar to those of meteor showers. Adonis is very difficult to observe from the ground, so we do not know much about it. Oljato, also a difficult target for a telescope, has intrigued scientists since it was first observed in 1979. The jury is still out on whether or not this asteroid is an extinct comet, but the evidence now seems to suggest that it is an asteroid. One thing is certain: the object is not normal even when considered an asteroid.

In 1983, Fred Whipple recognized the orbital elements of an asteroid found by an Earth-orbiting infrared telescope to be essentially the same as the Geminid meteor shower, which occurs in mid-December. There is little doubt that this asteroid, now named 3200 Phaethon, is the parent body of the Geminid meteors. But is Phaethon an extinct cometary nucleus? The supposition is yes, according to one line of thought based on similarities of orbital inclinations and the location of perihelion (longitude of perihelion relative to the ecliptic) of asteroids and comets compared to meteor showers. In 1981, Jack Drummond determined that a total of 15 comets had orbital characteristics that were similar to those of 15 of 48 meteor showers. However, in order to say yes definitively, we must find another, independent and distinguishing characteristic of Phaethon that indicates that it was derived from material from the distant solar system—that is from a comet—and not the inner solar system where the asteroids are derived. [See INFRARED VIEWS OF THE SOLAR SYSTEM FROM SPACE.]

E. DYNAMICAL HISTORY

Scientists have simulated the pathway that an object might take from different regions of the solar system to near-Earth orbits using computations of dynamical forces acting in the solar system. From the Main Belt, one possible pathway to near-Earth orbit is from an asteroid–asteroid collision within the Main Belt. Either the asteroid breaks up completely or sizable chunks are ejected from the impacted body upon collision. It is possible that large chunks then become small asteroids, being injected by the force of the collision into a chaotic region of space. Within these regions, small changes in motion can result in large, exponential changes in an orbital parameter that could change the orbit significantly on a short timescale. Thus the effects of chaotic regions is more than the sum of small changes in motion over long periods of time. These regions of chaotic motion are associated with resonances with Jupiter's orbit and are believed to play a significant role in directing meteorites to Earth, and presumably also many of the near-Earth asteroids (Fig. 7).

Other objects evolve from Jupiter-family comets or Halley-type short-period comets. Life in the Jupiter family is not long-lived, as resonances and Jupiter itself

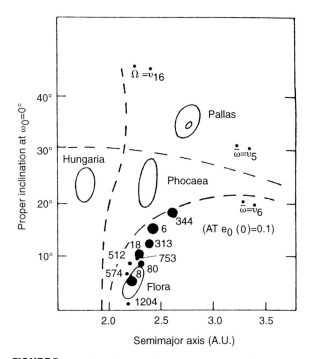

FIGURE 7 Graph of dynamical resonances, regions where gravitational interactions either deplete or protect asteroids from changes in their orbit. (From Jim Williams, Jet Propulsion Laboratory, Pasadena, Calif.)

impart changes to the orbits on time-scales of 10^4 yr. Leaving Jupiter's gravitational sphere of influence, the soon to be near-Earth asteroids may sometimes be perturbed by Mars and other terrestrial planets and also affected by the influences of nongravitational forces, such as volatile outgassing or splitting of the nucleus. These phenomena also contribute to orbital changes that result in planet-crossing orbits.

David Rabinowitz analyzed the distribution of orbits of the smallest near-Earth asteroids found with the Spacewatch Telescope. When he created a hypothetical population of thousands of near-Earth asteroids with the same distribution of orbital elements (semimajor axis, inclination, eccentricity, longitude of perihelion, and a time-fixing parameter called the mean-anomoly) as those found with earlier programs finding larger asteroids, he found an excess of small Earth-approaching asteroids. By comparing the hypothetical population to the orbital elements of the small asteroids actually discovered in the Spacewatch survey, he discovered what he called a Near-Earth Asteroid Belt. This population is characterized by small size, diameters < 50 m, low eccentricity (almost circular orbits), a wide range of inclinations, and possibly distinctive colors. The wide range of inclinations suggests that this is a cloud of Earth-approaching objects, not a belt, which is commonly conceived to occupy one primary plane and to have the shape of a disk. The origin of these small asteroids could be one of those considered before: the Main Asteroid Belt, short-period comets, or, because of their close proximity to the Earth–Moon system, the Moon itself. The population of near-Earth asteroids may indeed be diverse, contrary to the earlier statement that most of them probably are derived from the Main Asteroid Belt.

Following the discovery of an excess of small, Earth-approaching asteroids and meteorites from the Moon and Mars, dynamical studies of the lifetime of ejected fragments from neighboring satellites and planets have been carried out. This topic is covered in other chapters. [*See* SOLAR SYSTEM DYNAMICS.]

III. IMPACT HAZARDS

It does not take much imagination to envision an asteroid or comet hurtling through space that just happens to be on a collision course with Earth. Arthur C. Clarke used this scenario in his 1994 book "The Hammer of God." But consider this from the point of view of the physics of the solar system. Meteorites are seen to fall to Earth, and we see evidence of impact craters on Earth and plenty of them on the Moon. Remember, the largest lunar impact basins formed more than 3.8 billion years ago, and the largest impactors were swept up just after the solar system formed. Most objects colliding today with Earth and the Moon are small and harmless. There are two aspects of the collision hazard that we have to consider: the magnitude of the collision and their frequency in time.

A. MAGNITUDE

First consider the magnitude of a collision. The primary physical parameter that concerns us as inhabitants of Earth is the energy of the collision and particularly that which is transferred to Earth. The parameters controlling the energy of an impact are mass and velocity according to the relation

$$E = \tfrac{1}{2} mv^2$$

So we must consider two variables, mass and velocity. Clearly, if a massive body were to collide with Earth, the energy of impact would be proportional to its mass. Objects that are tens of kilometers to a kilometer in

size can cause significant damage to Earth as a whole by triggering changes in climate that will affect human systems such as agriculture. Objects less than a kilometer in size still pose a significant threat, having energies that rival the world's arsenal of nuclear weapons. Fragments that are a meter to a centimeter in size usually form local craters and great news stories.

Energy is also proportional to the square of the velocity, so a high-velocity object will make the energy larger by its value squared. Most near-Earth asteroids travel at approximately the same orbital velocity as Earth when nearby, about 30 km/s, hence the difference in the energy at impact is the critical factor. Objects in highly eccentric orbits, such as comets, can have tremendous energy.

To assess the damage that a particular impact will impart to Earth, we need to know how much energy any particular location on Earth can absorb and whether or not that location can recover from an impact. What range of energies can the Earth absorb? We know that meteoroids enter Earth's atmosphere with energies estimated in the 10^{14}–10^{15} J range and are crushed under the pressure of Earth's atmosphere (Fig. 8), leaving perhaps only scattered dust. On the other hand, damage from meteorites has been documented on various scales, from killing a dog in Egypt in 1911, to bruising a human thigh, to denting the trunk of a car. Holes of various sizes have been made in the ground, produced by impacts with energies on the order of 4.2×10^{16} J, or 10 megatons (MT) of TNT (Fig. 9). Impacts of greater energies by an order of magnitude or so can impart regional damage. Studies have shown that an impact of 4.2×10^{17} J or 100 MT can destroy areas within a 25-km radius. For example, an impact into the ocean could induce tsunamis that would destroy far-flung coastal areas. The Cretaceous–Tertiary Boundary event 65 million years ago has been estimated at $>4.2 \times 10^{23}$ J or 100,000,000 MT! Such large impacts occur very infrequently. But they do occur.

B. FREQUENCY

Knowing that kilometer-sized objects and larger pose a serious threat to civilization and that smaller objects can result in significant destruction on a regional scale is quite a sobering realization. A complete assessment of the situation requires knowledge of the frequency of collisions by objects of different sizes. We continue to make educated guesses of current collision rates as we improve knowledge of the near-Earth population

FIGURE 8   A daytime photograph of a bolide with possible energy on the order of 10^{13}–10^{14} J, seen over Grand Teton National Park, Wyoming, USA, 10 August, 1972. Courtesy Dennis Milon. Bolide mass is brightest over the highest peak "above" the cloud.

with our exploration tools of telescopes, spacecraft, and geochemical analysis of impact craters on Earth.

Objects in the range of hundreds of kilometers in diameter, with masses proportional to their size, were swept up and incorporated into the planets as the solar system was formed during the period called Late Heavy Bombardment. The lunar basins formed during this time, which ended ~3.8 billion years ago. We do not expect a collision from such a large object today. [*See* THE MOON.]

FIGURE 9   Meteor crater in Arizona, formed from a collision of approximately 10^{16} J.

An impact by an object <50 m in diameter with energies <4.2 × 10^{16} J (<10 MT) occurs about once per century. The frequency of impacts increases linearly with decreasing size and energy. Conversely, for larger objects the frequency decreases.

To assess the potential for any near-Earth object to collide with the Earth, it is imperative that we have an accurate assessment of the numbers and locations of this population. It is then important to know the nature of the orbit, as that bears a direct relation to the energy of the object relative to that of Earth in its motion around the Sun. It is the relative energy between the Earth and the impacting projectile that controls the damage resulting from a collision. A collision between a near-Earth asteroid and the Earth is very intriguing but is not a likely and/or frequent danger. The threat, however, is very real.

IV. POPULATION

A. SEARCH PROGRAMS AND TECHNIQUES

Organized, telescope-based search programs for near-Earth asteroids operate worldwide. The longevity of the programs is a complex function of availability of resources and science policy that governs the funding agencies in different countries. The search programs supported by the National Aeronautics and Space Administration (NASA) include the Near-Earth Asteroid Tracking (NEAT) system, Lowell Observatory's Near-Earth Object Search (LONEOS), and Spacewatch, the latter operated by the University of Arizona at Kitt Peak Observatory. Though the objectives of these programs are all similar, to inventory the objects in the vicinity of Earth, each has its own design and approach. A summary of each program follows.

1. Near-Earth Asteroid Tracking

The NEAT program began in 1996, is designed to study the dynamics and size distribution of near-Earth objects to further knowledge of our solar system, both its formation and processes. The objective is to discover and inventory NEAs larger than ~1 km by obtaining precise ephemerides that enable physical observations and follow-up studies. Eleanor Helin at the Jet Propulsion Lab (JPL) in Pasadena, California, heads the program with the assistance of three additional staff members. The program operates twelve nights a month centered around new moon, when the skies are darkest.

NEAT uses a 1-m telescope, operated by the U.S. Air Force at the Ground-Based Electro-Optical Deep Space Surveillance (GEODSS) site on the island of Maui, Hawaii. The telescope is automated and operated from the JPL; the coordinates for observing are uplinked to the telescope on Maui via modem. A charge-coupled device (CCD) camera is mounted on the back of the telescope and is commanded to make exposures. This program is capable of covering 110 square degrees of sky per night, but plans to improve the program will increase the coverage to 500 square degrees per night. Images are sent via telemetry to JPL, where they are monitored and scanned for fast-moving objects (Fig. 10). The supreme advantage of this approach is that scientists do not have to stay awake all night on a cold mountaintop and then try to stay awake the next day as well to scan their images. With this setup, the astonomer can sleep at night and come to work in the morning, ready to apply her mental acumen to the task of discovering asteroids. Any discoveries are reported to the Minor Planet Center in Cambridge, Massachusetts, where the discovery is announced via electronic mail to those subscribing to this service. Announcing new discoveries is important so that observers at other telescopes and amateur as-

FIGURE 10 A discovery image from the Spacewatch program. (Courtesy of Jim Scotti, University of Arizona.)

tronomers as well can follow up with additional observations. Unless there are enough positional measurements, the asteroid may not be found again, and it is considered lost. In the first nine months of 1996, NEAT discovered eight new near-Earth asteroids and covered 6114 square degrees of sky.

2. Lowell Observatory's Near-Earth Object Search

LONEOS has a primary objective of providing a near-complete inventory of large NEAs and comets and assessing their Earth-impact probability. Of secondary importance is an inventory and examination of the structure of the Main Asteroid Belt and other more distant and dynamically unusual objects such as Centaurs and large Kuiper Belt objects. Ted Bowell of Lowell Observatory heads this program with the assistance of six others. The program operates approximately three weeks of every month, weather cooperating. [See KUIPER BELT.]

The LONEOS program uses a 58-cm telescope of the Schmidt design, having excellent image quality and wide field of view. The camera is a large-format CCD, which, in combination with the telescope, provides a 10-square-degree field of view to a limiting magnitude through a V-band filter of $V = 19.4$ with a S/N of 3.2.

3. Spacewatch

The Spacewatch program has been operational since 1981. Its objective is more inclusive than inventorying the region around Earth, for this program plans to inventory the entire solar system! Tom Gehrels is the director, who operates with the assistance of about half a dozen others. The program has 18 nights for observing per month on the 0.91-m Newtonian reflector telescope at Kitt Peak National Observatory outside of Tucson, Arizona. A CCD camera and automated search routines scan the skies, covering 140 square degrees per month. Although the sky coverage of this program is less than those using Schmidt telescopes, it has the capability of detecting objects as faint as magnitude 21.5, hence their discovery rate is high because they are locating small objects, whose population is larger. Approximately three near-Earth asteroids a month are discovered by this program. Spacewatch has been most successful in extending the population estimates to smaller asteroids. Spacewatch scientists believe there are 135,000 objects that are 100 m or smaller.

4. Early Programs

The pioneering search program for near-Earth asteroids was initiated by Gene Shoemaker and Eleanor Helin at Mr. Palomar in the 1970s. They used photographic plates and the 0.46-m Schmidt telescope at Mt. Palomar during a week-long run every month. Over time, the program evolved and new photographic emulsioins were developed. Carolyn Shoemaker and others joined the teams and increased the discovery rate using stereo scanning techniques. Most of the first 200 or so of the near-Earth asteroid discoveries were made by these early search programs. As technology evolved with the availability of large-format CCD cameras and computer-assisted scanning algorithms, the photographic techniques have been replaced. The early search programs paved the way for the effective programs that operate today, finding fainter and smaller asteroids in the vicinity of Earth.

B. HOW MANY?

It is impossible to quote the definitive size of the near-Earth asteroid population. Search programs start up and are canceled, but they are constantly adding to the inventory. There are also inherent limitations in our search techniques. Consider setting out to count the number of near-Earth asteroids. First of all, you can only look for them at night. At any one time, you can only search half the sky. Then there are limitations in how much sky you can cover in one night, limited by

the field of view of your telescope and the recording instrumentation. These are just the limiting factors related to spatial coverage. The realities of weather and equipment failure further hinder the search. Then there are limitations related to measurable signal, for each instrument has a limiting magnitude beyond which it will not detect light from an object in a finite amount of time. The combination of spatial coverage, instrumental sensitivity, and the statistical probabilities of observing a moving target at any particular time forces us to consider the level of completeness of any search activity. This level represents an estimate of what fraction of an expected population has been found for a range of size and brightness.

We are more certain of knowing how many large and bright near-Earth objects exist than we are of the smaller ones, hence our estimates of these are more complete. We know this because in time we find fewer and fewer large asteroids, while at the same time the technology has improved, increasing the probability of finding larger asteroids. If we discover objects of a particular size and brightness more frequently, we suspect that there are many more of them. The LONEOS program is designed to maximize the discovery of objects between 500 m and 1 km across, but such a search covers less of the sky. As the discovery rate goes down with time, the completeness estimates derived from LONEOS will go up.

As of a few years ago, discovery was complete for objects with an absolute, H magnitude of 13.2. It was 35% complete to $H = 15.0$, 15% complete to $H = 16.0$, and 7% complete to $H = 17.0$. We are now discovering objects as faint as $H = 29$, which can be as small as 5 m in diameter! An object of magnitude 29 is 100 times fainter than an object of magnitude 24, and 10,000 times fainter than an object of magnitude 19! These are indeed faint and therefore small.

By converting brightness to size ranges, assuming the asteroids' inherent reflectivity or albedo, one can convert the completeness estimates in magnitude to completeness as a function of size (Fig. 11). Then by adding up the estimates across all size ranges one arrives at an estimate of the total population. The first estimates were for objects down to a diameter of 1 km, as photographic plates could not detect anything smaller than that. This approach was derived from the discovery rate in a specific fraction of the sky and projected to the entire sky. Based on a multiyear search program using the 0.46-m Schmidt telescope at Mt. Palomar Observatory in a program run by Eugene and Carolyn Shoemaker and David Levy, an estimate of approximately 2000 asteroids provides a lower limit on the population size. The real factors controlling the

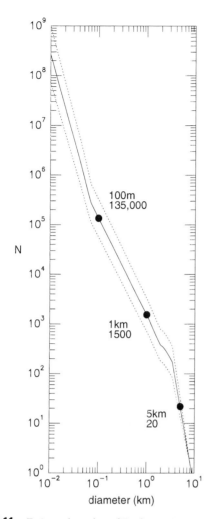

FIGURE 11 Estimated number of Earth-crossing asteroids larger than a given diameter. [Reprinted with permission from D. Rabinowitz *et al.* (1994). "Hazards Due to Comets and Asteroids." Univ. Arizona Press, Tucson.]

When converted to size ranges, this estimate assumes that there are equal components of bright asteroids (albedo ~0.15) and dark ones (albedo ~ 0.05).

That the size distribution of objects <1 km is the same among near-Earth asteroids as in the Main Belt was verified by data from the Spacewatch program, which has the capability to search for fainter objects than the 0.46-m Schmidt telescope. This program surveyed the size distribution of 1-km objects farther away in the Main Belt and found the size distribution to be similar to that seen among the near-Earth population. Accounting for the observational biases against dark objects among near-Earth asteroids, the estimate of equal probability of dark and bright objects is reasonable.

As a check on these estimates, a survey of the frequency of craters on both Earth and the Moon has been made and from that an estimate of the size of the cratering population derived. This method has its own set of inherent assumptions, and it may be just coincidence that the results agree to within the uncertainties of the technique.

With the advent of large-scale CCD cameras over the past decade, the population estimates have extended to smaller sizes. Rabinowitz now has evidence that there are at least 135,000 objects with diameters less than 100 m. To many, such a large number represents a swarm of small asteroids. However, that term is relative. One hundred thirty-five thousand objects in a space the size of the vicinity of Earth is not a very densely populated area.

discovery rate, in addition to the spatial and brightness factors, include a bias controlled by their orbits, there being certain sets of orbits that bring objects into the search volume more frequently than others. Hence this first-order estimate is a number that is not smaller than the real population. This value was derived by estimating the size and area of the detection volume given the limiting magnitude of the field (which varies from night to night depending on sky transparency conditions) and the motion and geometric circumstances at discovery. A scaling factor representing the fraction of time that a member of that population would spend in the observable window is included in the estimate, as is the assumption that the collision history, or size distribution, of asteroids in Earth-crossing orbits is the same as for the Main Belt population.

V. PHYSICAL PROPERTIES

The first measurement made after the position of a near-Earth asteroid is established is its brightness measured on the astronomical magnitude scale. The changing cross section of an asteroid as viewed from Earth affects its brightness and with time reflects the shape and rotation rate of the asteroid. Analysis of the changing brightness of an asteroid with time, accounting for its observation geometry, results in constraints on its shape and determination of its rotation rate and orientation in space. From analysis of reflected sunlight off asteroid surfaces at different wavelengths, we have binned the asteroids into different taxonomic types. Further analysis allows us to determine surface mineralogy, and from that we can place constraints on the temperatures at which they formed.

A. BRIGHTNESS

The standard asteroid photometric magnitude system compensates for the distance from which the asteroid is observed and the phase angle. The effect of distance on the magnitude is controlled by the inverse square law. As the distance increases, the brightness decreases by a factor equal to the square of the distance. The phase effect is removed by interpolating the magnitude to 0° phase after observing at different phase angles. This task can take years of observations. For comparison purposes, a magnitude measurement is converted to an absolute scale, H, which is defined as the brightness of an object at a distance of 1.0 AU and 0° phase angle. The phase function, that is, the change in brightness as a function of geometry, is controlled by the composition of the asteroid and its microscopic surface texture. For near-Earth asteroids, this is possible to observe at high phase angles, although the observing opportunities may not be very frequent. Measured phase coefficients, g, exist for some of the brighter near-Earth asteroids. Large phase coefficients indicate a very rough surface with significant effects due to shadowing, such that the magnitude changes significantly with changing phase angle. Low values of g indicate either a very dark surface, where the impact of shadows is not significant against a dark surface, or that few scattering centers exist and hence there is minimal shadowing.

B. SHAPE

Lightcurves are measurements of brightness as a function of time (Fig. 12). If the asteroid is perfectly spherical such that its cross section does not change with time, there will be no variation and the lightcurve would be flat. There are no such asteroids known, although there are lightcurves with very small amplitudes (not commonly found among near-Earth asteroids). Asteroid lightcurves most often show two or more maxima and minima, often with inflections embedded within. The shape of an asteroid is most often modeled as a triaxial ellipsoid of a smooth and uniform surface. Though there are some large Main Belt asteroids that do not deviate much from a sphere, for example, 1 Ceres, virtually all the near-Earth asteroids have lightcurves with pronounced magnitude variations from which three axes of triaxial ellipsoids are derived. Inflections in the lightcurves represent changes in the asteroid's cross section that may be the result of excavation from large-scale impacts.

With the availability of data at many geometric

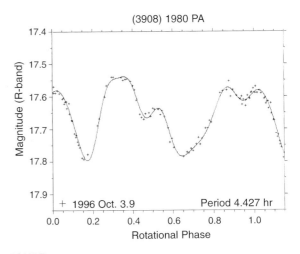

FIGURE 12 Lightcurve of Amor asteroid 3908 1980 PA indicating its irregular shape. (Courtesy of Petr Pravec, Astronomical Institute, Academy of Sciences of the Czech Republic, Ondrejov, Czech Republic.)

orientations, which we anticipate having after the *Near-Earth Asteroid Rendezvous* mission completes its year-long data acquisition at asteroid 433 Eros, we will prepare to analyze data that may not fit a triaxial ellipsoid very well. One method of extracting an irregular shape from lightcurve data was derived by Steve Ostro and Robert Connelly. Integral geometry combined with numerical analysis techniques can be used to invert lightcurves to a convex profile of the object, which estimates the average cross section, a two-dimensional average of the asteroid's three-dimensional shape. Assuming uniform and geometric scattering, that the Sun and Earth are in the asteroid's equatorial plane, that the phase angle is 0°, and that all the surfaces are convex and define a set of average cross sections cut by planes a fixed distance above the asteroid's equatorial plane, this method can produce a representation of the asteroid's shape. This technique has been used for a number of near-Earth asteroids, as their shapes are very irregular.

Radar measurements are also analyzed to produce images that reveal the shape of asteroids. Continuous-wave radar signals transmitted from Earth to an asteroid reflect back and are received as a radar echo. The power in the radar signal is reduced by the square of the distance to the target and again by the same factor on its return to receivers on Earth. The bandwidth of the echo power spectrum is proportional to the cross section of the asteroid presented to Earth and normal to the line of sight at the time of interaction with the surface. The signal can be built up as the asteroid rotates, producing an image that represents the shape

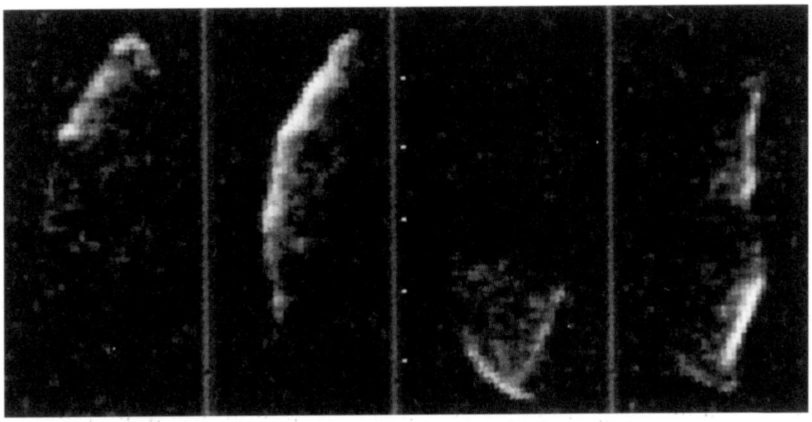

FIGURE 13 Radar images of 1620 Geographos at its cardinal points of rotation. (Courtesy of Steve Ostro, Jet Propulsion Laboratory, Pasadena, Calif.)

of the asteroid. For those asteroids that have approached Earth at close enough range to employ this technique, such as 4769 Castalia, 4179 Toutatis, 1627 Ivar, 1620 Geographos, and 433 Eros, the results show shapes varying from slightly noncircular to very irregular. 1620 Geographos (Fig. 13) is elongated in shape and exhibits contrast variations that can be attributed to regional topographic features, possibly large craters or a basin. [See PLANETARY RADAR.]

Knowledge of the asteroids' shape provides clues to the collisional history of this population. If all the asteroids are spherical, we would believe them to have formed from a viscous and rotating material that was not disturbed since formation. We know this is not the case, and so are in a position to make some inferences about the population's collisional history. The fact that many near-Earth asteroids are irregularly shaped implies that they are products of collsions that have knocked off significant chunks of material from a larger body.

Some near-Earth asteroid shapes have been interpreted as being two bodies stuck together and are referred to as a contact binary. This interpretation is intriguing because it leads one to speculate that the two components were brought together in a low-velocity collision and just stuck together instead of one or both being destroyed. An alternative interpretation is that the asteroid is so irregularly shaped that it appears to be two pieces, but really is continuous. Such a situation would imply a history of collisional fragmentation that kept the main body of the asteroid intact, albeit severely altering its shape, but not disrupting it totally. Measurements at different aspect angles are required to truly confirm the interpretation that some asteroids are contact binaries. The evidence that some of these objects are really two distinct entities close together

is only suggestive. Radar and lightcurve measurements are illustrative of how knowledge of asteroid shape provides information relating to their collisional history.

C. ROTATION RATES

As mentioned earlier, the rotation rate is derived from the relationship between brightness and time. The statistics of rotation rates show that near-Earth asteroids rotate faster than Main Belt asteroids of the same size. Of 32 near-Earth asteroids, the mean rotation rate is 4.94 ± 0.54 rev/day, whereas a sample of the same number of small, Main Belt asteroids has a mean rotation rate of 4.30 ± 0.46 rev/day. Because the standard deviation of these means overlaps, there is no great significance placed on these differences. The significant differences are in the size dependency of rotation rates, which is a problem in the study of Main Belt asteroids. The rotation rates of near-Earth asteroids were once compared with the same parameter for comets to seek differences or similarities indicative of the origin of near-Earth asteroids from either the Main Belt or the cometary reservoir. With time and additional data on rotation rates of comets, we have learned that the two populations are not distinguishable by this parameter alone.

D. SIZE

The size of near-Earth asteroids ranges from the largest, 1036 Ganymed at 41 km in diameter (Fig. 14), to the smallest seen by a telescope and estimated to be about 300 m across. Note the comparison of size

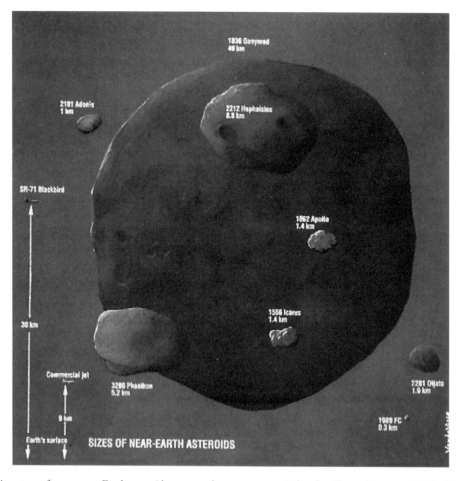

FIGURE 14   Diameters of some near-Earth asteroids compared on an aeronautical scale. [From *Astronomy* **20**(8), 34 (1992). Reprinted with permission.]

with various altitudes in Fig. 14. We are certain that we have found most of the asteroids at the larger end of the scale. Not many NEAs are larger than 10 km, roughly the distance across town (provided your town is 16 miles wide). As can be seen in the histogram of the number of asteroids in various size ranges (Fig. 15), there are only about 8 asteroids in the near-Earth population with diameters greater than or equal to 10 km. The occurrence of smaller ones increases, however, a presumed consequence of collisions. There are about 15 objects with diameters between 6 and 10 km, 86 with diameters between 1 and 5 km, and 106 known objects with diameters less than 1 km.

What does it take to determine an asteroid's size? Basically a lot of careful astronomical observing. For an object illuminated by the Sun alone, the sum of the reflected and emitted radiation from a nonradiating body in space is equal to the total incident solar radiation upon it. Hence, knowing where the body is, in terms of its distance from the Sun and the output of

the Sun, we can determine the amount of incident energy on an asteroid's surface. By measuring the reflected and reemitted component of this radiation, and with some rudimentary knowledge of the nature of the body's surface determined from the albedo, we can constrain its diameter. Usually, the two parameters diameter and albedo are derived in tandem, with the requirement that the sum of reflected and emitted components is equal to the incident flux. This can be expressed mathematically as

$$\pi R^2 (F/r^2)(1 - A) = 4\pi R^2 \varepsilon \sigma T^4$$

In this equation, R is the radius and F is solar flux, a constant. The distance from the Sun is r and A is a term called the bolometric Bond albedo. The emissivity of the asteroid, ε, is assumed to be 1 and the parameter σ is the Stefan–Boltzmann constant. The temperature, T, is derived from the radiated flux from the asteroid measured in a broadband pass in the thermal infrared

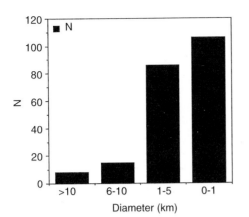

FIGURE 15 Frequency, *N*, of near-Earth asteroid diameters derived from radiometric measurements.

region of the spectrum. Its albedo is derived from the reflected magnitude at all wavelengths that reflect light. The measured parameter for albedo is usually brightness in a broadband pass filter covering the visible region of the spectrum.

Another method of estimating the size of small asteroids is from their measured brightness and an assumed albedo. This method is referred to as a photometric diameter. It is used when no thermal measurements and only visual magnitude are available, according to the equation

$$\log d = k - 0.5 \log p - 0.2 m(0°)$$

where *p*, geometric albedo, is assumed. Unfortunately, the range of asteroid albedos is large, producing considerable uncertainty in the photometric diameters, such that we often express a high albedo diameter and a low albedo one. Notice that an object with a lower albedo, reflecting the same amount of light, will be significantly larger than a bright object. For example, a 10th magnitude object (on the bright end of any near-Earth asteroid) with an albedo of 0.15, an average, "bright" asteroid, would have a diameter of 34 km, whereas an asteroid with a 0.06 albedo, at the high end of the range of dark asteroids, would be 1.5 times as large at 54 km.

Keep in mind that the plot showing the frequency of near-Earth asteroids as a function of brightness and size (see Fig. 11) provides only an estimate of the size and frequency of asteroids and, except at the large end of the magnitude scale, is an extrapolation and estimate of the size of the population. The histogram of Fig. 15 contains the data for asteroid diameters from radiometric measurements, which number far fewer than the anticipated total population.

E. MASS

Mass determinations of small asteroids are estimated from assumed densities and their estimated sizes. Thus the mass of any of these near-Earth asteroids is not directly determined and is poorly known. Most of the known near-Earth asteroids are a few kilometers in diameter and for purposes of calculations, assume they are spherical. Let's estimate the mass of the largest NEA, 1036 Ganymed. We do not know much about its shape, so we can assume it has a 41-km diameter. That yields a volume of 36,087 km^3 or 3.61×10^{13} m^3. We can make a reasonable guess that its density is either like that of stony meteorites, with a density of 3000 kg/m^3 (see Section V, G), or like that of the stony-iron meteorites, with somewhat higher densities of 5000–7000 kg/m^3. Hence, the mass of 1036 Ganymed is likely between 1.1×10^{17} kg for a stony material and 2.1×10^{17} kg for a material with more metallic iron. You can see from this estimate and limited knowledge that we arrive at two numbers varying by a factor of 2. For comparison, the mass of Earth is 6×10^{24} kg and that of our Moon is 7×10^{22} kg.

We will have a very precise mass for the Amor asteroid 433 Eros when the *NEAR* spacecraft goes into its orbit in early 1999. This mass will be derived from changes exerted on the spacecraft as it orbits Eros. $F = GMm/r^2$ expresses the gravitational force between the spacecraft and the asteroid, and we will know the mass of the spacecraft very precisely at any given moment, even accounting for burning of the spacecraft's fuel. We will also keep close track of the range between the spacecraft and the asteroid, *r*, making sure that number is always positive. Hence we expect a mass measurement that will be accurate to 0.1%. Eros is the only asteroid for which such a precise measurement will be available. No other asteroid has had a spacecraft orbit it for this type of precision measurement.

F. COLOR AND TAXONOMY

Since the early part of the twentieth century, astronomers have recognized that asteroids come in different colors. As observational techniques evolved and our ability to investigate asteroids improved, the number of observable characteristics increased. The sorting of characteristics into meaningful groups is the process of classification or taxonomy. The asteroid taxonomy developed in response to advances in observing techniques and new technology in the field of stellar photometric astronomy.

Photographic plates with different emulsions were first used to study the difference in radiation at different wavelengths. Photoelectric detectors followed and were used with filters of different bandpasses. Distinctive color differences were observed, and the first three categories of asteroids—C, M, and S—were identified. C-types are low albedo, between 0.03 and 0.06 with $B - V = 0.70$ and $U - B = 0.34$. M-types have a higher albedo, but are considered moderately bright with $B - V = 0.70$ and $U - B = -0.23$. The S-types are a little brighter than M's, with the center of their color domain falling at $B - V = 0.84$ and $U - B = 0.42$. In the 1970s, a standard set of filters was selected specifically to measure the spectral reflectance characteristics of asteroids, and Dave Tholen surveyed the colors of about 500 asteroids. From this survey, different taxa were created. Today, the alphabet soup of asteroid taxonomy extends to about 12 letters.

Near-Earth asteroids have representatives from all taxonomic types, indicating that no one location in the asteroid belt feeds the near-Earth population. Most of the NEAs are S-type asteroids, although after applying bias corrections, it appears that the near-Earth population is equally represented by bright, S-types and dark, C-types. The other groups are sparsely represented.

Current asteroid taxonomy is based on the application of statistical clustering techniques to the parameters of color and albedo. We believe this classification scheme reflects the compositional variations among the asteroids. However, one must always remember that the classification may in some cases be false, meaning that the color and albedo of a particular object may not directly correspond with a particular mineralogical composition and surface albedo. We are constantly attempting to test and refine the asteroid taxonomy by employing new statistical methods and extending the number of meaningful parameters that are included in the classification process while eliminating meaningless or redundant parameters.

Coupled to the shortcomings of any classification scheme are observational biases that exist when observing astronomical objects. There is an inherent bias toward observing brighter objects and to miss darker ones, since the ability to make an astronomical observation depends on the number of photons available. A bias correction factor can be applied to all known asteroids based on their proximity to Earth and the amount of time they are observable from Earth. These weights can be applied to those asteroids of known taxonomy and so a bias-corrected ratio of S : C type asteroids can be estimated. Whereas most of the observed near-Earth asteroids are bright, S-types, we believe there are an equal number of dark, C-types among the population when the observability factor is applied.

Other evidence in support of an equal representation of S- and C-type asteroids can be found in the 3 : 1 Kirkwood Gap, a likely source of near-Earth asteroids. NEAs from this region are equally represented in C-types and S-types. But among those C-types, some extinct comet nuclei could be lurking. There is little doubt that the S-type asteroids are derived from the inner regions of the Main Asteroid Belt, for they have spectral characteristics that are unique in the solar system. Only the spectra of some lunar regions are as red as those of the S-type asteroids, but usually lunar soils are redder than the S-types.

G. MINERALOGY

By measuring the percentage of reflected sunlight from the surface of an asteroid, it is possible to place some constraints on the mineralogy of that surface. This technique was pioneered by Tom McCord and his students and colleagues in the 1970s. It was first applied to the Moon, and the technique was given validity when lunar samples were brought to Earth and the interpretation of remote-sensing measurements agreed with direct chemical and mineral analysis of the lunar samples.

We find that most near-Earth asteroids contain two strong absorption bands, one in the ultraviolet and the other in the near infrared. Sometimes a second near-infrared band is observed at 2 μm, but the size of the asteroids and the sensitivity of instruments limit our opportunities to observe them in the 1- to 2-μm region. If the asteroids do not have prominent absorption bands, they are found to be featureless and flat. Most often these asteroids also have a low albedo.

Figure 16 shows some spectral reflectance measurements of near-Earth asteroids. The top spectrum is characteristic of a C-type asteroid in that there is only a hint of absorption bands in the ultraviolet and near-infrared spectral regions. This asteroid, 1580 Betulia, is also extremely dark, having an albedo of approximately 3%. The other three asteroids shown here have prominent ultraviolet absorption bands that are common in silicate minerals. Unfortunately, they are not diagnostic of a specific mineralogy as they are found in any mineral-bearing silicon and oxygen bound by a wide range of cations, including aluminum, magnesium, and iron. The band at 1 μm, however, is diagnostic of a mineral called pyroxene, which consists of silicon oxide tetrahedra bound in eight-fold symmetry by magnesium, calcium, and iron cations. Subtle differences in these spectra, such as the position of the center of the band, can constrain the chemistry of the pyroxene, which can accommodate a range of magnesium, iron,

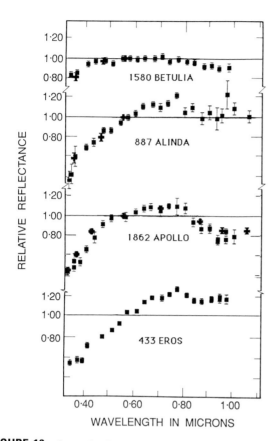

FIGURE 16   Spectral reflectances of some near-Earth asteroids.

are they alike in the Main Asteroid Belt. The range of variations in mineral composition reflects that seen in the Main Asteroid Belt and supports the supposition that near-Earth asteroids are derived from the Main Belt for the most part. I have looked for evidence that the NEAs are compositionally similar to any of the major planets, for they might be fragments from them. But they do not share any of the spectral reflectance characteristics of the major planets or the Moon.

From knowledge of the mineralogy of asteroids, we can constrain the temperature and pressure under which the asteroid formed. Analysis of the mineral composition of an asteroid reveals not only the chemistry but the structure that the elements have formed based on the temperature and pressure of their environment. Mineralogy is essential for determining the history of the formation of the rock. Reflectance studies beyond 1 μm have been made for a few NEAs. The S-type NEAs have a 2-μm absorption band, as expected.

VI. FUTURE STUDIES

A. NEAR

On February 16, 1996, the *Near-Earth Asteroid Rendezvous* spacecraft was launched from Cape Canaveral, Florida, on a three-year journey to the Amor asteroid 433 Eros. It will orbit Eros for a year in 1999, training its six scientific instruments on the asteroid's surface. This will be the first spacecraft mission dedicated to the scientific study of an asteroid, and the first to orbit an asteroid. We plan to study the chemical and physical properties of this asteroid to learn of the relationship between Eros and meteorites, the nature of the internal state of the asteroid, and its collisional and surface history.

The spacecraft carries a complement of instruments that cover the electromagnetic spectrum, illustrated in Fig. 17. The magnetometer will measure magnetic field strengths and orientations and determine whether the magnetic field is intrinsic or a remanent field. This knowledge will constrain the temperature at which the asteroid initially formed. The radio science team will monitor the position of the spacecraft as radioed to Earth relative to the asteroid to derive the asteroid's mass. Its shape will be determined from a laser range finder and the imaging camera, both of which will provide images with spatial resolution of approximately 6 km. With the near-infrared spectrometer,

and calcium abundances in the same mineralogical structure. There is also information tied up in the shape of the ultraviolet band, but we are not certain exactly what that is telling us.

One can see that the UV band in the spectrum of Alinda and Eros is more linear than that in the spectrum of Apollo. We suspect that the more linear band indicates the presence of an additional mineral component, either olivine, another silicate, or metallic iron. Data in spectral regions beyond 1 μm are needed to differentiate between these possible interpretations. In these spectra, the vertical bars extending above and below each data point indicate the 1-sigma uncertainty in the data. There is a 66% chance that the actual value is greater or less than the observed value to within the error bars. In cases where the error bars are large, either the asteroid was very faint when measured or the observing conditions were not ideal. The largest error bar in the spectrum of 887 Alinda falls in a region where water vapor absorptions occur in the terrestrial atmosphere. The effects of this were not completely removed in the calibration process, hence the value has a large uncertainty.

The first result of mineralogical studies of near-Earth asteroids revealed that they are not all alike. Nor

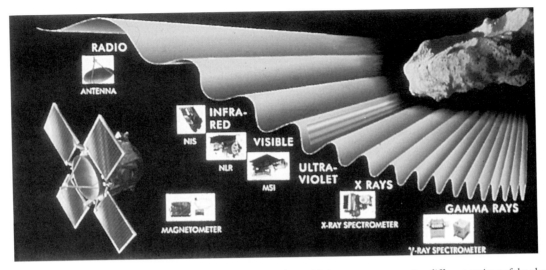

FIGURE 17   The *NEAR* spacecraft at 433 Eros with the complement of scientific instruments measuring different regions of the electromagnetic spectrum. The magnetometer measures magnetic fields that are perpendicular to the electromagnetic radiation. (Copyright 1996 University of Maryland, reprinted with permission. Design by Eric Johnson and Lucy McFadden. *NEAR* images provided by Applied Physics Lab.)

we will combine photometric measurements from the multispectral imaging (MSI) camera and extract information about the optical properties of the asteroid's surface and its mineralogy. The X-ray and γ-ray spectrometers will provide knowledge of the elemental abundance of about a dozen rock-forming elements, including Mg, Al, Si, S, Ca, Ti, and Fe. The γ-ray spectrometer provides additional abundance data for naturally occurring K, U, Th, and H. Analysis of the images from MSI will allow the ability to count the number of craters on the asteroid's surface and to estimate its cratering age, or how long it has been exposed in space to impacting projectiles. We will look for residual evidence of volcanic and/or tectonic activity, which would indicate that Eros is a fragment of a much larger body. An asteroid of merely ~20 km in diameter is not expected to show evidence of planetary processes. But is Eros a fragment of a once larger body? We eagerly await answers to these questions.

B. SAMPLE RETURN MISSION

No matter how much remote-sensing data we acquire from spacecraft missions, we will never be certain that our interpretations are correct without ground-truth measurements. The Apollo missions and returned lunar samples provided this important extension of lunar studies. We still find that we need information from areas of the Moon where no samples have been returned. There is hope that we will be convinced that

there are meteorites on Earth that reasonably came from Eros or at least are the same type as Eros. If we meet that primary objective, determining the elemental composition of the asteroid, and find that it coincides with a meteorite type, we will have established ground truth. If we find that the composition of Eros is not like that of any known meteorites, we will be frustrated by the uncertainties in our interpretations. Missions to return samples from a near-Earth asteroid have been proposed to NASA, but to date none has been selected for funding and implementation. However, such a mission is definitely the next step in the exploration of small solar system bodies.

VII. EXTRATERRESTRIAL RESOURCES

Whether near-Earth asteroids represent an enabling or a driving force in the development of extraterrestrial resources remains a matter of economic speculation until we have a better picture of the composition of these asteroids. Enabling products are those that will contribute to the infrastructure required for space activities and/or life support. Any solid material would fulfill this criterion. Whether other critical materials for life support, such as water or volatile-rich compounds, can be converted to a consumable form for humans requires additional compositional assessment that cannot be carried out by remote-sensing measurements alone. It is likely that multiple generations will pass before we have the ability to assess many near-Earth asteroids at the elemental level. Considering the engineering development required to utilize the materials and support human activities, use of near-Earth asteroids for space activities is pushed further into the future. However, the lobbying efforts of citizen-driven organizations such as the Planetary Society and the National Space Society promote the cause and objectives of space exploration.

The complement to enabling products is driving products. These have intrinsic value, independent of their usage. The primary products in this category are energy and precious metals, both of which have economic value on Earth. Again, the assertion that near-Earth asteroids contain materials that may serve as driving products for the economic development of space is entirely speculative at this point. However, the speculation is merited and based on sound reasoning, for any matter can be converted into energy. The question is whether or not it can be done efficiently and economically. That there are precious metals present among the near-Earth asteroids can be stated with certainty, as all material in the solar system has the same distribution of elements. The matter becomes an issue of accessibility. What the metallurgists and mining engineers like to hear is that in a subset of the meteorites, the platinum metal group of elements is present in abundances that are 3 to 10 times higher than in ore-grade material currently mined on Earth. What we do not know is what fractions of and which near-Earth asteroids specifically are the parent bodies of which meteorites. The ordinary chondrite meteornites have platinum-group elements with concentrations of 30–60 ppm, whereas some metallic meteorites have abundances as high as 100 ppm. Such abundances will grow in value when the available supply at Earth's crust becomes depleted. The time to develop techniques to assess the composition and the mechanism for extracting the needed materials is now.

SEE ALSO THE FOLLOWING ARTICLE

ASTEROIDS

BIBLIOGRAPHY

Barnes-Svarney, P. (1996). "Asteroid." Plenum, New York.

Gehrels, T. (ed.) (1994). "Hazards Due to Comets and Asteroids." Univ. Arizona Press, Tucson.

Lewis, J. S. (1996). "Mining the Sky, Untold Riches from the Asteroids, Comets, and Planets." Helix Books/Addison–Wesley, Reading, Mass.

Lewis, J. S., Matthews, M. S., and Guerrieri, M. L. (eds.) (1993). "Resources of Near-Earth Space." Univ. Arizona Press, Tucson.

METEORITES

I. Introduction
II. Meteorite Classification
III. Meteorites of Asteroidal Origin and Their Parent Bodies
IV. Chemical and Isotopic Constituents of Meteorites
V. Meteorite Chronometry

Michael E. Lipschutz[1]
Purdue University

Ludolf Schultz
Max-Planck-Institut für Chemie

GLOSSARY

Achondrite: Differentiated igneous stony meteorite, apparently solidified from a magma.

Age: Time elapsed since some event at a discrete time, t_0.

Breccia: Rock composed of fragments derived from previous generations of rocks.

Chondrite: Undifferentiated stony meteorite, usually containing chondrules or their fragments.

Chondrule: Approximately spherical droplet formed by partial or complete melting and quenching prior to incorporation in the meteorite.

Meteorite: Natural solid object of extraterrestrial origin that survives passage through Earth's atmosphere.

Mineral: Naturally occurring substance of specified chemical composition and physical properties having a characteristic atomic structure and/or crystalline form.

Regolith: Layer of impact-fragmented and comminuted material on the surface of planets having no atmosphere.

Meteorites, which have been called "The Poor Man's Space Probe," are of special interest because they contain the oldest solar system materials available for research and sample a wide range of parent bodies—exteriors and interiors—some primitive, some highly evolved. Meteorites carry decipherable records of certain solar and galactic effects and yield otherwise unobtainable data relevant to the genesis, evolution, and composition of Earth and other major planets, satellites, asteroids, and the Sun. Some contain inclusions that trace events from before the solar system formed; others contain organic matter produced by catalyzed reactions on grain boundaries in the early nebula and/or in giant interstellar clouds. Meteorites also provide an important body of "ground truth," in a chemical and physical sense, which is critical to interpreting planetary data obtained by remote sensing. It is especially advantageous that meteorites are delivered to Earth's surface, where the full spectrum of laboratory analytical techniques can be applied, ranging from the simplest to the most sophisticated. Truly, if one picture is worth 10,000 words, then one sample is worth 10,000 pictures. Although meteorites are only fragmentary pieces of their sources, proper

[1] Research supported by National Aeronautics and Space Administration Grant NAGW-3396.

arrangement of the information they contain can yield a more complete picture of their sources, just as a mosaic can be depicted from but a few tesserae.

I. INTRODUCTION

A. GENERAL

The quincentenary year 1492 marked not only the discovery of the Old World by the New, and the Spanish Expulsion, but also the oldest preserved and scientifically studied meteorite *fall* in the Western world, a 127-kg stone (LL6) that fell at 11:30 A.M. on 16 November 1492 at Ensisheim in Alsace, France. (A meteorite is named for the nearest post office or geographic feature. The chemical-petrologic classification scheme by which Ensisheim, for example, is classified as an LL6 chondrite is described in Section I, B.) The recordholder for Earth's oldest preserved meteorite fall is Nogata (Japan), another stone (L6), which fell in 861 and has been preserved since then in a Shinto shrine there. Recovered meteorites whose falls are unobserved are called *finds.* Some of these have been found, occasionally artificially reworked, in archaeological excavations in Old World locations such as Ur, Egypt, and Poland and in New World burial sites. Obviously, humans in prehistoric and early historic civilizations recognized meteorites as unusual, even venerable, objects.

Despite this, and direct evidence for the Ensisheim fall and subsequent ones, it was not until the beginning of the nineteenth century that scientists generally accepted that meteorites are genuine samples of other planetary bodies. Prior to that time, acceptance of meteorites as being extraterrestrial, and thus of great scientific interest, was spotty. One person might assemble a meteorite collection over some extended period, only to have this invaluable material disposed of by another person later on. Such a case occurred, for example, when the noted mineralogist Ignaz Edler von Born discarded the imperial collection in Vienna as "useless rubbish" in the latter part of the eighteenth century.

With the recognition that meteorites represent extraterrestrial planetary bodies, collections of them have proved to be particularly important. In 1943, as the Soviet army planned the invasion of Germany, the Russian government drew up plans to have "trophy brigades" accompany their advancing armies and collect artistic, scientific, and production materials in Ger-

many as restitution for Russian property seized or destroyed by Nazi armies during their occupation of parts of Russia. Meteorites that had fallen in Russia, portions of which had been acquired by and housed in German collections, were specifically identified as material to be seized. As this chapter was being revised at the end of 1996, the reserve auction price for meteorites from Mars was $2500 per gram; in contrast, the current price of gold is $9 per gram.

Apart from its distinction as the oldest observed fall, Ensisheim is typical. Outside of Antarctica, the recognition factor is important for discovery of finds—hence the high proportion of iron meteorites (with their "unearthly" high density) in the finds (Table I). For this reason, the makeup of observed falls is taken to best approximate the contemporary population of near-Earth extraterrestrial materials. Of course, the fall population may also be biased if, for example, highly friable meteoroids (whose existence is suggested by some data) are largely or totally disaggregated during atmospheric passage.

Meteorites' initial entry velocities, which can range from 11 to 70 km/s and average 15 km/s, are sufficient to cause surface material to melt and ablate by frictional heating during atmospheric passage. Since heat generation rates and ablation rates are rapid and very similar in magnitude, detectable heat effects rarely penetrate more than a few millimeters below the surface, so that the meteorite's interior is preserved in its cool, preterrestrial state. Atmospheric ablation and fragmentation—which can cause substantial (~90%) mass loss and deceleration, often to terminal velocity—leave a dark brown-to-black, sculpted fusion crust as the surface, which is diagnostic of a meteorite on Earth (Fig. 1a; see also color insert).

If the meteoroid is properly shaped, for example, by ablation, it can assume a quasi-stable orientation during the latter part of its atmospheric traversal. In such a case, material ablated from the front surface can be deposited as delicate droplets or streamlets on the sides and rear of the meteorite Lafayette (Fig. 1b). Since the delicate droplets on its fusion crust would have been erased after a few days of weathering, it must have been recovered almost immediately after it fell. Yet, when it was recognized as meteoritic during a 1931 visit to Purdue University by the prominent meteoriticist O. C. Farrington, the chemistry professor on whose desk it was found thought it was a terrestrial glacial artifact. Who actually recovered this Martian meteorite is a mystery. Meteorites derive from asteroids and, less commonly, from larger parent bodies: 16 individual samples representing 12 separate falls (all but 2 in Antarctica) come from Earth's Moon, and 12

TABLE I

**Numbers of Classified Non-Antarctic
Meteorite Falls and Finds Compared with
Antarctic Meteorite Recoveries by U.S. Teams**

| Meteorite type | Non-Antarctic | | Victoria Land and Transantarctic Mts., Antarctica[c] |
	Falls[a]	Finds[b]	
Chondrites, total	772	897	*1476* (7004)
CI1	5	0	1 (1)
CM2	18	15	48 (172)
C other	16	15	31 (91)
E	14	11	26 (72)
H	301	405	*741* (3059)
L	349	350	*564* (3341)
LL	70	30	*61* (263)
Other	2		4 (5)
Achondrites, total	73	22	118 (204)
Angrites	1	0	2 (2)
Aubrites	9	1	5 (33)
Howardites	19	3	11 (21)
Eucrites	24	7	49 (82)
Diogenites	9	0	13 (13)
Lunar	0	2	5 (7)
Martian			
Nakhlites	1	2	1 (1)
Shergottites	2	0	5 (5)
Ureilites	4	6	22 (31)
Other	4	2	5 (9)
Irons, total	47	683	35 (51)
Stony-irons, total	11	57	14 (32)
Lodranites	1	0	3 (3)
Mesosiderites	6	22	9 (25)
Pallasites	4	35	2 (4)

[a] Data from Dr. Monica Grady, Natural History Museum, London, as of September 1997. There also exist 70 unclassified stones and 7 unclassified meteorites among the 980 known falls.

[b] These do not include the many meteorites recently recovered in the hot deserts of Australia and the Sahara.

[c] Numbers of meteorite fragments recovered and classified as of May 1996 are listed in parentheses. Numbers listed before these entries are corrected for pairing and, in the case of ordinary chondrites, include an estimate of four fragments per fall event (values in italics). Several hundred additional ordinary chondrites from Antarctica remain unclassified. (Data for Antarctic meteorites courtesy of Dr. J. Grossman.)

others (half from Antarctica) almost certainly are from Mars. Some interplanetary dust particles may also come from these sources, as well as from comets. Meteorites are rocks and are therefore polymineralic (Table II), with each of the 350 known meteoritic minerals generally having some chemical compositional range, reflecting its formation conditions and/or subsequent alteration processes. Important episodes in the genesis of meteorites are shown in Fig. 2. [*See* MARS; THE MOON.]

B. FROM PARENT BODY TO EARTH

To be delivered to Earth, meteoritic matter must first be excavated and removed from the gravitational field of its parent body by an impact. This impact, which may generate short-lived but intense shocks, provides the impulse necessary for the meteoroid (the immediate source for a meteorite landing on Earth) to exceed the parent body's escape velocity. In general, the higher the shock pressure acting upon a material, the

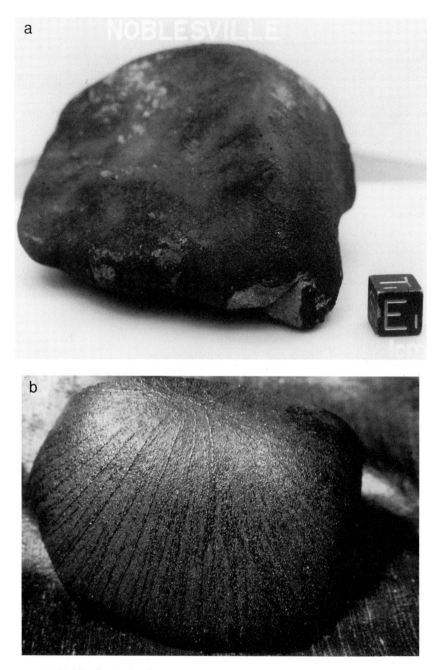

FIGURE 1 Fusion crusts: (a) Noblesville H chondrite (see also color insert) and (b) Lafayette Martian meteorite. The Noblesville H chondrite, which fell on 31 August 1991, has nearly complete fusion crust and exposed surface at lower right of a genomict (H6 in H4) breccia, next to the 1-cm cube. (Photo courtesy of NASA Johnson Space Center.) The Lafayette meteorite exhibits very delicate, redeposited droplets on its sides, indicating that it was oriented with its top pointing toward Earth in the last part of its atmospheric traversal. (Photo courtesy of the Smithsonian Institution.)

higher are its ejection velocity and its temperature—both the shock temperature derived from passage of the pressure wave and the postshock residual temperature (from compressional nonadiabatic heat) persisting after decompression. Residual temperatures as high as

1250°C, which have been recorded in recovered stony meteorites, correspond to pressures exceeding 57 gigapascals (GPa) or 570,000 atm (570,000 times the pressure at sea level on Earth). Significantly higher pressures would cause matter to vaporize, so there is a

TABLE II
Common Meteoritic or Cited Minerals

Mineral	Formula	Mineral	Formula
Anorthite	$CaAl_2Si_2O_8$	Oldhamite	CaS
Clinopyroxene	$(Ca,Mg,Fe)SiO_3$	Olivine	$(Mg,Fe)_2SiO_4$
Chromite	$FeCr_2O_4$	Olivine solid solution	
Cohenite	$(Fe,Ni)_3C$	Fayalite (Fa)	Fe_2SiO_4
Cristobalite	SiO_2	Forsterite (Fo)	Mg_2SiO_4
Diamond	C	Orthopyroxene	$(Mg,Fe)SiO_3$
Diopside	$CaMgSi_2O_6$	Pentlandite	$(Fe,Ni)_9S_8$
Enstatite	$MgSiO_3$	Plagioclase	
Epsomite	$MgSO_4 \cdot 7H_2O$	Albite (Ab)	$NaAl_2Si_2O_8$
Fayalite	Fe_2SiO_4	Anorthite (An)	$CaAl_2Si_2O_8$
Feldspar solid solution		Pyroxene solid solution	
Albite (Ab)	$NaAlSi_3O_8$	Enstatite (En)	$MgSiO_3$
Anorthite (An)	$CaAl_2Si_2O_8$	Ferrosilite (Fs)	$FeSiO_3$
Orthoclase (Or)	$KAlSi_3O_8$	Wollastonite (Wo)	$CaSiO_3$
Ferrosilite	$FeSiO_3$	Schreibersite	$(Fe,Ni)_3P$
Forsterite	Mg_2SiO_4	Serpentine (or chlorite)	$(Mg,Fe)_6Si_4O_{10}(OH)_8$
Gehlenite	$Ca_2Al_2SiO_7$	Spinel	$MgAl_2O_4$
Graphite	C	Spinel solid solution	
Hibonite	$CaAl_{12}O_{19}$	Spinel	$MgAl_2O_4$
Ilmenite	$FeTiO_3$	Hercynite	$FeAl_2O_4$
Kamacite	$\alpha\text{-}(Fe,Ni)$	Chromite	$FeCr_2O_4$
Lonsdaleite	C	Taenite	$\gamma\text{-}(Fe,Ni)$
Magnetite	Fe_3O_4	Tridymite	SiO_2
Melilite solid solution		Troilite	FeS
Åkermanite (Åk)	$Ca_2MgSi_2O_7$	Whitlockite	$Ca_3(PO_4)_2$
Gehlenite (Ge)	$Ca_2Al_2SiO_7$		

practical limit to the ejection velocity of survivable meteoritic material during an impact, ~5 km/s, that is, Mars's escape velocity.

In very special impact scenarios, it is possible for ejecta to be accelerated by impact-jetting—especially during oblique impacts—and, thus, to acquire a velocity higher than would be expected from the degree of shock-loading. Some of the Martian meteorites, the nakhlites, are not heavily shocked and may represent such an unusual case. In general, however, it is unlikely that we will receive meteorites on Earth from a parent body much larger than Mars.

The overwhelming majority of meteorites, those of asteroidal origin, seem not to be a random sampling of the thousands of asteroids but rather of a few hundred dominant ones. These may include the near-Earth asteroids (NEA) that are already in Earth-crossing or Earth-approaching orbits, ejected from Kirkwood Gap regions by chaotic motion and the gravitational effects of Jupiter. [See ASTEROIDS; NEAR-EARTH ASTEROIDS.] As will be discussed later, some types of meteorites and asteroids can be linked. The only four meteorite falls whose orbits were determined photographically seem to be NEA-like (Fig. 3). At present, some evidence suggests that co-orbital streams of meteorites and/or asteroids exist—perhaps arising from meteoroids' gentle disruption in space—but this is very controversial. Evidence from temperature-sensitive components indicates that, in their orbits about the Sun, some meteorites have perihelia within 0.8 AU, resulting in detectable solar heating.

Some meteorites derive from regolith material that was bombarded by particles having considerable energies. From the time that material is ejected by an impact from its parent body until it falls on Earth, meter-sized meteoroids are irradiated by cosmic rays (mainly protons) of solar or galactic origin. Solar cosmic rays are characterized by a power–law energy distribution

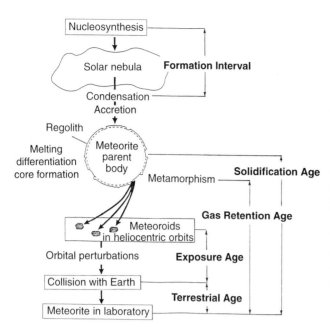

FIGURE 2 From nebula to meteorite: genetic processes and the corresponding age that can be determined for each process. Nuclides of nearly all elements were formed by nuclear reactions in the interiors of large stars, which then ejected them in very energetic events. The ejected nebular gas and dust subsequently condensed and accreted into primitive bodies. Bodies from which most meteorites derive were heated, causing solid-state metamorphism or, at higher temperatures, differentiation involving separation of solids, liquids, and gases. As the body evolved, it was battered by numerous impacts and, if it had no atmosphere, its surface was irradiated by solar and galactic particles that were embedded in the skins of small particles and/or caused nuclear reactions. During larger impacts, pieces of the body were ejected and traveled around the Sun as do larger bodies. Subsequently, orbital changes caused by the gravitational attraction of much larger bodies placed meteoroids into Earth-crossing orbits, from which they landed on Earth and were recovered immediately as a fall or later as a find. Each of these processes can alter elemental and/or isotopic contents. Which of these processes affected a given meteorite and the time elapsed since it occurred can be established by analysis of the meteorite in the laboratory.

in which the particle flux increases rapidly with decreasing energy: most solar particles have energies <1 MeV. Galactic and some solar particles have energies of hundreds of MeV to GeV and can induce nuclear reactions that produce cosmogenic radioactive or stable nuclides. In larger meteoroids, these cosmogenic nuclear reactions are confined to the meter-thick shell that cosmic ray primaries and secondaries penetrate. As will be discussed in Sections V, A and V, B, the amounts of nuclides produced during cosmic ray exposure (CRE) can establish the duration of energetic particle bombardment (the CRE age) and the time spent by a meteoritic find on Earth's surface (the terrestrial age) (Fig. 4).

C. IMPACT ON EARTH

If a meteoroid is small enough to be decelerated significantly by passage through the Earth's atmosphere, it may reach the ground as an individual or as a shower. A recovered individual can have a mass of a gram or less [e.g., the 31 March 1965 fall of the Revelstoke stone (CI1) in British Columbia] or up to 60 metric tons (e.g., the Hoba IVB iron meteorite found in 1920 on a farm in Namibia, where it remains). A meteorite shower results from a meteoroid that fragments high in the atmosphere, usually leaving a trail of particles down to dust size (Fig. 5b). Shower fragments striking the earth define an ellipse whose long axis, which can extend for tens of kilometers, is a projection of the original trajectory. Typically, the most massive shower fragments travel farthest along the trajectory and fall at the farthest end of the ellipse.

Some falls are signaled by both light and sound displays; others, like the Peekskill meteorite (Fig. 5d), exhibit a spectacular fireball trail observed over many states. Small falls, like Noblesville (see Fig. 1a), fall silently and unspectacularly, and when recovered immediately after fall have cold to slightly warm surfaces. Meteorites can fall anywhere at any time. The 500-g Borodino stone (H5) fell on 5 September 1812—two days before the famous battle there—and was recovered by a Russian sentry. The ultimate result of the Battle of Borodino was the devastating retreat of Napoleon's army from Russia, which was musically celebrated by Tchaikovsky's Overture of 1812.

Recently, the U.S. Department of Defense declassified data demonstrating that, since 1975, reconnaissance satellites have detected large explosions at seemingly random locations in Earth's atmosphere. On average, about nine of these mysterious explosions [which can involve energies of 1 megaton (Mt) equivalent of TNT] occur annually; no meteorite falls or fireballs have been associated with any of these events.

Large meteoroids—tens of meters or larger—are not decelerated much by atmospheric transit and, with an appropriate trajectory, may ricochet off Earth's atmosphere (Fig. 5a; see also color insert) or strike it at full geocentric velocity. (In this case, the distinction between a large meteoroid and a small asteroid is arbitrary.) Such impacts will be explosive, leaving a characteristic crater, and can do considerable damage. For example, the 1-km-diameter Meteor Crater (Fig. 5c; see also color insert) in northern Arizona was formed 50,000 years ago by the impact of a 25- to 86-m meteoroid, fragments of which survive as Canyon Diablo iron meteorites. At present, at least 40 craters have been identified on Earth as having features believed

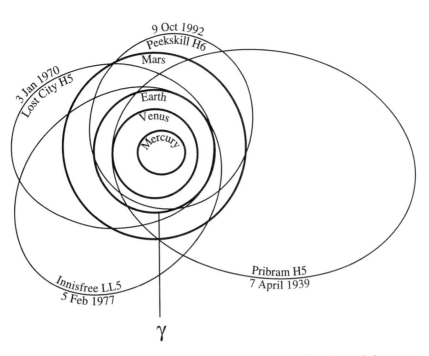

FIGURE 3 Orbits determined by overlapping photography of four ordinary chondrite falls. The symbol γ represents the vernal equinox. Each of the first three falls was detected by camera networks in Central Europe, the United States, and Canada. Only the first of these, the European Fireball Network covering 106 km² with 34 cameras, remains operational.

to be produced only from the intensive explosive impact of a large meteoroid, or a comet nucleus. The 30 June 1908 event at Tunguska, Siberia was explosive but left no crater. An additional 269 features on Earth are possibly of impact origin. One expert recently

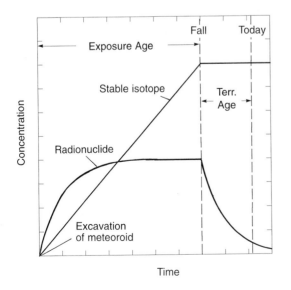

FIGURE 4 Concentrations of cosmic ray-produced radioactive and stable nuclides during cosmic ray exposure and after the meteorite's fall on Earth.

classed 130 of them as definite impact craters. One of these is the 180-km-diameter Chicxulub feature in Yucatan, Mexico, suspected to be the impact site of a 10-km-sized meteoroid. This impact is believed to have generated the climatic consequences responsible for the extinction of ~60% of then-known species of biota—including dinosaurs—ending the Cretaceous period on Earth and beginning the Tertiary, 65 million years ago.

A number of meteorites have struck man-made structures. The most recent of these is the Peekskill stone meteorite (H6), with a recovered mass of 12.4 kg, which ended its journey in the trunk of a car (Fig. 5e). Its descent on 9 October 1992 was observed and videotaped over a five-state area of the eastern United States by many individuals at high school football games on that Friday evening (Fig. 5d). These videotapes yielded a very well determined orbit (see Fig. 3).

Two authenticated reports of human injury by meteorite falls exist. The first involves a 3.9-kg (H4) stone (the larger of two fragments), which struck a recumbent woman on the thigh and badly bruised her, after passing through the roof of a house in Sylacauga, Alabama, on 30 November 1954. The second involved a 3.6-g piece of the Mbale (Uganda) L6 meteorite shower of 14 August 1992, which bounced off the

FIGURE 5   Large meteoroids: (a) fireball of 80-m object with estimated mass of 1 million metric tons moving left to right over Grand Teton National Park, Wyoming, on 10 August 1972, that apparently skipped out of the atmosphere (photo courtesy of Dennis Milon) (see also color insert); (b) Russian postage stamp illustrating dust trail during the fall of the Sikhote Alin IIB octahedrite (recovered mass of 23 metric tons) on 12 February 1947; (c) the 1-km-diameter Meteor Crater in Arizona formed by the explosive impact of the Canyon Diablo IA octahedrite meteoroid about 50,000 years ago (photo courtesy of Allan E. Morton) (see also color insert); (d) part of the videotape record of the Peekskill meteoroid during its atmospheric traverse on 9 October 1992. During fragmentation episodes such as the one pictured here (over Washington, D.C.), large amounts of material fell, but these were not recovered. (e) Landing site of Peekskill H6 chondrite in the right rear of the automobile (photo courtesy of Peter Brown, University of Western Ontario).

FIGURE 5 (*continued*)

leaves of a banana tree and hit a boy on the head. Chinese records from 616 and 1915 claim numerous human and animal casualties, including many killed, by meteorite falls. A number of unauthenticated reports of human injuries or deaths exist, and a dog is reported to have been killed by a piece of the 40-kg Nakhla meteorite shower of 28 June 1911 near Alexandria, Egypt; this, incidentally, is one of the 12 meteorites thought to come from Mars. Despite the small number of casualties to date, calculations indicate that the probability of dying in a meteoroid impact exceeds that of being killed in an airplane crash. This reflects the fact that the impact of a large meteoroid, small asteroid, or comet nucleus capable of causing devastating loss, indeed the total extinction of life, is fortunately rare.

As is already evident, meteorites may impact anywhere on Earth and the current numbers of known falls and non-Antarctic finds are 1103 and 1763, respectively (cf. Table I). In such cases, it can readily be established whether meteorite fragments found near each other derive from the same meteoroid. This is not so easy to do in the case of the very numerous (~17,000 to date) meteorite fragments found in various parts of Antarctica during the past three decades (Fig. 6). Starting in 1969, but mainly since 1976, annual Japanese and U.S. expeditions have recovered most of these; expeditions from other countries (and a European consortium) have contributed to the collection effort. The 14-million-km^2 Antarctic ice sheet seems to be such a meteorite trove because of the continent's possibly unique topography and its effect on ice motion, which promotes the collection, preservation (because of the climate), transportation, and concentration of the samples (Fig. 7). The number of fragments per meteoroid is estimated to be 4 ± 2, so that Antarctic meteorite fragments recovered thus far correspond to

about 4000 different impact events. In the past few years, many additional meteorite fragments have been found in arid regions, for example, the Sahara in the Northern Hemisphere and Australia in the Southern Hemisphere, but the "pairing" question for these has yet to be addressed.

II. METEORITE CLASSIFICATION

A. GENERAL

Meteorites, like essentially all matter in the solar system, ultimately derive from the most primitive materials that condensed and accreted from the gas- and dust-containing presolar disk. Most of these primitive materials were subsequently altered by postaccretionary processes—as in lunar, terrestrial, and Martian samples—but some have survived essentially intact as specific chondrites or inclusions in them. Some primitive materials can be recognized unambiguously (with considerable effort), usually from isotopic abundance peculiarities; others are merely conjectured to be unaltered primary materials. Postaccretionary processes also affected many meteorites, producing obvious characteristics that permit classification of the thousands of known meteorites into a much smaller number of types. Since many classification criteria contain genetic implications, it is important to summarize this information now.

At the coarsest level, meteorites may be considered as irons, stones, or stony-irons based on their predominant constituent (Figs. 8a and 9; Fig. 9-see also color

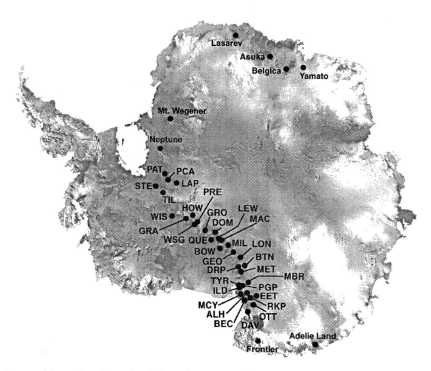

FIGURE 6 Schematic map of Antarctica with regions indicated. Yamato, Asuka, and Belgica are regions that have yielded very large numbers of meteorites collected by Japanese Antarctic Research Expedition Teams since 1969. Three-letter abbreviations designate areas along the Transantarctic Mountains that yielded large numbers of meteorites to U.S. teams operating annually since 1976 (from 1976 to 1979, these were joint U.S.–Japanese teams). In the Frontier Mountains, European meteorite recovery teams collected 350 samples during expeditions in 1990, 1993, and 1995. Other symbols indicate four other areas where individual meteorites were recovered accidently during the 1912–1964 span. (Courtesy of Anita Dodson, NASA Johnson Space Center.)

insert): each of these can then be classified by a scheme providing genetic information (Fig. 8b). The stones are subdivided into the very numerous chondrites (Figs. 9a and 9b), which are more or less primitive, and the achondrites (Fig. 9e), which are igneous in origin. Irons (Fig. 9d), stony-irons (Fig. 9c), and achondrites are differentiated meteorites, presumably formed by secondary processes in parent bodies (see Fig. 2) from melted chondritic precursors. During this process, physical (and chemical) separation occurred, with the high-density iron sinking to form pools or a core below the lower-density achondritic parent magma. Ultimately, these liquids crystallized as parent regions of the differentiated meteorites, the irons forming parent body cores or, in some cases perhaps, dispersed "raisins" within their parent body. Stony-iron meteorites are generally taken to represent metal–silicate interface regions. Pallasites (Fig. 9c), in which large (centimeter-sized) rounded olivine crystals are embedded in well-crystallized metal, seem to resemble an "equilibrium" assemblage that may have solidified within a few years but that cooled slowly at rates like those at which iron meteorites formed, a few degrees per million years

(Ma). Structures of mesosiderites, on the other hand, suggest more rapid and violent mixing of metal and silicate, possibly by impacts.

During differentiation, siderophilic elements are more easily reduced than iron to their metallic state. Hence, they follow metallic iron geochemically and are extracted into metallic melts. Such elements (e.g., Ga, Ge, Ni, and Ir) are thereby depleted in silicates and enriched in metal to concentrations substantially higher than those in precursor chondrites. Conversely, magmas become enriched in lithophilic elements—like Ca, Cr, Al, and Mg—above chondritic levels: concentrations of such elements approach zero in metallic iron. During any substantial heating, noble gases and other atmophile elements—like carbon and nitrogen—are vaporized and lost from metallic or siliceous regions. Chalcophilic elements that geochemically form sulfides like troilite (Table II) include Se, Te, Tl, and Bi. Chalcophiles and a few siderophiles and lithophiles are also quite easily mobilized (i.e., vaporized from condensed states of matter) so that they may be enriched in sulfides in the parent body or lost from it. Concentrations of these elements in specific meteor-

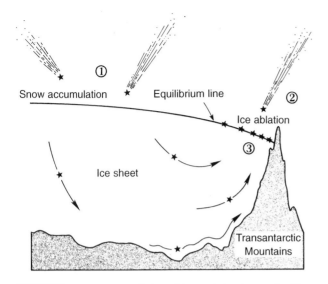

FIGURE 7 Cross-sectional view of Antarctic ice sheet and subice topography. Meteorites fall (1), are collected by the ice sheet and are buried (i.e., preserved), transported, and concentrated near a barrier that stops the ice sheet (2), and are exposed by strong winds from the South Pole that ablate the stagnant ice (3). [Reprinted from C. Bull and M. E. Lipschutz (eds.), "Workshop on Antarctic Glaciology and Meteorites," LPI Tech. Rept. No. 82-03. Copyright 1982 with kind permission from the Lunar and Planetary Institute, 3600 Bay Area Boulevard, Houston, TX 77058-1113.]

ites then depend in part on the fractionation histories of their parent materials and are, indeed, markers of thermal episodes.

B. CHARACTERISTICS OF SPECIFIC CLASSES

It is readily apparent, even to the naked eye, that most iron meteorites consist of large metallic iron crystals. These are usually single-crystal, bcc α-Fe (kamacite) lamellae that are 0.2–50 mm thick with decimeter (dm) to meter lengths (Fig. 9e). These relatively wide, Ni-poor lamellae are bounded by thin, Ni-rich fcc γ-Fe (taenite). The solid-state nucleation and diffusive growth process by which kamacite grew from taenite previously nucleated from melt is quite well understood and kamacite is interpreted as having formed by slow cooling in their parent bodies. Use of the 1-atm Fe–Ni phase diagram and measurement of Ni-partitioning between kamacite and taenite permit estimation of cooling rates between ~900 and 400°C. These typically are a few degrees or so per Ma, depending on iron meteorite group, consistent with formation in objects of asteroidal size. The concentration of Ni in the melt determines the temperature at which incipient crystallization occurs and this, in turn, estab-

lishes orientations of kamacite in the final meteorite. These orientations are revealed in iron meteorites by short etching (with nitric acid in alcohol) of highly polished cut surfaces: the discovery of this in the eighteenth century by Baron Alois von Widmanstätten led to the etched structure being called the "Widmanstätten pattern."

Meteorites containing <6% Ni are called hexahedrites because they yield a hexahedral etch pattern of large, single-crystal (centimeter-thick) kamacite (see Fig. 8a). Iron meteorites containing 6–16% Ni crystallize in an octahedral pattern and are called octahedrites. The lower-Ni meteorites have the thickest kamacite lamellae (>3.3 mm) and yield the very coarsest Widmanstätten pattern, whereas those highest in Ni are composed of very thin (<0.2 mm) kamacite lamellae and are called very fine octahedrites. Iron meteorites containing >16% Ni nucleate kamacite at such low temperatures that large single crystals could not form over times equivalent to solar system history. These therefore yield no obvious Widmanstätten pattern and are called Ni-rich ataxites (i.e., without structure). The Ni-poor ataxites are hexahedrites or octahedrites that were reheated in massive impacts, or artificially at some time after their fall on Earth.

As noted earlier, during differentiation of primitive parent body materials, siderophilic elements were extracted from chondritic precursors into molten metal. During crystallization of the melt, fractionation or separation of siderophiles can occur depending on local conditions. Almost 40 years ago, it was established that the Ga and Ge contents of iron meteorites are not continuous but are quantized, so they could be used to classify irons into groups denoted as I to IV. Originally, these Ga–Ge groups—which correlate well with Ni content and Widmanstätten pattern—were interpreted as samplings of core materials from a very few parent bodies. Subsequent extensive studies of many additional meteorites and some additional elements, especially Ni and Ir, modified this view. At present, the chemical groups (Fig. 10) suggest that the iron meteorites sample a number of parent bodies, perhaps a hundred or so, although many, if not most, derive from but five parents represented by the IAB, IIAB, IIIABCD, IVA, and IVB irons. (The earlier Roman numeral notation of the Ga–Ge groups was retained to indicate in a semiquantitative fashion the meteorite's Ga or Ge content. However, a letter suffix was added to indicate whether siderophiles were fractionated from each other or not.) In addition to the major minerals, kamacite and taenite and mixtures of them, minor amounts of other minerals like troilite (FeS) and graphite may be present. Also, silicates or other oxygen-

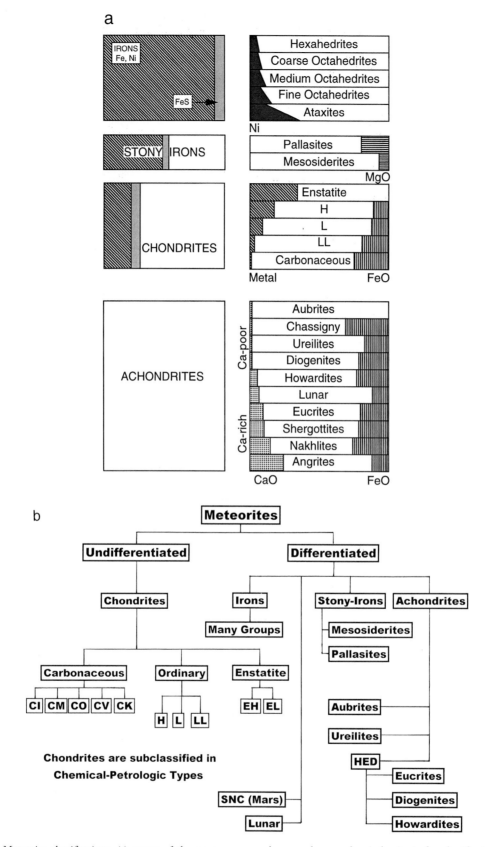

FIGURE 8 Meteorite classifications: (a) names of the most common classes and some chemical criteria for classification; (b) genetic relationships among meteorites.

FIGURE 9 Whole-rock samples of common meteorite types (with approximate longest dimension in cm): (a) Whitman, H5 (6 cm); (b) Allende, C3V (8 cm)—note the 1-cm chondrule in the center; (c) Springwater pallasite (18 cm); (d) Sioux County eucrite (8 cm); (e) Sanderson IIIB medium octahedrite (13 cm)—note the large FeS inclusions; and (f) the Widmanstätten pattern in the background of this stamp from Greenland is based on that of the Gibeon fine octahedrite. Many tons of this meteorite have been recovered, as have small artifacts made by Eskimos using the iron–nickel from it. (See also color insert.)

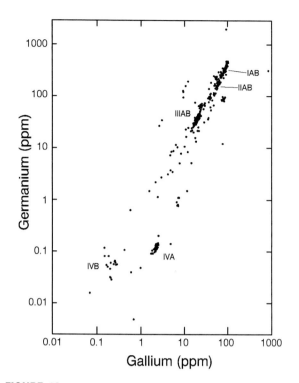

FIGURE 10 Contents of Ga and Ge in iron meteorites: some of the larger chemical groups are indicated. (Reprinted from D. W. G. Sears, "Nature and Origin of Meteorites." Copyright 1978 with kind permission from Oxford University Press, New York.)

containing inclusions are present in some iron meteorites.

In most cases, chondrites are distinguished by the presence of spherical, millimeter- to centimeter-sized chondrules or their fragments, silicates that were rapidly melted at temperatures of near 1600°C and rapidly cooled, some early in the solar system's history and immediately at about 1000°C/hr, others more slowly at 10–100°C/hr. Such rapid heating and cooling are relatively easy to perform in the laboratory but difficult to achieve on a solar system scale. Yet large volumes of chondrules must have been present in the solar system because the number of chondrites is large (see Table I). Chondrites (and many achondrites) date back to the solar system's formation—indeed provide chronometers for it (see Sections V, D and V, E)—and represent accumulated primary nebular condensate and accretionary products. A portion of this condensate formed from the hot nebula as millimeter-sized Ca- and Al-rich inclusions (CAI) that are aggregates of minerals predicted as vapor-deposition products by thermodynamic calculations. These CAI, found mainly in chondrites rich in carbonaceous (organic) material, exhibit many isotopic anomalies and contain atoms with distinct nucleosynthetic histories. Other inclu-

sions (like SiC and extremely fine diamond) represent relict presolar material. Other condensates formed at much lower temperatures. It should be noted, however, that some—perhaps even many—CAI may be refractory residues, not condensates.

Although most chondrites are composed of the same minerals, the proportions of these and their compositions differ in the six or so principal chondritic chemical groups. The primary bases for the classification of a given chondrite involve the proportions of iron as metal and silicate (in which oxidized iron, expressed as FeO, may be present) and the total iron (from Fe, FeO, and FeS) content (see Fig. 8a). The latter (Fig. 11) defines meteorites as having high and low total iron (H and L, respectively) or low total iron and low metal (LL). Numbers of H, L, and, to a lesser extent, LL chondrites are so large (see Table I) that these meteorites are often called the ordinary chondrites. Needless to say, chondrite compositions (typically, as in Table III, with elements apportioned by chemical form) do not form a continuum but are, rather, compositionally quantized. Table III lists major element ratios diagnostic of specific chondritic groups. The total amount of iron in some enstatite (E) chondrites is higher even than that in the H group of ordinary chondrites, and these are sometimes called EH chondrites. The reason for denoting EL chondrites should thus be self-evident.

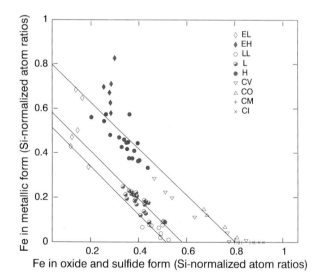

FIGURE 11 Silicon-normalized contents of Fe as metal (ordinate) versus Fe in FeS and ferromagnesian silicates (abscissa) in various chondritic groups: each diagonal is a line of constant total iron content. [Reprinted from D. W. G. Sears and R. T. Dodd, Overview and classification of meteorites. *In* "Meteorites and the Early Solar System" (J. F. Kerridge and M. S. Matthews, eds.), pp. 3–31. Copyright 1988 with kind permission from University of Arizona Press, Tucson.]

TABLE III
Average Chemical Compositions and Elemental Ratios
of Carbonaceous and Ordinary Chondrites and Eucrites

Species[a]	C1	C2M	C3V	H	L	LL	EUC
SiO_2	22.69	28.97	34.00	36.60	39.72	40.60	48.56
TiO_2	0.07	0.13	0.16	0.12	0.12	0.13	0.74
Al_2O_3	1.70	2.17	3.22	2.14	2.25	2.24	12.45
Cr_2O_3	0.32	0.43	0.50	0.52	0.53	0.54	0.36
Fe_2O_3	13.55						
FeO	4.63	22.14	26.83	10.30	14.46	17.39	19.07
MnO	0.21	0.25	0.19	0.31	0.34	0.35	0.45
MgO	15.87	19.88	24.58	23.26	24.73	25.22	7.12
CaO	1.36	1.89	2.62	1.74	1.85	1.92	10.33
Na_2O	0.76	0.43	0.49	0.86	0.95	0.95	0.29
K_2O	0.06	0.06	0.05	0.09	0.11	0.10	0.03
P_2O_5	0.22	0.24	0.25	0.27	0.22	0.22	0.05
H_2O^+	10.80	8.73	0.15	0.32	0.37	0.51	0.30
H_2O^-	6.10	1.67	0.10	0.12	0.09	0.20	0.08
Fe^0		0.14	0.16	15.98	7.03	2.44	0.13
Ni			0.29	1.74	1.24	1.07	0.01
Co			0.01	0.08	0.06	0.05	0.00
FeS	9.08	5.76	4.05	5.43	5.76	5.79	0.14
C	2.80	1.82	0.43	0.11	0.12	0.22	0.00
S (elem)	0.10						
NiO	1.33	1.71					
CoO	0.08	0.08					
NiS			1.72				
CoS			0.08				
SO_3	5.63	1.59					
CO_2	1.50	0.78					
Total	98.86	99.82	99.84	99.99	99.99	99.92	100.07
ΣFe	18.85	21.64	23.60	27.45	21.93	19.63	15.04
Ca/Al	1.08	1.18	1.10	1.11	1.12	1.16	1.12
Mg/Si	0.90	0.89	0.93	0.82	0.80	0.80	0.19
Al/Si	0.085	0.085	0.107	0.066	0.064	0.062	0.290
Ca/Si	0.092	0.100	0.118	0.073	0.071	0.072	0.325
Ti/Si	0.004	0.006	0.006	0.004	0.004	0.004	0.0019
ΣFe/Si	1.78	1.60	1.48	1.60	1.18	1.03	0.66
ΣFe/Ni	18.12	16.15	16.85	15.84	17.73	18.64	
Fe^0/Ni			9.21	5.67	2.29		
$Fe^0/\Sigma Fe$			0.58	0.32	0.12		

[a] ΣFe includes all iron in the meteorite whether existing in metal (Fe^0), FeS, iron silicates as Fe^{2+} (FeO), or Fe^{3+} (Fe_2O_3). The symbol H_2O^- indicates loosely bound (adsorbed?) water removable by heating up to 110°C; H_2O^+ indicates chemically bound water that can be lost only above this temperature. (Data courteously provided by Dr. E. Jarosewich, Smithsonian Institution.)

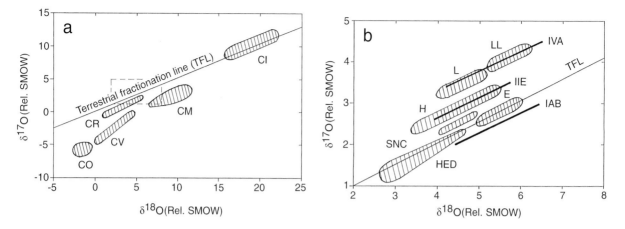

FIGURE 12 Relation between oxygen isotopic compositions in whole-rock and separated mineral samples from Earth, the Moon, and various meteorite classes. The units, $\delta^{17}O$ (%) and $\delta^{18}O$ (%), are those in use by mass spectrometrists and are, in effect, $^{17}O/^{16}O$ and $^{18}O/^{16}O$ ratios, respectively. The $\delta^{17}O$ (%) is defined in terms of a standard, SMOW (Standard Mean Ocean Water), as $[(^{17}O/^{16})_{sample}/(^{17}O/^{16}O)_{smow} -1] \times 1000$; $\delta^{18}O$ is defined in an analogous fashion. Oxygen isotopic compositions for carbonaceous chondrites (a) are much more variable than for other meteorite classes (b). The dashed box in (a) is expanded (b).

Achondrites, because of their high-temperature origins, contain essentially no metal or sulfide and are enriched in refractory lithophiles (cf. Table III), which, with their constituent minerals, serve to classify them into specific groups (see Fig. 8a). Most of these groups are named for a specific prototypical meteorite; others—howardites, eucrites, and diogenites (HED meteorites)—are named nonsystematically. At least 10 achondrite groups can be distinguished from their oxidized iron and calcium contents (FeO and CaO). Some of these apparently were associated in the same parent body but derive from different regions: the HED and the SNC (Shergottites-Nakhlites-Chassigny) associations. The HED meteorites are thought to come from 4 Vesta and/or the other V class asteroids believed to come from it. The consensus that the 12 SNC meteorites come from Mars is so strong that these are now denoted as Martian meteorites, not SNCs.

C. OXYGEN ISOTOPICS AND INTERPRETATION

Meteorites provide an unconventional map of the solar system in the form of the isotopic composition of oxygen (Fig. 12), a major element in all but the irons. Because of its high chemical reactivity, which causes oxygen to form numerous compounds, it is present in a variety of meteoritic minerals, even in silicate inclusions in iron meteorites. In standard references, such as the Chart of the Nuclides, the terrestrial composition of its three stable (i.e., nonradioactive) isotopes is given as 99.756% ^{16}O, 0.039% ^{17}O, and 0.205%

^{18}O. In fact, any physical or chemical reaction alters its isotopic composition slightly by mass-fractionation. Since the mass difference between ^{16}O and ^{18}O is twice that between ^{16}O and ^{17}O, a reaction that is mass-dependent, like physical changes and most chemical reactions, increases or decreases the $^{18}O/^{16}O$ ratio by a given amount and will alter the $^{17}O/^{16}O$ ratio in the same direction, but half as much. Accordingly, in a plot of $^{17}O/^{16}O$ versus $^{18}O/^{16}O$ or units derived from these ratios, that is, $\delta^{17}O$ and $\delta^{18}O$ (cf. Fig. 12 caption), all mass-fractionated samples derived by chemical or physical processes from an oxygen reservoir with a fixed initial isotopic composition will lie along a line of slope $\sim 1/2$.

Data derived from terrestrial samples delineate the Terrestrial Fractionation Line (TFL) in Fig. 12, whose axes are essentially those just described, but normalized to a terrestrial reference material, Standard Mean Ocean Water (SMOW). Not only do the oxygen isotopic compositions of all terrestrial samples lie along this line, but so too do the compositions of lunar samples, which occupy a small part of it. The existence of a single Earth–Moon line (defined by data covering the full length shown by the solid line) suggests that both bodies derive from a common oxygen isotopic reservoir, thus supporting the idea that the Moon's matter was spun off during a massive impact of a Mars-sized projectile with a proto-Earth.

One important feature of Fig. 12 is that many chondrite and achondrite groups defined by major element composition and mineralogy (e.g., Figs. 8a and 8b) occupy their own regions in oxygen isotope space. These data suggest that at least six major chondritic groups (H, L, LL, CI, CM, and E) and a few minor ones (R and K), acapulcoites and brachinites, the two achondrite associations [SNC and HED]. and the silicate inclusions in group IAB iron meteorites derive from different "batches" of nebular material. The HED region also encompasses oxygen isotope data for most pallasites and many mesosiderites, suggesting that these stony-irons derive from a common parent body. Extension of the HED region by a line with slope 1/2 passes through the oxygen isotopic region of the oxygen-containing silicate inclusions from the IIIAB irons, suggesting that they, too, may be related to the HED association. One possibility is that these irons represent the deeper interior of the HED parent body, but this would imply more complete disruption than that represented by V-class asteroids like 4 Vesta. The oxygen isotopic compositions of the rare angrites and brachinites are similar to those of the HED association, but differences in other properties weaken the connection. Other possible links based on limited oxygen isotopic data indicating common nebular reservoirs are: silicate inclusions in IIE irons with H chondrites; silicates in IVA irons with L or LL chondrites; winonaites (primitive meteorites modified at high temperatures) with silicates from IAB and IIICD irons; and the very rare, highly metamorphosed—even melted—primitive acapulcoites and lodranites.

One interpretation of Fig. 12 is that the solar system was isotopically inhomogeneous, since each "batch" of nebular matter seems to have its characteristic oxygen isotopic composition. Isotopic homogenization of gases is more facile than is chemical homogenization, so that the isotopic inhomogeneity demonstrated by Fig. 12 implies that the solar system condensed and accreted from a chemically inhomogeneous presolar nebular (see Fig. 2). We should add that the coincidence of the EH and EL chondrite (Fig. 12) region with that of enstatite achondrites, or aubrites (not included in that figure), is generally taken to imply an origin from a common batch of nebular matter, perhaps—but not necessarily—a single parent body. The position of the EH and EL region squarely on the Earth–Moon line in Fig. 12b may well be coincidental since, after all, these data must lie somewhere in the figure.

The other important feature to be noted from Fig. 12 is the "carbonaceous chondrite anhydrous minerals line," with slope near 1. One feature that distinguishes two sorts of carbonaceous chondrites (see Section IV, A)—C1 and C2—from all others (cf. Fig. 8b) is the evidence for preterrestrial aqueous alteration or hydrolysis of some phases in them. (Evidence for hydrous

alteration of minerals is also observed in some unequilibrated ordinary chondrites; see Section II, D, 2.) Other phases in carbonaceous chondrites, the anhydrous minerals (which include CAI), appear never to have been exposed to water, so that these chondrites are generally regarded as a mixture of materials with different histories. As can be seen from Fig. 12, the oxygen isotopic composition of anhydrous minerals in CM, CV, and CO chondrites are consistent with a line defined by CAI whose slope cannot reflect the mass-fractionation process indicated by a slope 1/2 line like that of TFL. Instead, the anhydrous minerals line is generally taken to represent a parental mixture of two end-members ("batches" of nebular material), whose oxygen isotopic compositions lay at or beyond the solid segment of the line, rich in ^{16}O. The fact that the oxygen isotopic compositions of ureilites lie on an extension of the anhydrous minerals line suggests a link, and these achondrites contain amounts of carbon (in the form of graphite–diamond mixtures) intermediate between those of CV or CO chondrites and CM. Data for ureilites indicate that they did not form by differentiation of material with a uniform oxygen isotopic composition. Rather, formation of ureilites may have involved carbonaceous chondritelike components in various proportions.

As originally interpreted, the anhydrous minerals line represents a mixture of nebular matter with high ^{17}O and ^{18}O and material containing oxygen free of these isotopes, that is, pure ^{16}O. If so, the latter would have had its own nucleosynthetic history, perhaps representing material condensed from the expanding, He- and C-burning shell of a supernova. Subsequently, it was shown that photochemical reactions of molecular oxygen with a given isotopic composition yield an ensemble of oxygen molecules whose isotopic compositions lie along a line of slope 1 as in Fig. 12.

Which of these processes—nebular or photochemical—was responsible for trends in Fig. 12 is unknown at present, but this does not affect the use of this figure to "fingerprint" parts of the solar system that produced various meteorites or groups of them. In using Fig. 12, it should be noted that the position of any sample(s) could reflect some combination of both sorts of line. For example, the primary material(s) that ultimately yielded L chondrites (or indeed any ordinary chondrite group) and HED meteorites could have had a single initial composition that was subsequently mass-fractionated and/or mixed or reacted photochemically to produce meteorite groups with very different oxygen isotopic compositions. However, suitable meteorites with intermediate oxygen isotopic compositions are unknown.

D. CHONDRITES

Most of the available data suggest that heat sources for the melting of primitive bodies, presumably chondritelike in composition, that formed differentiated meteorites were within parent bodies rather than external to them. Important sources no doubt include radioactive heating from radionuclides—both extant (^{40}K, ^{232}Th, ^{235}U, and ^{238}U) and extinct (e.g., ^{26}Al)—that were more abundant in the early solar system and impact heating. Calculations indicate that ^{26}Al was important in heating small (a few kilometers across) primitive parent bodies; the other heat sources were effective in differentiating larger ones. A possible heat source, whose importance cannot yet be properly judged, is that of electrical inductive heating driven by dense plasma outflow along strong magnetic lines of force associated with the very early, pre-main-sequence (T-Tauri stage) Sun.

1. Petrographic Properties

Major element and/or oxygen isotope data demonstrate that differences between parent materials of chondrites of the various chemical groups (e.g., H, CM, or EH) are of primary nebular—preaccretionary—origin. Parent body differentiation, on the other hand, is secondary—postaccretionary—in nature. Such heating does not necessarily cause melting of the entire parent body and, in such a case, it is quite reasonable to expect an intermediate region between the primitive surface and the molten interior, where differentiated matter forms. Properties of many chondrites are consistent with this expectation and suggest that solid-state alteration of primary chondritic parent material occurred during secondary heating. Eight characteristics obtained from petrographic study of optically thin sections (Fig. 13) can be used (with data for bulk carbon and water contents) to estimate the extent of thermal metamorphism experienced by a chondrite and to categorize it into one of six types (Table IV). The absence of chondrules and the presence of abnormally large (≥ 100 μm) feldspar characterize very rare type 7. However, these seven "pigeonholes" approximate a chondritic, thermal metamorphic continuum.

Two of these characteristics can be compared in the thin sections illustrated in Fig. 13: the opaque matrix and distinct chondrules of the petrographic type 3 chondrite Sharps (Fig. 13a) should be contrasted with the recrystallized matrix and poorly defined chondrules of the extensively metamorphosed type 6 chondrite Kernouve (Fig. 13b). Chemically, Fe^{2+} contents of the ferromagnesian silicates—olivine and pyroxene (see Table II)—are almostly completely random in a chondrite like Sharps and quite uniform in a chondrite like Kernouve. To a first approximation, chondrites of higher numerical types could acquire their petrographic characteristics (see Table IV) by extended heating (thermal metamorphism) of a more primitive (i.e., lower type) chondrite of the same chemical group. Temperature ranges estimated for formation of types 3 through 7 are 400–600°C, 600–700°C, 700–750°C, 750–950°C, and >950°, respectively.

Petrographic properties of achondrites, like the Martian meteorite Nakhla, clearly indicate igneous processes that occurred in parent bodies at temperatures $\geqslant 1000$°C. The resultant melting and differentiation erased all textural characteristics of the presumed chondritic precursor (Fig. 13c) so its nature can only be inferred.

Chemical changes involving loss of a constituent, like carbon or water in the case of chondrites, require an open system; other changes indicated in Table IV could have occurred in open or closed systems. It is important to note that thermal metamorphism can affect only secondary (parent body) characteristics—those listed horizontally in Table IV—not primary ones. Postaccretionary processes by which H chondritelike material can be formed from L or vice versa are unknown.

2. Chemical-Petrologic Classification

Since properties of a given chondrite reflect both its primary and subsequent histories, a classification scheme reflecting both has been adopted for chondrites. By this scheme, chondrites already mentioned are: Ensisheim, LL6; Nogata, L6; Sharps, H3 (see Fig. 13a); Sylacauga, H4; and Kernouve and Peekskill, H6 (see Fig. 13b). No ordinary (or enstatite) chondrites of types 1 or 2 are known. Type 3 ordinary chondrites, sometimes called the unequilibrated ordinary chondrites (UOC), as a type differ to the greatest extent among themselves and from chondrites of other petrographic types. Within UOC, a variety of properties—for example, the chemical heterogeneity of ferromagnesian silicates, concentrations of volatile trace elements (mainly Bi, Tl, In, and noble gases), and thermoluminescence sensitivity—permit subdivision of the UOC into subtypes 3.0 to 3.9. Sharps (see Fig. 13a) proves to be the most primitive H chondrite known, being an H3.0 or H3.4, depending on

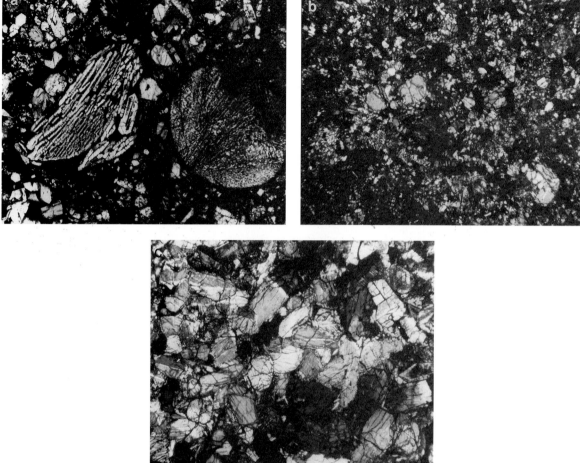

FIGURE 13 Petrographic (2.5-mm-wide) thin sections in polarized transmitted light. Partial large chondrules are easily recognizable in the H3 chondrite (a) but are barely recognizable in the H6 chondrite (b). (a) Sharps (H3); (b) Kernouve (H6); (c) Nakhla, believed to be of Martian origin. (Photos courtesy of Dr. Robert Hutchison, Natural History Museum, London.)

the classification criteria used. (A similar subclassification of C30 chondrites has also been proposed.)

Frequencies of occurrence of various chondritic groups are listed in Table I. Many properties of ordinary chondrites demonstrate that each chemical group has its own special history. This is evident even in something as simple as the numbers of each chemical-petrographic type. For example, proportions of H3 or L3 are low, 2–3% (8 of 276 H and 8 of 319 L chondrite falls), whereas 15% of LL chondrite falls (10 of 66) are LL3. Proportions of more evolved chondrites also differ (see Table I). The plurality of H chondrite falls are H5 (117 of 276 or 42%), whereas the largest numbers of L and LL chondrites are of type 6 (213 of 319 or 67% and 32 of 66 or 48%, respectively). Non-Antarctic chondrite finds generally exhibit similar trends, but stony-iron and, especially, iron finds are far more numerous because they are obviously "strange,"

hence more likely to be brought to a knowledgeable person who can identify them as meteoritic. By the same token, except for their fusion crust, achondrites grossly resemble terrestrial igneous rocks and are less likely to be picked up: hence, they are underrepresented among finds (see Table I).

3. Breccias

Though most chondrites are readily classified by chemical-petrographic type, a few consist of two or more meteorite types, each of which is readily identifiable in the resulting lithified breccia. Noblesville, for example, consists of light H6 clasts embedded in dark H4 matrix (see Fig. 1a). Such an assemblage—two petrographic types of the same chondritic chemical group—is called a genomict breccia. A polymict brec-

TABLE IV
Definitions of Chondrite Petrographic Types[a]

	Petrographic types					
	1	2	3	4	5	6
(i) Homogeneity of olivine and pyroxene compositions	—	>5% mean deviations		>5% mean deviations to uniform	Uniform	
(ii) Structural state of low-Ca pyroxene	—	Predominantly monoclinic		Monoclinic >20%	Monoclinic <20%	Orthorhombic
(iii) Degree of development of secondary feldspar	—	Absent		<2-μm grains	<50-μm grains	50- to 100-μm grains
(iv) Igneous glass	—	Clear and isotropic primary glass; variable abundance		Turbid if present	Absent	
(v) Metallic minerals (maximun Ni content)	—	(<20%) Taenite absent or very minor	Kamacite and taenite present (>20%)			
(vi) Sulfide minerals (average Ni content)	—	>0.5%	<0.5%			
(vii) Overall texture	No chondrules	Very sharply defined chondrules		Well-defined chondrules	Chondrules readily delineated	Poorly defined chondrules
(viii) Texture of matrix	All fine-grained, opaque	Much opaque matrix	Opaque matrix	Transparent micro-crystalline matrix	Recrystallized matrix	
(ix) Bulk carbon content	~3.5%	1.5–2.8%	0.1–1.1%	<0.2%		
(x) Bulk water content	~6%	3–11%	<2%			

[a] The strength of the vertical line is intended to reflect the sharpness of the type boundaries. A few ordinary chondrites of petrographic type 7 have abnormally large ($\geq$100 μm) feldspar and no chondrules: these have been interpreted as reflecting higher metamorphic temperatures than those associated with type 6. Water contents do not include loosely bound, that is, terrestrial water.

FIGURE 14 The Cumberland Falls meteorite: an extreme example of a polymict breccia in which chunks of a primitive chondrite (black) are embedded in an aubrite matrix (white). (Photo courtesy of the Smithsonian Institution.)

cia is an assemblage in which two or more chemically different sorts of meteorite types are present (Fig. 14), implying the mixing of materials from two (or more) parent bodies, each with its own histories.

Other sorts of breccias exist, with perhaps one of the most important being the regolith breccias. The matrix of such a meteorite, like Noblesville, is typically dark and fine-grained and contains large quantities of light noble gases (He and Ne) of solar origin (cf. Section IV, A). These gases are present, as is radiation damage in the form of solar-flare tracks (linear solid-state dislocations), in regolith breccias in a 10-nm-thick rim on a number of the myriad crystals of which the matrix is composed. However, solar gases and flare tracks are absent in the larger, lighter-colored clasts present in most regolith breccias. Clearly, the dark matrix represents lithified fine dust originally spread out on the very surface of the regolith or fragmental rocky debris layer, some meters thick, produced by repeated impacts on bodies lacking an appreciable protective atmosphere. (The regolith on the Moon is both thicker, ~1 km, and more mature and gardened or better mixed by impacts than are asteroidal regoliths.) This dust acquired its gas and track components from particles with energies of keV per nucleon streaming from the Sun as solar wind or solar flares with MeV energies, so that the dust sampled the composition of the solar photosphere. The irradiated dust, which is often quite rich in volatile trace elements from another source, was mixed with coarser, unirradiated pebblelike material and was formed into a breccia by some relatively mild impact that did not heat or degas the breccia to any great extent. Regolith breccias are found in many types of meteorites, but are especially frequently encountered as aubrites, howardites, and H (and R) chondrites. [*See* THE SOLAR WIND.]

4. Carbonaceous Chondrites

a. Composition

The only type 1 or 2 chondrites known are carbonaceous chondrites, nearly all non-Antarctic examples being observed falls. A dominant genetic process evident in these meteorites is preterrestrial hydrolysis, that is, the action of liquid water in the nebula or on their parent bodies that altered preexisting grains, producing various hydrated, claylike minerals. The preterrestrial nature of this hydrolysis is demonstrated both by the meteorites' petrographic properties and by the decidedly nonterrestrial hydrogen isotopic composition ($^2H/^1H$) of water from them. As noted earlier, oxygen isotopic compositions of hydrated minerals in these meteorites demonstrate that the two groups derive from different batches of nebular matter, so that

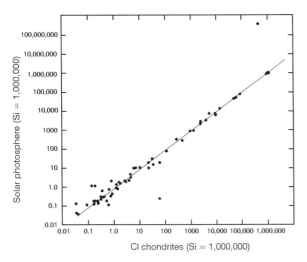

FIGURE 15 Correlation between abundances in the solar photosphere and C1 chondrites: the only disagreements involve gaseous elements and light elements easily destroyed by low-energy nuclear reactions in the Sun. (Reprinted from D. W. G. Sears, Chemical processes in the early solar system: A discussion of meteorites and astrophysical models. *Vistas in Astronomy,* **32** (1), 1–21. Copyright © 1988 with kind permission from Elsevier Science-NL, Amsterdam, Netherlands.)

C2 (or CM) could not have formed from C1 (or CI) by thermal metamorphism, nor could C1 have formed by hydrolysis of C2 parent material. It is for this reason that some specialists prefer to use the CM designation; others prefer to use a hybrid classification like C2M or CM2 since other C2-like meteorites exist.

It may seem peculiar that C1 chondrites contain *no* chondrules (see Table IV); however, their obvious compositional and mineralogic similarities to other chondrule-containing meteorites prompts their chondritic classification. The composition of C1 (or CI) chondrites is remarkably like that of the solar photosphere (Fig. 15), but there are a few obvious differences. Some elements are depleted in C1 chondrites relative to the Sun's surface: these elements, such as hydrogen, helium, and carbon, are gaseous or easily form volatile compounds that remained in the vapor state for the most part, in the nebular region where C1 chondrite parent material condensed and accreted. Other elements, such as lithium, beryllium, and boron, are easily destroyed by low-temperature nuclear reactions during pre-main-sequence evolution of stars like the Sun, so that they are depleted in the solar photosphere relative to C1 chondrites (see Fig. 15).

Since chemical analysis of C1 chondrites (or any planetary material) in the laboratory is more precise and accurate for most elements than is spectral analysis of the solar photosphere, chemical and isotopic abundances listed in "cosmic abundance" tables are derived mainly from analyses of C1 chondrites. It is generally these data that are used to estimate the composition of our solar system. Only in cases where reasons exist to suspect that processes such as incomplete nebular condensation are solar photospheric values adopted in such compilations. However, it is important to recall that chemical heterogeneity of the presolar nebula has been inferred (see Fig. 2), so cosmic abundances may not have been the same in all nebular regions.

b. Organic Constituents

Although chondrites are depleted in carbon, hydrogen, and nitrogen relative to the solar photosphere, a very important property of C1 and, to only a somewhat lesser extent, of C2 chondrites involves the large amount of organic matter in them (see Table III). Since the midnineteenth century, the existence of these organic compounds—now strongly believed to be of mainly abiogenic origin—in meteorites has been known. However, only during the past few decades have analytical techniques developed to the point that we can be assured of its preterrestrial origin. Evidence for the nonterrestrial nature of such organic matter is provided by properties such as: the presence in meteorites of compounds, such as certain amino acids not found naturally on Earth; the nonterrestrial carbon and hydrogen isotopic compositions of organic compounds in C1 and C2 chondrites; and the nature of chiral organic compounds in meteorites. Chiral compounds are structurally asymmetric and can exist in two "mirror-image" forms. When such compounds are biogenic, one "mirror-image" form dominates and the compound is optically active, rotating the plane-polarized light in a counterclockwise (i.e., "levorotatory") direction. These same compounds in meteorites have previously been found to consist of equal numbers of their "mirror-image" forms, thus being optically inactive or racemic. However, as this chapter was being finalized, an article appeared in the journal *Nature* reporting that a number of amino acids extracted from the freshly broken interior of a piece of the Murchison CM chondrite fall are optically active and levorotatory, and are enriched in ^{15}N to levels above those found terrestrially. It is premature to speculate on how these new findings will alter the debate on the biogenic versus abiogenic origin of organic compounds in meteorites.

Over 400 different organic compounds of extraterrestrial origin have been identified in C1 and C2M chondrites (Table V). These represent but a fraction of the kinds of organic molecules that exist in meteorites, since these compounds have been identified in less than 5% of the organic matter that is extractable with standard solvents. The types of molecules identified include: amino acids of various sorts, both cyclic and acyclic; aliphatic and aromatic hydrocarbons; mono-, di-, and hydroxycarboxylic acids; nitrogen-containing heterocyclic compounds; amines and amides; and alcohols, aldehydes, and ketones. Examples of some less familiar molecules identified in meteorites include:

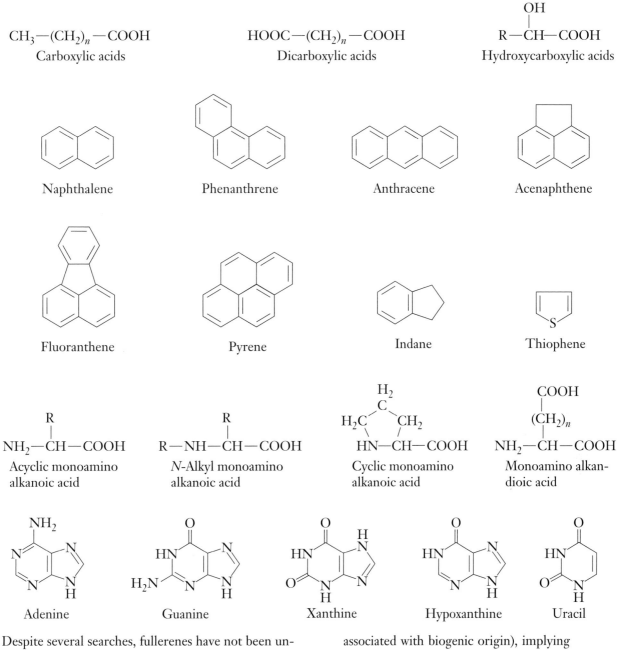

$CH_3-(CH_2)_n-COOH$
Carboxylic acids

$HOOC-(CH_2)_n-COOH$
Dicarboxylic acids

$R-\overset{\overset{\displaystyle OH}{|}}{CH}-COOH$
Hydroxycarboxylic acids

Naphthalene

Phenanthrene

Anthracene

Acenaphthene

Fluoranthene

Pyrene

Indane

Thiophene

$NH_2-\overset{\overset{\displaystyle R}{|}}{CH}-COOH$
Acyclic monoamino alkanoic acid

$R-NH-\overset{\overset{\displaystyle R}{|}}{CH}-COOH$
N-Alkyl monoamino alkanoic acid

Cyclic monoamino alkanoic acid

$NH_2-\overset{\overset{\displaystyle COOH}{\overset{|}{\overset{\displaystyle (CH_2)_n}{|}}}}{CH}-COOH$
Monoamino alkandioic acid

Adenine

Guanine

Xanthine

Hypoxanthine

Uracil

Despite several searches, fullerenes have not been unambiguously identified in any meteorite.

Important properties exhibited by these compounds include:

1. complete structural diversity (i.e., all possible structural isomers are found), implying that all compounds capable of forming will form;
2. no optical activity (the absence of a preferred enantiomer of chiral compounds, normally associated with biogenic origin), implying abiogenic genesis;
3. decreasing concentrations with increasing carbon number in homologous series, implying decreasing probability of carbon addition to existing molecules;
4. predominance of branched-chain structural isomers over straight-chain ones, implying generally random additions of carbon atoms to molecules.

TABLE V
Concentrations and Molecular Characteristics or Soluble Organic Compounds of Meteorites

Class	1 Concentration (ppm)	2 Compounds identified	3 Chain length	4 Homologous decline[a]	5 Branched- or straight-chain predominance[b]	6 Structural diversity[b]	7 Chirality[c]
Amino acids	60	74	C_2–C_7	Yes	Br	Yes	R
Aliphatic hydrocarbons	>35	140	C_1–$C_{\geq 23}$	?	$\begin{cases} <C_{10}Br \\ >C_{10}St \end{cases}$	Yes No	?
Aromatic hydrocarbons	15–28	87	C_6–C_{20}	NA[d]	(Br)	Yes	?
Carboxylic acids	>300	20	C_2–C_{12}	Yes		Br	Yes?
Dicarboxylic acids	>30	17	C_2–C_9	Yes		Br	Yes R
Hydroxycarboxylic acids	15	7	C_2–C_6	Yes		St	Yes R
Purines and pyrimidines	1.3	5	NA	NA		NA	NA
Basic N-heterocycles	7	32	NA	NA		NA	Yes?
Amines	8	10	C_1–C_4	Yes		Br	Yes?
Amides	55–70	>2	NA	NA		NA	Yes?
Alcohols	11	8	C_1–C_4	Yes		?	Yes?
Aldehydes and ketones	27	9	C_1–C_5	Yes		?	Yes?
Total	≥560	411					

[a] Implied progressive buildup of larger molecules from smaller ones.

[b] Br, branched; St, straight. The latter can result from a biogenic process.

[c] R, racemic; ?, unknown.

[d] NA, not applicable.

From "Organic Matter in Carbonaceous Chondrites," by J. R. Cronin *et al.*, pp. 819–857 of *Meteorites and the Early Solar System*, edited by John F. Kerridge and Mildred Shapley Matthews. Copyright © 1988 by the Arizona Board of Regents. Reprinted by permission of the University of Arizona Press.

Possible mechanisms for the formation of these organic compounds include Miller-Urey-type processes, Fischer-Tropsch reactions, and ion–molecule reactions. A Miller-Urey-type process involves the deposition of energy—usually electrical discharge but also thermal, shock, or radioactive—into a mixture of gaseous H_2O, NH_3, and CH_4. Fischer-Tropsch reactions involve the passage of gaseous CO, H_2, and NH_3 over heated minerals, including magnetite (Fe_3O_4), phyllosilicates (the hydrated, claylike matrix materials of CI and CM chondrites), and metallic iron–nickel. Currently gaining in popularity are gaseous ion–molecule reactions in dark interstellar clouds (in which well over 50 different organic molecules have been identified), followed by hydrolytic reactions on the surface of carbonaceous chondritelike material, for example, on or in primitive parent bodies. No absolute choice between these mechanisms is possible at this point and all may have occurred. Developing technology may help us choose between these alternatives as techniques permitting *in situ* analysis of attomole (10^{-18} mole) quantities of organic molecules are applied to C1 and C2 chondrites.

Since many organic compounds indigenous to meteorites can be altered or destroyed by even brief exposure to temperatures of 200–300°C, their presence in meteorites constitutes a thermometer for postaccretionary heating during metamorphism, shock, or atmospheric transit.

Polyaromatic hydrocarbons (PAH) have also been measured in the interiors of two Martian meteorites, but not near their surfaces. This may indicate that the PAH are not terrestrial contaminants but, rather, were present on Mars. Together with particles identified as microfossils in at least one Martian meteorite, these organic constituents raise the very controversial, but fascinating, possibility that life, in some form; existed on Mars. Currently, studies are under way to test this possibility, but some years of research may be needed to settle this matter.

5. Shock

It appears that a meteorite parent body cannot be disrupted by internal processes, but only by collision with

TABLE VI

**Degrees of Shock-Loading in Ordinary Chondrites
as a Function of Specific Chemical-Petrologic Type[a]**

Type	Total count	S1	S2	S3	S4	S5	S6
H3	46	15	19	10	—	2	—
H4	78	15	35	25	3	—	—
H5	216	23	82	88	21	1	1
H6	91	22	23	34	10	1	1
L3	29	2	11	8	6	1	1
L4	37	—	10	15	6	—	6
L5	81	1	20	36	10	7	7
L6	209	7	10	56	87	19	30
L7	4	—	—	1	2	1	—
LL3	11	1	6	3	1	—	—
LL4	10	—	5	5	—	—	—
LL5	27	1	16	8	2	—	—
LL6	31	1	10	12	6	1	1

[a] Data courtesy of Prof. D. Stöffler, Humboldt University, Berlin.

another substantial object. Hence, it is not surprising that evidence for exposure to a significant degree of shock is found in a large number of meteorites. A few decades ago, chondrites were qualitatively classed as "shocked" if the interior of the hand-specimen exhibited blackening, veining, or brecciation. More recent, detailed petrographic studies reveal mineralogic characteristics that can be used to semiquantitatively estimate the degree of shock exposure. Such characteristics result from changes induced directly, by the peak pressure wave, or indirectly, by the shock-associated, elevated residual temperature. Particular shock-pressure indicators ("shock barometers") have been calibrated against direct measurements of mineralogic characteristics produced by shock-loading experiments in the laboratory. Though classification criteria and schemes utilizing them have been developed, fewer than 20% of the known ordinary chondrites have been studied in sufficient detail to quantify their degree of shock-loading (Table VI). We think it valuable to summarize the very substantial range in shock exposure that is evident in various meteorites.

The currently popular scheme to classify equilibrated ordinary chondrites involves the addition of S1, S2, . . . , S6 to its chemical-petrographic classification. The peak shock pressures corresponding to the transitions are: <5 GPa (i.e., <50,000 atm or <50 kbar), S1/S2; 5–10 GPa, S2/S3; 15–20 GPa, S3/S4; 30–35 GPa, S4/S5; and 45–55 GPa, S5/S6. Whole-rock melting and formation of impact melt rocks or melt

breccias occurs at 75–90 GPa. According to this scheme, the Noblesville H4 regolith breccia (see Fig. 1a) as a whole is S1 with some H6 clasts being S2. As will be seen, other criteria (radiogenic gases and certain trace elements that are thermally mobile, i.e., easily volatilized and lost in an open system) also give information on chondritic shock histories (cf. Sections IV, C and V, C). Equilibrated L chondrites as a group show the highest proportion of heavily shocked chondrites, with almost half having been shocked above 20 GPa. Lesser, but still significant, proportions of H and LL chondrites show substantial degrees of shock-loading (see Table VI). The only C and E chondrites exhibiting evidence for unusually strong shock, that is, to S5, are a CK5 and an EL3.

The mineralogy (really, metallography) of iron meteorites is relatively simple and classification of such meteorites by degree of shock-loading at <13 GPa, 13–75 GPa, and >75 GPa is relatively easy. Laboratory studies also permit identification of iron meteorites shock-loaded at 13–75 GPa that were subsequently annealed at 400–500°C for days or weeks, presumably by contact with massive chunks of collisional debris having residual temperatures in this range or higher. About half of all iron meteorites studied are estimated to have been shocked at ≥13 GPa, nearly all during collisional impacts that disrupted their parent bodies. Only for large meteoroids that formed terrestrial explosion craters is it conceivable that pressures as high as 13 GPa were generated during impact with Earth.

The best-preserved, perhaps only, case of strong shock-loading during terrestrial impact involves the meteoroid that produced Meteor Crater in Arizona (see Fig. 5c). Some Canyon Diablo meteorite fragments of the impacting meteoroid contain millimeter- to centimeter-sized graphite–diamond aggregates, indicating partial transformation of graphite to diamond. [Highly unequilibrated chondrites contain very tiny (~0.002 μm) vapor-deposited diamond grains that do not have a high-pressure origin; see Section IV, B, 1.] These aggregates contain lonsdaleite, a hexagonal diamond polymorph produced, as far as is known, only by shock transformation of graphite, which also has a hexagonal crystal structure. Diamond-containing Canyon Diablo specimens always show metallographic evidence for exposure to shock ≥13 GPa, are found mainly on the crater rim and not in the surrounding plain, and contain cosmogenic stable nuclides and radionuclides in quantities indicating their derivation from deeper in the interior, closer to the front of the impacting meteoroid where explosion shock pressures would have been the greatest. The high mutual correlations between degree of shock-loading, depth in the impacting meteoroid, and geographic locations around Meteor Crater suggest that those Canyon Diablo specimens exhibiting evidence for exposure to strong shock acquired this during terrestrial impact.

The percentages of strongly shocked (i.e., ≥13 GPa) specimens in iron meteorite chemical groups differ widely. Virtually all IIIAB irons, which constitute the plurality of all known iron meteorites, have been shocked preterrestrially in the 13- to 75-GPa range. Nearly 60% of IVA irons, the next largest group, have been shocked to this extent. A similar proportion of IIB iron meteorites have been shocked at ≥13 GPa, but such meteorites are rare. No other chemical group of iron meteorites shows an unusually high proportion of shocked members. It has been shown by dynamic laboratory studies that shock pressures of 13 to 75 GPa acting on metallic Fe impart a free-surface velocity of 1–3 km/s. It has been suggested, therefore, that this shock impulse has been important, if not essential, in bringing strongly shocked meteorites to Earth in such high proportions. A possible scenario is that the parent bodies of the IIIAB and IVA irons were originally located in the Asteroid Belt and that the shock-impulse made ejecta orbits more elliptical, perhaps Mars-crossing, so that Mars could, with time, gravitationally perturb these fragments into Earth-crossing orbits.

Semiquantitative, petrographic shock indicators for basaltic achondrites, that is, mainly the HED association and shergottites, suggest a six-stage shock scale corresponding to pressures of ≤5, 5–20, 20–45, 45–60, 60–80, and ≥80 GPa. The full range of this scale is observed in samples from the HED association—primarily in clasts in the howardites, a meteorite group consisting mainly of polymict breccias with solar gases (see Section IV, A) representing mixtures of eucrite and diogenite fragments. These and other, mainly chemical compositional data suggest that howardites represent a shock-produced, near-surface mixture of two deeper igneous layers—from which eucrites and diogenites derive—in the HED parent body.

Nearly all ureilites show evidence for very substantial shock on petrographic study. Most also contain large graphite–diamond aggregates that are generally believed to have formed during preterrestrial impacts.

Shergottites also seem to have been heavily shocked, in keeping with their accepted derivation from a massive object, like Mars with its 5 km/s escape velocity. The nakhlites, linked to shergottites by oxygen isotopic compositions (see Fig. 12) and other properties, are less shocked.

Lunar meteoroids, which were ejected by impacts somewhere on the 95% of the Moon's surface not sampled by the Apollo or Luna programs, are breccias in most cases. Otherwise, they show no unusual evidence for shock greater than that evident in rocks returned by these programs.

Metallic portions of few stony-irons have been shocked at ≥13 GPa. Essentially no pallasites and only 3 of 18 mesosiderites studied show evidence of exposure to such strong shock. Somehow, parent bodies of the stony-irons and half of the known iron meteorites were disrupted and the meteoroids excavated from appreciable depth without subjecting them to major shocks. More puzzling is the fact that the silicate portion of mesosiderites contains a high proportion of shocked material. This implies that these stony-irons formed by intrusion of shock-loaded silicate into or onto preexisting, generally unshocked metal, possibly after its excavation from parent body interiors.

III. METEORITES OF ASTEROIDAL ORIGIN AND THEIR PARENT BODIES

A. THE METEORITE–ASTEROID CONNECTION

Two links have already been noted that suggest or imply an asteroidal origin for most meteorites. These are:

1. Photographically determined orbits (see Fig. 3) for four ordinary chondrites: Pribram (an H5 that fell in Czechoslovakia on 7 April 1959), Lost City (an H5 that fell in Oklahoma on 3 January 1970), Innisfree (an LL5 that fell in Alberta, Canada, on 5 February 1977), and Peekskill (an H6 that landed in the trunk of a car in New York state on 9 October 1992).

2. Mineralogic evidence indicating origin of most meteorites from objects of asteroidal size. (Some chondrites could come from much smaller primary objects.) This includes the cooling rates of iron meteorites, which imply formation depths consistent with asteroidal dimensions, the presence of minerals (e.g., tridymite) and phase relations (e.g., the Widmanstätten pattern) indicative of low-pressure ($\ll 1$ GPa) origin, and the absence of any mineral indicating high *lithostatic* (generated by the rocky overburden) rather than *shock* pressures.

One other property linking meteorites to certain asteroidal types, spectral reflectance, is a very active, on-going research area. The reflectivity (albedo) as a function of wavelength for an asteroid upon which white (solar) light is incident can be used to characterize its mineralogy and mineral chemistry to some extent. The spectral reflectance of an asteroid can be compared to that of possible meteoritic candidates, both as recovered or treated in the laboratory in some way to simulate effects of extraterrestrial processes, to uncover possible links.

The best matches (Fig. 16) exist between: the HED association and the rare V-class asteroids (4 Vesta and its smaller progeny); iron meteorites and the numerous M-class asteroids; CI and CM chondrites, thermally metamorphosed at temperatures up to 700°C, with the very numerous C-class and apparently related B-, F-, and G-class asteroids; aubrites with the somewhat unusual E-class asteroids; pallasites with a few of the very abundant and diverse S-class—which constitute a plurality of all classified asteroids—and/or rare A-type asteroids; and ordinary chondrites with the very rare Q-type asteroids, which happen to be near-earth asteroids, or 6 Hebe, an inner Asteroid Belt object belonging to the S(IV) subclass of S asteroids.

These resemblances constitute a typical good news/bad news situation. The good news is that specific meteorite types seem to be similar to (derive from) surface regions of identifiable asteroid types. The bad news is that relative frequencies with which meteorites of a given type, and asteroids of a supposedly similar type, are encountered do not agree. Specifically, there is the ordinary chondrite–S asteroid paradox (cf. Table I): Why are there so few asteroidal candidates for the

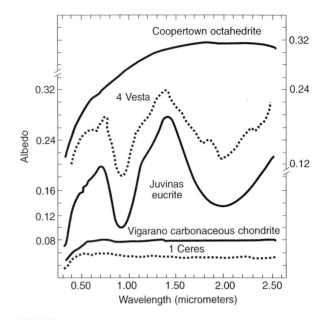

FIGURE 16 Spectral reflectances of the Coopertown IIIE coarse octahedrite, Juvinas eucrite and V-class asteroid 4 Vesta, and Vigarano C3V chondrite and G-class asteroid 1 Ceres. The albedo scale for all but Coopertown is on the left; that for Coopertown is on the right. Solid lines delineate meteorite spectra and dashed lines define asteroid spectra. (Courtesy of Dr. Lucy-Ann McFadden, University of Maryland.)

very numerous ordinary chondrites and so few olivine-dominated stony-irons from the very numerous S asteroids? The obvious answer is that Earth is collecting a biased meteorite population compared with the asteroid population in either near-Earth space or the Asteroid Belt. This explanation may also account for the absence of any known meteorites from the numerous D or P asteroids that are located at >3 AU from the Sun. Alternatively, ejecta from such asteroids might not survive passage through the Earth's atmosphere since surface materials of D and P asteroids are inferred to be very organic-rich and, presumably, very friable.

B. SAMPLING BIAS

The foregoing discussion indicates that the contemporary flux of meteorites is biased and unrepresentative of the meteoroid population in near-Earth space, let alone in the Asteroid Belt. Thus, generalizations about formation and evolutionary processes from studies of meteorites falling on Earth today may be incomplete.

The contemporary flux of meteorites is represented not only by the observed falls that occurred essentially during the past 200 years, but also by the finds (omitting all those from Antarctica, as well as the large numbers of samples recovered from arid desert regions). Those meteorites consisting of metallic iron, which should be readily oxidized on Earth, are surprisingly resistant to destructive oxidation, even in temperate climates. The smaller iron–nickel grains of chondrite finds are more readily oxidized. Non-Antarctic finds generally have terrestrial ages of up to 10,000 years or so, but the oldest one actually dated is the Tamarugal IIIA octahedrite, which has a terrestrial age of 3.6 Ma. A fossilized, almost unrecognizable H chondrite has been found in a block of limestone from a quarry near Brunflo, Sweden, which appears to be Ordovician, that is, about 450 Ma old. Thirteen other fossilized meteorites from about the same time were recently discovered in another quarry at Kinnekulle, Sweden.

The oldest Antarctic meteorite dated using cosmogenic radionuclides is Lazarev, an Antarctic octahedrite that belongs to no established chemical group of iron meteorites; its terrestrial age is 5 Ma. Two Antarctic chondrites have terrestrial ages of ~2 Ma, but more typical terrestrial ages for Antarctic meteorites are in the 0.1- to 1-Ma range (averaging 0.3 Ma for the population from Victoria Land), with a very few having been on Earth for up to 2 Ma. It could be that the meteorite population landing on Earth during that time window differs genetically from the contemporary flux. As can be seen from Table I, the populations of observed falls and finds from Victoria Land are roughly comparable, in terms of numbers of chondrites, achondrites, irons, and stony-irons, but there are some differences in detail. Another general, more easily quantified difference is that samples from Victoria Land seem to have, on average, smaller masses than do contemporary falls, perhaps because small samples are readily detected in Antarctica. Among Antarctic meteorites, rare specimens—like achondrites—are recognized rather easily, even in hand-specimen, and pieces of them can be readily paired with other pieces of the same fall. Hence, we can be quite certain of the identity of different Antarctic achondrites listed in Table I.

The numbers of aubrites and howardites seem smaller and that of ureilites larger in the Victoria Land population. Differences may also exist for nakhlites and C1 and LL chondrites, but for the first two of these, the numbers are small. The C1 chondrites are typically friable and it might be difficult for them to survive pulverization in the Antarctic ice sheet. The large number of still unclassified ordinary chondrites and pairing uncertainties that are always serious for ordinary chondrites that differ only subtly from each other—even in the case of falls—cloud the significance

of the LL chondrite difference and of a possible overabundance of H chondrites in the Victoria Land population. A number of studies of Antarctic meteorites reveal many differences of preterrestrial origin that exist between them and falls, but detailed interpretations of these differences remain controversial.

Many more fragments have been collected from Queen Maud Land, Antarctica, by Japanese meteorite recovery teams. Quite a few fragments of rare or unique types of meteorites have been identified by Japanese scientists, including four different lunar meteorites (six fragments), a Martian meteorite, three thermally metamorphosed C1 and C2 chondrites, and a unique C1M or C2I chondrite. In general, samples from Queen Maud Land seem to have terrestrial ages of up to 0.3 Ma, averaging 0.1 Ma (i.e., intermediate between those of contemporary falls and Antarctic samples from Victoria Land), and seem to have smaller masses, on average, than even those from Victoria Land. The Queen Maud Land population is less well characterized than the Victoria Land population, so we have not attempted to list any of its members in Table I.

IV. CHEMICAL AND ISOTOPIC CONSTITUENTS OF METEORITES

In earlier sections, meteorite compositions and genetic processes have been discussed to the extent necessary to understand general meteoritic properties. In this section, we will focus upon these topics in greater detail.

A. NOBLE GASES

The chemical inertness of noble gases makes it possible to separate them readily from all other chemical elements. This is why gas mass spectrometers are able to determine the very small concentrations of a noble gas in a meteorite and, in addition, measure its isotopic composition. Most analyses of meteorites are carried out on samples that weigh less than 100 mg. In total, about 5500 analyses of the light noble gases—He, Ne, and Ar—have been reported for all types of meteorites.

Noble (rare) gases in meteorites have different origins. Each component is characterized by a specific isotopic or elemental composition. Some components have been produced *in situ* in meteoritic material, like the radiogenic gases. Radiogenic ^{40}Ar is produced by spontaneous radioactive decay of long-lived, naturally occurring ^{40}K [half-life, 1.28 billion years (Ga)], whereas ^{4}He is produced in a similar manner from ^{232}Th, ^{235}U, and ^{238}U (14.1 Ga, 0.704 Ga, and 4.47 Ga, respectively). Fission components of Kr and Xe are produced by spontaneous or induced fission of heavy nuclei (e.g., the stable U isotopes), each having a characteristic isotopic distribution. In addition, decay products of extinct radionuclides (^{129}I and ^{244}Pu: 15.7 Ma and 81 Ma, respectively) are found in meteorites.

Other *in situ* produced gases include cosmogenic nuclides formed by nuclear reactions of high-energy particles of galactic or solar origin with meteoritic matter. The specific nuclear reaction process is mainly determined by the energy of the particles and by the chemical composition of the target meteoroid. Nuclear reactions of primary (GeV energies) particles involve the initiation of a cascade of secondary particles with smaller energies. The isotopic or elemental ratios of cosmogenic noble gas isotopes within a meteorite depend on its position within the meteoroid and on its size. Cosmogenic nuclides are limited to the surface regions (<1 m depth) of larger bodies and to meter-sized objects in space. Inert gases found in iron meteorites are mainly cosmogenic, but stony meteorites contain a mixture of many components.

The trapped gases include a whole family of noble gas components. This type of gas is not produced *in situ* but is incorporated in the meteoroid when it formed. Trapped gases are of three main varieties: solar, planetary, and "exotic." The elemental ratios of solar gases are similar to those observed in the Sun. Solar gases are introduced into meteoritic mineral grains by direct implantation of solar-wind ions or more energetic solar particles. This process occurs in the regolith of atmosphereless surfaces of parent bodies such as the Moon. The planetary noble gas pattern is characterized by a systematic fractionation in which the light noble gases—He and Ne—are depleted relative to Ar, Kr, and Xe. Different meteorite types or individual mineral separates have characteristic isotopic and elemental signatures that differ, for example, from those of the noble gases in the terrestrial atmosphere.

Each of the events depicted in Fig. 2 can, in principle, alter planetary matter mineralogically and/or chemically. To illustrate this, let us qualitatively consider an element like Ne, whose concentration in meteorites could be affected by any or all of the events shown in Fig. 2. As a noble gas, it does not form chemical bonds, so it must be physically bound in meteorites. The three stable Ne isotopes of mass 20, 21, and 22 were produced in some amounts by several stellar nucleosynthetic processes, and a mixture of these isotopes was introduced into the primitive nebula with other nucleosynthesized nuclides. Some proportion of this Ne (with its characteristic ^{20}Ne/^{22}Ne and ^{21}Ne/^{22}Ne ratios) was trapped in condensing and accreting nebular material. Presolar grains incorporated into the material and not subsequently destroyed contain another component—pure ^{22}Ne (Ne-E)—produced by decay of the short-lived, now extinct radionuclide ^{22}Na (2.60 years).

The interior of the primitive parent body was heated and transformed into a more evolved form, accompanied by partial or complete Ne degassing, with the Ne components escaping, possibly into space or perhaps being redeposited into cooler parent body material nearer the surface. Fine-grained matter on the parent body's surface could acquire solar-wind and solar-flare Ne, with their characteristic isotopic compositions, implanted as discussed earlier in connection with regolith breccias (see Section II, D, 3). Material in the topmost meter of the parent body would experience cosmogenic nuclear reactions, producing cosmogenic Ne with its own isotopic composition. The regolith would be repeatedly churned ("gardened") by impacts so that a mixture from a variety of sources could be present in any sample. Finally, an impact would occur that removed a meter-sized meteoroid, thereby starting the CRE "clock" measured by accumulation of a new batch of cosmogenic Ne and other nuclides, including radionuclides.

As can be imagined, a meteorite sample can contain Ne from any or all of these sources, so that its isotopic composition represents a weighted average of the isotopic compositions of the components present in it. These can be recognized on a three-isotope plot (Fig. 17a). A sample that consists essentially of just one component is represented by one point in such a diagram, and neon that is a mixture of two of these components will lie on a line connecting the isotopic compositions of these components. Included in Fig. 17a, as an example, is the Ne isotopic composition of samples of the meteoritic breccia ALH 85151, which contains solar and cosmogenic gas. Lunar soils also contain Ne that derives essentially from the Sun, but this is a mixture of Ne components from the low-energy solar wind and from more energetic solar particles, each differing in isotopic composition. The solar Ne isotopic composition extrapolated from analyses of ALH 85151 samples lies almost midway between the Ne isotopic composition determined by the Apollo Solar Wind Catcher foil measurements (SWC) and that of solar energetic particles (SEP).

Addition of Ne from other sources, like Ne-E, will

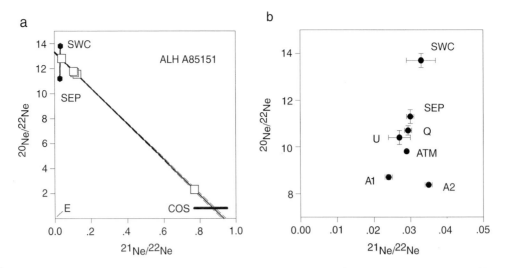

FIGURE 17   Three-isotope plot for stable Ne isotopes in meteorites. (a) ALH 85151 and its primary Ne constituents; (b) Ne components with low $^{21}Ne/^{22}Ne$ ratios. In (a), data points for mineral grains separated from ALH 85151 lie on lines (mean ± one standard deviation) that connect the composition of average solar Ne (keV/nucleon solar wind [SWC] and MeV solar energetic or flare particles [SEP]) with cosmogenic [COS] Ne produced by GeV galactic cosmic rays. The data points in the upper left represent grains exposed as fine dust on a regolith surface, so their constituent Ne is almost entirely of solar origin. The point at the lower right is from the interior of larger grains (with low surface-to-volume ratios) whose Ne is almost entirely of cosmogenic origin. Pure ^{22}Ne (designated by E, so-called Ne-E) is formed by radioactive decay of very short-lived (2.60 years) ^{22}Na in the protoplanetary nebula. If ALH 85151 contained substantial Ne-E, data points would be in the triangular region defined by E and Ne of solar and cosmogenic origin. In (b), we depict isotopic compositions of Ne from: solar wind, solar energetic particles, ureilites (U), Earth's atmosphere (ATM), two trapped planetary components (A1 and A2), and an absorbed presolar, planetary component that is released when mineral grains are etched with nitric acid (Q).

complicate this picture. A mixture of three Ne components will fall within a triangle whose apexes each have the Ne isotopic composition of one of the pure components. Furthermore, many chondrites contain one or more trapped Ne components, examples of which are shown in Fig. 17b.

A similar picture can be drawn for Kr and Xe, with many isotopes and several possibilities for three-isotope plots, but there are additional individual complications for He and Ar. For He, only two stable isotopes exist—^{3}He and ^{4}He—so that a three-isotope plot is not possible. Furthermore, an additional monoisotopic component can be present in meteorites in the form of radiogenic ^{4}He (see Section V, C). The Ar situation differs from that of He in that three stable isotopes exist. In most stony meteorites, one isotope, ^{40}Ar, is mainly radiogenic, deriving from decay of ^{40}K. This admixture of a monoisotopic component limits the utility of three-isotope plots for interpretation of the trapped Ar component. Krypton and Xe systematics are complicated for several reasons. The Kr and Xe isotopes are produced from a variety of nucleosynthetic sources. For our purposes, two Xe sources are especially important. One is now extinct ^{129}I, which decayed to produce ^{129}Xe—a process that can provide a relative, chronometric formation interval (cf. Section V, E).

Another Xe source of similar origin is fission of now extinct ^{244}Pu, which produced a Xe component with a characteristic fission–yield curve. In addition to induced and spontaneous fission products of the U isotopes, different trapped components are observed. Kr and Xe in presolar grains provide almost pure gas from individual nucleosynthetic events.

Each solar system body seems to have had its particular formation history and, therefore, contains its own noble gas isotopic "fingerprint." Gases on Earth, the Moon, Venus, and Mars can be distinguished from each other and from those in chondrites (Fig. 18). Data for samples from one Martian atmospheric meteorite, the EET A79001 shergottite, seem to lie on a mixing line between the terrestrial and Martian points. This is one compelling bit of evidence that is consistent with the idea that this shergottite (and the other 11 Martian meteorites) from Antarctica formed on Mars, later to be somewhat contaminated by the terrestrial atmosphere as it lay in or on the ice sheet.

B. NOBLE GAS COMPONENTS AND MINERAL SITES

In our brief discussion of Ne, we sketched out how, in principle, it is possible to disentangle several Ne

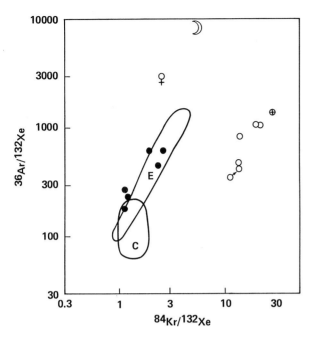

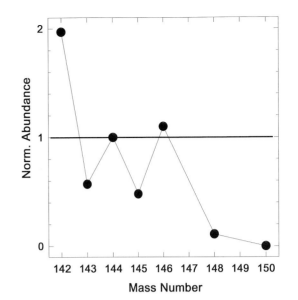

FIGURE 18 Noble gas nuclides in the Moon (crescent), enstatite (E), and carbonaceous (C) chondrites, and from Venus, Earth, and Mars (appropriate symbols). Gases collected from EET A79001 shergottite (open symbols) apparently lie on a mixing line between Mars and Earth. (Reprinted from D. D. Bogard, L. E. Nyquist, and P. Johnson, Noble gas contents of shergottites and implications for the Martian origin of SNC meteorites. *Geochimica et Cosmochimica Acta,* **48,** 1723–1739. Copyright 1984 with kind permission from Elsevier Science Ltd., Kidlington, England.)

FIGURE 19 Isotopic composition (normalized at 144) of neodymium isolated from SiC grains in Murchison CM meteorite. Pronounced anomalies are evident in the isotopes' departures from the horizontal line at 1.0.

components from the average value found in a meteorite. Actually, the situation is more complicated because each "component" may, in fact, be resolvable into constituents introduced from specific sources, each with reproducible isotopic patterns involving not only one noble gas but several. These can be disentangled by ingenious laboratory treatments that yield a phase enriched in one true gaseous constituent over others. Such enrichments can be carried out by investigating individual grains, selective acid dissolution of specific minerals, enrichment by mineral density using heavy liquids, stepwise heating in which gases evolved in a certain temperature interval are mass-analyzed, or some combination of these steps (and others).

1. Interstellar Grains in Meteorites

Until about 1970, the solar system was regarded as "isotopically homogeneous." All objects within it were thought to have formed from a well-mixed and homogenized primordial nebula that was chemically and isotopically uniform. (The later discovery of oxygen isotopic variations, as illustrated in Fig. 12, seems to have disproved this.) However, even at that time, a number of data indicated that some rare samples extracted from meteorites exhibit anomalous isotopic abundances, for example, in their contents of certain isotopes of neon or xenon. These anomalies could not and cannot be explained by well-established processes like decay of naturally occurring radionuclides, the interaction of cosmic rays with matter, or mass-dependent fractionation due to physical or chemical processes. These isotopic anomalies, usually orders of magnitude larger than observed in other solar system materials, are associated with very minor mineral phases of primitive chondrites distributed irregularly in unequilibrated meteorites. These minerals include diamond, graphite, silicon carbide, and aluminum oxide. Typical grains of these minerals are between 1 and 10 μm, although diamond is much smaller ($\sim$0.002 μm). Presolar SiC grains, at least, exhibit a power–law mass distribution dominated by submicron particles so that large ones are rare. The abundance of these minerals in meteorites is low (e.g., SiC in the CM meteorite Murchison is about 5 ppm by mass). An example of such an anomaly is shown in Fig. 19, which depicts the isotopic composition of Nd measured in presolar SiC grains separated from the Murchison meteorite. The data have been normalized to terrestrial values with ^{144}Nd = 1. The isotopic anomalies are evident in deviations from the horizontal line.

The isotopic compositions of these grains differ wildly from those of ordinary solar system matter, thus these rare mineral grains derive from outside our solar system. These presolar grains were incorporated into the solar nebula with intact memories of their individual sources of nucleosynthesis. The ones identified obviously survived all later metamorphic events connected with the history of meteoritic material.

The nature of the isotopic anomalies identified thus far point at specific genetic processes of their constituent elements. Most of the SiC grains were probably produced by stars found on the asymptotic giant branch (i.e., AGB stars) in the Hertzsprung–Russell diagram. It is thought that this is the source of isotopes produced by slow neutron capture processes (or so-called s-process nuclides). Supernovae seem to be required to explain the isotopic anomalies in tiny diamonds.

The wealth of information recorded by the isotopic anomalies of many trace elements in these presolar grains provides unique insights regarding the evolution of stars and processes of nucleosynthesis. This information is obtainable only by exhaustive, detailed, highly sensitive, and highly accurate analyses of rare interstellar grains from primitive samples in terrestrial laboratories. The identification and separation of these tiny mineral grains in meteorites require both inspiration and perspiration. Undoubtedly, much remains to be learned from isotopic anomalies in these rare meteoritic constituents about stellar formation and evolution, as well as the formation and early history of the solar system.

2. CAI

Not only low-temperature materials, like the matrix of C1 chondrites and presolar grains, record early solar history, but refractory grains like CAI also do so. The CAI are millimeter- to centimeter-sized refractory inclusions, especially recognizable in C2 and C3 chondrites but also identifiable in a few UOC and in an E3 chondrite. Typically, CAI consist of refractory silicate and oxide mineral assemblages rimmed by thin, multi-layered bands of minerals. Major-element compositions of CAI agree with calculations by equilibrium vapor-deposition evaporation models to represent the first 5% of condensable nebular matter solidifying at ≥ 1400 K from a gas of cosmic (solar photospheric) composition at a pressure of 10^{-3} atm. Most individual CAI contain tiny particles (usually $<50~\mu m$) very rich in refractory siderophiles (Re, W, Mo, Pt, Pd, Os, Ir, and Rh) and, occasionally, refractory lithophiles like

Zr and Sc. Sometimes, even smaller (micron-sized) refractory metal nuggets are found that consist of single-phase pure noble metals or their alloys.

The textural and mineralogic complexities of CAI are such that a variety of formation and alteration processes must have occurred in their history. Undoubtedly, CAI formed at high temperatures: properties of some suggest vapor condensation as crystalline solids, whereas others seemingly reflect liquid or amorphous intermediates. Volatilization, melting, solid-state metamorphism, and/or alteration in the nebula or after accretion may also have affected properties of some or many CAI. Clearly, CAI had complicated histories that obscured their primary textural properties but left their chemical and isotopic properties relatively unaltered.

Volumetrically, fine-grained CAI are encountered more often than coarse-grained ones, but the latter are studied more easily. Coarse-grained CAIs can be grouped into four types, defined mainly by their mineralogy: type A, dominated by melilite, compositionally Åkermanite (Åk) 0–70; type B, a mixture of melilite, fassaitic pyroxene, spinel, and minor anorthite; anorthite-dominated type C; and forsterite-bearing inclusions. Type A CAI may be the most diverse, having apparently condensed as solid from vapor, and many are heavily altered, so that reconstruction of their original composition is very difficult. The other three types formed from partly molten mixtures to melt droplets, respectively. Type B CAI are mineralogically the most complex and host a much wider array of isotopic anomalies. Compositionally, CAIs reflect their high-temperature origin and are refractory-rich: refractory lithophilic trace elements like the rare earth elements (REE) are generally enriched relative to C1 compositions by 20-fold or more, although considerable variability may be evident in individual CAI owing to thermal history and oxygen fugacity variations. The oxygen isotopic compositions of the CAIs help define the anhydrous minerals line (with slope 1) in Fig. 12.

Most interest has been devoted to centimeter-sized type B CAI in C3V chondrites. Individual minerals in CAI have been probed by a wide variety of very sophisticated instruments that put to shame conventional chemical microanalytical techniques. Many CAI exhibit isotopic anomalies (in the positive and/or negative directions) for O, Ca, Ti, and Cr. A few CAI, mineralogically and texturally indistinguishable from other CAI, are called FUN inclusions because they exhibit Fractionated and Unidentified Nuclear isotopic effects involving not only Kr and Xe but also elements like Mg, Si, Sr, Ba, Nd, and Sm. Six FUN inclusions contain mass-fractionated oxygen (i.e., follow slope

1/2 lines in Fig. 12) and the two type B inclusions of these six exhibit isotopic anomalies for every element thus far studied.

Although a great deal of information has been gathered from CAI, in general, and FUN inclusions, in particular, we do not yet understand why isotopic anomalies appear in some CAI but not in others and why some elements in a given sample exhibit specific anomalies whereas others do not. The CAI apparently formed early in the solar system's history, by processes analogous to those that formed chondrules from unhomogenized matter.

C. ELEMENTS OTHER THAN NOBLE GASES

Having briefly touched upon some important but complicated meteoritic constituents, let us return to the whole rocks themselves to consider information conveyed by trace elements in them. Most elements in the periodic table are present in a meteorite at very low levels: microgram/gram (ppm), nanogram/gram (ppb), or picogram/gram (ppt) concentrations. This occurs because the constituent stable isotopes of trace elements were produced only in relatively small amounts during nucleosynthesis, and because their geochemical and/or physical properties prevent enrichment—indeed, may cause significant depletion—during primary or secondary genetic episodes. Because of their geochemical properties, some trace elements may be sited in specific hosts in particular meteorites [e.g., siderophiles like Ir, Ga, and Ge are enriched (relative to Cl levels) in a metal phase of iron meteorites (cf. Fig. 10)] whereas others are dispersed among a variety of minerals. The same element may be a dispersed element in some meteorite class and be sited in a particular host in another class. For example, REE are found in phosphates in achondrites, but some REE are dispersed elements in chondrites. They are concentrated in whitlockite in eucrites and are even more enriched in CAI. Trace elements convey important genetic information because a small absolute concentration change induced by a genetic process will cause a large, and therefore significant, relative effect.

This improvement in "signal-to-noise" is perhaps best illustrated by explosive meteorite impacts. Whatever the initial composition of the material from which Earth derived, it appears certain that a goodly proportion of the proto-Earth's initial complement of refractory siderophiles has been extracted into the core, so that the crust is generally depleted in them. The fall of a massive chondrite or, even better, of an iron meteorite enriched in siderophiles, followed by an explo-

sion, will disseminate mixed projectile and target ejecta over a considerable distance and allow the ejecta to be redeposited in a thin layer. Subsequent chemical analysis of a vertical slice that includes the deposition layer will reveal enrichment of the siderophiles in that layer. In this fashion, siderophile enrichments—especially refractory Ir—in the K-T boundary layer around Earth suggested that dinosaurs (and many other biota) died off as the result of a meteoroid/asteroid impact 65 Ma ago. This idea was initially controversial but now is more accepted. In many instances, enrichments by several siderophiles in impact breccias at a meteorite explosion crater on Earth or the Moon provide a "fingerprint" that can identify the nature of the meteoroid that created the crater.

As discussed in Section II, D, 4, a, the compositional similarity of C1 chondrites and the solar photosphere is marred mainly by volatile elements condensible only at very low temperatures. To put meteorite compositions on the common footing of readily condensible material, the normalization standard for meteorites is a refractory lithophile—most commonly Si (as in Fig. 15), but Mg or Al also serve—rather than hydrogen as in the solar photosphere. For meteorites, these ratios can be on a weight or atom basis; in the latter case, trace element contents are usually referred to as atomic abundances. Very often, data are normalized to C1 values since compositions of the most primitive chondrites—EH or UOC—typically approach C1 levels. It is on such a basis that we can say that moderately to highly refractory siderophiles are enriched in iron meteorites, or that refractory lithophiles are enriched in achondrites. Contents of the more refractory trace elements are characteristic of, and hence can be used to classify, achondrite associations (Fig. 20).

A priori identification of a trace element as refractory or volatile is impossible since the chemical form in which it is bound in a meteorite is generally unknown. To illustrate, macroscopic amounts of indium metal or gaseous oxygen are each quite volatile, but when chemically bonded in InO, the compound is quite refractory. Since In is present only at ppb levels in even the most volatile-rich meteorites, neither InO nor any other In compound is observable. Several approaches have been taken to obtain at least a qualitative elemental volatility order, and the lists obtained generally agree, with some minor differences. The criteria used include: calculation of theoretical condensation temperatures in a nebular gas of solar composition at pressures of 10^{-3}–10^{-6} atm; determination of C1-normalized atomic abundances in equilibrated (petrographic types 5 and 6) ordinary chondrites; and laboratory studies of elemental mobility (ease of vaporization

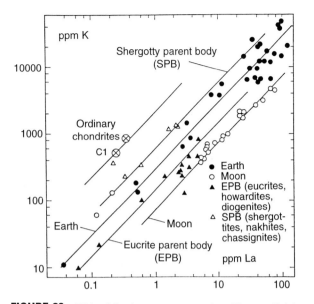

FIGURE 20 Lithophile element concentrations (K versus La) in ordinary chondrites, Cl chondrites, and samples from evolved bodies: lunar samples from various Apollo missions and lunar meteorites; the eucrite parent body (EPB) consisting of HED achondrites (eucrites, howardites, diogenites); terrestrial rocks; and Martian meteorites (shergottites, nakhlites, and Chassigny). (Reprinted from G. Dreibus and H. Wänke, Volatiles on Earth and Mars: A comparison. *Icarus,* **71.** Copyright 1987 with kind permission from Academic Press, San Diego.)

and loss) during weeklong heating of primitive chondrites under conditions simulating metamorphism in a parent body (400–1000°C, 10^{-4} atm H_2). By these criteria, elements considered as moderately volatile include (in increasing order) Ni, Co, Au, Mn, As, P, Rb, Cu, K, Na, Ga, and Sb, whereas strongly volatile ones include Ag, Se, Cs, Te, Zn, Cd, Bi, Tl, and In.

Small but real ($\leq$ factor of 2) differences exist in the contents of the more refractory trace elements of the various chondritic groups. Contents of siderophiles are higher in EH than in EL chondrites and decrease in ordinary chondrites in the order H > L > LL, as expected from their total iron contents. Achondrites are, of course, enriched in refractory lithophiles, and depleted by orders of magnitude in siderophiles (whether refractory or volatile) and volatile elements of whatever geochemical character. In some achondrites (mainly HED meteorites and at least one lunar meteorite), high levels of volatiles are present, having been deposited by late volcanic emanations on their parent bodies. As might be expected from the formation processes of iron meteorites, these meteorites are rich only in the more refractory of the siderophiles; they contain essentially no volatiles, or lithophiles except in silicate inclusions.

The more volatile elements exhibit much greater

variability in stony meteorites. Concentrations of the three or four most volatile elements are several orders of magnitude higher in UOC than in their equilibrated analogues and decrease by one or two orders of magnitude with increasing UOC subtype. Contents of most strongly volatile elements in ordinary chondrites of petrographic types 4–6 are highly variable and do not vary to any great extent with petrographic type. However, in H chondrites, concentrations of a number of moderately volatile elements vary as H4 > H5 > H6, consistent with, say, loss at progressively higher metamorphic temperatures in a stratified parent or parents. As will be discussed in Section V, D, chronometric data also are consistent with this theory for the H chondrite parent(s). Such a model cannot be established for the L chondrites, because the late shock evident in the petrographic properties of many of them affected other thermometric properties, thereby obscuring their earlier histories. In addition to the petrographic evidence, strongly shocked L4–L6 chondrites exhibit loss of some noble gases and highly mobile elements.

Mean contents of Ag, Te, Zn, Cd, Bi, Tl, and In decrease with increasing amount of shock-loading (and, therefore, residual temperature) estimated from petrographic shock indicators. Trace element contents of H chondrites show no such dependence with shock and, in unshocked chondrites, volatile element contents are significantly lower in H than in L chondrites. The latter suggests that L chondrite parent material formed from the nebula at lower temperatures than did H chondrite parent material. Apparently, nebular temperatures during formation of H chondrite parent material were so high ($\sim$700 K) that only a very small complement of volatile trace elements could condense. Hence, essentially none was present to be lost later at high, shock-induced residual temperatures.

The H chondrite regolith breccias, like Noblesville (see Fig. 1a), are clear exceptions to "normal" H chondrites in that the dark, gas-rich portions of such breccias are quite rich in volatile trace elements, sometimes containing levels exceeding those of C1. These volatiles, which are very heterogeneously distributed in the dark matrix, were apparently not implanted by the solar wind but rather derive from identifiable black clasts. These black clasts represent either volatile-rich nebular condensate or a sink for material degassed from deeper in the parent body interior. During exposure on the asteroidal surface, these dark clasts and light clasts (containing "normal" levels of volatiles) were apparently comminuted by repeated impacts and mixed, ultimately forming the matrix of the regolith breccia. Insufficient data exist for equilibrated LL chondrites, so we do not yet know if they have a unique thermal history or one that resembles that of H or L chondrites.

In contrast to the situation in most ordinary chondrites, volatile trace elements in carbonaceous chondrites are very homogeneously distributed. These elements are unfractionated from each other in nearly all carbonaceous chondrites, implying that the parent material of these chondrites incorporated greater or lesser amounts of C1-like matter during accretion, the proportions defining a continuum ranging from pure C1 down to about 20% in C5 or C6. As in enstatite chondrites, the more volatile-rich samples seem to have higher proportions of the more siderophile trace elements. These trends accord with oxygen isotope data, implying a continuum of formation conditions for parent materials of carbonaceous chondrites. At least three Antarctic C1 and C2 chondrites show unambiguous mobile trace element, noble gas, and petrographic evidence for open-system thermal metamorphic alteration in their parent bodies at temperatures of 400–700°C and they do not seem to have been rehydrated during their residence in Antarctica. In all cases, petrographic properties of C1–C6 chondrites were established during nebular condensation and accretion, or by closed-system thermal processing.

Trends in enstatite chondrites lead to dichotomous views of their genesis. On the one hand, contents of major elements and the more refractory trace elements indicate that EH (mainly E3 and E4) and EL (mainly E5 and E6) chondrites each derive from separate batches of compositionally different nebular materials. On the other hand, contents of moderately to strongly mobile trace elements decline from ~C1 levels in some EH chondrites by ~2 orders of magnitude in EH6 or EL6 chondrites in a manner consistent with open-system metamorphism. These elemental trends suggest further that enstatite achondrites derived from E6 chondritelike material that previously experienced loss of FeS–Fe eutectic (formation temperature, 988°C). Oxygen isotopic data for all enstatite meteorites (i.e., chondrites and achondrites) are similar (see Fig. 12). It may be that enstatite meteorites derive from a single parent body, stratified in refractories but not volatiles during primary accretion, which subsequently lost these volatiles by secondary thermal processes. Alternatively, Mother Nature may have been particularly perverse in providing samples of two parent bodies (EH and EL) with similar oxygen isotopic compositions, with primitive material coming essentially from the former and evolved portions mainly from the latter.

Though there is little doubt that for meteorites of less common types, the Antarctic collection constitutes a broader sampling of extraterrestrial materials than do contemporary falls of the same type (see Section III, B), systematic and reproducible differences involving moderately to highly volatile elements suggest that this extends even to ordinary chondrites. This suggestion, initially disbelieved, acquired some credence with the observation of meteorite and asteroid streams, the formation of a comet stream by differential tidal disruption of comet Shoemaker-Levy 9, failures of alternatives to explain Antarctic meteorite/fall compositional differences, and identification of population differences of unambiguous preterrestrial origin. Differences have been suggested to reflect flux variations with time.

V. METEORITE CHRONOMETRY

How old are meteorites? An "age" is a time interval between two events and is measured by specific chronometers. Any accurate chronometer must contain a mechanism that operates on a predictable, although not necessarily constant rate. The chronometer is started by some special event at the beginning of a time interval and its end must be clearly and sharply recorded. Chronometers used in modern geo- and cosmochronology usually involve radioactive isotopes that are long-lived and naturally occurring. Examples are the U isotopes, ^{87}Rb, and ^{40}K. The process of radioactive decay allows calculation of an age if the concentrations of both parent and daughter nuclide are determined, the beginning of the time interval is defined, and the system was not disturbed during the time interval (so that the sample constitutes a "closed system"). Some meteorite ages involve the production of particular stable or radioactive nuclides, or the decay of the latter. Typically, the half-life of the chronometer should be of the same magnitude as the time interval being measured.

Meteoritic materials yield a variety of ages, each corresponding to a specific episode in its history. Some of these events are shown in Fig. 2: the end of nucleosynthesis in stars; the first formation of solid material in the solar system; formation of "rocks" on parent bodies; excavation of meteoroids from their parent bodies; and the meteorite's fall to Earth. It is also possible to date other events, like volcanism or metamorphism on parent objects. It is further possible to establish a formation interval (based on extinct radionuclides) that measures the time between the last production of new nucleosynthetic material and the formation of minerals in the early solar system materials.

Cosmic ray exposure ages date the duration of the flight of a meteoroid as a small body (≤1 m) in interplanetary space: the meteorite's terrestrial age is the time elapsed since it landed on Earth's surface. In the following sections, we will discuss some of these meteorite ages.

A. TERRESTRIAL AGES

Terrestrial ages are established from comparison of amounts of cosmogenic radionuclides found in meteorite falls and finds. The principles of the method are depicted in Fig. 4, with ^{14}C ($t_{1/2} = 5.73$ thousand years, or ka), ^{81}Kr ($t_{1/2} = 200$ ka), ^{36}Cl ($t_{1/2} = 301$ ka), and ^{26}Al ($t_{1/2} = 730$ ka) being the nuclides most frequently employed. In Section II, B, we summarized the most important conclusions obtained from determination of meteorites' terrestrial ages. The survival time of a meteorite during its terrestrial residence is determined by the weathering conditions in the environment in which the meteorite is stored. Survival times (and hence terrestrial ages) for meteorites are much lower for warm and/or wet areas than for arid regions. Typical chondrites from the southwestern part of the United States have survival times of about 10 to 15 ka, whereas some stone meteorites in the drier and colder Antarctic "deserts," which preserve them very well, reach 2 Ma (Fig. 21).

B. CRE AGES

In principle, the determination of amounts of a cosmogenic radionuclide (or its decay product) and of a stable nuclide is required to establish cosmic ray exposure ages. In practice, however, production rates of stable cosmogenic noble gas nuclides in stony meteorites are well known and it is usually sufficient to measure their concentrations. In the absence of evidence to the contrary, it is usually assumed that the irradiation by cosmic rays of solar and galactic origin is simple, that is, that the meteoroid was completely shielded (buried in a parent body) until an impact ejected it as a meter-sized object that remained essentially undisturbed until it collided with Earth. Some larger stones (e.g., the H4 chondrite Jilin) and irons (e.g., Canyon Diablo) exhibit complex irradiation histories involving preirradiation on the surface of the parent body or secondary collisions in space that fractured the meteoroid and exposed new surfaces of CRE. In such cases, different samples of a meteorite having a complex irradiation history then exhibit different CRE ages. As noted in

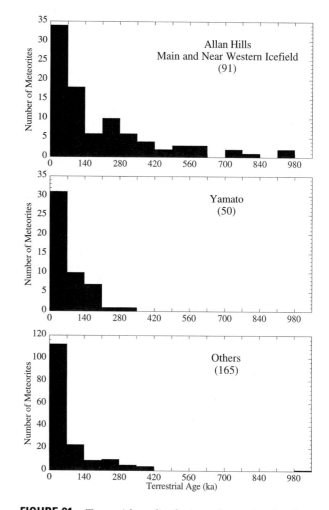

FIGURE 21   Terrestrial age distributions of meteorites (numbers analyzed in parentheses) from: (a) part of the Allan Hills region of Victoria Land (West Antarctica); (b) the Yamato Mountains region of Queen Maud Land (East Antarctica); and (c) various Antarctic regions other than the Allan Hills or Yamato Mountains. (Courtesy of Dr. Kunihiko Nishiizumi, University of California, Berkeley.)

Section I, B, meteoroids approaching the Sun to within 0.8 AU are warmed, causing diffusive loss of gases, especially ^3H (a contributor to ^3He production) from iron; such cases can be readily recognized by low isotopic ratios, particularly $^3He/^4He$ or $^3He/^{21}Ne$, or low natural TL.

The CRE data for ordinary chondrites are illustrated in Fig. 22. Though all groups have CRE ages that range up to 90 Ma, it is clear that the distributions differ markedly. For H chondrites, there is at least one major peak in the neighborhood of 7 Ma (which may, in fact, include two closely spaced ones at 6 and 8 Ma) and a smaller one at 33 Ma (Fig. 22a). For L chondrites, major peaks are not evident, but some clustering of ages is observed at 20–30 Ma and 40 Ma, with a smaller one at 5 Ma (Fig. 22b). For LL chondrites, the major peak at 15 Ma includes ~30% of all measured samples (Fig. 22c). Major peaks correspond, of course, to major collisional breakup events on the chondrite parent bodies. Contrary to the situation among H chondrites, nearly two-thirds of the L chondrites have CRE ages exceeding 10 Ma.

Current data suggest that the presence of major CRE peaks differs with petrographic grade for each ordinary chondrite group, but of course poor statistics cloud the situation in some cases. For example, among L chondrites, the 40-Ma event produced mainly L5 and L6 chondrites. The 7-Ma CRE age peak is particularly evident among H4 and H5 chondrites and a cluster at 4 Ma is evident for H5 and H6. The 10-Ma peak for LL chondrites includes mainly LL6. It has long been recognized that the fall frequency of most ordinary chondrites (except H5) is twice as great between noon and midnight (i.e., P.M. falls) as between midnight and noon (A.M. falls). A social bias for this observation can be ruled out, so the difference must reflect the meteoroids' orbits. Meteoroids with perihelia ~1 AU will be predominantly P.M. falls, whereas those having aphelia of ~1 AU will be A.M. falls. These A.M. falls result from Earth's overtaking meteoroids or involve meteoroids that narrowly miss Earth, and subsequently experience gravitational attractions causing them to land on Earth's forward hemisphere. Fall frequencies for H5 chondrites are significantly different, with A.M. and P.M. falls being about equal. Clearly, there is a fundamental difference between the orbital elements of H5 and other ordinary chondrites.

The number of CRE ages for carbonaceous and enstatite chondrites are too sparse, at present, to exhibit significant peaks. Carbonaceous chondrites tend to have short CRE ages (<20 Ma). Lunar meteorites contain mostly cosmogenic nuclides produced on the Moon's surface: the Moon–Earth transit time is generally short compared to the exposure ages of other stone meteorites. For Martian meteorites, exposure ages range from 0.5 to 16 Ma, with some clustering being apparent.

Clustering of exposure ages in also observed for meteorite groups from the HED parent body. In this case, two diogenite clusters (at about 22 and 39 Ma) coincide with clusters in the eucrite and howardite CRE distributions. Major impacts on 4 Vesta or its daughters are believed to produce meteorites of these three different achondritic classes.

Although attempts have been made to develop a reliable CRE age method for iron meteorites, results have generally not proven to be satisfactory except in one case, a difficult, tedious, and no longer practiced

technique involving ^{40}K ($t_{1/2} = 1.28$ Ga) and stable ^{39}K and ^{41}K. About 70 iron meteorites have been dated by the ^{40}K/^{41}K method and the resulting ages range from 100 Ma to 1.2 Ga (Fig. 23). The fact that CRE ages for iron meteorites greatly exceed those of stones is attributed to the greater resistance of iron meteoroids to destructive collisions (so-called "space erosion") while in space. Peaks of CRE exposure ages are evident for a few iron meteorite chemical groups. In group IIAB, 13 of 14 meteorites have a CRE age of 650 ± 60 Ma; this age is also exhibited by 3 of 4 measured IIICD meteorites, suggesting a major collisional event involving the parent of the chemical group III iron meteorites (see Fig. 23). The only other group exhibiting a peak in CRE ages is the IVA irons, 7 of the 9 dated samples exhibiting an age of 400 ± 60 Ma (see Fig. 23). It should be recalled (cf. Section II, D, 5) that these iron meteorite groups are the two most numerous ones and contain the highest proportions of strongly shocked members. Either the parent asteroids of these groups were unusually large (thus requiring unusually large and violent breakup events) and/or Earth preferentially sampled collisional fragments that had been strongly shocked (thus acquiring a significant shock-induced impulse).

C. GAS RETENTION AGE

As discussed in Section IV,A, the decay series initiated by the long-lived radioactive nuclides ^{232}Th, ^{235}U, and ^{238}U yield six, seven, and eight α particles, respectively, whereas long-lived ^{40}K produces ^{40}Ar. Thus, from measurements of U, Th, and radiogenic ^{4}He, or of ^{40}K and radiogenic ^{40}Ar, it is possible to calculate a *gas retention* age that quantifies the time elapsed since a meteorite sample cooled to a temperature low enough to retain these noble gases, provided that the system was closed during this period. This radiogenic age could record the primary formation of the meteorite's parent material, but in most cases it is disturbed by some subsequent events (metamorphic and/or shock) accompanied by substantial heating that partially or completely degassed the primary material. A variant of the K/Ar age, the ^{40}Ar–^{39}Ar method, involves conversion of some stable ^{39}K to ^{39}Ar by fast-neutron bombardment, that is, ^{39}K(n, p)^{39}Ar, followed by stepwise heating and mass-spectrometric analysis. From the ^{39}Ar/^{40}Ar ratio in each temperature step, it is possible to correct for later gas loss. This variant even permits analysis of small, inhomogeneous samples by use of a pulsed laser heat source.

Gas retention ages of many chondrites, achondrites, and even silicate inclusions in iron meteorites range

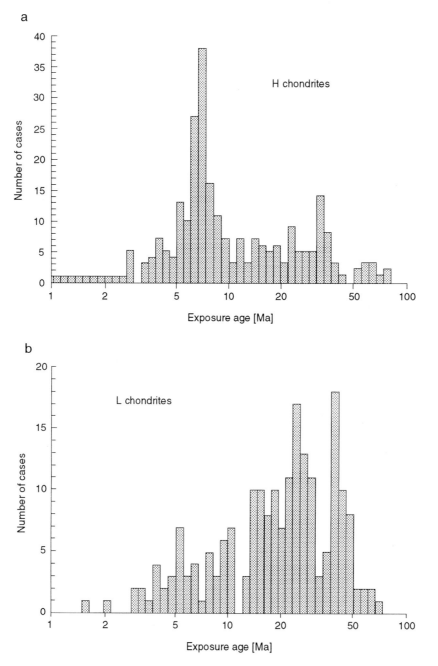

FIGURE 22 Cosmic ray exposure ages for ordinary (a) H, (b) L, and (c) LL chondrites. Peaks in the histograms indicate major collisional events on parent bodies that generated substantial proportions of meter-sized fragments (see Section V, B).

up to about 4.6 Ga. A large number of meteorites, particularly L chondrites, have young gas retention ages in the neighborhood of 500 Ma, whereas H chondrites cluster at higher ages (Figs. 24a and 24b). Meteorites with young gas retention ages generally exhibit petrographic evidence for substantial exposure to strong shock, so diffusive gas loss took place from material having quite high residual temperatures gen-

erated in major destructive collisions. Almost always, meteorites having young K–Ar or ^{40}Ar–^{39}Ar ages have lower U, Th–He ages compared with ^{40}K–^{40}Ar results, since He is less easily retained in most minerals than is Ar because of the lower activation energy for He diffusion and loss. Diffusive loss of ^{40}Ar, incidentally, is much more facile than is loss of trapped ^{36}Ar or ^{38}Ar, enhancing its value as a chronometer. Preferential ^{40}Ar

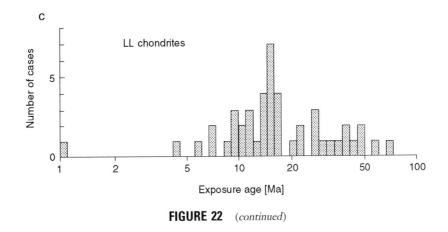

FIGURE 22 *(continued)*

loss occurs because most of it is sited in feldspars and in minerals, where ^{40}K was and is associated with radiation damage that provides a ready diffusive escape path. Highly mobile trace elements are lost more readily than is even ^{40}Ar, so L chondrites with young gas retention ages have lower contents of such elements than do those with old ages.

As will be seen, solidification ages for most Martian meteorites (or at least the shergottites and nakhlites) are quite young, ~1.3 Ga, or less—although one (ALH 84001) has an age of ~3.7 Ga—implying that parent magmas were present as recently as 1.3 Ga ago. The $^{40}Ar-^{39}Ar$ ages of essentially unshocked nakhlites accord with the 1.3-Ga age, but shergottites, which are

heavily shocked, have gas retention ages ranging upward from 250 Ma. It seems, then, that gas retention ages for shergottites represent partial degassing of their parent material ≲250 Ma ago, probably 180 Ma ago.

D. SOLIDIFICATION AGE

As with gas retention ages, solidification ages establish the time elapsed since the last homogenization of parent and daughter nuclides, normally by crystallization of a rock or mineral phase. Nuclides used to establish solidification ages are isotopes of nongaseous elements that are insensitive to events that might have affected

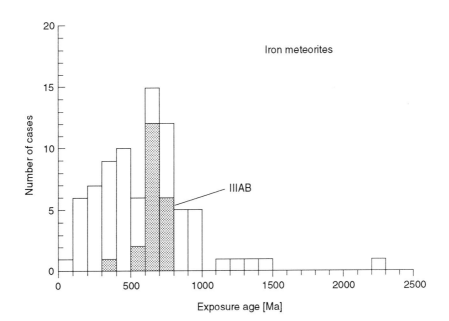

FIGURE 23 Cosmic ray exposure ages determined by the $^{40}K-^{41}K$ method for iron meteorites. The group IIIAB irons, nearly all of which were apparently produced by a single, massive collision of their parent body 650 ± 60 Ma ago, are depicted in the shaded region.

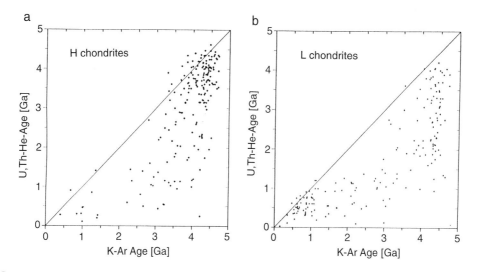

FIGURE 24 Gas retention ages of H and L chondrites. Data obtained from the U, Th–He and K–Ar methods are plotted against each other. The 45° line represents concordant ages. The very different trends indicate that the thermal histories of the two types of ordinary chondrites differ. The concordant long ages of H chondrites suggest that, in general, their parent body or bodies have remained thermally unaltered since they formed 4–4.5 Ga ago. The concordant short ages of L chondrites suggest that they were shock-heated in a major collision(s) 0.1–1.0 Ga ago. Nearly all discordant meteorites lie below the concordance lines because radiogenic ^{4}He is lost far more readily than is radiogenic ^{40}Ar.

gas retention. Some techniques, such as the Pb/Pb method, which involves the ultimate decay products of ^{235}U, ^{238}U, and ^{232}Th (^{207}Pb, ^{206}Pb, and ^{208}Pb, respectively), involve a relatively mobile element, Pb, that should be more easily redistributed than would be the ^{147}Sm–^{143}Nd dating pair. Hence, in principle, a sample dated by several techniques might yield slightly different ages depending on its postformation thermal history.

Common techniques that yield useful solidification ages include: the Pb–Pb method mentioned here; ^{147}Sm ($t_{1/2} = 106$ Ga)–^{143}Nd; ^{87}Rb ($t_{1/2} = 48$ Ga)–^{87}Sr; and ^{187}Re ($t_{1/2} = 41$ Ga)–^{187}Os. Generally, methods used to determine solidification ages depend on data depicted in isochron diagrams, such as that shown in Fig. 25, in which the enrichment of radiogenic ^{87}Sr is proportional to the amount of ^{87}Rb, and ^{86}Sr is taken for normalization. The slope of such a line yields an "internal isochron" for a meteorite or a single inclusion of a meteorite, if minerals having various ^{87}Rb/^{86}Sr ratios are measured. The y intercept provides the initial ^{87}Sr/^{86}Sr ratio—a relative measure of the time that nucleosynthetic products were present in the system prior to solidification—that is, how "primitive" the system is. Clearly, the lower the ^{87}Sr/^{86}Sr ratio in Fig. 25, the less radiogenic (or evolved) was the source material. For some time, basaltic achondrites (e.g., HED meteorites) and the angrite, Angra dos Reis, competed as the source containing the most prim-

itive (least radiogenic) Sr, but more recently, Rb-poor CAI inclusions in the C3V chondrite, Allende, have become "champions" in this category, with ^{87}Sr/^{86}Sr = 0.69877 ± 2.

Solidification ages for most meteoritic samples are "old," that is, close to 4.56 to 4.57 Ga. The results

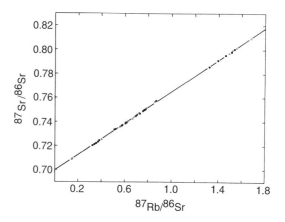

FIGURE 25 Three-isotope plot (Rb–Sr method) for whole-rock samples of various sorts of chondrites (squares, H chondrites; triangles, LL chondrites; circles, E chondrites): model age, 4.498 ± 0.015 Ga, initial ^{87}Sr/^{86}Sr = 0.69885 ± 0.00010. (Reprinted with permission from J.-F. Minster, J.-L. Birck, and C. J. Allégre, Absolute age of formation of chondrites by the ^{87}Rb–^{87}Sr method. *Nature*, **300**, 414–419. Copyright 1982 from Macmillan Magazines, Limited, New York.)

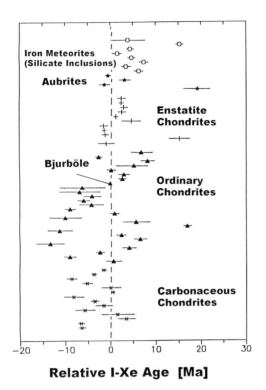

FIGURE 26 ^{129}I–^{129}Xe formation ages for various sorts of chondrites, aubrites, and silicate portions of iron meteorites, relative to that of Bjurböle (older ages to the left and more recent ones to the right).

Figure labels: Iron Meteorites (Silicate Inclusions); Aubrites; Enstatite Chondrites; Bjurböle; Ordinary Chondrites; Carbonaceous Chondrites; **Relative I-Xe Age [Ma]**; x-axis: −20, −10, 0, 10, 20, 30

obtained by different methods agree quite well, although some "fine structures" can be detected. A large number of chondrites have been studied by the Pb–Pb, Rb–Sr, and Nd–Sm techniques and results for them are consistent with an age of about 4.56 Ga. Recently, H6 chondrites have been found to exhibit small but significant differences in Pb–Pb ages from H4 and H5 chondrites, with the H6 population being younger by 30–50 Ma. If correct, this is consistent with a stratified ("onion-shell") model for the H chondrite parent body, and is suggestive of a simple, progressive metamorphic alteration with increasing depth in it.

Though most meteorites have solidification ages around 4.56 Ga, there is clear evidence of more recent disturbances of chronometric systems—particularly Pb–Pb and Rb–Sr—in many meteorites. For example, Rb–Sr internal isochrons for E chondrites (believed by some to have experienced open-system thermal metamorphism, as discussed in Section IV, C) were disturbed 4.3–4.45 Ga ago. Of course, chronometers in heavily shocked L chondrites show clear evidence for late disturbance.

Four techniques (^{40}Ar–^{39}Ar, Rb–Sr, Pb–Pb, and Sm–Nd) yield an age for nakhlites of 1.3 Ga, which

has long been interpreted as indicating their derivation from a large planet, like Mars. Objects of asteroidal dimension cannot attain as high internal temperatures as can larger planets, nor can they retain magma-forming temperatures to recent times. The heavily shocked shergottites seem to have derived from several magma reservoirs and Rb–Sb internal isochrons suggest a major shock-induced disturbance 180 Ma ago, before the Martian meteoroids were ejected from their parent planet.

E. FORMATION INTERVALS

Measurements of decay products of an extinct radionuclide do not provide absolute dates in the sense discussed in earlier sections, but they do permit relative chronologies on timescales comparable with the half-life of the radionuclide. Thus far, clear positive evidence has been found in meteorites or their constituent minerals for the presence in the early solar system of the following nuclides: ^{41}Ca ($t_{1/2} = 110$ ka), ^{26}Al ($t_{1/2} = 730$ ka), ^{60}Fe ($t_{1/2} = 1.5$ Ma), ^{53}Mn ($t_{1/2} = 3.7$ Ma), ^{107}Pd ($t_{1/2} = 6.5$ Ma), ^{129}I ($t_{1/2} = 15.7$ Ma), ^{244}Pu ($t_{1/2} = 82$ Ma), and ^{146}Sm ($t_{1/2} = 103$ Ma). In most cases, relative ages are calculated from three-isotope plots involving decay products of the extinct radionuclide. However, in some cases, the relative chronologic information can be combined with data for absolute ages, allowing small time differences in the early solar system to be established. For example, combining the $^{53}Mn/^{55}Mn$ ratio measured in the Omolon pallasite with the absolute Pb–Pb age of the LEW 86010 angrite yields an absolute age of 4557.8 ± 0.4 Ma for Omolon.

In recent years, considerable effort has gone into this area, so we should focus on one set of results in concluding this chapter. The oldest technique used is that of I–Xe dating, which depends on the decay of ^{129}I into ^{129}Xe. In this technique, a meteorite on Earth is bombarded with neutrons in a nuclear reactor as in ^{40}Ar–^{39}Ar dating (see Section V, B) to convert some stable ^{127}I into short-lived ^{128}I ($t_{1/2} = 25$ m), which decays into stable ^{128}Xe. Stepwise heating releases Xe: a linear array with slope > 0 on a three-isotope plot of $^{129}Xe/^{132}Xe$ versus $^{128}Xe/^{132}Xe$ indicates an iodine-correlated ^{129}Xe release, whose slope is proportional to $^{129}I/^{127}I$ at the last time ^{129}I and ^{129}Xe were in equilibrium. This ratio is a measure of the formation interval. Absolute age values, however, can be obtained only if the ratio $^{129}I/^{127}I$ at the time of the closure of the solar nebula is known. Because this number is not available, only relative ages can be given.

FIGURE 27 A recent stamp of the French Southern and Antarctic Territories illustrating a micrometeorite (left) or cosmic dust particle collected by melting Antarctic ice cores, the coring drill being shown at the right. Representations of trails from meteors (of cometary origin) and a fireball appear at the top of the stamp.

The I–Xe "clock" proves to be remarkably resistant to resetting by heating: the principal effect is to degrade the linearity, but not to destroy it completely. Shock seems quite effective in resetting this clock, and hydrolysis—which affected C1 and C2 chondrites—even more so.

Figure 26 summarizes data for 79 chondrites, aubrites, and silicate inclusions in iron meteorites, relative to the Bjurböle L4 chondrite, which gives highly reproducible I–Xe intervals and is therefore arbitrarily assumed to have an age of zero. As illustrated in Fig. 26, each meteorite class spans an I–Xe interval ≥ 10 Ma, whereas all meteoritic materials possessing isochrons span ~55 Ma. Apparently, the only systematic variation of the I–Xe formation interval with chondritic petrographic type involves E chondrites: EH chondrite parent material formed earlier than did EL. Clearly, while the nuclide ^{129}I was still alive, that is, during or shortly after nucleosynthesis, primitive nebular matter condensed and evolved into essentially the materials that we now receive as meteorites. The conclusion is supported by other isotopic and charged-particle track evidence (see Sections IV, A, IV, B, V, C, and V, D).

As we have seen from the foregoing summary, the meteoritic record can be read best in an interdisciplinary light. Results of one type of study—say, trace element chemical analysis—provide insight to another—orbital dynamics, for example. Early experience gained from meteorite studies provided guidance for proper handling, preservation, and analysis of Apollo lunar samples. In turn, studies of these samples led to the development of extremely sensitive techniques that are now being used to analyze meteorites and microgram-sized interplanetary dust particles of probable cometary origin collected by high-altitude aircraft and from sea sediments, and most recently in Antarctica (Fig. 27). Undoubtedly, this experience will prove invaluable when samples from other planets, their satellites, and small solar system bodies are returned to Earth for study. [See INTERPLANETARY DUST AND THE ZODIACAL CLOUD.]

Previous studies of meteorites have provided an enormous amount of knowledge about the solar system, and there is no indication that the scientific growth curve in this area is beginning to level off. Indeed, the first version of this chapter was prepared when the publication date of this Encyclopedia was scheduled for 1992. Work on the present version began late in 1996, and we were amazed to see how much had been learned about meteorites in the intervening five years. Predictions about future developments are very hazardous, but we can expect surprises, probably from Antarctic meteorites, which seem to include so many peculiar objects. As has been said in another connection, we who work with meteorites don't pray for miracles, we absolutely rely on them.

BIBLIOGRAPHY

Binzel, R. P., Gehrels, T., and Matthews, M. S. (eds.) (1989). "Asteroids II." Univ. Arizona Press, Tucson.

Buchwald, V. F. (1975). "Handbook of Iron Meteorites." Univ. California Press, Berkeley.

Burke, J. G. (1986). "Cosmic Debris." Univ. California Press, Berkeley/Los Angeles.

Hewins, R. H., Jones, R. H., and Scott, E. R. D. (eds.) (1996). "Chondrules and the Protoplanetary Disk." Cambridge Univ. Press, Cambridge, England.

Kerridge, J. F., and Matthews, M. S. (eds.) (1988). "Meteorites and the Early Solar System." Univ. Arizona Press, Tucson.

Lewis, J. S. (1996). "Rain of Iron and Ice." Addison–Wesley, New York.

Melosh, H. J. (1989). "Impact Cratering—A Geologic Process." Oxford Univ. Press, Oxford, England.

INTERPLANETARY DUST AND THE ZODIACAL CLOUD

I. Introduction

II. Observations

III. Dynamics and Evolution

IV. Future Studies

Eberhard Grün

Max-Planck-Institut für Kernphysik

GLOSSARY

Beta-meteoroid: Small meteoroid for which the repulsive radiation pressure force is comparable to solar gravitational attraction.

Carbonaceous material: Material similar to that found in carbonaceous meteorites, which are believed to be among the most primitive (unaltered since their formation in the solar nebula) objects found in the solar system. They contain complex carbon compounds (hydrocarbons, amino acids, etc.) made up mostly from the elements C, H, O, and N.

Ecliptic plane: Plane of Earth's orbit around the Sun. The orbit planes of all other planets are close to this plane, therefore, it is considered the principal plane of the planetary system.

Lorentz force: Force exerted by a magnetic field on a moving charged particle. This force is always perpendicular to the motion of the particle.

Meteor: Light phenomenon that results from the entry of a meteoroid from space into Earth's atmosphere.

Meteorite: Meteoroid that has reached the surface of Earth without being completely vaporized.

Meteoroid: Solid object moving in interplanetary space of a size smaller than an asteroid and larger than a molecule.

Micrometeoroid: Meteoroid smaller than about 0.1 mm in size.

Refractories: Materials not deformed or damaged by high temperatures. Classic refractories are high-melting oxides, like silica and alumina, but also carbides, nitrides, sulfides, and pure carbon. In our terminology, refractories are materials that are not modified by space conditions (temperature and vacuum) in the inner solar system. The opposite are volatile materials, for example, ices that rapidly sublimate close to the Sun.

Space debris: Man-made particulates littered in space.

Spallation zone: Fracture zone around an impact crater in brittle materials from which large fragments have been chipped off.

Interplanetary dust is finely divided particulate matter that exists between the planets. Interplanetary dust particles are also often called micrometeoroids, and range in size from assemblages of a few molecules to tenth-millimeter-sized grains, above which size they are called meteoroids. The boundary between interplanetary dust and larger meteoroids is semantic and, hence, it will be ignored in the following. Sources for interplanetary dust are larger meteoroids, comets, asteroids, the planets, and their satellites, and there is interstellar dust sweeping through the solar system. The zodiacal cloud is seen as a diffuse glow in the west after twilight and in the east before dawn; it appears to be wedge-shaped and lies along the ecliptic. It is widest in the parts near the Sun, and is caused by the reflection of sunlight from the myriads of interplanetary dust particles that are concentrated in the ecliptic plane.

I. INTRODUCTION

Zodiacal light is a prominent light phenomenon to the human eye in the morning and evening sky in nonpolluted areas (Fig. 1). Already in 1683, Giovanni Domenico Cassini presented the correct explanation of this phenomenon: it is sunlight scattered by dust particles orbiting the Sun. The relation to other "dusty" interplanetary phenomena, like comets, was soon suspected. Comets shed large amounts of dust, visible as dust tails, during their passage through the inner solar system. The genetic relation between meteors and comets was already known in the last century. Meteoroids became the link between interplanetary dust and the larger objects: meteorites, asteroids, and comets.

It is not the individual properties of dust particles that are important, but rather, the global properties of interplanetary particles that matter. Interplanetary dust is described not by one definitive set of parameters, but by distributions of parameters describing different populations of dust. Interplanetary dust can have different appearances in different regions of the solar system. It consists not only of refractory rocky or metallic material as in stony and iron meteorites, but also of carbonaceous material; dust in the outer solar system can even be ice particles.

Individual dust particles in interplanetary space have much shorter lifetimes than the age of the solar system. Several dynamic effects disperse the material in space and in size (generally going from bigger to smaller particles). Therefore, interplanetary dust must have contemporary sources, namely, bigger objects like meteoroids, comets, and asteroids.

Dust is often the synonym for dirt to human perception, which is difficult to quantify. This is also true for interplanetary dust. Astronomers who want to observe extra-solar system objects have to fight the foreground from the zodiacal light. Theoreticians who want to model interplanetary dust have the difficulty of representing these particles by simplified models, for example, a spherical particle of uniform composition and optical properties of a pure material. True interplanetary dust particles are far different from these simple models (Fig. 2).

Another aspect of dust is its danger to technical systems. A serious concern of the first spaceflights was the hazard from meteoroid impacts. Among the first instruments flown were simple dust detectors, many of which were unreliable devices that responded not only to impacts but also to mechanical, thermal, or electrical interferences. A dust belt around Earth was initially suggested, which was dismissed only years later when instruments had developed enough to suppress this noise by several orders of magnitude. Modern dust detectors are able to reliably measure dust impact rates from a single impact in one month up to a thousand impacts per second.

In the early days of spaceflight, measures were developed to protect spacecraft against the heavy bombardment by meteoroids. The bumper shield concept found its ultimate verification in the European Space Agency's *Giotto* mission to comet Halley. This spacecraft was designed to survive impacts of particles of up to 1 g mass at an impact speed of 70 km/s. These grains carry energies comparable to cannon balls that are 1000 times more massive. Heavy metal armor was prohibited since spacecraft are notoriously lightweight. The *Giotto* bumper shield combined a 1-mm-thick aluminum sheet positioned 23 cm in front of a 7-cm-thick lightweight composite rear shield. A dust particle that struck the thin front sheet was completely vaporized. The vapor cloud then expanded into the empty space between the two sheets and struck the rear shield, where its energy was absorbed by being distributed over a large area. In this way, the 2.7-m² front surface of the spacecraft was effectively protected by armor that weighed only 50 kg.

FIGURE 1   Cone of zodiacal light seen in the west one hour after sunset. The ecliptic plane is delineated by Venus at the top of the cone and the crescent moon just above the horizon. (Courtesy of C. Leinert.)

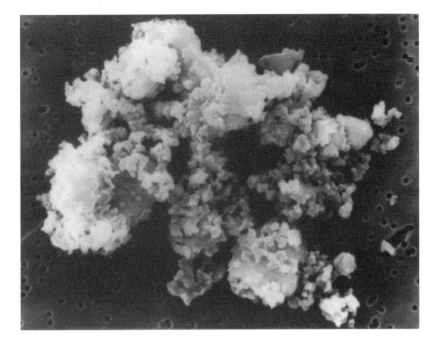

FIGURE 2 Interplanetary dust particle collected in the atmosphere. The size of the particle is a few micrometers in diameter.

Only recently has the dust hazard become important again, because of man-made space debris in Earth orbit. Each piece of equipment carried into space becomes, after disruption by an impact, the source of small projectiles, which endanger other satellites. Some estimates indicate that, in 50 years, the continuous increase in man-made space activity will lead to a runaway effect that will make the near-Earth space environment unhabitable to humans and equipment.

However, it is not this aspect of interplanetary dust that we are concerned with; the topic of this chapter is interplanetary dust as an exciting object of astrophysical research. Through its wide distribution over the solar system, interplanetary dust can tell stories about its parents (comets, asteroids, even interstellar matter) that otherwise are not easily readable. This view, however, requires that dust particles be traced back to their origins. To do this, we must understand its dynamics. Dust particles follow not only the gravitational pull of the Sun and the planets, but also feel the interplanetary magnetic field and the electromagnetic radiation that fills interplanetary space. In addition, they interact with the corpuscular emissions from the Sun and with other dust particles that they encounter in space, generally at high speeds. These collisions lead to erosion or to disruption of both particles, thus generating many more smaller particles. The dynamics of interplanetary dust cannot be described solely in position and velocity space, but also size or mass dimension must be added to this consideration.

II. OBSERVATIONS

Different methods are available to study interplanetary dust (Fig. 3). They are distinguished by the size or mass range of particles that can be studied. The earliest

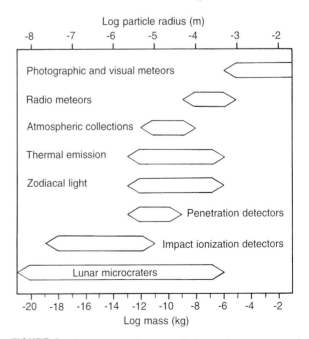

FIGURE 3 Comparison of meteoroid sizes and masses covered by different observational methods.

methods were ground-based zodiacal light and meteor observations. Fifty years ago, radar observations of meteor trails became available. With the onset of space-flight, *in situ* detection by space instrumentation provided new information on small interplanetary dust particles. Among the first reliable instruments were simple penetration detectors; modern impact ionization detectors allow not only the detection but also the chemical analysis of micrometeoroids. Deep space probes have identified micrometeroids in interplanetary space from 0.3 AU out to 18 AU from the Sun. Natural (e.g., lunar samples) and artificial surfaces exposed to micrometeoroid impacts have been returned from space and analyzed. High-flying aircraft have collected dust in the stratosphere that was identified as extraterrestrial material and that was analyzed by the most advanced microanalytic tools. Modern space-based infrared observatories now allow the observation of the thermal emission from interplanetary dust in the outer solar system.

A. METEORS

Looking up at the clear night sky, one can record about 10 faint meteors (or shooting stars in colloquial language) per hour. Once in a while, a brighter streak or trail of light or "fireball" will appear. Around the year 1800, the extraterrestrial nature of meteors was established when triangulation was used to deduce their height and speed. This technique is still used in modern meteor research by employing specifically equipped cameras and telescopes. About 50 years ago, radar techniques were also developed to observe faint meteor trains even during daylight.

Visible meteors result when centimeter-sized meteoroids enter the Earth's atmosphere at a speed greater than 10 km/s. At this speed, the energy of motion, which is converted to heat, is sufficient to totally vaporize the meteoroid. During the deceleration of the meteoroid in the atmosphere at about 100 km in height, the meteoroid will heat up and atoms from its outer surface will be ablated. These atoms, still traveling at high speed, collide with gas molecules from the ambient atmosphere, exciting and ionizing them. A luminating train of several kilometers in length follows the meteoroid. It is this ionized and luminating atmospheric gas that is visible and that scatters radar signals. The incident meteoroid may start to ablate at about 115 km in height, and complete evaporation is achieved at about 85 km. The meteoroid undergoes little deceleration before it is completely evaporated. This occurs after it has intercepted an air mass of about 1% of its own mass. From triangulation of the meteor train by ground stations (several cameras or a radar), the preatmospheric meteoroid orbit is obtained with high accuracy. Table I lists some classes of meteor orbits that were identified by these methods.

During the atmospheric entry of objects larger than several tens of kilograms or about 10 cm in diameter, a surface layer of several centimeters in thickness will burn away and the object will be decelerated. The rest that reaches Earth's surface is called a meteorite. Meteorites of 1 kg to several tons are sufficiently decelerated and fall on Earth with the interior little altered by atmospheric entry. These meteorites are the source of our knowledge about extraterrestrial material. However, even meter-sized meteoroids may not reach Earth's surface if its material does not have sufficient strength to withstand the enormous forces during de-

TABLE I
Classification of Meteor Orbits
Observed with Methods Sensitive to Different Meteoroid Masses[a]

Observational method:	Radar	Super Schmidt cameras	Small-camera meteors	Prairie Network fireballs
Typical mass range:	10^{-5}–10^{-3} g	10^{-3}–10^{-1} g	10^{-1}–10^3 g	10^2–10^6 g
Asteroidal		1%	5%	32%
Short-period cometary		69%	60%	46%
Long-period cometary	30%	31%	55%	22%

[a] Asteroidal meteors have orbits with medium eccentricities (0.6) and low inclinations (10°), and are of high densities (3.7 g/cm³); short-period cometlike meteors have oribts with higher eccentricities and their densities are lower (1–2 g/cm³); long-period cometlike meteors have almost parabolic orbits and random inclinations, and are of very low densities (0.2–0.6 g/cm³). After Z. Ceplecha (1977). *In* "Comets, Asteroids, Meteorites, Interrelations, Evolution and Origin" (A. H. Delsemme, ed.), pp. 143–152. Univ. Toledo Press, Toledo.

celeration. An outstanding example of this type is the Tunguska event of 1908, when an object of several tens of thousands of tons disintegrated about 10 km above the ground in Siberia. This caused massive destruction on the ground through detonation waves in the air, but left only fine dust of the original object. These observations indicate that there is material in interplanetary space of which we do not have any record in our meteorite collections. Of the 10,000 meteorites that we do have in our collections, only four falls have been observed with adequate equipment to derive their interplanetary orbits. These meteorites are of the ordinary chondrite type and have aphelia in the asteroid belt, directly indicating a genetic relationship between these meteorites and asteroids. Less fragile, big meteoroids of tens of meters in size or bigger will hit the ground at close to original speed and produce an impact crater. [*See* METEORITES; NEAR-EARTH ASTEROIDS; PLANETARY IMPACTS.]

Much of the ablated material from a meteor will condense again into small droplets, which will cool down and form cosmic spherules that subsequently rain down to Earth. These cosmic spherules can be found and identified in abundance in deep-sea sediments and on the large ice masses of Greenland, the Arctic, and Antarctica. An average of 40 tons of extraterrestrial material per day in the form of fine dust falls onto the surface of Earth.

FIGURE 4 An unusually strong meteor shower (Leonid) was observed on 17 November 1966. The meteor trails seem to radiate from the constellation Leo.

At certain times, meteor showers can be observed at a rate that is a hundred (and more) times higher than the average sporadic meteor rate. Figure 4 shows several meteors in a photograph of the night sky taken on Nobember 17, 1966. The visible rate was about one meteor per second. Because all of these meteoroids travel on parallel trajectories, they seem for an observer to arrive from a common point in the sky (the radiant), which in this case lies in the constellation Leo. Therefore, this meteor shower is called the Leonid shower. A common feature of all meteor showers is that the meteors in one shower have a common radiant; actually, that is how they are defined as members of a meteor stream. Many meteor showers reoccur each year on the same days at about the same rate (Table II).

The explanation for the yearly occurrence of meteor showers is that all meteoroids in one stream closely follow a common elliptic orbit around the Sun but are spread out all along the orbit. Each year when the Earth crosses this orbit on the same day, some meteoroids of the stream hit the atmosphere and cause the shower. If the meteoroids are not spread out over the full orbit but are concentrated in one segment, a meteor shower is observed only when Earth crosses this segment. This meteor stream will show an intensity variation with the orbit period of the meteoroids around the Sun.

Many meteor streams have orbits similar to those of known comets (cf. Table II). It is a generally accepted view that meteor streams are derived from comets. Centimeter- to decimeter-sized particles that are emitted from comets are not visible in the normal comet tail but form so-called comet trails along a short segment of the comet's orbit. Their different speeds will slowly spread the particles out over the full orbit. Gravitational interactions with planets will scatter meteoroids out of the stream and they will become part of the sporadic background cloud of meteoroids. The fact that some meteor showers display strong variations of their intensities indicates that they are young streams that are still concentrated in a small segment of the parent's orbit. The parent comet of the Leonids, the periodic comet Tempel-Tuttle, has the same periodicity of 33.3 years. The parent object of one of the strongest yearly meteor showers, the Geminids, is Phaeton, which had been previously classified as an asteroid because it shows no cometary activity. However, its association with a meteor stream indicates that it is an inactive, dead comet that at some time in the past emitted large quantities of meteoroids. [*See* PHYSIC AND CHEMISTRY OF COMETS; COMETARY DYNAMICS.]

TABLE II

**Major Meteor Showers, Date of Shower Maximum, Radiant in Celestial Coordinates
(Right Ascension, RA, and Declination, DEC, in Degrees), Geocentric Speed (km/s),
Maximum Hourly Rate, and Parent Objects (If Known, Short-Period Comets Are Indicated by P/)[a]**

Name	Date	Radiant RA	Radiant DEC	Speed	Rate	Parent object
Quadrantids	Jan. 3	230	+49	42	140	
April Lyrids	Apr. 22	271	+34	48	10	Comet 1861 I Thatcher
Eta Aquarids	May 3	336	−2	66	30	P/Halley
June Lyrids	June 16	278	+35	31	10	
S. Delta Aquarids	July 29	333	−17	41	30	
Alpha Capricornids	July 30	307	−10	23	30	P/Honda-Mrkos-Pajdusakova
S. Iota Aquarids	Aug. 5	333	−15	34	15	
N. Delta Aquarids	Aug. 12	339	−5	42	20	
Perseids	Aug. 12	46	+57	59	400 (1993)	P/Swift-Tuttle
N. Iota Aquarids	Aug. 20	327	−6	31	15	
Aurigids	Sept. 1	84	+42	66	30	Comet 1911 II Kiess
Giacobinids	Oct. 9	262	+54	20	10	P/Giacobini-Zinner
Orionids	Oct. 21	95	+16	66	30	P/Halley
Taurids	Nov. 3	51	+14	27	10	P/Encke
Taurids	Nov. 13	58	+22	29	10	P/Encke
Leonids	Nov. 17	152	+22	71	3000 (1966)	P/Tempel-Tuttle
Geminids	Dec. 14	112	+33	34	70	Phaeton
Ursids	Dec. 22	217	+76	33	20	P/Tuttle

After A. F. Cook (1973). *In* "Evolutionary and Physical Properties of Meteoroids" (C. L. Hemenway, P. M. Millman, and A. F. Cook, eds.), pp. 183–192. NASA SP-319, National Aeronautics and Space Administration, Washington, D.C.

B. INTERPLANETARY DUST PARTICLES

There is another "window" through which extraterrestrial material reaches the surface in a more or less undisturbed state. Small interplanetary dust particles (IDPs) of a few to 50 μm in diameter are decelerated in the tenuous atmosphere above 100 km. At this height, the deceleration is so gentle that the grains will not reach the temperature of substantial evaporation (T < 800°C), especially since these small particles have a high surface area-to-mass ratio that enables them to effectively radiate away excessive heat. These dust particles subsequently sediment through the atmosphere and become accessible to collection and scientific examination. The abbreviation IDP (or "Brownlee particle" after Don Brownlee, who first reliably identified their extraterrestrial nature) is often used for such extraterrestrial particles that are collected in Earth's atmosphere.

Early attempts to collect IDPs by rockets above about 60 km were not successful because of the very low influx of micrometeoroids into the atmosphere and the short residence times of IDPs at these altitudes. More successful were airplane collections in the stratosphere at or above 20 km. At this height, the concentration of 10-μm-diameter particles is about 10^6 times higher than in space and terrestrial contamination of this sized particles is still low. Only micron- (1 millionth of a meter) and submicron-sized terrestrial particles (e.g., from volcanic eruptions) can reach these altitudes in significant amounts. Another type of interference is caused by man-made contamination: about 90% of all collected particles in the 3- to 8-μm size range are aluminum oxide spheres, which are products of solid fuel rocket exhausts. Because of this overwhelming contamination problem for small particles, the lower size limit of IDPs collected by airplanes is a few micrometers in diameter.

The typical size range of IDPs is about 5 to 50 μm in diameter. The upper limit is caused by the low abundance of bigger particles (e.g., only about 10 IDPs of more than 10 μm in size are collected during one hour of aircraft flight). Since 1981, IDP collection by airplanes has been routinely performed by NASA using high-flying aircraft like the ER-2 plane (Fig. 5), which can cruise at 20 km for many hours. It carries a 300-cm^2 flat plate dust collector below its wings, which sweeps huge amounts of air because of the high speed of the airplane. After deployment in the stratosphere, dust particles stick to the collector surface, which is coated with silicone oil. After several hours of exposure, the collector is retracted into a sealed storage container and returned to the laboratory. There, all particles are removed from the collector plate, the silicone oil is washed off, and the particles are preliminarily examined.

After microscopic inspection, the collected particles are chemically characterized and catalogued. From cosmic dust catalogues issued by the NASA Johnson Space Center in Houston (Fig. 6 shows a sample of IDPs from such a catalogue), individual IDPs can be ordered for further scientific investigation. The widest variety of microanalytic tools has been applied to examine and analyze IDPs. Some of the methods have been specifically developed or modified for optimum use in IDP research.

The first goal of IDP research was the proof that these particles indeed are of extraterrestrial nature. The first step was to determine their chemical composition, which often resembles that of chondritic meteorites. The ratios of the elements magnesium, silicon, sulfur, iron, and nickel are characteristic of the terrestrial or extraterrestrial nature of the collected particles. The final proof for the extraterrestrial origin of some classes of collected particles came after traces of solar-wind helium and tracks from the exposure to high-energetic ions in space had been identified in the minerals.

According to their elemental composition, IDPs come in three major types: chondritic, 60% (cf. Table III); iron–sulfur–nickel, 30%; and mafic silicates (iron–magnesium-rich silicates, i.e., olivine and pyroxene), 10%. Most chondritic IDPs are porous aggregates, but some smooth chondritic particles are found as well. Chondritic aggregates may contain varying amounts of carbonaceous material of unspecified composition. Table III shows a significant enrichment in volatile (low condensation temperature) elements if compared to C1 chondrites. This observation is being used to support the argument that these particles consist of some very primitive solar system material that had never seen temperatures above about 500°C, as is the case for cometary material. This and some compositional similarity with comets argue for a genetic relation between comets and IDPs. Although recently some volatile components have been suspected to be terrestrial contaminants, IDPs are an important source of extraterrestrial material that at the present time is not available in any other way.

A remarkable feature of IDPs is their large variability in isotopic composition. Extreme isotopic anomalies have been found in some IDPs (e.g., factors of 1000 of the solar hydrogen isotope ratio). Under typical solar system conditions, only fractions of a percent

FIGURE 5 The high-flying NASA ER-2 airplane is used to collect interplanetary dust particles sedimenting through the atmosphere. The dust collector is mounted below the wings. (Courtesy of NASA.)

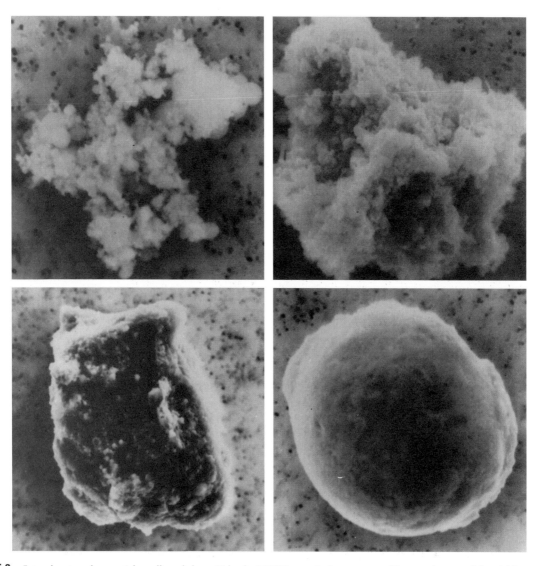

FIGURE 6 Interplanetary dust particles collected above 20 km by NASA's cosmic dust program. Three grains are of chondritic composition and of various degrees of compactness and there is one Fe–S–Ni sphere (lower right). The widths of the photographs are 15 μm (first and third photos, clockwise from upper left) and 30 μm (second and fourth photos). (Courtesy of NASA.)

of isotopic variations can occur. These huge isotopic variations indicate that some grains are not homogenized with other solar system material but have preserved much of their presolar character.

C. ZODIACAL LIGHT

The wedge-shaped appearance of the zodiacal light (see Fig. 1) demonstrates the concentration of zodiacal light in the ecliptic plane. For an observer on Earth, the zodiacal light extends in the ecliptic all the way around to the antisolar direction, however, at strongly reduced intensities (Fig. 7). In the direction opposite

to the Sun, this light forms a hazy area of a few degrees in dimension known as the gegenschein, or counterglow. If seen from the outside, the zodiacal dust cloud would have a flattened, lenticular shape that extends along the ecliptic plane about seven times farther from the Sun than perpendicular to the ecliptic plane.

The brightness of zodiacal light arrives from the light scattered by a huge number of particles in the direction of observation. The scattering angle, that is, the Sun–particle–observer angle, varies systematically along the line of sight. This angle is biggest closest to the observer and can approach 180°. For one particle, the scattered light intensity is a strong function of the scattering angle. For particles larger than the wave-

TABLE III
Average Elemental Composition (All Major and Selected Minor and Trace Elements) of Several Chondritic IDPs Is Compared with C1 Chondrite Composition[a]

Element	C1	IDP	Variation	T_c
Mg	1,071,000	0.9	0.6–1.1	1067
Si	1,000,000	1.2	0.8–1.7	1311
Fe	900,000	1	1	1336
S	515,000	0.8	0.6–1.1	648
Al	84,900	1.4	0.8–2.3	1650
Ca	61,100	0.4	0.3–0.6	1518
Ni	49,300	1.3	1.0–1.7	1354
Cr	13,500	1.1	0.9–1.4	1277
Mn	9,550	1.1	0.8–1.6	1190
Cl	5,240	3.6	2.8–4.6	863
K	3,770	2.2	2.0–2.5	1000
Ti	2,400	1.5	1.3–1.7	1549
Co	2,250	1.9	1.2–2.9	1351
Zn	1,260	1.4	1.1–1.8	660
Cu	522	2.8	1.9–4.2	1037
Ge	119	2.3	1.6–3.4	825
Se	62	2.2	1.6–3.0	684
Ga	38	2.9	2.1–3.9	918
Br	12	34	23–50	690

[a] The IDP abundances are normalized to iron (Fe) and to C1. C1 abundance is normalized to Si = 1,000,000 condensation temperatures T_c (°C). From E. K. Jessberger *et al.* (1992). *Earth Planet. Sci. Lett.* **112**, 91–99.

length of the scattered light, this scattering function is strongly peaked in the forward direction (scattering angle = 180°). For particles much smaller than the wavelength, the scattering function is more uniform. Variable particle structure and composition affect the scattering function as well. Therefore, the observed zodiacal brightness is a mean value, averaged over all sizes, compositions, and structures of particles along the line of sight. The increased brightness toward the Sun is in part the effect of the enhancement in the scattering function.

At visible wavelengths the spectrum of the zodiacal light closely follows the spectrum of the Sun. A slight reddening (i.e., the ratio of red and blue intensities is larger for zodiacal light than for the Sun) indicates that the majority of particles are larger than the mean

visible wavelength of 0.54 μm. In fact, most of the zodiacal light is scattered by 10- to 100-μm-sized particles. Therefore, the dust seen by zodiacal light is only a subset of the interplanetary dust cloud. Submicron- and micron-sized particles, as well as millimeter and bigger particles, are not well represented by the zodiacal light.

Above about 1 μm in wavelength the intensities in the solar spectrum rapidly decrease. The zodiacal light spectrum follows this decrease until about 5 μm, above which the thermal emission of the dust particles prevails. The maximum of the thermal infrared emission from the zodiacal dust cloud lies between 10 and 20 μm. From the thermal emission observed by the *IRAS* and *COBE* satellites, an average dust temperature at 1 AU distance from the Sun between 20°C and 0°C has been derived. [*See* INFRARED VIEWS OF THE SOLAR SYSTEM FROM SPACE.]

The large-scale distribution of the zodiacal dust cloud is determined from zodiacal light measurements. However, because of the strong influence of the average scattering function (the detailed function is not known), observations from Earth are not sufficient to derive uniquely the spatial density of interplanetary dust. The *Helios* spacecraft went inward from Earth orbit to 0.3 AU on an elliptic orbit, carrying instruments to characterize the interplanetary environs. Measurements of the zodiacal light photometer are

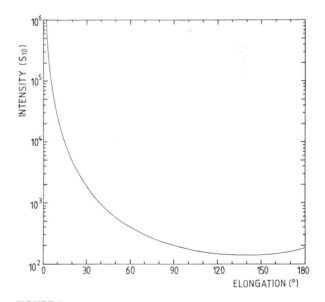

FIGURE 7 Zodiacal light brightness along the ecliptic. The unit S_{10} is the equivalent of one 10th magnitude solar-type star per square degree. At the ecliptic poles the brightness is about $60S_{10}$. [After C. Leinert and E. Grün (1990). *In* "Physics of the Inner Heliosphere" (R. Schwenn and E. Marsch, eds.), pp. 207–275. Springer-Verlag, Berlin. Copyright 1990 Springer-Verlag.]

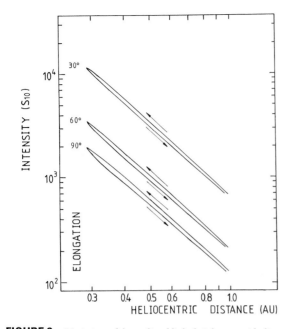

FIGURE 8 Variation of the zodiacal light brightness with distance from the Sun as observed by *Helios*. The profiles are similar for all elongations (angle between line of sight and Sun direction), except for their absolute brightness. The split of the intensities between the inbound (upper trace) and outbound legs (lower trace) of the orbit is due to the small inclination of the plane of symmetry with respect to the orbit plane of *Helios*. [From C. Leinert and E. Grün (1990). *In* "Physics of the Inner Heliosphere" (R. Schwenn and E. Marsch, eds.), pp. 207–275. Springer-Verlag, Berlin. Copyright 1990 Springer-Verlag.]

red observations indicate that this inclination changes somewhat outside the Earth orbit.

Zodiacal light can be traced clearly into the solar corona. However, most of this dust is foreground dust close to the observer because of a favorable scattering function. Nevertheless, the vicinity of the Sun is of considerable interest for zodiacal light measurements because it is expected that close to the Sun the temperature of the dust rises and that inside a few solar radii distance all dust should sublimate. Unfortunately, no clear picture has arisen from such observations so far, since no consistent results on the edge of the dust-free zone have been obtained by different investigators, which may indicate a time-variable inner edge of the zodiacal cloud.

Optical and infrared observations of other extraterrestrial dusty phenomena have provided important insights into the zodiacal complex. Cometary dust is considered to be an important source of the zodiacal cloud. The study of circumplanetary dust and rings has stimulated much research in the dynamics of dust clouds. Interstellar dust is believed to be the ultimate source of all refractory material in the solar system. Circumstellar dust clouds, like the one around β-Pictoris, are "zodiacal clouds" in their own right, the study of which may eventually provide information on extrasolar planetary systems. [*See* EXTRA-SOLAR PLANETS: SEARCHING FOR OTHER PLANETARY SYSTEMS.]

displayed in Fig. 8. The intensities observed in the same direction (as specified by elongation) but at different distances to the Sun show a steep gradient going to the Sun. This gradient can be directly converted to the radial dependence of the spatial density (number of particles per volume element) of zodiacal dust particles. Though the intensity increases from 1 to 0.3 AU by a factor of 16, the spatial density of dust needs only to increase by a factor of 5. The radial dependence of the number density is slightly steeper than the inverse distance dependence.

The split seen in the radial intensity profiles in Fig. 8 is a consequence of a slight inclination of the symmetry plane of zodiacal light with respect to the ecliptic plane. While *Helios* traveled in the ecliptic plane, one zodiacal light photometer was directed 16° to the north. On the inbound leg (upper trace), the symmetry plane was above *Helios*, so the instrument recorded light from more particles. On the outbound leg, *Helios* was above the symmetry plane and hence a lower intensity was observed. The inclination of this symmetry plane is only 3° inside the Earth orbit. Infra-

D. LUNAR MICROCRATERS AND THE NEAR-EARTH DUST ENVIRONMENT

The size distribution of interplanetary dust particles is represented by the lunar microcrater record. Microcraters on lunar rocks have been found ranging from 0.02 μm to millimeters in diameter (Fig. 9). Craters on fragile materials like rock display a central crater that is surrounded by a spallation zone from which large chips have been removed. The ratio of the spallation zone diameter to the central pit diameter is quite variable. Micron-sized lunar craters show a narrow spallation zone, if at all, whereas the spallation zone of millimeter-sized craters cover two-thirds of the crater diameter. Still bigger craters are hard to identify because they consist mostly of an irregular spallation zone.

Laboratory simulations of high-velocity impacts on lunarlike materials have been performed to calibrate crater sizes with projectile sizes and impact speeds. Submicron- to centimeter-sized projectiles have been used with speeds above several kilometers per second. The typical impact speed of interplanetary meteoroids on the Moon is about 20 km/s. For the low-mass

FIGURE 9 Microcraters on the glassy surface of a lunar sample. Bright spallation zones surround circular central pits.

end, electrostatic dust accelerators were used that reach speeds up to 100 km/s. The high-mass projectiles were accelerated with light-gas guns, which reached speeds up to about 10 km/s. For the intermediate-mass range, plasma drag accelerators reached impact speeds of 20 km/s. The crater diameter to projectile diameter varies from 2 for the smallest microcrater to about 10 for centimeter-sized projectiles.

The difficulty in deriving the impact rate from a crater count on the Moon is the generally unknown exposure geometry (e.g., shielding by other rocks) and exposure time of any surface on a rock. Therefore, the crater size or meteoroid distribution has to be normalized with the help of an impact rate or meteoroid flux measurement obtained by other means. *In situ* detectors or recent analyses of impact plates that were exposed to the meteoroid flux in a controlled way provide this flux calibration (Fig. 10). One finding of such comparisons was that the flux of micron-sized particles on the Moon is larger by two orders of magnitude compared to the interplanetary dust flux. This is due to secondary high-speed ejecta particles that are emitted from larger primary meteoroid impacts on the Moon.

From the size distribution of meteoroids, it is possible to determine which particles are most effective in scattering sunlight and hence producing the zodiacal light. It has been found that the main contribution comes from particles ranging from 10 to 100 μm in radius. The mass distribution of meteoroids peaks at 10^{-5} g. The total mass density of interplanetary dust at 1 AU is 10^{-16} g/m³ and the total mass of the zodiacal

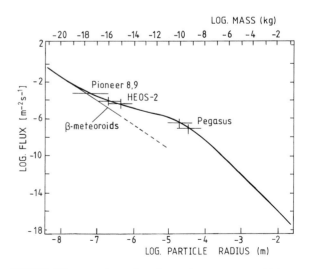

FIGURE 10 Cumulative flux of interplanetary meteoroids on a spinning flat plate at 1 AU from the Sun. The solid line has been derived from lunar microcrater statistics and it is compared with satellite and spaceprobe measurements.

FIGURE 11   Release of the *Long Duration Exposure Facility* (*LDEF*) from the space shuttle in 1984. Various trays containing collectors and materials have been exposed to the space environment for 6 years. (Courtesy of NASA.)

cloud inside Earth's orbit is 10^{16} to 10^{17} kg, which corresponds to the mass of a single object (comet or asteroid) of about 20 km in diameter.

In 1984, NASA released the *Long Duration Exposure Facility* (*LDEF*) (Fig. 11) into near-Earth space at about 450 km to study the effects on materials during prolonged exposure to the space environment. Of primary interest was the effect of meteoroids. Six years after launch *LDEF* was retrieved by the space shuttle and brought back to Earth. Near Earth, the meteoroid flux is about a factor of two higher than in deep space because of gravitational concentration by Earth. Micron-sized natural meteoroids are outnumbered (by a factor of three) by man-made space debris. Craters produced by space debris particles are identified by chemical analyses of residues in the craters. Residues have been found from space materials and signs of human activities in space, such as paint flakes, plastics, aluminum, titanium, and human excretions.

E. *IN SITU* DUST MEASUREMENTS

Complementary to ground-based and astronomical dust observations are *in situ* observations by dust im-

pact detectors onboard interplanetary spacecraft. *In situ* measurements of interplanetary dust have been performed in the heliocentric distance range from 0.3 AU out to 18 AU (Table IV). Included in the list of interplanetary dust detectors are the two Earth-orbiting satellites, the European *HEOS 2* and the Japanese *Hiten* satellites, because they performed significant portions of their measurements in interplanetary space outside the near-Earth's environment and provided important insights into the interplanetary dust flux.

Two types of impact detectors were mainly used for interplanetary dust measurements: penetration detectors (where the micrometeoroid has to penetrate 25- to 50-μm-thick metal films) with detection thresholds of 10^{-9} and 10^{-8} g and impact ionization detectors, which detect the charge released upon impact, with detection thresholds of 10^{-16} to 10^{-13} g. These detection thresholds refer to a typical impact speed of 20 km/s. At lower impact speeds, the minimum detectable particle mass is bigger and vice versa.

Figure 12 shows a photo of the dust detector flown on the *Galileo* and *Ulysses* spacecraft. The detector has a sensitive area of 0.1 m² and is based on the impact

TABLE IV

In Situ Dust Detectors in Interplanetary Space: Distance of Operation, Mass Sensitivity, and Sensitive Area

Spacecraft	Launch year	Distances (AU)	Mass threshold (g)	Area (m²)
Pioneer 8	1967	0.97–1.09	2×10^{-13}	0.0094
Pioneer 9	1968	0.75–0.99	2×10^{-13}	0.0074
HEOS 2	1972	1	2×10^{-16}	0.01
Pioneer 10	1972	1–18	2×10^{-9}	0.26
Pioneer 11	1973	1–10	10^{-8}	0.26
Helios 1/2	1974/1976	0.3–1	10^{-14}	0.012
Galileo	1989	0.7–5.3	10^{-15}	0.1
Hiten	1990	1	10^{-15}	0.01
Ulysses	1990	1.0–5.4	10^{-15}	0.1
Cassini	1997	1–10	10^{-15}	0.1

ionization effect: a dust particle that enters the detector and hits the gold target at a speed above 1 km/s will produce an impact crater and free electrons and ions are generated. At impact speeds in excess of 15 km/s, part or all of the projectile's material will vaporize. Because of the high temperature at the impact site, some electrons are stripped off atoms and molecules and the generated vapor is partially ionized. These

FIGURE 12 The *Galileo* micrometeoroid detector (sensor and electronics box); the detector onboard the *Ulysses* spacecraft is of the same kind. The cylindrical sensor has a diameter of 43 cm. The bottom of the sensor contains the hemispherical impact target; in the center are charge-collecting electrodes. The detector records impacts of submicron- and micron-sized dust particles above 1 km/s impact speed.

ions and electrons are separated in an electric field within the detector and collected by electrodes. Coincident electric pulses on these electrodes signal the impact of a high-velocity dust particle. The strength and the waveform of the signal are measures of the mass and speed of the impacting particle. Combinations of impact ionization detectors with mass spectrometers give the chemical composition of the produced ions. Detectors of this latter type have been flown on the *Helios* spacecraft, the *Giotto* and *VEGA* missions to comet Halley, and the *Cassini* spacecraft to Saturn. Electrostatic dust accelerators are used to calibrate these detectors with micron- and submicron-sized projectiles at impact speeds of up to about 100 km/s.

The radial profile of the dust flux in the inner solar system between 1 and 0.3 AU from the Sun has been determined by the *Helios 1* and *2* spaceprobes. Three dynamically different interplanetary dust populations have been identified in the inner solar system. First, particles that orbit the Sun in low-eccentric orbits had already been detected by the *Pioneer 8/9* and *HEOS 2* dust experiments. They relate to particles probably originating in the asteroid belt and spiraling under the Poynting–Robertson effect toward the Sun. Second, there are particles on highly eccentric orbits that have, in addition, large semimajor axes and that probably derive from a cometary source. Third, the *Pioneer 8/9* dust experiments detected a significant flux of small particles from approximately the solar direction (Fig. 13). Existence of these particles was recently confirmed by *Hiten* measurements. With lunar microcrater data and *Pioneer 8/9* and *HEOS 2* measurements it was possible to determine the flux of small ($<10^{-12}$ g) interplanetary particles at 1 AU (cf. Fig. 10).

In the outer solar system the penetration detectors onboard *Pioneers 10* and *11* measured the flux of 10-μm-sized interplanetary dust particles. The flux decreased from 1 AU going outward. No sign of a flux enhancement in the asteroid belt was detected. Except for a strong increase of the flux near Jupiter, outside about 3 AU *Pioneer 10* recorded a flat flux profile (Fig. 14), which indicates a constant spatial density of dust in the outer solar system. The *Pioneer 11* spacecraft traversed the region between 4 and 5 AU three times and these data are best explained by meteroids moving on highly eccentric or highly inclined orbits.

Recently, the *Galileo* and *Ulysses* spacecraft carried dust detectors through interplanetary space between the orbits of Venus and Jupiter and above the ecliptic plane (Fig. 15). Swingbys of Venus and Earth (twice) were necessary to give the heavy *Galileo* spacecraft (mass of 2700 kg) the necessary boost to bring it to

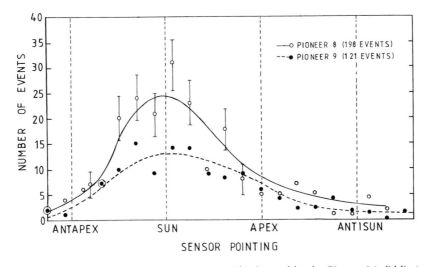

FIGURE 13 Angular distribution of small impact events (beta-meteoroids) detected by the *Pioneer 8* (solid line) and *Pioneer 9* (dashed line) dust detectors. While the spacecraft spins, the detector scans the ecliptic plane; "Sun" stands for the detector pointing toward the Sun, "Apex" is in the direction of the spacecraft orbital motion, and "Antisun" and "Antapex" are the opposite directions. [After O. E. Berg and E. Grün (1973). *Space Res.* **13**, 1047–1055.]

Jupiter within 6 years of flight time, where it became the first man-made satellite of this giant planet. The *Ulysses* spacecraft, being much lighter (mass of 375 kg), made the trajectory to Jupiter within 1.5 years. By a swingby of Jupiter the *Ulysses* spacecraft was brought into an orbit almost perpendicular to the ecliptic plane that carried it under the south pole, through the ecliptic plane, and over the north pole of the Sun.

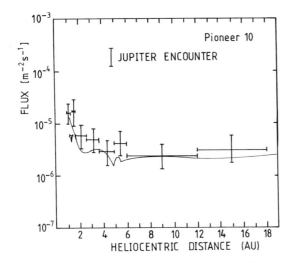

FIGURE 14 Flux of meteoroids with masses $> 8 \times 10^{-10}$ kg (about 10 μm in size) in the outer solar system measured by the *Pioneer 10* penetration detector. At 18 AU from the Sun, the instrument quit operation. The measurements are in agreement with a model of constant spatial dust density in the outer planetary system. [From D. H. Humes (1980). *J. Geophys. Res.* **85**, 5841–5852.]

Dust measurements by the *Galileo* spacecraft obtained in the ecliptic plane between Venus's orbit and the Asteroid Belt are displayed in Fig. 16. The dust impact rate was generally higher closer to the Sun than it was farther away. After all planetary flybys (Venus, V, and two times Earth, E1 and E2) the spacecraft moved away from the Sun. At these times the impact rate was more than an order of magnitude higher than before the flyby when the spacecraft moved toward the Sun. This observation is explained by the fact that interplanetary dust inside the asteroid belt orbits the Sun on low inclination ($<30°$) and in low-eccentric-bound orbits. Thus, the detector that looks all the time away from the Sun detects more dust impacts when the spacecraft moves in the same direction (outward) than in the opposite case when the spacecraft moves inward. The spatial dust density follows roughly an inverse radial distance dependence. Close passages of the asteroids Gaspra (G) and Ida (I) did not exhibit increased dust impact rates. Also, an interstellar dust population did not contribute significantly to this data set.

Most of the particles recorded outside 3 AU from the Sun have been identified as interstellar particles: their arrival direction coincides with that of interstellar gas and their speed is so high (>26 km/s) that they are not bound by the solar gravitational field. The mass flux in these particles is comparable to the expected mass flux of interstellar particles in the local interstellar medium. During the out-of-ecliptic portion of *Ulysses'* orbit, it detected a roughly constant dust flux arriving only from one side of *Ulysses'* orbit plane, which is

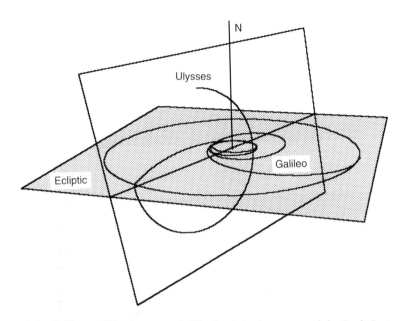

FIGURE 15 Trajectories of the *Galileo* and *Ulysses* spacecraft. The Sun is in the center, and the Earth, Jupiter, and *Galileo* trajectories are in the ecliptic plane (shaded). The initial trajectory of *Ulysses* from Earth to Jupiter was also in the ecliptic plane. Subsequently, *Ulysses* was thrown onto an orbit plane inclined 79° to the ecliptic. This orbit carried *Ulysses* below the south pole of the Sun through the ecliptic plane and above the north pole (N).

the signature of interstellar dust particles passing through the planetary system on hyperbolic orbits. This flux also persisted over the poles of the Sun. Figure 17 shows the measurements during the South-to-North traverse through the ecliptic plane. Above 40° ecliptic latitude, that is, outside 1.5 AU from the Sun, interstellar dust dominates. Closer to the ecliptic plane, interplanetary dust on bound orbits is the dominant dust population.

Inside a distance of 2 AU from Jupiter, both *Ulysses* and *Galileo* spacecraft detected an unexpected phenomenon: swarms of submicron-sized dust particles arrived from the direction of Jupiter. Figure 18 shows the strongly time-variable dust flux observed by *Ulys-*

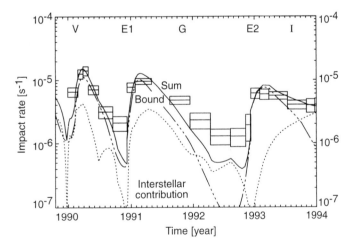

FIGURE 16 *Galileo* impact rate during the first 4 years of the mission. The boxes indicate the mean impact rates and their uncertainties. V, E1, E2, G, and I are flybys of Venus (10, February 1990), Earth (8, December 1990 and 8, December 1992), and the asteroids Gaspra (29, October 1991) and Ida (28, August 1993). At the end of 1993, *Galileo* had reached a distance of 3.7 AU. Model calculation of the impact rate during the first 4 years of the *Galileo* mission is shown. Both individual contributions from interplanetary dust on *bound* orbits and *interstellar* dust on hyperbolic trajectories and the sum of both (solid line) are displayed.

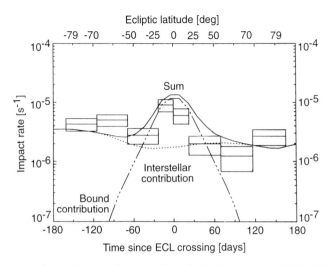

FIGURE 17 *Ulysses* dust impact rate observed around the time of its ecliptic plane crossing (ECL). ECL occurred on 12 March 1995 at a distance of 1.3 AU from the Sun. The boxes indicate the mean impact rates and their uncertainties. The top scale gives the spacecraft latitude. Model calculation of the impact rate during *Ulysses'* south to north traverse through the ecliptic plane is shown by the lines. Contributions from interplanetary dust on *bound* orbits and *interstellar* dust on hyperbolic trajectories and the sum of both are displayed. From these measurements it is concluded that interstellar dust is not depleted down to a distance of 1.3 AU from the Sun.

ses during its flyby of Jupiter. About one month after its closest approach to Jupiter, *Ulysses* encountered the most intense dust burst at a distance to Jupiter of about 40 million km. For about 10 hours the impact rate of submicron-sized particles increased by a factor of 1000 above the background rate. The similarity of the impact signals and the sensor pointing directions indicated that the particles in the burst were moving in collimated streams at speeds of several hundred kilometers per second. Even stronger and longer-lasting dust streams were observed in 1995 by the *Galileo* dust detector during its approach to Jupiter. Dust measurements inside the Jovian magnetosphere showed a modulation of the small particle impact rate with a period of 10 hours, which is the rotation period of Jupiter and its magnetic field. Positively charged dust particles in the 10-nm size range couple to the magnetic field and are thrown out of Jupiter's magnetosphere in the form of a warped dust sheet. Potential sources of these dust particles are the volcanoes on Io, Jupiter's ring, or any other dust electromagnetically trapped in the magnetosphere of Jupiter. Some of the bigger dust impacts that were observed close to Jupiter (at a distance of several hundreds of thousands of kilometers) are dust grains that are in orbit about Jupiter. The remainder of big particles are interplanetary and interstellar meteoroids passing through the Jovian system. [*See* PLANETARY RINGS.]

In interplanetary space, the highest dust fluxes have been observed near comets. So far three comets have

been visited by spacecraft: comets P/Giacobini-Zinner, P/Halley, and P/Grigg-Skjellerup. Specially optimized dust analyzers have been used to study comet Halley's dust. Chemical analyses showed that besides the expected dust particles consisting of silicates, a large fraction of cometary dust consists of carbonaceous materials. Extreme isotopic anomalies have been found to exist in some of these particles. Similar compositions are expected for interplanetary dust.

III. DYNAMICS AND EVOLUTION

A. GRAVITY AND KEPLERIAN ORBITS

In the planetary system, solar gravitation determines the orbits (Keplerian orbits) of all bodies larger than dust particles for which other forces become important as well. But even for dust, gravity is an important factor. Near planets, planetary gravitation takes over. However, the basic orbit characteristics remain the same. Two types of orbits are possible: bound and unbound orbits around the central body. Circular and elliptical orbits are bound to the Sun; the planets exert only small disturbances to these orbits. Planets, asteroids, and comets move on such orbits. Objects on unbound orbits will eventually leave the solar system.

Typically, interstellar dust particles move on unbound, hyperbolic orbits through the solar system. Similarly, interplanetary particles are unbound to any planetary system and traverse it on hyperbolic orbits with respect to the planet. [*See* SOLAR SYSTEM DYNAMICS.]

A Keplerian orbit is a conic section that is characterized by its semimajor axis a, eccentricity e, and inclination i. The Sun (or a planet) is in one focus. The perihelion distance (closest to the Sun) is given by $q = a(1 - e)$. Circular orbits have eccentricity $e = 0$, elliptical orbits have $0 < e < 1$, and hyperbolic orbits have $e > 1$ and a is taken negative. The aphelion distances (farthest from the Sun) are finite only for circular and elliptical orbits. The inclination is the angle between the orbit plane and the ecliptic, that is, the orbit plane of Earth.

Dust particles in interplanetary space move on very different orbits, and several classes of orbits with similar characteristics have been identified. One class of meteoroids moves on orbits that are similar to those of asteroids, which peak in the asteroid belt (Fig. 19). Another class of orbits that represents the majority of zodiacal light particles has a strong concentration toward the Sun. Both orbit populations have low to intermediate eccentricities ($0 \le e < 0.6$) and low inclinations ($i < 40°$). These asteroidal and zodiacal core populations satisfactorily describe meteors, the lunar crater size distribution, and a major portion of zodiacal light observations. Also, *Galileo* measurements inside 2 AU (See Fig. 16) are well represented by the core population. [*See* ASTEROIDS.]

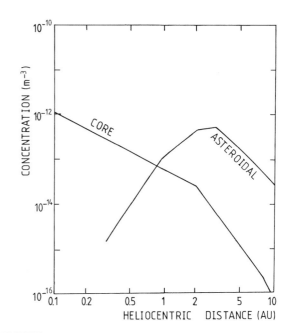

FIGURE 19 Radial dependence of meteoroid concentrations for two main populations in interplanetary space according to Divine (1993). The values given refer to particles with masses $> 10^{-4}$ g. The zodiacal core population comprises particles of all sizes, whereas the asteroidal population comprises only big ($>10^{-6}$ g) particles.

B. RADIATION PRESSURE AND THE POYNTING–ROBERTSON EFFECT

Electromagnetic radiation from the Sun (most intensity is in the visible wavelength range, $\lambda_{max} = 0.5 \ \mu$m) being absorbed, scattered, or diffracted by any particulate exerts pressure on a dust particle. Since solar radiation is directed outward from the Sun, radiation pressure is also directed away from the Sun and depends on the intensity of the radiation. Thus gravitational attraction is reduced by the radiation pressure force. Both radiation pressure and gravitational forces have an inverse square dependence on the distance from the Sun. Radiation pressure depends on the cross section of the particle and gravity on the mass, therefore, for the same particle, the ratio β of radiation pressure, F_R, over gravitational force, F_G, is constant everywhere in the solar system and depends only on particle properties: $\beta = F_R/F_G \sim Q_{pr}/s\rho$, where Q_{pr} is the efficiency factor for radiation pressure, s is the particle radius, and ρ is its density.

Figure 20 shows the dependence of β on the particle size for different materials (for simplicity we show results obtained for homogeneous spheres). For big particles ($s \gg \lambda_{max}$), radiation pressure force is proportional to their geometric cross section, giving rise to the $1/s$ dependence of β. At particle sizes comparable to

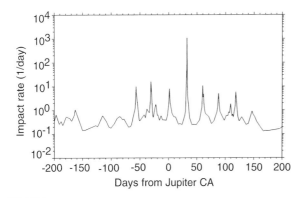

FIGURE 18 Dust impact rate observed by the *Ulysses* dust detector during 400 days around the closest approach to Jupiter (CA, 8 February 1992). At the beginning and end of the period shown, *Ulysses* was 240 million km (1.6 AU) from Jupiter, while at CA the distance was only 450,000 km. Except for the flux peak at CA, when bigger particles were detected, the peaks at other times consisted of submicron-sized dust particles.

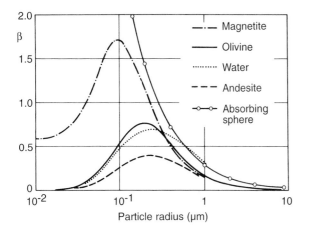

FIGURE 20 Ratio β of the radiation pressure force over solar gravity as a function of particle radius. Values are given for particles made of different materials and for a totally light-absorbing particle. [From G. Schwehm and M. Rhode (1997). *J. Geophys.* **42**, 727–735.]

the wavelength of sunlight, $s \approx \lambda_{max}$, β values peak and decline for smaller particles as their interaction with light decreases.

A consequence of the radiation pressure force is that particles with $\beta > 1$ are not attracted by the Sun but rather are repelled by it. If such particles are generated in interplanetary space either by a collision or by release from a comet, they are expelled from the solar system on hyperbolic orbits. But even particles with β values smaller than 1 will leave the solar system on hyperbolic orbits if their speed at formation is high enough so that the reduced solar attraction can no longer keep the particle on a bound orbit. Figure 21 depicts such a scenario. If a particle that is released from a parent body moving on a circular orbit has $\beta > 1/2$, then it will leave on a hyperbolic orbit. These particles are termed beta-meteoroids.

Because of the finite speed of light ($c \approx 300,000$ km/s), radiation pressure acts not perfectly radial but has an aberration in the direction of motion of the particle around the Sun. Thus a small component (approximately proportional to v/c, where v is the speed of the particle) of the radiation pressure force always acts against the orbital motion, reducing its orbital energy. This effect is called the Poynting–Robertson effect. As a consequence of this drag force the particle is decelerated. This deceleration is largest at its perihelion distance because both the light pressure and the velocity peak. Consequently, the eccentricity (aphelion distance) is reduced and the orbit is circularized. Subsequently, the particle spirals toward the Sun, where it finally sublimates.

The lifetime τ_{PR} of a particle on a circular orbit that

spirals slowly to the Sun is given by $\tau_{PR} = 7 \times 10^5$ $s\rho r^2/Q_{pr}$, where τ_{PR} is in years, r is given in AU, and all other quantities are in SI units. Even a centimeter-sized ($s = 0.01$ m), stony ($\rho = 3000$ kg/m^3, $Q_{pr} \approx 1$) particle requires only 21 million years to spiral to the Sun. This example shows that all interplanetary dust had to be recently generated; no dust particles remain from the times of the formation of the solar system. The dust we find today had to be stored in bigger objects (asteroids and comets), which have sufficient lifetimes.

The effect of solar wind impingement on particulates is similar to that of radiation pressure and the Poynting–Robertson effect. Although direct particle pressure can be neglected with respect to radiation pressure, solar-wind drag is about 30% of Poynting–Robertson drag.

Particle orbits that evolve under Poynting–Robertson drag will eventually cross the orbits of the inner planets and, thereby, will be affected by planetary gravitation. Figure 22 shows the evolution of particles that have been released in the asteroid belt. Even if the orbit periods of the particle and the planet are not the same but form a simple integer ratio, a resonance effect will occur. These effects are largest for big particles, the orbits of which evolve slower and which spend more time near the resonance position. Density enhancements of interplanetary dust have been found,

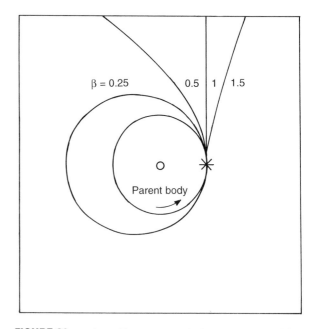

FIGURE 21 Orbits of beta-meteoroids that were generated from a parent body at the position indicated by the asterisk. β values of differently sized fragments are indicated; big β values refer to small particles.

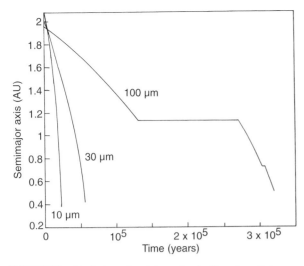

FIGURE 22 Evolution of particles under the Poynting–Robertson effect. Dust particles are assumed to be generated in the asteroid belt at about 2 AU distance from the Sun. Big particles (>100 μm) are temporarily trapped in resonant orbit with the Earth (1 AU) and Venus (0.7 AU). [After A. A. Jackson and H. A. Zook (1992). *Icarus* **97**, 70–84. With permission from Academic Press, San Diego.]

that is, the Earth resonant ring was identified in *IRAS* data and later confirmed by data from the *Cosmic Background Explorer* (*COBE*) satellite.

Dust near other stars will also evolve under the Poynting–Robertson effect and form a dust disk around this star. Such a disk has been found around the star β-Pictoris. A search is on to find resonance enhancements in this disk, which would indicate planets around this star. [*See* EXTRA-SOLAR PLANETS: SEARCHING FOR OTHER PLANETARY SYSTEMS.]

C. COLLISIONS

Mutual high-speed ($v > 1$ km/s) collisions among dust particles lead to grain destruction and fragment generation. By these effects, dust grains are modified or destroyed and many new fragment particles are generated in interplanetary space (Fig. 23). From impact studies in stony material, we know that at a typical collision speed of 10 km/s an impact crater is formed on the surface of the target particle if it is more than 50,000 times more massive than the projectile. This mass ratio is strongly speed and material dependent. A typical impact crater in brittle stony material (see Fig. 9) consists of a central hemispherical pit surrounded by a shallow spallation zone. The largest ejecta particle (from the spallation zone) can be many times bigger than the projectile, however, it is emitted at a very low speed on the order of meters per second. The total

mass ejected from an impact crater at 10 km/s impact speed is about 500 times the projectile mass.

However, if the target particle is smaller than the stated limit, the target will be catastrophically destroyed. The material of both colliding particles will be transformed into a huge number of fragment particles (Fig. 24). Thus, catastrophic collisions are a very effective process for generating small particles in interplanetary space. Figure 25 compares the lifetimes of interplanetary dust particles at 1 AU with respect to collisions and Poynting–Robertson drag. It has been found that interplanetary particles bigger than about 0.1 mm in diameter will be destroyed by a catastrophic collision rather than transported to the Sun by Poynting–Robertson drag.

D. CHARGING OF DUST AND INTERACTION WITH THE INTERPLANETARY MAGNETIC FIELD

Any meteoroid in interplanetary space will be electrically charged, and several competing charging processes determine the actual charge of a meteoroid (Fig. 26). Irradiation by solar UV light frees photoelectrons, which leave the grain. Electrons and ions are collected from the ambient solar-wind plasma. Energetic ions and electrons then cause the emission of secondary

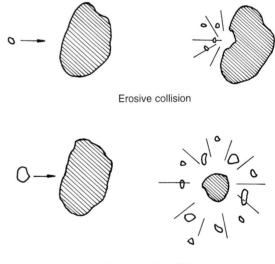

Erosive collision

Catastrophic collision

FIGURE 23 Schematics of meteoroid collisions in space. If the projectile is very small compared to the target particle, only a crater is formed in the bigger one. If the projectile exceeds a certain size limit the bigger particle is also shattered into many fragments. The transition from one type to the other is abrupt.

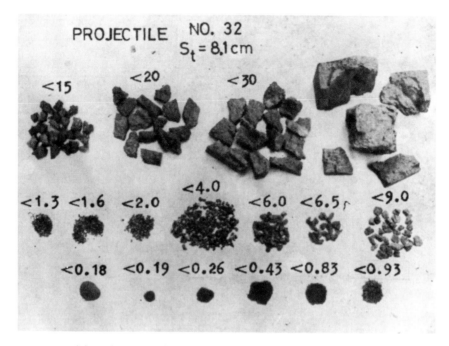

FIGURE 24 Fragments recovered from the catastrophic collision between a 0.3-g projectile and a 1.4-kg basalt cube of 8.1 cm at about 2.7 km/s impact speed. The fragments are sorted according to their sizes. [From A. Fujiwara *et al.* (1977). *Icarus* **31**, 277–288. With permission from Academic Press, San Diego.]

electrons. Whether electrons or ions can reach or leave the grain depends on their energy and on the polarity and electrical potential of the grain. Because of the predominance of the photoelectric effect in interplanetary space, meteoroids are mostly charged positive at a potential of a few volts. Only at times of very high solar-wind densities does the electron flux to the particle dominate and the particle gets charged negatively. The final charging state is reached when all currents to and from the meteoroid cancel. The timescale for charging is seconds to hours depending on the size of the particle; small particles charge slower.

The outward- (away from the Sun) streaming solar wind carries a magnetic field away from the Sun. The polarity of the magnetic field can be positive or negative depending on the polarity at the base of the field line in the solar corona, which varies spatially and temporally. Because of the rotation of the Sun (at a

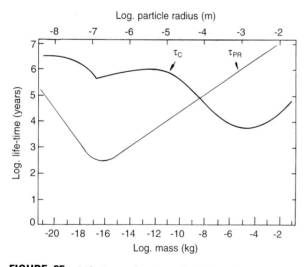

FIGURE 25 Life-times of meteoroids in interplanetaray space with respect to destruction by collisions τ_C and transport to the Sun by the Poynting–Robertson effect τ_{PR} as a function of particle mass. The shorter the lifetime, the more effective is the process of removing particles out of the zodiacal cloud.

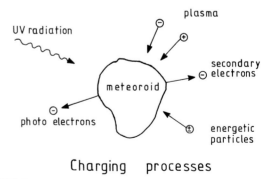

Charging processes

FIGURE 26 Charging processes of meteoroids in interplanetary space. UV radiation releases photoelectrons, electrons, and ions from the solar-wind plasma and they are collected; the impact of energetic particle radiation releases secondary electrons.

period of 25.7 days), magnetic field lines are drawn in a spiral like water from a rotating lawn sprinkler. For an observer or a meteoroid in interplanetary space, the magnetic field sweeps outward at the speed of the solar wind (400 to 600 km/s). In the magnetic reference frame the meteoroid moves inward at about the same speed, since its orbital speed is comparatively small. The Lorentz force on a charged dust particle near the ecliptic plane is mostly either upward or downward depending on the polarity of the magnetic field. Near the ecliptic plane, the polarity of the magnetic field changes at periods (days to weeks) that are much faster than the orbital period of an interplanetary dust particle, and the net effect of the Lorentz force on micron-sized particles is small. Only secular effects on the bigger zodiacal particles are expected to occur, which could have an effect on the symmetry plane of the zodiacal cloud close to the Sun. [See SOLAR WIND.]

The overall polarity of the solar magnetic field changes with the solar cycle of 11 years. For one solar cycle, positive magnetic polarity prevails away from the ecliptic in the northern hemisphere and negative polarity in the southern hemisphere. Submicron-sized interstellar particles that enter the solar system are deflected either toward the ecliptic plane or away from it (Fig. 27) depending on the overall polarity of the magnetic field. Interstellar particles entering the heliosphere from one direction at a speed of 26 km/s need about 20 years (two solar cycles) to get close to the Sun. Therefore, trajectories of small interstellar particles (0.1 μm in radius) are strongly diverted: in some regions of space their density is strongly increased, in others they are depleted. At the time of the *Ulysses* and *Galileo* measurements (1992 to 1996), the overall solar magnetic field had changed to the unfavorable configuration and, therefore, only big (micron-sized) interstellar particles reached the positions of *Ulysses* and *Galileo*.

E. EVOLUTION OF DUST IN INTERPLANETARY SPACE

Forces acting on interplanetary particles are compared in Table V. For large particles the dominating force is the solar gravity. This force depends on the mass of the particle, therefore it depends on the size as $F_G \sim s^3$. Radiation pressure depends on the cross section of the particle, hence $F_R \sim s^2$. The electric charge on a dust grain depends on the size directly, as does the Lorentz force $F_L \sim s$. Therefore, these latter forces become more dominating at smaller dust sizes. At a size of the wavelength of visible light ($s \sim 0.5$ μm),

radiation pressure is dominating gravity, and below that size the Lorentz force dominates the particles' dynamics. Though gravity is attractive to the Sun, radiation pressure is repulsive. The Lorentz force depends on the time-variable interplanetary magnetic field, but the net effect on small particles is that they are convected out of the solar system by the solar wind.

Besides energy-conserving forces, there are also dissipative forces: the Poynting–Robertson effect and the ion drag from the solar wind. Although small, these forces have a significant effect on the secular evolution of interplanetary dust orbits. They cause a loss of orbital energy and force particles to slowly spiral to the Sun, where they eventually evaporate. These atoms and molecules become ionized and are flushed out of the solar system by the solar wind.

Figure 28 shows the flow of meteoritic matter through the solar system as a function of the meteoroid size. There is a constant input of mass from comets and asteroids. From the intensity enhancement of zodiacal light toward the Sun it was deduced that, inside 1 AU, significant amounts of mass have to be injected into the zodiacal cloud, probably by comets. Comets shed their debris over a large range of heliocentric distances but preferentially close to the Sun, asteroid debris is mostly generated in the asteroid belt, between 2 and 4 AU from the Sun. Collisions dominate the fate of big particles and are a constant source of smaller fragments. Meteoroids in the range of 1 to 100 μm are dragged by the Poynting–Robertson drag to the Sun. Smaller fragments are driven out of the solar system by radiation pressure and Lorentz force.

Estimates of the mass loss from the zodiacal cloud inside 1 AU give the following numbers. About 10 tons per second are lost by collisions from the big (meteor-sized) particle population. A similar amount (on the average) has to be replenished by cometary and asteroidal debris. Nine tons per second of the collisional fragments are lost as small particles to interstellar space, and the remainder of 1 ton per second is carried by the Poynting–Robertson drag to the Sun, evaporates, and eventually becomes part of the solar wind. Interstellar dust transiting the solar system becomes increasingly important farther away from the Sun. At 3 AU from the Sun, the interstellar dust flux seems to already dominate the flux of submicron- and micron-sized interplanetary meteoroids.

IV. FUTURE STUDIES

New techniques will generate new insights. These techniques will include innovative observational meth-

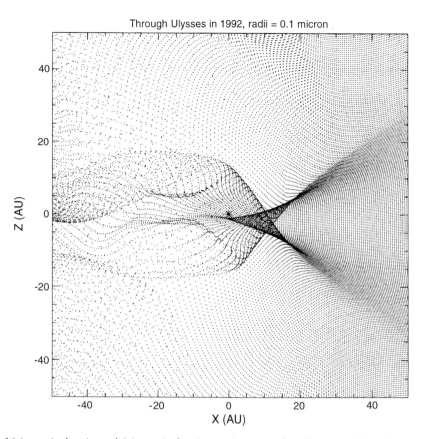

FIGURE 27   Grid of 0.1-μm-sized grains and 0.4-μm-sized grains moving across the solar system from the right side. The electrically charged grains interact with the solar gravitational field, with the solar radiation pressure, and with the interplanetary magnetic field. Positions refer to 1992 when *Ulysses* first identified interstellar particles out at Jupiter's distance. The magnetic polarity switched in 1979 and 1990. The heliospheric boundary, outside of which no solar effects are felt by the dust grains, was assumed to be at 100 AU upstream. The 0.1-μm-sized grains are much more affected by the interplanetary magnetic field than are the 0.4-μm-sized grains. [After E. Grün *et al.* (1994). *Astron. Astrophys.* **286**, 915–924.]

ods, new space missions to unexplored territory, and new experimental and theoretical methods to study the processes effecting interplanetary dust. Questions to address are: the composition (elemental, molecular, and isotopic) and spatial distribution of interplanetary dust; the quantitative understanding of effects or processes affecting dust in interplanetary space; and the quantitative determination of the contributions from different sources (asteroids, comets, planetary environs, interstellar dust).

Analyses of brightness measurements at infrared wavelengths up to 200 μm by the *COBE* satellite will result in refined models of the distribution of dust mostly outside 1 AU. Spectrally resolved observations of asteroids, comets, and zodiacal dust by the *Infrared Space Observatory (ISO)* are expected to show the genetic relation between these larger bodies and interplanetary dust. Improved observations of the inner

zodiacal light and the edge of the dust-free zone around the Sun will provide some clues to the composition of zodiacal dust. Optical and infrared observations of extrasolar systems will bring new insights to zodiacal clouds around other stars.

Interplanetary space missions presently under way that carry dust detectors are the *Ulysses* and *Galileo* missions. *Ulysses* has probed space above the poles of the Sun and outside 1 AU and continues its study of the interplanetary dust cloud at times of high solar activity. *Galileo* has become the first man-made satellite of Jupiter and has started to study its dust environment. With the newly launched *Cassini* mission, an in-depth study of the Saturnian system and its surrounding interplanetary space is planned. *Cassini* carries a complex dust analyzer (similar to the ones flown on the previous Halley missions) that measures the chemical composition of dust particles impacting

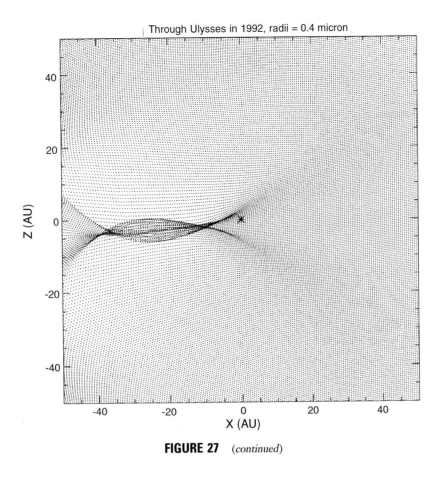

FIGURE 27   (*continued*)

the detector. The Japanese Mars probe *Planet B* will carry a simple dust detector to study the near-mars environment. The detailed study of cometary and interstellar dust is the goal of NASA's *Stardust* mission and the European Space Agency's *Rosetta* mission, which will follow a comet through its perihelion and investigate its release of dust to interplanetary space.

In near-Earth space, ambitious new techniques will be applied to collect meteoritic material that is not accessible by other methods. High-speed meteoroid

TABLE V

Comparison of Various Forces (Dominating Forces in Bold) Acting on Dust Particles of Size *s* under Typical Interplanetary Conditions at 1 AU from the Sun[a]

s (μm)	F_G (N)	F_R (N)	F_L (N)	F_{PR} (N)	F_{ID} (N)
0.01	9×10^{-23}	1.4×10^{21}	$\mathbf{1.5 \times 10^{-20}}$	1.4×10^{-25}	4×10^{-26}
0.1	9×10^{-20}	$\mathbf{1.4 \times 10^{-19}}$	$\mathbf{1.5 \times 10^{-19}}$	1.4×10^{-23}	4×10^{-24}
1	$\mathbf{9 \times 10^{-17}}$	1.4×10^{-17}	1.5×10^{-18}	1.4×10^{-21}	4×10^{-22}
10	$\mathbf{9 \times 1^{-14}}$	1.4×1^{-15}	1.5×10^{-17}	1.4×10^{-19}	4×10^{-20}
100	$\mathbf{9 \times 10^{-11}}$	1.4×10^{-13}	1.5×10^{-16}	1.4×10^{-17}	4×10^{-18}

[a] Subscripts for the *F* variables refer to gravity, radiation pressure, Lorentz force, Poynting–Robertson drag, and ion drag.

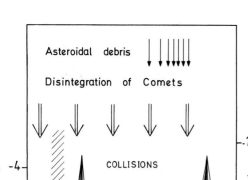

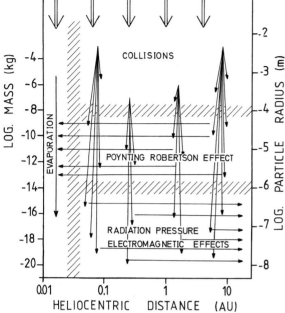

FIGURE 28 Mass flow of meteoritic matter through the solar system. Most of the interplanetary dust is produced by collisions of larger meteoroids, which represent a reservoir continually being replenished by disintegration of comets or asteroids. Most of it is blown out of the solar system as submicron-sized grains. The remainder is lost by evaporation after being driven close to the Sun by the Poynting–Robertson effect. In addition to the flow of interplanetary matter shown, there is a flow of interstellar grains through the planetary system.

catchers are under development that will permit the determination of the trajectory as well as the recovery of material for analysis in ground laboratories. A cosmic dust collector is currently being flown on the Russian *Mir* station.

Laboratory studies are instrumental in improving our understanding of planetary and interplanetary processes in which dust plays a major role. The study of dust–plasma interactions is a new and expanding field that is attracting considerable attention. New phenom-

ena are expected to occur when plasma is loaded with large amounts of dust. Processes of this type are suspected to play a significant role in cometary environments, in planetary rings, and in proto planetary disks.

BIBLIOGRAPHY

General Reading

Giese, R. H., and Lamy, P. (eds.) (1985). "Properties and Interactions of Interplanetary Dust." Reidel, Dordrecht/Boston.

Gustafson, B. A. S., and Hanner, M. S. (eds.) (1996). "Physics, Chemistry, and Dynamics of Interplanetary Dust," Conference Series Vol. 104. Astronomical Society of the Pacific, San Francisco.

Leinert, C., and Grün, E. (1990). *In* "Physics of the Inner Heliosphere" (R. Schwenn and E. Marsch, eds.), pp. 207–275. Springer-Verlag, Berlin.

Levasseur-Regourd, A. C., and Hasegawa, H. (eds.) (1991). "Origin and Evolution of Interplanetary Dust." Kluwer, Dordrect.

McDonnell, J. A. M. (ed.) (1978). "Cosmic Dust." John Wiley & Sons, Chichester, England.

McDonnell, J. A. M. (ed.) (1992). "Hypervelocity Impacts in Space." Univ. Kent Press, Canterbury, England.

Specialized Papers

Beg, O. E., and Grün, E. (1973). *Space Res.* **13**, 1047–1055.

Ceplecha, Z. (1977). *In* "Comets, Asteroids, Meteorites. Interrelations, Evolution and Origin" (A. H. Delsemme, ed.), pp. 143–152. Univ. Toledo Press, Toledo.

Dermott, S., *et al.* (1994). *Nature* **369**, 719–723.

Divine, N. (1993). *J. Geophys. Res.* **98**, 17029–17048.

Fujiwara, A., Kamimoto, G., and Tsukamoto, A. (1977). *Icarus* **31**, 277–288.

Grün, E., Zook, H. A., Fechtig, H., and Giese, R. H. (1985). *Icarus* **62**, 244–272.

Grün, E., *et al.* (1993). *Nature* **362**, 428–430.

Grün, E., *et al.* (1994). *Astron. Astrophys.* **286**, 915–924.

Humes, D. H. (1980). *J. Geophys. Res.* **85**, 5841–5852.

Jackson, A. A., and Zook, H. A. (1992). *Icarus* **97**, 70–84.

Jessberger, E. K., BohSung, J., Chakaveh, S., and Traxel, K. (1992). *Earth Planet. Sci. Lett.* **112**, 91–99.

Morfill, G. E., Grün, E., and Leinert, C. (1986). *In* "The Sun and the Heliosphere in Three Dimensions" (R. G. Marsden, ed.), pp. 455–474. Reidel, Dordrecht.

Schwehm, G., and Rhode, M. (1977). *J. Geophys.* **42**, 727–735.

Zook, H. A., and Berg, O. E. (1975). *Planet. Space Sci.* **23**, 183–203.

THE SOLAR SYSTEM AT ULTRAVIOLET WAVELENGTHS

I. A Brief History of Ultraviolet Astronomy

II. Nature of Solar System Astronomical Observations

III. Observations of Planetary Atmospheres

IV. Observations of Solid Surfaces

V. Conclusions

VI. Appendix: Calculation of the Bond Albedo

Robert M. Nelson
Jet Propulsion Laboratory,
California Institute of Technology

Deborah L. Domingue
Applied Physics Laboratory,
California Institute of Technology

GLOSSARY

Bond albedo: Ratio of the total radiation reflected in all directions from a solar system object to the total incident flux.

Column density: Integral over the number of molecules above a column of unit area in an atmosphere.

Electromagnetic spectrum: Division of electromagnetic radiation according to wavelength of the radiation.

Geometric albedo: Ratio of the brightness of a solar system object to the brightness of a perfectly diffusing disk at the same distance from the Sun.

Mixing ratio: Fractional mass of a particular component of an intimate mixture.

Orbital phase angle: Angular position of a planetary satellite in orbit about its primary object as measured from the point of superior geocentric conjunction.

Phase integral: Integral over all directions of the function that describes the directional scattering properties of a surface.

Solar phase angle: Angular distance between the Sun, the object under observation, and an Earth-based observer.

Superior geocentric conjunction: Point in a planetary satellite's orbit where it is directly opposite Earth, such that the satellite lies on a straight line connecting Earth, the planet, and the satellite.

I. A BRIEF HISTORY OF ULTRAVIOLET ASTRONOMY

The ultraviolet spectral region is important to the entire community of astronomers, from those who study nearby objects such as Earth's Moon to those who study objects at the edge of the observable universe. From the perspective of a planetary astronomer, the spectral information is important for determining the composition of, and understanding the physical processes that are occurring on, the surfaces and atmospheres of solar system objects.

Spectrophotometry of solar system objects at wavelengths shorter than ~3000 Å has long been desired in order to complement observations made by ground-based telescopes at longer wavelengths. However, the presence in Earth's atmosphere of ozone, a strong absorber of ultraviolet light between 2000 and 3000 Å, and molecular oxygen (O_2), which is the dominant ultraviolet absorber below 2000 Å, prevented astronomers of the 1950s and earlier from observing the universe in this important spectral region.

The ultraviolet wavelengths were not observed until the space age, when astronomical instruments could be deployed above Earth's atmosphere. A spacecraft provides a platform from which astronomical observations can be made where the light being collected has not been subjected to absorption from Earth's atmospheric gases. Thus, the space revolution dramatically enhanced the ability of astronomers to access the full spectrum of electromagnetic radiation emitted by celestial objects.

In the 1950s, a series of rocket-flown instruments began slowly to reveal the secrets of the ultraviolet universe. The first photometers and spectrometers were flown on unstabilized Aerobe rockets. They remained above the ozone layer for several tens of minutes while they scanned the sky at ultraviolet wavelengths. By the early 1960s, spectrometers on three-axis-stabilized platforms launched by rockets on suborbital trajectories were able to undertake observations with sufficient resolution such that individual spectral lines could be resolved in the target bodies.

Shortly thereafter, the military spacecraft designated *1964-83C* carried an ultraviolet spectrometer into Earth orbit. This was followed closely by NASA's launch of the first *Orbiting Astronomical Observatory* (*OAO*) satellite in 1966. These space platforms permitted long-duration observations compared to what was possible from a rocket launch on a suborbital trajectory. By 1972, the third spacecraft of the *OAO* series was launched. It was designated the *Copernicus* spacecraft and was an outstanding success.

In Europe, a parallel development pattern for exploring the ultraviolet sky was under way using sounding rockets followed by orbiting spacecraft. By 1972 the European Space Research Organization had launched a spacecraft dedicated to ultraviolet astronomy. These parallel developments set the stage in the 1970s for a joint U.S.–European collaboration: the *International Ultraviolet Explorer* satellite (*IUE*). Additional Earth-orbiting satellites with ultraviolet-observing capabilities would not be launched until the early 1990s. These include the joint U.S.–European project known as the *Hubble Space Telescope* (*HST*) and NASA's *Extreme Ultraviolet Explorer satellite* (*EUVE*). In 1990 and 1995, the *Hopkins Ultraviolet Telescope* (*HUT*) was flown aboard the U.S. space shuttle as part of the Astro Observatory.

During this time frame, several interplanetary spacecraft missions were launched that also included ultraviolet instruments in their payloads. *Pioneers 10* and *11*, which were launched in 1970 and 1973, respectively, included ultraviolet photometers among their scientific instruments. [See PLANETARY EXPLORATION MISSIONS.]

These two spacecraft were the first to safely pass through the asteroid belt and fly by Jupiter and Saturn. *Pioneer Venus*, which was launched in 1978, was the first U.S. mission dedicated to the exploration of the planet Venus. It included an ultraviolet spectrometer among its instrument package. Russian spacecraft missions *Vega 1* and *Vega 2*, launched in 1985, dropped two descent probes into Venus's atmosphere, which included the French–Russian ISAV ultraviolet spectroscopy experiment. The Voyager project sent two spacecraft to the outer solar system that included ultraviolet spectrometers (UVS) within their instrument payloads. *Voyager 2* was the first spacecraft to fly by all four of the Jovian planets (Jupiter, Saturn, Uranus, and Neptune). In 1989 the *Galileo* spacecraft was launched. This spacecraft was the first dedicated mission to the Jupiter system and it included within its scientific instrument payload two ultraviolet spectrometers, the EUV (extreme ultraviolet spectrometer, which operates between 500 and 1400 Å) and the UVS (an ultraviolet spectrometer that operates between 1150 and 4300 Å). The *Galileo* spacecraft collected ultraviolet spectra as it flew by Venus. Observations of Mars were undertaken in 1976 by instruments aboard the *Viking* orbiter spacecraft.

II. NATURE OF SOLAR SYSTEM ASTRONOMICAL OBSERVATIONS

Most astronomers observe objects that have their own intrinsic energy source, such as stars and galaxies. However, most of the observations undertaken by planetary astronomers are of targets that do not emit their own radiation but are observable principally because they reflect the sunlight that falls on them or emit energy as a result of various physical processes. Measuring reflected light at ultraviolet wavelengths can pose some interesting problems for instrument designers.

First, instrument spectral responses become weaker with decreasing wavelength and so does the Sun's energy output. The energy output of the Sun will change by a factor of 30 over the spectral range of most ultraviolet instruments. This exceeds the dynamic range of the ultraviolet detectors and therefore several spectra must be taken at increasing exposure levels in order that the entire spectral range can be covered at an adequate signal-to-noise ratio.

The contribution of the observed solar spectrum to the spectrum observed must be removed to understand the spectrum of the object. The acquisition of a good solar spectrum in the ultraviolet range is no easy task. In addition, below 1800 Å, the spectrum of the Sun is variable. Therefore, a simultaneous spectrum of the Sun (or the reflection spectrum from an object whose spectrum is well understood) must be gathered at the same time that any ultraviolet observations are undertaken.

Lastly, solar system objects change positions against the background of stars in the course of an individual observation. In most cases, special tracking rates must be calculated prior to each observing run in order to know the change of the position of the target with time. Inaccurately calculated tracking rates can cause the observed target to drift from the instrument's field of view, thus adding noise and uncertainty to a measurement.

Every planet in the solar system except Mercury has been observed in the ultraviolet by Earth-orbiting telescopes. Many of the larger planetary satellites, selected asteroids, and comets have also been observed. This data set has provided important information regarding the atmospheres and surfaces of solar system objects and the processes shaping their compositions.

A major objective of a solar system astronomical observing program is to measure the energy balance of a solar system object. This is done by calculating the Bond albedo. The Bond albedo of a nonluminous object is defined as the total radiative flux reflected in all directions to the total incident flux. The calculation of the Bond albedo is shown in the Appendix. In addition to the energy balance, the variation of the geometric albedo as a function of wavelength is used to measure the strength of mineral absorption features, from which the abundance of spectrally active species can be estimated. Lastly, the reflectance spectrum, the geo-

metric albedo variation normalized at a particular wavelength, is useful in locating the wavelengths at which significant absorptions exist.

III. OBSERVATIONS OF PLANETARY ATMOSPHERES

With the exception of the innermost planet Mercury, all of the planets in the solar system (and a few planetary satellites) are surrounded by substantial atmospheres. In some cases the surface of the object is seen through its atmosphere and in others it is not. All the planets with atmospheres absorb ultraviolet light, and as a result ultraviolet observations provide information on the composition of, and processes that are occurring in, the object's atmosphere.

In general, the atmospheres of the terrestrial planets (Mercury, Venus, Earth, and Mars) are thought to be secondary atmospheres that evolved after the primordial atmospheres were lost. However, the atmospheres of the four giant or Jovian planets (Jupiter, Saturn, Uranus, and Neptune), because of their strong gravitational attraction and comparatively low temperatures, retained the primordial, cosmically very abundant, but very low atomic mass elements, particularly hydrogen and helium. From ground-based observations, methane (CH_4) and ammonia (NH_3) were identified in the atmospheres of the giant planets and therefore atmospheric processes were suspected of producing a host of daughter products that can be detected at ultraviolet wavelengths. [See ATMOSPHERES OF THE GIANT PLANETS.]

Sunlight entering a planetary atmosphere can experience or initiate a wide variety of processes that contribute to the total energy emitted by the object and observed by an astronomical facility. The objects described previously all possess atmospheres that contribute significantly to their spectral behavior. Astronomical observations of such bodies are used to search for and measure the depths of absorption bands in the spectrum or emission bands due to atmospheric interactions with energetic particles that originate from the solar wind or the planet's magnetosphere. These bands are unique to specific gases, thus it is possible to identify or eliminate particular gases as candidate materials in the atmospheres of these objects. The interpretation of an ultraviolet spectrogram can be an arduous task, given that the bands and lines observed in the spectrum may arise from a combination of processes. These include:

1. Single and multiple scattering of photons by aerosols (haze and dust) in the planetary atmosphere.
2. Absorption of the ultraviolet light by atmospheric species.
3. Stimulation of an atmospheric gas by incident sunlight and emission by fluorescence, chemiluminescence, or resonant scattering.
4. Photoionization and photodissociation reactions that produce a reaction product in an excited state.
5. Excitation of gas by precipitation of magnetospheric particles.

Each of these processes is associated with a well-understood physical mechanism, the details of which are beyond the scope of this chapter. The reader is referred to the Bibliography and other chapters in this volume.

Limited-wavelength facilities can identify some but not all of the constituents present and processes ongoing in a planetary atmosphere. The ultraviolet data from Earth-orbiting satellites have been used in combination with ground-based observations at other wavelengths and with observations by other spacecraft (including flyby missions) to develop an understanding of the atmospheres of planetary objects. The following discussion summarizes the results of those bodies in the solar system that contain atmospheres.

A. VENUS

For more than half a century, the very dense Venus atmosphere has been known to be composed principally of carbon dioxide (CO_2) based on the existence of strong spectral absorption features in the near-infrared spectrum. Within a few years of launch, *IUE* identified several important trace constituents, including nitrogen oxide (NO), and confirmed the presence of several others, such as sulfur dioxide (SO_2). The *Vega 1* and *2* probes measured local ultraviolet absorptions due primarily to SO_2 and aerosols. Ultraviolet reflectance spectra obtained during two sounding rocket observations in 1988 and 1991 found that SO_2 is the primary spectral absorber between 1900 and 2300 Å and that sulfur monoxide (SO) is also present in Venus's atmosphere. Strong spectral absorptions prevent the observation of the Venus surface at ultraviolet wavelengths. The *Galileo* spacecraft EUV observed Venus in the extreme ultraviolet wavelength range (550 to 1250 Å) during its flyby. It detected emissions due to helium, ionized atomic oxygen, atomic hydrogen, and an atomic hydrogen–atomic oxygen blend. In 1994,

an extreme ultraviolet spectrograph (EUVS) was launched aboard a sounding rocket to observe the Venusian atmosphere from 825 to 1110 Å. The EUVS identified about ten spectral features, including N I, N II, N_2, H I, O I, O II, and a probable CO (C-X) band. The results of the *EUVE* observations are in alignment with earlier observations by *IUE*, *Pioneer Venus*, *Venera 11* and *12*, and the *Galileo* EUV spectrometer.

Spectra of the Venus dayside and nightside have been obtained while Venus was near elongation. More SO_2 absorptions were found at 2080–2180 Å, which when combined with the column densities reported by the *Pioneer Venus* orbiter and with ground-based observations provide a measure of the SO_2 mixing ratio with altitude and its variation at the top of the cloud deck. This provides information on its variation in spatial distribution and permits models to be constructed of the planet's atmospheric dynamics. Observations of the Venus nightside have led to the identification of the Venus nightglow, which is caused by the emission bands of nitric oxide (NO). Because of the short lifetime of NO on the nightside, this finding implies the rapid dayside–nightside transport of material in the Venus atmosphere. Observations of the Venus dayside have led to the discovery that the dayglow emission is carbon monoxide fluorescence, probably due to fluorescent scattering of solar Lyman alpha radiation.

B. MARS

The atmosphere of Mars, like that of Venus, is also dominated by carbon dioxide, which was first detected based on absorptions in its near-infrared spectrum. The Venus atmosphere is very thick and the surface is not visible to a remote observer in the ultraviolet to infrared wavelengths. However, on Mars the atmosphere is much less dense, and in comparison to Venus the Martian atmosphere is relatively transparent at these wavelengths. Therefore ultraviolet to infrared obsevations of Mars reveal information about both its atmosphere and its surface. Shortly after launch, *IUE* detected ozone over the southern region, and subsequent observations by *HST* have studied the seasonal variation of atmospheric ozone on Mars. *HST* imaged the Martian surface in the ultraviolet to infrared wavelengths in an attempt to measure diurnal, seasonal, and latitudinal variations in atmospheric ozone content. These same images are also used in an effort to monitor dust storms within the Martian atmosphere. In general, the ozone is in greatest abundance where the atmo-

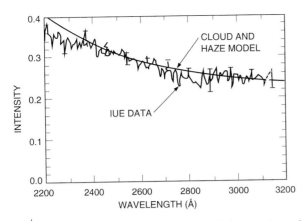

FIGURE 1 The spectral geometric albedo of Jupiter as obtained from *IUE*. The smooth solid line is the best fit from a computer model that assumed a layer of haze particles with single scattering albedo of 0.42 that overlie a cloud deck of geometric albedo of 0.25.

sphere is dry and cold. *EUVE* has provided the first measurements of helium within the Martian atmosphere. These helium observations have been used to set constraints on outgassing processes.

The Martian surface has been found to be spectrally featureless at the *IUE* and *HST* wavelengths, and the planet's albedo in the ultraviolet is about what it is at visual wavelengths (~0.1).

C. JUPITER

The spectral geometric albedo of Jupiter as measured by *IUE* is shown in Fig. 1. Most of this spectral behavior is attributable to hazes that are high above the cloud deck. The solid line in the figure shows the change in geometric albedo as a function of wavelength as predicted by a computer model that simulates the astronomical observations. The best-fit model to the data occurs for a Jovian cloud deck with a geometric albedo of 0.25 and for a haze composed of particles with a single scattering albedo of 0.42. Though such a result may not be able to provide an unambiguous identification of the materials that compose the haze, it can constrain the eligible candidate materials that are suggested by other observations.

In addition to the major atmospheric species that have been identified in the Jovian atmosphere from ground-based observation (hydrogen, helium, methane, and ammonia), *IUE* discovered ultraviolet absorptions that led to the identification of acetylene (C_2H_2).

Figure 2 shows the *IUE* spectrum of Jupiter in the range 1700–1800 Å. A solar spectrum is also shown for comparison. The strong absorption features at 1715, 1735, 1755, and 1775 Å (indicated by the arrows) are considered to be positive and unambiguous evidence for the presence of C_2H_2 in the Jovian atmosphere. The mixing ratio of acetylene to hydrogen has been found to be 2.2×10^{-9}.

IUE observations have also permitted a mixing ratio for ammonia to hydrogen to be calculated, and it is found to be 5×10^{-7}. The fact that *IUE* is able to observe the absorption features of these species indicates that they are above the Jovian tropopause, where the clouds create an opaque barrier to light emitted from the material underneath and hence make spectral identification of the underlying material impossible. It was not until the impact of the fragmented comet Shoemaker-Levy 9 (SL-9) that studies of this underlying material became possible.

In July 1994, the comet SL-9 collided with Jupiter. The *EUVE* satellite observed Jupiter before, during, and after this event. *EUVE* found that 2 to 4 hours after the impact of several of the larger fragments, the amount of neutral helium temporarily increased by a factor of ~10. This transient increase is attributed to the interaction of sunlight with the widespread high-altitude remnants of the plumes from the larger impacts. The *HST* also observed this event with the Goddard High-Resolution Spectrograph (GHRS) and the Faint Object Camera (FOC). The ultraviolet spectra obtained by *HST* of Jupiter after the collision of SL-9 identified approximately 10 molecules and atoms in the perturbed atmosphere, many of which had never been detected before in Jupiter's atmosphere. Among

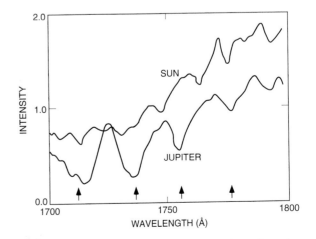

FIGURE 2 The *IUE* identification of acetylene absorptions (arrows) in the atmosphere of Jupiter.

these were S_2, CS_2, CS, H_2S, and S^+, which are believed to be derived from a sulfur-bearing parent molecule native to Jupiter. These observations also detected stratospheric ammonia (NH_3). Neutral and ionized metals, including Mg II, Mg I, Si I, Fe I, and Fe II, were also observed in emission. The surprising observation was the absence of absorptions due to oxygen-containing molecules.

Ultraviolet emissions of the hydrogen Lyman alpha line and bands of molecular hydrogen have been detected in the Jupiter atmosphere. They are thought to be associated with polar auroral activity originating from particle impact excitation processes. Jupiter's auroral displays are the most energetic in the solar system. Synoptic observations of these ultraviolet emissions using *IUE* have shown that they vary with Jupiter's magnetic (not planetary) longitude, and hence these emissions are magnetospheric and not atmospheric phenomena. *IUE* observations have been used to construct a spatial map of the Lyman alpha emission and the data indicate that the emitting material is upwelling at about 50 m/sec relative to the surrounding material. More intensive ultraviolet observations with the *HST* GHRS and FOC have measured the temporal variability within the aurora and temperature variations within the auroral ovals seen at both poles. These variations are reflections of possible distortions in the magnetic field of Jupiter. *HST* has measured the first detection of reversed Lyman alpha emissions, which are linked to variable atomic hydrogen. Estimates of vertical column densities ($1-5 \times 10^{16}$ cm^{-2}) of atomic hydrogen above the auroral source have been made. *HST* has also detected ultraviolet emission from a superthermal hydrogen population. The *Galileo* spacecraft EUV and UVS spectrometers have also observed Jupiter's aurora. These simultaneous observations have placed constraints on the vertical distribution of methane (CH_4) in Jupiter's atmosphere. Slant methane column abundances are estimated to be 2×10^{16} cm^{-2} in the north and 5×10^{16} cm^{-2} in the south based on the *Galileo* observations.

An equatorial bulge has been observed in the Lyman alpha emission that has been continually present but has changed in shape over many years. This bulge has been associated with an anomaly in the Jovian magnetosphere that is not well understood.

The *IUE* discovery of emissions in the vicinity of Jupiter's moon Io has led to the identification of several ionized species of sulfur and oxygen. The origin of the sulfur and oxygen is the surface of Io itself. Io is the most volcanically active body in the solar system and these gases are ejected by the volcanoes. The process by which material is transported from the volcanoes

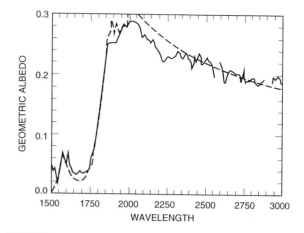

FIGURE 3 The spectral geometric albedo of Saturn as observed by *IUE*.

to the atmosphere has been a lively topic of debate in the planetary science community for several decades.

The *Voyager* spacecraft detected sulfur dioxide gas in regions localized over Io's hot spots. This led to the incorrect assumption that the SO_2 gas was distributed planetwide in the same abundance and that Io had an atmosphere of some significance. However, subsequent ultraviolet observations with both *IUE* and *HST* of the entire Io disk were unable to detect substantive amounts of SO_2 gas. If the gas were uniformly distributed, then these instruments should have been able to detect it. Ultraviolet observations are an important piece of evidence in support of Io having only a very tenuous atmosphere that is only of significance over Io's hot spots or volcanoes. [*See* Io.]

Associated with Jupiter's satellite Io is a plasma torus, or a donut-shaped ion cloud centered at Io's orbital radius. This torus has been studied by *Voyager*, *IUE*, *HST*, *EUVE*, *HUT*, and *Galileo*. Oxygen, sulfur, and sodium ions are the major constituents of the torus. The torus is not uniform, and the density of ions shows various asymmetries dependent on Io's position and dawn–dusk timings, in addition to temporal variations.

Observations with the *HST* GHRS have detected atomic oxygen emission from Jupiter's satellite Europa, which has been interpreted as evidence for a tenuous O_2 atmosphere about this satellite. Study of the nature and extent of this emission species will be an active research topic in the future.

D. SATURN

The spectral geometric albedo of Saturn as determined by *IUE* is shown as a solid line in Fig. 3. The broken line is a best-fit model to the data, which assumes a hydrogen atmosphere above a homogeneous cloud deck with a reflectivity of about 0.2. The column density of the hydrogen above the clouds is about 3 km atm.

IUE has discovered absorption features in the ultraviolet spectrum of Saturn that have been associated with acetylene in the upper atmosphere. The mixing ratio of the acetylene is about 1×10^{-7}. Although acetylene is a well-known strong absorber of ultraviolet radiation, it alone cannot explain the low UV spectral geometric albedo of Saturn that has been reported by *IUE*. Other ultraviolet-absorbing materials must be present. Comparisons of laboratory spectra of C_2H_2, PH_3, AsH_3, and GeH_4 with the *IUE* observations show that the best-fit model for Saturn's atmospheric ultraviolet spectrum includes absorptions by C_2H_2, H_2O, CH_4, C_2H_6, PH_3, and GeH_4. The distribution of PH_3 and GeH_4 decreases with increasing altitude in these models, suggesting that ultraviolet photolysis is an important process occurring at higher altitudes.

Pole-to-pole mapping studies of the hydrogen Lyman alpha emission across Saturn's disk led to the discovery of pronounced spatial asymmetries in the emission. Other observations of hydrogen do not find a variation in intensity with rotational period as with Jupiter. There is no rotational bulge in the Lyman alpha emission as seen on Jupiter. This is probably due to the fact that Saturn's magnetic pole is coincident with the rotational pole, whereas in Jupiter's case the poles are offset.

Like Jupiter, Saturn also displays auroral activity. On both planets this auroral activity is related to the formation of aerosols that are detectable in the ultraviolet as dark-absorbing regions. *HST* FOC ultraviolet observations discovered a dark oval encircling the north magnetic pole that is spatially coincident with the aurora detected by the *Voyager* UVS. *Voyager 2* ultraviolet Photopolarimeter Subsystem (PPS) measurements also demonstrate a geographical correlation between the auroral zones of Jupiter and Saturn with UV-dark polar regions. Additional ultraviolet observations with the *HST* FOC of Saturn's northern ultraviolet aurora and polar haze support the hypothesis that the polar haze particles are composed of hydrocarbon aerosols produced during H_2^+ auroral activity.

E. URANUS

Uranus presents a unique observational circumstance to the inner solar system observer because of the fact that its pole is inclined 89° to the ecliptic and that at the present position in its 84-year orbit about the Sun it presents its pole to Earth. This unusual inclination, combined with its great distance from Earth, makes it impossible to use an Earth-based instrument to undertake pole-to-pole comparisons as was done with Jupiter and Saturn. Uranus has a geometric albedo at *IUE* wavelengths of about 0.5, more than twice that of Jupiter and Saturn. This is consistent with the hypothesis that additional absorbers are present in the Jovian and Saturanian atmospheres that are not present in the atmosphere of Uranus.

Voyager 2 spacecraft observations of Uranus found a very small internal heat source compared to the large internal heat sources found at Jupiter and Saturn. This means that there is very little atmospheric mixing by eddy current in the Uranian atmosphere. Thus, ultraviolet observations are able to sense a deeper region of the atmosphere.

The ultraviolet emissions from Uranus's atmosphere have been measured by *IUE* and the *Voyager* UVS. To increase the signal-to-noise ratio, *IUE* observers have used principally low-resolution observations and have binned broad-wavelength regions together to search for broadband absorbers at ultraviolet wavelengths. Analysis of the *IUE* observations has detected acetylene absorptions, which have also been detected on Jupiter and Saturn. Based on these observations, the mixing ratio of the acetylene is estimated to be 3×10^{-8}. Analysis of the *Voyager* UVS observations of H_2 band ultraviolet airglow emissions shows aurora at both magnetic poles. The auroral emissions on Uranus are very localized in magnetic longitude and do not form complete auroral ovals as are seen on Jupiter and Saturn. A sharp increase in measured reflectance intensity above 1500 Å is indicative of C_2H_2 present in the atmosphere.

F. NEPTUNE

Neptune is so distant that only broadband ultraviolet measurements are possible. The geometric albedo of Neptune measured by *IUE* is 0.5, which, like that of Uranus, is twice that of Jupiter and Saturn. Most of the important data for Neptune at ultraviolet wavelengths have come from the UVS onboard the *Voyager 2* spacecraft and, more recently, the *HST* satellite. *Voyager 2* UVS measurements tentatively identified weak auroral emissions on Neptune's nightside. CH_4 and C_2H_6 abundances inferred from the Voyager UVS solar occultation experiment are between 0.0006 and 0.005 mole fraction for CH_4 in the lower stratosphere (with a mixing ratio of $5-100 \times 10^{-5}$) and a density of C_2H_6 estimated to reach 3×10^9 per cm^3.

In 1994, *HST* imaged Neptune in six broadband filters, one of which was in the ultraviolet. The goal of these observations was to study the cloud structure on Neptune and compare the measurements with the observations made by *Voyager 2*. The *HST* images showed that the Great Dark Spot seen by *Voyager* no longer existed, but a new large dark feature of comparable size had appeared in the northern latitudes.

G. PLUTO

Pluto and its large satellite Charon are at a great distance and are quite small compared to the four gas-giant planets that populate the outer solar system. Nevertheless, *IUE* has been able to obtain a few spectra of these objects and it has been observed that the ultraviolet albedo of these objects changes as they rotate. The size of the rotational variation as measured at ultraviolet wavelengths by *IUE* is larger than the rotational variation that is measured at longer wavelengths by Earth-based observers. This is consistent with the presence of an absorbing material being present on the surface that is spectrally active in this wavelength range. The composition of the absorbing material is unknown. [*See* PLUTO AND CHARON.]

Observations made with the FOS on *HST* measured Pluto's atmospheric haze. These measurements found that at visual wavelengths the upper limit of the optical depth is 0.26. These observations also allowed upper limits to be set on the column abundances of CO, OH, and NO. The ultraviolet spectrum of Pluto's satellite, Charon, was also measured by the *HST* FOS.

H. TITAN AND TRITON

Saturn's satellite Titan and Neptune's satellite Triton are among the largest satellites in the solar system. In addition they are far from the Sun and therefore the reduced solar energy keeps the atmospheric gases cold enough that they cannot easily escape by thermal processes. [*See* TITAN; TRITON.]

Ground-based and *Voyager* spacecraft observations have identified methane as a major constituent of Titan's atmosphere. Nitrogen has also been suggested as an atmospheric constituent, but it is difficult to detect although its presence is inferred. The *Voyager* UVS observations also measured an atomic hydrogen torus about Titan that shows some temporal variability. Analysis of *IUE* observations of Titan have placed

constraints on the properties of Titan's high-altitude haze and the abundances of simple organic compounds.

A few *IUE* observations of Triton indicate that the satellite's ultraviolet brightness varies as Triton rotates. Its albedo varies from about 0.42 to 0.58 from one hemisphere to another. This observation has not been confirmed by the photopolarimeter on *Voyager 2, which measures the albedo at the same wavelengths as IUE*. The photopolarimeter on *Voyager* measured an albedo of 0.59 on all sides of Triton. Later analyses of Triton with *IUE* measured a geometric albedo of 0.28 at 2700 Å, which increases monotonically from 2600 to 3200 Å with a slope of 0.13 per 1000 Å. This difference may indicate that observations of faint objects such as Triton test the limits of *IUE*'s sensitivity. Mixing ratio upper limits for atmospheric constituents of OH, NO, and CO of 3×10^{-6}, 8×10^{-5}, and 1.5×10^{-2}, respectively, are derived from the analysis of *HST* FOC observations.

IV. OBSERVATIONS OF SOLID SURFACES

Many solid-state materials that make up the surfaces of solar system objects exhibit spectral absorption features, and thus it is possible to identify or constrain the abundance of solid components on the surfaces of these objects. This is accomplished by comparing the spectral geometric albedo of the object with the reflection spectrum of the solid-state materials as measured in the laboratory. The following discussion will focus on ultraviolet observations of solid surfaces throughout the solar system.

A. GALILEAN SATELLITES

Many Earth-orbiting satellites have observed the bright satellites of Jupiter and also the satellites of Saturn and Uranus. By far, the most extensive study has been of Jupiter's Galilean satellites Io, Europa, Ganymede, and Callisto. Several hundred usable spectra of these objects have been taken over the lifetime of *IUE*. *HST* has also focused its attention on these moons of Jupiter and has made some interesting new discoveries, such as the tenuous oxygen atmosphere about Europa mentioned earlier.

The very high spatial resolution provided by *Voyager* and *Galileo* images shows that the Galilean satellites, particularly Io, are variegated in color on continental scales. Compositional information may be

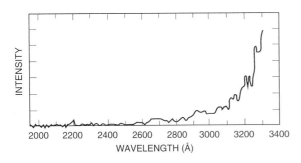

FIGURE 4 Spectral geometric albedo of Io as observed by *IUE*. The strong increase in reflectance longward of 3000 Å is consistent with the presence of sulfur dioxide frost on Io's surface.

derived from high spectral resolution studies from Earth or near Earth orbit because the satellite's synchronous rotation permits any given full-disk observation of a satellite to be associated with a uniquely defined hemisphere of that particular object. The extension of the available spectral range to shorter wavelengths with ultraviolet telescopes enhances this data set by permitting the identification of more absorption features, thereby providing further constraints on the compositional models that have been developed. [*See* OUTER PLANET ICY SATELLITES.]

One of the first set of ultraviolet observations of the Galilean satellites at wavelengths shorter than 3000 Å was done from space using *IUE*'s precursor, the *OAO-2* spacecraft. Io was found to have an extremely low ultraviolet geometric albedo (3% at 2590 Å), in marked contrast to its very high albedo at longer wavelengths (70% at visible light wavelengths). This result was so unusual that it remained in doubt until *IUE* confirmed it by measuring Io's spectrum in this spectral range. Figure 4 shows an *IUE* spectrum of Io that dramatically confirms that Io's albedo rises by an order of magnitude between 2000 and 3300 Å. This albedo change with wavelength is caused by the presence of a very absorbing material, sulfur dioxide frost, on Io's surface.

The first in-depth studies of the Galilean satellites in the ultraviolet were accomplished with the use of the *IUE* satellite. Subsequent observations with *HST* have supported the initial findings of *IUE*, in addition to adding to our knowledge of the composition of the surfaces of these satellites. To calculate a geometric albedo, the 7 Å resolution *IUE* information was averaged over three wide bandpasses to reduce noise. These bandpasses are hereafter referred to as bands 1, 2, and 3. The wavelength ranges for these bandpasses are 2400–2700 Å (band 1), 2800–3000 Å (band 2), and 3000–3200 Å (band 3). The geometric albedos of the Galilean satellites change as a function of the satellites'

orbital phase angle at the time of observation, and the albedos derived from the three bandpasses have been grouped according to the specific orbital phase angles at which the observations were made. The term "leading side" and "trailing side" refer to orbital phase angles at or near 90° and 270°, respectively. For this purpose, the leading side includes observations between orbital phase 45° and 135° and the trailing side includes observations between 235° and 315°.

The data in Table I show that, at ultraviolet wavelengths, Io's trailing side has a higher albedo than it's leading side, just the opposite of what is seen when Io is observed at longer wavelengths. This reversal in brightness associated with orbital phase behavior is more pronounced for Io than for any other object in the solar system, and proves to be important in efforts to determine the surface composition variation in longitude across Io's surface.

For Io, shortward of the *IUE* bandpass 3 (~3200 Å), Io's leading hemisphere is less reflective than its trailing hemisphere. This absorption on Io's leading side is somewhat stronger in band 2 (~2900 Å) and still stronger at band 3 (~2500 Å). Therefore, it can be directly inferred from the Io data that there is a longitudinally asymmetric distribution of a spectrally active surface component on Io's surface. The material is strongly absorbing shortward of ~3200 Å and is strongly reflecting longward of that wavelength. It is in greatest abundance on the leading hemisphere of Io, and it is in least abundance on the trailing hemisphere. This is consistent with the expected behavior of SO_2 frost being present in greater abundance on Io's leading hemisphere. [*See* IO.]

Europa and Ganymede exhibit a variation in brightness at *IUE* wavelengths with orbital phase that is in the same sense as the variation reported at the visible wavelengths; at all wavelengths, these objects are brighter on their leading sides than on their trailing sides. A gradual decrease in brightness toward shorter wavelengths occurs on both hemispheres of both objects.

Ground-based observations of Callisto have found that its albedo varies with orbital phase angle in the opposite sense to that of Europa and Ganymede (i.e., its trailing side has a higher albedo than its leading side). This is also true at the three *IUE* wavelength bands. The albedo of Callisto decreases shortward of 5500 Å and continues to decrease throughout the *IUE* bandpasses. Its albedo at all wavelengths is lower than the albedo of Europa and Ganymede.

Gross UV albedo changes are apparent on all four objects, and these contrasts are most pronounced when approximately opposite hemispheres centered at or-

TABLE I
Ultraviolet Geometric Albedos (%) of the Galilean Satellites

	Band 1	Band 2	Band 3
Io (L)[a]	1.5 ± 0.1 ($N = 34$)	1.7 ± 0.1 ($N = 35$)	4.2 ± 0.1 ($N = 27$)
Io (T)	2.8 ± 0.2 ($N = 28$)	3.0 ± 0.5 ($N = 24$)	3.8 ± 0.3 ($N = 27$)
Europa (L)	18.0 ± 0.4 ($N = 22$)	26.0 ± 1.0 ($N = 18$)	37.0 ± 2.0 ($N = 24$)
Europa (T)	9.6 ± 0.2 ($N = 31$)	12.9 ± 0.4 ($N = 33$)	17.1 ± 0.6 ($N = 28$)
Ganymede (L)	12.8 ± 0.7 ($N = 26$)	16.8 ± 0.9 ($N = 20$)	20.0 ± 0.1 ($N = 20$)
Ganymede (T)	7.0 ± 0.3 ($N = 16$)	7.5 ± 0.04 ($N = 16$)	10.5 ± 0.8 ($N = 17$)
Callisto (L)	4.0 ± 0.8 ($N = 15$)	4.9 ± 0.1 ($N = 13$)	6.6 ± 0.02 ($N = 14$)
Callisto (T)	5.6 ± 0.2 ($N = 26$)	6.4 ± 0.02 ($N = 21$)	10.5 ± 0.8 ($N = 26$)

[a] L = Leading side ($45° < 0° < 135°$); T = trailing side ($235° < 0° < 315°$).

bital phase angles at or near eastern and western elongation (90° and 270° orbital phase angle) are compared for a given object. These albedo differences between individual hemispheres of each object imply that differences in chemical composition exist across the surfaces of each satellite. [See OUTER PLANET ICY SATELLITES.]

The ultraviolet spectral signature of materials distributed nonuniformly on a satellite's surface becomes apparent when the spectra of the opposite hemispheres of each object are ratioed. The individual spectra taken at or near one hemisphere of a given object are co-added and ratioed to similarly co-added spectra of the opposite hemisphere. The resulting opposite hemisphere ratio spectra are shown in Fig. 5 for the four satellites. All the ratio spectra have been normalized at 2700 Å. Identical ratio characteristics are also seen in similar *HST* observations.

The leading/trailing side UV spectral ratio for Io is shown in Fig. 5A. The ratio spectrum indicates that a strong spectrally active UV absorber is asymmetrically distributed across Io's surface. The absorbing material (SO_2 frost) is characterized by a strong absorption shortward of 3300 Å, and is distributed in greater abundance on Io's leading side than its trailing side. Images from *HST* and *Galileo* have made it possible to associate specific volcanic features with absorbing species [See IO.]

The trailing/leading side UV spectral ratio for Europa is shown in Fig. 5B. This ratio spectrum shows that Europa's opposite hemispheric UV ratio spectrum is characterized by a broad, weak absorption centered at approximately 2800 Å. The spectral absorption is relatively stronger on Europa's trailing hemisphere compared to its leading hemisphere. This same broad, weak absorption is also detected in *HST* ultraviolet spectra of Europa. This feature has been attributed to

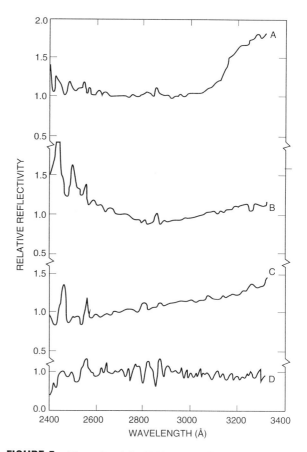

FIGURE 5 The ratio of the *IUE* spectra of the opposite hemispheres of the Galilean satellites. From top to bottom, they are Io (leading side to trailing side), Europa (trailing side to leading side), Ganymede (trailing side to leading side), and Callisto (leading side to trailing side). The differences in the spectra of the opposite hemispheres of these objects are due to material being asymmetrically distributed in longitude on the surfaces of these objects. All the ratios have been normalized at 2700 Å.

the presence of sulfur atoms from Jupiter's magnetosphere embedded in Europa's water ice surface.

The trailing/leading side ratio spectrum of Ganymede is shown in Fig. 5C. The opposite hemispheric UV ratio spectrum of Ganymede is characterized by a weak spectral absorption beginning at ~3200 Å that gradually increases in absorption down to 2600 Å. The absorption, attributed to ozone (O_3), is more pronounced on Ganymede's trailing hemisphere than on its leading hemisphere.

The leading/trailing side ratio spectrum of Callisto is shown in Fig. 5D. The ratio spectrum indicates that Callisto has no UV spectral asymmetry. Inspection of Callisto's opposite hemisphere UV ratio spectrum indicates that there are no spectrally active UV absorbers on Callisto. Given that the hemispheric UV albedo differences are in the same sense as the visual albedo hemispheric differences, then one material may be responsible for both absorptions. Water ice contaminated by trace amounts of elemental sulfur is not inconsistent with this observation, although a host of other possibilities exist. Reanalysis of many of the *IUE* spectra, in addition to analysis of recent *HST* spectra, shows a broad, weak absorption at 2800 Å similar to that seen on Europa. These observations indicate the presence of SO_2 in a few leading hemisphere regions. The source of this SO_2 is not well understood.

Water ice is reflective at ultraviolet wavelengths. Therefore, another material must be responsible for the ultraviolet absorption of the icy Galilean satellites. The most likely darkening agents are elemental sulfur and sulfur-bearing compounds originating from the very young and active surface of Io, which are transported as ions outward from Io's orbit by Jovian magnetospheric processes. These energetic ions and neutrals interact with the icy surfaces of Europa, Ganymede, and Callisto and cause the ices to become darkened at UV wavelengths. This process competes with other processes of surface modification such as infall of interplanetary debris.

B. SATURNIAN SATELLITES

Quantitative spectrophotometric studies of most Saturnian satellites using ground-based telescopes were undertaken concurrent with the launch of *IUE*. Up until then, the only Saturnian satellites that had been subjected to extensive observation were the two most unusual planetary satellites, Iapetus and Titan. Iapetus has a pronounced leading-to-trailing hemisphere albedo asymmetry that is the largest by far of any satellite in the solar system, and it was first reported by Cassini,

the satellite's discoverer, in 1671. Titan is also unusual and was known to have a thick methane atmosphere.

The ground-based observations confirm that all of these satellites, like the Galilean satellites and Earth's Moon, are in synchronous rotation. At visual wavelengths, Tethys, Dione, and Rhea all have leading side albedos that are 10–20% higher than those of their trailing sides, which suggests that there are longitudinal differences in chemical/mineralogical abundance and/or composition in the optically active regoliths of these objects.

The hemispheric albedo asymmetry of Iapetus at visual wavelengths is extremely large (the trailing side is brighter by a factor of 5) and the cycle-to-cycle repeatability is variable, which has been interpreted as being caused by the effect of different scattering properties of the optically active regoliths of the two very different hemispheres.

Infrared observations of the large satellites of Saturn have identified water ice as the principal absorbing species of the optically active surface of Tethys, Dione, and Rhea and the trailing (bright) hemisphere of Iapetus. The leading (dark) hemisphere of Iapetus does not show spectral features consistent with water ice and has an infrared spectrum that is featureless.

The albedo of water ice alone is too high for the surfaces of the satellites to be covered only by this material. Other materials must be present in varying amounts to explain the albedos of all the Saturnian satellites. In the case of the dark hemisphere of Iapetus, the darkening material is most probably the dominant species on the surface. Observations at improved spectral resolution and extended spectal range are required to identify these absorbers on the surfaces of the Saturnian satellites.

A limited number of observations of the Saturnian satellites with *IUE* were undertaken. The ultraviolet geometric albedos for the Saturnian satellites were calculated for the three *IUE* wavelength bandpasses using the same method as was used for Galilean satellites. These are shown in Table II.

The UV albedo of Tethys (~60%) is the highest of the Saturnian satellites and is comparable to the high visual albedo reported by *Voyager* and ground-based visual observations. The leading side of Dione is ~10% brighter than its trailing side, which is somewhat less than the ~30% hemispheric albedo brightness variation reported from ground-based visual wavelength observations. The leading side of Rhea is 40% brighter than its trailing side at the *IUE* wavelengths. This is more than the ~20% observed from the ground at visual wavelengths.

The UV albedo of Iapetus is consistent with the

TABLE II
Ultraviolet Geometric Albedos (%) of the Saturnian Satellites

	Band 1	Band 2	Band 3
Tethys (L)[a]	61 ($N = 1$)	61 ($N = 1$)	62 ($N = 1$)
Dione (L)	27 ± 0.04 ($N = 5$)	27 ± 0.03 ($N = 5$)	29 ± 0.04 ($N = 5$)
Dione (T)	22 ± 0.03 ($N = 4$)	27 ± 0.02 ($N = 2$)	26 ± 0.04 ($N = 4$)
Rhea (L)	26 ± 0.05 ($N = 10$)	27 ± 0.05 ($N = 8$)	30 ± 0.06 ($N = 9$)
Rhea (T)	16 ± 0.09 ($N = 4$)	19 ± 0.09 ($N = 4$)	22 ± 0.12 ($N = 4$)
Iapetus (L)	3 ± 0.009 ($N = 5$)	3 ± 0.001 ($N = 2$)	3 ± 0.002 ($N = 5$)
Iapetus (T)	21 ± 0.01 ($N = 4$)	24 ± 0.01 ($N = 2$)	25 ± 0.02 ($N = 4$)

[a] L = Leading side ($45° < 0° < 135°$); T = trailing side ($235° < 0° < 315°$).

albedos reported at longer wavelengths from ground-based observations and the *Voyager* spacecraft. In the UV, as in the visual, the leading side of Iapetus is extremely absorbing, and the trailing side is comparable to the trailing side albedos of other Saturnian satellites. The leading side albedo is at least 7 times less than the trailing side albedo at the *IUE* wavelengths. This is greater than the 5 times darker reported at visual wavelengths. The spectral absorber that darkens the leading hemisphere of Iapetus is more absorbing toward shorter wavelengths. Efforts to identify this absorber should focus on a similar decrease in reflectance in the laboratory spectrum of any candidate absorber.

The broadband UV albedos reported by *IUE* observations of the Saturnian satellites confirm the suggested differences in chemical/mineralogical composition on Dione, Rhea, and Iapetus that the longer-wavelength observations imply. In the case of Dione, these observations indicate that there are no strong UV absorptions in the unidentified materials on the satellite's surface. In the case of Rhea and Iapetus, the UV absorption becomes greater toward longer wavelengths. This may be due to a gradual decrease in reflectance or may be the effect of an absorption band. If such absorptions occur at the *IUE* wavelengths, they should be detectable in the opposite hemispheric spectral ratios as occurred in the case of the Galilean satellites, most notably on Io.

The individual spectra for each object are few in number and quite noisy when compared to spectra of the much brighter Galilean satellites. For the sake of comparison, the spectra ratios (normalized at 2700 Å) are presented (unsmoothed) in Figs. 6A, 6B, and 6C for Dione, Rhea, and Iapetus, respectively.

Figure 6A is the ratio spectrum of Dione's leading side to its trailing side. The ratio spectrum suggests the presence of a very slight absorption feature

shortward of ~2600 Å. This absorption is not significantly above the noise to permit a positive confirmation of its existence in the *IUE* data set. However, recent *HST* observations of Dione have detected this absorption feature at a significantly high signal-to-noise ratio to confirm its existence. This absorption feature is similar to the 2600-Å absorption feature detected by

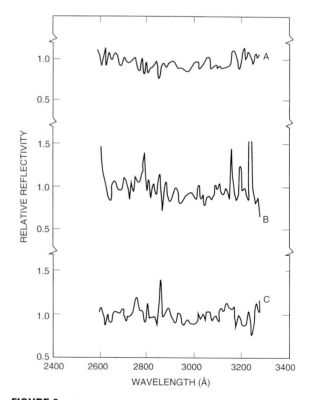

FIGURE 6 The ratio of the *IUE* spectra from the opposite hemispheres of the Saturnian satellites. From top to bottom, they are Dione (leading side to trailing side), Rhea (leading side to trailing side), and Iapetus (leading side to trailing side). Although water ice is present on the surfaces of all of these objects, other unidentified spectral absorbers are present as well.

HST on Ganymede and has been attributed to the presence of ozone on both satellites.

The ratio spectra of Rhea's leading side to its trailing side is shown in Fig. 6B. The noise level in the data is unusually high given that Rhea is brighter than Dione. A number of absorption features can be inferred, in spite of the fact that the albedo asymmetries reported for Rhea at the *IUE* wavelengths are different from those reported at visual wavelengths. This suggests that an absorption feature is present somewhere in the 2600, to 5600-Å range. More recent *HST* ultraviolet observations have detected an absorption feature centered at 2600 Å that is similar to the absorption feature seen on Dione and Ganymede. Thus, Rhea also appears to have ozone present within its surface. Like Ganymede, Dione and Rhea orbit within the magnetosphere of their planet. It has been suggested that the ozone on these satellites may be the product of magnetospheric–surface interactions.

The ratio spectrum of the leading hemisphere of Iapetus to its trailing hemisphere spectra is shown in Fig. 6C. Although there is a very great difference in albedo between the two hemispheres (a factor of 7 at the *IUE* wavelengths), the differences in spectral reflectance are very slight. A possible absorption feature at ~3000 Å is suggested. However, given the noise level of the data, the existance of this feature should by no means be considered certain. Since the hemispheric albedo ratio does increase at the *IUE* wavelengths compared to the visual wavelengths, the possibility of an absorption feature somewhere between 2400 and 5600 Å can be inferred.

C. URANIAN SATELLITES

The five major satellites of Uranus—Miranda, Ariel, Umbriel, Titania, and Oberon—are a suite of icy satellites that are situated at about the limit at which *IUE* can confidently return spectral information. They are so faint that it is not even possible to divide the *IUE* wavelength range into several bands, as was done with the Jovian and Saturnian satellites. All the spectral information is integrated into one wavelength range and a geometric albedo can be determined.

The Uranian satellites are in an orbital plane that is parallel with the Uranian equator, and the pole of Uranus's orbit is tilted such that, at the present time, it is pointed toward Earth. Therefore, only the poles of one hemisphere of the satellites of Uranus are observable with *IUE*, and hence it is not possible to construct orbital phase curves and leading/trailing side ratio spectra.

IUE was able to observe Oberon, Uranus's brightest satellite. The *IUE* result proved to be an important and independent confirmation of results from the *Voyager 2* photopolarimeter experiment. The ultraviolet geometric albedo of Oberon was found to be 0.19 ± 0.025, an excellent confirmation of the earlier *Voyager 2 PPS* result of 0.17.

Spectra from 2200 to 4800 Å were obtained with the *HST* FOC for the Uranian satellites Ariel, Titania, and Oberon. The inner Uranian satellites Miranda and Puck were also observed from 2500 to 8000 Å with the *HST* FOC. The geometric albedos for Ariel, Titania, and Oberon display a broad, weak absorption at 2800 Å, similar to the feature seen on Europa and Callisto. Although this absorption feature on the Galilean satellites has been attributed to SO_2, it has been attributed to OH on the Uranian satellites. Both SO_2 and OH produce an absorption near 2800 Å, however, the molecule OH (a by-product of the photolysis and radiolysis of water) is unstable at the surface temperatures of the Galilean satellites but is stable at the colder surface temperatures of the Uranian satellites. No detection of the 2600 Å ozone feature seen on Ganymede, Dione, and Rhea has been detected in any of the Uranian satellite spectra.

D. DISCUSSION

The ultraviolet observations of planetary satellites can be integrated with the results of observations at longer wavelengths to provide a comparative assessment of the families of large planetary satellites in the solar system. The *IUE*- and PPS-determined photometric properties of the larger planetary satellites of Jupiter, Saturn, and Uranus are shown in Fig. 7 as a plot of ultraviolet-to-infrared color ratio versus ultraviolet geometric albedo. The geometric albedos of the Saturnian satellites indicate that in this system there is a wide variation in UV geometric albedo. The Galilean and Uranian satellites have photometric properties that are common within each group. This is consistent with the hypothesis that the surface modification processes that have occurred are similar within the Galilean and Uranian satellite systems, but the two systems have surface modification processes that are distinct from each other. The diverse nature of the photometric properties of the Saturnian satellites suggests that there is no common surface modification process that is altering the surfaces of the satellites or, alternatively, that the entire satellite system was recently modified and the slow process of space weathering has not had

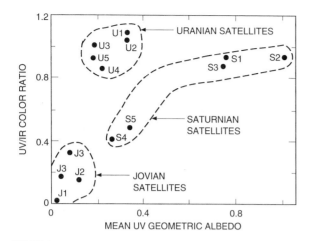

FIGURE 7 The color ratio versus mean ultraviolet geometric albedo of the satellites of Jupiter, Saturn, and Uranus. The clustering of the Jovian and Uranian satellites is consistent with their having similar compositions or surface modification processes. The Saturnian satellites are not entirely alike. This suggests that they have diverse surface composition or surface morphology.

time to restore the surfaces to a common photometric state.

E. PLUTO AND CHARON

The first spatially resolvable images of Pluto and its satellite Charon were obtained by the *HST* FOC in the visible and ultraviolet. The image resolution is sufficient to show the presence of large, longitudinally asymmetric polar cap regions in addition to a variety of albedo markings.

F. ASTEROIDS AND COMETS

The *IUE* satellite obtained ultraviolet observations of several (~45) main belt asteroids in the wavelength range between 2300 and 3250 Å. The geometric albedos for these objects are consistently low, and three major asteroid taxonomic classes seen in the visible persists into the ultraviolet. [See ASTEROIDS.]

Analysis of the *IUE* asteroid data shows that the asteroids observed have ultraviolet albedos that range from 0.02 for C-class asteroids to 0.08 for M-class asteroids. Both *HST* and *IUE* observed the Centaur asteroid 2060 Chiron, a possible former resident of the Kuiper Belt. Neither instrument detected emission from gaseous species at ultraviolet wavelengths, in contrast to CN emissions that have been reported at visible

wavelengths. The UV albedo of Chiron is similar to that of some of the Saturnian and Uranian satellites. Chiron's UV/IR color and ultraviolet albedo are very similar to those of Dione. [See THE KUIPER BELT.]

Observations of comets at ultraviolet wavelengths were first accomplished by sounding rockets and the *OAO* satellite prior to the launch of *IUE*. These observations established the emission of hydroxyl ion at near the limit of ground-based observations, 3085 Å. This is consistent with a principally water ice cometary composition, of which hydroxyl ion is a daughter product of exposure to solar radiation.

Since its launch, *IUE* has observed several dozen comets (~400 individual spectra). *IUE*'s photometric constancy has provided the ability to compare observations of comets that have appeared several years apart. Those observed range from short-period comets with aphelion near Jupiter to long-period comets that may be first-time visitors to our solar system.

All the comets observed by *IUE* have shown the 3085 Å hydroxyl line, which is consistent with water ice being a major part of cometary composition. Although all comets appear to have in common principal compositional components (water), each has different trace components and also different dust-to-gas ratios. *IUE* and *HST* observations have been able to distinguish these differences. Several comets that were observed over a long period of time exhibited differences in their dust-to-gas ratios from one observation to the next, consistent with a variation as a function of heliocentric distance. Though water ice is the major cometary constituent, smaller but detectable amounts of other ices are present and common to all comets, including carbon dioxide, ammonia, and methane. Gas production rates have been derived for species such as H_2O, CS_2, and NH_3.

The first detection of diatomic elemental sulfur in a comet was seen in comet IRAS–Araki–Alcock. Figure 8 shows an *IUE* spectrum of the comet and the positions of the S_2 emission lines are shown. The upper part of the figure shows the spectrum of the area near, but not centered on, the comet and the emission lines of sulfur are absent. This indicates that the lifetime of the diatomic sulfur in the cometary atmosphere is quite short. Based on this information, it is estimated to be about 500 seconds. This makes sulfur a useful tracer of the dynamics of the tenuous cometary atmosphere, which appears during the short time that the comet is near the Sun. Analysis of the S I triplet emission band near 1814 Å in cometary comae spectra taken with *IUE* and the *HST* FOS shows that cometary sulfur, which is present and stored in a variety of volatile species, is depleted in abundance compared to solar

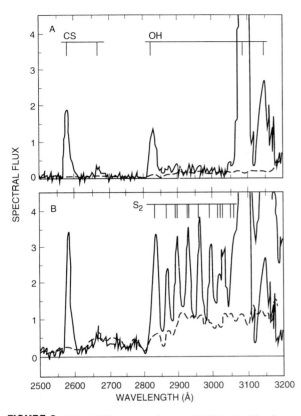

FIGURE 8 Two *IUE* spectra of comet IRAS-Araki-Alcock near perihelion passage. The lower spectrum has the comet directly centered in the *IUE* aperture and shows the strong and unusual emission from diatomic sulfur. The upper spectrum was taken with the comet slightly displaced in the aperture, and it shows the emission of CS and OH in the comet's coma.

abundances. The detection of CS at ultraviolet wavelengths in comae is attributed to the presence of CS_2 in the comet. Sulfur detected in the comae in excess of the sulfur attributable to CS_2 is assumed to originate from H_2S and nuclear S in the comet. Using this assumption, models have been used to measure total sulfur versus water abundances, which range between ~0.001 and ~0.01. [*See* PHYSICS AND CHEMISTRY OF COMETS.]

HST and *IUE* cometary observations have also detected ultraviolet emission bands from CO, which is useful for measuring the CO_2 production rate. This rate derived from *IUE* observations of comet 1P/Halley agrees with the rate measured *in situ* by the spacecraft *Giotto*. These *HST* and *IUE* observations suggest that the level of activity of a comet may be linked to its CO abundance, however, this is based on a small sample of the comet population.

V. CONCLUSIONS

The *IUE* spacecraft provided the astronomical community with the first stable long-term (spanning nearly two decades) observing platform in space, from which astronomers have been able to study regions of the spectrum that are inaccessible from telescopes on Earth's surface. *IUE* was able to observe every solar system object except the Sun and Mercury. This foundation, with the support of ultraviolet spectrometers incorporated into the payloads of deep space missions, filled an observational void that had existed since the dawn of astronomy. These observations have led to important new discoveries and have provided tests of physical models that have been developed based on ground-based observations. Yet much of *IUE*'s observing capability has been surpassed by *HST*. *HST* has provided the astronomical community with the opportunity to look at fainter and more distant solar system objects, which has led to new discoveries in the ultraviolet spectrum.

APPENDIX:
CALCULATION OF THE BOND ALBEDO

If S is the solar energy flux at Earth's orbit, then the amount of energy impinging on a solar system object of radius r at a distance R (in astronomical units) from the Sun is

$$\frac{\pi S r^2}{R^2} \tag{1}$$

If an observer were exactly on the Sun–object line (solar phase angle = 0), a fully illuminated disk would be seen when observing the object. The intensity measured in this case is defined to be $C(0)$. In general, however, an observer is not exactly on the Sun–object line, and therefore a different intensity $C(\alpha)$ will be observed, where α is the solar phase angle. The phase function, $B(\alpha)$, is defined as the ratio of the intensity at zero phase angle to the intensity at phase angle α. In this more general observational circumstance, the rate at which the object reflects radiation in all directions is then

$$2\pi d^2 C(0) \int_0^\pi B(\alpha) \sin(\alpha)\, d\alpha \tag{2}$$

where d is the distance between the observer and the object.

The Bond albedo, A, is then

$$A = 2 \left(\frac{d^2 R^2 C(0)}{S r^2} \right) \int_0^\pi B(\alpha) \sin(\alpha) d\alpha \qquad (3)$$

The term on the right side is often divided into two parts, the phase integral, q, and the geometric albedo, p, where

$$q = 2 \int_0^\pi B(\alpha) \sin(\alpha) d\alpha \qquad (4)$$

and

$$p = \left(\frac{d^2 R^2 C(0)}{S r^2} \right) \qquad (5)$$

The geometric albedo is the ratio of the brightness of the object to the brightness of a perfectly diffusing disk of the same radius at the same distance from the Sun, and is often used to make compositional identifications of the materials in the atmosphere or on the surface of a solar system object.

For the outer planets and their satellites, Earth-based observations can only be made over a narrow range of solar phase angles. Thus, the phase integral, q, cannot be determined and therefore neither can the Bond albedo. However, the geometric albedo as a function of wavelength (i.e., the spectral geometric albedo) is obtainable and proves to be of great use in making comparisons to laboratory spectra of hypothesized atmospheric or surface constituents in order to attempt a compositional identification.

ACKNOWLEDGMENTS

We would like to thank the *IUE* staff astronomers and the director of the *IUE* observatory, Dr. Yoji Kondo, for the many discussions that provided the background for this manuscript. We value highly the many stimulating conversations with Bruce W. Hapke and Arthur L. Lane. We would also like to acknowledge the information obtained from the *HST*, *HUT*, and *EUVE* websites. R.M.N. gratefully appreciates the patience and advice of editor Lucy-Ann A. McFadden in assisting with the final revisions of this manuscript. This work was carried out at the Jet Propulsion Laboratory and the Lunar and Planetary Institute, both under contract with NASA.

BIBLIOGRAPHY

A'Hearn, M. F., Schleicher, Donn, B., and Jackson, M. (1981). *In* "Exploring the Universe with the IUE Satellite" (Y. Kondo, ed.). NASA CP-2171, Washington, D.C.

Ballaster, G. E., Clarke, J. T., Rego, D., Combi, M., Larsenn, N., Ajello, J., Strobel, D. F., Schneider, N. M., and McGrath, M. (1996). *Bull. Am. Ast. Soc.* **28**, 1156.

Budzien, S. A., Scheicher, D. G., and Dymond, K. F. (1997). *Bull. Am. Ast. Soc.* **29**, 1031.

Buie, M. W., and Stern, S. A. (1996). *Bull. Am. Ast. Soc.* **28**, 1079.

Chen, F., Judge, D. L., Wu, C. Y. R., Caldwell, J., White, H. P., and Wagener, R. (1991). *Conf. Lab. Res. Planetary Atmospheres* **96**, 17519–17527.

Clarke, J. T., Ben Jaffel, L., Vidal-Madjar, A., Gladstone, G. R., Waite, J. H., Prange, R., Gerard, J-C., Ajello, J., and James, G. (1994). *Astrophys. J.* **430**, L73–L76.

Courtin, R., Wagener, R., McKay, C. P., Caldwell, J., Fricke, K. H., Raulin, F., and Bruston, P. (1991). *Icarus* **90**, 43–56.

Davis, R. J., Deutschman, W. A., Lindquist, C. A., Nozawa, Y., and Bass, S. D. (1972). *In* "The Scientific Results from the Orbiting Astronomical Observatory OAO-2" (A. D. Code, ed.), NASA SPP-310-1. NASA, Washington, D.C.

Dolls, V., Gerard, J. C., Paresce, F., Prange, R., and Vidal-Madjar, A. (1992). *Geophys. Res. Lett.* **19**, 1803–1806.

Falker, F., Gorder, F., and Sandford, M. C. W. (1987). *In* "Exploring the Universe with the IUE Satellite" (Y. Kondo, ed.). D. Reidel, Dordrecht.

Feldman, P. D., Festou, M. G., Tozzi, G. P., and Weaver, H. A. (1997). *Astrophys. J.* **475**, 829.

Feldman, P. D., Weaver, H. A., and Festou, M. C. (1984). *Icarus* **60**, 455.

Festou, M. C., and Feldman, P. D. (1987). *In* "Exploring the Universe with the IUE Satellite" (Y. Kondo, ed.). D. Reidel, Dordrecht.

Gerard, J. C., Dolls, V., Prange, R., and Paresce, F. (1994). *Planetary Space Sci.* **42**, 905–917.

Ghormley, J., and Hochanadel, C. (1971). *J. Chem. Phys.* **75**, 40.

Gladstone, G. R., Hall, D. T., and Waite, J. H. (1995). *Science* **268**, 1595.

Hall, D. T., Gladstone, G. R., Moos, H. W., Bagenal, F., Clarke, J. T., Feldman, P. D., McGrath, M. A., Schneider, N. M., Shemansky, D. E., Strobel, D. F., and Waite, J. H. (1994). *Astrophys. J. Lett.* **426**, L51.

Hall, D. T., Strobel, D. F., Feldman, P. D., McGrath, M. A., and Weaver, H. A. (1995). *Nature* **373**, 676.

Hammel, H. B., Lockwood, G. W., Millis, J. R. and Barnet, C. D. (1995). *Science* **268**, 1740.

Herbert, F., and Sandel, B. R. (1994). *J. Geophys. Res.* **99**, 4143–4160.

Hord, C. W., Barth, C. A., Esposito, L. W., McClintock, W. E., Pryor, W. R., Simmons, K. E., Stewart, A. I. F., Thomas, G. E., Ajello, J. M., Lane, A. L., West, R. W., Sandel, B. R., Broadfoot, A. L., Hunten, D. M., and Shemansky, D. E. (1991). *Science* **253**, 1548–1549.

Ip, W. H. (1996). *Astrophys. J.* **457**, 922.

James, P. B., Clancy, R. T., Lee, S. W., Martin, L. J., Singer, R. B., Smith, E., Khan, R. A., and Zurek, R. W. (1994). *Icarus* **109**, 79.

Johnson, R. E., and Jesser, W. A. (1997). *Astrophys. J. Lett.* **480**, L79.

Kondo, Y. (1990). *In* "Observatories in Earth Orbit and Beyond" (Y. Kondo, ed.). Kluwer, Dordrecht.

Krasnopolsky, V. A., Bowyer, S., Chakrabarti, S., Gladstone, G. R., and McDonald, J. S. (1994). *Icarus* **109**, 337.

Lane, A., and Domingue, D. (1997). *Geophys. Res. Lett.* **25**, 1143.

Lane, A., Nelson, R., and Matson, D. (1981). *Nature* **292**, 38.

Livengood, T. A. (1992). Ph.D. thesis, Johns Hopkins University, Baltimore, Md.

McClintock, W. W., Barth, C. A., and Kohnert, R. A. (1994). *Icarus* **112**, 382–388.

Meier, R., and A'Hearn, M. F. (1997). *Icarus* **125**, 161.

Moos, H. W., and Encrenas, Th. (1987). *In* "Exploring the Universe with the IUE Satellite" (Y. Kondo, ed.). D. Reidel, Dordrecht.

Na, C. Y., Esposito, L. W., McClintock, W. E., and Barth, C. A. (1994). *Icarus* **112**, 389–395.

Nelson, R. M. (1982). *In* "Proceedings, Conference on the Universe at Ultraviolet Wavelengths—Four Years of IUE," NASA CP-2238. NASA, Washington, D.C.

Nelson, R. M., and Lane, A. L. (1987). *In* "Exploring the Universe with the IUE Satellite" (Y. Kondo, ed.). D. Reidel, Dordrecht.

Nelson, R. M., Lane, A. L., Matson, D. L., Fanale, F. P., Nash, D. B., and Johnson, T. V. (1980). *Science* **210**, 784.

Noll, K., Johnson, R., Lane, A., Domingue, D., and Weaver, H. (1996). *Science* **273**, 341.

Noll, K., Johnson, R., McGrath, M., and Caldwell, J. (1997). *Geophys. Res. Lett.* **24**, 1139.

Noll, K. S., McGrath, M. A., Trafton, L. M., Atreya, S. K., Caldwell, J. J., Weaver, H. A., Yelle, R. V., Carnet, C., and Edgington, S. (1995). *Science* **267**, 1307–1313.

Noll, K. S., Roush, T. L., Cruikshank, D. P., Johnson, R. E., and Pendleton, Y. J. (1997). *Geophys. Res. Lett.*

Noll, K., Weaver, H., and Gonnella, A. (1995). *J. Geophys. Res.* **100**, 19057.

Parker, J. Wm., Stern, S. A., A'Hearn, M. F., Bertaux, J. L., Feldman, P. D., Festou, M. C., Schultz, R., and Weintraub, D. A. (1996). *Bull. Am. Ast. Soc.* **28**, 1083.

Prange, R., Rego, D., Pallier, L., Ben Jaffel, L., Emerich, C., Ajello, J., Clarke, J. T., and Ballaster, G. E. (1997). *Astrophys. J. Lett.* **484**, 1169.

Pryor, W. R., and Hord, C. W. (1991). *Icarus* **91**, 161–172.

Roettger, E. E., and Buratti, B. J. (1994). *Icarus* **112**, 496.

Roush, T. L., Noll, K., Cruikshank, D. P., and Pendleton, Y. J. (1997). *Bull. Am. Ast. Soc.* **29**, 1010.

Skinner, T. E., Durrance, S. T., Feldman, P. D., and Moos, H. W. (1983). *Astrophys. J. Lett.* **25**, L23.

Stern, S. A., Skinner, T. E., Brosch, N., Van Santvoort, J., and Trafton, L. M. (1989). *Astrophys. J. Lett.* **341**, L107–L110.

Stern, S. A., Slater, D. C., Gladstone, G. R., Wilkenson, E., Cash, W. C., Green, J. C., Hunten, D. M., Owen, T. C., and Paxton, L. (1996). *Icarus* **122**, 200.

Stern, S. A., Trafton, L. M., and Flynne, B. (1995). *Astron. J.* **109**, 2855.

Strobel, D. F., Yelle, R. V., Shemansky, D. E., and Atreya, S. K. (1991). *In* "Uranus" (J. T. Bergstralh, E. D. Miner, and M. S. Matthews, eds.). Univ. Arizona Press, Tucson.

Trafton, L. M., and Stern, S. A. (1996). *Bull. Am. Ast. Soc.* **28**, 1080.

Yelle, R. V., Herbert, F., Sandel, B. R., Vervack, R. J., Jr., and Wentzel, T. M. (1993). *Icarus* **104**, 38–59.

INFRARED VIEWS OF THE SOLAR SYSTEM FROM SPACE

Mark V. Sykes
University of Arizona

I. Introduction
II. The *Infrared Astronomical Satellite*
III. Observing Strategies
IV. The Zodiacal Dust Cloud and Its Sources
V. Comets and Their Tails
VI. Asteroids
VII. Pluto
VIII. Other Planets
IX. Earth-Orbiting Debris
X. The Legacy of *IRAS*

GLOSSARY

Azimuthal: Around an axis. An azimuthally symmetric disk would look the same in all directions when viewed from its central axis.

Beta-meteoroids: Particles, usually around 1 μm in diameter, that are so sensitive to radiation pressure that they travel along escape trajectories from the solar system.

Cryogenic cooling: Method of cooling something by bringing it into thermal contact with a liquified gas such as helium, which has a temperature of only a few degrees above absolute zero.

Devolatilize: To remove, usually through heat-

ing and sublimation, any volatiles, such as water, methane, and certain organic compounds.

Ecliptic: Mean plane of Earth's orbit.

Eos family: One of the prominent Hirayama asteroid families, named after its largest asteroid, Eos. Family members have an average semimajor axis of 3.0 AU and orbital inclination near 10°.

Fluence: Rate at which something (mass, numbers of particles, energy) passes through a unit area normal to the direction of the flow.

Geometric albedo: Ratio of the brightness of an object to the brightness of a perfectly diffusing disk of unit reflectivity having the same projected area as the object observed.

Graybody: Sphere that absorbs a fraction of all light incident on it and reemits that energy in all directions as thermal radiation. A graybody that absorbs all light incident on it is considered a "blackbody."

High-pass filtering: Method by which spatial structures that appear narrow in a given direction (e.g., latitude) are segregated and displayed separately from spatial structures that appear to be broad in the same direction.

Hirayama asteroid families: Clusters of asteroids having orbits with similar semi-major axes, eccentricities, and inclinations. The most prominent of these were discovered by Kiyotsugu Hirayama in 1914.

Hyperbolic orbit: Description of the path of a body not bound by the gravity of the Sun. "Orbit" is something of a misnomer, since such a body never returns to the Sun, but rather escapes the solar system.

Koronis family: One of the prominent Hirayama asteroid families, named after its largest asteroid, Koronis. Family members have an average semimajor axis of 2.9 AU and orbital inclination near 2.1°.

Optically thin: Being somewhat transparent, as when the amount of gas or dust blocking a given area being viewed is a small fraction of that area. When something is optically thick, one cannot see through it.

Poynting–Robertson drag: As a consequence of its motion, a particle orbiting the Sun will absorb solar photons that appear to be coming from a direction slightly displaced from the Sun, in the direction of the particle's motion. Solar radiation thus acts to slow down the particle, causing it to spiral in toward the Sun.

Raster-scanned: Pattern of sweeping a detector or detector array back and forth over an area of interest.

Resonance: Effect arising when the orbital periods of two bodies allow them to gravitationally interact at regular periodic intervals of time.

Solar phase angle: Angle between the Sun and Earth as viewed from the object in question.

Themis family: Most prominent Hirayama asteroid family, named after its largest member, Themis. Family members have an average semimajor axis of 3.1 AU and orbital inclination near 1.4°.

I. INTRODUCTION

All objects in the universe radiate heat. The energy distribution of this radiation with wavelength is a function of the temperature of the source. The Sun, at a temperature of more than 5000 K, radiates primarily at visual wavelengths and appears yellow. Colder sources radiate at longer wavelengths. Thus, the heating element of an electric stove appears orange red. Objects in space, such as asteroids, comets, and planets, also radiate, but at wavelengths much longer than can be detected by the human eye. This region of the spectrum is referred to as the thermal infrared. Analysis of the thermal radiation from an object can tell us much about its composition and other physical properties. Observing this radiation from ground-based telescopes is complicated by thermal emission from the telescope itself and the atmosphere, which are much brighter than the astrophysical sources at which they are looking. This has been compared to observing a star in the daytime with the telescope on fire.

Our large-scale view of the sky at thermal wavelengths was obtained primarily by the *Infrared Astronomical Satellite* (*IRAS*), which was launched on January 26, 1983, into Earth orbit. It carried a cryogenically cooled telescope to begin a sensitive and unbiased survey of the sky (Fig. 1). From orbit, *IRAS* measured the thermal radiation of the solar system and beyond, while avoiding the enormous background emission of the atmosphere and ground-based telescopes. Previously, only tiny patches of sky had been observed by different ground-based and aircraft-borne telescopes, and a small sampling of the sky had been made by small-aperture rocket-borne telescopes. *IRAS* successfully surveyed more than 96% of the sky before it ceased operating on November 23, 1983, when its cryogens were depleted. Over the course of its mission, it opened up a new vista on the universe: one dominated not by stars and fuzzy galaxies, but filled with clouds and bands and contrail-like structures—phenomena never before observed (Fig. 2; see also color insert). *IRAS* gathered a wealth of information about astrophysical and solar system objects, which has subsequently been added to on large spatial scales by the *Cosmic Background Explorer* (*COBE*) and on targeted objects by the *Infrared Space Observatory* (*ISO*). No spacecraft, however, has matched the detailed coverage of the sky and the surprises revealed by *IRAS*.

FIGURE 1 An artist's conception of the *Infrared Astronomical Satellite* in orbit.

II. THE INFRARED ASTRONOMICAL SATELLITE

IRAS contained a helium-cooled, 60-cm telescope whose surfaces were maintained at temperatures between 2 and 5 K. Since the objects to be observed by *IRAS* were much warmer than this, contamination by thermal radiation from the telescope was thus minimized. Light passed through the entrance pupil to the telescope, was reflected off the primary to the secondary mirror, and was then reflected back by the secondary through a hole in the primary to an array of detectors mounted at the focal plane of the telescope. The primary mapping mission of *IRAS* was carried out using these detectors, sensitive to infrared radiation over broad passbands centered at 12, 25, 60, and 100 μm. Though the mission was designed primarily with astrophysical sources in mind, the *IRAS* passbands were ideally suited for the study of thermal emission from solar system objects. Peak thermal emission from inner solar system objects near or inside the orbit of Jupiter were well covered by the 12- and 25-μm passbands, whereas emission from outer solar system objects from Uranus to beyond Pluto were well sampled by the 60- and 100-μm passbands (Fig. 3).

The focal plane array consisted of duplicate sets of detectors arranged in parallel (Fig. 4). Consequently, as the image of a source passed over the detector array, it would be sampled twice by detectors in each passband. This allowed real sources to be distinguished from spurious sources such as cosmic rays and Earth-orbiting debris passing near the telescope. This also allowed for more reliable determination of the brightness of a given source.

III. OBSERVING STRATEGIES

IRAS was placed into a near-polar orbit at an altitude of 900 km whose plane was perpendicular to the Earth–Sun line (Fig. 5). As it orbited Earth every 103 minutes, its array of detectors swept out a strip 0.5° wide on the sky. The orbit of the satellite precessed at a rate of about 1° a day, thus maintaining the satellite–Earth–Sun angle (solar elongation) of about 90°. "Looking" straight out from Earth, *IRAS* would have mapped out the entire sky in 6 months as Earth orbited the Sun. To increase this rate of coverage, the orientation of the satellite was shifted to allow it to scan over a range of solar elongation angles. From one orbit to the next, the solar elongation was changed enough to shift the longitude of observation at the ecliptic by about a quarter degree. Consequently, over two consecutive orbits the same location of sky was observed twice (by

FIGURE 2 The sky as seen from space in the thermal infrared is filled with interstellar cirrus (the reddish clouds), rings of dust around the solar system arising from asteroid collisions (one of which is seen as the broad band extending diagonally across the top of the image), and contrail-like structures consisting of cometary debris—the birth of a meteor stream (see below the band). The image is 12° in width. By comparison, the Moon has an apparent diameter of only one-half a degree. This picture was constructed from scans at 12, 60, and 100 μm made by *IRAS*. (See also color insert.)

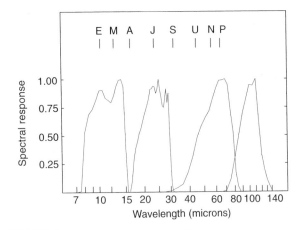

FIGURE 3 The *IRAS* passbands. Relative system response is shown as a function of wavelength for the different detectors. For comparison, the wavelength of peak thermal emission is shown for a graybody at the mean heliocentric distances of Earth (E), Mars (M), the asteroid belt (A), Jupiter (J), Saturn (S), Uranus (U), Neptune (N), and Pluto (P). The graybody is assumed to have a Bond albedo of 0.05 and an emissivity of 0.9.

different halves of the detector array), which allowed real astrophysical sources to be confirmed. By making confirming observations of a given field, moving sources such as asteroids and comets could be more easily identified, as could spurious sources due to the stimulation of detectors by cosmic rays.

Solar elongations were restricted to between 60° and 120°. This prevented sunlight and thermal radiation from Earth from heating any telescope component in the field of view of the detectors, while at the same time allowing a Sun sensor to function, which maintained satellite orientation. Scanning too close to either the Sun or Earth's limb would put a huge thermal load on the cryogenic system, resulting in a more rapid loss of helium and thus a shorter mission lifetime. In addition to the Sun and Earth, *IRAS* also avoided scanning too closely to the Moon, Jupiter, and Mars.

As the satellite scanned the sky, it stored the accumulating data onboard. Every 10 to 14 hours, the satellite would pass over a receiving station at Chilton, England, transmit its data, and receive a new set of observing plans.

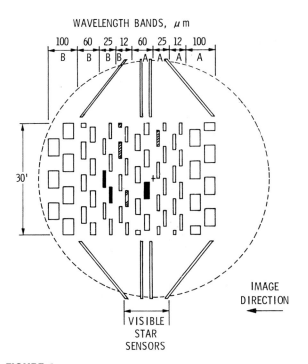

FIGURE 4 The *IRAS* focal plane array. The rectangles indicate the effective field of view of each detector. Solid detectors were inoperative, and cross-hatched detectors showed degraded performance during the mission. Images of sources crossed the field of view along the short dimension of the detectors. An asteroid image would cross a 12-μm detector in region A, then cross another such detector in region B, resulting in a "seconds-confirmed" detection. A spurious source, such as a cosmic ray, would strike only one detector.

The mission was structured so that increasing ecliptic longitudes were observed over a period of about a week, and then the same region of the sky was re-scanned the following week. This pattern was maintained over the first 7 months of the mission, so that 95% of the sky was observed a total of four times. Over the remaining three months of the mission, 76% of the sky was observed another two times.

In addition to the "survey-mode" observations that utilized the orbital motion of the satellite and a fixed solar elongation angle, *IRAS* also conducted over 10,000 "pointed" observations in which a small segment of sky, generally 0.5° × 1.5°, was raster-scanned at slower than survey rates. As with the survey-mode observations, this allowed for more sensitive measurements to be made of specific sources of interest. Fewer than 100 of these pointed observations were made of solar system objects, most of which were comets and asteroids. Fortunately for solar system studies, all pointed observation fields were scanned at least twice, generally an hour and a half (one orbit) apart, for purposes of confirmation. This allowed for the identi-

fication of moving solar system objects that were not specifically targeted in all of the pointed observation fields, as well as the identification of variable astrophysical sources.

IV. THE ZODIACAL DUST CLOUD AND ITS SOURCES

When we think of the solar system, the image that often comes to mind is the textbook picture of planets orbiting the Sun on concentric orbits with the addition of asteroids and the occasional comet flying by. However, in the inner solar system we are immersed in a cloud of dust that we see sometimes in the morning or evening sky as the zodiacal light and sometimes in the direction opposite the Sun as the gegenschein. A variety of parametric models of the shape of this cloud have been proposed—fans, disks, even a "sombrero"—based on its brightness distribution as seen from Earth.

The thermal infrared sky is dominated by emission from warm interplanetary dust at all but the longest wavelengths observed by *IRAS*. At these longer wavelengths, large amounts of cold dust in the galactic plane become prominent. The zodiacal cloud is optically thin, so the *IRAS* observations represent an integration along the line of sight. At large solar elongations, *IRAS* scanned away from the Sun and saw dust extending from Earth's orbit through the asteroid belt. At its smallest solar elongations, *IRAS* scanned through a portion of the cloud that is actually interior to Earth's orbit. Since the satellite is scanning dust that is closer to the Sun, and therefore warmer, the cloud appears brighter at the lower solar elongations (Fig. 6). [See INTERPLANETARY DUST AND THE ZODIACAL CLOUD.]

The origin of the zodiacal cloud has long been thought to be primarily cometary, though estimates of dust production by short-period comets fell far short of that needed to maintain the cloud against losses as the particle orbits inexorably spiral in toward the Sun as a consequence of Poynting–Robertson drag. A cometary cloud would be replenished by the occasional capture of "new," highly active comets into short-period orbits. Comet Encke was suggested as one such possible source in the past. Asteroid collisions have also been considered to be a source of interplanetary dust, but there were few observational constraints on estimates of their contribution.

Our knowledge of the relationship between comets, asteroids, and the interplanetary dust complex is

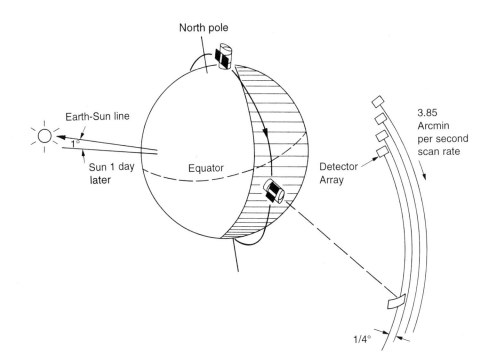

FIGURE 5 *IRAS* orbital geometry. The satellite sweeps out a one-half-degree swath of sky as it orbits Earth, roughly along the day–night terminator.

changing as a result of *IRAS*. Within the broad emission from the zodiacal dust, *IRAS* discovered extensive structures that represent the "missing link" between the small bodies of the inner solar system and zodiacal dust. These are the asteroid dust bands and cometary dust trails.

The asteroid dust bands appear as parallel rings of material extending completely around the sky (Fig. 7). When they were first discovered during the IRAS mission, it was expected that the rings above and below the ecliptic would at some point cross, because the upper and lower bands were thought to be composed of separate groups of material that would have to cross

the midplane given the Sun's central force. When this did not happen, mission scientists were perplexed, until it was realized that the upper and lower bands were part of the same distribution of material having similar orbits that shared a common inclination, but whose nodes (the points where the orbits crossed the ecliptic plane) were spread over all longitudes. In such a case, particles would be distributed within a torus having a somewhat boxy cross section. Particle number densities would be maximum near the corners (Fig. 8), giving rise to the appearance of parallel rings above and below the ecliptic when viewed from Earth.

Initial analysis of the dust bands indicated a location

FIGURE 6 The last scan of the ecliptic plane made by *IRAS* was only 76% complete, but it illustrates the effects of changing the solar elongation angle from scan to scan in order to maximize coverage of the sky over the remaining life of the satellite's mission. The zodiacal cloud is seen extending from 0° to 360° in ecliptic longitude from right to left. Ecliptic latitudes between 30° (top) and −30° are shown. The diagonal structure crossing the ecliptic plane near 90° and 270° longitude is the galactic plane. Where the cloud is bright and wide (in latitude), the sky is being scanned at lower solar elongations, picking up the brighter thermal emissions of the warmer dust that lies closer to the Sun. As the satellite scans farther away from the Sun at higher solar elongations, it is looking through less dust near Earth and seeing a greater fraction of colder, fainter dust.

FIGURE 7 The zodiacal dust bands are seen as parallel bands above and below the ecliptic, encircling the inner solar system. This image is constructed from the same scans used in constructing Fig. 6, high-pass-filtered in ecliptic latitude to reveal structures in the zodiacal cloud. The dust bands lie out in the asteroid belt, so when they are scanned at lower solar elongations we are looking at them from a greater distance and their parallactic separation above and below the ecliptic is smaller than where we scan them at higher solar elongations, at which they are closer and have a greater parallactic separation. This effect and its relation to solar elongation are the opposite of what we see for the zodiacal dust in Fig. 6.

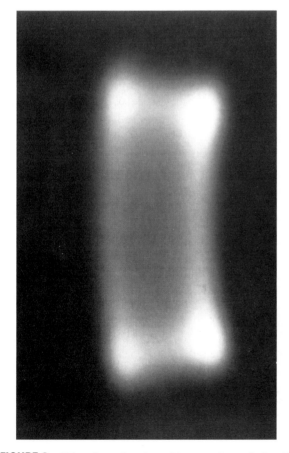

FIGURE 8 When the nodes of an orbit are randomized, they fill a torus. This cross section corresponds to the torus associated with members of the Eos asteroid family. Particle number densities are maximum near the outer surface and are highest near the corners of the squarish cross section. When viewed from near the center of the torus, these particle concentrations appear as parallel rings.

in the asteroid belt, specifically associating them with the most prominent Hirayama asteroid families. The central band pairs are associated with the Themis and Koronis families, and the outer pair have been associated with the Eos family. Each of the asteroid families was formed when a large (100–200 km) asteroid was catastrophically disrupted in a collision, leaving a large number of fragments having somewhat similar orbits. However, collisions continue to break these fragments up into smaller pieces and those smaller pieces into dust. Dust production is continuous down to sizes at which they are finally removed from the production region by radiation forces (Fig. 9). When the fragments are around 1 μm in size, they may be immediately ejected from the solar system along hyperbolic orbits. These are known as beta-meteoroids. Otherwise the solar radiation field and solar wind act as a friction to the particle's orbital motion and it will slowly spiral into the Sun. It is thought that the dust ultimately vaporizes and is incorporated into the Sun or recondenses into small particles that are then lost to the solar system as beta-meteoroids.

The dust bands indicate the primary locations of collisional dust production in the asteroid belt today. That production, together with the gradual orbital decay of the dust, creates much of the zodiacal dust cloud that extends from the outer asteroid belt, where the major asteroid families are located, inward past the orbit of Earth (Fig. 10). Dust production in a given family will slowly decline as more and more of its mass is ground up and removed up radiation forces. However, the zodiacal cloud would not be expected to fade away anytime soon. The more frequent disruption of small asteroids, 10 km in size, is expected to contribute to the dust complex—though such a contribution would be shorter-lived than that from disruption of an object 10 times larger. [See NEAR-EARTH AS-TEROIDS.]

FIGURE 9   A view from outside the solar system of the principal dust production regions in the asteroid belt, associated with the Themis, Koronis, and Eos asteroid families. Earth's orbit is shown for scale.

Although they are significant contributors to the zodiacal dust cloud, the major Hirayama asteroid families do not account for all of it. General dust production from collisional activity in the rest of the asteroid belt likely contributes. In addition, there is still some contribution by comets. The latter is evidenced by interplanetary dust particles collected in Earth's atmosphere and on the surface having low densities and anhydrous mineralogies thought to indicate a cometary origin.

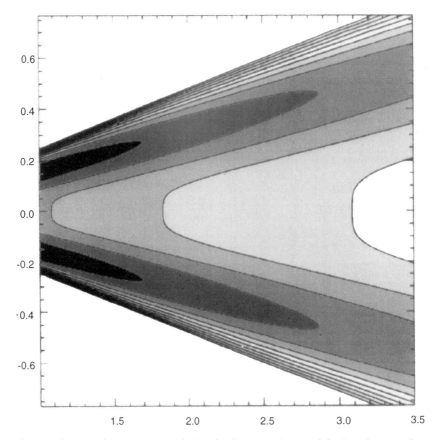

FIGURE 10   As interplanetary dust particles originating in the Eos family migrate in toward the Sun, they contribute to the overall zodiacal cloud. The density contours of their contribution are shown. As the particles evolve to smaller heliocentric distances, the number density increases and the extrema near the upper and lower edges of the cloud component are maintained. (Figure provided by William Reach.)

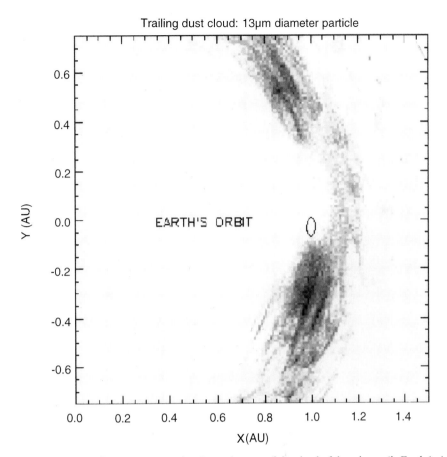

Trailing dust cloud: 13µm diameter particle

FIGURE 11 A simulated image of Earth's resonant ring showing a close-up of the cloud of dust that trails Earth in its orbit through the year. The ellipse shows the orbit of Earth in a rotating reference frame. The resolution of the image is 0.01 AU in the X and Y directions. The proximity of the clump of dust behind Earth near its orbit explains why scans of the sky in this direction were always brighter than scans in front of Earth at the same solar elongation angle. (Figure provided by S. Jayaraman.)

The main problem with a significant cometary component to the zodiacal cloud observed by *IRAS* is the difficulty in smoothly distributing cometary particles over all ecliptic longitudes on timescales smaller than their collisional lifetimes. A cometary cloud may be more azimuthally "bumpy" than is observed. Numerical simulations, however, have indicated that small cometary particles may be put into Jupiter-crossing orbits by radiation pressure and have their orbital nodes rapidly distributed over all longitudes as a consequence of gravitational interactions with the giant planet.

IRAS observations of the zodiacal cloud did have a mysterious aspect to them: scans of the cloud trailing Earth tended to be brighter than scans of the cloud forward of Earth—always. It was thought to be some inexplicable calibration problem down at the 5% level. A solution presented itself when researchers simulated dust slowly spiraling in from the asteroid belt past Earth and found that some of it would be trapped in resonance with Earth's motion in such a way that its distribution relative to Earth was not symmetric: there was a clump that would have the appearance of trailing Earth as it moved around the Sun (Fig. 11). The brighter, trailing cloud can now be explained by this phenomenon.

V. COMETS AND THEIR TRAILS

IRAS has provided new insights into the nature, evolution, and origin of comets. Five months into its mission, *IRAS* discovered its first comet, IRAS-Araki-Alcock (1983 H1), which passed within 0.031 AU of Earth. From the ground the comet appeared to be a typical "gassy" comet, faintly visible to the naked eye, exhibiting a sunward fan of gas and dust. In the infra-

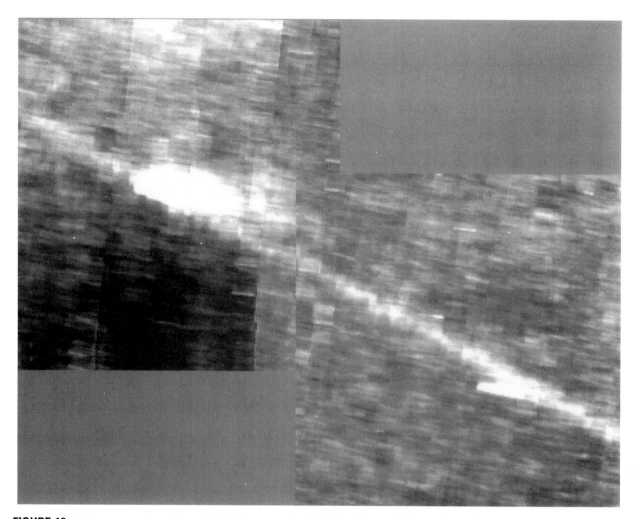

FIGURE 12   The most prominent dust trail in 1983 was associated with the short-period comet Tempel 2. The dust coma and tail appear as the fish to the dust trail's stream. Trails are characteristically narrow (as a consequence of the small relative velocities of the constituent dust relative to the nucleus of the parent comet) and trace out a portion of the comet's orbit. The particles ahead of the comet (to the left) are preferentially larger than those following the comet.

red, however, it presented a far different picture: a huge coma and tail of dust extending behind the comet away from the Sun as much as half a million kilometers. Over the remainder of the mission, *IRAS* discovered five more comets—many more than had been expected before the mission. This was because comets were "dustier" than had been inferred from visual wavelength observations. [*See* PHYSICS AND CHEMISTRY OF COMETS.]

This was further evidenced by the discovery of cometary dust trails. Dust trails derive their name from their long, narrow appearance (Fig. 12). They consist of large (millimeter to centimeter) particles escaping from the comet at low (meters per second) velocities. Radiation pressure increases their orbit sizes slightly, so these particles will have somewhat lower orbital

velocities than their parents and tend to spread out behind it along its orbital path. The largest trail particles are so insensitive to radiation pressure that their motions are determined primarily by their small ejection velocities. In this case, they end up with orbital velocities higher or lower (by a small amount) than their parent comet, and consequently spread along the orbital path both ahead of and behind the comet.

Trails detected by *IRAS* were associated preferentially with those short-period comets having the lowest perihelion distances and that happened to be near perihelion at the time of the *IRAS* mission. These particles were therefore the hottest and intrinsically brightest at thermal wavelengths. The nature of the selection effect for trail detection suggests that trails are a phenomenon associated with all short-period comets, and

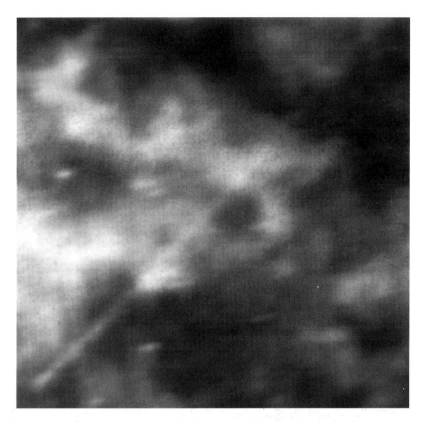

FIGURE 13 The brightest of the "orphan trails" detected by *IRAS*, seen against a background of interstellar clouds. Elongated sources are stars distorted by the rectangular shape of the detectors. Orphan trails are probably associated with comets never before detected. (See also color insert.)

that if *IRAS* were launched today it would likely detect a different ensemble of trails associated with other short-period comets at perihelion.

In addition to trails associated with known short-period comets, *IRAS* also detected trails associated with unknown comets (Fig. 13; see also color insert). Unfortunately, since these were discovered in the data long after the mission had ended, it was not possible to follow up the *IRAS* observations with observations from the ground in order to determine their orbits. So these objects are now lost. However, the numbers of these "orphan trails" suggest that there may be twice as many short-period comets as previously recorded, with the majority of these being less active and hence more difficult to detect by traditional means.

Trails tend to consist of particles emitted over a timespan of many orbits, consistent with the fact that Jupiter gravitational perturbations tend to shift these orbits on the same timescales. The most stable orbits have the oldest trails. The extreme example is Tempel 2, whose trail consists of particles emitted over centuries.

Trails represent the principal mechanism by which short-period comets lose mass. Since these particles quickly devolatilize after leaving the comet nucleus, this means that most of the comet's mass loss is in refractory particles.

The discovery of cometary dust trails is changing the picture of comet nuclei from being primarily icy bodies to objects more akin to "frozen mudballs," because of their much higher than expected fraction of refractory dust. The fraction of "dust to gas" in comet nuclei provides important information about where they formed and how they evolve, once captured into short-period orbits.

Dust-to-gas mass ratios corresponding to the canonical "dirty snowball" model range between 0.1 and 1. If we were to compress comet nucleus material so that refractories have a density of 3 g/cm^3 and volatiles had a density of 1 g/cm^3, this would give us a nucleus in which 3 to 33% of the volume consisted of refractory material. This picture is based largely on ground-based observations of dust at visual wavelengths, sensitive to particles within a decade or so of 1 μm in size. These observations underestimate the mass fluence of dust from comets. Most of the cometary mass loss appears to be in much larger (and dark) particles that are difficult to detect at these wavelengths. This was also the

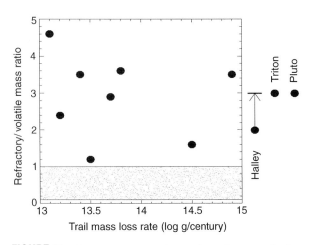

FIGURE 14 Dust-to-gas mass ratios are shown for comets having detected dust trails. For comparison, values are shown for Halley, Triton, and Pluto. The shaded area spans the "canonical" ratios between 0.1 and 1.

conclusion after the European *Giotto* spacecraft flew by comets Halley and Grigg-Skellerup. Analysis of the *IRAS* observations of eight trails indicates that short-period comets lose their mass primarily in refractory particles in the millimeter to centimeter diameter size range. An average dust-to-gas mass ratio of 3 was calculated (Fig. 14). This was the upper limit inferred for Halley by *Giotto* (with a nominal value of 2). Assuming the same densities as here, this corresponds to a comet nucleus that is 75% refractive by mass and 50% by volume (Fig. 15). A simple backyard experiment demonstrates the apt description of such a mixture as a "mudball." [*See* Physics and Chemistry of Comets.]

These dust-to-gas ratios also provide insight into the formation location of short-period comets. Dynamical considerations have led investigators to focus on the proto-Uranus and proto-Neptune regions as

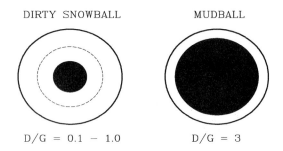

FIGURE 15 The canonical "dirty snowball" model of comets, inferred from ground-based observations at visible wavelengths, is compared to the "frozen mudball" model, inferred from space-based observations in the thermal infrared. All refractories are collected at the center in both cases for the purpose of comparison.

that location. Significant amounts of ice have long suggested the outer solar system as the source of short-period comets. Consideration has also been given to their formation beyond the solar system, for instance, in molecular clouds. Models of comets forming in such interstellar locations yield comets dominated by their volatile components, contradicting inferences drawn from *IRAS* observations. On the other hand, it is very interesting that both Pluto and Triton have effective dust-to-gas ratios that are identical to the average comet values determined from *IRAS* and *Giotto* (see Fig. 14). This is not unexpected if Pluto and Triton accumulated from protocomets in the vicinity of Neptune's orbit. [*See* The Origin of the Solar System.]

The existence of dust trails indicates that short-period comets are losing mass more rapidly than previously thought. Hence, their lifetime against sublimation may be shorter. A greater fraction of refractory material, however, would allow for the rapid formation of a nonvolatile mantle that is difficult to blow off, progressively choking off cometary activity. Such a mantle was apparent in the *Giotto* images of the Halley nucleus, which was near perihelion at the time. When activity is choked off, the comet would look like an asteroid until such time as sufficient pressure built up from subsurface ices to break through the crust in a burst of resumed activity. The discovery in August 1992 that asteroid 4015 was actually comet P/Wilson-Harrington (last seen in outburst in 1949) provided the first incontrovertible evidence of such "dormant" comets in the inner solar system.

VI. ASTEROIDS

The *IRAS* observations constitute the single largest, most complete, and least biased asteroid survey ever made at thermal wavelengths. Comparison with the orbits of asteroids known in 1983 resulted in the identification of observations of 1811 asteroids, which led to the detection of a large number of asteroids with unknown orbits. [*See* Asteroids.]

Although *IRAS* scanned 96% of the celestial sphere, it scanned a much smaller fraction of the asteroid population. This was due to a gap in the coverage of the asteroid belt in combination with asteroid motion in the direction that the satellite was precessing, thus staying away from its scan path. Figure 16 shows the positions of more than 15,000 asteroids on the day that *IRAS* was launched. The overlying concentric rings (at heliocentric intervals of 0.2 AU) show locations where the satellite scanned the asteroid belt in a rotat-

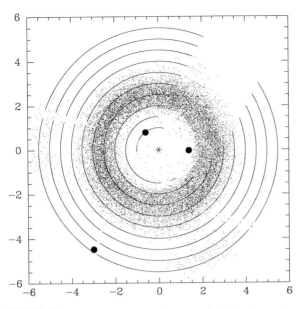

FIGURE 16 *IRAS* sampling of the asteroid belt is shown in a dynamical reference frame. Scan positions at a particular heliocentric distance, indicated by concentric rings spaced by 0.2 AU, are compensated for the motion of an object in a circular orbit since the mission start time. Positions for 15000 asteroids are shown. Those that fall in the "gap" areas were missed in the survey. The positions of Earth, Mars, and Jupiter are given as reference. Jupiter was deliberately avoided as a target because of its brightness at thermal wavelengths.

ing reference frame that compensates for the motions of objects in circular orbits. Generally speaking, the asteroids in the scan gaps were not observed by the satellite. These include a preferentially large number of inner main belt objects, as well as the Hildas (some of which are clustered on the other side of the Sun from Jupiter at the outer edge of the main belt owing to their 2 : 3 resonance with the planet). As a consequence of their geometry and large heliocentric distance, the Jupiter Trojans (which are in a 1 : 1 resonance with the planet) were the most thoroughly scanned population. For the same reason, however, they are colder and less visible at thermal wavelengths and only their larger members (>50 km) were detectable by *IRAS*.

When thermal observations of asteroids are combined with their visual magnitudes it is possible to determine their diameters and albedos. The models used for this purpose presume the asteroids to be spherical with uniform surface properties. Several observations over a range of solar phase angles can also be used to constrain other properties, such as the thermal inertia of the surface. The large number of observations by *IRAS* has allowed for the distribution of these properties to be determined for various asteroid populations.

Known asteroid diameters inferred from each detection by *IRAS* are shown in Fig. 17 as a function of heliocentric distance. The absence of small asteroids with increasing heliocentric distance is a consequence of the limits of detector sensitivity. The absence of larger asteroids beyond 4 AU, however, is real.

Albedo provides some insight into composition. Meteorite studies show that very dark surfaces arise from largely carbonaceous materials, whereas high-albedo surfaces are associated with silicic compositions lacking such carbonaceous material. *IRAS* confirmed that most C-type asteroids (thought to be carbonaceous) are indeed dark compared to the "stony" S-type asteroids, and that there is a trend toward darker asteroid surfaces with increasing heliocentric distance (Fig. 18). This is consistent with the view that not only is there a primordial composition gradient through the asteroid belt, but that inner belt asteroids (predominantly S-type) were significantly processed by heating in the early solar system, which melted them, whereas the outermost asteroids experienced little heating and have retained a more "primitive" mineralogy.

Only the largest asteroids are thought to have survived since the formation of the solar system. There are so many asteroids in crisscrossing orbits that a significant amount of collisional evolution has occurred throughout the asteroid belt since that time. Most of the asteroids we observe today, particularly the smaller ones, are the fragments of larger asteroids that were destroyed in a catastrophic disruption. For some period, these fragments will have similar orbits and would be expected to form clusters when asteroid orbital elements are plotted together. Such clusters were first noticed by Kiyotsugu Hirayama in 1914, and many investigators since then have identified many more "families" of asteroids as more asteroids have been found. There has been some controversy as to whether all such families mark the site of a past catastrophic disruption or whether in some cases asteroids might be clumped together because of dynamical forces such as gravitational perturbations on their orbits by Jupiter. Albedo distributions can also provide clues to the origin of some asteroid families. Assuming the parent to have been compositionally homogeneous, the fragments should exhibit similar spectral properties. On the other hand, members of purely dynamical clusters would not be expected to have similar compositions. *IRAS* scanned enough of the members of the largest families to show that family members had albedos more similar to each other than to the background

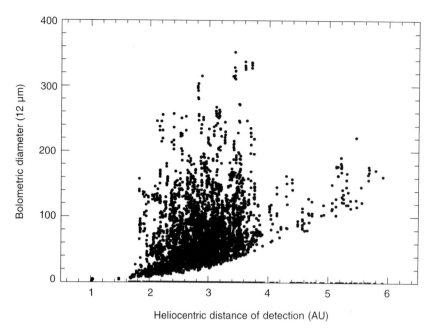

FIGURE 17 Asteroid diameters versus heliocentric distance of detection. The lower limit of detected asteroid sizes reflects the sensitivity limits of the *IRAS* detectors. The outer asteroid belt is shown to have few large asteroids compared to the inner belt.

asteroids nearby, giving support, in those cases, to the asteroid breakup hypothesis (Fig. 19).

New observations often result in as many new questions as new answers, and the *IRAS* asteroid observations are no exception. Prior to *IRAS*, ground-based thermal observations had been preferentially made of the largest asteroids. It was noticed that there was a bimodal distribution in the inferred albedos, which was consistent with the main belt asteroid population being dominated by dark C-type and bright S-type

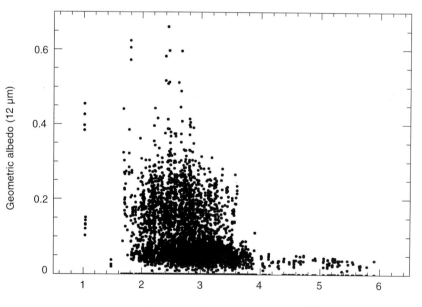

FIGURE 18 Geometric albedo versus heliocentric distance of known asteroids detected by *IRAS*. The high-albedo asteroids are located almost exclusively in the inner portion of the main asteroid belt.

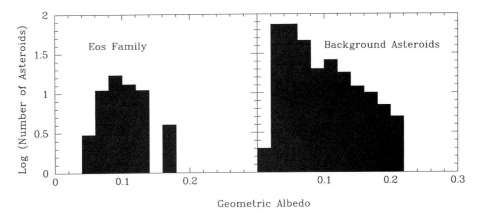

FIGURE 19 A comparison of the albedo distributions of the Eos asteroid family and of nonfamily members near the same location in the asteroid belt supports the hypothesis that the family members derived from a single parent body.

asteroids. *IRAS* added large numbers of observations of smaller asteroids and it was found that they had an albedo distribution quite different from that of the larger asteroids. Small asteroids have a unimodal distribution spanning the total range of albedos inferred for the large asteroids (Fig. 20). Since the small asteroids are fragments of larger asteroids, this might imply that surface mineralogies are not representative of interior mineralogies or that significant "space weathering" may affect the surface spectra of the larger bodies.

VII. PLUTO

Thermal radiation from Pluto and its moon Charon was first detected by *IRAS* (Fig. 21). The thermal flux of the system was consistent with that of a rapidly rotating graybody having an equatorial temperature of

~60 K. This information, in combination with ground-based spectroscopic measurements and albedo maps derived from the mutual eclipses between Pluto and its moon between 1984 and 1990, has provided important insights into the nature and dynamics of the surface of Pluto. [*See* PLUTO AND CHARON.]

When methane was first detected on the surface of these objects, it was thought that Pluto must be completely covered by the frozen ice, and would be isothermal because of the transport of heat as high-insolation locations would be cooled by sublimation and less insolated locations would be warmed by the condensation of atmospheric methane. Charon was thought to be a less likely location of such a coating of methane frost because of its lower gravity, from which methane would be expected to escape over time.

The detection of an extended atmosphere from a stellar occultation in 1988 and the subsequent detection of nitrogen ice on Pluto's surface required that

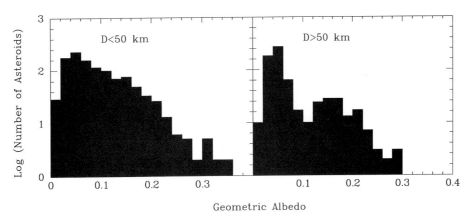

FIGURE 20 Large and small asteroids show different albedo distributions.

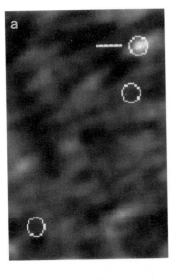

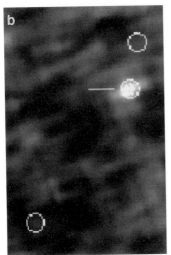

FIGURE 21 Pluto–Charon were detected moving across the infrared sky by *IRAS*. These images were constructed from 60-μm scans for (a) July 13, (b) July 23–24, and (c) August 16 in 1983. The predicted positions of Pluto–Charon at each of these times are indicated by circles. The August 16 position is the lower left circle in (a) and (b).

the volatile surface ices be dominated by nitrogen with a small fraction of the more spectroscopically active methane, and that these surface ices must be very cold, ~35 K.

The spectroscopic and *IRAS*-derived temperatures appear to contradict each other. Nitrogen ice at the warmer *IRAS* value would produce an enormous atmosphere that would have been evident in the occultation observations. The surface albedo maps, however, show that Pluto's surface is segregated into bright and dark regions with bright ices generally at higher latitudes. A high-albedo surface is bright at visible wavelengths (reflected light) and faint at thermal wavelengths, whereas a dark surface is faint at visible wavelengths and bright in the thermal. On Pluto, visible-wavelength spectroscopy samples primarily the icy polar regions of the planet, and *IRAS* detected emission dominated by the dark equatorial region.

IRAS tells us that the volatile nitrogen ice—from which the atmosphere derives—is segregated on the surface of Pluto, away from the warmer regions, giving rise to the thermal emission detected by the satellite. These warmer regions are probably a mixture of water ice and carbonaceous residue resulting from the radiation processing of methane ice over the age of the solar system. The dark regions are not contributing significantly to the atmospheric gases, which is consistent with a water/organic composition that would have negligible vapor pressure at the temperature inferred.

VIII. OTHER PLANETS

Jupiter and Mars were avoided by *IRAS* because of their thermal brightness. Even though the satellite avoided directly scanning Jupiter, the planet was bright enough to cause observable diffraction spikes (from secondary mirror struts) several degrees away. When observed in the early *IRAS* image products, they looked very similar to cometary dust trails (and confused the author, who identified one as a possible trail associated with P/Shoemaker 2). Mercury and Venus were too close to the Sun to be within the range of solar elongations scanned by *IRAS*. Saturn was scanned, but it saturated the detectors, appearing as a large spiky smudge in the *IRAS* image products (Fig. 22). Diffraction spikes are also apparent in the vicinity of the planet. Uranus and Neptune were used as sources to provide some verification of the transmission functions of the 60- and 100-μm bandpasses.

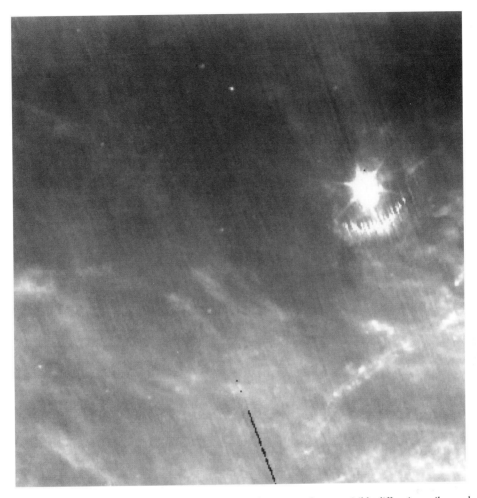

FIGURE 22 Saturn at 100 μm is so bright that it saturates the *IRAS* detectors and causes visible diffraction spikes and other artifacts in this image constructed from *IRAS* scans. Other sources in the background are the infrared cirrus and galaxies (appearing as point sources).

IX. EARTH-ORBITING DEBRIS

Several decades of human activity (manned and unmanned) in Earth orbit have resulted in a rapidly growing cloud of debris around Earth, which is an increasing hazard to operations there (Fig. 23). Though *IRAS* was not affected by the debris environment in which it flew, it did detect a number of fragments that passed within its field of view. Most appeared as "non-seconds-confirming" sources (see Fig. 4) rapidly moving across the sky kilometers to hundreds of kilometers away. Others came so close that they illuminated the entire focal plane array, sometimes saturating the detectors. Near-field sources were identified in *IRAS* image products as an image of the detector array on the sky (Fig. 24).

X. THE LEGACY OF *IRAS*

The unbiased nature of the *IRAS* survey allows the data to be continually reexamined as we learn more and more about the diverse phenomena in the universe and as new questions arise. Whenever a new asteroid or short-period comet is discovered, we can look to see if *IRAS* detected it. Searches for new classes of objects, such as Earth Trojans, have been conducted by reprocessing the data with a different hours-confirmation strategy that takes specific kinds of motion into account. Debris in the orbits of newly discovered dormant comets can be sought.

Even if the *IRAS* mission were completely redone with more sensitive detectors, the old data would still retain their value owing to the dynamic nature of the

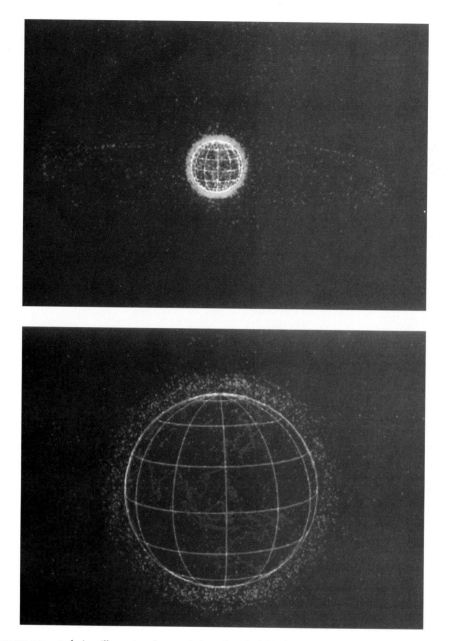

FIGURE 23   Computer-generated view illustrating the population of tracked Earth orbital debris on December 14, 1990; it is typical of such a view at any time. In the lower image are those in low Earth orbit predominantly below 2000 km. Most debris is either at very high inclination, nearly crossing the poles, or at relatively low inclination, rarely going above 30° latitude. In the upper image, the view is from far out in space; one can see the arc of geostationary objects over the equator and the highly inclined Molnia orbits used by the Russians for communication at very high latitudes. (Reproduced from Interagency Report on Orbital Debris, 1995, November 1995.)

universe. Solar system phenomena in particular are far from static. Bodies change in temperature with changing heliocentric distance and orientation. Their positions change. Any new survey would undoubtedly see many asteroids missed by *IRAS*, and miss many that were scanned. Just as the Palomar Sky Survey continues to be an extensively used tool by the astronomical community decades after it was conducted,

the *IRAS* data stream and resultant data products will continue to be an inexhaustible resource on which researchers can draw.

Future spacecraft, such as the *Wide-field Infrared Explorer* (*WIRE*) and the *Space Infrared Telescope Facility* (*SIRTF*), will add depth of understanding to the breadth of information provided by *IRAS* as they conduct detailed investigations of thousands of solar

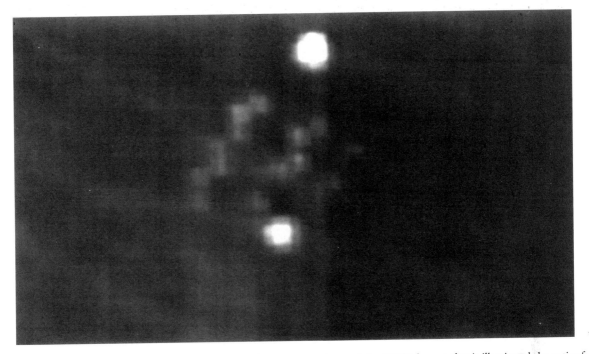

FIGURE 24 Some orbital debris passed close enough to the entrance aperture of the *IRAS* telescope that it illuminated the entire focal plane array. This resulted in a characteristic signature: an image of the array on the sky. By comparison, nearby stars appear pointlike.

system and extra-solar system objects. These missions will examine them over a wider range of infrared wavelengths, as well as in greater spectral and spatial detail.

We have now seen what the sky looks like were we to travel into space with infrared eyes. It is filled with giant clouds. Thick bands of dust encircle it. Narrow trails extend tens of degrees across it. Asteroids, stars, distant galaxies, and other objects pepper it. The galactic plane shouts brightly. At visible wavelengths, the sky is black, deep, cool, and filled with stars. In the infrared, it roils. It is a beautiful sight.

BIBLIOGRAPHY

Neugebauer, G., Habing, H. J., Clegg, P. E., and Chester, T. (eds.). (1988). "IRAS Explanatory Supplement." U.S. Government Printing Office, Washington, D.C.

Sykes, M., and Walker, R. (1992). Cometary dust trails. *Icarus* **95**, 180–210.

Sykes, M., *et al.* (1989). Dust bands in the asteroid belt. *In* "Asteroids II" (R. Binzel, T. Gehrels, and M. Matthews, eds.). Arizona Press, Tucson.

Tedesco, E. (ed.) (1992). "IRAS Minor Planet Survey," Phillips Laboratory Report No. PL-TR-92-2049.

THE SOLAR SYSTEM AT RADIO WAVELENGTHS

I. Introduction

II. Terrestrial Planets and the Moon

III. Giant Planets

IV. Small Bodies

V. Nonthermal Radiation

VI. Future Directions

Imke de Pater

University of California, Berkeley

GLOSSARY

Adiabatic temperature structure: For an atmosphere that is marginally unstable to convection, and where there is no heat transfer between the rising and sinking parcels of air with the environment, the temperature profile with altitude follows a so-called adiabat. The dry adiabatic lapse rate on Earth is roughly 10 K/km.

Blackbody radiation: Continuous spectrum of electromagnetic radiation emitted by the thermal activity of the electrons in an object that absorbs all radiation incident on it.

Brightness temperature: Temperature that a body would have if it were a black-body emitting the same amount of radiation as received at the observing wavelength. It can also be defined as the *radiant intensity* scaled to units of temperature by $\lambda^2/2k$ (λ = wavelength, k = Boltzmann's constant).

Dipole magnetic field: Magnetic field structure similar to that of a bar magnet.

Dynamical spectrum: Graph of the emission intensity as a function of frequency and time. Usually the intensity is shown on a gray scale ranging from black to white, with frequency plotted along the vertical y axis and time along the horizontal x axis.

Flux density: Power per unit area and per unit frequency interval received from an object. The units of flux density are in Jansky; 1 Jy = 10^{-26} W/m²Hz.

Nonthermal radio emission: Radio emission produced by processes other than those that produce thermal emission. In particular, in planetary science we are concerned with *cyclotron* and *synchrotron* radiation. Cyclotron or auroral radiation is emitted by (nonrelativistic) electrons in the auroral (near-polar) regions of a planet's magnetic field at the frequency of gyration around the magnetic field lines (*cyclotron frequency*). The emission resembles a hollow cone pattern. Synchrotron radiation is produced by relativistic (i.e., particle velocity approaching the speed of light) electrons. This radiation is strongly beamed in the direction in which the particle is moving.

Radio spectrum: Graph of the brightness temperature as a function of wavelength or frequency.

Thermal radio emission: Continuous radio emission from objects resulting from the thermal activity of their electrons.

Van Allen belts: Also referred to as radiation belts; a region in Earth's magnetic field, inside of ~4 Earth radii, filled with energetic particles.

Ground-based radio astronomical observations of planets, satellites, asteroids, and comets provide information that is complementary to that obtained at visual and infrared wavelengths. One typically receives thermal emission from planetary bodies and nonthermal emissions from charged particles in a planet's magnetosphere. The thermal emission provides information regarding a body's atmosphere and surface layers, whereas the nonthermal radiation can be used to retrieve information regarding the planet's magnetic field and charged particle distributions.

I. INTRODUCTION

A. BLACKBODY RADIATION

Any object with a temperature above absolute zero emits a continuous spectrum of electromagnetic radiation at all wavelengths, which results from the thermal activity of its electrons. This emission is referred to as thermal or "blackbody" radiation. A blackbody radiator is defined as an object that absorbs all radiation that falls on it at all frequencies and all angles of incidence; none of the radiation is reflected. Blackbody radiation can be described by Planck's radiation law, which, at radio wavelengths, can usually be approximated by the Rayleigh–Jeans law:

$$B_\nu(T) = \frac{2\nu^2}{c^2} kT \qquad (1)$$

where $B_\nu(T)$ is the brightness (W/m²/Hz/sr), ν is the frequency (Hz), T is the temperature (K), k is Boltzmann's constant (1.38×10^{-23} J/deg K), and c is the velocity of light (3×10^8 m/s). Using radio telescopes, radio astronomers measure the power flux density emitted by the object. A common unit is the flux unit or Jansky, where 1 Jy = 10^{-26} W/m²Hz. If a planet emits blackbody radiation and has a typical size of $2a \times 2b$ arcsec, the flux density can be related to the temperature of the object:

$$S = \frac{abT}{4.9 \times 10^6 \, \lambda^2} \qquad (2)$$

with λ the observing wavelength (in m) and T the temperature (in K). Usually, planets do not behave like a blackbody, and the temperature T in Eq. (2) is called the brightness temperature, defined as the temperature of an equivalent blackbody of the same brightness.

B. RADIO ANTENNAS AND INTERFEROMETRY

The radio power emitted by an object is received by a radio telescope, which basically consists of an antenna and a receiver. The sensitivity of the antenna depends on many factors, but the most important are the effective aperture and system temperature. The effective aperture depends on the size of the dish and the aperture efficiency. The sensitivity of the telescope increases when the effective aperture increases and/or the system temperature decreases. Figure 1a shows a photograph of one of the BIMA (Berkeley–Illinois–Maryland Array) antennas, an array of telescopes that operates at millimeter wavelengths.

The response of an antenna as a function of direction is given by its antenna pattern, which consists of a "main" lobe and a number of smaller "side" lobes, as depicted in Fig. 2a. The resolution of the telescope depends on the angular size of the main lobe. It is common to express the main lobe width as the angle between the directions for which the power is half that at lobe maximum; this is referred to as the half-power beamwidth. This angle depends on the size of the dish and the observing wavelength: for a uniform illumination, the beamwidth is approximately λ/D radians, with D the dish diameter in the same units as the wavelength λ. In practice, the illumination is tapered off toward the edge, and the resolution is roughly $(1.2\lambda/D)$ radians. At 6-cm wavelength, an antenna with a diameter of 25 m has a resolution of approximately 10 arcmin. A large planet has a diameter of 0.5–1 arcmin and would be unresolved.

The resolution of a radio telescope can be improved by connecting the outputs of two antennas that are separated by a distance S at the input of a radio receiver. Such a system is called a radio interferometer. The response to an unresolved radio source is an interference pattern, as sketched in Fig. 2b, where the maxima are separated by an angle λ/S' radians, where S' is the baseline length as projected on the sky. As shown in Fig. 2b, the single-antenna output is essentially modulated on a scale λ/S'. This angle is the resolving power of the interferometer in the direction of the projected baseline S'. Assume that the interferometer is built along the east–west direction, and that you observe a radio source X. Because of Earth's rotation, the baseline projected onto the sky at X will trace out an ellipse during the course of the day. The coordinates of this

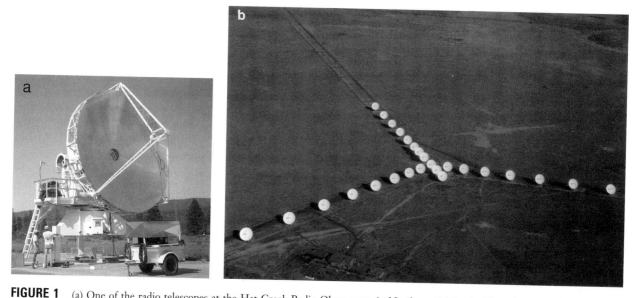

FIGURE 1 (a) One of the radio telescopes at the Hat Creek Radio Observatory in Northern California. The telescope has a diameter of 6 m and receives emission at millimeter wavelengths. It is part of an array of telescopes, operated by the University of California at Berkeley, the University of Illinois, and the University of Maryland. (b) Aerial photograph of the Very Large Array of radio telescopes in New Mexico. (Courtesy of the National Radio Astronomy Observatory, operated by the Associated Universities, Inc., under contract with the National Science Foundation.)

ellipse are generally referred to as the u (east–west on the sky) and v (north–south on the sky) coordinates in the (u, v)-plane. The parameters of the ellipse depend on the declination of the radio source, the length and orientation of the baseline, and the latitude of the center of the baseline. The ellipse determines the angular resolution on the radio source.

The VLA (Very Large Array) in Socorro, New Mexico, consists of a Y-shaped track with nine antennas along each of the arms. An aerial photograph is

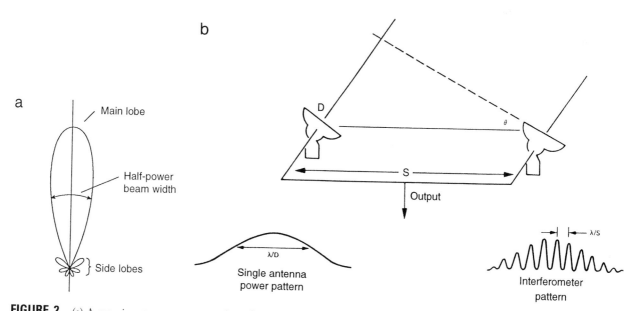

FIGURE 2 (a) A generic antenna pattern consists of a "main" lobe and a number of smaller "side" lobes. The half-power beamwidth is the full width at half power (FWHP). [After J. D. Kraus (1986). "Radio Astronomy." Cygnus Quasar Books, Powell, Ohio.] (b) Top: Geometry of a two-element interferometer. Bottom: Antenna response for a single element of the interferometer (left) and response of the interferometer (right) to an unresolved radio source. [From S. Gulkis and I. de Pater (1992). Radio astronomy, planetary. *In* "Encyclopedia of Physical Science and Technology," Vol. 14, pp. 141–164. Academic Press, San Diego.]

shown in Fig. 1b. The antennas operate at centimeter wavelengths, and each antenna is connected with each of the others into an interferometer. With this instrument one can thus gather data from 351 individual interferometer pairs, which each trace out their own unique ellipse in the (u, v)-plane; in other words, each interferometer pair has its own instantaneous resolution along its projected baseline S'. With such an array of antennas one can build up an image that shows both the large- and small-scale structure of a radio source. At short spacings the entire object can be "seen," but details on the planet are washed out owing to the low resolution of such baselines. At longer baselines, details on the planet can be distinguished, but the large-scale structure of the object gets resolved out, and hence would be invisible on the image unless short spacing data are included as well. Hence, arrays of antennas are crucial in imaging an object at radio wavelengths.

C. IMAGING

Figure 2b showed a sketch of the fringe pattern received from an unresolved radio source. If there were two unresolved radio sources, separated by one-half the fringe period, $\lambda/2S'$, the fringe patterns of the two sources would be in antiphase and cancel each other. If the source is resolved with an arbitrary brightness profile, the responses from different parts of the source correspond to different fringe phases and amplitudes and combine vectorially. In general, to recover the spatial brightness distribution of an object, one needs to measure both the amplitude and phase of fringes received (called the complex visibilities) by many interferometer pairs. This measurement forms the basis of mapping by Fourier synthesis in radio astronomy.

In practice, to construct an image of the object, the responses of all the individual interferometer pairs, the visibility data, are gridded into cells having uniform intervals in projected antenna-pair separation on the sky, or the (u, v)-plane. This grid of data is then Fourier-transformed to give a map of the brightness distribution on the sky. Before this image can be analyzed, however, one needs to remove the response of the synthesized antenna pattern from the image. The synthesized antenna pattern is essentially the response of the instrument to a point source in the sky. Since the antenna pattern (see Fig. 2a) consists of a main lobe and a large number of side lobes, the image can be severely distorted. The antenna response can be modeled to high accuracy and can be removed from the observed image by using a deconvolution technique.

The most widely used technique is the CLEAN algorithm.

CLEAN deconvolves an image by an iterative procedure in which a fraction of the interferometer response to the brightest point remaining on the map is successively subtracted. For planets, one usually first subtracts a model of a uniform planet, which consists of a set of equal brightness points that uniformly fills an oblate disk, with a size and total flux density equal to that of the planet itself. The CLEAN process is repeated until the entire source is removed from the map. All CLEAN components (including the disk model) are then convolved with a Gaussian beam that best fits the central part of the antenna pattern and are restored to the map. At this point, one has a reasonable image of the planet that can be used for research, although the quality of the image (dynamic range) can still be improved by applying "self-calibration" techniques or different deconvolution or cleaning algorithms.

D. SPECTRA

A spectrum of a planet can be obtained by observing the object at different frequencies, for example, in a series of channels where each channel has a finite width in frequency (or wavelength). Radio spectra can vary from a broadband quasi continuum to narrow line spectra. Broadband quasi-continuum spectra consist of observations at discrete wavelengths covering a large fraction of the radio band; for example, thermal spectra of the giant planets typically cover the wavelength range between 0.01 and 50 cm. Narrow line spectra have resolutions varying from a few kHz up to many MHz. They are centered at the frequency of a particular energy transition in a molecule, for example, the $J = 1-0$ or $2-1$ rotational transition of the carbon monoxide (CO) molecule.

Conventional spectra are obtained using single antennas, where the total flux density of the object is recorded as a function of frequency. One can also image an object in the spectral line mode; the result is a data cube, in which a third dimension is added to the two-dimensional brightness distribution of the object in units of frequency (wavelength or velocity).

The low-frequency radio emissions detected by spacecraft are usually represented as a dynamical spectrum, which displays the emission intensity as a function of frequency (y axis) and time (x axis).

In Sections II–IV, ground-based observations of the thermal emission of planets, satellites, asteroids, and comets are discussed. Nonthermal emissions from the giant planets and Earth, usually observed only by spacecraft, are summarized in Section V.

II. TERRESTRIAL PLANETS AND THE MOON

Radio observations of the terrestrial planets and the Moon can be used to extract information on the (sub)surface layers of these objects and, for Venus and Mars, on their atmospheres. The temperature structure of the (sub)surface layers of airless bodies depends on a balance between solar insolation, heat transport within the crust, and reradiation outward. The fraction of the solar flux absorbed by the surface depends on the object's albedo, whereas the energy radiated by the surface (at a given temperature) depends on its emissivity. During the day, a planet's surface will heat up and reach its peak temperature at noon or early afternoon (the exact time depends on the body's thermal inertia, see the following), while at night the object cools off. Its lowest temperature is reached just before sunrise. Since it takes time for the heat to be carried downward, there will be a phase lag in the diurnal heating pattern of the subsurface layers with respect to that at the surface, and the amplitude of the variation will be suppressed. At night, heat is carried upward and radiated away from the surface. Hence, during the day the surface is hotter than the subsurface layers, and at night the opposite is true.

The amplitude and phase of the diurnal temperature variations and the temperature gradient with depth in the crust are largely determined by the thermal inertia and the thermal skin depth of the material. The thermal inertia measures the ability of the surface layers to store energy and is defined by

$$\gamma = \sqrt{K\rho C} \tag{3}$$

The amplitude of diurnal temperature variations is largest at the surface and decreases exponentially into the subsurface, with an e-folding scale length equal to the thermal skin depth:

$$L_t = \sqrt{\frac{2K}{\Omega\rho C}} \tag{4}$$

Variations in γ are dominated by variations in the thermal conductivity, K. The parameters γ and L_t further depend on the density of the material (ρ), the specific heat (C) or heat capacity (ρC), and Ω, the average angular velocity of the Sun as viewed from the surface. Temperature variations are always largest at the surface, where the temperature is determined directly by the solar radiation. Note that, averaged over the day, the surface must be in radiative equilibrium. If the thermal conductivity is low, the amplitude of the temperature wave is large and it does not penetrate deeply into the crust. Alternatively, if K is high, temperature variations are smaller, but penetrate to greater depths in the subsurface layers.

Radio waves typically probe a depth of ~10 wavelengths into the crust. Hence, by observing at different wavelengths, one can determine the diurnal heating pattern of the Sun in the subsurface layers. Such observations can be used to constrain thermal and electrical properties of the crustal layers. The thermal properties relate to the physical state of the crust (e.g., rock versus dust), and the electrical properties are related to the mineralogy of the surface layers (e.g., metallicity).

Venus and Mars have atmospheres that consist of over 95% carbon dioxide gas (CO_2). The surface pressure of Venus's atmosphere is approximately 90 atm (the surface pressure on Earth is 1 atm), and the microwave opacity (i.e., absorptivity at radio wavelengths) of this much CO_2 gas is substantial. The opacity decreases with wavelength, and Venus's surface can be probed at wavelengths longward of ~4 cm. The surface pressure of Mars's atmosphere is approximately 7 mbar, and the atmosphere is transparent throughout most of the microwave region. CO_2 gas is photodissociated (molecules are broken up) by sunlight into carbon monoxide (CO) and oxygen (O). CO gas has strong rotational transitions at millimeter wavelengths, which can be utilized to determine the atmospheric temperature profile and the CO abundance on Venus and Mars in the altitude regions probed. [See VENUS: ATMOSPHERES; MARS: ATMOSPHERE AND VOLATILE HISTORY.]

A. THE MOON

At radio wavelengths, one can probe the Moon's surface and subsurface layers. The microwave emission from a surface element on the Moon depends on the temperature structure in and radio emissivity of the (sub)surface layers. These are determined by the Sun–Moon geometry, the Moon's orbit and thermal history, the angle of emission, surface roughness, and physical properties with depth (complex dielectric constant, density, heat capacity, and thermal conductivity). Measurements at infrared and radio wavelengths of the variations in brightness temperature with lunar phase, and the change in temperature over time when the Moon goes into an eclipse, contain information to constrain the thermal and electromagnetic properties of the upper meter of the lunar regolith, or "soil" in analogy to terrestrial soil (although it does not contain organic matter). Planetary regoliths are formed by meteorite impacts, which pulverize the bedrock. Since the cumulative number of meteoritic impacts increases with time, the thickness of the lunar regolith depends on the age of the underlying bedrock. The lunar soil is typically a few meters thick in the maria and over 10 m deep in the 4.5-billion-year-old lunar highlands. The ejecta from larger impacts have formed a 2- to 3-km-thick mega-regolith, which consists of (many) meter-sized boulders. The mega-regolith may be bonded at depth owing to the high temperatures and pressures.

Lunar radio astronomy dates back to the early 1960s, well before the first Apollo landing on the Moon. Since the mid-1970s, after a decade of "neglect," there was renewed interest in lunar radio astronomy since radio receivers had improved substantially and laboratory measurements of Apollo samples provided a ground truth for several sites on the Moon. Lunar core samples have been used to derive a density profile with depth near the landing sites: the regolith is very porous (density of ~1 g/cm^3 or 1000 kg/m^3) in the upper ~2 cm, below which it rises very rapidly, maybe abruptly, to a density of 1.8 g/cm^3 in the next few centimeters. This structure is likely caused by the bombardment of small meteorites, which maintains the top layer at a low density while compacting deeper layers.

Rock samples allowed determination of the complex dielectric constant of lunar rocks and powders and the derivation of an empirical relation ("mixing formula") between the dielectric constant and density of the regolith material. In addition, the microwave opacity, absorptivity, or loss tangent (ratio between the imaginary and real parts of the dielectric constant) could be determined over a range of wavelengths. Although losses due to both scattering and absorption are included in the loss tangent measurements of lunar samples at short wavelengths ($\lesssim$ sub-millimeter wavelengths), the scattering component in longer-wavelength samples is likely underestimated or absent, since rock chips that would produce significant scattering at centimeter wavelengths were removed from the sample prior to the measurements.

Remote microwave observations of the Moon at wavelengths from a few up to 30 cm, when compared to sophisticated thermophysical models, suggest the presence of microwave opacity due to scattering in addition to, and of the same magnitude as, that due to pure absorption. The scattering could be from isotropic scattering off buried rock chips, or from backscattering in a stratified medium. As mentioned earlier, the density of the regolith increases with depth. If this increase occurs over a distance similar to or smaller than the observing wavelength, the radio emissivity will be less than that at the surface. This will translate into a depressed brightness temperature. Hence, since one probes deeper layers at longer wavelengths, the observed brightness temperature may be relatively low at the longer wavelengths. This effect, together with the opacity caused by microwave scattering, makes it very difficult to interpret the Moon's microwave spectrum. This is unfortunate, since in principle a microwave spectrum of the Moon from a few up to 30 cm contains information on its heat flow. At the present time, it is very difficult to distinguish between gradients in brightness temperature caused by changes in radio emissivity or by an internal heat source.

Analysis of microwave and infrared measurements showed that the Apollo-based thermophysical properties are representative of a large portion of the lunar nearside hemisphere. However, radio images of the Moon reveal differences in brightness temperature between the maria and highlands that have not yet been fully explained. A microwave image at full Moon, at a wavelength of 3.55 cm is shown in Fig. 3. The maria are typically 5 K warmer than the highlands, which may, in part, be due to the difference in Bond albedo or reflectivity between maria and highlands (maria, ~7–10%; highlands, ~11–18%). Any remaining difference is suggestive either of a difference in radio emissivity between the two regions or a difference in microwave opacity. There is evidence from lunar samples that the microwave opacity in the highlands is somewhat (by a factor of ~2) lower than that in the maria. If this is true, deeper, cooler layers will be probed in the lunar highlands during full Moon (as observed); at new Moon, the temperature contrast should be reversed (no observations have yet been reported), since the temperature increases with depth at night.

The loss tangents of lunar fines increase with the abundance of ilmenite, which is the most common titanium-bearing mineral on the Moon (FeTiO$_3$). Ilmenite is also opaque at optical wavelengths and is largely responsible for the dark appearance of maria compared to the highlands. [See THE MOON.]

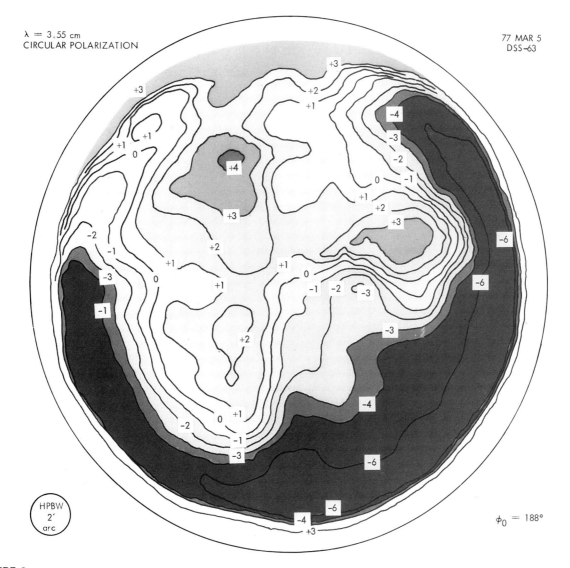

FIGURE 3 Microwave difference map of the full Moon at a wavelength of 3.55 cm. This image shows the residuals after subtracting a model map from the observed data. Subtracting a model map from the data will "enhance" any anomalies in the observations. The regions of enhanced brightness temperature correspond to lunar maria, and the highlands typically show depressed temperatures. The contours are marked in degrees kelvin. [From S. J. Keihm and B. L. Gary (1979). Comparison of theoretical and observed λ3.55 cm wavelength brightness temperature maps of the full Moon. *Proc. 10th Lunar Sci. Conf. Geochim. Cosmochim. Acta,* **Suppl. 10,** 2311–2319.]

B. MERCURY

Temperature variations within Mercury's regolith depend on a balance between solar insolation, heat transport within the regolith, and reradiation outward. Because of Mercury's slow rotation, the surface temperature reaches its peak value at noon. The highest subsurface temperature is reached in the afternoon, however, since there is a phase lag between solar insolation and heat transport downward. As a result of the 3/2 resonance between Mercury's rotational and orbital periods, in combination with Mercury's large or-

bital eccentricity, the average diurnal insolation varies significantly with longitude on the planet. Regions along Mercury's equator near longitudes $\phi = 0°$ and 180° (the subsolar longitudes when the planet is at perihelion) receive roughly 2.5 times more sunlight than do longitudes 90° away from it. As a result of this nonuniform heating, the diurnally averaged surface and subsurface temperatures vary by ~100 K with longitude. The diurnal temperature variations are superimposed on this hot–cold longitude pattern. The nighttime equatorial surface temperature is approximately 100 K, independent of longitude, but the peak

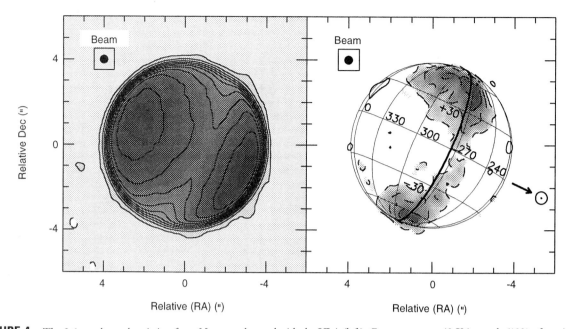

FIGURE 4   The 3.6-cm thermal emission from Mercury observed with the VLA (left). Contours are at 42 K intervals (10% of maximum), except for the lowest contour, which is at 8 K (dashed contours are negative). The beamsize is 0.4″ or one-tenth of a Mercurian radius. The geometry of Mercury during the observation, the direction to the Sun, and the morning terminator (dashed line) are superimposed on the image on the right. This image shows the residuals after subtracting a model image from the observed map. Note that the grayscaling shows peaks in emission in the image on the left and peaks in depression on the right. The image on the right clearly shows the thermal depressions at both poles and along the sunlit side of the morning terminator. Contour intervals are in steps of 10 K, which is roughly three times the rms noise in the image. [From D. L. Mitchell and I. de Pater (1994). Microwave imaging of Mercury's thermal emission: Observations and models. *Icarus* **110**, 2–32.]

(noon) surface temperature varies between 700 K at $\phi = 0°$ and 180° to 570 K at $\phi = 90°$ and 270°.

Whereas the subsurface temperature is below that of the surface during the day, it is above the surface temperature at night. Thus, heat is transported downward during the day and upward at night. Since the thermal conductivity and heat capacity depend on the temperature of the crust, downward conduction is very effective during the day, when the surface temperature is high. At night the surface cools rapidly and acts like an insulator, so that heat is trapped in the subsurface layers. This results in an ~100 K increase in temperature with depth.

Radio images of the planet show a brightness variation across the disk, which displays the history of the solar insolation. At short wavelengths, where shallow layers are probed, the dayside·temperature is usually highest. However, when deeper layers are probed, the diurnal heating pattern is less obvious, and one can distinguish two hot regions, one at longitude 0° and one at 180°. Figure 4 shows a radio image at 3.6 cm. An image of the total intensity is shown on the left side. The image on the right side shows the difference between the observed map and a thermal model. The viewing geometry is superimposed on the latter image.

At this wavelength, one probes ~70 cm into the crust. The hot region at longitude 0° is clearly visible on the nightside (left image), whereas the high temperature on the dayside is caused by both the 180° hot longitude and solar insolation. These regions have been modeled quite well and hence are removed from the image on the right side. Most remarkable on the latter image are the negative temperatures near the poles and along the terminator. These regions show that the poles and terminator are colder than predicted in the model. This is, likely caused by surface topography, which causes a permanent shadowing effect at high latitudes and transient effects in the equatorial regions, where crater floors and hillsides are alternatively in shadow and sunlight as the day progresses. Some crater floors near the poles are permanently shadowed, and radar observations have revealed evidence for the existence of water ice at such crater floors.

Radio spectra and images, together with *Mariner 10* infrared data of the planet, have been used to derive Mercury's surface properties, which appear to be quite similar to those of the Moon. The top ~2 cm consist of a thermally insulating layer of low-density powder ($\rho \sim 1.0$ g/cm³), with dielectric constant $\varepsilon \sim 2$. Below this layer, the thermal conductivity is much higher and the regolith is more compact ($\rho = 1.9$ g/cm^3; $\varepsilon \sim 3$). Like on the Moon, this structure is expected to result from bombardment by small meteorites, which maintains the top layer at a very low density while compacting deeper layers.

Although the thermophysical properties of Mercury's regolith are quite similar to those of the Moon, the loss tangent, or microwave opacity, appears to be roughly a factor of two to three smaller than that of most lunar samples. The actual difference may be even larger since the lunar samples were sieved prior to the laboratory measurements; thus scattering losses in the lunar samples are underestimated. This suggests that the ilmenite content on Mercury is low compared to that on the Moon. As mentioned before, the ilmenite content is largely responsible for the albedo difference between the lunar highlands and maria; in fact, Mercury's surface is much brighter than that of the Moon and exhibits much smaller brightness contrasts between its smooth plains and cratered terrain. Iron and titanium are major constituents of basalt, common in volcanic areas on Earth. Basalt is the first rock to form from a cooling magma. The absence of iron and titanium from Mercury's surface suggests that the planet is devoid of basalt. Differentiation of heavy elements such as Fe and Ti could have extended to deeper levels on Mercury than on the Moon, due to a more extensive lithospheric melting and a larger surface gravity. Volcanic regions on Mercury might then be depleted in Fe and Ti relative to the Moon. Alternatively, several models have been put forward to explain the high average density of the planet, the most popular being that a giant impact might have stripped off Mercury's mantle. Although it is not clear how such an event could result in a low Fe and Ti abundance on Mercury's surface, no theories have yet been developed to explain the consequences of the mineralogy on Mercury's surface under such circumstances. [See MERCURY.]

C. VENUS

The first radio astronomical observations of Venus were carried out in the mid-1950s at a wavelength of 3 cm. These measurements indicated a surface temperature of over 560 K, which seemed too high compared to the expected terrestrial analogue of 300 K. Subsequent measurements led to Venus's radio spectrum, a graph of the disk-averaged brightness temperature of the planet as a function of wavelength. This radio spectrum showed a steep increase in temperature from ~ 230 K at short millimeter wavelengths up to ~ 700 K near a wavelength of 10 cm. These observations could

be reconciled with models of a strong greenhouse effect in the Venusian atmosphere, as proposed by Carl Sagan in the early 1960s.

The atmosphere of Venus is opaque at millimeter and short centimeter wavelengths, but gets gradually more transparent at longer wavelengths. Hence, the planet's brightness temperature at millimeter wavelengths is close to the value expected for a planet heated up by the Sun, and at longer wavelengths deeper layers of the atmosphere are probed, where it is warmer thanks to the greenhouse effect. Longward of ~ 4 cm, the planet's surface can be probed. Owing to the adiabatic temperature structure in Venus's atmosphere, the radio spectrum shows a steep temperature increase from millimeter to centimeter wavelengths. There are no diurnal temperature variations, because of the atmosphere's large heat capacity.

1. Surface

Radio observations (single dish and interferometric) together with radar data have been used to study the Venusian surface, in particular to determine the emissivity, which can be related to the dielectric constant, ε. Venus's emissivity averaged over the disk is 0.86 ± 0.04, which translates into a dielectric constant of 5 ± 0.9. The emissivity of the highland regions is lower, 0.5–0.6; results from the *Magellan* spacecraft indicate values as low as ~ 0.3 near the mountaintops. These low values may imply large dielectric constants; the average value of 0.5–0.6 suggests dielectric constants of over 30, whereas emissivities closer to 0.3 suggest $\varepsilon \sim 80$, if the low emissivities are indeed caused by high dielectric permittivities and not from some kind of subsurface scattering phenomenon. All of these values are much higher than the dielectric constants measured for the Moon, Mercury, and Mars, which are typically close to 2. An $\varepsilon \sim 2$ implies porous surface materials; $\varepsilon \approx 5-9$ is typical for solid rocks (granite–basalt), and much higher dielectric constants can be caused by the inclusion of metallic and sulfide material. Hence, Venus's surface is overlain, at most, by only a few centimeters of soil or dust, and likely consists of dry solid rock. The highlands may contain substantial amounts of minerals and sulfides close to the surface, materials one might expect to find in volcanic areas.

2. Atmosphere

Roughly half the microwave opacity in the Venusian atmosphere is attributed to CO_2 gas, and prime sus-

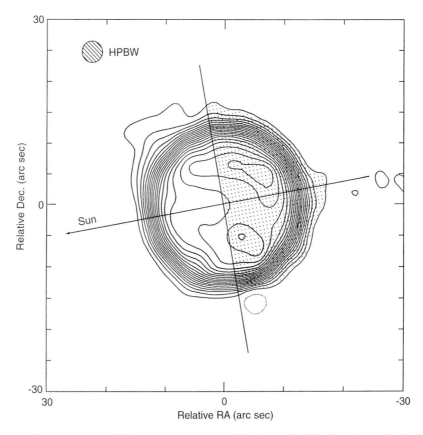

FIGURE 5   Radio image of Venus at 3 mm. The beamsize is 3.5″ or 0.3 Venusian radii. The direction to the Sun and the terminator are indicated. The contour levels are in 5% intervals with a maximum of ~365 K [From I. de Pater, F. P. Schloerb, and A. Rudolph (1991). CO on Venus imaged with the Hat Creek radio interferometer. *Icarus* **90**, 282–298.]

pects for the other half are gaseous sulfuric acid (H_2SO_4) and sulfur dioxide (SO_2) gas. An image of Venus at 3-mm wavelength, taken in January 1987, shows that Venus's nightside is approximately 10% brighter than its dayside (Fig. 5), which suggests that there is probably less microwave opacity at the nightside compared to the dayside, so deeper, warmer regions can be probed at night. At 3-mm wavelength, one probes altitudes around 50 km, well within Venus's cloud layers. Spatial variations in opacity in this region might be caused by local changes in cloud humidity or gaseous sulfuric acid (note that the clouds of Venus consist of liquid sulfuric acid particles).

At millimeter wavelengths, in the $J = 1$–0 and $J = 2$–1 transitions of the CO line, one probes altitudes between 70 and 120 km, the mesosphere of Venus, a region not well studied at other wavelengths. CO is produced upon photodissociation of CO_2 by solar UV radiation, in the 70- to 120-km altitude region. This is a transition region between the massive lower atmosphere (altitudes $\lesssim 70$ km),

in which the radiative time constant is much greater than a solar day, and the upper atmosphere (altitudes $\gtrsim 120$ km), which has a low heat capacity. The temperature structure in the lower atmosphere follows an adiabatic curve, whereas we find a strong day-to-night gradient in temperature in the upper atmosphere. This temperature gradient drives strong winds to blow from the dayside to the nightside, a pattern very different from the retrograde zonal winds observed in the visible cloud layers. Both wind patterns have approximate velocities of 100 m/s.

The ground and first excited rotational transitions of CO have been observed routinely at wavelengths of 3 and 1 mm, respectively. Since CO is formed in the upper part of the atmosphere, the line is seen in absorption against the warm continuum background. Examples of CO spectra averaged over Venus's dayside and nightside hemispheres, respectively, are shown in Fig. 6. It has been possible to determine the planet's temperature structure and the altitude profile of the CO abundance (volume mixing ratio) from such obser-

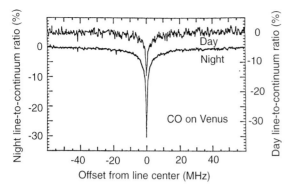

FIGURE 6 Spectra of Venus in the $J = 1-0$ line: the upper curve is for the dayside hemisphere (when Venus is near superior conjunction) and the lower curve is for the nightside hemisphere (when Venus is near inferior conjunction) [From F. P. Schloerb (1985). Millimeter-wave spectroscopy of solar system objects: Present and future. *In* "ESO Proceedings 22, ESO–IRAM–Onsala Workshop on (Sub)millimeter Astronomy," Aspenas, Sweden, June 17–20, 1985 (P. A. Shaver and K. Kjar, eds.), pp. 603–616, Garching bei Munich.]

vations. The spectra in Fig. 6 show the nightside line to be approximately three times deeper and also narrower than the dayside line, which is suggestive of a larger concentration of CO at high altitudes on the nightside of the planet than the dayside. Since CO is presumably formed at the dayside upon photodissociation of CO_2, this is just opposite to what one would expect. This suggests that CO is likely blown from the dayside to the nightside by the day-to-night winds in Venus's upper atmosphere.

As mentioned earlier, the mesosphere is a transition region between the zonal winds in the main cloud deck and the solar-to-antisolar winds in the upper atmosphere. This region is probed in the CO lines. By measuring the Doppler shift of these lines at different locations on Venus's disk, the mesospheric wind pattern can be studied in detail. This project is now feasible with the recent development of sensitive arrays of telescopes at millimeter wavelengths.

D. MARS

1. Surface

Because of its variation in heliocentric distance, Mars's surface brightness temperature varies approximately as $1/\sqrt{r}$ (r = heliocentric distance). The temperature of the crust may be slightly increased above that expected from solar illumination alone owing to the at-

mospheric greenhouse effect. In addition, although atmospheric dust storms are transparent at radio wavelengths, these storms influence the Martian radio brightness directly by reducing the amount of solar radiation to the surface and thus lowering the surface brightness temperature.

As described earlier, sunlight heats Mars during the day, and the heat is transported downward mainly by conduction. The amplitude and phase of the diurnal temperature variations and the temperature gradient with depth in the crust are largely determined by the thermal inertia and skin depth of the material. Temperature variations are largest at the surface, where the temperature is determined directly by the solar radiation; there is a phase lag in the diurnal temperature variation deeper in the crust.

In addition to the diurnal heating cycle, the disk-averaged radio brightness temperature of Mars varies as a function of central meridian longitude of the planet, by up to 5–10 K. These variations are suggestive of a nonuniformity in the Martian surface properties, an observation that has not yet been fully explained. Radio images of the planet at millimeter wavelengths suggest that Mars is hottest in the afternoon, as expected from the phase lag in the diurnal heating pattern between the surface and subsurface layers. These and images at centimeter wavelengths

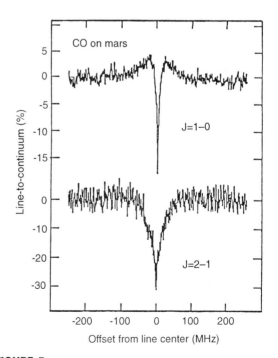

FIGURE 7 Spectra of Mars in the CO $J = 1-0$ and $J = 2-1$ transitions, during the 1984 opposition. [From F. P. Schloerb (1985).]

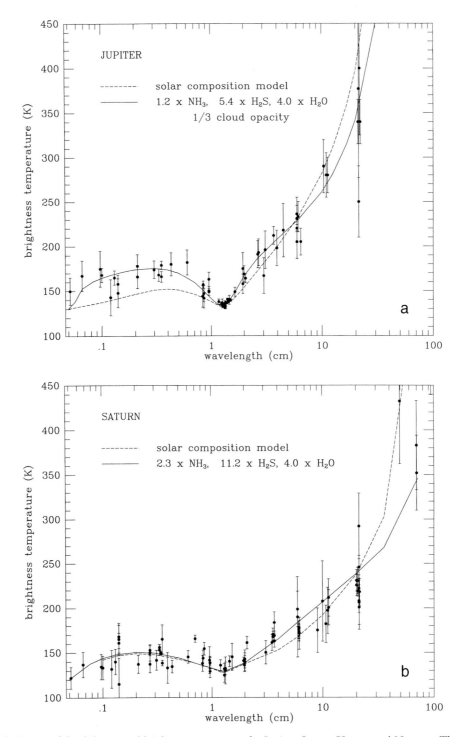

FIGURE 8 (a–d) Spectra of the disk-averaged brightness temperature for Jupiter, Saturn, Uranus, and Neptune. The data are indicated by dots. The dashed line shows a model spectrum expected for the planet if it would have a solar composition, that is, the abundances of NH_3, CH_4, H_2S, and H_2O would be equal to the solar elemental N, C, S, and O abundances. The solid line shows a model calculation for an atmosphere that fits the observed spectra. The enhancement in NH_3, CH_4, H_2S, and H_2O above the solar elemental values is indicated for each planet. [After I. de Pater and J. J. Lissauer (2000). "The Solar System: An Introduction to Planetary Sciences." Cambridge Univ. Press, Cambridge, England.]

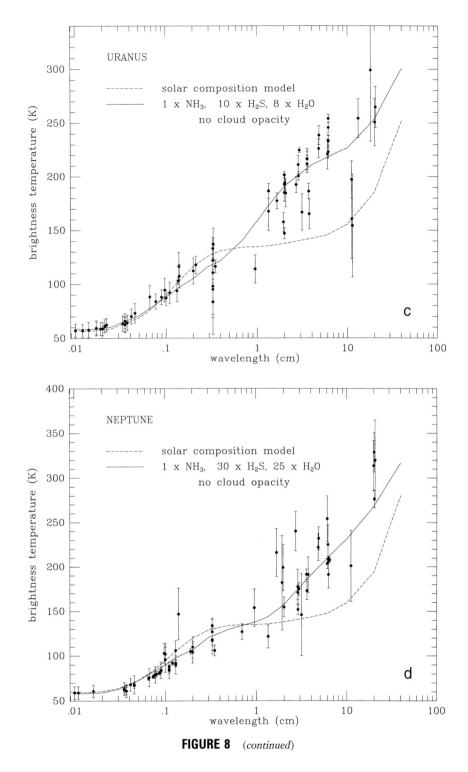

FIGURE 8 (*continued*)

also reveal a reduced brightness temperature in the polar regions, by ~20% compared to the average brightness temperature. This is likely caused by a thick layer of CO_2 frost covering most of the Martian polar regions.

2. Atmosphere

Both the ground and first excited state of CO have been observed in the Martian atmosphere. Examples of CO spectra are shown in Fig. 7. Whether the line is seen in emission, absorption, or a combination thereof depends mainly on the temperature–pressure profile in the atmosphere, and how this compares to the brightness temperature of the surface. The temperature profiles measured by the Viking spacecraft show a roughly constant temperature of 140 K down to an altitude of about 60 km, below which it increases to ~210–220 K at the surface. The CO lines are optically thick; thus the core of the line is formed high up in the atmosphere where it is cold, and hence is seen in absorption against the continuum background from the planet's surface. The wings of the line are formed in the lower atmosphere, just above the surface. As a consequence of the surface emissivity, the brightness temperature of the surface is somewhat less than the kinetic temperature in the atmosphere just above it. The wings of the line are therefore seen in emission against the continuum background. Since the CO abundance and the temperature structure in the Martian atmosphere both influence the line profiles, a unique solution of the CO abundance can be found only if measurements are made of both the optically thick ^{12}CO lines and the optically thin lines of its isotope ^{13}CO. From such measurements, it has been found that the CO abundance, averaged over the disk, is quite stable in time, whereas the temperature structure in the atmosphere varies considerably.

One of the most intriguing problems is the apparent low CO mixing ratio: $CO/CO_2 \sim 10^{-3}$, much less than expected from theories on photolysis of CO_2 and recombination of CO and O. Some kind of catalysis plays a role. Although various processes have been suggested, no firm explanation yet exists. Observations of the spatial distribution of the CO abundance and maps of the wind velocity fields on the planet may help unravel this problem. Such endeavors can now be undertaken, with the completion of new sensitive millimeter arrays.

Water vapor in the Martian atmosphere has been imaged at 1.3 cm with the VLA: the water vapor shows up in a partial ring around the limb of the planet. There is no emission above the poles, owing to the low atmospheric temperature. The strength and shape of the spectrum have been used to determine the water abundance as a function of height in the atmosphere. The atmosphere is very dry; the VLA measurements in 1990 indicated a column density less than 5 precipitable μm over the equator, with even smaller amounts at higher latitudes. The vapor, however, is well mixed vertically up to an altitude of at least 50 km. The water abundance has been determined previously from *Viking* observations (1977) and infrared-reflectance spectra (1988). A comparison of all data sets suggests strong (a factor ~ 4) variations in the water abundance over time.

III. GIANT PLANETS

At radio wavelengths in the millimeter to centimeter regime, one generally probes regions in the atmospheres of the giant planets Jupiter, Saturn, Uranus, and Neptune, which are inaccessible to optical or infrared wavelengths. One typically probes pressure levels of ~0.5–10 bars in Jupiter's and Saturn's atmospheres, and down to 50–100 bars on Uranus and Neptune. Much information is contained in the planet's radio spectrum: a graph of the disk-averaged brightness temperature of the planet as a function of wavelength. An example is shown in Fig. 8a, where observations of Jupiter's brightness temperature are plotted as a function of wavelength (the curves on the figure will be discussed in Section III, A). These spectra generally show an increase in brightness temperature with increasing wavelength beyond 1.3 cm, due to the combined effect of a decrease in opacity at longer wavelengths and an increase in temperature at increasing depth in the planet. At millimeter to centimeter wavelengths, the main source of opacity is ammonia (NH_3) gas, which has a broad absorption band at 1.3 cm. At longer wavelengths (typically >10 cm), absorption by water vapor (H_2O) and droplets becomes important, whereas at short millimeter wavelengths the contribution of collision-induced absorption by hydrogen gas (H_2) becomes noticeable. On Uranus and Neptune there may be additional absorption by hydrogen sulfide (H_2S) gas. [See ATMOSPHERES OF THE GIANT PLANETS.]

Radio spectra of the planets can be interpreted by comparing observed spectra with synthetic spectra, which are obtained by integrating the equation of radiative transfer through a model atmosphere:

$$B_\nu(T_D) = 2 \int_0^1 \int_0^\infty B_\nu(T) e^{(-\tau/\mu)} d(\tau/\mu) \, d\mu \qquad (5)$$

where $B_\nu(T_D)$ can be compared to the observed disk-averaged brightness temperature. The brightness $B_\nu(T)$ is given by the Planck function, and the optical depth $\tau_\nu(z)$ is the integral of the total absorption coefficient over the altitude range z at frequency ν. The parameter μ is the cosine of the angle between the line of sight and the local vertical. By integrating over μ, one obtains the disk-averaged brightness temperature, to be compared to the brightness temperatures shown in Fig. 8.

However, before the integration in Eq. (5) can be carried out, the atmospheric structure, as composition and temperature–pressure profile, needs to be defined. The temperature, pressure, and composition of an atmosphere are related to one another through the equation of state (ideal gas law) and the fact that condensable gases will form clouds when the temperature gets cold enough. The temperature in the lower atmosphere (below the tropopause, or at $P \gtrsim 0.4$ bar) will probably follow a wet adiabat. In addition to H_2 and He, we find the condensable gases CH_4, NH_3, H_2S, and H_2O in the atmospheres of the giant planets, and therefore expect the following cloud layers to form: an aqueous ammonia solution cloud (H_2O–NH_3–H_2S) at relatively deep levels in the atmosphere ($T > 270$ K), and then stepping up in altitude we find water ice (base level at 273 K), a cloud of ammonium hydrosulfide particles ($NH_3 + H_2S \rightarrow NH_4SH$ around 250 K), ammonia ice (~140 K), hydrogen sulfide ice (~150 K), and methane ice (~80 K; only on Uranus and Neptune is the temperature cold enough for methane ice to form). The "visible" cloud layers on Jupiter and Saturn consist of ammonia ice, whereas methane haze is visible on Uranus and Neptune.

At first approximation the spectra of both Jupiter and Saturn resemble those expected for a solar composition[1] atmosphere, and the spectra of Uranus and Neptune indicate a depletion of ammonia gas compared to the solar value by about two orders of magnitude. Resolved images of both Jupiter and Saturn show bands of enhanced brightness temperature on their disks. This implies latitudinal variations in the opacity, or precise ammonia abundance. When the opacity is lower, deeper, warmer regions will be probed in the planet's atmosphere. Uranus shows a brightening toward the visible (south) pole. Images of Neptune suggest that the brightness temperature is depressed in a wide latitude range around the equator.

---

[1] We use the following mixing ratios: CH_4/H_2, 7.96×10^{-4}; NH_3/H_2, 2.24×10^{-4}; H_2O/H_2, 1.70×10^{-3}; H_2S/H_2, 3.70×10^{-5}.

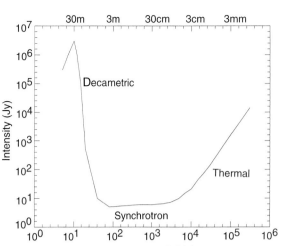

FIGURE 9 A sketch of Jupiter's radio spectrum from 1 MHz up to 300 GHz.

A. JUPITER

Radio signals from Jupiter were first detected in 1955 at a frequency of 22.2 MHz. This emission was sporadic in character and confined to frequencies less than 40 MHz. It is commonly referred to as decametric radiation and is discussed in more detail in Section V. In subsequent years the planet's thermal emission was detected at short centimeter wavelengths and its synchrotron radiation at wavelengths between ~2 cm and a few meters. The latter radiation is emitted by high-energy electrons in a Jovian Van Allen belt (see Section V). Figure 9 gives a schematic representation of Jupiter's spectrum, showing the relative intensities and the frequency ranges at which atmospheric, synchrotron, and decametric emissions dominate.

1. Thermal Emission

A spectrum of Jupiter's emission was shown in Fig. 8, along with model spectra. The dashed line shows the spectrum for a solar composition atmosphere. Although the line agrees to first order with the data, a better fit can be obtained by changing the altitude profile of ammonia gas. An ammonia mixing ratio slightly enhanced above the solar nitrogen (N) value in Jupiter's deep atmosphere provides a better fit to data longward of ~6 cm, whereas ammonia gas must be depleted at higher altitudes to match the spectrum at short centimeter wavelengths. An example of a reasonable synthetic spectrum is shown by the solid line, which is for an atmosphere in which the ammonia mixing ratio, NH_3/H_2, in the planet's deep atmosphere

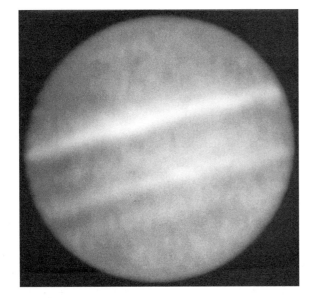

FIGURE 10.   A radio photo of Jupiter at a wavelength of 2 cm. The resolution is 1.2″ and size of the disk 32″. The image was obtained from data taken with the VLA in 1983. [From I. de Pater and J. R. Dickel (1986). Jupiter's zone-belt structure at radio wavelengths. I. Observations. *Astrophys. J.* **308,** 459–471.]

is enhanced by a factor of approximately 1.2 compared to the solar nitrogen value; CH_4 is enhanced by a factor of 2.3 above solar C, and H_2S and H_2O are likely enhanced by a factor of 5–7 above solar S (see the following) and 4–5 above solar O, respectively. At higher altitudes, all species follow the saturated vapor curves wherever appropriate. Ammonia gas is decreased owing to the formation of an ammonium hydrosulfide (NH_4SH) cloud layer around 230 K and an ammonia ice cloud at higher altitudes, around 140 K. The NH_4SH cloud is formed upon condensation of ammonia gas together with hydrogen sulfide gas. Despite searches at infrared and radio wavelengths, no H_2S gas has been detected above the NH_4SH cloud layer. This fact, together with the "observed" decrease in the abundance of NH_3 gas, has been used to determine an H_2S mixing ratio of 5–7 times the solar sulfur value in Jupiter's deep atmosphere (as used in the spectrum above).

A radio image of the planet at 2-cm wavelength is shown in Fig. 10. The resolution is 1.2″, and the disk diameter is 32″ at the time of the observations. The image shows bright horizontal bands across the disk, which coincide with the brown belts seen at visible and infrared wavelengths. These bands have a higher brightness temperature, likely due to a lower opacity in the belts relative to the zonal regions, so deeper, warmer layers are probed in the belts. This phenome-

non is suggestive of gas rising up in the zones; when the temperature drops below ∼140 K, ammonia gas will condense out. In the belt regions, the air, now depleted in ammonia gas (so-called dry air), descends. This general picture is in agreement with that suggested by other researchers from analyses of visible and infrared data. We note here, however, that the radio data probe the gas from which the clouds condense, and the observations at visible and infrared wavelengths are sensitive primarily to the cloud particles. Thus the base level of the clouds is determined through radio observations, whereas the altitude of the cloud tops is defined by measurements at optical and infrared wavelengths. The location of the base level is needed to develop dynamical models of an atmosphere. In thermochemical equilibrium one usually expects the base level of a cloud at the altitude where the temperature drops below freezing. However, in analogy with Earth's atmosphere, the gas may have to be supercooled before condensation takes place, which may raise the base level of the cloud.

B. SATURN

1. Atmosphere

We receive thermal radio emission from Saturn's atmosphere as well as from its rings. In analogy to Jupiter, radio spectra of the atmospheric emission (Fig. 8b) can be interpreted in terms of its ammonia abundance and local variations therein with altitude and latitude. The ammonia and hydrogen sulfide abundances on Saturn are slightly more enhanced than on Jupiter: NH_3 gas may be enhanced by a factor of ∼3 compared to the solar nitrogen value, whereas H_2S gas may be enhanced by a factor of 10–15 compared to solar sulfur. Radio images of the planet at different wavelengths and ring inclination angles are shown in Fig. 11. At 2-cm wavelength the disk is very smooth: no latitudinal structure can be distinguished. At 6-cm wavelength, however, a bright band is visible at mid northern latitudes, indicative of an average lack of NH_3 gas over the altitude region probed at this wavelength. The region at midlatitudes is likely a region of subsiding gas, just like the bright belts seen on Jupiter. A similar conclusion was reached from an analysis of infrared data, where the cloud opacity at midlatitudes was found to be less than that at neighboring latitudes.

Interesting enough, the radio brightness distribution as shown in Fig. 11 changed drastically in the 1990s. Bands appeared at a wavelength of 2 cm, the

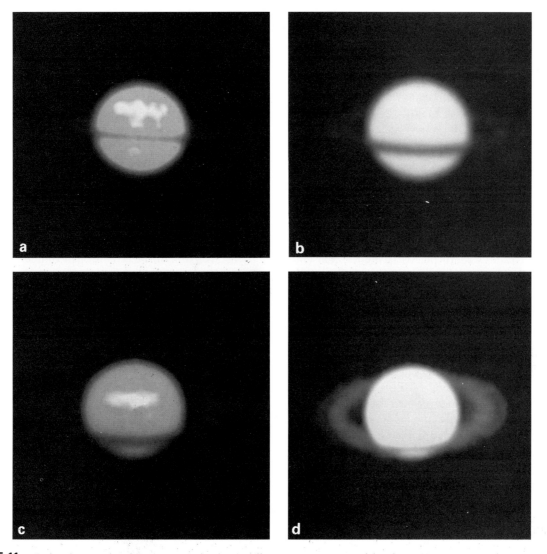

FIGURE 11. Radio photographs of Saturn at 2 and 6 cm, at different viewing aspects of the planet. (a) 6 cm, ring inclination angle $B = 5.8°$; (b) 2 cm, $B = 12.5°$; (c) 6 cm, $B = 20°$; (d) 2 cm, $B = 26°$. The resolution is 1.5″, or approximately one-sixth of a Saturnian radius. [From I. de Pater and J. R. Dickel (1991). Multi-frequency VLA observations of Saturn at ring inclination angles between 5° and 26°. *Icarus* **94**, 474–492.]

bright northern band at 6 cm disappeared, and other bright bands became visible at this wavelength in both hemispheres. Neither the timescales nor the causes for these drastic changes are known or understood.

2. Rings

As shown in Fig. 11, in addition to atmospheric emission, we receive radiation from Saturn's main rings, the A-, B-, and C-rings. This emission consists primarily of Saturn's radiation reflected off the ring particles. The ring brightness temperature can be measured at either side of the planet; on the average, the temperature is

roughly 6 K in the centimeter range, rising at shorter (millimeter) wavelengths. In front of the planet the rings block out part of Saturn's radio emission, resulting in an absorption feature. From this feature one can determine the optical depth of the rings. The optical depth is largest in the B-ring, approximately 1–1.5; it is about half that value in the A-ring, and only ∼0.03 in the C-ring. Variations in the optical depth as a function of wavelength and ring inclination angle, together with a radio spectrum of the ring brightness temperature, have been used to determine the scattering properties of the ring particles. Such properties are unique for certain size distributions and compositions of the particles. The rings consist primarily of dirty ice particles, with sizes between a few millimeters up to ~5 m. The number of ring particles at a particular particle radius r varies approximately as r^{-2}; thus, although there are more small than large particles, the area covered by large and small particles is roughly equal. A particle distribution like this can be explained from a collisionally evolved distribution of particles.

C. URANUS AND NEPTUNE

Uranus is unique among the planets in having its rotation axis closely aligned with the plane in which the planet orbits the Sun. With its orbital period of 84 years, the seasons on Uranus last for 21 years. At the moment, Uranus's south pole is facing the Sun (and us), while the pole is slowly moving out of sight. In less than 40 years from now, the opposite pole will be visible. This geometry must have a pronounced effect on the large-scale circulation of the Uranian atmosphere.

Even though Neptune orbits the Sun at a considerably larger distance than Uranus (roughly 30 versus 20 AU), the temperature at each planet's tropopause, just below the inversion layer (i.e., the altitude where the temperature is a minimum), is remarkably similar: 53 K for Uranus versus 52 K for Neptune. This has been explained by the presence of a large internal heat source on Neptune. With similar atmospheric composition and temperature structure, one would expect rather similar radio spectra. However, even though there are similarities, we also find large differences.

The opacity in the Uranian and Neptunian atmospheres is dominated by collision-induced absorption due to hydrogen gas at millimeter wavelengths, and by ammonia and possibly hydrogen sulfide gases at centimeter wavelengths. Water vapor and cloud particles may provide opacity as well. The radio spectra of Uranus and Neptune (Figs. 8c and 8d) suggest an overall depletion of ammonia gas in their upper atmospheres, by roughly two orders of magnitude compared to the solar nitrogen value. This apparent depletion is likely caused by a nearly complete removal of NH_3 gas in the upper atmosphere through the formation of NH_4SH. In contrast to Jupiter and Saturn, the H_2S abundance on these two outer giants is likely larger than the NH_3 abundance, leading to a rather effective removal of ammonia gas, whereas H_2S gas is merely decreased in abundance in the upper atmosphere compared to lower layers. Model atmosphere calculations suggest that H_2S is enhanced above solar S by a factor of ~10 on Uranus and ~30 on Neptune. Ammonia gas may be close to the solar N value in Uranus's

deep atmosphere, but seems depleted on Neptune. A subsolar value for nitrogen gas is inconsistent with theories on planet formation. If a planet forms directly from the primordial solar nebula, the elemental abundances must be equal to that measured on the Sun. Condensable materials accrete as solids, sublime in the protoplanet's atmosphere, and therefore enhance the elemental abundances above solar values. Carbon, sulfur, and oxygen compounds, present as CH_4, H_2S, and H_2O in the giant planets, are therefore enhanced above solar values. Similarly, nitrogen is expected to be enhanced above solar N. A subsolar value on Neptune is simply not possible. As we will see in the next section, nitrogen is likely not depleted on Neptune, but may be present in the form of both NH_3 and N_2 gases.

Just like ammonia gas, hydrogen sulfide condenses out in its own ice cloud at temperatures $\lesssim$ 140–150 K. Since the temperature on Uranus and Neptune is much less than that on Jupiter and Saturn, the H_2S ice cloud is not the upper cloud deck. Methane ice freezes out at higher altitudes, at a temperature of ~80 K. The latter cloud forms a haze seen at visible wavelengths.

Uranus's brightness temperature has been monitored since 1966, and a pronounced increase was noticed when the south pole came into view. This was the first indication that the Uranian pole is considerably warmer than the equatorial region, a theory later confirmed by radio images obtained with the VLA. Figure 12 shows an image of Uranus at 6-cm wavelength. The brightest point on the disk is displaced toward the pole, and the temperature contrast between the pole and the equator is roughly 60 K. As in the case of Jupiter and Saturn, the pole is likely warmer than the equator owing to a relative lack of absorbing gases. Such a scenario implies descending gas in the polar region and rising gas at other latitudes. A similar pattern seems to exist on Neptune, where the equatorial region appears to be ~15 K colder than at higher latitudes.

These same rising and subsiding motions have been derived from visible/infrared data at higher altitudes. Hence, these vertical motions happen over large distances, extending from the cloud tops at levels of a few hundred millibars down to at least the bottom of the NH_4SH clouds, if not deeper, which on Uranus and Neptune is close to 40 atm.

1. CO and HCN on Neptune

Despite many searches at radio wavelengths for emission and/or absorption lines in giant planet atmospheres, only carbon monoxide (CO) and hydrogen cyanide (HCN) have been detected in emission in

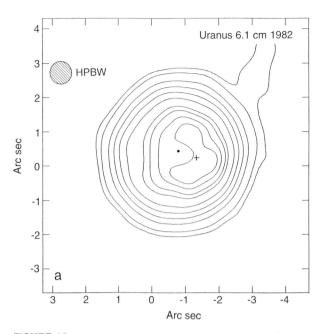

Uranus 6.1 cm 1982

HPBW

FIGURE 12. Contour map of Uranus at a wavelength of 6 cm. The cross indicates the position of the pole and the dot that of the subsolar point. The resolution of the image is 0.65″, and contour values are 14, 28, 69, 110, 151, 193, 248, 261, and 270 K. [From I. de Pater and S. Gulkis (1988). VLA observations of Uranus at 1.3–20 cm. *Icarus* **75**, 306–323.]

Neptune's atmosphere, in the 1- and 2-mm-wavelength bands. CO has also been seen in absorption. Whether the line is seen in emission or absorption depends on the temperature structure in the atmosphere and the altitudes probed. If a line is seen in emission, it must be formed above the tropopause, where the temperature is increasing with increasing altitude; absorption lines are seen in the troposphere, where the temperature is decreasing with increasing altitude. The observations suggest that CO is present both below and above the tropopause; it may be brought up from deep depths by a fast upward convection. HCN is seen only in the stratosphere, above the tropopause. It cannot have been brought up from below, since the gas would condense out before reaching the tropopause. Hence, HCN must be formed in the stratosphere, from nitrogen and carbon products. Although carbon compounds are readily produced in Neptune's stratosphere through photolysis of methane gas, the presence of nitrogen atoms is harder to explain because of the relative lack of ammonia gas. Nitrogen gas may fall in from outside (e.g., from the satellite Triton), or it may be brought up by fast convection, in analogy to CO. If the latter scenario is right, the relative lack of ammonia gas in Neptune's atmosphere

can be reconciled with formation theories if nitrogen exists in the form of both N_2 and NH_3 gas.

Even though Uranus is very similar to Neptune in many respects, no CO or HCN have been detected on Uranus. This is attributed to the lack of a significant internal heat source in Uranus, and the consequent suppression of strong vertical convection in the deeper layers of the planet.

IV. SMALL BODIES

A. ASTEROIDS

By analogy to the terrestrial planets, radio spectra of small airless bodies provide information on the (sub)-surface properties of the material as composition and compactness. A small number of asteroids have been observed from 350 μm up to 20 cm. An example is shown in Fig. 13, where Ceres's spectrum is plotted between 10 μm and 20 cm. Unfortunately, the error bars are still rather large. Most asteroids show a pronounced decrease in brightness temperature between $\sim$10 μm and centimeter wavelengths. This is because at longer wavelengths one probes deeper, colder layers in the body's crust (note that we can observe only the dayside of objects that orbit the Sun at distances larger than 1 AU). The precise shape of the spectrum can be interpreted in terms of composition and compactness of the material. The interpretation of the spectra of

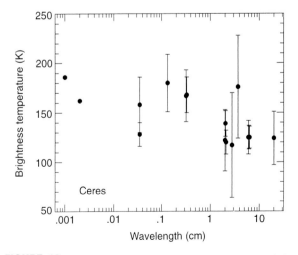

FIGURE 13 A spectrum of the brightness temperature of the asteroid Ceres, between 10 μm and 30 cm. [After I. de Pater and J. J. Lissauer (2000).]

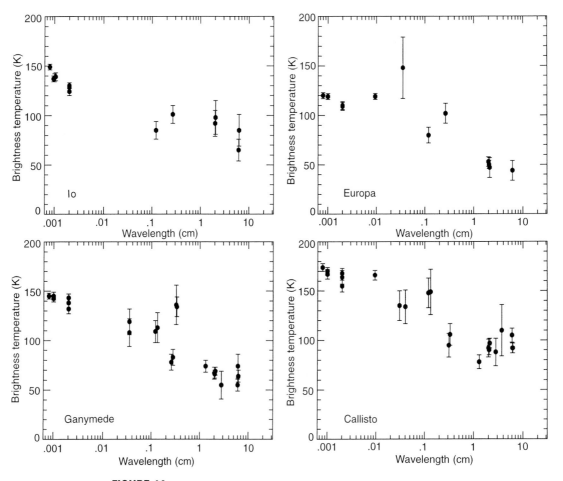

FIGURE 14   Spectra of the Galilean satellites, between 10 μm and 30 cm.

most asteroids indicates a 1- to 10-cm layer of fluffy dust overlying a more compacted regolith. [*See* ASTEROIDS.]

B. GALILEAN SATELLITES

The radio spectra of the Galilean satellites are quite diverse and interesting. The spectra of Io, Europa, Ganymede, and Callisto are shown in Fig. 14. The brightness temperature at infrared wavelengths can be related directly to the satellite's Bond albedo (the ratio between reflected and incident sunlight). Callisto, with its relatively low albedo, 0.13, is warmer than Io and Europa. The brightness temperature at radio wavelengths is determined by the physical temperature of the subsurface layers and the emissivity of the material. The radio emissivity is approximately equal to $1 - \alpha$, with α being the radar geometric albedo. If α is close to zero, the radio brightness temperature is in close

agreement with the physical temperature at depth in the satellite's crust. This is seen on Io. The observed brightness temperatures for Ganymede and, in particular, Europa are well below the physical temperature of the subsurface layers, since the radar albedo for these objects is very large: $\alpha = 0.33$ for Ganymede and 0.65 for Europa. These high albedos and consequently low emissivities and radio brightness temperatures are likely caused by coherent backscattering in fractured ice. [*See* OUTER PLANET ICY SATELLITES.]

1. IO

Since the detection of an ionosphere around Io by the *Pioneer 10* spacecraft in 1973, this satellite is known to possess a tenuous atmosphere. The first detection of a global atmosphere was obtained in 1990, where a rotational line of sulfur dioxide gas was measured at 222 GHz (near a wavelength of 1.35 mm). Since that time, more SO_2 lines have been observed, which allowed researchers to derive Io's atmospheric structure. SO_2 gas is likely the major constituent of Io's atmosphere. It appears to have a global temporal stability and a surface pressure of 3–40 nano-atm (1 natm = 10^{-9} atm), and it covers only a limited fraction (5–20%) of Io's surface. The atmosphere appears to be relatively hot, 500–600 K at 40 km altitude. It is not clear whether the atmosphere is of volcanic origin or in vapor pressure equilibrium with SO_2 frost on its surface. The data favor a volcanic source atmosphere, however. [See Io.]

C. TITAN

Titan's atmosphere consists primarily of nitrogen gas, with traces of methane gas, argon, carbon monoxide, and a variety of hydrocarbons and nitriles. CO, HCN, and HC_3N (cyanoacetylene) have been detected at millimeter wavelengths. Whereas the altitude profile for CO could not be determined unambiguously, both HCN and HC_3N appear to be most abundant at high altitudes, in the stratosphere, which agrees with photochemical models in which HCN is formed from CH_3 and N radicals at altitudes above 800 km, in the thermosphere. Small amounts of HCN are expected to convert into more complex nitriles, under which HC_3N, at lower levels in the atmosphere forms. Downward diffusion would bring the gases down to the stratosphere, where they can be observed; below an altitude of ~80 km, the gases are lost owing to condensation.

Titan's surface can be probed at radio wavelengths and at a few selected infrared wavelengths. The surface temperature is 94 K and the pressure is 1.5 atm. At this temperature, ethane will form a liquid, and it has been suggested in the past that Titan might be covered by a kilometer-thick ocean of liquid ethane. However, more recent radio, radar, and infrared measurements do not show evidence of a deep global ethane ocean. The radio emissivity is about 0.88, which suggests a dielectric constant $\varepsilon \sim 3$, which is much higher than the value of 1.6–1.8 expected for a global ethane ocean. It is entirely possible that ethane lakes, pools, and small oceans cover Titan's surface, and several research groups are studying the satellite in detail to probe its secrets. [See TITAN.]

D. PLUTO

Until recently, Pluto was the only planet from which we had not yet been able to measure thermal emission, owing to its small angular extent and low surface temperature. Assuming the planet is in radiative equilibrium with an albedo of approximately 0.6, its surface temperature should be approximately 53 K. At radio wavelengths one probes depths below the surface, approximately 10 wavelengths deep. Pluto's physical temperature at depths is expected to be ~40 K. New large millimeter telescopes enabled the detection of Pluto at a wavelength of 1.2 mm. The observed brightness temperature is approximately 30 K, which is considerably colder than the temperature expected. This may suggest a low radio emissivity, as seen on many icy bodies; the emissivity is probably similar to that measured for Ganymede.

Since Charon orbits Pluto at a radius of about 19,000 km, or 17 Pluto radii, the combined angular extent of Pluto and Charon is less than 1″; thus it is hard to resolve the two objects ("see" them separately). Thus far, Pluto has been observed with a single-element radio telescope (30-m IRAM telescope in Spain), which allowed detection of the radio flux density from the total system. Since the albedo and size of Charon are well known, its radio intensity can be calculated and has been subtracted from the observed flux density to allow determination of Pluto's brightness temperature. With new sensitive millimeter arrays, the two objects can be imaged, and more reliable measurements of the flux density of each object can be obtained. [See PLUTO AND CHARON.]

E. COMETS

Radio observations of comets provide information that complements studies at other wavelengths. Continuum measurements are sensitive to the thermal emission from the cometary nucleus and any ≳(sub)millimeter-sized material around the comet. Typical spectra at submillimeter to millimeter wavelengths show the radio flux density $S \propto \nu^q$, with $2 < q < 3$, that is, a slope steeper than that of blackbody thermal emission ($q = 2$), but smaller than that expected from Rayleigh scattering from small particles ($q = 4$). Such spectra suggest the presence of large [≳(sub)millimeter sized] grains in a halo around the comet. Upper limits to the radio continuum emission of a few comets have been used to suggest that the temperature gradient in the nucleus may be very steep, or that the radio brightness temperature at depth may be substantially depressed by subsurface scattering.

The most significant advances in cometary radio research have been obtained from spectroscopic studies. The cometary nucleus consists primarily of water ice, which sublimates off the surface when the comet approaches the Sun. It is difficult to detect water directly from the ground; the first definitive detection of water had not been obtained until the apparition of comet Halley in 1986. The detection was made at infrared wavelengths, using the *Kuiper Airborne Observatory*. After about a day, H_2O dissociates into OH and H. It is relatively easy to observe the hydroxyl (OH) molecule at a wavelength of 18 cm. Since the early 1970s, OH has been observed in many comets and monitored in several of them. The line can be seen in either emission or absorption against the galactic background. The OH emission is maser emission, which is stimulated emission from molecules in which the population of the various energy levels is inverted, so that the higher energy level is overpopulated compared to the lower energy level. This population inversion is caused by fluorescence or absorption of solar photons at UV wavelengths, which acts as a pumping mechanism. Photons from the galactic background or 3 K cosmic background can induce a maser action or stimulated emission if the population of energy levels is inverted.

Whether the 18-cm OH line appears in emission or absorption depends on the comet's velocity with respect to the Sun (heliocentric velocity), an effect that is generally referred to as the Swings effect. The OH radical is excited by solar UV photons; however, when the heliocentric velocity is such that solar Fraunhofer (absorption) lines are Doppler-shifted into the excitation frequency, the radical is not excited. In that case, OH will absorb 18-cm photons from the galactic background and be seen in absorption against the galactic background. If the line is excited, background radiation at the same wavelength (18 cm) will trigger its de-excitation, and the line is seen in emission. A detailed study of the 18-cm OH line as a function of heliocentric velocity has revealed many clues regarding the OH excitation mechanism and its production rate in comets. Since OH is produced upon dissociation of water molecules, observations of the OH line yield indirect information on the production rate of water, a comet's primary constituent. The OH production rate appears to vary drastically on short timescales, sometimes in less than a day, and the emission is usually anisotropic, often in the form of jets. Using radio interferometers, the OH emission can be imaged, and one can determine its spatial brightness distribution. Because of the nature of interferometers, one usually detects only 10–20% of the comet's total OH emission; one obtains, in essence, an image of "the top of the iceberg." An image of comet Halley in the OH line is shown in Fig. 15. The brightness distribution is very

irregular and changes on short time scales in spatial as well as velocity coordinates.

Radio observations have been carried out over the entire wavelength range, from submillimeter wavelengths up to at least 20 cm. Many (potential) parent molecules in comets have rotational transitions in this region. Hence, one of the strengths of radio astronomy is the detection of parent molecules in a cometary coma. Such detections are crucial for our understanding of the cometary composition. Since comets have presumably not been altered by excessive heating or high pressures, they are pristine objects, and any compositional information yields clues to the conditions in the early solar nebula from which our planetary system formed. To date, in addition to the OH molecule, a growing number of other molecular species have been detected at radio wavelengths, for example, water, ammonia, hydrogen cyanide, formaldehyde, hydrogen sulfide, and methanol. A large number of molecular lines have been observed at submillimeter wavelengths from comets Hyakutake (1996) and Hale-Bopp (1996–1997). Whereas most molecules may originate from the cometary nucleus, the source of formaldehyde, as for carbon monoxide, seems to be distributed over a larger space and may originate from dust grains in the coma. With the advent of new powerful millimeter arrays, the distribution of parent molecules in the cometary coma can be, and have been (Hale-Bopp), imaged directly. This will provide additional information to constrain cometary models and theories regarding the formation and evolution of our solar system. [*See* PHYSICS AND CHEMISTRY OF COMETS; COMETARY DYNAMICS.]

V. NONTHERMAL RADIATION

Before the era of spacecraft missions, we had only received nonthermal radio emissions from the planet Jupiter. As mentioned earlier (see Section III, A), strong radio bursts were observed from Jupiter at frequencies below 40 MHz, which were attributed to electron cyclotron emissions. Synchrotron radiation was detected at centimeter wavelengths. Despite several searches, no positive detections of nonthermal radio emissions from any of the other three giant planets were made until the *Voyager* spacecraft approached these objects. Now we know that all four giant planets, as well as Earth, are strong radio sources at low frequencies (kilometric wavelengths). In most cases the mode of propagation is in the extra-ordinary (X) sense, and the polarization depends on the direction of the

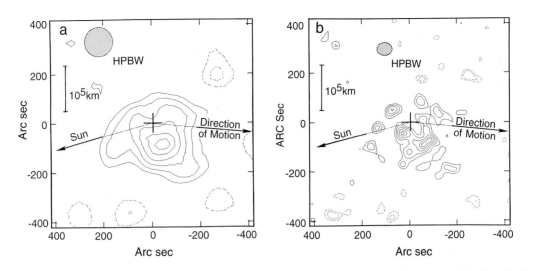

FIGURE 15 Contour plots of comet Halley, Nov. 13–16, 1985. The image is taken at the peak flux density of the line (0.0 km/s in the reference frame of the comet). The left side shows a low-resolution image (3′) and the right side shows a high-resolution image (1′), after the data for both dates were combined. Contour levels for the low-resolution image are 4.9, 7.8, 10.8, 13.7, 16.7, and 18.6 mJy/beam. For the high-resolution image they are 4.4, 6.0, 7.7, 9.3, and 10.4 mJy/beam. Dashed contours indicate negative values. The beamsize, a linear scale, the direction of motion, and the direction of the Sun are indicated in the figures. The cross indicates the position of the nucleus at the time of the observations. [From I. de Pater, P. Palmer, and L. E. Snyder (1986). The brightness distribution of OH around comet Halley. *Astrophys. J. Lett.* **304**, L33–L36.]

magnetic field. The emission is right-handed circularly polarized (RH) if the field is directed toward the observer, and left-handed circularly polarized (LH) if the field points away from the observer.[2] The emission is most likely electron cyclotron radiation, emitted by keV electrons in the planet's magnetic field. In some cases the radio emissions are identified to propagate in the ordinary (O) magneto-ionic mode of propagation. In this mode the polarization is reversed. However, since this mode is quickly damped in the surrounding medium, it is less common. The emission mechanism for this radiation is much more complicated.

A. EARTH

The terrestrial kilometric radiation (TKR) has been studied at both close range and larger distances by many Earth-orbiting satellites. The radiation is very intense; the total power is 10^7 W, sometimes up to 10^9 W. The intensity is highly correlated with the presence of geomagnetic substorms, thus it is indirectly modu-

lated by the solar wind. It originates in the nightside auroral regions and in the dayside polar cusps at low altitudes and high frequencies, and spreads to higher altitudes and lower frequencies. Typical frequencies are between 100 and 600 kHz. The mode of propagation is in the extra-ordinary sense, and emissions usually occur in regions where the local cyclotron frequency is larger than the local electron plasma frequency, ω_e:

$$\omega_e = \left(\frac{4\pi N_e q^2}{m_e}\right)^{1/2} \tag{6}$$

with q the elemental charge, N_e the electron density, and m_e the electron mass.

TKR, as well as the low-frequency auroral emissions from the giant planets, is generally attributed to electron cyclotron radiation, emitted by nonrelativistic particles in a magnetic field, at the frequency of gyration around the magnetic field lines. The radiation can escape its region of origin only if the local cyclotron frequency is larger than the electron plasma frequency. If this condition is not met, the waves are locally trapped and amplified, until it reaches a region from where it can escape.

Cyclotron radiation is emitted in a dipole pattern, where the lobes are bent in the forward direction. The resulting emission is like a hollow cone pattern, as

---

[2] Circular polarization is in the RH sense when the electric vector in a plane perpendicular to the magnetic field direction rotates in the same sense as a RH screw advancing in the direction of propagation. Thus, rotation is counterclockwise when propagation is toward and viewed by the observer. RH polarization is defined as positive and LH as negative.

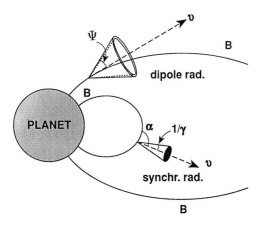

FIGURE 16 Radiation patterns in a magnetic field. Indicated are the hollow cone patterns caused by cyclotron (dipole) radiation from nonrelativistic electrons in the auroral zone. The electrons move outward along the planet's magnetic field lines. The hollow cone opening half-angle is given by Ψ. At low magnetic latitudes, in the Van Allen belts, the filled radiation cone of a relativistic electron is indicated. The angle between the particle's direction of motion and the magnetic field, commonly referred to as the particle's pitch angle, α, is indicated on the sketch. The emission is radiated into a narrow cone with a half-width of $1/\gamma$. [From I. de Pater and J. J. Lissauer (2000).]

displayed in Fig. 16. The radiation intensity is zero along the axis of the cone, in the direction of the particle's motion, and reaches a maximum at an angle Ψ. Theoretical calculations show that Ψ is very close to 90°. Observed opening angles, however, can be much smaller, down to ~50°, which has been attributed to refraction of the electromagnetic waves as they depart from the source region.

In the following, the phenomenology of the non-thermal radio emissions from each of the giant planets is discussed, as they are observed from the ground and by spacecraft (*Voyager* and *Ulysses*). A graph of the average normalized spectra of the auroral radio emissions from the four giant planets and Earth is displayed in Fig. 17. All data are adjusted to a distance of 1 AU. Jupiter is the strongest low-frequency radio source, followed by Saturn, Earth, Uranus, and Neptune.

B. JUPITER

1. Synchrotron Radiation

Jupiter is the only planet from which we receive synchrotron radiation in addition to low-frequency radio emissions. Synchrotron radiation is emitted by relativistic electrons gyrating around magnetic field lines.

The emission is received at decimetric wavelengths, and is commonly referred to as DIM radiation. The emission is beamed in the forward direction (see Fig. 16) within a cone $1/\gamma$:

$$\frac{1}{\gamma} = \sqrt{1 - \frac{v^2}{c^2}} \qquad (7)$$

with v the particle's velocity and c the speed of light. The relativistic beaming factor $\gamma = 2E$, with E the energy in MeV. The radiation is emitted over a wide range of frequencies, but shows a maximum at $0.29\nu_c$, with the critical frequency, ν_c, in MHz:

$$\nu_c = 16.08E^2B \qquad (8)$$

where the energy E is in MeV and the field strength B is in gauss. For emission that is received at 20 cm, we require $E^2B = 320$. If B is 1 G, the typical energy of electrons emitting at 20 cm is close to 25 MeV. At lower strengths and/or higher observing frequencies, the typical energy increases. Hence one probes a different electron population when observing at different frequencies. Furthermore, since the magnetic field strength decreases with planetary distance r, approximately as r^{-3} for a dipole magnetic field, we also observe different electron distributions at different distances from the planet.

Synchrotron emission is generally polarized, and we express the observed quantities in terms of the Stokes parameters I, Q, U, and V. The degree of linear polarization is given by $\sqrt{Q^2 + U^2}/I$, with the position angle of the electric vector $PA = 0.5\text{atan}(U/Q)$. The direction of the projected magnetic field can be found by rotating PA over 90°. Note, however, that the emission is integrated along the entire line of sight, weighted most heavily by the regions that emit the most radiation. The degree of circular polarization, V/I, is a measure of the strength of the component of the magnetic field directed along the line of sight. In general, one expects zero circular polarization for a dipole field if the observer is in the magnetic equatorial plane, and maxima (with opposite sign) when the magnetic poles are facing the observer.

Following the "discovery" of Jupiter's synchrotron radiation, this component of the planet's microwave emission was studied in detail. The variation of the total nonthermal intensity and polarization characteristics during one Jovian rotation (so-called beaming curves) is indicated in Fig. 18 for the total intensity S, the position angle of the electric vector, PA, and the linearly and circularly polarized flux densities, P_L and P_C. The bottom curve shows the variation in the mag-

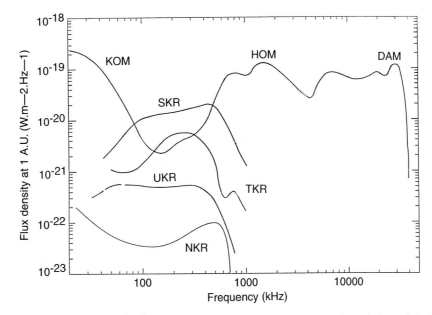

FIGURE 17 A comparison of the peak flux density spectrum of the kilometric continuum radio emissions of the four giant planets and Earth. All emissions were scaled such that the planets appear to be at a distance of 1 AU. [Adapted from P. Zarka (1992). The auroral radio emissions from planetary magnetospheres: What do we know, what don't we know, what do we learn from them. *Adv. Space Res.* **12**(8), 99–115.]

netic latitude of Earth, ϕ_m, with respect to Jupiter. The orientation of Jupiter's magnetosphere is indicated at the top. The maxima and minima in S and P_L occur approximately at $\phi_m = 0$ and $|\phi_m| = $ max, respectively. The circularly polarized flux density is zero where S and P_L show maxima, and P_C shows a positive or negative maximum where S and P_L show minima. These curves indicate that Jupiter's magnetic field is approximately dipolar in shape, offset from the planet by roughly one-tenth of a planetary radius toward a longitude of 140°, and inclined by ~10° with respect to the rotation axis. Most electrons are confined to the magnetic equatorial plane, as visualized in Fig. 19. The magnetic north pole is in the northern hemisphere, tipped toward a longitude of 200°. The 40-MHz cutoff at decametric wavelengths (see the following) implied a field strength of ~10 G at the surface.

Figures 20a and 20b display images of Jupiter's synchrotron radiation at 20 cm. Figure 20a displays contour maps at different rotational aspects of the planet; Jupiter's central meridian longitude is indicated in the upper left corner of each row. From left to right, the contour plots indicate the total intensity, circularly polarized flux density, linearly polarized flux density, and projected direction of the magnetic field, obtained by rotating the position angle of the electric vector by 90°. The effective beamsize (HPFW) or resolution of the images is shown at the center. The images were constructed from data obtained with the Westerbork array in the Netherlands. Figure 20b shows radio pho-

tographs obtained with the VLA in 1994, at a resolution of ~6" or $0.3R_J$. Consider for now the images on the left side only. These images reveal more details than those displayed in Fig. 20a.

Both sets of images show that the emission is confined to the magnetic equatorial plane out to a distance of ~$4R_J$. Several intriguing features are visible. The main radiation peaks (the brightness peaks, one at each side of the planet) are usually asymmetric, in that one of the peaks appears to be brighter than the other peak, as when a hot region in the torus of radiating electrons rotates around with the planet (see Fig. 19). The asymmetry of the main radiation peaks, as well as the asymmetry in the polarized emission, in particular in the circularly polarized flux density (see Fig. 20a), is entirely caused by deviations in Jupiter's magnetic field from a pure dipole configuration. If Jupiter's field were a dipole field, the radiation peaks would always be equal in intensity and they would be minimal when one of the magnetic poles is directed toward Earth. The peaks in circularly polarized intensity would also be equal, and be maximal (with positive or negative sign) when the peaks in total intensity are minimal (see Fig. 18). The asymmetric nature of the peaks reveals the existence of large asymmetries in Jupiter's magnetic field, very similar to what is found in Earth's magnetic field. A detailed comparison between models and observations of the emission can be used to refine models of Jupiter's magnetic field configuration.

Other intriguing features in the images displayed

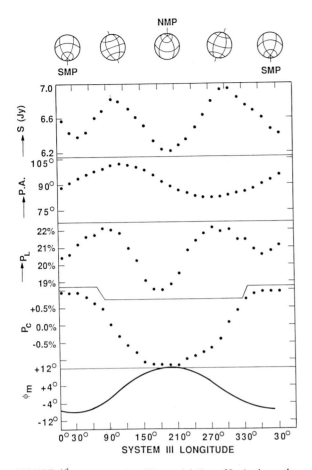

FIGURE 18 An example of the modulation of Jupiter's synchrotron radiation due to the planet's rotation. The orientation of the planet is indicated at the top; the different panels show subsequently the total flux density, S, the position angle, P.A., of the electric vector, the degree of linear and circular polarization, P_L and P_C, and the magnetic latitude of the Earth, ϕ_m, with respect to Jupiter. [From I. de Pater and M. J. Klein (1989). Time variability in Jupiter's synchrotron radiation. *In* "NASA Proceedings SP-494, Time Variable Phenomena in the Jovian System," Flagstaff, Arizona, Aug. 25–27, 1987, pp. 139–150, Washington, D.C.; after I. de Pater (1980). 21 cm maps of Jupiter's radiation belts from all rotational aspects. *Astron. Astrophys.* **88**, 175–183.]

in Fig. 20b are the secondary emission peaks just north and south of the main peaks. These are produced by electrons at their mirror points, and they reveal the presence of a rather large number of particles that bounce up and down the field lines at a Jovian distance of 2 to $3R_J$. It is believed that these electrons may have been scattered (pitch angle scattered) out of the magnetic equatorial plane by the moon Amalthea, which orbits Jupiter at a distance of $2.5R_J$.

The total radio intensity of Jupiter varies significantly in time (Fig. 21). Variations on timescales of years appear to be correlated with solar wind parameters, in particular the solar wind ram pressure, suggesting that the solar wind is influencing the supply and/or loss of electrons into Jupiter's inner magnetosphere. The impact of comet D/Shoemaker-Levy 9 with Jupiter in July 1994 caused a sudden sharp increase in Jupiter's total flux density, by ~40% at 6 cm and 10–15% at 70–90 cm (see Fig. 21). At the same time, the brightness distribution of the radio flux density changed drastically, as displayed by images in Fig. 20b. The images on the left side, which were just discussed, were taken in June of 1994, a few weeks before the comet crashed into the planet. Those on the right side were taken during the week of cometary impacts, at the same viewing aspects. The data show that the impacts triggered large changes in the total flux density as well as the brightness distribution. These changes are not yet understood.

2. Decametric Radio Emissions

At frequencies below 40 MHz, or wavelengths longward of 7.5 m, Jupiter is a strong emitter of sporadic nonthermal emissions. From the ground we can observe emission at decametric (DAM) wavelengths. The

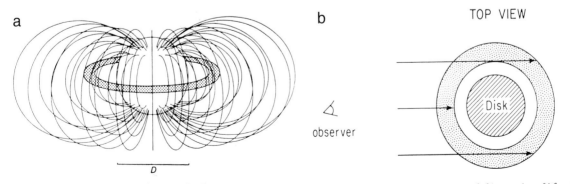

FIGURE 19   A schematic representation of the energetic electrons in Jupiter's magnetic field: (a) front view and (b) top view. [After I. de Pater (1981). Radio maps of Jupiter's radiation belts and planetary disk at λ = 6 cm. *Astron. Astrophys.* **93**, 370–381.]

emission is characterized by a complex, highly organized structure in the frequency–time domain and depends on the observer's position relative to Jupiter. Such decametric observations made over a long period of time, together with long-term observations of the synchrotron emission, made possible an accurate determination of Jupiter's rotation period, $9^h55^m29\overset{s}{.}71$. In addition, the satellite Io strongly modulates the emission; both the intensity and the probability of occurrence increase when Io is at certain locations in its orbit with respect to Jupiter and the observer. Decametric radiation is probably electron cyclotron emission originating close to Jupiter's ionosphere. Using the cutoff frequency of 40 MHz and the definition for the cyclotron or Larmor frequency,

$$\nu_L = \frac{qB}{2\pi m_e c} \qquad (9)$$

one can determine Jupiter's "surface" magnetic field strength. With $\nu_L = 40$ MHz, q the elemental charge (4.8×10^{-10} esu), m_e the electron mass (9.11×10^{-28} g), and c the speed of light (3×10^{10} cm/s), the field strength $B = 14.3$ G.

From the ground, the emission has occasionally been observed down to 4 MHz; at lower frequencies, the radiation cannot propagate through Earth's ionosphere. With radio experiments onboard the *Voyager* spacecraft, the existence of sporadic emissions has been detected as low as a few kHz. DAM exhibits a broad peak in emission near 8 MHz. The dynamic spectra in the frequency–time domain are extremely complex, but well ordered. On timescales of minutes, the emission displays a series of arcs, like "open" or "closed parentheses," shown in Fig. 22. We distinguish greater arcs between 1 and 40 MHz and lesser arcs at frequencies below 20 MHz. Within one storm, the arcs are all oriented the same way.

The intensity of DAM as observed on Earth shows a modulation with a period of 11.9 years, Jupiter's orbital period. This is likely caused by the changing declination of Earth with respect to Jupiter and suggests that the DAM source is strongly beamed latitudinally. Even more striking is the modulation in DAM intensity and its activity level with Jovian longitude and the position of Io with respect to the observer (Io phase). Figure 23 shows a graph of the dependence of DAM activity on Jupiter's central meridian longitude (λ_{cml}) and Io phase. The latter is counted counterclockwise from 0° when Io is behind Jupiter, via 90° at east elongation, to 270° at west elongation. The modulation by Io and the dependence on λ_{cml} is obvious. Not all storms are related to Io's position relative to the observer, but the most intense storms are. The upper and lower panels show the shape of the storm's dynamical spectra.

The decametric emission is strongly polarized. At frequencies above 20 MHz, the radiation is primarily right-handed circularly polarized; below 20 MHz, the sense of polarization depends on the planet's central meridian longitude. In the vicinity of 10 MHz, it tends to be left-handed circularly polarized at $0° < \lambda_{cml} < 135°$, when the south pole is tipped toward the observer, and RH at $135° < \lambda_{cml} < 300°$, when the north pole (at $\lambda_{cml} \approx 200°$) is turned toward the observer. Both types of polarization are seen at $300° < \lambda_{cml} < 360°$. Using the polarization characteristics and the fact that the radiation propagates in the extra-ordinary mode, the sources of the various Io and non-Io related storms can be determined. The RH sense must correspond to a source in the northern hemisphere, whereas the LH emission suggests a southern source. The cutoff frequencies of 40 MHz for RH and 20 MHz for LH polarization correspond to surface field strengths of ~14 and 7 G in the north and south polar regions, respectively. The extent of the source region has been determined using observations at very long baselines (VLBI); the sources are typically less than a few hundred kilometers in size.

Modulations of the emission appear on timescales of seconds. The emission drifts in frequency at a rate between −150 and +150 kHz/s. In addition to these L-bursts (long-lived bursts), there are also S-bursts (short-lived), emission features of short duration ($\lesssim 0.01$ s) that drift rapidly in frequency at a rate between −5 and −45 MHz/s (always negative). Although the emission is generally attributed to electron cyclotron radiation, the detailed emission mechanism is not known.

3. Kilometric Radio Emissions

Between a few kHz up to 1 MHz, the *Voyager* and *Ulysses* spacecraft detected kilometric (KOM) radiation from Jupiter. This radiation consists of broadband (bKOM) and narrowband (nKOM) components. The bKOM emission typically lasts for over an hour and covers a frequency range of a few hundred kHz (Fig. 24). The events usually last longer at the lower frequencies, and details of the emission (e.g., upper- and lower-frequency cutoffs) vary from event to event. The emission is very bursty. On occasion the low-frequency cutoff can be as low as 5 kHz, although often it is over 20 kHz. The cutoff frequency is likely set by

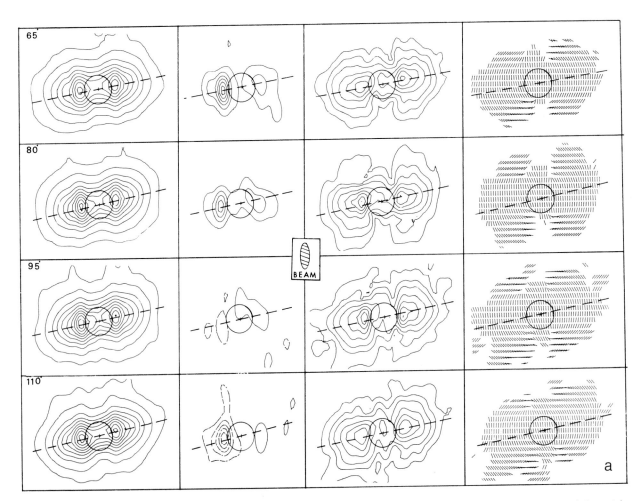

FIGURE 20 (a) Images of Jupiter's radio brightness distribution as obtained with the Westerbork Radio Telescope. From left to right, maps of the total intensity, *I*, the circularly and linearly polarized flux densities, *V* and *P*, and a vector diagram of the magnetic field of the planet, obtained by rotating the position angle of the electric vector, P.A., over 90°. For the circular polarization, dashed contours indicate left-handed and solid contours indicate right-handed circular polarization. The contours belonging to the three highest values in all three maps are drawn with heavy lines. The central meridian longitude is indicated in the top left corner of each row of pictures. Contour values in Kelvin are: *I*: 9.5 K, 65–1065 K in steps of 125 K; *V*: 1.9–21.5 K in steps of 2.8 K; *P*: 9.5, 32, 65–325 K in steps of 65 K. [From I. de Pater (1980).] (b) False color radio photographs of Jupiter's emission at longitudes (from top to bottom) $\lambda_{cml} \approx 110°$, 290°, and 350°. The images were taken with the VLA. Those shown on the left were taken in June 1994 and those on the right on July 19, 1994. The latter images were taken during the time that comet D/Shoemaker-Levy 9 crashed into the planet. The angular resolution (HPFW) of the data is ~6″ or 0.3R_J. [From I. de Pater *et al.* (1997). Time evolution of the east–west asymmetry in Jupiter's radiation belts after the impact of comet D/Shoemaker-Levy 9. *Icarus,* **129,** 21–47.]

propagation of the radiation through the Io plasma torus.

On the dayside the emission peaks at a longitude of 200°, the λ_{cml} of the north magnetic pole, whereas on the nightside there is a "bite" out of the emission at this longitude; instead, we see two bKOM periods centered around $\lambda_{cml} \approx 150°$ and $\approx 240°$. The intensity of bKOM is generally higher when observed at the dayside and when the observer is at higher latitudes, indicative of a source at high northern latitudes. A second, weaker source of bKOM emission is seen at $\lambda_{cml} \approx 20°$, when the south magnetic pole is tipped toward the observer.

The emission is strongly polarized over its full frequency range. At $\lambda_{cml} \approx 200°$, its sense of polarization was LH when the spacecraft was above the dayside hemisphere and RH when above the nightside. There are indications that the reverse is true at $\lambda_{cml} \approx 20°$. These data suggest a radio source that is fixed in local time rather than at a particular longitude rotating with the planet. The forward lobe near the north magnetic pole is of opposite polarization than a "back lobe" of the same source.

Narrow band emission peaks at 100 kHz and drops off sharply to zero ~50 and ~180 kHz. The events typically last a few hours or less and show a smooth

Pre-Impact

July, 19 1994

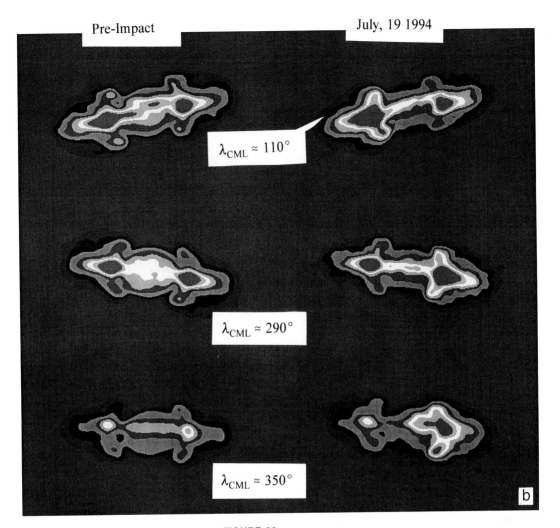

$\lambda_{CML} \approx 110°$

$\lambda_{CML} \approx 290°$

$\lambda_{CML} \approx 350°$

b

FIGURE 20 (*continued*)

rise and fall in intensity. The emission is usually observed to be LH if the observer is in the northern hemisphere and RH when in the south. The recurrence period for nKOM events is a little longer than expected from strict corotation with the planet: the source seems to lag behind by 3–5%. Similar departures from corotation are expected at a distance of 8–9 Jovian radii and have been observed in the outer plasma torus. This gave a first hint that the source of nKOM emission may be near the outer edge of the Io plasma torus. The radio experiment aboard the *Ulysses* spacecraft confirmed this suspicion and detected several discrete sources of nKOM in the outer plasma torus.

Continuum emission has been observed by the *Ulysses* spacecraft below ~25 kHz, in both its escaping and trapped forms. The latter form of radiation is trapped inside the magnetic cavity and cannot propagate through the high plasma density just inside the magnetopause. Inside the magnetosphere, the contin-

uum emission was often observed from a few hundred Hz up to ~5 kHz, but at times the emission could be detected up to 25 kHz, suggesting a compression of the magnetosphere caused by an increased solar wind ram pressure. Outside the magnetosphere, the lower-frequency cutoff of the freely propagating radiation appears to be well correlated with the solar wind ram pressure. At times of increased solar wind activity, Jupiter's continuum emission was sometimes absent owing to an upward shift in its lower-frequency cutoff.

C. SATURN

Saturn's nonthermal radio spectrum consists of three components: (a) Saturn kilometric radiation at frequencies from 3 to 1200 kHz, with a peak intensity near 175 kHz; (b) low-frequency narrowband emissions from 300 Hz to 100 kHz; and (c) Saturn electro-

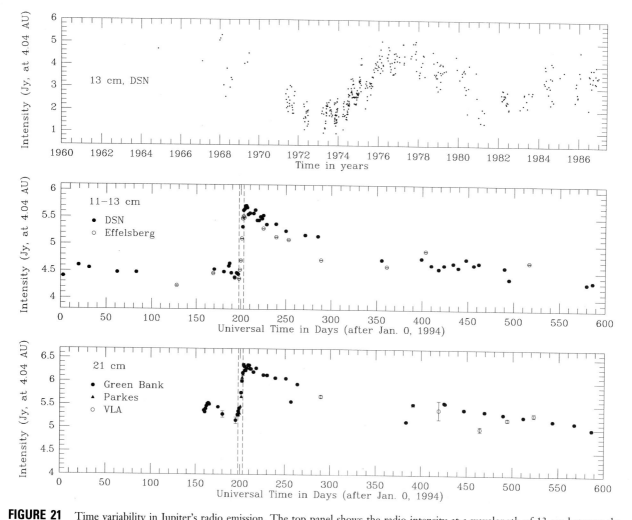

FIGURE 21 Time variability in Jupiter's radio emission. The top panel shows the radio intensity at a wavelength of 13 cm between the years 1960 and 1987. The data were taken by M. J. Klein, during his Jupiter Patrol Observing program with the NASA Deep Space Antennas. [From M. J. Klein *et al.* (1989). Systematic observation and correlation studies of variations in the synchrotron radio emission from Jupiter. *In* "Proceedings, Time Variable Phenomena in the Jovian System," Flagstaff, Arizona, Aug. 25–27, 1987.] The middle and bottom panels show Jupiter's radio intensity at 11–13 and 21 cm, respectively, during 1994 up to the summer of 1995. The impact of comet D/Shoemaker-Levy 9 with Jupiter occurred in July of 1994 (indicated by the vertical dashed lines). The data are from M. J. Klein, S. Gulkis, and S. J. Bolton (1995). Changes in Jupiter's 13-cm synchrotron radio emission following the impacts of comet Shoemaker/Levy 9. *Geophys. Res. Lett.* **22**, 1797–1800; M. K. Bird *et al.* (1996). Multi-frequency radio observations of Jupiter at Effelsberg during the SL-9 impact. *Icarus* **121**, 450–456; and M. H. Wong *et al.* (1996, updated and extended). Observations of Jupiter's 20-cm synchrotron emission during the impacts of comet P/Shoemaker-Levy 9. *Icarus* **121**, 457–468. [After I. de Pater and J. J. Lissauer (2000).]

static discharges, short bursts of radio emission at frequencies from ~20 kHz, to at least 40 MHz, the cutoff frequency of the radio instrument aboard the *Voyager* spacecraft.

1. Saturn Kilometric Radio Emissions

Saturn kilometric radio emissions (SKR) could be detected as far away as 3 AU from the planet. An example of the emission is shown in Fig. 25. A broad band of emission extends from 20 kHz up to ~1 MHz. No SKR has ever been observed above 1200 kHz. The 20-kHz lower-frequency cutoff is caused by the lower-frequency cutoff of the planetary radio astronomy (PRA) instrument onboard *Voyager*. Using Eq. (9), the upper cutoff of 1200 kHz suggests that the radiation is confined to magnetic field lines where the field strength is less than 0.43 G. Noting that the surface field strength is ~0.83 G in the north polar region and ~0.69 G near the south pole, the radiation must originate quite close to Saturn's ionosphere. The SKR

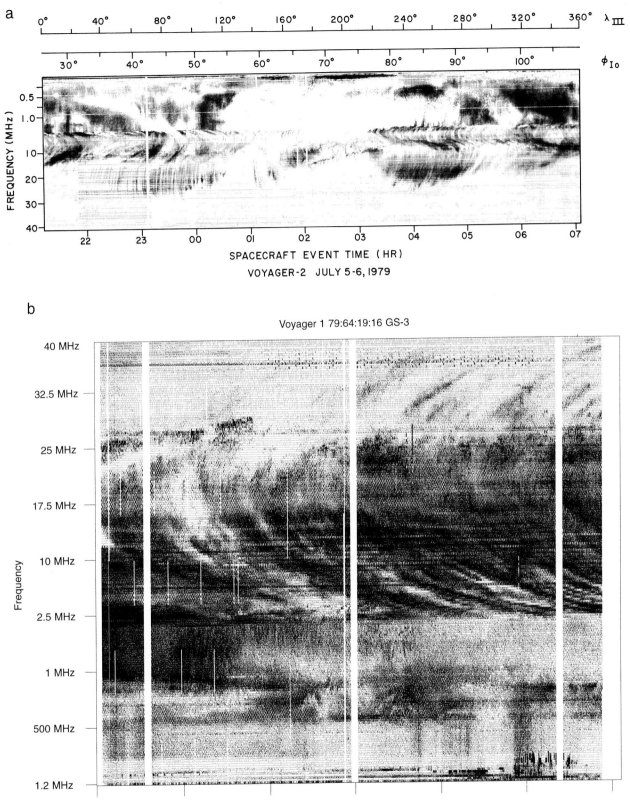

0°   40°   80°   120°   160°   200°   240°   280°   320°   360°   λ III

30°   40°   50°   60°   70°   80°   90°   100°   φ Io

FREQUENCY (MHz)

0.5
1.0
10
20
30
40

22   23   00   01   02   03   04   05   06   07

SPACECRAFT EVENT TIME (HR)

VOYAGER-2 JULY 5-6, 1979

b

Voyager 1 79:64:19:16 GS-3

40 MHz

32.5 MHz

25 MHz

17.5 MHz

Frequency

10 MHz

2.5 MHz

1 MHz

500 MHz

1.2 MHz

10 minute tick marks   Jupiter

FIGURE 22 (a) A representative dynamic spectrum of Jupiter's low-frequency radio emissions during one complete rotation of the planet. The darkness of the gray shading is proportional to the intensity of the emission. The emission is plotted as a function of frequency (along y axis) and time (along x axis). The pattern of vertex-late lesser arcs at longitudes less than 180° and vertex-early lesser arcs at longitudes above 180° is typical of the data before encounter. [From T. D. Carr, M. D. Desch, and J. K. Alexander (1983). Phenomenology of magnetospheric radio emissions. *In* "Physics of the Jovian Magnetosphere." (A. J. Dessler, ed.), pp. 226–284. Cambridge Univ. Press, Cambridge, England.] (b) A dynamical spectrum of Jupiter taken around closest approach. In contrast to (a), the timespan along the x axis is only 1 h. (Courtesy of D. Evans.)

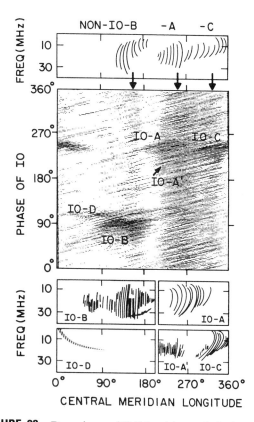

FIGURE 23 Dependence of DAM activity on Jupiter's central meridian longitude (CML) and Io's phase, at a frequency of about 20 MHz. The Io-related sources are labeled in the intensity plot. Schematic illustrations of the various plots are shown in the lower (Io-related events) and upper (non-Io-related events) panels. [From T. D. Carr, M. D. Desch, and J. K. Alexander (1983).]

emission is sometimes organized in arclike structures, reminiscent of the decametric arcs seen in Jupiter's radio emission. The emission is strongly circularly polarized, right-handed when the spacecraft was in the northern hemisphere and left-handed when in the southern hemisphere. The occurrence of SKR emission appears to be periodic, with a period of $10^h39^m24^s \pm 7^s$. Since the emission is tied to Saturn's magnetic field, the radio period is likely the true rotation period of the planet. In addition to SKR modulations due to the planet's rotation, there have been reports of a possible modulation by the satellite Dione (66-hour periodicity), which is not completely understood at the present time.

The polarization characteristics and the fact that the emission source appeared fixed relative to the Sun rather than to the planet suggest that the radio source is located in the north and south polar regions. This also agrees with the high magnetic field strength derived from the upper cutoff in frequency. A more precise determination of the source locations shows that

they correspond to active regions in Saturn's magnetosphere, where the field lines tend to open up into the interplanetary medium and intense particle precipitation and auroral activity occur. It is therefore not too surprising that the intensity of SKR appears to be correlated with the solar wind ram pressure, indicative of a continuous mass transfer of the solar wind into Saturn's low-altitude polar cusps.

2. Low-Frequency Emissions

Around a few kHz, Saturn appears to emit a low-level nonthermal continuum emission. At frequencies below 2–3 kHz, the radiation is trapped inside the magnetosphere. At higher frequencies the emission can escape. It is concentrated in narrow frequency bands and is therefore referred to as narrowband radio emissions. Although it is believed that both the "trapped" and narrowband radio emissions are generated by the same mechanism, the mechanism itself is still not resolved.

3. Saturn Electrostatic Discharges

Saturn electrostatic discharges (SEDs) are strong, unpolarized, impulsive events, which last for a few tens of milliseconds over the entire frequency range of the PRA experiment (20.4 kHz to 40.2 MHz) onboard *Voyager*. An example of SEDs is shown in Fig. 26. Structure in individual bursts can be seen down to the time resolution limit of 140 μs, which suggests a source size less than 40 km. Episodes of SED emissions occur approximately every 10^h10^m, distinctly different from the periodicity in SKR. In contrast to SKR, the SED source is fixed relative to the planet–observer line. The emissions are likely electrostatic discharge events, however, the source of these events is still controversial. They might be due to electrostatic discharges in Saturn's rings; the SED periodicity corresponds to the Keplerian orbital period of a particle in the planet's B-ring. Most researchers, however, attribute SED to the radio counterpart of lightning flashes in Saturn's atmosphere.

D. URANUS

Both smooth and bursty components are apparent in Uranus's emissions, which is reminiscent of Saturn's emissions. However, in contrast to Saturn, the emission is dominant on the nightside hemisphere, and detailed morphology is much more complicated than

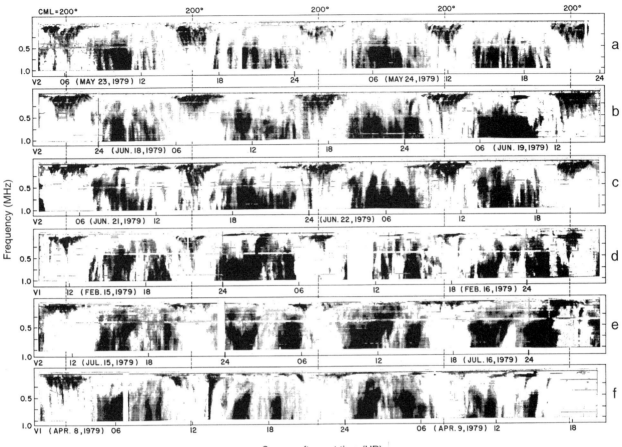

FIGURE 24 Dynamic spectra of Jupiter's HOM and KOM emissions, between 20 kHz and 1 MHz. The panels are aligned with respect to Jupiter's central meridian longitude. [From T. D. Carr, M. D. Desch, and J. K. Alexander (1983).]

that for Saturn. Many of the components are visible in the spectrogram shown in Fig. 27.

N-bursts are bursts of emission that occur in a narrow frequency band, with a typical width of 5 kHz, at frequencies between 15 and 120 kHz. They are right-handed circularly polarized, have timescales of about 500 ms, and have a greater probability of occurrence if the magnetic north pole is pointed toward the spacecraft.

The N-smooth component exhibits a very smoothly varying emission component, confined to a very narrow frequency range. It reminds one of the Jovian narrowband kilometric radiation, but the emission mode is likely different. The emission peaks at ~60 kHz, but has been observed over a broad frequency range, 20–350 kHz. The emission was left-handed circularly polarized before the spacecraft crossed the magnetic equator, and right-handed circularly polarized afterward. It has been postulated that the emission arises from a source in the magnetic equator, although

the north magnetic pole has also been suggested as a possible source location.

The dayside component is a component of Uranus's emission observed exclusively on the dayside hemisphere. It differs from the N-smooth component in its frequency coverage and polarization characteristics. It has been seen only from 100 to 250 kHz, and peaks at ~155 kHz. Although it is left-handed circularly polarized like the N-smooth component, it is substantially more linearly polarized. Its source location is on the dayside hemisphere, possibly near the magnetic north pole.

The B-smooth and B-burst emission components consist of a smooth and a bursty broadband component, respectively. They are the most intense emissions and extend over a broad range in frequency. A few rotations of the emission are shown in Fig. 28. We note, in particular, the "bite-out" between 4 and 6 h of the (arbitrary) rotation period. Its stability permitted an accurate determination of the Uranian rotation pe-

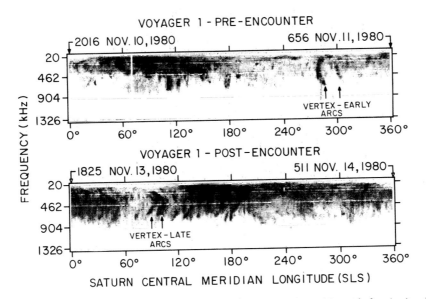

FIGURE 25 Dynamic spectra of Saturn's SKR emission. Data are shown from one rotation of Saturn before (top) and after (bottom) closest approach. The beginning and ending spacecraft acquisition times are indicated, as well as two vertex-early and two vertex-later arcs. [From J. R. Thieman and M. L. Goldstein (1981). Arcs in Saturn's radio spectra. *Nature* 292, 728–730.]

riod of 17.24 ± 0.01 h. The bite-out is likely caused by the sweeping motion of the radiation pattern past the spacecraft. The pattern can be described as a hollow cone pattern (see Section V, A), and every time the spacecraft is aligned with the beam axis of emission, the intensity diminishes to zero. The B-bursts are very impulsive, on timescales less than tens of milliseconds. Both components appear to come from a source near the south magnetic pole, likely from the auroral region. The bursty emission is probably beamed more perpendicular to the magnetic field than is the smooth component.

In addition to the emission components mentioned here, Uranus emits a bursty narrowband component centered around 5 kHz, referred to as 5-kHz noise. A continuum "trapped" radiation component, similar to that seen on Saturn, has been detected between 1 and 3 kHz.

Similar to the SED events from Saturn, impulsive bursts of radio emission have also been detected from Uranus, referred to as UED or Uranian electrostatic discharge events. They generally were fewer in number and less intensive than the SEDs. The UEDs were detected during a 24-h period centered around closest approach, have a typical duration of 60 ms, are unpolarized, and show a systematic change in the low-frequency cutoff from about 7 MHz before closest approach to 900 kHz when the spacecraft entered the nightside hemisphere. If these radio emissions are caused by lightning discharges, the lower-frequency cutoff can be related to the peak ionospheric electron

density, which should be ~6 × 10^5/cm^3 over the dayside hemisphere and ~10^4/cm^3 above the nightside. These numbers agree with ionospheric densities determined by other experiments onboard the *Voyager* spacecraft.

E. NEPTUNE

Neptune's radio emissions appear to be very similar to those received from Uranus. Both smoothly varying and bursty emission components have been detected, as displayed on a dynamical spectrum in Fig. 29 (only the smooth component is visible on this timescale). The bursts appeared throughout most of the low-frequency band of the PRA receiver, from 450 kHz up to 1.326 MHz. The threshold level for detection at higher frequencies is much higher, which makes scientists suspect that the upper-frequency cutoff may be an instrumental rather than a natural cutoff. The lower-frequency cutoff is probably natural. The bursts usually occur in about 1-h episodes, in which a few to a few tens of individual events are recorded, with a collective bandwidth of ~200 kHz. The individual events last less than 30 ms; they reoccur with a well-defined period of 16.11 ± 0.02 h. Using Eq. (9), the cutoff frequency of 1326 kHz corresponds to a magnetic field strength of 0.46 G; thus the surface field strength must be at least 0.46 G, well above the surface field strength of 0.14 G "measured" at Neptune's equator by the magnetometer aboard *Voyager*, assuming a dipole mag-

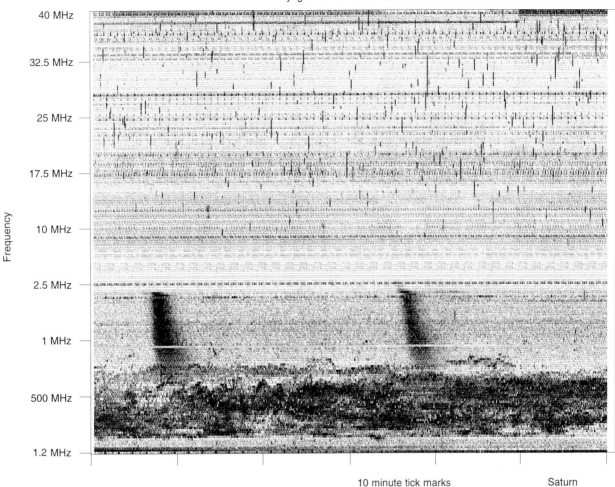

FIGURE 26 A dynamic spectrum of Saturn during 1 h near closest approach. The SEDs are clearly visible, in particular between 10 and 40 MHz. They appear as short streaks parallel to the frequency scale. At lower frequencies, the emission is dominated by SKR emission; the two prominent bursts between about 0.7 and 2 MHz are solar type III bursts. (Courtesy of D. Evans.)

netic field configuration. This immediately suggests that Neptune's magnetic field is complex. More detailed modeling of the *Voyager* magnetometer data shows minimum and maximum field strengths of 0.1 and 0.9 G, respectively, with the high field strength in the southern hemisphere near the south magnetic pole. The radio bursts are left-handed circularly polarized. The source of these impulsive emissions is likely located just above the south magnetic pole, within the auroral region. The radiation is emitted into a thin hollow cone with a wide opening angle (~80°). Some researchers believe the source to be a flickering searchlight rotating with the planet; others advocate the source to be fixed in local time, like Saturn's kilometric emission. In the latter case, the source will "turn on" or intensify when a particular (active) longitude sector passes by the planet's dawn terminator.

Within the smooth emission, three distinct components can be seen. The main emission occurs at frequencies between 20 and 600 kHz and is generally referred to as Neptunian kilometric radiation (NKR). This component was observed during 10 days around closest approach. A second component was observed for a few rotations around closest approach, at frequencies between 600 and 870 kHz (referred to as HF on Fig. 29, or higher frequency). A third component was seen only when the spacecraft was at low magnetic latitudes, from a few tens of Hz up to 56 kHz. This emission is referred to as low-frequency continuum radiation. It is characteristic of the low-frequency continuum emissions from Earth, Jupiter, and Uranus. The NKR emission is clearly periodic, lasting for about 8 h in each rotation. Together with the bursty emissions, this emission has been used to determine Nep-

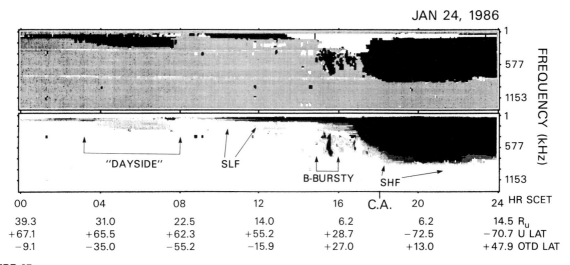

							HR SCET
00	04	08	12	16	20	24	
39.3	31.0	22.5	14.0	6.2	6.2	14.5	R$_u$
+67.1	+65.5	+62.3	+55.2	+28.7	−72.5	−70.7	U LAT
−9.1	−35.0	−55.2	−15.9	+27.0	+13.0	+47.9	OTD LAT

FIGURE 27 Dynamic spectra of Uranus's low-frequency radio emissions. The various components in the emission are labeled. The top panel shows the emission polarization, with white (black) corresponding to right (left)-hand circularly polarized signals. The bottom panel shows the intensities of the emissions. [From M. D. Desch *et al.* (1991). Uranus as a radio source. *In* "Uranus" (J. T. Bergstrahl *et al.*, eds.), pp. 894–925. Univ. Arizona Press, Tucson.]

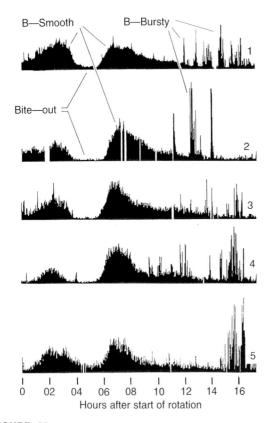

FIGURE 28 Five consecutive rotations of Uranus's nightside emission at a frequency of 481 kHz. The B-smooth and B-bursty components are indicated, as well as the "bite-out" from the B-smooth emissions. [From M. D. Desch *et al.* (1991).]

tune's rotation period of 16.108 ± 0.006 h. The main smooth radiation is sometimes left-handed, at other times right-handed circularly polarized, which is indicative of two independently radiating sources, one in each hemisphere. Each of the sources is located at relatively high magnetic latitudes and resembles auroral emissions from other planets. The second smooth component between 600 and 870 kHz is primarily RH polarized and appears to originate from a region near the magnetic equator.

When the spacecraft crossed Neptune's ring plane, many impacts by dust particles were detected, much more than during the ring plane crossings at Saturn and Uranus. They typically last several milliseconds and are seen from a few Hz up to 1.3 MHz.

VI. FUTURE DIRECTIONS

The research outlined here shows the value of radio observations in atmospheric research (composition, dynamics), surface composition and structure, cometary research (parent molecules, source of material, outgassing), and magnetospheric research (magnetic field configurations and strength, particle distributions). With the recent completion of millimeter arrays in California, France, and Japan, the high angular resolution one can obtain with one of the instruments (BIMA), and the upcoming (Smithsonian) submillimeter array, a new window for planetary research has opened up, in particular with regard to atmospheric

Voyager 2 PRA - Neptune

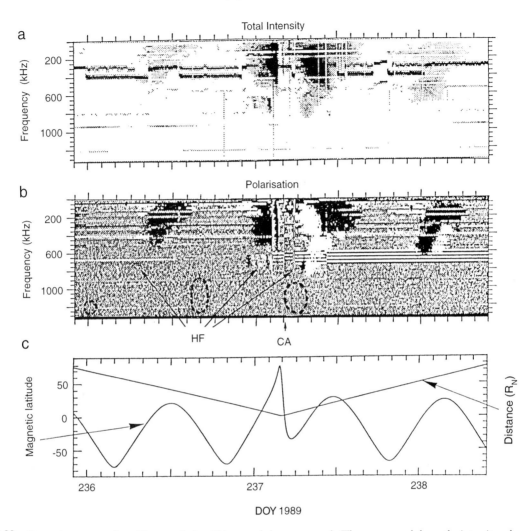

FIGURE 29 Dynamic spectrum from Neptune during 60 h around closest approach. The upper panel shows the intensity, where increasing darkness represents increasing intensity. The middle panel shows the polarization, where white indicates LH and black RH polarization. The bottom panel shows the magnetic distance latitude and distance of the spacecraft at the time of the observations. The abbreviation CA stands for closest approach and HF for higher-frequency emissions (see text). [From P. Zarka *et al.* (1995). Radio emissions from Neptune. *In* "Neptune" (D. P. Cruikshank, ed.), pp. 341–387. Univ. Arizona Press, Tucson.]

and cometary research. The campaign carried out with the various interferometers during the apparition of comet Hale-Bopp, with its fantastic results (all very preliminary at the time of this writing), shows the potential of planetary research with these relatively new millimeter arrays. Several countries (United States, Europe, and Japan) are planning the construction of very large millimeter arrays (possibly in Chile), either individually and/or in possible collaborations. These arrays will consist of many antennas, on the order of (at least) 40–60. They will work at both millimeter and submillimeter wavelengths. In addition to being ideal instruments for conventional planetary research, such instruments will allow the detection of hundreds of asteroids, "bare" cometary nuclei, impulsive emissions from molecular "jets" from comets with high time resolution, and "proto-Jupiters" in nearby stellar systems. Besides simple detection experiments, we expect to actually carry out scientific research in these areas, such as to determine the mass and chemical composition of protoplanets. [*See* ASTEROIDS.]

BIBLIOGRAPHY

Berge, G. L., and Gulkis, S. (1976). Earth based radio observations of Jupiter: Millimeter to meter wavelengths. *In* "Jupiter" (T. Gehrels, ed.), pp. 621–692. Univ. Arizona Press, Tucson.

Carr, T. D., Desch, M. D., and Alexander, J. K. (1983). Phenomenology of magnetospheric radio emissions. *In* "Physics of the Jovian Magnetosphere" (A. J. Dessler, ed.), pp. 226–284. Cambridge Univ. Press, Cambridge, England.

Crovisier, J., and Schloerb, F. P. (1991). The study of comets at radio wavelengths. *In* "Comets in the Post-Halley Era" (R. L. Newburn and J. Rahe, eds.), pp. 149–174. Kluwer Acad. Publ., Dordrecht, Netherlands.

de Pater, I. (1990). Radio images of planets. *Annu. Rev. Astron. Astrophys.* **28,** 347–399.

de Pater, I. (1991). The significance of radio observations for planets. *Phys. Rep.* **200,** 1.

de Pater, I., Palmer, P., and Snyder, L. E. (1991). Review of interferometric imaging of comets. *In* "Comets in the Post-Halley Era" (R. L. Newburn and J. Rahe, eds.), pp. 175–207. Kluwer Acad. Publ., Dordrecht, Netherlands.

de Pater, I., *et al.* (1995). The outburst of Jupiter's synchrotron radiation following the impact of comet P/Shoemaker-Levy 9. *Science* **268,** 1879–1883.

Desch, M. D., Kaiser, M. L., Zarka, P., Lecacheux, A., LeBlanc, Y., Aubier, M., and Ortega-Molina, A. (1991). Uranus as a radio source. *In* "Uranus" (J. T. Bergstrahl, A. D. Miner, and M. S. Matthews, eds.), pp. 894–925. Univ. Arizona Press, Tucson.

Kaiser, M. L., Desch, M. D., Kurth, W. S., Lecacheux, A., Genova, F., Pederson, B. M., and Evans, D. R. (1984). Saturn as a radio source. *In* "Saturn" (T. Gehrels and M. S. Matthews, eds.), pp. 378–415. Univ. Arizona Press, Tucson.

Kraus, J. D. (1986). "Radio Astronomy." Cygnus Quasar Books, Powell, Ohio.

Perley, R. A., Schwab, F. R., and Bridle, A. H. (1989). "Synthesis Imaging in Radio Astronomy," NRAO Workshop No. 21. Astronomical Society of the Pacific, San Francisco.

Thompson, A. R., Moran, J. M., and Swenson, G. W., Jr. (1986). "Interferometry and Synthesis in Radio Astronomy." John Wiley & Sons, New York.

Zarka, P., Pederson, B. M., Lecacheux, A., Kaiser, M. L., Desch, M. D., Farrell, W. M., and Kurth, W. S. (1995). Radio emissions from Neptune. *In* "Neptune" (D. Cruikshank, ed.), pp. 341–388. Univ. Arizona Press, Tucson.

PLANETARY RADAR

I. Introduction

II. Techniques and Instrumentation

III. Radar Measurements and Target
Properties

IV. Prospects for Planetary Radar

Steven J. Ostro
*Jet Propulsion Laboratory,
California Institute of Technology*

GLOSSARY

Aliasing: Overlapping of echo at different frequencies or at different time delays.

Antenna gain: Ratio of an antenna's sensitivity in the direction toward which it is pointed to its average sensitivity in all directions.

Circular polarization ratio: Ratio of echo power received in the same sense of circular polarization as transmitted (the SC sense) to that received in the opposite (OC) sense.

Doppler shift: Difference between the frequencies of the radar echo and the transmission, caused by the relative velocity of the target with respect to the radar.

Echo bandwidth: Dispersion in Doppler frequency of an echo, that is, the width of the echo power spectrum.

Ephemeris: Table of planetary positions as a function of time (plural: ephemerides).

Klystron: Vacuum-tube amplifier used in planetary radar transmitters.

Radar albedo: Ratio of a target's radar cross section in a specified polarization to its projected area; hence, a measure of the target's radar reflectivity.

Radar cross section: Most common measure of a target's scattering efficiency, equal to the projected area of that perfect metal sphere that would give the same echo power as the target if observed at the target's location.

Scattering law: Function giving the dependence of a surface element's radar cross section on viewing angle.

Synodic rotation period: Apparent rotation period of a target that is moving relative to the observer, to be distinguished from the "sidereal" rotation period measured with respect to the fixed stars.

Time delay: Time between transmission of a radar signal and reception of the echo.

Planetary radar astronomy is the study of solar system entities (the Moon, asteroids, and comets, as well as the major planets and their satellites and ring systems) by transmitting a radio signal toward the target and then receiving and analyzing the echo. This field of research has primarily involved observations with Earth-based radar telescopes, but also includes certain experiments with the transmitter and/or the receiver on board a spacecraft orbiting or passing near a planetary object. However, radar studies of Earth's surface, atmosphere, or ionosphere from spacecraft, aircraft, or the ground are not considered part of planetary radar astronomy. Radar studies of the Sun involve such distinctly individual methodologies and physical considerations that solar radar astronomy is considered a field separate from planetary radar astronomy.

I. INTRODUCTION
· ·

A. SCIENTIFIC CONTEXT

Planetary radar astronomy is a field of science at the intersection of planetology, radio astronomy, and radar engineering. A radar telescope is essentially a radio telescope equipped with a high-power radio transmitter and specialized electronic instrumentation designed to link transmitter, receiver, data-acquisition, and telescope-pointing components together in an integrated radar system. The principles underlying operation of this system are not fundamentally very different from those involved in radars used, for example, in marine and aircraft navigation, measurement of automobile speeds, and satellite surveillance. However, planetary radars must detect echoes from targets at interplanetary distances ($\sim 10^5$–10^9 km) and therefore are the largest and most powerful radar systems in existence.

The advantages of radar observations in astronomy stem from the high degree of control exercised by the observer on the transmitted signal used to illuminate the target. Whereas virtually every other astronomical technique relies on passive measurement of reflected sunlight or naturally emitted radiation, the radar astronomer controls all the properties of the illumination, including its intensity, direction, polarization, and time/frequency structure. The properties of the transmitted waveform are selected to achieve particular scientific objectives. By comparing the properties of the echo to the very well known properties of the transmission, some of the target's properties can be deduced. Hence, the observer is intimately involved in an active astronomical observation and, in a very real sense, performs a controlled laboratory experiment on the planetary target.

Radar delay-Doppler and interferometric techniques can spatially resolve a target whose angular extent is dwarfed by the antenna beamwidth, thereby bestowing a considerable advantage on radar over optical techniques in the study of asteroids, which appear like "point sources" through ground-based optical telescopes. Furthermore, by virtue of the centimeter-to-meter wavelengths employed, radar is sensitive to scales of surface structure many orders of magnitude larger than those probed in visible or infrared regions of the spectrum. Radar is also unique in its ability to "see through" the dense clouds that enshroud Venus and the glowing gaseous coma that conceals the nucleus of a comet. Because of its unique capabilities,

radar astronomy has made essential contributions to planetary exploration for a third of a century.

B. HISTORY

Radar technology was developed rapidly to meet military needs during World War II. In 1946, soon after the war's conclusion, groups in the United States and Hungary obtained echoes from the Moon, giving birth to planetary radar astronomy. These early postwar efforts were motivated primarily by interest in electromagnetic propagation through the ionosphere and the possibility for using the Moon as a "relay" for radio communication.

During the next two decades, the development of nuclear weaponry and the need for ballistic missile warning systems prompted enormous improvements in radar capabilities. This period also saw rapid growth in radio astronomy and the construction of huge radio telescopes. In 1957, the Soviet Union launched *Sputnik* and with it the space age, and in 1958, with the formation by the U.S. Congress of the National Aeronautics and Space Administration (NASA), a great deal of scientific attention turned to the Moon and to planetary exploration in general. During the ensuring years, exhaustive radar investigations of the Moon were conducted at wavelengths from 0.9 cm to 20 m, and the results generated theories of radar scattering from natural surfaces that still see wide application.

By 1963, improvements in the sensitivity of planetary radars in both the United States and the U.S.S.R. had permitted the initial detections of echoes from the terrestrial planets (Venus, Mercury, and Mars). During this period, radar investigations provided the first accurate determinations of the rotations of Venus and Mercury and the earliest indications for the extreme geologic diversity of Mars. Radar images of Venus have revealed small portions of that planet's surface at increasingly fine resolution since the late 1960s, and in 1979 the Pioneer Venus Spacecraft Radar Experiment gave us our first look at Venus's global distributions of topography, radar reflectivity, and surface slopes. During the 1980s, maps having sparse coverage but resolution down to ~ 1 km were obtained from the Soviet *Venera 15* and *16* orbiters and from ground-based observations with improved systems. Much more recently, the *Magellan* spacecraft radar revealed most of the planet's surface with unprecedented clarity, revealing a rich assortment of volcanic, tectonic, and impact features.

The first echoes from a near-Earth asteroid (1566 Icarus) were detected in 1968; it would be nearly another decade before the first radar detection of a main belt asteroid (1 Ceres in 1977), to be followed in 1980 by the first detection of echoes from a comet (Encke). During 1972 and 1973, detection of 13-cm-wavelength radar echoes from Saturn's rings shattered prevailing notions that typical ring particles were 0.1 to 1.0 mm in size—the fact that decimeter-scale radio waves are backscattered efficiently requires that a large fraction of the particles be larger than a centimeter. Observations by the Voyager spacecraft confirmed this fact and further suggested that particle sizes extend to at least 10 m.

In the mid-1970s, echoes from Jupiter's Galilean satellites Europa, Ganymede, and Callisto revealed the manner in which these icy moons backscatter circularly polarized waves to be extraordinarily strange, and totally outside the realm of previous radar experience. We now understand that those echoes were due to high-order multiple scattering from within the top few decameters of the satellites' regoliths, but there remain important questions about the geologic structures involved and the nature of electromagnetic interactions with those structures.

The late 1980s saw the initial detections of Phobos and Titan; the accurate measurement of Io's radar properties; the discovery of large-particle clouds accompanying comets; the dual-polarization mapping of Mars and the icy Galilean satellites; and radar imaging of asteroids that revealed an extraordinary assortment of radar signatures and several highly irregular shapes, including a "contact-binary" near-Earth asteroid.

During the 1990s, the novel use of instrumentation and waveforms has yielded the first full-disk radar images of the terrestrial planets, revealing the global diversity of small-scale morphology on these objects and the surprising presence of radar-bright polar anomalies on Mercury as well as Mars. Similarities between the polarization and albedo signatures of these features and those of the icy Galilean satellites argue persuasively that Mercury's polar anomalies are deposits of water ice in the floors of craters that are perpetually shaded from sunlight by Mercury's low obliquity. The first time-delay-resolved ("ranging") measurements to Ganymede and Callisto were carried out in 1992. That same year, delay-Doppler images of the closely approaching asteroid 4179 Toutatis revealed this strange object to be in a very slow, non-principal-axis spin state and provided the first geologically detailed pictures of an Earth-orbit-crossing asteroid. This decade also saw the first intercontinental radar observations and the beginning of planetary radar in Germany and Japan. By 1997, the list of small planetary objects detected by radar included 6 comets, 37 main belt asteroids, and 47 near-Earth asteroids (Table I).

Perhaps the most far-reaching recent development is the upgrading of the Arecibo telescope's sensitivity by over an order of magnitude. At this writing, Arecibo is about to return to operation, and radar astronomy is poised to begin a new era of major contributions to planetary science.

II. TECHNIQUES AND INSTRUMENTATION

A. ECHO DETECTABILITY

How close must a planetary target be for its radar echo to be detectable? For a given transmitted power P_T and antenna gain G, the power flux a distance R from the radar will be $P_T G/4\pi R^2$. We define the target's radar cross section, σ, as 4π times the backscattered power per unit of solid angle per unit of flux incident at the target. Then, letting λ be the radar wavelength and defining the antenna's effective aperture as $A = G\lambda^2/4\pi$, we have the received power

$$P_R = P_T GA\sigma/(4\pi)^2 R^4 \qquad (1)$$

This power might be much less than the receiver noise power, $P_N = kT_S \Delta f$, where k is Boltzmann's constant, T_S is the receiver system temperature, and Δf is the frequency resolution of the data. However, the mean level of P_N constitutes a background that can be determined and removed, so P_R will be detectable as long as it is at least several times larger than the standard deviation of the random fluctuations in P_N. These fluctuations can be shown to have a distribution that, for usual values of Δf and the integration time Δt, is nearly Gaussian with standard deviation $\Delta P_N = P_N/(\Delta f \Delta t)^{1/2}$. The highest signal-to-noise ratio, or SNR $= P_R/\Delta P_N$, will be achieved for a frequency resolution equal to the effective bandwidth of the echo. As discussed in the following, that bandwidth is proportional to $D/\lambda P$, where D is the target's diameter and P is the target's rotation period, so let us assume that $\Delta f \sim D/\lambda P$. By writing $\sigma = \hat{\sigma}\pi D^2/4$, where the radar albedo $\hat{\sigma}$ is a measure of the target's radar reflectivity, we arrive at the following expression for the echo's signal-to-noise ratio:

$$\text{SNR} \sim (\text{system factor})(\text{target factor}(\Delta t)^{1/2} \qquad (2)$$

TABLE I
Radar-Detected Planetary Targets

Year of first detection	Planets, satellites, rings	Main belt asteroids	Near-Earth asteroids	Comets
1946	Moon			
1961	Venus			
1962	Mercury			
1963	Mars			
1968			1566 Icarus	
1972			1685 Toro	
1973	Saturn's rings			
1974	Ganymede			
1975	Callisto Europa		433 Eros	
1976	Io		1580 Betulia	
1977		1 Ceres		
1979		4 Vesta		
1980		7 Iris 16 Psyche	1862 Apollo	Encke
1981		97 Klotho 8 Flora	1915 Quetzalcoatl 2100 Ra-Shalom	
1982		2 Pallas 12 Victoria 19 Fortuna 46 Hestia		Grigg-Skjellerup
1983		5 Astraea 139 Juewa 356 Liguria 80 Sappho 694 Ekard	1620 Geographos 2201 Oljato	IRAS-Araki-Alcock Sugano-Saigusa-Fujikawa
1984		9 Metis 554 Peraga 144 Vibilia	2101 Adonis	
1985		6 Hebe 41 Daphne 21 Lutetia 33 Polyhymnia 84 Klio 192 Nausikaa 230 Athamantis 216 Kleopatra 18 Melpolmene	1627 Ivar 1036 Ganymed 1866 Sisyphus	Halley
1986		393 Lampetia 27 Euterpe	6178 1986DA 1986JK 3103 Eger (1982BB) 3199 Nefertiti	
1987		532 Herculina 20 Massalia	1981 Midas 3757 1982XB	
1988	Phobos	654 Zelinda 105 Artemis	3908 1980PA	
1989	Titan		4034 1986PA 1989JA 4769 Castalia (1989PB) 1917 Cuyo	

continues

continued

Year of first detection	Planets, satel-lites, rings	Main belt asteroids	Near-Earth asteroids	Comets
1990		78 Diana 194 Prokne	1990MF 1990OS 4544 Xanthus (1989FB)	
1991		324 Bamberga 796 Sarita	1991AQ 6489 Golevka (1991JX) 1991EE	
1992			5189 1990UQ 4179 Toutatis	
1994			4953 1990MU	
1995			2062 Aten (1976AA)	
1996			1992QN 1993QA 2063 Bacchus 1996JG 1991CS 4197 1982TA	Hyakutake (C/1996 B2)
1997			7341 1991VK 7482 1994PC1 1997BR	
1998			1998BY7 6037 1988EG 4183 Cuno 1998KY26	

where

$$\text{system factor} \sim P_T A^2 / \lambda^{3/2} T_S$$
$$\sim P_T G^2 \lambda^{5/2} / T_S \qquad (3)$$

and

$$\text{target factor} \sim \hat{\sigma} D^{3/2} P^{1/2} / R^4 \qquad (4)$$

The inverse-fourth-power dependence of SNR on target distance is a severe limitation in ground-based observations, but it can be overcome by constructing very powerful radar systems.

B. RADAR SYSTEMS

The world has two active planetary radar facilities: the Arecibo Observatory (part of the National Astronomy and Ionosphere Center) in Puerto Rico and the Goldstone Solar System Radar in California. Radar wavelengths are 13 cm and 70 cm for Arecibo and 3.5 cm and 13 cm for Goldstone; with each instrument, enormously more sensitivity is achievable with the shorter wavelength. The upgraded Arecibo telescope has twice the range and will see three times the volume of Goldstone, whereas Goldstone sees twice as much sky as Arecibo and can track targets at least three times

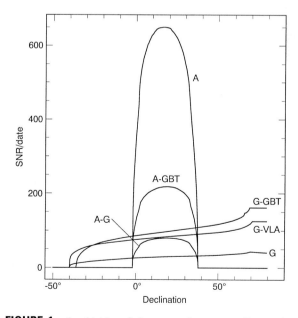

FIGURE 1 Sensitivities of planetary radar systems. Curves plot the single-date, signal-to-noise ratio of echoes from a typical 1-km asteroid ($\hat{\sigma} = 0.1$, $P = 5$ hr) at a distance of 0.1 AU for the upgraded Arecibo telescope (A), Goldstone (G), and bistatic configurations using those instruments and the Very Large Array (VLA) or the Greenbank Telescope (GBT).

FIGURE 2 The Arecibo Observatory in Puerto Rico. (a) Aerial view prior to the recent upgrade. The diameter of the spherical reflector is 305 m. (b) A close-up view of the structure suspended above the reflector, showing the old 430-MHz line feed and the radome that encloses the new Gregorian secondary and tertiary subreflectors.

longer. Figure 1 shows the relative sensitivities of planetary radar systems as a function of target declination.

The Arecibo telescope (Fig. 2) consists of a 305-m-diameter, fixed reflector whose surface is a 51-m-deep section of a 265-m-radius sphere. Moveable feeds designed to correct for spherical aberration are suspended from a triangular platform 137 m above the reflector and can be aimed toward various positions on the reflector, enabling the telescope to point within about 20 degrees of the overhead direction (declination 18.3°N). Components of the recent upgrade have included a megawatt transmitter, a ground screen to reduce noise generated by radiation from the ground, and replacement of most of the old single-frequency line feeds with a Gregorian reflector system (named after the seventeenth-century mathematician James Gregory) that employs 22-m secondary and 8-m tertiary subreflectors enclosed inside a 26-m dome.

The Goldstone main antenna, DSS-14 (DSS stands for Deep Space Station), is part of the NASA Deep Space Network, which is run by the Jet Propulsion Laboratory (JPL). It is a fully steerable, 70-m, parabolic reflector (Fig. 3). Bistatic (two-station) experiments employing transmission from DSS-14 and reception of echoes at the 27-antenna Very Large Array (VLA) in New Mexico have synthesized a beamwidth as small as 0.24 seconds of arc, versus 2 minutes of arc for single-dish observations with Arecibo or Goldstone. Bistatic experiments using DSS-14 transmissions and reception of echoes at DSS-13, a 34-m antenna 22 km away, have been conducted on several very close targets. In coming years, bistatic observations between Arecibo and Goldstone, or using transmission from Arecibo or Goldstone and reception at the 100-m Greenbank Telescope, now under construction in West Virginia, should prove advantageous for outer planet satellites and nearby asteroids and comets.

Figure 4 is a simplified block diagram of a planetary radar system. A waveguide switch, a moveable subreflector, or a moveable mirror system is used to place the antenna in a transmitting or receiving configuration. The heart of the transmitter is one or two klystron vacuum-tube amplifiers. In these tubes, electrons accelerated by a potential drop of some 60 kV are magnetically focused as they enter the first of five or six cavities. In this first cavity, an oscillating electric field at a certain radio frequency (RF, e.g., 2380 MHz for Arecibo) modulates the electrons' velocities and hence

b

FIGURE 2 (*continued*)

their density and energy flux. Subsequent resonant cavities enhance this velocity bunching (they constitute what is called a "cascade amplifier") and about half of the input DC power is converted to RF power and sent out through a waveguide to the antenna feed system and radiated toward the target. The other half of the input power is waste heat and must be transported away from the klystron by cooling water. The impact of the electrons on the collector anode generates dangerous X rays that must be contained by heavy metal shielding surrounding the tube, a requirement that further boosts the weight, complexity, and hence cost of a high-power transmitter.

In most single-antenna observations, one transmits for a duration near the roundtrip propagation time to the target (i.e., until the echo from the beginning of the transmission is about to arrive), and then receives for a similar duration. In the "front end" of the receiving system, the echo signal is amplified by a maser and converted from RF frequencies (e.g., 2380 MHz for Arecibo at 13 cm) down to intermediate frequencies (IF, e.g., 30 MHz), for which transmission line losses

FIGURE 3 The 70-m Goldstone Solar System Radar antenna, DSS-14, in California.

are small, and passed from the proximity of the antenna feed to a remote control room containing additional stages of signal-processing equipment, computers, and digital tape recorders. The signal is filtered, amplified, and converted to frequencies low enough for analog voltage samples to be digitized and recorded. The frequency down-conversion can be done in several stages using analog devices called superheterodyne mixers, but in recent years it has become possible to do this digitally, at increasingly higher frequencies. The nature of the final processing prior to recording of data on a hard disk or magnetic tape depends on the nature of the radar experiment and particularly on the time/frequency structure of the transmitted waveform. Each year, systems for reducing and displaying echoes in "real time" and techniques for processing recorded data are becoming more ambitious as computers get faster.

C. ECHO TIME DELAY AND DOPPLER FREQUENCY

The time between transmission of a radar signal and reception of the echo is called the echo's roundtrip time delay, τ, and is of order $2R/c$, where c is the speed of light, $299{,}792{,}458$ m sec^{-1}. Since planetary targets are not points, even an infinitesimally short transmitted pulse would be dispersed in time delay, and the total extent $\Delta \tau_{TARGET}$ of the distribution $\sigma(\tau)$ of echo power (in units of radar cross section) would be D/c for a sphere of diameter D and in general depends on the target's size and shape.

The translational motion of the target with respect to the radar introduces a Doppler shift, ν, in the frequency of the transmission. Both the time delay and the Doppler shift of the echo can be predicted in advance from the target's ephemeris, which is calculated using the geodetic position of the radar and the orbital elements of Earth and the target. The predicted Doppler shift can be removed electronically by continuously tuning the local oscillator used for RF-to-IF frequency conversion (see Fig. 4). The predicted Doppler must be accurate enough to avoid smearing out the echo in delay, and this requirement places stringent demands on the quality of the observing ephemeris. Time and frequency measurements are critical, because the delay/Doppler distribution of echo power is the source of the finest spatial resolution, and also because delay and Doppler are fundamental dynamical observables. Reliable, precise time/frequency measurements are made possible by high-speed data acquisition systems and stable, accurate clocks and frequency standards.

Because different parts of the rotating target will have different velocities relative to the radar, the echo will be dispersed in Doppler frequency as well as in time delay. The basic strategy of any radar experiment always involves measurement of some characteristic(s) of the function $\sigma(\tau, \nu)$, perhaps as a function of time and perhaps using more than one combination of transmitted and received polarizations. Ideally, one would like to obtain $\sigma(\tau, \nu)$ with very fine resolution, sampling that function within intervals whose dimensions, $\Delta \tau \times \Delta \nu$, are minute compared to the echo dispersions $\Delta \tau_{TARGET}$ and $\Delta \nu_{TARGET}$. However, one's ability to resolve $\sigma(\tau, \nu)$ is necessarily limited by the available echo strength. Furthermore, as described in the next section, an intrinsic upper bound on the product $\Delta \tau \Delta \nu$ forces a trade-off between delay resolution and Doppler resolution for the most efficient waveforms used. Under these constraints, many planetary radar experiments employ waveforms aimed at providing estimates of one of the marginal distributions, $\sigma(\tau)$ or $\sigma(\nu)$. Figure 5 shows the geometry of delay-resolution cells and Doppler-resolution cells for a spherical target and sketches the relation between these cells and $\sigma(\tau)$ and $\sigma(\nu)$. Delay-Doppler measurements are explored further in the following.

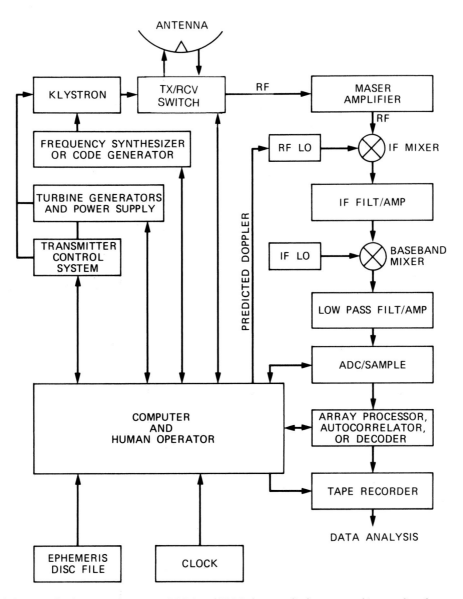

FIGURE 4 Block diagram of a planetary radar system. RF LO and IF LO denote radio frequency and intermediate frequency local oscillators, and ADC denotes analog-to-digital converter.

D. RADAR WAVEFORMS

In the simplest radar experiment, the transmitted signal is a highly monochromatic, unmodulated continuous-wave (cw) signal. Analysis of the received signal comprises Fourier transformation of a series of time samples and yields an estimate of the echo power spectrum $\sigma(\nu)$, but contains no information about the distance to the target or $\sigma(\tau)$. To avoid aliasing, the sampling rate must be at least as large as the bandwidth of the low-pass filter (see Fig. 4) and usually is comparable to or larger than the echo's intrinsic dispersion

$\Delta\nu_{\text{TARGET}}$ from Doppler broadening. Fast-Fourier transform (FFT) algorithms, implemented via software or hard-wired (e.g., in an array processor), greatly speed the calculation of discrete spectra from time series, and are ubiquitous in radar astronomy. In a single FFT operation, a string of N time samples taken at intervals of Δt seconds is transformed into a string of N spectral elements with frequency resolution $\Delta\nu = 1/N\,\Delta t$. Most planetary radar targets are sufficiently narrowband for power spectra to be computed, accumulated, and recorded directly on magnetic tape at convenient intervals. In some situations, it is desirable

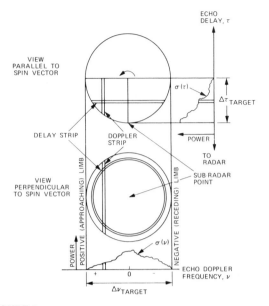

FIGURE 5 Time-delay and Doppler-frequency resolution of the radar echo from a rotating spherical target.

to record time samples on tape and Fourier-analyze the data later, perhaps using FFTs of different lengths to obtain spectra at a variety of frequency resolutions. In others, it is convenient to pass the signal through an autocorrelator to record autocorrelation functions, and then apply FFTs to extract spectra.

To obtain delay resolution, one must apply some sort of time modulation to the transmitted waveform. For example, a short-duration pulse of cw signal lasting 1 μsec would provide delay resolution of 150 m. However, the echo would have to compete with the noise power in a bandwidth of order 1 MHz so the echo power from many consecutive pulses would probably have to be summed to yield a detection. One would not want these pulses to be too close together, however, or there would be more than one pulse incident on the target at once and interpretation of echoes would be insufferably ambiguous. Thus, one arranges the pulse repetition period t_{PRP} to exceed the target's intrinsic delay dispersion $\Delta\tau_{TARGET}$, ensuring that the echo will consist of successive, nonoverlapping "replicas" of $\sigma(\tau)$ separated from each other by t_{PRP}. To generate this "pulsed cw" waveform, the transmitter is switched on and off while the frequency synthesizer (see Fig. 4) maintains phase coherence from pulse to pulse. Then Fourier transformation of time samples taken at the same position within each of N successive replicas of $\sigma(\tau)$ yields the power spectrum of echo from a certain delay resolution cell on the target. This spectrum has an unaliased bandwidth of $1/t_{PRP}$ and a frequency resolution of $1/Nt_{PRP}$. Repeating this process for a different position within each replica of $\sigma(\tau)$ yields the power spectrum for echo from a different delay resolution

cell, and in this manner one obtains the delay-Doppler image $\sigma(\tau, \nu)$.

In practice, instead of pulsing the transmitter, one usually codes a cw signal with a sequence of 180° phase reversals and cross-correlates the echo with a representation of the code (e.g., using the decoder in Fig. 4), thereby synthesizing a pulse train with the desired values of Δt and t_{PRP}. With this approach, one optimizes SNR, because it is much cheaper to transmit the same average power continuously than by pulsing the transmitter. Most modern ground-based radar astronomy observations employ cw or repetitive, phase-coded cw waveforms.

A limitation of coherent-pulsed or repetitive, binary-phase-coded cw waveforms follows from combining the requirement that there never be more than one echo received from the target at any instant (i.e., that $t_{PRP} > \Delta\tau_{TARGET}$) with the antialiasing frequency requirement that the rate $(1/t_{PRP})$ at which echo from a given delay resolution cell is sampled be no less than the target bandwidth $\Delta\nu_{TARGET}$. Therefore, a target must satisfy $\Delta\tau_{TARGET} \Delta\nu_{TARGET} \leq 1$ or it is "overspread" (Table II) and cannot be investigated completely and simultaneously in delay and Doppler without aliasing, at least with the waveforms discussed so far. Various degrees of aliasing may be "acceptable" for overspread factors less than about 10, depending on the precise experimental objectives and the exact properties of the echo.

How can the full delay-Doppler distribution be obtained for overspread targets? Frequency-swept and frequency-stepped waveforms have seen limited use in planetary radar; the latter approach was applied to Saturn's rings. A new technique uses a nonrepeating, binary-phase-coded cw waveform. The received signal for any given delay cell is decoded by multiplying it by a suitably lagged replica of the code. Developed for observations of the highly overspread ionosphere, this "coded-long-pulse" or "random-code" waveform redistributes delay-aliased echo power into an additive white-noise background. The SNR is reduced accordingly, but this penalty is acceptable for strong targets.

III. RADAR MEASUREMENTS AND TARGET PROPERTIES

A. ALBEDO AND POLARIZATION RATIO

A primary goal of the initial radar investigation of any planetary target is estimation of the target's radar cross

TABLE II
Characteristics of Selected Planetary Radar Targets[a]

Target	Minimum echo delay[b] (min)	Radar cross section (km²)	Radar albedo $\hat{\sigma}_0$	Circular polarization ratio, μ_C	Maximum dispersions[c]		
					Delay (msec)	Doppler (Hz)	Product
Moon	0.04	6.6×10^5	0.07	0.1	12	60	0.7
Mercury	9.1	1.1×10^6	0.06	0.1	16	110	2
Venus	4.5	1.3×10^7	0.11	0.1	40	110	4
Mars	6.2	2.9×10^6	0.08	0.3	23	7600	170
Phobos	6.2	22	0.06	0.1	0.1	100	10^{-2}
1 Ceres	26	2.7×10^4	0.05	0.0	3	3100	9
2 Pallas	25	1.7×10^4	0.08	0.0	2	2000	4
12 Victoria	15	2.3×10^3	0.22	0.1	0.5	590	3
16 Psyche	28	1.4×10^4	0.31	0.1	0.8	2200	2
216 Kleopatra	20	7.1×10^3	0.44	0.0	?	750	?
324 Bamberga	13	2.9×10^3	0.06	0.1	0.8	230	0.2
1685 Toro	2.3	1.7	0.1	0.2	0.02	14	10^{-4}
1862 Apollo	0.9	0.2	0.1	0.4	0.01	16	10^{-4}
2100 Ra-Shalom	3.0	1.0	0.1	0.2	0.01	5	10^{-4}
2101 Adonis	1.5	0.02	<0.3	1.0	?	2	?
4179 Toutatis	0.4	1.3	0.24	0.3	0.01	1	10^{-5}
4769 Castalia	0.6	0.2	0.15	0.3	0.01	10	10^{-4}
6178 1986DA	3.4	2.4	0.6	0.1	12	15	0.2
IAA[d] nucleus	0.5	2.4	0.04?	0.1	?	4	?
IAA coma	0.5	0.8	?	0.01	?	600	?
HYA[e] nucleus	1.7	0.11	?	0.5	?	12	?
HYA coma	1.7	1.3	?	<1	?	3000	?
Io	66	2×10^6	0.2	0.5	12	2400	29
Europa	66	8×10^6	1.0	1.5	10	1000	11
Ganymede	66	1×10^7	0.6	1.4	18	850	15
Callisto	66	5×10^6	0.3	1.2	16	330	5
Saturn's rings	134	10^8–10^9	0.7	0.5	1600	6×10^5	10^6

[a] Typical 3.5- to 13-cm values. Question marks denote absence of radar data or of prior information about target dimensions.

[b] For asteroids and comets, this is the minimum echo time delay for radar observations to date.

[c] Doppler dispersion for transmitter frequency of 2380 MHz (λ13 cm). The product of the dispersions in delay and Doppler is the overspread factor at 2380 MHz.

[d] IAA denotes comet IRAS-Araki-Alcock.

[e] HYA denotes comet Hyakutake (C/1996 B2).

section, σ, and its normalized radar cross section or "radar albedo," $\hat{\sigma} = \sigma/A_p$, where A_p is the target's geometric projected area. Since the radar astronomer selects the transmitted and received polarizations, any estimate of σ or $\hat{\sigma}$ must be identified accordingly. The most common approach is to transmit a circularly polarized wave and to use separate receiving systems for simultaneous reception of the same sense of circular polarization as transmitted (i.e., the SC sense) and the opposite (OC) sense. The handedness of a circularly polarized wave is reversed on normal reflection from a smooth dielectric interface, so the OC sense dominates echoes from targets that look smooth at the radar wavelength. In this context, a surface with minimum radius of curvature very much larger than λ would "look smooth." SC echo power can arise from single scattering from rough surfaces, multiple scattering from smooth surfaces or subsurface heterogeneities (e.g., particles or voids), or certain subsurface refraction effects. The circular polarization ratio, $\mu_C = \sigma_{SC}/\sigma_{OC}$, is thus a useful measure of near-surface structural complexity or "roughness." When linear polarizations are used, it is convenient to define the ratio $\mu_L = \sigma_{OL}/\sigma_{SL}$, which would be close to zero for normal reflection from a smooth dielectric interface. For all radar-detected planetary targets, $\mu_L < 1$ and $\mu_L < \mu_C$. Although the OC radar albedo, $\hat{\sigma}_{OC}$, is the most widely used gauge of radar reflectivity, some radar measurements are reported in terms of the total power (OC + SC = OL + SL) radar albedo $\hat{\sigma}_T$, which is four times the geometric albedo used in optical planetary astronomy. A smooth metallic sphere would have $\hat{\sigma}_{OC} = \hat{\sigma}_{SL} = 1$, a geometric albedo of 0.25, and $\mu_C = \mu_L = 0$.

If μ_C is close to zero (see Table II), its physical interpretation is unique, as the surface must be smooth at all scales within about an order of magnitude of λ and there can be no subsurface structure at those scales within several $1/e$ power absorption lengths, L, of the surface proper. In this special situation, we may interpret the radar albedo as the product $g\rho$, where ρ is the Fresnel power-reflection coefficient at normal incidence and the backscatter gain g depends on target shape, the distribution of surface slopes with respect to that shape, and target orientation. For most applications to date, g is <10% larger than unity, so the radar albedo provides a reasonable first approximation to ρ. Both ρ and L depend on very interesting characteristics of the surface material, including bulk density, porosity, particle size distribution, and metal abundance.

If μ_C is ~0.3 (e.g., Mars and some near-Earth asteroids), then much of the echo arises from some backscattering mechanism other than single, coherent reflections from large, smooth surface elements. Possibilities include multiple scattering from buried rocks or from the interiors of concave surface features such as craters, or reflections from very jagged surfaces with radii of curvature much less than a wavelength. Most planetary targets have values of μ_C <0.3 at decimeter wavelengths, so their surfaces are dominated by a component that is smooth at centimeter to meter scales.

The observables $\hat{\sigma}_{OC}$ and μ_C are disk-integrated quantities, derived from integrals of $\sigma(\nu)$ or $\sigma(\tau)$ in specific polarizations. Later, we will see how their physical interpretation profits from knowledge of the functional forms of $\sigma(\nu)$ and $\sigma(\tau)$.

B. DYNAMICAL PROPERTIES FROM DELAY/DOPPLER MEASUREMENTS

Consider radar observation of a point target a distance R from the radar. As noted earlier, the "roundtrip time delay" between transmission of a pulse toward the target and reception of the echo would be $\tau = 2R/c$. It is possible to measure time delays to within 10^{-7} sec. Actual delays encountered range from $2\frac{1}{2}$ s for the Moon to $2\frac{1}{2}$ hr for Saturn's rings. For a typical target distance ~1 astronomical unit (AU), the time delay is ~1000 sec and can be measured with a fractional timing uncertainty of 10^{-9}, that is, with the same fractional precision as the definition of the speed of light.

If the target is in motion and has a line-of-sight component of velocity toward the radar of v_{LOS}, the target will "see" a frequency that, to first order in v_{LOS}/c, equals $f_{TX} + (v_{LOS}/c)f_{TX}$, where f_{TX} is the transmitter frequency. The target reradiates the Doppler-shifted signal, and the radar receives echo whose frequency is, again to first order, given by

$$f_{TX} + 2(v_{LOS}/c)f_{TX}$$

That is, the total Doppler shift in the received echo is

$$2v_{LOS}f_{TX}/c = v_{LOS}/(\lambda/2)$$

so a 1-Hz Doppler shift corresponds to a velocity of half a wavelength per second (e.g., 6.3 cm sec^{-1} for λ12.6 cm). It is not difficult to measure echo frequencies to within 0.01 Hz, so v_{LOS} can be estimated with a precision finer than 1 mm sec^{-1}. Actual values of v_{LOS} for planetary radar targets can be as large as several tens of kilometers per second, so radar velocity measurements have fractional errors as low as 10^{-8}. At this level, the second-order (special relativistic) contribution to the Doppler shift becomes measurable; in fact, planetary radar observations have provided the initial experimental verification of the second-order term.

By virtue of their high precision, radar measurements of time delay and Doppler frequency are very useful in refining our knowledge of various dynamical quantities. The first delay-resolved radar observations of Venus, during 1961–1962, yielded an estimate of the light-second equivalent of the astronomical unit that was accurate to one part in 10^6, constituting a thousandfold improvement in the best results achieved with optical observations alone. Subsequent radar observations provided additional refinements of nearly two more orders of magnitude. In addition to determining the scale of the solar system precisely, these observations greatly improved our knowledge of the orbits of Earth, Venus, Mercury, and Mars, and were essential for the success of the first interplanetary missions. Radar observations still contribute to maintaining the accuracy of planetary ephemerides for objects in the inner solar system, and have played an important role in dynamical studies of Jupiter's Galilean satellites. For newly discovered near-Earth asteroids, whose orbits must be estimated from optical astrometry that spans short arcs, a few radar observations can mean the difference between successfully recovering the object during its next close approach and losing it entirely. Even for near-Earth asteroids with secure orbits, delay-Doppler measurements can shrink the positional error ellipsoid significantly for decades or even centuries.

Precise interplanetary time-delay measurements have allowed increasingly decisive tests of physical theories for light, gravitational fields, and their interactions with matter and each other. For example, radar observations verify general relativity theory's prediction that for radar waves passing nearby the Sun, echo time delays are increased because of the distortion of space by the Sun's gravity. The extra delay would be ~100 μsec if the angular separation of the target from the Sun were several degrees. (The Sun's angular diameter is about half a degree.) Since planets are not point targets, their echoes are dispersed in delay and Doppler, and the refinement of dynamical quantities and the testing of physical theories are tightly coupled to estimation of the mean radii, the topographic relief, and the radar scattering behavior of the targets. The key to this entire process is resolution of the distributions of echo power in delay and Doppler. In the next section, we will consider inferences about a target's dimensions and spin vector from measurements of the dispersions ($\Delta\tau_{TARGET}$, $\Delta\nu_{TARGET}$) of the echo in delay and Doppler. Then we will examine the physical information contained in the functional forms of the distributions $\sigma(\tau)$, $\sigma(\nu)$, and $\sigma(\tau, \nu)$.

C. DISPERSION OF ECHO POWER IN DELAY AND DOPPLER

Each backscattering element on a target's surface returns echo with a certain time delay and Doppler frequency (see Fig. 5). Since parallax effects and the curvature of the incident wave front are negligible for most ground-based observations (but not necessarily for observations with spacecraft), contours of constant delay are intersections of the surface with planes perpendicular to the line of sight. The point on the surface with the shortest echo time delay is called the subradar

point; the longest delays generally correspond to echoes from the planetary limbs. As noted already, the difference between these extreme delays is called the dispersion, $\Delta\tau_{TARGET}$, in $\sigma(\tau)$ or simply the "delay depth" of the target.

If the target appears to be rotating, the echo will be dispersed in Doppler frequency. For example, if the radar has an equatorial view of a spherical target with diameter D and rotation period P, then the difference between the line-of-sight velocities of points on the equator at the approaching and receding limbs would be $2\pi D/P$. Thus the dispersion of $\sigma(\nu)$ would be $\Delta\nu_{TARGET} = 4\pi D/\lambda P$. This quantity is called the bandwidth, B, of the echo power spectrum. If the view is not equatorial, the bandwidth is simply $(4\pi D \sin \alpha)/\lambda P$, where the "aspect angle" α is the acute angle between the instantaneous spin vector and the line of sight. Thus, a radar bandwidth measurement furnishes a joint constraint on the target's size, rotation period, and pole direction.

In principle, echo bandwidth measurements obtained for a sufficiently wide variety of directions can yield all three scalar coordinates of the target's intrinsic (i.e., sidereal) spin vector $\mathbf{W}$. This capability follows from the fact that the apparent spin vector $\mathbf{W}_{app}$ is the vector sum of $\mathbf{W}$ and the contribution ($\mathbf{W}_{sky} = \dot{\mathbf{e}} \times \mathbf{e}$, where the unit vector $\mathbf{e}$ points from the target to the radar) from the target's plane-of-sky motion. Variations in $\mathbf{e}$, $\dot{\mathbf{e}}$, and hence $\mathbf{W}_{sky}$, all of which are known, lead to measurement of different values of $\mathbf{W}_{app} = \mathbf{W} + \mathbf{W}_{sky}$, permitting unique determination of all three scalar components of $\mathbf{W}$.

These principles were applied in the early 1960s to yield the first accurate determination of the rotations of Venus and Mercury (Fig. 6). Venus's rotation is retrograde with a 243-day sidereal period that is close to the value (243.16 days) characterizing a resonance with the relative orbits of Earth and Venus, wherein Venus would appear from Earth to rotate exactly four times between successive inferior conjunctions with the Sun. However, two decades of ground-based observations and ultimately images obtained by the *Magellan* spacecraft while in orbit around Venus have conclusively demonstrated nonresonance rotation: the period is 243.0185 ± 0.0001 days. To date, a satisfactory explanation for Venus's curious spin state is lacking.

For Mercury, long imagined on the basis of optical observations to rotate once per 88-day revolution around the Sun, radar bandwidth measurements (see Fig. 6) demonstrated direct rotation with a period (59 days) equal to two-thirds of the orbital period. This spin-orbit coupling is such that during two Mercury years, the planet rotates three times with respect to

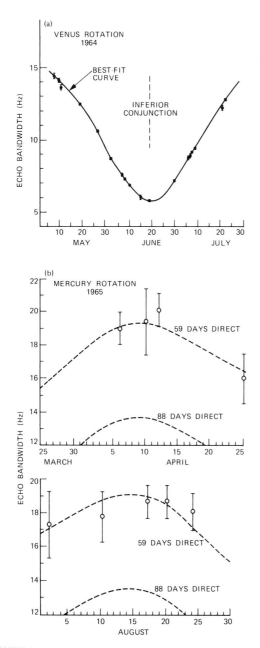

FIGURE 6 Measurements of echo bandwidth (i.e., the dispersion of echo power in Doppler frequency) used to determine the rotations of (a) Venus and (b) Mercury. [From R. B. Dyce, G. H. Pettengill, and I. I. Shapiro (1967). *Astron. J.* **72**, 351–359.]

the stars but only once with respect to the Sun, so a Mercury-bound observer would experience alternating years of daylight and darkness.

What if the target is not a sphere but instead is irregular and nonconvex? In this situation, which is most applicable to small asteroids and cometary nuclei, the relationship between the echo power spectrum and

the target's shape is shown in Fig. 7. We must interpret D as the sum of the distances r_+ and r_- from the plane ψ_o containing the line of sight and the spin vector to the surface elements with the greatest positive (approaching) and negative (receding) line-of-sight velocities. In different words, if the planes ψ_+ and ψ_- are defined as being parallel to ψ_o and tangent to the target's approaching and receding limbs, then ψ_+ and ψ_- are at distances r_+ and r_- from ψ_o. Letting f_o, f_+, and f_- be the frequencies of echoes from portions of the target intersecting ψ_o, ψ_+, and ψ_-, we have $B = f_+ - f_-$. Note that f_o is the Doppler frequency of hypothetical echoes from the target's center of mass and that any constant-Doppler contour lies in a plane parallel to ψ_o.

It is useful to imagine looking along the target's pole at the target's projected shape, that is, its pole-on silhouette S. D is simply the width, or "breadth," of this silhouette (or, equivalently, of the silhouette's convex envelope or "hull," H) measured normal to the line of sight (see Fig. 7). In general, r_+ and r_- are

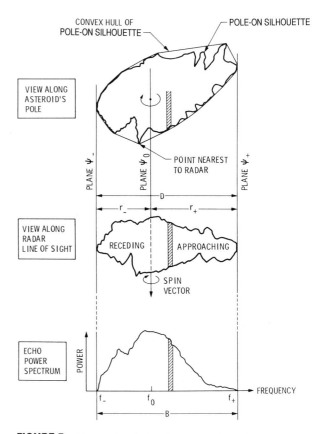

FIGURE 7 Geometric relations between an irregular, nonconvex rotating asteroid and its echo power spectrum. The plane ψ_o contains the asteroid's spin vector and the asteroid–radar line. The cross-hatched strip of power in the spectrum corresponds to echoes from the cross-hatched strip on the asteroid.

periodic functions of rotation phase ϕ, and depend on the shape of H as well as on the projected location of the target's center of mass, about which H rotates. If the radar data thoroughly sample at least 180° of rotational phase, then in principle one can determine $f_+(\phi)$ and $f_-(\phi)$ completely, and can recover H as well as the astrometrically useful quantity f_o. For many small, near-Earth asteroids, pronounced variations in $B(\phi)$ reveal highly noncircular pole-on silhouettes (see Fig. 8 and Section III, J).

D. TOPOGRAPHY ON THE MOON AND INNER PLANETS

For the Moon, Mercury, Mars, and Venus, topography along the subradar track superimposes a modulation on the echo delay above or below that predicted by ephemerides, which generally are calculated for a sphere with the object's a priori mean radius. Prior to spacecraft exploration of these objects, there were radar-detectable errors in the radii estimates as well as in the target's predicted orbit. These circumstances required that an extended series of measurements of the time delay of the echo's leading edge be folded into a computer program designed to estimate simultaneously parameters describing the target's orbit, mean radius, and topography. These programs also contain parameters from models of wave propagation through the interplanetary medium or the solar corona, as well as parameters used to test general relativity, as noted earlier.

Radar has been used to measure topography on the Moon and on the inner planets. For example, Fig. 9 shows a three-dimensional reconstruction of topography derived from altimetric profiles obtained for Mars in the vicinity of the giant shield volcano Arsia Mons. The altimetric resolution of the profiles is about 150 m (1 μsec in delay), but the surface resolution, or footprint, is very coarse ($\sim$75 km). Figure 10 shows altitude profiles across impact basins on Mercury. The *Magellan* radar altimeter, with a footprint typically 20 km across and vertical resolution on the order of tens of meters, has produced detailed topographic maps of most of Venus.

E. ANGULAR SCATTERING LAW

The functional forms of the distributions $\sigma(\tau)$ and $\sigma(\nu)$ contain information about the radar scattering process and about the structural characteristics of the target's surface. Suppose the target is a large, smooth, spherical

planet. Then echoes from the subradar region (near the center of the visible disk; see Fig. 5), where the surface elements are nearly perpendicular to the line of sight, would be much stronger than those from the limb regions (near the disk's periphery). This effect is seen visually when one shines a flashlight on a smooth, shiny ball—a bright glint appears where the geometry is right for backscattering. If the ball is roughened, the glint is spread out over a wider area and, in the case of extreme roughness, the scattering would be described as "diffuse" rather than "specular."

For a specular target, $\sigma(\tau)$ would have a steep leading edge followed by a rapid drop. The power spectrum $\sigma(\nu)$ would be sharply peaked at central frequencies, falling off rapidly toward the spectral edges. If, instead, the spectrum were very broad, severe roughness at some scale(s) comparable to or larger than λ would be indicated. In this case, knowledge of the echo's polarization properties would help to ascertain the particular roughness scale(s) responsible for the absence of the sharply peaked spectral signature of specular scattering.

By inverting the delay or Doppler distribution of echo power, one can estimate the target's average angular scattering law, $\sigma_o(\theta) = d\sigma/dA$, where dA is an element of surface area and θ is the "incidence angle" between the line of sight and the normal to dA. For the portion of the echo's "polarized" (i.e., OC or SL) component that is specularly scattered, $\sigma_o(\theta)$ can be related to statistics describing the probability distribution for the slopes of surface elements. Examples of scattering laws applied in planetary radar astronomy are the Hagfors law,

$$\sigma_o(\theta) \sim C(\cos^4\theta + C\sin^2\theta)^{-3/2} \tag{5}$$

the Gaussian law,

$$\sigma_o(\theta) \sim [C\exp(-C\tan^2\theta)]/\cos^4\theta \tag{6}$$

and the Cosine law,

$$\sigma_o(\theta) \sim (C + 1)\cos^{2C}\theta \tag{7}$$

where $c^{-1/2} = S_o = \langle\tan^2\theta\rangle^{1/2}$ is the adirectional rms slope.

Echoes from the Moon, Mercury, Venus, and Mars are characterized by sharply peaked OC echo spectra (Fig. 11). Although these objects are collectively referred to as "quasi-specular" radar targets, their echoes also contain a diffusely scattered component and have full-disk circular polarization ratios averaging about

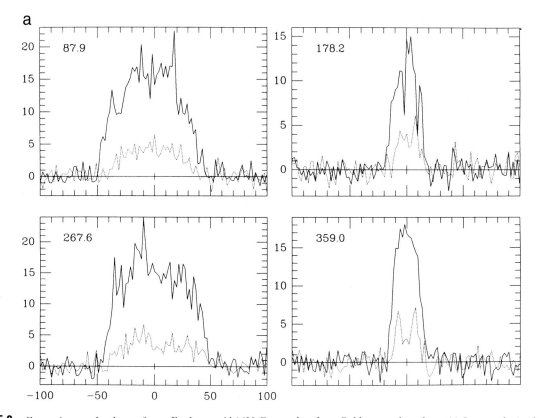

FIGURE 8   Constraints on the shape of near-Earth asteroid 1620 Geographos from Goldstone radar echoes. (a) Spectra obtained at phases of bandwidth extrema. OC (solid curve) and SC (dotted curve) echo power is plotted versus Doppler frequency. (b) Comparison of an estimate of the hull (*H*) on the asteroid's pole-on silhouette (*S*) with an estimate of *S* itself. The white curve is the cw estimate of *H* and the X marks the projected position of the asteroid's center of mass (COM) with respect to *H*. That curve and the X are superposed on an estimate of *S* from delay-Doppler images; the bright pixel is the projection of the COM determined from analysis of those images. The absolute scales and relative rotational orientations of the two figures are known: border ticks are 1 km apart. The offset between the X and the bright pixel is a measure of the uncertainty in our knowledge of the COM's delay-Doppler trajectory during the experiment. In the diagram at right, the arrows point to the observer at phases of lightcurve maxima (*M1*, *M2*) and minima (*m1*, *m2*). [From S. J. Ostro *et al.* (1996). *Icarus* **121**, 46–66.]

0.07 for the Moon, Mercury, and Venus, but ranging from 0.1 to 0.4 for Mars, as discussed next.

Typical rms slopes obtained at decimeter wavelengths for these four quasi-specular targets are around 7° and consequently these objects' surfaces have been described as "gently undulating." As might be expected, values estimated for S_o increase as the observing wavelength decreases. For instance, for the Moon, S_o increases from ~4° at 20 m to ~8° at 10 cm, to ~33° at 1 cm. At optical wavelengths, the Moon shows no trace of a central glint, that is, the scattering is entirely diffuse. This phenomenon arises because the lunar surface (Fig. 12) consists of a regolith (an unconsolidated layer of fine-grained particles) with much intricate structure at the scale of visible wavelengths. At decimeter wavelengths, the ratio of diffusely scattered power to quasi-specularly scattered power is about one-third for the Moon, Mercury, and Venus, but two to three times higher for Mars. This ratio can be determined

by assuming that all the SC echo is diffuse and then calculating the diffusely scattered fraction (*x*) of OC echo by fitting to the OC spectrum a model based on a "composite" scattering law, for example, $S_o(\theta) = x \, \sigma_{DIF}(\theta) + (1 - x)\sigma_{QS}(\theta)$. Here $\sigma_{QS}(\theta)$ might be the Hagfors law and usually $\sigma_{DIF}(\theta) \sim \cos^m\theta$; when this is done, estimated values of *m* usually fall between unity (geometric scattering, which describes the optical appearance of the full Moon) and two (Lambert scattering).

For the large, nearly spherical asteroids 1 Ceres and 2 Pallas (see Section III, J), the closeness of μ_C to zero indicates quasi-specular scattering, but the OC spectra, rather than being sharply peaked, are fit quite well using a Cosine law with *C* between 2 and 3, or a Gaussian law with *C* between 3 and 5. Here we can safely interpret the diffuse echo as due to the distribution of surface slopes, with S_o between 20° and 50°. OC echo spectra obtained from asteroid 4 Vesta and

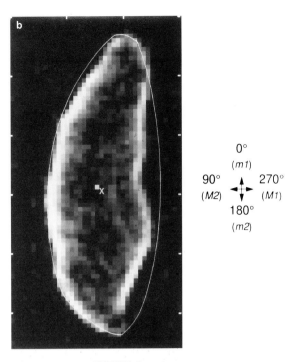

0°
(m1)

90°   270°
(M2)   (M1)

180°
(m2)

FIGURE 8   (continued)

Jupiter's satellite Io have similar shapes, but these objects' substantial polarization ratios (μ_C ~0.3 and ~0.5, respectively) suggest that small-scale roughness is at least partially responsible for the diffuse echoes. Circular polarization ratios between 0.5 and 1.0 have been measured for several asteroids (see Table II) and parts of Mars and Venus, implying extreme decimeter-scale roughness, perhaps analogous to terrestrial lava flows (Fig. 13). Physical interpretations of the diffusely scattered echo employ information about albedo, scattering law, and polarization to constrain the size distributions, spatial densities, and electrical properties of wavelength-scale rocks near the surface, occasionally using the same theory of multiple light scattering applied to radiative transfer problems in other astrophysical contexts.

F. RADAR MAPPING OF SPHERICAL TARGETS

The term "radar image" usually refers to a measured distribution of echo power in delay, Doppler, and/or up to two angular coordinates. The term "radar map" usually refers to a display in suitable target-centered coordinates of the residuals with respect to a model that parameterizes the target's size, shape, rotation, average scattering properties, and possibly its motion

with respect to the delay-Doppler ephemerides. Knowledge of the dimensions of the Moon and inner planets has long permitted conversion of radar images to maps of these targets. For small asteroids, the primary use of images is to constrain the target's shape (see Section III, J).

As illustrated in Fig. 5, intersections between constant-delay contours and constant-Doppler contours on a sphere constitute a "two-to-one" transformation from the target's surface to delay-Doppler space. For any point in the northern hemisphere, there is a conjugate point in the southern hemisphere at the same delay and Doppler. Therefore, the source of echo in any delay-Doppler resolution cell can be located only to within a twofold ambiguity. This north–south ambiguity can be avoided completely if the radar beamwidth (~2 arcmin for Arecibo at 13 cm or Goldstone at 3.5 cm) is comparable to or smaller than the target's apparent angular radius, as in the case of observations of the Moon (angular radius ~ 15 arcmin). Similarly, no such ambiguity arises in the case of side-looking radar observations from spacecraft (e.g., the *Magellan* radar) for which the geometry of delay-Doppler surface contours differs somewhat from that in Fig. 5. For ground-based observations of Venus and Mercury, whose angular radii never exceed a few tens of arcseconds, the separation of conjugate points is achievable by either (1) offsetting the pointing to place a null of the illumination pattern on the undesired hemisphere or (2) interferometrically, using two receiving antennas, as follows.

The echo waveform received at either antenna from one conjugate point will be highly correlated with the echo waveform received at the other antenna from the same conjugate point. However, echo waveforms from the two conjugate points will be largely uncorrelated with each other, no matter where they are received. Thus, echoes from two conjugate points can, in principle, be distinguished by cross-correlating echoes received at the two antennas with themselves and with each other, and performing algebraic manipulations on long time averages of the cross product and the two self products.

The echo waveform from a single conjugate point will experience slightly different delays in reaching the two antennas, so there will be a phase difference between the two received signals, and this phase difference will depend only on the geometrical positions of the antennas and the target. This geometry will change as the Earth rotates, but very slowly and in a predictable manner. The antennas are best positioned so contours of constant phase difference on the target disk are as orthogonal as possible to the constant-Doppler con-

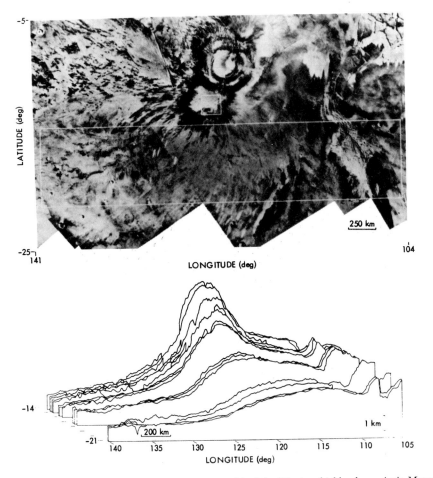

FIGURE 9 Topographic contours for the southern flank (large rectangle) of the Martian shield volcano Arsia Mons, obtained from radar altimetry. [From L. Roth, G. S. Downs, R. S. Saunders, and G. Schubert (1980). *Icarus* **42**, 287–316.]

tours, which connect conjugate points. Phase difference hence becomes a measure of north–south position, and echoes from conjugate points can be distinguished on the basis of their phase relation.

The total number of "fringes," or cycles of phase shift, spanned by the disk of a planet with diameter D and a distance R from the radar is approximately $(D/R)(b_{PROJ}/\lambda)$, where b_{PROJ} is the projection of the interferometer baseline normal to the mean line of sight. For example, Arecibo interferometry linked the main antenna to a 30.5-m antenna about 11 km farther north. It placed about seven fringes on Venus, quite adequate for separation of the north–south ambiguity. The Goldstone main antenna (see Fig. 3) has been linked to smaller antennas to perform three-element as well as two-element interferometry. Tristatic observations permit one to solve so precisely for the north–south location of a given conjugate region that one can obtain the region's elevation relative to the mean

planetary radius. Altimetric information can be extracted also from bistatic observations using the time history of the phase information, but only if the variations in the projected baseline vector are large enough.

In constructing a radar map, the unambiguous delay-Doppler distribution of echo power is transformed to planetocentric coordinates, and a model is fit to the data, using a maximum-likelihood or weighted-least-squares estimator. The model contains parameters for quasi-specular and diffuse scattering as well as prior information about the target's dimensions and spin vector. For Venus, effects of the dense atmosphere on radar wave propagation must also be modeled. Residuals between the data and the best-fit model constitute a radar reflectivity map of the planet. Variations in radar reflectivities evident in radar maps can be caused by many different physical phenomena, and their proper interpretation demands due attention to the radar wavelength, echo polarization, viewing geome-

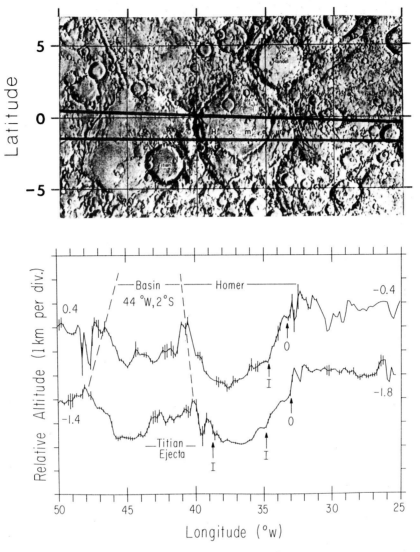

FIGURE 10 Mercury altitude profiles (bottom) showing topography across Homer Basin and a large, unnamed basin to the west, estimated from observations whose subradar tracks are shown on the USGS shaded-relief map (top). Broken lines indicate approximate locations of the basin rims as seen in *Mariner 10* images. Arrows locate Homer's inner/outer (I/O) basin rings. [From J. K. Harmon, D. B. Campbell, D. L. Bindschadler, J. W. Head, and I. I. Shaprio (1986). *J. Geophys. Res.* **91**, 385–401.]

try, prior knowledge about surface properties, and the nature of the target's mean scattering behavior. Similar considerations apply to inferences based on disk-integrated radar albedos.

Delay-Doppler interferometry is not currently feasible for targets like the Galilean satellites and the largest asteroids, which are low-SNR and overspread (see Section II, D). A different, "Doppler mapping" technique, developed for these spherical targets, reconstructs the global albedo distribution from cw echo spectra acquired as a function of rotation phase and at an arbitrary number of subradar latitudes. To visualize

how Doppler mapping works, note that a target's reflectivity distribution can be expanded as a truncated spherical harmonic series, and that the distribution of echo power in rotation phase and Doppler frequency can be obtained as a linear, analytic function of the series coefficients. Estimation of those coefficients from an observed phase-Doppler distribution can be cast as a least-squares problem to form a linear imaging system. Doppler mapping works best when the limb-darkening is minimal and is ideally suited to overspread targets whose echoes are too weak for the random-code method. Removal of the north–south ambiguity

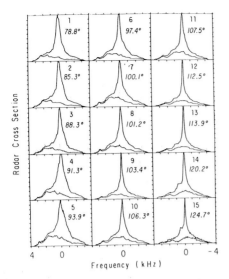

FIGURE 11 Mars 13-cm radar echo spectra for subradar points along 22° north latitude at the indicated west longitudes, obtained in the OC (upper curves) and SC (lower curves) polarizations. In each box, spectra are normalized to the peak OC cross section. The echo bandwidth is 7.1 kHz. Very rough regions on the planet are revealed as bumps in the SC spectra, which move from positive to negative Doppler frequencies as Mars rotates. [From J. K. Harmon, D. B. Campbell, and S. J. Ostro (1982). *Icarus* **52**, 171–187.]

in Doppler images (or in single-antenna delay-Doppler images) is possible if the data sample nonequatorial subradar latitudes, because then the Doppler (or delay-Doppler) versus rotation-phase trajectories of conjugate points are different.

Recently, a new kind of radar observation using Goldstone as a 3.5-cm cw transmitter and the Very Large Array in New Mexico as a synthetic aperture receiver has been used for direct measurement of the angular distribution of echo power. The 27-antenna VLA constitutes a 351-baseline interferometer whose synthesized beamwidth is as narrow as 0.24 arcsec at the Goldstone transmitter frequency, providing resolutions of roughly 80, 40, and 70 km for Mercury, Venus, and Mars at closest approach. (The finest resolutions in published ground-based delay-Doppler maps of those planets are approximately 15, 1, and 40 km, but the maps cover small fractions of the surface at such fine resolutions.) Radar aperture synthesis has produced full-disk images of those planets and Saturn's rings that are free from north–south ambiguities, avoid problems related to overspreading, and permit direct measurement of local albedo, polarization ratio, and scattering law. The Goldstone–VLA system has achieved marginal angular resolution of main belt asteroid echoes and could readily image any large-particle comet clouds as radar detectable as the one around

comet IRAS-Araki-Alcock. Goldstone–VLA resolution of asteroids and comet nuclei is impaired by the VLA's coarse spectral resolution (>380 Hz) and inability to accommodate time-modulated waveforms. Figures 14–18 show examples of radar maps constructed using various techniques.

G. RADAR EVIDENCE FOR ICE DEPOSITS AT MERCURY'S POLES

The first full-disk (Goldstone–VLA) radar portraits of Mercury surprisingly revealed anomalously bright polar features with $\mu_C > 1$, and subsequent delay-Doppler imagery from Arecibo established that the anomalous radar echoes originate from interiors of craters that are perpetually shaded from sunlight because of Mercury's near-zero obliquity (see Fig. 15). The angle between the orbital planes of Mercury and Earth is 7°, so portions of the permanently shadowed regions are visible to Earth-based radars. Most of the south pole anomalies are confined to the floor of the 155-km crater Chao Meng-Fu. At each pole, bright radar features in regions imaged by *Mariner 10* correlate exactly with craters; numerous radar features lie in the hemisphere not imaged by that spacecraft.

Similarities between the radar scattering properties of the Mars and Mercury polar anomalies and those of the icy Galilean satellites (see Section III, K) support the inference that the radar anomalies are deposits of water ice. Temperatures below 120 K in the permanent shadows are expected and are low enough for ice to be stable against sublimation for billions of years. Temperatures several tens of kelvins lower may exist inside high-latitude craters and perhaps also beneath at least 10 cm of optically bright regolith. Plausible sources of water on Mercury include comet impacts and outgassing from the interior. It has been noted that most water vapor near the surface is photodissociated, but that some molecules will random-walk to polar cold traps. Ices of other volatiles, including CO_2, NH_3, HCN, and SO_2, might also be present.

H. VENUS REVEALED BY *MAGELLAN*

The *Magellan* spacecraft entered Venus orbit in August 1990 and during the next two years explored the planet with a single scientific instrument operating as a radar imager, an altimeter, and a thermal radiometer. *Magellan*'s imaging resolution (~100 m) and altimetric resolution (5 to 100 m) improves upon the best previous spacecraft and ground-based measurements by an

FIGURE 12 Structure on the lunar surface near the *Apollo 17* landing site. Most of the surface is smooth and gently undulating at scales much larger than a centimeter. This smooth component of the surface is responsible for the predominantly quasi-specular character of the Moon's radar echo at $\lambda \geqslant 1$ cm. Wavelength-scale structure produces a diffuse contribution to the echo. Wavelength-sized rocks are much more abundant at $\lambda \sim 4$ cm than at $\lambda \sim 10$ m (the scale of the boulder being inspected by astronaut H. Schmitt), and hence diffuse echo is more substantial at shorter wavelengths.

FIGURE 13 This lava flow near Sunset Crater in Arizona is an example of an extremely rough surface at decimeter scales and is similar to terrestrial flows yielding large circular polarization ratios at decimeter wavelengths.

order of magnitude, and does so with nearly global coverage. Analysis of *Magellan*'s detailed, comprehensive radar reconnaissance of Venus's surface topography, morphology, and electrical properties ultimately will revolutionize our understanding not only of that planet, but of planetary geology itself.

Venus's surface contains a plethora of diverse tectonic and impact features, but its formation and evolution have clearly been dominated by widespread volcanism, whose legacy includes pervasive volcanic planes, thousands of tiny shield volcanoes, monoumental edifices, sinuous lava flow channels, pyroclastic deposits, and pancakelike domes. The superposition of volcanic signatures and elaborate, complex tectonic forms records a history of episodic crustal deformation. The paucity of impact craters smaller than 25 km and the lack of any as small as a few kilometers attests to the protective effect of the dense atmosphere. The multilobed, asymmetrical appearance of many large craters presumably results from atmospheric breakup of projectiles before impact. Atmospheric entrainment and transport of ejecta are evident in very elongated

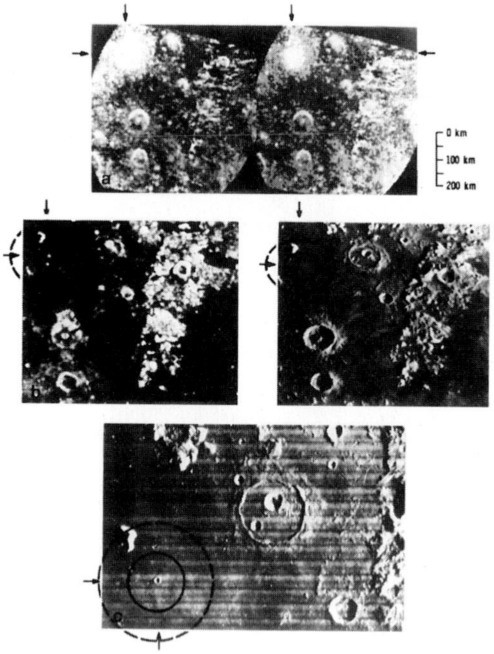

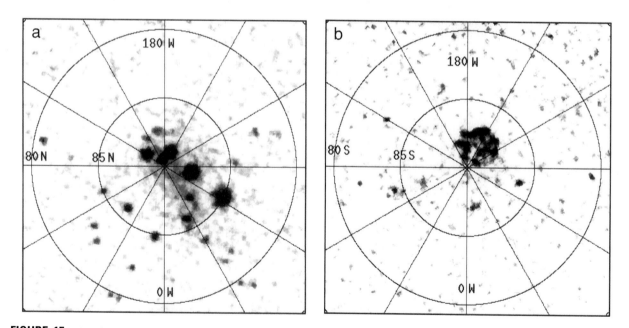

FIGURE 14 The lunar crater Piton B (located by arrows) is surrounded by an ejecta blanket that is conspicuous in the 3.8-cm-radar image (displayed in both continuous- and incremental-tone maps) but invisible in Earth-based and Lunar Orbiter photographs. The sketched 50- and 100-km-diameter circles are concentric to the crater and coplanar with the local mean surface. (Courtesy of T. W. Thompson.)

ejecta blankets. Numerous craters are surrounded by radar-dark zones, perhaps the outcome of atmospheric pressure-wave pulverization and elevation of surface material that upon resettling deposited a tenuous and hence unreflective "impact regolith." Figure 17 shows examples of *Magellan* radar images.

I. THE RADAR HETEROGENEITY OF MARS

Ground-based investigations of Mars have achieved more global coverage than those of the other terrestrial targets, because the motion in longitude of the subradar point on Mars (whose rotation period is only 24.6 hr) is rapid compared to that on the Moon, Venus, or Mercury, and because the geometry of Mars's orbit and spin vector permits subradar tracks throughout the Martian tropics. The existing body of Mars radar data reveals extraordinary diversity in the degree of small-scale roughness as well as in the rms slope of smooth surface elements. For example, Fig. 19 shows the variation in OC echo spectral shape as a function of longitude for a subradar track along ~16°S latitude. Slopes on Mars have rms values from less than 0.5° to more than 10°. Chryse Planitia, site of the first Viking Lander, has fairly shallow slopes (4°–5°) and, in fact, radar rms slope estimates were utilized in selection of the Viking Lander (and Mars Pathfinder) sites.

Diffuse scattering from Mars is much more substan-

FIGURE 15 Arecibo 13-cm delay-Doppler images of the (a) north and (b) south poles of Mercury, taken in the SC polarization. The resolution is 15 km. The radar-brightness regions are shown here as dark. [From J. K. Harmon, M. A. Slade, R. A. Velez, A. Crespo, M. J. Dryer, and J. M. Johnson (1994). *Nature* **369**, 213–215. Copyright 1994 Macmillan Magazines Limited.]

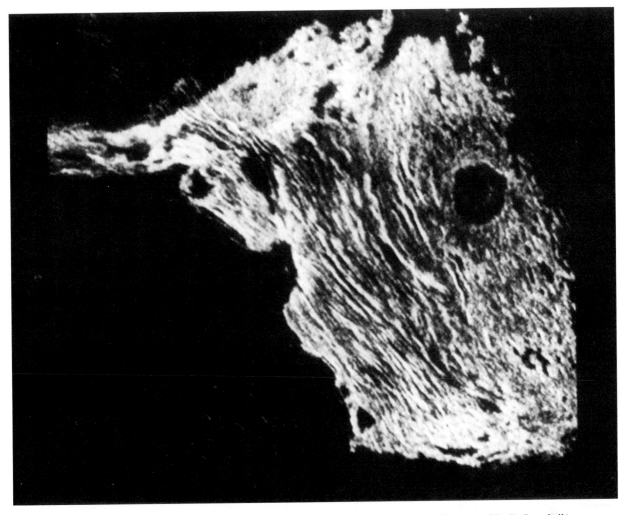

FIGURE 16   Arecibo delay-Doppler OC radar map of Maxwell Montes on Venus. (Courtesy of D. B. Campbell.)

tial than for the other quasi-specular targets, and often accounts for most of the echo power, so the average near-surface abundance of centimeter-to-meter-scale rocks presumably is much greater on Mars than on the Moon, Mercury, or Venus. Features in Mars SC spectra first revealed the existence of regions of extremely small-scale roughness (see Fig. 11), and the trajectory of these features' Doppler positions versus rotation phase suggested that their primary sources are the Tharsis and Elysium volcanic regions. The best terrestrial analog for this extremely rough terrain might be young lava flows (see Fig. 13). Goldstone–VLA images of Mars at longitudes that cover the Tharsis volcanic region (see Fig. 18) confirm that this

area is the predominant source of strong SC echoes and that localized features are associated with individual volcanoes. A 2000-km-long band with an extremely low albedo cuts across Tharsis; the radar darkness of this "Stealth" feature probably arises from an underdense, unconsolidated blanket of pyroclastic deposits ~1 m deep. The strongest SC feature in the Goldstone–VLA images is the residual south polar ice cap, whose scattering behavior is similar to that of the icy Galilean satellites (Section III, K). Arecibo observations of Mars, including Doppler-only and random-code delay-Doppler mapping, have charted the detailed locations and fine structure of Mars SC features (see Fig. 18b).

FIGURE 17 *Magellan* radar maps of Venus. (a) Northern-hemisphere projection of mosaics. The north pole is at the center of the image, with 0° and 90°E longitudes at the six and three o'clock positions. Gaps use Pioneer Venus data or interpolations. The bright, porkchop-shaped feature is Maxwell Montes, a tectonically produced mountain range first seen in ground-based images. (b) 120-m-resolution map of the crater Cleopatra on the eastern slopes of Maxwell Montes. Cleopatra is a double-ringed impact basin that resembles such features seen on the Moon, Mercury, and Mars. (Courtesy of JPL/NASA.)

J. ASTEROIDS

Echoes from 37 main belt asteroids (MBAs) and 47 near-Earth asteroids (NEAs) have provided a wealth of new information about these objects' sizes, shapes, spin vectors, and surface characteristics such as decimeter-scale morphology, topographic relief, regolith porosity, and metal concentration. During the past de-

cade, radar has been established as the most powerful Earth-based technique for determining the physical properties of asteroids that come close enough to yield strong echoes.

The polarization signatures of some of the largest MBAs (e.g., 1 Ceres and 2 Pallas) reveal surfaces that are smoother than that of the Moon at decimeter scales but much rougher at some much larger scale. For ex-

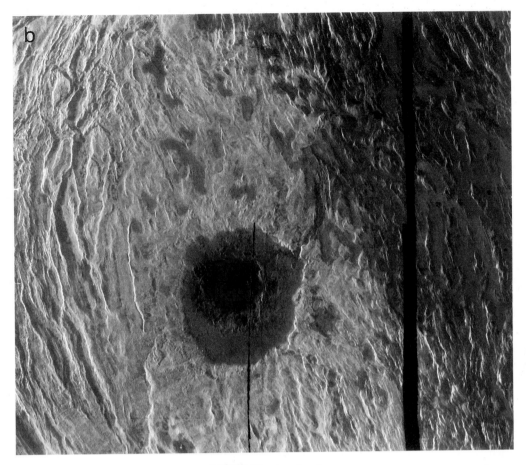

FIGURE 17 (*continued*)

ample, for Pallas, μ_c is only ~0.05 and, as noted earlier, surface slopes exceed 20°. For asteroids in the 200-km-diameter range, the echoes provide evidence for large-scale topographic irregularities. For example, brightness spikes within narrow ranges of rotation phase suggest large, flat regions on 7 Iris (Fig. 20), 9 Metis, and 654 Zelinda, whereas bimodal spectra imply nonconvex, possibly bifurcated shapes for 12 Victoria and 216 Kleopatra.

There is a 10-fold variation in the radar albedos of MBAs, implying substantial variations in these objects' surface porosities or metal concentrations, or both. The lowest MBA albedo estimate, 0.04 for Ceres, indicates a lower surface bulk density than that on the Moon. The highest MBA albedo estimates, 0.31 for 16 Psyche and 0.44 for Kleopatra, are consistent with metal concentrations near unity and lunar porosities. These objects might be the collisionally stripped cores of differentiated asteroids and by far the largest pieces of refined metal in the solar system.

The diversity of NEA radar signatures is extreme (see Table II). Some small NEAs are much rougher at decimeter scales than MBAs, comets, or the terrestrial planets. The radar albedo of the 2-km object 6178 (1986DA), 0.58, strongly suggests that this Earth-approacher is a regolith-free metallic fragment, presumably derived from the interior of a much larger object that melted, differentiated, cooled, and subsequently was disrupted in a catastrophic collision. This asteroid, which appears extremely irregular at 10- to 100-m scales and shows hints of being bifurcated, might be (or have been a part of) the parent body of some iron meteorites. At the other extreme, an interval estimate for 1986JK's radar albedo (0.005 to 0.07) suggests a surface bulk density within a factor of 2 of 0.9 g cm^{-3}. Similarly, the distribution of NEA circular polarization ratios runs from near zero to near unity. The highest values, for 2101 Adonis, 1992QN, 3103 Eger, and 3980 1980PA, indicate extreme near-surface structural complexity, but we cannot distinguish between multiple scattering from subsurface heterogeneities (see Section III, K) and single scattering from complex structure on the surface.

The MBAs 951 Gaspra and 243 Ida, imaged by the

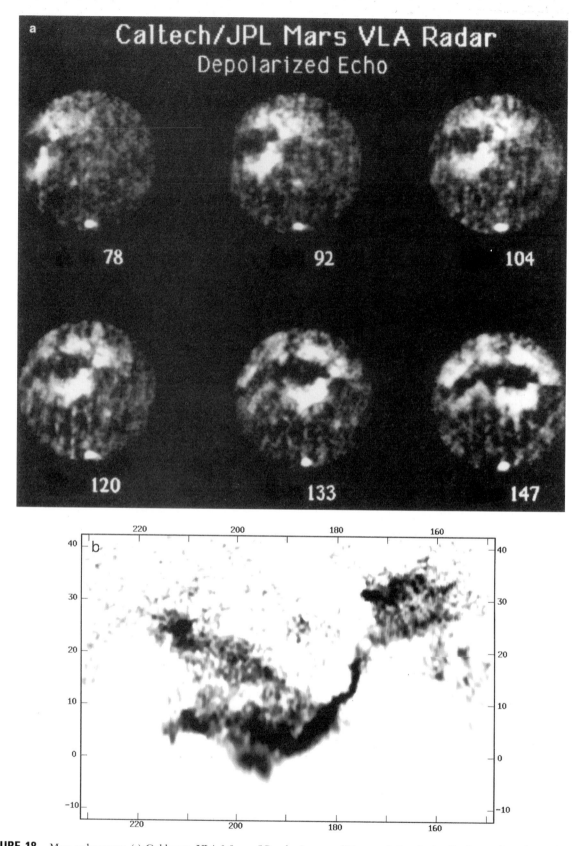

FIGURE 18 Mars radar maps. (a) Goldstone–VLA 3.5-cm, SC radar images of Mars at six longitudes. In the northern hemisphere, the brightest features are in Tharsis, which is traversed by the low-albedo Stealth region. Note the very bright residual south polar ice cap. [From D. O. Muhleman, B. J. Butler, A. W. Grossman, and M. A. Slade (1991). *Science* **253**, 1508–1513. Copyright 1991 American Association for the Advancement of Science.] (b) Arecibo 13-cm, SC reflectivity map of the Elysium region of Mars, obtained from random-code observations made with the subradar latitude 10°S. The map is north–south ambiguous, but more northerly observations confirm that all of the strong features come from the north. The radar-bright regions (shown here as dark) correspond to Elysium Mons (at ~214°W longitude, 25°N latitude) and the Elysium flood basin and outflow channel. The SC brightness of these regions is probably caused by extremely rough lava flows. [From J. K. Harmon, M. P. Sulzer, P. J. Perillat, and J. F. Chandler (1992). *Icarus* **95**, 153–156.]

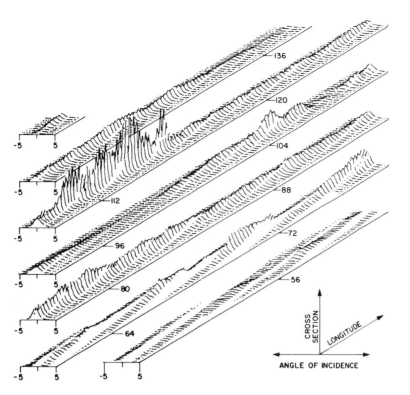

FIGURE 19 Mars echo power spectra as a function of longitude obtained along a subradar track at 16° south latitude. The most sharply peaked spectra correspond to the smoothest regions (i.e., the smallest rms slopes). (Courtesy of G. S. Downs, P. E. Reichley, and R. R. Green.)

Galileo spacecraft, probably are marginally detectable with the upgraded Arecibo. Both Goldstone and Arecibo have investigated the Gaspra-sized Martian moon Phobos, whose radar properties differ from those of most small, Earth-approaching objects but resemble those of large (~100-km), C-class, main belt asteroids. Phobos's surface characteristics may be more representative of Ceres and Pallas than most NEAs. The upper limit on the radar cross section of Deimos, which has defied radar detection, argues for a surface bulk density no greater than about 1 g cm^{-3}.

During the past decade, delay-Doppler imaging of asteroids has produced spatial resolutions as fine as a few decameters. The images generally can be "north–south" ambiguous, that is, they constitute a two-to-one (or even many-to-one) mapping from the surface to the image. However, if the radar is not in the target's equatorial plane, then the delay-Doppler trajectory of any surface point is unique. Hence images that provide adequate orientational coverage can be inverted, and in principle one can reconstruct the target's three-dimensional shape as well as its spin state, the radar-scattering properties of the surface, and the motion of the center of mass through the delay-Doppler ephemerides.

The first asteroid radar data set suitable for recon-struction of the target's shape was a 2.5-hr sequence of 64 delay-Doppler images of 4769 Castalia (1989PB) (Fig. 21a), obtained two weeks after its August 1989 discovery. The images, which were taken at a subradar latitude of about 35°, show a bimodal distribution of echo power over the full range of sampled rotation phases, and least-squares estimation of Castalia's three-dimensional shape (Fig. 21b) reveals it to consist of two kilometer-sized lobes in contact. Castalia apparently is a contact-binary asteroid formed from a gentle collision of the two lobes.

If the radar view is equatorial, unique reconstruction the asteroid's three-dimensional shape is ruled out, but a sequence of images that thoroughly samples rotation phase can allow unambiguous reconstruction of the asteroid's pole-on silhouette. For example, observations of 1620 Geographos yield ~400 images with ~100-m resolution. The pole-on silhouette's extreme dimensions are in a ratio, 2.76 ± 0.21, that establishes Geographos as the most elongated solar system object imaged so far (see Fig. 8). The images show craters as well as indications of other sorts of large-scale topographic relief, including a prominent central indentation. Protuberances at the asteroid's ends may be related to the pattern of ejecta removal and deposition caused by the asteroid's gravity field.

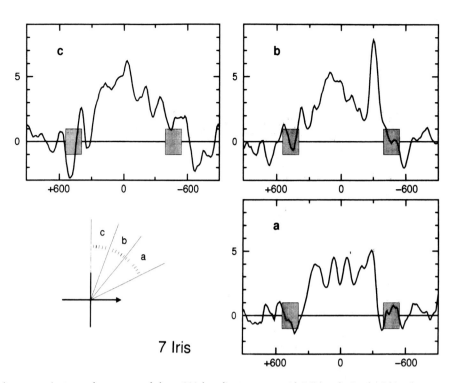

FIGURE 20   Thirteen-centimeter echo spectra of the ~200-km-diameter asteroid 7 Iris, obtained within three narrow rotation-phase intervals. OC echo power in standard deviations is plotted versus Doppler frequency. The shaded boxes show frequency intervals thought to contain the echo edges. A radar "spike" appears in (b) at −305 Hz, but not in spectra at adjacent phases, so it is probably not due to a reflectivity feature but rather to a temporary surge in radar-facing surface area, perhaps a flat facet ~20 km wide. [From D. L. Mitchell *et al.* (1995). *Icarus* **118**, 105–131.]

Delay-Doppler imaging of 4179 Toutatis in December 1992 achieved resolutions as fine as 125 nsec (19 m in range) and 8.3 mHz (0.15 mm sec^{-1} in radial velocity), placing hundreds to thousands of pixels on the asteroid. This data set provides physical and dynamical information that is unprecedented for an Earth-crossing object. The images (Fig. 22) reveal this asteroid to be in a highly unusual, non-principal-axis (NPA) spin state with several-day characteristic timescales. Extraction of the information in this imaging data set required inversion with a much more comprehensive physical model than in the analysis of Castalia images; free parameters included the asteroid's shape and inertia matrix, initial conditions for the asteroid's spin and orientation, the radar scattering properties of the surface, and the delay-Doppler trajectory of the center of mass. The shape (Fig. 23) reconstructed from the low-resolution images of Toutatis has shallow craters, linear ridges, and a deep topographic "neck" whose geologic origin is not known. It may have been sculpted by impacts into a single, coherent body, or Toutatis might actually consist of two separate objects that came together in a gentle collision. Toutatis is rotating in a long-axis mode (see Fig. 23) characterized

by periods of 5.4 days (rotation about the long axis) and 7.4 days (average for long-axis precession about the angular momentum vector). The asteroid's principal moments of inertia are in ratios within 1% of 3.22 and 3.09, and the inertia matrix is indistinguishable from that of a homogeneous body. Such information has yet to be determined for any other asteroid or comet, and probably is impossible to acquire in a fast spacecraft flyby. Higher-resolution images (e.g., Fig. 22b) from the 1992 and 1996 experiments are now being used to refine the Toutatis model.

Accurate shape models of near-Earth asteroids open the door to a wide variety of theoretical investigations that previously have been impossible or have used simplistic models (spheres or ellipsoids). For example, the Castalia and Toutatis models are being used to explore the stability and evolution of close orbits, with direct application to the design of robotic and piloted spacecraft missions, to studies of retention and redistribution of impact ejecta, and to questions about plausible origins and lifetimes of asteroidal satellites. Accurate models also allow realistic investigations of the effects of collisions in various energy regimes on the object's rotation state, surface topography, regolith, and inter-

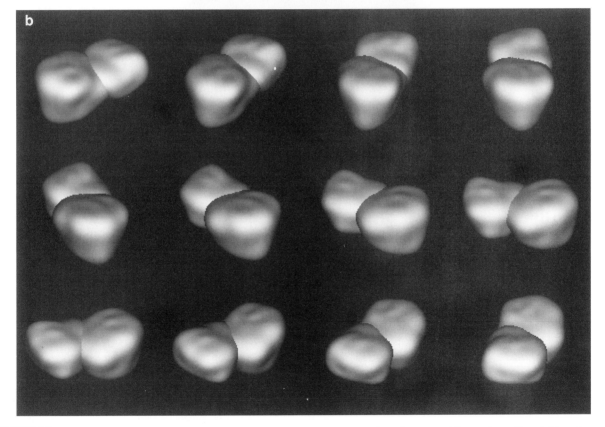

FIGURE 21 Radar results for near-Earth asteroid 4769 Castalia (1989PB). (a) Arecibo radar images. This 64-frame "movie" is to be read like a book (left to right in the top row, etc.). The radar lies toward the top of the page, in the image plane, which probably is about 35° from the asteroid's equatorial plane. In each frame, OC echo power (i.e., the brightness seen by the radar) is plotted versus time delay (increasing from top to bottom) and frequency (increasing from left to right). The object is seen rotating through about 220° during the 2.5-hr sequence. [From S. J. Ostro, J. F. Chandler, A. A. Hine, I. I. Shapiro, K. D. Rosema, and D. K. Yeomans (1990). *Science* **248,** 1523–1528. Copyright 1990 AAAS.] (b) Three-dimensional computer model of Castalia, from inversion of the images in (a). The reconstruction uses 167 shape parameters and has a resolution of about 100 m. This contact-binary asteroid is about 1.8 km long. [From R. S. Hudson and S. J. Ostro (1994). *Science* **263,** 940–943. Copyright 1994 AAAS.]

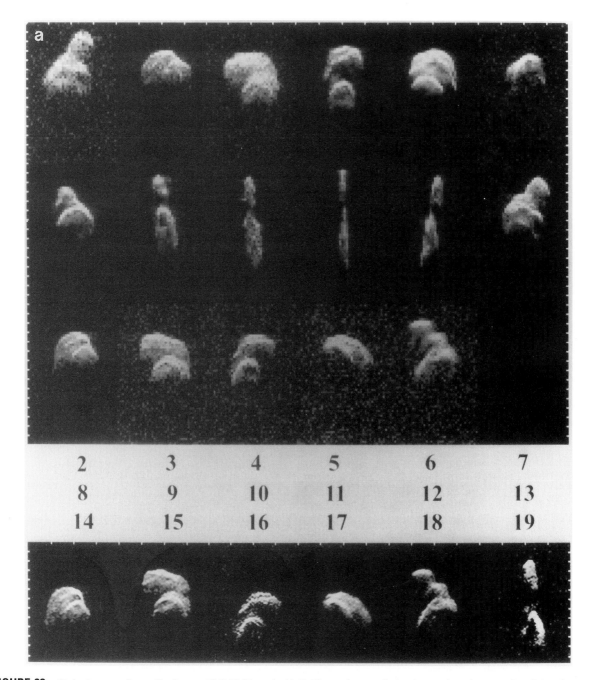

FIGURE 22 Radar images of near-Earth asteroid 4179 Toutatis. (a) Goldstone low-resolution images (top three rows) and Arecibo images (bottom row) obtained on the indicated dates in December 1992, plotted with time delay increasing toward the bottom and Doppler frequency increasing toward the left. On the vertical sides, ticks are 2 μsec (300 m) apart. Two horizontal sides have ticks separated by 1 Hz for Goldstone and 0.28 Hz for Arecibo; those intervals correspond to a radial velocity difference of 18 mm sec^{-1}. (b) A high-resolution (125 nsec × 33 mHz) Goldstone image obtained with Toutatis 3.6 million km (10 lunar distances) from Earth. The spatial resolution is 19 × 46 m. [From S. J. Ostro *et al.* (1995). *Science* **270**, 80–83. Copyright 1995 AAAS.]

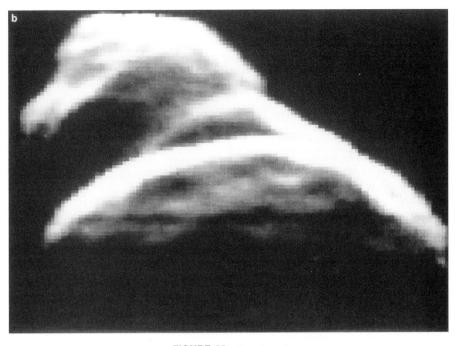

FIGURE 22 (*continued*)

nal structure. Simulations of impacts into Castalia using smooth-particle hydrodynamics code have begun to suggest how surface and interior damage depends on impact energy, impact location, and the equation of state of the asteroidal material. These computer investigations have clear ramifications for our understanding of asteroid collisional history, for exploitation of asteroid resources, and eventually for deflection/destruction of objects found to be on a collision course with Earth.

K. JUPITER'S ICY GALILEAN SATELLITES

Among all the radar-detected planetary bodies in the solar system, Europa, Ganymede, and Callisto have the most bizarre radar properties. Their reflectivities are enormous compared with those of the Moon and inner planets (see Table II). Europa is the extreme example (Fig. 24), with an OC radar albedo (1.0) as high as that of a metal sphere. Since the radar and optical albedos and estimates of fractional water frost coverage increase by satellite in the order Callisto–Ganymede–Europa, the presence of water ice has long been suspected of playing a critical role in determining the unusually high reflectivities even though ice is less radar-reflective than silicates. In spite of the satellites' smooth appearances at the several-kilometer scales of Voyager high-resolution images, a diffuse scattering

process and hence a high degree of near-surface structure at centimeter to meter scales is indicated by broad spectral shapes and large linear polarization ratios ($\mu_L \sim 0.5$).

The most peculiar aspect of the satellites' echoes is their circular polarization ratios, which exceed unity. That is, in contrast to the situation with other planetary targets, the scattering largely preserves the handedness, or helicity, of the transmission. Mean values of μ_C for Europa, Ganymede, and Callisto are about 1.5, 1.4, and 1.2, respectively. Wavelength dependence is negligible from 3.5 to 13 cm, but dramatic from 13 to 70 cm (Fig. 25). Significant polarization and/or albedo features are present in the echo spectra and in a few cases correspond to geologic features in Voyager images.

The icy satellites' echoes are due not to external surface reflections but to subsurface "volume" scattering. The high radar transparency of ice compared with that of silicates permits deeper radar sounding, longer photon path lengths, and higher-order scattering from regolith heterogeneities—radar is seeing Europa, Ganymede, and Callisto in a manner that the Moon has never been seen. The satellites' radar behavior apparently involves the coherent backscatter effect, which accompanies any multiple-scattering process; occurs for particles of any size, shape, and refractive index; and was first discovered in laboratory studies of the scattering of electrons and of light. Coherent

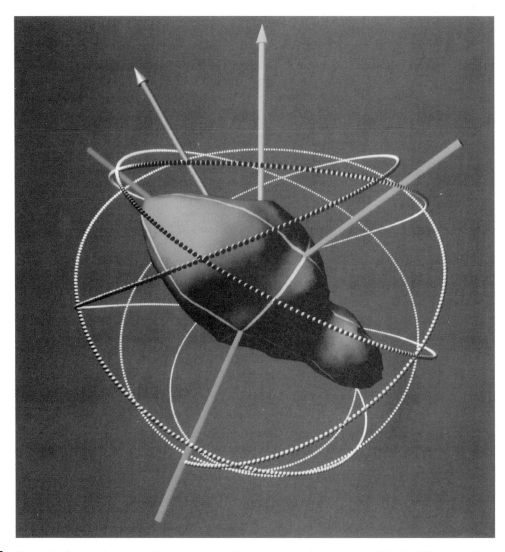

FIGURE 23 Toutatis's shape and non-principal-axis spin state from inversion of the images in Fig. 21a. The axes with no arrow tips are the asteroid's principal axes of inertia and the vertical arrow is its angular momentum vector; the direction of the spin vector (the arrow pointing toward eleven o'clock) relative to the principal axes is a (5.41-day) periodic function. A flashlamp attached to the short axis of inertia and flashed every 15 minutes for 20 days would trace out the intricate path indicated by the small spheres stacked end-to-end; the path never repeats. Toutatis's spin state differs radically from those of the vast majority of solar system bodies that have been studied, which are in principal-axis spin states. For those objects, the spin vector and angular momentum vector point in the same direction and the flashlamp's path would be a circle.

backscatter yields strong echoes and $\mu_C > 1$ because the incident, circularly polarized wave's direction is randomized before its helicity is randomized and also before its power is absorbed. The vector-wave theory of coherent backscatter accounts for the unusual radar signatures in terms of high-order, multiple anisotropic scattering from within the upper few decameters of the regoliths, which the radar sees as an extremely low-loss, disordered random medium. Inter- and intrasatellite albedo variations show much more dynamic range than μ_C variations, and probably are due to variations in ice purity.

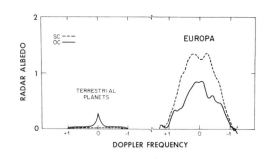

FIGURE 24 Typical 13-cm echo spectra for the terrestrial planets are compared to echo spectra for Jupiter's icy moon Europa. The abscissa has units of half the echo bandwidth.

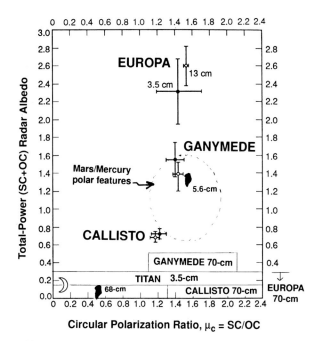

FIGURE 25 Radar properties of Europa, Ganymede, and Callisto compared to those of some other targets. The icy Galilean satellites' total-power radar albedos do not depend on wavelength between 3.5 and 13 cm, but plummet at 70 cm. There are large uncertainties in those objects' μ_C at 70 cm and in Titan's μ_C at 3.5 cm (the only wavelength at which it has been detected by radar). Solid symbols shaped like Greenland indicate properties of that island's percolation zone at 5.6 and 68 cm. The domain of most of the bright polar features on Mars and Mercury is sketched.

As sketched in Fig. 25, there are similarities between the icy Galilean satellites' radar properties and those of the radar-bright polar caps on Mars, features inside perpetually shadowed craters at the poles of Mercury (see Fig. 15), and the percolation zone in the Greenland ice sheet. However, the subsurface configuration in the Greenland zone, where the scattering heterogeneities are "ice pipes" produced by seasonal melting and refreezing, are unlikely to resemble those on the satellites. Therefore, unique models of subsurface structure cannot be deduced from the radar signatures of any of these terrains.

L. COMETS

Since a cometary coma is nearly transparent at radio wavelengths, radar is much more capable of unambiguous detection of a cometary nucleus than are optical and infrared methods, and radar observations of several comets (see Table I) have provided useful constraints on nuclear dimensions. The radar signature of one particular comet (IRAS-Araki-Alcock, which came within 0.03 AU of Earth in May 1983) revolutionized our concepts of the physical nature of these intriguing objects. Echoes obtained at both Arecibo (Fig. 26) and Goldstone have a narrowband component from the nucleus as well as a much weaker broadband component from large particles ejected mostly from the sunlit side of the nucleus. Models of the echoes suggest that the nucleus is very rough on scales larger than a meter, that its maximum overall dimension is within a factor of two of 10 km, and that its spin period is 2–3 days. The particles are probably several centimeters in size and account for a significant fraction of the particulate mass loss from the nucleus. Most of them appear to be distributed within ~1000 km of the nucleus, that is, in the volume filled by particles ejected at several meters per second over a few days. The typical particle lifetime may have been this short, or the particle ejection rate may have been highly variable.

In late 1985, radar observations of comet Halley, which was much more active than IRAS-Araki-Alcock, yielded echoes with a substantial broadband component presumed to be from a large-particle swarm, but no narrowband component, a negative result consistent with the hypothesis that the surface of the nucleus has an extremely low bulk density. In 1996, Goldstone obtained 3.5-cm echoes from the nucleus and coma of comet Hyakutake (C/1996 B2). The coma-to-nucleus ratio of radar cross section is about 12 for Hyakutake versus about 0.3 for IAA. The radar signatures of these three comets strengthen impressions about the diversity, and unpredictability, of comet physical properties

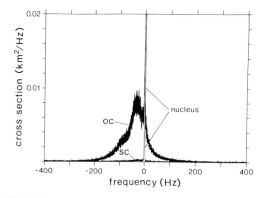

FIGURE 26 OC and SC echo spectra obtained at 13 cm for comet IRAS-Araki-Alcock, truncated at 2% of the maximum OC amplitude. The narrowband echo from the nucleus is flanked by a broadband echo from large (1-cm) particles in a 1000-km-radius cloud surrounding the nucleus. [From J. K. Harmon, D. B. Campbell, A. A. Hine, I. I. Shapiro, and B. G. Marsden (1989). *Astrophys. J.* **338**, 1071–1093.]

and have obvious implications for spacecraft operations close to comets.

M. SATURN'S RINGS AND TITAN

The only radar-detected ring system is quite unlike other planetary targets in terms of both the experimental techniques employed and the physical considerations involved. For example, the relation between ring-plane location and delay-Doppler coordinates for a system of particles traveling in Keplerian orbits is different from the geometry portrayed in Fig. 5. The rings are grossly overspread (see Table II), requiring the use of frequency-stepped waveforms in delay-Doppler mapping experiments.

Radar determinations of the rings' backscattering properties complement results of the Voyager spacecraft radio occultation experiment (which measured the rings' forward scattering efficiency at identical wavelengths) in constraining the size and spatial distributions of ring particles. The rings' circular polarization ratio is ~1.0 at 3.5 cm and ~0.5 at 13 cm, more or less independent of the inclination angle δ between the ring plane and the line of sight. Whereas multiple scattering between particles might cause some of the depolarization, the lack of strong dependence of μ_C on δ suggests that the particles are intrinsically rougher at the scale of the smaller wavelength. The rings' total-power radar albedo shows only modest dependence on δ, a result that seems to favor many-particle-thick models of the rings over monolayer models. Delay-Doppler resolution of ring echoes indicates that the portions of the ring system that are brightest optically (the A and B rings) also return most of the radar echoes. The C ring has a very low radar reflectivity, presumably because of either a low particle density in that region or compositions or particle sizes that lead to inefficient scattering.

Apart from landing a spacecraft on Titan, radar provides the most direct means to study the cloud-covered surface of Saturn's largest moon. Voyager and ground-based data indicate a surface temperature and pressure of 94 K and 1.5 bar, and show that the atmosphere is mostly N_2 with traces of hydrocarbons and nitriles. Thermodynamic considerations imply a near-surface reservoir of liquid hydrocarbons, possibly consisting of a kilometer-deep global ocean. However, that configuration, which would give a radar albedo of only several percent, has been ruled out by Goldstone–VLA detections that yield OC albedo estimates that average about 0.13. More work is needed to elucidate Titan's radar properties.

IV. PROSPECTS FOR PLANETARY RADAR

The 1993–1997 upgrading of the Arecibo telescope has increased the instrument's average radar sensitivity by a factor of 20, more than doubling its range and reducing by nearly an order of magnitude the diameter of the smallest object detectable at any given distance. The impact of the Arecibo upgrade on planetary science is expected to be fundamental and far-reaching, especially for studies of small bodies and planetary satellites. The quality, in terms of signal-to-noise ratio and spatial resolution, of radar measurements will jump by more than an order of magnitude. Several short-period comets will become easy targets. The pre-upgrade Arecibo could barely skim the inner edge of the main asteroid belt, but the upgraded telescope will have access to asteroids throughout the belt. The instrument is expected to provide high-resolution images of dozens of asteroids per year.

Efforts are under way to increase the near-Earth asteroid discovery rate by one to two orders of magnitude. Most of the optically discoverable NEAs traverse the detectability windows of the upgraded Arecibo and/or Goldstone telescopes at least once during any given several-decade interval. In view of the utility of radar observations for orbit refinement and physical characterization, there is considerable motivation to do radar observations of newly discovered NEAs whenever possible. The initial radar reconnaissance of a new NEA might eventually become an almost daily opportunity.

Radar investigations of natural satellites will reap enormous benefits from ground-based and spaceborne radar reconnaissance. The near-surface physical properties of Deimos, Io, and Titan will be readily discernible with Arecibo. Doppler images of Titan may furnish a coarse-resolution, nearly global albedo map, while the *Cassini* spacecraft, with its high-resolution, 13.8-GHz (2.2-cm) radar instrument, is journeying toward its arrival at Saturn in 2004. That instrument, which will function as a synthetic-aperture radar imager, an altimeter, and a passive radiometer, is designed to determine whether oceans exist on Titan, and, if so, to determine their distribution. There is growing interest in the possibility of a subsurface ocean on Europa, and in the feasibility of using an orbiting, long-wavelength (~6-m) radar sounder to probe many kilometers below that object's fractured crust, perhaps within a decade. In summary, planetary radar astronomy appears to be on the verge of producing an enormously valuable body of new information about asteroids, comets, and the satellites of Mars, Jupiter, and Saturn.

BIBLIOGRAPHY

Butrica, A. J. (1996). "To See the Unseen: A History of Planetary Radar Astronomy," NASA History Series No. SP-4218. NASA, Houston.

Harmon, J. K., Slade, M. A., Velez, R. A., Crespo, A., Dryer, M. J., and Johnson, J. M. (1994). Radar mapping of Mercury's polar anomalies. *Nature* **369**, 213–215.

Hudson, R. S., and Ostro, S. J. (1995). Shape and non-principal axis spin state of asteroid 4179 Toutatis. *Science* **270**, 84–86.

Mitchell, D. L., Ostro, S. J., Hudson, R. S., Rosema, K. D., Campbell, D. B., Velez, R., Chandler, J. F., Shapiro, I. I., Giorgini, J. D., and Yeomans, D. K. (1996). Radar observations of asteroids 1 Ceres, 2 Pallas, and 4 Vesta. *Icarus* **124**, 113–133.

Muhleman, D. O., Grossman, A. W., and Butler, B. J. (1995). Radar investigation of Mars, Mercury, and Titan. *Annu. Rev. Earth Planet Sci.* **23**, 337–374.

Ostro, S. J. (1993). Planetary radar astronomy. *Rev. Modern Physics* **65**, 1235–1279.

Pettengill, G. H., Ford, P. G., Johnson, W. T. K., Raney, R. K., and Soderblom, L. A. (1991). Magellan: Radar performance and data products. *Science* **252**, 260–265.

Shapiro, I. I., Chandler, J. F., Campbell, D. B., Hine, A. A., and Stacy, N. J. S. (1990). The spin vector of Venus. *Astron. J.* **100**, 1363–1368.

Simpson, R. A., Harmon, J. K., Zisk, S. H., Thompson, T. W., and Muhleman, D. O. (1992). Radar determination of Mars radar properties. *In* "Mars" (H. Kieffer, B. Jakosky, C. Snyder, and M. Matthews, eds.), pp. 652–685. Univ. Arizona Press, Tucson.

Slade, M. A., Butler, B. J., and Muhleman, D. O. (1992). Mercury radar imaging: Evidence for polar ice. *Science* **258**, 635–640.

Stacy, N. J. S., Campbell, D. B., and Ford, P. G. (1997). Arecibo radar mapping of the lunar poles: A search for ice deposits. *Science* **276**, 1527–1530.

Tyler, G. L., Ford, P. G., Campbell, D. B., Elachi, C., Pettengill, G. H., and Simpson, R. A. (1991). Magellan: Electrical and physical properties of Venus' surface. *Science* **252**, 265–270.

Yeomans, D. K., Chodas, P. W., Keesey, M. S., Ostro, S. J., Chandler, J. F., and Shapiro, I. I. (1992). Asteroid and comet orbits using radar data. *Astron. J.* **103**, 303–317.

SOLAR SYSTEM DYNAMICS

I. Introduction: Keplerian Motion

II. The Two-Body Problem

III. Planetary Perturbations and the Orbits of Small Bodies

IV. Dissipative Forces and the Orbits of Small Bodies

V. Long-Term Stability of Planetary Orbits

Martin J. Duncan
Queen's University

Jack J. Lissauer
NASA Ames Research Center

GLOSSARY

Apoapse: Point on an orbit farthest from the center of gravity (called aphelion for orbits about the Sun and apogee for orbits about Earth).

Apsides: The two points in an orbit that lie closest to (*periapse*) and farthest from (*apoapse*) the center of gravity. The line of apsides is the straight line connecting the two apsides and is the major axis of an elliptical orbit.

Chaotic motion: Motion that is extremely sensitive to initial conditions in the sense that tiny changes in the starting state can lead to dramatically different evolution. In a chaotic region, nearby orbits diverge exponentially with time.

Disturbing function: Mathematical expression for the gravitational potential of bodies that perturb an otherwise Keplerian orbit about a central body.

Eccentricity: Measure of the departure of an orbit from circularity, represented by e. For elliptical orbits, $0 < e = (1 - b^2/a^2)^{1/2} < 1$, where $2a$ is the major axis of the ellipse and $2b$ is the minor axis. For circular orbits, $e = 0$; for parabolic orbits, $e = 1$; and for hyperbolic orbits, $e > 1$.

Ecliptic: Plane of Earth's orbit about the Sun.

Escape velocity: Minimum speed required to escape to infinity.

Hill sphere: Region around a secondary in which the secondary's gravity is more influential for the motion of a particle about the secondary than is the tidal influence of the primary.

Horseshoe orbits: Orbits that librate encircling the L_3, L_4, and L_5 Lagrangian points in the circular restricted three-body problem. These orbits appear to be shaped like horseshoes in the frame rotating with the mean motion of the system.

Kepler's laws: Three rules that describe the motion of planets about the Sun (and of moons about planets) with good accuracy: (1) Planets move on elliptical paths with the Sun at one focus. (2) An imaginary line from the Sun to a planet sweeps out area at a constant rate. (3) The square of a planet's orbital period varies as the cube of the semimajor axis of its orbit.

Lagrangian points: The five locations in the restricted circular three-body problem at which the net gravitational and centrifugal forces in the frame rotating with the massive bodies is zero.

The first three Lagrangian points, L_1, L_2, and L_3, lie on the line connecting the massive bodies; all three colinear Lagrangian points are unstable. The L_4 and L_5 Lagrangian points each make equilateral triangles with the two massive bodies: orbits about the triangular Lagrangian points are stable to small perturbations provided the ratio between the masses of the two bodies is ≥ 27.

Libration: Oscillation in time of an angle with respect to its mean value.

Newton's laws: Three laws of motion and one of gravitation that describe aspects of the physical world: (1) A body remains at rest or in uniform motion unless it is acted upon by an external force. (2) The acceleration of a body is directly proportional to the force acting upon it and inversely proportional to its mass. (3) For every action, there exists an equal and opposite reaction. (Gravitation) The gravitational attraction between any two spherically symmetric objects is proportional to the product of their masses and inversely proportional to the square of the distance between their centers.

Periapse: Point on an orbit closest to the center of gravity (called perihelion for orbits about the Sun and perigee for orbits about Earth).

Phase space: Multidimensional space in which the coordinates are the spatial positions together with the corresponding momenta.

Poynting–Robertson drag: Force produced as a result of the absorption and reradiation of photons that causes the orbits of micron-sized grains to decay toward the Sun.

Reynolds number: Dimensionless number that governs the conditions for the occurrence of turbulence in fluids.

Roche's limit: Minimum distance from the center of a massive primary at which a fluid satellite of a given density can remain stable (assuming it is held together exclusively by its self-gravity, is traveling on a circular orbit, and is in synchronous rotation).

Secular: Continuing or changing over a long period of time.

Secular resonance: Near-commensurability among the frequencies associated with the precessions of the line of nodes and/or apsides.

Tadpole orbits: Orbits that librate about the stable L_4 or L_5 triangular Lagrangian points in the restricted three-body problem. These orbits appear to be shaped like tadpoles in the frame rotating with the mean motion of the massive bodies.

Three-body problem: Study of the trajectories of three point masses attracted to each other by mutual gravitation. When the mass of one of the bodies is negligible compared to the masses of the other two bodies, it is referred to as the restricted three-body problem. If, in addition, the two massive bodies move on circular orbits about their common center of mass, the problem becomes the circular restricted three-body problem.

I. INTRODUCTION: KEPLERIAN MOTION

In 1687, Isaac Newton showed that the motion of two spherically symmetric bodies resulting from their mutual gravitational attraction is described by simple conic sections (see Section II, D). However, the introduction of additional gravitating bodies produces a rich variety of dynamical phenomena, even though the basic interactions between pairs of objects can be straightforwardly described. Indeed, one of the most profound results in the last two decades in physics has been the realization that even few-body systems governed by apparently simple nonlinear interactions can display remarkably complex behavior, which has come to be known collectively as chaos. Perhaps with hindsight, it should come as no surprise that on sufficiently long timescales, the apparently regular orbital motion of many bodies in the solar system can exhibit symptoms of this chaotic behavior.

In this chapter we describe the basic orbital properties of solar system objects (planets, moons, minor bodies, and dust) and their mutual interactions. We also provide several examples of important dynamical processes that occur in the solar system and lay the groundwork for describing some of the phenomena that are discussed in more detail in other chapters of this book.

A. KEPLER'S LAWS OF PLANETARY MOTION

By analyzing Tycho Brahe's careful observations of the orbits of the planets, Johannes Kepler deduced the following three laws of planetary motion:

(1) All planets move along elliptical paths with the Sun at one focus. We can express the heliocentric distance r (i.e., the planet's distance from the Sun) as

$$r = \frac{a(1 - e^2)}{1 + e \cos f} \tag{1}$$

with a the semimajor axis (average of the minimum and maximum heliocentric distances) and e (the eccentricity of the orbit) $\equiv (1 - b^2/a^2)^{1/2}$, where $2b$ is the minor axis of an ellipse. The true anomaly, f, is the angle between the planet's perihelion (closest heliocentric distance) and its instantaneous position (Fig. 1).

(2) A line connecting a planet and the Sun sweeps out equal areas ΔA in equal periods of time Δt:

$$\frac{\Delta A}{\Delta t} = \text{constant} \tag{2}$$

Note that the value of this constant differs from one planet to the next.

(3) The square of a planet's orbital period P about the Sun (in years) is equal to the cube of its semimajor axis a (in AU):

$$P^2 = a^3 \tag{3}$$

B. ELLIPTICAL MOTION, ORBITAL ELEMENTS, AND THE ORBIT IN SPACE

The Sun contains more than 99.8% of the mass of the known solar system. We shall see that the gravitational force exerted by a body is proportional to its mass, so to an excellent first approximation we can regard the

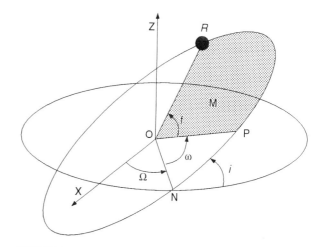

FIGURE 2 Geometry of an orbit in three dimensions. The Sun is at one focus of the ellipse (O) and the planet is instantaneously at location R. The location of the perihelion of the orbit is P. The intersection of the orbital plane (X–Y) and the reference plane is along the line ON (where N is the ascending node). The various angles shown are described in the text. The mean anomaly M is an angle proportional to the area OPR swept out by the radius vector **OR** (Kepler's third law).

motion of the planets and many other bodies as being solely due to the influence of a fixed central pointlike mass. For objects like the planets, which are bound to the Sun and hence cannot go arbitrarily far from the central mass, the general solution for the orbit is the ellipse described by Eq. (1). The orbital plane, although fixed in space, can be arbitrarily oriented with respect to whatever reference plane we have chosen (such as Earth's orbital plane about the Sun, which is called the ecliptic). The inclination, i, of the orbital plane is the angle between the reference plane and the orbital plane and can range from 0 to 180°. Conventionally, bodies orbiting in a counterclockwise (clockwise) sense as viewed from the north are defined to have inclinations from 0° to 90° (90° to 180°) and are said to be on prograde (retrograde) orbits. The two planes intersect in a line called the line of nodes and the orbit pierces the reference plane at two locations— one as the body passes upward through the plane (the ascending node) and one as it descends (the descending node). A fixed direction in the reference plane is chosen and the angle to the direction of the orbit's ascending node is called the longitude of the ascending node, Ω. Finally, the angle between the line to the ascending node and the line to the direction of periapse (perihelion for orbits about the Sun, perigee for orbits about Earth) is called the argument of periapse ω. Thus the six orbital elements a, e, i, Ω, ω, and f uniquely specify the location of the object in space (Fig. 2). The first

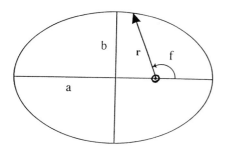

FIGURE 1 Geometry of an elliptic orbit. The Sun is at one focus and the vector **r** denotes the instantaneous heliocentric location of the planet (i.e., r is the planet's distance from the Sun). a is the semimajor axis (average heliocentric distance) and b is the semiminor axis of the ellipse. The true anomaly, f, is the angle between the planet's perihelion (closest heliocentric distance) and its instantaneous position.

three quantities (a, e, and i) are often referred to as the principal orbital elements, as they describe the orbit's size, shape, and tilt, respectively.

II. THE TWO-BODY PROBLEM

In this section we discuss the general solution to the problem of the motion of two otherwise isolated objects in which the only force acting on each body is the mutual gravitational interaction.

A. NEWTON'S LAWS OF MOTION AND THE UNIVERSAL LAW OF GRAVITATION

Although Kepler's laws were originally found from careful observation of planetary motion, they were subsequently shown to be derivable from Newton's laws of motion together with his universal law of gravity. Consider a body of mass m_1 at instantaneous location $\mathbf{r}_1$ with instantaneous velocity $\mathbf{v}_1 \equiv d\mathbf{r}_1/dt$ and hence momentum $\mathbf{p}_1 \equiv m_1\mathbf{v}_1$. The acceleration $d\mathbf{v}_1/dt$ produced by a net force $\mathbf{F}_1$ is given by Newton's second law of motion:

$$\mathbf{F}_1 = \frac{d(m_1\mathbf{v}_1)}{dt} \qquad (4)$$

Newton's universal law of gravity states that a second body of mass m_2 at position $\mathbf{r}_2$ exerts an attractive force on the first body given by

$$\mathbf{F}_1 = -\frac{Gm_1m_2}{r_{12}^3} \mathbf{r}_{12} \qquad (5)$$

where $\mathbf{r}_{12} \equiv \mathbf{r}_1 - \mathbf{r}_2$ is the location of particle 1 with respect to particle 2, and G is the gravitational constant. Newton's third law states that for every action there is an equal and opposite reaction; thus, the force on each object of a pair is equal in magnitude but opposite in direction. We will use these facts to reduce the two-body problem to an equivalent one-body case in the next section.

B. REDUCTION TO THE ONE-BODY CASE

From the foregoing discussion of Newton's laws, we see that for the two-body problem the force exerted by body 1 *on* body 2 is

$$\frac{d(m_2\mathbf{v}_2)}{dt} = \mathbf{F}_2 = -\mathbf{F}_1 = \frac{Gm_1m_2}{r_{12}^3} \mathbf{r}_{12} \qquad (6)$$

Thus, from Eqs. (4) and (6) we see that

$$\frac{d(m_1\mathbf{v}_1 + m_2\mathbf{v}_2)}{dt} = \mathbf{F}_1 + \mathbf{F}_2 = 0 \qquad (7)$$

This is of course a statement that the total linear momentum of the system is conserved, which means that the center of mass of the system moves with constant velocity.

Multiplying Eq. (6) by m_1 and Eq. (5) by m_2 and subtracting, we find that the equation for the relative motion of the bodies can be cast in the form

$$\mu_r \frac{d^2\mathbf{r}_{12}}{dt^2} = \mu_r \frac{d^2(\mathbf{r}_1 - \mathbf{r}_2)}{dt^2} = -\frac{G\mu_r M}{r_{12}^3} \mathbf{r}_{12} \qquad (8)$$

where $\mu_r \equiv m_1m_2/(m_1 + m_2)$ is called the reduced mass and $M \equiv m_1 + m_2$ is the total mass. Thus, the relative motion is completely equivalent to that of a particle of reduced mass μ_r orbiting a *fixed* central mass M. For known masses, specifying the elements of the relative orbit and the positions and velocities of the center of mass is completely equivalent to specifying the positions and velocities of both bodies. A detailed solution of the equation of motion (8) is discussed in any elementary text on orbital mechanics and in most general classical mechanics books. In the remainder of Section II, we simply give a few key results.

C. ENERGY, CIRCULAR VELOCITY, AND ESCAPE VELOCITY

The centripetal force necessary to keep an object of mass μ_r in a circular orbit of radius r with speed v_c is $\mu_r v_c^2/r$. Equating this to the gravitational force exerted by the central body of mass M, we find that the circular velocity is

$$v_c = \sqrt{\frac{GM}{r}} \qquad (9)$$

Thus the orbital period (the time to move once around the circle) is

$$P = 2\pi r/v_c = 2\pi \sqrt{\frac{r^3}{GM}} \qquad (10)$$

The total (kinetic plus potential) energy E of the system is a conserved quantity:

$$E = T + V = \frac{1}{2} \mu_r v^2 - \frac{GM\mu_r}{r} \qquad (11)$$

where the first term on the right is the kinetic energy of the system, T, and the second term is the potential energy of the system, V. If $E < 0$, the absolute value of the potential energy of the system is larger than its kinetic energy and the system is bound. The body will orbit the central mass on an elliptical path. If $E > 0$, the kinetic energy is larger than the absolute value of the potential energy, and the system is unbound. The relative orbit is then described mathematically as a hyperbola. If $E = 0$, the kinetic and potential energies are equal in magnitude, and the relative orbit is a parabola. By setting the total energy equal to zero, we can calculate the escape velocity at any separation:

$$v_e = \sqrt{\frac{2GM}{r}} = \sqrt{2} v_c \qquad (12)$$

For circular orbits it is easy to show [using Eqs. (9) and (11)] that both the kinetic energy and the total energy of the system are equal in magnitude to half the potential energy:

$$T = -\frac{1}{2} V \qquad (13)$$

$$E = -\frac{GM\mu_r}{2r} \qquad (14)$$

For an elliptical orbit, Eq. (14) holds if the radius r is replaced by the semimajor axis a:

$$E = -\frac{GM\mu_r}{2a} \qquad (15)$$

Similarly, for an elliptical orbit, Eq. (10) becomes Newton's generalization of Kepler's third law:

$$P^2 = \frac{4\pi^2 a^3}{G(m_1 + m_2)} \qquad (16)$$

It can be shown that Kepler's second law follows immediately from the conservation of angular momentum, **L**:

$$\frac{d\mathbf{L}}{dt} = \frac{d(\mu_r \mathbf{r} \times \mathbf{v})}{dt} = 0 \qquad (17)$$

D. ORBITAL ELEMENTS: ELLIPTICAL, PARABOLIC, AND HYPERBOLIC ORBITS

As we noted earlier, the relative orbit in the two-body problem is either an ellipse, parabola, or hyperbola depending on whether the energy is negative, zero, or positive, respectively. These curves are known collectively as conic sections and the generalization of Eq. (1) is

$$r = \frac{p}{1 + e \cos f} \qquad (18)$$

where r and f have the same meaning as in Eq. (1), e is the generalized eccentricity, and p is a constant. For a parabola, $e = 1$ and $p = 2q$, where q is the pericentric separation (distance of closest approach). For a hyperbola, $e > 1$ and $p = q(1 + e)$, where q is again the pericentric separation. For all orbits, the three orientation angles i, Ω, and ω are defined as in the elliptical case.

III. PLANETARY PERTURBATIONS AND THE ORBITS OF SMALL BODIES

Gravity is not restricted to interactions between the Sun and the planets or individual planets and their satellites, but rather all bodies feel the gravitational force of one another. Within the solar system, one body typically produces the dominant force on any given body, and the resultant motion can be thought of as a Keplerian orbit about a primary, subject to small perturbations by other bodies. In this section we consider some important examples of the effects of these perturbations on the orbital motion.

A. FORMULATION OF THE PROBLEM

Classically, much of the discussion of the evolution of orbits in the solar system used perturbation theory as its foundation. Essentially, the method involves writing the equations of motion as the sum of a part that describes the independent Keplerian motion of the bodies about the Sun plus a part (called the disturbing function) that contains terms due to the pairwise interactions among the planets and minor bodies and the indirect terms associated with the back-reaction of the planets on the Sun. In general, one can then expand the disturbing function in terms of the small parameters of the problem (such as the ratio of the planetary masses to the solar mass, the eccentricities and inclinations, etc.), as well as the other orbital elements of the bodies, including the mean longitudes (i.e., the location of the bodies in their orbits), and attempt to solve the resulting equations for the time-dependence of the orbital elements. However, Poincaré showed roughly a century ago that these perturbation series are often divergent and have validity only over finite time spans. The full significance of Poincaré's work has become apparent only in the last two decades, in part because of long-term direct integrations on computers of the trajectories of bodies in the solar system. What is often found in practice is that for some initial conditions the trajectories are "regular" with variations in their orbital elements that seem to be well described by the perturbation series, whereas for other initial conditions the trajectories are found to be "irregular" (or in current parlance "chaotic") and are not as confined in their motions.

There is a key feature of the irregular orbits that we will use here as a definition of "chaos": two trajectories that begin arbitrarily close in phase space in a chaotic region will typically diverge exponentially with time until they are separated by macroscopically large distances. Ironically, the timescale for that divergence within a given chaotic region does not typically depend on the precise values of the initial conditions, provided that close approaches between bodies do not occur! Thus, if one computes the distance $d(t)$ between two particles having an initially small separation, it can be shown that for regular orbits, $d(t) - d(t_0)$ grows as a power of time t (typically linearly), whereas for irregular orbits $d(t)$ grows exponentially as $d(t_0)e^{\gamma(t-t_0)}$ where γ is conventionally called the Lyapunov exponent and γ^{-1} is called the Lyapunov timescale. Examples of both sorts of behavior are illustrated in Fig. 3.

From this definition of chaos, we see that chaotic orbits show such a sensitive dependence on initial conditions that the detailed long-term behavior of the orbits is lost within several Lyapunov timescales. Even a perturbation as small as 10^{-10} in the initial conditions will result in a 100% discrepancy (e.g., in the longitudes) in about 20 Lyapunov times. However, one of the interesting features of much of the chaotic behavior seen in simulations of the orbital evolution of bodies in the solar system is that the timescale for large changes in the principal orbital elements a, e, and i is often many orders of magnitude longer than the Lyapunov timescale.

In dynamical systems like the solar system, chaotic regions do not appear randomly, but rather they are

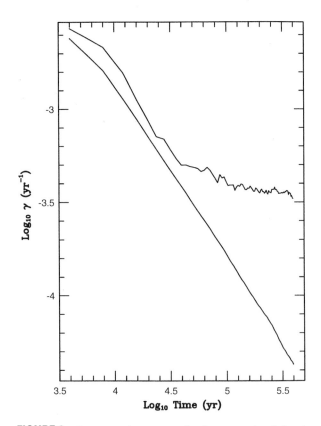

FIGURE 3 Distinction between regular (lower curve) and chaotic (upper curve) trajectories as characterized by the Lyapunov exponent discussed in the text. In this example, the trajectories are those of two massless particles near the 3 : 1 mean-motion resonance with Jupiter in the restricted three-body problem. Jupiter is assumed to move on a fixed elliptical orbit with eccentricity of 0.05. For chaotic trajectories, a plot of log γ versus log t eventually levels off at a value of γ that is the inverse of the Lyapunov timescale for the divergence of initially adjacent trajectories.

associated with trajectories in which the ratios of characteristic frequencies of the original problem are nearly commensurable, that is, sufficiently well approximated by rational numbers. The system is then said to be near a resonance, the simplest of these to visualize being so-called mean-motion resonances, in which the orbital periods of two bodies are commensurable. In the next two subsections, we give some examples of the consequences of this type of resonance. In Section III, E we describe resonances that are important for particles in planetary rings, and in Section V we define secular resonances and indicate their relation to the stability of the planets. Chaotic behavior in the solar system is discussed in greater detail in C. Murray's chapter in this encyclopedia. [See CHAOTIC MOTION IN THE SOLAR SYSTEM.]

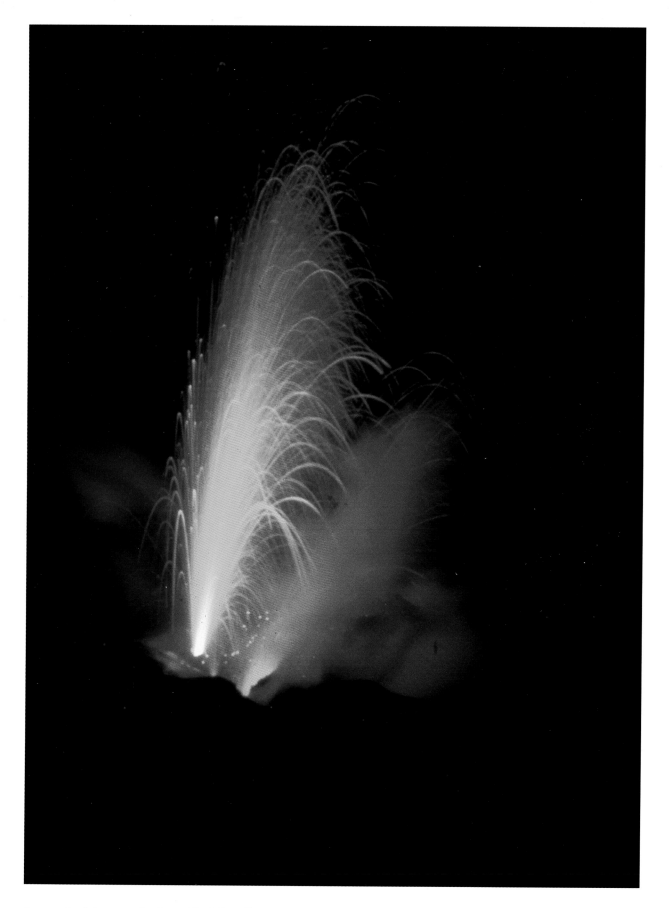

An example of planetary volcanism on Earth. Jets of hot gas and entrained incandescent basaltic pyroclasts ejected from a transient strombolian explosion on the volcano Stromboli, which is located on an island in the Tyrrhenian Sea off the northeastern coast of Sicily. Stromboli is one of the few constantly active volcanoes in Europe. (Photo by Lionel Wilson)

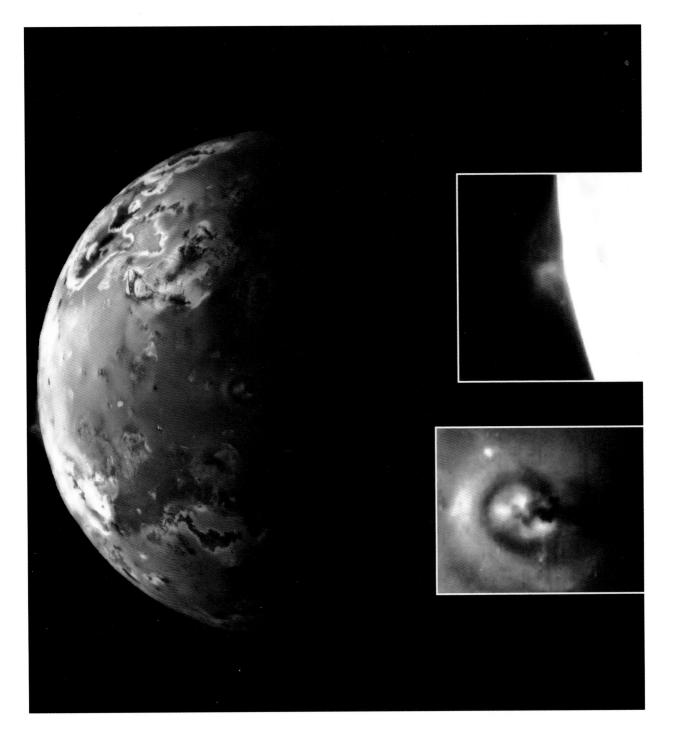

Two volcanic plumes on the satellite Io, the innermost of the four Galilean moons of Jupiter, viewed by the spacecraft Galileo on June 28, 1997. One plume is on the bright limb of the satellite (at ~8:30 o'clock) and is shown enlarged in the upper inset. It comes from a volcanic depression or caldera named Pillan Patera (after a South American god of thunder, fire, and volcanoes). This picture is the first showing activity at Pillan Patera. This plume is ~140 km high (~86 miles) and was also detected by the Hubble Space Telescope. The second plume is near the terminator (the boundary where day meets night). It is enlarged and shown in the lower inset. The location of this eruption is at a volcano called Prometheus, named after the Greek fire god. Note the shadow cast by this plume. From measurement of the shadow, we know that this plume has a height of ~75km (~45 miles). Prometheus was active at the times of the flybys of Voyagers 1 and 2.

Above, the Space Shuttle Columbia gets a flyby visit from Comet Hale-Bopp (shown as the streak at left center; note arrow) while awaiting launch on the STS-83 mission. This photo was taken the night before the planned liftoff on April 4, 1997. (NASA) Below, Comet Hale-Bopp photographed on April 5, 1997 by Dennis di Ciccio. (Courtesy Dennis di Ciccio, Sky and Telescope Magazine.)

Top, photograph taken by Eugene Cernan, commander of Apollo 17, showing lunar module pilot Harrison Schmitt standing in front of a large split boulder on the Moon. The lunar rover is in the foreground at left. Apollo 17 was launched on December 7, 1972 and landed on the Moon December 11. It was the last of the Apollo Moon landing missions. Bottom, view of the surface of Mars from Viking lander 1. (NASA photos)

B. PERTURBED KEPLERIAN MOTION AND RESONANCES

Although perturbations on a body's orbit are often small, they cannot always be ignored. They must be included in short-term calculations if high accuracy is required, for example, for predicting stellar occultations or targeting spacecraft. Most long-term perturbations are periodic in nature, their directions oscillating with the relative longitudes of the bodies or with some more complicated function of the bodies' orbital elements.

Small perturbations can produce large effects if the forcing frequency is commensurate or nearly commensurate with the natural frequency of oscillation of the responding elements. Under such circumstances, perturbations add coherently, and the effects of many small tugs can build up over time to create a large-amplitude, long-period response. This is an example of resonance forcing, which occurs in a wide range of physical systems.

An elementary example of resonance forcing is given by the simple one-dimensional harmonic oscillator, for which the equation of motion is

$$m \frac{d^2x}{dt^2} + m\Omega^2 x = F_0 \cos \omega t \qquad (19)$$

In Eq. (19), m is the mass of the oscillating particle, F_0 is the amplitude of the driving force, Ω is the natural frequency of the oscillator, and ω is the forcing or resonance frequency. The solution to Eq. (19) is

$$x = x_0 \cos \omega t + A \cos \Omega t + B \sin \Omega t \qquad (20a)$$

where

$$x_0 \equiv F_0/[m(\Omega^2 - \omega^2)] \qquad (20b)$$

and A and B are constants determined by the initial conditions. Note that if $\omega \approx \Omega$, a large-amplitude, long-period response can occur even if F_0 is small. Moreover, if $\Omega = \omega$, this solution to Eq. (19) is invalid. In this case the solution is given by

$$x = \frac{F_0}{2m\Omega} t \sin \Omega t + A \cos \Omega t + B \sin \Omega t \qquad (21)$$

The t in front of the first term at the right-hand side of Eq. (21) leads to secular growth. Often this linear growth is moderated by the effects of nonlinear terms that are not included in the simple example provided

here. However, some perturbations have a secular component.

Nearly exact orbital commensurabilities exist at many places in the solar system. Io orbits Jupiter twice as frequently as Europa does, which in turn orbits Jupiter twice as frequently as Ganymede does. Conjunctions (at which the bodies have the same longitude) always occur at the same position of Io's orbit (its perijove). How can such commensurabilities exist? After all, the rational numbers form a set of measure zero on the real line (which means that the probability of randomly picking a rational from the real number line is 0), and the number of small integer ratios is infinitely smaller still! The answer lies in the fact that orbital resonances may be held in place as stable locks, which result from nonlinear effects not represented in the foregoing simple mathematical example. For example, differential tidal recession (cf. Section IV, E) brings moons into resonance, and nonlinear interactions among the moons can keep them there.

Other examples of resonance locks include the Hilda asteroids, the Trojan asteroids, Neptune–Pluto, and several pairs of co-orbital moons about Saturn, including Mimas–Tethys and Enceladus–Dione. Resonant perturbation can also force material into highly eccentric orbits that may lead to collisions with other bodies; this is believed to be the dominant mechanism for clearing the Kirkwood gaps in the asteroid belt (see Section III, D). Spiral density waves can result from resonant perturbations of a collective disk of particles by an orbiting satellite. Density waves are seen at many resonances in Saturn's rings; they explain most of the structure seen in Saturn's A ring. The vertical analog of density waves, bending waves, are caused by resonant perturbations perpendicular to the ring plane due to a satellite in an orbit that is inclined to the ring. Spatial bending waves excited by the moons Mimas and Titan have been seen in Saturn's rings. In the next few subsections we discuss these manifestations of resonance effects in more detail.

C. EXAMPLES OF RESONANCES: LAGRANGIAN POINTS AND TADPOLE AND HORSESHOE ORBITS

We can understand many features of the orbits considered in this section by examining an idealized system in which two massive bodies move on circular orbits about their common center of mass. If a third body is introduced that is much less massive than either of the first two, we can follow its motion by assuming that its gravitational force has no effect on the orbits of the other bodies. By considering the motion in a frame corotating with the massive pair (so that the pair remain fixed on a line that we can take to be the x axis), Lagrange found that there are five points where particles placed at rest would feel no net force in the rotating frame. Three of the so-called Lagrange points (L_1, L_2, and L_3) lie along a line joining the two masses m_1 and m_2. The other two Lagrange points (L_4 and L_5) form equilateral triangles with the two massive bodies.

Particles displaced slightly from the first three Lagrangian points will continue to move away and hence these locations are unstable. The triangular Lagrangian points are potential energy maxima, which are stable for sufficiently large primary to secondary mass ratio due to the Coriolis force. Provided that the most massive body has at least 27 times the mass of the secondary (which is the case for all known examples in the solar system except the Pluto–Charon system), the Lagrangian points L_4 and L_5 are stable points. Thus, a particle at L_4 or L_5 that is perturbed slightly will start to "orbit" these points. Lagrangian points 4 and 5 are important in the solar system. For example, we see the Trojan asteroids in Jupiter's Lagrangian points and a recently discovered body called Eureka appears to be a Martian Trojan. There are also small moons in the triangular Lagrangian points of Tethys and Dione, in the Saturnian system. The L_4 and L_5 points in the Earth–Moon system have been suggested as possible locations for space stations.

1. Horseshoe and Tadpole Orbits

Consider a moon on a circular orbit about a planet. Figure 4 shows some important dynamical features in the frame corotating with the moon. All five Lagrangian points are indicated in the picture. A particle just interior to the moon's orbit has a higher angular velocity than the moon in the stationary frame, and thus moves with respect to the moon in the direction of corotation. A particle just outside the moon's orbit has a smaller angular velocity, and moves away from the moon in the opposite direction. When, in the stationary frame, the outer particle approaches the moon, the particle is slowed down (loses angular momentum) and, provided the initial difference in semimajor axis is not too large, the particle drops to an orbit lower than that of the moon. The particle then recedes in the forward direction. Similarly, the particle at the lower orbit is accelerated as it catches up with the moon, resulting in an outward motion toward the higher, slower orbit. Orbits like these encircle the L_3, L_4, and L_5 points and are called horseshoe orbits. Sa-

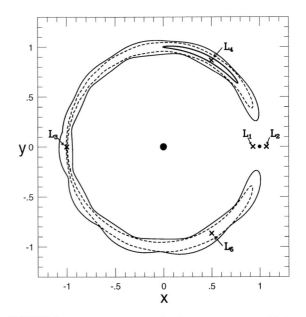

FIGURE 4 Diagram showing the five Lagrangian equilibrium points (denoted by crosses) and three representative orbits near these points for the circular restricted three-body problem. In this example, the secondary's mass is 0.001 times the total mass. The coordinate frame has its origin at the barycenter and corotates with the pair of bodies, thereby keeping the primary (large solid circle) and secondary (small solid circle) fixed on the x axis. Tadpole orbits remain near one or other of the L_4 and L_5 points. An example is shown near the L_4 point on the diagram. Horseshoe orbits enclose all three of L_3, L_4, and L_5 but do not reach L_1 or L_2. The outermost orbit on the diagram illustrates this behavior. There is a critical curve dividing tadpole and horseshoe orbits that encloses L_4 and L_5 and passes through L_3. A horseshoe orbit near this dividing line is shown as the dashed curve in the diagram.

turn's small moons Janus and Epimetheus execute just such a dance, changing orbits every 4 years.

Since the Lagrangian points L_4 and L_5 are stable, material can librate about these points individually: such orbits are called tadpole orbits. The tadpole libration width at L_4 and L_5 is roughly equal to $(m/M)^{1/2}r$, and the horseshoe width is $(m/M)^{1/3}r$, where M is the mass of the planet, m the mass of the satellite, and r the distance between the two objects. For a planet of Saturn's mass, $M = 5.7 \times 10^{29}$ g, and a typical small moon of mass $m = 10^{20}$ g (i.e., an object with a 30-km radius, with density of ~1 g/cm³), at a distance of 2.5 Saturnian radii, the tadpole libration half-width is about 3 km and the horseshoe halfwidth about 60 km.

2. Hill Sphere

The approximate limit to a planet's gravitational dominance is given by the extent of its Hill sphere,

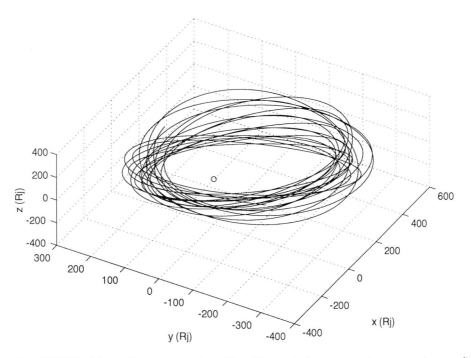

FIGURE 5 The orbit of J VIII Pasiphae, a distant retrograde satellite of Jupiter, is shown as seen in a nonrotating coordinate system with Jupiter at the origin (open circle). The satellite was integrated as a massless test particle in the context of the circular restricted three-body problem for approximately 38 years. The unit of distance is Jupiter's radius, R_J. During the course of this integration, the distance to Jupiter varied from 122 to 548R_J. Note how the large solar perturbations produce significant deviations from a Keplerian orbit. [Figure reprinted with permission from Jose Alvarellos (1996). "Orbital Stability of Distant Satellites of Jovian Planets," M.Sc. thesis, San Jose State University.]

$$R_H = \left[\frac{m}{3(M + m)}\right]^{1/3} a \qquad (22)$$

where m is the mass of the planet and M is the Sun's mass. A test body located at the boundary of a planet's Hill sphere is subjected to a gravitational force from the planet comparable to the tidal difference between the force of the Sun on the planet and that on the test body. The Hill sphere essentially stretches out to the L_1 point and is roughly the limit of the Roche lobe of a body with $m \ll M$. Planetocentric orbits that are stable over long periods of time are those well within the boundary of a planet's Hill sphere; all known natural satellites lie in this region. The trajectories of the outermost planetary satellites, which lie closest to the boundary of the Hill sphere, show large variations in planetocentric orbital paths (Fig. 5). Stable heliocentric orbits are those that are always well outside the Hill sphere of any planet.

D. EXAMPLES OF RESONANCES: KIRKWOOD GAPS

There are obvious patterns in the distribution of asteroidal semimajor axes that seem to be associated with mean-motion resonances with Jupiter, in which a particle's period of revolution about the Sun is an integer ratio times Jupiter's period. The Trojan asteroids travel in a 1:1 mean-motion resonance with Jupiter as described earlier. These asteroids execute small-amplitude (tadpole) librations about the L_4 and L_5 points 60° behind or ahead of Jupiter and therefore never suffer a close approach to Jupiter. Another example of a protection mechanism provided by a resonance is the Hilda group of asteroids at Jupiter's 3:2 mean-motion resonance and the asteroid 279 Thule at the 4:3 resonance. The Hilda asteroids have a libration about 0° of their critical argument, $\sigma \equiv 3\lambda' - 2\lambda - \varpi$, where λ' is Jupiter's longitude, λ is the asteroid's longitude, and $\varpi = \omega + \Omega$ is the asteroid's longitude of perihelion. In this way, whenever the asteroid is in conjunction with Jupiter ($\lambda = \lambda'$), the asteroid is close to perihelion ($\lambda \approx \varpi$), and well away from Jupiter.

Using resonances to explain the *gaps* in the main asteroid belt and the general depletion of the outer belt proves to be more difficult than understanding the protection mechanisms at other resonances. A feature subject to much investigation has been the gap at the 3:1 mean-motion resonance. Early investigations found that most orbits starting at small eccentricity were regular and showed very little variation in eccentricity or semimajor axis over timescales of 50,000 yr. Investigation of more realistic models was limited because of the amount of computer time needed to follow orbits over longer periods of time. A breakthrough occurred when Jack Wisdom devised an algebraic mapping of phase space onto itself with a resonant structure similar to the 3 : 1 commensurability. His surprising result was that an orbit near the resonance could maintain a low eccentricity ($e < 0.1$) for nearly a million years and then have a sudden increase in eccentricity to $e > 0.3$.

Any doubts about this apparent chaotic behavior were later dispelled by numerical integrations and the measurement of a nonzero maximum Lyapanov exponent (as defined in Section III, A) and by a semianalytic perturbation theory that explained many of the features found in the mappings. We simply note here a result that illustrates an important feature that often occurs in simulations to be discussed later—a particle can remain in a low-eccentricity state for hundreds of Lyapunov times before "jumping" relatively quickly to high eccentricity.

The outer boundaries of the chaotic zone as determined by Wisdom's work coincide well with the boundaries of the 3 : 1 Kirkwood gap as shown in the numbered minor planets and the Palomar–Leiden survey. Since asteroids that begin on near-circular orbits in the gap acquire sufficient eccentricities to cross the orbit of Mars and in some cases Earth, the perturbative effects of these two planets are believed capable of clearing out the 3 : 1 gap over the age of the solar system. Moreover, collisions and chaotic diffusion near the edge of the 3 : 1 gap provide a source of meteorites arriving near Earth at the present epoch. [*See* ASTEROIDS; METEORITES.]

E. EXAMPLES OF RESONANCES: RING PARTICLES AND SHEPHERDING

In our discussions in Section II, we described the gravitational force produced by a spherically symmetric body. In this section we must include the effects of deviations from spherical symmetry in computing the force. This is most conveniently done by introducing the gravitational potential $\Phi(\mathbf{r})$, which is defined such that the acceleration $d^2\mathbf{r}/dt^2$ of a particle in the gravitational field is

$$d^2\mathbf{r}/dt^2 = -\nabla\Phi \tag{23}$$

In empty space, the Newtonian gravitational potential $\Phi(\mathbf{r})$ always satisfies Laplace's equation

$$\nabla^2\Phi = 0 \tag{24}$$

Most planets are very nearly axisymmetric, with the major departure from sphericity being due to rotationally induced polar flattening. Thus, the gravitational potential can be expanded in terms of Legendre polynomials instead of the complete spherical harmonic expansion, which would be required for the potential of a body of arbitrary shape:

$$\Phi(r, \phi, \theta) = -\frac{Gm}{r}\left[1 - \sum_{n=2}^{\infty} J_n P_n(\cos\theta)(R/r)^n\right] \tag{25}$$

In this equation we use standard spherical coordinates, so that θ is the angle between the planet's symmetry axis and the vector to the particle. The terms $P_n(\cos\theta)$ are the Legendre polynomials, and J_n are the gravitational moments determined by the planet's mass distribution. If the planet's distribution is symmetrical about the planet's equator, the J_n are zero for odd n. For large bodies, J_2 is generally significantly larger than the other gravitational moments.

Let us consider a particle in Saturn's rings, which revolves around the planet on a circular orbit in the equatorial plane ($\theta = 90°$) at a distance r from the center of the planet. The centripetal force must be provided by the radial component of the planet's gravitational force [cf. Eq. (9)], so the particle's angular velocity $\tilde{\Omega}$ satisfies

$$r\tilde{\Omega}^2(r) = \left[\frac{\partial\Phi}{\partial r}\right]_{\theta=90°} \tag{26}$$

If the particle suffers an infinitesimal displacement from its circular orbit, it will oscillate freely in the horizontal and vertical directions about the reference circular orbit with radial (epicyclic) frequency $\kappa(r)$ and vertical frequency $\mu(r)$, respectively, given by

$$\kappa^2(r) = r^{-3}\frac{d}{dr}[(r^2\tilde{\Omega})^2] \tag{27}$$

$$\mu^2(r) = \left[\frac{\partial^2\Phi}{\partial\theta^2}\right]_{\theta=90°} \tag{28}$$

From Eqs. (24)–(28) we find the following relation between the three frequencies for a particle in the equatorial plane:

$$\mu^2 = 2\tilde{\Omega}^2 - \kappa^2 \tag{29}$$

For a perfectly spherically symmetric planet, $\mu = \kappa = \tilde{\Omega}$. Since Saturn and the other ringed planets are oblate, μ is slightly larger and κ is slightly smaller than the orbital frequency $\tilde{\Omega}$. Resonances occur where the radial (or vertical) frequency of the ring particles is equal to the frequency of a component of a satellite's horizontal (or vertical) forcing, as sensed in the rotating frame of the particle. In this case the resonating particle is always near the same phase in its radial (vertical) oscillation when it experiences a particular phase of the satellite's forcing. This situation enables continued coherent "kicks" from the satellite to build up the particle's radial (vertical) motion, and significant forced oscillations may thus result. The location and strengths of resonances with any given moon can be determined by decomposing the gravitational potential of the moon into its Fourier components. The disturbance frequency, $\tilde{\omega}$, can be written as the sum of integer multiples of the satellite's angular, vertical, and radial frequencies:

$$\tilde{\omega} = l\tilde{\Omega}_s + n\mu_s + p\kappa_s \qquad (30)$$

where the azimuthal symmetry number, l, is a nonnegative integer, and n and p are integers, with n being even for horizontal forcing and odd for vertical forcing. The subscript s refers to the satellite. A particle placed at distance $r = r_L$ will undergo horizontal (Lindblad) resonance if r_L satisfies

$$\tilde{\omega} - l\tilde{\Omega}(r_L) = \pm\kappa(r_L) \qquad (31)$$

It will undergo vertical resonance if its radial position r_v satisfies

$$\tilde{\omega} - l\tilde{\Omega}(r_v) = \pm\mu(r_v) \qquad (32)$$

When Eq. (31) is valid for the lower (upper) sign, we refer to r_L as the inner (outer) Lindblad or horizontal resonance. The distance r_v is called an inner (outer) vertical resonance if Eq. (32) is valid for the lower (upper) sign. Since all of Saturn's large satellites orbit the planet well outside the main ring system, the satellite's angular frequency $\tilde{\Omega}_s$ is less than the angular frequency of the particle, and inner resonances are more important than outer ones. When $m \neq 1$, the approximation $\mu \approx \tilde{\Omega} \approx \kappa$ may be used to obtain the ratio

$$\frac{\tilde{\Omega}(r_{L,v})}{\tilde{\Omega}_s} = \frac{l + n + p}{l - 1} \qquad (33)$$

The notation $(l + n + p)/(l - 1)$ or $(l + n + p):(l - 1)$ is commonly used to identify a given resonance.

The strength of the forcing by the satellite depends, to lowest order, on the satellite's eccentricity, e, and

inclination, i, as $e^{|p|}[\sin i]^{|n|}$. The strongest horizontal resonances have $n = p = 0$, and are of the form $l:(l - 1)$. The strongest vertical resonances have $n = 1$, $p = 0$, and are of the form $(l + 1):(l - 1)$. The location and strengths of such orbital resonances can be calculated from known satellite masses and orbital parameters and Saturn's gravity field. Most strong resonances in the Saturnian system lie in the outer A ring near the orbits of the moons responsible for them. If $\tilde{\Omega} = \mu = \kappa$, the inner horizontal and vertical resonances would consider: $r_L = r_v$. Since, owing to Saturn's oblateness, $\mu > \tilde{\Omega} > \kappa$, the positions r_L and r_v do not coincide: $r_v < r_L$. A detailed discussion of spiral density waves, spiral bending waves, and gaps at resonances produced by moons is presented elsewhere in this encyclopedia. [See PLANETARY RINGS.]

1. Precession of Particle Orbits

Using Eqs. (24)–(29), one can show that the orbital and epicyclic frequencies can be written as

$$\tilde{\Omega}^2 = \frac{GM}{r^3}\left[1 + \frac{3}{2}J_2\left(\frac{R}{r}\right)^2 - \frac{15}{8}J_4\left(\frac{R}{r}\right)^4\right.$$
$$\left. + \frac{35}{16}J_6\left(\frac{R}{r}\right)^6 + \cdots\right] \qquad (34)$$

$$\kappa^2 = \frac{GM}{r^3}\left[1 - \frac{3}{2}J_2\left(\frac{R}{r}\right)^2 + \frac{45}{8}J_4\left(\frac{R}{r}\right)^4\right.$$
$$\left. - \frac{175}{16}J_6\left(\frac{R}{r}\right)^6 + \cdots\right] \qquad (35)$$

$$\mu^2 = \frac{GM}{r^3}\left[1 + \frac{9}{2}J_2\left(\frac{R}{r}\right)^2 - \frac{75}{8}J_4\left(\frac{R}{r}\right)^4\right.$$
$$\left. + \frac{245}{16}J_6\left(\frac{R}{r}\right)^6 + \cdots\right] \qquad (36)$$

Thus, the oblateness of a planet causes apsides of particle orbits in and near the equatorial plane to precess in the direction of the orbit and lines of nodes of nearly equatorial orbits to regress.

IV. DISSIPATIVE FORCES AND THE ORBITS OF SMALL BODIES

The foregoing sections describe the gravitational interactions between the Sun, planets, and moons. We have ignored solar radiation, which is an important force for small particles in the solar system. We can distinguish three effects: (1) the radiation pressure, which pushes particles primarily outward from the Sun (micron-sized dust); (2) the Poynting–Robertson drag, which causes centimeter-sized particles to spiral inward toward the Sun; and (3) the Yarkovski effect, which changes the orbits of meter- to kilometer-sized objects owing to uneven temperature distributions at their surfaces. We discuss each of these effects in the next three subsections and then examine the effect of gas drag. In the final subsection we discuss the influence of tidal interactions, which (in contrast to the other dissipative effects described in this section) is most important for larger bodies such as moons and planets. [See INTERPLANETARY DUST AND THE ZODIACAL CLOUD.]

A. RADIATION FORCE (MICRON-SIZED PARTICLES)

The Sun's radiation exerts a force, F_r, on all other bodies of the solar system. The magnitude of this force is

$$F_r = \frac{L_\odot A}{4\pi c r^2} Q_{pr} \qquad (37)$$

where A is the particle's geometric cross section, $L_\odot$ is the solar luminosity, c is the speed of light, r is the heliocentric distance, and Q_{pr} is the radiation pressure coefficient, which is equal to unity for a perfectly absorbing particle and is of order unity unless the particle is small compared to the wavelength of the radiation. We define the parameter β as the ratio between the forces due to the radiation pressure and the Sun's gravity:

$$\beta \equiv \frac{F_r}{F_g} = 5.7 \times 10^{-5} \frac{Q_{pr}}{\rho R} \qquad (38)$$

where the radius, R, and the density, ρ, of the particle are in c.g.s. units. Note that β is independent of heliocentric distance and that the solar radiation force is important only for micron- and submicron-sized particles. Using the parameter β, we can write a more general expression for the effective gravitational attraction:

$$F_{geff} = \frac{-(1 - \beta)GmM_\odot}{r^2} \qquad (39)$$

that is, the small particles "see" a Sun of mass $(1 - \beta)M_\odot$. It is thus clear that small particles with $\beta > 1$ are in effect repelled by the Sun, and thus quickly escape the solar system, unless they are gravitationally bound to one of the planets. Dust released at the Keplerian velocity from bodies on circular orbits is ejected from the solar system if $\beta > 0.5$.

The importance of solar radiation pressure can be seen, for example, in comets: cometary tails always point in the antisolar direction due to the Sun's radiation pressure. The dust tails are curved rather than straight as a result of the continuous ejection of dust grains from the comet, which itself is on an elliptical orbit around the Sun. [See COMETARY DYNAMICS; PHYSICS AND CHEMISTRY OF COMETS.]

B. POYNTING–ROBERTSON DRAG (CENTIMETER-SIZED GRAINS)

A small particle in orbit around the Sun absorbs solar radiation and reradiates the energy isotropically in its own frame. The particle thereby preferentially radiates (and loses momentum) in the forward direction in the inertial frame of the Sun. This leads to a decrease in the particle's energy and angular momentum and causes dust in bound orbits to spiral sunward. This effect is called the Poynting–Robertson drag.

The net force on a rapidly rotating dust grain is given by

$$\mathbf{F}_{rad} \approx \frac{L_\odot Q_{pr} A}{4\pi c r^2}\left[\left(1 - \frac{2v_r}{c}\right)\hat{\mathbf{r}} - \frac{v_\theta}{c}\hat{\theta}\right] \qquad (40)$$

The first term in Eq. (40) is that due to radiation pressure and the second and third terms (those involving the velocity of the particle) represent the Poynting–Robertson drag.

From this discussion, it is clear that small-sized dust grains in the interplanetary medium disappear: (sub)-micron sized grains are blown out of the solar system, whereas larger particles spiral inward toward the Sun. Typical decay times (in years) for circular orbits are given by

$$\tau_{P-R} \approx 400 \frac{r^2}{\beta} \qquad (41)$$

with the distance r in AU. However, particles with radii greater than about 50 μm are destroyed by mutual collisions before they spiral close enough to the Sun for them to evaporate.

Particles that produce the bulk of the zodiacal light (at infrared and visible wavelengths) are between 20 and 200 μm, so their lifetimes at Earth orbit are on the order of 10^5 yr, which is much less than the age of the solar system. A possible source for the dust grains is the asteroid belt, where numerous collisions occur between countless small asteroids.

C. YARKOVSKI EFFECT (METER-SIZED OBJECTS)

Consider a rotating body heated by the Sun. The evening hemisphere is typically warmer than the morning hemisphere, by an amount $\Delta T \ll T$. Let us assume that the temperature of the morning hemisphere is $T - \Delta T/2$, and that of the evening hemisphere $T + \Delta T/2$. The radiation reaction upon a surface element dA, normal to its surface, is $dF = 2\sigma T^4 dA/3c$. For a spherical particle of radius R, the Yarkovski force in the orbit plane due to the excess emission on the evening side is

$$F_Y = \frac{8}{3}\pi R^2 \frac{\sigma T^4}{c}\frac{\Delta T}{T}\cos\psi \qquad (42)$$

where σ is the Stefan–Boltzmann constant and ψ is the particle's obliquity, that is, the angle between its rotation axis and orbit pole. The reaction force is positive for an object that rotates in the prograde direction, $0 < \psi < 90°$, and negative for an object with retrograde rotation, $90° < \psi < 180°$. In the latter case, the force enhances the Poynting–Robertson drag.

The Yarkovski force is important for bodies in the meter-sized range. Asymmetric outgassing from comets produces a nongravitational force similar in form to the Yarkovski force. This effect is discussed in greater detail in the chapter by Julio Fernandez. [See COMETARY DYNAMICS.]

D. GAS DRAG

Although interplanetary space generally can be considered an excellent vacuum, there are certain situations in planetary dynamics where interactions with gas can significantly alter the motion of solid particles. Two prominent examples of this process are planetesimal interactions with the gaseous component of the protoplanetary disk and orbital decay of ring particles as a result of drag caused by extended planetary atmospheres.

In the laboratory, gas drag slows solid objects down until their positions remain fixed relative to the gas. In the planetary dynamics case, the situation is more complicated. For example, a body on a circular orbit about a planet loses mechanical energy as a result of drag with a static atmosphere, but this energy loss leads to a decrease in semimajor axis of the orbit, which implies that the body actually speeds up! Other, more intuitive effects of gas drag are the damping of eccentricities and, in the case where there is a preferred plane in which the gas density is the greatest, the damping of inclinations relative to this plane.

Objects whose dimensions are larger than the mean free path of the gas molecules experience Stokes' drag,

$$F_D = -C_D A\rho v^2/2 \qquad (43)$$

where v is the relative velocity of the gas and the body, ρ is the gas density, A is the projected surface area of the body, and C_D is a dimensionless drag coefficient, which is of order unity unless the Reynolds number is very small. Smaller bodies are subject to Epstein drag,

$$F_D = -A\rho v v' \qquad (44)$$

where v' is the mean thermal velocity of the gas. Note that as the drag force is proportional to surface area and the gravitational force is proportional to volume (for constant particle density), gas drag is usually most important for the dynamics of small bodies.

The gaseous component of the protoplanetary disk is believed to have been partially supported against the gravity of the Sun by a negative pressure gradient in the radial direction. Thus, less centrifugal force was required to complete the balance, and consequently the gas orbited less rapidly than the Keplerian velocity. The "effective gravity" felt by the gas is

$$g_{eff} = -\frac{GM_\odot}{r^2} - (1/\rho_{gas})\frac{dP}{dr} \qquad (45)$$

To maintain a circular orbit, the effective gravity must be balanced by centrifugal acceleration, $r\tilde{\Omega}^2$. For estimated protoplanetary disk parameters, the gas rotates $\approx 0.5\%$ slower than the Keplerian speed.

Large particles moving at (nearly) the Keplerian speed thus encountered a headwind, which removed part of their angular momentum and caused them to spiral inward toward the Sun. Inward drift was greatest for mid-sized particles, which have large ratios of surface area to mass yet still orbit with nearly Keplerian velocities. The effect diminishes for very small particles, which are so strongly coupled to the gas that the headwind they encounter is very slow. Peak rates of inward drift occur for particles that collide with roughly their own mass of gas in one orbital period. Meter-sized bodies in the inner solar nebula drift inward at a rate of up to $\approx 10^6$ km/yr! Thus, the material that survives to form the planets must complete the transition from centimeter to kilometer size rather quickly, unless it is confined to a thin dust-dominated subdisk in which the gas is dragged along at essentially the Keplerian velocity.

Drag induced by a planetary atmosphere is even more effective for a given density, as atmospheres are almost entirely pressure supported so the relative velocity between the gas and particles is high. As atmospheric densities drop rapidly with height, particles decay slowly at first, but as they reach lower altitudes, their decay can become very rapid. Gas drag is the principal cause of orbital decay of artificial satellites in low Earth orbit.

E. TIDAL INTERACTIONS AND PLANETARY SATELLITES

Tidal forces are important to many aspects of the structure and evolution of planetary bodies:

(1) On short timescales, temporal variations in tides (as seen in the frame rotating with the body under consideration) cause stresses that can move fluids with respect to more rigid parts of the planet (e.g., the ocean tides with which we are familiar) and even cause seismic disturbances (though the evidence that the Moon causes some earthquakes is weak and disputable, it is clear that the tides raised by Earth are a major cause of moonquakes).

(2) On long timescales, tides cause changes in the orbital and spin properties of planets and moons. Tides also determine the equilibrium shape of a body located near any massive body; note that many materials that behave as solids on human timescales are effectively fluids on very long geological timescales (e.g., Earth's mantle).

The gravitational attraction of the Moon and Earth on each other causes tidal bulges that rise in a direction close to the line joining the centers of the two bodies. Particles on the nearside of the body experience gravitational forces that exceed the centripetal force, whereas particles on the far side experience gravitational forces that are less than the centripetal forces needed for motion in a circle. It is the gradient of the gravitational force across the body that gives rise to the double tidal bulge.

The Moon spins once per orbit, so that the same face of the Moon always points toward Earth and the Moon is always elongated in that direction. Earth, however, rotates much faster than the Earth–Moon orbital period. Thus, different parts of Earth point toward the Moon and are tidally stretched. If Earth were perfectly fluid, the tidal bulges would respond immediately to the varying force, but the finite response time of Earth's figure causes the tidal bulge to lag behind, at the point where the Moon was overhead slightly earlier. Since Earth rotates faster than the Moon orbits, this "tidal lag" on Earth leads the position of the Moon in inertial space. As a result, the tidal bulge of Earth accelerates the Moon in its orbit. This causes the Moon to slowly spiral outward. The Moon slows down Earth's rotation by pulling back on the tidal bulge, so the angular momentum in the system is conserved. This same phenomenon has caused most, if not all, major moons to be in synchronous rotation: the rotation and orbital periods of these bodies are equal. In the case of the Pluto–Charon system, the entire system is locked in a synchronous rotation and revolution of 6.4 days. Satellites in retrograde orbits (e.g., Triton) or satellites whose orbital periods are less than the planet's rotation period (e.g., Phobos) spiral inward toward the planet as a result of tidal forces.

Mercury orbits the Sun in 88 days and rotates around its axis in 59 days, a 3 : 2 spin–orbit resonance. Hence, at every perihelion one of two locations is pointed at the Sun: the subsolar longitude is either 0° or 180°. This configuration is stable because Mercury has both a large orbital eccentricity and a significant permanent deformation that is aligned with the solar direction at perihelion. Indeed, at 0° longitude we find a large impact crater, Caloris Planitia.

(3) Under special circumstances, strong tides can have significant effects on the physical structure of bodies. Generally, the strongest tidal forces felt by solar system bodies (other than Sun-grazing or planet-grazing comets) are those caused by planets on their closest satellites. Near a planet, tides are so strong that they rip a fluid (or weakly aggregated solid) body apart. In such a region, large moons are unstable, and even small moons, which could be held together by inertial strength, are unable to accrete because of tides. The boundary of this region is known as *Roche's limit*. Inside Roche's limit, solid material remains in the form of small bodies and we see rings instead of large moons.

The closer a moon is to a planet, the stronger is the tidal force to which it will be subjected. Let us consider Roche's limit for a spherical satellite in synchronous rotation at a distance r from a planet. This is the distance at which a loose particle on an equatorial subplanet point just remains gravitationally bound to the satellite. At the center of the satellite of mass m and radius R_s, a particle would be in equilibrium and so

$$\frac{GM}{r^2} = \tilde{\Omega}^2 r \qquad (46)$$

where $M\,(\geqslant m)$ is the mass of the planet. However, at the equator, the particle will experience (i) an excess gravitational or centrifugal force due to the planet, (ii) a centrifugal force due to rotation, and (iii) a gravitational force due to the satellite. If the equatorial particle is *just* in equilibrium, these forces will balance and

$$-\frac{d}{dr}\left(\frac{GM}{r^2}\right)R_s + \tilde{\Omega}^2 r = \frac{Gm}{R_s^2} \qquad (47)$$

In this case, Roche's limit r_{Roche} is given by

$$r_{\text{Roche}} = 3^{1/3}\left(\frac{\rho_{\text{planet}}}{\rho_s}\right)^{1/3} R_{\text{planet}} \qquad (48)$$

with ρ the density for the planet and satellite, and R_{planet} the planetary radius. When a fluid moon is considered and flattening of the object due to gravity is taken into account, the correct result for a liquid moon (no internal strength) is

$$r_{\text{Roche}} = 2.45\left(\frac{\rho_{\text{planet}}}{\rho_s}\right)^{1/3} R_{\text{planet}} \qquad (49)$$

Most bodies have significant internal strength, which allows bodies with sizes $\lesssim 100$ km to be stable somewhat inside Roche's limit. Mars's satellite Phobos is well inside Roche's limit; it is subjected to a tidal force equivalent to that in Saturn's B ring.

(4) Internal stresses caused by variations in tides on a body in an eccentric orbit or not rotating synchronously with its orbital period can result in significant tidal heating of some bodies, most notably in Jupiter's moon Io. If no other forces were present, this would lead to a decay of Io's orbital eccentricity. By analogy to the Earth–Moon system, the tide raised on Jupiter by Io will cause Io to spiral outward and its orbital eccentricity to decrease. However, there exists a $2:1$ mean-motion resonant lock between Io and Europa. Io passes on some of the orbital energy and angular momentum it receives from Jupiter to Europa, and Io's eccentricity is increased as a result of this transfer. This forced eccentricity maintains a high tidal dissipation rate and large internal heating in Io, which displays itself in the form of active volcanism.

V. LONG-TERM STABILITY OF PLANETARY ORBITS

We turn now to one of the oldest problems in dynamical astronomy: whether or not the planets will continue indefinitely in nearly circular, nearly coplanar orbits.

A. SECULAR PERTURBATION THEORY

In Section III, we briefly described how celestial mechanics classically attempted to study the time-dependence of planetary orbits by introducing a disturbing function and employing perturbation theory. For very long term behavior, a fruitful approach (due to Lagrange and Laplace) involves averaging the disturbing function over the mean motions of the planets, resulting in what is known as the secular part of the disturbing function. If the disturbing function is further limited to terms of lowest order, the equations of motion of the orbital elements of the planets can be expressed as a coupled set of first-order linear differential equations. This system can then be diagonalized to find the proper modes, which are sinusoids, and the corresponding eigenfrequencies. The evolution of a given planet's orbital elements is, therefore, a sum of the proper modes. With the addition of higher-order terms, the equations are no longer linear; however, it is sometimes possible to find a solution of a form similar to the linear solution, except with shifted proper mode frequencies and terms involving combinations of the proper mode frequencies. We discuss the long-term validity of this method in the next section.

B. CHAOS AND PLANETARY MOTIONS

As we have described, Laplace and Lagrange showed that if the mutual planetary perturbations were calculated to first order in the masses, inclinations, and eccentricities, the orbits could indeed be described by a sum of periodic terms, indicating stability. Successive workers have shown that this is still the case if the perturbations are expanded to somewhat higher orders. It can be shown that such stable orbits would describe the solar system if the masses, eccentricities, and inclinations were sufficiently small. The real solar system, however, is far from satisfying these stringent requirements, so the question of its stability is unresolved. Indeed, the work of Poincaré in the late 1800s casts doubt on the long-term convergence of the various perturbation schemes. The problem with the perturbation expansion is that although the expansion is done in powers of small parameters, the existence of resonances between the planets introduces small divisors into the expansion terms. Such small divisors make high-order terms in the power series unexpectedly large and destroy the convergence of the series.

There are two separate points in the construction of the secular system at which resonances can cause nonconvergence of the expansion. The first is in averaging over mean motions. Mean-motion resonances between the planets can introduce small divisors, leading to divergences when forming the secular disturbing function. Second, there can be resonances between the proper mode frequencies, leading to problems when trying to solve the secular system using an expansion approach.

The analytical complexity of the perturbation techniques and the development of ever faster computers has led others to the investigation of stability by purely numerical models. Early integrations of the orbits of the four outer planets on million-year timescales compared well with perturbation calculations, showing quasi-periodic behavior for the four major outer planets. Pluto's behavior, however, was sufficiently different to inspire further study. It was found that the angle $3\lambda - 2\lambda_N - \varpi$ is in libration with a period of 20,000 yr, where λ and λ_N are the mean longitudes of Pluto and Neptune, respectively, and ϖ is the longitude of perihelion of Pluto. It was also shown that the argument of perihelion of Pluto librates with a period of 4 Myr and that the angles $\Omega-\Omega_N$ and $\varpi-\varpi_N$ seemed to be in resonance with this libration. All of these resonances act to prevent close encounters of Pluto with Neptune and hence protect the orbit of Pluto. However, numerical integrations performed by Gerald Sussman and Jack Wisdom of MIT show that Pluto's orbit is not quasi-periodic. There is evidence for the existence of very long period changes in Pluto's orbital elements and Sussman and Wisdom calculate a Lyapunov exponent of $(1/20\,\mathrm{Myr}^{-1})$. Data from the European LONGSTOP project suggests that Pluto is locked in a complicated system of three resonances and that the value of the Lyapunov exponent for its motion could be sensitive to the assumed initial conditions and planetary masses. Thus, a detailed understanding of Pluto's behavior is likely to be obtained only with the next generation of simulations.

Returning to the question of the stability of the planets other than Pluto, Jacques Laskar in Paris recently performed a critical test of the quasi-periodic hypothesis by numerically integrating the perturbations calculated to second order in mass and fifth order in eccentricities and inclinations. Such an expansion consists of about 150,000 polynomial terms. By numerically integrating the secular system, he avoids the small divisor problem caused by resonances between proper modes. He found the surprisingly high value of $(1/5\,\mathrm{Myr}^{-1})$. He argued that the exponential divergence is due to the transition from libration to circulation of the critical argument of a secular resonance related to the motions of perihelia and nodes of Earth and Mars. He also argued from his results that the chaotic nature of the inner solar system is robust against small variations in the initial conditions or in the model.

Laskar's important conclusions have recently been confirmed by other researchers using direct numerical calculations, although the underlying dynamical mechanism for the chaos has not been uniquely identified. The confirmation of such large Lyapunov exponents certainly suggests chaotic behavior. However, the apparent regularity of the motion of Earth and Pluto, and indeed the fact that the solar system has survived for 4.5 billion years, implies that the chaotic regions must be narrow. What the chaotic motion does mean is that there is a horizon of predictability for the detailed motions of the planets. Thus, the exponential divergence of orbits with a 4- to 5-Myr timescale shown by the calculations implies that an error as small as 10^{-10} in the initial conditions will lead to a 100% discrepancy in the longitudes of the planets in 100 Myr, although the principal orbital elements remain tightly bounded on that timescale. It is also worth bearing in mind the lessons learned from integration of test particle trajectories, namely, that the timescale for macroscopic changes in the system can be many orders of magnitude longer than the Lyapunov timescales. Thus the apparent stability of the current planetary system on billion-year timescales may simply be a manifestation of the fact that the solar system is in the chaotic sense a dynamically young system.

BIBLIOGRAPHY

Burns, J. A. (1987). The motion of interplanetary dust. *In* "The Evolution of the Small Bodies of the Solar System," pp. 252–275. Soc. Italiana di Fisica, Bologna, Italy.

Danby, J. M. A. (1992). "Fundamentals of Celestial Mechanics." Willmann–Bell, Richmond, Virginia.

Duncan, M., and Quinn, T. (1993). The long-term dynamical evolution of the solar system. *Annu. Rev. Astron. Astrophys.* **31**, 265–295.

Lissauer, J. J. (1993). Planet formation. *Annu. Rev. Astron. Astrophys.* **31**, 129–174.

Murray, C. D., and Dermott, S. F. (1998). "Solar System Dynamics." Cambridge University Press.

Peale, S. J. (1976). Orbital resonances in the solar system. *Annu. Rev. Astron. Astrophys.* **14**, 215–246.

CHAOTIC MOTION IN THE SOLAR SYSTEM

I. Chaotic Motion

II. Chaotic Rotation

III. Orbital Evolution of Minor Bodies

IV. Chaotic Behavior of the Planets

Carl D. Murray

Queen Mary and Westfield College,
University of London

GLOSSARY

Algebraic mapping: Mathematical technique for greatly increasing the speed with which gravitational interactions can be modeled on a computer. A mapping is usually derived by replacing the continuous gravitational effect of a perturbing body by a series of discrete impulses.

Chaotic motion: Motion that is sensitively dependent on initial conditions such that small changes in the starting position or velocity produce a dramatically different final state. Nearby orbits diverge exponentially in chaotic regions.

Deterministic system: Dynamical system in which the individual bodies move according to fixed laws described mathematically in the form of equations of motion. A deterministic system can still give rise to chaotic, unpredictable motion because of the finite precision with which any physical measurement or numerical computation can be made.

Equilibrium points: Those points in a dynamical system (usually considered in a rotating reference frame) where the velocity and acceleration are zero. Such points can be stable or unstable to small displacements.

Lyapunov exponent: Measure of the rate of divergence of two nearby trajectories in a system. A positive Lyapunov exponent is associated with

chaotic motion and its inverse gives an estimate of the timescale for exponential separation of nearby orbits.

Moment of inertia: Measure of the resistance of a body to changes in its rotational state. The moment of inertia increases with the mass and radius of an object.

Obliquity: Angle between a planet's equator and its orbital plane. Earth's current obliquity of 23.5° is sufficient to cause seasons.

Orbit–orbit resonance: Condition in which two solar system objects have orbital periods in the ratio of small integers. Orbit–orbit resonances are common in the satellite systems of Jupiter and Saturn.

Phase space: Higher-dimensional space in which the coordinates of a point are the components of the position and velocity of an object at a given time. The coordinates in phase space define the dynamical state of the object. A body moving in two or three dimensions of physical space has a phase space that is four or six dimensional, respectively.

Secular perturbations: Long-period variations in the orbital elements of a solar system body caused by the gravitational perturbations from other bodies. Secular perturbations can be studied analytically provided the eccentricities and inclinations of all the bodies are sufficiently small.

825

Separatrix: Boundary of a resonance separating resonant or "librating" motion inside the resonance from nonresonant or "circulating" motion outside.

Spin–orbit resonance: Simple numerical relationship between the spin period of a planet or satellite and its orbital period. Most natural satellites in the solar system are in the 1:1 spin–orbit resonance, also called the synchronous spin state.

Surface of section: Means of studying the regular or chaotic nature of an orbit by plotting a sequence of points in two dimensions that can represent all or part of the coordinates of the point in phase space.

Three-body problem: Problem of the motion of three bodies moving under their mutual gravitational attraction. In the restricted three-body problem, the third body is considered to have negligible mass such that it does not affect the motion of the other two bodies.

A n object in the solar system exhibits chaotic behavior in its orbit or rotation if the motion is sensitively dependent on the starting conditions, such that small changes in its initial state produce different final states. Possible examples of chaotic motion in the solar system include the rotation of the Saturnian satellite Hyperion, the orbital evolution of numerous asteroids and comets, and the orbit of Pluto. Numerical investigations suggest that the motion of the planetary system is chaotic although there are no signs of any gross instability in the orbits of the planets. Chaotic motion has probably played an important role in determining the dynamical structure of the solar system.

I. CHAOTIC MOTION
. . . .

A. CONCEPTS OF CHAOS

The equations of motion that describe the gravitational forces acting on the orbits of solar system bodies are relatively simple. In the Newtonian approximation, a point mass m_1 is attracted toward another point mass m_2 at a distance r by a force of attraction, with a magnitude, F, given by

$$F = \frac{Gm_1m_2}{r^2} \tag{1}$$

where G is the universal gravitational constant. If the initial positions and velocities of each body are known, then previous or future values can be determined by solving the resulting equations of motion. In the seventeenth century, Isaac Newton showed that in the case of a planet moving under the effect of the Sun's gravity, the force of attraction results in the planet describing an elliptical path about the Sun. The basic properties of the two-body problem are described in the chapter by Duncan and Lissauer. [*See* Solar System Dynamics.]

Since each planet in the solar system will also be affected by the gravitational attraction of other planets, all according to Eq. (1), a more complicated set of equations is required. Such a system is still deterministic, since the nonlinear differential equations governing its evolution can be written down explicitly and the current values of the various positions and velocities determine the past and future values. However, in the general case of three or more bodies, an analytical solution is either impractical or unobtainable.

In the nineteenth century, Henri Poincaré studied the mathematics of the circular restricted three-body problem. In this problem, one mass (the secondary) moves in a fixed, circular orbit about a central mass (the primary), while a test particle moves under the gravitational effect of both masses but does not perturb their orbits. From this work, Poincaré realized that despite the simplicity of the equations of motion, some solutions to the problem exhibit complicated behavior.

Poincaré's work in celestial mechanics provided the framework for the modern theory of nonlinear dynamics and ultimately led to a deeper understanding of the phenomenon of chaos, whereby dynamical systems described by simple equations can give rise to unpredictable behavior. One characteristic of such motion is that small changes in the starting conditions can produce vastly different final outcomes. Since all measurements of positions and velocities of objects in the solar system have finite accuracy, relatively small uncertainties in the initial state of the system can lead to large errors in the final state, for initial conditions that lie in chaotic regions in phase space.

Figure 1 is a schematic illustration of this phenomenon in the context of the three-body problem. Three chaotic trajectories are shown for the motion of a test particle in the vicinity of a planet. The motion of the particle and the planet are dominated by the gravitational attraction of a central star, but the region of space close to the planet is one where chaotic motion

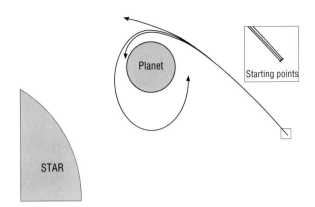

FIGURE 1 An illustration of the possible consequences that a small displacement in starting position could have for an object moving around a central star and being perturbed by a planet. The object could either impact the planet, go into temporary orbit around the planet, or have its orbit deflected slightly by the encounter. Such a dependence on initial conditions is one characteristic of chaotic motion. The diagram is schematic; in reality the actual orbits of the object are more complicated.

of the particle might be expected. Although all three particle trajectories are started close to one another, they each have dramatically different outcomes: one ends up in orbit about the planet, one impacts it, and the third is only slightly perturbed and continues to orbit the star. This is an example of what has become known as the "butterfly effect," first mentioned in the context of chaotic weather systems. It has been suggested that under the right conditions, a small atmospheric disturbance (such as the flapping of a butterfly's wings) in one part of the world could ultimately lead to a hurricane in another part of the world.

Not all examples of chaotic motion in the solar system are so dramatic; the changes in an orbit that reveal it to be chaotic may take place over several thousands or even millions of years. Although there have been a number of significant mathematical advances in the study of nonlinear dynamics since Poincaré's time, the digital computer has proved to be the most important tool in investigating chaotic motion in the solar system. This is particularly true in studies of the gravitational interaction of all the planets, where there are few analytical results. The ability to carry out long-term numerical investigations of orbital and rotational motion has led to a number of important discoveries in the last decade. Although chaotic behavior had been detected by a number of researchers in numerical experiments, the concept of chaotic motion was not widely appreciated and most had not realized that their simple sets of equations could give rise to such unusual behavior.

B. THE THREE-BODY PROBLEM AS A PARADIGM

The characteristics of chaotic motion are common to a wide variety of dynamical systems. In the context of the solar system, the general properties are best described by considering the circular restricted three-body problem, consisting of a massless test particle and two bodies of masses m_1 and m_2 moving in circular orbits about their common center of mass at constant separation with all bodies moving in the same plane. The test particle is attracted to each mass under the influence of the inverse square law of force given in Eq. (1). In Eq. (2), a is the constant separation of the two masses and n is their constant angular velocity about the center of mass. Kepler's third law gives

$$n^2 a^3 = G(m_1 + m_2) \tag{2}$$

where G is the universal gravitational constant. Using x and y as components of the position vector of the test particle referred to the center of mass of the system (Fig. 2), the equations of motion of the particle in a reference frame rotating at angular velocity n are

$$\ddot{x} - 2n\dot{y} - n^2 x = -G\left(m_1 \frac{x + \mu_2}{r_1^3} - m_2 \frac{x - \mu_1}{r_2^3} \right) \tag{3}$$

$$\ddot{y} + 2n\dot{x} - n^2 y = -G\left(\frac{m_1}{r_1^3} + \frac{m_2}{r_2^3} \right) y \tag{4}$$

where $\mu_1 = m_1 a/(m_1 + m_2)$, $\mu_2 = m_2 a/(m_1 + m_2)$ are constants and

$$r_1^2 = (x + \mu_2)^2 + y^2 \tag{5}$$

$$r_2^2 = (x - \mu_1)^2 + y^2 \tag{6}$$

where r_1 and r_2 are the distances of the test particle from the masses m_1 and m_2, respectively.

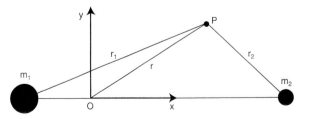

FIGURE 2 The rotating coordinate system used in the circular restricted three-body problem. The masses are at a fixed distance from one another and this is taken to be the unit of length. The position and velocity vectors of the test particle (at point P) are referred to the center of mass of the system at O.

These two second-order, coupled, nonlinear differential equations can be solved numerically provided the initial position (x_0, y_0) and velocity $(\dot{x}_0, \dot{y}_0)$ of the particle are known. Therefore the system is deterministic and at any given time the orbital elements of the particle (such as its semimajor axis and eccentricity) can be calculated from its instantaneous position and velocity.

Although the reality of the problem is slightly different because circular, coplanar orbits of natural bodies do not exist in the solar system, nevertheless it is a reasonable approximation for the study of some systems. For example, observed hierarchies of three bodies such as a Sun–planet–asteroid system or a planet–satellite–ring particle system are all cases where the smallest body (the asteroid or the ring particle) has negligible effect on the other two and the orbital eccentricity of the orbiting body (the planet or the satellite) is small.

The equations of motion given in Eqs. (3) and (4) differ from those of the full three-body problem in that the orbital energy and angular momentum of the system are no longer conserved. However, the system is still constrained by the existence of a constant of the motion called the Jacobi constant, C, given by

$$C = n^2(x^2 + y^2) + 2G\left(\frac{m_1}{r_1} + \frac{m_2}{r_2}\right) - \dot{x}^2 - \dot{y}^2 \quad (7)$$

The values of (x_0, y_0) and $(\dot{x}_0, \dot{y}_0)$ determine the value of C for the system and this value has to be preserved for all subsequent motion. At any instant the particle is at some position on the two-dimensional (x, y) plane. However, since the actual orbit is also determined by the components of the velocity $(\dot{x}, \dot{y})$, the particle can also be thought of as being at a particular position in a four-dimensional $(x, y, \dot{x}, \dot{y})$ space. This is referred to as the phase space of the system. Note that the use of four dimensions rather than the customary two is simply a means of representing the position *and* the velocity of the particle at a particular instant in time, and its motion is always restricted to the x–y plane. The existence of the Jacobi constant implies that the particle is not free to wander over the entire 4-D phase space, but rather that its motion is restricted to the 3-D "surface" defined by Eq. (7). This has an important consequence for studying the evolution of orbits in the problem.

The usual method is to solve the equations of motion, convert x, y, $\dot{x}$, and $\dot{y}$ into orbital elements such as semimajor axis, eccentricity, longitude of periapse, and mean longitude, and then plot the variation of these quantities as a function of time. However, another method is to produce a surface of section, also called a Poincaré map. This makes use of the fact that the orbit is always subject to Eq. (7), where C is determined by the initial position and velocity. Therefore if any three of the four quantities x, y, $\dot{x}$, and $\dot{y}$ are known, the fourth can always be determined by solving Eq. (7). One common surface of section that can be obtained for the circular restricted three-body problem is a plot of values of x and $\dot{x}$ whenever $y = 0$ and $\dot{y}$ is positive. The actual value of $\dot{y}$ can always be determined uniquely from Eq. (7), and so the two-dimensional $(x, \dot{x})$ plot implicitly contains all the information about the particle's location in the four-dimensional phase space. Although surfaces of section make it more difficult to study the evolution of the orbital elements, they have the advantage of revealing the characteristic motion of the particle (regular or chaotic) and a number of orbits can be displayed on the same diagram.

As an illustration of the different types of orbits that can arise, the results of integrating a number of orbits using a mass $m_2/(m_1 + m_2) = 10^{-3}$ and a value of the Jacobi constant $C = 3.07$ are described next. In each case the particle was started with the initial longitude of periapse $\varpi_0 = 0$ and initial mean longitude $\lambda_0 = 0$. This corresponds to $\dot{x} = 0$ and $y = 0$. Since the chosen mass ratio is comparable to that of the Sun–Jupiter system, and Jupiter's eccentricity is small, we will use this as a good approximation to the motion of fictitious asteroids moving around the Sun under the effect of gravitational perturbations from Jupiter. The asteroid is assumed to be moving in the same plane as Jupiter's orbit.

1. Regular Orbits

The first asteroid has starting values $x = 0.55$, $y = 0$, $\dot{x} = 0$, with $\dot{y} = 0.9290$ determined from the solution of Eq. (7). Here we use a set of dimensionless coordinates in which $n = 1$, $G = 1$, and $m_1 + m_2 = 1$. In these units the orbit of m_2 is a circle at distance $a = 1$ with uniform speed $v = 1$. The corresponding initial values of the heliocentric semimajor axis and eccentricity are $a_0 = 0.6944$ and $e_0 = 0.2065$. Since the semimajor axis of Jupiter's orbit is 5.202 AU, this value of a_0 would correspond to an asteroid at 3.612 AU.

Figure 3 shows the evolution of e as a function of time. The plot shows a regular behavior with the eccentricity varying from 0.206 to 0.248 over the course of the integration. In fact, an asteroid at this location would be close to an orbit–orbit resonance

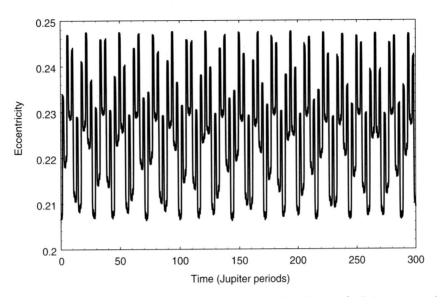

FIGURE 3   The eccentricity as a function of time for an object moving in a regular orbit near the 7 : 4 resonance with Jupiter. The plot was obtained by solving the circular restricted three-body problem numerically using initial values of 0.6944 and 0.2065 for the semimajor axis and eccentricity, respectively. The corresponding position and velocity in the rotating frame were $x_0 = 0.55$, $y_0 = 0$, $\dot{x} = 0$, and $\dot{y} = 0.9290$.

with Jupiter, where the ratio of the orbital period of the asteroid, T, to Jupiter's period, T_J, is close to a rational number. From Kepler's third law of planetary motion, $T^2 \propto a^3$. In this case, $T/T_J = (a/a_J)^{3/2} = 0.564 \approx 4/7$ and the asteroid orbit is close to a 7 : 4 resonance with Jupiter. Figure 4 shows the variation of the semimajor axis of the asteroid, a, over the same time interval as shown in Fig. 3. Although the changes in a are correlated with those in e, they are smaller in

amplitude and a appears to oscillate close to the location of the exact resonance at $a = (4/7)^{2/3} \approx 0.689$. An asteroid in resonance experiences enhanced gravitational perturbations from Jupiter, which can cause regular variations in its orbital elements. The extent of these variations depends on the asteroid's location within the resonance, which is, in turn, determined by the starting conditions.

The equations of motion can be integrated with

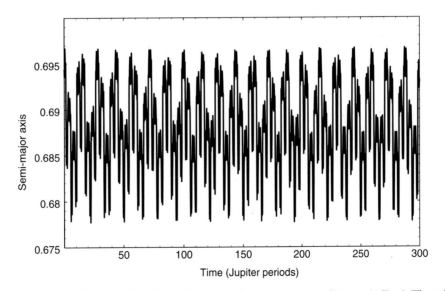

FIGURE 4   The semimajor axis as a function of time for an object using the same starting conditions as in Fig. 3. The units of the semimajor axis are such that Jupiter's semimajor axis (5.202 AU) is taken to be unity.

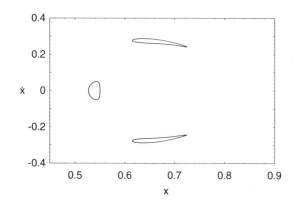

FIGURE 5 A surface of section plot for the same regular orbit as shown in Figs. 3 and 4. The 2000 points were generated by plotting the values of x and $\dot{x}$ whenever $y = 0$ with positive $\dot{y}$. The three "islands" in the plot are due to the third-order $7:4$ resonance.

the same starting conditions to generate a surface of section by plotting the values of x and $\dot{x}$ whenever $y = 0$ with $\dot{y} > 0$ (Fig. 5). The pattern of three distorted curves or "islands" that emerges is a characteristic of resonant motion when displayed in such plots. If a resonance is of the form $(p + q):p$, where p and q are integers, then q is said to be the order of the resonance. The number of islands seen in a surface of section plot of a given resonant trajectory is equal to q. In this case, $p = 4$, $q = 3$ and three islands are visible.

The center of each island would correspond to a starting condition that placed the asteroid at exact resonance where the variation in e and a would be minimal. Such points are said to be fixed points of the Poincaré map. If the starting location was moved farther away from the center, the subsequent variations in e and a would get larger, until eventually some starting values would lead to trajectories that were not in resonant motion.

2. Chaotic Orbits

Figures 6 and 7 show the plots of e and a as a function of time for an asteroid orbit with starting values $x_0 = 0.56$, $y_0 = 0$, $\dot{x}_0 = 0$, and $\dot{y}$ determined from Eq. (7) with $C = 3.07$. The corresponding orbital elements are $a_0 = 0.6984$ and $e_0 = 0.1967$. These values are only slightly different from those used earlier, indeed the initial behavior of the plots is quite similar to that seen in Figs. 3 and 4. However, subsequent variations in e and a are strikingly different. The eccentricity varies from 0.188 to 0.328 in an irregular manner and the value of a is not always close to the value associated with exact resonance. This is an example of a chaotic trajectory where the variations in the orbital elements have no obvious periodic or quasi-periodic structure.

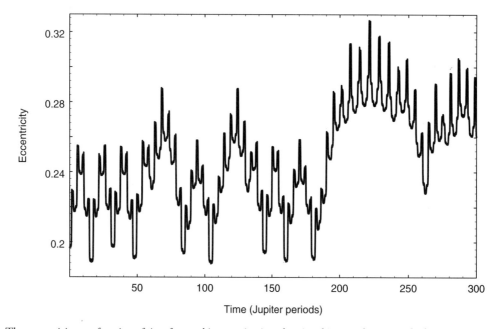

FIGURE 6 The eccentricity as a function of time for an object moving in a chaotic orbit started just outside the $7:4$ resonance with Jupiter. The plot was obtained by solving the circular restricted three-body problem numerically using initial values of 0.6984 and 0.1967 for the semimajor axis and eccentricity, respectively. The corresponding position and velocity in the rotating frame were $x_0 = 0.56$, $y_0 = 0$, $\dot{x} = 0$, and $\dot{y} = 0.8998$.

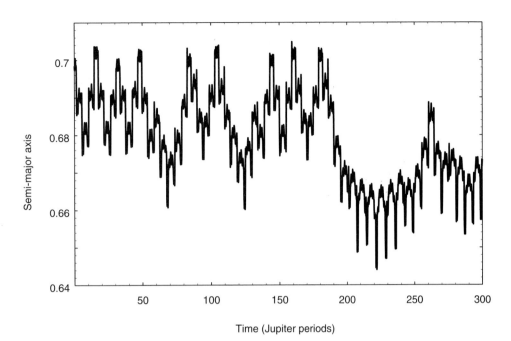

FIGURE 7 The semimajor axis as a function of time for an object using the same starting conditions as in Fig. 6. The units of the semimajor axis are such that Jupiter's semimajor axis (5.202 AU) is taken to be unity.

The anticorrelation of *a* and *e* can be explained in terms of the Jacobi constant.

The identification of this orbit as chaotic becomes apparent from a study of its surface of section (Fig. 8). Here it is clear that the orbit covers a much larger region of phase space than the previous example. Furthermore, the orbit does not lie on a smooth curve, but is beginning to fill an area of the phase space. The points also help to define a number of empty regions, one of which is clearly associated with the 7:4 resonance seen in the regular trajectory. There is also a tendency for the points to "stick" near the edges of the islands; this gives the impression of regular motion for short periods of time.

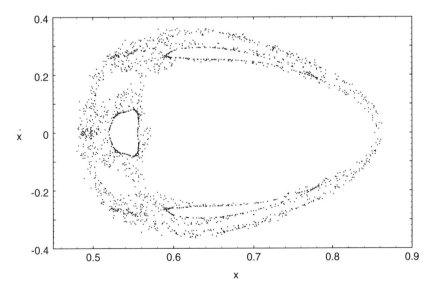

FIGURE 8 A surface of section plot for the same chaotic orbit as shown in Figs. 6 and 7. The 2000 points were generated by plotting the values of *x* and $\dot{x}$ whenever $y = 0$ with positive $\dot{y}$. The points are distributed over a much wider region of the $(x, \dot{x})$ plane than the points for the regular orbit shown in Fig. 5, and they help to define the edges of the regular regions associated with the 7:4 and other resonances.

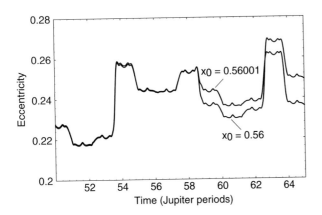

FIGURE 9 The variation in the eccentricity for two chaotic orbits started close to one another. One plot is part of Fig. 6 using the chaotic orbit started with $x_0 = 0.56$, and the other is for an orbit with $x_0 = 0.56001$. Although the divergence of the two orbits is exponential, the effect becomes noticeable only after 60 Jupiter periods.

Chaotic orbits have the additional characteristic that they are sensitively dependent on initial conditions. This is illustrated in Fig. 9, where the variation in e as a function of time is shown for two trajectories; the first corresponds to Fig. 6 (where $x_0 = 0.56$) and the second has $x_0 = 0.56001$. The initial value of $\dot{y}$ was chosen so that the same value of C was obtained. Although both trajectories show comparable initial variations in e, after 60 Jupiter periods it is clear that the orbits have drifted apart. Such a divergence would not occur for nearby orbits in a regular part of the phase space.

The rate of divergence of nearby trajectories in such numerical experiments can be quantified by monitoring the evolution of two orbits that are started close together. In a dynamical system such as the three-body problem, there are a number of quantities called the Lyapunov characteristic exponents. A measurement of the local divergence of nearby trajectories leads to an estimate of the largest of these exponents and this can be used to determine whether or not the system is chaotic. If two orbits are separated in phase space by a distance d_0 at time t_0, and d is their separation at time t, then the orbit is chaotic if

$$d = d_0 \exp \gamma(t - t_0) \qquad (8)$$

where γ is a positive quantity equal to the maximum Lyapunov characteristic exponent. In practice, γ can be estimated from the results of a numerical integration by writing

$$\gamma = \lim_{t \to \infty} \frac{\ln (d/d_0)}{t - t_0} \qquad (9)$$

and monitoring the behavior of γ with time. A plot of γ as a function of time on a log–log scale reveals a striking difference between regular and chaotic trajectories. For regular orbits, $d \approx d_0$ and a log–log plot has a slope of -1. However, if the orbit is chaotic, then γ tends to a positive value. This method may not always work because γ is defined only in the limit as $t \to \infty$ and sometimes chaotic orbits may give the appearance of being regular orbits for long periods of time by sticking close to the edges of the islands. If the nearby trajectory drifts too far from the original one, then γ is no longer a measure of the local divergence of the orbits. To overcome this problem, it helps to rescale the separation of the nearby trajectory at fixed intervals. Figure 10 shows $\log \gamma$ as a function of $\log t$ calculated using this method for the regular and chaotic orbits described here. This leads to an estimate of $\gamma = 10^{-0.77}$ (Jupiter periods)$^{-1}$ for the maximum Lyapunov characteristic exponent of the chaotic orbit. The corresponding Lyapunov time is given by $1/\gamma$, or in this case ~6 Jupiter periods. This indicates that for this starting condition the chaotic nature of the orbit quickly becomes apparent.

3. Location of Regular and Chaotic Regions

The extent of the chaotic regions of the phase space of a dynamical system can depend on a number of factors. In the case of the circular restricted three-

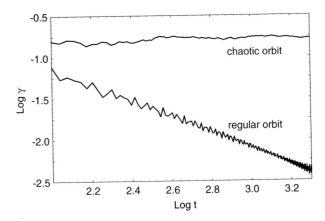

FIGURE 10 The evolution of the quantity γ [defined in Eq. (9)] as a function of time (in Jupiter periods) for a regular ($x_0 = 0.55$) and chaotic ($x_0 = 0.56$) orbit. In this log–log plot, the regular orbit shows a characteristic slope of -1 with no indication of $\log \gamma$ tending toward a finite value. However, in the case of the chaotic orbit, $\log \gamma$ tends to a limiting value close to -0.77.

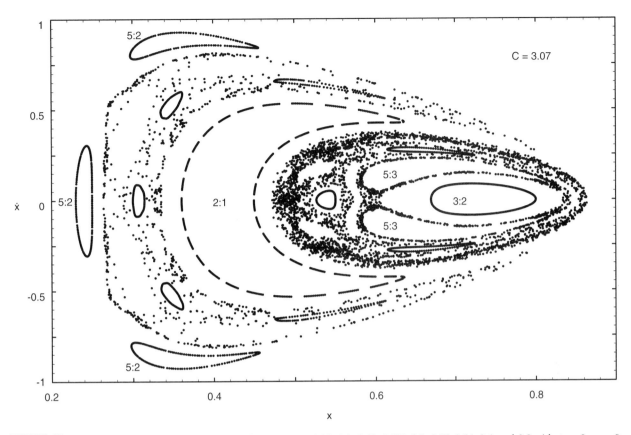

FIGURE 11 Representative surface of section plots for $x_0 = 0.25, 0.29, 0.3, 0.45, 0.475, 0.5, 0.55, 0.56, 0.6$, and 0.8 with $\dot{x}_0 = 0$, $y_0 = 0$, and Jacobi constant $C = 3.07$. Each trajectory was followed for a minimum of 500 crossing points. The plot uses the points shown in Figs. 5 and 8 (although the scales are different), as well as points from other regular and chaotic orbits. The major resonances are identified.

body problem, the critical quantities are the values of the Jacobi constant and the mass ratio μ_2. In Figs. 11 and 12, ten trajectories are shown for each of two different values of the Jacobi constant. In the first case (Fig. 11), the value is $C = 3.07$ (the same as the value used in Figs. 5 and 8), whereas in Fig. 12 it is $C = 3.13$. It is clear that the extent of the chaos is reduced in Fig. 12. The value of C in the circular restricted problem determines how close the asteroid can get to Jupiter. Larger values of C correspond to orbits with greater minimum distances from Jupiter. For the case $\mu_2 = 0.001$ and $C > 3.04$, it is impossible for their orbits to intersect, although the perturbations can still be significant.

Close inspection of the separatrices in Figs. 11 and 12 reveals that they consist of chaotic regions with regular regions on either side. As the value of the Jacobi constant decreases, the extent of the chaotic separatrices increases until the regular curves separating adjacent resonances are broken down and neighboring chaotic regions begin to merge. This can be thought of as the overlap of adjacent resonances giving

rise to chaotic motion. It is this process that permits chaotic orbits to explore regions of the phase space that are inaccessible to the regular orbits. In the context of the Sun–Jupiter–asteroid problem, this observation implies that asteroids in certain orbits are capable of large excursions in their orbital elements.

II. CHAOTIC ROTATION

A. SPIN–ORBIT RESONANCE

One of the dissipative effects of the tide raised on a natural satellite by a planet is to cause the satellite to evolve toward a state of synchronous rotation, where the rotational period of the satellite is approximately equal to its orbital period. Such a state is one example of a spin–orbit resonance, where the ratio of the spin period to the orbital period is close to a rational num-

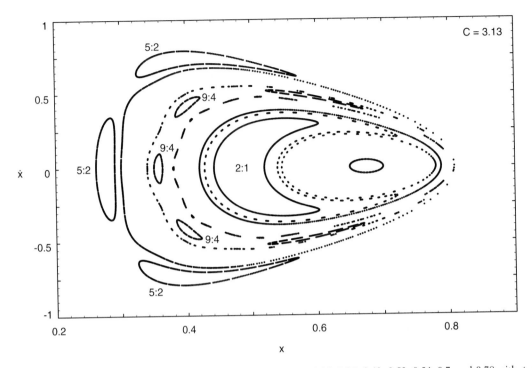

FIGURE 12   Representative surface of section plots for $x_0 = 0.262, 0.3, 0.34, 0.35, 0.38, 0.42, 0.52, 0.54, 0.7$, and 0.78 with $\dot{x}_0 = 0$, $y_0 = 0$, and Jacobi constant $C = 3.13$. Each trajectory was followed for a minimum of 500 crossing points. It is clear from a comparison with Fig. 11 that the phase space is more regular; chaotic orbits still exist for this value of C, but they are more difficult to find. The major resonances are identified.

ber. The time needed for a near-spherical satellite to achieve this state depends on its mass and orbital distance from the planet. Small, distant satellites take a longer time to evolve into the synchronous state than do large satellites that orbit close to the planet. Observations by spacecraft and ground-based instruments suggest that most natural satellites are in the synchronous spin state, in agreement with theoretical predictions.

A satellite in synchronous rotation will adopt a configuration such that its longest axis points in the approximate direction of the planet–satellite line. Let θ denote the angle between the long axis and the planet–satellite line in the planar case of a rotating satellite (Fig. 13). The variation of θ with time can be described by equating the time variation of the rotational angular momentum with the restoring torque. The resulting differential equation is

$$\ddot{\theta} + \frac{\omega_0^2}{2r^3}\sin 2(\theta - f) = 0 \qquad (10)$$

where ω_0 is a function of the principal moments of inertia of the satellite, r is the radial distance of the satellite from the planet, and f is the true anomaly (or

angular position) of the satellite in its orbit. The radius is an implicit function of time and is related to the true anomaly by the equation

$$r = \frac{a(1 - e^2)}{1 + e\cos f} \qquad (11)$$

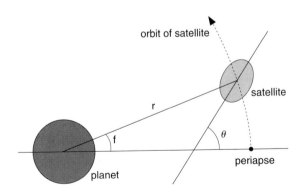

FIGURE 13   The geometry used to define the orientation of a satellite in orbit about a planet. The planet–satellite line makes an angle f (the true anomaly) with a reference line, which is taken to be the periapse direction of the satellite's orbit. The orientation angle, θ, of the satellite is the angle between its long axis and the reference direction.

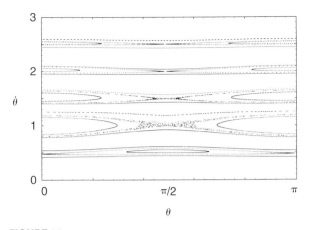

FIGURE 14 Representative surface of section plots of the orientation angle, θ, and its time derivative, $\dot{\theta}$, obtained from the numerical solution of Eq. (10) using $e = 0.1$ and $\omega_0 = 0.2$. The values of θ and $\dot{\theta}$ were obtained at every periapse passage of the satellite. Four starting conditions were integrated for each of the $1:2$, $1:1$, $3:2$, $2:1$, and $5:2$ spin–orbit resonances in order to illustrate motion inside, at the separatrix, and on either side of each resonance. The thickest "island" is associated with the strong $1:1$ spin–orbit state at $\dot{\theta} = 1$, whereas the thinnest is associated with the weak $5:2$ resonance at $\dot{\theta} = 2.5$.

where a and e are the constant semimajor axis and the eccentricity of the satellite's orbit, respectively, and the orbit is taken to be fixed in space. The relationship between r and the time t involves the solution of Kepler's equation. In Eq. (10) we have taken the unit of distance to be the semimajor axis of the satellite's orbit; in such a system the satellite's mean motion is unity and its orbital period is 2π time units.

Equation (10) defines a deterministic system where the initial values of θ and $\dot{\theta}$ determine the subsequent rotation of the satellite. Since θ and $\dot{\theta}$ define a unique spin position of the satellite, a surface of section plot of $(\theta, \dot{\theta})$ once every orbital period, say at every periapse passage, produces a picture of the phase space. Figure 14 shows the resulting surface of section plots for a number of starting conditions using $e = 0.1$ and $\omega_0 = 0.2$. The chosen values of ω_0 and e are larger than those that are typical for natural satellites, but they serve to illustrate the structure of the surface of section; large values of e are unusual since tidal forces also act to dampen eccentricity. The surface of section shows large, regular regions surrounding narrow islands associated with the $1:2$, $1:1$, $3:2$, $2:1$, and $5:2$ spin–orbit resonances at $\dot{\theta} = 0.5$, 1, 1.5, 2, and 2.5, respectively. The largest island is associated with the strong $1:1$ resonance and, although other spin states are possible, most regular satellites are observed to be in this state. Note the presence of diffuse collections of points associated with small chaotic regions at the separatrices of

the resonances. These are particularly obvious at the $1:1$ spin–orbit state at $\theta = \pi/2$, $\dot{\theta} = 1$. Although this is a completely different dynamical system compared to the circular restricted three-body problem, there are distinct similarities in the types of behavior visible in Fig. 14 and parts of Figs. 11 and 12.

In the case of near-spherical objects it is possible to investigate the dynamics of spin–orbit coupling using analytical techniques. The sizes of the islands shown in Fig. 14 can be estimated by expanding the second term in Eq. (10) and isolating the terms that will dominate at each resonance. Using such a method, each resonance can be treated in isolation and the gravitational effects of nearby resonances can be neglected. However, if a satellite is distinctly nonspherical, ω_0 can be large and this approximation is no longer valid. In such cases it is necessary to investigate the motion of the satellite using numerical techniques.

B. HYPERION

Hyperion is a satellite of Saturn that has an unusual shape (Fig. 15). It has approximate radial dimensions of $175 \times 120 \times 100$ km, an orbital eccentricity of 0.1, a semimajor axis of 24.55 Saturn radii, and a corresponding orbital period of 21.3 days. Such a small object at this distance from Saturn has a large tidal despinning timescale, but the unusual shape implies an estimated value of $\omega_0 = 0.89$.

The surface of section for a *single* trajectory is shown in Fig. 16 using the same scale as Fig. 14. It is clear that there is a large chaotic zone that encompasses most of the spin–orbit resonances. The islands associated with the synchronous and other resonances survive but in a much reduced form. Although this work assumes that Hyperion's spin axis remains perpendicular to its orbital plane, studies have shown that the satellite should also be undergoing a tumbling motion, such that its axis of rotation is not fixed in space.

Voyager observations of Hyperion indicated a spin period of 13 days, which suggested that the satellite was not in synchronous rotation. However, the standard techniques that are used to determine the period are not applicable if it varies on a timescale that is short compared with the timespan of the observations. In principle, the rotational period can be deduced from ground-based observations by looking for periodicities in plots of the brightness of the object as a function of time (the lightcurve of the object). The results of one such study for Hyperion are shown in Fig. 17. Since there is no recognizable periodicity, the lightcurve is consistent with that of an object undergoing

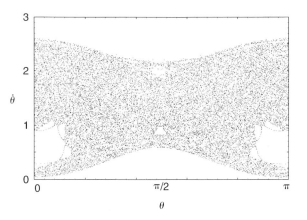

FIGURE 16 A single surface of section plot of the orientation angle, θ, and its time derivative, $\dot{\theta}$, obtained from the numerical solution of Eq. (10) using the values $e = 0.1$ and $\omega_0 = 0.89$, which are appropriate for Hyperion. The points cover a much larger region of the phase space than any of those shown in Fig. 14, and although there are some remaining islands of stability, most of the phase space is chaotic.

FIGURE 15 Three *Voyager 2* images of the Saturnian satellite Hyperion taken from distances of 1,200,000 km (top), 700,000 km (middle), and 500,000 km (bottom). The different aspects show the unusual shape of the satellite, which is one cause of its chaotic rotation. [Courtesy of National Space Science Data Center, team leader Dr. Bradford A. Smith (77-076A-01Q).]

orbital motion inside the resonance results, in part, in the extent of the chaos in its rotational motion. [*See* OUTER PLANET ICY SATELLITES.]

C. OTHER SATELLITES

Although there is no evidence that other natural satellites are undergoing chaotic rotation at the present time, it is possible that several irregularly shaped satellites did experience chaotic rotation at some time in

chaotic rotation. It is likely that Hyperion is the first natural satellite that has been observed to have a chaotic spin state. Numerical studies of Hyperion's rotation in three dimensions suggest that its spin axis does not point in a fixed direction. Therefore the satellite also undergoes a tumbling motion in addition to its chaotic rotation. An improved model of the shape of Hyperion and a more detailed investigation of its rotational dynamics have shown that while its spin is formally chaotic, it can give the appearance of being in a regular state for long periods of time. This may be consistent with the *Voyager* observations.

The dynamics of Hyperion's motion is complicated by the fact that it is in a 4 : 3 orbit–orbit resonance with the larger Saturnian satellite Titan. Although tides act to decrease the eccentricities of satellite orbits, Hyperion's eccentricity is maintained at 0.104 by means of the resonance. Titan effectively forces Hyperion to have this large value of e and so the apparently regular

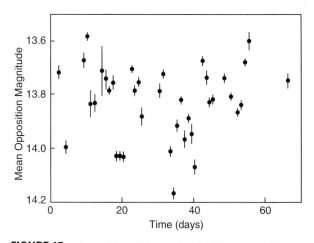

FIGURE 17 Ground-based observations by J. Klavetter of Hyperion's lightcurve obtained over 13 weeks (4.5 orbital periods) in 1987. The fact that there is no obvious curve through the data points is convincing evidence that the rotation of Hyperion is chaotic. (Courtesy of the American Astronomical Society.)

their histories. In particular, since satellites have to cross chaotic separatrices before capture into synchronous rotation can occur, they must have experienced some episode of chaotic rotation. This may also have occurred if the satellite suffered a large impact that affected its rotation. Such episodes could have induced significant internal heating and resurfacing events in some satellites. The Martian moon Phobos and the Uranian moon Miranda have been mentioned as possible candidates for this process. If this happened early in the history of the solar system, then the evidence may well have been obliterated by subsequent cratering events. There is also the possibility that tidal heating associated with capture in orbit–orbit resonances could lead to significant resurfacing. This process is currently taking place on Io, one of the Galilean moons of Jupiter. [*See* Io; Phobos and Deimos.]

III. ORBITAL EVOLUTION OF MINOR BODIES

A. ASTEROIDS

With more than 8500 accurately determined orbits and one major perturber (the planet Jupiter), the asteroids provide a natural laboratory in which to study the consequences of regular and chaotic motion. Using suitable approximations, asteroid motion can be studied analytically in some special cases. However, it is frequently necessary to resort to numerical integration. [*See* Asteroids.]

Investigations have shown that a number of asteroids have orbits that result in close approaches to planets. Of particular interest are asteroids such as 433 Eros, 1033 Ganymed, and 4179 Toutatis, because they are on orbits that bring them close to Earth. One of the most striking examples of the butterfly effect (see Section I, A) in the context of orbital evolution is the orbit of asteroid 2060 Chiron, which has a perihelion inside Saturn's orbit and an aphelion close to Uranus's orbit. Numerical integrations based on the best available orbital elements show that it is impossible to determine its past or future orbit with any degree of certainty since it frequently suffers close approaches to Saturn and Uranus. In such circumstances, the outcome is strongly dependent on the initial conditions as well as the accuracy of the numerical method. These are the characteristic signs of a chaotic orbit. By inte-

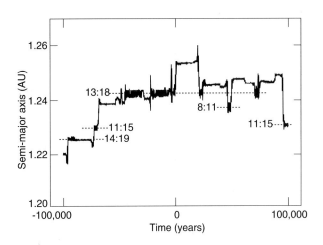

FIGURE 18 A plot of the semimajor axis of the near-Earth asteroid 1620 Geographos over a backward and forward integration of 100,000 years starting in 1986. Under perturbations from the planets, Geographos moves in a chaotic orbit and gets temporarily trapped in a number of high-order, orbit–orbit resonances (indicated in the diagram) with Earth. The data are taken from a numerical study of planet-crossing asteroids undertaken by A. Milani and co-workers. (Courtesy of Academic Press.)

grating several orbits with initial conditions close to the nominal values, it is possible to carry out a statistical analysis of the orbital evolution. Studies suggest that there is a 1 in 8 chance that Saturn will eject Chiron from the solar system on a hyperbolic orbit, while there is a 7 in 8 chance that it will evolve toward the inner solar system and come under strong perturbations from Jupiter. Telescopic observations of a faint coma surrounding Chiron suggest that it may be a comet rather than an asteroid and that perhaps its future orbit will resemble that of a short-period comet of the Jupiter family.

Numerical studies of the orbital evolution of planet-crossing asteroids under the effects of perturbations from all the planets have shown a remarkable complexity of motion for some objects. For example, the Earth-crossing asteroid 1620 Geographos gets trapped temporarily in a number of resonances with Earth in the course of its chaotic evolution (Fig. 18).

A histogram of the number distribution of asteroid orbits in semimajor axis (Fig. 19) shows that apart from a clustering of asteroids near Jupiter's semimajor axis at 5.2 AU, there is an absence of objects within 0.75 AU of the orbit of Jupiter. The objects in the orbit of Jupiter are called the Trojan asteroids, which move around the Lagrangian equilibrium points, which are located 60° ahead of and behind Jupiter. These asteroids are in a stable 1 : 1 resonance with Jupiter and are protected from having close approaches to the planet.

The cleared region near Jupiter's orbit can be un-

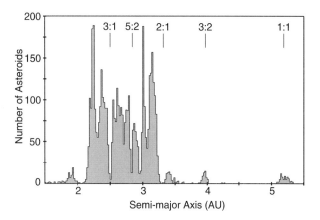

FIGURE 19 A histogram of the distribution of the numbered asteroids with semimajor axis together with the locations of the major Jovian resonances. Most objects lie in the main belt between 2.0 and 3.3 AU, where the outer edge is defined by the location of the 2 : 1 resonance with Jupiter. As well as gaps (the Kirkwood gaps) at the 3 : 1, 5 : 2, 2 : 1, and other resonances in the main belt, there are small concentrations of asteroids at the 3 : 2 and 1 : 1 resonances (the Hilda and Trojan groups, respectively).

derstood in terms of chaotic motion due to the overlap of adjacent resonances. In the context of the Sun–Jupiter–asteroid restricted three-body problem, the perturber (Jupiter) has an infinite sequence of first-order resonances that lie closer together as its semimajor axis is approached. For example, the 2 : 1, 3 : 2, 4 : 3, and 5 : 4 resonances with Jupiter lie at 3.3, 4.0, 4.3, and 4.5 AU, respectively. Since each $(p + 1):p$ resonance (where p is a positive integer) has a finite width in semimajor axis that is almost independent of p, adjacent resonances will always overlap for some value of p greater than a critical value, p_{crit}. This value is given by

$$p_{crit} \approx 0.51 \left(\frac{m}{m + M} \right)^{-2/7} \qquad (12)$$

where, in this case, m is the mass of Jupiter and M is the mass of the Sun. This equation can be used to predict that resonance overlap and chaotic motion should occur for p values greater than 4; this corresponds to a semimajor axis near 4.5 AU. Therefore chaos may have played a significant role in the depletion of the outer asteroid belt.

The histogram in Fig. 19 also shows a number of regions in the main belt where there are few asteroids. The gaps at 2.5 and 3.3 AU were first detected in 1867 by Daniel Kirkwood using a total sample of less than 100 asteroids; these are now known as the Kirkwood gaps. Their locations coincide with prominent Jovian

resonances (indicated in Fig. 19), and this led to the hypothesis that they were created by the gravitational effect of Jupiter on asteroids that had orbited at these semimajor axes. The exact removal mechanism was unclear until the 1980s, when several numerical and analytical studies showed that the central regions of these resonances contained large chaotic zones.

Early work concentrated on the 3 : 1 resonance at 2.5 AU since the lack of nearby strong resonances permitted the development of special techniques, including the use of algebraic mappings, to study asteroid motion. It was realized that the Kirkwood gaps cannot be understood using the model of the circular restricted three-body problem described in Section I, B. The eccentricity of Jupiter's orbit, although small (0.048), plays a crucial role in producing the large chaotic zones that help to determine the orbital evolution of asteroids. On timescales of several hundreds of thousands of years, the mutual perturbations of the planets act to change their orbital elements and Jupiter's eccentricity can vary from 0.025 to 0.061. This means that the extent of the chaotic zones will also vary with time. Research on the removal mechanism at the 3 : 1 resonance showed that an asteroid in the chaotic zone would undergo large, essentially unpredictable changes in its orbital elements. In particular, the eccentricity of the asteroid could become large enough for it to cross the orbit of Mars. This is illustrated in Fig. 20 for a fictitious asteroid with an initial eccentricity of 0.15 moving in a chaotic region of the phase space at the resonance. Although the asteroid can have periods of relatively low eccentricity, there are large deviations and e can reach values in excess of 0.3. Allowing for the fact that the eccentricity of Mars's orbit can reach 0.14, this implies that there will be times when the orbits could intersect (Fig. 21). In this case, the asteroid orbit would be unstable since it is likely to either impact the surface of Mars or suffer a close approach that would drastically alter its semimajor axis. Although Jupiter provides the perturbations, it is Mars that ultimately removes the asteroids from the 3 : 1 resonance. Figure 22 shows the excellent correspondence between the distribution of asteroids close to the 3 : 1 resonance and the maximum extent of the chaotic region determined from numerical experiments.

The situation is less clear for other resonances, although there is good evidence for large chaotic zones at the 2 : 1 and 5 : 2 resonances. In the outer part of the main belt, large changes in eccentricity will cause the asteroid to cross the orbit of Jupiter before it gets close to Mars. There may also be perturbing effects from other planets. In fact, it is now known that secular

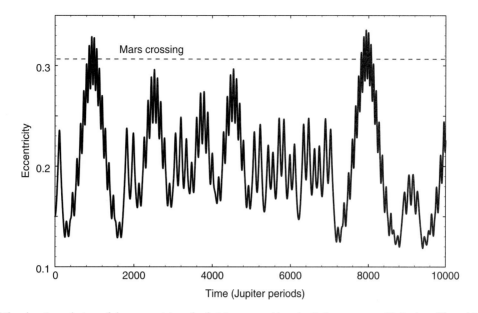

FIGURE 20 The chaotic evolution of the eccentricity of a fictitious asteroid at the 3:1 resonance with Jupiter. The orbit was integrated using an algebraic mapping technique developed by J. Wisdom. The line close to $e = 0.3$ denotes the value of the asteroid's eccentricity, above which it would cross the orbit of Mars. It is believed that the 3:1 Kirkwood gap was created when asteroids in chaotic zones at the 3:1 resonance reached high eccentricities and were removed by direct encounters with Mars.

perturbations have an important role to play in the clearing of the Kirkwood gaps, including the one at the 3:1 resonance. Once again, chaos is involved. Studies of asteroid motion at the 3:2 Jovian resonance indicate that the motion is regular, at least for low values of the eccentricity. This may help to explain why there is a local concentration of asteroids (the Hilda group) at this resonance, whereas others are associated with an absence of material.

The motion of an asteroid in a chaotic orbit can still be bounded in the sense that there may be regions of space that it cannot reach. An example of this has

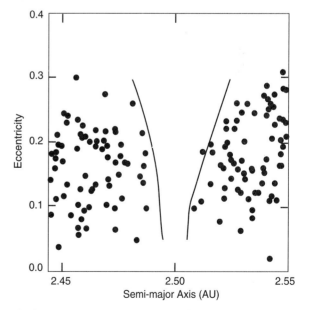

FIGURE 22 The eccentricity and semimajor axes of asteroids in the vicinity of the 3:1 Jovian resonance; the Kirkwood gap is centered close to 2.5 AU. The two curves denote the maximum extent of the chaotic zone determined from numerical experiments, and there is excellent agreement between these lines and the edges of the 3:1 gap.

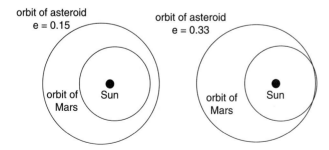

FIGURE 21 The effect of an increase in the orbital eccentricity of an asteroid at the 3:1 Jovian resonance on the closest approach between the asteroid and Mars. For $e = 0.15$, the orbits do not cross. However, for $e = 0.33$, a typical maximum value for asteroids in chaotic orbits, there is a clear intersection of the orbits and the asteroid could have an encounter with Mars.

been detected in the orbital evolution of the asteroid 522 Helga, which is close to a 12 : 7 resonance with Jupiter. Although the Lyapunov exponent has been measured as $1/6600$ yr^{-1}, numerical integrations of Helga's orbit for 6 million years show no signs of gross instability and it never has a close approach to Jupiter. Therefore chaotic motion does not necessarily correspond to unstable motion. This is the phenomenon of bounded chaos.

Since the dynamical structure of the asteroid belt has been determined by the perturbative effects of nearby planets, it seems likely that the original population was much larger and more widely dispersed. Therefore the current distribution of asteroids may represent objects that are either recent collision products or that have survived in relatively stable orbits over the age of the solar system.

B. METEORITES

Most meteorites are thought to be the fragments of material produced from collisions in the asteroid belt, and the reflectance properties of certain meteorites are known to be similar to those of common types of asteroids. Since most collisions take place in the asteroid belt, the fragments have to evolve into Earth-crossing orbits before they can be collected as samples when they eventually hit Earth.

An estimate of the time taken for a given meteorite to reach Earth after the collisional event that produced it can be obtained from a measure of its cosmic ray exposure age. Prior to the collisions, the fragment may have been part of the mantle or core of a much larger body, and as such it would have been shielded from all but the most energetic cosmic rays. However, after a collision the exposed fragment would be subjected to cosmic ray bombardment in interplanetary space. A detailed analysis of meteorite samples allows these exposure ages to be measured.

In the case of one common class of meteorites called the ordinary chondrites, the cosmic ray exposure ages are typically less than 20 million years and the samples show little evidence of having been exposed to high pressure, or "shocking." Prior to the application of chaos theory to the origin of the Kirkwood gaps, there was no plausible mechanism that could explain delivery to Earth within the exposure age constraints and without shocking. However, small increments in the velocity of the fragments as a result of the initial collision could easily cause them to enter a chaotic zone near a given resonance. Numerical integrations of such orbits near the 3 : 1 resonance showed that it was possible for

them to achieve eccentricities large enough for them to cross the orbit of Earth. This result complemented previous research that had established that this part of the asteroid belt was a source region for the ordinary chondrites. [See METEORITES.]

C. COMETS

Typical cometary orbits have large eccentricities and therefore planet-crossing trajectories are commonplace. Comets are thought to originate in the Oort Cloud at several tens of thousands of AU from the Sun. Those that have been detected from Earth are classified as either long period (most of which have made single apparitions and have periods > 200 yr) or short period (those that have made several apparitions). It is believed that the short-period comets are objects that experienced a close approach to Jupiter or one of the other giant planets; it is now believed that there may be a reservoir of such objects in the outer solar system. By their very nature, the orbits of comets are chaotic, since the outcome of any planetary encounter will be sensitively dependent on the initial conditions.

Studies of the orbital evolution of the short-period comet P/Lexell highlight the possible effects of close approaches. A numerical integration has shown that prior to 1767 it was a short-period comet with a semimajor axis of 4.4 AU and an eccentricity of 0.35. In 1767 and 1779, it suffered close approaches to Jupiter. The first encounter led to its discovery and its only apparition in 1770, whereas the second was at a distance of ~3 Jovian radii. This changed its semimajor axis to 45 AU with an eccentricity of 0.88.

A more recent example is the orbital history of comet Shoemaker–Levy 9 prior to its spectacular collision with Jupiter in 1994. Orbit computations suggest that the comet was first captured by Jupiter at some time during a 9-year interval centered on 1929. Prior to its capture, it is likely that it was orbiting in the outer part of the asteroid belt close to the 3 : 2 resonance with Jupiter. However, the chaotic nature of its orbit means that it is impossible to derive a more accurate history unless prediscovery images of the comet are obtained. [See PHYSICS AND CHEMISTRY OF COMETS; COMETARY DYNAMICS.]

D. SMALL SATELLITES AND RINGS

Chaos is also involved in the dynamics of a satellite embedded in a planetary ring system. The processes differ from those discussed in Section III, A because there is a near-continuous supply of ring material and direct scattering by the perturber is now important. In this case, the key quantity is the Hill's sphere of the satellite. Ring particles on near-circular orbits passing close to the satellite exhibit chaotic behavior due to the significant perturbations they receive at close approach. This causes them to collide with surrounding ring material, thereby forming a gap. Studies have shown that for small satellites, the expression for the width of the cleared gap is

$$W \approx 4.8 \left(\frac{m}{M}\right)^{1/3} a \qquad (13)$$

where m and a are the mass and semimajor axis of the satellite and M is the mass of the planet. Thus, an icy satellite with a radius of 10 km and a density of 1 g cm^{-3} orbiting in Saturn's A ring at a radial distance of 135,000 km would create a gap approximately 140 km wide.

Since such a gap is wider than the satellite that creates it, this provides an indirect method for the detection of small satellites in ring systems. There are two prominent gaps in Saturn's A ring: the ~35-km-wide Keeler gap at 135,800 km and the 320-km-wide Encke gap at 133,600 km. The predicted radii of the icy satellites required to produce these gaps are ~2.5 and ~24 km, respectively. In 1991, an analysis of *Voyager* images revealed a small satellite, Pan, with a radius of ~10 km orbiting in the Encke gap. *Voyager 2* images of the dust rings of Uranus show pronounced gaps at certain locations. Because most of the proposed shepherding satellites needed to maintain the narrow rings have yet to be discovered, these gaps may provide indirect evidence of their orbital locations.

E. PLANETARY SATELLITES

All the foregoing examples described in Sections III, A–III, D involve dynamical systems where the objects are acted upon by gravitational forces alone. However, in the case of satellites orbiting a planet, the tides raised by the satellites cause dissipation in the planet and the total orbital energy of the system is not conserved. Objects in prograde orbits that lie outside the synchronous orbit can evolve outward at different rates and so there may have been occasions in the past when pairs of satellites evolved toward an orbit–orbit resonance. The outcome of such a resonant encounter depends on the direction from which the resonance is approached. For example, capture into resonance is possible only if the satellites are approaching one another. If the satellites are receding, then capture is not possible, but the resonance passage can lead to an increase in the eccentricity and inclination. In certain circumstances it is possible to study the process using a simple mathematical model. However, this model breaks down near the chaotic separatrices of resonances and in regions of resonance overlap.

It is likely that the major satellites of Jupiter, Saturn, and Uranus have undergone significant tidal evolution and that the numerous resonances in the Jovian and Saturnian systems are a result of resonant capture. The absence of orbit–orbit resonances in the Uranian system is thought to be related to the fact that the oblateness of Uranus is significantly less than that of Jupiter or Saturn. In these circumstances, there can be large chaotic regions associated with resonances and stable capture may be impossible. Howerver, temporary capture into some resonances can produce large changes in eccentricity or inclination. For example, the Uranian satellite Miranda has an anomalously large inclination of 4°, which is thought to be the result of a chaotic passage through the 3 : 1 resonance with Umbriel at some time in its orbital history. Under tidal forces, a satellite's eccentricity is reduced on a shorter timescale than its inclination, and Miranda's current inclination agrees with estimates derived from a chaotic evolution. [See OUTER PLANET ICY SATELLITES.]

IV. CHAOTIC BEHAVIOR OF THE PLANETS

A. THE *N*-BODY PROBLEM

The entire solar system can be approximated by a system of nine planets orbiting the Sun. In a center of mass frame, the vector equation of motion for planet i moving under the Newtonian gravitational effect of the Sun and the remaining 8 planets is given by

$$\ddot{\mathbf{r}}_i = G \sum_{j=0}^{9} m_j \frac{\mathbf{r}_j - \mathbf{r}_i}{r_{ij}^3} \qquad (j \neq i) \qquad (14)$$

where $\mathbf{r}_i$ and m_i are the position vector and mass of planet i ($i = 1, 2, \ldots, 9$), respectively, $\mathbf{r}_{ij} = \mathbf{r}_j - \mathbf{r}_i$, and the subscript 0 refers to the Sun. These are the equations of the N-body problem for the case where $N = 10$, and although they have a surprisingly simple form, they have no general, analytical solution. However, as in the case of the three-body problem, it is possible to tackle this problem mathematically by making some simplifying assumptions.

Provided the eccentricities and incinations of the N bodies are small and there are no resonant interactions between the planets, it is possible to derive an analytical solution that describes the evolution of all the eccentricities, inclinations, perihelia, and nodes of the planets. This solution, called Laplace–Lagrange secular perturbation theory, gives no positional information about the planets yet it demonstrates that there are long-period variations in the planetary orbital elements that arise from mutual perturbations. The secular periods involved are typically tens or hundreds of thousands of years and the evolving system always exhibits a regular behavior. In the case of Earth's orbit, such periods may be correlated with climatic change, and large variations in the eccentricity of Mars are thought to have had important consequences for its climate.

In the early nineteenth century, Pierre Simon de Laplace claimed that he had demonstrated the long-term stability of the solar system using the results of his secular perturbation theory. Although the actual planetary system violates some of the assumed conditions (e.g., Jupiter and Saturn are close to a $5:2$ resonance), the Laplace–Lagrange theory can be modified to account for some of these effects. However, such analytical approaches always involve the neglect of potentially important interactions between planets. The problem becomes even more difficult when the possibility of near-resonances between some of the secular periods of the system is considered. However, nowadays it is always possible to carry out numerical investigations of long-term stability.

B. NUMERICAL EXPERIMENTS

The availability of low-cost, fast, digital computers means that studies of the long-term behavior of the planetary orbits can be easily undertaken by numerical integration. Purpose-built machines, such as the Digital Orrery and the Supercomputer Toolkit at the Massachusetts Institute of Technology, and a variety of supercomputers have been used to integrate the orbits of the planets for timescales that approach the age of the solar system. Since most numerical methods required a fixed time step, the period of the innermost planet in the integration tends to determine the value of the time step. Hence initial investigations concerned the motion of the five outer planets (Jupiter, Saturn, Uranus, Neptune, and Pluto), since integrations of all

nine planets require a time step that is determined by Mercury, the fastest-moving planet.

There is a possible compromise between carrying out a numerical integration of the full equations of motion of the N-body problem and attempting a mathematical treatment. By getting the computer to manipulate algebraic expressions rather than numbers, it is possible to produce a system of averaged equations of motion. These are an approximation to the full equations and their production can involve the manipulation of several hundred thousand symbolic terms. Although the resulting averaged equations are more complicated than Eq. (14), they have the advantage that they can be integrated numerically using larger time steps and for longer periods. For example, a typical time step in a full integration of the outer planets is 40 days, whereas that of an averaged system for all the planets is 500 years. Where direct comparisons have been made, the results of each method show good agreement, although it is recognized that the full integration is always a better approximation to the real system. Both methods have been used to study the long-term behavior of the planetary orbits over timescales that approach or even exceed the age of the solar system.

C. STABILITY OF THE SOLAR SYSTEM

The results of numerical integrations have shown that the orbits of the planets are chaotic, although there is no indication of gross instability in their motion provided that the integrations are restricted to durations of 5 billion years (the age of the solar system). The planets remain more or less in their current orbits with small, nearly periodic variations in their eccentricities and inclinations; close approaches never seem to occur. Early investigations of the long-term behavior of Pluto's orbit showed that it was chaotic, partly as a result of its $3:2$ resonance with the planet Neptune, although the perturbing effects of other planets are also important. Despite the fact that the timescale for exponential divergence of nearby trajectories (the inverse of the Lyapunov exponent) is about 20 million years, none of the studies has shown evidence for Pluto leaving the resonance.

Chaos has also been observed in the motion of other planets, and it appears that the solar system as a whole is chaotic with a timescale for exponential divergence of 4 or 5 million years. The effect is most apparent in the orbits of the inner planets. Though there are no dramatic consequences of this chaos, it does mean that the use of the deterministic equations of celestial mechanics to predict the future positions of the planets will always be limited by the accuracy with which their orbits can be measured. For example, if the position of Earth along its orbit is known to within 1 cm today, then the exponential propagation of errors that is characteristic of chaotic motion implies that we have no knowledge of Earth's orbital position 200 million years in the future.

The solar system appears to be "stable" in the sense that all numerical integrations show that the planets remain close to their current orbits for timescales approaching a billion years or more. Therefore the planetary system appears to be another example of bounded chaos, where the motion is chaotic but always takes place within certain limits. Although an analytical proof of this numerical result and a detailed understanding of how the chaos has arisen have yet to be achieved, the solar system seems to be chaotic yet stable. When the planetary orbits are integrated forward for timescales as long as 10 billion years using the averaged equations of motion, there is some evidence that the orbit of Mercury can become unstable and intersect the orbit of Venus. However, the dynamical model becomes physically meaningless in such circumstances, since by that stage the innermost planets would have been engulfed in the outer envelope of an expanding Sun.

D. CHAOTIC OBLIQUITY

The fact that a planet is not a perfect sphere means that it experiences additional perturbing effects due to the gravitational forces exerted by its satellites and the Sun, and these can cause long-term evolution in its obliquity (the angle between the planet's equator and its orbit plane). Numerical investigations have shown that chaotic changes in obliquity are particularly common in the inner solar system. For example, it is now known that the stabilizing effect of the Moon results in a variation of $\pm 1.3°$ in Earth's obliquity around a mean value of $23.3°$. Without the Moon, Earth's obliquity would undergo large, chaotic variations from $0°$ to $85°$. In the case of Mars there is no stabilizing

factor and the obliquity varies chaotically from $0°$ to $60°$ on a timescale of 50 million years. Therefore an understanding of the long-term changes in a planet's climate can be achieved only by an appreciation of the role of chaos in its dynamical evolution.

It is clear that nonlinear dynamics has provided us with a deeper understanding of the dynamical processes that have helped to shape the solar system. We now realize that chaotic motion is a natural consequence of even the simplest systems of three or more interacting bodies. The realization that chaos has played a fundamental role in the dynamical evolution of the solar system came about because of contemporary and complementary advances in mathematical techniques and digital computers. For example, just as algebraic mappings and new averaging methods were being derived, low-cost computers were becoming fast enough to undertake extensive, long-term numerical integrations. This coincided with an explosion in our knowledge of the solar system and its major and minor members. If we can understand how a random system of planets, satellites, ring and dust particles, asteroids, and comets interacts and evolves under a variety of chaotic processes and timescales, ultimately we can apply our knowledge to trace the history and predict the fate of other planetary systems.

BIBLIOGRAPHY

Diacu, F., and Holmes, P. (1996). "Celestial Encounters. The Origins of Chaos and Stability." Princeton Univ. Press, Princeton, N.J.

Duncan, M. J., and Quinn, T. (1993). The long-term dynamical evolution of the solar system. *Annu. Rev. Astron. Astrophys.* **31**, 265–295.

Dvorak, R., and Henrard, J. (eds.) (1988). "Long Term Evolution of Planetary Systems." Kluwer, Dordrecht, Holland.

Ferraz-Mello, S. (ed.) (1992). "Chaos, Resonance and Collective Dynamical Phenomena in the Solar System." Kluwer, Dordrecht, Holland.

Peterson, I. (1993). "Newton's Clock. Chaos in the Solar System." W. H. Freeman, New York.

Roy, A. E., and Steves, B. A. (eds.) (1995). "From Newton to Chaos: Modern Techniques for Understanding and Coping with Chaos in *N*-Body Dynamical Systems." Plenum, New York.

PLANETARY IMPACTS

I. Impact Craters

II. Impact Process

III. Craters as Planetary Probes

IV. Impacts and Planetary Evolution

Richard A.F. Grieve
Geological Survey of Canada

Mark J. Cintala
NASA Johnson Space Center

GLOSSARY

Allochthonous breccia: Rock composed of broken, angular fragments set in a finer-grained matrix, which has been moved into its present position.

Autochthonous breccia: Rock composed of broken, angular fragments set in a finer-grained matrix, which has formed essentially in place.

Clastic: Composed of fragments of preexisting rock. A clastic rock represents at least a second-generation product, consisting of pieces of other rocks.

Crater morphometry: Measurement and mathematical expression of the dimensions of various attributes (e.g., depth, diameter, rim height, rim width) of impact craters.

Cratering flow-field: Movement of target materials in an impact event in response to the passage of the shock and rarefaction, or decompression, waves.

Diaplectic glass: Amorphous phase produced from minerals by the destruction of internal structural order, without melting, by the passage of a shock wave.

Graben: Low, elongated region, resembling a valley, caused by a downdropped block of ground; German for "ditch" or "trench." Graben are characteristic of tensile stresses in the planet's upper layers.

Hugoniot elastic limit: Stress at which a rock or mineral's response to shock changes from elastic to plastic. Stresses over the Hugoniot elastic limit cause the rock or mineral to deform plastically.

Photic zone: Uppermost layer of the ocean that receives enough sunlight to permit its use by organisms, particularly those using photosynthesis.

Polymorph: Crystal form of a mineral that displays a form different from that of the original mineral.

Regolith: Layer of broken and loose surface materials that overlies solid bedrock on planetary surfaces.

Scarp: Abrupt transition in elevation on a planetary surface; a cliff or steep slope at the edge of a plateau or other elevated region.

Shock metamorphism: Permanent physical, chemical, and mineralogic changes in rocks resulting from the passage of a shock wave.

Shock wave: Compressional wave that travels at supersonic velocities and has an amplitude greater than the elastic limit of the medium it is traversing.

I mpact has been a common process throughout solar system history. Small bodies, such as asteroids and comets and their derivatives, meteoroids, and cosmic dust, can have their orbits disturbed by gravitational forces and collisions. Some of these disturbed orbits have a finite probability of colliding or impacting another body. Because these interplanetary bodies have cosmic velocities that range from approximately 10 to 70 km per second, they contain considerable kinetic energy. On impact, this kinetic energy is transferred to the target body by means of an approximately hemispherical propagating shock wave, resulting in the formation of an impact crater much larger in diameter than the impacting body and so-called shock metamorphic effects in the target materials. The collision of bodies was a fundamental process during planetary accretion, and in its final stages impacts were taking place involving planetesimal-sized objects.

The impact of a Mars-sized body with the proto-Earth may have been responsible for the formation of Earth's moon. As the solar system stabilized, the impact rate decreased, but it was still sufficient at ~4.0 billion years ago to produce 1000-km-sized impact basins on planetary surfaces. Although impacts producing craters 100–200 km in diameter are relatively rare in more recent geologic time, they do occur on timescales of approximately 100 million years. One such event 65 million years ago produced the boundary between the Cretaceous and Tertiary geologic periods and the mass extinction of approximately 75% of the species living on Earth at that time. The recently discovered 180-km-diameter Chicxulub impact crater in the Yucatan, Mexico, is the impact site for this global extinction event.

I. IMPACT CRATERS

A. CRATER FORM

Impact craters on the solid surfaces of solar system bodies are the most obvious manifestation of the impact process. On a small body that has no atmosphere, even the smallest pieces of interplanetary material can produce impact craters down to micron-sized cavities on individual mineral grains. On larger bodies, atmospheric breakup and deceleration serve to slow smaller impacting objects. On Earth, for example, impacting bodies with masses of less than 10,000 kg can lose up to 90% of their impact velocity while penetrating the atmosphere. At these reduced velocities, the resultant impact pit is only slightly larger than the impacting body. At masses >100,000 kg, however, atmospheric effects are less and the projectile impacts with relatively undiminished velocity, producing an impact crater that is considerably larger than the impacting body.

The basic shape of an impact crater is a depression with an upraised rim. Detailed appearance, however, varies with crater diameter. With increasing diameter, impact craters become proportionately shallower and develop more complicated rims and floors, including the appearance of central peaks and interior rings. Craters are divided into three basic morphologic subdivisions: simple craters, complex craters, and basins.

Virtually all small impact structures have the form of a bowl-shaped depression with an upraised rim and are known as simple craters (Figs. 1 and 2). The exposed rim, walls, and floor define the so-called apparent crater. At the rim, there is an overturned flap of ejected target materials, which displays inverted stratigraphy with respect to the original target materials. Beneath the floor is a lens of brecciated target material that is roughly parabolic in cross section (see Fig. 2). This breccia lens is polymict, that is, a mixture of different rocks, with fractured blocks of target materials set in a finer-grained clastic matrix. These are allochthonous materials, having been moved into their present position. In places, near the top and the base, for example, the breccia lens may contain regions of highly shocked, including melted, target material. Beneath the breccia lens, relatively in-place, or autochthonous, fractured target rocks define the walls and floor of what is known as the true crater. In the case of terrestrial simple craters, the depth to the base of the breccia lens (i.e., the base of the true crater) is roughly twice that of the depth to the top of the breccia lens (i.e., the base of the apparent crater). Shocked rocks in the autochthonous materials of the true crater floor are confined to a small central volume at the base (see Fig. 2).

With increasing diameter, simple craters show increasing evidence of wall and rim collapse and evolve into complex craters. The transition diameter varies between planetary bodies and is, to a first approximation, an inverse function of planetary gravity (Table 1). Other variables, such as target material and possibly projectile type and velocity, play a lesser role, so that the transition diameter is actually over a small range. For example, simple craters can reach diameters of 25 km and complex craters can have diameters as small as 15 km on the Moon. Perhaps the most obvious effect of secondary variables appears on Earth, where there are major areas of both sedimentary and crystalline rocks on the surface. Complex craters on Earth occur at diameters greater than 2 km in sedimentary target rocks, but not until diameters of 4 km or greater in stronger igneous or metamorphic crystalline target rocks.

FIGURE 1 Alfraganus C, which is 10 km in diameter, is an example of a simple lunar crater. Though this type of crater is often described as bowl-shaped, the freshest examples show a small flat floor. (*Apollo 16* Panoramic Camera Photograph 4615.)

Complex craters are a highly modified craterform compared to simple craters. They are characterized by a central topographic peak or peaks, a broad, flat floor, and terraced, inwardly slumped rim areas (Fig. 3). The rim of a typical complex crater is a structural feature corresponding to a series of fault terraces. An annular trough lies interior to the rim, within which there has been little or no excavation of material in the cratering process. The trough is partially filled by a sheet of impact melt rock and/or polymict breccia (Fig. 4). Only in the central area is there evidence of substantial excavation of material. This region is structurally complex and, in large part, occupied by a central peak, which is the topographic manifestation of a much broader and extensive area of uplifted rocks that occurs beneath the center of complex craters. With increasing diameter, a fragmentary ring of interior peaks appears,

marking the transition from craters to basins. While a single interior ring is required to define a basin, they can be subdivided further into central-peak basins with both a peak and ring; peak ring basins (Fig. 5), with only a ring; and multiring basins, with two or more interior rings (Fig. 6). The transition from central-peak basins to multiring basins also represents a sequence with increasing diameter. As with the simple to complex transition, there is a small amount of overlap in basin form near the transition diameter, the definition of which varies slightly between workers, depending on the specific characteristics examined.

Most detailed models of impact cratering indicate that about 50% at most of a simple crater's final volume is ejected. The remainder is permanently displaced, much of which is involved in structural uplift of the crater's rim. In many cases, however, even this rela-

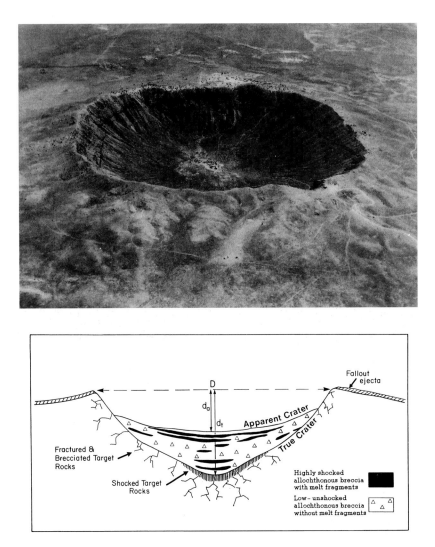

FIGURE 2   Top: Oblique aerial photograph of the 1.2-km-diameter, 50,000-year-old Barringer or Meteor Crater, in Arizona, U.S.A., which is a well-preserved simple crater. It consists of a bowl-shaped depression with an upraised rim. The rim is overlain by ejecta, which shows reverse stratigraphy. The hummocky terrain surrounding the rim is the remnant of the original ejecta blanket. Bottom: Schematic cross section of a simple crater based on terrestrial data. Below the apparent crater, which is pictured in the top frame, there is a subsurface breccia lens of shocked and unshocked target rocks, which is confined by the fractured and brecciated target rocks of the walls and floor of the so-called true crater. The depth of the true crater (d_t) is approximately twice the depth of the apparent crater (d_a). Shocked, in-place, target rocks are confined to a small volume at the base of the true crater.

tively small amount of ejecta can have major effects on the terrain around the crater, often to great distances. One of the principal features examined at the *Apollo 17* site, for example, was a landslide that was probably caused by secondary projectiles from the impact that formed the lunar crater Tycho (see Fig. 3), over 2200 km away.

In the coarsest sense, ejecta deposits can be subdivided simply into continuous and discontinuous facies (Fig. 7). The continuous deposits are those closest to the crater, being thickest at the rim crest. These materials are ejected late in the excavation stage of the

event, are the most weakly shocked, and leave the cavity at the lowest velocities of all the ejecta. In the case of simple craters, the net effect of such ejection is to invert the stratigraphy at the rim. The case is not so clear-cut for large, complex craters, as it is conceivable that rim failure could occur contemporaneous with excavation.

As the distance from the crater rim increases, the ejecta is emplaced at higher velocities and, therefore, with higher kinetic energies. Though these energies may not be sufficient to create secondary craters, they will favor mixing of ejecta with local material. Thus,

TABLE I
**Summary of Simple
to Complex Crater Transition
Diameters for the Terrestrial Planets[a]**

Planetary body	D_t (km)	Gravity (cm s^{-1})
Moon		
Highlands	27	162
Mare	19	162
Mars	6	372
Mercury	13	378
Earth		
Crystalline	4	981
Sedimentary	2	981

[a] D_t is the diameter at which 50% of the craters are complex. Venus has no observed simple craters owing to atmospheric effects. See text.

at increasing distance from the crater, the final ejecta blanket on the ground includes increasing amounts of local material. In addition to the impact energy, the formation of secondary craters (see Fig. 7) also depends on the state of the ejecta. If it is relatively fine-grained, secondaries are more difficult to form than if it is in the form of clods or more coherent, isolated objects.

Secondary crater fields surrounding fresh craters, particularly those on the Moon (see Fig. 7), Mercury, and other airless bodies, are often associated with typically bright or high-albedo "rays" that define an overall pattern radial to the crater. Various mechanisms have been suggested to account for the rays. They have been described as thin layers of ejecta from the responsible crater, as well as material mobilized by the impact of secondary projectiles. Their relatively high albedo can often be explained by the higher albedo of the material composing the primary ejecta, although this cannot be the case when bright rays from impacts into dark mare material occur in light highland terrain. Excavation of younger regolith or freshly crushed rock might be part of the explanation of the albedo, since older materials are known to darken with exposure to micrometeoroid bombardment.

The size of the crater is important when ejecta energies are considered. Ejecta impacting two crater radii from a 10-m crater will be traveling much more slowly than ejecta impacting at the same scaled distance from a 10-km crater. Assuming identical distributions of ejection angles, reduction of the ballistic-range equations for a small and a large crater yields the ratio

$$\frac{v}{V} = \sqrt{\frac{r}{R}}$$

where v and V are the ejection velocities for the small and large craters, respectively, and r and R are the ballistic ranges of each over a flat surface, respectively. If the scaled ranges are identical for comparative purposes, the variables r and R can be replaced by their respective crater radii. Comparing the 10-m and 10-km craters, it is easy to determine that $v \approx 0.03\,V$. This implies that the corresponding ratio of specific kinetic energy (i.e., energy per unit mass) for a piece of ejecta from each of the two craters is $ke \approx 0.001\,KE$, where ke and KE are the specific kinetic energies of the 10-m and 10-km craters, respectively. That is, there is a thousand times more specific kinetic energy in the ejecta from the 10-km-diameter crater. Clearly, differences in the effects of ejecta emplacement at a given scaled range are highly dependent on the size of the crater.

Some Martian craters were considered to have a unique form in that they possessed examples of fluidized ejecta (Fig. 8). They are most often called "fluidized-ejecta," "rampart," or "pedestal" craters. These ejecta deposits possess many indications of having been emplaced as a ground-hugging flow. Typically, they extend well past the limits of the continuous ejecta deposits of craters on other bodies, such as the Moon and Mercury. Most hypotheses treating the origin of these features invoke the presence of ground ice (or water), which, upon heating by impact, is incorporated into the ejecta in either liquid or vapor form. This, then, provides lubrication for the mobilized material.

Recently, other hypotheses have been put forward, most of which invoke the effects of the Martian atmosphere on the expanding plume of ejecta. Laboratory experiments under different atmospheric pressures appear to duplicate many of the features of the fluidized-ejecta deposits. On the other hand, *Voyager* photographs of Ganymede, an icy satellite of Jupiter, show craters that appear to resemble many fluidized-ejecta craters on Mars. The resolution of those photographs, unfortunately, is insufficient for the detection of diagnostic flow features and, therefore, do not permit detailed comparison with their Martian counterparts. Given that Ganymede possesses no atmosphere but has a crust whose major component is H_2O ice, these craters are targetted for imaging at the 100 m resolution available through the *Galileo* spacecraft. *Magellan* images of Venus, however, which has a dry crust and a thick atmosphere, provide numerous examples of rampart-type craters over a range of sizes. [*See* OUTER PLANET ICY SATELLITES.]

FIGURE 3   The lunar complex crater Tycho, which is 85 km in diameter, has well-developed wall terraces, a large flat floor, and multiple central peaks. The deposits on the crater floor are interpreted as impact melt, as are the smooth pools on the eastern and southern rim and terrace areas. (*Lunar Orbiter V* Photograph 125M.)

Impact craters on Venus have come into focus through the recent radar imagery supplied by the *Magellan* orbiter. Craters more than 15 km in diameter exhibit central peaks and/or peak rings (Fig. 9) and appear, for the most part, to be similar to complex craters and basins on the other terrestrial planets. Many of the craters smaller than 15 km have multiple floors or occur as clusters. This is attributed to the effects of the dense atmosphere of Venus, which has a surface pressure of ~90 bars. During atmospheric passage, smaller impacting bodies are effectively crushed and broken up, so that they form clusters of craters. In many cases, craters on Venus have ejecta deposits out to greater distances than expected from simple ballistic emplacement and, in some cases, the distal deposits are clearly lobate (see Fig. 9). It has been suggested that these deposits owe their origin to atmospheric entrainment or the high proportion of impact melt that would be produced on a relatively high gravity planetary body such as Venus. The relatively abundant impact melt produced in cratering events on Venus versus the Moon, for example, is also attributable, in part, to the initially higher temperature of the target rocks. Because of the greenhouse effect

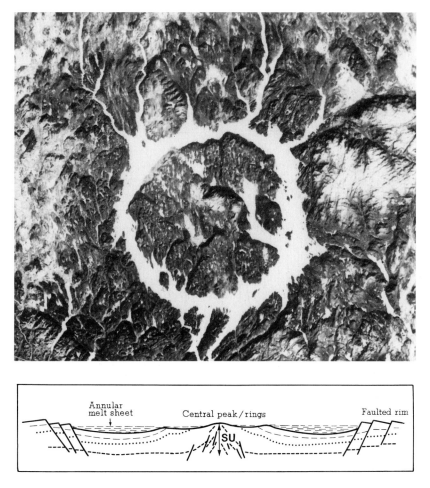

FIGURE 4   Top: LANDSAT image of the 212-m.y.-old Manicougan peak ring basin, in Quebec, Canada. The annular lake (frozen) is 65 km in diameter and represents the erosionally overdeepened inner edge of the original rim area. The original diameter of the rim is estimated to be 100 km. Bottom: Schematic cross section of the subsurface of a complex crater based on terrestrial data. Note the faulted rim, annular trough, and complex uplifted central area, and that target rocks are missing, due to excavation and melting, only in the central area. SU represents the amount of structural uplift of originally deeper rocks in the center.

of the dense CO_2 atmosphere of Venus, the surface temperature is 480°C.

Another unusual feature on Venus is a radar-dark zone surrounding some craters. Approximately half the craters imaged by *Magellan* are partially or wholly surrounded by these zones, which extend three to four crater diameters from the crater center. It has been suggested that they are due to the destruction of surface roughness by an atmospheric shock wave produced by the impacting body. No crater smaller than 3 km occurs on Venus. Small crater clusters, however, have dark haloes and dark circular areas are observed with no central crater form. In these latter cases, it has been suggested that the impacting body did not survive passage through the atmosphere. The accompanying atmospheric shock wave, however, had sufficient energy to interact with the surface to create a dark, radar-smooth area. The situation is somewhat analogous to the Tunguska event in 1908, when a relatively small body exploded over Siberia at an altitude of 10 km and the resultant atmospheric pressure wave leveled some 2000 km^2 of forest. [*See* VENUS: SURFACE AND INTERIOR.]

Remarkable ring structures occur on the Galilean satellites of Jupiter, Callisto and Ganymede. The largest is Valhalla on Callisto (Fig. 10), which has a diameter of 4000 km and consists of a bright central area up to 800 km in diameter, surrounded by darker terrain with bright ridges 20–30 km apart. This zone is about 300 km wide and gives way to an outer zone with riftlike features 50–100 km apart. These (very) multiring basins are generally considered to be of impact origin, although the original crater is believed to be confined to the central area. Models suggest that the rings formed as a result of the

FIGURE 5   Schrödinger, which is 320 km in diameter, is an example of a lunar peak ring basin. It has an almost complete interior peak ring of massifs, an expansive flat floor, and a terraced, faulted rim area. (*Lunar Orbiter IV* Photograph 9M.)

original crater puncturing the outer, strong shell, or lithosphere, of these bodies. This permitted the weaker, underlying layer, the asthenosphere, to flow toward the crater, setting up stresses at the base of the lithosphere and leading to fracturing and the formation of scarps and graben.

In such basins on the icy satellites, the original crater has essentially disappeared. Craters that no longer have an obvious craterform but appear as bright or high-albedo spots on the surfaces of these bodies are known as palimpsests. They are believed to have begun as complex craters but have had their topography relaxed by the slow, viscous creep of the target's icy crust over time. An alternative explanation for the occurrence of palimpsests is that, in larger impacts into icy materials, such a high volume of the target material is heated by the impact that the target turns into a semiliquid slush.

If this is the case, then no permanent crater can form, as the target has no strength and topography cannot be supported.

Another type of anomalous crater occurs on Mars but is best developed on Ganymede and Callisto. On these latter icy satellites, some craters between 30 and 120 km in diameter have a central pit. Pit craters (Fig. 11) are shallower than other craters of comparable size. They are also generally smaller than palimpsests, which have diameters >70 km. It has been suggested that pit craters are an intermediate stage between normal craters and that palimpsests and the pits are due to impact melting of the centers of the craters. The fact that some craters on these icy bodies are anomalous and others are not has been ascribed to a velocity effect, as higher impact velocities result in greater melting of the target. Interpretations of the origin of the various

FIGURE 6 Multiring basin Orientale, which is ~900 km in diameter as defined by the outer ring, the Cordillera Mountains. Orientale is the youngest and, thus, the best-preserved multiring basin on the Moon. The interior is filled with impact melt deposits (light, smooth) and some younger basaltic mare rocks (dark, smooth). (*Lunar Orbiter IV* Photograph 194M.)

anomalous crater forms on the icy satellites, however, must be regarded as relatively poorly constrained.

B. CRATER MORPHOMETRY

Whereas crater morphology involves the qualitative description of crater-related features, morphometry treats these features and their interrelationships quantitatively. The depth–diameter relations for craters on the terrestrial planets are given in Table II. Other relations involving parameters such as rim height, rim width, central peak diameter, central peak height can be found in the literature. Because of the abundant imagery and low rate of erosion, these morphometric relations for impact craters are best defined for the Moon.

Simple craters have essentially the same apparent depth–diameter relationship on all the terrestrial planets (Table II and Fig. 12), indicating that planetary gravity and variations in target material properties are not important factors in the formation of simple craters. At first glance, terrestrial craters appear to be shallower than their planetary counterparts. This, however, is more apparent than real. Erosive processes are most severe on Earth compared to the other terrestrial planets, and the crater rim is rapidly affected by erosion. Few terrestrial craters have well-preserved rims and it is common to measure terrestrial crater depths with respect to the ground surface, which is known and is assumed to erode more slowly. In the case of other planetary bodies, depths are measured most often by the shadow that the rim casts on the crater floor. That is, the topographic measure is a

FIGURE 7 The change in ejecta facies with distance from the point of impact is apparent in this photograph of the lunar crater Aristarchus, which is 40 km in diameter. Fine, somewhat random texturing near the rim crest is due to flows of impact melt downslope both into and away from the crater. These flows are superimposed on a deposit that exhibits a more regular pattern on a somewhat larger scale. This yields to a facies that gives a strong impression of radial outward motion, gradually merging into irregular, but closely packed, secondary craters. The rille cutting across the center of the frame acts as a demarcation between the appearance of discontinuous and the more continuous deposits closer to the crater. The fields of secondary craters then become better defined in chains, clusters, and other groupings. Although this sort of trend is representative of a variety of crater sizes, the relative distances at which the various facies appear change with crater size. The bottom of the figure is about 90 km from the center of Aristarchus in this foreshortened view. (*Apollo 15* Panoramic Camera Photograph 0326.)

relative one between the rim crest and the floor. In most cases, the elevation of the surrounding ground surface is not well known. Thus, the measurements of depth for Earth and for other planetary bodies are not the same. In the very few cases where the rim is well preserved in terrestrial craters, depths from the top of the rim to the crater floor are comparable to those of similar-sized simple craters on the other terrestrial planets.

Unlike simple craters, the depth of complex craters with respect to diameter does vary between the terrestrial planets (see Fig. 12). The sense of variation is that increasing planetary gravity shallows final crater depths. This is not a strict relationship, as Martian craters are shallower than equivalent-sized Mercurian craters (see Table II). This is probably a function of differences between target material, with the trapped volatiles and relatively abundant sedimentary deposits on Mars making it, in general, a weaker target. The secondary effect of target strength is also well illustrated by the shallower depth–diameter relationship for terrestrial complex craters in sedimentary targets compared to those in crystalline targets (see Table II and Fig. 12).

The depth–diameter relationships for craters on the icy satellites of the outer gas-giant planets have the same general trends as those on the rocky terrestrial planets. They are, however, shallower. In fact, even simple craters are shallower, for a given diameter, by 20–40%. This is a reflection of the extreme differences in material properties between icy and rocky worlds. Yet it is difficult to make many generalizations about craters on icy bodies, as the resolution of the currently available imagery is low. This will change as the results of the *Galileo* mission continue to be received and interpreted. [*See* OUTER PLANET ICY SATELLITES.]

II. IMPACT PROCESS

Impact is an extremely transient process and involves energies, velocities, pressures, and temperatures not normally encountered in other geologic processes. It is, therefore, inherently difficult to study. Small-scale impacts can be produced in the laboratory by firing small projectiles at high velocity (<6 km s^{-1}) at various targets. Some insights can also be gained from observing the results of the explosion of nuclear devices and from complex numerical model calculations. Finally, the planetary record of impact provides constraints on the process. In particular, the terrestrial record is a source of important ground truth data,

FIGURE 8   Many characteristics of fluidized-ejecta deposits are apparent in this photograph of the Martian crater Arandas, which is 28 km in diameter. Note the striated, radial texture on the uppermost surface, the layered nature of the underlying deposits, and the "upstream" buildup of material apparently caused by the smaller crater near the center of the photograph. (*Viking Orbiter 1* Photograph 32A28.)

especially with regard to the subsurface nature and spatial relations at impact craters, and the effects of impact on rocks.

When an interplanetary body impacts a planetary surface, it transfers the bulk of its kinetic energy to the target. The kinetic energy of such bodies is extremely high, as they impact generally with a velocity that is a combination of their cosmic approach velocity plus escape velocity of the target body, such that

$$V_i = \sqrt{V_\infty^2 + V_{esc}^2}$$

where V_i is impact velocity, V_∞ is the approach velocity of the body in space, and V_{esc} is the escape velocity of the target body. A listing of the mean impact velocity on the terrestrial planets for asteroidal bodies is given in Table III. The impact velocity of comets is even

higher. For example, long-period comets (comets with an orbital period greater than 200 years) have an average impact velocity with Earth of $\sim$55 km s^{-1}; whereas short-period comets have a somewhat lower average impact velocity of $\sim$40 km s^{-1}. [*See* PHYSICS AND CHEMISTRY OF COMETS.]

A. CRATERING MECHANICS

The bulk of the impacting body's sizable amount of kinetic energy is transferred to the target by means of a shock wave, which propagates into the target with the approximate shape of a hemisphere. The shock wave compresses and heats the target, increasing its internal energy. It also accelerates the target materials in the direction of shock propagation, imparting ki-

FIGURE 9   A group of complex craters on Venus. The craters are 37 to 50 km in diameter and display the features common to complex central-peak craters. In this radar image, the rough exterior ejecta appears bright and shows lobate flowlike features at their distal ends. The flat interior floors of the craters are radar dark, indicating that they are smooth. (*Magellan* Photograph P-36711.)

netic energy to the target. It is this kinetic energy in the target that produces the crater.

As the shock wave propagates into the target, it sets the target materials in motion (Fig. 13). The direction of acceleration is perpendicular to the shock front and material is accelerated downward and outward. Since a state of stress cannot be maintained at a free surface, such as the original ground surface or the edges and rear of the impacting body, a series of secondary release waves or rarefaction waves are generated, which bring the shock-compressed target materials back to ambient pressure. The rarefaction waves tend to have a shallower profile than the hemispherical shock wave. Thus, as the rarefaction wave interacts with the target material, it changes the direction of some of the material

set in motion by the shock wave. In particular, it changes some of the outward and downward motions in the relatively near-surface materials to outward and upward, leading to the ejection of target material from the growing cavity. Directly below the impacting body, however, the two wave fronts are more nearly parallel and material is still driven downward (see Fig. 13).

This motion, generated by the combination of shock and rarefaction waves, is known as the cratering flow-field and the cavity grows by a combination of upward ejection and downward displacement of target materials. This "transient cavity" reaches its maximum depth before its maximum radial dimensions are established, but it is usually depicted in illustrations at its maximum growth in all directions (see Fig. 13). At this

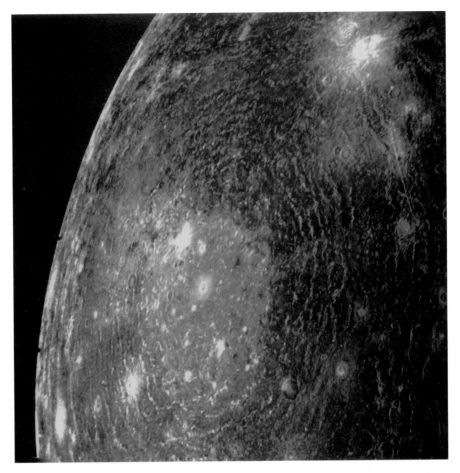

FIGURE 10   The Valhalla Basin and its related structure dominate this image of Callisto. The relatively bright central region is 600–800 km across. Its brightness could be due to segregation by settling of dense silicates from relatively pure ice when it was liquid H_2O impact melt. Note the well-defined concentric structure. (*Voyager 1* Photograph 2622.11.)

point, it is generally parabolic in cross section and, at least for the terrestrial case, has a depth-to-diameter ratio of about 1 to 3.

The time required for transient cavity formation depends on the size of the impacting body and to a lesser extent on the impact velocity and physical properties of the impacting body and the target. For the case of an asteroidal body of density 3 g cm^{-3} impacting crystalline target rocks at 25 km s^{-1}, initial shock velocities in the target are over 20 km s^{-1}, with corresponding initial particle velocities over 10 km s^{-1} for the materials set in motion by the shock wave. The rarefaction wave has initial velocities similar to the shock wave but, since the target materials are compressed by the shock wave, it has a smaller distance to cover and can rapidly overtake the moving material, altering its direction of movement. Although shock pressures and attendant particle velocities attenuate

radially, as increasing volumes of the target materials are involved in the impact process, transient cavity growth is an extremely rapid event. The time (t) to achieve maximum depth is on the order of

$$t \approx (2\,d_{\mathrm{t}}/g)^{1/2}$$

where d_{t} is transient cavity depth and g is gravitational acceleration.

For convenience, the cratering process is often divided into stages: initial contact and compression, excavation, and modification. However, it is a continuum and, in reality, different volumes of the target undergo different stages at the same instant of time. For example, excavation by the cratering flow-field is taking place close to the impact point, while materials more distant from the impact point are experiencing only compression by the shock wave (see Fig. 13). The

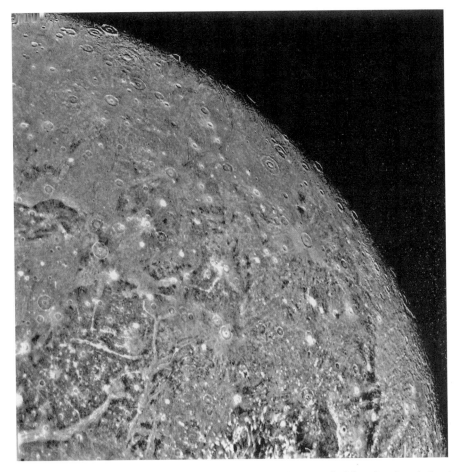

FIGURE 11 A multitude of craters with central pits are evident in this image of Ganymede. Note that the relative size of the central pit is not constant, even in craters that are similar in size. The crater on the terminator at the top of the photograph with the prominent central pit is about 80 km in diameter. (*Voyager 2* Photograph 20631.33.)

formation of the transient cavity, by compression and excavation due to the impact-induced cratering flow-field, marks the end of the excavation stage.

As the excavation stage draws to a close, a major change in the direction of movement of target material occurs. The direction of movement is no longer outward but is inward and the transient cavity collapses, marking the beginning of the modification stage. Collapse is due to gravity and, in larger structures, also to the rebound of the compressed materials of the transient cavity floor. It ranges in style from minor landslides on the cavity walls of the smallest simple craters to complete modification, involving the uplift of the center and collapse of the rim area to form central peaks and terraced, structural rims in large complex craters.

In simple craters, the interior breccia lens is the result of collapse. Toward the end of the excavation stage, shocked and melted debris is moving upward along the transient cavity walls. As the cratering flow ceases, the fractured and oversteepened transient cavity walls become unstable and collapse inward, carrying with them the lining of shocked material. As the walls collapse into the cavity, they undergo fracturing and mixing and come to rest as the bowl-shaped allochthonous breccia lens of mixed unshocked and shocked target materials (see Fig. 13). The collapse of the cavity walls increases the diameter, such that the final crater diameter is about 20% larger than that of the transient cavity. This is offset by the shallowing of the cavity by the production of the breccia lens, with the final apparent crater about half the depth of the original transient cavity. The collapse process is rapid and probably takes place on time-scales comparable to those of transient cavity formation.

The simple-to-complex transition on the terrestrial planets appears to be a function of collapse under gravity (Fig. 14), once some threshold diameter has been

TABLE II

Summary of Apparent Depth–Diameter Relations for Craters on the Terrestrial Planets[a]

Planetary body	Slope	Intercept	Gravity (cm s⁻¹)
	Simple Craters		
Moon			
Highlands	1.013	0.195	162
Mare	1.022	0.195	162
Mars	1.019	0.204	372
Mercury	0.995	0.199	378
Earth	1.06	0.13	981
	Complex Craters		
Moon			
Highlands	0.313	1.088	162
Mare	0.332	0.841	162
Mars	0.395	0.415	372
Mercury	0.415	0.492	378
Venus	0.33	0.20	891
Earth			
Sedimentary	0.30	0.12	981
Crystalline	0.43	0.15	981

[a] The equation for the depth–diameter relation is of the form log (depth) = log (intercept) + slope × log (diameter), where depth and diameter are in km. Values are obtained by different methods on different bodies and have variable precision. For all bodies except Earth, slopes are within ±0.035 or better. Intercepts are within ±0.090 or better. The terrestrial uncertainties are higher owing to the very small number (<15) of well-preserved craters available for measurement. The Venus relation is based on the results of the Russian Venera missions.

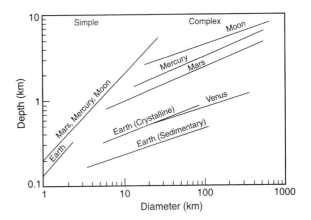

FIGURE 12 Apparent depth versus diameter for craters on the terrestrial planets. Both scales are logarithmic. Note the relatively shallower depth–diameter relationship for complex compared to simple craters and the different transition diameters and relations for the various planetary bodies, which are ascribed to the effects of planetary gravity. Note also the effect of target strength, resulting in shallower craters in sedimentary compared to crystalline target rocks on Earth. The depth–diameter relations for simple craters appear constant for all bodies. Lower values for the terrestrial case are attributed to erosion of the rim. In general, the relationships for terrestrial craters are less well defined because of erosional effects.

exceeded. Most of our understanding of complex crater formation comes from the interpretation of observations at terrestrial craters. In craters formed in sedimentary strata on Earth, it has been possible to trace the movement of beds and to show that central peaks are the result of the uplift of rocks from depth (see Fig. 4). Shocked autochthonous target rocks, analogous to those found in the floor of the true crater in simple craters, are present in the centers of complex structures. It is clear, therefore, that the central structure represents the uplifted floor of the original transient cavity. The amount of uplift based on terrestrial data corresponds to

$$SU = 0.086D^{1.03}$$

where SU is the maximum amount of uplift and D is final crater diameter. Further observations at terrestrial craters in the tens of kilometers size range indicate

that excavation is also limited to the central area and that the transient cavity diameter was about half the diameter of the final crater. Beyond this, original near-surface units are preserved in the down-dropped annular trough. The rim area is a series of fault terraces, progressively stepping down from the rim to the floor (see Figs. 3 and 4).

Although models for the formation of complex craters are less constrained than those for the formation of simple craters, there is a general consensus that, in their initial stages, complex craters were not unlike simple craters. It is more difficult to reconstruct the

TABLE III

Mean Impact Velocity of Asteroidal Bodies with the Terrestrial Planets

Planetary body	Impact velocity (km s⁻¹)
Moon	16
Mars	12
Mercury	24
Venus	21
Earth	18

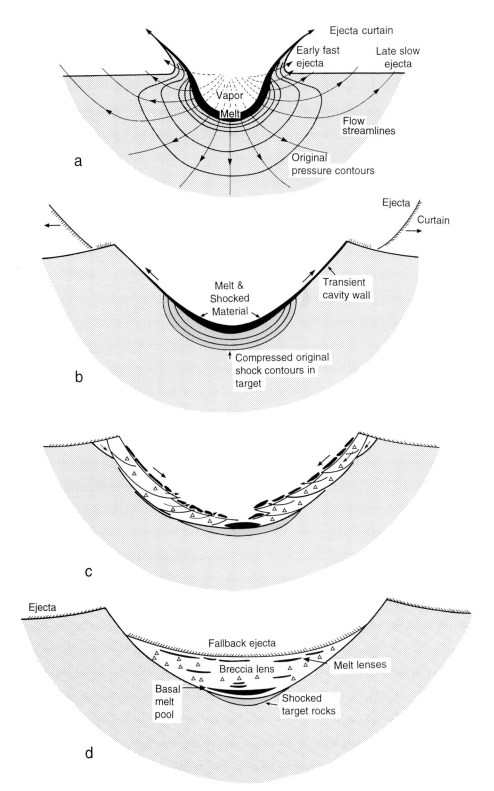

transient cavity at a given complex crater but, again, the study of terrestrial craters in sedimentary rocks with good marker horizons suggests that at the point of its maximum growth its shape was comparable to that of simple craters. The main difference is that the downward displacements in the transient cavity floor observed in simple craters are not locked in at complex craters and the cavity floor rebounds upward (see Fig. 14). Because the maximum depth of the transient cavity is reached before maximum transient cavity diameter, it is likely that this rebound and reversal of the flow-field in the center occur while the diameter of the transient cavity is still growing by excavation. With the upward movement of material of the transient cavity floor, the entire rim area outside the transient cavity collapses down and inward (see Fig. 14). This greatly enlarges the final crater diameter. Based on various structural and stratigraphic reconstructions, terrestrial data indicate that for central-peak craters

$$D_t = 0.5 - 0.65 D$$

where D_t is the diameter of the transient cavity and D is the final rim diameter. Data from other planetary bodies are very limited. There have been a number of reconstructions for large lunar craters, where the rim terraces are restored to their original, preimpact positions. In these cases,

$$D_t \sim 1.5 D^{0.85}$$

It is clear that uplift and collapse in the modification of complex craters are extremely rapid and that the target material behaves as if it had very low strength. Why the target materials are weak is not known. A number of mechanisms, including fluidization, by impact melt, interstitial water, or so-called acoustic fluidization, have been suggested. Although the details of cavity modification are not well known, the sense of movement during cavity collapse is well established

and there is no doubt that central peaks expose uplifted transient cavity floor.

There is less consensus on the formation of rings within impact basins. The most popular hypothesis is that the rings represent uplifted material in excess of what can be accommodated in a central peak. This may explain the occurrence of both peaks and rings in central-peak basins but offers little explanation for the absence of peaks and the occurrence of only rings in peak ring and multiring basins. A number of analogies have been drawn with the formation of "craters" in liquids and semiconsolidated materials such as muds, where the initial uplifted peak of material has no strength and collapses completely, sometimes oscillating up and down several times. At some time in the formation of ringed basins, however, the target rocks must regain their strength, so as to preserve the interior rings. An alternative recent explanation is that the uplift process proceeds, as in central-peak craters, but the uplifted material in the very center is essentially fluid due to impact melting. This hypothesis stems from the observation that impact melt volumes increase at a faster rate than transient cavity volumes with increasing event size. Thus, in large impact events, the depth of impact melting may reach and even exceed the depth of the transient cavity floor. When the transient cavity is uplifted in such events, the central, melted part has no strength and, therefore, cannot form a positive topographic feature, such as a central peak. Only rings from the unmelted portion of the uplifted transient cavity floor can form some distance out from the center (see Fig. 14).

B. SHOCK METAMORPHISM

Some of the impacting body's kinetic energy is partitioned into increasing the internal energy of the target rocks. The target rocks are first highly compressed by the passage of the shock wave and transformed into

FIGURE 13 Formation of a simple crater. (a) Shortly after impact, the shock wave has propagated into target rocks. Close to the point of impact, material has been set in motion by the cratering flow-field due to the shock and rarefaction waves. A cavity has started to grow by excavation and displacement, while melted and vaporized materials are being driven down and line the expanding cavity. Ultimately, target materials will follow the paths outlined by the flow streamlines. (b) Final geometry of the cavity formed by the cratering flow-field. The cavity, known at this point as the transient cavity, is formed by a combination of excavation and displacement of target rocks. It is lined with impact melt and shocked materials and is floored in the center by compressed and shocked target rocks. (c) Collapse of the transient cavity as the cratering flow-field ceases. Transient cavity walls collapse downward and inward, carrying with them the lining of impact melt and high-shocked debris. During cavity collapse, the transient cavity diameter is enlarged. (d) Final form of simple crater with interior allochthonous breccia lens formed from collapse of transient cavity. The breccia lens is a mixture of unshocked material from the transient cavity wall and shocked material from the original lining of the transient cavity. The diameter of the final crater has been enlarged over that of the transient cavity and the final apparent depth reduced by about a half by the formation of the breccia lens.

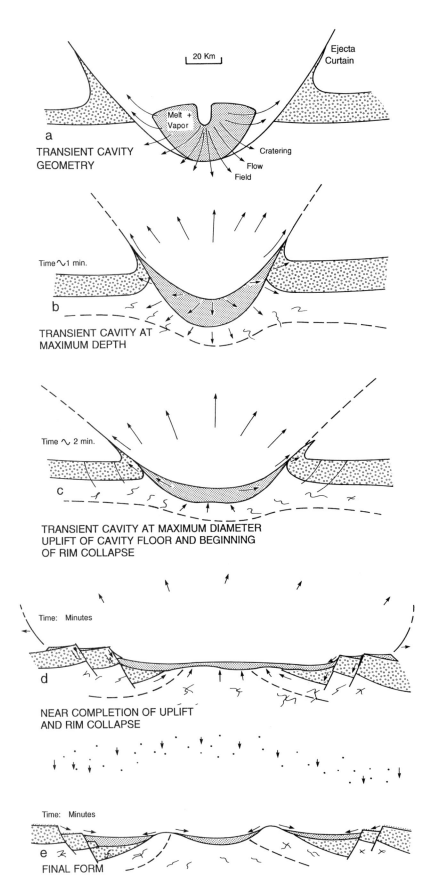

a
TRANSIENT CAVITY
GEOMETRY

Melt + Vapor

Ejecta Curtain

20 Km

Cratering

Flow Field

b
TRANSIENT CAVITY AT
MAXIMUM DEPTH

Time ～1 min.

c
TRANSIENT CAVITY AT MAXIMUM DIAMETER
UPLIFT OF CAVITY FLOOR AND BEGINNING
OF RIM COLLAPSE

Time ～ 2 min.

d
NEAR COMPLETION OF UPLIFT
AND RIM COLLAPSE

Time: Minutes

e
FINAL FORM

Time: Minutes

high-density phases, and are then rapidly decompressed by the rarefaction wave and converted back toward their original state. They do not, however, recover fully to their preshock state. They are of slightly lower density than their original form and the entropy of the constituent minerals is increased. The collective term for these shock-induced irreversible changes in minerals and rocks is "shock metamorphism." Shock metamorphic effects are found naturally in many lunar samples and meteorites and at terrestrial impact craters. They have also been produced in nuclear explosions and in laboratory shock-recovery experiments, in which a high-velocity projectile is fired at geologic materials. No other geologic process is capable of producing the high transient pressures and temperatures required for shock metamorphism (Fig. 15). Therefore, the occurrence of shock metamorphism is considered to be diagnostic of the impact process.

Metamorphism of rocks normally occurs in planetary bodies as a consequence of thermal and tectonic events originating within the body. The maximum pressures and temperatures recorded in surface rocks by such metamorphic events in planetary crusts are generally on the order of 1 GPa (10 kb) and 1000°C. In shock metamorphism, materials deform along their so-called Hugoniot curves, which describe the locus of pressure–volume states achieved by the material due to the shock process. Shock metamorphic effects do not appear until the material has exceeded its Hugoniot elastic limit (HEL), which is on the order of 5–10 GPa for most geologic materials (see Fig. 15). This is the pressure–volume point beyond which the shocked material no longer deforms elastically and permanent changes are recorded on removal of the shock pressure. The upper limit is controlled by the peak pressures generated on impact. These vary with the type of impacting body and target material, but are principally a function of impact velocity, reaching into the hundreds to thousands of GPa. Table IV gives estimates

of peak pressures for various impact velocities for iron, stony, and icy bodies impacting a common lunar rock type, gabbroic anorthosite. Shock metamorphism is also characterized by strain rates that are orders of magnitude higher than those produced by internal geologic processes. For example, the duration of so-called regional metamorphism associated with tectonism on Earth is generally considered to be in the millions of years. In contrast, the pressures associated with shock metamorphism produced during the formation of, for example, a 20-km crater occur on a timescale of much less than a second.

1. Subsolidus Effects

At pressures below the HEL, minerals and rocks respond to the shock wave by brittle deformation, which is manifested as fracturing, shattering, and brecciation. These features are not readily distinguished from those produced by, for example, tectonism. There is, however, a unique shock wave effect in the development of unusual, striated, and horse-tailed conical fractures, known as shatter cones (Fig. 16). Shatter cones are best developed at relatively low shock pressures (5–10 GPa) and in fine-grained, structurally homogeneous rocks, such as carbonates, quartzites, and basalts.

Apart from shatter cones, all diagnostic subsolidus shock effects are microscopic in character. The most obvious are so-called planar deformation features and diaplectic glasses. Planar deformation features are a few microns in width and are arranged in parallel sets, with specific crystallographic orientations (see Fig. 16). They develop in many minerals but are best known from the common silicate minerals, quartz and feldspar, for which shock-recovery experiments have calibrated the onset shock pressures of particular orientations. They develop initially at ~10 GPa and continue to 20 to 30 GPa (see Fig. 15), by which pressure multiple sets occur, with differing orientations in individual

---

FIGURE 14 Formation of a large complex crater. (a) Schematic cross section of transient cavity geometry. Note the similarity of transient cavity geometry to that of a simple crater. It contains, however, considerably more melt due to decreased cratering efficiency at larger cavity diameters. This geometry does not exist at any one instant in time, as maximum depth and diameter are not reached at the same time (see below). (b) Transient cavity at maximum depth. Cavity floor is covered by a thick lining of impact melt and highly shocked debris. Note time of ~1 minute to achieve a depth of ~30 km. (c) Transient cavity at maximum diameter. Though excavation has been proceeding and the diameter being enlarged, the transient cavity floor has begun to rebound upward toward the surface. As the transient cavity floor rises upward, the area exterior to the cavity rim begins to collapse downward and inward. (d) Near completion of collapse of rim area and uplift of center. Transient cavity diameter has been enlarged considerably by faulting in the rim area and the interior impact melt has overridden the cavity to line the floor of the final crater. (e) Final form of ~200-km-diameter complex crater form. The rim is a structural feature, with down-faulted blocks of target materials. The crater floor is relatively flat with central uplifted structures, in this case a ring. Local readjustments take place for some time after the event as the crater form stabilizes and material is shed from topographic highs.

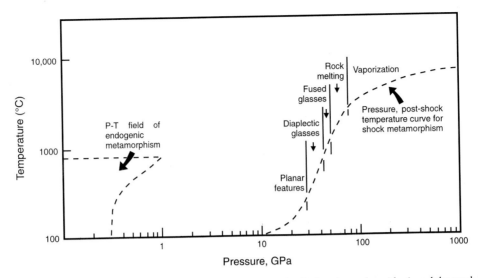

FIGURE 15 Pressure–temperature range of shock metamorphic effects. Note that both scales are logarithmic and the much higher pressures and temperatures reached during shock metamorphism compared to normal metamorphism generated by internal mechanisms in planetary crusts. The pressure–temperature range of a number of diagnostic shock metamorphic effects are indicated (1 GPa = 10 kilobars).

grains. The increasing effects of shock pressure, as observed optically through the development of planar deformation features, are mirrored by changes in X-ray characteristics, indicative of the increasing breakdown of individual mineral structure to smaller and smaller domains.

By ~30–40 GPa (see Fig. 15), quartz and feldspar are converted to so-called diaplectic glass. These are solid–state glasses, with no evidence of flow, that exhibit the same outline as the original crystal (see Fig. 16). For this reason, they are sometimes referred to as thetomorphic ("same shape") glasses. The variety produced from plagioclase, known as maskelynite, was originally discovered in 1872 in the Shergotty meteorite. Diaplectic glasses differ from silicate glasses

quenched from melts in having higher densities and refractive indices.

At shock pressures of 40–50 GPa (see Fig. 15), the waste heat trapped in minerals, due to the incomplete recovery of the pressure–volume work done on them by shock compression, is sufficient to initiate melting in some minerals (see Fig. 16). For example, feldspar grains show incipient melting and flow at shock pressures of ~45 GPa. Melting tends initially to be mineral specific, favoring mineral phases with the highest compressibilities. It also tends to be concentrated at grain boundaries, where there are reverberations of the shock wave. As a result, highly localized melts of mixed mineral compositions can arise.

The effects of shock reverberations on melting are most obvious when comparing the pressures required for melting of particulate materials, such as those that make up parts of the lunar surface and solid rock of similar composition. Shock-recovery experiments indicate that intergranular melts can occur at pressures as low as 30 GPa in particulate basaltic material, compared to 45 GPa in solid basalt. This is due to reverberations of the shock wave at the free surfaces of the individual grains and possibly to very violent shearing between grains. The density contrast between minerals and voids leads to highly localized pressure concentrations and to the heterogeneous deposition of waste heat.

Most minerals undergo phase transitions to dense, high-pressure phases during shock compression. Little is known, however, about the mineralogy of the high-pressure phases, as they generally revert back to low-

TABLE IV
Peak Shock Pressures for Various Bodies Impacting Gabbroic Anorthosite

Impact velocity (km s⁻¹)	Impacting body		
	Iron (GPa)	Stony (GPa)	Icy (GPa)
5.0	82	62	29
7.5	153	99	57
15.0	481	304	185
30.0	1590	1010	633
40.0	3360	2130	1330

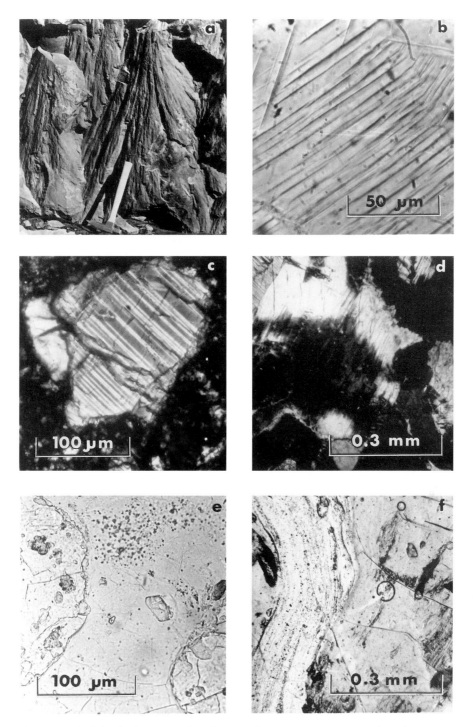

FIGURE 16 Some diagnostic subsolidus shock metamorphic effects. (a) Shatter cones in quartzite at the Sudbury impact structure, Ontario, Canada. (b) Photomicrograph of planar deformation features in the mineral quartz. (c) Photomicrograph of planar deformation features in the mineral pyroxene. (d) Photomicrograph of the mineral plagioclase partly converted to diaplectic glass (black areas in large central white grain). (e) Photomicrograph of coesite, a high pressure polymorph of quartz (cluster of grains in the top third of the image) set in diaplectic quartz glass. (f) Photomicrograph of mixed glasses produced by the melting and flow of individual minerals.

pressure phases during decompression by the rarefaction wave. Where metastable high-pressure phases are preserved, they may be either high-pressure polymorphs of preexisting low-pressure phases or high-pressure assemblages due to mineral breakdown. Some known high-pressure phases, such as diamond from carbon or stishovite from quartz (SiO_2), form during shock compression. Others, such as coesite (SiO_2), form by reversion of such minerals during pressure release (see Fig. 16). Several high-pressure phases have been noted in shocked meteorites, however, they are relatively rare at terrestrial craters. This may be due to postshock thermal effects, which are sufficiently prolonged in a large impact event to inhibit preservation of metastable phases.

2. Impact Melting

Shock-recovery experiments indicate that the waste heat trapped in shocked rocks is sufficient to cause whole-rock melting above ~60 GPa. Thus a zone is produced during an impact event, in which the target rocks are melted and even vaporized (see Fig. 14). Ultimately, these liquids cool to form impact melt rocks. These occur as glassy bodies in ejecta and breccias, as dikes in the crater floor, as pools and lenses within the breccia lenses of simple craters, and as annular sheets surrounding the central structures and lining the floors of complex craters and basins (see Figs. 3, 4, and 17).

Although some terrestrial impact melt rocks were initially identified as having volcanic origins, they were compositionally distinct from usual volcanic rocks. Impact melt rocks have compositions determined by the target rocks, whereas volcanic rocks have compositions determined by internal melting processes that involve partial melting of more mafic and refractory progenitors within Earth's mantle or crust. In addition, impact melt rocks have unusual isotopic compositions, with initial ratios in systems such as $^{87}Sr/^{86}Sr$ and $^{144}Nd/^{143}Nd$ that are equivalent to those of the target rocks. On the other hand, when the same systems involving daughter products (or other systems such as $^{40}Ar/^{39}Ar$) are used for isotopic dating, they indicate younger crystallization ages, reflecting the time of impact.

Impact melt rocks can also contain shocked and unshocked lithic and mineral clastic debris. During the cratering event, as the melt is driven down into the expanding transient cavity (see Fig. 14), it overtakes and incorporates less-shocked materials as clasts ranging in size from small grains to large blocks. Impact melt rocks that cool quickly generally contain large fractions of clasts. In melt rocks that cool more slowly, there is evidence of melting and resorption of the clastic debris. This is possible because impact melt rocks are initially superheated, being a mixture of melt and vapor. This is another characteristic that sets impact melt rocks apart from volcanic rocks, which generally have temperatures equivalent to their melting temperature and no higher. In larger impacts that produce thick, annular impact melt sheets within complex craters and basins, recognizable clastic debris is limited, therefore, to the fast-cooling units at the bottoms and tops of the melt sheets.

Enrichments above target rock levels in so-called siderophile (iron-favoring) elements, and sometimes chromium, have been identified in some impact melt rocks. This is the result of an admixture of a small amount of material from the impacting body. The shock pressures in the impacting body typically are sufficient to vaporize or even ionize most of it. Most of the projectile material is removed from the immediate area in high-speed, early-time ejecta, but a few percent can become mixed with the melted target rock and driven down into the cavity. Identification of the projectile type can sometimes be made through the relative abundances of siderophile elements, which are relatively enriched, compared to planetary crustal rocks, in chondritic impactors whose parent bodies did not undergo a core-forming event. The formation of an iron core in a planetary or meteorite parent body scavenges siderophile elements and locks them in the core. Impactors from meteorite parent bodies that did undergo such an event, for example, those that produced the basaltic achondrites, do not have enriched siderophiles and cannot be readily identified in impact melt rocks.

III. CRATERS AS PLANETARY PROBES

A. SAMPLING TOOLS

As large impacts transport material from within the target planet's crust (and possibly the underlying mantle), they act as excavation tools for planetary geologists for both sampling and remote sensing studies. One of the fundamental tenets of cratering mechanics is that the deepest material excavated and ejected is emplaced closest to the crater and at the lowest velocities. In applying this to the lunar sample collection, many of the deepest-seated rocks (e.g., so-called norites, troctolites, and dunites) were, in fact, found at the *Apollo 17* landing site, which is on the rim of the ~700-km-diameter Serenitatis Basin. However, our understanding of the excavation depth of such large lunar basins is poor, complicating efforts to derive the depth of origin of such samples. This uncertainty is heightened by the processes of mixing and redistribution at the lunar surface and its components by the ceaseless bombardment of the Moon throughout geologic time, which further confuses the question of origin of any given sample.

The effects of large impacts are also important in that the uplift of material in the centers of the complex structures also brings deep-seated material to the surface. Remotely sensed reflectance spectra of the 100-km-diameter lunar crater Copernicus indicate that the dominant mafic mineral in the central peaks is olivine, which is probably mixed with feldspar. Use of the equation $SU = 0.086D^{1.03}$ for structural uplift for Earth yields a value of about 10 km of uplift for the rocks at the center of a 100-km crater. Although the relevance of this value for a lunar crater is somewhat problematical, it nevertheless serves to illustrate that the formation of a large crater can expose material that otherwise would be inaccessible to the planetary geologist.

Samples have been obtained on Earth from other bodies in the solar system by virtue of the impact process. All meteorites in the terrestrial collection have been shocked to some extent, implying that they have been subjected to at least one impact event. Ejection of material from asteroids during collisions, for example, can place it in heliocentric orbits that often can be perturbed by the gravitational fields of other bodies, particularly Jupiter. These orbits can then evolve into Earth-crossing trajectories. Those that collide with Earth and survive atmospheric entry become meteorites. Along with those recovered elsewhere, the meteorites found in Antarctica have yielded a wide variety of specimens, including those confirmed to have had a planetary origin. These include samples of lunar highland breccias (Fig. 18) and examples of the Shergotty, Nakhla, and Chassigny group of meteorites, the so-called SNC meteorites, which have come from Mars. The scientific advantages of having such samples are obvious. However, the source regions for the lunar and Martian meteorites on their respective planets can only be guessed. [*See* METEORITES.]

B. GEOLOGICAL GAUGES

On many planetary surfaces, craters can be completely or partially buried by later units, such as volcanic flows

FIGURE 17 Edge of the eroded impact melt sheet that lines the floor of the Manicouagan impact crater, in Quebec, Canada (see Fig. 4). The original melt volume for this 100-km crater is estimated to be ~1000 km³.

or sedimentary deposits. Because craters have defined morphometric relations, they can be used to determine the thickness of the infilling unit within the crater. For example, the morphometric relations in Table II and the observation that a 20-km lunar crater is all but buried by mare volcanics indicate that the mare fill inside the crater is ~3 km thick. However, this does not mean that the lunar mare in question is 3 km thick. This would only be the case if the filled crater had formed on the original mare basin floor before any subsequent lava had erupted. This is likely not the case, with the flooded crater in question having been formed sometime during the eruption history of the mare. As a result, depths of mare thickness calculated by this method tend to be minima. Nevertheless, this is one of the few photogeologic techniques to estimate parameters such as thicknesses of volcanic units from planetary imagery. In the case of the Moon, its use gives estimates of basalt thickness of 0.5–1 km at the edge of the mare basins, rising to a few kilometers in the interiors.

The populations of craters on a given planet are sufficiently similar that they can also be used as indicators of various deformational processes. For example, very long, lobate scarps occur on Mercury. In analyses of the characteristics of these scarps, it was found that they cut craters and that the distortions in the shapes of these craters could be used to estimate the amount of displacement induced by the formation of the scarps

FIGURE 18 "Lunar meteorite" ALHA 81005,0 was found in 1981 near the Allan Hills in Antarctica. Judging from its clast content, this regolith breccia was ejected from somewhere in the lunar highlands during a cratering event. The glassy coating visible on the top, right, and bottom right is part of the fusion crust formed during the rock's passage through Earth's atmosphere. The cube is 1 cm on a side. (NASA Photograph S82-35865.)

(Fig. 19). On the basis of such deformations and other evidence, it was determined that these scarps are thrust features and can be explained by a decrease in Mercury's radius of 1 to 2 km. This crustal shortening event was in all likelihood caused by the cooling and contraction of Mercury's interior. [See MERCURY.]

Another application of crater morphometry and morphology is in constraining estimates of the strengths or viscosities of planetary interiors. A topography of a given shape imposed on a viscous medium will change its configuration with time, as the medium flows in response to the stress field generated by that shape. Impact craters have relatively well-defined topographies and shapes as a function of size and thus are well suited as gauges of the viscosities of planetary interiors. When the viscosity of the target is sufficiently low, models show that viscous relaxation will cause the crater's floor to bow upward, its rim to subside, and a depressed annulus to form outside the rim crest.

There is abundant evidence from *Voyager* photography of Ganymede that large craters degrade viscously and that small ones persist for longer times. This indicates that the viscosity of the outer layer of the planet is or was less viscous than that of the Moon, in keeping with the high H_2O ice content of Ganymede. The temperature gradient inside the planet, however, is unknown, hindering quantitative estimates of the viscosity of the outer layer. Further study of crater relaxation on Ganymede could help to constrain the viscosity with depth and, perhaps, the evolution of subsurface temperature with time.

C. PLANETARY CHRONOLOGIES

The longer a surface is exposed to the impacting flux of interplanetary objects, the greater the areal density of craters on that surface. If the impacting flux (i.e., the number of impacts of a given size with time) were known, it then would be a simple matter of counting craters on the surface to establish its absolute age. The procedure used in age-dating planetary surfaces is straightforward. All craters greater than a predetermined size are measured, recorded, and sorted with respect to diameter. The number of craters greater than a given diameter are then plotted against that diameter, resulting in a cumulative size–frequency diagram (Fig. 20). This is repeated for other surfaces.

FIGURE 19 One of a number of compressional scarps on the surface of Mercury. This scarp is about 500 km long. Average relief to the west (left) of the scarp is about 2 km higher than that to the east. The larger of the two craters cut by the scarp is about 60 km in diameter. Note the deformation induced by the scarp in the shape of the crater near the top of the photograph. (*Mariner 10* Photograph FDS 27399.)

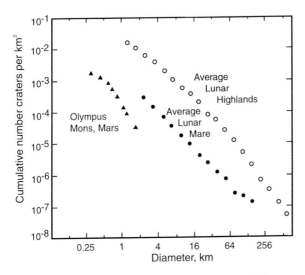

FIGURE 20 Cumulative number of craters per square kilometer versus crater diameter for various planetary surfaces. Note that both scales are logarithmic. The higher number of craters at any given diameter for the lunar highlands indicates that this is the oldest surface. This technique of crater counting is the basic method for determining the relative ages of planetary surfaces.

Interpretation of these plots is also straightforward. The surfaces with the greatest density of craters are the oldest.

This approach is normally accurate when large craters are used. Unfortunately, many surfaces of interest are so young that small craters must be used to obtain a sufficiently large statistical base for interpretation. The danger in using smaller craters lies in the growing population of secondary craters at smaller diameters. Secondary craters are those created by the impact of ejecta from a distant, larger crater, and they can be virtually indistinguishable from primary impacts. A number of secondary craters counted as primaries will result in an overestimate of the age of a surface.

Another difficulty occurs at the other end of the age spectrum. Craters on a very old surface can be obliterated if the density of impacts reaches a certain level. When craters of a given size are buried by ejecta or destroyed by impact erosion as fast as they are formed, that surface is said to be at equilibrium at that size. The identification of equilibrium surfaces, however, is a highly contentious issue. Computer simulations of the extended bombardment of idealized surfaces by idealized populations of impacting bodies imply that equilibrium occurs at crater densities much higher than those observed on any terrestrial planet. Some workers, however, claim that equilibrium has been observed in a number of cases and that the computer simulations are too simplified. At this time, all that can be stated is that crater counts can give relative age information, provided they are interpreted with care. In the case where supporting calibration ages are available, such as for lunar samples, absolute ages of some surfaces can be obtainable from crater statistics.

Crater densities often provide the only basis for assigning relative ages to various geological units on planetary surfaces. For example, because Venus is similar in size to Earth, there has been considerable interest in determining the age of its surface units. Crater counts from recently acquired *Magellan* data, coupled with reasoned assumptions about Venus's impact flux, indicate that its average surface age is ~500 million years, not unlike the average age of Earth's land surface. In addition, some areas on Venus, such as the 6×10^6-km^2 Sappho area, have no impact craters, indicating that they are young, probably volcanic, areas. Conversely, there are no areas with a high density of craters. Thus, like Earth, Venus has preserved little if any of its early crust. [*See* VENUS: SURFACE AND INTERIOR.]

Crater counts can be used to provide some information on surface history. The production, or primary, population of craters on the terrestrial planets (Fig. 21) has a size–frequency distribution that approximates

$$N \propto D^{-2}$$

where N is the cumulative number of craters with diameters greater than D. If some resurfacing event occurs, small craters may be removed from the population by burial. As a result, the size–frequency distribution of craters will deviate from the production relationship. There will be a deficit of small craters and the slope of the size–frequency distribution will be reduced at sizes below some critical diameter. The diameter at which the deviation begins gives some indication of the magnitude and thickness of the resurfacing event. On some Martian surfaces, for example, it is possible to observe a dip in the size–frequency slope, then a return to a production slope at smaller diameters. This is indicative of an initial, stable surface with a production population of craters, a later resurfacing event, followed by stabilization of this new surface with the accumulation of a new production population of, in this case, smaller craters.

Not all deviations from a production population size–frequency relation are due to resurfacing events. For example, Earth and Venus show very similar size–frequency distributions (see Fig. 21), which deviate from a production distribution at diameters less than 20–25 km. In the case of Earth, the relative absence of craters at smaller diameters is due to erosion, which can be considered a resurfacing event. In the case of Venus, however, the relative absence of craters at smaller diameters is due largely to atmospheric effects, with the retardation and crushing of smaller impacting bodies by the dense atmosphere of Venus.

IV. IMPACT AND PLANETARY EVOLUTION

As the impact flux has varied through geologic time, so has the potential for impact to act as an evolutionary agent. The Moon has recorded almost the complete record of cratering since planetary crustal formation. Crater counts combined with isotopic ages on returned lunar samples have established an estimate of the cratering rate and its variation with time. Terrestrial data have been used to extend knowledge of the crater rate, at least in the Earth–Moon system, to more recent geologic time. The lunar data are generally interpreted as indicating an exponential decrease in the rate until ~4.0 billion years (b.y.) ago, further slower decline for an additional billion years, and a relatively constant rate, within a factor of two, since ~3.0 b.y. ago (Fig. 22). The actual rate much prior to ~4.0 b.y. ago is imprecisely known, as there is the question of whether crater counts from the most ancient lunar highlands reflect all the craters that were produced (i.e., they are a production population) or only those that have not been obliterated by subsequent impacts (i.e., they are an equilibrium population). Thus, the oldest lunar surfaces may give a minimum estimate of the ancient cratering rate. Similarly, there is some question as to whether the largest recorded events, represented by the major multiring basins, occurred over the relatively short time period of 4.2–3.8 b.y. ago or were spread more evenly with time, as expected if this early high cratering rate represented the impacts by remnants of the process of planetary accretion.

One of the fundamental hypotheses for the growth of planetary bodies from the early solar nebula is their accretion by collision. Models of planetary accretion involve the collision of progressively larger objects until the solar system is populated by asteroid-sized bodies, then embryonic planets (planetesimals), and finally planets at their current size. At first, the relative veloci-

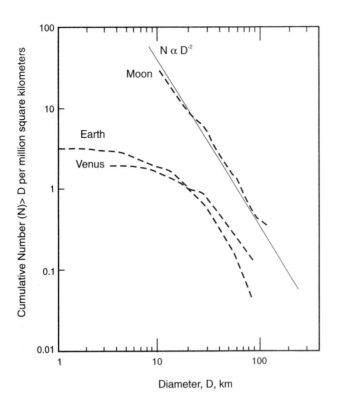

FIGURE 21 Cumulative number of craters (N) against crater diameter (D) for the Moon, Earth, and Venus. Note that both scales are logarithmic. The lunar data are for the average lunar mare and the size–frequency distribution of craters approximates $N \propto D^{-2}$, which is the primary production population of craters. The terrestrial and Venusian crater populations are similar and deviate from $N \propto D^{-2}$ at $D \leq 20$ km. There is a shortage of small craters, which is due to planetary processes, principally erosion, in the case of Earth, and atmospheric breakup and destruction of smaller impacting bodies in the case of Venus.

ties between these bodies are low, but they are gradually increased by gravitational interactions until they reach kilometers per second. The assembly of the final planets by massive collisions is believed to have happened very rapidly. During this period of runaway growth, some of the growing bodies may have been completely disrupted by collisions and reaccreted under their own self-gravitation. This is believed to be the case for several of the moons of Uranus, where widely varying juxtaposed surface units are explained by repeated breakup by impact and reassembly.

A. IMPACT ORIGIN OF EARTH'S MOON

The impacts of the greatest magnitude dominate the cumulative effects of the much more abundant impacts in terms of affecting planetary evolution. In the case of Earth, this may have been a massive impact that produced the Moon. Earth is unique among the terrestrial planets in having a large satellite, and the origin of the Moon has always presented a problem.

Previous hypotheses for the origin of the Moon fell into three general categories: capture of an existing body, fission from Earth, or coaccretion with Earth. None of these hypotheses, however, satisfies the constraints of post-Apollo knowledge of the chemistry and physics of the moon and the Earth–Moon system. The suggestion that the Moon formed from a massive impact with Earth was originally proposed over fifteen years ago. With the development of complex code calculations and more efficient computers, it has been possible recently to model such an event. Most models involve the impact of a Mars-sized object, about half the diameter of Earth, which would produce an Earth-orbiting disk of impact-produced vapor, consisting mostly of mantle material from Earth and the impacting body. This disk, depleted in volatile and enriched in refractory elements, would cool, condense, and accrete to form the Moon. In the computer simulations, very little material from the iron core of the impacting body goes into the accretionary disk, accounting for the low iron and, ultimately, the small core of the Moon. In addition to the formation of the

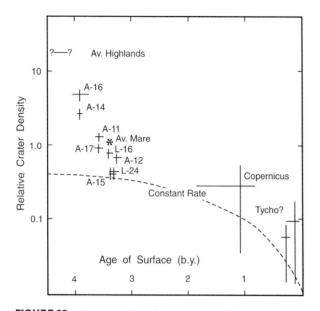

FIGURE 22 Crater density, relative to average lunar mare, against time for various lunar surfaces and Earth. Note that the relative crater density scale is logarithmic. Apollo sites and Luna sites on the Moon are designated A and L, respectively. Cratering rate has fallen sharply in the first 1.5 b.y. of planetary history and can be modeled as an approximately constant rate from ~3.0 b.y. till the present. Uncertainties for the various estimates are indicated by vertical and horizontal bars.

Moon, effects of such a massive impact on the earliest Earth itself would have been extremely severe, leading to massive remelting of Earth and loss of the original atmosphere. [*See* THE MOON.]

B. EARLY CRUSTAL EVOLUTION

Following planetary formation, the subsequent high rate of bombardment by the remaining tail of accretionary debris is recorded on the Moon (see Fig. 22), and on some of the other terrestrial planets and icy satellites of the outer solar system that have preserved portions of their earliest crust. Because of the age of its crust, the relatively large number of planetary missions, and the availability of samples, the Moon is the source of most interpretations of the effects of such an early, high flux on planetary geologic evolution. In the case of the Moon, a minimum of 6000 craters with diameters greater than 20 km are known to have been formed during this early period. In addition, there were ~45 impacts producing impact basins, ranging in diameter from Bailly at 300 km to the South Pole Aitken Basin at 2600 km and the putative Procellarum Basin at

3500 km. The results of the Apollo missions demonstrate clearly the dominance of impact in the nature of the samples from the lunar highlands. Over 90% of the returned samples are impact lithologies, with 30% by volume being impact melt rocks. The dominance of impact as a process for change is also reflected in the age of the lunar highland samples. The bulk of the near-surface rocks, which are impact products, have ages in the 4.0–3.8 b.y. range. Only a few, so-called pristine, igneous rocks with ages ≥3.9 b.y. from the early lunar crust occur in the Apollo collection. Computer simulations indicate that the cumulative thickness of materials ejected from major lunar craters in the lunar highlands is 2–10 km. Beneath this, the lunar crust is believed to be brecciated and fractured by impacts, but essentially autochthonous or in place, to a depth of 20–25 km.

The large multiring basins on the Moon define the basic surface dichotomy between the ancient anorthositic highlands and younger, infilling, basaltic maria. They also define the major topographic features of the Moon. For example, the topography associated with the Orientale Basin (see Fig. 6), the youngest multiring basin at ~3.8 b.y. and, therefore, the basin with the least topographic relaxation, is over 8 km, somewhat less than Mt. Everest at ~9 km. The impact energies released in the formation of impact basins in the 1000-km size range are on the order of 10^{27}–10^{28} J, a million to ten million times the present annual output of internal energy of Earth. The volume of crust melted in basin-forming events of this size is on the order of a million cubic kilometers. Although crater ejecta are generally confined to within <2.5 diameters of the source crater, this still represents essentially hemispheric redistribution of materials in the case of an Orientale-sized impact on the Moon.

Following formation, these impact basins localize subsequent endogenic geologic activity in the form of tectonism and volcanism. As a consequence of such a large impact, the uplift of originally deep-seated isotherms results in a thermal anomaly. The subsequent tectonic evolution of the basin and its immediate environs is a function of the gradual loss of this thermal anomaly, which may take as long as a billion years to completely dissipate. Cooling leads to stresses, crustal fracturing, and basin subsidence. In addition to thermal subsidence, the basins may be loaded by later mare volcanics, leading to further subsidence and stress.

All the terrestrial planets experienced the formation of large impact basins early in their histories. Earth and Venus, however, do not retain any record of this massive bombardment, because subsequent geologic activity has removed and renewed their earliest surfaces. There has been considerable speculation as to the potential effects of basin-sized impacts on Earth, which actually received a higher cratering flux than the Moon because of its greater gravitational attraction. From the lunar cratering rate, and assuming relatively low, early cosmic approach velocities of 6–10 km s^{-1} at 4.0 b.y., it is estimated that as many as 200 impact basins greater than 1000 km may have been formed on Earth from 4.5 to 3.8 b.y. The cumulative effect of such a bombardment on the surface and upper crustal rocks of Earth is unknown.

Basin-sized impacts will have also affected any existing atmosphere, hydrosphere, and potential biosphere. The vapor plume created by high-velocity ejecta in sufficiently large impacts can blow off the atmospheric mass above the tangent plane through the point of impact. Considering the cumulative impact flux from the lunar record, it has been estimated that the primordial atmosphere on early Earth could have been reduced in density by a factor of five by such a process. Similar calculations for Mars indicate that atmospheric erosion by impact would be even more severe, with a reduction in density between the primordial atmosphere at 4.5 b.y. and the atmosphere at ~3.5 b.y. by a factor of ~100. This high rate of atmospheric erosion is due to the relatively small size and low gravity of Mars. This reduction in atmospheric pressure with time due to impact blowoff may account for the cessation of the period of ancient fluvial activity and enhanced erosion by surface water early in martian history.

The impact on Earth of a body in the 500-km size range, similar to the present-day asteroids Pallas and Vesta, would be sufficient to evaporate the world's present oceans, if 25% of the impact energy were used in vaporizing the water. Such an event would have effectively sterilized the surface of Earth. It would have been enveloped by an atmosphere of hot rock and water vapor that would radiate heat downward onto the surface, with an effective temperature of a few thousand degrees. It would take some 3000 years for the water-saturated atmosphere to rain out and reform the oceans. Smaller impacts by bodies in the 200-km size range would be sufficient to evaporate the photic zone of the oceans. Models of the potential of impact to frustrate the development of life on Earth indicate that life could have survived in a deep marine setting at 4.2–4.0 b.y. Smaller impacts, however, would continue to make the surface inhospitable until 4.0–3.7 b.y. Similarly, photosynthesis would have been difficult before 3.7 b.y., because of relatively frequent and violent, impact-induced climatic excursions.

C. SURFACE EVOLUTION

The comminution (or fracturing) effects of impact range from the fragmentation of submicron mineral grains to the disruption of large asteroids. The pulverization of surfaces by impact is most severe on bodies with no atmosphere to reduce the velocity and flux of smaller impacting bodies. It is probable that the Moon never had a surface of solid rock, in the sense that much of Earth possesses one. Continuous impact events, from the time of accretion to the present, have fractured, ejected, mixed, and repeatedly pulverized the lunar surface to a degree that is dependent on depth. This layer is known as the regolith. Even the mare basalts, which were erupted and solidified over relatively short periods of time, have what are believed to be regolith surfaces between flow units and now are covered with a highly disaggregated, impact-generated layer. The frequency of impacts is highly dependent on the sizes of the impacting objects, with small particles being much more abundant than large ones. Only the upper layers of the regolith are affected by these small impacts. The net result is that the uppermost layers become finer-grained with time, suffer impact melting at a higher rate, and are mixed more rapidly than the lower portions of the regolith. Occasionally, a large impact will occur, exhuming previously buried material and burying the former surface. This process occurred at all scales, even to the size of impact basins.

There are important differences, however, between the debris layer generated at the scale of impact basins and that of the much thinner surface regolith. The overall conditions of the generation and evolution of such a layer are comparable with the surface regolith, so this larger counterpart is called the megaregolith. A typical mare regolith is on the order of 5 to 10 m deep and is irregularly bounded at its base by a fractured but nevertheless distinct layer of its parent basalt. The megaregolith, on the other hand, occurs in the ancient lunar highlands and probably underlies the mare basalts, extending to local depths perhaps greater than 10 km. The nature of its lower boundary is problematic, in that the depth of fracturing by large craters is poorly understood, due both to our general lack of knowledge of the effects of large-scale cratering at extreme depth and to the poorly understood behavior of the lower lunar crust and even the upper mantle during such deep "fracturing" events.

Just as the evolution of the regolith and megaregolith is dominated by impact, so the results of impact events are affected by the regolith and megaregolith. For example, the amount of impact energy partitioned into heating is higher when the target is porous and impacts into regoliths typically melt and vaporize more material than those into solid rock. The morphologies of relatively small craters are affected to a great degree by the thickness of the regolith in the target area. Indeed, a progression of morphologies from small, cone-shaped craters through those with a central mound or a flat floor to those with a concentric (or "nested") morphology (Fig. 23) has been described as a function of regolith thickness. This morphological sequence has also been duplicated in the laboratory, using targets that simulate unconsolidated regolith overlying a solid substrate. This spectrum of morphologies, when observed, can be used to map regolith thicknesses from planetary images.

D. BIOSPHERE EVOLUTION

As the residual planetesimals were removed by collisions with planetary bodies and ejection from the solar system, the impact record was increasingly dominated by impacts of material perturbed from the more stable asteroid belt between Mars and Jupiter and by cometary material from the outer reaches of the solar system. By ~3.0 b.y., evidence from the Earth-Moon system suggests that the cratering rate had essentially stabilized to something approaching a constant value (see Fig. 22). While major basin-forming impacts were no longer occurring, there were still occasional impacts resulting in craters in the size range of 100 km or greater. For example, the terrestrial record contains remnants of the Sudbury, Canada, and Vredefort, South Africa, structures, which have estimated original crater diameters of ~250 km and ~300 km, respectively, and ages of ~2 b.y. Events of this size are unlikely to have caused significant long-term changes in the geosphere, but likely affected the biosphere of Earth.

In more recent Phanerozoic time, a number of large craters are preserved that are temporally close to transitions in the biostratigraphic record on Earth. However, this may be coincidental. At present, there is only one case where there is a direct physical or chemical link between a large impact event and changes in the biostratigraphic record—this is at the so-called Cretaceous–Tertiary (K/T) boundary, which occurred ~65 million years (m.y.) ago. The physical evidence for impact contained in K/T boundary deposits includes: shock-produced, microscopic planar deformation features in quartz and other minerals; the occurrence of stishovite (a high-pressure polymorph of quartz); high-temperature minerals believed to be vapor condensates; and various, generally altered, spherules. The

chemical evidence consists primarily of a siderophile anomaly in the K/T boundary deposits, indicative of an admixture of meteoritic material. In undisturbed North American K/T sections, which were laid down in swamps and pools on land, the K/T boundary consists of two units: a lower one, linked to ballistic ejecta, and an upper one, linked to atmospheric dispersal in the impact fireball and subsequent gentle fallout over a period of time. This fireball layer occurs worldwide. However, the ejecta horizon is known only in North America.

The K/T boundary marks a mass extinction in the biostratigraphic record of the Earth. In this case, ~75% of the various types of life at the time became extinct. With discovery of evidence for a large impact, a cause–effect relationship has been suggested. Originally, it was suggested that dust in the atmosphere from the impact led to global darkening, the cessation of photosynthesis, and cooling. Models indicated that the residence time for the dust in the atmosphere would be on the order of 6 months, with no photosynthesis for 3 months and below freezing temperatures on land for 6 months. These models are heavily dependent on initial assumptions, the most critical being the particle size of the dust. In these cases, it was assumed to be very small (about half a micron) and similar to windblown and volcanic dust in the upper atmosphere.

Other potential killing mechanisms have been suggested. They include a period of initial cooling due to suspended dust, followed by global warming as the result of an oceanic impact and injection of water into the atmosphere, producing an enhanced greenhouse effect and a transient increase in global temperatures on the order of 10°C. Calculations of shock-wave heating of the atmosphere and conversion of atmospheric nitrogen to oxides of nitrogen indicate that a K/T-sized impact would have created global acid rains. This would result in the defoliation of plants, the asphyxiation of lung-respiring animals, and the killing of calcareous organisms in the top 150 m of the world's oceans. Soot has also been identified in K/T boundary deposits and its origin has been ascribed to global wildfires. Soot in the atmosphere may have enhanced or even overwhelmed the effects produced by global dust clouds.

During the study of the K/T impact, more and more emphasis has been placed on understanding the effects on the atmosphere by the vaporized and melted ejecta. It is this hot, expanding vapor cloud that may provide a mechanism for igniting global wildfires. Models of the thermal radiation produced by the ballistic reentry of ejecta condensed from the vapor plume of the K/T impact indicate a thermal radiation pulse

FIGURE 23 The morphologies of relatively small craters can be used as probes of the lunar regolith, and a wide range is displayed by this portion of an *Apollo 15* panoramic camera photograph. The inner bench of the concentric crater at A is almost square, perhaps reflecting jointing in the basalt underlying the regolith in that area. A smaller, flat-floored crater at B is just a few hundred meters from crater A. The depth of the regolith at that location is very nearly equal to the depth of the crater. The two concentric craters at C also possess mounds on their floors. This multiplicity of features is probably caused by multiple sequences of regolith and more competent parent basalts. The outermost rim delineates the crater formed in the uppermost regolith, whereas the inner rim outlines the crater formed in the stronger basalt. Beneath that basalt, however, is another, older layer of regolith, which was not completely penetrated during formation of either of these two craters. The lower crater at C is about 250 m across. (*Apollo 15* Panoramic Camera Photograph 9331.)

on Earth's surface. This thermal radiation may have been directly or indirectly responsible, through the global killing and drying of vegetation that is later ignited by lightning strikes, for the global wildfires. Calculations also suggest that the thermal radiation produced by the K/T event was at the lower end of that required to ignite living plant materials. If global wildfires are a requirement to produce global mass extinction effects, in addition to the deleterious effect of atmospheric dust loadings and acid rain, this may explain why impacts that result in 50- to 100-km-sized craters do not result in mass extinctions. They produce sufficient dust, according to the models, but the accom-

panying thermal radiation is insufficient to result in the global ignition of living plant materials.

Although the record in the K/T boundary deposits is consistent with the impact interpretation, it is clear that many of the details are not known at the present time, particularly with respect to the potential killing mechanism and associated mass extinctions. One of the arguments of opponents of the impact interpretation was that there is no known K/T crater. Without this "smoking gun," they have argued, the physical and chemical evidence are circumstantial and open to various interpretations.

This is no longer the case. The "killer crater" has

been identified as the ~180-km-diameter structure, known as Chicxulub, buried under 1 km of sediments on the Yucatan peninsula, Mexico. Evidence for impact appears in the form of shock-produced planar deformation features in the minerals quartz and feldspar in deposits both interior and exterior to the structure, as well as the occurrence of impact melt rocks. Isotopic dating of impact melt rocks from Chicxulub indicates a K/T boundary age. There is evidence from variations in the concentration and size of shocked quartz grains and the thickness of K/T boundary deposits, particularly the ejecta layer, that points toward a source crater in the area of Central America.

If Chicxulub is the K/T "killer crater," then it provides evidence of potential extinction mechanisms. The Chicxulub target contains beds of anhydrite ($CaSO_4$). At ~1500°C, anhydrite decomposes to release SO_2. Calculations for the Chicxulub impact indicate that the SO_2 released would have produced over 600 billion tons of sulfuric acid in the atmosphere. Studies have shown that the lowering of temperatures following large volcanic eruptions is mainly due to sulfuric acid aerosols. Scaling these known effects to the mass of sulfuric acid from the Chicxulub impact leads to estimates of global temperature decreases of up to 11°C. The sulfuric acid would eventually return to Earth as acid rain and, depending on the timescale of return, could have had an additional disastrous effect on any surviving biota.

The frequency of K/T-sized events on Earth is on the order of one every ~100 m.y. Smaller, but still significant, impact events occur on shorter timescales. Although such events may not be sufficient to produce mass extinctions, they will affect the terrestrial climate and biosphere to varying degrees. Model calculations suggest that dust loadings from the formation of impact craters as small as 20 km could produce light reductions and temperature disruptions, leading to the equivalent of a "nuclear winter." Such impacts occur on Earth with a frequency of approximately two or three every million years. If these impacts were to land in the ocean, they would produce tsunamis or tidal waves, which would be ocean basin to hemispheric in scale. The most recent known impact crater in this size range is Zhamanshin, Kazakhstan, with a diameter of 15 km and an age of ~1 m.y. Impacts of this scale are not likely to have a serious affect upon the biosphere. The most fragile component of the present environment, however, is human civilization, which is highly dependent on an organized and technologically complex infrastructure for its survival. Though we seldom think

of civilization in terms of millions of years, there is little doubt that if civilization lasts long enough, it will suffer severely or may even be destroyed by an impact event.

Yet impact events can occur on human timescales. For example, the Tunguska event in Russia in 1908 was due to the atmospheric explosion of a relatively small body at an altitude of ≤10 km. The energy released, based on that required to produce the observed seismic disturbances, has been estimated to be equivalent to the explosion of ~10 megatons of TNT. Although the air blast resulted in the devastation of ~2000 km² of Siberian forest, there was no loss of human life because of the very sparse population in the area. Events such as Tunguska occur on a timescale of hundreds of years. At the rate at which the human population is expanding, it makes it all the more likely that loss of life will occur in the next Tunguska-sized event to occur over land.

Humanity can take some collective comfort in the fact that civilization-threatening impact events occur only every half million years or so, yet it must be remembered that impact is a random process not only in space but also in time. The next large impact with Earth could be an "impact-winter"-producing event or even a K/T-sized global catastrophe. In March 1989, an asteroidal body named 1989 FC passed within 700,000 km of Earth. Although 700,000 km is a considerable distance, it translates to a miss of Earth by only a few hours, when relative orbital velocities are considered. This Earth-crossing body was not discovered until it had passed by. Its diameter is estimated to have been in the 0.5-km range, and its impact would have had hemispherical, if not global, climatic effects.

BIBLIOGRAPHY

Dressler, B. O., Grieve, R. A. F., and Sharpton, V. L. (eds.) (1994). "Large Meteorite Impacts and Planetary Evolution," Special Paper 293. Geological Society of America, Boulder, Colo.

French, B. M., and Short, N. M. (eds.) (1968). "Shock Metamorphism of Natural Materials." Mono Book Corp., Baltimore.

Melosh, H. J. (1989). "Impact Cratering." Oxford Univ. Press, New York.

Roddy, D. J., Pepin, R. O., and Merrill, R. B. (eds.) (1977). "Impact and Explosion Cratering." Pergamon, New York.

Schultz, P. H., and Merrill, R. B. (eds.) (1981). "Multi-Ring Basins." Pergamon, New York.

Silver, L. T., and Schultz, P. H. (1980). "Geological Implications of Large Asteroids and Comets on the Earth," Special Paper 190. Geological Society of America, Boulder, Colo.

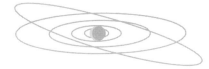

PLANETARY VOLCANISM

I. Summary of Planetary Volcanic Features

II. Classification of Eruptive Processes

III. Effusive Eruptions and Lava Flows

IV. Explosive Eruptions

V. Inferences about Planetary Interiors

Lionel Wilson
Lancaster University, United Kingdom

GLOSSARY

Andesite: Rock type with intermediate metal and silica content, formed on Earth by the remelting in the presence of water of basaltic rocks carried down into the planetary interior.

Basalt: Rock formed as a result of the eruption and subsequent solidification of basaltic magma.

Basaltic magma: Relatively metal-rich, silica- and volatile-poor liquid rock formed as a result of the partial melting of the materials making up planetary mantles.

Convection: Process whereby material heated in a gravitational field expands and moves upward to be replaced by cooler, denser material moving downward.

Effusive eruption: Eruption in which magma emerges from a vent without significant disruption to form a lava flow.

Eruption: Any emergence of magma at the surface of a planet as a lava flow or as pyroclastic material.

Eruption cloud: Mixture of pyroclasts and gas that has a bulk density less than that of the surrounding atmosphere, so that it convects upward into the atmosphere above the vent.

Explosive eruption: Eruption in which magma is disrupted into fragments (called pyroclasts) by the expansion of exsolved gases.

Exsolution: Process whereby volatile compounds dissolved in magmas become saturated due to decreasing pressure and are forced out of solution to create small gas bubbles in the liquid.

Halogen: One of the elements in the family that includes chlorine and fluorine. These gases are released in small amounts from erupting magmas along with larger amounts of water, carbon dioxide, and sulfur compounds.

Hawaiian eruption: Explosive eruption in which a relatively steady discharge of volcanic gas and incandescent pyroclasts forms a fountain above the vent.

Intrusion: Emplacement of magma at a shallow depth below the surface without any surface eruption.

Lava: Completely or partially molten rock erupting onto the surface of a planet. The liquid may contain bubbles of gas produced by exsolution of volatile compounds, mineral crystals produced by cooling of the liquid after its formation, and larger pieces of solid rock removed from the walls of the fissures through which the parent magma reached the surface.

Lava flow: Coherent body of rock emplaced on the surface of a planet as a result of the eruption of lava from a vent and the flowage of the lava downslope away from the vent under gravity.

Magma: Molten or partially molten rock formed in some region of the interior of a planet as a result of a local decrease in pressure or increase in temperature.

Phreato-magmatic eruption: Explosive eruption involving the vigorous interaction of erupting magma with surface water (or any other liquid volatile).

Plinian eruption: Vigorous explosive eruption in which a relatively steady discharge of volcanic gas and hot pyroclasts entrains and heats gas from the surrounding atmosphere to form a high convecting eruption cloud above the vent.

Pyroclast: Any piece of rock resulting from the tearing apart of magma as it nears the surface, usually as a result of the rapid (explosive) expansion of bubbles of gas in the magma. Pyroclasts are generally irregular in shape and commonly contain trapped gas bubbles.

Pyroclastic flow: Mixture of hot pyroclasts and volcanic gas forming a dense, ground-hugging, fluidlike mixture flowing downslope away from a vent.

Rheology: Study of the way fluids deform and flow as a result of applied stresses.

Rhyolite: Rock type formed on Earth mainly by the extensive crystallization of a more basaltic magma.

Strain: Fractional amount of deformation induced by the application of a stress to a body of solid or liquid material.

Stress: Any force applied to a body of solid or liquid material that is potentially able to alter its shape.

Strombolian eruption: Eruption in which large, pressurized bubbles of volcanic gas emerge intermittently from a vent and burst, disrupting the magma surface into a spray of pyroclasts.

Viscosity: Response of a fluid to the application of a stress, defined as the ratio of the applied stress to the rate of strain (i.e., the rate of deformation). For a Newtonian fluid, the viscosity is a constant; for all other materials, it is a function of the amount of stress.

Vulcanian eruption: Form of abrupt, intermittent explosive eruption in which gas pressure builds up beneath a retaining cover of cooled lava until the strength of the covering layer is exceeded.

Volcanism is one of the major processes whereby a planet transfers heat produced in its interior outward to the surface. Volcanic activity has been directly responsible for forming at least three quarters of the surface rocks of Earth and Venus, and for forming extensive parts of the surfaces of Mars, Earth's Moon, Jupiter's satellite Io, and probably Mercury. Investigation of the styles of volcanic activity (e.g., explosive or effusive) on a planet's surface provides clues to the composition of the erupted magma and hence, indirectly, to the chemical composition of the interior. The observation of the products of volcanic eruptive processes formed under a wide range of environmental conditions (gravity, atmospheric pressure, etc.) has been an important spur to the development of an understanding of volcanic processes on Earth.

I. SUMMARY OF PLANETARY VOLCANIC FEATURES

A. EARTH

Only in the middle part of the twentieth century did it become entirely clear that much of Earth's surface, specifically the crust forming the ocean floor, consists of geologically very young rocks. These erupted from long lines of volcanoes, generally located along ridges near the centers of ocean basins, within the last 300 Ma (million years). Along with this realization came the development of the theory of plate tectonics, which explained the locations and distributions of volcanoes over Earth's surface. Volcanoes erupting relatively metal-rich, silica- and volatile-poor magmas (basalts) tend to concentrate along the midocean ridges, which mark the constructional margins of Earth's rigid crust plates. These magmas represent the products of the partial melting of the mantle at the tops of convection cells in which temperature variations cause the solid mantle to deform and flow on very long timescales. Their compositions are very closely related to the bulk composition of the mantle, which makes up most of Earth's volume outside of the iron-dominated core. The volcanic edifices produced by ocean-floor volcanism consist mainly of relatively fluid (low-viscosity) lava flows with lengths from a few kilometers to a few tens of kilometers. Lava flows erupted along the midocean ridges simply add to the topography of the edges of the growing plates as they move slowly (~10

FIGURE 1 A hawaiian-style lava fountain feeding a lava flow and building a cinder cone (Pu'u 'O'o on the flank of Kilauea volcano in Hawai'i). Steaming ground marks the location of the axis of the rift zone along which a dike propagated laterally to feed the vent. (Photograph by P.J. Mouginis-Mark.)

mm/year) away from the ridge crest. [*See* EARTH AS A PLANET: ATMOSPHERE AND OCEANS; EARTH AS A PLANET: SURFACE AND INTERIOR.]

Lavas erupted from vents located some distance away from the ridge crest build up roughly symmetrical edifices that generally have convex-upward shapes and are described, depending on their height-to-width ratio, as shields (having relatively shallow flank slopes) or domes (having relatively steeper flanks). Some of these vent systems are not related to the spreading ridges at all, but instead mark the locations of "hot spots" in the underlying mantle, in particular, vigorously rising plumes of mantle material from which magmas migrate through the overlying plate. Because the plate moves over the hot spot, a chain of shield volcanoes can be built up in this way, marking the trace of the relative motion. The largest shield volcanoes on Earth form such a line of volcanoes, the Hawaiian Islands, and the two largest of these edifices, Mauna Loa and Mauna Kea, rise ~10 km above the ocean floor and have basal diameters of about 200 km.

Eruptive activity on shield volcanoes tends to be concentrated either at the summit or along linear or arcuate zones radiating away from the summit, called rift zones (Fig. 1). It appears to be very common for a long-lived reservoir of magma, a magma chamber, to exist at a depth of a few to several kilometers below the summit. This reservoir, which is roughly equant

in shape and may be up to 1 to 3 km in diameter, intermittently feeds surface eruptions, either when magma ascends vertically from it in the volcano summit region or when magma flows laterally in a subsurface fracture called a dike, which most commonly follows an established rift zone, to erupt at some distance from the summit. In many cases, magma fails to reach the surface and instead freezes within the fracture it was following, thus forming an intrusion. The summit reservoir is fed, perhaps episodically, from partial melt zones in the mantle beneath. Rare but important events in which a large volume of magma leaves such a reservoir lead to the collapse of the rocks overlying it, and a characteristically steep-sided crater called a caldera is formed, with a width similar to that of the underlying reservoir.

Volcanoes erupting relatively silica- and volatile-rich magma (andesite or, less commonly, rhyolite) mark the destructive margins of plates, where the plates bend downward to be subducted into the interior and at least partly remelted. These volcanoes tend to form an arcuate pattern (called an island arc when the volcanoes rise from the sea floor) marking the trace on the surface of the zone where the melting is taking place, at depths on the order of 100 to 150 km. The andesitic magmas thus produced represent the products of melting of a mixture of subducted ocean floor basalt, sedimentary material washed onto the ocean floor from

FIGURE 2 The upper three layers of gray, dark, and bright material are air-fall pyroclastic deposits from the 1875 plinian eruption of Askja volcano in Iceland. They clearly mantle earlier, dark, more nearly horizontal pyroclastic deposits. (Photograph by L. Wilson.)

the continents (which are themselves an older, silica-rich product of the chemical differentiation of Earth), seawater trapped in the sediments, and the primary mantle materials into which the plates are subducted. As a result, the andesites are much less representative of the current composition of the mantle. Because they are rich in volatiles (mainly water, carbon dioxide, and sulfur compounds), andesitic volcanoes often erupt explosively, producing localized pyroclastic deposits with a range of grain sizes; alternatively, they produce relatively viscous lava flows that travel only short distances (a few kilometers) from the vent. The combination of short flows and localized ash deposits tends to produce steep-sided, roughly conical volcanic edifices.

When large bodies of very silica- and volatile-rich magma (rhyolite) accumulate—in subduction zones or, in some cases, where hot spots exist under continental areas, leading to extensive melting of the continental crustal rocks—the potential exists for very large scale explosive eruptions, in which finely fragmented magma is blasted at high speed from the vent to form a convecting eruption cloud in the atmosphere. These clouds may reach heights up to 50 km, from which pyroclastic fragments fall to create a characteristic deposit spreading downwind from the vent area (Fig. 2). Under certain circumstances, the cloud cannot convect in a stable fashion and collapses to form a fountainlike structure over the vent, which feeds a series of pyroclastic flows—mixtures of incandescent pyroclastic fragments, volcanic gas, and entrained air—that can travel for at least tens of kilometers from the vent at speeds

in excess of 100 m/s, eventually coming to rest to form a rock body called ignimbrite. These fall and flow deposits may be so widespread around the vent that no appreciable volcanic edifice is recognizable; however, there may be a caldera, or at least a depression, at the vent site due to the collapse of the surface rocks to replace the large volume of material erupted from depth.

B. THE MOON

During the 1970s, continuing analyses of the samples collected from the Moon by the Apollo missions showed that there were two major rock types on the lunar surface. The relatively bright rocks forming the old, heavily cratered highlands of the Moon were recognized as being a primitive crust that formed about 4.5 Ga (billion years) ago by the accumulation of solid minerals at the cooling top of an extensively melted region referred to as a magma ocean. This early crust was extensively modified prior to about 3.9 Ga ago by the impacts of meteoroids and asteroids with a wide range of sizes to form impact craters and basins. Some of the larger craters and basins (the mare basins) were later flooded episodically by extensive lava flows, many more than 100 km long, to form the darker rocks visible on the lunar surface. [See THE MOON.]

Radiometric sample dating and photogeologically established relative ages of flow units show that these mare lavas were mostly erupted between 3 and 4 Ga ago, forming extensive, relatively flat deposits inside the basins. Individual flow units, or at least groups of flows, can commonly be distinguished using multispectral remote-sensing imagery on the basis of their differing chemical compositions, which give them differing reflectivities in the visible and near-infrared parts of the spectrum. In composition these lavas are basaltic, and their detailed mineralogy shows that they are the products of partial melting of the lunar mantle at depths between 150 and more than 400 km, the depth of origin increasing with time as the lunar interior cooled. Mapping of individual flow units (Fig. 3) shows that many of them must have had vents near the edges of the interiors of the basins they occupy, probably associated with the arcuate rilles found in similar positions. These are arcuate depressions, the boundaries of which are controlled by faults (fractures) produced by tensional forces in the crust. Individual vents for these flows are hard to find, however. This is consistent with the idea, supported by theoretical calculations based on the mineralogy and by melting experiments on samples, that these lavas were extremely fluid (i.e., had very low viscosities, at least a

FIGURE 3 Lava flows in southwest Mare Imbrium on the Moon. The source vents are off the image to the lower left and the ~300-km-long flows extend down a gentle slope toward the center of the mare basin beyond the upper right edge of the frame. (NASA *Apollo* photograph.)

factor of 3 to 10 less than those of typical basalts on Earth) when they were erupted, and thus had a tendency to flow back into, and cover up, their vents at the ends of the eruptions.

A second class of lunar volcanic features associated with the edges of large basins is the sinuous rilles. These are meandering depressions, commonly hundreds of meters wide, tens of meters deep, and tens of kilometers long, which occur almost entirely within the mare basalts. Some are discontinuous, giving the impression of an underground tube that has been partly revealed by partial collapse of its roof, and these are almost certainly the equivalent of lava tube systems (lava flows whose top surface has completely solidified) on Earth. Other sinuous rilles are continuous open channels all along their length; these generally have origins in source depressions two or three times wider than the rille itself, and become narrower and shallower with increasing downslope distance from the

source. At least some of these sinuous rilles appear to have been caused by long-duration, turbulent lava flows heating up and eventually thermally eroding the preexisting surface.

In contrast to the lava flows and lava channels, two types of pyroclastic deposit are recognized on the Moon. There are numerous regions, often roughly circular and up to at least 200 km in diameter, where the fragmental lunar surface regolith is darker than usual, and spectroscopic evidence shows that it contains a component of small volcanic particles in addition to the locally derived rock fragments. The centers of these regions, called dark mantles, are commonly near the edges of mare basins, suggesting that the dark mantle deposits are produced by the same (or similar) source vents as the lava flows. Chemical analysis of the Apollo lava samples suggests that the main gas released from mare lava vents was carbon monoxide, produced in amounts up to a few hundred parts per million by weight as a result of a chemical reaction between free carbon and oxides, mainly iron oxide, in the magma as it neared the surface.

Several more localized, dark, fragmental deposits occur on the floor of the old, 90-km-diameter impact crater Alphonsus. These patches, called dark haloes, extend for a few kilometers from the rims of subdued craters that are centered on, and elongated along, linear depressions (called linear rilles) on the crater floor. It is inferred that these are the sites of less energetic volcanic explosions.

Localized volcanic constructs such as shield volcanoes and domes are not common on the Moon, though more than 200 low shieldlike features with diameters mainly in the range 3 to 10 km are found in the Marius region within Oceanus Procellarum, in northeast Mare Tranquillitatis, and in the region between the craters Kepler and Copernicus. Conspicuously absent are edifices with substantial summit calderas, implying the presence of large, shallow magma reservoirs. However, some collapse pits with diameters up to 3 km do occur, often located near the tops of domes or aligned along the fault-bounded depressions called linear rilles.

C. MARS

The first spacecraft to fly past, and later orbit, Mars showed impact-generated craters and basins modifying an early crust covering about 60% of the surface area of the planet. However, the other 40% was found to consist mainly of relatively young, flat, plains-forming units. The identification within the plains of many lobate features with all the morphological characteris-

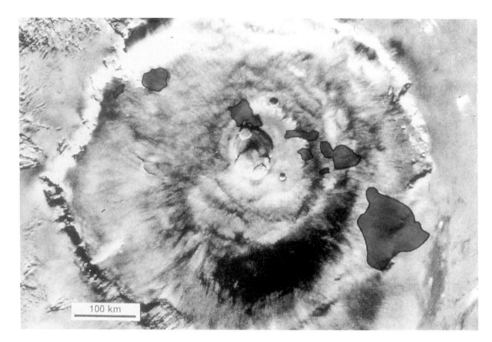

FIGURE 4   The Olympus Mons shield volcano on Mars with the Hawaiian islands superimposed for scale. (NASA image with overlay by P.J. Mouginis-Mark. Reproduced by permission of the Lunar and Planetary Institute.)

tics of basaltic lava flows on Earth soon led to the interpretation that the plains were largely of volcanic origin, in many areas with aeolian or in some cases even fluvial modification. This view was reinforced by the data from the two *Viking* lander craft, which, in the course of making chemical measurements designed to detect living organisms, found indirect evidence for the surface rocks having an essentially basaltic composition. [*See* MARS: SURFACE AND INTERIOR.]

More obvious evidence for basaltic volcanism was the recognition, in the orbiter images, of four extremely large (~500 km diameter, heights up to ~25 km) shield volcanoes with the same general morphology as those found on Earth (Fig. 4). There are also about 20 smaller shields on Mars in various stages of preservation. Some of these volcanoes are large enough to allow meaningful statistics to be collected on the numbers of small impact craters on their flanks, which can be used to estimate the ages of the surface units. It seems that activity on at least some of these volcanoes has extended from more than 3 Ga ago to as recently as ~300 Ma. Complex systems of nested and intersecting calderas are found on the larger shields, implying protracted evolution of the internal plumbing of the volcano. Individual caldera depressions are up to at least 30 km in diameter, much larger in absolute size than any found on Earth.

Most shields appear to have flanks dominated by lava flows, though some, especially some of the more eroded edifices like Tyrrhena Patera, appear to contain a high proportion of relatively weak, presumably pyroclastic, rocks. This is also true of the ~500-km-diameter volcano Alba Patera, which, though not very heavily eroded, has an anomalously low height (~3 km) for its size. There is a hint, from the relative ages of the volcanoes and the stratigraphic positions of the mechanically weaker layers within them, that pyroclastic eruptions were commoner in the early part of Mars's history. More contentious is the suggestion that some of the volcanic plains units, generally interpreted as modified lava flows, in fact consist of pyroclastic fall or flow deposits.

D. VENUS

Because of its dense, optically opaque atmosphere, the only synoptic imaging of the Venus surface comes from Earth-based and various orbiting satellite-based radar systems. Despite the differences between conventional optical images and radar images (radar is sensitive to both the dielectric constant and the roughness of the surface on a scale similar to the radar wavelength), numerous kinds of volcanic features have been unambiguously detected on Venus. Large parts of the planet are covered with plains-forming units that appear to consist of lava flows: certainly, in many places, the well-defined lobate shapes of the edges of the features

FIGURE 5 A variety of radar-bright lava flows radiate from the summit area down the flanks of a shield volcano on Venus. (NASA *Magellan* image.)

and the clear control of topography on their direction of movement leave little room for doubt as to their origin (Fig. 5). The lengths (which can be up to several hundred kilometers) and thicknesses (generally significantly less than 30 m, since they are not resolvable in the radar altimetry data) of these flows suggest that they are basaltic in composition. This interpretation is supported by the (admittedly small) amounts of major-element chemical data obtained from six of the Soviet probes that soft-landed on the Venus surface. Some areas show concentrations of particularly long flows called fluctus (Latin for floods). Most of the lava plains, judging by the numbers of superimposed impact craters, were emplaced within the last 700 Ma. [*See* VENUS: SURFACE AND INTERIOR].

Many areas within the plains and other geological units contain groupings (dozens to hundreds) of small volcanic edifices, from less than one to several kilometers in diameter, with profiles that lead to them being classified as shields or domes. These groupings are called shield fields, and at least 500 have been identified. Some of the individual volcanoes have small summit depressions, apparently due to magma withdrawal and collapse, and others are seen to feed lava flows. Quite distinct from these presumably basaltic shields and domes is a class of larger, steep-sided domes (Fig. 6) with diameters of a few tens of kilometers and heights up to ~1 km. The surface morphologies of these domes suggest that most were emplaced in a single episode, and current theoretical modeling shows

that their height-to-width ratio is similar to that expected for highly viscous silicic (perhaps rhyolitic) lavas on Earth.

Many much larger volcanic constructs occur on Venus. About 300 of these are classed as intermediate volcanoes and have a variety of morphologies, not all including extensive lava flows. A further 150, with diameters between 100 and about 600 km, are classed as large volcanoes. These are generally broad shield volcanoes with extensive systems of lava flows, and have heights above the surrounding plains of up to about 3 km.

Large numbers of collapse-depressions occur, ranging in size from a few kilometers for summit calderas on some shield volcanoes to a few tens of kilometers for more isolated features. There are two particularly large volcano-related depressions, called Sacajawea and Colette, located on the upland plateau Lakshmi Planum. With diameters on the order of 200 km and depths of ~2 km, these features appear to represent the downward sagging of the crust over some deep-seated site of magma withdrawal.

Finally, there are a series of large, roughly circular features on Venus, which, though intimately linked with the large-scale tectonic stresses acting on the crust (they range from a few hundred to a few thousand kilometers in diameter), also have very strong volcanic associations. These are the coronae, novae, and arachnoids. Though defined in terms of the morphology of circumferential, moatlike depressions, radial fracture systems, and so on, these features commonly contain small volcanic edifices (fields of small shields or

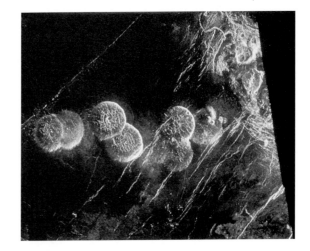

FIGURE 6 A cluster of ~25-km-diameter "pancake" domes on Venus. These domes are evidence of the eruption of lava which is more viscous than that forming the majority of flows on Venus. (NASA *Magellan* image.)

domes), small calderas, or lava flows, the latter often apparently fed from elongate vents coincident with the distal parts of radial fractures. In such cases, it is strongly inferred that the main feature is underlain by some kind of magma reservoir, which feeds the more distant eruption sites via lateral dike systems.

E. MERCURY

Much of the surface of Mercury is a heavily cratered ancient terrain like that of the Moon. There are some relatively flat plains-forming units dispersed among the craters, and it is tempting to speculate that these contain lava flows. Half of the surface of the planet was imaged by the flyby probe *Mariner 10*, but the resolution of the images is too poor to allow the lobate edges of any flow units to be identified unambiguously. Earth-based spectroscopic measurements suggest that many of the surface rocks are close to basaltic in composition, but this does not in itself guarantee that they were emplaced volcanically after the era of early intense bombardment. [*See* MERCURY.]

F. IO

The bulk density of Io is about the same as that of Earth's Moon, suggesting that it has a silicate composition, similar to that of the inner, Earth-like planets. Io and the Moon also have similar sizes and masses, and it might therefore be expected by analogy with the Moon's thermal history that any volcanic activity on Io would have been confined to the first one or two billion years of its life. However, as the innermost satellite of the gas-giant Jupiter, Io is subjected to strong tidal forces. An orbital period resonance driven by the mutual gravitational interactions of Io, Europa, and Ganymede causes the orbit of Io to be slightly elliptical. This, coupled with the fact that it rotates synchronously (i.e., the orbital period is the same as that of the axial rotation), means that the interior of Io is subjected to a periodic tidal flexing. The inelastic part of this deformation generates heat in the interior on a scale that far outweighs any heat source due to the decay of naturally radioactive elements. As a result, Io is currently the most volcanically active body in the solar system. The most obvious manifestation of the activity is the presence, at any one time, of several eruption clouds above the surface, marking sites from which gases and fragmental materials are being ejected

at speeds of up to 1000 m/s to reach heights up to 300 km (Fig. 7). [*See* Io.]

At first sight the materials involved in the eruptive activity are strange: only sulfur and sulfur dioxide are detected as vapors in the eruption clouds, and much of the surface is coated with highly colored deposits of liquid or solid sulfur and solid sulfur compounds. However, it seems very likely, based on the fluid dynamic and thermodynamic analysis of the eruption clouds, that the underlying cause of the activity is the ascent of basaltic magmas from the interior of Io. These interact with, and mobilize, sulfur compounds that have been degassed from the interior of the body over solar system history and are now concentrated in the near-surface layers. Some features that resemble lava flows are visible on the surface, often associated with calderas located at the centers of low shieldlike features, but there is still uncertainty as to whether these are flows of silicate (probably basaltic) lava or of sulfur. Many of these uncertainties should be resolved by data from the *Galileo* spacecraft, which is active at the time of writing.

G. THE ICY BODIES

Many of the satellites of the gas-giant planets have bulk densities indicating that their interiors are mixtures of silicate rocks and the ices of the common volatiles (mainly water, with varying amounts of ammonia and methane). On some of these bodies (e.g., Ganymede, Ariel, and Triton), flowlike features are seen that have many of the morphological attributes of very viscous lava flows. However, there is no spectroscopic evidence for silicate magmas having been erupted onto the surfaces of these bodies.

These features have forced us to recognize that there is a more general definition of volcanism than that employed so far. Volcanism is the generation of partial melts from the internal materials of a body and the transport out onto the surface of some fraction of that melt. In these bodies, it is the generation of liquid water that plays the role of partial melting of rocks, and the ability of the water to erupt at the surface is controlled in part by the amounts of ammonia and methane that it contains. Since the surface temperatures of most of these satellites are very much less than the freezing temperature of water, and since they do not have appreciable atmospheres, the fate of any liquid water erupting at the surface is complex. Cooling will produce ice crystals at all boundaries of the flow and, being less dense than liquid water, the crystals will rise toward the flow surface. Because of the low external

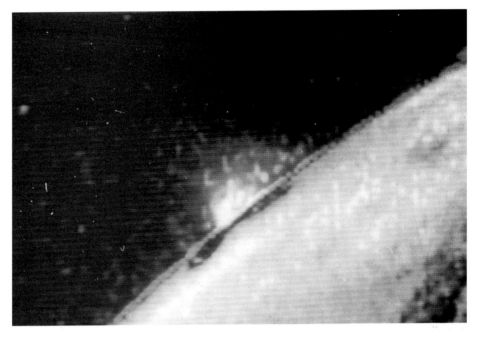

FIGURE 7   An explosive eruption plume on Io. The great height of the plume, more than 100 km, implies that magma is mixing with and evaporating volatile materials (sulfur or sulfur dioxide) on the surface as it erupts. (NASA *Voyager* image.)

pressure, evaporation (boiling) will take place within the region up to a few hundred millimeters below the upper surface of the flow. The vapor produced will freeze as it expands, to settle out as a frost or snow on the surrounding surface. The boiling process extracts heat from the liquid and adds to the rate of ice crystal formation. If enough ice crystals collect at the surface of a flow, they will impede the boiling process, and if a stable ice raft several hundreds of millimeters thick forms, it will suppress further boiling. Thus, if it is thick enough, a water flow may be able to travel a significant distance from its eruption site.

II. CLASSIFICATION OF ERUPTIVE PROCESSES

Volcanic eruption styles on Earth have traditionally been classified partly in terms of the physics of the processes occurring at the vent and partly in terms of the observed dispersal of the eruption products. A similar scheme can be adopted for all planetary bodies, though differences in surface gravity and atmospheric pressure mean that care must be taken in relating the deposits seen (or anticipated) to the mechanisms forming them.

Eruptive processes are classified as either explosive or effusive. An effusive eruption is one in which lava spreads steadily away from a vent to form one or more lava flows, whereas an explosive eruption is one in which the magma emerging through the vent is disrupted, as a result of gas expansion, into clots of liquid that are widely dispersed. The clots cool while in flight above the ground and may be partly or completely solid by the time they land to form a layer of pyroclasts. There is a certain amount of ambiguity concerning this basic distinction between effusive and explosive activity, because many lava flows form from the coalescence, near the vent, of large clots of liquid that have been partly disrupted by gas expansion but that have not been thrown high enough or far enough to cool appreciably.

There is also some ambiguity about the use of the word explosive in a volcanic context. Conventionally, an explosion involves the sudden release of a quantity of material that has been confined in some way at a high pressure. Most often the expansion of trapped gas drives the explosion process. In volcanology, the term explosive is used not only for this kind of abrupt release of pressurized material, but also for any eruption in which magma is torn apart into pyroclasts that are accelerated by gas expansion, even if the magma is being erupted in a steady stream over a long time period. Eruption styles falling into the first category include strombolian, vulcanian, and phreato-magmatic activity, whereas those falling into the second include hawaiian and plinian activity. All of these styles are discussed in detail later.

III. EFFUSIVE ERUPTIONS AND LAVA FLOWS

Whatever the complications associated with prior gas loss, an effusive eruption is regarded as taking place once lava leaves the vicinity of a vent as a continuous flow. The morphology of a lava flow, both while it is moving and after it has come to rest as a solid rock body, is an important source of information about the rheology (the deformation properties) of the lava, which is determined largely by its chemical composition, and about the rate at which the lava is being delivered to the surface through the vent. Because lava flows basically similar to those seen on Earth are so well exposed on Mars, Venus, and the Moon, a great deal of effort has been made to understand lava emplacement mechanisms.

In general, a lava contains a small proportion of solid crystals of various minerals. Above a certain temperature called the liquidus temperature, however, all the crystals will have melted and a given lava will be completely liquid. Under these circumstances, lavas have almost perfectly Newtonian rheologies, which means that the rate at which the lava deforms, the strain rate, is directly proportional to the stress applied to it under all conditions. This constant ratio of the stress to the strain rate is called the Newtonian viscosity of the lava (Fig. 8). At temperatures below the liquidus but above the solidus (the temperature at which all the components of the lava are completely solid), the lava in general has a non-Newtonian rheology. The ratio of stress to strain rate is now a function of the stress, and is called the apparent viscosity. At high crystal contents, the lava may develop a finite strength, called the yield strength, which must be exceeded by the stress before any flowage of the lava can occur. The simplest kind of non-Newtonian rheology is that in which the increase in stress, once the yield strength is exceeded, is proportional to the increase in strain rate: the ratio of the two is then called the Bingham viscosity and the lava is described as a Bingham plastic. Figure 8 shows examples of these and other, more complex rheologies found for lavas.

The earliest theoretical models of lava flows treated them as Newtonian fluids. Such a fluid released on an

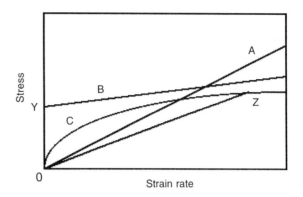

FIGURE 8   Curves showing the relationship between stress and strain rate for fluids with various rheological properties. The straight line A represents a Newtonian fluid and its slope is the (constant) viscosity of the fluid; line B represents a Bingham plastic fluid with yield strength Y; curve C is a fluid with nonlinear rheology. The slope of the line O–Z is the apparent viscosity, now no longer constant, of the fluid C under any particular stress–strain rate conditions.

inclined plane will spread both downslope and sideways indefinitely (unless surface tension stops it, a negligible factor on the scale of lava flows). Some lavas are clearly channeled by preexisting topography, and so it is understandable that they have not spread sideways. However, lavas do not flow downhill indefinitely once the magma supply from the vent ceases: they always stop moving quite soon afterward, often while the front of the flow is on a region with an appreciable slope and almost all of the lava is still at least partly liquid. Also, liquid lava present in a channel at the end of an eruption does not drain completely out of the channel: a significant thickness of lava is left in the channel floor. These observations led to the suggestion that no lavas are Newtonian. As a result, analyses of lava flow morphology, like the simple example that follows, generally assume that the lava rheology is at least as complex as that of a Bingham plastic.

If a Bingham plastic fluid with bulk density ρ has a yield strength τ, then on a planet where the acceleration due to gravity is g, it will not spread any farther, on ground of slope α, than is necessary to reduce the flow depth to a value B, where

$$B = \tau/(\rho \, g \, \sin \alpha) \qquad (1)$$

Thus, as a flow forms it will spread sideways at first to form lateral banks, called levées, which have a maximum thickness B. The width, W_b, of the levée on each side of the flow will be

$$W_b = B/(2 \sin \alpha) \qquad (2)$$

Between these levees will be a lava-filled channel of width W_c in which the mean downslope flow speed U and average thickness D of the lava are just enough to accommodate the eruption rate, V (the volume of lava per second being erupted from the vent), that is,

$$V = U \, D \, W_c \qquad (3)$$

D, the mean lava thickness in the channel, will be somewhat greater than the levée thickness B: the lava surface will have a curved profile that is convex upward. Mathematical relationships can be found between these variables and the plastic viscosity of the lava, η, such that, for the majority of planetary lava flows, which have central channels wider than their retaining levées,

$$W_c = \{[(24 \, V \, \eta)^4 \, \rho \, g]/[\tau^5 \, (\sin \alpha)^6]\}^{1/11} \qquad (4)$$

When the central channel is relatively narrow, the relationship becomes

$$W_c = \{(24 \, V \, \eta)/[\tau \, (\sin \alpha)^2]\}^{1/3} \qquad (5)$$

When high enough resolution spacecraft images are available, the linear dimensions of lava flow features can easily be measured. The thicknesses of flows can be found using stereoscopic images, shadow lengths, or photoclinometry (the interpretation of small brightness changes across an image in terms of changes in the local tilt of the surface), and regional slopes can be found from stereoscopy or radar altimetry. Thus, if B and α are measured for a lava flow deposit seen in a spacecraft image, and if the density ρ of the lava is estimated, the yield strength of the lava can be found from Eq. (1). If the central channel width W_c is then measured, Eq. (4) or (5) can be used to find the product of V and η. Although unique values of V and η cannot be obtained in this way, enough is known about the likely trends of pairs of values of τ and η for various compositions of lavas on Earth that reasonable approximations to both quantities can be obtained.

There is a possible alternative way to estimate V if it can be assumed that the flow unit being examined has come to rest because of cooling. An empirical relationship has been established for cooling-limited flows on Earth between the length, L, of a flow unit and the effusion rate, V, that fed it, such that

$$V = (125 \, \kappa \, W_c \, L)/D \qquad (6)$$

where κ is the thermal diffusivity of lava, a measure of its heat transport ability ($\sim 10^{-6} \ m^2/s$), and the numerical factor 125 applies to channels that are about three times as wide as they are deep. If a flow is treated as cooling-limited when in fact it was not (the alternative being that it was volume-limited, meaning that it came to rest because the magma supply from the vent ceased at the end of the eruption), the effusion rate will inevitably be an underestimate by an unknown amount. Cooling-limited flows can sometimes be recognized because they have breakouts from their sides where lava was forced to form a new flow unit when the original flow front came to rest. Once V has been obtained, by whatever method, the lava flow speed U can be estimated from Eq. (3) either using the measured value of D or taking B as a rough approximation to it.

Lava rheologies and effusion rates have been estimated in this way for lava flows on Mars, the Moon, and Venus: the values found suggest that essentially all the lavas studied so far have properties similar to those of basaltic to intermediate (andesitic) lavas on Earth. Many of these lavas have lengths up to several hundred kilometers, to be compared with basaltic flow lengths up to a few tens of kilometers on Earth in geologically recent times, and this implies that they were erupted at much higher volume fluxes than is now common on Earth. There is a possibility, however, that some of these flow lengths have been overestimated. If a flow comes to rest so that its surface cools, but the eruption that fed it continues and forms other flow units alongside it, a breakout may eventually occur at the front of the original flow. A new flow unit is fed through the interior of the old flow, and the cooled top of the old flow, which has now become a lava tube, acts as an excellent insulator. As a result, the breakout flow can form a new unit almost as long as the original flow, and a large, complex compound flow field may eventually form in this way. Unless spacecraft images of the area have sufficiently high resolution for the compound nature of the flows to be clear, the total length of the group of flows will be interpreted as the length of a single flow, and the effusion rate will be greatly overestimated.

There are, however, certain volcanic features on the Moon and Mars that may be more unambiguous indicators of high effusion rates: the sinuous rilles. The geometric properties of these meandering channels—widths and depths that decrease away from the source, lengths of tens to a few hundred kilometers—are consistent with the channels being the result of the eruption of a very fluid lava at a very high volume flux for a long time. The turbulent motion of the initial flow, meandering downhill away from the vent, led to efficient heating of the ground on which it flowed, and it can be shown theoretically that both mechanical and thermal erosion of the ground surface are expected to have occurred on a timescale from weeks to months. The flow, which may have been ~10 m deep and moving at ~10 m/s, slowly subsided into the much deeper channel that it was excavating. Beyond a certain distance, the lava would have cooled to the point where it could no longer erode the ground, and it would have continued as an ordinary surface lava flow. The volume eruption rates deduced from the longer sinuous rille channel lengths are very similar to those found for the longest conventional lava flow units; modeling studies show that the turbulence leading to efficient thermal erosion was probably encouraged by a combination of unusually steep slope and unusually low lava viscosity. Recently, sinuous channels associated with lava plains have been identified on Venus, but the lengths of some of the Venus channels are several to ten times as great as those seen on the Moon and Mars. It is not yet clear if the thermal erosion process is capable of explaining these channels by the eruption of low-viscosity basalts, or whether some more exotic volcanic fluid (or some other process) must be assumed.

There are numerous uncertainties in using the foregoing relationships to estimate lava eruption conditions. Thus, there have been many studies of the way heat is transported out of lava flows, taking account of the porosity of the lava generated by gas bubbles, the effects of deep cracks extending inward from the lava surface, and the external environmental conditions—the ability of the planetary atmosphere to remove heat lost by the flow by conduction, convection, and radiation. However, none of these has yet dealt in sufficient detail with turbulent flows, or with the fact that cooling must make the rheological properties of a lava flow a function of distance inward from its outer surface, so that any bulk properties estimated in the ways described earlier can only be approximations to the detailed behavior of the interior of the lava flow. There is clearly some feedback between the way a flow advances and its internal pattern of shear stresses. For example, lava flows on Earth have two basic surface textures. Basaltic flows erupted at low effusion rates or while still hot near their vents have smooth, folded surfaces with a texture called pahoehoe (a hawaiian word), the result of plastic stretching of the outer skin as the lava advances; at higher effusion rates, or at lower temperatures farther from the vent, the surface fractures in a more brittle fashion to produce a very rough texture called 'a'a. A similar but coarser, rough, blocky texture is seen on the surfaces of more andesitic flows. Because there is a possibility of relating effusion rate and composition to the surface roughness of a

flow in this way, there is a growing interest in obtaining relatively high resolution radar images of planetary surfaces (and Earth's surface) in which, as in the *Magellan* images of Venus, the returned signal intensity is a function of the small-scale roughness.

IV. EXPLOSIVE ERUPTIONS

A. BASIC CONSIDERATIONS

Magmas ascending from the mantle on Earth commonly contain volatiles, mainly water and carbon dioxide together with sulfur compounds and halogens. All of these have solubilities in the melt that are both pressure and temperature dependent. The temperature of a melt does not change greatly if it ascends rapidly enough toward the surface, but the pressure to which it is subjected changes enormously. As a result, the magma generally becomes saturated in one or more of the volatile compounds before it reaches the surface. Only a small degree of supersaturation is needed before the magma begins to exsolve the appropriate volatile mixture into nucleating gas bubbles, which are commonly only a few tens of microns in diameter. As a magma ascends to shallower levels, existing bubbles grow by decompression and new ones nucleate. It is found empirically that once the volume fraction of the magma occupied by the bubbles exceeds about 75%, the foamlike fluid can no longer deform in response to the shear stresses applied to it and disintegrates into a mixture of released gas and entrained pyroclastic clots and droplets. The eruption is then, by definition, explosive. The pyroclasts have a range of sizes dictated by the viscosity of the magmatic liquid, in turn a function of its composition and temperature, and the rate at which the decompression is taking place, essentially proportional to the rise speed of the magma.

It is not a trivial matter for the volume fraction of gas in a magma to become high enough to cause disruption into pyroclasts. The lowest pressure to which a magma is ever exposed is the planetary surface atmospheric pressure. This ranges from 4 to 10 MPa on Venus, depending on the height of the vent relative to mean planetary radius; it is about 0.1 MPa on Earth (but higher, up to 60 MPa, on the deep ocean floor); it ranges from 50 to 500 Pa on Mars, again depending on the volcano height; and it is essentially zero on the Moon and Io. If the magma volatile content is low enough, then even at atmospheric pressure no gas will be exsolved—or at least too little will be exsolved to cause magma fragmentation. Using the solubilities of common volatiles in magmas, calculations show that explosive eruptions can occur on Earth as long as the water content exceeds 0.07 wt.% in a basalt. On Mars the critical level is 0.01 wt.%. On Venus, however, a basalt would have to contain about 2 wt.% water before explosive activity could occur, even at highland sites; this is greater than is common in basalts on Earth by a factor of 10, and leads to the suggestion that explosive activity may never happen on Venus, at least at lowland sites, or may happen only when some process leads to the local concentration of volatiles within a magma. Examples of this are discussed later.

Finally, the present discussion assumes that exsolving magmatic vulatiles drive the explosive activity. However, many vulcanian and all phreato-magmatic explosive eruptions involve interaction of erupting magma with solid or liquid volatiles present at the surface. The weight fraction of gas in the eruption products in such cases will depend on the detailed nature of the interaction as well as the composition and inherent volatile content of the magma.

B. STROMBOLIAN ACTIVITY

A strombolian eruption is an excellent example of how the rise speed, gas content, and viscosity of a magma are critical in determining the style of explosive activity that occurs. While the magma as a whole is ascending through a fracture in the planetary crust, bubbles of exsolved gas are rising through the liquid at a finite speed determined by the liquid viscosity and the bubble sizes. If the magma rise speed is negligible, for example, when magma is trapped in a shallow reservoir or a shallow intrusion, and if its viscosity is low, as in the case of a basalt, there may be enough time for gas bubbles to rise completely through the magma and escape into overlying fractures that convey the gas to the surface, where it escapes or is added to the atmosphere if there is one. Subsequent eruption of the residual liquid will be essentially perfectly effusive. If a low-viscosity magma is rising to the surface at a slow enough speed, most of the gas will still escape as bubbles rise to the liquid surface and burst. Because relatively large bubbles (those that nucleated first and have decompressed most) will rise faster through the liquid than very small bubbles, it is common in some magmas, especially basalts, for large bubbles to overtake and coalesce with small ones. The even larger bubbles produced in this way rise even faster and overtake more smaller bubbles. In many cases a runaway

FIGURE 9 Jets of hot gas and entrained incandescent basaltic pyroclasts ejected from a transient strombolian explosion on the volcano Stromboli in Italy. (Photograph by L. Wilson.) (See also color insert.)

situation develops in which a single large bubble completely fills the diameter of the vent system. As this emerges at the surface of the slowly rising liquid column, it bursts, and a discrete layer of magma forming the "skin" of the bubble disintegrates into clots and droplets that are blown outward by the expanding gas in what is described as a strombolian explosion (Fig. 9; see also color insert). The pyroclasts produced accumulate around the vent to form a cinder cone. If the largest rising gas bubble does not completely fill the vent, continuous overflow of a lava lake in the vent may take place to form one or more lava flows at the same time that intermittent explosive activity is occurring, resulting in a simultaneously effusive and explosive eruption.

As long as any volatiles are exsolved from a low-viscosity magma rising sufficiently slowly to the surface, some kind of strombolian explosive activity, however feeble, should occur at the vent on any planet, even on Venus or Earth's ocean floors. Strombolian eruptions commonly involve excess pressures in the bursting bubbles of only a few tenths of a MPa, so that the amount of gas expansion that drives the dispersal of pyroclasts is small. Pyroclast ranges in air on Earth of several tens to at most a few hundred meters are possible, and ranges would be much smaller in submarine strombolian events on the ocean floor or on Venus because of the higher ambient pressure. Subaerial strombolian eruptions on Mars would eject pyroclasts to distances about three times greater than on Earth because of the lower gravity, but as a result the deposits formed would have a 10-fold lower relief than on Earth, and no examples have been unambiguously identified in spacecraft images so far.

C. VULCANIAN ACTIVITY

At the other extreme of a slowly rising viscous magma, it would be relatively difficult for gas bubbles to escape from the melt. Particularly if the magma stalls as a shallow intrusion, slow diffusion of gas through the liquid and rise of bubbles in the liquid concentrate gas in the upper part of the intrusion and the gas pressure in this region rises. The pressure rise is greatly enhanced if any volatiles existing near the surface (groundwater on Earth; ground ice on Mars; sulfur or sulfur dioxide on Io) are evaporated. Eventually the rocks overlying the zone of high pressure break under the stress and the rapid expansion of the trapped gas drives a sudden, discrete explosion in which fragments of the overlying rock and of the disrupted magma are scattered around the explosion source: this is called vulcanian activity (Fig. 10), named for the Italian volcanic island Vulcano. Again, as long as any volatiles are released from a magma or are present in the near-surface layers of the planet, activity of this kind can occur.

Several vulcanian events on Earth involving fairly viscous magmas have been analyzed in enough detail to provide estimates of typical pressures and gas concentrations. Observed ejecta ranges up to 5 km imply pressures as high as a few MPa in regions that are tens of meters in size and that have gas mass fractions in the explosion products up to 10%. On Mars, with the same initial conditions, the lower atmospheric pressure would cause much more gas expansion to accelerate the ejected fragments, and the lower atmospheric density would exert much less drag on them; also the lower gravity would allow them to travel farther for a given initial velocity. The result is that the largest clasts could travel up to 50 km, and the deposit from the explosion would be spread over an area 100 times greater than on Earth, being on average 100 times thinner. Apart from the possibility that the pattern of small craters

FIGURE 10 A dense cloud of large and small pyroclasts and gas ejected to a height of a few hundred meters in a transient vulcanian explosion by the volcano Ngauruhoe in New Zealand. (Image courtesy of the University of Colorado in Boulder, Colorado, and the National Oceanic and Atmospheric Administration, National Geophysical Data Center.)

produced by the impact of the largest boulders on the surface might be recognized, such a deposit, with almost no vertical relief and having very little influence on the preexisting surface, would almost certainly go unnoticed in spacecraft images.

A similar explosion on Venus would also be very different from its equivalent on Earth. In this case, the high atmospheric pressure would suppress gas expansion and lead to a low initial velocity for the ejecta, and the atmospheric drag would also be high. Pyroclasts that would have reached a range of 5 km on Earth would travel less than 200 m on Venus. On the one hand, this should concentrate the eruption products around the vent and make the deposit more obvious; however, the resolution of the best radar images from *Magellan* is only ~75 m, and so such a deposit would represent only three or four adjacent pixels, which again would probably not be recognized.

FIGURE 11   A hawaiian eruption from the Pu'u 'O'o vent in Hawai'i showing a convecting cloud of gas and small particles in the atmosphere above the 300-m-high lava fountain (commonly termed fire fountain) of coarser basaltic pyroclasts. (Photograph by P.J. Mouginis-Mark.)

Only on the Moon have vulcanian explosion products been identified. The dark halo craters on the floor of Alphonsus have ejecta deposits with ranges up to 5 km. Since the Moon has a much lower atmospheric pressure than Mars (essentially zero), the preceding analysis suggests at first sight that lunar volcanian explosions should eject material to very great ranges. However, the Alphonsus event seems to have involved the intrusion of basaltic magma into the ~10-m-thick layer of fragmental material forming the regolith in this area, and the strength of the resulting mixture of partly welded regolith and chilled basalt was quite low. Thus only a small amount of pressure buildup occurred before the retaining rock layer fractured. As a result, the initial speeds of the ejected pyroclasts were low and their ranges were unusually small.

D. HAWAIIAN ACTIVITY

The commonest situation in low-viscosity basaltic magmas rising at appreciable rates (more than about 1 m/s) is for some gas bubble coalescence to occur, but for the magma release at the vent to be continuously explosive. A lava fountain, more commonly called a fire fountain, forms over the vent, consisting of pyroclastic clots and droplets of liquid entrained in a magmatic gas stream that fluctuates in its upward velocity on a timescale of a few seconds (Fig. 11). The largest clots

of liquid rise some way up the fountain and fall back around the vent to coalesce into a lava pond that overflows to feed a lava flow—the effusive part of the eruption—whereas smaller clasts travel to greater heights in the fountain. Some of the intermediate-sized pyroclasts cool as they fall from the outer parts of the fountain and collect around the lava pond in the vent to build up a roughly conical edifice called an ash cone, cinder cone, or scoria cone, the term used depending on the sizes of the pyroclasts involved. Such pyroclastic cones are commonly asymmetric owing to the influence of the prevailing wind.

Atmospheric gases are entrained into the edge of the fire fountain and heated by contact with the hot pyroclasts and mixing with the hot magmatic gas. In this way, a convecting gas cloud is formed over the upper part of the fountain, and this gas entrains the smallest pyroclasts so that they take part fully in the convective motion. The whole cloud spreads downwind and cools, and eventually the pyroclasts are released again to form a layer on the ground that consists of finer particles at greater distances from the vent. This whole process, involving formation of lava flows and pyroclastic deposits at the same time, is called hawaiian eruptive activity (see Fig. 11). This style of activity is expected to have occurred on Mars, but may be suppressed in basaltic magmas on Venus, especially in lowland areas, unless, as noted earlier, magma volatile contents are several times higher than is common on Earth.

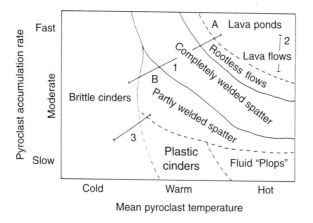

FIGURE 12 Schematic indication of the relative influences of the volatile content and the volume eruption rate of magma on the dispersal and thermal state of pyroclastic material produced in explosive eruptions. (Reprinted from Fig. 5 in the *Journal of Volcanology and Geothermal Research*, Vol. 37, J.W. Head and L. Wilson, Basaltic pyroclastic eruptions: Influence of gas-release patterns and volume fluxes on fountain structure, and the formation of cinder cones, spatter cones, rootless flows, lava ponds and lava flows, pp. 261-271, 1989, with kind permission of Elsevier Science–NL, Sara Burgerhartstraat 25, 1055 KV Amsterdam, The Netherlands.)

Figure 12 shows qualitatively how the combination of erupting mass flux and magma gas content in a hawaiian eruption on Earth determines the nature and size of the possible products: a liquid lava pond at the vent that directly feeds lava flows; a pile of slightly cooled pyroclasts accumulating fast enough to weld together and form a "rootless" lava flow; a cone in which almost all of the pyroclasts are welded together; or a cone formed from pyroclasts that have had time to cool while in flight so that none, or only a few, weld on landing. Attempts have been made to quantify the results in Fig. 12 and extend them to other planetary environments. These results confirm that hot lava ponds around vents on Earth are expected to be no more than a few tens of meters wide even at very high mass eruption rates. On the Moon, the greater gas expansion due to the lack of an atmosphere causes very thorough disruption of the magma (even at the low gas contents implied by analysis of the *Apollo* samples) and gives the released volcanic gas a high speed. This, together with the lower gravity, allows greater dispersal of pyroclasts of all sizes, and allows an explanation of the 100- to 300-km-wide dark mantle deposits as the products of extreme dispersal of the smallest, 30- to 100-μm-sized particles.

Nevertheless, it appears that hot lava ponds up to ~5 km in diameter could have formed around basaltic vents on the Moon if the eruption rates were high enough—as high as those postulated to explain the

long lava flows and sinuous rilles. The motion of the lava in such ponds would have been thoroughly turbulent, thus encouraging thermal erosion of the base of the pond, and this presumably explains why the circular to oval depressions seen surrounding the sources of many sinuous rilles have just these sizes. Similar calculations for the Mars environment show that, as long as eruption rates are high enough, the atmospheric pressure and gravity are low enough on Mars to allow similar hot lava source ponds to have formed there, again in agreement with the observed sizes of some depressions of this type.

E. PLINIAN ACTIVITY

In the case of a basaltic magma that is very rich in volatiles, or (much more commonly on Earth) in the case of a volatile-rich andesitic or rhyolitic magma, the magma fragmentation process in a steadily erupting magma is very efficient, and most of the pyroclasts formed are small enough to be thoroughly entrained by the gas stream. Furthermore, the speed of the mixture emerging from the vent, which is proportional to the square root of the amount of gas exsolved from the magma, will be much higher (perhaps up to 500 m/s) than in the case of a basaltic hawaiian eruption (where speeds are commonly less than 100 m/s). The fire fountain in the vent now entrains so much atmospheric gas that it develops into a very strongly convecting eruption cloud in which the heat content of the pyroclasts is converted in the buoyancy of the entrained gas. The resulting cloud rises to a height that is proportional to the fourth root of the magma eruption rate (and hence the heat supply rate) and that may reach several tens of kilometers on Earth. Only the very coarsest pyroclasts fall out near the vent, and almost all of the erupted material is dispersed over a wide area from the higher parts of the eruption cloud (Fig. 13). This activity is termed plinian, after Pliny's description of the A.D. 79 eruption of Vesuvius. Not all eruptions of this type produce stable convection clouds. If the vent is too wide or the eruption speed of the magma is too low, insufficient atmospheric gas may be entrained to provide the necessary buoyancy for convection, and a collapsed fountain forms over the vent, feeding pyroclastic flows or smaller, more episodic pyroclastic surges.

Figure 14a shows the results of some calculations that relate the rise heights of plinian eruption clouds on Earth to the mass eruption rate of magma from the vent. Also shown on the left side of the graph are the speeds of the gas and small pyroclasts in the vent for

FIGURE 13 The plinian phase of the explosive eruption of Pinatubo volcano in 1991. A dense cloud of large and small pyroclasts and volcanic gases is ejected at high speed from the vent and entrains and heats the surrounding air. Convection then drives the resulting cloud to a height of tens of kilometers, where it drifts downwind, progressively releasing the entrained pyroclasts. (Photo credit: R.S. Culbreth, U.S. Air Force. Photo courtesy of the National Oceanic and Atmospheric Administration, National Geophysical Data Center.)

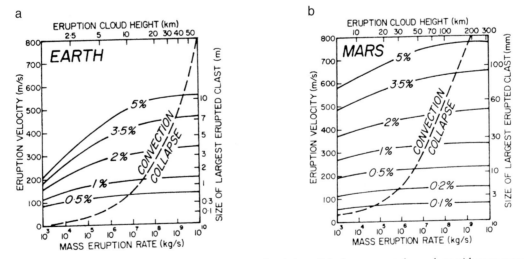

FIGURE 14 Diagrams showing the variations of gas eruption speed and sizes of the largest erupted pyroclasts with mass eruption rate of magma from the vent and magma volatile content for plinian eruptions on (a) Earth and (b) Mars. [Reprinted from Fig. 1 in L. Wilson, J.W. Head, and P.J. Mouginis-Mark (1982). "Theoretical Analysis of Martian Volcanic Eruption Mechanisms," ESA Special Publication 185, pp. 107–113. Reproduced by permission of the authors and The European Space Agency Scientific and Technical Publications Branch.]

various magma gas contents (assuming the volatile is water, the commonest on Earth); on the right side of the graph are the sizes of the largest pyroclasts that can just be transported out of the vent by the drag force of the gas stream. Finally, the dashed line on this figure marks the boundary between combinations of eruption rate and released gas content for which a convecting eruption cloud can remain stable and those for which instead a low fountain forms over the vent, feeding pyroclastic flows.

Mars is the obvious place to look for explosive eruption products other than on Earth, since the low atmospheric pressure encourages explosive eruptions to occur and the atmospheric density is high enough to allow convecting eruption clouds to form. Figure 14b shows the results of the equivalent calculations to those given in Fig. 14a for plinian eruption clouds on Mars. Note the much greater cloud rise heights (by a factor of more than 4), the higher eruption velocities (by a factor of more than 2), and the reduced ability to transport large clasts out of the vent (by a factor of more than 100), all due directly or indirectly to the lower atmospheric pressure. In fact, only one potential fall deposit has yet been identified on Mars with any certainty. This is a region on the flank of the shield volcano Hecates Tholus, where, in contrast to the rest of the volcano, small impact craters appear to be hidden by a blanket of fine material in a region about 50 km wide and at least 70 km long. The sizes of the hidden craters suggest that the deposit is ~100 m thick, giving it a volume of ~65 km^3 and a mass of 7×10^{13} kg if its bulk density is similar to that of terrestrial pyroclasts, say, 1000 kg/m^3. Using the width of this deposit as an indicator of the height of the cloud that fed it, the height of up to 70 km implies an eruption rate of ~3×10^7 kg/s. To erupt the estimated mass, an eruption duration of 20 to 30 days is then needed. The erupted volume of 65 km^3 would be equivalent to a dense rock volume of 23 km^3. It is interesting that this value is comparable to the volumes of the four summit depressions on Hecates Tholus, which range from ~10 to ~30 km^3, suggesting that they may be calderas produced by collapse of the summit to compensate for the volume removed from a fairly shallow magma storage reservoir in each of a series of eruptions like the one documented here.

The fact that a fall deposit (rather than a pyroclastic flow) was able to form at the eruption rate of ~3 × 10^7 kg/s requires that the magma volatile content must have been greater than about 1 wt.% if the volatile was water. To allow this amount of water to be dissolved in the magma, it must have had an initial pressure in excess of 8 MPa, implying a depth of origin in the

Martian interior greater than about 1 km. These figures would also be consistent with the magma having been stored in a relatively shallow reservoir. If the volatile had been carbon dioxide, another likely candidate for Mars given that this gas dominates the composition of the atmosphere, calculations similar to those used to produce Fig. 14b show that the minimum required amount in the magma is about 2 wt.%, implying a much greater minimum depth of origin, between 50 and 100 km. Unfortunately, there is little hope of deciding which of these volatiles was involved in the Hecates Tholus eruption unless rock samples are returned from the deposit by some future lander mission. For it turns out that the only other useful parameter of the erupted deposit that might be measured by high-resolution remote sensing, the maximum pyroclast size near the vent, does not depend on the chemical composition of the volatile phase. Nevertheless, pyroclastic fall deposits give enough information about the eruption conditions that produced them that they will be sought carefully in images from future Mars orbiter missions.

Although the high magma gas content needed suggests that large-scale, steady (plinian) explosive eruptions are rare on Venus, it is possible to calculate the heights to which their eruption clouds would rise. The high density and temperature of the atmosphere lead to rise heights about a factor of two lower than on Earth for the same eruption rate, and very large (at least a few tens of meters) clasts may be transported into near-vent deposits. At distances greater than a few kilometers from the vent, pyroclastic fall deposits will not be very different from those on Earth. A few examples of elongate markings on the Venus surface have been proposed as fall deposits, but no detailed analysis of them has yet been carried out.

The conditions that cause a steady explosive eruption to generate pyroclastic flows instead of feeding a stable, convecting eruption cloud are fairly well understood (see Fig. 14). If the eruption rate exceeds a critical value (which increases with increasing gas content of the mixture emerging through the vent and decreases with increasing vent diameter), stable convection is not possible. Since pyroclastic flow formation is linked automatically to high eruption rate and, in general, to high eruption speed, which will encourage a great travel distance, it would not be surprising if large-scale pyroclastic flow deposits distributed radially around a vent were the products of high discharge rate eruptions of gas-rich magmas. Many of the flanking deposits of the large Martian volcanoes, especially Alba Patera, may be viewed in this way, and their thermal inertia properties are consistent with them being fragmental deposits having a small average grain size. However, there is currently no way of being certain of this interpretation.

Theoretical work has shown that pyroclastic flows on Mars may be able to transport quite large blocks of rock (up to several meters in size, similar to those found on Earth) out of the vent and into nearby deposits. These pyroclast sizes are much greater than those expected in fall deposits on Mars (compare with the right side of Fig. 14b), thus making it potentially possible to distinguish flow and fall deposits in future, high-resolution spacecraft images of Martian vents. No equivalent work has yet been carried out for Venus, again mainly because of the expectation that voluminous explosive eruptions may be rare under the high atmospheric pressure conditions.

Short-lived or intermittent explosive eruptions (e.g., vulcanian explosions, phreato-magmatic explosions, or events in which a gas-rich, high-viscosity lava flow or dome disintegrates into released gas and pyroclasts as a result of excessive gas pressure) can also produce small-scale pyroclastic flows. Because these are shorter lived and have characteristically different grain size distributions, they are called surges. The least well understood aspect of these phenomena is the way in which the magmatic material interacts with the atmosphere. As a result, it is currently almost impossible to predict in detail what the results of this kind of activity on Mars or Venus would look like. Such deposits, by the nature of the way they are generated, would not be very voluminous, however, and so would be spread very thinly, and might not be recognized if they were able to travel far from the vent.

F. PHREATO-MAGMATIC ACTIVITY

Finally, some types of eruption on Earth are controlled by the vigorous interaction of magma with water. If an intrusion into water-rich ground causes steam explosions, these are called phreatic events (from the Greek word for a well). If some magma also reaches the surface, the term used is phreato-magmatic, as distinct from normal, purely magmatic eruptions. When the equivalents of strombolian or hawaiian explosive events take place from eruption sites located in shallow water, they lead to much greater fragmentation of the magma than usual because of the stresses induced as pyroclasts are chilled by contact with the water. This activity is usually called surtseyan, named after a classic eruption that formed the island of Surtsey off the south coast of Iceland. A much more vigorous eruption under similar circumstances leads to a pyro-

clastic fall deposit similar to that of a plinian event, but again involving greater fragmentation of magma: the result is called phreato-plinian activity. Since the word phreatic does not specifically refer to water as the nonmagmatic volatile involved in these kinds of explosive eruption, it seems safe to apply these terms, as appropriate, to the various kinds of interactions between magma and liquid sulfur compounds forming the plumes seen on Io. Such eruptions should also have occurred on Mars in the distant past when the atmospheric pressure was high enough to allow liquid water to exist on the surface.

G. DISPERSAL OF PYROCLASTS INTO A VACUUM

The conditions in the region above the vent in an explosive eruption on a planet with an appreciable atmosphere (e.g., Venus, Earth, or Mars) are very different from those when the atmospheric pressure is very small (much less than about 1 Pa), as on the Moon or Io. If the mass of atmospheric gas displaced from the region occupied by the eruption products after the magmatic gas has decompressed to the local pressure is much less than the mass of the magmatic gas, there is no possibility of a convecting eruption cloud forming in eruptions that would have been classed as hawaiian or plinian on Earth. In the region immediately above the vent, the gas expansion must be quite complex, and will involve a series of shock waves. Relatively large pyroclasts will pass through these shocks with only minor deviations in their trajectories, but intermediate-sized particles may follow very complex paths, and few studies have yet been made of these conditions. The magmatic gas eventually expands radially into space, accelerating as it expands and reaching a limiting velocity that depends on its temperature. As the density of the gas decreases, its ability to exert a drag force on pyroclasts also decreases, and even the smallest particles eventually decouple from the gas and fall back to the planetary surface.

These are the conditions that led to the formation of the dark mantle deposits on the Moon, with ultimate gas speeds on the order of 500 m/s, leading to ranges up to 150 km for small pyroclasts 30 to 100 μm in size. They are also the conditions that exist now in the eruption plumes on Io, though there is an added complication. The driving volatiles in the Io plumes appear to be mainly sulfur and sulfur dioxide, evaporated from the solid or liquid state by intimate mixing with rising basaltic magma in what are effectively phreato-magmatic eruptions. The plume heights imply gas speeds just above the vent of ~1000 m/s, and these speeds are consistent with the plume materials being roughly equal mixtures of basaltic pyroclasts and volatiles. However, as the gas ,phase expands to very low pressures, both sulfur and sulfur dioxide will begin to condense again, forming small solid particles that rain back onto the surface along with the silicate particles to be potentially recycled again in future eruptions.

A final point concerns pyroclastic eruptions on very small atmosphereless bodies. There is strong evidence that many asteroids were subjected to significant heating events early in their lifetimes, probably due to the decay of the short-lived radioactive isotopes. Basaltic partial melts formed within these bodies were erupted at the surface at speeds that depended on the released volatile content. This is estimated to have been as much as 0.2 to 0.3 wt.%, leading to speeds up to 150 m/s. These speeds are greater than the escape velocities from asteroids with diameters less than about 200 km, and so instead of falling back to the surface, pyroclasts would have been expelled into space, eventually to spiral into the Sun. This process explains the otherwise puzzling fact that we have meteorites representing samples of the residual material left in the mantle of at least one asteroid after a partial melting event, but have no meteorites from this asteroid with the expected partial melt composition.

V. INFERENCES ABOUT PLANETARY INTERIORS

The presence of the collapse depressions called calderas at or near the summits of many volcanoes on Earth, Mars, Venus, and Io suggests that it is common on all of these bodies for large volumes of magma to accumulate in reservoirs at relatively shallow depths. Theories of magma accumulation suggest that the magma in these reservoirs must have an internal pressure greater than the stress produced in the surrounding rocks by the weight of the overlying crust. This excess pressure may be due to the formation of bubbles by gas exsolution, or to the fact that heat loss from the magma to its cooler surroundings causes the growth of crystals that are less dense than the magmatic liquid and so occupy a larger volume. Most commonly, a pressure increase leads to fracturing of the wall of the reservoir and to the propagation of a magma-filled crack, called a dike, as an intrusion into the surrounding rocks. If the dike reaches the surface, an eruption occurs, and removal of magma from the reservoir allows the wall rocks to relax inward elastically as the pressure decreases. If magma does not reach the surface, the dike propagates underground until either the magma within it chills and comes to rest as its viscosity becomes extremely high, or the pressure within the reservoir falls to the point where there is no longer a great enough stress at the dike tip for rock fracturing to continue.

Under certain circumstances, an unusually large volume of magma may be removed from a shallow reservoir, reducing the internal pressure beyond the point where the reservoir walls behave elastically. Collapse of the overlying rocks may then occur to fill the potential void left by the magma, and a caldera (or, on a smaller scale, a pit crater) will form. The circumstances causing large-volume eruptions on Earth include the rapid eruption to the surface immediately above the reservoir of large volumes of low-density, gas-rich silicic (rhyolitic) magma, and the drainage of magma through extensive lateral dike systems extending along rift zones to distant flank eruption sites on basaltic volcanoes. This latter process appears to have been associated with caldera formation on Kilauea volcano in Hawai'i, and it is tempting to speculate that the calderas on some of the Martian basaltic shield volcanoes (especially Pavonis Mons and Arsia Mons) are directly associated with the large-volume eruptions seen on the distal parts of their rift zones. In contrast, we saw earlier that, at the Martian volcano Hecates Tholus, a large explosive summit eruption is implicated in the formation of at least one of its calderas.

The size of a caldera must be related to the volume of the underlying magma reservoir, or more exactly to the volume of magma removed from it in the caldera-forming event. If the reservoir is shallow enough, the diameter of the caldera is probably similar to that of the reservoir. Diameters from 1 to 3 km are common on basaltic volcanoes on Earth and on Venus, with depths up to a few hundred meters implying magma volumes less than about 10 km^3. In contrast, caldera diameters up to at least 30 km occur on several volcanoes on Mars and, coupled with caldera depths up to 3 km, imply volumes ranging up to as much as 10,000 km^3. The stresses implied by the patterns of fractures on the floors and near the edges of some of these Martian calderas suggest that the reservoirs beneath them are centered on depths on the order of 10 to 15 km, about three to four times greater than the known depths to the centers of shallow basaltic reservoirs on Earth. The simplest models of the internal structures of volcanoes suggest that, due to the progressive closing of gas cavities in rocks as the pressure increases, the density of the rocks forming a volcanic edifice should increase, at first quickly and then more slowly, with depth. Rising magma from deep partial melt zones may stall when its density is similar to that of the rocks around it, so that it is neither positively nor negatively buoyant, and a reservoir may develop in this way. Since the pressure at a given depth inside a volcano is proportional to the acceleration due to gravity, and since Martian gravity is about three times less than that on Earth or Venus, the finding that Martian magma reservoirs are centered three to four times deeper than on Earth is not surprising. However, these simple models do not address the reason for the Martian calderas being much more than three times wider than those on Earth or Venus. Much is still not understood about the formation and stability of shallow magma bodies.

Evidence for significant shallow magma storage is conspicuously absent from the Moon. The large volumes observed for the great majority of eruptions in the later part of lunar volcanic history, and the high effusion rates inferred for them, imply that almost all of the eruptions took place directly from large bodies of magma stored at great depth—at least at the base of the crust and possibly in partial melting zones in the lunar mantle. Not all the dikes propagating up from these depths will have reached the surface, however, and some shallow intrusions almost certainly exist. Recent work suggests that many of the linear rilles on the Moon represent the surface deformation resulting from the emplacement of such dikes, the dikes having thicknesses of at least 100 m, horizontal and vertical extents on the order of 100 km, and tops extending to within 1 or 2 km of the surface. Minor volcanic activity associated with some of these features would then be the result of gas loss and small-scale magma redistribution as the main body of the dike cooled.

The emplacement of very large dike systems extending most or all of the way from mantle magma source zones to the surface is not confined to the Moon. It has long been assumed that such structures must have existed to feed the high-volume basaltic lava flow sequences called flood basalts that occur on Earth every few tens of millions of years. These kinds of feature are probably closely related to the systems of giant dikes, tens to hundreds of meters wide and traceable laterally for hundreds to more than 1000 km, that are found exposed in very ancient rocks on the Earth. The radial patterns of these ancient dike swarms suggest that they are associated with major areas of mantle upwelling and partial melting, with magma migrating vertically above the mantle plume to depths of a few tens of kilometers and then traveling laterally to form the longest dikes. Some of the radial surface fracture patterns associated with the novae and coronae on Venus are almost certainly similar features that have been formed more recently in that planet's geologic history.

BIBLIOGRAPHY

Cattermole, P. (1989). "Planetary Volcanism: A Study of Volcanic Activity in the Solar System." Ellis Horwood Limited, Chichester, England.

Fagents, S. A., and Wilson, L. (1995). *J. Geophys. Res.* **100**, 26327–26338.

Frankel, C. (1996). "Volcanoes of the Solar System." Cambridge Univ. Press, Cambridge, England.

Head, J. W., and Wilson, L. (1991). *Geophys. Res. Lett.* **18**, 2121–2124.

Head, J. W., and Wilson, L. (1992). *Geochim. Cosmochim. Acta* **56**, 2155–2175.

Head, J. W., and Wilson, L. (1992). *J. Geophys. Res.* **97**, 3877–3903.

Head, J. W., and Wilson, L. (1993). *Planet. Space Sci.* **41**, 719–727.

Mouginis-Mark, P. J., Wilson, L., and Zuber, M. T. (1992). *In* "Mars" (H. H. Kieffer, B. M., Jakosky, C. W. Snyder, and M. S. Mathews, Eds.), pp. 424–452. Univ. Arizona Press, Tucson.

Pinkerton, H., and Wilson, L. (1994). *Bull. Volcanol.* **56**, 108–120.

Wilson, L., and Head, J. W. (1994). *Rev. Geophys.* **32**, 221–264.

Wilson, L., and Keil, K. (1996). *J. Geophys. Res.* **101**, 18927–18940.

PLANETS AND
THE ORIGIN OF LIFE

Christopher P. McKay
and Wanda L. Davis
NASA Ames Research Center

I. Introduction
II. What Is Life?
III. History of Life on Earth
IV. Origin of Life
V. Limits to Life
VI. Life in the Solar System
VII. Mercury and the Moon
VIII. Venus
IX. Mars
X. Viking Results
XI. Early Mars
XII. Subsurface Life on Mars
XIII. Meteorites from Mars
XIV. Giant Planets
XV. Europa
XVI. Titan
XVII. Asteroids
XVIII. Comets
XIX. Life about other Stars
XX. Conclusion

GLOSSARY

Algae: Any of a large group of mostly aquatic organisms that contain chlorophyll and other pigments and can carry on photosynthesis, but that lack true roots, stems, or leaves; they range from microscopic unicellular organisms to very large multicellar structures.

Amino acid: Any organic compound containing an amino acid ($-NH_2$) and a carboxyl ($-COOH$) group; specifically, one of the so-called building blocks of life, a group of 20 such compounds from which proteins are synthesized during ribosomal translation of messenger RNA.

Autotrophy: Literally, self-feeding; the capacity of an organism to obtain its essential nutrients by synthesizing nonorganic materials from the environment, rather than by consuming organic materials; photosynthetic green plants and chemosynthetic bacteria are examples of autotrophic organisms.

Chemoautotrophy: The capacity of an autotrophic (self-feeding) organism to derive the energy required for its growth from certain chemical reactions (e.g., methanogenesis) rather than from photosynthesis; some bacterial forms are chemoautotrophic organisms.

Entropy: Broadly, the degree of disorder, or randomness in a system; in thermodynamics, a measure of the amount of heat energy in a closed system that is not available to do work. In a condition of low entropy (high efficiency), the system will convert to energy a large portion of the heat transferred to it from an external source (no actual system can utilize 100% of the heat it receives).

Gas chromatography: A chemical technique for separating gas mixtures, in which the gas is passed through a long column containing a fixed absorbent phase that separates the gas into its component parts.

Heterotrophy: Literally, other-feeding; the condition of an organism that is not able to obtain nutrients by synthesizing nonorganic materials from the environment, and that therefore must consume other life forms to obtain the organic products necessary for life; e.g., animals, fungi, most bacteria.

Miller-Urey experiments: Noted studies by the U. S. biochemists Stanley Miller and Harold Clayton Urey, indicating that it is possible to synthesize amino acids by circulating a mixture of simple gases through a closed system while discharging a spark into the system; this process is regarded as a simulation of the production of organic molecules from inorganic materials under the conditions of the primitive Earth, and thus a support for the theory that life arose spontaneously on this planet.

Panspermia: The theory that life exists, or did exist, elsewhere in the universe and that the ori-

gin of life on Earth occurred when life forms were transported to this planet from an extraterrestrial source, rather than life arising on Earth itself.

Photoautotrophy: The capacity of an autotrophic (self-feeding) organism to derive the energy required for its growth from sunlight, by means of photosynthesis; green plants are photoautotrophic.

Phototaxis: The movement of an organism in response to light, either toward or away from the source; e.g., certain microorganisms are phototactic and will migrate in the direction of sunlight.

Polysaccharide: Any of a group of carbohydrates consisting of long chains of simple sugars; e.g., starch, glycogen.

Stromatolite: A geological feature formed by the conversion of loose, unconsolidated sediment into a coherent layer, as a result of the growth, movement, or activity of microorganisms; e.g., blue-green algae. Microfossils associated with stromatolite formation are an important form of evidence for early life on Earth, and thus a search for stromatolites could be undertaken on other planets in sites where liquid water might have accumulated.

I. INTRODUCTION

Life is widespread on the Earth and appears to have been present on the planet since early in its history. Biochemically, all life on Earth is similar and seems to share a common origin. Throughout geological history, life has significantly altered the environment of the Earth while at the same time adapting to this environment. It would not be possible to understand the Earth as a planet without the consideration of life. Thus life is a planetary phenomenon and is arguably the most interesting phenomenon observed on planetary surfaces.

Everything we know about life is based on the example of life on Earth. Generalization to other areas or extended forms of life must proceed with this caveat. Although uncertain of the process or the time for its origin, the advent of life on Earth was established within 1 billion years after the formation of the planet. The geological/fossil record of life on Earth indicates that life exists in a planetary environment where liquid water is present. A liquid water environment is currently the best predictor we have for extraterrestrial life. Liquid water environments are not expected on any of the recently discovered large extrasolar planets because they are too distant from their sun. Looking out into our own solar system we do not see liquid water surface environments that resemble Earth. Europa may have liquid water underneath a global ice surface, but the present evidence is inconclusive. We do, however, see the past remnants of liquid water surfaces on Mars. Direct images from spacecraft show fluvial features on the surface of Mars. Our understanding of life, albeit limited to one example and one planet, would suggest that life is possible on other planets whenever conditions allow for environments like those on Earth — essentially liquid water. This predicts early microbial life on Mars and forms the basis for a search for Earth-like planets orbiting other stars. Studies of a second example of life — a second genesis — to which we can compare and contrast terrestrial biochemistry will be the beginning of a more general understanding of life as a process in the universe. This implies a search for more than just fossils but a search for the biochemical remains of organisms, dead or alive.

II. WHAT IS LIFE?

Our understanding of life as a phenomenon is currently based only on the study of life on Earth. One of the profound results of biology is the basic biochemical similarity of living things. The impression of vast diversity that we experience in nature is a result of manifold variations on a single fundamental biochemistry. All life on Earth shares a similar biochemistry based on 20 amino acids and 5 nucleotide bases. Added to this are the few sugars, from which are made the polysaccharides, and the simple alcohols and fatty acids that are the building blocks of lipids. This simple collection of primordial biomolecules (Fig. 1) represents the set from which the rest of biochemistry derives.

The basic building blocks of life are the amino acids from which proteins are made. Proteins are the basic structural units of life. The 20 common amino acids used in living systems are shown in Fig. 1. Enzymes such as nitrogenase, used by bacteia to obtain nitrogen; transport molecules such as hemoglobin, which carry oxygen in the blood; and energy transducers such as chlorophyll, the molecule of photosynthesis, are all made from amino acids. An important molecular aspect

of amino acids is that they come in two symmetries. Except for glycine, the simplest which is symmetrical, amino acids can have either a left handed (L-) or a right handed (D-) form. Figure 2 shows the two versions known as enantiomers (from the Greek *enantios* meaning opposite) for alanine. Life predominantly uses the L enantiomer, although there are some bacteria that use certain D forms and many others have enzymes that can convert the D form to the L form. In addition, amino acids other than the 20 listed in Fig. 1 are used occasionally in proteins and are sometimes used directly, for example, as toxins by fungi and plants. How and why life acquired a preference for L amino acids over D amino acids is one of the observations that theories for the origins of life seek to explain.

There is also commonality in the genetic material that records the blueprints for all life forms. The two types of genetic material are DNA (deoxyribonucleic acid, which uses the sugar deoxyribose) and RNA (ribonucleic acid, which uses the sugar ribose). These nucleic acids are both constructed from sugars to which are attached four nucleotide bases. These bases form the information alphabet of life. In DNA the nucleotide bases are adenine (A), thymine (T), cytosine (C), and guanine (G). When strung in the double helix pattern characteristic of DNA, A always pairs with T and C always pairs with G. In RNA, thymine is replaced by uracil (U), which always binds with adenine. The nucleic acids each provide a four-letter alphabet in which the codes for the construction of proteins are based. This same information recording system is found in all living systems.

The biochemical unity of life, particularly genetic unity, strongly suggests that all living things on Earth are descendants from a common ancestor. This is the phylogenic unity of life shown in Fig. 3. These genetic trees are obtained by comparing the ribosomal RNA within each organism. Sections within the RNA are remarkably similar within all life forms. These conserved sections show only random point changes and not evolutionary trends. Thus the similarity between the sequences of any two organisms is a measure of their evolutionary distance or, more precisely, the time since they shared a common ancestor. When viewed in this way, life on Earth is divided into three main groups: eucarya, bacteria, and archaea. Eucarya include the multicellular life forms encompassing all plants and animals. Bacteria are the familiar bacteria including intestinal bacteria, common soil bacteria, and pathogens. Archaea are a class of microorganisms that are found in unusual and often harsh environments such as hypersaline ponds and H_2-rich anaerobic sediments.

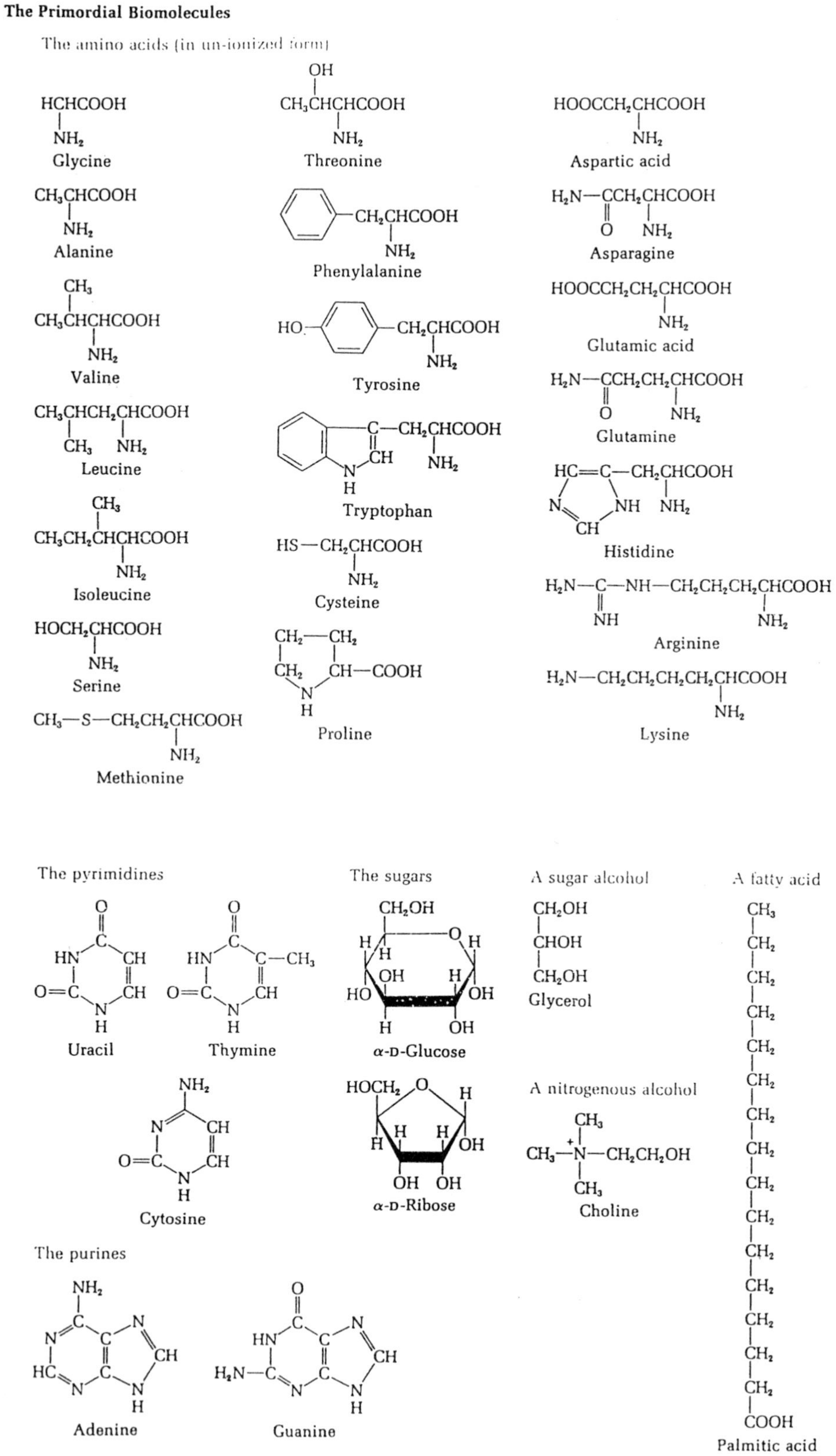

The Primordial Biomolecules

The amino acids (in un-ionized form)

HCHCOOH
|
NH₂
Glycine

OH
|
CH₃CHCHCOOH
|
NH₂
Threonine

HOOCCH₂CHCOOH
|
NH₂
Aspartic acid

CH₃CHCOOH
|
NH₂
Alanine

[phenyl ring]—CH₂CHCOOH
|
NH₂
Phenylalanine

H₂N—CCH₂CHCOOH
‖ |
O   NH₂
Asparagine

CH₃
|
CH₃CHCHCOOH
|
NH₂
Valine

HO—[phenyl ring]—CH₂CHCOOH
|
NH₂
Tyrosine

HOOCCH₂CH₂CHCOOH
|
NH₂
Glutamic acid

CH₃CHCH₂CHCOOH
| |
CH₃   NH₂
Leucine

[indole ring]—C—CH₂CHCOOH
‖ |
CH   NH₂
Tryptophan

H₂N—CCH₂CH₂CHCOOH
‖ |
O   NH₂
Glutamine

CH₃
|
CH₃CH₂CHCHCOOH
|
NH₂
Isoleucine

HS—CH₂CHCOOH
|
NH₂
Cysteine

HC=C—CH₂CHCOOH
| | |
N   NH   NH₂
‖
CH
Histidine

HOCH₂CHCOOH
|
NH₂
Serine

CH₂—CH₂
| |
CH₂   CH—COOH
\   /
N
|
H
Proline

H₂N—C—NH—CH₂CH₂CH₂CHCOOH
‖ |
NH   NH₂
Arginine

CH₃—S—CH₂CH₂CHCOOH
|
NH₂
Methionine

H₂N—CH₂CH₂CH₂CH₂CHCOOH
|
NH₂
Lysine

The pyrimidines

O
‖
C
/   \
HN   CH
| ‖
O=C   CH
\   /
N
|
H
Uracil

O
‖
C
/   \
HN   C—CH₃
| ‖
O=C   CH
\   /
N
|
H
Thymine

The sugars

α-D-Glucose

α-D-Ribose

A sugar alcohol

CH₂OH
|
CHOH
|
CH₂OH
Glycerol

A fatty acid

CH₃
|
CH₂
|
CH₂
|
CH₂
|
CH₂
|
CH₂
|
CH₂
|
CH₂
|
CH₂
|
CH₂
|
CH₂
|
CH₂
|
CH₂
|
CH₂
|
CH₂
|
COOH
Palmitic acid

NH₂
|
C
/   \
N   CH
‖ |
O=C   CH
\   /
N
|
H
Cytosine

A nitrogenous alcohol

CH₃
|
CH₃—N⁺—CH₂CH₂OH
|
CH₃
Choline

The purines

NH₂
|
C
/   \
N   C—N
‖ | \
HC   C   CH
\   / \   /
N   N
|
H
Adenine

O
‖
C
/   \
HN   C—N
| | \
H₂N—C   C   CH
\   / \   /
N   N
|
H
Guanine

FIGURE 1   The basic molecules of life. (From Lehninger, 1975.)

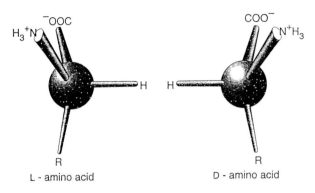

FIGURE 2 The L and D form of the amino acid alanine.

TABLE I
Ecological Requirements for Life

Requirement	Occurrence in the solar system
Energy	Common
Predominately sunlight	Photosynthesis at 100 AU light levels
Chemical	e.g., $H_2 + CO_2 \rightarrow CH_4 + H_2O$
Carbon	Common as CO_2 and CH_4
Liquid water	Rare, only on Earth for certain
N, P, S, and other elements	Common

A. THE ECOLOGY OF LIFE: LIQUID WATER

In addition to describing the building blocks of life, it is instructive to consider what life does. In this regard it is possible to define a set of ecological or functional requirements for life. There are four fundamental requirements for life on Earth: energy, carbon, liquid water, and a few other elements. These are listed in Table I along with the occurrence of these environmental factors in the solar system.

Energy is required for life from basic thermodynamic considerations. Typically, on Earth, this energy is provided by sunlight, which is a thermodynamically efficient (low entropy) energy source. Some limited systems on Earth are capable of deriving their energy from chemical reactions (e.g., methanogenesis, $CO_2 + 4H_2 \rightarrow CH_4 + 2H_2O$) and do not depend on photosynthesis. On Earth these systems are confined to locations where the more typical photosynthetic organisms are not able to grow, and it is not clear if an ecosystem that was planetary in scale or survived over billions of years could be based solely on chemical energy. No known organisms on Earth make use of temperature gradients to derive energy. These organisms would be analogous to a Carnot heat engine. Table II lists some of the most important metabolic reactions by which living systems generate energy. This list includes autotrophs (which derive energy from nonbiological sources) as well as heterotrophs (which derive energy by the consumption of organic material, usually other life forms).

Elemental material is required for life, and on Earth carbon has the dominant role as the backbone molecule of biochemistry. Life almost certainly requires other elements as well. Life on Earth utilizes a vast array of the elements available on the surface. However, this does not prove that these elements are absolute requirements for life. Among the other elements, N, S, and P are probably the leading candidates for the status of required elements. Table III lists the distribution of elements in the cosmos and on the Earth and compares these with the common elements in life.

As indicated in Table I, sunlight and the elements required for life are common in the solar system. What appears to be ecologically limiting for life in the solar system is the stability of liquid water. Liquid water is a necessary requirement for life on Earth. Liquid water is key to biochemistry because it acts as the solvent in which biochemical reactions take place and, furthermore, it interacts with many biochemicals in ways that influence their properties. For example, water forms hydrogen bonds with some parts of a large molecule, the hydrophilic groups, and repels other parts, the hydrophobic groups, thereby forcing these molecules to curl up with their hydrophobic groups in the interior and the hydrophilic groups on the exterior in contact with the water. Certain organsims, notably lichen and some algae, are able to utilize water in the vapor phase if the relative humidity is high enough. Many organisms can continue to metabolize at temperatures well below the freezing point of water because

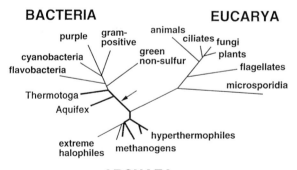

FIGURE 3 A phylogenetic tree showing the relatedness of all life on Earth.

TABLE II
Metabolic Pathways

Heterotrophy
1. Fermentation $C_6H_{12}O_6 \rightarrow 2CO_2 + 2C_2H_5OH$
2. Anaerobic respiration $C_6H_{12}O_6 + 12NO_3^- \rightarrow 6CO_2 + 6H_2O + 12NO_2^-$
3. Aerobic respiration $C_6H_{12}O_6 + 6O_2 \rightarrow 6CO_2 + 6H_2O$

Photoautotrophy
1. Anoxic photosynthesis $12CO_2 + 12H_2S + h\nu \rightarrow 2C_6H_{12}O_6 + 9S + 3SO_4$
2. Oxygenic photosynthesis $6CO_2 + 6H_2O + h\nu \rightarrow C_6H_{12}O_6 + 3O_2$

Chemoautotrophy
  Anaerobic
1. Methanogens $CO_2 + 4H_2 \rightarrow CH_4 + 2H_2O$
                  $CO + 3H_2 \rightarrow CH_4 + H_2O$
                  $4CO + 2H_2O \rightarrow CH_4 + 3CO_2$
2. Acetogens $2CO_2 + 4H_2 \rightarrow CH_3COOH + 2H_2O$
3. Sulfate reducers $H_2SO_4 + 4H_2 \rightarrow H_2S + 4H_2O$
4. Sulfur reducers $S^\circ + H_2 \rightarrow H_2S$
5. Thionic denitrifiers $H_2S + 2NO_3^- \rightarrow SO_4^{2-} + H_2O + N_2O$
                  $3S^\circ + 4NO_3^- + H_2 \rightarrow 3SO_4^{2-} + 2N_2 + 2H^+$
6. Iron reducers $2Fe^{3+} + H_2 \rightarrow 2Fe^{2+} + 2H^+$
  Aerobic
1. Sulfide oxidizers $2H_2S + 3O_2 \rightarrow 2SO_4S + 2H_2O$
2. Iron oxidizers $4FeO + O_2 \rightarrow 2Fe_2O_3$

their intracellular material contains salts and other solutes that lower the freezing point. No microorganism known is currently able to obtain water directly from ice. Many organisms, such as the snow algae *Chlamydomonas nivalis,* thrive in liquid water associated with ice, but in these circumstances the organisms are the beneficiaries of external processes that melt the ice. There is no known occurrence of an organism using metabolic methods to overcome the latent heat of fusion of ice, thereby liquifying it.

Because liquid water is universally required for life and because it appears to be rare in our solar system, the search for life may be, for all intents and purposes, equated with a search for liquid water.

B. GENERALIZED THEORIES FOR LIFE

If we could provide a precise definition of life, this would aid in our investigation for life on other planets and help unravel the origin of life on Earth. This would also be the first step toward a fundamental theory of life that could be generalized to life elsewhere. Despite the fundamental unity of biochemistry and the universality of the genetic code, no single definition has proven adequate in describing life. Many of the attributes that we would associate with life; e.g., self-replication, self-ordering, and response to environmental stimuli, can be found in nonliving systems; fire, crystals, and bimetallic thermostats, respectively. Furthermore, there are various and peculiar life forms such as viruses and giant cell-less slime molds that defy even a biological definition of life in terms of the cell or the separation of internal and external environments. In attempting a resolution of this problem, the most useful definition of life is that it is a system that develops Darwinian evolution: reproduction, mutation, and

TABLE III
Elemental Abundances by Mass

	Cosmos		Earth's crust		Humans		Bacteria	
1	H	70.7%	O	46.6%	O	64%	O	68%
2	He	27.4	Si	27.7	C	19	C	15
3	O	0.958	Al	8.13	H	9	H	10.2
4	C	0.304	Fe	5.00	N	5	N	4.2
5	Ne	0.174	Ca	3.63	Ca	1.5	P	0.83
6	Fe	0.126	Na	2.83	P	0.8	K	0.45
7	N	0.110	K	2.59	S	0.6	Na	0.40
8	Si	0.0706	Mg	2.09	K	0.3	S	0.30
9	Mg	0.0656	Ti	0.44	Na	0.15	Ca	0.25
10	S	0.0414	H	0.14	Cl	0.15	Cl	0.12

TABLE IV
Functional Properties of Life

Mutation
Selection
Reproduction

selection (Table IV). This is an answer to the question what does life do?

We are able to answer the questions, what does life need and what does life do, but we are unable to answer the question what is life (and where is life)? Thus, the requirements for life listed in Table I and the functions of life in Table IV are therefore very general and it is probably unwise to apply more restrictive criteria. For example, for evolution to occur, some sort of information storage mechanism is required. However, it is not certain that this information mechanism needs to be a DNA/RNA-based system or even that it be expressed in structures dedicated solely for replication. Whereas on the present Earth all life uses dedicated DNA and RNA systems for genetic coding, evidence shows that genetic and structural coding were combined at one time into one molecule, RNA. In this so-called RNA world there would have been no distinction between genotype (genetic) and phenotype (structural) molecular replicating systems, both processes would have been performed by an RNA replicating molecule. In present biology the phenotype is composed of proteins for the most part. This example illustrates the difficulty in determining which aspects of biochemistry are fundamental and which are the result of the peculiarities of life's history on Earth.

In basing our consideration of life on the distribution we observe here on Earth as a general phenomenon, we suffer simultaneously from the problem that there is only one kind of life on this planet whereas the variety of that life is too complex to allow for precise definitions or characterization. Thus we can neither extrapolate nor be specific in our theories for life.

It has been suggested that living systems extend far beyond terrestrial biology and anticipate extremely different life forms. Some common examples of this type of alternate life propose the substitution of ammonia for water or silicon for carbon. Certainly ammonia is an excellent solvent — in some respects it is better than water. The range of temperatures over which it is liquid is prevalent in the universe (melting point: $-78°C$, normal boiling point: $-33°C$, liquid at room temperature when mixed with water) and the elements that compose it are abundant in the cosmos. Silicon

has been suggested as a substitute for carbon in alien life forms. However, silicon does not form polymeric chains either as readily or as long as carbon does and its bonds with oxygen (SiO_2) are much stronger than carbon bonds (CO_2), rendering its oxide essentially inert.

Although speculations of alien life capable of using silicon and ammonia are intriguing, no specific experiments directed toward alternate biochemistry have been designed. Thus we have no strategies for where or how to search for such alternate life or its fossils. More significantly, these speculations have not contributed to our understanding of life. One can only conclude that our unique understanding of terrestrial life is based on Earth systems and the wide range speculation of alternate chemistry is too unconstrained to be fruitful. Perhaps some day we will develop general theories for life or, more likely, have many sources of life to compare, thereby allowing for complete theories. Basing our theories on Earth-like life should be considered a practical consequence of our discipline and not a limitation on our approach.

III. HISTORY OF LIFE ON EARTH

There are several sources of information about the origin of life on Earth. These are the physical record, the genetic record, the metabolic record, and laboratory simulations. The physical record includes the collection of sedimentary and fossil evidence of life. This record is augmented by models of the Earth and the solar system, all of which provide clues to conditions billions of years ago when the origin of life is thought to have occurred. There is also the record stored in the genomes of living systems that comprises the collective gene pool of our planet. Genetic information tells us the path of evolution as shaped by environmental pressures, biological constraints, and random events that connect the earliest genomic organism, through the common ancestor to the present tree of life (Fig. 3). There is also the record of metabolic pathways in the biochemistry of organisms that has evolved in response to changes in the environment as well as causing changes in that environment. All of these records are palimpsest in that they have been overwritten, often repeatedly, over time. Laboratory simulations of prebiotic chemistry — the chemistry assumed present before life — can provide clues to the conditions and chemical solutions leading up to the origin of life. Experiments of DNA/RNA replication sequences can provide clues to the selection process that optimizes

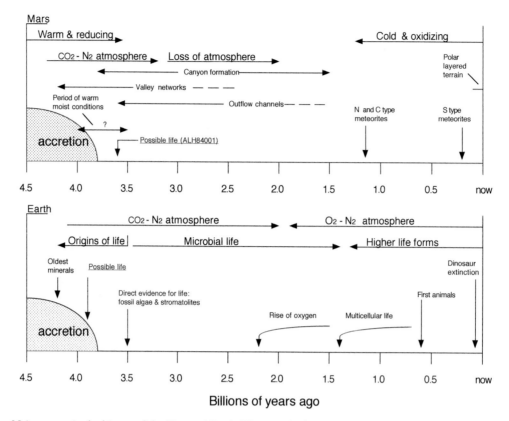

FIGURE 4   Major events in the history of the Mars and Earth. The period of moist surface conditions on Mars may correspond to the time during which life originated on Earth. Similarities between the two planets at this time raise the possibility for the origin of life on Mars.

mutations as well as provide a basic understanding of reproduction. Perhaps one day the process that initiates life will be studied in the laboratory or discovered on another planet.

Major events in the history of life are shown in Fig. 4. As the Earth was forming about 4.5 Gyr ago its surface would have been inhospitable to life. The gravitational energy released by the formation of the planet would have kept surface temperatures too high for liquid water to exist. Eventually, as the heat flow subsided, rain would have fallen for the first time and life could be sustained in liquid water. However, it is possible that subsequent impacts could have been large enough to sterilize the Earth by melting, excavating, and vaporizing the planetary surface, removing all liquid water. Thus, life may have been frustrated in its early starts. For a sufficiently large impact the entire upper crust of the Earth would be ejected into outer space and any remnant left as a magma ocean. Barring these catastrophic events, however, sterilizing the Earth is a difficult task as it is not sufficient to merely heat the surface to high temperatures. At the present time, microorganisms survive at the bottom of the

ocean and even kilometers below the surface of the planet. An Earth-sterilizing impact must not only completely evaporate the oceans, but must then heat the surface and subsurface of the Earth such that the temperature does not fall anywhere below about 200°C, which is the requirement for the heat sterilization of dry, dormant organisms. This is a difficult requirement since the time it takes to diffuse down a given distance scales as the square of the distance. Thus, heat must be applied a million times longer to sterilize to a depth of 1 km compared to a depth of 1 m.

It is not known when the last life-threatening impact occurred. As shown schematically in Fig. 4, the rate of impact, extrapolated from the record on the moon, rises steeply before 3.8 Gyr ago. Thus, it is likely that the Earth was not suitable for life much before 3.8 Gyr ago. There is direct evidence that microbial life was present on the Earth as early as 3.5 Gyr ago. This evidence includes microbial fossils, but it primarily consists of stromatolites. Stromatolites are large features, often many meters in size, that are formed by the lithification of laminated microbial mats (see Fig. 5). Phototactic microorganisms living on the bottom

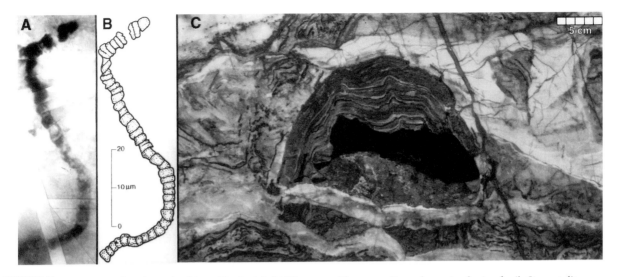

FIGURE 5 Earliest fossil evidence for life on Earth. A 3.5 billion-year-old stromatolite and associated microfossil. Stromatolites are an important form of fossil evidence of life because they form macroscopic structures that could be found on Mars. It is therefore possible that a search for stromatolites near the shores of an ancient martian lake or bay could be conducted in the near future. Expecting microbial communities to have formed stromatolites on Mars is not entirely misplaced geocentricism. Properties of a microbial mat community that result in stromatolite formation need only be those associated with photosynthetic uptake of CO_2. There are broad ecological properties that we expect to hold on Mars even if details of the biochemistry and community structure of martian microbial mats were quite alien compared to their terrestrial counterparts. Trace microfossils can sometimes be found within stromatolites.

of a shallow lake or ocean shore may be covered periodically with sediment carried in by spring runoff, for example. To retain access to sunlight the organism must move up through this sediment layer and establish a new microbial zone. After repeated cycles, a mat formed of laminated sediments and organic material is formed. One characteristic of these biogenic mats that distinguishes them from nonbiologically caused layering is that the response is phototactic, not gravitational. Thus the layering is not usually flat but is more often domed shaped because microorganisms on the periphery would move more toward the side to reach light. In this way stromatolites can be distinguished from similar but nonbiological laminae. Stromatolites often contain microfossils, further testimony to their biological origin.

Comparative studies of the morphology of the earliest microfossils as well as the phototaxis evident in stromatolites imply that life present at 3.5 Gyr ago was based on photosynthesis. Further evidence for this comes from the analysis of the carbon isotope ratios in organic material in these early sediments. Photosynthetic systems preferentially uptake the lighter isotope of carbon (^{12}C) by about 2%. Interestingly, this characteristic shift is found in the earliest sedimentary material (3.9 Gyr old) found on Earth, in Greenland. Unfortunately, these old sediments have experienced extensive reworking and heating over time that has destroyed any morphological clues to their origin.

However, isotopic evidence suggests the presence of photosynthetic life even at this earlier date.

Sophisticated microbial life, clearly capable of photosynthesis and mobility, originated early in the history of the Earth, possibly before the end of the late bombardment 3.8 Gyr ago and certainly not later than 3.5 Gyr ago. This suggests that the time required for this onset of life was brief. If the Greenland sediments are taken as evidence for life, it suggests that, within the resolution of the geological record, life arose on Earth as soon as a suitable habitat was provided. The stromatolites at 3.5 Gyr ago put an upper limit of 300 million years on the length of time it took for the first record of life, after clement conditions.

It is possible, in principle, to determine which organism on the Earth is the most similar to the common ancestor. To do so one must determine which organism has changed the least compared to all other organisms. For example, if some taxon of organism contains a certain mutation but many do not, it is possible to trace the mutation to an ancestor common to all organisms in that taxon. Within this related group of organisms the most primitive traits can be established based on how widespread they are. Traits that are found in all or most of the major groupings should be primitive, particularly if these traits are found in groups that diverged early. Traits found in only a few recently related groups are probably younger traits. This line of reasoning applied to the entire phylogenetic tree would indicate which organism extant today has the most primitive set of traits. This organism would therefore be most similar to the common ancestor. Studies of this type have indicated that the organisms alive today that are most similar genetically, and hence presumably ecologically, to the common ancestor are the thermophilic hydrogen-metabolizing bacteria and maybe also sulfur-metabolizing bacteria. The arrow in Fig. 3 represents the suggested position of the last common ancestor.

It is important to note here that the common ancestor is not necessarily the first organisms on Earth but was merely the last organism from which all life forms today were descendant. The common ancestor may have existed within a world of multiple lineages, none of which are in evidence today. If all life on Earth has indeed descended from a sulfur bacterium living in a hot springs environment, this could be the result of at least three possibilities. First, it may be the case that hot sulfurous environments are important in the origin of life and the common ancestor may represent this primal cell. Second, the common ancestor may have been a survivor of a catastrophe that destroyed all other life forms. The survival of the common ancestor may have been the result of its ability to live deep within a hydrothermal system. Third, the nature of the common ancestor may be serendipitous with no implications as to origin or evolution of the biosphere.

For over 2 Gyr after the earliest evidence for life, life on the Earth was composed of microorganisms only. There were certainly bacteria and possibly one-celled eukaryotes as well. There seemed to be a major change in the environment of the Earth with the rise of photosynthetically produced oxygen beginning at 2.2 Gyr ago, reaching significant levels about 1 Gyr ago and culminating about 600 Myr ago. (Figure 4 shows a time line of Earth's history with these events.) Soon after the development of high levels of oxygen in the atmosphere multicellular life forms appeared. These radiated rapidly into the major phylum known today (as well as many that have no living representative). In time, organisms colonized the land and plants and animals developed.

IV. ORIGIN OF LIFE

Numerous and diverse theories exist for the origin of life presently under serious consideration within the scientific community. A diagram, and classification of current theories for the origin of life on Earth, is shown

in Fig. 6. At the most fundamental level, theories may be characterized within two broad categories: theories that suggest that life originated on Earth (terrestrial in Fig. 6) and those that suggest that the origin took place elsewhere (extraterrestrial in Fig. 6). Extraterrestrial or panspermia theories suggest that life existed in outer space and was transported by meteorites, asteroids, or comets to a receptive Earth. Along similar lines, life may have been ejected by impacts from another planet in our solar system and jettisoned to Earth or vise versa. Furthermore, the scientific literature has suggested that life may have been purposely directed to Earth (directed panspermia in Fig. 6) by an intelligent species from another planet.

Terrestrial theories are further subdivided into organic origins (carbon based) and inorganic origins (mineral based). Mineral-based theories suggest that life's first components were mineral substrates that organized and synthesized clay organisms. These organisms have evolved via natural selection into the organic-based life forms visible on Earth today. The majority of theories that do not invoke an extraterrestrial origin require an organic origin for life on Earth. Theories postulating an organic origin suggest that the initial life forms were composed of the same basic building blocks present in biochemistry today — organic material. If life arose in organic form, then there must have been a prebiological source of organics. The Miller–Urey experiments and their successors have demonstrated how organic material may have been produced naturally in the primordial environment of Earth (endogenous production in Fig. 6). An alternative to the endogenous production of organics on early Earth is the importation of organic material by celestial impacts and debris: comets, meteorites, interstellar dust particles, and comet dust particles. A comparison of these sources is shown in Table V. Table VI lists the organics found in the Murchison meteorite and compares these with the organics produced in a Miller–Urey abiotic synthesis. Organic origins differ mainly in the type of primal energy sources: photosynthetic, chemosynthetic, or heterotrophic. Chemotrophs and phototrophs (collectively called autotrophs) use energy sources that are inorganic, chemical energy and sunlight, respectively, whereas heterotrophs acquire their energy by consuming organics (Table II).

Hydrothermal vent environments have been suggested for the subsurface origin of chemotrophic life. In the absence of sunlight, these organisms must utilize chemical energy (e.g., $CO_2 + 4H_2 \rightarrow CH_4 + 2H_2O +$ energy). Alternatively, phototrophic life utilizes solar radiation from the surface for prebiotic synthesis. With their ability to chemosynthesize and photosynthesize,

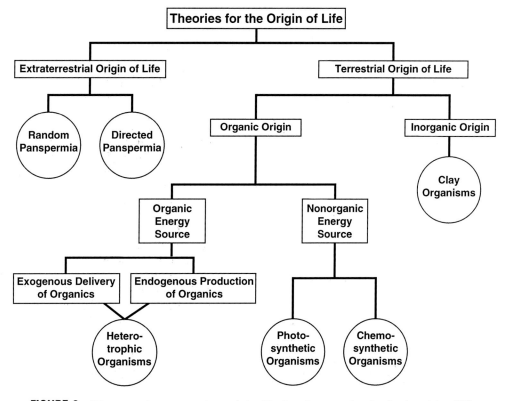

FIGURE 6 Diagrammatic representation and classification of current theories for the origin of life.

these organisms can assimilate their own energy from materials in their environment. One feature that the various theories for the origin of life have in common is the requirement for liquid water. This is because the chemistry of earliest life requires a liquid water medium. This is true if the primal organism appears fully developed (panspermia) and if it engages in organic chemistry, as well as for the clay inorganic theories.

V. LIMITS TO LIFE

In considering life beyond the Earth it is useful to quantitatively determine the limits that life has been able to reach on this planet with respect to environmental conditions. Life is not everywhere. There are environments on Earth in which life has not been able to effectively colonize even though these environments could be suitable for life. Perhaps the largest life-free zone on Earth is the polar ice sheets. Here there is abundant energy, carbon, and nutrients (from atmospheric deposition) to support life. However, water is available only in the solid form. No organism on Earth has adapted to use metabolic energy to liberate water from ice, even though the energy required per molecule is only $\simeq 1\%$ of the energy produced by photosynthesis per molecule. Table VII lists the limits to life as currently known. The lower temperature limit clearly ties to the presence of liquid water whereas the

TABLE V
Sources of Prebiotic Organics on Early Earth

Source	Energy dissipation (J year^{-1})	Organic production (in a reducing atmosphere) (kg year^{-1})
Lightning	1×10^{18}	3×10^{9}
Coronal discharge	5×10^{17}	2×10^{8}
Ultraviolet light ($\lambda <270$ nm)	1×10^{22}	2×10^{11}
Ultraviolet light ($\lambda <200$ nm)	6×10^{20}	3×10^{9}
Meteor entry shocks	1×10^{17}	1×10^{9}
Meteor postimpact plumes	1×10^{20}	2×10^{10}
Interplanetary dust	—	6×10^{7}

TABLE VI
Comparison of Amino Acids in Murchison Meteorite and in Electric Discharge Synthesis, Normalized to Glycine

Amino acid	Murchison meteorite	Electric discharge
Glycine	100	100
Alanine	>50	>50
α-Amino-n-butyric acid	>50	>50
α-Aminoisobutyric acid	10	>50
Valine	10	1
Norvaline	10	10
Isovaline	1	1
Proline	10	0.1
Pipecolic acid	0.1	<1
Aspartic acid	10	10
Glutamic acid	10	1
β-Alanine	1	1
β-Amino-n-butyric acid	0.1	0.1
δ-Aminoisobutyric acid	0.1	0.1
γ-Aminobutyric acid	0.1	1
Sarcosine	1	10
N-Ethylglycine	1	10
N-Methylalanine	1	1

higher temperature limit seems to be determined by the stability of proteins, also in liquid water. Life can survive at extremely low light levels, corresponding to 100 AU, roughly three times the distance between Pluto and the Sun. Salinity and pH also allow for a wide range. Water activity, effectively a measure of the relative humidity of a solution or vapor, can support life only for values above 0.6 for yeasts, lichens, and molds. Bacteria requires levels above 0.8. Radiation-resistant organism such as *Deinococcus radiodurans* can easily survive radiation doses of 1–2 Mrad and maybe higher when in a dehydrated or frozen state.

VI. LIFE IN THE SOLAR SYSTEM

Because the knowledge of life is restricted to the unique but varied case found here on Earth, the most practical approach to the search for life on other planets has been to proceed by way of analogy with life on Earth. The argument for the origin of life on another world would then be based on the similarity of the environments there to the environments on early Earth. Whatever process led to the establishment of life in one of these environments on Earth could then be logically expected to have led to the origin of life on this comparable world. The more exact the comparison between the early Earth and another planet the more compelling is the argument by analogy. This comparative process should be valid for all the theories for the origin of life listed in Fig. 6, ranging from panspermia to the standard theory. (The standard theory for the origin of life posits a terrestrial organic origin requiring the endogenous production of organics leading to the development of heterotrophic organisms.)

Following this line of reasoning further we can conclude that if similar environments existed on two worlds and life arose in both of them then these life forms should be comparable in their broad ecological characteristics. If sunlight was the available energy source, CO_2 the available carbon source, and liquid water the solvent then one could expect phototrophic autotrophs using sunlight to fix carbon dioxide with water as the medium for chemical reactions. Our knowledge of the solar system suggests such an environment could have existed on Mars early in its history as well as Earth early in its history. Thus, whereas life forms originating independently on these two planets would have different biochemical details, they would be recognizably similar in many fundamental attributes. This approach — by analogy to Earth life and the early Earth — provides a specific search strategy for life elsewhere in the solar system. The key element of that strategy is the search for liquid water habitats.

Spacecraft have now visited or flown past comets,

TABLE VII
Limits to Life

Parameter	Limit	Note
Lower temperature	$\sim -15°C$	Liquid water
Upper temperature	113°C	Thermal denaturing of proteins
Low light	$\sim 10^{-4} S_\odot$	Algae under ice and deep sea
pH	1–11	
Salinity	Saturated NaCl	Depends on the salt
Water activity	0.6	Yeasts and molds
	0.8	Bacteria
Radiation	1–2 Mrad	May be higher for dry or frozen state

asteroids, and all of the planets in our solar system except Pluto. Observatory missions have studied all of the major celestial objects in our solar system as well. A preliminary assessment of the occurrence of liquid water habitats, and indirectly life, in the solar system can be done.

VII. MERCURY AND THE MOON

Mercury and the Moon appear to have little prospects for liquid water, now or anytime in the past. These virtually airless worlds have negligible amounts of volatiles (such as water and carbon dioxide) essential for life. There are no geomorphological features that indicate fluid flow. There is speculation that permanently shaded regions of the polar areas on Mercury and the Moon can act as traps for water ice. Recent radar data support this hypothesis for Mercury. However, there is no indication that the pressure and temperature were ever high enough for liquid water to exist at the surface. [See MERCURY.]

VIII. VENUS

Venus currently has a surface that is clearly inhospitable to life. There is no liquid water on the surface and the temperature is over 450°C at an atmospheric pressure of 92 times the Earth's. There is water on Venus but only in the form of vapor and clouds in the atmosphere. The most habitable zone on Venus is the level in the atmosphere where the pressure is about half of the sea level on Earth. At that location there are clouds composed of about 25% water and 75% sulfuric acid and the temperature is about 25°C; these might be reasonable conditions for life. It is possible therefore to speculate that life can be found, or survive if implanted, in the clouds of Venus. What argues against this possibility is the fact that clouds on Earth, at similar pressures and temperatures, do not harbor life. No life forms are known that exist in cloud environments. Perhaps the essential elements are there but a stable environment is required. [See VENUS, ATMOSPHERE.]

Theoretical considerations suggest that Venus and Earth may have initially had comparable levels of water. In this case Venus may have had a liquid water surface early in its history when it was cooler, (3.5 billion years ago) due to the reduced brightness of the fainter early sun. Unfortunately, all record of this early epoch has been erased on Venus and the question of the origin of life during such a liquid water period remains untestable. [See VENUS, SURFACE AND INTERIOR.]

IX. MARS

Of all the extraterrestrial planets and smaller objects in our solar system, Mars is the one that has held the most fascination in terms of life. Early telescopic observations revealed Earth-like seasonal patterns on Mars. Large white polar caps, that grew in the winter and shrunk in the summer were clearly visible. Regions of the planet's surface appeared to darken beginning near the polar cap that was at the start of its spring season and spreading toward the equator. It was natural that these changes, familar to patterns on the Earth, would be attributed to similar causes. Hence, the polar caps were thought to be water ice and the wave of darkening was believed to be caused by the growth of vegetation. The 19th century arguments for the existence of life, and even intelligent life, on Mars culminated in the book *Mars as the Abode of Life* by Percival Lowell in 1908 and the investigations of the celebrated canals. The Mars revealed by spacecraft exploration is decidedly less alive than Lowell's anticipation, but its standing as the most interesting object for biology outside Earth still remains.

X. VIKING RESULTS

In 1976 the Viking landers successfully reached the Martian surface while the two orbiters circled the planet repeatedly photographing and monitoring the surface. The primary objective of the Viking mission was the search for microbial life. Previous reconnaissance of Mars by the *Mariner* flyby spacecraft and the photographs returned from the *Mariner 9* orbiter had already indicated that Mars was a cold, dry world with a thin atmosphere. There were intriguing features indicative of past fluvial erosion, but there was no evidence for current liquid water. It was thought that any life to be found on Mars would be microbial. The Viking biology package consisted of three experiments shown schematically in Fig. 7.

The pyrolytic release (PR) experiment searched for photosynthesis as a sign of life. The PR was designed to see if martian microorganisms could incorporate

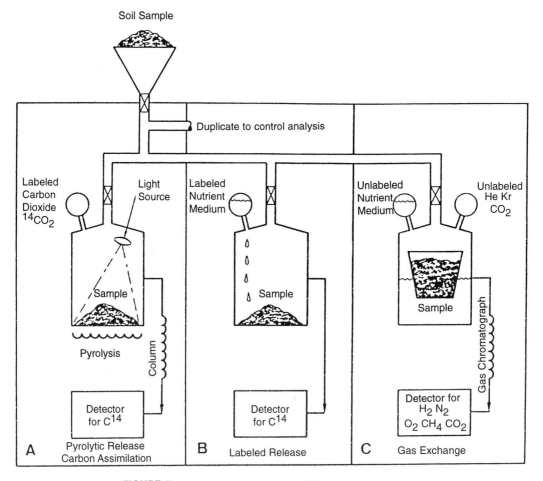

FIGURE 7   Schematic diagram of the *Viking* biology experiments.

CO_2 under illumination. The experiment could be performed under dry conditions, similar to the martian surface, or it could be run in a humidified mode. The CO_2 in the chamber was labeled with radioactive carbon, which could then be detected in any organic material synthesized during the experiment. The very first run of the labeled release experiment produced a significant response. It was well below the typical response observed when biotic soils from Earth had been tested in the experiment, but it was much larger than the noise level. Subsequent trials did not reproduce this high result and this initial response was attributed to a startup anomaly, possibly some small prelaunch contamination.

The gas exchange (GEx) experiment searched for heterotrophs, which are microorganisms capable of consuming organic material. The GEx was designed to detect any gases that the organisms released as a by-product of their metabolism — bacterial flatulence. After a sample was placed in the chamber the soil was first equilibrated with water vapor and then combined with a nutrient solution. At prescribed intervals, a sample of the gas above the sample was removed and analyzed by a gas chromatograph.

The GEx results were startling. When the martian soil was merely exposed to water vapor it released oxygen gas at levels of 70–700 nmol per gram soil, much larger than could be explained by absorption of the gas onto the soil grains. The GEx results are summarized in Table VIII. It was clear that some chemical or biological reaction was responsible for the oxygen release. A biological explanation was deemed unlikely as the reactivity of the soil persisted even after it had been heat sterilized to temperatures of over 160°C. Furthermore, adding the nutrient solution did not change the result, some chemical in the soil was highly reactive with water.

The labeled release (LR) experiment also searched for heterotrophic microorganisms. In the LR experiment, a solution of water containing seven organics

TABLE VIII
Comparison of GEx O₂ and LR ¹⁴C Results[a]

Sample	GEx O$_2$ (nmol cm^{-3})	Oxidant[b] (KO$_2$ → O$_2$)	LR CO$_2$ (nmol cm^{-3})	Oxidant[b] (H$_2$O$_2$ → O)
Viking 1 (surface)	770	35 ppm/m	~30	1 ppm/m
Viking 2 (surface)	194	10	~30	1
Viking 2 (subrock)	70	3	~30	1

[a] After Klein (1979).
[b] Assuming a soil density of 1.5 g cm^{-3}.

was added to the soil. The carbon atoms in each organic were radioactive. A radiation detector in the headspace detected the presence of radioactive CO_2 released during the experiment. Any organisms in the soil would be detected as they broke down the organics and released radioactive CO_2.

When the LR experiment was performed on Mars there was a steady release of radioactive CO_2; the results are summarized in Table VIII. When the soil sample was heat sterilized before exposure to the nutrient solution, no radioactive CO_2 was detected. Results of the LR experiment were precisely that expected if there were microorganisms in the soil sample. Taken alone the LR results would have been a strong positive indication for life on Mars.

In addition to the three biology experiments, there was another instrument that gave information pivotal to the interpretation of the biological results. This was a combination of a gas chromatograph and a mass spectrometer (GCMS). This instrument received martian soil samples from the same sampling arm that provided soil to the biology experiments. The sample was then heated to release any organics. The decomposed organics were carried through the gas chromatograph and identified by the mass spectrometer. No martian organics were detected. The only signal was due to cleaning agents used on the spacecraft before launch. The limit on the concentration of organics that would remain undetectable by the GCMS was one part per billion. A part per billion of organic material in a soil sample represents over a million individual bacteria, each the size of a typical *Escherichia coli*. This may not seem to rule out a biological explanation for the LR results. However, all life is composed of organic material and it is constantly exuded and processed in the biosphere. On Earth, it is difficult to imagine life without a concomitant matrix of organic material. This apparent absence of organic material is the main argument against a biological interpretation of the positive LR results.

The prevailing explanation for the reactivity of the martian soil relies on the presence of reactive chemicals in the martian atmosphere. In particular hydrogen peroxide (H_2O_2) is thought to be produced by ultraviolet light in the atmosphere and deposited onto the soil surface. Hydrogen peroxide itself could explain many of the LR results, including the loss of reactivity with heating, but it cannot explain the thermally stable results of the GEx. However, peroxide, possibly abetted by ultraviolet radiation, could somehow result in the production of the stable reactive chemicals responsible for the release of oxygen on humidification and the breakdown of organics in the LR experiment. In addition, these reactive chemicals would have cleansed the surface of any naturally occurring organic material or any material carried in by a meteorite. Table VIII also lists the concentration of oxidant necessary to explain the Viking results for typical models of the chemistry of the oxidants.

Amplifying the apparently negative results of the Viking biology experiments, the environment of Mars appears to be inhospitable to life, although the atmosphere contains many of the elements necessary for life; it is composed of 95% CO_2 with a few percent N_2 and argon and trace levels of water. However, the mean surface pressure is less than 1% of sea level pressure on the Earth, and the mean temperature is −60°C. The pressure is close to the triple point pressure of water. This is the minimum pressure at which a liquid state can exist. The low pressures and low temperatures make it unlikely that water will exist as a liquid on Mars. In the rare locations at low elevation where the pressure and temperature are sufficient to support liquid water, the surface is desiccated. Due to seasonal transport, the available water is trapped as ice in the polar regions. Even saturated brine solutions cannot exist in equilibrium with the atmosphere near the equator. The absence of liquid water on the surface of Mars is probably the most serious argument against the presence of life anywhere at the surface of the planet. A second significant hazard to life on the martian surface is the presence of solar ultraviolet light in wavelengths between 190 and 300 nm. This radiation, which is largely shielded from Earth's surface by the ozone layer, is highly effective at destroying terrestrial organisms. Wavelengths below 190 nm are absorbed even by the present thin martian CO_2 atmosphere. Compounding the effects of UV irradiation, and perhaps caused by it, are possible chemical oxidants that are thought to exist in the martian soil. Such strong oxidants have been suggested as the causative agent for the chemical reactivity observed at the Viking sites. [*See* MARS: ATMOSPHERE AND VOLATILE HISTORY.]

XI. EARLY MARS

There is considerable evidence that early in its history Mars did have liquid water on its surface. Images from the *Viking* orbiters show complex dendritic valley networks that are believed to have been carved by liquid water. One of these is shown in Fig. 8. These valleys are found predominantly in the heavily cratered, hence ancient, terrains in the southern hemisphere. This would suggest that the period of liquid water on Mars occurred contemporaneously with the end of the last stages of heavy cratering, about 3.8 Gyr ago. This epoch is the same at which life is thought to have originated on Earth (Fig. 4). [*See* MARS, SURFACE AND INTERIOR.]

The presence of liquid water habitats on early Mars at approximately the time that life is first evident on Earth suggests that life may have originated on Mars during the same time period. Liquid water is the most critical environmental requirement for life on Earth, and the general similarity between Earth and Mars leads us to assume that life on Mars would be similar in this basic environmental requirement. More exotic approaches to life on Mars cannot be ruled out, but they are not supported by any available evidence.

It is interesting therefore to consider how evolution may have progressed on Mars by comparison with the Earth. The history of Earth and Mars is compared in Fig. 4. Figure 4 shows that the period between 4.0 and 3.5 Gyr ago is the time when life is most likely to have evolved on both planets. On Earth life persists and remains essentially unchanged for several billion years until the cumulative effects of O_2 production induced profound changes on the atmosphere of that planet. On Mars, conditions become unsuitable for life (no liquid water) in a billion years or less. Thus, it is likely that if there was any life on early Mars it remained microbial.

The evidence of liquid water on early Mars, particularly that provided by the valley networks, suggests that the climate on early Mars may have been quite different than the present. It is generally thought that the surface temperature must have been close to freezing, much warmer than the present $-60°C$. These warmer temperatures are thought to have occurred as a result of a greatly enhanced greenhouse due to a thick (~ 1–5 atm) CO_2 atmosphere. However, CO_2 condensation may have limited the efficacy of the CO_2 greenhouse, but recent work indicates that CO_2 clouds can enhance the greenhouse and maintain warm temperatures.

If Mars did have a thick CO_2 atmosphere this strengthens the comparison to the Earth, which is thought to also have had a thick CO_2 atmosphere early in its history. The duration of a thick atmosphere on Mars and the concomitant warm, wet surface conditions are unknown, but simple climate models suggest that significant liquid water habitats could have existed on Mars for ~ 0.5 Gyr after the mean surface temperature reached freezing. This model is based on the presence of deep lakes (over 30 m), such as those in the dry valleys of the Antarctic where mean annual temperatures are $-20°C$.

If we divide the possible scenario for the history of water on the surface of Mars into four epochs, the first epoch would have warm surface conditions and liquid water. As Mars gradually loses its thick CO_2 atmosphere, the second and third epochs would be characterized by low temperatures but still relatively high atmospheric pressures. This is because the temperature would drop rapidly as the pressure decreased. During the second epoch, temperatures would rise above freezing during some of the year and liquid water habitats would require a perennial ice cover. However, by epoch three the temperature would never rise above freezing and the only liquid water would be found in porous rocks with favorable exposures to sunlight. In epoch four the pressure would fall too low for the presence of liquid water.

A point worth emphasizing here is that the biological requirement is for liquid water per se. Current difficulties in understanding the composition and pressure of the atmosphere need not lessen the biological importance of the direct evidence for the presence of liquid water. In fact, as observed in Antarctic dry valleys, ecosystems can exist when mean temperatures are well below freezing. Mars need not have ever been above freezing for life to persist.

The particular environment on the early Earth in

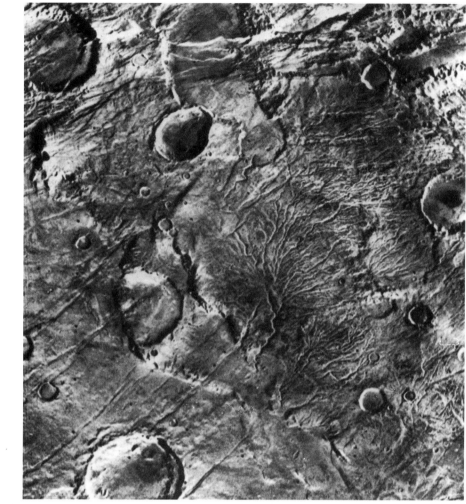

N

25 km

FIGURE 8   A well-developed valley network in the ancient cratered terrain. These valleys were probably formed under a significantly warmer climate, presumably caused by a denser atmosphere, than the present Mars. This evidence for the stability of liquid water in the martian surface 3.8 Gyr ago is the primary motivation for considering the possible origin of life on Mars.

which life originated is not known. However, this does not pose as serious a problem to the question of the origin of life on Mars as might be expected. The reason is that all of the environments found on the early Earth would be expected to be found on Mars, including hydrothermal sites, hot springs, lakes, oceans (i.e., planetary scale water reservoirs), volcanos, tidal pools (solar tides only), marshes, salt flats, and others. Thus, whatever environment or combination of environments that was needed for life to get started on Earth should have been present on Mars as well, and at the same time.

Because the rationale for life on Mars early in its history is based on analogs with fossil evidence for life on the early Earth it is natural to look to the fossil record on Earth as a guide to how relics of early martian

life might be found. The most persuasive evidence for microbial life on the early Earth comes from stromatolites, as discussed earlier. The resulting structures can be quite large, as they are macroscopic fossils generated by microorganisms.

Sites have been identified on Mars that are prime targets for a search for stromatolites or microfossils. These are locations where liquid water would have accumulated — lakes — which are now dry. Most of these locations have been identified from orbital images as basins into which valley networks flow. In one location a possible paleolake has been identified based on studies of laminated sediments forming a plateau within a canyon that has no obvious inlets or outlets. The canyon, shown in Fig. 9, is Hebes chasma and is part of the Valles Marineris system. It is thought that

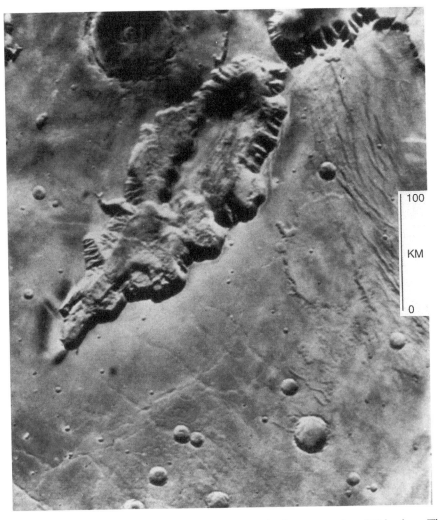

FIGURE 9 *Viking* orbiter image of Hebes Chasma (0°S, 75°W). This canyon is a box canyon about 280 km long. The mesa in the center of the canyon shows layered sediments that are believed to have been deposited in a lake, possibly an ice-covered one like those in the Antarctic dry valleys. Such lakes are of interest as a possible habitat for life after ambient conditions on Mars had become inimical to life.

the sediments represent carbonate deposits formed from the CO_2 in the martian atmosphere.

XII. SUBSURFACE LIFE ON MARS

Although there is currently no direct evidence to support speculations about extant life on Mars, there are several interesting possibilities that cannot be ruled out at this time. Protected subsurface niches associated with hydrothermal activity could have continued to support life even after surface conditions became inhospitable. Liquid water could be provided by the heat of geothermal or volcanic activity melting permafrost or other subsurface water sources. Gases from volcanic

activity deep in the planet could provide reducing power (as CH_4, H_2, or H_2S) percolating up from below and enabling the development of a microbial community based on chemolithoautotrophy. An example is a methanogen (or acetogen) that uses H_2 and CO_2 in the production of CH_4. Such ecosystems have been found deep underground on the Earth. However, their existence is neither supported nor excluded by current observations of Mars. Tests for such a subsurface system involve locating active geothermal areas associated with ground ice or detecting trace quantitites of reduced atmospheric gases that would leak from such a system.

While it certainly seems clear that volcanic activity on Mars has diminished over geological time, intriguing evidence for recent (on the geological time scale) volcanic activity comes from the young crystalization ages (all less than 1 Gyr) of the Shergotty meteorite (and other similar meteorites thought to have come from Mars). Volcanic activity by itself does not provide a suitable habitat for life, as liquid water, which may be derived from the melting of ground ice, is also required. Presumably, the volcanic source in the equatorial region would have depleted any initial reservoir of ground ice and there would be no mechanism for renewal, although there are indications of geologically recent volcano/ground ice interactions at equatorial regions. Closer to the poles ground ice is stable. It is conceivable that a geothermal heat source could result in cycling of water through the frozen ice-rich surface layers. The heat source would be melting and drawing in water from any underlying reservoir of groundwater or ice that might exist. [*See* METEORITES.]

Another line of reasoning also supports the possibility of subsurface liquid water. There are outflow channels on Mars that appear to be the result of the catastrophic discharge of subsurface aquifers of enormous sizes. Evidence based on craters and stratigraphic relations shows that these have occurred throughout martian history. If this is the case then it is possible that intact aquifers remain. This would have profound implications for exobiology (as well as human exploration). Furthermore, it suggests that the debris field and outwash regions associated with the outflow channel may hold evidence of life that existed within the subsurface aquifer just prior to its catastrophic release.

The collection of available water on Mars in the polar regions naturally suggests that summer warming at the edges of the permanent water ice cap may be a source of meltwater, even if short lived. In the polar regions of Earth there are complex microbial ecosystems that survive in transient summer meltwater. However, on Mars the temperature and pressures remain too low for liquid water to form. Any energy available is lost due to sublimation of the ice before any liquid is produced. It is unlikely that there is even seasonal habitats at the edge of the polar caps. This situation may be different over longer time scales. Changes in the obliquity axis of Mars can significantly increase the amount of insolation reaching the polar caps in summer. If the obliquity increases to over about 50°, then increased temperatures, increased atmospheric pressures, and the increased polar insolation that result may cause summer liquid water meltstreams and ponds at the edge of the polar cap.

The polar regions may harbor remnants of life in another way. Tens of meters beneath the surface the temperature is well below freezing ($< -70°C$) and does not change from summer to winter. It is likely that these permafrost zones have remained frozen, particularly in the southern hemisphere, since the end of the intense crater formation period. In this case there may be microorganisms frozen into the permafrost that date back to the time when liquid water was common on Mars, over 3.5 Gyr ago. On Earth, permafrost of such an age does not exist, but there are sediments in the polar regions that have been frozen for many millions of years. When these sediments are exhumed and samples are extracted using sterile techniques, viable bacteria are recovered. The sediments on Mars have been frozen much longer (1000 times) but the temperatures are also much colder. Thus, it may be possible that intact microorganism could be recovered from the martian permafrost. Natural radiation from U, Th, and K in the soil would be expected to have killed any organism, but their biochemical remains would be available for study. The southern polar region seems like the best site as the terrain there contains ancient heavily cratered regions.

XIII. METEORITES FROM MARS

Of the thousands of meteorites known, there are 12 that are thought to have come from Mars. It is certain that these meteorites came from a single source because they all have similar ratios of the oxygen isotopes — values distinct from terrestrial, lunar, or asteroidal ratios. These meteorites can be grouped into four classes. Three of these classes contain 11 of the 12 samples and are known by the name of the type specimen: the S (Shergotty), N (Nakhla), and C (Chassigny) class meteorites. The S, N, and C meteorites are relatively young, having crystallized from lava flows between 200 and 1300 million years ago. Gas inclusions in two of the S-type meteorites contain gases similar to the present martian atmosphere as measured by the *Viking* landers, proving that this meteorite, and by inference the other 10 as well, came from Mars. The fourth class of martian meteorite is represented by the single specimen known as ALH84001. Studies of this meteorite indicate that it formed on Mars about 4.5 Gyr ago in warm, reducing conditions. There are even indications that it contains martian organic material and appears to have experienced aqueous alteration after formation. This rock formed during the time period when Mars is thought to have had a warm, wet climate capable of supporting life.

Researchers have suggested that ALH84001 contains plausible evidence for life on Mars. They base their conclusion on four observations. (1) Complex

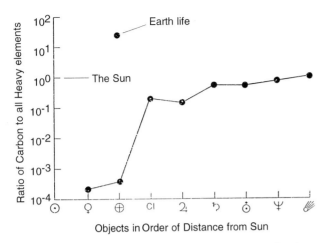

FIGURE 10   Ratio of carbon atoms to total heavy atoms (heavier than He) for various solar system objects illustrating the depletion of carbon in the inner solar system. The *x* axis is not a true distance scale, but the objects are ordered by increasing distance from the sun. Mars is not shown because the size of its carbon reservoir is unknown.

organic material (polycyclic aromatic hydrocarbons) consistent with, but not proof of, a biogenic source are present inside ALH84001. (2) Carbonate globules indigenous to the meteorite that are enriched in ^{12}C over ^{13}C. The isotopic shift is in the range that, on Earth, results only from biogenic activity. (3) Magnetite and iron-sulfide particles are present, and the distribution and shape of these particles are most naturally explained by microbial activity. (4) Microfossils are seen that could be actual traces of microbial life.

ALH84001 does not provide convincing evidence of past life on Mars when compared to the convincing fossil evidence of life on Earth 3.5 Gyr ago (Fig. 3). However, ALH84001 results do provide strong support to the suggestion that conditions suitable for life were present on Mars early in its history. When compared to the SNC meteorites, ALH84001 indicates that Mars experienced a transition from a warm reducing environment with organic material present to a cold oxidizing environment in which organic material was unstable.

XIV. GIANT PLANETS

The inner solar system has the temperature conditions necessary for liquid water, but the outer solar system is richer in the organic material from which life is made. This comparison is shown in Fig. 10, which shows the ratio of carbon to heavy elements (all elements other than H and He) for various objects in the solar system. Earth is in fact depleted with respect to the solar system value of carbon by a factor of about 10^4. It may be interesting then to consider life in the organic-rich outer solar system.

The giant planets Jupiter, Saturn, Uranus, and Neptune do not have firm surfaces on which water could accumulate and form a reservoir for life. Here the only clement zone would be that region of the clouds in which temperatures were in the range suitable for life. Cloud droplets would provide the only source of liquid water. Such an environment would have all the elements needed for life as well as sunlight. [*See* ATMOSPHERES OF THE GIANT PLANETS.]

There have been speculations that life could exist in such an environment, including advanced multicellular creatures. However, such speculations are not supported by considerations of the biological state of clouds on Earth. No organisms have adapted themselves to live exclusively in clouds on Earth, even in locations where clouds are virtually always present. This niche remains unfilled on Earth and by analogy is probably unfilled elsewhere in the solar system.

Following this line of thought leads us to search for environments suitable for life on planetary bodies with surfaces. In the outer solar system this focuses on the moons of the giant planets. Of particular interest are Europa and Titan.

XV. EUROPA

Europa, one of the moons of Jupiter, appears to be an airless, ice-shrouded world. However, theoretical calculations suggest that there may be a layer of liquid water under the ice surface of Europa. The surface of Europa is crisscrossed by streaks that are slightly darker than the rest of the icy surface. If there is an ocean beneath a relatively thin ice layer, then these streaks may represent cracks where the water has come to the surface. [*See* ICY SATELLITES.]

Many ecosystems on Earth thrive and grow in water that is continuously covered by ice. These are found in both Arctic and Antarctic regions. In addition to the polar oceans where sea ice diatoms perform photosynthesis under the ice cover, there are perennially ice-covered lakes in the Antarctic continent in which microbial mats based on photosynthesis are found in the water beneath a 4-m ice cover. The light penetrating these thick ice covers is minimal, about 1% of the incident light. Using these Earth-based systems as a guide, it is possible that sunlight penetrating through the cracks (the observed streaks) in the ice of Europa could support a transient photosynthetic community. Otherwise, if there were hydrothermal sites on the bottom of the Europa ocean it may be possible that chemosynthetic life could survive there, by analogy to life at hydrothermal vent sites at the bottom of the Earth's oceans. The biochemistry of hydrothermal sites on Earth depends on O_2 produced at the Earth's surface. On Europa, a chemical scheme such as that suggested for subsurface life on Mars would be appropriate ($H_2 + CO_2$).

The main problem with life on Europa is the question of its origin. Lacking a complete theory for the origin of life, and lacking any laboratory synthesis of life, we have to base our understanding of the origin of life on other planets on analogy with the Earth. It has been suggested that hydrothermal vents may have been the site for the origin of life on Earth and, in this case, the prospects for life in a putative ocean on Europa are improved. However, the early Earth contained many environments other than hydrothermal vents, such as surface hot springs, volcanos, lake and ocean shores, tidal pools, and salt flats. If any of these environments were the locale for the origin of the first life on Earth, then the case for an origin on Europa is weakened considerably.

XVI. TITAN

Titan, the largest moon of the planet Saturn, has a substantial atmosphere composed primarily of N_2 and CH_4 with many other organic molecules present. The temperature at the surface is close to 94 K and the surface pressure is 1.5 times the pressure of Earth at sea level. The conditions at the surface are uncertain but there may be lakes or oceans of liquid CH_4 and C_2H_6. [*See* TITAN.]

Voyager images and spectral data, as well as previous ground-based studies, indicated that there is an optically thick haze in the upper atmosphere. The haze is thought to be composed of organic material based on the following considerations. (1) During the *Voyager 1* flyby of Titan the spectral signature of nine organic molecules heavier than CH_4 was detected in Titan's stratosphere. (2) photochemical models suggest that CH_4 and N_2 are being destroyed by chemical reactions driven by solar photons and by magnetospheric electrons. The observed organic species and even heavier organic molecules are predicted to result from these chemical transformations. (3) Laboratory simulations of organic reactions in Titan-like gas mixtures produce solid refractory organic substances (tholin), and similar processes are expected to occur in Titan's atmosphere.

Conditions on Titan are much too cold (although the pressure is in an acceptable regime) for liquid water to exist. For this reason it is unlikely that life could originate there. However, Titan remains interesting because it is a naturally occurring Miller–Urey experiment in which simple compounds are transformed into more complex organics. A detailed study of this process may yield valuable insight into how such a mechanism might have operated on the early Earth.

There is also some speculation that Titan may have liquid water on or near the surface under unusual conditions. This could have occurred early in its formation when the gravitational energy released by the formation of Titan would have heated it to high temperatures. More recently, impacts could conceivably melt local regions generating warm subsurface temperatures that could last for thousands of years. Whether such brief episodes of liquid water could have lead to life remains to be tested.

XVII. ASTEROIDS

Asteroids seem like unlikely locations for life to have originated. Certainly they are too small to support an atmosphere sufficient to allow for the presence of liquid water at the present time. However, asteroids, particularly the so-called carbonaceous type, are thought to contain organic material. Thus they might have played a role in the delivery of organics to the prebiotic Earth. A more intriguing aspect of some asteroids is the presence of hydrothermally altered materials, which seems to indicate that asteroids were once part of a larger parent body. Furthermore, conditions on this larger parent body were such that liquid water was present, at least in thin films. Containing both organic material and liquid water, the parent body of the asteroids is thus an interesting target for a life search. However, a thorough assessment of this will require a more detailed study of carbonaceous asteroids in the asteroid belt. Meteorites found on the Earth provide only a glimpse of small fragments of these objects and no signs of extraterrestrial life have been found, but the samples are small and the potential for contamination by Earth life is great.

XVIII. COMETS

Comets, like asteroids, are known to be rich in organic material. However, unlike asteroids, comets also con-

TABLE IX
Drake Equation[a]

$N = R_* \times f_p \times n_e \times f_l \times f_i \times f_c \times L$	
N	Number of civilizations in the galaxy
R_*	Number of stars forming each year in the galaxy
f_p	Fraction of stars possessing planetary systems
n_e	Average number of habitable planets in a planetary system
f_l	Fraction of habitable planets on which life originates
f_i	Fraction of life forms that develop intelligence
f_c	Fraction of intelligent life forms that develop advanced technology
L	Length of time, in years, that a civilization survives

[a] After Drake (1965).

tain a large fraction of water. In their typical state this water is frozen as ice, which is unsuitable for life processes. As a comet approaches the sun, its surface is warmed considerably, but this leads only to the sublimation of the water ice. Liquid does not form because the pressure at the surface of the comet is much too low.

There has been the suggestion that soon after their formation the interior of large comets would have been heated by short-lived radioactive elements (^{26}Al) to such an extent that the core would have melted. In this case there would have been a subsurface liquid water environment similar to that postulated for the present-day Europa. Again the question of the origin of life in such an environment rests on the assumption that life can originate in an isolated deep dark underwater setting.

XIX. LIFE ABOUT OTHER STARS

In our solar system we find only one planet with clear signs of life and one other planet with past evidence of liquid water and the possibility of past life. Our understanding of life as a planetary phenomenon would clearly benefit from finding another planet, around another star, that was rich in life.

One way of formulating the probability of life, and intelligent life, elsewhere in the galaxy is known as the Drake equation, after Frank Drake, a pioneer in the search for extraterrestrial intelligence. The equation and the terms that comprise it are listed in Table IX. The most accurately determined variable in the Drake equation at this time is R_*, the number of stars forming

in the galaxy each year. Because we know that there are about 10^{11} stars in our galaxy and that their average lifetime is about 10^{10} years, then $R_* \simeq 10$ stars per year. All the other terms are uncertain and can be only estimated by extrapolating from what has occurred on Earth. Estimates by different authors for N, the number of civilizations in the galaxy capable of communicating by radios waves, range from 1 to millions. Perhaps the most uncertain term is L, the length of time that a technologically advanced civilization can survive.

The criterion for determining whether a planet can support life is the availability of water in the liquid state. This in turn depends on the surface temperature of the planet, which is controlled primarily by the distance to a central star. Life appeared so rapidly on Earth after its formation that it is likely that other planets may only have had to sustain liquid water for a short period of time for life to originate. Planets orbiting a variety of star types could satisfy this criteria at some time in their evolution. The development of advanced life on Earth, particularly intelligence, took much longer, almost 4 billion years. Earth maintained habitable conditions for the entire period of time. Locations about stars in which temperatures are conducive to liquid water for such a long period of time have been called continuously habitable zones (CHZ). Calculations of the CHZ about main sequence stars indicate that the mass of the star must be less than 1.5 times the mass of our sun for the CHZ to persist for more than 2 billion years.

An interesting result of this calculation is that the current habitable zone for the sun has an inner limit at about 0.8 AU and extends out to between 1.3 and 1.6 AU, depending on the way clouds are modeled. Thus, while Venus is not in the habitable zone, Earth and Mars both are. This calculation would suggest that Mars is currently habitable, but we see no indication of life. This is due to the fact that the distance from the sun is not the only determinant for the presence of liquid water on a planet's surface. The presence of a thick atmosphere, and the resultant greenhouse effect, is required as well. On Earth the natural greenhouse effect is responsible for warming the Earth by 30°C; without the greenhouse effect the temperature would average −15°C. Mars does not have an appreciable greenhouse effect and hence its temperature averages −60°C. If Earth were at the same distance from the sun that Mars is, it would probably be habitable. The reason is the thermostatic effect of the long-term carbon cycle. This cycle is driven by the burial of carbon in seafloor sediments as organic material and carbonates. The formation of carbonates is due to chemical erosion of the surface rocks. Subduction car-

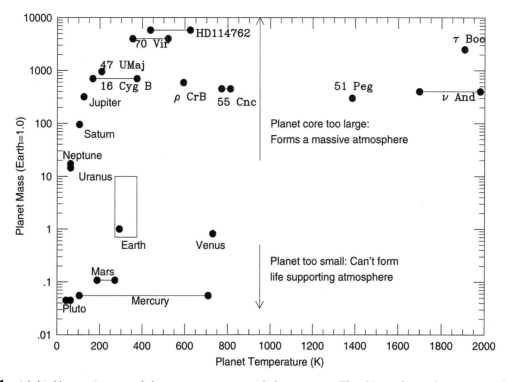

FIGURE 11 A habitable zone in terms of planetary temperatures and planetary mass. The objects of our solar system are shown as well as stars with evidence suggesting newly discovered extrasolar planets.

ries this material to depths where the high temperatures release the sedimentary CO_2 gas. These gases escape to the surface in volcanos that lie on the boundary arc of subduction zones. The thermostatic action of this cycle results because the erosion rate is strongly dependent on temperature. If the temperature were to drop, erosion would slow down. Meanwhile the outgassing of CO_2 would result in a buildup of this greenhouse gas and the temperature would rise. Conversely, higher temperatures would result in higher erosion rates and a lowering of CO_2, again stabilizing the temperature.

Mars became uninhabitable because it lacks plate tectonics and hence has no means of recycling carbon-containing sediments. As a result, the initial thick atmosphere that kept Mars warm has dissipated, presumably into carbonate rocks located on the floor of ancient lake and ocean basins on Mars. Mars lacks plate tectonics because it is too small (10 times smaller than the Earth) to maintain the active heat flows that drive tectonic activity. Hence, planetary size and its effect on geological activity also play a role in determining the surface temperature and thereby the presence of liquid water and life. Figure 11 shows the habitable zone for a planet in terms of its surface temperature and mass. Planets of the solar systems are shown, as are stars with evidence for extrasolar planets.

XX. CONCLUSION

Life is a planetary phenomenon. We see its profound influences on the surface of one planet: the Earth. Its origin, history, and present state and global scale interactions remain a mystery primarily because we have only one datum. The many questions about life await the discovery of another life form with which to compare. Mars early in its history is probably the best prospective case, although Europa may be a promising candidate if it has liquid water under its surface ice. In any case, it is likely that our true understanding of life will be found in the exploration of other worlds, those both with and without life-forms. We have only just begun to search.

BIBLIOGRAPHY

Davis, W. L., and McKay, C. P. (1996). Origins of life: A comparison of theories and application to Mars. *Orig. Life Evol. Biosph.* **26**, 61–73.

Drake, F. D. (1965). The radio search for intelligent extraterrestrial life. *In* "Current Aspects of Exobiology" (G. Mamikunian and M. H. Briggs, eds.), pp. 323–345. Pergamon Press, New York.

Goldsmith, D. (1997). "The Hunt for Life on Mars." Penguin, Baltimore.

Klein, H. P. (1979). The Viking mission and the search for life on Mars. *Rev. Geophys. Space Phys.* **17,** 1655–1662.

Lederberg, J. (1960). Exobiology: Approaches to life beyond the Earth, *Science* **132,** 393–400.

Lehninger, A. L. (1975). "Biochemistry." Worth, New York.

McKay, C. P. (1997). The search for life on Mars. *Orig. Life Evol. Biosph.* **27,** 263–289.

Miller, S. L. (1992). The prebiotic synthesis of organic compounds as a step toward the origin of life. *In* "Major Events in the History of Life" (J. W. Schopf, ed.), pp. 1–28. Jones and Bartlett, Boston, MA.

Shapiro, R. (1986). "Origins: A Skeptics Guide to the Creation of Life on Earth." Summit Books, New York.

PLANETARY
EXPLORATION MISSIONS

I. Developing a Planetary Program

II. Moon Missions

III. Interplanetary Missions

IV. Venus (and Mercury) Missions

V. Mars Missions

VI. Outer Planet Missions

VII. Comet and Asteroid Missions

VIII. Future Planetary Missions

Louis Friedman and Robert Kraemer
The Planetary Society

I. DEVELOPING A PLANETARY PROGRAM

During the 1950s the Soviet Union developed large ICBM boosters and started to employ them to send spacecraft to the Moon and, without early success, to Venus and Mars. Americans advanced beyond the German V-2 technology to develop the relatively small but very efficient Atlas ICBM. In the 1960s the Atlas was mated first to Agena and later to Centaur upper stages to send spacecraft to the Moon, Venus, and Mars (Table I).

Reliability in both launch vehicles and spacecraft was gradually and torturously achieved by both the United States and USSR during the 1960s, providing excellent reconnaissance of the Moon and paving the way for the successful *Apollo* manned landing on July 20, 1969. At the planets, the Soviet Union first focused on Mars and later on Venus. The U.S. strategy was to strive for a "balanced" program of reconnaissance of the entire solar system while simultaneously pursuing the in-depth study of the highest priority bodies. The 1970s, with finally reliable launch vehicles and spacecraft and high-gain antennas around the world, produced a golden era of planetary exploration, sending highly capable space-craft to every planet but Pluto and successfully landing payloads on both Mars and Venus. Sadly this was followed by an 11-year drought in the American program, with no launches from 1978 to 1989. During this time, the Soviet Union produced a string of successful Venus missions and the first fly through of the coma of Halley's comet. The 1970s and 80s also saw the beginnings of active solar system exploration programs by the European and Japanese Space Agencies. The 1990s have seen a renewal and reinvigoration of planetary exploration by the United States.

A. THE 1960s: IN THE SHADOW OF APOLLO

Early in the 1960s launch vehicles were only about 50% reliable and primitive spacecraft were no better. By unofficial count the Soviets had 19 scientific failures in a row. The United States failed with the first six *Ranger* lunar missions but succeeded in 1964 when it reached the Moon with *Ranger 7*. In 1962 *Mariner 2* conducted a successful flyby of Venus. Success was achieved at the Moon first with Soviet *Lunas* and later with America's *Rangers, Surveyors,* and *Lunar Orbiters,* as well as increasingly more sophisticated *Lunas,*

TABLE I
Planetary Missions[a]

Spacecraft	Country	Launch date	Mission type
		Moon	
Luna 2	USSR	Sept. 12, 1959	Impacter
Luna 3	USSR	Oct. 4, 1959	Circumlunar
Ranger 7	USA	July 28, 1964	Impacter
Ranger 8	USA	Feb. 17, 1965	Impacter
Ranger 9	USA	Mar. 21, 1965	Impacter
Zond 3	USSR	July 18, 1965	Flyby
Luna 9	USSR	Jan. 31, 1966	Lander
Luna 10	USSR	Mar. 31, 1966	Orbiter
Surveyor 1	USA	May 30, 1966	Lander
Lunar Orbiter 1	USSR	Aug. 10, 1966	Orbiter
Luna 11	USSR	Aug. 24, 1966	Orbiter
Luna 12	USSR	Oct. 22, 1966	Orbiter
Lunar Orbiter 2	USSR	Nov. 6, 1966	Orbiter
Luna 13	USSR	Dec. 21, 1966	Lander
Lunar Orbiter 3	USSR	Feb. 4, 1967	Orbiter
Surveyor 3	USA	Apr. 17, 1967	Lander
Lunar Orbiter 4	USSR	May 4, 1967	Orbiter
Lunar Orbiter 5	USSR	Aug. 1, 1967	Orbiter
Surveyor 5	USA	Sept. 8, 1967	Lander
Surveyor 6	USA	Nov. 7, 1967	Lander
Surveyor 7	USA	Jan. 6, 1968	Lander
Luna 14	USSR	Apr. 7, 1968	Orbiter
Zond 5	USSR	Sept. 15, 1968	Circumlunar
Zond 6	USSR	Nov. 10, 1968	Circumlunar
Apollo 8	USA	Dec. 21, 1968	Human orbiter
Apollo 10	USA	May 18, 1969	Human orbiter
Apollo 11	USA	July 16, 1969	Human orbiter/lander
Zond 7	USSR	Aug. 7, 1969	Circumlunar
Apollo 12	USA	Nov. 14, 1969	Human orbiter/lander
Luna 16	USSR	Sept. 12, 1970	Sample return
Zond 8	USSR	Oct. 20, 1979	Circumlunar
Luna 17	USSR	Nov. 10, 1970	Lander/rover
Apollo 14	USA	Jan. 31, 1971	Human orbiter/lander
Apollo 15	USA	July 26, 1971	Human orbiter/lander/rover
Luna 19	USSR	Sept. 28, 1971	Orbiter
Luna 20	USSR	Feb. 14, 1972	Sample return
Apollo 16	USA	Apr. 16, 1972	Human orbiter/lander/rover
Apollo 17	USA	Dec. 7, 1972	Human orbiter/lander/rover
Luna 21	USSR	Jan. 8, 1973	Lander/rover
Luna 22	USSR	May 29, 1974	Orbiter

continues

continued

Spacecraft	Country	Launch date	Mission type
		Moon	
Luna 23	USSR	Oct. 28, 1974	Sample return
Luna 24	USSR	Aug. 9, 1976	Sample return
Clementine	USA	Jan. 25, 1994	Orbiter
Lunar Prospector	USA	Jan. 6, 1998	Orbiter
		Venus and Mercury	
Mariner 2	USA	Aug. 27, 1962	Flyby
Venera 4	USSR	June 12, 1967	Probe
Mariner 5	USA	June 14, 1967	Flyby
Venera 5	USSR	Jan. 5, 1969	Lander
Venera 6	USSR	Jan. 10, 1969	Lander
Venera 7	USSR	Aug. 17, 1970	Lander
Venera 8	USSR	Mar. 27, 1972	Lander
Mariner 10	USA	Nov. 3, 1973	Flybys of Venus and Mercury
Venera 9	USSR	June 8, 1975	Orbiter/lander
Venera 10	USSR	June 14, 1975	Orbiter/lander
Pioneer Venus 1	USA	May 20, 1978	Orbiter
Pioneer Venus 2	USA	Aug. 8, 1978	Multiple probes
Venera 11	USSR	Sept. 9, 1978	Flyby/lander
Venera 12	USSR	Sept. 14, 1978	Flyby/lander
Venera 13	USSR	Oct. 30, 1981	Lander and orbiter
Venera 14	USSR	Nov. 4, 1981	Lander
VEGA 1	USSR	Dec. 15, 1984	Balloon, lander (also comet mission)
VEGA 2	USSR	Dec. 21, 1984	Balloon, lander (also comet mission)
Venera 15	USSR	Oct. 10, 1983	Orbiter (radar images)
Venera 16	USSR	Oct. 16, 1983	Orbiter (radar images)
Magellan	USA	May 4, 1989	Orbiter (radar images)
		Mars	
Mariner 4	USA	Nov. 28, 1964	Flyby
Mariner 6	USA	Feb. 24, 1969	Flyby
Mariner 7	USA	Mar. 27, 1969	Flyby
Mars 2	USSR	May 19, 1971	Orbiter
Mars 3	USSR	May 28, 1971	Orbiter and lander
Mariner 9	USA	May 30, 1971	Orbiter
Mars 4	USSR	July 21, 1973	Flyby
Mars 5	USSR	July 25, 1973	Orbiter and lander
Viking 1	USA	Aug. 20, 1975	Orbiter and lander
Viking 2	USA	Sept. 9, 1975	Orbiter and lander
Phobos 2	USSR	July 12, 1988	Orbiter
Mars Pathfinder	USA	Dec. 4, 1996	Lander/rover
Mars Global Surveyor	USA	Nov. 7, 1996	Orbiter

continues

continued

Spacecraft	Country	Launch date	Mission type
		Outer planets	
Pioneer 10	USA	Mar. 2, 1972	Jupiter flyby
Pioneer 11	USA	Apr. 5, 1973	Jupiter and Saturn flybys
Voyager 1	USA	Sept. 5, 1977	Jupiter and Saturn flybys
Voyager 2	USA	Aug. 20, 1977	Jupiter, Saturn, Uranus, and Neptune flybys
Galileo	USA	Oct. 18, 1989	Jupiter orbiter and probe (also Venus, Earth, Moon, two asteroid flybys)
Cassini	USA	Oct. 15, 1997	Saturn orbiter and Titan (en route)
		Comets and asteroids	
Sakigake	Japan	Jan. 8, 1985	Comet flyby
Suisei	Japan	Aug. 18, 1985	Comet flyby
VEGA 1	USSR	Dec. 15, 1984	Comet flyby
VEGA 2	USSR	Dec. 21, 1984	Comet flyby
Giotto	ESA	July 2, 1985	Comet flyby
NEAR	USA	Feb. 17, 1996	Asteroid flyby/orbiter (en route)

[a] Only missions that returned data consistent with their objectives are listed here, i.e., successful missions that reached their destination and returned data.

providing excellent reconnaissance for the successful lunar landing of *Apollo 11*. Hardware and software problems in the United States doomed *Mariners 1* and *3*, but *Mariners 4, 6,* and *7* made highly productive flybys of Mars, while *Mariner 5* made another visit to Venus. The USSR failed at Mars but succeeded late in the decade with the first entry and landing at Venus. The technology was ready for a major burst of exploration in the 1970s.

B. THE 1970s: THE LEGACY OF APOLLO

In 1975, the Soviet *Veneras* made another successful landing on Venus, returning surprisingly clear, first images of a rugged surface. In the United States, NASA tried to ride the coattails of *Apollo* with a highly ambitious program to land and rove on Mars. It was eliminated as too expensive by the Congress in 1967, leaving no American planetary program approved or even planned for the 1970s. NASA was able to rebuild a balanced program that sent *Mariner* spacecraft to orbit and map the highly varied surface of Mars and image the clouds of Venus and half the cratered surface of Mercury; *Viking* spacecraft to search for life on the surface of Mars; *Pioneer* spacecraft to make the first close flybys of awesome Jupiter and ringed Saturn and to characterize the atmosphere and surface features of

Venus; and new *Voyager* spacecraft to make a Jupiter–Saturn–Uranus–Neptune Grand Tour of the outer solar system. Meanwhile, the Russians tried a series of Mars missions, achieving only limited success with *Mars 4* and *5* in 1973. They were much more successful at Venus, however, beginning a series of successful *Veneras* in 1967. Thanks to, at least partly, the international competition between the United States and the Soviet Union, the 1970s produced a golden era of planetary exploration.

C. THE 1980s: DARKNESS

In 1972 James Fletcher, the NASA administrator, reached an agreement with President Nixon that NASA would undertake the challenging development of a reusable space shuttle within a continuing NASA budget that would rise each year by at least the going rate of inflation. That "golden handshake" agreement did not hold up for even 6 months and the NASA budget began a continuing slide. Soon, sufficient funds were not available for space science missions. In addition, a late 1970s decision to make the space shuttle, as yet unbuilt and unflown, the sole launch vehicle for the civil space program effectively led to the cessation of space exploration missions by the U.S. for the next decade. *Pioneer Venus* in 1978 was the last U.S. planetary exploration mission launched until *Magellan* in 1989, a gap of 11 years in the American program. In 1985 it was possible to redirect a previously launched Explorer spacecraft, *ISEE 3*, through the ingenious use of lunar flybys to achieve the first comet tail fly through. The Soviets continued their success with the *Veneras* throughout the 1980s. The most spectacular new mission of the 1980s was the encounter with Halley's comet in 1986 by an armada of international spacecraft: two Russian *Vegas*, the European *Giotto*, and the Japanese *Suisei* and *Sakigake* spacecraft. There was no American entry, emphasizing a decade of darkness in the American planetary exploration program.

II. MOON MISSIONS

A. *LUNAS*

Luna 1 was launched January 2, 1959, and flew past the Moon, the first spacecraft to escape Earth's gravity. *Luna 2,* launched on September 12, 1959, impacted the Moon, and *Luna 3,* launched October 4, 1959, circumnavigated the Moon and sent back the first images of the far side. The Soviet Union launched a number of other robotic missions to the Moon throughout the 1960s: *Zond 3*, a flyby, on July 18, 1965; *Luna 9,* a lander, on Janurary 31, 1966: *Luna 16,* the first automated sample return on September 12, 1970; and *Luna 17,* a lander with the first automated rover, on November 10, 1970.

The Soviet *Luna 16, 20, 23,* and *24* missions, launched on September 12, 1970; February 14, 1972; October 28, 1974, and August 9, 1976, respectively, carried out successful automated sample returns from the lunar surface. To this date these remain the only automated sample return missions from another world. The accomplishments of automated rovers and sample returns stand as crowning achievements of the Soviet planetary program, remarkable in that the first U.S. automated rover was not attempted until 1996, and there has still been no robotic sample return mission other than *Luna*.

B. *RANGERS*

After six failed attempts (two nonlunar test flights, three aimed at rough lunar landings and one television carrying impacter), *Ranger 7*, on July 31, 1964, re-turned thousands of high-resolution images before impacting in the region now called Mare Cognitum. *Ranger 8* followed on February 21, 1965, impacting in Mare Tranquillitatis. *Ranger 9* was then targeted to the great crater Alphonsus and impacted northeast of the 100-km crater's central peak. The television images of *Ranger 9* gave close-up views of the small dark-halo craters in Alphonsus that are thought to be ancient volcanic vents. The three successful *Rangers* provided strong evidence that *Apollo* landings would be practical, given the smoothly eroded appearance of the lunar surface at fine scale.

C. *SURVEYORS*

After a long, difficult, and costly development of both the soft-landing spacecraft and its hydrogen-fueled Centaur launch vehicle, *Surveyor 1* landed successfully on June 2, 1966, near the crater Flamsteed. It returned large mosaics of television images of its surroundings, proving beyond a doubt that suitable landing sites could be found for Apollo. *Surveyor 2* experienced an in-flight failure, but *Surveyor 3* made a bouncy successful landing in Oceanus Procellarum. (Thirty-one months later *Apollo 12* landed near the *Surveyor* and the astronauts walked over to it, cut off the soil sampler and TV camera, and brought them back to Earth.) Contact with *Surveyor 4* was lost during the final seconds of its landing sequence. *Surveyor 5,* despite a helium leak that required risky replanning in flight, landed successfully in Mare Tranquillitatis, not far from the future landing point of *Apollo 11,* and sent not only television mosaics, but also data from an α-particle backscatter instrument that gave chemical composition measurements of the lunar soil. *Surveyor 6* landed near the center of the Moon's Earth-facing side and added more images and chemical data before being commanded to hop with a short firing of its small rocket engines, providing stereo views and more soil-mechanics observations. Finally, *Surveyor 7* was targeted to a site near the huge crater Tycho, with the intention of learning more about the Moon's ancient highlands and also sampling materials excavated from depth by the impact that made the crater. With this highly successful scientific mission in January 1968, the *Surveyor* series came to an end.

D. *LUNAR ORBITERS*

The *Rangers* and *Surveyors,* initially conceived as scientific missions, were pressed into service as pathfinders for *Apollo.* In contrast, the *Lunar Orbiters* of the same period were planned from the beginning to survey the Moon for *Apollo,* but in the process they yielded a rich scientific harvest. Using a photo system with on-board film development and scanning to produce telemetry-enabling photographic reproduction on Earth, the *Orbiters* produced hundreds of excellent views of the Moon's surface features. Five *Orbiters* were launched between August 1966 and August 1967. The first three were so successful that they fulfilled *Apollo* requirements for landing-site surveys, so *Lunar Orbiter 4* was sent into a high polar orbit enabling it to map the entire Moon. *Lunar Orbiter 5* was targeted to sites of particular scientific interest. The resulting image collection, supplemented by Apollo coverage and then, many years later, by data from the multispectral surveys of *Clementine* and *Galileo,* provided basic knowledge for lunar science in the decades following the great contest of the 1960s.

E. *APOLLOS*

Describing the Apollo program is beyond the scope of this review of planetary missions. The *Apollo* series of missions were politically motivated, part of the contest between the United States and Soviet Union to prove their technical superiority to the rest of the world (and to each other and themselves). However, it is important to remember that 12 men landed on and explored the Moon, 6 used mobile vehicles (rovers) to extend their exploration, and 15 more men observed the Moon from orbit during 1968–1972. Samples were returned from the six landing missions (in addition to samples from Soviet *Luna* missions). This exploration, coupled with the dozens of automated robotic vehicles, have provided extensive exploration of Earth's satellite. Almost a generation passed from Apollo before lunar exploration would continue.

F. *CLEMENTINE*

The United States Ballistic Missile Defense Organization, part of the Department of Defense, was set up to carry out research and development for military space activities. The original goal was the "Star Wars" defense shield, but as that goal was discredited, the organization took on a broader role of developing advanced technology for military space operations. The *Clementine* spacecraft was developed to test and prove some instrumentation technologies as well as a new concept of smaller and lower cost satellites employing

microelectronics and other miniature components. The test plan included sending the spacecraft to the Moon, and after its primary technological aims were complete, the mission was dedicated to scientific measurements at the Moon. The spacecraft mapped the lunar surface extensively from orbit, returning images and multispectral maps. It was the first U.S. mission to the Moon in 22 years.

G. *LUNAR PROSPECTOR*

Launched on January 6, 1998, the *Lunar Prospector* orbited the Moon with a suite of instruments to measure the Moon's surface composition, gas emissions, electromagnetic properties, and gravity field. Early data include evidence of hydrogen (possibly in water ice) trapped in cold, dark regions near the lunar poles.

III. INTERPLANETARY MISSIONS

Many spacecraft have journeyed into interplanetary space, thereby qualifying perhaps as "planetary" missions. Many were investigating phenomena of the particles and fields environment of the solar system or the Sun itself, including solar flares. Some were errant vehicles from planetary misses. In the section on planetary exploration, we omit almost all of the interplanetary or Sun observing missions. The notable *Pioneer* series launched starting in 1959 and continued in the late 1960s. The first international mission, *Helios,* a U.S.–German cooperative mission, which had interplanetary spacecraft observing the solar wind and radiation, was launched in the mid-1970s. One went as close as 0.29 AU from the Sun. The *Ulysses* mission flew past Jupiter for a gravity assist out of the ecliptic.

A. *PIONEERS 1–4*

Pioneers 1 and *2* were built by the USAF Space Technology Laboratories (now TRW) and *Pioneers 3* and *4* by JPL. Launched in 1958 and 1959 and intended to fly past the Moon and go into orbit around the Sun, the first three were considered to be failures, although they did return useful data showing radiation beyond the Van Allen Belts to be low enough to permit manned flights to the Moon.

B. *PIONEERS 6–9*

These four small spinning spacecraft were built for NASA's Ames Research Center by TRW. Launched from 1965 to 1968 and placed around the Sun at 1 AU (Earth's orbital path), they produced a wealth of data on the Sun and the interplanetary environment. With lifetimes exceeding 20 years they provided the Apollo astronauts with early warnings of dangerous solar flares.

C. *HELIOS*

This solar observing mission was the first international interplanetary mission—a cooperative effort of the United States and West Germany. Launched in 1974, *Helios 1* went as close to the Sun as 0.31 AU, and *Helios 2,* launched in 1976, passed even closer at 0.29 AU, surviving solar heating 11 times than that experienced at Earth's orbit. The spacecraft operated from solar minimum through solar maximum and provided fundamental data on solar dynamics and the characteristics of cosmic rays. For the United States, *Helios* provided an opportunity for its space physicists to participate in vital science and also provided a test flight of the new high-performance Titan III E/Centaur before its 1975 launch of the *Viking* mission to Mars.

D. *ULYSSES*

Launched in 1990, *Ulysses* flew by Jupiter in February 1992 to receive a gravity assist, enabling the spacecraft to fly into a solar polar orbit—over the Sun's south pole in September 1994 and north pole in September 1995. *Ulysses* measured the high latitude properties of the solar corona, the solar wind, and the Sun's magnetic field. [*See* THE SUN.]

IV VENUS (AND MERCURY) MISSIONS
· ·

A. *MARINER 2*

In the early 1960s the U.S. space program was in turmoil. The Soviet Union was achieving remarkable success with Earth-orbiting Sputniks and shortly thereafter with exploratory *Luna* flights to the Moon. The United States was experiencing nothing but failures

with its *Pioneer* and *Ranger* attempts to reach the Moon. Further funding for JPL was being threatened. The savior was *Mariner 2,* a hastily assembled derivative of the *Ranger* spacecraft, which encountered one crisis after another from launch and beyond but managed to survive to make its way to Venus and return the first-ever close-up observations of another planet. Infrared scans confirmed very high day and night-time surface temperatures in excess of 800°F, later attributed to a CO_2 greenhouse effect. There were no detectable magnetic fields or radiation belts. [*See* VENUS: ATMOSPHERE.]

B. *VENERAS 4–6*

The first Soviet success at a planet came in 1967 with the *Venera 4* atmospheric probe. It transmitted data during its parachute descent and provided the first *in situ* measurements of the atmosphere. The mission presaged a remarkable 17-year era of Venus exploration by the Soviet Union. *Venera 5* and *6* in 1969 also provided atmospheric sampling of Venus in 1969.

C. *VENERAS 7–14*

Throughout the 1970s, the Soviet Union carried out eight landings, returning information from Venus. Contrary to the United States, the Soviets used the same spacecraft design repeatedly, making only minor changes for each mission. This gave them the ability to fly frequently (every opportunity) at low cost. The lander provided our first views of another planetary surface and showed us the hostile, hot world of Venus' reality, countering the old view of a habitable Earth's twin.

D. *MARINER 5*

The very low-cost *Mariner 5,* assembled from *Mariner 4* spares, had a much smoother flight than *Mariner 2* and passed within 4100 km of Venus in 1967. Two days earlier, the attempt of *Venera 4* to land on Venus had gone off the air at an atmospheric pressure of 18 Earth atmospheres, leading the Russians to conclude that they had impacted the surface. Radio occultation data of *Mariner 5,* combined with Earth-based radar returns, disproved the *Venera* conclusion and confirmed a surface pressure closer to a crushing 90 Earth atmospheres. No attempt was made to image the surface through the thick venusian clouds.

E. *MARINER 10* (*MARINER* VENUS–MERCURY)

Launched in 1973, *Mariner 10* was the first spacecraft to utilize gravity assist. In this application of planetary gravity assist, the gravitational attraction of Venus was used to slow the spacecraft's velocity around the Sun so that it "fell" in toward the Sun to reach the planet Mercury. During its flight past Venus, vidicon cameras viewing through ultraviolet filters returned over 3500 images. The upper cloud layers were found to rotate about the planet 60 times faster than the planet rotates on its axis.

Intricate swirl patterns in the clouds conformed to Hadley cell circulation, a much-sought confirmation of a theory first advanced in 1735 (to explain Earth's trade winds) and used since then by meteorologists but never before proved. After a first close flyby of Mercury, *Mariner 10* went round the Sun to make a second and third pass at Mercury. Twin television cameras with telescopic lenses returned a total of 12,000 images, resulting in a mosaic map of Mercury's sun-lit hemisphere at excellent resolution. The surface was ancient and Moon-like, heavily cratered, and with evidence of volcanism. A magnetic field was detected, a surprise for a slowly rotating planet. [*See* MERCURY.]

F. *PIONEER VENUS 1 AND 2*

Venus lacked some of the public appeal of the other planets. There were no telescope images of surface details, there were no prospects for finding life, and launch opportunities were not rare as in the Grand Tour of the outer planets. But undaunted, a small group of dedicated scientists persisted and helped to get NASA funding for a two-spacecraft mission, one to study Venus from orbit and the other to probe the depths of the atmosphere. The small spin-stabilized spacecraft, built by Hughes Aircraft under the management of NASA Ames, were launched on two Atlas/ Centaurs in 1978, arriving at Venus in December of that year. The *Pioneer Venus 1* orbiter studied the upper atmosphere and used a surface radar altimeter to discover great rift valleys, mountains, continents, and volcanoes. Over the next few years in orbit it provided a detailed map of the local gravity variations over the planet. *Pioneer Venus 2* sent four probes into the atmosphere. They returned detailed measurements all the way to the surface, confirming that the clouds were composed of droplets of concentrated sulfuric acid and measuring winds both globally and vertically. The atmosphere was found to be 96% CO_2, with the remaining 4% mostly nitrogen, but also with interesting amounts of sulfur dioxide, argon, and neon. In confirming that a CO_2 greenhouse effect was indeed what produced such extreme surface temperatures (over 700°K), *Pioneer Venus* helped elevate Venus to a high priority for its relevance to how to keep its sister planet Earth from following the same fate.

G. *VENERAS 15 AND 16*

The Soviet Union surprised American scientists by carrying out a sophisticated imaging radar orbiter mission on two orbiters of Venus in 1983–1984. The imaging radar technology—an adaptation of military technology used for all-weather "spy" flights—was provided by the Institute for Radio Electronics for their first planetary mission. The images supplemented and extended the first Earth-based radar images that were being taken at this time and confirmed the solid surface and many diverse features of Venus geology. [*See* VENUS: SURFACE AND INTERIOR.]

H. *MAGELLAN*

The original mission was called the *Venus Orbital Imaging Radar*. It was canceled for budgetary reasons in 1980, but resurrected with a lower-cost mission design the following year. It was to have launched in 1984, but delays caused by difficulties in the space shuttle program delayed its launch to 1989, the first planetary spacecraft launched on the shuttle. The mission was a huge success, doing for Venus what *Mariner 9* had done for Mars, providing a global mapping of nearly the entire surface. As with the *Venera 15 and 16* missions, imaging radar turned out to be a powerful technique for seeing cloud-shrouded planetary surfaces.

V. MARS MISSIONS

A. SOVIET MARS

The Soviet Union had much more difficulty at Mars than at Venus. Although *Mars 2, 3,* and *4* returned data, it was very little and very poor, and only *Mars 5* in 1973 gave enough to be called even a limited success. Following the U.S. *Viking* mission in 1976, the Soviets turned their attention strictly to Venus and only tried Mars again in 1988 with launches of orbiters to Mars and its small moon Phobos.

B. *MARINER 4*

This spacecraft was a derivative of the lunar *Ranger* spacecraft, augmented with four solar panels to power it at Mars' distance from the Sun. Launched in 1964 after a shroud failure doomed *Mariner 3*, *Mariner 4* made the first successful flyby of Mars, returning 22 vidicon images. Even though the communication system was advanced from *Ranger's* L-band to S-band, recorded images had to be transmitted at the agonizingly slow rate of just 8-1/3bits/seconds (bps). The images, mostly in the southern hemisphere of Mars, showed a heavily cratered Moon-like surface with no sign of water erosion and nothing that looked like the linear features seen by Schiaparelli in the 1870s and popularized as water-bearing canals by the dedicated astronomer Percival Lowell. These images somewhat dampened any expectations of finding life on Mars. The Martian atmosphere was found to be very thin, only 0.5 to 1% that of Earth. No magnetic field was found.

C. *MARINERS 6 AND 7*

Launched in 1969, these spacecraft were substantial advances over prior *Mariner* spacecraft. S-band transmitters, improved coding, and upgrades in NASA's worldwide tracking stations raised the data rate from 8-1/3 bps all the way to 16,200 bps. The payload included not only television cameras, but also an infrared radiometer and both infrared and ultraviolet spectrometers. As a bonus, radio occulation sciences and celestial mechanics required no additional weight. Just as *Mariner 6* was about to start its near-encounter sequence, the trailing *Mariner 7* spacecraft went off the air. JPL engineers were sorely stretched but managed to determine that a battery on *Mariner 7* had exploded and the spacecraft was reactivated. Again, as with *Mariner 4,* celestial mechanics dictated that all of the high-resolution images were concentrated in the southern hemisphere of Mars. They again confirmed an ancient cratered Moon-like surface. The only features remotely related to dark "canals" were the dark floors of some impact craters. The atmosphere was confirmed to be predominantly CO_2 with any nitrogen concentration determined to be less than 1%. None of the results increased the prospects for finding life.

D. *MARINER 9*

A Centaur stage failure doomed the launch of *Mariner 8* in 1971, but *Mariner 9* was launched successfully within the window. Meanwhile, the Russians had launched *Mars 2* and *3,* due to arrive at Mars after *Mariner 9*. An agreement was negotiated at a joint meeting in Moscow to provide quick imaging data from *Mariner 9* to the Russians to enhance their landing attempts. Unfortunately, when *Mariner 9* went into orbit around Mars, a giant dust storm totally obscured the surface of Mars. These high surface winds may have contributed to the failures of the two Russian landers. The dust storm eventually subsided to reveal a Martian surface startling in its variety. The southern hemisphere was indeed heavily cratered and ancient appearing, but the northern hemisphere was maked by huge volcanoes rising up to 21,500 m (70,000 feet), massive lava flows, and a 7-km-deep by 5000-km-long "Grand Canyon" appropriately named Valles Marineris after *Mariner 9*. The only apparent explanation for the origin of this canyon was massive erosion from water flow, perhaps triggered by volcanic eruptions melting a frozen subsurface. *Mariner 9* also revealed that Mars probably had ancient flooding and was shaped by fluid processes. Prospects for finding Martian microbial life, or at least its fossil remains, suddenly soared. [*See* PLANETS AND THE ORIGIN OF LIFE.]

E. *VIKINGS 1 AND 2*

The *Viking* mission, as initially funded, featured a pair of orbiters and two modest capsules to be hard landed on Mars. The mission, managed by NASA Langley, was subsequently escalated to feature two orbiters (upgraded from *Mariner 9* by JPL) deploying two highly sophisticated sterilized vehicles (built by Martin Marietta) to soft land on the surface. The *Viking* missions were launched on August 20 and September 9, 1975. The orbiters played a key role in providing images and other data for lander site selection and for relaying lander data to Earth. Orbiter science instruments also provided years of data about Mars, emphasizing its warmer and wetter past and therefore its relevance to questions of the evolution of life in the solar system.

Viking 1 landed at a "safe" site (Chryse Planitia) on July 20, 1976, whereas *Viking 2* was sent to a more risky northern site. Both areas proved to be more rocky than expected and the landers were lucky to survive. The rugged terrain was surprisingly like some high desert areas on Earth, but with reddish iron-rich soil. Essentially all instruments and subsystems worked as designed. The orbiters mapped Mars at high resolution. The landers began moving rocks and digging soil samples as directed by operators on Earth. Two of the three life detection experiments gave positive results, but were contradicted by the gas chromatograph–mass spectrometer instrument, which was sensitive to one part per billion and yet found no trace of organics or even any complex molecules. The science team consensus was that intense UV radiation had created highly reactive hydroxyl radicals on the Martian surface, which produced the confusing life detection data. The lander seismometer readings were negated by the vibrations of high winds, but all other instruments returned excellent data. The soil composition was analyzed and found to be rich in iron, as expected, but also surprisingly having 100 times the sulfur content compared to Earth and the Moon.

The *Viking 1* meteorology instrument package continued to send daily weather reports from Mars for several years, with cameras showing a layer of frost forming on the surface during Martian winter. *Viking* was a spectacular technical success, but it was still a disappointment to many not to have found life, at least not on the surface and at two different sites on Mars.

F. *PHOBOS*

The Soviets sent a new design interplanetary spacecraft to Mars with launches on July 7 and 12, 1988. These were the only Mars missions from Earth in the 1980s. *Phobos-1* was lost during its interplanetary cruise when the spacecraft began tumbling, and communications and power were lost. *Phobos-2* entered Mars orbit successfully in January 1989 and began remote sensing of Mars and the spacecraft's ultimate destination, the Martian moon Phobos. Just prior to its rendezvous with Phobos, on March 27, 1989, where it was supposed to deploy two landers on the moon, this second spacecraft of the mission was also lost. *Phobos-2* made a number of useful measurements of Mars and Phobos but failed to reach its exciting rendezvous and deployment of landers on Phobos. The missions included experiments from more than a dozen European countries.

G. *MARS PATHFINDER*

This mission consisted of a lander and rover, named *Sojourner*, which landed on Mars on July 4, 1997, and operated on the surface for about 3 months—2 months longer than its planned lifetime. The spacecraft was

launched on December 4, 1996. *Pathfinder* was the second mission of the NASA Discovery program. Following a novel landing in which parachutes, retrorockets, and air bags were used, *Sojourner* drove down a ramp from the lander and began exploring the environment and surface around the landing spot. It logged more than 100 m of traveling while staying within a 15-m radius from the lander. Spectacular panoramic stereo images from the lander showed an area very different from the two *Viking* sites (the only other two spots we have seen close-up on Mars)—a debris basin from an ancient flood. Any doubts about liquid water having been an important part of Mars' evolution were dispelled by *Pathfinder* data. [*See* MARS: SURFACE AND INTERIOR.]

H. *MARS GLOBAL SURVEYOR*

Mars Global Surveyor (MGS) was launched on November 7, 1996, and inserted into Martian orbit on September 12, 1997. The spacecraft was inserted into a highly elliptical orbit to be circularized slowly at a low altitude by repeated aerobraking maneuvers—cautious passages through Mars' atmosphere which dissipate some of the orbital energy. This mission is the first of the *Mars Surveyor* program (see later) born out of the failure of *Mars Observer* and was designed to recapture that mission's mapping and remote-sensing objectives in a series of three flights: orbiters launched in 1996, 1998, and 2001. The science instruments on *MGS* are a high resolution CCD camera, a magnetometer, a thermal emission spectrometer, and a laser altimeter.

As we prepare this chapter we can report only on the beginning of the mission. The first success is the discovery of a Martian magnetic field. The orbiting spacecraft discovered an irregular field around the planet, indicating not an intrinsic field but the remnants of one still buried in the rocks and surface material of the planet. As the spacecraft concluded its first few weeks of aerobraking, a severe anomaly in the mechanical structure of the solar panels was noted. It was known immediately after launch in 1996 that one of the solar panels had apparently not fully deployed and that the mission was going to have to be conducted with the panels in a partially deployed configuration. Because the panels were the main contributors to the spacecraft drag in the atmosphere, this impacts the aerobraking mission plan. The aerobraking has proceeded, but more slowly than planned and achievement of the final mapping orbit will be delayed many months. The mapping strategy was changed so that data, including remarkable 1- to 2-m resolutions images, are now being received even during the aerobraking and, with luck, all of the scientific objectives will still be met.

VI. OUTER PLANET MISSIONS

A. *PIONEERS 10 AND 11*

These *Pioneers,* as built by TRW under the direction of NASA Ames Research Center, were relatively small and simple spinning spacecraft. Yet simplicity led to exceptional reliability and long life. The penalties were in the need for frequent pointing commands from Earth and a limitation to relatively simple instruments. For example, imaging came from a scanning photopolarimeter rather than from any dedicated camera. *Pioneer 10* was launched in 1972 on an Atlas/Centaur, setting a record for the highest launch velocity yet achieved. It passed through the dreaded asteroid belt beyond the orbit of Mars with only minor particle impacts. Close-up images of Jupiter and its Giant Red Spot were fascinating if not full of scientific surprises. The big surprises came from the magnetic fields and particles around Jupiter. A strong magnetic field was found to be tilted 11° from the axis of rotation, which caused surrounding belts of concentrated high-energy electrons to wobble up and down as the planet rotated. *Pioneer 10* escaped being fried only by the sheer luck of zooming past just when the radiation belt was dipping. A twin *Pioneer 11* spacecraft was launched in 1973. It was targeted to pass low on the leading edge of Jupiter, escaping the worst of the radiation and sending it up and over the ecliptic plane to an encounter with Saturn. *Pioneer 11* revealed new details about the rings of Saturn and discovered a magnetic field, trapped radiation belts moderated by the rings, and a complex magnetosphere. Perhaps as important, it mapped a clear path for the *Voyager* spacecraft to follow.

B. *VOYAGERS 1 AND 2*

A rare alignment (every 176 years) of the outer planets in 1977 would permit multiplanet Grand Tours. NASA planned missions with two advanced highly autonomous spacecraft, one to fly from Jupiter to Saturn and out of the ecliptic to Pluto, and the other to pass from Jupiter to Uranus to Neptune. Faced with a con-

strained budget, the NASA administrator canceled the Grand Tours rather than slow down development of the space shuttle. However sufficient funds were put in the budget to permit development of a pair of *Mariner*-class spacecraft to visit just Jupiter and Saturn. Trying to still perform a Grand Tour, JPL engineers added some autonomous capability to the spacecraft central computer. The spacecraft, retitled with the resurrected name of *Voyager,* were launched in 1977. The sophisticated computer was confused by data it saw and kept shutting down the spacecraft, requiring a major rewrite of its software on the way to Jupiter. *Voyager 1* produced highly detailed images of the swirling multicolored clouds of Jupiter, but even more spectacular images of the Galilean satellites. Io amazed everyone with its multiple active sulfur-spewing volcanoes. Europa's surface, as viewed by *Voyager 2,* was so smooth that it had to be relatively recent frozen ice, promising a potential ocean of water beneath the ice. The *Voyagers* looked back at Jupiter and discovered faint rings. *Voyager 1* reached Saturn and found there were actually hundreds of delicate shining rings. Specifically targeted at the large satellite Titan, possessor of the only thick nitrogen-rich atmosphere other than Earth's, *Voyager 1* also found a substantial presence of methane, making Titan a promising future objective for the *Cassini* spacecraft and probe to study the evolution of organic compounds as a precursor of life. [*See* TITAN; OUTER PLANET ICY SATELLITES.]

Achieving the scientific objectives at Titan with *Voyager 1* permitted retargeting *Voyager 2* to stay in the ecliptic plane, passing safely outside the rings of Saturn, and continuing on its way to Uranus. This was done in an extended mission as *Voyager's* nominal mission ended at Saturn. In 1981 there was a short-lived political effort to turn off the working spacecraft as a budget-cutting device; fortunately it was beaten back. Arriving in January 1986, *Voyager* viewed the only planet in the solar system to be lying on its side with its spin axis pointed at the Sun, producing a totally unique meteorology. Its magnetic pole was tilted a surprising 60° from the spin axis. New rings and satellites were discovered. If Uranus' bland blue-green cloud surface was a bit disappointing, the satellites certainly were not. Miranda has a violently tortured surface of deep rifts and sharp angles unlike anything ever seen before. [*See* ATMOSPHERES OF THE GIANT PLANETS.]

Voyager 2 reached remote Neptune in August 1989, still functioning well and taking beautifully sharp pictures after 12 arduous years, thanks to tender loving care and a multitude of ingenious solutions from JPL. Back on Earth all available antennas were strung together to produce a very large effective receiving aperture that enabled the signals from the spacecraft to be received and processed despite the enormous distance from Neptune. Detailed images were returned showing a bluish planet with a definite cloud structure, including large circulating dark and light spots to confound scientists who compared them to the Giant Red Spot on Jupiter. Rings were found to exist in segmented arcs, only thinly connected as complete circles. Once again, the biggest surprise came from the satellite Triton, with its unique retrograde orbit. Triton was revealed to have a variegated white/gray/pink/brown surface with active geysers or plumes of dark material (perhaps organics) erupting from its surface of frozen nitrogen and methane. *Voyager 2* joined its companion spacecraft in heading out of the solar system to search for the heliopause boundary between the solar wind and the interstellar wind, continuing what has been acclaimed as the greatest single voyage of exploration of all time.

C. *GALILEO*

A long and torturous journey was made by this project from the time of its conception (1975) to its arrival at Jupiter some 20 years later on December 7, 1995. The project became a metaphor for both what was wrong and what is right about the space program since the mid-1970s.

Galileo's launch in October 1989 came after years of delay due to problems and changes in the U.S. shuttle and expendable launch vehicle program. It was launched on the shuttle on a trajectory that had to make Venus and Earth flybys for gravitational assist. Only this way could enough energy be obtained to send the 2380 kg spacecraft on its trajectory to Jupiter. In 1990 the main antenna failed to open and for 2 years spacecraft engineers tried to come up with ways to "unstick" the antenna. Those attempts failed, but by ingenious software and mission-sequencing techniques, the *Galileo* mission was still able to fulfill nearly all of its scientific requirements and return a rich quantity of scientific data through the low-gain antenna on the spacecraft. The spacecraft obtained fine images of the Moon, Earth, and Venus during these flybys and, on the way out to Jupiter, of asteroids Ida and Gaspra. It also discovered a moon of Ida, subsequently named Dactyl, and observed the impacts of the fragments of Comet Shoemaker-Levy 9 on Jupiter. [*See* AS-TEROIDS.]

On December 7, 1995, *Galileo* finally reached Jupiter, made a close flyby of Io, the innermost of the large

Galilean satellites, and was inserted into Jupiter orbit. An atmospheric probe targeted and separated several days earlier, simultaneously entered the jovian atmosphere, and sent back (through the orbiter) 58 min of high-quality data about the composition and structure of the atmosphere to a depth of 160 km below the cloud tops.

Throughout 1996 and 1997 *Galileo* conducted a rich and successful exploration of the large moons: Europa, Ganymede, and Callisto. (Although the orbit did not pass near Io, excellent distant images of Io were obtained as well.) Most interesting is Europa, an ice-covered moon that looks as if it may be harboring a sea of liquid water, heated by tidal friction, under the ice. Speculation about Europa being a place to support extraterrestrial life led NASA to use the still working *Galileo* orbiter beyond its lifetime to perform a *Galileo Europa Mission (GEM)*. The spacecraft's orbit was modified for repeated flybys of Europa, and will close out its mission in 1999 with two close flybys of Io. [*See* IO; OUTER PLANET ICY SATELLITES.]

D. *CASSINI*

Cassini was launched on September 15, 1997, on a Titan 4/Centaur rocket on its intended 6.5-year trajectory to Saturn. The trajectory will take it around the inner solar system twice, with gravitational assists from Venus and Earth, then outward past Jupiter for another gravitational assist, and then to Saturn. There, with its European Space Agency probe, *Huygens,* an extensive exploration of the planet, moons, rings, and magnetosphere will begin. The *Cassini* spacecraft will orbit Saturn with repeated close flybys of many of the 21 or so moons and observations of the rings. *Huygens* will probe into the organic-rich atmosphere of Titan and land on its icy surface. Surviving the landing in this unknown environment is not a mission design requirement, but may occur.

VII. COMET AND ASTEROID MISSIONS
· ·

A. *ICE*

In 1978, the United States launched an *International Sun Earth Explorer* (*ISEE,* pronounced "I-See") into a high orbit to measure solar phenomena before the solar particles arrived at Earth. The spacecraft was renamed *ICE* (pronounced "Ice") and was redirected using lunar flybys to fly through the tail of Comet Giaccobini-Zinner in September 1985. The mission was cooperative with ESA. The spacecraft did not come close to the comet's nucleus or provide imaging.

B. *VEGA*

The Soviet *VEnera-GAlley (VEGA)*, Venus-Halley, mission launched two spacecraft on June 10 and 14, 1985, to Venus, where they deposited balloons in the upper atmosphere and where they then used a gravity-assisted flyby of Venus to shape their trajectories to permit fly throughs of Halley's Comet on March 6 and March 9, 1986. Showing boldness and flexibility, the Soviets did not decide until 1981 or 1982 to redesign their planned 1985 *Venera* mission to perform the Halley's Comet exploration.

The mission provided the first close-up images of a comet nucleus as well as many other scientific measurements of the environment near the comet. An international team contributed some 14 experiments. The two spacecraft passed within 9000 km of the nucleus. The mission was also a keystone of an international cooperative effort among the Soviet Union, Japan, Europe (who all conducted missions to the comet), and the United States (who did not). Among other results from this effort was the use of the American Deep Space Network and the Soviet images of the comet to help target the European *Giotto* spacecraft for the closest approach to the nucleus.

VEGA was also politically significant in that it was the first Soviet mission to be carried out openly and publicly, a presage to the era of *glasnost* (openness) being introduced into the political system. Prior to this mission, Soviet missions were generally carried out in secret, with details only emerging long after their completion.

C. *GIOTTO*

The European Space Agency carried out its first planetary mission with the fly through of Comet Halley by *Giotto*. The mission was launched on July 2, 1985, and made its 596-km flyby of the nucleus on March 14, 1986. The close-up images from *Giotto* remain our best pictures of an active comet nucleus, a remarkable achievement given the flyby speed of nearly 70 km/sec (288,000 km/hr). The spacecraft also carried 11 other instruments for observations at the comet.

Giotto surprisingly survived the comet fly through,

despite being hit by many particles outflowing from the comet. Data from the spacecraft damage, including the destruction of the imaging system, have been valuable in putting together models of comet activity.

D. *SUISEI AND SAKIGAKE*

These spacecraft were the first Japanese interplanetary missions. *Sakigake* was more of a precursor test mission, launched on January 8, 1985, and flying 7 million km from Halley's Comet. *Suisei*, launched on August 18, 1985, flew within 151,000 km of the nucleus of the comet on March 8, 1986. Both spacecraft sent back fields and particles data about the cometary environment, and both were part of the international Halley armada, which included the two Soviet spacecraft and the European Space Agency mission. [*See* PHYSICS AND CHEMISTRY OF COMETS.]

E. *NEAR*

NEAR, the *Near Earth Asteroid Rendezvous*, was launched on Febraury 17, 1996, on a 3-year mission to rendezvous (match position and velocity) with the asteroid Eros. It is the first of the Discovery missions (see later) and is being managed by the John Hopkins Applied Physics Laboratory. *NEAR* will reach the vicinity of Eros in late 1998 and conduct its close-up mission observing and sensing the asteroid for several months. In June 1997 the *NEAR* spacecraft flew by the main belt asteroid Mathilde, sending back the first images of a dark C-type asteroid. [*See* ASTEROIDS.]

VIII. FUTURE PLANETARY MISSIONS
• •

For this article the future starts after the launch of *Cassini*. It is the beginning of the "Goldin" era of planetary exploration. From here on the American missions were conceived and developed during Dan Goldin's administration of NASA. His plan, enunciated during his first year as administrator, of seeing spacecraft fly into the solar system every month, is not so far from realization when we consider the number and variety of planned missions over the next decade. The missions are being developed in the framework of "programs"—broad multimission activities in which technology and exploration goals are added to that of space science. The missions are thus not defined by

TABLE II
Future Planetary Missions

Current missions
  Galileo
  Ulysses
  Near Earth Asteroid Rendezvous
  Mars Global Surveyor
  Lunar Prospector
  Cassini/Huygens

1998
  Mars Climate Orbiter
  Planet B
  New Millennium Deep Space 1 (comet/asteroid)

1999
  Mars Polar Lander (with *Deep Space 2*)
  Stardust
  Lunar A

2001
  Mars Surveyor Orbiter
  Mars Surveyor Lander
  Genesis

2002
  Muses-C (asteroid rover/sample return)
  Selene
  Contour

2003
  Mars Surveyor Orbiter and Lander
  Mars Express
  Rosetta (comet/asteroid)
  New Millennium Deep Space 4 (comet)
  Europa Orbitor

2004
  Pluto/Kuiper Belt

2005
  Mars sample return

National Academy of Science and NASA Science committees, as was the case in the United States in previous years, but by various considerations of which science is one, albeit a major one. Outside the United States a number of missions are being planned as well, a few in Russia, more in Japan, and some also by the European Space Agency. These still tend to be defined and dominated by space science considerations in those countries.

The programs in which the future U.S. missions are being implemented are Mars Surveyor, Discovery, New Millennium, and a new one now just being initiated in 1998: Outer Planets. The following section describes each of the missions planned in these programs and then lists them (Table II) by launch year in order to provide an overall sense of the pace and diversity of solar system exploration. (Another catego-

rization is that of the earlier sections in this article, viz. by target body. However, as will soon be noted, with the exception of *Mars Surveyor*, the programs and the mission rationale are less characterized by the target body and more by the technical goals.

One mission under study not fitting into any of these categories is a proposed privately funded mission, *Near Earth Asteroid Prospector (NEAP)* for launch in 1999 or 2000. Funds are being solicited for investment and for services provided to potential science customers, but no large commitments have yet been made.

A. MARS SURVEYOR

American national policy (signed in a memorandum by President Clinton in September 1996) dictates "a sustained program to support a robotic presence on Mars by the year 2000." This was begun with the Mars *Pathfinder* and *Mars Global Surveyor* missions launched in 1996. It is to be continued with two launches (nominally an orbiter and a lander) at every Mars launch opportunity (approximately every 26 months) for the foreseeable future, culminating in the return of samples from the Martian surface in the middle of the next decade. As is to be expected, near term missions in the program (and in all programs) are well defined, whereas later missions are defined only vaguely. The current missions of the Mars Surveyor program include:

- The *Mars Climate Orbiter* launched in December 1998, continuing the reflight of *Mars Observer* instruments to meet that mission's original objectives. Included are color imaging and a pressure-modulator infrared radiometer to provide vertical profiles of temperature, dust, and water vapor.
- The *Mars Polar Lander* to be launched in January 1999. The spacecraft has a stationary lander with a highly capable robotic arm for sampling the polar area for the first time. The lander includes the robotic arm for soil investigation, a surface stereo camera, and meteorology instruments. The goals are to provide a measure of the volatiles and climate history of the planet. Also included is the first Russian instrument to ever fly on a U.S. planetary spacecraft, a lidar instrument to measure atmospheric opacity. In this instrument the first microphone, developed and funded by the Planetary Society, will be placed on the surface of Mars to listen for whatever sounds might exist

there, from the planet, or from the spacecraft on the planet.

- Two "microprobes" or penetrators will be carried as "piggybacks" on the 1999 lander launch. These are ultrasmall devices that are deployed from orbit and stick into the Martian surface, penetrating to a depth of about 1–2 m, depending on the type of soil they hit. They will measure the temperature and water content of the subsurface. These have been developed by the New Millennium Program and are known as the *Deep Space 2* mission.

- An orbiter launched in March 2001 will complete the failed *Mars Observer* (launched in 1992) instrument reflight. This mission will fly the gamma ray spectrometer and a thermal emission imaging system in a 400-km circular orbit. It will image and map the surface for sample return site selection and serve as a communications relay.

- The 2001 lander will consist of at least two vehicles: a science rover of the *Sojourner* design and a stationary lander that will carry the rover to the surface. The lander includes a test of *in situ* propellant production and a radiation and soil/ dust experiment to investigate the conditions on the surface for eventual human exploration there.

- The 2003 orbiter and lander missions are not yet defined. However, the 2003 lander will include a rover, larger than the 2001 lander. The *Athena* rover will carry a number of instruments for *in situ* analysis and practice collection of samples for future sample return missions in 2005 and 2007.

- *Mars Sample Return* is being planned for the 2005 launch opportunity. Actually one of the many mission options being considered would provide for a 2004 launch with an extra loop around the inner solar system in order to adjust the landing date to meet the requirements for the collection of cached samples and return to Earth. No mission design has yet been defined, but many different options are under study, including international participation. Russian interest in the mission is long-standing and remains a major goal in their program. Another possibility is cooperation with France, or more broadly in Europe, with use of the Ariane 5 launch vehicle for the mission.

B. DISCOVERY MISSIONS

The purpose of the Discovery program is to encourage innovative scientist-led mission designs and developments for exploration of the solar system. Breaking from the tradition of NASA defining missions and

scientists proposing experiments for them, in this program scientists propose the mission, including the target, goals, ojgectives, management, and organization, as well as, of course, the experiments. The first Discovery mission is *NEAR*, which was discussed earlier in Section VII. The second mission in the Discovery program was *Mars Pathfinder,* also discussed earlier, as part of the Mars program.

Discovery missions in development and approved for future flight are *Lunar Prospector, Stardust, Contour,* and *Genesis. Lunar Prospector* was launched in January 1998. The mission sent a polar orbiter to the moon, with a neutron energy spectrometer and other instruments, in the hopes of determining the presence or absence of lunar ice in the polar regions.

Stardust is a comet coma sample return mission that will send a spacecraft through the coma of comet Wild 2, collect samples of the material streaming from the nucleus, and return to Earth with the capsule of material. The capsule will reenter Earth's atmosphere and be recovered by an airplane, snatching it as it comes in by parachute over Utah. The mission launch is scheduled for early 1999.

Genesis is also a sample return mission with a capsule that will be recovered similarly over Utah. In this mission the samples are charged particles from the solar wind. It is scheduled for launch in January 2001. The return is scheduled for August 2003.

Contour is a three-comet flyby mission. It will image the nucleus, make spectral maps, and analyze dust in the comets' coma. It is scheduled for launch in July 2002, with its first comet flyby to occur in November 2003. This flyby of Comet Encke at a distance of about 60 miles (100 km) will be followed by similar encounters with Comet Schwassmann-Wachmann-3 in June 2006 and Comet d'Arrest in August 2008.

C. NEW MILLENNIUM MISSIONS

New Millennium is an advanced technology program, with the goal of enabling space science missions to be truly "cheaper, faster, and better" and smaller. Emphasis is placed on micro- (and nano-) miniaturization in sensors, data processing, spacecraft subsystems, and even the system design. Science goals are only added to the mission after technology goals are defined and a preliminary mission design to meet those goals is completed. The missions in this program are known as *Deep Space X*, where X is the mission number of the series.

Deep Space 1 is to be the first interplanetary spacecraft to use solar electric propulsion (ion drive). The mission is scheduled to launch in October 1998 and fly by asteroid 1992 KD in 1999 and possibly Comet Borrelly in 2000. *Deep Space 2* is the *Mars Microprobe* mission flying piggyback on the *Mars Surveyor 1998* mission (described earlier). *Deep Space 3* and *Deep Space 4* are not officially defined yet, but are understood to be an earth-orbiting interferometer and a comet sample return. The former is a technology precursor for the development of deep space interferometers to be used to observe planets in distant solar systems, an important goal of the Origins program in NASA. Origins is a broad space science initiative to search the universe for information helping us to understand the origin and evolution of planets and of processes related to life, including the possibilities for extraterrestrial life. The Deep Space 4 comet sample return is scheduled for launch in 2003 to Comet Tempel 1. Trip time to the comet is about 2.7 years. The main spacecraft will orbit the comet for several months, then deploy a lander to drill and collect samples and bring them back to the orbiter for their return to Earth in 2010.

D. OUTER PLANET PROGRAMS

The nascent program for exploration of the outer planets following *Galileo* and *Cassini* is just beginning to be defined. Discoveries at the Galilean satellites such as hints of liquid water under the ice at Europa, the magnetosphere of Ganymede (and possibly of Europa as well), and the continuous volcanic activity on Io have whetted the appetites of scientists to learn more about the planets and objects of the outer solar system. Such objects may even be relevant to the search for extraterrestrial life and/or to Origins because of the interaction of the many forces and dynamics of solar system evolution.

In a way, the first mission is defined: it is the *GEM*, which began on December 8, 1997. It is an extension of the *Galileo* mission made possible by the remarkable success of that orbiter to continue working and sending valuable data to Earth. After 2 years in orbit around Jupiter, it will now be targeted for repeated close flybys of Europa for intensive study of the moon. Ultimately it will be targeted for a close pass or even an impact on Io.

The two candidate new missions now receiving the most study for outer planet exploration are a Europa orbiter with a possible sample return and a Pluto fast flyby/Kuiper belt explorer. The Kuiper belt is the region beyond Neptune of many small bodies, some of the larger of which have now been observed with Earth-based telescopes. As we go to press, it appears that the Europa orbiter will be the first mission. Following Europa and Pluto in the "Outer Planets" plan is a mission into the Sun (solar probe) using a gravity assist flyby of Jupiter.

E. EUROPEAN MISSIONS

The Europeans have made a major contribution to *Cassini* with its *Huygens* probe. They are now developing *Rosetta*, a comet rendezvous mission for launch in January 2003. *Rosetta* will make several flybys of Earth and Mars before reaching its target: Comet Wirtanen in 2012. The mission includes remote sensing, dust and gas instruments, fields and particles experiments for detailed observations, and a lander to observe the nucleus of the comet.

The European Space Agency also is now planning *Mars Express* for a 2003 launch. This will be an orbiter to principally refly the several European instruments intended for the Russian *Mars 96* mission that failed on launch. Landers may also be included on this mission. In addition to the European Space Agency, national space agencies in Europe, principally in France, Germany, and Italy, have investments and programs for planetary exploration. These usually do not take the form of independent missions, but cooperative arrangements with others. As new innovations in the "cheaper, faster, better" category are defined, we can anticipate small mission suggestions from national space agencies as well. One category of such missions might be piggyback launches on the *Ariane 5* rocket.

F. JAPANESE MISSIONS

Japan has three planetary missions in development, each under the aegis of the very small Institute for Space and Astronautical Sciences (ISAS). These are *Lunar A*, *Planet B*, and *MUSES-C*. *Lunar A* is scheduled for launch in 1998 or 1999 and will send three penetrators into the lunar surface. *Planet B* is a Mars aeronomy orbiter scheduled for launch in July 1998. *MUSES-C* is a near-Earth asteroid sample return scheduled for launch to Nereus in 2003. It will include a NASA/JPL nanorover to be placed on the asteroid surface. Also under study in Japan by the main space agency, NASDA, in cooperation with ISAS, is a lunar mission (*Selene*) for potential launch in 2002 or 2003, a mission still to be defined.

G. THE RUSSIAN PLANETARY PROGRAM

The loss of *Mars 96* coupled with the precipitous decline in government support for space science has crippled the Russian planetary program. Although *Mars Together*, a U.S.–Russian government initiative, and the Russian national Mars program remain official commitments, the lack of funding has prevented their implementation. Studies continue for a Russian Mars 2003 mission, with either a rover (*Marsokhod*) or a small station (as was launched on *Mars 96*), and for a Russian Mars or Phobos sample return mission. Phobos in particular continues to hold special interest in the Russian science and space mission community. Russian cooperation with ESA on Mars Express and in "Mars Together" with the United States in 2003 and sample return missions (possibly to the martian moon, Phobos) are also being discussed.

In addition, new interest in a lunar orbiter mission for 2000 or 2001 has been shown, and the Russian Academy of Sciences has endorsed plans for study of such a mission. No plans, however, are now approved, for either the Moon or Mars.

H. CONCLUSION

The pace of missions described in Section VIII and shown in Table II is breathtaking. Smaller missions with lower cost launch vehicles and shorter development times permit frequent missions and rapid adaptation to new knowledge and discoveries. Albeit they do not permit comprehensive surveys in a single mission, it does appear that there will be a significant net increase to planetary science from this approach, certainly over the previous strategy of planned big missions being delayed continually. The question of life in the universe is dominant in the space science program, and the missions now in development seem to be using the strategy of following water, with evidence on Mars and Europa being crucial at this time. The robotic search for life and searching for Earth analogs in the solar system will undoubtedly lead to human exploration of the planets, specifically Mars. Whether this exploration is presaged by human return to the Moon, by side trips to asteroids, or followed by settlement on another world remains to be learned. What is exciting is that it will probably be learned in the next decade or two from now (1998).

ACKNOWLEDGMENT

The authors thank James Burke for his contributions.

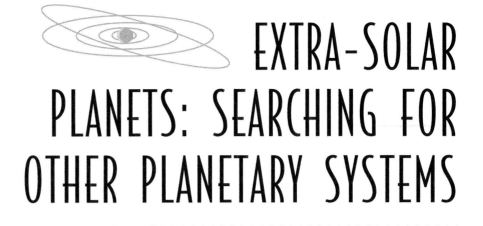

EXTRA-SOLAR PLANETS: SEARCHING FOR OTHER PLANETARY SYSTEMS

I. Introduction

II. Some Key Concepts

III. Results to Date

IV. Future Directions and Opportunities

David C. Black
Lunar and Planetary Institute

GLOSSARY

Astrometry: Observations of the position of celestial objects and how they vary over time. Measurements of the position of a "target" star relative to the positions of several other stars can be used to determine if the target star has companions. The presence of companions is inferred from "wobbles" in the apparent motion of the target star.

Brown dwarf: Object formed by the same process(es) that form stars but of insufficient mass to burn hydrogen into helium (roughly 80 times the mass of Jupiter). The lower limit to the mass of a brown dwarf is unknown.

Planet: Object formed by accretion of dust and gas. This process is different from the process that forms stars and brown dwarfs.

Planetary system: System consisting of at least two planets in Keplerian orbit around a star.

Pulsar: Magnetic neutron star that remains after a star has undergone a supernova. Pulsars rotate, in some cases as fast as a thousand revolutions per second. This gives rise to pulsed radio

emission from charged particles present in the pulsar's magnetic field.

Spectroscopy: Observational technique that breaks the light from an object into discrete wavelength bins. Comparison of the wavelength of known emission or absorption lines in the light from a star with the wavelength of those lines from a source that is at rest with respect to an observer provides a measure of the velocity of the star along the line of sight to the observer. The velocity is determined by the extent of the Doppler shift in the observed wavelength.

Ever since Copernicus proposed that Earth was part of a planetary system that revolved around the Sun, philosophers and scientists have long sought evidence of the existence of other planetary systems. It has only recently been realized fully that we will never understand the origin of our planetary system if we do not have information concerning others. For example, what fraction of stars have planetary companions, and what are the detailed characteristics of other planetary systems? We have developed a paradigm that describes in general terms the formation of the Sun and the solar system from its beginnings as part of a molecular cloud. For the first time in the history of science, we find ourselves in the position of being able to test our view of how the solar system formed and to modify that view if necessary in light of data concerning the existence and nature of other planetary systems. The search for and study of other planetary systems involve a variety of astronomical observational techniques and tools in an effort to detect the presence of relatively dim and low-mass objects (planets) in the presence of relatively bright, high-mass objects (the central star or stars in a planetary system).

I. INTRODUCTION

The presence of a chapter dealing with the topic of other planetary systems in an encyclopedia of this type is remarkable in at least two ways. It is remarkable because what might be considered a cousin of a planetary system, planets revolving around a pulsar, has been recently discovered, and there have been discoveries of what may turn out to be planetary companions to other stars. However, until the nature of these companions is clear, there is at present no unequivocal observational evidence of any other *planetary system.* The distinction between all of these is discussed later in the chapter. This chapter is also remarkable because it is explicit recognition of the fact that information from a comprehensive search for other planetary systems is essential to understanding the origin and early evolution of our own planetary system. [*See* THE SOLAR SYSTEM AND ITS PLACE IN THE GALAXY.]

We have been able to construct a paradigm as to how the solar system, and by inference other planetary systems, formed. That paradigm is in large measure built upon a foundation of data from the study of objects in the solar system, as well as upon general features of the solar system. These features include, but are not limited to, the orbital architecture of the solar system and the general trend in compositions of the planets. A scientific test of that paradigm must come by using it to predict both the frequency of occurrence of planetary systems and their general features.

Viewed from a scientific perspective, the rationale for studying other planetary systems has three major components. The first component has been noted, namely, that the results from a comprehensive search for and characterization of other planetary systems are essential to understanding the origin of our own

planetary system. The second component is related to the first in that it associates planetary systems with the process of star formation. The structure (e.g., orbits and masses) of other planetary systems can be considered as a fossil signature of key events and processes that took place during their formation, events and processes that occur over relatively short timescales and in difficult to observe phases in the formation of a star. Therefore, the detailed nature of other planetary systems can provide us with valuable, perhaps unique, insight into the process of star formation. The last component of the rationale involves the possibility of life, particularly intelligent life, existing beyond the solar system. Planets are like "cosmic petri dishes." They provide a stable thermal environment for emerging life to develop and evolve, perhaps to an intelligent state. Discovery of other planetary systems will bolster the assumption that sites for life abound in the galaxy, and it will also provide a set of specific targets for efforts to detect extraterrestrial intelligence. [*See* PLANETS AND THE ORIGIN OF LIFE.]

The first objective of any effort involving other planetary systems is to obtain unambiguous observational evidence for the existence of another planetary system. As noted earlier, although there have been several claims of discovery of extrasolar planets over the years, none of those claims has yet been substantiated and most have been shown to be incorrect. That does not mean that some or many of these objects will not turn out to be members of a planetary system, it simply means that there is insufficient data at this time to reach that conclusion with certainty.

The second objective should be to gather statistical evidence concerning the nature of planetary systems. For example, what is the distribution of planetary systems as a function of stellar spectral type? Are planetary systems found in association with binary stars? What are the masses of the largest planets and how do those masses vary with spectral type? The reader will note that whereas the first objective could be accomplished in principle with existing observational systems, the second objective will require a far more comprehensive search and also new observational tools. It is also worth remarking that while the confirmed discovery of another planetary system will be an exciting event, significant scientific return will begin with realization of the second objective.

The third objective involves detailed study of specific planetary systems. Emphasis is placed upon determining the masses and orbits of planets in specific planetary systems, and, where possible, observations should be made that provide information regarding the composition of planets in these systems. If we are able to determine the composition of the atmospheres of planets revolving about other stars, there is a chance of inferring whether life exists on those planets. The composition of Earth's atmosphere is not in chemical equilibrium. Something is acting to maintain the relatively high abundance of molecular oxygen (O_2) and ozone (O_3). That something is the abundant life on the planet. The presence of detectable amounts of methane (CH_4) would also herald the presence of life on a planet (the bulk of the CH_4 inventory on Earth is due principally to cows and termites).

II. SOME KEY CONCEPTS

The focus in this chapter is on *other planetary systems.* Though it is clear that the definition of "planetary system" must include the notion of a planet, it might come as a surprise to many readers to find that there is no generally agreed to definition of what is meant by the term "planet." It is almost like good art, the observer believes that they will know it when they see it. Although this is probably the case for objects similar to Earth, the situation is far less clear when the objects in question are like Jupiter.

This is a sufficiently important issue in terms of interpreting some of the exciting recent observational discoveries that are discussed later in this chapter, so I take a small detour at this point to consider the definition of a planetary system.

A. SCIENTIFIC DEFINITIONS

A definition, to be useful, should be unambiguous. It should, according to Webster, "set apart in a class by identifying marks or characteristics." It should also provide the basis for distinguishing between superficially similar, but different objects. For example, defining an elephant as any object that is gray and weighs several thousand pounds is not a good definition. One could not distinguish between a small tank and an elephant on the basis of such a definition.

Defining a planet on the basis of its mass alone is likely to be a similarly flawed definition. Yet it is one that is used by many scientists in spite of the fact that there is no indication of any break in the mass of objects ranging from the most massive star to the smallest asteroid.

Mass can be used as a defining parameter if it is related to some other defining characteristic that is

secondarily clearly related to mass. For example, a star is defined as an object that at some time in its life is self-luminous through nuclear reactions in its interior. The specific nuclear reaction is not critical here as long as it is one of the reaction chains that converts hydrogen into helium. The physics of stellar structure coupled with fundamental nuclear physics indicates that any object with a mass greater than about 80 times the mass of Jupiter is able to generate the required nuclear reactions to qualify as a star. Objects less massive than this cannot qualify as a star based on the definition.

Note that this definition says nothing about how a star is formed. A star is defined purely in terms of an intrinsic process—nuclear conversion of hydrogen into helium. This can be related to a minimum mass. We know that there are objects that are formed as stars are formed, but they happen to have masses below the critical value of 80 Jupiter masses. These objects are called "brown dwarfs." From the point of view of how they are formed, there is no distinction between them and a star.

B. THE REALM OF THE BROWN DWARF

The upper limit to the mass of a brown dwarf is well defined and quantitatively well known. It is the same as the lower limit to the mass of a star. Is there a lower limit to the mass of a brown dwarf, and if so, what is it?

To answer these questions, one needs first to define a brown dwarf. Two approaches are in use at the present time. One is modeled after the definition of a star, relying on a process that is intrinsic to the object for its definition. The other defines a brown dwarf in terms of its process of formation.

The former approach defines a brown dwarf as any object that is capable of burning deuterium, a heavy isotope of hydrogen, and the most easily burned. Using fairly well understood physics of objects like brown dwarfs, it can be shown that any object more massive than about 12 times the mass of Jupiter is able to burn deuterium. This would imply that a brown dwarf is any object with a mass between about 12 and 80 Jupiter masses. This definition, like that of a star, says nothing about the way in which a brown dwarf is formed. It would also imply that any object less massive than 12 Jupiter masses is something else, even if such an object is formed by the same process that makes stars and brown dwarfs.

An alternative approach recognizes that planets, at least the ones that we know something about, are formed by a different process than the process responsible for star formation. Accordingly, it defines brown dwarfs and planets in terms of their formation process. Specifically, a brown dwarf is defined to be any object that is formed by the same process that a star is formed by, but has insufficient mass to burn hydrogen into helium. Note that this definition has the same upper limit to the mass of a brown dwarf as did the previous definition, but the lower limit to its mass is less obvious.

It is generally thought that a process referred to as fragmentation of molecular clouds is responsible for the formation of stars, and therefore brown dwarfs. Fragmentation is a catchall concept to some extent, encompassing gravitational collapse as a key ingredient. Most studies of fragmentation are very idealized, as it is an intrinsically difficult process to study. It is too complex to deal with on a purely theoretical basis, and even sophisticated numerical models emulate the process only approximately.

If we take what is currently known about fragmentation as a guide, one of the key parameters that determines the smallest mass that can be formed by this process is the opacity of the material that is undergoing fragmentation. The mass, M_J, that can be formed by fragmentation of a gas is roughly given by $M_J \approx$ const. $(T^3/\rho)^{1/2}$, where T and ρ are the temperature and density of the gas, respectively.

It is known that during the early phases of collapse to form a star, the gas and dust that are collapsing are relatively transparent to radiation of the energy that is created by the collapse, so the collapse is isothermal. In light of the increase in density during this phase, the value of M_J will get progressively smaller until the temperature begins to increase because the opacity of the material is sufficiently high to trap radiation and thereby heat the collapsing fragment (Fig. 1). When that happens, there is a minimum mass, and that mass is often referred to in the literature as the "opacity-limited fragmentation mass." The numerical value for this minimum M_J is not known as well as the nuclear burning masses, but the number of $M_J \approx 10$ Jupiter masses, about 0.01 times the mass of the Sun, is often quoted. Values as low as $M_J \approx 5$ Jupiter masses have appeared in the literature on this subject. It is not even clear that this simple model of star formation applies as described when we are considering the formation of binary stellar systems, which includes the case of a star with a brown dwarf companion. As can be seen, there is some uncertainty about how small an object can form by the process of fragmentation. It could be comparable to the mass of Jupiter, or it could be an order of magnitude greater.

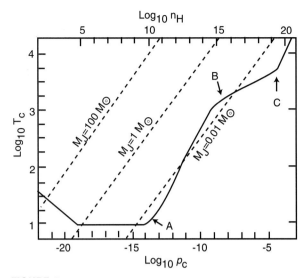

FIGURE 1 Plot of the logarithm of central temperature, T_c, in degrees kelvin as a function of the logarithm of the central density, ρ_c, in g/cm³ of a collapsing protostar. The solid line is the run of (T_c, ρ_c) for a roughy solar-mass protostar undergoing idealized spherical collapse. The dashed lines are lines of constant Jeans' mass with the indicated values being in units of solar masses. The points labeled A, B, and C correspond, respectively, to the formation of a central, 10 Jupiter-mass core due to opacity effects, the onset of significant dissociation of molecular hydrogen leading to a second collapse phase, and the formation of the final stellar core with a mass of roughly a few Jupiter masses.

C. WHAT IS A PLANET?

We come now to the question of defining a planet. If we adopt the nuclear definition of a brown dwarf, then an upper limit to the mass of a planet would be the lower limit to the mass of a brown dwarf. That is, a planet cannot be more massive than about 12 times the mass of Jupiter. This approach does not tell us whether a planet would ever really be that massive, as the process of formation is not an aspect of the definition. Indeed, if a planet did manage to get more massive as it was forming, it would suddenly no longer be a planet, but a brown dwarf.

If we adopt a definition of a planet that is based on its means of formation, we must first realize that, based on the nine planets that we know something about, it seems clear that planets are formed by a different process than the process or processes responsible for star formation. Planets appear to form from the bottom up via a process of accretion. We see ample evidence of this in the modern asteroid belt, as well as in the record of the early solar system embodied in ancient craters on the Moon.

Is there any phenomenon inherent in the process of planetary formation that would limit its mass? The "good news" is that the answer to this question is "yes." The "bad news" is that there is more than one phenomenon that will limit the growth, and we do not understand any of them well enough to quantify the upper limit. One process is simply that a growing planet runs out of gas to accrete from its surroundings. It is generally thought that this was the case for Neptune and perhaps Uranus. Another process is that as a planet gets massive, it tends to open a gap in the disk in which it formed. The torques exerted by the massive planet on the disk prevent material from getting to the growing planet, thereby limiting its growth. Because we do not understand these limiting processes well enough to determine a unique upper limit to the mass of a planet, we cannot say whether there may be planets revolving around other stars that managed to achieve masses slightly or even substantially greater than that of Jupiter, perhaps even more massive than the limit of 12 Jupiter masses that is required to initiate burning of deuterium. [*See* THE ORIGIN OF THE SOLAR SYSTEM.]

It is argued in some quarters that the foregoing discussion is simply a matter of semantics. If an object has a mass near that of Jupiter, call it a planet and be done with it. The problem is that we could end up mistaking a simple binary system, one with a star and a low-mass brown dwarf, for a planetary system. Such a mistake would lead to totally erroneous conclusions regarding all aspects of planetary systems.

The potential for such confusion is high. Nature prefers to make stars in binary systems. There is no generally accepted model, and certainly none that is quantitatively definitive, for the formation of binary systems. Thus, we have no firm basis for predicting either the frequency of occurrence of brown dwarf binary systems or the range of masses of brown dwarfs that might form in such systems. This should be borne in mind when the recent detections of substellar-mass binary companions to stars are discussed later in the chapter.

Which definition of a brown dwarf is better, and correspondingly, which definition of a planet should be used? There is no firm rule on this, neither is wrong nor right. The definition that will be used in this chapter is that *a planet is any object, irrespective of its mass, that is formed as we believe planets in this planetary system were formed,* that is, by the process of accretion. Brown dwarfs will be taken to be any object formed by the same process that forms a star, but with a mass

less than roughly 80 Jupiter masses. The lower limit to the mass of a brown dwarf is not clear by this definition, but is likely to be on the order of a few Jupiter masses. Similarly, the upper limit to the mass of a planet is not prescribed, but could exceed the lower limit to the mass of a brown dwarf.

Is there a difference between a brown dwarf of a given mass and a planet of the same mass? The answer is "yes." The planet, by virtue of the way in which a planet is formed, would have a higher proportion of heavy elements, relative say to the abundance of hydrogen, than would its equal mass, but differently formed brown dwarf counterpart. The latter would have a composition that would be similar to, if not identical with, that of the star that it revolved around. Another difference is that the brown dwarf is likely to appear as a single companion to a star, whereas the process of planet formation is far more likely to lead to multiple companions, just as is seen in the solar system, and in the regular satellite systems of the giant planets Jupiter, Saturn, and Uranus. For this reason, I here use the following definition of a planetary system: *a planetary system is any system in which at least two planets revolve around a star in Keplerian orbits.*

The reader is cautioned that these definitions are not uniformly accepted. However, I think that in the spirit of what a definition should do, and given the focus on planetary *systems* and not extrasolar planets, this definition and its companion definition of a planet are more likely to be scientifically useful than the other definitions that have been proposed. Armed with these definitions, we next consider results from searches for other planetary systems.

III. RESULTS TO DATE

A discussion of search results to date must be divided into two eras and must include a few words about search techniques. Techniques to search for other planetary systems have been grouped into one of two general approaches, direct and indirect. *Direct searches* seek photons that are either scattered (e.g., visible light) or emitted (e.g., thermal infrared radiation) by a companion to a star. *Indirect searches* focus on a star in an effort to detect some observable effect (e.g., an apparent dimming due to a transit of the star by the companion, or reflex motion of the star due to gravitational interaction with a companion) that companions have on that star. Each of these approaches has its strengths and weaknesses.

TABLE I

**Information That Can
Be Gained from Direct and
Indirect Planetary Detection Techniques**

	Detection techniques	
	Direct	Indirect
Orbital period	*	*
Orbital structure (inclination, etc.)		*
Planet temperature	*	
Planet atmospheric composition	*	
Planet mass		*

A. SEARCH TECHNIQUES

Direct detection is likely to provide detection in a shorter time than indirect detection, and depending on the observation wavelength it potentially provides different information about companions than is provided by indirect observation. For example, direct detection at infrared wavelengths can say something about the temperature, or perhaps the composition, of the atmosphere of the companion.

Indirect detection typically takes longer to provide a clean detection, as one must observe the system for a reasonable fraction (≥one-half) of the orbital period of the companion to be certain about detection. Indirect detection does provide more information about the orbit of any discovered companions. A comparison of the information content of the two general approaches is given in Table I.

Most of the search results to date that are discussed in the following have come from indirect searches, primarily by astrometric and spectroscopic/radial velocity observations. This is more a comment on the inherent difficulty of direct detection because of the tremendous brightness contrast ratio between a star and likely planetary companions, and on the associated state of technology for direct detection, than it is on the power of the technique. Indeed, as will be discussed near the end of the chapter, direct detection in the infrared is the long-term objective of large-scale search programs.

B. HISTORICAL ERA OF SEARCHES

The historical era of searches covers most of this century and ends in the mid-1980s. Searches during this time were few and far between, and the bulk of them were conducted at what most astronomers would consider quiet backwaters of modern astronomy, at small astrometric observatories.

While astrometric searches were conducted at the U.S. Naval Observatory and other astrometric observatories, far and away the most concerted effort was being conducted at the Sproul Observatory in Pennsylvania. The staff at Sproul had conducted much of the pioneering astrometry of this century, and during that time they had been gathering observations on a small number of stars and looking for perturbations in the motion of those stars that could be attributed to the presence of planetary companions. Because of the small aperture of the Sproul telescope, it was limited to looking at relatively nearby stars when looking for the very small perturbations that a planetary mass object would cause.

One of the stars monitored very carefully since the early part of this century was Barnard's star. This star, which is about one-seventh the mass of the Sun, has the highest proper motion of any star. This makes it a good candidate for astrometric study as it optimizes the visibility of any perturbation in the motion of the star. Also, although Barnard's star is intrinsically fairly faint, it is the second closest stellar system to the Sun, and so its apparent brightness was adequate to conduct astrometric observations. After examining several decades of photographic plates on Barnard's star, Peter van de Kamp at Sproul Observatory concluded that there was evidence for one and perhaps two Jupiter-mass companions with orbits in the range of roughly 12 to 22 years. These observations were near the performance limit of the Sproul astrometric telescope.

Subsequent and more accurate observations of Barnard's star, primarily by Robert Harrington at the U.S. Naval Observatory and by George Gatewood at the Allegheny Observatory, revealed that van de Kamp's interpretation of the data could not be correct. They were unable to corroborate the predicted motion of the star if the putative planetary companions were really present. It was later discovered that there were systematic errors at the Sproul facility that probably influenced the data used by van de Kamp.

Two other search efforts in this early era are noteworthy. One involves the pulsar that powers the Crab Nebula. The other involves the nearby star Van Biesbroeck 8. Shortly after pulsars were discovered in the late 1960s, there was some evidence for periodicity in the pulse structure other than the remarkable periodic pulses that gave rise to the name of this unusual class of object. One interpretation of the extra signal was that it was due to a low-mass companion, possibly a planet, to the pulsar. That example was subsequently shown not to be the case, but it was a harbinger of what was to come.

The case of the star Van Biesbroeck 8, or VB 8 as it is often called, was interesting in that it marked the first time in which a direct detection of a possible substellar-mass companion was claimed. VB 8 had been studied by astrometrists who also thought that they had evidence that there might be a companion to the star. The technique of infrared speckle interferometry, a powerful method for minimizing deleterious effects on astronomical studies that arise from Earth's atmosphere, was used to study VB 8. There was evidence that there was a companion to the star, a companion that appeared to be consistent with the companion claimed by the astrometrists.

So exciting was this result that a conference was held in 1985 on the subject of brown dwarfs, but the major topic of interest was the companion to VB 8, a companion that came to be called "VB 8B." Several excellent papers were presented at the conference, many of them concerned with trying to interpret the data to get a better estimate of the nature of VB 8B. There was a general consensus as to the mass of the object, something in the range of a few to several tens of Jupiter masses. The excitement, however, was not sustained as subsequent efforts to observe VB 8B failed to even detect it. It appears that what astronomers thought was a companion was an artifact of some kind.

The failure of these early efforts to detect other planetary systems began to introduce an air of urgency among those who had devoted a significant portion of their research careers to looking for these systems. However, it is now clear that the failure to find substellar-mass companions was in the end more a comment about the sensitivity or accuracy of the observational tools than of the existence of such companions.

C. MODERN ERA OF SEARCHES

Though there is not at this time an unambiguous detection of another planetary system, a few observational studies stand out as seminal in this field of observation. One concerns yet another study of a pulsar, and the other two relate to two long-term programs to monitor the radial velocities of stars.

The first convincing evidence of the detection of a companion that was clearly substellar in mass, although it was debated for some time before being accepted, came in 1989. David Latham and his colleagues at Harvard University had been conducting a moderate-accuracy radial velocity survey of a large number of

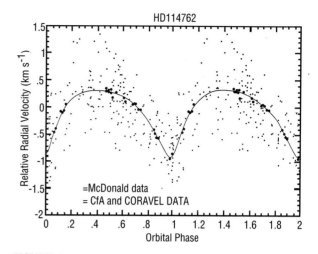

FIGURE 2 Radial velocity variations for the star HD 114762 shown as a function of orbital phase over two phase cycles. The smaller dots are from the less accurate studies of Latham and co-workers, and the larger dots represent data from the more accurate study by Cochran and co-workers. The solid line is the best fit to the data and indicates that the orbit is highly eccentric.

stars for several years. They noticed that one of the standard stars, HD 114762, displayed evidence of periodic changes in its spectrum (Fig. 2).

After analyzing data from many years of observing, they concluded that there was a companion in an orbit similar to that of the planet Mercury. The companion has a minimum mass of about 10 times the mass of Jupiter. One of the interpretational ambiguities associated with radial velocity detections of companions is that the amplitude of the detected signal is related to the true amplitude via the sine of the angle between the observer's line of sight to the star and the orbital plane of the companion about the star. As the angle is in general unknown, the data provide only a lower limit to the mass of the companion. The observations also showed that this companion was in a highly eccentric orbit, very much different than the nearly circular orbits of Jupiter and Saturn, but typical of binary systems.

The next major discovery was the astronomical equivalent of "back to the future" as it involved observations of a pulsar. This observation was also part of an ongoing survey program to monitor pulses from pulsars, particularly the rapidly spinning millisecond pulsars. Alex Wolszczan and David Frail used the Arecibo radio telescope to monitor precisely the time of arrival of pulses from a number of pulsars. They noticed that one of these pulsars, PSR 1257+12, showed

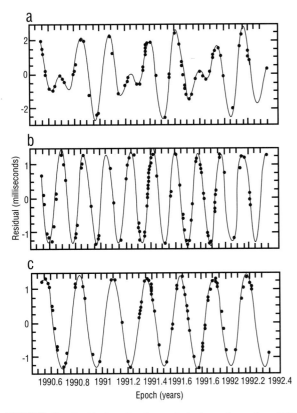

FIGURE 3 Residuals of pulse arrival times for pulsar PSR 1257+12 as a function of observing epoch. Residuals from a fit using only positional and rotational parameters are shown in panel (a). Panels (b) and (c) show the effects of removing perturbations due to a 98.2-day or a 66.5-day Keplerian orbit, respectively.

signs of periodic behavior in the arrival times of pulses (Fig. 3). At times the pulses arrived earlier, and then later than expected. They interpreted this as being due to gravitational perturbations in the position of the pulsar arising from the presence of at least two companions revolving around the pulsar.

The remarkable precision of pulsar observations meant that Wolszczan and Frail were able to determine that the lower limits to the masses of the unseen companions (pulsar timing is basically a radial velocity observation and so suffers the same ambiguity with regard to the unknown angle) were both about three times the mass of the Earth! The history of pulsar observations had been checkered in the area of detection of companions, so there was concern as to whether this interpretation would stand the test of time.

The fact that there appeared to be at least two companions, in orbits roughly equivalent to the orbits of Venus and the Earth as far as distance from the pulsar, meant that the gravitational interaction between the two putative companions would provide a

direct test of Wolszczan and Frail's interpretation. It became clear within a little more than a year that the pulsar arrival time pattern was indeed showing the type of shift that it would if the variations were due to orbiting companions. This same gravitational interaction provided a means to place an upper limit to the mass of the companions—they were less than roughly 10 Earth masses.

Pulsars are neutron stars, relics of a violent supernova event. It is highly unlikely that any planetary system that might have been present prior to the supernova would have survived. Where then did the purported planets come from?

The answer appears to lie in the mechanism that forms a millisecond pulsar. It is thought that these rapidly spinning systems are created as a normal pulsar accretes high angular momentum material from a disk, a disk created after the supernova event. The assumption at this time is that the planetary companions to PSR 1257+12 were created in a manner analogous to how it is believed that planets formed around the Sun, via accretion in a disk. Thus, although the physical and chemical conditions in the disk that was associated with PSR 1257+12 were much different that the conditions that attend the formation of a star, the process of accretion seems to have occurred.

The system of companions to PSR 1257+12 is not considered here as a "planetary system" because the system was formed by a distinctly different, albeit related, set of processes than those involved in the formation of a star. The pulsar system is thus similar to the regular satellite systems of the giant planets in the solar system. Those systems display all the orbital regularities and features that are seen in the solar system and the pulsar system, and it is thought that they too formed via accretion in a disk that surrounded these large planets as they were forming.

Both the pulsar and satellite systems provide support for the notion of accretion as the mode of formation of planetary objects, but they cannot be used to draw inferences regarding either the frequency or the characteristics of planetary systems in general. The possible exceptions to this, and they are important exceptions, are that companion systems formed via accretion from a disk having low-eccentricity orbits and they are multiple-companion systems, not binaries.

Before leaving the story of PSR 1257+12, which is mainly one of success, there is an aspect of this story that continues to remind those who search for very small effects how nature can fool even the most careful observers. In addition to the two signals that Wolszczan and Frail found with periods of 98.22

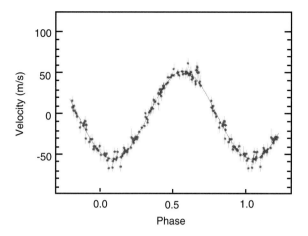

FIGURE 4 Radial velocity (m/sec) as a function of phase for the star 51 Pegasi. The best fit period is 4.231 days, with an orbital eccentricity of 0.014.

and 66.54 days, they also reported evidence of a period of 25.34 days in their data. The former two periods are the ones that are associated with the two terrestrial-mass companions discussed earlier. Wolszczan and Frail associated the third period with a lunar-mass "planet" orbiting the pulsar. The reality of this inner companion has not been as widely accepted as that of the two longer-period companions. There was no firm independent demonstration of its existence, as was the case with the mutual perturbations that characterize the two longer-period companions, but there was no firm evidence indicating that the interpretation was flawed.

Recent studies of signals from the *Pioneer 10* spacecraft from 1987 to 1997 as it moved from 40 to 65 AU heliocentric distance, signals that pass through the heliosphere in almost the same path that signals from the pulsar PSR 1257+12 pass, reveal the same periodicity as found in the pulsar timing data. This suggests that although the periodicity in the data gathered by Wolszczan and Frail is real, the interpretation of the periodicity as evidence for a third companion to the pulsar is probably incorrect. The pulsar observations appear to have simply detected the rotation of the Sun! This demonstrates again that the search for companions takes observers into new realms of phenomena, and that it is all too easy to misinterpret periodic signals in terms of companions to familiar celestial objects such as stars and pulsars.

The third of the major discoveries came in 1994 and was announced in 1995. Didier Queloz and Michel Mayor, two Swiss astronomers observing at the Haute-Provence Observatory, detected a companion to the star 51 Pegasi (Fig. 4). The companion was in an orbit

with a period of just over 4 days! Using the estimated mass of the star and Kepler's laws of motion, this period implies that the orbit is only about 5% as large as that of Earth. The detection technique used was radial velocity observations, and this companion had a lower limit to its mass of about 0.5 times the mass of Jupiter.

Because the lower limit to the mass is so close to the mass of Jupiter, the immediate reaction and detection statements declared this companion to be an "extrasolar planet," the first of its kind to be discovered. The discovery was quickly confirmed by two other groups, so there was no doubt that a signal was present. As we shall see, this discovery was rapidly followed by a number of others that were similar in many respects. It was as though the Queloz and Mayor discovery had broken some form of observational log jam. However, the true nature of these companions remains unclear.

It is worthwhile noting that the seminal observations to date were the product of long-term routine observational surveys. Though such programs are often viewed with mild enthusiasm at best, this is but one example of how such efforts often provide discoveries that were not anticipated at the inception of the program.

D. THE ZOO OF SUBSTELLAR-MASS COMPANIONS

Since the discovery of the companion to 51 Pegasi, roughly 20 additional detections of substellar-mass companions to "normal" stars have been announced. There have also been a small number of additional claims of detection of planetary-mass companions to pulsars, but the focus for the remainder of this chapter will be on companions to normal stars.

The vast majority of the newly discovered companions were detected by radial velocity observations. The exceptions are one or possibly two companions to the star Lalande 21185 that were detected by astrometric observations. This means that the masses that are associated with these companions are lower limits to their masses. It should be noted, however, that if the orientation of orbital planes is random in space, then the uncertainty in knowing the angle between the line of sight for an Earth-bound observer and the orbital plane is not likely to lead, on average, to the lower-mass limit being too far from the true mass. Errors in mass estimates using the lower limit will typically be less than a factor of two. It is possible that a small percentage of the companions have true masses that are a factor of 10 larger than their lower limit, but those

TABLE II
Properties of Newly
Discovered Substellar Companions

Star	Mass[a]	Period (days)	Eccentricity
51 Peg[1]	0.45	4.23	0.00
υ And[2]	0.65	4.61	0.11
55 Cnc[2]	0.84	14.65	0.05
Rho Cr[3]	1.1	39.64	0.03
16 Cyg[4]	1.6	804	0.65
47 UMa[2]	2.3	1090	0.03
τ Boo[2]	3.9	3.31	0.00
70 Vir[2]	7.4	116.7	0.37
HD 114762[5]	9.0	84.02	0.33
HD 110833[6]	17	270	0.69
BD −04:782[6]	21	240.92	0.28
HD 112758[6]	35	103.22	0.16
HD 98230[7]	37	3.98	0.00
HD 18445[6]	39	554.67	0.54
HD 29587[6]	40	1471.70	0.37
HD 140913[6]	46	147.94	0.61
BD +26:730[6]	50	1.79	0.02
HD 89707[6]	54	298.25	0.95
HD 217580[6]	60	454.66	0.52

[a] Mass is expressed in Jupiter masses.

cases should be rare if the assumption of random orientation is valid.

A list of the discovered companions is given in Table II. They are arranged in order of increasing value of their minimum mass expressed in units of Jupiter masses. Also given in the table are the values of the companions' orbital period in days and the eccentricity of its orbit. Table II contains only those companions that have been discovered by radial velocity searches. The remarkable push to high accuracy in radial velocity efforts owes much to Krzysztof Serkowski, who died before his dream of detecting planetary systems with high-accuracy radial velocity observations could be realized, and to Bruce Campbell and Gordon Walker, whose pioneering use of hydrogen fluoride to provide a precise calibration of incoming starlight blazed the path now followed by the "planet hunters."

One important aspect of the searches is that they have been conducted with instruments of widely differing accuracy. There are currently at least three systems that are capable of high-accuracy radial velocity observations, with an accuracy of 10 m/sec or better.

Some of the observations were made with less accurate systems, ranging in accuracy from a few tens of meters per second to a few hundred meters per second. The question of the accuracy of the system used to detect a companion will be important when we consider the statistics of these systems.

The normal procedure in science, when a new class of objects is discovered, is to observe patterns emerging in comparisons of all measured variables. In the case of the companions, the three basic parameters determined to date are orbital period, eccentricity, and minimum mass. Let us first consider the mass of the companions.

Nine of the companions in Table II have masses below 10 Jupiter masses. Assuming that orbital planes are randomly distributed in space, it is possible, even likely, that the true masses of the companions to 70 Vir and HD 114762 are slightly above 10 Jupiter masses. Recalling the discussion about the masses of brown dwarfs and planets, the value of 10 Jupiter masses was significant for two reasons. One is that it is near the cutoff mass, 12 Jupiter masses, for the burning of deuterium. Another is that it is the oft-quoted lower limit to the formation of an object by the process of fragmentation. For this reason, these nine companions are the ones that are most often referred to as "extrasolar planets" by their discoverers and in the literature on the subject.

A plot of the number of companions, $N(M)$, as a function of their mass, M, can provide an indication of possible trends. As there are so few companions, it is common practice to display the data as a histogram, where the number of companions $N(M)dM$ with a mass between M and $M + dM$ is displayed. Such a plot is shown in Fig. 5.

It is important to recognize that the appearance of the data in a plot such as Fig. 5 is heavily influenced by how the data are binned and counted. The format used is one that is used often, but it can be very misleading, especially from the perspective of making it seem as though the population of objects with low mass is disproportionately greater than it really is. For example, instead of plotting the data in bins of 10 Jupiter masses, and there is no scientifically based reason to use such a binning, we could plot the logarithm of the number of objects per logarithmic mass unit. This type of display is common when one assumes that the true distribution might be some form of power law as is so often the case in nature. Figure 6 shows the revised plot.

Note that Fig. 6 gives a totally different impression of the data, namely that there is a relative *underabundance* of objects in the lower-mass bins when compared

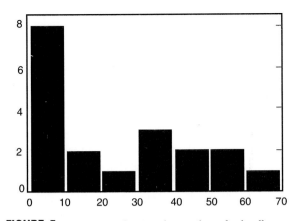

FIGURE 5 Histogram showing the number of substellar-mass companions discovered by radial velocity searchers as a function of their minimum mass. The mass binning is linear with a width of 10 Jupiter masses. These data are uncorrected for selection effects arising from differences in the various search accuracies and number of stars surveyed.

to the number in the upper end of the brown dwarf mass range. If incompleteness is taken into account, at least qualitatively, one would expect that the harder to detect low-mass companions will increase in number, and perhaps the number per logarithmic mass unit will be a constant, a result not inconsistent with what is known about the lower end of the stellar-mass function. One possible interpretation of this is that most, if not all, of these companions come from the same population of objects, which in this case could be brown dwarfs.

The moral here is that one needs to take care in

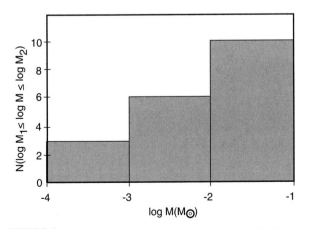

FIGURE 6 Histogram showing the same data as in Fig. 5, but with the mass binning logarithmic, with each bin being a decade in mass. Note the major change in the inferred nature of the underlying population depending on the manner in which the masses are binned. The same effect is present if one uses data corrected for accuracy and sample-sized differences among the searches.

interpreting data when there is possible ambiguity in the manner in which it is presented. The true nature of the mass distribution of these substellar-mass companions must await more observations.

There is one other aspect of the mass distribution that should be mentioned, and that deals with how the results of searches vary depending on the accuracy of the instruments. Three of the groups currently engaged in spectroscopic searches have high-accuracy instruments that have been used to search a smaller number of stars for the presence of companions. The interesting aspect of these searches is that only companions with minimum masses less than 10 Jupiter masses have been found with these instruments. They have not yet detected objects with a minimum mass above 10 Jupiter masses, objects that would be easier to detect than are the lower-mass companions.

All of the massive companions, those with masses greater than 10 Jupiter masses, listed in Table II were found by less accurate systems that surveyed a larger population of stars. If one takes the results from the more accurate systems as the baseline, it would appear that companions with masses less than 10 Jupiter masses are more abundant than companions with masses above 10 Jupiter masses. This is a key aspect of arguments that are made that the objects with masses less than 10 Jupiter masses are truly a separate class of objects than the more massive, but still substellar, companions listed in Table II.

Some caution is in order in interpreting this interesting result. First, the number of stars searched is still small enough that the possibility of selection effects cannot be ruled out. It may be that the next sample of comparable size will have no, or very few, low-mass companions and be characterized predominately by companions with masses in excess of 10 Jupiter masses. It is also worth noting that many of the companions with low mass also have short orbital periods (they are the easiest to detect because of the higher stellar velocity that is associated with short-period companions). It is possible, indeed likely, that evolutionary effects have altered the mass distribution of systems that are only a few stellar radii away from their central star.

Examination of the other two variables for these companions on which observations provide information, orbital period and eccentricity, is also interesting. If orbital eccentricity is plotted as a function of the logarithm of the orbital period in days, there is highly suggestive evidence of a trend, one in which there is a systematic increase in eccentricity with increasing orbital period (Fig. 7). Two aspects of this trend are particularly interesting. First, the trend is present for

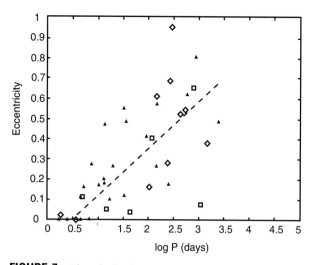

FIGURE 7 Plot of orbital eccentricity as a function of the logarithm of orbital period P, expressed in days, for three separate classes of objects. Solid triangles are data from pre-main-sequence binary systems. Open diamonds are from substellar-mass companion radial velocity searches with minimum masses above 10 Jupiter masses (see Table II), and open squares are for similar companions but with minimum masses less than 10 Jupiter masses (see Table II). The line is a least-squares fit to the stellar data. With the exception of two points, all the substellar systems could have come from the same (e, $\log P$) distribution that describes pre-main-sequence binary systems. This distribution is sharply distinct from that evidenced by systems where accretion is thought to be the formation process for companions.

all of the substellar-mass companions with two possible exceptions and, second, the trend is distinctly different from that for systems where the companions are presumed to have formed by accretion.

The possible significance of this trend lies in the fact that it is indistinguishable from that displayed by pre-main-sequence binary star systems. It is generally believed that the increase in eccentricity with increasing orbital period in the case of binaries is a consequence of their formation mechanism. If so, the fact that the distribution in eccentricity and period for most of the substellar-mass companions is so similar to that for stars suggests that these systems were formed by a similar, perhaps identical, process as were the binary stars. This would make their identification as planets debatable, while making their identification as brown dwarfs at least reasonable. Support for the notion that the majority of these companions may not be members of a planetary system lies in the fact that planetary systems, as well as other systems where there are multiple companions formed by accretion, have essentially zero eccentricity independent of the period of the orbit. The companions to Rho Cr and 47 Ursa Major are possible exceptions. They appear to have eccentricities

that are very low given their orbital period, and appear in this regard like objects formed by accretion.

The search for other planetary systems has brought us to a remarkable place. We now have growing evidence that substellar-mass companions exist in reasonable, but not great abundance. It is clear that the majority of the objects discovered to date are brown dwarfs, but the situation is less clear for those objects for which the lower limit to their masses is ≤10 Jupiter masses. It may turn out to be the case that these companions are not alone, that there are other companions to their central star. If that is the case, they qualify as members of a planetary system. If that is not the case, if there are no other companions of mass comparable to the masses of planets in the solar system, it is likely that they are not planets, but are either brown dwarfs or a new class of object.

IV. FUTURE DIRECTIONS AND OPPORTUNITIES

The search efforts to date and the technology available for searches in the near future may permit the discovery of planetary systems around nearby stars. However, they are unlikely to be able to conduct an extensive survey that would provide a firm statistical basis for establishing even the frequency of planetary systems, and they will do little to illuminate the issue of whether planets exist that either might be, or in fact are, habitable. The key observational demands are to be able to detect systems where one component (the star) is much brighter than the other components (the planets).

There are plans for a next generation of telescopes that should be able to attack these problems. The key wavelength regime for this next generation is in the near to mid infrared, the wavelengths ranging from several to a few tens of microns. Why is this wavelength regime so favorable? The answer lies both in the likely intrinsic brightness ratio between a typical star and a planetary companion, as well as in the fact that many of the key molecules that tell us something about the nature of a planetary atmosphere have lines in this portion of the infrared.

A. DIRECT DETECTION

Direct detection as a means to discover planetary systems was discussed earlier in this chapter. Although direct detection offers many advantages over indirect

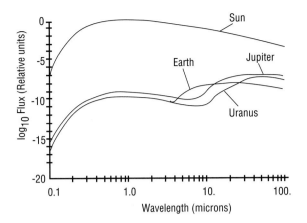

FIGURE 8 Plot of the power as a function of wavelength as radiated by the Sun and several planets. The curve for the Sun is essentially a blackbody spectrum, whereas those for the planets are a combination of a blackbody spectrum at longer wavelengths plus reflected solar radiation. Note that the near-infrared region of the spectrum is a favorable one for detecting Earth-like planets.

detection, it is intrinsically a more difficult task observationally and technically because planets are in general so much dimmer than their stellar companions (Fig. 8). Detecting systems where the contrast ratios are as large as those shown in Fig. 8, and where the distance between star and companion is small enough to provide surface temperatures that may be conducive to life, is impossible at the current time.

Use of direct detection to discover substellar-mass companions has had one resounding success, the detection of the brown dwarf companion to the star GL 229. This companion is relatively massive as brown dwarfs go and is located more than 40 AU away from its stellar companion. This makes it possible to detect directly. If this companion were located a few AU from its stellar companion, it would have escaped detection.

As can be seen from Fig. 8, detection of planets located in orbits similar to that of Jupiter is best accomplished at wavelengths in the vicinity of a few tens of microns where the planet has the peak in its blackbody curve. Planets located in the terrestrial planet region are best detected in the wavelength range from a little less than 10 μm out to perhaps 20 μm.

The argument for the infrared goes beyond contrast ratios. If we are interested in finding planets that are habitable, we must be able to gather spectral information that points to the presence of key molecules in the atmosphere of discovered planets. The definition of habitability is somewhat debatable, but it generally involves the presence of liquid water. This suggests that evidence of water in the atmosphere of a planet is a key step in establishing whether a planet is habitable.

There are a number of molecules that tell us something about the nature of any atmosphere associated with a planet. The presence of CO_2 is generally regarded as a signpost that a significant atmosphere is present. This molecule has lines in the wavelength band between 7 and 20 μm. The presence of ozone is an indicator that the atmosphere may have a significant abundance of O_2, another key molecule for life. Ozone also has lines in this wavelength band. Finally, there is the molecule H_2O itself, also with lines in this region of the spectrum, and a major indicator that a planet may be habitable. Some of the molecular lines in this region of the spectrum are shown in Fig. 9. [*See* PLANETS AND THE ORIGIN OF LIFE.]

B. THE PLANET FINDER—THE SEARCH TOOL OF THE FUTURE

A key consideration with any telescope is how large it must be to accomplish the observational task for which it is designed. In the case of a system that must detect as weak a signal, relative to that of its stellar companion,

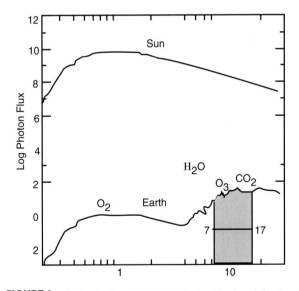

FIGURE 9 A plot similar to Fig. 8, but for just Earth and the Sun. The location of key molecular lines, such as CO_2, O_2, O_3, and H_2O, are indicated. Presence of the former is a strong indicator that a substantial atmosphere is present. The presence of oxygen is a strong indicator that photosynthesis is occurring, and the presence of substantial water in the atmosphere could imply the presence of large reservoirs of liquid water on the planet. Note that although the line for molecular oxygen is not in a favorable region of the spectrum, lines for ozone are. Ozone is a reasonably good surrogate for oxygen. The shaded region lies in the wavelength range 7–17 μm, a range that includes all the key spectral signatures that are diagnostic of life.

as that expected from planets revolving around other stars, and a system that will operate at wavelengths ranging from 7 to 20 μm, the effective aperture must be several tens of meters!

Construction of a filled aperture of the scale that is required is a daunting task. However, the aperture need not be filled, leading to the notion of an infrared interferometer. Several independent considerations help to drive the design concept. One is that the observing system needs to be able to suppress the strong stellar signal in order to detect the planetary signal. An innovative scheme, first conceived in the 1970s by Ron Bracewell of Stanford University, and subsequently refined by Roger Angel of the University of Arizona, uses an interferometer with four small- to modest-aperture infrared telescopes mounted on a linear structure. In principle, this system is capable of reducing the contribution of the central star by a factor of a million by combining the signal from the four apertures to null the signal from the star. The size of the individual apertures is set according to how much of the zodiacal cloud that is present in our planetary system the interferometer must see through. If the system operates in an orbit that is roughly 1 AU from the Sun, then the individual apertures on the interferometer must be at least 4 m in size. That is, each element in the interferometer must be almost twice the size of the *Hubble Space Telescope*. If, on the other hand, the interferometer is placed in an orbit that is 5 AU from the Sun, the zodiacal dust cloud is less significant and smaller apertures may be used. Estimates for this case indicate that apertures of a little more than a meter would suffice.

Whatever the aperture sizes for this interferometer, the baseline of the system must remain around 70 m or even larger. This is clearly a major step in space-based observing systems, and it is being approached by NASA in small steps. The first step is to gain experience on the ground using a variety of infrared interferometers. This would be followed by development of a Space Interferometer Mission that would be the first space-based interferometer. This device would probably operate at optical wavelengths rather than infrared. However, it would permit a number of critical tests of how such systems operate in space.

A major consideration in the design of the large infrared interferometer, which is currently referred to as "The Planet Finder," is that once it discovers planets in a location where they could be habitable, the instrument will "stare" at that planet for a sufficiently long time to gather spectral evidence concerning the nature of its atmosphere. Emphasis will be placed on detecting the lines that were discussed earlier, but any lines will

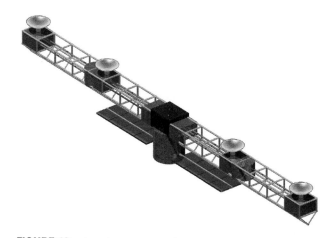

FIGURE 10 Artist's conception of the Planet Finder, an infrared interferometer that is being considered by NASA for development in the first decade of the new millennium. Planet Finder will be a large facility, something in the range of 75–100 m in the current model. The interferometer has four separate telescopes that will operate in the 7- to 17-μm region of the spectrum, which is rich in diagnostic information about possible life on planets revolving around other stars (see Fig. 9). This is also the region of the spectrum that is most favorable from the perspective of contrast ratio between star and planet with a strong signal from the planet (see Fig. 8). By carefully combining the infrared radiation detected by each of the four telescopes, the light from the central star can be reduced by a factor of roughly one million, making detection of the dimmer planet much easier.

provide information regarding the planet. As more information is gained, the path to a system like the Planet Finder will remain flexible. At the present, however, the system depicted in Fig. 10 appears to be a viable device for the detection and characterization of habitable planets.

C. CONCLUDING REMARKS

This chapter is concerned with searches for other planetary systems, both those ongoing and those that will come in the future. The focus on this topic, which has taken place in the last few decades, has led to an intensification of effort to detect other planetary systems. A consequence of that intensification has been the discovery of a number of substellar-mass companions to nearby stars. Whether those companions are members of planetary systems or are brown dwarfs, objects more closely related to stars in binary systems, will only be known when more observational evidence is in hand.

We now stand in a position to test our current ideas

of how the Sun and its retinue of planets, as well as other such systems, are formed, and to gain insight on how those ideas need to be changed if the results of comprehensive surveys take us in that direction. This program will also lead to the completion of the revolution in thought that was started by the astronomer Copernicus some 500 years ago. It is reasonable to suppose that after the first few decades of the new millennium have passed, we will know with certainty whether habitable planets exist around other stars, and we may even have gathered spectral evidence that life is present on some of those planets.

APPENDIX

Planetary Exploration Missions   958

Selected Astronomical Constants   961

Physical and Orbital Properties of
the Sun and Planets   961

Physical and Orbital Properties of
the Satellites   962

PLANETARY EXPLORATION MISSIONS

Spacecraft	Source	Launch	Target	Mission	Notes
Pioneer 1	USA	Oct. 11, 1958	Moon	flyby	Reached 70,700 km altitude-missed Moon
Pioneer 2	USA	Nov. 8, 1958	Moon	flyby	Third stage ignition failure
Pioneer 3	USA	Dec. 6, 1958	Moon	flyby	Reached 63,600 km altitude-missed Moon
Luna 1	USSR	Jan. 2, 1959	Moon	flyby	Flew by Moon at 5998 km distance
Pioneer 4	USA	Mar. 3, 1959	Moon	flyby	Missed Moon by 60,030 km
Luna 2	USSR	Sept. 12, 1959	Moon	impact	Impacted on Moon
Luna 3	USSR	Oct. 4, 1959	Moon	flyby	Photographed far side of Moon
(Pioneer)	USA	Nov. 26, 1959	Moon	flyby	Payload shroud failed during launch
(Pioneer)	USA	Sept. 25, 1960	Moon	flyby	Second stage malfunction
(unnamed)	USSR	Oct. 10, 1960	Mars	flyby	Failed to achieve Earth orbit
(unnamed)	USSR	Oct. 14, 1960	Mars	flyby	Failed to achieve Earth orbit
(Pioneer)	USA	Dec. 15, 1960	Moon	flyby	First stage exploded
Sputnik 7	USSR	Feb. 4, 1961	Venus	flyby	Failed to depart from low Earth orbit
Venera 1	USSR	Feb. 12, 1961	Venus	flyby	Communications failure in transit
Ranger 1	USA	Aug. 23, 1961	Moon	flyby	Failed to depart from low Earth orbit
Ranger 2	USA	Nov. 18, 1961	Moon	flyby	Failed to depart from low Earth orbit
Ranger 3	USA	Jan. 26, 1962	Moon	impact	Missed Moon by 36,745 km
OSO 1	USA	Mar. 7, 1962	Sun	orbiter	Solar observatory in Earth orbit
Ranger 4	USA	Apr. 23, 1962	Moon	impact	Impacted Moon with experiments inoperative
Mariner 1	USA	July 22, 1962	Venus	flyby	Launch failure
Sputnik 19	USSR	Aug. 25, 1962	Venus	flyby	Failed to depart from low Earth orbit
Mariner 2	USA	Aug. 27, 1962	Venus	flyby	Flew by Venus at 34,745 km range
Sputnik 20	USSR	Sept. 1, 1962	Venus	flyby	Failed to depart from low Earth orbit
Sputnik 21	USSR	Sept. 12, 1962	Venus	flyby	Failed to depart from low Earth orbit
Ranger 5	USA	Oct. 18, 1962	Moon	impact	Missed Moon by 724 km
Sputnik 22	USSR	Oct. 24, 1962	Mars	flyby	Failed to depart from low Earth orbit
Mars 1	USSR	Nov. 1, 1962	Mars	flyby	Communications failure in transit
Sputnik 24	USSR	Nov. 4, 1962	Mars	flyby	Failed to depart from low Earth orbit
Sputnik 25	USSR	Jan. 4, 1963	Moon	impact	Failed to depart from low Earth orbit
Luna 4	USSR	Apr. 2, 1963	Moon	impact	Missed Moon by 8499 km
Kosmos 21	USSR	Nov. 11, 1963	Venus	test	Failed to depart from low Earth orbit
Ranger 6	USA	Jan. 30, 1964	Moon	impact	Impacted Moon with TV inoperative
Kosmos 27	USSR	Mar. 27, 1964	Venus	flyby	Failed to depart from low Earth orbit
Zond 1	USSR	Apr. 2, 1964	Venus	flyby	Communications failure in transit
Ranger 7	USA	July 28, 1964	Moon	impact	Impacted moon-returned 4308 photos
Mariner C	USA	Nov. 5, 1964	Mars	flyby	Shroud failed to separate after launch
Mariner 4	USA	Nov. 28, 1964	Mars	flyby	Flew by Mars: July 15, 1965-returned 21 photos
Zond 2	USSR	Nov. 30, 1964	Mars	flyby	Communications failure in transit
OSO 2	USA	Feb. 3, 1965	Sun	orbiter	Solar observatory in Earth orbit
Ranger 8	USA	Feb. 17, 1965	Moon	impact	Impacted moon-returned 7137 photos
Kosmos 60	USSR	Mar. 12, 1965	Moon	lander	Failed to depart from low Earth orbit
Ranger 9	USA	Mar. 21, 1965	Moon	impact	Impacted moon-returned 5814 photos
Luna 5	USSR	May 9, 1965	Moon	lander	Landing attempt failed-crashed on Moon
Luna 6	USSR	June 8, 1965	Moon	lander	Missed Moon by 160,935 km
Zond 3	USSR	July 18, 1965	Mars	test	Flew by Moon as test of Mars spacecraft
OSO C	USA	Aug. 23, 1965	Sun	orbiter	Solar observatory-failed to orbit
Luna 7	USSR	Oct. 4, 1965	Moon	lander	Retros fired early-crashed on Moon
Venera 2	USSR	Nov. 12, 1965	Venus	probe	Communication failure just before Venus arrival
Venera 3	USSR	Nov. 16, 1965	Venus	probe	Communication failure before Venus entry
Kosmos 96	USSR	Nov. 23, 1965	Venus	probe	Failed to depart from low Earth orbit
Luna 8	USSR	Dec. 3, 1965	Moon	lander	Retros fired late-crashed on Moon
Luna 9	USSR	Jan. 31, 1966	Moon	lander	First lunar soft landing-returned photos
Kosmos 111	USSR	Mar. 1, 1966	Moon	lander	Failed to depart from low Earth orbit
Luna 10	USSR	Mar. 31, 1966	Moon	orbiter	First successful lunar orbiter
Surveyor 1	USA	May 30, 1966	Moon	lander	Lunar soft landing-returned 11,150 photos
Lunar Orbiter 1	USA	Aug. 10, 1966	Moon	orbiter	Lunar photographic mapping
Pioneer 7	USA	Aug. 17, 1966	Solar wind	interplanetary	Monitored solar wind
Luna 11	USSR	Aug. 24, 1966	Moon	orbiter	Lunar orbit science mission
Surveyor 2	USA	Sept. 20, 1966	Moon	lander	Crashed on Moon attempting landing
Luna 12	USSR	Oct. 22, 1966	Moon	orbiter	Lunar photographic mapping
Lunar Orbiter 2	USA	Nov. 6, 1966	Moon	orbiter	Lunar photographic mapping
Luna 13	USSR	Dec. 21, 1966	Moon	lander	Soft lander science mission
Lunar Orbiter 3	USA	Feb. 4, 1967	Moon	orbiter	Lunar photographic mapping
Surveyor 3	USA	Apr. 17, 1967	Moon	lander	Surface science mission

continues

Spacecraft	Source	Launch	Target	Mission	Notes
Lunar Orbiter 4	USA	May 4, 1967	Moon	orbiter	Photographic mapping
Kosmos 159	USSR	May 17, 1967	Venus	probe	Possible Venera or Molniya failure in Earth orbit
Venera 4	USSR	June 12, 1967	Venus	probe	Successful atmospheric probe: Oct. 18, 1967
Mariner 5	USA	June 14, 1967	Venus	flyby	Flew by Venus at 3990 km distance
Kosmos 167	USSR	June 17, 1967	Venus	probe	Failed to depart from low Earth orbit
Surveyor 4	USA	July 14, 1967	Moon	lander	Communications ceased before landing
Lunar Orbiter 5	USA	Aug. 1, 1967	Moon	orbiter	Photographic mapping
Surveyor 5	USA	Sept. 8, 1967	Moon	lander	Lunar surface science
OSO 4	USA	Oct. 18, 1967	Sun	orbiter	Solar observatory in Earth orbit
Surveyor 6	USA	Nov. 7, 1967	Moon	lander	Lunar surface science
Pioneer 8	USA	Dec. 13, 1967	Solar wind	interplanetary	Monitored solar wind
Surveyor 7	USA	Jan. 7, 1968	Moon	lander	Lunar surface science
Zond 4	USSR	Mar. 2, 1968	Moon	test	Unmanned test of Soyuz lunar craft
Luna 14	USSR	Apr. 7, 1968	Moon	orbiter	Mapped lunar gravity field
Zond 5	USSR	Sept. 14, 1968	Moon	test	Circumlunar flyby-spacecraft recovered
Pioneer 9	USA	Nov. 8, 1968	Solar wind	interplanetary	Monitored solar wind
Zond 6	USSR	Nov. 10, 1968	Moon	test	Lunar flyby-precursor of manned flight
Apollo 8	USA	Dec. 21, 1968	Moon	manned	Manned lunar orbiter and return
Venera 5	USSR	Jan. 5, 1969	Venus	probe	Atmospheric entry probe: May 16, 1969
Venera 6	USSR	Jan. 10, 1969	Venus	probe	Atmospheric entry probe: May 17, 1969
OSO 5	USA	Jan. 22, 1969	Sun	orbiter	Solar observatory in Earth orbit
Mariner 6	USA	Feb. 24, 1969	Mars	flyby	Flew by July 31, 1969-returned 75 pictures
Mariner 7	USA	Mar. 27, 1969	Mars	flyby	Flew by Aug 5, 1969-returned 126 pictures
Apollo 10	USA	May 18, 1969	Moon	manned	Manned lunar orbit test-precursor to landing
Luna 15	USSR	July 13, 1969	Moon	orbiter	Impacted on Moon after Apollo 11 landing
Apollo 11	USA	July 16, 1969	Moon	manned	First manned lunar landing and return
Zond 7	USSR	Aug. 8, 1969	Moon	test	Unmanned circumlunar flight and return
OSO 6	USA	Aug. 9, 1969	Sun	orbiter	Solar observatory in Earth orbit
Kosmos 300	USSR	Sept. 23, 1969	Moon	test	Possible test of Earth-orbit lunar equipment
OSO 7	USA	Sep. 29, 1971	Sun	orbiter	Solar observatory in Earth orbit
Kosmos 305	USSR	Oct. 22, 1969	Moon	test	Aborted lunar landing mission
Apollo 12	USA	Nov. 14, 1969	Moon	manned	Manned lunar landing and return
Apollo 13	USA	Apr. 11, 1970	Moon	manned	Aborted lunar landing-crew returned safely
Venera 7	USSR	Aug. 17, 1970	Venus	lander	First successful lander: Dec. 15, 1970
Kosmos 359	USSR	Aug. 22, 1970	Venus	lander	Failed to depart from low Earth orbit
Luna 16	USSR	Sept. 12, 1970	Moon	return	Returned lunar surface samples to Earth
Zond 8	USSR	Oct. 20, 1970	Moon	test	Unmanned circumlunar flight and return
Luna 17	USSR	Nov. 10, 1970	Moon	rover	Unmanned lunar rover: Lunokhod
Apollo 14	USA	Jan. 31, 1971	Moon	lander	Manned lunar landing and return
Mariner H	USA	May 8, 1971	Mars	orbiter	Second stage failure at launch
Kosmos 419	USSR	May 10, 1971	Mars	lander	Failed to depart from low Earth orbit
Mars 2	USSR	May 19, 1971	Mars	orbiter	First Mars orbiter: Nov. 27, 1971
Mars 2 lander	USSR	May 19, 1971	Mars	lander	First Mars lander: Dec. 2, 1971. No data returned
Mars 3	USSR	May 28, 1971	Mars	orbiter	Orbited Mars: Dec. 2, 1971
Mars 3 lander	USSR	May 28, 1971	Mars	lander	Failed immediately after landing
Mariner 9	USA	May 30, 1971	Mars	orbiter	Orbital photographic mapper
Apollo 15	USA	July 26, 1971	Moon	lander	Manned lunar landing and return
P&F satellite	USA	Aug. 4, 1971	Moon	orbiter	Subsatellite deployed by Apollo 15
Luna 18	USSR	Sept. 2, 1971	Moon	return	Crashed on Moon
Luna 19	USSR	Sept. 28, 1971	Moon	orbiter	Photographic mapping mission
Luna 20	USSR	Feb. 14, 1972	Moon	return	Lunar surface sample return
Pioneer 10	USA	Mar. 3, 1972	Jupiter	flyby	Jupiter flyby on Dec. 3, 1973
Venera 8	USSR	Mar. 27, 1972	Venus	lander	Landed on July 22, 1972
Kosmos 482	USSR	Mar. 31, 1972	Venus	lander	Failed to depart from low Earth orbit
Apollo 16	USA	Apr. 16, 1972	Moon	manned	Manned lunar landing and return
P&F satellite	USA	Apr. 19, 1972	Moon	orbiter	Subsatellite deployed by Apollo 15
Apollo 17	USA	Dec. 7, 1972	Moon	manned	Manned lunar landing and return
Luna 21	USSR	Jan. 8, 1973	Moon	rover	Unmanned lunar rover: Lunokhod 2
Pioneer 11	USA	Apr. 6, 1973	Jupiter	flyby	Jupiter flyby on Dec. 4, 1974
			Saturn	flyby	Saturn flyby on Sept. 1, 1979
Explorer 49	USA	June 10, 1973	Moon	orbiter	Solar and galactic radio science experiment
Mars 4	USSR	July 21, 1973	Mars	orbiter	Failed to achieve Mars orbit
Mars 5	USSR	July 25, 1973	Mars	orbiter	Orbited Mars on Feb. 12, 1974
Mars 6	USSR	Aug. 5, 1973	Mars	lander	Communications lost just before landing
Mars 7	USSR	Aug. 9, 1973	Mars	lander	Engine failure-missed Mars

continues

Spacecraft	Source	Launch	Target	Mission	Notes
Mariner 10	USA	Nov. 3, 1973	Venus	flyby	Flew by Feb. 5, 1974 en route to Mercury
			Mercury	flyby	Flew by Mercury three times in 1974
Luna 22	USSR	May 29, 1974	Moon	orbiter	Photographic mapper
Luna 23	USSR	Oct. 28, 1974	Moon	return	Drill arm damaged-no return attempt
Helios 1	Germany	Dec. 10, 1974	Solar wind	interplanetary	Monitored solar wind and dust
Venera 9	USSR	June 8, 1975	Venus	lander	Landed Oct. 22, 1975
Venera 9 orbiter	USSR	June 8, 1975	Venus	orbiter	Orbited Venus Oct. 22, 1975
Venera 10	USSR	June 14, 1975	Venus	lander	Landed Oct. 25, 1975
Venera 10 orbiter	USSR	June 14, 1975	Venus	orbiter	Orbited Venus Oct. 25, 1975
OSO 8	USA	Jun. 21, 1975	Sun	orbiter	Solar observatory in Earth orbit
Viking 1	USA	Aug. 20, 1975	Mars	orbiter	Orbited Mars June 19, 1976
Viking 1 lander	USA	Aug. 20, 1975	Mars	lander	Landed on Mars July 20, 1976-surface science
Viking 2	USA	Sept. 9, 1975	Mars	orbiter	Orbited Mars Aug. 7, 1976
Viking 2 lander	USA	Sept. 9, 1975	Mars	lander	Landed on Mars Sept. 3, 1976-surface science
Helios 2	Germany	Jan. 15, 1976	Solar wind	interplanetary	Monitored solar wind and dust
Luna 24	USSR	Aug. 9, 1976	Moon	return	Lunar surface sample return
Voyager 2	USA	Aug. 20, 1977	Jupiter	flyby	Flew by Jupiter July 9, 1979
			Saturn	flyby	Flew by Saturn Aug. 26, 1981
			Uranus	flyby	Flew by Uranus Jan. 24, 1986
			Neptune	flyby	Flew by Neptune Aug. 25, 1989
Voyager 1	USA	Sept. 5, 1977	Jupiter	flyby	Flew by Jupiter Mar. 5, 1979
			Saturn	flyby	Flew by Saturn Nov. 12, 1980
IUE	USA/ESA	Jan. 26, 1978	All	orbiter	Ultraviolet observatory in Earth orbit
Pioneer 12	USA	May 20, 1978	Venus	orbiter	Orbited Venus Dec. 8, 1978
Pioneer 13	USA	Aug. 8, 1978	Venus	probes	Four atmospheric entry probes: Dec. 9, 1978
Venera 11	USSR	Sept. 9, 1978	Venus	lander	Landed Dec. 25, 1978
Venera 12	USSR	Sept. 14, 1978	Venus	lander	Landed Dec. 21, 1978
ISEE 3	USA	Aug. 12, 1978	Solar wind	inteplanetary	Monitored solar wind; flew through tail of Comet Giacobini-Zinner: Sept. 11, 1985
Solar Max	USA	Feb. 14, 1980	Sun	Orbiter	Solar observatory in Earth orbit
Venera 13	USSR	Oct. 30, 1981	Venus	lander	Landed Feb. 27, 1982
Venera 14	USSR	Nov. 4, 1981	Venus	lander	Landed Mar. 5, 1982
IRAS	USA/UK/ Netherlands	Jan. 26, 1983	All	orbiter	Infrared observatory in Earth orbit
Venera 15	USSR	June 2, 1983	Venus	orbiter	Orbited Venus Oct. 10, 1983-radar mapper
Venera 16	USSR	June 7, 1983	Venus	orbiter	Orbited Venus Oct. 14, 1983-radar mapper
Vega 1	USSR	Dec. 15, 1984	Venus	lander	Landed June 11, 1985
	USSR	Dec. 15, 1984	Venus	balloon	Deployed in Venus atmosphere
	USSR	Dec. 15, 1984	Halley	flyby	Flew by Comet Halley at 8890 km: Mar. 6, 1986
Vega 2	USSR	Dec. 21, 1984	Venus	lander	Landed June 15, 1985
	USSR	Dec. 21, 1984	Venus	balloon	Deployed in Venus atmosphere
	USSR	Dec. 21, 1984	Halley	flyby	Flew by Comet Halley at 8030 km: Mar. 9, 1986
Sakigake	Japan	Jan. 7, 1985	Halley	flyby	Distant flyby of Comet Halley: Mar. 8, 1986
Suisei	Japan	Aug. 18, 1985	Halley	flyby	Flew by Comet Halley at 151,000 km: Mar. 14, 1986
Giotto	ESA	July 2, 1985	Halley	flyby	Flew by Comet Halley at 596 km: Mar. 14, 1986
Phobos 1	USSR	July 7, 1988	Phobos	orbiter	Communications lost en route
Phobos 2	USSR	July 12, 1988	Phobos	orbiter	Orbited Mars: Jan. 29, 1989-failed prior to landing
Magellan	USA	May 5, 1989	Venus	orbiter	Orbited Venus: Aug. 10, 1990-radar mapper
Galileo Orbiter	USA	Oct. 18, 1989	Jupiter	orbiter	Flew by Venus: Feb. 10,1990
					Flew by Earth: Dec. 8, 1990, Dec. 8, 1992
					Flew by asteroid Gaspra: Oct. 29, 1991
					Flew by asteroid Ida: Aug. 28,1993
					Orbited Jupiter: Dec. 7, 1995
Galileo Probe	USA	Oct. 18, 1989	Jupiter	probe	Entered Jupiter's atmosphere: Dec. 7, 1995
Hiten	Japan	Jan. 24, 1990	Moon	flyby	Flew by moon
Hagormo	Japan	Jan. 24, 1990	Moon	orbiter	Deployed into lunar orbit by Hiten: Mar. 1990
HST	USA/ESA	Apr. 25, 1990	All	orbiter	2.4 meter telescope in Earth orbit
Ulysses	ESA/USA	Oct. 6, 1990	Sun	orbiter	Injected into solar polar orbit by Jupiter flyby: Feb. 8, 1992
Yokhoh	Japan	Aug. 30, 1991	Sun	orbiter	Solar observatory in Earth orbit
Mars Observer	USA	Sept. 25, 1992	Mars	orbiter	Communications lost en route to Mars
Clementine	USA	Jan. 25, 1994	Moon	orbiter	Orbited Moon on Feb. 19, 1994. Failed after lunar departure
ISO	ESA	Nov. 17, 1995	All	orbiter	Infrared observatory in Earth orbit
SOHO	ESA	Dec. 2, 1995	Sun	orbiter	Solar observatory in Earth orbit-lost June, 1998

continues

Spacecraft	Source	Launch	Target	Mission	Notes
NEAR	USA	Feb. 17, 1996	Eros	orbiter	Flew by asteroid Mathilde: June 27, 1997. Will orbit asteroid Eros: Jan. 10, 1999
Mars Global Surveyor	USA	Nov. 7, 1996	Mars	orbiter	Orbited Mars: Sept. 12, 1997
Mars 96	CIS	Nov. 16, 1996	Mars	orbiter	Failed to depart low Earth orbit
Mars Pathfinder	USA	Dec. 2, 1996	Mars	lander/rover	Landed Mars: July 4, 1997. Deployed rover Sojourner
Cassini	USA	Oct. 15, 1997	Saturn	orbiter/probe	En route to Saturn. Arrival: July 1, 2004
Lunar Prospector	USA	Jan. 6, 1998	Moon	orbiter	Orbited Moon: Jan. 11, 1998

SELECTED ASTRONOMICAL CONSTANTS

Astronomical unit, AU	$1.4959787066 \times 10^{11}$ meters
Speed of light, c	2.99792458×10^{8} meters second^{-1}
AU in light time	499.00478384 seconds
Gravitational constant, G	6.67259×10^{-11} meters^3 kg^{-1} sec^{-2}
Mass of the Sun	1.9891×10^{30} kilograms
Mass of the Earth	5.9742×10^{24} kilograms
Solar constant	1368 Watts meters^{-2}
Sun-Jupiter mass ratio	1047.3486
Earth-Moon mass ratio	81.30059
Equatorial radius of the Earth	6378.136 kilometers
Obliquity of the ecliptic (J2000)	$23^0\ 26'\ 21.412''$
Earth sidereal day	23 hours 56 minutes 4.09054 seconds
Sidereal year	365.25636 days
Semimajor axis of the Earth's orbit	1.00000003 AU
Parsec, pc	206,264.806 AU
Age of the solar system	4.55×10^{9} years
Age of the galaxy	$10-15 \times 10^{9}$ years

PHYSICAL AND ORBITAL PROPERTIES OF THE SUN AND PLANETS

Name	Mass kg	Equatorial Radius km	Density g cm^{-3}	Rotation Period	Obliquity degrees	Escape Velocity km sec^{-1}	Semimajor Axis AU	Eccentricity	Inclination degrees	Period years
Sun	1.989×10^{30}	696,000	1.41	24.65–34 d.	7.25*	617.7				
Mercury	3.302×10^{23}	2,439	5.43	58.646 d.	0.	4.43	0.38710	0.205631	7.0048	0.2408
Venus	4.868×10^{24}	6,051	5.20	243.018 d.	177.33	10.36	0.72333	0.006773	3.3947	0.6152
Earth	5.974×10^{24}	6,378	5.52	23.934 h.	23.45	11.19	1.00000	0.016710	0.0000	1.0000
Mars	6.418×10^{23}	3,396	3.93	24.623 h.	25.19	5.03	1.52366	0.093412	1.8506	1.8807
Jupiter	1.899×10^{27}	71,492	1.33	9.925 h.	3.08	59.54	5.20336	0.048393	1.3053	11.856
Saturn	5.685×10^{26}	60,268	0.69	10.656 h.	26.73	35.49	9.53707	0.054151	2.4845	29.424
Uranus	8.683×10^{25}	25,559	1.32	17.24 h.	97.92	21.33	19.1913	0.047168	0.7699	83.747
Neptune	1.024×10^{26}	24,764	1.64	16.11 h.	28.80	23.61	30.0690	0.008586	1.7692	163.723
Pluto	1.32×10^{22}	1,170	2.0	6.387 d.	119.6	1.25	39.4817	0.248808	17.1417	248.02

* Solar obliquity relative to the ecliptic plane
Orbital data for January 1, 2000

PHYSICAL AND ORBITAL PROPERTIES OF THE SATELLITES

Name	Semimajor Axis 10³ km	Orbital Eccentricity	Orbital Inclination degrees	Orbital Period days	Mean Radius km	Mass (10²³ g)	Density g cm⁻³	Year of Discovery	Discovered By
Earth									
Moon	384.40	0.0549	18–29	27.3216	1,737.4	734.9	3.34	—	—
Mars									
Phobos	9.38	0.0151	1.08	0.319	$13 \times 11 \times 9.2$	0.000106	~2.	1877	Hall
Deimos	23.46	0.0003	1.79	1.262	$7.5 \times 6.1 \times 5.2$	0.000024	~2.	1877	Hall
Jupiter									
J16 Metis	128.0	0.0	0.0	0.295	20	—	—	1979	*Voyager 1/2*
J15 Adrastea	129.0	0.0	0.0	0.298	10	—	—	1979	*Voyager 1/2*
J5 Amalthea	181.3	0.003	0.45	0.498	$131 \times 73 \times 67$	0.15	—	1892	Barnard
J14 Thebe	221.9	0.015	0.8	0.674	50	—	—	1979	*Voyager 1/2*
J1 Io	421.6	0.004	0.04	1.769	1,818	893.2	3.55	1610	Galileo
J2 Europa	670.9	0.010	0.47	3.552	1,560	480.0	3.02	1610	Galileo
J3 Ganymede	1,070	0.002	0.21	7.154	2,634	1481.9	1.94	1610	Galileo
J4 Callisto	1,883	0.007	0.51	16.69	2,409	1075.9	1.84	1610	Galileo
J13 Leda	11,094	0.148	26.70	238.7	5	—	—	1974	Kowal
J6 Himalia	11,480	0.163	27.63	250.6	85	—	—	1904	Perrine
J10 Lysithea	11,720	0.107	29.02	259.2	12	—	—	1938	Nicholson
J7 Elara	11,737	0.207	24.77	259.6	40	—	—	1904	Perrine
J12 Ananke	21,200	0.169	147.	631	10	—	—	1951	Nicholson
J11 Carme	22,600	0.207	163.	692	15	—	—	1938	Nicholson
J8 Pasiphae	23,500	0.378	145.	735	18	—	—	1908	Melotte
J9 Sinope	23,700	0.275	153.	758	14	—	—	1914	Nicholson
Saturn									
S18 Pan	133.6	0.0	0.0	0.575	10	—	—	1980	*Voyager 1*
S15 Atlas	137.6	0.0	0.0	0.602	$19 \times 17 \times 14$	—	—	1980	*Voyager 1*
S16 Prometheus	139.3	0.002	0.0	0.613	$74 \times 50 \times 34$	—	—	1980	*Voyager 1*
S17 Pandora	141.7	0.004	0.05	0.629	$55 \times 44 \times 31$	—	—	1980	*Voyager 1*
S11 Epimetheus	151.4	0.009	0.14	0.695	$69 \times 55 \times 55$	—	—	1979	*Pioneer 11*
S10 Janus	151.5	0.007	0.34	0.695	$97 \times 95 \times 77$	—	—	1966	Dollfus
S1 Mimas	185.5	0.020	1.53	0.942	199	0.37	1.1	1789	Herschel
S2 Enceladus	238.0	0.004	0.0	1.370	249	0.73	1.1	1789	Herschel
S3 Tethys	294.7	0.000	1.0	1.888	530	6.3	1.0	1684	Cassini
S14 Calypso	294.7	0.0	1.10	1.888	$15 \times 8 \times 8$	—	—	1980	Pascu
S13 Telesto	294.7	0.0	1.0	1.888	$15 \times 12 \times 8$	—	—	1980	Reitsema
S4 Dione	377.4	0.002	0.02	2.737	560	11.0	1.5	1684	Cassini
S12 Helene	377.4	0.005	0.15	2.737	16	—	—	1980	Lecacheux
S5 Rhea	527.0	0.001	0.35	4.518	764	23.1	1.2	1672	Cassini
S6 Titan	1,222	0.029	0.33	15.945	2,575	1346.	1.88	1655	Huygens

S7 Hyperion	1,481	0.104	0.4	21,277	180 × 140 × 112	15.9	1.0	1848	Bond
S8 Iapetus	3,561	0.028	14.72	79.330	718	18.8	1.21	1671	Cassini
S9 Phoebe	12,952	0.163	150.	550.48	110	0.1	—	1898	Pickering
Uranus									
U6 Cordelia	49.75	0.000	0.14	0.335	13	—	—	1986	*Voyager 2*
U7 Ophelia	53.76	0.010	0.09	0.376	15	—	—	1986	*Voyager 2*
U8 Bianca	59.16	0.001	0.16	0.435	21	—	—	1986	*Voyager 2*
U9 Cressida	61.78	0.000	0.04	0.464	31	—	—	1986	*Voyager 2*
U10 Desdemona	62.66	0.000	0.16	0.474	27	—	—	1986	*Voyager 2*
U11 Juliet	64.36	0.001	0.06	0.493	42	—	—	1986	*Voyager 2*
U12 Portia	66.10	0.000	0.09	0.513	54	—	—	1986	*Voyager 2*
U13 Rosalind	69.93	0.000	0.28	0.558	27	—	—	1986	*Voyager 2*
U14 Belinda	75.26	0.000	0.03	0.624	33	—	—	1986	*Voyager 2*
U15 Puck	86.00	0.000	0.31	0.762	77	—	—	1985	*Voyager 2*
U5 Miranda	129.8	0.003	3.40	1.413	236	0.61	1.1	1948	Kuiper
U1 Ariel	191.2	0.003	0.0	2.520	579	13.5	1.7	1851	Lassell
U2 Umbriel	266.0	0.005	0.0	4.144	585	11.7	1.4	1851	Lassell
U3 Titania	435.8	0.002	0.0	8.706	789	35.3	1.7	1787	Herschel
U4 Oberon	582.6	0.001	0.0	13.46	761	30.1	1.6	1787	Herschel
S/1997 U1	7169.	0.082	140.	579.	30	—	—	1997	Nicholson et al.
S/1997 U2	12214.	0.509	153.	1289.	60	—	—	1997	Nicholson et al.
Neptune									
N3 Naiad	48.23	0.000	0.0	0.294	29	—	—	1989	*Voyager 2*
N4 Thalassa	50.08	0.000	4.5	0.311	40	—	—	1989	*Voyager 2*
N5 Despina	52.53	0.000	0.0	0.335	74	—	—	1989	*Voyager 2*
N6 Galatea	61.95	0.000	0.0	0.429	79	—	—	1989	*Voyager 2*
N7 Larissa	73.55	0.000	0.0	0.555	104 × 89	—	—	1989	*Voyager 2*
N8 Proteus	117.6	0.000	0.0	1.122	208	—	—	1989	*Voyager 2*
N1 Triton	354.8	0.000	157.	5.877	1,353	214.0	2.06	1846	Lassell
N2 Nereid	5,513.	0.751	29.	360.14	170	—	—	1949	Kuiper
Pluto									
P1 Charon	19.40	0.0076	96.16	6.387	593	16.	1.8	1978	Christy

INDEX

· ·

A

Absorption bands
    infrared, carbonate in Mars atmospheric dust, 282
    near-Earth asteroids, 624–625
    Titan, 380
Absorption features, Saturnian satellites, 708–709
Abundances
    carbon
        Comet P/Halley, 534
        Jovian planet atmosphere, 351
    CO_2, Titan atmosphere, 386–387
    elemental, comets, 533–534
    helium
        within high-speed solar wind streams, 100
        Jupiter and Saturn atmospheres, 350–351
        in solar wind, 120
    iron
        Earth and Moon, 271
        Mercury, 141, 144
    potassium and uranium, terrestrial planets, 61
    solar, Jovian atmospheres, 344
    sulfur, Mercury core, 128–129
Acceleration
    Coriolis, 206
    seed particles, 120–121
    slowest coronal mass ejections, 109–110
Accretion, see also Runaway accretion
    cross section for, in planetesimals, 26
    gas, outer planets, 50–51, 61
    planetary, and debris clearing from planetary zones, 27–28
    planets formed by, 944–945
    Pluto and Charon, 579
    selective, model of Mercury origin, 141, 144
    terrestrial planets, 198–199, 273–275
    vertical crustal, Venus, 188
Accretional remnants, Mercurian craters, 130
Accretion disk
    in solar nebula, 26–27
    stellar, 583

Accretion zone, planetesimal, 47
Accuracy, instruments for searching for planetary systems, 950–951
Acetylene, giant planet stratospheres, photochemical formation, 325–326
Achondrites
    graphite–diamond mixtures, 645
    petrographic properties, 646
    refractory lithophile-enriched, 644, 661–662
Acid rain, global, following K/T impact, 874–875
Active Cavity Radiometer Irradiance Monitor, 91–92
Active regions, solar atmosphere, 81
    occurrence of solar flares, 87–88
Adiabat, Jovian planet interior temperature distribution, 343, 347
Aerosols
    giant planet atmospheres, 321–328
    layer enshrouding Titan, 381, 383
    Pluto atmospheric layer, 512
Age
    cratering, 433 Eros, 626
    mare basalt, 264–265
    Mars, 56–57
    meteorites
        gas retention, 665–667
        solidification, 667–669
        terrestrial and cosmic ray exposure, 664–665
    solar system, 54–57
    Sun, 4
    Venus surface, and implications, 178–179
Albedo, see also Bond albedo
    asteroidal, 727–729, 797
    correlation with thermal emission on Io, 364
    dark visual, Uranus satellites, 451
    Deimos, 310–311
    Galilean satellites, 705–707
    geometric
        near-Earth asteroids, 622–623
        in near-infrared, Titan, 391–392
        satellites, 441
        Triton, 413–414
    Iapetus, hemispheric asymmetry, 20, 398–399, 707–708
    Mercury, 124, 135–137

Moon, 251
Nereid, 454
Pluto, 505, 507
    radar measurements, 782–784
Saturnian satellites, 707–709
and temperature, correlation on Jupiter, 322–323
Triton surface, 429
Alfvén speed, 118, 479, 494
Alfvén waves
    detected in wind near Earth, 85
    heating of upper chromosphere, 77
887 Alinda, spectral error bar, 625
Alpha Centauri, double-star system, 30
Alpha particles, solar wind, 97–98, 119–120
Altimetry
    radar, Pioneer Venus, 163, 165
    Venus, Magellan data, 167–169
Aluminum
    ^{26}Al
        nucleosynthetic event, 56
        spallation-produced, 58
    magma ocean enriched in, 270–271
Amino acids
    extracted from Murchison CM chondrite, 650
    symmetries, 901
Ammonia
    condensing to clouds, 340
    in early Mars atmosphere, 287
    giant planet atmospheres, 320
    ice, giant planet clouds, 324, 327–328
    and water, viscous lava composition, 418–419
Ammonium hydrosulfide
    atmospheric
        giant planets, 320–321
        Jupiter, 324–325
    cloud formation, 323
Ancestor, common, life on Earth, 908
Angular momentum
    conservation during contraction, 39
    global atmospheric, 196
    lost by gas inside planet orbit, 51
    Moon, 252
    transport in nebula, 62
Angular scattering law, 787–789
Anhydrous minerals line, carbonaceous chondrites, 645, 660

Antarctica
  chondrites, thermal metamorphism, 663
  meteorite fragments, 637, 656
  SSI imaging by *Galileo,* 194
Anticyclonic spots, Jupiter, 327, 332–333
Aphelion distribution, new comets,
      549–551
Apollo 11, lunar landing, 923
Apollo program, Moon missions, 928
Archaea, main group of life on Earth, 901
Architecture, solar system, 4–26
Arecibo Observatory, Puerto Rico, 162,
      777–780, 790, 792, 805–806,
      947–948
Argon
  ^{37}Ar, conversion of ^{37}Cl to, 69
  ^{40}Ar, calculation of gas retention age
        from, 665–667
  escaping from lunar surface rocks, 19
  radiogenic, in meteorites, 656–658
  in Titan atmosphere, observation
        attempts, 385–386
Ariel, regoliths and cratered terrain,
      451–452
Asteroid belt, *see also* Main Asteroid Belt
  compositional gradient, 13
  Kirkwood gaps, 590, 817–818
  localization of dust bands, 720–722
  theory of origin, 38–39
Asteroid collisions
  dust production from, 721–722
  in Main Belt, 612, 614
  providing material for zodiacal cloud,
        21, 721–722
Asteroids
  Amors, 550–551
    approaching Earth orbit, 608
  Apollos, 550–551
    Earth-crossing orbit, 608
  Aten–Apollo–Amor objects, 196, 591
  Atens, orbit inside Earth orbit, 608
  composition, 597–601
    telescopic observations, 594–595
  density and porosity and rotation rates,
        601–602
  description, 586
  distinction from comets, 587
  dust bands, 719–721
  exploration missions, 934–935
  and extinction, 603
  families and satellites, 591–592
  interactions with solar wind, 482
  IRAS observations, 726–729
  large, Doppler mapping, 791–792
  main belt
    echoes, 796–802
    radar-detected, 776–777
  naming system, 3–4, 587–588
  near-Earth, *see* Near-Earth asteroids
  observations at ultraviolet wavelengths,
        710–711
  orbital evolution, 837–840

orbit–orbit resonance with Jupiter,
      828–829
  orbits and zones, 590–591
  parent body, target for life search, 919
  planet-crossing, 7, 611
  at radio wavelengths, 753–754
  relationship to
    comets, 594
    meteorites, 592–593
      and meteorite delivery, 593–594
    resource potential, 604–605
    sizes and shapes, 588–589
    spacecraft missions to, 604
    surface, 602–603
    taxonomy and distribution of classes,
          595–597
  thermally processed, 13
Asthenosphere, Earth upper mantle, 239
Astronomy
  planetary radar
    measurements and target properties,
          782–806
    prospects for, 806
    scientific context and history,
          774–775
    techniques and instrumentation,
          775–782
  Triton, pre-*Voyager,* 408–412
  ultraviolet, history, 697–698
Aten Patera, plume deposits on Io, 362
Atlas, shepherd satellite of Saturn, 451
Atmosphere
  Earth, *see* Earth atmosphere
  gas-giant planets, *see* Jovian planet at-
        mospheres
  Io, *see* Io atmosphere
  Mars, *see* Mars atmosphere
  Mercury, 126
  Moon
    elemental composition, 251
    transient, 17
  planetary
    observations, 699–704
    vertical structure, 201
  Pluto, 501, 510–513
  Saturn
    bright bands, 750–751
    helium depletion, 319
  Sun, *see* Solar atmosphere
  terrestrial planets, 9–10
  Triton, 423–429
    thermospheric temperature, 417
  Venus, *see* Venus atmosphere
Auroral emissions
  generation, 489
  Jovian, 336–337
Auroras
  giant planets, 335–337, 702–703
  polar, Ganymede, 445
  signature of magnetospheric dynamics,
        491–493
Australia, SSI imaging by *Galileo,* 194

B

Babcock's empirical model, sunspot cycle,
      86
Bacteria
  main group of life on Earth, 901
  source of methane and N_2O, 194
Ballerina skirt model, heliospheric current
      sheet, 102
Barnard's star, astrometric observations,
      946
Barred spirals, description, 30
Barringer Crater, breccia, 197
Basalt
  KREEP, as impact melt, 269
  lunar maria
    age and composition, 264–265
    origin, 265–266
    thickness, 258–259
  Mercury surface, 137–138
  SNC meteorites, 292–293
Basins, *see also specific basins*
  drainage
    inputs from major storms, 228
    topography, 218–219
  impact, *see* Impact basins
  Mercury, 129–132
  multiring, Moon, 260–261
  oceanic
    opening and closing, 224
    submarine, 216
    transition from crater to, 847
Beta–Atla–Themis region, Venus, 171–
      172, 174
Binary system, *see also* Double planet
      system
  Pluto–Charon, origin, 514–515
Bingham plastic, lava, 886
Biogenic material, in Martian meteorites,
      3
Biogenic mats, stromatolites on Earth,
      906–907
Biological weathering, rock, 232
Biosphere, evolution, 874–876
Bipoles, solar atmospheric, 78–79
  active regions, 81
Bistatic experiments, planetary radar sys-
      tems, 778
Blackbody
  photospheric continuum, 75–76
  radiation, definition, 736
Black dwarf, observed through micro-lens-
      ing events, 33
Blocking highs, Earth atmosphere, 202
Bode's law, describing orbital regular spac-
      ing, 6–7
Bombardment, *see also* Late Heavy Bom-
      bardment
  extended, of idealized surfaces, com-
        puter simulations, 869–870
  meteoritic
    icy bodies, 440
    Venus, 182

meteoroid, into rings, 467–468
regolithic material, by solar particles, 633–634
Bond albedo
calculation, 711–712
near-Earth asteroids, 622–623
satellites, 441
Bouguer anomaly, lunar gravity, 253–254
Boundary conditions, restrictive, on terrestrial planets, 205
Bow shock
beyond heliopause, 25
in interstellar plasma, 115
magnetospheric, 208, 496
Brachinites, meteorites, analog to A class asteroids, 599
Breccia
Barringer Crater, 197
chondrites
lithified, 647
regolith, 649, 653, 662
lens, as result of impact crater collapse, 858
lunar highlands, 269–270
meteoritic, ALH 85151, 657
Bremsstrahlung, X rays, 89–90
Brightness
asteroid, converting to size ranges, 618–619
aurora, 491
comet nucleus, 538
coronal streamer, 82
magnetic region types, 92
Mercury, 742–743
terrains, 124
near-Earth asteroids, 620
Nereid, 453–454
oscillations, in photosphere, 75
Pluto, 502
Titan, hemispheric asymmetry, 389, 393–395
umbra, 80–81
zodiacal light, 680–681
Brightness temperature
giant planets, 748–749, 751–752
Mars, 745, 748
Broken power laws, Kuiper Belt, 577–578
Brown dwarf
definition, 943–944
objects identified as, 952
Brownlee particles, see Interplanetary dust particles
Bulge
central, Milky Way, 28
triaxial shape, 30
tidal, outer planet icy satellites, 438

C

Calcium, magma ocean enriched in, 270–271

Calderas
Earth, 418
formation, 879–880
Io, 360–362
Mars, 896
Callisto
albedo, 705–707
impact records, 20
pit craters, 852–853
remnant structures of impacts, 446
Caloris Basin, Mercury
exospheric enhancement coinciding with, 126
floor structure, 132
Canyon Diablo
iron asteroid, 605
meteorite fragments, 653–654
Canyons, Mars, 301–303
Capture
Halley-type comets, into inner solar system, 532
Moon, from independent orbit, 272
Phobos and Deimos, 313–314
Triton
by collision, 431–432
tidal heating after, 406–407, 432–434
Carbon
abundance
Comet P/Halley, 534
Jovian planet atmosphere, 61, 351
C IV, 75, 81
isotope ratios, in early Earth sediments, 907
Carbonaceous chondrites
anhydrous minerals line, 645
asteroid analogs, 597–600
CAI
igneous inclusions, 55
isotope abundance ratios, 58
isotopic anomalies, 642, 660–661
meteoritic inclusions, 62
CI
chemical composition, 54–55
unequilibrated material, 59
composition, 649–650
organic constituents, 650–652
volatile-rich, 14–15
Carbonate
infrared absorption bands, Mars atmospheric dust, 282
Mars atmospheric CO_2 removed as, 286, 289
Carbon dioxide
Mars
greenhouse effect, 285–287, 304, 920–921
augmentation mechanisms, 287
outgassing, 921
volatile inventory, 278–283
terrestrial planets, microwave opacity, 739
Titan atmosphere, abundance, 386–387
Venus atmosphere, major gas, 150–151

Carbon monoxide
chemical recombination in Venus atmosphere, 154–155
ice, Pluto surface, 506
Neptune
atmosphere, 752–753
stratosphere, 321
Titan atmosphere, 386–387
tracer of outer solar nebula chemistry, 416
volatile, 412
Cassini spacecraft
dust analyzer, 694–695
future planetary missions, 938
outer planet missions, 934
satellites, 443
Saturn's rings, 471–472
Titan mission, 401–404
4769 Castalia, delay-Doppler images, 799–800
Cataclysm, lunar, 261
Catastrophic theories, solar system origin, 39
Cavity
heliospheric, solar wind confined within, 479
solar wind-blown, in interstellar plasma, 115
transient, in cratering process, 856–859, 861
Cavity radiometers, aboard NASA satellites, 75
Celestial mechanics
application of chaos to, 5–6
Henri Poincaré, 826
Centaurs
brightest of ecliptic comets, 560
colors, 566–568
found between Saturn and Uranus, 590
maximum inclination, 564
stray bodies in outer planetary region, 554
Centrifugal force, on unit mass of gas, 42
Ceres
quasi-specular scattering, 788–789
surface composition, 13
Chalcophiles, sulfide-enriched, 638–639
Chaos
application to celestial mechanics, 5–6
concepts of, 826–827
defined, 810, 814
large-scale, Kuiper Belt inner edge, 553
and planetary motions, 823–824
three-body problem as paradigm, 827–833
Chaotic behavior, planets, 841–843
Chaotic orbits, and three-body problem, 830–832
Chaotic region, small asteroids injected into, 614
Chaotic rotation, spin–orbit resonance, 833–835

Charge-coupled device camera, near-Earth asteroid tracking, 617–619
Charged particles
  continuous outflow from Sun, 96–97
  energetic
    accelerated near termination shock, 480
    impacts on planet surfaces, 479
  escape mechanisms, 204
  interaction with interplanetary magnetic field, 691–693
Charon
  discovery, 501–502
  IRAS observations, 729–730
  orbit, 503–504
  as plasma source, 484
  and Pluto
    mutual events, 504–505
    origin of binary, 514–515
    tidally evolved system, 21
  spatially resolvable images, 710
  surface composition, constraints, 513
  water frost covering, 12
Chassignites, *see* SNC meteorites
Chemical composition
  atmosphere
    giant planets, 317–321
    Titan, 385–387
  chondrites and eucrites, 643
  CI carbonaceous chondrites, 54–55
  IDPs, 679
  meteorites, 656–663
Chemical recombination, in Venus atmosphere, 154–155
Chemical species, detected in comets, 528–530
Chemical weathering, rock, 231
Chicxulub, Yucatan
  evidence of extinction mechanisms, 875–876
  impact basin, 198
  impact site, 635
Chiral compounds, carbonaceous chondrites, 650
2060 Chiron, first Centaur, 568–570
Chlorine
  ^{37}Cl, conversion to ^{37}Ar, 69
  single atom, effect on Venus atmosphere, 150
Chlorofluorocarbons, phase-out, 203
Chondrites
  breccias, 647, 649
  carbonaceous, *see* Carbonaceous chondrites
  chemical-petrologic classification, 646–647
  enstatite, 642, 663
  meteorite stones, 638
  ordinary
    asteroidal source, 600
    evidence of nebular environment, 59
    unequilibrated, 646, 662
  origin, 642

  petrographic properties, 646
  shocked, 652–654
Chondrules
  found in chondritic meteorites, 59, 642
  origin unknown, 61
Chromophore, candidate materials, outer planet atmospheres, 324–325
Chromosphere
  optical spectrum, 75–76
  spicules, 81
Chronologies, planetary, craters as probes, 868–870
Chronometry, meteorite, 663–670
Circulation
  stratospheric, giant planets, 331–332
  Titan atmosphere, 388
  Venus atmosphere, 158
Circumstellar disk
  direct imaging by *Hubble,* 39–40
  in formation of solar system, 38
  T Tauri stars, 52–53
Classes
  asteroids, distribution, 595–597
  Venetian structures: coronae, 177
Classification
  chondrites, chemical-petrologic, 646–647
  eruptive processes, 885–886
  meteorites, 637–639
  S class asteroids, Gaffey scheme, 600
Clasts, black, in regolith breccias, 662
Clementine spacecraft, Moon missions, 928
Climate
  change on Mars
    early epochal, 283–285
    long-term, 285–287
    ongoing periodic, 287–288
  evolution on Titan, 400
  and polar cap, Triton, 429–430
Close encounters, Sun and other stars, 30–31
Cloud core
  collapse
    gravitational, 26
    timescale, 42
  density increase, 38
Clouds
  ammonia and methane condensing to, 340
  ammonium hydrosulfide, formation, 323
  convecting eruption
    collapse, 880
    formation, 892, 894
  giant molecular, comet encounters with, 547–548
  giant planet atmospheres, 321–328
  interstellar
    star formation in, 36–38
    Sun and planets formed from, 3
  Venus atmosphere
    appearance and motions, 155
    early measurements, 149

  layers and chemistry, 155–156
  lightning, 157
  water–ammonia, Jupiter, 326
Clusterings, anomalous, aphelion points of young comets, 549–550
C/1995 O1, *see* Hale–Bopp comet
Cold-trapping, sulfur, at Mercury poles, 127
Collapse
  cloud core, 26, 38, 42
  convecting eruption cloud, 880
  gravitational, in creation of protoplanets, 439
  impact crater
    breccia lens as result, 858
    wall and rim, 846–847
  isothermal, in star formation, 944
  protostellar, 41–43
Collisions
  asteroid, *see* Asteroid collisions
  continental plates, 223
  destructive, iron meteorites, 665–666
  between dust particles, 675, 691
  hypervelocity, in planet-building stage, 60–61
  inelastic
    among ring particles, 468–470
    Pluto–Charon binary originating from, 515
  in Kuiper Belt, 579
  near-Earth asteroids with Earth, 615–616
  planetesimals, cross section, 46–47
  plasmas dominated by, 488–489
  protoplanets, violent inelastic, 48
  in solar wind plasma, 116–117
  Triton capture by, 431–432
Color
  Centaurs and Kuiper Belt objects, 566–568
  near-Earth asteroids, 623–624
  Pluto surface, 505
  Triton surface, 409–410
    change, 415–416
Cometary coma
  detection of parent molecules in, 756
  Halley's comet, dust from, 59
  outflow models, 526
Cometary nucleus
  compositional differences, 530–531
  density and mass, 524–525
  extinct, near-Earth asteroids made of, 612, 614
  frozen gas composition, 520
  as frozen mudballs, 725–726
  gas and dust overflow, 522–523
  Halley's comet, 15, 523–524
  surface
    dust particles left on, 550–551
    sandblasted, 525
  VEGA mission, 935
Cometesimals, unaccreted, 554
Comets, *see also* Scattered disk

causing solar wind field to drape, 496
contribution to zodiacal cloud, 21, 722–723
distinction from asteroids, 587
dust tails, grains in, 527
ecliptic, 580–581
exploration missions, 934–935
formation, 51–52
impacts, hazards, 535–536
interactions with solar wind, 482–483
Jupiter family, *see* Jupiter family comets
long-period, *see* Long-period comets
meteor streams derived from, 677
molecular and elemental abundances, 527–534
most primitive bodies in solar system, 15
naming system, 4, 522
new, injection in Oort Cloud, 544–545
temporal variations, 548–550
observations at ultraviolet wavelengths, 710–711
orbital evolution, 840
orbits
planet-crossing, 7
statistics, 538–540
origin
dynamical aspects, 554–555
theory, 38–39
plasma tails
CO$^+$, 532
description, 526–527
manifestation of solar wind, 25
1950s observations, 97
radar-detected, 776–777
radar observations, 804–805
at radio wavelengths, 755–756
relationship to
asteroids, 594
near-Earth asteroids, 612–613
short-period
computer simulations of orbits, 613
Kuiper Belt, 51–52
sources of meteor showers, 534–535
trails, *IRAS* information, 723–726
water-containing, 919–920
Comet showers, 548–550
Commensurability, nearly exact orbital, 815
Companions
asteroids, 592
to PSR 1257+12, 948–949
substellar-mass, 947, 949–953
Compression regions
bounded by shock waves, 106
followed by rarefaction regions, 115
Compressive stress, Mercurian lithosphere, 140
Computer simulations
extended bombardment of idealized surfaces, 869–870
planetary spacing, 6–7
short-period comet orbits, 613

Condensates, in Titan stratosphere, 389–390
Condensation
gas, from Io plumes, 372
Triton atmosphere onto surface, 423–424
Condensation nuclei, for comet gases, dust particles as, 532
Contact binary, near-Earth asteroid shape, 621
Continental margins, Atlantic style, 216
Continental plates
collisions, 223
and oceanic plates, tectonics, 213–215
stable interior cratons, 215
Continents
geomorphology, and plate tectonics, 217–218
surficial, Titan, 393–395
tear in, Gulf of California resulting from, 224
Convection
continued, in liquid outer core, 484
Jupiter interior
energy transport by, 341, 343
and magnetic field generation, 350–351
mantle, Earth, 199–201
Convection zone, solar, 66–67, 71–72
Convective cells
Jupiter interior, 341
solar, spatial scales, 78
Co-orbital satellites, in Saturnian system, 451
Core
cloud, *see* Cloud core
convection continuance, 484
Earth
inner
evolution, 241
solid, 233
outer
magnetic dynamo, 240–241
P-wave speed, 233
giant planet
gravitational harmonics, 349–350
growth, 27
inner
Mercury, 128, 139–140
Oort Cloud, 555–556
Io, 364
Mars, 307–308
Moon, 260
Sun, hydrogen fusion at, 32–33
Core–mantle boundary
Earth, 196, 200, 233, 244
S asteroids representing, 600
Corona
class of structure on Venus, 177
consisting of dark matter, 28–29
heating
by magnetohydrodynamic waves, 77
models, 98

Miranda, 452
rotation, 77–78
streamers, 82–83
structure
and magnetic field, 99–100
relationship to solar wind stream structure, 101–102
Coronal holes
energy loss, 84–85
place of solar wind expansion, 99, 101–102
plasma expansion, 67
Coronal mass ejections
characteristics, 109
in solar wind at 1 AU, 111–113
as disruption of helmet streamer, 91
fast
interplanetary disturbances driven by, 110–111
and nonrecurrent geomagnetic storms, 114
frequency of occurrence, 110
Corotating interaction regions, solar wind stream, 106–109
Cosmic ray exposure, meteorites, 634
ages, 664–665
Coulomb interactions, solar wind, 117
Cratering
impact
Mars, 297–298
Venus, 177–178
lunar, history, 261
mechanics, 855–861
outer planet icy satellites, 440
terrestrial planets, 9
Craters, *see also* Microcraters
Callisto and Ganymede, 445–446
Ganymede, 849
anomalous, 852–853
impact, *see* Impact craters
Mimas, 447
produced by space debris particles, 684
rayed
description, 849
Mercury, 135–137
simple
and complex, 197–198
final volume ejecta, 847–849
Titan surface, density, 394
Cratons, continental plate stable interior, 215
Cretaceous–Tertiary boundary, 198, 232, 603, 615, 635, 846, 874–876
Crust, *see also* Lunar highland crust
early evolution on Moon, 872–873
Earth
continental and oceanic, 238–239
shock waves through, 197–198
fusion, meteorite, 630, 647
Mars, 307–308
thin ice, on Europa, 20

Crust (continued)
  Venus
    and interior compositional structure, 184–185
    thickness variations, 168
    vertical accretion, 188
Cryovolcanic activity, Titan, 394–395
Crystallization
  magma ocean on Moon, 266–271
  melt, siderophile fractionation during, 639
Cycles
  seasonal, on Triton, 411
  sunspot, 85–87
  volcanic activity on Earth, 418

D

Dark material, coating one side of Iapetus, 449
Dark matter, description, 28
Data
  IRAS, reexamined, 731–732
  Neptune and Uranus interiors, from Voyager 2, 351
Day length, Earth, 195–196
Debris clearing
  cometary, and Late Heavy Bombardment, 17
  in formation of solar system, 38
  from orbital zones, after planet formation, 3
  from planetary zones, late in accretion of planets, 27–28
Debris flows, Mars, 305
Decametric emission, Jupiter, 760–761
Decay, zonal wind, with height, 331
Deceleration
  entry, Huygens probe, 403
  interplanetary dust particle, 690
  meteoroid, 676–677
Deep Space spacecraft, future planetary missions, 937–938
Deformation
  planar, minerals, 863–864
  tessera on Venus, 176–177
Deimos
  composition, 311–312
  exploration, 309
  orbit, and gravitational environment, 309–310
  origin and history, 313–314
  surface bulk density, 799
Density
  asteroid, in relation to porosity and rotation rate, 601–602
  average, Pluto and Charon, 504–505
  cloud core, increase in, 38
  cometary nucleus, 524
  discontinuities, lacking in Jovian interior, 330

function of temperature, 241
Moon, 252
Phobos and Deimos, 312
photosphere, and radial temperature, 76
Pluto, 508–509
spatial, interplanetary dust, 681–682, 686
variation with radius, for Jupiter interior model, 350
Deposition, and erosion, on Mars, 303–306
Depressions, collapse-produced, on Venus, 883
Deuterium
  giant planet atmospheres, 319
  Titan atmosphere, 386
D/H ratio
  Comet P/Halley, 534
  original planetary water, 205
Diacetylene, giant planet stratospheres, photochemical formation, 325–326
Diamond
  Canyon Diablo meteorites containing, 654
  mixture with graphite, in achondrites, 645
Diapirism, Triton surface, 422
Differentiated body, geochemical evolution, 600–601
Diffusion equation, comets, 542
Dike system, exposed on ancient Earth rocks, 897
Diogenites, see Meteorites, HED association
Dione
  heavily cratered, 449
  ultraviolet albedos, 707–709
Direct detection, in discovery of planetary systems, 952–953
Dirty snowball, cometary nucleus model, 725–726
Discontinuities
  density, Jovian interior lacking, 330
  Earth transition zone, 239–240
Discovery
  asteroids, 587–588
  Charon, 501–502
  Kuiper Belt, 559
  magnetic field, Mars Global Surveyor, 932–933
  near-Earth asteroids, Spacewatch program, 617–619
  outer planet satellites, 436–438
  planetary systems, direct detection in, 952–953
  Pluto, 500
  solar wind, 96–98
  Titan, 379–380
  Triton, 408
Discovery program, future planetary missions, 937
Dissipative forces, and orbits of small bodies, 819–823

Distortion, surface of constant total potential, 342
Distribution function, 117–118
Disturbance
  orbital, from planetary impacts, 845–846
  transient, and coronal mass ejections, 109–114
DNA, nucleotide bases, 901
Doppler frequency, radar echo, 780
  dispersion of echo power in, 785–787
  dynamical properties from, 784–785
Doppler Wind Experiment, Titan atmosphere, 403–404
Double planet system
  hypothesis of lunar origin, 272–273
  Pluto–Charon, 504–505
Drag
  gas
    particle effects, 46
    small bodies, 821–822
  Poynting–Robertson, 820–821
Drake equation, extraterrestrial intelligence, 920
Draping
  field line, about fast coronal mass ejections, 113–114
  solar wind field, caused by comets, 496
Driving products, near-Earth asteroids representing, 627
Dust
  atmospheric
    following K/T impact, 874–876
    Mars, carbonate in, 282
  bands, asteroid, 719–721
  comae, cometary, evidence for, 613
  flux, inner solar system, 685–688
  interplanetary, see Interplanetary dust
  lithified, regolith breccias, 649
Dust cloud, zodiacal, see Zodiacal cloud
Dust devils, Tritonian, 425, 427–428
Dust grains
  asteroid collision-produced, 7–8
  charging, and interaction with interplanetary magnetic field, 691–693
  formation of planetesimals, 44–46
  infalling and settling, 26
  Poynting–Robertson drag, 820–821
Dust particles
  appearing as meteors upon burning up, 21
  condensation nuclei for comet gases, 532
  eddy transport effects, 332
  interplanetary, see Interplanetary dust particles
  Io as source, 358
  left on comet nucleus surface, 550–551
Dynamical history, near-Earth asteroids, 614–615
Dynamics
  interplanetary dust and zodiacal cloud, 688–693

orbital, in relation to Kuiper Belt,
    560–564
planetary magnetosphere, 490–494
solar system, 809–824
Dynamo, internal magnetic, planetary,
    484–485, 496
Dynamo models, sunspot cycle, 86–87

E

Earth
    auroras, 335
    based observations of ring systems,
        460–461
    collision with protoplanet, 15–16
    cometary impact hazards, 535–536
    day length, 195–196
    debris orbiting, 731
    dust environment near, and lunar micro-
        craters, 682–684
    Earth–Moon line, in relation to oxygen
        isotopics, 644–645
    erosive action of water, 292
    explosive eruptions, 888–896
    fossil record of liquid environment,
        900–901
    geomorphology, role in planetary explo-
        ration, 210–212
    grazing impact late in accretion,
        273–274
    life
        history, 905–908
        origin, 908–909
            comet role, 522
    magnetosphere, 207–208
        hybrid nature, 496
        solar wind transferred into, 108–109
    meteorites landing on, 13–14
        after removal from parent body,
            631–634
        SNC, from Mars, 192
        terrestrial ages, 664
    and Moon
        double planet system hypothesis,
            272–273
        tidal interaction, 822
    near-Earth asteroids, proximity, 609
    near-Earth objects, 196
    nonthermal radiation, 757–758
    orbit, IRAS sky survey, 716–719
    radio emissions, 193
    reconnection process, 493
    as Rosetta stone, 245
    structure, spherically symmetric model,
        237–238
    three-dimensional model, 241–245
    Titan compared to, 397–398
Earth atmosphere
    composition, 9–10
    exosphere and ionosphere, 204
    greenhouse effect, 201–202, 286

mesosphere, 203
oxygen and methane, 194–195
stratosphere, 203
thermosphere, 203–204
thunderstorms, 157, 202
troposphere, 202–203
unpredictability, 206–207
vertical structure, 201
volatile inventories, 204–205
weather, 205–206
Earth interior
    deep, tools for studying, 233–235
    gravity field, 200–201
    inner core (5251–6371 km), 241
    mantle convection, 199–200
    outer core (2981–5151 km), 240–241
    seismic sources, 235–237
    seismology, 200
    and Venus interior, differences, 184
Earthquake
    deep, 237
    global seismic tomography, 242
    stress release on fault plane, 235
Earth surface
    cratering, 9
    crust, 238–239
    features, 193–194
    geomorphic processes
        constructive, 222–226
        destructive, 227–233
    impact craters, 196–199, 634–635
    lower mantle (660–2890 km), 240
    physiographic provinces
        basic divisions, 212–216
        landforms, 216–221
    plate tectonics, 199
    transition zone (400–660 km depth),
        239–240
    upper mantle (25–40 km depth), 239
    volcanic activity, cycles, 418
    volcanic events, 890
    volcanic features, 878–880
Eccentricity
    Hyperion, 836
    increase with increasing orbital period,
        951–952
    Mercury, 124–125
    near-Earth asteroids, 614
    Neptune, 579
    object moving in chaotic orbit, 830–831
    sudden increase due to chaos, 6
Echo
    icy satellite, 803
    radar
        asteroids, 796–802
        detectability, 775–777
        time delay and Doppler shift, 780
        waveform, from conjugate points,
            789–790
Eclipse
    lunar, 251
    mutual, see Mutual events
    solar, coronal observations, 99–100

Ecliptic plane
    not inertially fixed in space, 4
    solar wind statistical properties at 1 AU,
        98
    zodiacal light, 680, 682
Ecology, of life, requirements, 903–904
Eddy mixing, horizontal, Jupiter, 332
Edgeworth–Kuiper Belt, see Kuiper Belt
Ejecta
    atmospheric entrainment on Venus,
        793–794
    lobate, Mars craters, 298
    reaccretion back into Mercury, 144
    simple crater final volume, 847–849
Ejecta blankets
    Moon, 260–261, 264–265
    outer planet icy satellites, 440
    Phobos and Deimos, 310–311
Elastic limit, Hugoniot, for shocked mate-
    rial, 863
Elastic sphere, dissipative, model for Tri-
    ton, 432–433
Electrical conductivity, fluids in Neptune
    and Uranus interiors, 352–353
Electric fields, aligned with magnetic
    fields in high atmosphere, 334–335
Electron neutrinos, 70
Electrons
    delocalization at high pressure, 347
    precipitation in high atmosphere, 335
    solar wind, kinetics, 119
    velocity modulation, 778–779
Electrostatic discharge, Saturn, 766
Elemental abundances, comets, 533–534
Ellipsoid, triaxial, asteroid shape, 589
Elliptical motion, and orbital elements,
    811
El Niño–Southern Oscillation, 196
Elysium province, Mars, volcanism,
    298–301
Emission lines
    collisionally excited, 489
    condensed ethane on Titan, 389–390
    in quiet Sun transition zone, 75–76
    red-shifted, from stellar wind, 52
    ultraviolet, Jupiter atmosphere, 702
Enabling products, near-Earth asteroids
    representing, 626–627
Enceladus
    extensive resurfacing, 448
    orbital ring of material, 20
Encounters
    catastrophic, Oort Cloud with giant mo-
        lecular clouds, 547–548
    close, Sun and other stars, 30–31
    Earth, by Galileo, 193–194, 207–208
    Jupiter family comets with Jupiter, 551
    stellar, effect on Kuiper Belt, 31
    Triton, by Voyager 2, 412–413
Energetic particles
    effect on cometary nucleus, 524
    magnetospheric plasmas, 489–490
    solar wind, 120–121

Energetic particles (*continued*)
  sputtering, on satellite surface, 487
Energetic processes, high atmosphere of
    giant planets, 333–337
Energy
  auroral, 337
  and circular velocity and escape veloc-
      ity, 812–813
  free mechanical, rings, 468–470
  geomorphic, Earth surface, 222–233
  magnetospheric plasmas, 487–489
  orbital, comet, 538
  radiant, solar, 66, 68
  spectral, T Tauri stars, distributions,
      52–53
  transport
    in Jupiter interior, 341
    subsurface, Triton, 428–429
Energy balance, solar atmosphere, 76–77
Energy space, random walk in, by long-pe-
    riod comets, 541–542, 550
Ensisheim, France, meteorite fall, 630
Environment
  change, impact-generated, 232
  dust, near-Earth, and lunar microcra-
      ters, 682–684
  gravitational, Deimos and Phobos,
      309–310
  liquid
    Earth fossil record, 900–901
    Mars fossil record, 915
  nebular, evidence from ordinary chon-
      drites, 59
  space, asteroids exposed to, 602–603
Eolian processes
  Earth arid regions, 219, 229–230
  Venus surface, 178
Epimetheus, and Janus, co-orbital satel-
    lites of Saturn, 451
Equation of state
  giant planet interior
    general considerations, 344–345
    hydrogen and helium, 345–346
    ices and rock, 346
  giant planet interior components, 354
  mixtures, 347
  solar interior model, 68
433 Eros
  Amor asteroid, 623
  cratering age estimates, 626
  UV band linearity, 625
Erosion
  asteroids kept small by, 610
  collisional, Kuiper Belt, 578–579
  collisional and dynamical, 30- to 50-AU
      zone, 517
  fluvial, 227–229
  gravitational, comets from Kuiper Belt,
      7
  magnetospheric, affecting satellites,
      440–441
  on Mars, 9, 283–285, 298
  sediment, and epirogenic uplift, 224

Eruptions
  effusive, and lava flows, 886–888
  explosive, magma volatile content,
      888–889
  massive, in written human history, 233
Eruptive processes, classification, 885–886
Escape
  charged particles, mechanisms, 204
  hydrodynamic, Pluto atmosphere, 512
  ionospheric plasma, 486
Ethane
  condensed, emission lines on Titan,
      389–390
  giant planet stratospheres, photochemi-
      cal formation, 325–326
Eucaryotae, main group of life on Earth,
    901
Eucrites, *see* Meteorites, HED association,
    see also achondrites
Europa
  liquid water under ice surface, 918–919
  network of fractures and dark features,
      443–444
  orbital resonance with Io, 358
  smooth icy surface, 192
  thin ice crust, 20
  ultraviolet spectrum, 705–707
European Space Agency
  future planetary missions, 938
  Giotto comet mission, 935
Evolution
  dust, in interplanetary space, 693
  dynamical
    long-period comets, 540–544
    trans-Neptunian objects, 552–554
  Earth inner core, 241
  interplanetary dust and zodiacal cloud,
      688–693
  Jovian planet, 353–354
  Mercury, 129–141
  Neptune satellites, 454–455
  orbital, *see* Orbital evolution
  outer planet icy satellites, 440–441
  planetary, and impact, 870–876
  solar nebula, 43–44
  stream structure with distance from
      Sun, 104–109
  thermal, Venus interior, 185
    uniformitarian, 186–187
  tidal, *see* Tidal evolution
  Titan, 396–397
    unanswered questions, 400–401
  Triton, 430–434
  Venus, geological and geophysical,
      186–188
  Venus atmosphere, 158, 204
Excavation, by cratering flow-field,
    857–858
Exosphere
  Earth, 204
  Mercury, 126
  Venus, 153
Exploration missions, *see also* Spacecraft

comet and asteroid, 934–935
interplanetary, 928–929
Mars, 292–293, 930–933
Moon, 927–928
outer planet icy satellites, 442–443
outer planets, 933–934
Phobos and Deimos, 309
planetary
  development of, 923, 926–927
  Earth geomorphology role, 210–212
  future, 935–939
  Titan, 401–404
Venus
  history, 162–167
  key scientific results, 185–186
  and Mercury, 929–930
Extinction
  from asteroid impact, 603
  life on Mars, 284
  mass, at K/T boundary, 198
  mechanisms, provided by Chicxulub cra-
      ter, 876
Extreme Ultraviolet Explorer satellite,
    701–702

F

Faint Object Camera, of *Hubble Space
    Telescope,* 507, 701–704, 709–711
Faulting
  Mars canyons, 302
  thrust and strike-slip, Earth, 236–237
Ferroan anorthosite, lunar highland crust,
    267
Ferrous oxide, content of Mercury mantle,
    136–137
Filaments
  activation during preflare phase, 88
  dextral and sinistral, 79
  quiescent, magnetic fields, 83
  radial bright and dark, penumbra, 80
Fischer–Tropsch reaction, organic com-
    pound formation, 652
Fission hypothesis, lunar origin, 273
Flaring magnetic loop, *Yohkoh* images, 89
Flash phase, solar flare, 88–89
Flood
  continental eruptions, 223
  forming Martian channel, 296
  Mars, size estimation, 303–304
  volcanic, Venus, 183
Flux
  comet, background to comet shower,
      548–550
  dust, inner solar system, 685–688
  Earth temperature, peak, 201–202
  galactic cosmic rays, 60
  hot solar wind electrons, 119
  magnetic, *see* Magnetic flux
  meteorites, contemporary, 655–656
  neutrino, high-energy, 69
  transfer events, 493

Flux rope
  magnetic, 482
  plasmoid, 207–208
Flux tube, magnetic, 492–494
Formation
  calderas, on Earth, 879–880, 896–897
  comets, 51–52
  convecting eruption clouds, 892, 894
  crater, rock uplift in, 859
  Galilean satellite rings, 851–852
  Hevelius, Moon, 259
  Imbrium Basin, Moon, 261
  kamacite, in iron meteorites, 639
  Kuiper Belt and scattered disk, 581–583
  Mercury inner core, 139–140
  Moon, from giant impact, 15–16, 49,
    273–275, 871–872
  oldest solids, and nucleosynthesis,
    55–56
  organic compounds, Fischer–Tropsch re-
    action, 652
  outer planet icy satellites, theoretical
    models, 439–440
  photochemical, stratospheric aerosols,
    325–326
  planetary satellites, Roche limit in, 439
  planetesimals, 44–46, 158
  planets, 46–51, 944–945
    from interstellar cloud, 3
  Saturn E-ring, 448
  shock, in outer heliosphere, 104–106
  solar prominences, 79
  solar system
    circumstellar disk in, 38
    and cometary origin, 554–555
    theory, 36–52
  stars
    in interstellar cloud, 36–38
    isothermal collapse in, 944
  stromatolites, on Earth, 906–907
  terrestrial planets, 48–49
Formation intervals, meteorites, 669–670
Forward shock
  coronal mass ejections, 110–111
  solar wind stream, 105–107
Fossil record, liquid environment
  Earth, 900–901
  Mars, 915
Fractionation
  chemical
    chondrules, 59
    primitive meteorites, 62
  isotope, 58
Fractures
  network on Europa, 443–444
  Venus surface, 176
Fragmentation
  atmospheric, effect on meteor surface,
    630
  micron-sized particles, 45
  molecular clouds, 944
Friction, destructive geomorphic process
  on Earth, 227

Frost deposits
  H₂O, Charon, 12
  N₂, Triton, 429–430
  SO₂, Io, 373–375
1993 FW, trans-Neptunian object, 564,
  566

G

Galactic cosmic rays
  anomalous, seed particles for, 121
  flux, 60
  lunar surface exposed to, 255
  1930s observations, 96
Galactic disk
  arranged in multiarm spiral, 28–29
  tidal force, on Oort Cloud comets, 547
  warp, 29–30
Galatea, gravitational resonance with Neptune, 466
Galaxy, solar system place in, 28–32
Galilean satellites, see also Callisto; Europa; Ganymede; Io
  competing dynamical processes, 6
  discovery, 436
  Doppler mapping, 791–792
  gravitational interactions, 20
  historical survey, 443
  radar properties, 802–804
  at radio wavelengths, 754
  ring structures, 851–852
  ultraviolet observations, 704–707
Galileo
  dust detectors, 684–687
  Earth encounters, 193–194, 207–208
  future planetary missions, 938
  Io intensive phase, 375–376
  Jovian ring system, 458
  Jupiter mission, 443
  outer planet missions, 934
Galileo Probe
  giant planet cloud stratigraphy, 323
  Jupiter
    cloud regions, 326–327
    sinkhole, 317
Gamma rays, appearance in flash phase of
  solar flare, 89
Ganymede
  craters, 849
    anomalous, 852–853
  geologic activity, 444–445
  internal magnetic field, 495
  partially resurfaced, 20
  ultraviolet spectrum, 705–707
Gas, see also Noble gases
  accretion, outer planets, 50–51, 61
  collapsing, inward supersonic motion,
    42
  condensation, from Io plumes, 372
  disk, planet embedded in, 51
  frozen, cometary nuclei made of, 520

  in magma, explosive eruptions, 888–
    889, 894–895
  nebular, 62
  parcels, vertical mixing on Jupiter, 318
  particle speed distribution in, 117–118
  radial pressure gradient, 44
  rare, degassed in Mars atmosphere, 281
  released by comets, 483
  solar, hydrostatic equilibrium, 68
  Titan atmosphere, latitudinal variations,
    387–388
  trapped, in meteorites, 657
  unit mass of, centrifugal force on, 42
Gas drag
  particle effects, 46
  small bodies, 821–822
Gas exchange experiment, Mars, 912
Gas giants, see Jovian planets; Jupiter;
  Saturn
Gas–liquid phase boundary, Jovian planet
  atmospheres, 340
Gas nebula, around solar-mass protostars,
  27
Gas retention age, meteorites, 665–667
G2 dwarf, Sun classified as, 4
Geochemical models, Mars volatile inven-
  tory, 280–281
1620 Geographos, imaging, 799
Geological features, Venus, 170–178
Geological gauges, craters as, 867–868
Geology
  Mercury, 129–141
  Triton, Voyager encounter, 416–423
  Venus
    history, 179–183
    Magellan data, 167–169
Geomagnetic activity
  governed by solar wind speed, 491
  recurrent, and quasi-stationary streams,
    108–109
Geomagnetic storms, nonrecurrent, and
  fast coronal mass ejections, 114
Geomorphology, Earth, role in planetary
  exploration, 210–212
Giant impacts
  in final stages of planetary accretion, 27,
    273–275
  Moon formation from, 15–16, 49, 273–
    275, 871–872
  Pluto–Charon binary originating from,
    515
  proto-Mercury, 141, 144
Giant planets, see Jovian planets; Jupiter;
  Neptune; Saturn; Uranus
Giotto spacecraft
  bumper shield, 674
  comet trail observations, 726
  surviving comet fly-through, 935
Glaciers, Earth, 226
Glasses, thetomorphic, 864
Global Oscillation Network Group, helio-
  seismological studies, 73
Global seismic tomography, 241–242

Global warming, following K/T impact, 874
Goldstone Solar System Radar, California, 391, 777–780, 790, 792, 795–796, 805–806
Gondwana, 199
Grabens
    Mars canyons, 302
    pattern, on floor of Caloris, 130
    tensional, postdating ridges, 132
Grains
    in cometary dust tails, 527
    dust, see Dust grains
    interstellar, in meteorites, 659–660
    silicate sand, saltation, 230
Granulation, solar convective cell, 78
Graphite, mixture with diamond, in achondrites, 645
Gravitational field
    constraint on giant planet interior, 341–343
    Earth interior, 200–201
Gravitational harmonics
    calculation, 347
    giant planets, 342–343, 354
    in modeling of giant planet interiors, 349–350
Gravitational interactions
    Galilean satellites, 20
    mutual, in multisatellite systems, 6
Gravity
    interplanetary dust and zodiacal cloud, 688–689
    lunar, 253–254
    Newton's law, and law of motion, 812
Gravity assist, utilized by Mariner 10, 930
Gravity field, Earth interior, 200–201
Gravity waves, solar deep interior, 73–75
Great Dark Spot, Neptune, 328, 332–333
Great Red Spot, Jupiter, 327, 331–332
Greenhouse effect
    Earth, 201–202, 286
    Io, solid-state, 374
    Mars, 285–287, 304, 920–921
        augmentation mechanisms, 287
    Titan, 400
    Triton
        solid-state, 428
        by tidal heating, 433
    Venus, 148, 150–151, 158, 850–851
Greenstone belts, Canadian Shield, 215
Grooves
    on Enceladus and Tethys, 448–449
    surface features of Phobos and Deimos, 310–311
Ground truth
    near-Earth asteroid samples, establishment, 626
    provided by meteorites, 629
Groundwater, sapping, networks generated by, 229
Gulf of California, result of continental tear, 224

Gulf Stream rings, Earth, 202

H

Habitability
    continuously habitable zones, 920–921
    Mars, 913–914
    requirements, 953
Hadley Rille, sinuous, 257, 264
Hale–Bopp comet, 539
Halley's comet
    arrival at perihelion, 543
    coma dust, 59
    dust-to-gas ratio, 531–532
    elemental discrepancies, 533
    Giotto images, 726
    nucleus, 15, 523–524
    VEGA mission, 935
Halley-type comets
    captured into inner solar system, 532
    periodic, 538–539
Halo
    Milky Way, oblate spheroid shape, 28
    radar-bright, Mercurian anomalies, 135
    radar-dark, surrounding Venus craters, 851
    solar wind electrons, energies, 119
Hawaiian activity, pyroclastic deposits, 891–892
Hazards
    cometary impacts, 535–536
    dust, and man-made space debris, 675
    near-Earth asteroids
        assessment, 609
        impact magnitude and frequency, 615–616
Hazes
    aerosol, Pluto atmosphere, 512
    particle, giant planet stratospheres, 325
    Titan atmosphere
        banded appearance, 388
        precipitation onto surface, 390–391
    Venus atmosphere, 155–157
Heat flow
    from Io, 367–368
    Jovian planets, 353–354
    lunar, 254–255
    Triton, during tidal heating, 433
Heating
    corona, possible mechanisms, 77, 98
    internal, Mars, 283–284
    Joule, 333–334
    tidal, after Triton capture, 406–407, 432–434
Heat transport, in atmospheres of giant planets, 318–319
Hectometric emission, Jupiter, 760–761
522 Helga, asteroid, 12:7 resonance with Jupiter, 840

Heliocentric distance
    temperature gradient with, exhibited by nebula, 8
    time-average, Oort Cloud comet at, 547–548
    trans-Neptunian objects, 566
    trans-Neptunian population, 553
    where ices condense, Jovian planets formed at, 10
Heliopause
    defining boundary of solar system, 25–26
    separating interstellar plasma from solar wind plasma, 115–116
Helioseismology
    acoustic oscillations, 71–72
    detection schemes, 72–73
    gravity waves, 73–75
    oscillations of brightness, 75
    solar interior rotation, 73
Heliosphere
    beyond giant planet orbits, 480
    inner, stream evolution in, 106–107
    outer
        shock formation in, 104–106
        two-dimensional stream structure in, 107–108
    solar wind speed, 479
Heliospheric current sheet
    ballerina skirt model, 102
    coronal streamers surrounding, 104
    relationship to solar magnetic field, 102
Helios spacecraft
    dust detectors, 685
    interplanetary missions, 929
    zodiacal light measurements, 681–682
Helium
    burning, during red giant phase, 33
    depletion hypothesis, 353
    giant planet atmosphere, 319
    giant planet interior, equation of state, 345–346
    ³He, implanted in grains during solar wind exposure, 24
    solar wind component in form of alpha particles, 97–98
Helium abundance
    within high-speed solar wind streams, 100
    Jupiter and Saturn atmospheres, 350–351
    in solar wind, 120
Hemispheric asymmetry
    albedo, Iapetus, 20, 398–399, 707–708
    brightness, Titan, 389, 393–395
Heterogeneity
    isotopic, mass-independent meteorites, 62
    radar, Mars, 794–796
Hevelius formation, Moon, 259
Hexahedrites, meteorites, 639
Hildas, asteroid population, 590
Hill sphere, planetary, 816–817

Hill sphere radius, 47
Hilly and lineated terrain, Mercury, 132–133
Himalayan Range, uplift rate, 214–215
Hipparcos astrometry satellite, 31
History
    collisional, near-Earth asteroid populations, 621
    Deimos and Phobos, 313–314
    dynamical, near-Earth asteroids, 614–615
    exploration of Venus, 162–167
    geological
        Mercury, 140–141
        Triton, 423
        Venus surface, 179–183
    life on Earth, 905–908
    lunar cratering, 261
    planetary radar, 774–775
    Russian spacecraft, 1960s-1980s, 923, 926–927
    thermal, Mercury, 139–140
    ultraviolet astronomy, 697–698
    water, on early Mars, 914–916
    written, of massive eruptions, 233
Howardites, *see* Meteorites, HED association
Hubble Space Telescope
    Charon orbit, 503–504
    circumstellar disk imaging, 39–40, 52
    Ganymede atmospheric polar auroras, 445
    observations at ultraviolet wavelengths, 700–711
    Pluto surface imaging, 507
    SO_2 gas measurements on Io, 374
    Titan imaging, 381
        surface, 393
Hugoniot elastic limit, for shocked material, 863
Humans
    ancient, identification of stars, 2–3
    development, Moon role, 248, 251
    early records of comets, 521
    injury from meteorites, 635–637
    written history, of massive eruptions, 233
Hurricanes
    lifting-condensation level, 202
    tracking, 206–207
Huygens probe, Titan mission, 401–404
Hydrogen, *see also* Metallic hydrogen
    atmospheric
        giant planets, 318–319
        Uranus and Neptune, 352
    depletion, in Comet P/Halley, 533
    fusion, at Sun's core, 32–33
Hydrogen cyanide
    Neptune atmosphere, 752–753
    stratospheric, Jupiter and Neptune, 321
Hydrogen sulfide, atmospheric, giant planets, 320–321

Hydrolysis, preterrestrial, carbonaceous chondrites, 649–650
Hydromagnetic dynamo process, 344
Hydrostatic equilibrium, for rotating planet, 347
Hyperion
    chaotic rotation, 835–836
    ice-covered, 450
    nonspherical body, 20
    surficial water ice, 399

I

Iapetus
    dark side, D-type primordial matter, 438
    hemispheric albedo asymmetry, 20, 398–399, 707–708
    reflectivity, 449
Ice
    covering Hyperion, 450
    deposits at Mercury poles, radar evidence, 792
    diacetylene, formation on Uranus, 326
    giant planet interior, equation of state, 346
    Mars, 305–306
        regolith filled with, 281–282
    methane, Triton surface, 409–410
    methane and ammonia, giant planet clouds, 324
    N_2
        Kuiper Belt object surfaces, 569–570
        Triton surface, 414–416
    N_2 and CO, Pluto surface, 506
    Triton surface, 407, 419–420
    volcanoes, on Triton, 21
ICE, cometary exploration mission, 934
Ice dwarfs, Pluto–Charon and Triton as, 516
Ice giants, *see* Jovian planets; Neptune; Uranus
Ida/Dactyl asteroid system, 592
IDPs, *see* Interplanetary dust particles
Imaging
    CLEAN algorithm for, 738
    measurements of Venus atmosphere, 149
    multispectral, near-Earth asteroids, 626
    Titan, surficial continents, 393–395
    Venus, by *Magellan*, 792–794
Imbrium Basin, Moon
    Bouguer gravity, 253
    comparison with Caloris Basin, 133
    formation, 261
Impact basins
    Chicxulub, 198
    Earth, 215–216
    tectonic evolution, 872–873
    Titan, 394

Impact craters
    Earth, 196–199, 634–635
    form, 846–853
    formed by iron meteorites, 197
    impact
        and planetary evolution, 870–876
        process, 854–866
    Mars, morphology, 298
    Mercury, 129–132
    Moon, 260–261
    morphometry, 853–854
    Phobos and Deimos, 310–311
    as planetary probes, 866–870
    record, evidence of comet showers, 549
    Tethys, 448
    Venus, 177–178
        areal distribution, 174
Impact events
    occurring on human timescales, 876
    SNC meteorites, 867
Impacts
    asteroidal, extinction from, 603
    Callisto
        records of, 20
        remnant structures, 446
    Chicxulub, Yucatan, 198, 635, 875–876
    cometary, hazards, 535–536
    detectors, for interplanetary dust measurements, 684–686
    Earth
        crustal shock waves during, 197–198
        grazing, late in accretion, 273–274
        life-threatening, 906
        by meteorites, 634–637
    energetic charged particles on planet surfaces, 479
    environmental change generated by, 232
    giant, *see* Giant impacts
    high-velocity, laboratory simulations, 682–683
    K/T, global acid rain following, 874–875
    large, terrestrial planets, unusual characteristics from, 144
    magnitude and frequency, near-Earth asteroids, 615–616
    melt
        basalt KREEP as, 269
        rocks, 866
    meteoroid, danger to spacecraft, 674–675
    planetary, *see* Planetary impacts
    process, 854–866
    Saturn's rings, by micrometeoroids, 468
Inclination, wide range, for near-Earth asteroids, 614–615
Inclusions
    CAI carbonaceous chondrites
        FUN, 660–661
        igneous, 55
        meteoritic, 62
    silicate, in iron meteorites, 670
Indium, in meteorites, 661–662

Infrared Astronomical Satellite
  asteroid observations, 726–729
  comets and comet trails, 723–726
  Earth-orbiting debris, 731
  imaging other planets, 730
  legacy, 731–733
  observing strategies, 717–719
  Pluto
    surface dynamics, 729–730
    surface temperature, 506–507
  search for planet X, 12–13
  sky survey from Earth orbit, 716
  telescope, 716–717
  zodiacal dust cloud and its sources,
    719–723
Infrared Space Observatory
  Titan mission, 401
  zodiacal dust observations, 694
Inner planets, *see* Earth; Mars; Mercury;
  Terrestrial planets; Venus
Instability
  current-driven, in triggering flare, 90
  gravitational, 38
  massive, 232
Instrumentation
  on *Cassini/Huygens* missions to Titan,
    402–404
  planetary radar, 775–782
  planetary systems searches, accuracy,
    950–951
  for radio wavelength observations,
    770–771
  Soviet *Lidar,* on *Mars Polar Lander,*
    936
Intelligence
  extraterrestrial, Drake equation, 920
  life exhibiting, importance of studying
    planetary systems, 942–943
Intensity, zodiacal light, 681–682
Interferometry, and radio antennas,
  736–738
Interior
  deep, containing electrically conducting
    fluids, 484
  Earth, *see* Earth interior
  giant planet, *see* Jovian planet interiors
  Io, 364
  Mars, 306–308
  Mercury, 127–129
  planetary, inferences from volcanism,
    896–897
  Pluto, 508–510
  Sun, 67–75
  Triton, freezing, 433
  Venus, *see* Venus interior
International Ultraviolet Explorer satellite,
  698, 700–705, 707–711
Interplanetary dust
  charging, and interaction with interplan-
    etary magnetic field, 691–693
  evolution in interplanetary space, 693
  gravity and Keplerian orbits, 688–689
  IRAS discoveries, 719

  observations, 675–688
    in situ measurements, 684–688
    meteors, 676–677
    near-Earth environment, 682–684
    zodiacal light, 680–682
  radiation pressure and Poynting–
    Robertson effect, 689–691
  sources, 673–674
Interplanetary dust particles
  anhydrous mineralogy, 722–723
  collisions, 675, 691
  composition, 22–24
  isotopic anomalies, 679–680
  size range, 678–679
Interplanetary space
  dust evolution in, 693
  energetic particle acceleration in, 121
  solar wind permeating, 96
Interstellar cloud
  star formation in, 36–38
  Sun and planets formed from, 3
Interstellar molecules, in comets, 529–530
Interstellar space, long-period comets
  ejected to, 542, 544
Inventory, *see also* Mars volatile inventory
  initial, water on terrestrial planets,
    204–205
  near-Earth asteroids, 609
Io
  heat flow, 367–368
  hot lava eruptions, 364–366
  information on World Wide Web, 375
  interior, 364
  plume eruptions, 368–373
  at radio wavelengths, 754–755
  as source of
    plasma to Jupiter magnetosphere,
      494–495
    sulfur and oxygen, 337
  surface, *see* Io surface
  thermal emission, 358–359
Io atmosphere
  SO_2 sources, 373–374
  stealth plumes, 375
  transport processes, 375
  ultraviolet observations, 702
  vapor pressure, 374–375
  volcanic, 374
Iodine, I–Xe dating, meteorites, 669–670
Ionosphere
  Earth, 204
  plasma
    escape, 486
    temperature, 487
  unmagnetized planets, 481–482
  Venus, 153–154
Ions
  chemistry, importance at high altitudes,
    333
  interaction with Earth magnetic field,
    204
  lost to space from Mercury, 126
  pick-up, 483, 488

solar wind
  ionization states, 119–120
  kinetics, 117–119
Io surface
  physiography, 360–364
  sulfur and silicate composition, 359
  volcanic features, 884
  volcanism, 20
IRAS, see Infrared Astronomical Satellite
IRAS-Araki-Alcock comet, 805
Iron
  Canyon Diablo asteroid, 605
  meteorites, 638–639
  oxidized, in relation to Mars volatile in-
    ventory, 280
Iron abundance
  Earth and Moon, 271
  Mercury, 141, 144
Irradiance, solar, variations, 91–92
Isochron, internal, for meteorite, 668–669
Isotopes
  duration of nebular phase derived from,
    57–59
  mare basalts, 266
  rare gas, lost to space from Mars, 281
  uranium, CAI carbonaceous chondrites,
    55
Isotopic anomalies
  CAI chondrites, 642
  in IDPs, 679–680
  meteorites, 659–661
  oxygen, 58

J

Janus, and Epimetheus, co-orbital satellites of Saturn, 451
Japan
  cometary exploration missions, 935
  future planetary missions, 938
Jeans instability, 38, 42–44
Jets
  Tritonian plumes as, 428
  zonal, outer planet atmospheres,
    329–330
Jovian planet atmospheres
  carbon enhancement, 61
  chemical composition, 317–321
  clouds and aerosols, 321–328
  constraint on interior, 343–344
  gas–liquid phase boundary, 340
  high, energetic processes, 333–337
  thermosphere structure, 204
  troposphere and stratosphere, meteorol-
    ogy, 328–333
  vertical structure, 201
  weather, 205–206
Jovian planet interiors
  atmospheric constraint, 343–344
  general considerations, 340–341
  gravitational field constraint, 341–343

magnetic field constraint, 344
modeling, 347–348
models, mass–radius calculations,
348–349
Jovian planets
accounting for majority of angular momentum, 39–40
attraction of hydrogen and helium from
nebula, 44
composition, 10, 61
cores, growth, 27
evolution, 353–354
gas accretion, 50–51, 61
magnetospheres, 485–486
masses, 62
migration, 28
oscillations, detection, 354–355
at radio wavelengths, 748–753
removal of Oort Cloud comets, 545
rich in organic material, 918
ring systems, 21
satellites, *see* Outer planet icy satellites
Jupiter
asteroid in orbit–orbit resonance with,
828–829
atmosphere, ultraviolet wavelengths,
701–702
atmospheric conditions, 316–317
aurora, 336–337
avoided by *IRAS,* 730
gas layer, 318
Great Red Spot, 327, 331–332
heat flux, 349
interior
general considerations, 340–341
models, 350–351
inward migration, 51
magnetosphere, 335–337, 688
magnetospheric plasma pressure, 485
perturbations, cometesimals under, 554
ring structure, 462
satellites
large, *see* Galilean satellites
small, 446–447
satellite system, 20
synchrotron radiation, 758–760
thermal emission, 749–750
visible albedo and temperature, correlation, 322–323
Jupiter family comets
Kuiper Belt as source, 554, 560
periodic, 538–539
short-lived, 570–571, 614
subject to physical loss mechanisms,
550–551

K

Kamacite, formation in iron meteorites,
639
Kelvin–Helmholtz cooling, 353–354

Kepler's laws, planetary motion, 4, 810–
811, 814–815
Kilometric emission
Jupiter, 761–763
Saturn, 764–766
Kinetics, solar wind plasma, 116–119
Kirkwood gaps
coinciding with Jovian resonances,
838–839
location, 590
resonance, 817–818
3:1, 593
Klystron, in planetary radar system,
778–779
KREEP, lunar highland crust component,
266–269
Kreutz family comets, sungrazers, 538,
544
Krypton isotopes, in meteorites, 658
K/T boundary, *see* Cretaceous–Tertiary
boundary
Kuiper Belt
asteroids, 590–591
comets
gravitational erosion, 7
in low eccentricity, 522
dynamical evolution in, 552–554
effect of stellar encounter, 31
historical perspective, 558–560
long-term stability of orbits in, 570–
576
mass, 574
observations, 564–566
orbital dynamics, 560–564
short-period comets, 51–52
100- to 500-km-diameter objects in,
516–518
total mass, and sizes of trans-Neptunian
objects, 576–580
true population, 62

L

Labeled release experiment, Mars,
912–913
Lagrangian points
resonance, 815–816
Trojan asteroids, 590
Lagrangian satellites, in Saturnian system,
451
Landforms, Earth
formed by water, 226
produced by volcanism, 219, 222–223
submarine and subaerial, 216–219
Landscape
Earth
constructive processes, 222–226
destructive processes, 227–233
Titan, speculations, 395–396
Landslides, subaerial, 232

Lapse rate, adiabatic, convective part of at-
mosphere, 328–329
Late Heavy Bombardment
Mercury, 130, 133
Moon, 17
objects incorporated into planets dur-
ing, 616
terrestrial planets, 28
Latitude, solar, effect on solar wind speed,
102–104
Latitudinal variations, Titan atmosphere,
387–388
Lava eruptions
Io
infrared radiometric detection,
364–365
1990 outburst, 365–366
low viscosity, Earth, 223
Lava flows
Earth, 878–880
and effusive eruptions, 886–888
Io, heat radiated by, 358–359
Moon, 262–264, 880–881
Venus, 170–171
Layered terrain
creation, relationship to pressure variation, 288
cyclic process-produced, 290
Martian water reservoir, 279
Libration
Pluto with Neptune, 2:3 resonance, 502,
516
Trojans, 13
Life
Earth
history, 905–908
origin, 908–909
ecology of, 903–904
evidence from interplanetary spacecraft,
192–195
generalized theories, 904–905
giant planets and Europa, 918–919
intelligent
importance of studying planetary systems, 942–943
about other stars, 920–921
limits to, 909–910
Mars, subsurface, 916–917
need for liquid environment, 901
in solar system, 910–911
Titan, drawbacks, 397
Lightcurve
asteroid, 620
mutual event, 504–505, 507
Pluto, 502–503
steepening, 511–512
symmetric, for comet, 543
Lightning
on giant planets, 328
Venus atmosphere, 157
Lineae, on Europa, 444
Lithophiles, refractory, achondrites enriched in, 644, 661

Lithosphere
  Earth, 192, 199
    upper mantle, 239
  Mercury, compressive stresses, 140
Local Standard of Rest, 30
Loki Patera, thermal anomaly on Io,
    367–368
Long Duration Exposure Facility, 684
Long-period comets
  computed orbits, 538–540
  nearly parabolic orbits, 520
  nongravitational forces, 543
  Oort Cloud, 7, 51
  osculating and original orbits, 540–541
  planetary perturbations, 541–542
  sungrazers, 538, 544
Lorentz force, 487–488, 492, 693
Loss cones, Oort Cloud, 548–549
Love waves, 233
Lowell Observatory, near-Earth object
    search, 617–618
Low-frequency emission, Saturn, 766
Luminosity
  intrinsic, T Tauri stars, 52–53
  solar
    increased during red giant phase, 32
    and Mars greenhouse effect, 286, 289
Luna, Soviet Moon missions, 927
Lunar highland crust
  breccias, polymict predominance, 266,
      269–270
  feldspar, 248
  ferroan anorthosite, 267
  KREEP and KREEP basalt, 268–269
  magma ocean, 270–271
  Mg suite, 267–268
  modified by meteoroids, 880
  seismic velocities, 258–259
Lunar Orbiters, Moon missions in 1960s,
    927–928
Lunar Prospector, 1998 Moon mission, 928
Lunar surface
  cratering history, 261
  craters and multiring basins, 260–261
  magma ocean, crystallization, 266–271
  microcraters, and near-Earth dust envi-
      ronment, 682–684
  oldest rocks from, 56
  regolith, 256–257
    and megaregolith, 873–874
  stratigraphy, 257–258
  tectonics, 257
  temperatures, 255
  volcanian explosion products, 891
  volcanic features, 880–881
Lunation, longer than sidereal month, 249
Lyapunov exponent, 814, 818, 824, 832
Lyman alpha line, ultraviolet emissions, Ju-
    piter atmosphere, 702

M

Magellan spacecraft, Venus surface
  global geology and altimetry, 167–169
  global mapping, 930
  radar reconnaissance, 792–794
  scientific goals, 165–166
  tectonics, 175
Magma
  cryomagma, Triton mantle, 433
  eruptions on Venus, 170–171
  leaving reservoir, subsequent caldera for-
      mation, 879–880, 896–897
  lithiophilic element-enriched, 638
  source, Mars volcanoes stationary over,
      300
  volatile content, 888–889
Magma ocean, lunar, crystallization,
    266–271
Magnesium suite, lunar highland crust,
    267
Magnetic dipole
  Mercury, 127–128
  planetary, 485–486
  Saturn and Jupiter, 325
  tilted, giant planets, 335–336
Magnetic field
  aligned with electric field in high atmo-
      sphere, 334–335
  in and around quiescent filaments, 83
  coronal, 99–100
  coronal mass ejections, topology, 113
  dipole, *see* Magnetic dipole
  draping about fast coronal mass ejec-
      tions, 113–114
  Ganymede, 445
  giant planets
    interior constraint, 344
    rotation rate of, 342–343
  interplanetary
    basic nature, 98–99
    interaction with charged dust parti-
        cles, 691–693
  Jupiter, 758–760
  Moon, 255
  near-Earth asteroids, measurements,
      625
  Neptune, 768–770
  planetary, 484–486
  role in gravitational collapse, 42
  solar, 67, 78–79
  unmagnetized planets, 481–482
Magnetic flux
  balance, in relation to coronal mass ejec-
      tions, 113, 122
  filling factor, 78–79
  sunspot as, 80
Magnetohydrodynamics, Earth, 207–208
Magnetohydrodynamic waves, corona pos-
    sibly heated by, 77
Magnetopause
  dayside of Earth, 492–493
  not plasma-tight, 486
Magnetosphere, *see also* Planetary magne-
    tosphere
  Earth, 207–208
    hybrid nature, 496

  solar wind transferred into, 108–109
  Jupiter, 335–337, 688
  Mercury, 127–128
  Saturn, Titan interaction with, 385, 404
  terrestrial, 478
  types, 479–484
Main Asteroid Belt
  collisions in, 612, 614
  near-Earth asteroids derived from, 608
  relationship of near-Earth asteroids,
      610–611
Manicouagan structure, Quebec, 197
Mantle
  Earth
    convection, 199–201
    lower, flow models, 240
    upper, seismic velocities, 239
  Europa, liquid, 444
  Mercury, FeO content, 136–137
  Moon, seismic activity, 259–260
  Triton, cryomagmas in, 433
Mantle waves, 234
Mapping
  delay-Doppler, Saturn rings and Titan,
      805–806
  global, Venus, 166, 930
  radar, spherical targets, 789–792
  submarine trenches of Earth, 213
Maps
  Pluto surface, 507
  solar system, provided by meteorites,
      644
Maria
  basalt
    age and composition, 264–265
    origin, 265–266
  dark mantle deposits, 263–264
  lava vents, 881
  Mercurian smooth plains similar to,
      134–135
  result of massive lava eruptions, 17
  subcircular in form, 262
Mariner 2
  1962 plasma experiment, 97–98
  Venus mission, 929
Mariner 4, Mars surface, 292
Mariner 5, Venus surface pressure, 929
Mariner 10
  Mercury surface
    composition, 136
    imaging, 124, 129
  utilizing gravity assist, 930
Mariner spacecraft
  Mars missions, 211, 931
  in 1960s and 1970s, 923, 926
Mars
  age, 56–57
  avoided by *IRAS,* 730
  early, water history, 914–916
  exploration missions, 292–293, 930–933
  explosive eruptions, 890–895
  interior, 306–308
  magnetic field, 481–482

internal, 485
orbit and rotation, 293
Planet B probe, 695
radar heterogeneity, 794–796
removal of asteroids from 3:1 resonance, 838
satellites, *see also* Deimos; Phobos
surface composition, 19–20
SNC meteorites, 644
ALH84001, 917–918
found on Earth, 192
role in Mars volatile inventory, 280
subsurface life, 916–917
Viking landers, search for microbial life, 911–914
Mars atmosphere
climate change
early epochal, 283–285
ongoing period, 287–288
CO₂ greenhouse, 285–287, 304, 920–921
augmentation mechanisms, 287
isotopic evidence, 10
at radio wavelengths, 748
at ultraviolet wavelengths, 700–701
Mars Global Surveyor
future program missions, 936–937
magnetic field discovery, 932–933
Mars spacecraft, Soviet, 930–931
Mars surface
calderas, 896–897
canyons, 301–303
cratering rates, 297–298
craters
forms, 849
morphology, 298
early climate, 283–284
erosion, 9, 283–285, 298
and deposition, 303–306
global physiography and topography, 294–295
polar caps, 911
poles, 306
at radio wavelengths, 745, 748
sand dunes, 231
tectonics, 301
temperature ranges, 293–294
Viking landers measurements, 295–296
volcanic features, 881–882
volcanism, 298–301
water on, *see* Water on Mars
Mars volatile inventory
geochemical models, 280–281
ice-filled regoliths, 282
morphological evidence, 278–280
thermokarst features, 281
Mass
cometary, loss, 725–726
cometary nucleus, 524–525
companions, 950–952
Earth, redistribution by near-Earth objects, 196
Kuiper Belt, 574

and sizes of trans-Neptunian objects, 576–580
Moon, 252
near-Earth asteroids, 623
Neptune satellites, 454
Pluto–Charon system, 503–504
Sun, 62, 67
Triton, 408–409
Mass ejections
coronal, *see* Coronal mass ejections
stellar, expanding nebula produced by, 33
Mass extinction, at K/T boundary, 198
Mass–radius calculations, in giant planet interior modeling, 348–349
Mass wasting, in landscape alteration, 232
Maunder minimum, 85
Maxwellian distribution, 118, 204, 489
Mean motion resonance, in Kuiper Belt, 561–563, 572–573
Mechanical weathering, rock, 231–232
Mechanics
celestial, *see* Celestial mechanics
cratering, 855–861
secular resonance, 563–564
Mediterranean salt lenses, as long-lived vortices, 202–203
Megaregolith, lunar, 257
in ancient highlands, 873–874
Melting
rock, by impact, 866
runaway, Triton, 432–433
shock reverberation effects, 864
Mercury
albedo, 124
atmosphere, 126
compared with Moon, 124
cratering, 9
exploration missions, 929–930
geologic history, 140–141
geologic surface units, 129–135
giant impact, 49
interior and magnetic field, 127–129
liquid water, prospects, 911
lobate scarps, 867–868
motion and temperature, 124–126
orbit, perihelion precession, 5
origin, 141–144
polar deposits, 127, 792
at radio wavelengths, 741–743
surface composition, 135–138
tectonic framework, 138–139
thermal history, 139–140
volcanic features, 884
Mesogranulation, solar convective cell, 78
Mesosiderites, meteorites, 638
Mesosphere, Earth, 203
Metallic hydrogen
density, 348
equation of state, 345
Jupiter interior, convection in, 350–351
Jupiter mass, 340–341

transition from molecular hydrogen, 355
Metamorphism
shock, planetary impacts, 861–866
thermal, chondrites, 646, 663, 669
Meteor Crater, Arizona, 634, 653–654
Meteorites, *see also* Chondrites
asteroidal origin, 593–594, 654–656
chronometry, 663–670
formation intervals, 669–670
gas retention age, 665–667
solidification age, 667–669
terrestrial and cosmic ray exposure ages, 664–665
cometary, fragility, 14
co-orbital streams, 633
deceleration, 676–677
on Earth, 13–14, 192, 664
falls and finds, 630, 647
general classification, 637–639
HED association, 597, 599, 645, 654
iron
characteristics, 639–644
impact craters formed by, 197
resource potential, 605
shock-loading, 653–654
isotopic heterogeneity, mass-independent, 62
lunar, 272
regolith gardened by, 256–257
Martian, biogenic material in, 3
noble gas components and mineral sites, 658–661
noble gases, 656–658
orbital evolution, 840
oxygen isotopics, 644–645
from parent body to Earth, 631–634
relationship to
asteroids, 592–593
near-Earth asteroids, 611–612
SNC, *see* SNC meteorites
trace elements, 661–663
Meteoroids, *see also* Micrometeoroids
electrical charging, 691–693
flux, 683–684
high-speed, catchers, 695–696
impacts, danger to spacecraft, 674–675
Moon, 654
streams, cometary, 534–535
Meteorology, giant planet troposphere and stratosphere, 328–333
Meteors
atmospheric deceleration, 676–677
dust particles appearing as, 21
Meteor showers
associations with near-Earth asteroids, 613–614
comets as sources, 534–535
Leonid, 677
Methane
absorptions, Pluto surface, 506
Earth atmosphere, NIMS detection, 194
frost, Triton atmosphere, 427

Methane (*continued*)
  giant planet atmospheres, 319–320
  ice
    giant planet clouds, 324
    Triton surface, 409–410
  Pluto and Charon surfaces, 729–730
  presence in early Mars atmosphere, 287
  Titan atmosphere
    saturation, 383–385
    thermal emission spectrum, 380–381
Microcraters, lunar, and near-Earth dust
    environment, 682–684
Micro-lensing events, black dwarf ob-
    served via, 33
Micrometeoroids
  identification by deep space probes, 676
  IDPs as, 673
  impacts onto Saturn's rings, 468
  low influx into atmosphere, 678
Microwave emission
  flare flash phase, 89
  Moon, 739–740
Midoceanic ridges
  associated anomalies, 243
  Earth, 213–214
    faulting, 236–237
  magnetic anomalies along, 199
  volcanism on Venus, 183
Migration
  giant planets, 28
  outward, Uranus and Neptune orbits,
    582
  radial, protoplanets, 48
Milky Way, solar system place in, 28–32
Miller–Urey experiment
  naturally occurring on Titan, 919
  organic material production, 908
Miller–Urey-type process, organic com-
    pound formation, 652
Mimas, craters, 447
Mineralogy
  anhydrous, IDPs, 722–723
  near-Earth asteroids, 624–625
  S class asteroids, 596
Minerals
  lunar, 272
  Mercury surface
    fracture-associated regions, 137–138
    sources for exosphere, 126
  in meteorites, sites, 658–661
  phase transitions, 864, 866
  planar deformation features, 863–864
  rock and sulfurous, on Io, 359–360
  shock-metamorphosed, 197–198
Miranda
  complex terrains, 20–21
  coronae, 452
Mixtures
  giant planet interior components, equa-
    tion of state, 347
  graphite–diamond, in achondrites, 645
Modeling, giant planet interiors, 347–
    350

Moment of inertia
  Earth, 195
  Moon, 252
Moon
  absence of shallow magma storage, 897
  composition
    minerals and meteorites, 272
    refractory element enrichment, 271,
      275
    tektites, 272
  core, 260
  crust, *see* Lunar highland crust
  early crustal evolutions, 872–873
  and Earth, tidal interaction, 16, 822
  Earth–Moon line, in relation to oxygen
    isotopics, 644–645
  exploration missions, 927–928
  formation via giant impact, 15–16, 49,
    273–275, 871–872
  geophysics, 253–255
  Imbrium Basin, 133, 253, 261
  Late Heavy Bombardment, 17
  liquid water, prospects, 911
  lunar landing and orbital missions, 250
  mantle, 259–260
  mare, *see* Maria
  Mercury compared with, 124
  meteoroids, 654
  motion, theories predicting, 195
  origin
    categories of hypotheses, 272–273
    single-impact, 273–275
  physical properties, 249–252
  at radio wavelengths, 739–740
  surface, *see* Lunar surface
  topography along subradar track, 787
Moons
  interactions with planetary magneto-
    sphere, 494–495
  large, *Galileo* explorations, 934
Morphology, *see also* Geomorphology
  Mars, evidence for volatile inventory,
    278–280
  Venus surface, and global processes,
    167–178
Morphometry, impact crater, 853–854
Mount Palomar Observatory, search pro-
    grams for near-Earth asteroids,
    618–619
Muon neutrinos, 70
Mutual events, Pluto and Charon, 501,
    504–505

N

Nakhlites, *see also* SNC meteorites
  age, 669
  ejection velocity, 633
Narrowband emissions, from Earth, 193
N-body problem, *see also* Three-body
    problem; Two-body problem
  planetary chaotic behavior, 841–842

Near-Earth Asteroid Rendezvous mission,
    608, 620, 623, 625, 935
Near-Earth asteroids
  future studies, 625–626
  impact hazards
    magnitude and frequency, 615–616
    shock waves, 232
  inventory and hazard assessment, 609
  origin, 609–615
  physical properties, 619–625
  population size, 591, 618–619
  radar-detected, 776–777
  remnants of early solar system, 608–
    609
  role in development of extraterrestrial
    resources, 626–627
  search programs and techniques,
    616–618
  shape models, accurate, 800
  telescopic searches and exploration, 603
Near-Earth objects, redistribution of
    Earth mass, 196
Near-infrared mapping spectrometer,
    Earth atmosphere studies, 194
Near-infrared sounding, Venus atmo-
    sphere, 151
Nebula, *see also* Orion nebula
  expanding, produced by mass ejections
    from star, 33
  solar, *see* Solar nebula
  turbulence, effect on particle accumula-
    tion, 45–46
Nebular hypothesis, explaining planetary
    systems, 62
Nebular models, solar system origin,
    39–41
Neon
  concentration in meteorites, 657–658
  ^{21}Ne, spallation-produced, 60
Neptune
  atmosphere, CO and hydrogen cyanide,
    752–753
  broadband ultraviolet measurements,
    703–704
  crossed by Pluto orbit, 515–516
  differential rotation, 330
  eccentricity, 579
  Great Dark Spot, 328, 332–333
  interior
    models, 351–353
    water ocean, 318
  Kuiper Belt objects crossing, 552–554
  mean motion resonances, 572–573
  nonthermal radiation, 768–770
  and Pluto, 2:3 resonance, 562
  rings, future monitoring, 472, 475
  ring system, 463–464
  satellites
    appearance and composition, 454
    general observations, 452–453
    orbital and bulk properties, 453–454
    origins and evolution, 454–455
  satellite system, 21

stratosphere
    CO detection, 321
    methane abundance, 320
Nereid, eccentric orbit, 452–453
Networks
    fractures and dark features, Europa,
        443–444
    generated by groundwater sapping, 229
    valley, Mars
        branching, 304–305
        erosion, 283–285
        similar to sapping channels on Earth,
            289
Neutral-plasma sheet, Earth magneto-
    sphere, 208
Neutrinos
    detectors, 69–70
    production, 68–69
New Millennium program, future plane-
    tary missions, 937–938
Newton's laws, motion and gravity, 812
Nickel, in iron meteorites, 639
NIMS, see Near-infrared mapping spec-
    trometer
Nitrogen
    depletion in Comet P/Halley, 533
    frost, Triton atmosphere, 429–430
    ices
        Kuiper Belt object surfaces, 569–570
        Pluto surface, 506
        Triton surface, 414–416
    liquid, seas on Triton, 410–411
    Pluto atmosphere, 511
Noble gases
    depletion in Comet P/Halley, 533
    in meteorites
        components and mineral sites,
            658–661
        origins, 656–658
Noblesville assemblage, genomict breccia,
    647, 653
Noctis Labyrinthus, Mars, 301
Nomenclature
    asteroids, 587–588
    comets, 522
    natural planetary satellites, 438
    planetary, 3–4
Nongravitational forces, long-period com-
    ets, 543
Nonthermal radiation
    discoveries, 756
    Earth, 757–758
    Jupiter, 758–763
    Neptune, 768–770
    Saturn, 763–766
    Uranus, 766–768
Nucleosynthesis
    and formation of oldest solids, 55–56
    in stars, 58–59
Nuclides, see also Radionuclides
    cosmogenic, in meteorites, 657
    isotopic, in establishment of solidifica-
        tion ages, 667–668

Numerical integration, planetary orbits,
    842–843
Numerical studies, ring systems, 461–462

O

Oberon, cratered terrain, 452
Oblateness, Titan, 389
Obliquity
    chaotic, 843
    Mars, 288–290, 293
    Mercury, 125, 127
    polar, Pluto, 502
    Uranus, 486
Observations
    astrometric, Barnard's star, 946
    astronomical
        Io thermal emissions, 367
        and nature of solar system, 698–699
        attempts: argon in Titan atmosphere,
            385–386
    coronal, solar eclipse, 99–100
    early/historic
        1930s: galactic cosmic rays, 96
        1950s: comet plasma tails, 97
        solar wind, 96–98
        trans-Neptunian objects, 564, 566
    Earth-based, ring systems, 460–461
    Giotto spacecraft, comet trail, 726
    interplanetary dust, 675–688
    IRAS
        asteroids, 726–729
        Charon and Pluto, 729–730
    IRAS strategies, 717–719
    Kuiper Belt and related structures,
        564–566
    Neptune satellites, 452–453
    Oort Cloud, 544
    physical, trans-Neptunian objects,
        566–570
    radar
        comets, 804–805
        satellites, 442
    at radio wavelength, instrumentation
        for, 770–771
    remote, Moon spectra, 252
    rings, 457–459
    telescopic
        asteroids, 594–595
        outer planet icy satellites, 441–442
    T Tauri star, evidence for solar system,
        52–54
    Yohkoh satellite
        coronal mass ejection, 91
        solar flare, 88–90
    zodiacal dust, spectrally resolved, 694
Observations at ultraviolet wavelengths
    asteroids and comets, 710–711
    Galilean satellites, 704–707
    planetary atmospheres, 699–704
    Pluto and Charon, 710

Saturnian satellites, 707–709
Uranian satellites, 709
Observatories
    Arecibo, Puerto Rico, 162, 777–780,
        790, 792, 805–806, 947–948
    Infrared Space Observatory, 401, 694
    Lowell Observatory, 617–618
    Mount Palomar Observatory, Califor-
        nia, 618–619
    Solar Oscillation and Heliospheric Ob-
        servatory, 73
    Sproul Observatory, Pennsylvania,
        946
Occultation
    star/asteroid, 589
    stellar, by planetary rings, 460
    12th magnitude star, by Pluto, 510
Ocean floor, strombolian activity, 889–890
Oceanic basins, submarine, dominant land-
    form on Earth, 216
Oceanus Procellarium, Moon, 262
Offset, lunar center of mass from center
    of figure, 252
Olivine, A class asteroids, 599
Olympus Mons, Martian volcano, 299
Oort Cloud
    asteroids ejected to, 28
    catastrophic encounters with giant mo-
        lecular clouds, 547–548
    comets ejected to, 15
    frozen-storage comets, 522
    galactic tidal forces, 547
    injection of new comets, 544–545
        temporal variations, 548–550
    long-period comets, 7, 51
    observations, 544
    stellar perturbations, 546–547
    three regions, 555–556
    true population, 62
Opposition effect, satellites, 441, 443
Opposition surge, satellites of Uranus,
    451–452
Orbit
    Charon, 503–504
    circular, gas in disk, 43
    comets, 520–521, 538–540
        Jupiter-crossing, 542, 545
        short-period, computer simulations,
            613
    Earth, characteristics, 195–196
    eccentric
        Nereid, 452–454
        Pluto, 500, 502
    highly inclined
        Iapetus, 449
        small satellites of Jupiter, 447
    horseshoe and tadpole, 816
    Keplerian, 560–561, 688–689
    long-term stability in Kuiper Belt,
        570–576
    Moon, 249, 251
    original, long-period comets, 540–541
    Phobos and Deimos, 309–310, 313–314

Orbit (*continued*)
  planet-crossing
    comets and asteroids, 7
    from gravitational perturbations, 48
  planets
    long-term stability, 823–824
    relativistic effects, 5
  regular and chaotic, 828–832
  retrograde
    Phoebe, 20, 450
    Triton, 21, 406, 408, 434
  small bodies
    and dissipative forces, 819–823
    and planetary perturbations, 813–819
  in space, and elliptical motion, 811
Orbital dynamics, in relation to Kuiper
    Belt, 560–564
Orbital elements
  distribution, predicted, 573
  and elliptical motion, 811
  two-body problem, 813
Orbital evolution
  asteroids, 837–840
  meteorites and comets, 840
  planetary satellites, 841
  small satellites and rings, 840–841
  Triton, 431–432
Orbital period, increasing, eccentricity in-
    crease with, 951–952
Orbital velocity
  comet dust trails, 724
  long-period comet, 546–547
  Mercury, 125–126
Organic constituents, carbonaceous chon-
    drites, 650–652
Orientation, planetary magnetic field, 486
Origin
  comets
    dynamical aspects, 554–555
    theory, 38–39
  IDPs, extraterrestrial, 679–680
  life on Earth, 908–909
  Mercury, 141–144
  Moon, 272–275
    by impact, 871–872
    mare basalts, 265–266
  near-Earth asteroids, 609–615
  Neptune satellites, 454–455
  noble gases in meteorites, 656–658
  Phobos and Deimos, 313–314
  Pluto and Charon, 513–518
  rings, 470–471
  solar system, 26–28
    nebular models, 39–41
    T Tauri star role, 52–54
    unresolved questions, 61–63
  tessera terrain on Venus, 186–188
  Titan, 396–397
    unanswered questions, 400–401
  Triton, 430–434
  Venus atmosphere, 158
  zodiacal cloud, 719
Orion nebula, circumstellar disks, 52

Oscillations
  acoustic
    helioseismology, 71–72
    ring features produced by, 467
  brightness of, 75
  free, Earth, 234–235
  Jovian temperature, 332
  Jupiter, detection, 354–355
Outer planet icy satellites
  craters, 852–854
  evolution, 440–441
  physical and dynamical properties,
    438–439
  spacecraft exploration, 442–443
  telescopic observations, 441–442
  theoretical models of formation,
    439–440
  viscous lava composition, 418–419
  volcanic features, 884–885
  water ice composition, 10
Outflow channels
  Martian water inventory, 279, 281, 287
  trapped water, eruptions, 303–304
Outgassing
  carbon dioxide, Mars, 921
  cometary, 613–614
Oxygen, NIMS detection in Earth atmo-
    sphere, 194
Oxygen isotopes
  anomalies, 58
  meteoritic, 644–645
Ozone, Earth stratosphere, 203

P

Pallasites, meteorites, 638
Pandora, shepherd satellite of Saturn, 451
Pangaea, 199
Parabola, comet orbit classified as, 541
Parent body
  asteroid, 591–593, 597
    target for life search, 919
  meteorite, 596
  meteoritic material removal from,
    631–634
Parker's model, solar wind, 97–99
Particles, *see also* Alpha particles; Inter-
    planetary dust particles
  accumulation, in formation of planetesi-
    mals, 44–46
  charged, *see* Charged particles
  cloud, Venus, sedimentation velocity,
    156
  dust, *see* Dust particles
  energetic, *see* Energetic particles
  haze, giant planet stratospheres, 325
  high-speed ejecta, 683
  interstellar, mass flux, 686–687
  micron-sized, radiation force, 820
  orbits, precession, 819

ring
    inelastic collisions among, 468–470
    resonance, 818–819
  solar, regolithic material bombarded by,
    633–634
  space debris, craters produced by, 684
  test, orbital evolution, 571
Particulates
  extraterrestrial, 24
  and gas, components of Io plumes, 372
Patches, surficial, Venus, 178
Pathfinder, Mars landing, 295–296
Peekskill meteorite, 634–635
Pele, plume deposits on Io, 360–362, 369
Penumbra, radial bright and dark fila-
    ments, 80
Perihelion
  long-period comets, 539–540
  Mercury orbit, 125–126
    precession, 5
  Oort Cloud comet, 544–545
  Pluto, 502
  secular resonance with Neptune, 563
Perturbations
  external, on ring particles, 465–467
  gravitational, planet-crossing orbits
    from, 48
  Jupiter, cometesimals under, 554
  planetary
    long-period comet orbit, 541–542
    and orbits of small bodies, 813–819
  secular
    effect on orbital perihelion, 5
    Laplace–Lagrange theory, 842
    long-term planetary, 544, 823–824
  stellar, long-period comets, 546–548
Petrographic properties, chondrites, 646
Phaethon
  association with meteor stream, 677
  comet-derived, 613–614
1P/Halley, *see* Halley's comet
Phase function, near-Earth asteroids, 620
Phase space, chaotic regions of, 832–833
Phobos
  composition, 311–312
  exploration, 309
  orbit, and gravitational environment,
    309–310
  origin and history, 313–314
Phobos-2, Mars missions, 932
Phoebe, retrograde orbit, 20, 450
Phosphine, stratospheric, Jupiter and Sat-
    urn, 321
Phosphorus, and rare earth elements and
    potassium, *see* KREEP
Photometry, satellites, 441, 709–710
Photons, ultraviolet, emissions from Jupi-
    ter aurora, 336–337
Photosphere
  magnetic fields, 78–79
  rotation, 77–78
  solar, CI chondrites resembling, 650
Phreatomagmatic activity, surtseyan, 895

Physics
    solar flare, 90–91
    vortices: study of idealized fluid flows,
        333
Physiography
    Earth provinces
        basic divisions, 212–216
        landforms, 216–221
    Io surface, 360–364
    Mars surface, 294–295
Pioneer 11, approaching termination
        shock region, 116
Pioneers 10 and *11*, outer planet missions,
        933
Pioneer spacecraft
    interplanetary missions, 928–929
    in 1970s, 926
Pioneer Venus
    atmospheric probes, 930
    radar altimetry, 163, 165
Plagioclase, crystallization, forming lunar
        highlands, 270–271
Plains
    Earth, abyssal, 216
    Io, volcanic, 360
    lunar, *see* Maria
    Mars, various elevations, 295
    Mercury, intercrater and smooth, 133–
        135, 140
    Triton
        undulating high, 418–420
        walled and terraced and smooth, 420
    Venus
        forming varieties, 172–175
        lava, 883
        material composing, 180
        surface age, 179
Planar deformation features, minerals,
        863–864
Planetary dynamo current theory, 334
Planetary impacts
    cratering mechanics, 855–861
    craters
        form, 846–853
        morphometry, 853–854
        as planetary probes, 866–870
    disturbed orbits, 845–846
    and planetary evolution, 870–876
    shock metamorphism, 861–866
    as transient process, 854–855
Planetary magnetosphere
    description, 478–479
    dust grains trapped in, 8
    dynamics, 490–494
    interactions with moons, 494–495
    interface with solar wind, 25
    magnetized planets, 484
    plasma
        energetic particles, 489–490
        energy, 487–489
        sources, 486–487
    satellites embedded in, 440–441
    unmagnetized planets, 480–482

Planetary nebula, 33
Planetary systems
    defined, 945
    future directions and opportunities,
        952–955
    searches
        historical era, 946–947
        modern era, 947–949
        study rationale, 942–943
        substellar-mass companions, 949–952
Planet B probe, Mars, 695
Planetesimals
    asteroids as, 610
    containing water ice, 50
    cross section for accretion, 26
    disintegration, in lunar origin, 273
    formation, 44–46, 158
    icy, Pluto as, 10
    scattered, 582–583
        into different orbit, 27–28
    trans-Neptunian, 551–552
Planet Finder, search tool of future,
        953–954
Planets
    atmospheres
        observations, 699–704
        vertical structure, 201
    chaotic behavior, 841–843
    compositions, 60–61
    definition, 943–945
    formation, 46–51, 944–945
        from interstellar cloud, 3
    gravitational effects on each other, 561
    ice dwarf, 516
    interaction with solar wind, 24–25
    Keplerian motion, 4, 810–811, 814–815
    magnetized, magnetospheres, 484
    motions, and chaos, 823–824
    prograde rotation, 41
    radar-detected, 776
    unmagnetized, magnetospheres,
        480–482
Planet X, searches for, 12–13
Plane wave, tracing series of arcs, 72
Plasma
    expanding hot, bubble, solar system on
        edge of, 31–32
    flare, expansion, 89–90
    interactions with dust, 696
    interstellar, solar wind-blown cavity in,
        115
    magnetospheric
        energetic particles, 489–490
        energy, 487–489
        sources, 486–487
    solar wind
        entering magnetosphere, 494
        nearly collisionless, 116–117
        role in magnetosphere, 478–479
    solar wind stream
        coronal mass ejections identified in,
            111–113
        low- and high-speed, 105–107

Plasma tails, cometary, manifestation of so-
        lar wind, 25
Plate tectonics
    and continental geomorphology,
        217–218
    Earth, 199
    Mars, 301, 308
    model of Venus geological evolution, 187
    oceanic and continental crusts, differ-
        ence between, 238–239
    oceanic and continental plates, 213–215
Platinum
    in Canyon Diablo iron, 605
    near-Earth asteroids as source, 627
Plinian activity, pyroclastic flows, 892–895
Plume deposits, on Io, 360–363
Plume eruptions
    Io, 368–373
    Triton, velocity and temperature, 428
Plumes
    eruptive, ejection velocity, 369
    gas condensation from, 372
    SO_2, temperature, 370–372
    stealth, in Io atmosphere, 375
    Triton, models, 425–429
Pluto
    albedo
        and color, 505
        ultraviolet, 704
    atmosphere
        composition, 510–511
        structure, 511–513
    bulk composition and internal structure,
        509–510
    in context of outer solar system,
        516–518
    density, 508–509
    designation as planet, debate, 10, 12
    interactions with solar wind, 483–484
    IRAS observations, 729–730
    lightcurve and pole direction, 502–503
    as member of Kuiper Belt, 522
    as oddity, 559
    orbit
        eccentric and highly inclined, 500
        heliocentric, 502
        Neptune-crossing, 515–516
        resonances, 824
        unique aspects, 561–562
    origin, 515–516
    similarities with Triton, 411–412
    solar phase curve, 505–506
    spatially resolvable images, 710
    surface
        appearance and markings, 507
        composition, 506
        temperature, 506–507
    Titan compared to, 399–400
Pluto–Charon system
    mutual events, 504–505
    origin, 514–515
    tidally evolved, 21
    as trans-Neptunian object, 552

Poincaré, Henri, celestial mechanics, 826
Polar cap
    and climate, Triton, 429–430
    Mars, 911
Polarimetry, satellites, 442
Polarity
    magnetic, sunspot cycle during, 85–86
    reversals, in interplanetary magnetic
        field, 102
    solar magnetic field, 692–693
Polarization ratio, radar measurements,
    782–784
Polarization signatures, large asteroids,
    796–797
Poles
    Mars, CO$_2$ cap, 306
    Mercury
        ice deposits, 792
        sulfur cold-trapping at, 127
        warm, 125
    Pluto, direction, 502–503
    pole-on silhouette, 786–787, 799
Poloidal field, conversion to toroidal field,
    86
Polyaromatic hydrocarbons, in Martian
    meteorite interior, 652
Populations
    asteroids, 590–591
    Kuiper Belt and Oort Cloud, 62
    near-Earth asteroids
        collisional history, 621
        size, 618–619
    related to Kuiper Belt, 560
    trans-Neptunian, heliocentric distance,
        553
Pores, appearance as umbra, 80
Porosity
    asteroid, in relation to density and rota-
        tion rate, 601–602
    Phobos and Deimos, 312
Postaccretion vaporization hypothesis,
    Mercury origin, 141, 144
Postflare phase, relaxation process, 90
Potassium
    and rare earth elements and phospho-
        rus, see KREEP
    and uranium, abundances in terrestrial
        planets, 61
Poynting–Robertson drag, dust grains,
    820–821
Poynting–Robertson effect, 8
    interplanetary dust and zodiacal cloud,
        689–691
Precession
    particle orbits, 819
    perihelion of Mercury orbit, 5
Preflare phase, filament activation during, 88
Pressure
    atmospheric, Mars, and layered terrain
        creation, 288
    radiation
        dust grains blown out of solar system
            by, 7–8

interplanetary dust and zodiacal
    cloud, 689–691
shock, experienced by meteorites,
    631–633
and temperature, increase in giant
    planet interiors, 340–341
vapor, Io atmosphere driven by,
    374–375
Primordial matter, D-type, Iapetus dark
    side, 438
Probes
    deep-space, micrometeoroid identifica-
        tion, 676
    Galileo
        giant planet cloud stratigraphy, 323
        Jupiter
            cloud regions, 326–327
            sinkhole, 317
    Huygens, Titan missions, 401–404
    planetary, craters as, 866–870
    Planet B, Mars, 695
    Venus atmosphere, 149, 152
Products
    driving and enabling, near-Earth aster-
        oids representing, 626–627
    volcanian explosion, Moon, 891
Prometheus, shepherd satellite of Saturn,
    451
Prominences
    quiescent, 83–84
    solar, formation, 79
Proteus, tidally despun, 454
Proton beams, secondary, in solar wind,
    118–119
Protostars
    collapse, 41–43
    solar-mass, nebula dust and gas around,
        27
Puck, nonspherical, 452
Pulsars
    planets revolving around, 942
    PSR 1257+12, periodic behavior,
        947–949
1996 PW, on long-period comet orbit, 28
Pyroclastic deposits
    from Hawaiian activity, 891–892
    on Io, 362
    lunar, 263
    from strombolian activity, 889–890
Pyroclasts, dispersal into vacuum, 895–896
Pyroxene, absorption band, on near-Earth
    asteroids, 624–625

Q

Questions
    unanswered, Titan evolution, 400–401
    unresolved, origin of solar system,
        61–63
Quiet Sun, spectrum, 75–76

R

Radar
    measurements, producing images of as-
        teroid shape, 620–621
    planetary
        measurements and target properties,
            782–806
        prospects for, 806
        scientific context and history,
            774–775
        techniques and instrumentation,
            775–782
    satellite observations, 442
Radial pressure gradient, gas, 44
Radiant energy, solar, 66, 68
Radiation
    blackbody, 736
    emission from solar flare, 87
    nonthermal, see Nonthermal radiation
    synchrotron, Jupiter, 758–760
    terrestrial kilometric, 757
    thermal
        adsorbed, giant planets, 330
        following K/T impact, 874–875
Radiation belt, Uranus and Neptune, 490
Radiation force, micron-sized particles,
    820
Radiation pressure
    dust grains blown out of solar system
        by, 7–8
    interplanetary dust and zodiacal cloud,
        689–691
Radio antennas, and interferometry,
    736–738
Radiochronometers, ambiguous results
    from, 58
Radio emissions
    Earth, 193
    Jupiter
        decametric and hectometric, 760–761
        kilometric, 761–763
    Saturn, 449
        kilometric, 764–766
Radiometry
    measurements of Venus atmosphere,
        149
    satellites, 441–442
Radionuclides
    extinct, 58, 669
    short-lived, in chondrites, 55–56
Radio wavelengths
    asteroids, 753–754
    comets, 755–756
    Galilean satellites, 754
    Io and Titan, 754–755
    Jupiter, 749–750
    Mars, 745, 748
    Mercury, 741–743
    Moon, 739–740
    Pluto, 755
    Saturn, 750–752
    Uranus and Neptune, 752–753

Venus, 743–745
Radius
    Pluto and Charon, 504–505
    Triton, 408–409
Rainfall
    surface runoff from, 227–228
    terrestrial drainage patterns from, 218–219
Random walk
    in energy space, by long-period comets, 541–542, 550
    quasi-, terrestrial Markovian process, 228
Ranger, United States Moon missions, 927
Rare earth elements
    dispersed, in chondrites, 661
    enriched relative to CI composition, 660
    and potassium and phosphorus, *see* KREEP
Rarefaction waves, in cratering process, 856–857
Rayed craters
    description, 849
    Mercury, 135–137
Rayleigh number, 199–200
Rayleigh waves, 233
Reconnection, magnetic
    and coronal mass ejections, 113
    role in magnetospheric dynamics, 492–494
    and solar wind streams, 108–109
Red giant, fate of Sun, 32–33
Reflectivity
    belts and zones, giant planets, 322–323
    Enceladus, 448
    Galilean satellites, 802–803
    Iapetus, 449
    Pluto surface, 510
Regolith
    breccias, 649, 653, 662
    Mars
        adsorbed CO_2, 278, 289–290
        ice-filled, 281–282
    material, meteorites derived from, 633–634
    Mercury, 127
        temperature variations, 741–743
    Moon, 256–257
    satellites of Uranus, 451–452
Resonance
    2:3, Pluto with Neptune, 502, 516, 518
    3:1, asteroid, removal by Mars, 838
    3:2
        Kuiper Belt objects, 553
        Mercury, 125–126
    forcing, 815
    gravitational, Galatea with Neptune, 466
    Kirkwood gaps, 817–818
    Kosai, 544

in Kuiper Belt
    mean motion, 561–563, 572–573
    secular, 563–564, 571–572
Lagrangian points, 815–816
orbital, Io with Europa, 358
orbit–orbit, asteroid with Jupiter, 828–829
ring particles and shepherding, 818–819
spin–orbit, 833–835
Resource potential
    asteroids, 604–605
    near-Earth asteroids, 609, 626–627
Resurfacing
    Enceladus and Tethys, 448–449
    Ganymede, partial, 20
    Venus, catastrophic, 187
Reverse shock
    coronal mass ejections, 110–111
    solar wind stream, 105–107
Rhea, crater size dichotomy, 449
Rheology, Newtonian and non-Newtonian, 886–887
Ridges
    midoceanic, *see* Midoceanic ridges
    Triton, quasi-linear, 422
    Venus plains, 172–175, 180
    wrinkle, lunar, 257
Rifts, three-cornered, 223–224
Rilles
    geometric properties, 887–888
    lunar, 257, 881
    sinuous and arcuate, 264
Ring current
    energy increase during magnetic storm, 491
    magnetic field produced by, 490
Ring of Fire, Pacific Ocean, 220
Rings
    dark, above Titan north pole, 381
    future exploration missions, 471–472
    on Galilean satellites, formation, 851–852
    Neptune, ring arcs confined by Galatea, 455
    origins, 470–471
    particles
        inelastic collisions among, 468–470
        resonance, 818–819
    plane crossings, 460–461
    radar-detected, 776
    Saturn
        brightness temperature, 751–752
        delay-Doppler mapping, 805–806
        E-ring, formation, 448
    spacecraft observations, 457–459
    structure
        external causes, 464–468
        internal causes, 468–470
        overview, 462–464
Ring systems
    governed by dynamical laws of motion, 6
    Jovian planets, 21

numerical studies, 461–462
orbital evolution, 840–841
RNA
    replicating molecule, 905
    ribosomal, 901
Roche limit
    in planetary satellite formation, 439
    planetary tidal interactions, 822–823
    ring systems, 21
Rock
    ancient Earth, dike system exposed on, 897
    giant planet interior, equation of state, 346
    and ice, ratio on Pluto, 412
    impact melt, terrestrial, 866
    lunar surface
        argon escaping from, 19
        microwave opacity, 740
    Martian, silica contents, 296–297
    molten, on Io, 359, 365–366, 368
    oldest, from Moon, 56
    Pluto bulk density consisting of, 509–510, 514
    shock metamorphism, 861, 863
    uplift, in crater formation, 859
    weathering, 231–232
Rosetta stone, Earth as, 245
Rotation
    chaotic, 833–837
    coronal holes, 85
    differential, Neptune, 330
    giant planets, and gravitational field, 341–343
    heliocentric, planetesimals, 46
    Moon, 249, 251
        synchronous, 16
    nonsynchronous, Hyperion, 450
    solar, large-scale motions, 77–78
    solar interior, 73
    super-rotation, zonal, Titan stratosphere, 388–389
    variable, Earth, 195
    winds relative to, 205–206
Rotational effects, in solar nebula, 43
Rotation rates
    asteroids, in relation to density and porosity, 601–602
    giant planet magnetic field, 342–343
    near-Earth asteroids, 621
Ruach Planitia, Triton, 420
Rubble piles, asteroids as, 13, 602
Rubidium, and strontium, internal isochron in chondrites, 668–669
Runaway accretion
    in accumulation of planetary cores, 50–51
    termination in terrestrial planet region, 47–48
Runaway melting, Triton, 432–433
Russia
    Mars missions, 930–932
    Martian meteorites, 630

Russia (*continued*)
  Moon missions, 927
  planetary program, 938–939
  spacecraft history
    1960s, 923, 926
    1970s and 1980s, 926–927
  Venus missions, 929–930

S

Sagittarius, center of galaxy, 29–30
Sakigake, cometary exploration missions,
    935
Saltation, silicate sand grains, 230
Salt lenses, Mediterranean, as long-lived
    vortices, 202–203
Samples, near-Earth asteroids, establish-
    ment of ground truth, 626
Sampling bias, contemporary meteorite
    flux, 655–656
Sampling tools, craters as, 866–867
Sand dunes, star and seif, 230–231
San Gorgonio Pass, California, windmills,
    230
Sapping, groundwater, networks generated
    by, 229
Satellites, *see also* Moons; *specific satel-
    lites*
  to asteroids, 592
  chaotic rotation, 836–837
  embedded in ring system, 840–841
  governed by dynamical laws of motion,
      6
  icy, *see* Outer planet icy satellites
  planetary
    orbital evolution, 841
    tidal interactions, 822–823
  radar-detected, 776
  spin–orbit resonance, 833–835
  surface, energetic particle sputtering,
      487
Satellite systems
  multi-, mutual gravitational interactions,
      6
  Saturn, 20
    discovery, 436
    unusual, 451
  as test of origin theories, 41
Saturn
  ammonia-ice cloud, 327–328
  atmosphere
    bright bands, 750–751
    upper, helium depletion, 319
  hydrogen bulk composition, 349
  interior
    helium rains, 346
    models, 351
  magnetic dipole, 325
  magnetosphere, Titan interaction with,
      385, 404
  nonthermal radiation, 763–766

rings
  brightness temperature, 751–752
  delay-Doppler mapping, 805–806
  E-ring, formation, 448
  structure, 464
satellites
  icy, medium-sized, 447–450
  small, 450–451
  ultraviolet observations, 707–709
satellite system, 20
  discovery, 436
  Titan as member, 398–399
scanned by *IRAS,* 730
ultraviolet spectrum, 702–703
Scarps, lobate, Mercury, 138, 867–868
Scattered disk
  and ecliptic comets, 580–581
  and formation of Kuiper Belt, 581–583
Scattering
  angular scattering law, 787–789
  gravitational, 7, 46
  high-order, from Galilean satellite rego-
      lith, 803
  planetesimals, 27–28, 582–583
Scattering angle, zodiacal light, 680–681
Scorpius–Centarus OB association, 31–32
Search programs, for near-Earth asteroids,
    603, 616–618
Search techniques, for planetary systems,
    direct, 952–953
  and indirect, 945–946
Secular resonance, in Kuiper Belt, 563–
    564, 571–572
Sedimentation velocity, cloud particle of
    Venus, 156
Seismic waves, compressional and shear,
    233
Seismology
  Earth interior, 200
  Jovian, 354–355
  Moon, 254
  seismic sources on Earth, 235–237
Selective accretion model, Mercury origin,
    141, 144
Shallow-atmosphere models, outer planets,
    329–330
Shape
  asteroids, 588–589
  impact crater, 846–847
  near-Earth asteroids, 620
    accurate models, 800
  oblate spheroid, Milky Way halo, 28
  triaxial, Milky Way central bulge, 30
Shear waves, Earth interior, 200
Shepherding, example of resonance,
    818–819
Shepherding satellites
  effect on ring particle orbits, 466
  in Saturnian system, 451
Shergottites, *see also* SNC meteorites
  EETA79001, Martian, 658
  shocked, 654, 669

Shock
  bow, *see* Bow shock
  chondrites, 652–654, 662
  experienced by meteorites on removal
      from parent body, 631–633
  formation in outer heliosphere, and
      stream damping, 104–106
  metamorphism, in cratering, 861
    impact melting, 866
    subsolidus effects, 863–866
  in nearly collisionless plasma, 121
  reverse
    coronal mass ejections, 110–111
    solar wind stream, 105–107
  termination, *see* Termination shock
Shock waves
  compression regions bounded by, 106
  in cratering process, 855–858
  experimental, in constructing equations
      of state, 344–346
  near-Earth asteroid impact, 232
  through Earth crust during impact,
      197–198
Shoemaker–Levy 9 comet
  collision with Jupiter, 701–702
  disruption, 524
Siderophile elements
  in CAI, 660
  iron meteorites enriched in, 661–662
  low abundance on Moon, 271
  meteorites, 638–639
Signal strength, radio emissions from
    Earth, 193
Silhouette, pole-on, 786–787, 799
Silica
  mare basalts containing, 265
  Martian rocks, 296–297
Silicate
  composition, Mercury, 141, 144
  inclusions in iron meteorites, 670
Silicon carbide, presolar grains, in meteor-
    ites, 659–660
Sinkhole, downwelling, Jupiter, 317
Size
  asteroids, 588–589
  craters on Rhea, dichotomy, 449
  earthquake, 235–236
  flood on Mars, estimation, 303–304
  IDPs, 678–679
  near-Earth asteroid population,
      618–619
  near-Earth asteroids, 621–622
  planetary, role in presence of life,
      921
  trans-Neptunian objects, and total mass
      of Kuiper Belt, 576–580
  Triton, 406–407
Sky, scanned by *IRAS* in Earth orbit,
    716–719
Skylab, polar hole measurements, 84
Small bodies
  atmosphereless, pyroclastic eruptions,
      896
  mean motion resonance, 562–563

orbits
  and dissipative forces, 819–823
  and planetary perturbations, 813–819
SNC meteorites
  ALH84001, 192, 917–918
  basaltic, 292–293
  impact events, 867
  isotopic studies, 306–307
  Martian, 644
  role in Mars volatile inventory, 280
SOHO, see Solar and Heliospheric Observatory satellite
Soil
  lunar, components, 256–257
  Martian, search for microorganisms, 912–914
Solar abundances, Jovian atmospheres, 344
Solar activity cycle, effect on corona, 102
Solar and Heliospheric Observatory, 73
Solar and Heliospheric Observatory satellite, 82, 91, 93
Solar atmosphere
  continual expansion into interplanetary space, 97
  energy balance, 76–77
  large-scale motions, 77–78
  magnetic fields, 78–79
  radial temperature and density profiles, 76
  spectrum of quiet Sun, 75–76
  structures in, 80–85
Solar elongation, maintained by IRAS, 717–718
Solar flares
  gradual, thermal phase, 89–90
  models and physics, 90–91
  occurrence with coronal mass ejections, 110
  postflare phase, 90
  preflare and flash phases, 88–89
  radiation and particle emissions, 87
Solar nebula
  accretion disk, 26–27
  chondrite-forming region, 59–60
  evolution, 43–44
  strong temperature gradient with heliocentric distance, 8
  thermal state, 38
Solar phase curve, Pluto, 505–506
Solar system
  age and composition, 54–57
  architecture, 4–26
  astronomical observations, 698–699
  dynamics, 809–824
  early
    near-Earth asteroids as remnants, 608–609
    physical conditions, 59–60
  evidence from T Tauri star observations, 52–54
  fate, 32–33

formation
  and cometary origin, 554–555
  theory, 36–52
life in, 910–911
map, provided by meteorites, 644
origin, 26–28
  nebular models, 39–41
  T Tauri star role in understanding, 52–54
  unresolved questions, 61–63
outer, Pluto in context of, 516–518
place in galaxy, 28–32
stability, 842–843
timescales, 57–59
Solar wind
  ballerina skirt model, 102
  composition and interaction with planets, 24–25
  confined within heliospheric cavity, 479
  coronal mass ejections
    at 1 AU, 111–113
    and transient disturbances, 109–114
  early observations
    direct, 97–98
    indirect, 96–97
  effect on particulates, 690, 693
  electrons, kinetics, 119
  energetic particles, 120–121
  heavy ion content, 119–120
  interactions
    asteroids and comets, 482–483
    Pluto, 483–484
  and interplanetary magnetic field, 98–99
  ions
    implantation, 60
    kinetics, 117–119
  Parker's model, 97
  plasma
    entering magnetosphere, 494
    nearly collisionless, 116–117
    role in magnetosphere, 478–479
  relationship to solar magnetic field, 102
  solar latitude effects, 102–104
  speed, variations with distance from Sun, 114–115
  statistical properties in ecliptic plane at 1 AU, 98
  supersonic, interaction with unmagnetized planet, 481–482
  termination, 115–116
  traced back to coronal hole, 84
Solar wind streams
  high-speed, 100
  quasi-stationary, 108–109
  structure
    evolution with distance from Sun, 104–109
    relationship to corona structure, 101–102
Solidification age, meteorites, 667–669
Solid-state imaging system, Galileo, Earth surface, 193–194

Sound waves
  propagation, 73
  in solar convection zone, 71–72
South Pole, helioseismological detection schemes, 72–73
South Pole–Aitken Basin
  excavated into lunar mantle, 259
  not flooded with lava, 263
Space, see also Energy space; Interplanetary space
  deep, probes in, micrometeoroid identification, 676
  ecliptic plane not inertially fixed in, 4
  environment, asteroids exposed to, 602–603
  interstellar, long-period comets ejected to, 542, 544
  ions lost to, from Mercury, 126
  orbit in, and elliptical motion, 811
  phase, chaotic regions of, 832–833
  rare gas isotopes lost to, from Mars, 281
  weather, forecasting by spacecraft, 494
Spacecraft, see also Exploration missions; specific spacecraft
  bumper shield, 674
  exploring outer planet icy satellites, 442–443
  forecasting space weather, 494
  future, adding to IRAS information, 732–733
  interplanetary
    dust detectors, 684–688
    evidence for life, 192–195
  missions to asteroids, 604
  observations of rings, 457–459
  sampling planetary gravitational field, 342
  studying Venus atmosphere, 148–149
  tracking Neptune inner satellites, 453–454
  ultraviolet instruments, 698
Space debris
  man-made, and dust hazards, 675
  particles, craters produced by, 684
Spacewatch program, discovery of near-Earth asteroids, 617–619
Space weathering, asteroids, 602–603
Spallation
  ^{26}Al and ^{53}Mn produced by, 58
  ^{21}Ne produced by, 60
Spectral characteristics, Phobos and Deimos, 311–312
Spectral energy distributions, T Tauri stars, 52–53
Spectral reflectance
  asteroid, 597–600
  Io, 358, 363–364
  linking meteorites to asteroids, 655
  near-Earth asteroids, 624
Spectrometers
  near-infrared mapping, Earth atmosphere studies, 194
  on Titan missions, 402–404

Spectroscopy
  measurements of Venus atmosphere,
    149
  satellites, 441
Spectrum
  infrared and visible, Triton, 409–410
  Moon, remote observations, 252
  planetary, at radio wavelengths,
    738–739
  quiet Sun, 75–76
  Titan atmosphere, 380–381
  Titan surface, 392
  ultraviolet, importance to astronomers,
    697–698
Spherical harmonics, 72, 201, 244
Spherically symmetric Earth model,
  237–241
Spherical targets, radar mapping, 789–792
Spicules, chromospheric, 81
Spin axis, Hyperion, 835–836
Spiral
  barred, description, 30
  interplanetary, and solar wind streams,
    100
  magnetic field, in equatorial plane, 106
Spiral structure, long-wavelength, 43
Sproul Observatory, Pennsylvania, 946
SSI, see Solid-state imaging system
Stability
  long-term
    Kuiper Belt orbits, 570–576
    planetary orbits, 823–824
  solar system, 842–843
  water, importance for Mars geology,
    294
Stars, see also Protostars; specific stars
  birthplace, 26
  defined, 943
  formation
    in interstellar cloud, 36–38
    isothermal collapse in, 944
  in galactic disk, 30
  identification by ancient humans, 2–3
  life around, probability, 920–921
  perturbations of long-period comets,
    546–548
  supernova explosions, 31–32
  T Tauri, in understanding solar system
    origin, 52–54
Static equilibrium models, solar interior,
  67–68
Statistical properties, solar wind, in ecliptic plane at 1 AU, 98
Stealth plumes, Io atmosphere, 375
Steepening
  kinematic, solar wind stream, 104
  lightcurve, 511–512
Stickney crater, Phobos, 310, 312
Stones, meteorite classification, 638–639
Stony-iron meteorites, 638, 654
Storms, see also Substorms
  geomagnetic, nonrecurrent, and fast co-
    ronal mass ejections, 114

high-pressure, Earth, 157, 202–203
magnetic, ring current energy increase
  during, 491
major, drainage basin inputs from, 228
Stratigraphy
  giant planet cloud, Galileo Probe, 323
  Moon, 257–258
  Venus surface, 179–183
Stratosphere
  Earth, 203
  giant planets, meteorology, 328–333
  Titan
    condensates in, 389–390
    zonal super-rotation, 388–389
Streaks
  crescent, Triton atmosphere, 424–425
  Venus surface, 178
Streamers
  coronal, 82–83
  helmet, coronal mass ejections as disrup-
    tion of, 91
Streams
  co-orbital, meteorites, 633
  meteor, 677
  meteoroid, cometary, 534–535
  quasi-stationary, and recurrent geomag-
    netic activity, 108–109
  solar wind, see Solar wind streams
Strength
  meteorites, 593
  signal, radio emissions from Earth, 193
Stress release, earthquake, 235
Stromatolites, formation on Earth,
  906–907
Strombolian activity, ocean floor, 889–890
Strontium, and rubidium, internal isochron in chondrites, 668–669
S/1997 U1, eccentric orbit around Uranus, 21
S/1997 U2, eccentric orbit around Uranus, 21
Subduction zones, thrust faulting at,
  236–237
Subsidence, on Io, 363
Subsolidus effects, in shock metamorphism, 863–866
Substellar-mass, companions, 947,
  949–953
Substorms, signatures, 491–494
Sudden Electrostatic Discharge events,
  Saturn, 328
Suisei, cometary exploration missions, 935
Sulfur
  abundance in Mercury core, 128–129
  cold-trapping, at Mercury poles, 127
  explaining Io spectrum, 364
  lava, 366
  role in Io surficial processes, 359–360
Sulfur dioxide
  issuing from vents on Io, 362–363
  plume eruptions on Io, 370–373
  presence in early Mars atmosphere, 287

Sulfuric acid, Venus atmospheric clouds,
  156
Sun
  atmosphere, see Solar atmosphere
  composition and age, 4
  coronal mass ejections, 91
  distance from
    asteroid composition related to, 611
    giant planet atmospheres determined
      by, 317
    solar wind speed variations with,
      114–115
  episodic evolution, 38
  gravitational sphere of influence, 26
  interior
    helioseismology, 71–75
    models, 67–68
    neutrinos, 68–71
  irradiance variations, 91–92
  mass, 62, 67
  properties and chemical composition, 66
  red giant and white dwarf phases, 32–33
  solar flares, 87–91
  sunspot cycle, 85–87
  T Tauri star phase, 53–54
  velocity vector, 30–31
  vicinity, zodiacal light
    measurements, 682
Sungrazers, long-period comets, 538, 544
Sunspots
  cycle, magnetic polarity during, 85–86
  growing, 80–81
Supergranulation, solar convective cell, 78,
  80–81, 86
Supernova
  explosions, 31–32
  pulsars as relicts of, 948
Surface
  asteroid, 602–603
    incident energy on, 622
  Ceres, 13
  cometary nucleus, sandblasted, 525
  Earth, see Earth surface
  Europa, smooth icy, 192
  Iapetus, dichotomous albedos, 20
  impacting flux of interplanetary objects,
    868–870
  Io, see Io surface
  Kuiper Belt object, N_2 ices, 569–570
  Mars, see Mars surface
  Mercury, 124, 129–138, 742–743
  Moon, see Lunar surface
  Phobos and Deimos, 310–311
  Pluto, 505–507
  Triton
    color change, 415–416
    geology, 417–423
    ices, 407
Surt Patera, plume deposits on Io, 362
Surtseyan activity, phreatomagmatic, 895
Surveyor spacecraft, Moon missions, 927
Swarm
  planetesimal, 46–47

submicron-sized dust particles, 687–688
subparallel and anastomosing faults,
  Venus surface, 176
Synchronicity
  Moon rotation, 16
  spin–orbit, Pluto–Charon system, 503
Synchrotron emission, Jupiter, 758–760

T

Taenite, iron meteorites, 639
Tau-lepton neutrinos, 70
Taxonomy
  asteroids, 595–597
  near-Earth asteroids, 623–624
Tectonic framework, Mercury, 138–139
Tectonics
  Moon, 257
  plate, see Plate tectonics
  Tharsis province, Mars, 301
  Venus, 175–177
Tektites, lunar, 272
Telescopes
  IRAS, 716–717
  observations
    asteroids, 594–595
    outer planet icy satellites, 441–442
  Planet Finder, 953–954
  radio, 736–738
  study of Venus atmosphere, 148
Temperature
  and albedo, correlation on Jupiter,
    322–323
  asteroid composition change as function
    of, 611
  atmospheric
    giant planets, 318–319
    Pluto, 510–512
    Titan, 387
    Venus, 151–153
  brightness, see Brightness temperature
  density as function of, 241
  ionization state, 119–120
  ionospheric plasma, 487
  lava eruptions on Io, 364–366
  lunar, profile, 254–255
  Mercury, 126
    diurnal variations, 741–742
  plume eruption on Triton, 428
  and pressure, increase in giant planet in-
    teriors, 340–341
  radial, solar atmosphere, 76
  shock, experienced by meteorites, 632
  solar wind, 115
  SO$_2$ plumes on Io, 370–372
  structure, Titan atmosphere, 384–385
  surface
    Mars, 293–294
    Pluto, 506–507
T Tauri star disk, 52–53

Temporal variations, Titan atmosphere,
    387–388
Termination shock
  solar wind slowing down across, 480
  sunward of heliopause, 115–116
Terranes
  karstic, 229
  subaerial, Earth, 216–219
Terrestrial ages, meteorites, 664
Terrestrial Fractionation Line, 644
Terrestrial planets
  accretion progression, 198–199,
    273–275
  boundary conditions, 205
  composition and atmosphere, 8–10
  formation, 48–49
  full-disk radar images, 775
  at radio wavelengths, 739–748
  termination of runaway accretion,
    47–48
  topography along subradar track, 787
  unusual characteristics from large im-
    pacts, 144
  volatile element abundances, 61
  volatile inventories, 204–205
  volcanian activity, 890–891
Tessera terrain, Venus, 164, 175–177
  crater density on, 179
  material composing, 180
  origin, 186–188
Tethys
  companion satellites, 20
  impact craters and resurfacing, 448–449
Tharsis province, Mars
  radar heterogeneity, 795–796
  tectonics, 301
  volcanism, 298–300
Thermal anomalies
  Io
    data plot, 367–368
    while in Jupiter's shadow, 365
  isotherm uplift resulting in, 872
Thermal emission
  Io, 358–359
    astronomical observations, 367
    correlation with albedo, 364
  Jupiter, 749–750
  spectrum, methane in Titan atmo-
    sphere, 380–381
  from zodiacal dust cloud, 681
Thermal evolution, Venus interior, 185
  uniformitarian, 186–187
Thermal history models, Mercury core,
    128, 139–140
Thermal processing
  asteroids, 13
  recovered meteorites, 14–15
Thermal structure, internal
  Pluto, 510
  Venus, 185
Thermochemical equilibrium models,
    cloud base temperature, 323

Thermodynamic equilibrium hypothesis,
    194
Thermodynamics, governing eruption of
    SO$_2$ plumes on Io, 371–372
Thermosphere
  Earth, 203–204
  Titan, thermal balance, 385
  Triton, temperature, 417
Thickness
  lunar maria basalt, 258–259
  optical, Uranus clouds, 328
  Venus crust, variations, 168
Three-body problem, see also N-body
    problem; Two-body problem
  as paradigm of chaotic motion, 827–
    833
Thrust faults
  Mercurian system, 138
  at subduction zones, 236–237
Thunderstorms, Earth, 157, 202
Tidal dissipation, powering Io heat flow,
    359, 368
Tidal evolution
  Charon, 503–504
  outer planet icy satellites, 438
  satellite systems subject to, 41
Tidal forces, galactic, on Oort Cloud com-
    ets, 547
Tidal heating, Triton, 406–407, 432–434
Tidal interaction
  Earth and Moon, 16, 822
  and planetary satellites, 822–823
  Triton and Neptune, 21
Time delay, radar echo, 780
  dispersion of echo power in, 785–787
  dynamical properties from, 784–785
Time-dependent models, solar interior,
    67–68
Timescales
  cloud core collapse, 42
  human, impact events occurring on, 876
  Lyapunov, 814, 824
  solar system, 57–59, 842–843
Tisserand invariant
  for Jupiter family comets, 550–551
  in relationship of comets to asteroids,
    594, 612–613
Titan
  absorption bands, 380
  climate evolution, 400
  comparison with
    Earth and Venus, 397–398
    Triton and Pluto, 399–400
  delay-Doppler mapping, 805–806
  discovery, 379–380
  exploration missions, 401–404
  as member of Saturnian system,
    398–399
  as naturally occurring Miller–Urey ex-
    periment, 919
  origin and evolution, 396–397
    open questions, 400–401
  at radio wavelengths, 755

Titan (*continued*)
  Voyager mission
    atmospheric composition, 383–384
    images, 381–383
    magnetospheric interaction, 385
    temperature structure, 384–385
Titan atmosphere
  chemical composition, 385–387
  general circulation, 388
  historical context, 380–381
  latitudinal and temporal variations, 387–388
  organic residue clouds, 20
  stratosphere
    condensates, 389–390
    zonal super-rotation, 388–389
  thermal profiles, 387
  ultraviolet observations, 704
Titania, cratered terrain, 452
Titanium, mare basalts containing, 265
Titan surface
  geometric albedo in near infrared, 391–392
  imaging, 393–395
  landscape speculations, 395–396
  nature of, 390–391
  spectrum, 392
Tomography, global seismic, 241–242
Topography
  drainage basins, 218–219
  Mars surface, 294–295
  Moon and inner planets, along subradar track, 787
  Triton, 417
Topology, magnetic field of coronal mass ejections, 113
Toroidal field, poloidal field conversion to, 86
Torque, persistent, acting on Earth, 195–196
4179 Toutatis, delay-Doppler imaging, 799–800
Trace elements, stable isotopes, in meteorites, 661–663
Tracking
  hurricanes, 206–207
  near-Earth asteroids, 617–619
  Neptune inner satellites, by spacecraft, 453–454
Traits, found in Earth major groupings, 907–908
Trajectory, double-lobed, Pluto–Neptune resonance, 562
Transform faults, Earth, 199
Transition zone
  Earth, seismic models, 239–240
  quiet Sun, emission lines in, 75–76
  shear velocity anomalies in, 243–244
  solar atmosphere, 81–82
Trans-Neptunian objects
  disk of, historical perspective, 559–560
  dynamical evolution, 552–554
  historical observations, 564, 566

physical observations, 566–570
sizes, and total mass of Kuiper Belt, 576–580
Transport processes, Io atmosphere, 375
Trenches, Earth, 213–214
Triangulation, meteor height and speed deduced by, 676
Triton
  color change, 415–416
  discovery, 408
  geological history, 423
  geometric albedo, 413–414
  infrared and visible spectra, 409–410
  origin and evolution, 430–434
  polar cap and climate, 429–430
  radius and mass, 408–409
  retrograde orbit, 21, 406, 408, 434
  similarities with Pluto, 411–412
  tides, 406
  Titan compared to, 399–400
  ultraviolet brightness, 704
  Voyager 2 encounter, 412–413
Triton atmosphere
  condensation onto surface, 423–424
  crescent streaks, 424–425
  plume models, 425–429
  thermospheric temperature, 417
Triton surface
  bright polar terrains, 422–423
  cantaloupe terrain, 420–422
  liquid nitrogen seas, 410–411
  N_2 ices, 414–416
  ridges and fissures, 422
  rock-rich, 414
  smooth plains and zoned maculae, 420
  undulating, high plains, 418–420
  walled and terraced plains, 420
Troilite, meteorites, 599
Trojans
  asteroid population in orbit of Jupiter, 590
  librations, 13
Troposphere
  Earth, 202–203
  giant planets, meteorology, 328–333
  Titan, 390
Tsunami, ocean-wide, 232
Tunguska event, Russia, 233, 851, 876
Turbulence
  nebular, effect on particle accumulation, 45–46
  supersonic, 335
Two-body problem, *see also* N-body problem; Three-body problem
  circular and escape velocity, 812–813
  orbital elements, 813
  reduction to one-body case, 812

U

Ultraviolet emissions
  Jupiter atmosphere, Lyman alpha line, 702

Jupiter aurora, 336–337
  Uranus atmosphere, 703
Ultraviolet light
  absorption band, near-Earth asteroids, 624–625
  aerosols absorbing, Jupiter and Saturn, 325
Ultraviolet radiation, extreme, in high atmosphere of giant planets, 333
Ultraviolet wavelengths
  planetary atmospheres, 699–704
  solid surfaces
    asteroids and comets, 710–711
    Galilean satellites, 704–707
    Pluto and Charon, 710
    Saturnian satellites, 707–709
    Uranian satellites, 709
  spacecraft observations at, 698
Ulysses spacecraft
  dust detectors, 684–688
  solar polar orbit, 929
Umbriel, dark visual albedo, 451
Unequilibrated ordinary chondrites, 646, 662
United States
  comet and asteroid missions, 934–935
  future planetary missions, 935–939
  interplanetary missions, 928–929
  Mars missions, 930–933
  Moon missions, 927–928
  outer planet missions, 933–934
  planetary exploration
    1960s, 923, 926
    1970s and 1980s, 926–927
  Venus and Mercury missions, 929–930
Uplift
  epirogenic, and sediment erosion, 224
  isotherm, resulting in thermal anomalies, 872
  rate, Himalayan Range, 214–215
  rock, in crater formation, 859
Upwelling, and sinking, zonal pattern, 332–333
Uranium, and potassium, abundances in terrestrial planets, 61
Uranus
  atmospheric ultraviolet emissions, 703
  cloud optical thickness, 328
  diacetylene ice formation, 326
  giant impact, 49
  interior
    models, 351–353
    water ocean, 318
  nonthermal radiation, 766–768
  obliquity, 486
  orbit, outward migration, 582
  at radio wavelengths, 752
  ring structure, 463
  satellites
    medium-sized and small, 451–452
    ultraviolet spectral information, 709
  satellite system, 20–21
  stratospheric circulation, 331–332

synthetic, 346, 352
Ureilites, shocked, 654

V

Vacuum, pyroclast dispersal into, 895–896
Valley networks, Mars
   branching, 304–305
   erosion, 283–285
   similar to sapping channels on Earth, 289
Van Biesbroeck 8, astrometric studies, 947
Vapor pressure, Io atmosphere driven by, 374–375
VEGA, cometary exploration mission, 935
Velocity
   circular and escape, 812–813
   distributions
      planetesimal swarm, 47
      plasma particles, 489
   ejection
      Io eruptive plumes, 369
      meteoritic material, 633
   electrons, bunching, 778–779
   orbital
      comet dust trails, 724
      long-period comet, 546–547
      Mercury, 125–126
   plume eruption on Triton, 428
   radial, detections of companions, 947, 949–950
   sedimentation, cloud particles of Venus, 156
   seismic, in lunar crust, 258–259
   shear, anomalies in transition zone, 243–244
   supersonic, collapsing cloud core material, 42
Venera spacecraft, Venus missions, 163–165, 184–185, 929–930
Vents, hydrothermal, origin of chemotrophic life, 908–909
Venus
   evolution, geological and geophysical models, 186–188
   exploration missions, 929–930
   explosive eruptions, 889–891, 894–895
   magnetic field, 481–482
   revealed by *Magellan,* 792–794
   Titan compared to, 397–398
   water
      forms of, 911
      initial inventory, 204–205
Venus atmosphere
   clouds and hazes, 155–157
   composition, 150–151
   general circulation, 158
   greenhouse effect, 148, 150–151, 158, 850–851

historical studies, 148–149
   lower, temperatures and water vapor, 151–152
   measuring techniques, 149
   middle and upper
      chemical recombination, 154–155
      ionosphere, 153–154
      temperatures, 152–153
      winds, 154
   near-infrared sounding, 151
   origin and evolution, 158
   at radio wavelengths, 743–745
   at ultraviolet wavelengths, 700
Venus interior
   crustal and interior composition, 184–185
   exploration, key scientific results, 185–186
   internal thermal structure and thermal evolution, 185
   perspectives on viewing structure, 183–184
Venus surface
   age, and implications, 178–179
   cratering, 9
      impact, 177–178
   eolian processes, 178
   exploration
      history, 162–167
      key scientific results, 185–186
   impact craters, 850–851
   Magellan global geology and altimetry, 167–169
   at radio wavelengths, 743
   stratigraphy and geologic history, 179–183
   tectonism, 175–177
   volcanic features, 882–884
   volcanism, 170–174
Vertical motion, growing particle in nebula, 44–45
Very Large Array, New Mexico, 391, 737–738, 778, 792, 795–796
Vesta, thermally processed asteroid, 13
Viking landers
   Mars surface, 292, 295
   search for microbial life on Mars, 911–914
Viking spacecraft
   Mars missions, 931–932
   in 1970s, 926
Volatile inventories
   Mars, 278–283
   original, degassing, 211–212
   terrestrial planets, 204–205
Volcanic activity, *see also* Cryovolcanic activity
   Earth
      cycles, 418
      Hawaii, 216
   Io, volcanic system, 370–373
   terrestrial planets, 890–891
   Venus, 170–174

Volcanic features
   Earth, 878–880
   icy bodies, 884–885
   Io, 884
   Mars, 881–882
   Mercury, 884
   Moon, 880–881
   Venus, 882–884
Volcanism
   Earth landforms produced by, 219, 222–223
   effusive eruptions and lava flows, 886–888
   eruptive processes, classification, 885–886
   explosive eruptions, 888–896
   ice, on Enceladus, 448
   Io, 358–359
   Mars surface, 298–301
   planetary interiors, inferences about, 896–897
   planetary volcanic features, 878–885
   Venus, 171–174
Volcanoes
   ice, Triton, 21
   shield
      Earth, 223
      Mars, 298–300, 882–883
      Venus, 168, 882–883
Vortices
   coherent, Earth, 202–203
   long-lived, gas-giant planets, 206
   physics: study of idealized fluid flows, 333
Voyager 2
   Neptune and Uranus, data on interiors, 351
   Triton encounter, 412–413
Voyagers 1 and *2*
   approaching termination shock region, 116
   Io, Pele's plume fallout zone, 361–362
   outer planet missions, 933–934
Voyager spacecraft
   Io atmosphere, 374–375
   outer planet satellite images, 443
   planetary ring systems, 457–459
   in 1970s, 926
   Titan
      atmospheric composition, 383–384
      images, 381–383
      temperature structure, 384–385

W

Water
   atmospheric, giant planets, 319
   destructive geomorphic processes on Earth, 227–229
   forming constructive landforms on Earth, 226

Water (*continued*)
    forms, discrimination by infrared filters, 194
    liquid, requirement for life, 903–904
    outer planets, major constituent, 61
    stability, importance for Mars geology, 294
    terrestrial planets
        initial inventory, 204–205
        presence, 911
Water–ammonia cloud, Jupiter, 326
Water ice
    Charon surface, 513
    dirty, Mars north cap, 279
    Hyperion surface, 399
    Mars permafrost, 294
    Mercury polar radar features, 127
    partial melts, outer planet icy satellites, 438–439
    planetesimals containing, 50
    Pluto interior, 509–510
Water on Mars
    branching valley networks, 304–305
    early history, 914–916
    outflow channels, 303–304
    subsurface, 916–917
    volatile inventory, 278–283
Water vapor
    Earth troposphere, 202
    lower Venus atmosphere, 151–152
Waveforms
    echo, from conjugate points, 789–790
    radar, 781–782
Weak early Sun paradox, 286

Weather
    Earth and Jovian planet atmospheres, 205–206
    near Titan surface, 390
    space, forecasting by spacecraft, 494
Weathering
    asteroids, by space, 602–603
    rock, 231–232
West Germany, and United States, cooperative *Helios* missions, 929
White dwarf, fate of Sun, 33
Widmanstätten pattern, Ni-meteorites, 639, 655
Wildfires, global, following K/T impact, 874–875
Windmills, San Gorgonio Pass, California, 230
Winds
    east–west flow, *see* Zonal wind
    produced by young stars, 42–43
    relative to rotation period, 205–206
    solar, *see* Solar wind
    source of Mars erosion, 306
    supersonic, Jupiter high atmosphere, 335
    T Tauri stars, 62
        emission lines, 52
    Venus atmosphere, 154
World Wide Web, information on Io, 375

X

Xenon, I–Xe dating, meteorites, 669–670
Xenon isotopes, in meteorites, 658

Y

X rays
    appearance in flash phase of solar flare, 89
    Jupiter aurora, 336

Yarkovsky effect, 8
    meter-sized objects, 821
Yohkoh satellite
    coronal mass ejection observations, 91
    solar flare observations, 88–90
    study of active regions, 81

Z

Zodiacal cloud
    gravity and Keplerian orbits, 688–689
    material from asteroid collisions and comets, 21, 721–723
    origin, 719
    radiation pressure and Poynting–Robertson effect, 689–691
Zodiacal light
    brightness, 680–681
    described by Cassini, 674
    measurements by *Helios* spacecraft, 681–682
Zonal wind, diversity among giant planets, 329–331

ISBN 0-12-226805-9